Golf Course Management & Construction

ENVIRONMENTAL ISSUES

EDITED BY

James C. Balogh, B.A., M.S., Ph.D.
Soil, Water, and Environmental Research
Spectrum Research, Inc.
Duluth, MN

William J. Walker, Ph.D.
Environmental Chemistry
Sacramento, CA

CRC Press is an imprint of the
Taylor & Francis Group, an informa business

CRC Press
Taylor & Francis Group
6000 Broken Sound Parkway NW, Suite 300
Boca Raton, FL 33487-2742

First issued in paperback 2019

ISBN 13: 978-0-367-45030-4 (pbk)
ISBN 13: 978-0-87371-742-7 (hbk)

Visit the Taylor & Francis Web site at
http://www.taylorandfrancis.com

and the CRC Press Web site at
http://www.crcpress.com

Library of Congress Card Number 91-46754

Cover photo courtesy of James T. Snow, USGA.

Library of Congress Cataloging-in-Publication Data

Golf course management and construction: environmental issues.
James C. Balogh and William J. Walker, editors.
p. cm.
Includes bibliographical references and index.
ISBN 0-87371-742-2
1. Golf courses—Environmental aspects. 2. Golf courses—Maintenance and repair. 3. Turf management. I. Balogh, James C. II. Walker, William J.
GV975.G62 1992
796.352'06'8—dc20 91-46754
CIP

EDITORS

James C. Balogh, a Senior Research Soil Scientist and Chief Executive Officer of Spectrum Research, Inc., received his B.A. degree from St. John's University, Collegeville, Minnesota and his M.S. and Ph.D. degrees from the Department of Soil Science at the University of Minnesota. Spectrum Research, Inc. is a research and development company with offices located in Duluth, Minnesota and Tallahassee, Florida.

Dr. Balogh specializes in assessment of the environmental effects of agricultural, forest, and turfgrass systems; modeling and field evaluation of soil hydrology and chemical transport; development of strategies for reduction of nonpoint source pollution; evaluating the fate and partitioning of chemicals in soil, water, and sediment; remediation of petroleum contaminated soils; and the development of computer software for chemical fate and toxicology databases.

Since 1978 Dr. Balogh has conducted both academic and contract research for federal, state, and private agencies. Participating agencies include the National Science Foundation; the Department of Energy; the USDA Forest Service, Soil Conservation Service, and the Agricultural Research Service; the U.S. Environmental Protection Agency; the United States Golf Association; state departments of conservation and natural resources; and various private natural resource companies.

Dr. Balogh has been active in developing strategies to mitigate nonpoint source pollution from fertilizer and pesticide management as well as bioremediation of soils contaminated with petroleum products. Many of these projects include literature review, field and laboratory research, multivariate statistical analyses, and computer simulation. Development of research protocols, sampling designs, and environmental database software are all important components of Dr. Balogh's work at Spectrum Research, Inc.

Dr. Balogh is an active member in the Soil Science Society of America, American Society of Agronomy, Soil and Water Conservation Society, and Minnesota Association of Professional Soil Scientists. He also is an ARCPACS Certified Professional Soil Scientist.

William J. Walker is a Senior Inorganic Geochemist with James P. Walsh and Associates, Inc., Sacramento, California and an Associate Research Scientist at the University of California-Davis. Dr. Walker received a B.A. degree in Chemistry from St. Louis University, St. Louis, Missouri, his M.S. degree from the Department of Soil Science at the University of Minnesota, and his Ph.D. degree in Soil and Water Chemistry from the Department of Land, Air, and Water Resources at the University of California, Davis.

Dr. Walker's research specialization includes studies of aluminum solubility in acidified soils and natural waters, development of chemical speciation techniques for determining the sources of metal pollutants in soil and water, evaluation and implementation of quality assurance/quality control practices for laboratories conducting trace element analyses in a wide range of environmental matrices, development of remediation strategies for mine-waste contaminated soils, and modeling of aqueous chemical species and associated transport.

Over the past decade, Dr. Walker has conducted studies for the California Department of Health Services, the U.S Bureau of Reclamation, the San Joaquin Valley Drainage Program, the Kesterson Containment and Cleanup Program, the U.S. Environmental Protection Agency, the United States Golf Association, and a variety of private agencies and companies.

Dr. Walker is an active member in the American Chemical Society, the American Association for the Advancement of Science, and the Soil Science Society of America.

PREFACE

The establishment and culture of turfgrass for golf courses, lawns, parks, and roadsides is a well-accepted practice in the United States. Turfgrass management also is an important soil and water conservation practice in both urban and rural areas. In most areas of the United States, maintenance of quality turfgrass at the levels currently demanded by the public requires management of water, nutrients, and pests. As a result of these intensive management practices, potential environmental and water quality effects have become a public concern.

This book is a summary and assessment of the technical and scientific research on the environmental effects of construction and management of turfgrass systems. Although the focus of this series of interrelated articles is on golf courses, the concept of turfgrass management for residential and commercial lawns, parks, and greenways has been included. This book is intended as an introduction to the concepts of the nonpoint source environmental impacts of turfgrass management for turfgrass scientists and specialists, landscape and golf course architects, developers of turfgrass systems and golf courses, golf course superintendents, environmental scientists, and land use regulators. The information in this book includes sources of additional information on environmental effects for those interested in greater detail on the selected topics.

Assessment of the scientific and technical issues on the environmental effects of golf course construction and turfgrass management was originally developed by Spectrum Research, Inc., Duluth, MN, for the United States Golf Association: Green Section (USGA). With growing public concern over environmental, water use, water quality, and health-related issues, the USGA recognized the need for accurate and unbiased information on the potential effects of certain turfgrass management practices. The USGA has an active commitment to resource conservation and mitigation of potential adverse environmental impacts of turfgrass systems. The USGA currently has an active research program to evaluate the effects of turfgrass management and golf course development on the environment. Development of practices that maintain high quality turfgrass, preserve our environment, and conserve our water resources is the goal of integrated management of turfgrass.

Turfgrass systems have considerable environmental benefits. In addition to the recreational and aesthetic value of turfgrass in urban and rural environments, the positive effects of turfgrass include erosion control, adsorption of atmospheric pollutants, mitigation of water pollution problems, dust control, cooling effects, noise abatement, enhancement of real estate value, and provision of wildlife habitat. In contrast to the benefits, certain turfgrass management practices may potentially increase detrimental environmental impacts. Some of the major potential nonpoint source environmental effects of turfgrass management on golf courses include the following:

1. Leaching and runoff losses of nutrients and pesticides from established turfgrass sites
2. Soil erosion and runoff losses of sediment and nutrients during construction and losses from disturbed riparian zones
3. Exposure of beneficial nontarget soil organisms, wildlife, and aquatic systems to pesticides
4. Development and resurgence of insect and disease populations resistant to current chemical management strategies
5. Excessive use of water resources for irrigation during drought conditions and in semiarid and arid climatic zones
6. Degradation of stream and lake quality resulting from sediment, chemical, and thermal pollution
7. Disturbance or loss of wetlands which is a serious public concern especially regarding the development of new golf courses
8. Disturbance and toxicity impacts on wildlife

Many of these issues are not limited to turfgrass systems. The potential and documented adverse effects of intensive landscape management are a general concern in urban, industrial, and rural environments. Intensive management of cropland, forests, turfgrass, and urban areas has a history of controversy in the United States. Many of these multiple use issues involve a core of emotional, philosophical, and technical questions. As examples, public concerns about the increasing reliance on chemical fertilizers and pesticides is directly related to perceived effects on water quality, wildlife, and humans. Also, competition for use of limited water resources remains a sensitive issue in regions with perennial and periodic drought conditions. Development of new golf courses in areas with limited space for land development often confronts concerns and regulations surrounding wetland conservation. Therefore, the social and environmental benefits of golf courses and turfgrass management must be balanced with concerns regarding health and environmental quality.

The concepts originally developed for integrated crop management (ICM) in agriculture are suitable for ecologically sound management of turfgrass. The use of integrated management for turfgrass would be very effective for reducing detrimental nonpoint source environmental and water quality impacts. These practices would coordinate all management factors required for long-term sustained productivity and quality of turfgrass, golf course profitability, and ecological soundness of selected management options. Critical components of integrated management systems include proper design and construction of golf courses; selection of appropriate turfgrass species and cultivars; soil practices; clipping and cultivation practices; nutrient management; irrigation and drainage management; chemical, biological, and cultural pest management; and soil, water, energy, and natural resource conservation during construction and maintenance of lawns and golf courses.

Integrated pest management (IPM) programs developed for turfgrass would be a significant, *but not a singular,* component of integrated management systems for turfgrass. The goal of an integrated systems approach is to manage golf courses and turfgrass, while balancing turfgrass quality, costs, benefits, public health, and environmental quality. Use of integrated water, pest, and nutrient management strategies will ultimately resolve the issue of maintaining high quality turfgrass with a minimum of environmental disturbance.

This review and assessment is organized into eight chapters. The introduction provides an overview and historical perspective regarding turfgrass management and environmental quality. Chapter 2 discusses the relationship of turfgrass management to the critical issues of water resources. Chapter 2 focuses on the issues of water use, water quality, soil and water conservation, and movement within the water cycle. Chapters 3 and 4 provide a state-of-the-art scientific review and assessment of the literature regarding the environmental effects of nutrient and pest management practices. Chapter 5 provides an introduction to concepts necessary for development of integrated management systems for turfgrass. Chapter 6 covers the direct and indirect effects of golf course management and construction on wildlife and aquatic organisms. Chapter 7 is an introduction to the issues of conservation and protection of wetlands which is emerging as a critical environmental concern of the 1990s. An appendix, Chapter 8, contains tables of toxicity tests related to the effect of chemicals used for turfgrass management. Each of the chapters includes a section on research and information needed to resolve the issues surrounding the positive and potentially adverse effects of turfgrass management. It is not the intention of this book to suggest or recommend site-specific turfgrass management practices. However, the information presented in this publication may be used as an introduction to the general principles for evaluating the potential environmental effects of turfgrass management.

Golf course and turfgrass managers are caught in the dilemma of maintaining cost effective operations, while sharing public concern over potential hazards to the environment and limited water supplies. Many land use managers perceive few alternatives to chemical and water management options, adopted on the recommendation of both university and industry scientists. There currently is a consensus among many environmental regulators, turfgrass specialists, soil scientists, entomologists, weed scientists, and pathologists that scientific research regarding the ecological impacts of intensively managed turfgrass is in its initial stages.

Development of consistent practices, perspectives, and public policy regarding construction and maintenance of turfgrass systems is needed. Given the prohibitive expense of mitigating environmental problems, it has been recognized that prevention of adverse effects is a preferred strategy. Although it is relatively easy to identify potential problems, development of economically feasible systems to mitigate or resolve adverse effects related to golf course management still requires significant basic and applied research efforts. The challenge for turfgrass and landscape managers is to develop management strategies that optimize the benefit of intensive management, while minimizing the risk of environmental adversity. The challenge to those who might oppose golf courses and sound turfgrass management is to objectively evaluate the facts, as they are known, and to be thoroughly educated on all of the issues.

James C. Balogh
William J. Walker

CONTRIBUTORS

James L. Anderson, Ph.D.
Professor and Director
Center for Agricultural Impacts on Water Quality
Department of Soil Science
University of Minnesota
St. Paul, MN

James C. Balogh, Ph.D.
Senior Research Soil Scientist and Chief Executive Officer
Soil, Water, and Environmental Research
Spectrum Research, Inc.
Duluth, MN

Mary E. Balogh
Senior Scientist
Spatial Analysis Laboratory
Environmental Programs Office
Lockheed Engineering & Sciences Company
Las Vegas, NV

Bruce Branham, Ph.D.
Associate Professor
Crop and Soil Sciences Department
Michigan State University
East Lansing, MI

Victor A. Gibeault, Ph.D.
Professor
Department of Botany and Plant Sciences
University of California, Riverside
Riverside, CA

Michael P. Kenna, Ph.D.
Director
Green Section Research
United States Golf Association
Stillwater, OK

Patricia A. Kosian
Research Chemist and Vice President
Soil, Water, and Environmental Research
Spectrum Research, Inc.
Duluth, MN

Anne R. Leslie
Chemist
Office of Pesticide Programs
United States Environmental Protection Agency
Washington, D.C.

Sheila R. Murphy
Director
Software Research and Development
Spectrum Research, Inc.
Duluth, MN

James T. Snow
National Director
Green Section
United States Golf Association
Far Hills, NJ

Roberta M. Tietge
Research Biologist
Soil, Water, and Environmental Research
Spectrum Research, Inc.
Duluth, MN

William J. Walker, Ph.D.
Senior Inorganic Chemist
Walsh & Associates, Inc.
Sacramento, CA

James R. Watson, Jr., Ph.D.
Vice President
The Toro Company
Minneapolis, MN

TABLE OF CONTENTS

CHAPTER 1

Background and Overview of Environmental Issues

James C. Balogh,
Victor A. Gibeault,
William J. Walker,
Michael P. Kenna, and
James T. Snow

1.1 INTRODUCTION

The public demands high-quality turfgrass on golf courses, parks and urban greenways, sports fields, residential and institutional lawns, and conservation systems. As a result of this demand, the intensity of turfgrass management is rising steadily in the United States (Cockerham and Gibeault 1985; Walker et al. 1990). To meet the requirements for facilities with high-quality turfgrass, managers use numerous cultural, physical, and chemical management strategies as cost-effective tools (Beard 1982; Carrow et al. 1990; Madison 1982).

Construction and management of golf courses and turfgrass for other facilities often require (1) exposure of bare soil surfaces; (2) construction of areas with disturbed soil structure; (3) turfgrass planting, upkeep, and clipping; (4) intensive irrigation; (5) pest management; and (6) fertilization (Beard 1982). These practices may intensify potentially adverse environmental effects, such as soil erosion, sediment and runoff losses of applied chemicals, contamination of groundwater with leaching losses of pesticides and fertilizers, disturbance of adjacent ecosystems, and impacts on nontarget plants and animals.

Photo courtesy of Michael French.

In contrast to potential negative effects on the environment, golf courses and turfgrass systems have considerable environmental benefits. In urban and rural environments, turfgrass has recreational and aesthetic benefits. In addition to these obvious values, turfgrass benefits include erosion control, mitigation of nonpoint atmospheric and water pollution problems, dust control, cooling effects, noise abatement, and enhancement of real estate value (Leslie and Knoop 1989). The golf course and turfgrass industry constantly balances the potential adverse effects of turfgrass management in relation to public health concerns, effects on nontarget organisms, potential degradation of water resources, and effects on wetlands.

1.2 ENVIRONMENTAL ISSUES AND TURFGRASS MANAGEMENT: A HISTORICAL PERSPECTIVE

Increasing demands for high-quality agricultural products, wood and paper products from forests, and turfgrass throughout the United States have resulted in intensified management of many ecosystems. Some of the environmental and water quality impacts resulting from intensive management of agriculture and forests systems have been reviewed by Anderson et al. (1989), Balogh et al. (1990), Koehler et al. (1982a, 1982b, 1982c), Stewart et al. (1975, 1976), Smolen et al. (1984), and Walker et al. (1990). These reviews provide a perspective on the historical development of issues regarding soil and water conservation, water quality, nutrient and pesticide management, and impacts of management on the quality of the environment.

The major environmental concerns from the 1930s through the 1950s involved soil and water conservation on agricultural land (Stewart et al. 1975). Loss of valuable topsoil from water and wind erosion was compounded by the need to conserve soil water during droughts. Use of forage grasses and maintenance of turfgrass was one of the major soil and water conservation practices (SWCPs) developed by the USDA Soil Conservation Service (SCS) to reduce soil erosion (Anderson et al. 1989; Bennett 1979). Under most conditions, forage grass and turfgrass are still considered best management practices (BMPs) for protection of surface water quality (Anderson et al. 1989; Walker et al. 1990).

By the 1960s, the primary issues regarding nonpoint source pollution centered on surface losses of soil, nutrients, and chlorinated organic pesticides. Agricultural, forestry, and urban areas all were identified as potential sources of chemical and sediment losses (Daniel et al. 1982). Eutrophication of surface water, sedimentation of streams and lakes, introduction of persistent pesticides in the global food chain, and identification of high levels of nitrates in drinking water were important environmental issues (Keeney 1982, 1983; Larson et al. 1983; Sharpley 1980; Smolen et al. 1984; Taylor and Kilmer 1980). In the 1970s, the focus of environmental research shifted from identification of pollution problems to development of best management practices (BMPs) to mitigate nonpoint pollution from agricultural and silvicultural sources (Anderson et al. 1989; Moore et al. 1980; Warrington et al. 1980).

The need to protect surface water quality, soil resources and productivity, and groundwater resources remains a serious environmental issue (Foster and Lane 1987; Larson et al. 1983). By the 1980s, and most likely throughout the 1990s, one of the major environmental issues and public concerns is contamination of both surface water and groundwater with nutrients and pesticides used in intensive management of agricultural and turfgrass systems (EPA 1986, 1988a; Keeney 1986; Petrovic 1990; Pratt 1985; Pye et al. 1983). The research and regulatory goals of the federal government are currently focusing on the issue of contamination of groundwater with nitrates and pesticides (EPA 1988a; Onstad et al. 1991; USDA ARS 1989a, 1989b). In a recent survey of drinking water wells in the United States, the EPA identified nitrates as the leading chemical contaminant (EPA 1990). Atrazine, bentazon, simazine, and DCPA were found in some of the samples. These pesticides are registered for use on turfgrass. A metabolite, or degradation product of DCPA, was the most commonly detected residue. Contamination of groundwater has emerged as one of the major environmental issues of the 1990s (EPA 1987).

Several articles in the turfgrass trade literature suggest that chemical management of turfgrass will not be a source of surface water or groundwater contamination if appropriate management practices are followed (e.g., Cooper 1987; Watschke et al. 1988, 1989). Similar claims were made for agricultural chemicals in the 1960s and 1970s (e.g., Caro 1976; Frere 1976). Unfortunately, contamination of susceptible surface water and groundwater resources with agricultural chemicals has occurred throughout the United States (Fairchild 1987; Keeney 1986; Smith et al. 1987). Protection of both the quantity and quality of drinking water supplies is a major environmental concern of the public, regulators, and ecosystem managers in the United States (EPA 1986). Hopefully, these issues will be resolved from the standpoint of golf courses by use of environmentally sound and integrated management practices for turfgrass.

One of the principal goals of the U.S. Environmental Protection Agency's (EPA) water resources strategy is to protect the quality of surface water and groundwater (EPA 1986). The EPA is currently developing and implementing a state level pesticide registration program to protect local and regional groundwater resources (EPA 1987). The EPA also has a direct research interest and regulatory control of chemical management for turfgrass and ornamentals in urban areas (Leslie and Metcalf 1989). The EPA and many members of the golf course industry recognize the need for cooperative development of ecologically sound management of fertility and pests. These practices are essential for management of quality turfgrass in an era of increasingly complex technical and regulatory issues. Cooperative research and development should be coordinated with the technical expertise of the golf course industry; regulatory and environmental perspectives of the EPA; and the research capabilities of academia, industry, and the private sector.

The development of strategies by government agencies, academia, and the private sector to mitigate adverse impacts resulting from management of biological systems reflects the increasing sophistication and integration of interdisciplinary skills. In agricultural systems, SWCPs initially were utilized to alleviate movement of sediment and sediment-bound chemicals to surface water (Bennett 1979; Stewart et al. 1975, 1976). As transport of soluble chemicals in runoff water and leaching of soil solution was identified as a potential threat to water quality, the concept of BMPs evolved (Krivak 1978). BMPs attempt to mitigate adverse environmental impacts, especially for surface water and groundwater quality.

BMPs are used in an attempt to reduce the adverse water quality and environmental effects of agricultural and forestry management systems. The goals of BMPs are

1. To reduce offsite transport of sediment, nutrients, and pesticides
2. To control the rate, method, and type of chemical being applied
3. To reduce total chemical loads by use of IPM, economic thresholds, alternate pest control options, and fertility testing
4. To use both biological and mechanical SWCPs
5. To educate managers and the public on the relationship of environmental issues and systems management

There is a tendency for BMPs to be single or limited options within existing management systems. Therefore, the use of BMPs has not necessarily been effective in mitigating all negative environmental impacts. For example, attempts to improve surface water quality using no-tillage systems in agriculture have been frustrated by potential increases in movement of soluble chemicals in both surface runoff and soil leachate (Hinkle 1983; Logan et al. 1987).

Integrated systems approaches are the latest and most complete attempt to solve the adverse environmental effects of agricultural, silvicultural, and turfgrass systems (Leslie and Metcalf 1989; USDA-ARS 1988a, 1988b, 1989a, 1989b; Walker et al. 1990). The integrated systems approach combines the concepts of systems analysis, use of IPM and economic thresholds, and use of an appropriate combination of SWCPs and BMPs. This approach incorporates the use of simulation models to identify potential problems prior to implementation of land management plans. Successful adoption of management systems to reduce detrimental environmental impacts could be implemented through a combination of voluntary and incentive programs. Development of integrated programs requires the interdisciplinary efforts of turfgrass scientists, breeders, agronomists, designers, and managers; environmental regulators; soil scientists; entomologists; hydrologists; and the concerned public.

Wetlands are an important natural resource in the United States. They serve as a vital and direct link between land and water resources. Wetlands are the dynamic areas that develop between open water and land. Marshes, swamps, and bogs are examples of the wetlands. In the past, these wet landscape features were considered wastelands that required elimination to become productive areas (Mitsch and Gosselink 1986). Over half of the original wetlands in the United States have been destroyed or degraded due to the early negative perception of wetlands (EPA 1988b; Salvesen 1990). This attitude changed significantly in the 1970s and 1980s. In the 1990s, protection of wetland resources promises to be a highly visible and sensitive environmental issue.

With an increased understanding of ecological processes, scientists and land use managers now recognize the vital importance of wetlands as a valuable natural resource. In their natural or restored state, wetlands provide essential fish and wildlife habitat, a source for water quality improvement, flood control, shoreline erosion control, natural products, and aesthetic and recreation areas (EPA 1988b; Godfrey et al. 1985; Kusler and Kentula 1990; Mitsch and Gosselink 1986). Many golf course developers and managers recognize the importance of wetlands as natural land features (Peacock et al. 1990; Salvesen 1990). Development of golf courses in popular urban and coastal zones requires the turfgrass and golf course industry to understand the ecological and regulatory implication of wetland management.

1.3 SCOPE OF THE TURFGRASS INDUSTRY

Excellent reviews on the scope and value of the turfgrass and golf course industry have been prepared by Cockerham and Gibeault (1985) and Watson et al. (1990). Turfgrass management for lawns, golf courses, and conservation has been extensively reviewed by Beard (1973, 1982) and Madison (1982). Although a review of the turfgrass industry is beyond the scope of this article, several factors concerning the turfgrass industry directly affect the impacts of turfgrass on the environment. These factors are summarized from the previously mentioned sources and from surveys conducted by the National Golf Foundation (1989, 1991).

There has been a tremendous growth in the turfgrass industry since 1945 (Cockerham and Gibeault 1985; Nutter and Watson 1969; Watson et al. 1990). By 1982, turfgrass had become a $25 billion industry (Roberts and Roberts 1989). California, Florida, Michigan, New York, Pennsylvania, Ohio, Illinois, and Texas currently dominate the turfgrass industry (Table 1). Maintenance of home lawns accounts for 68% of all

Table 1. Annual Cost of Turfgrass Maintenance by State in 1982

State	Home lawns	General lawns	Golf courses	Parks	Highways	Schools	Airports	Cemeteries	State total
				in millions of dollars					
AL	197.6	34.3	15.6	1.3	10.8	6.7	6.4	5.5	278.2
AK	15.1	2.5	0.5	0.9	0.8	0.5	0.5	0.4	21.2
AZ	350.6	60.1	21.6	22.4	18.9	11.7	11.2	9.7	506.2
AR	169.2	29.3	11.9	10.9	9.2	5.7	5.5	4.7	246.4
CA	648.9	128.0	124.2	47.7	40.1	24.9	23.9	20.6	1,058.3
CO	477.0	80.3	9.9	30.0	25.2	15.7	15.0	12.9	666.0
CT	467.5	81.8	39.4	30.5	25.6	15.9	15.2	13.2	689.1
DE	40.0	7.4	5.6	2.7	2.3	1.4	1.4	1.2	62.0
DC	40.0	6.7	1.7	2.5	2.1	1.3	1.2	1.1	56.6
FL	723.2	132.5	75.1	49.4	41.6	25.8	24.7	21.3	1,093.6
GA	331.7	61.4	45.2	22.9	19.2	12.0	11.4	9.9	513.7
HI	200.7	34.9	16.5	13.0	10.9	6.8	6.5	5.6	294.9
ID	15.1	3.3	4.3	1.2	1.0	0.6	0.6	0.5	26.6
IL	600.1	114.5	99.5	42.7	35.9	22.3	21.3	18.4	954.7
IN	552.8	64.6	37.4	35.5	28.3	18.6	16.2	15.3	768.7
IA	467.5	62.0	37.7	30.4	20.7	15.9	15.2	13.0	662.4
KS	458.1	78.7	29.2	29.4	24.7	15.3	14.7	12.7	662.8
KY	297.6	51.8	24.5	19.3	16.3	10.1	9.7	8.3	437.6
LA	303.2	50.7	11.5	18.9	15.9	9.9	9.5	8.2	427.8
ME	20.1	5.6	14.9	2.1	1.8	1.1	1.0	0.9	47.5
MD	514.9	86.3	18.3	32.2	27.1	16.8	16.1	13.9	725.6
MA	486.5	84.2	36.9	31.4	26.4	16.4	15.7	13.6	711.1
MI	675.9	119.7	60.8	44.6	37.5	23.3	22.3	19.3	1,003.4
MN	505.4	88.3	35.3	32.9	27.7	17.2	16.5	14.2	737.5
MS	269.2	45.4	11.2	16.9	14.2	8.8	8.5	7.3	381.5
MO	325.0	57.3	29.4	21.3	18.0	11.2	10.7	9.2	482.1
MT	15.1	3.1	4.1	1.2	1.0	0.6	0.6	0.5	26.2
NE	166.6	28.0	22.6	10.4	8.8	5.5	5.2	4.5	251.6
NV	60.0	10.9	7.6	4.0	3.4	2.1	2.0	1.7	91.7
NH	101.5	18.0	10.5	6.7	5.6	3.5	3.4	2.9	152.1
NJ	514.9	91.9	50.7	34.2	28.8	17.9	17.1	14.8	770.3
NM	107.0	18.5	6.8	6.9	5.8	3.6	3.4	3.0	155.0
NY	732.7	139.5	130.4	52.0	43.7	27.2	26.0	22.5	1,174.0
NC	337.3	60.6	39.2	22.6	19.0	11.8	11.3	9.8	511.6
ND	19.5	5.5	14.9	2.1	1.7	1.1	1.0	0.9	46.7
OH	590.6	106.0	63.8	39.5	33.2	20.7	19.8	17.1	890.7
OK	357.0	59.8	13.1	22.3	18.7	11.6	11.1	9.6	503.2
OR	467.5	77.4	10.6	28.9	24.3	15.1	14.4	12.5	650.7
PA	675.9	119.7	66.8	44.6	37.5	23.3	22.3	19.3	1,009.4
RI	200.7	33.9	7.4	12.6	10.6	6.6	6.3	5.5	283.6
SC	280.5	48.6	21.7	18.1	15.2	9.5	9.1	7.8	410.5
SD	19.5	5.6	15.1	2.1	1.8	1.1	1.0	0.9	47.1
TN	314.6	54.6	24.6	20.3	17.1	10.6	10.2	8.8	460.8

Table 1. Annual Cost of Turfgrass Maintenance by State in 1982 (continued)

State	Home lawns	General lawns	Golf courses	Parks	Highways	Schools	Airports	Cemeteries	State total
				in millions of dollars					
TX	647.5	112.0	40.6	41.8	35.1	21.8	20.9	18.0	937.7
UT	41.7	7.7	4.6	2.9	2.4	1.5	1.4	1.2	63.4
VT	19.5	4.4	7.8	1.6	1.4	0.9	0.8	0.7	37.1
VA	552.8	93.4	26.9	34.8	29.3	18.2	17.4	15.0	787.8
WA	514.9	85.8	15.1	32.0	26.9	16.7	16.0	13.8	721.2
WV	163.5	28.8	15.8	10.8	9.0	5.6	5.4	9.0	247.9
WI	533.8	93.5	39.5	34.9	29.3	18.2	17.4	15.1	781.7
WY	15.1	2.9	3.2	1.1	0.9	0.6	0.6	0.5	24.9
Total	16,602.6	2,881.7	1,481.5	1,081.4	912.7	571.2	545.0	476.3	24,552.4

Adapted from Cockerham and Gibeault 1985.

turfgrass expenditures in the United States. Maintenance of turfgrass in parks, schools, highways, cemeteries, and other general lawns accounts for another 26% of all expenditures for turfgrass. Only 6% of total maintenance costs are allocated to turfgrass on golf courses (Cockerham and Gibeault 1985).

Although turfgrass is typically the most intensively managed component of urban landscapes (Feucht and Butler 1988; Madison 1982), the total extent of lawns and sports turfgrass sites is difficult to quantify. Tashiro (1987) estimated that turfgrass covers 8.1–10.1 million ha (20–25 million ac), while Roberts and Roberts (1989) estimated that lawns occupy an area between 10.1–12.1 million ha (25–30 million ac) in the United States. In a typical city in the United States, the percent of total turfgrass acreage has been estimated as 70% for residential lawns; 10% for city parks and sports facilities, 9% for golf courses; 9 percent for educational facilities, 2% for churches and cemeteries, and 1% for industrial purposes (Cockerham and Gibeault 1985).

Some of the factors associated with the increased demand for turfgrass in the United States are the increase in time for leisure activities, a substantial increase in discretionary income, a desire for increased recreational and aesthetic facilities in urban environments, and an increase of population in urban environments. As the population continues to concentrate in urban areas, the extent and intensity of lawn care is expected to increase (Roberts and Roberts 1989). Increasing public demand for dense, uniform, and pest-free turfgrass has resulted in increased dependence on intensive chemical and irrigation management strategies on lawns and golf courses (Potter et al. 1989).

The popularity of golf has increased steadily in the United States since 1946 (Cockerham and Gibeault 1985; Watson et al. 1990) (Tables 2 and 3). From 1985 to 1990, the number of golfers in the United States increased from 17.5 to 25 million.

Table 2. The Increase in the Number of Golfers and Rounds of Golf Played in United States from 1958 to 1990

Item	Year				
	1958[a]	1963[a]	1968[a]	1985[b]	1990[c]
			in millions		
Total number of golfers	4.0	7.3	11.3	17.5	25.0
Number of rounds played	—	—	194	415	488

[a] *Nutter and Watson 1969.*
[b] *National Golf Foundation 1989.*
[c] *National Golf Foundation 1991.*

Table 3. The Increase in the Total Number of Golf Facilities in the United States from 1958 to 1990

Item	Year				
	1946[a]	1958[a]	1963[a]	1968[a]	1990[b]
			in thousands		
Total of golf facilities	4.8	5.7	7.5	9.6	12.8[b]
Total estimated acreage (ha)[c]	—	—	299.6	417.4	532.7[d]

[a] *, Nutter and Watson 1969; Heuber 1984.*
[b] *National Golf Foundation 1991, estimate includes Pacific region.*
[c] *1 ha = 2.471 ac.*
[d] *Estimate from Table 7.*

In the same time period, the number of rounds of golf played annually increased by 17.6% (Table 2). Between 1968 and 1990, the number of golf courses in the United States increased from approximately 9.6 to 14,000 (Table 3). The National Golf Foundation estimates that over 12% of the population of the United States over the age of 12 participates in the game of golf (Table 4). California leads the country in the number of golfers with 2.8 million (Table 4). Several other states including Florida, Illinois, Michigan, New York, Ohio, Pennsylvania, and Texas have over 1 million golfers. Despite the impression that golfers and golf courses are concentrated in California, the Southwest, and eastern coastal regions, half of the golfers in the United States live in the Northeast and Midwest (Tables 4 and 5). The greatest density of golfers and golf course facilities are located in the Southern and Middle Atlantic, East north central, and Southern Pacific areas (Figure 1, Table 5).

Table 4. Distribution of Golf Courses, Number of Rounds Played, Golfers, and Rounds of Golf Played by State in 1990

State[a]	Golf courses	Total annual rounds played (in thousands)	Golfers	Percent of total population[b]
AL	211	5,254	264	7.9
AZ	236	8,736	416	13.7
AR	151	4,404	109	5.6
CA	853	56,556	2,842	11.8
CO	187	7,567	409	14.5
CT	168	7,823	354	12.8
DE	28	832	52	9.3
DC	10	459	23	4.0
FL	1,011	44,518	1,374	12.3
GA	324	11,577	510	9.5
ID	82	1,963	130	17.0
IL	606	22,950	1,530	16.0
IN	387	13,355	639	13.8
IA	367	8,235	394	17.4
KS	243	5,270	288	13.8
KY	228	8,280	345	11.1
LA	148	5,105	201	5.6
ME	121	2,038	109	10.8
MD	146	7,134	392	10.0
MA	329	12,960	720	14.3
MI	749	22,532	1,273	16.9
MN	394	11,123	727	21.2
MS	145	3,047	110	5.3
MO	280	8,283	499	11.7
MT	75	1,047	88	13.6
NE	173	3,899	194	15.0
NV	55	2,348	118	13.0
NH	99	1,629	90	9.5
NJ	263	11,680	730	11.0
NM	82	3,779	171	13.6
NY	781	25,631	1,643	10.9
NC	474	12,119	609	11.0
ND	103	1,634	99	19.3
OH	704	25,713	1,375	18.7
OK	172	6,528	257	9.6
OR	159	4,452	291	15.3
PA	643	18,850	1,071	10.8
RI	48	1,962	93	10.9
SC	316	9,222	290	10.0
SD	110	2,349	77	13.0
TN	234	7,482	358	8.5

Table 4. Distribution of Golf Courses, Number of Rounds Played, Golfers, and Rounds of Golf Played by State in 1990 (continued)

State[a]	Golf courses	Total annual rounds played (in thousands)	Golfers	Percent of total population[b]
TX	756	32,036	1,497	10.5
UT	94	3,257	267	21.0
VT	60	720	55	11.6
VA	256	8,348	469	9.4
WA	240	6,834	510	13.4
WV	109	3,213	135	8.8
WI	417	11,766	769	19.3
WY	49	1,668	67	17.4
Total[a]	13,876	488,167	25,033	12.7[c]

From National Golf Foundation 1991.

[a] *Lower 48 States.*
[b] *Age 12+.*
[c] *Average.*

The extent of golf courses and the number of golfers using these facilities are concentrated in the major urban centers in the United States (Table 5, Figure 1). Expansion of golf courses in regions with dense urban populations often overlaps with areas with a history of problems concerning land use, water resources, and environmental quality. Concentration of golf courses in areas requiring extensive use of water resources and chemical management practices will increase the exposure of golf course developers and superintendents to environmental and water allocation concerns.

The number of golf courses in the United States has grown to accommodate the increased demand for facilities since 1945 (Nutter and Watson 1969) (Table 3). Currently, the National Golf Foundation estimates there are over 13,951 golf courses and 12,846 golf facilities in the United States (National Golf Foundation 1991). Assuming an average size of 50–73 ha (124–180 ac) per facility (Beard 1982; Watson et al. 1990), there are from 642 to 938 thousand ha (1.6–2.3 million ac) of land dedicated to golf facilities. This estimate includes all areas of golf course facilities, such as club houses and parking lots, that are not covered by turfgrass. Actual acreage of land managed for turfgrass on golf courses is less than the total area of the facility.

The average dimensions of turfgrass on greens, tees, fairways, and rough have been reported by Beard (1982) and the National Golf Foundation (1985) (Table 6). Estimates of actual turfgrass acreage were calculated on the basis of the median turfgrass dimensions (Table 6) and estimates of the number of golf holes by state (National Golf Foundation 1991). Approximately 533,000 ha (1.3 million ac) of

Table 5. Regional Distribution of Golf Courses in the United States

Region/states	Number of golfers[a]	Number of golf holes[b]	Number of golf courses[a]
New England/ CT, MA, ME, RI, NH, VT	1,421,000	11,538	825
Middle Atlantic/ NJ, NY, PA	3,444,000	25,083	1,687
South Atlantic/ DC, DE, FL, GA, MD, NC, SC, VA, WV	3,854,000	42,516	2,674
East south central/ AL, KY, MS, TN	1,077,000	11,448	818
West south central/ AR, LA, OK, TX	2,064,000	17,217	1,227
East north central/ IL, IN, MI, OH, WI	5,586,000	41,742	2,863
West north central/ IA, KS, MN, MO, NE, ND, SD	2,278,000	19,737	1,670
Mountain/ AZ, CO, ID, MT, NV, NM, UT, WY	1,666,000	12,402	860
Pacific/ CA, OR, WA	3,643,000	18,342	1,252
Alaska	—	99	8
Hawaii	—	1,089	67

[a] *Summarized from Table 4.*
[b] *Summarized from Table 7.*

land are managed for golf course turfgrass in the United States (Table 7). On an area basis, Florida and California lead the nation in the extent of turfgrass used for golf courses with 42.8 and 33.2 thousand ha (106,000 and 82,000 ac), respectively. Illinois, Michigan, New York, North Carolina, Ohio, Pennsylvania, and Texas each have dedicated over 20,000 ha (50,000 ac) of land for management of golf course turfgrass.

The intensity of construction and maintenance of golf course turfgrass decreases in order from the putting greens, tees, fairways, and the rough. Total annual use and the number of applications of fertilizers and pesticides decreases from the greens to the rough (Beard 1982). The proportion of golf course turfgrass subjected to the intensive management of green and tee areas is relatively small (6%) compared to the

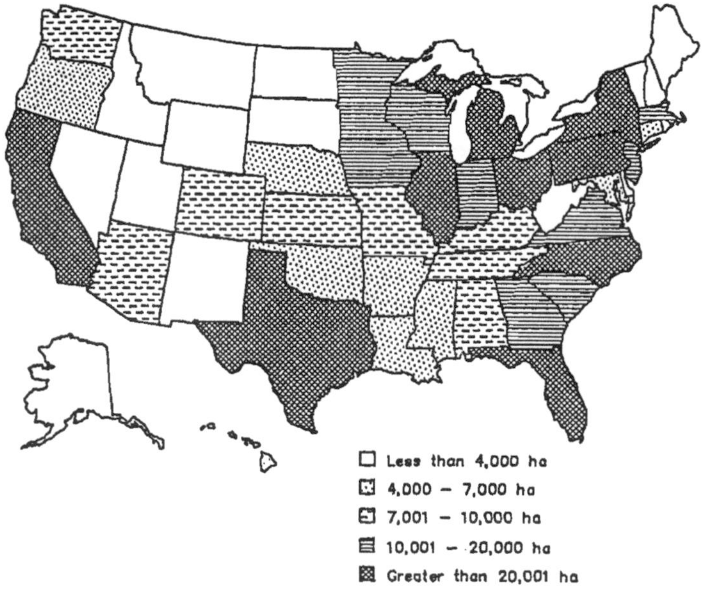

Fig. 1. Estimated distribution of turfgrass acreage on golf courses in the United States.

total acreage of golf course turfgrass (Table 7). This will have a direct effect on the extent of the environmental effects of turfgrass management.

In several scientific reviews, investigators have suggested that the extent of acreage subject to intensive water and chemical management is a better indicator of nonpoint source pollution and environmental problems than total amounts of chemicals or water used on specific sites (Anderson et al. 1989; Smolen et al. 1984). The spatial extent of fertilizer and pesticide application is more important than the total amounts of chemicals applied on a single site when evaluating the local and regional impacts of chemical management on water and environmental quality. The distribution of golf course turfgrass is concentrated in several regions of the United States (Figure 1, Table 5). These regions would be at greater risk of potential nonpoint impacts of golf course management than areas without heavy concentrations of golf courses. The impact of very intensive management, single heavy chemical applications, and accidental chemical spills on turfgrass could constitute a point source of contamination under certain conditions.

1.4 ENVIRONMENTAL BENEFITS OF TURFGRASS SYSTEMS

Although potential environmental risks have been identified for turfgrass management of lawns and golf courses, the overall benefits of turfgrass also should

Table 6. Median Dimensions of Turfgrass on Golf Courses

Turfgrass component	Range[a] (ha)	Median dimension by type of hole: Regulation (ha)	Executive[b] (ha)	Par-3[b] (ha)
Greens	0.011–0.255 (0.050–0.075)[d]	0.052[c]	0.045	0.035
Green surrounds (collar, bunkers, and grassy surrounds)	0.056–0.111	0.083[a]	0.060	0.060
Tee	0.010–0.067	0.020[c]	0.020	0.020
Fairways	0.667–1.333[e]	1.133[a]	0.800	0.667
Rough	0.889–2.000	1.444[a]	0.900	0.900
Average overall acreage[b,c,f]		9 holes — 25 ha 18 holes — 54 ha 1 ha = 2.471 ac		

[a] *Beard 1982.*
[b] *Estimates only.*
[c] *National Golf Foundation 1985.*
[d] *Most popular range.*
[e] *Range for 18-hole courses only.*
[f] *Total dimension includes nonturfgrass facilities.*

be considered (Klein 1990; Walker et al. 1990). Healthy turfgrass probably provides the greatest benefit to land surfaces in urban and suburban environments. Healthy turfgrass has the greatest resistance to disease, insect, and weed infestations (Beard 1982). Strong root systems use applied nutrients and water efficiently. Efficient use of nutrients and water limits the need for unnecessary irrigation, fertilizer, and pesticide applications. The benefits of turfgrass, in addition to the obvious values for homeowners, golf, and recreation, have been reviewed by Beard (1989), Roberts and Roberts (1989), and Walker et al. (1990).

One of greatest influences of turfgrass is on water resources. Research on grasslands, pastures, and turfgrass indicates that sediment and nutrient losses from these systems is considerably less than losses from agricultural systems (Tables 8–12). Turfgrass is capable of (1) entrapping and retaining large quantities of precipitation; (2) reducing the rate and volume of surface runoff; (3) reducing sediment losses and surface losses of residual pesticides and nutrients in the soil; and (4) scavenging nitrate and increasing infiltration, therefore acting as a nitrate filter and source of groundwater recharge. However, much of the research and monitoring studies on the

Table 7. Estimate of Turfgrass Acreage by Golf Courses by State[a]

State	Total holes[b]			Estimated acreage of turfgrass in hectares[c]					
	Reg.	Exec.	Par-3	Green	Green surrounds	Tee	Fairway	Rough	Total of turfgrass
AL	2,943	54	54	157.4	250.7	61.0	3,413.6	4,346.9	8,229.6
AK	90	9	0	5.1	8.0	2.0	109.2	138.1	262.4
AZ	3,006	621	117	188.4	293.8	74.9	3,980.6	5,004.9	9,542.6
AR	1,962	27	27	104.2	166.1	40.3	2,262.6	2,881.7	5,454.9
CA	10,710	1,341	891	648.5	1,022.9	258.8	13,801.5	17,474.0	33,205.7
CO	2,475	126	135	139.1	221.1	54.7	2,995.0	3,808.8	7,218.7
CT	2,412	63	81	131.1	208.8	51.1	2,837.2	3,612.5	6,840.7
DE	432	18	18	23.9	38.0	9.4	515.9	656.2	1,243.4
DC	126	27	0	7.8	12.1	3.1	164.4	206.2	393.6
FL	13,851	2,016	747	837.1	1,315.4	332.3	17,804.2	22,487.5	42,776.5
GA	4,626	63	90	246.5	393.1	95.6	5,351.7	6,817.6	12,904.5
HI	1,026	27	36	55.8	88.9	21.8	1,208.1	1,538.2	2,912.8
ID	1,035	36	45	57.0	90.8	22.3	1,231.5	1,567.4	2,969.0
IL	8,019	387	288	444.5	706.1	173.9	9,587.2	12,186.9	23,098.6
IN	5,148	189	360	288.8	460.2	113.9	6,224.0	7,927.8	15,014.7
IA	3,843	117	99	208.6	331.9	81.2	4,513.8	5,743.7	10,879.2
KS	2,736	81	45	147.5	234.6	57.2	3,194.7	4,064.2	7,698.2
KY	2,826	99	135	156.1	248.6	61.2	3,371.1	4,291.3	8,128.3
LA	2,043	9	18	107.3	171.2	41.4	2,323.9	3,494.2	6,138.0
ME	1,332	36	81	73.7	117.6	29.0	1,592.0	2,028.7	3,841.0
MD	2,169	90	54	118.7	188.7	46.3	2,565.5	3,261.6	6,180.8
MA	4,293	126	342	240.9	384.4	95.2	5,192.9	6,620.3	12,533.7
MI	10,098	522	396	562.4	893.2	220.3	12,122.8	15,407.7	29,206.4
MN	4,212	450	351	251.6	397.7	100.3	5,366.3	6,803.0	12,918.9

MS	1,881	0	54	99.7	159.4	38.7	2,167.2	2,764.8	5,229.8
MO	3,456	63	117	186.6	297.6	72.7	4,044.1	5,152.5	9,753.5
MT	864	0	45	46.5	74.4	18.2	1,008.9	1,288.1	2,436.1
NE	1,710	144	117	99.5	157.6	39.4	2,130.7	2,704.1	5,131.3
NV	801	45	9	44.0	69.7	17.1	949.5	1,205.2	2,285.5
NH	1,089	45	126	63.1	100.6	25.2	1,353.9	1,726.4	3,269.2
NJ	3,834	162	171	212.6	338.2	83.3	4,587.6	5,836.0	11,057.7
NM	1,053	27	36	57.2	91.2	22.3	1,238.7	1,577.2	2,986.6
NY	10,170	576	468	571.1	906.8	224.3	12,295.6	15,625.1	29,622.9
NC	7,326	144	189	394.0	628.0	153.2	8,541.6	10,878.4	20,595.2
ND	981	45	27	54.0	85.7	21.1	1,165.5	1,481.4	2,807.7
OH	9,702	567	405	544.2	863.6	213.5	11,716.1	14,884.5	28,221.9
OK	2,268	27	27	120.1	191.5	46.4	2,609.3	3,323.6	6,290.9
OR	1,764	234	99	105.7	166.4	41.9	2,251.8	2,846.9	5,412.7
PA	8,910	315	477	494.2	787.1	194.0	10,665.2	13,578.8	25,719.3
RI	675	0	27	36.0	57.6	14.0	782.8	999.0	1,889.4
SC	4,833	63	171	260.1	415.2	101.3	5,640.2	7,189.5	13,606.3
SD	1,089	36	18	58.9	93.6	22.9	1,274.6	1,621.1	3,071.1
TN	3,312	36	54	175.7	280.3	68.0	3,817.3	4,863.5	9,204.8
TX	10,350	162	297	555.9	886.6	216.2	12,054.2	15,358.5	29,071.4
UT	1,152	54	81	65.2	103.7	25.7	1,402.4	1,785.0	3,382.0
VT	765	27	18	41.6	66.2	16.2	900.4	1,145.2	2,169.6
VA	3,807	72	99	204.7	326.2	79.6	4,437.0	5,651.2	10,698.7
WA	2,871	198	234	166.4	264.2	66.1	3,567.3	4,534.5	8,598.5
WV	1,341	45	99	75.2	119.9	29.7	1,621.4	2,066.0	3,912.2
WI	5,121	252	288	287.7	457.4	113.2	6,195.8	7,880.7	14,934.8
WY	630	9	0	33.2	52.8	12.8	721.0	917.8	1,737.6
Total	183,168	9,882	8,163	10,255.1	16,285.4	4,024.2	220,869.8	281,254.4	532,688.9

[a] *Estimates of acreage based on the median dimensions of turfgrass areas on golf courses (Table 6) and the distribution of the types of golf course holes by state.*
[b] *The distribution of golf course holes is available from estimates of the number of regulation (Reg.), executive (Exec.), and par-3 holes (National Golf Foundation 1991).*
[c] *1 ha = 2.471 ac.*

Table 8. Estimated Pollutant Contributions to Surface Waters from Selected Nonpoint Sources in the Contiguous 48 States

Source	Sediment	BOD[a]	Nitrogen	Phosphorus	Acids	Salinit
	Average load (million tons $year^{-1}$)					
Cropland	1870	9	4.3	1.56	—	57.3
Pasture and range	1220	5	2.5	1.08	—	—
Forest	256	0.8	0.39	0.089	—	—
Construction	197	—	—	—	—	—
Mining	59	—	—	—	3.1	—
Urban runoff	20	0.5	0.15	0.019	—	—
Rural roadways	2	0.004	0.0005	0.001	—	—
Small feedlots	2	0.05	0.17	0.032	—	—
Landfills	—	0.3	0.026	—	—	—
Subtotal	3626	15.8	7.4	2.8	3.1	57.3
Natural background	1260	5.0	2.5	1.1	—	—
Total	4886	20.8	10.0	3.8	3.1	57.3

From Koehler et al. 1982a.

[a] *Biological oxygen demand.*

Table 9. Comparative Magnitude of Nonpoint Sources

Source in surface runoff	Total nitrogen[a]		Total phosphorus[a]	
	mg L^{-1}	kg ha^{-1} $year^{-1}$	mg L^{-1}	kg ha^{-1} $year^{-}$
U.S. precipitation	0.73–1.27	5.6–10	—	0.05–0.10
Lower limit for algal blooms	—	—	0.025	—
Maximum level— domestic water supply	10	—	—	—
Precipitation — Ohio	2.0–2.8	12.8	—	—
Forest — Ohio	0.54–0.89	2.1	0.011–0.020	0.04
Farmland — Ohio	0.91–3.11	5.1	0.020–0.023	0.06
Precipitation — coastal Delaware	—	44.6–45.4	—	1.45–1.48
Agricultural watersheds — coastal Delaware	—	14.4–15.7	—	0.39–0.46
Precipitation — Minnesota	—	—	0.011–0.042	0.10
Forest — Minnesota	—	—	0.04–1.2	0.08

Table 9. Comparative Magnitude of Nonpoint Sources (continued)

Source in surface runoff	Total nitrogen[a]		Total phosphorus[a]	
	mg L^{-1}	kg ha^{-1} $year^{-1}$	mg L^{-1}	kg ha^{-1} $year^{-1}$
Upland Native Prairie — Minnesota	—	1.0	—	0.13
Grassland (112 kg N ha^{-1}) North Carolina	—	2.3	—	—
Grassland (44 kg N ha^{-1}) North Carolina	—	8.4	—	—
Grassland (rotate graze) Oklahoma	1.52–1.64	1.5	0.56–0.83	0.89
Grassland (continuous graze) Oklahoma	2.58–3.25	6.8	1.29–1.32	3.24
Corn (204 kg N ha^{-1}) coastal Georgia	0.17–0.43[b]	0.1–0.2[b]	—	—
Corn (204 kg N ha^{-1})[c] coastal Georgia	7.07–10.31[b]	12.4–25.8[b]	—	—
Silvicultural piedmont — Virginia	1.1–1.8	2.7	0.12–0.19	0.28
Agricultural piedmont — Virginia	1.1–3.2	4.4	0.10–0.60	0.54
Poorly drained coastal plain — Virginia	1.7–2.3	1.6	0.19–0.31	0.21
Well–drained coastal plain — Virginia	1.5–4.1	4.9	0.41–0.65	0.88

From Koehler et al. 1982a.

[a] *Normalized to precipitation of 76 cm $year^{-1}$.*
[b] *Nitrate-nitrogen.*
[c] *Subsurface flow.*

Table 10. Range of Nitrogen Concentrations Observed from Nonpoint Sources

Source	Range in concentration (mg L^{-1})	
U.S. Precipitation	1–3	(Total N)
Forest surface runoff	<0.5–2	(Total N)
Grassland surface runoff	1–8	(Total N)
Surface runoff from fertilized cropland	<0.5–92	(NO_3^-)
Tile drain effluent from fertilized cropland	2–40	(NO_3^-)

From Koehler et al. 1982a.

Table 11. Range of Total Phosphorus Concentrations Observed from Nonpoint Sources

Source	Range in concentration (mg L^{-1})
U.S. Precipitation	<0.025–0.05
Forest surface runoff	0.030–1.10
Grassland surface runoff	0.030–1.55
Surface runoff from fertilized cropland	<0.025–1.40
Tile drainage from fertilized cropland	<0.025–1.55

From Koehler et al. 1982a.

Table 12. Average Annual Sediment Losses from Selected Soils Under Different Management Systems

	Management system				
Soil texture/	Fallow	Continuous cropping	Rotation	Grass	Forest
Percent slope	Annual sediment loss (tons of soil ac^{-1})				
Loam/4	41.6	19.7	2.7	0.3	—
Loam/8	—	—	—	—	<0.1–0.4
Silt loam/8	112.8	85.5	11.4	0.3	—
Silt loam/16	151.9	84.1	25.3	<0.1	—
Fine sandy loam/8	20.3	28.1	5.5	<0.1	—
Sandy loam/4	—	—	—	—	<0.1
Sandy clay loam/10	64.7	25.8	10.8	<0.1	—

From Bennett 1939; Watson 1985; Balogh et al. 1990.

ability of turfgrass to shield groundwater from contamination should be examined with extreme care. Conclusive data to support all claims for water quality improvement by turfgrass are not necessarily available, especially concerning nutrient and pesticide transport. Research on turfgrass, erosion control, and water quality benefits are discussed in Chapters 2–4 of this book.

Irrigation of turfgrass with effluent water is practiced on at least 10% of the golf courses in the United States (Roberts and Roberts 1989; Payne 1987). Turfgrass utilizes eutrophying nutrients such as nitrogen and phosphorus, filters sediments, and absorbs heavy metals (Schuler 1987; Harivandi 1982). These contaminants would otherwise require expensive effluent treatment or ultimately be transported to surface water and groundwater systems. Use of effluent water reduces the need to use potable water for irrigation of golf courses. This is a critical benefit for maintenance of golf courses in semiarid and arid climatic zones.

Recycling and disposal of sewage sludge on turfgrass systems is another beneficial conservation practice (Sommers and Giordano 1984). The sludge provides fertilizer elements for golf courses, while reducing potential groundwater contamination at waste disposal sites (Leslie 1989). Sewage sludge may be contaminated with heavy metals and other contaminants; therefore, it is important to mention that no food should be grown in the vicinity of turfgrass systems receiving sludge applications.

In addition to the water quality benefits, turfgrass beneficially reflects solar radiation and heat in comparison to other urban land surfaces (Beard and Johns 1985; Johns and Beard 1985). Turfgrass has a positive influence on the dissipation of solar energy in urban environments. Concrete and asphalt absorb high amounts of solar radiation, which is dissipated to the surrounding area as infrared radiation (Rosenberg 1974). This infrared radiation increases surface air temperatures. In contrast, turfgrass has lower reflectance and dissipates a greater amount of absorbed energy through transpiration (Beard 1973). This results in reduced heating of objects and air in the vicinity of turfgrass. Actively growing turfgrass reduces surface air temperatures by as much as 30–40°F compared to bare soil. Temperatures in the vicinity of Palm Springs, CA, are 2°F lower than adjacent areas (Anonymous 1988a). One theory attributes this beneficial phenomenon to the substantial acreage of golf courses in the area compared to the desert sand and concrete present in other areas.

In a small lysimeter study of irrigated Kentucky bluegrass, Feldhake et al. (1984) observed an increased canopy temperature of 1.7°C for each 10% decrease in soil moisture status up to 70% of field capacity. Decreased irrigation of urban lawns could have a negative effect on microclimate cooling by turfgrasses. Reduced irrigation of residential lawns may result in increased energy expenses for air conditioning.

Beard and Johns (1985) observed that during peak afternoon temperatures turfgrass surfaces were 9–12°F cooler than concrete and 36°F cooler than asphalt. Although this temperature differential dissipates with height, significant temperatures differences are still apparent 100 cm above the surface. Based on temperature profiles above a bermudagrass surface and bare soil, Johns and Beard (1985) calculated the energy savings resulting from reduced cost of residential air conditioning. An average savings of $4.00 per month was estimated for a 223 m^2 house surrounded on all sides by 3 m of irrigated turfgrass compared to a similar structure surrounded by bare soil. This energy savings is quite substantial when an entire community is considered. However, the cost-benefit analysis conducted by Johns and Beard (1985) did not include the indirect energy costs for mowing, irrigation, fertilization, and pest control.

In an energy audit of residential landscape features, Parker (1982) observed that on an area basis a lawn is less energy intensive than either shrubs or a 5-year-old tree.

It has been a common assumption in design of energy efficient landscapes that turfgrass is the most energy intensive sector. Parker (1982) analyzed the total energy cost of fertilization, pest control, irrigation, pruning or clipping, and use of tools on a residential landscape. He calculated that replacement of lawns with trees and shrubs would significantly increase the indirect fossil fuel inputs required for the converted vegetation. The high energy inputs associated with preventative pest control of shrubs or small trees is the major expense factor. The major energy costs associated with maintenance of residential turfgrass are for irrigation, mowing, pest control, and fertilization (Falk 1976; Parker 1982).

To a limited extent, grass absorbs atmospheric contaminants without sustaining permanent injury. Turfgrass absorbs ozone, carbon dioxide, and hydrogen fluoride and gives off oxygen (Beard 1973; Roberts and Roberts 1989). There have been claims that turfgrass filters other components of smog including smoke, dust, and fly ash (Beard 1973). Although respiration of turfgrass aids in "cleaning" various atmospheric pollutants from the air, absorption of these powerful chemical free radicals, reducing agents, peroxacetyl nitrate, and atmospheric sulfur dioxide may result in varying degrees of injury to turfgrass (Amthor and Beard 1982; Beard 1973; Richards et al. 1980; Youngner and Nudge 1980). Warm-season grasses exhibit greater tolerance to ozone than many of the cool-season grasses (Richards et al. 1980; Youngner and Nudge 1980). Considering only the cool-season grasses, "Merion" Kentucky bluegrass exhibited the greatest tolerance to ozone damage and annual bluegrass sustained the greatest damage (Richards et al. 1980). Most turfgrass acreage in the United States is located in the large metropolitan areas. The gases associated with these "smoggy" regions can destroy chlorophyll and result in the interruption of photosynthesis, interference with respiration, and restriction of growth (Beard 1973).

There has been considerable attention recently on the effects of increased atmospheric CO_2 on regional ecosystems, carbon storage in relation to global climate warming, and associated responses of plants (Allen 1990; Harmon et al. 1990). Research based on simulated integration of energy budgets, extent of production, and CO_2 absorption and storage characteristics could be used to assess the potential benefits of turfgrass on the global carbon balance. Assimilation of CO_2 by turfgrass has been studied in relation to photosynthesis and carbohydrate production (Mehall et al. 1984), mowing (Mehall et al. 1984), and irrigation (Peacock and Dudeck 1984). Although the rates of CO_2 absorption are known for some turfgrasses, research has not been published on the impact of turfgrass production, on CO_2 absorption, CO_2 storage, and global climate warming.

Turfgrass also provides excellent recreational surfaces in urban landscapes. On recreational surfaces, such as playgrounds and parks, turfgrass can minimize injury and provide a resilient playing surface. Glare and noise reduction are other positive attributes of turfgrass. On expressway banks, turfgrass can reduce traffic noise twice as much as cement embankments (Roberts and Roberts 1989).

Well-maintained lawns increase the property value of homes and neighborhoods. Property surrounding golf courses often has a higher appraised value than other areas. Of the resorts and year-round living areas built, about 50% of them are incorporating golf courses in their development. The homeowners who later sell their property find that it has increased significantly in value (Anonymous 1988b).

Turfgrass creates an "oasis" effect in urban areas, providing living greenery and aesthetic value to communities and residences (Roberts and Roberts 1989; Ulrich 1986). Turfgrass provides an excellent aesthetic alternative to blacktop (asphalt), concrete, and other artificial surfaces. Golf courses attract wildlife and provide a sense of serenity and relaxation to urban and suburban people. The benefit of golf courses as wildlife habitats is discussed in greater detail in Chapter 6. Urban and roadside turfgrass also reduces muddy areas and sources of dust. Turfgrass provides a green carpet which forms the basis for landscaping homes, office buildings, and industrial complexes.

Ulrich (1984, 1986) observed that vegetated landscapes have a positive psychological and physiological effect on inhabitants of urban and suburban areas. Ulrich (1986) reviewed the aesthetic, emotional, and physiological response of humans to vegetated landscapes in urban environments. In a comparison of urban and unspectacular vegetated natural views, North American and European adults exhibited a strong preference for natural vistas. Views of landscaped areas in urban settings have a positive effect on emotional and physiological status of these adults. The greatest effect of vegetation in urban environments is on individuals experiencing stress or anxiety (Ulrich 1981). Patients recovering from surgery had fewer complaints and required less postoperative analgesics when they had a view of a small natural landscape as opposed to a view of a brick wall (Ulrich 1984).

Ulrich (1986) found that preferences for unspectacular natural scenes should be comparatively high if

1. The complexity or number of vegetated elements in the scene is moderate to high.
2. The complexity of the scene is structured with a focal point or other organized pattern.
3. The scene has a moderate to high level of clearly defined depth.
4. A deflecting or curving sightline conveys the impression of a new landscape beyond the immediate visual boundary.
5. The ground surface has an even or uniform texture that is relatively smooth and gives an impression favorable to movement.

These high-preference views are described as park-like or savanna-like with a "civilized" assembly of different vegetation. Urban landscapes with a combination of smooth ground cover, such as turfgrass, scattered trees and shrubs, and open scenes, are considered the most highly preferred view (Hayward and Weitzer 1984; Ulrich 1986). The addition of water bodies to this park-like setting greatly enhances the preference of landscaped areas. The criteria for the most highly desirable landscapes in urban areas are all met by well-designed golf courses, parks, and lawns. A combination of turfgrass, trees, shrubs, and aquatic zones (streams, ponds, and water hazards) will improve the aesthetic quality, energy budgets, and maintenance costs for urban landscapes.

Kaplan and Kaplan (1989) recently summarized research on the psychological and sociological benefits of "nearby nature." Nearby nature is defined as parks, wooded areas, and large landscaped sites. Kaplan and Kaplan (1989) point out that in urban areas, nearby natural areas are used both for an appreciation of natural beauty as well as an opportunity to actively use the facility. The extent, or size of an individual facility, did not appear to be as important as the proximity of the study subjects to nearby nature. The authors observed an increased sense of residential satisfaction and increased sense of general well-being when nearby nature facilities were available. Higher rental and home values reflect that preference. They found that neighborhood and life satisfaction was higher if nature was actively used, such as gardening. Also, they found that local residential satisfaction was second, after marital role satisfaction, in predicting life satisfaction. Access to nature at the workplace was associated with lowered levels of perceived job stress. Employees were more satisfied, had fewer general ailments, and fewer headaches if they had views of "nearby nature."

1.5 IDENTIFICATION OF ENVIRONMENTAL IMPACTS AND NONPOINT POLLUTION PROBLEMS IN TURFGRASS SYSTEMS

The potential detrimental effects associated with golf course construction and maintenance have been identified as the following:

1. Contamination of surface water with sediment and nutrients during turfgrass construction
2. Potential contamination of runoff water and groundwater with applied nutrients and pesticides
3. Development of pest populations with increasing resistance to chemical control
4. Potentially negative impacts of chemical management on beneficial soil and nontarget organisms

5. Potentially toxic effects of applied chemicals to nontarget plants and animals
6. Excessive use of water resources during drought conditions and in semiarid and arid climatic zones
7. Loss or degradation of wetland resources during construction and turfgrass maintenance

One of the primary perceived hazards of construction and turfgrass management of golf courses is the risk of adverse effects on nontarget organisms through exposure to both sediment and chemicals. Regulatory and environmental issues related to golf course construction and management is another major problem for golf course managers and developers (Anonymous 1990). There currently is a consensus among many land use regulators, environmental scientists, and turfgrass specialists that scientific and technical research regarding the ecological impacts of intensively managed turfgrass is in its initial stages (Gold et al. 1989; Potter et al. 1989; Petrovic 1990).

The controversy concerning the environmental and water quality impacts of intensively managed turfgrass has emotional, philosophical, and technical elements (e.g., Cooper 1987; Erickson and Kuhlman 1987). The emotional concerns evolved from past and current environmental issues include the following:

1. Massive accelerated erosion in agricultural, forest, and urban environments (Daniel et al. 1982; Sweeten and Reddel 1978; Walter et al. 1979)
2. Eutrophication of surface waters (Uttormark et al. 1974)
3. Occurrence of persistent pesticides at toxic, subacute, and chronic levels in birds and fish (Madhun and Freed 1990)
4. Past use of chemical defoliants during the Vietnam War (CAST 1975)
5. Environmental release of phenoxy herbicides contaminated with dioxin (Bovey and Young 1980)
6. Potentially adverse effects on nontarget organisms resulting from intentional and accidental release of chemicals into the environment (Albert 1987; Cantor et al. 1988; Cookson 1979; Fiore 1987)
7. Suspected impacts of agricultural and industrial chemicals on drinking water supplies and human health (Anderson et al. 1989; Madhun and Freed 1990)
8. Loss of wetlands from drainage of agricultural land, urban development, industrial development, and discharge of pollutants (Salvesen 1990)

Many of these issues are not directly related to turfgrass management. However, these issues have a direct impact on public perception and reaction to all forms of intensive management including turfgrass culture on golf courses and lawns.

The second element of the controversy, closely interwoven with the emotional issue, is the philosophical or policy consideration. This aspect of environmental issues involves questions regarding the following:

1. What are the appropriate measures of risk and safety?
2. When does science know enough about a particular practice in order to proceed safely?
3. Who will make judgments regarding risk and safety?
4. How is absolute safety or "no-risk" practices to be proven?
5. What constitutes an adequate ratio of risk to benefit?
6. How should "imposed" vs "accepted" risks be evaluated?

The question of imposed vs accepted risk is particularly troublesome to those concerned with public policy and assessment of environmental risk (Starr and Whipple 1980; Zeckhauser and Viscusi 1990). Some individuals will "accept" substantial personal risks, such as smoking. However, these same individuals may choose to reject exposure to a risk situation over which they have no control, if the risk is perceived to be high enough (Zeckhauser and Viscusi 1990).

The technical aspects of the environmental controversy involve the techniques and practices used for intensive management of turfgrass. Some of the technical issues concerning intensive management of turfgrass include (1) extent of nutrient and pesticide applications, (2) chronic and acute toxicity to nontarget organisms, (3) potential for exposure of nontarget organisms to applied chemicals, (4) use of increasingly scarce water resources for irrigation, (5) potential off-site movement of fertilizers and pesticides, (6) effects of maintenance and storage facilities on soil and water quality, and (7) potential loss and effects on wetlands resulting from construction and turfgrass maintenance.

Technical information on the environmental impacts of turfgrass management has been used by both opponents and proponents of intensive turfgrass management to substantiate an emotional or policy position (Cooper 1987; Grant 1987; Klein 1990; Raupp et al. 1989; Wilkinson 1989). The primary goal of this book is to present a concise review of the technical issues in an unbiased manner. Previous experience with public controversy over intensive management of agricultural and silvicultural systems has shown that land managers need to respond promptly and objectively to all elements of public concern (Anderson et al. 1989; Balogh et al. 1990). A balanced perspective on potential risks and benefits to the environment will promote successful resolution to perceived and real issues.

Water resource issues in relation to irrigation, pest management, and fertilization are a particular concern for turfgrass managers, the public, and environmental regulators. As previously mentioned, nonpoint pollution of water resources and the perceived consumptive use of water on golf courses is under close public scrutiny. Nonpoint pollution refers to contamination of surface water and groundwater from diffuse sources. The following criteria define nonpoint source pollution (Daniel and Schneider 1979).

1. Occurrence is over extensive areas and transport is by surface runoff (solution and sediment) and subsurface percolation
2. Entrance to surface water and groundwater is diffuse and usually by intermittent flow
3. Monitoring the point of origin is difficult and pollutants are usually not traceable to their exact source
4. Magnitude of contamination is related to variable climatic processes

The extent of consumptive use and nonpoint pollution of water resources is spatially and temporally variable (van der Leeden et al. 1990). Variations depend on soil distribution, climate, topography, subsurface geology, land use, land management strategies, and intensity of chemical applications. In general, the intensity of land use is directly proportional to the potential for water quantity and quality problems. The major mechanisms for nonpoint pollution are surface runoff water (solution and particulate) and subsurface flow (Anderson et al. 1989; Daniel and Schneider 1979; Petrovic 1990; Pye and Patrick 1983). The greater the volume of water flux, either annual or individual event, the greater potential for nonpoint source loading (Gold et al. 1989; Welterlen et al. 1989). Major precipitation events within a year may be responsible for the greatest portion of contaminant transport (Anderson et al. 1989).

Although pollution of surface water and groundwater has numerous sources, recent studies suggest that both urban and rural land management practices make a significant contribution to water quality problems (Tables 8–12) (Daniel et al. 1982; Canter 1987; Leonard 1990; Schuler 1987). Major contributors to pollution in runoff not attributed to a single point source, the nonpoint sources, are sediments, nutrients, and pesticides (Erickson and Kuhlman 1987; Leonard 1990; Stewart et al. 1976; Watschke et al. 1989). Phosphorus and nitrogen are the fertilizer elements most often linked with nonpoint source pollution. Eutrophication and subsequent deterioration of surface water quality is associated with phosphorus contamination of surface waters (Uttormark et al. 1974). Pesticides, nitrate-nitrogen, and hydrocarbon-contaminated soils are the chemicals of concern in regard to potential contamination of groundwater (Daugherty 1991; Grant 1987; Petrovic 1990; Pratt 1985). All of these chemicals are associated with management of turfgrass on golf courses.

Public concern over urban and rural nonpoint source pollution of water resources has intensified with increased reliance on use of chemicals and irrigation in agricultural, silvicultural, and turfgrass areas (Anderson et al. 1989; Cooper 1987). Unmistakable links have been made between deterioration of environmental quality and intensive management practices in agriculture (EPA 1988a; Leonard 1988, 1990; Pratt 1985; Stewart et al. 1975, 1976). Sediment and nutrient transport resulting from urban construction sites also has been identified as a serious source of nonpoint source pollution (Daniel et al. 1979; Kuo et al. 1988; Schuler 1987). Heavy applications of organic and inorganic fertilizers and pesticides have the potential for movement to surface waters, as well as subsurface movement to groundwater (Anderson et al. 1989; Keeney 1982, 1986; Morton et al.1988; Uttormark et al. 1974). A complex array of agricultural and industrial compounds have been identified in surface and groundwater. These chemicals include fertilizers, pesticides, petroleum products, and industrial solvents.

The public, including urban and rural communities, is increasingly aware of water resource and environmental quality issues. Although the extent of surface water and groundwater deterioration has not been well characterized, it is currently apparent that water resources are increasingly under threat of both depletion and contamination (Canter 1987; Keeney 1982, 1986; Leonard 1988; Pye and Patrick 1983; van der Leeden et al. 1990; Watson 1985). Research efforts on the environmental impacts of turfgrass management have been only recently initiated in realistic turfgrass systems (Leslie and Metcalf 1989; Petrovic 1990; Walker et al. 1990). Past research on the environmental impacts of intensely managed systems has focused on agriculture and silviculture (e.g., Anderson et al. 1989; Balogh et al. 1990; Keeney 1986). Research directed specifically to identification of environmental impacts and water quality issues related to turfgrass is not as extensive (Leslie 1989).

Loss of valuable wetland resources is recognized as a serious threat to environmental quality in the United States. Wetlands have been drained to create additional farmland, filled for urban and industrial development, and used for disposal of urban and industrial waste. Dredging, diking, alteration of streams and rivers, overgrazing, pollution discharges, and mining are also responsible for wetland losses. Natural losses of wetlands may result from erosion, subsidence sea level rise, droughts, storms, and overgrazing by wildlife. Agricultural drainage of wetlands is responsible for 87% of wetland losses between the 1950s and 1970s (Salveson 1990). Urban development was responsible for 5% of the losses in the similar period. However, urban development in coastal zones is responsible for considerable losses of coastal and riparian wetlands (Salveson 1990).

Protection of wetland resources by many state and federal agencies combines the approaches of acquisition, economic incentives, and regulation. The current national trend in relation to land development is no net loss of wetlands. Section 404 of the Clean Water Act establishes many of the federal regulatory programs. Golf course developers with wetlands on or adjacent to a proposed site will be subject to wetland permit processes and environmental criteria (Salveson 1990). At the federal level, the

EPA, Army Corps of Engineers, and the Fish and Wildlife Service all have established wetland regulatory criteria and responsibilities. A recent survey conducted by the American Society of Golf Course Architects reports that many companies and courses have had difficulty obtaining permits for projects resulting from environmental concerns by regulatory agencies (Anonymous 1990). Fifty-six percent of the firms in the survey reported wetlands issues as the primary problem. Other difficulties in the permit process included habitat loss, chemical use and contamination, and groundwater protection. Understanding the importance of wetland, water quantity, and water quality protection are vital for successful construction and maintenance of turfgrass and golf course facilities (Dye 1989; Peacock et al. 1990). Published research on wetlands in relation to turfgrass and the golf course industry is very limited.

1.6 BENEFITS OF IMPLEMENTING INTEGRATED TURFGRASS MANAGEMENT STRATEGIES

Golf course and turfgrass managers are caught in the dilemma of maintaining cost-effective operations and meeting user demands, while sharing public concern over environmental effects, effects on water supplies, and potential health hazards. Many land use and turfgrass managers perceive limited alternatives to chemical management options, adopted on the recommendation of both university and industry scientists. Regulation of chemical management practices may temporarily frustrate an industry faced with increasing financial risks (EPA 1987; Wilkinson 1989).

The use of fertilizers (organic and inorganic), pesticides, and irrigation is considered necessary for economically successful golf courses and turfgrass systems (Beard 1982; Watson et al. 1990). Sound management of water, nutrients, and pests involves practices designed to retain the applied chemicals onsite and within the soil root zone. Protection of surface water, groundwater, and wetland resources has been the primary issue in recent research and regulatory reports (EPA 1987; Lee and Nielsen 1987; Pratt 1985; Salveson 1990).

The cost of remedial cleanup of major water resources and restoration of disturbed wetland is prohibitive. Prevention of adverse environmental impacts, such as surface water and groundwater contamination, by implementation of rational water, nutrient, and pesticide practices is a cost-effective measure. Avoidance or protection of wetlands prior to construction and continued protection during turfgrass maintenance will eliminate the need for expensive wetland mitigation or restoration. Avoiding protracted permit processes or litigation prevents losses of revenues and natural resources. Voluntary adoption of sound turfgrass management practices is a superior alternative to regulatory mandates. Establishing a record regarding prevention of adverse environmental impacts and chemical losses not only reduces operational costs, but minimizes exposure to loss of revenues from litigation and unfavorable publicity.

Most economic analyses of practices designed to protect water quality have focused on the cost of implementation in return for social, health, and environmental benefits (Lee and Nielsen 1987). Turfgrass and agricultural managers should be aware of the economic benefits of management practices that reduce the use of water and applied chemicals (Severn and Ballard 1990). Retention of applied chemicals within the target or turfgrass root zone increases the efficacy of treatments while reducing the threat of nonpoint source pollution. Runoff, leaching, and volatile losses of applied chemicals that degrade water and wetland resources represent production input losses. Fertilization beyond turfgrass requirements and excessive application of preventive pesticides result in leaching, surface, and volatile losses of applied chemicals. These practices are cost ineffective. Chemicals removed by runoff and leaching are not available for plant growth and pest control. Export of nutrients and pesticides to wetlands may involve expensive litigation and cleanup activities. Use of sound design, irrigation, nutrient, pesticide, and wetland "conservation" practices will have both short-term and long-term economic benefits.

Use of integrated pest, water, and nutrient management strategies will ultimately resolve the issue of maintaining high quality turfgrass with a minimum of environmental disturbance. Given the prohibitive expense of mitigating environmental problems, there has been an increasing recognition that prevention of adverse impacts is a viable strategy for mitigation of environmental quality problems (Lee and Nielsen 1987; Leslie and Knoop 1989). For example, incorporating wetlands into golf course designs or avoiding construction on sites with wetlands will enhance the value of golf courses while protecting a valuable national resource (Dye 1989; Peacock et al. 1990).

The challenge to those who might oppose golf courses and turfgrass management is to objectively evaluate the facts, as they are known, and to be as thoroughly educated on the issues as possible. Only in an objective and educated atmosphere will useful research and policy decisions be made concerning the competing demands on our turfgrass, water, and environmental resources.

1.7 OBJECTIVES AND SOURCES OF INFORMATION

This book was developed by Spectrum Research, Inc., Duluth, MN, from a state-of-the-art literature review and assessment for the United States Golf Association (USGA), Green Section (Walker et al. 1990). The scientific and technical staff has reviewed and summarized information on the environmental impacts of turfgrass management, consulted government and academic experts, and surveyed existing on-line computer databases for additional information on the impacts of turfgrass management.

The primary goal of this book is to review and evaluate the scientific research on potential environmental impacts of golf course construction and turfgrass management. The review of the scientific and documented technical literature will insure development of an objective and accurate appraisal of current environmental research status and needs. Peer reviewed scientific papers and technical reports based on actual data have been selected from academic, government, and private research sources. When appropriate and available, scientific and technical information from private sector sources was included in the research assessment.

Scientific and technical literature used in this appraisal of environmental impacts was initially identified from computer-based literature searches of DIALOG data bases (Lockheed Corporation, Palo Alto, CA), the USGA Turfgrass Information File (USGA TGIF) (Chapin et al. 1989), and several published literature searches of the National Technical Information Service (NTIS). Literature citations and abstracts were obtained from Agricola, Biosis, CAB Abstracts, and USGA TGIF.

Additional sources of information on environmental impacts of golf course and turfgrass management were obtained from recent literature reviews, monographs, and symposium publications (e.g., Anderson et al. 1989; Beard 1973, 1982; Buckner and Bush 1979; Cheng 1990; Gibeault and Cockerham 1985; Godfrey et al. 1985; Grover 1988; Grover and Cessna 1991; Hanson and Juska 1969; Leslie and Metcalf 1989; Petrovic 1990; Sawhney and Brown 1989; Schuler 1987; Sleper et al. 1989). These literature reviews are based on the evaluation of individual technical and scientific studies. Although the primary focus of this book is on golf courses and turfgrass, where appropriate, research literature on environmental impacts in agricultural and grassland systems has been used. Technical information on physical and chemical characteristics of applied chemicals was acquired from several literature sources and chemical company fact sheets (e.g., Anderson et al. 1989; CPCR 1991; Farm Chemicals Handbook 1991; Humburg et al. 1989).

Toxicity information on nontarget organisms was obtained from literature reviews (Atkinson 1985; Bovey and Young 1980; Kline et al. 1987; Murty 1986; Pickering et al. 1989; Pilli et al. 1988; Sheedy et al. 1991; Stephan et al. 1986) and computerized toxicity databases (TERRE-TOX, Meyers and Shiller 1985; AQUIRE, Murphy 1990; Pilli et al. 1989). These databases were originally developed by the U.S. Environmental Protection Agency and The University of Montana, Missoula.

Additional sources of information and scientific literature were solicited directly from academic, industry, and private sector scientists. Scientific and technical papers were solicited by direct contact with the USGA Research Committee, telephone contacts with government and university scientists, and a general telephone and mail survey of turfgrass scientists and appropriate environmental specialists.

REFERENCES

Albert, R. E. 1987. Health concerns and chronic exposures to pesticides assessing the risk from carcinogens. in Pesticides and Groundwater: A Health Concern for the Midwest. Proceedings of a Conference Oct. 16–17, 1986, Freshwater Foundation, St. Paul, MN, pp. 163–168.

Allen, L. H., Jr. 1990. Plant responses to rising carbon dioxide and potential interactions with air pollutants. *J. Environ. Qual.* 19(1):15–34.

Amthor, J. S. and Beard, J. B. 1982. Effects of atmospheric pollutant sulfur dioxide on four St. Augustine cultivars. in Texas Turfgrass Research — 1982. PR-4042. The Texas Agric. Exp. Stn., The Texas A&M Univ. Sys., College Station, TX, p. 38.

Anderson, J. L., Balogh, J. C. and Waggoner, M. 1989. Soil Conservation Service Procedure Manual: Development of Standards and Specifications for Nutrient and Pesticide Management. Section I & II. USDA Soil Conservation Service. State Nutrient and Pest Management Standards and Specification Workshop, St. Paul, MN, July 1989, 453 pp.

Anonymous. 1988a. Palm Springs turf benefits from reuse project. *SportsTURF* 5(2):28.

Anonymous. 1988b. Golfsites found on half of all new real estate projects. *Greenmaster* 24(1):43.

Anonymous. 1990. Survey indicates environmental standardization needed. *Golf & SportsTURF* 6(7):7.

Atkinson, D. 1985. Toxicological properties of glyphosate — A summary. in Grossbard, E. and Atkinson, D. (Eds.). *The Herbicide Glyphosate.* Butterworth & Co., Boston, MA, pp. 127–133.

Balogh, J. C., Gordon, G. A., Murphy, S. R. and Tietge, R. M. 1990. The Use and Potential Impacts of Forestry Herbicides in Maine. 89-12070101. Final Report Maine Dept. of Cons., 237 pp.

Beard, J. B. 1973. *Turfgrass Science and Culture.* Prentice-Hall, Inc., Englewood Cliffs, NJ, 658 pp.

Beard, J. B. 1982. *Turf Management for Golf Courses.* Macmillan Publishing Co., New York, 642 pp.

Beard, J. B. 1989. Science shows turf can save water resources. *Turf. Environ.* 1(1):5.

Beard, J. B. and Johns, D. 1985. The comparative heat dissipation from three typical urban surfaces: asphalt, concrete, and a Bermuda grass turf. in Texas Turfgrass Research — 1985. PR-4329. The Texas Agric. Exp. Stn., The Texas A&M Univ. Sys., College Station, TX, pp. 59–62.

Bennett, O. L. 1979. Conservation. in Buckner, R. C. and Bush, L. P. (Eds.). *Tall Fescue.* Agron. Monogr. 20:319–340. Am. Soc. Agron., Crop Sci. Soc. Am., Soil Sci. Soc. Am., Madison, WI.

Bennett, H. H. 1939. Rates of erosion and runoff. in *Soil Conservation.* McGraw-Hill Co., New York, pp. 125–168. [cited in Watson (1985)].

Bovey, R. W. and Young, A. L. 1980. *The Science of 2,4,5-T and Associated Phenoxy Herbicides.* John Wiley & Sons, New York, 462 pp.

Buckner, R. C. and Bush, L. P. (Eds.). 1979. *Tall Fescue.* Agron. Monogr. 20. Am. Soc. Agron. Madison, WI, 351 pp.

Canter, L. W. 1987. Nitrates and pesticides in ground water: An analysis of a computer-based literature search. in Fairchild, D. M. (Ed.). *Ground Water Quality and Agricultural Practices.* Lewis Publishers, Chelsea, MI, pp. 153–174.

Cantor, K. P., Blair, A. and Zahn, S. H. 1988. Health effects of Agrichemicals in groundwater: What do we know? in Agricultural Chemicals and Groundwater Protection: Emerging Management and Policy. Proceedings of a Conference, Oct. 22–23, 1987, Freshwater Foundation, St. Paul, MN, pp. 27–42.

Caro, J. H. 1976. Pesticides in agricultural runoff. in Stewart, B. A., Woolhiser, D. A., Wischmeier, W. H., Caro, J. H. and Frere, M. H. (Eds.). Control of water pollution from cropland: An overview. Vol. 2. EPA 600/2-75-026b. U.S. Environmental Protection Agency and USDA Agricultural Research Service, pp. 91–119.

Carrow, R. N., Shearman, R. C. and Watson, J. R. 1990. Turfgrass. in Stewart, B. A. and Nielsen, D. R. (Eds.). *Irrigation of Agricultural Crops.* Agron. Monogr. 30:889–919. Am. Soc. Agron., Crop Sci. Soc. Am., Soil Sci. Soc. Am., Madison, WI.

CAST. 1975. Effects of Herbicides in Vietnam and Their Relation to Herbicide Use in the United States. Council Agricult. Sci. Tech. Report. No. 46, 17 pp.

Chapin, R. E., Armstrong, C. S., Cookingham, P. O. and Ray, P. A. 1989. USGA Turfgrass Information File Dial-Up User's Manual. Turfgrass Information Center, Mich. State Univ. Library. USGA Turfgrass Res. Comm., O.J. Noer Foundation, East Lansing, MI.

Cheng, H. H. (Ed.). 1990. *Pesticides in the Soil Environment: Processes, Impacts, and Modeling.* SSSA Book Series No. 2. Soil Science Society of America, Inc., Madison, WI, 530 pp.

Cockerham, S. T. and Gibeault, V. A. 1985. The size, scope, and importance of the turfgrass industry. in Gibeault, V. A. and Cockerham, S. T. (Eds.). *Turfgrass Water Conservation.* University of California, Riverside, Division of Agriculture and Natural Resources, pp. 7–12.

Cookson, C. 1979. Emergency ban on 2,4,5-T herbicide in the US. *Nature (London)* 278(5700):108–109.

Cooper, R. J. 1987. Turfgrass pesticide risk in perspective. *Golf Course Manage.* 55(11): 50–60.

CPCR. 1991. *Crop Protection Chemicals Reference.* Chemical and Pharmaceutical Press, John Wiley & Sons. New York, 2170 pp.

Daniel, T. C. and Schneider, R. R. 1979. Nonpoint pollution: Problem assessment and remedial measures; economic and planning considerations for designing control methods. in *Planning the Uses and Management of Land.* American Society of Agronomy, Madison, WI, pp. 829–851.

Daniel, T. C., McGuire, P. E., Stoffel, D. and Miller, B. 1979. Sediment and nutrient yield from residential construction sites. *J. Environ. Qual.* 8(3):304–308.

Daniel, T. C., Wendt, R. C., McGuire, P. E. and Stoffel, D. 1982. Nonpoint source loading rates from selected land uses. *Water Res. Bull.* 18(1):117–120.

Daugherty, S. J. 1991. Regulatory approaches to hydrocarbon contamination from underground storage tanks. in Kostecki, P. T. and Calabrese, E. J. (Eds.). *Hydrocarbon Contaminated Soils and Groundwater.* Lewis Publishers, Chelsea, MI, pp. 23–63.

Dye, P. 1989. Golf course architects adapt to environmental challenges. *SportsTURF* 5(5):29.

Environmental Protection Agency. 1986. Pesticides in Ground Water: Background Document. EPA-WH550G, Office of Ground-Water, U.S. Environmental Protection Agency, Washington, D.C., 72 pp.

Environmental Protection Agency. 1987. Agricultural Chemicals in Groundwater: Proposed Pesticide Strategy. Proposed Strategy Document, U.S. Environmental Protection Agency, Office of Pesticides and Toxic Substances, Washington, D.C., 149 pp.

Environmental Protection Agency. 1988a. Protecting Ground Water: Pesticides and Agricultural Practices. EPA-440/6-88-001. U.S. Environmental Protection Agency, Office of Ground-Water Protection, Washington, D.C., 53 pp.

Environmental Protection Agency. 1988b. America's Wetlands; Our Vital Link Between Land and Water. OPA-87-016. U.S. Environmental Protection Agency, Office of Wetlands Protection, Washington, D.C., 9 pp.

Environmental Protection Agency. 1990. National Pesticide Survey: Summary Results of EPA's National Survey of Drinking Water Wells. U.S. Environmental Protection Agency, Office of Pesticides and Toxic Substances, Washington, D.C., 16 pp.

Erickson, L. E. and Kuhlman, D. 1987. What you need to know about groundwater concerns. *Grounds Maintenance* 22:74–75,128.

Falk, J. H. 1976. Energetics of a suburban lawn ecosystem. *Ecology* 57(1):141–150.

Fairchild, D. M. 1987. A national assessment of groundwater contamination from pesticides and fertilizers. in Fairchild, D. M. (Ed.). *Ground Water Quality and Agricultural Practices.* Lewis Publishers, Chelsea, MI, pp. 273–294.

Farm Chemicals Handbook. 1991. *Pesticide and Fertilizer Dictionary.* Meister Publishing Company, Willoughby, OH, 770 pp.

Feldhake, C. M, Danielson, R. E. and Butler, J. D. 1984. Turfgrass evapotransperation. II. Responses to deficit irrigation. *Agron. J.* 76(1):85–89.

Feucht, J. R. and Butler, J. D. 1988. *Landscape Management: Planting and Maintenance of Trees, Shrubs, and Turfgrasses.* Van Nostrand Reinhold Co., New York, 179 pp.

Fiore, M. 1987. Chronic exposure to aldicarb contaminated groundwater and human immune function. in Pesticides and Groundwater: A Health Concern for the Midwest. Proceedings of a Conference Oct. 16–17, 1986, Freshwater Foundation, St. Paul, MN, pp. 199–203.

Foster, G. R. and Lane, L. J. 1987. Beyond the USLE: Advancements in soil erosion prediction. in Boersma, L. L. (Ed. chair.). *Future Developments in Soil Science Research.* Soil Sci. Soc. Am., Madison, WI, pp. 315–326.

Frere, M. H. 1976. Nutrient aspects of pollution from cropland. in Stewart, B. A., Woolhiser, D. A., Wischmeier, W. H., Caro, J. H. and Frere, M. H. (Eds.). Control of Water Pollution From Cropland: An Overview. Vol. 2. EPA 600/2-75-026b. U.S. Environmental Protection Agency and USDA Agricultural Research Service, pp. 59–90.

Gibeault, V. A. and Cockerham, S. T. (Eds.). 1985. *Turfgrass Water Conservation.* University of California, Riverside. Division of Agriculture and Natural Resources. 155 pp.

Godfrey, P. J., Kaynor, E. R., Pelczarski, S. and Benforado, J. (Eds.). 1985. *Ecological Considerations in Wetland Treatment of Municipal Wastewaters.* Van Nostrand Reinhold Co., New York, 473 pp.

Gold, A. J., Sullivan, W. M. and Hull, R. J. 1989. Influence of fertilization and irrigation practices on waterborne nitrogen losses from turfgrass. in Leslie, A. R. and Metcalf, R. L. (Eds.) Integrated Pest Management for Turfgrass and Ornamentals. Office of Pesticide Programs, 1989-625-030. U.S. Environmental Protection Agency, Washington, D.C., pp. 143–150.

Grant, Z. 1987. One aspect of government relations: groundwater issues in pesticide regulation. *Golf Course Manage.* 55(6):14–20.

Grover, R. (Ed.). 1988. *Environmental Chemistry of Herbicides.* Vol. 1. CRC Press, Boca Raton, FL, 207 pp.

Grover, R. and Cessna, A. J. (Eds.). 1991. *Environmental Chemistry of Herbicides.* Vol 2. CRC Press, Boca Raton, FL, 302 pp.

Hanson, A. A. and Juska, F. V. (Eds.). 1969. *Turfgrass Science*. Agron. Monogr. 14. Am. Soc. Agron. Madison, WI, 715 pp.

Harivandi, M. A. 1982. The use of effluent water for turfgrass irrigation. *Calif. Turf. Cult.* 32(3–4):1–4.

Harmon, M. E., Ferrel, W. K. and Franklin, J. F. 1990. Effects on carbon storage of conversion of old-growth forests to young forests. *Science* 247(4943):699–702.

Hayward, D. G. and Weitzer, W. H. 1984. The public's image of urban parks: Past amenity, present ambivalence, uncertain future. *Urban Ecol.* 8:243–268.

Heuber, D. B. (Ed.). 1984. Statistical Profile of Golf in the United States. National Golf Foundation, North Palm Beach, FL.

Hinkle, M. K. 1983. Problems with conservation tillage. *J. Soil Water Cons.* 38:201–206.

Humburg, N. E., Colby, S. R., Hill, E. R., Kitchen, L. M., Lym, R. G., McAvoy, W. J. and Prasak, R. 1989. *Herbicide Handbook of the Weed Science Society of America*. 6th ed., Weed Science Society of America, Champaign, IL, 301 pp.

Johns, D. and Beard, J. B. 1985. A Quantitative Assessment of the Benefits from Irrigated Turf on Environmental Colling and Energy Savings in Urban Areas. in Texas Turfgrass Research — 1985. PR-4330. The Texas Agric. Exp. Stn., The Texas A&M Univ. Sys., College Station, TX. pp. 63–65.

Kaplan, R. and Kaplan, S. 1989. *The Experience of Nature — A Psychological Perspective*. Cambridge University Press, New York, 340 pp.

Keeney, D. R. 1982. Nitrogen management for maximum efficiency and minimum pollution. in Stevenson, F. J. (Ed.). *Nitrogen in Agricultural Soils*. Agron. Monogr. 22:605–649. Am. Soc. of Agron., Crop Sci. Soc. Am., Soil Sci. Soc. Am., Madison, WI.

Keeney, D. R. 1983. Transformations and transport of nitrogen. in Schaller, F. (Ed.). Proc. Natl. Conf. Agric Mgmt. and Water Quality, Ames, Iowa, 26–29 May, 1981, pp. 48–63.

Keeney, D. R. 1986. Sources of nitrate to groundwater. *CRC Crit. Rev. Environ. Control* 16:257–304.

Klein, R. D. 1990. *Protecting the Aquatic Environment from the Effects of Golf Courses*. Community & Environmental Defense Associates, Maryland Line, MD, p. 59.

Kline, E. R., Mattson, V. R., Pickering, Q. H., Spehar, D. L. and Stephan, C. E. 1987. Effects of pollution on freshwater organisms. *J. Water Pollut. Control Fed.* 59(6):539–572.

Koehler, F. A., Humenik, F. J., Johnson, D. D., Kreglow, J. M., Dressing, R. P. and Maas, R. P. 1982a. Best Management Practices for Agricultural Nonpoint Source Control. II. Commercial Fertilizer. USDA Coop. Agree. 12-05-300-472, EPA Interagency Agree. AD-12-F-0-037-0. North Carolina Agricult. Ext. Serv., 55 pp.

Koehler, F. A., Humenik, F. J., Johnson, D. D., Kreglow, J. M., Dressing, R. P. and Maas, R. P. 1982b. Best Management Practices for Agricultural Nonpoint Source Control. III. Sediment. USDA Coop. Agree. 12-05-300-472, EPA Interagency Agree. AD-12-F-0-037-0. North Carolina Agricult. Ext. Serv., 49 pp.

Koehler, F. A., Humenik, F. J., Johnson, D. D., Kreglow, J. M., Dressing, R. P. and Maas, R. P. 1982c. Best Management Practices for Agricultural Nonpoint Source Control. I. Animal waste. USDA Coop. Agree. 12-05-300-472, EPA Interagency Agree. AD-12-F-0-037-0. North Carolina Agricult. Ext. Serv., 67 pp.

Krivak, J. A. 1978. Best management practices to control nonpoint-source pollution from agriculture. *J. Soil Water Cons.* 33(4):161–166.

Kuo, Chin Y., Cave, K. A. and Loganathan, G. V. 1988. Planning of urban best management practices. *Water Res. Bull.* 24(1):125–132.

Kusler, J. A. and Kentula, M. E. (Eds.). 1990. *Wetland Creation and Restoration: The Status of the Science*. Island Press, Covelo, CA, 591 pp.

Larson, W. E., Pierce, F. J. and Dowdy, R. H. 1983. The threat of soil erosion to long-term crop production. *Science* 219:458–465.

Lee, L. K. and Nielsen, E. G. 1987. The extent and costs of groundwater contamination by agriculture. *J. Soil Water Cons.* 42(4):243–248.

Leonard, R. A. 1988. Herbicides in surface waters. in Grover, R. (Ed.). *Environmental Chemistry of Herbicides*. Vol. 1. CRC Press, Boca Raton, FL, pp. 45–87.

Leonard, R. A. 1990. Movement of pesticides into surface waters. in Cheng, H. H. (Ed.). *Pesticides in the Soil Environment: Processes, Impacts, and Modeling*. SSSA Book Series No. 2. Soil Science Society of America, Inc., Madison, WI, pp. 303-349.

Leslie, A. R. 1989. Societal problems associated with pesticide use in the urban sector. in Leslie, A. R. and Metcalf, R. L. (Eds.). Integrated Pest Management for Turfgrass and Ornamentals. Office of Pesticide Programs, 1989-625-030. U.S. Environmental Protection Agency, Washington, D.C., pp. 93–96.

Leslie, A. R. and Knoop, W. 1989. Societal benefits of conservation oriented management of turfgrass in home lawns. in Leslie, A. R. and Metcalf, R. L. (Eds.). Integrated Pest Management for Turfgrass and Ornamentals. Office of Pesticide Programs, 1989-625-030. U.S. Environmental Protection Agency, Washington, D.C., pp. 93–96.

Leslie, A. R. and Metcalf, R. L. (Eds.). 1989. Integrated Pest Management for Turfgrass and Ornamentals. Office of Pesticide Programs, 1989-625-030. U.S. Environmental Protection Agency, Washington, D.C., 337 pp.

Logan, T. J., Davidson, J. M., Baker, J. L. and Overcash, M. R. (Eds.). 1987. *Effects of Conservation Tillage on Groundwater Quality*. Lewis Publishers Chelsea, MI, 292 pp.

Madhun, Y. A. and Freed, V. H. 1990. Impact of pesticides on the environment. in Cheng, H. H. (Ed.). *Pesticides in the Soil Environment: Processes, Impacts, and Modeling*. SSSA Book Series No. 2. Soil Science Society of America, Inc., Madison, WI, pp. 429–466.

Madison, J. H. 1982. *Principles of Turfgrass Culture*. Robert E. Krieger Publishing Co., Melbourne, FL, 431 pp.

Mehall, B. J., Hull, R. J. and Skogley, C. R. 1984. Turf quality of Kentucky bluegrass cultivars and energy relations. *Agron. J.* 76(1):47–50.

Meyers, S. M. and Schiller, S. M. 1985. TERRE-TOX: a data base for effects of anthropogenic substances on terrestrial animals. *J. Chem. Inf. Comput. Sci.* 26(1):33–36.

Mitsch, W. J. and Gosselink, J. G. 1986. *Wetlands*. Van Nostrand Reinhold Co., New York, 539 pp.

Moore, D. G., Currier, J. B. and Norris, L. A. 1980. Introduced chemicals. in USDA Forest Service. An Approach to Water Resources Evaluation of Non-Point Silvicultural Sources (A Procedural Handbook). USDA Forest Service, Washington, D.C., EPA-600/8-012. p. XI. 1–XI.62.

Morton, T. G., Gold, A. J. and Sullivan, W. M. 1988. Influence of overwatering and fertilization on nitrogen losses from home lawns. *J. Environ. Qual.* 17(1):124–130.

Murphy, S. R. 1990. Aquatic Toxicity Information Retrieval (AQUIRE): A Microcomputer Database. System Documentation. Spectrum Research, Inc., Duluth, MN, 37 pp.

Murty, A. S. 1986. *Toxicity of Pesticides to Fish*. Vol. 1. CRC Press, Inc., Boca Raton, FL, 178 pp.

National Golf Foundation. 1985. *Golf Course Maintenance Report. 1985 Biennial Review*. Golf Course Superintendents Association of America and National Golf Foundation, North Palm Beach, FL.

National Golf Foundation. 1989. Golf's popularity continues to soar. *Golf Market Today* 28(4):1,4–8.

National Golf Foundation. 1991. *Golf Facilities in the United States*. National Golf Foundation, Jupiter, FL.

Nutter, G. C. and Watson, J. R., Jr. 1969. The turfgrass industry. in Hanson, A. A. and Juska, F. V. (Eds.). *Turfgrass Science*. Agron. Monogr. 14:9–26. Am. Soc. Agron., Madison, WI.

Onstad, C. A., Burkart, M. R. and Bubenzer, G. D. 1991. Agricultural research to improve water quality. *J. Soil Water Cons.* 46(3):184–188.

Parker, J. H. 1982. An energy and ecological analysis of alternate residential landscapes. *J. Environ. Syst.* 11(3):271–288.

Payne, R. A. 1987. Resource conservation: The golf course role. *Golf Course Manage.* 55(10):50–58.

Peacock, C. H. and Dudeck, A. E. 1984. Physiological response of St. Augustinegrass to irrigation scheduling. *Agron. J.* 76(2):275–279.

Peacock, C. H., Bruneau, A. H. and Spak, S. P. 1990. Protecting a valuable resource: Preservation of wetlands from a technical perspective. *Golf Course Manage.* 58(11):6–7, 10, 12, 16.

Petrovic, A. M. 1990. The fate of nitrogenous fertilizers applied to turfgrass. *J. Environ. Qual.* 19(1):1–14.

Pickering, Q., Carle, D. O., Pilli, A., Willingham, T. and Lazorchak, J. M. 1989. Effects of pollution on freshwater organisms. *J. Water Pollut. Control Fed.* 61(6):998–1042.

Pilli, A., Carle, D. O., Kline, E., Pickering, Q. and Lazorchak, J. 1988. Effects of pollution on freshwater organisms. *J. Water Pollut. Control Fed.* 60(6):994–1065.

Pilli, A., Carle, D. O. and Sheedy, B. A. 1989. AQUIRE: Aquatic Toxicity and Retrieval Data Base. Contract No. 68-01-7176. U.S. Environmental Protection Agency, ERL-Duluth, MN, 18 pp.

Potter, D. A., Cockfield, S. D. and Morris, T. A. 1989. Ecological side effects of pesticide and fertilizer use on turfgrass. in Leslie, A. R. and Metcalf, R. L. (Eds.). Integrated Pest Management for Turfgrass and Ornamentals. Office of Pesticide Programs, 1989-625-030. U.S. Environmental Protection Agency, Washington, D.C., pp. 33–44.

Pratt, P. F. (Chairman). 1985. Agriculture and Groundwater Quality. Council for Agricultural Science and Technology. CAST Report No. 103, 62 pp.

Pye, V. I. and Patrick. R. 1983. Ground water contamination in the United States. *Science* 221:713–221.

Pye, V. I., Patrick, R. and Quarles, J. 1983. *Groundwater Contamination in the United States.* University of Pennsylvania Press, Philadelphia, 315 pp.

Raupp, M. J., Smith, M. F. and Davidson, J. A. 1989. Educational, environmental and economic impacts of integrated pest management programs for landscape plants. in Leslie, A. R. and Metcalf, R. L. (Eds.). Integrated Pest Management for Turfgrass and Ornamentals. Office of Pesticide Programs, 1989-625-030. U.S. Environmental Protection Agency, Washington, D.C., pp. 77–83.

Richards, G. A., Mulchi, C. L. and Hall, J. R. 1980. Influence of plant maturity on the sensitivity of turfgrass species to ozone. *J. Envir. Qual.* 9(1):49–53.

Roberts, E. C. and Roberts, B. C. 1989. *Lawn and Sports Turf Benefits.* The Lawn Institute, Pleasant Hill, TN, 31 pp.

Rosenberg, N. J. 1974. *Microclimate: The Biological Environment.* John Wiley & Sons, New York, 315 pp.

Salvesen, D. 1990. *Wetlands: Mitigating and Regulating Development Impacts.* The Urban Land Institute. Washington, D.C., 117 pp.

Sawhney, B. L. and Brown, K. (Eds.). 1989. *Reactions and Movement of Organic Chemicals in Soils*. SSSA Special Publication No. 22. Soil Sci. Soc. Am., Am. Soc. Agron., Madison, WI, 474 pp.

Schuler, T. R. 1987. Controlling Urban Runoff: A practical manual for planning and designing urban BMPs. Dept. of Environ. Programs. Washington Metropolitan Water Resources Planning Board, Washington, D.C.

Severn, D. J. and Ballard, G. 1990. Risk/benefit and regulations. in Cheng, H. H. (Ed.). *Pesticides in the Soil Environment: Processes, Impacts, and Modeling*. SSSA Book Series No. 2. Soil Science Society of America, Inc., Madison, WI, pp. 467–503.

Sharpley, A. N. 1980. The enrichment of soil phosphorus in runoff sediments. *J. Environ. Qual.* 9(3):521–526.

Sheedy, B. A., Lazorchak, J. M., Grunwald, D. J., Pickering, Q. H., Pilli, A., Hall, D. and Webb, R. 1991. Effects of pollution on freshwater organisms. *J. Water Pollut. Control Fed.* 63(4):619–696.

Sleper, D. A., Asay, K. H. and Pedersen, J. F. 1989. Contributions from Breeding Forage and Turf Grasses. CSSA Spec. Publ. No. 15, Crop Sci. Soc. Am. Madison, WI, 122 pp.

Smith, R. A., Alexander, R. B. and Wolman, M. G. 1987. Water-quality trends in the nation's rivers. *Science* 235:1607–1615.

Smolen, M. D., Humenik, F. J., Spooner, J., Dressing, S. A. and Maas, R. P. 1984. Best Management Practices for Agricultural Nonpoint Source Control. IV. Pesticides. USDA Coop. Agree. 12-05-300- 472, EPA Interagency Agree. AD-12-F-0-037-0. North Carolina Agricult. Ext. Serv., 87 pp.

Sommers, L. E. and Giordano, P. M. 1984. Use of nitrogen from agricultural, industrial, and municipal wastes. in Hauck, R. D. (Ed.). *Nitrogen in Crop Production*. Am. Soc. Agron., Crop Sci. Soc. Am., Soil Sci. Soc. Am., Madison, WI, pp. 208–220.

Starr, C. and Whipple, C. 1980. Risks of risk decisions. *Science* 208:1114–1119.

Stephan, C. E., Spehar, D. L., Roush, T. H., Phipps, G. L. and Pickering, Q. H. 1986. Effects of pollution on freshwater organisms. *J. Water Pollut. Control Fed.* 58(6):645–671.

Stewart, B. A., Woolhiser, D. A., Wischmeier, W. H., Caro, J. H. and Frere, M. H. 1975. Control of Water Pollution from Cropland: A Manual for Guideline Development. Vol. 1. EPA 600/2-75-026a. U.S. Environmental Protection Agency and USDA Agricultural Research Service, 111 pp.

Stewart, B. A., Woolhiser, D. A., Wischmeier, W. H., Caro, J. H. and Frere, M. H. 1976. Control of Water Pollution from Cropland: An Overview. Vol. 2. EPA 600/2-75-026b. U.S. Environmental Protection Agency and USDA Agricultural Research Service, 187 pp.

Sweeten, J. M. and Reddel, D. L. 1978. Nonpoint sources: State-of-the-Art Overview. *Trans. ASAE* 21(3):474–483.

Tashiro, H. 1987. *Turfgrass Insects of the United States and Canada*. Cornell University Press, Ithaca, NY, 391 pp.

Taylor, A. W. and Kilmer, V. J. 1980. Agricultural phosporus in the environment. in Khasawneh, F. E. (Ed. chair.). *The Role of Phosphorus in Agriculture*. Am. Soc. Agron., Crop Sci. Soc. Am., Soil Sci. Soc. Am., pp. 545–557.

Ulrich, R. S. 1981. Natural versus urban scenes: Some psychophysiological effects. *Environ. Behav.* 13:523–556.

Ulrich, R. S. 1984. View through a window may influence recovery from surgery. *Science* 224(4647):420–421.

Ulrich, R. S. 1986. Human responses to vegetation and landscapes. *Landscape Urban Plan.* 13:29–44.

USDA Agricultural Research Service. 1988a. ARS Strategic Groundwater Plan. 1. Pesticides. Washington, D.C.

USDA Agricultural Research Service. 1988b. ARS Strategic Groundwater Plan. 1. Nitrates. Washington, D.C.

USDA Agricultural Research Service. 1989a. USDA Research Plan for Water Quality. United States Dept. of Agric., Coop. State Res. Serv., State Agric. Exp. Stns., 14 pp.

USDA Agricultural Research Service. 1989b. Midwest Water Quality Research Initiative Management Systems Evaluation Areas. FY 1990 Guidelines for Proposal Preparation and Submission. Agricultural Chemical Management Research: A multiagency focus on ground water quality. U.S. Dept. of Agric., Coop. State Res. Serv., State Agric. Exp. Stns., 17 pp.

Uttormark, P. D., Chapin, J. D. and Green, K. M. 1974. Estimating Nutrient Loading of Lakes from Nonpoint Sources. EPA-660/3-74-020. U.S. Environmental Protection Agency, Washington, D.C.

van der Leeden, F., Troise, F. L. and Todd, D. K. 1990. *The Water Encyclopedia*. Lewis Publishers, Inc., Chelsea, MI, 808 pp.

Walker, W. J., Balogh, J. C., Tietge, R. M. and Murphy, S. R. 1990. *Environmental Issues Related to Golf Course Construction and Management*. United States Golf Association, Green Section, Far Hills, NJ, 378 pp.

Walter, M. F., Steenhuis, T. S. and Haith, D. A. 1979. Nonpoint source pollution control by soil and water conservation practices. *Trans. ASAE* 22(4):834–840.

Warrington, G. E., Knapp, K. L., Klockm, G. O., Foster, G. R. and Beasley, R. S. 1980. in USDA Forest Service. An Approach to Water Resources Evaluation of Non-Point Silvicultural Sources (A Procedural Handbook). USDA Forest Service, Washington, D.C., EPA-600/8-012. p. IV.1–IV.81.

Watschke, T. L., Hamilton, G. and Harrison, S. 1988. Is pesticide runoff from turf increasing. *ALA* 9(7):43–44.

Watschke, T. L., Harrison, S. and Hamilton, G. W. 1989. Does Fertilizer/Pesticide Use on a Golf Course Put Water Resources in Peril? *USGA Green Sect. Record* May/June, pp. 5–8.

Watson, J. R., Jr. 1985. Water resources in the United States. in Gibeault, V. A. and Cockerham, S. T. (Eds.). *Turfgrass Water Conservation*. University of California, Riverside, Division of Agriculture and Natural Resources, pp. 19–36.

Watson, J. R., Jr., Kaerwer, H. E. and Martin, D. P. 1990. The turfgrass industry. in *Turfgrass Science*. ASA-CSSA Monogr. 14 (Revised). Am. Soc. Agron., Crop Sci. Soc. Am., Madison, WI, 106 pp. (in press).

Welterlen, M. S., Gross, C. M., Angle, J. S. and Hill, R. L. 1989. Surface runoff from turf. in Leslie, A. R. and Metcalf, R. L. (Eds.). Integrated Pest Management for Turfgrass and Ornamentals. Office of Pesticide Programs, 1989-625-030. U.S. Environmental Protection Agency. Washington, D. C. pp. 153–160.

Wilkinson, J. F. 1989. Current and future regulatory concerns for lawn care operators. in Leslie, A. R. and Metcalf, R. L. (Eds.). Integrated Pest Management for Turfgrass and Ornamentals. Office of Pesticide Programs, 1989-625-030. U.S. Environmental Protection Agency. Washington, D.C., pp. 45–50.

Youngner, V. B. and Nudge, F. J. 1980. Air pollutant oxidant effects on cool-season and warm-season turfgrasses. *Agron. J.* 72:169–170.

Zeckhauser, R. J. and Viscusi, W. K. 1990. Risk within reason. *Science* 248(4955):559–564.

CHAPTER 2

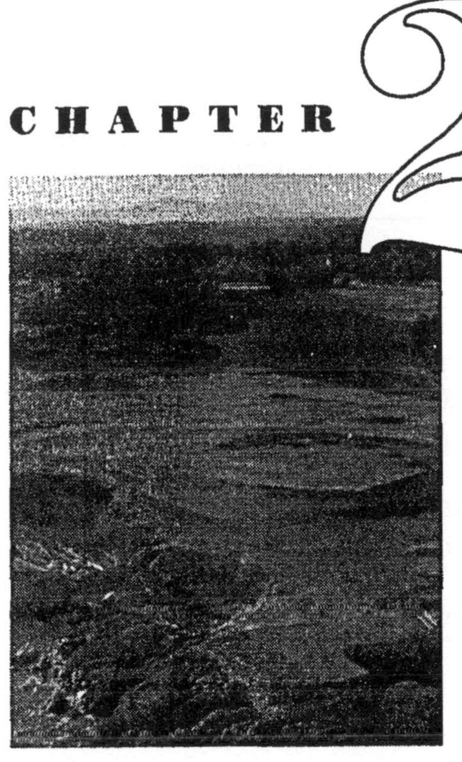

Role and Conservation of Water Resources

James C. Balogh and James R. Watson, Jr.

2.1 INTRODUCTION: WATER RESOURCES AND TURFGRASS

Groundwater, streams, rivers, lakes, and reservoirs are invaluable natural resources in the United States (Table 1). They supply drinking water, supply water for irrigation and industry, are a source of natural beauty and recreation, and are a mode of transportation (Table 2). The use of water resources and the establishment of water use priorities have always been controversial in the United States. Both urban and rural development in the United States, especially in the West, have been shaped by the distribution and location of water. Water and the lack of it has often been the critical source of success and failure of agricultural, forestry, and turfgrass projects. In the past, water use has been the basis of water rights and law. However, within the last two decades, the basis of water policy in the United States has been changing from the assumption of an expanding natural resource to that of water as a limited commodity (Matheson 1991). Recent extended droughts in the southeastern, midwestern, and western United States highlight the need to develop a rational water resource policy. Today, allocation of both groundwater and surface water resources can no longer rely on water surpluses stored during wet years to meet the ever-increasing demands of many urban and rural constituents.

Photo courtesy of British Columbia Ministry of Tourism.

Table 1. Water Resources in the Continental United States

Water source	Volume (10^9 m^3)	Volume (%)	Annual circulation (10^9 m^3 $year^{-1}$)	Replacement period (year)
Groundwater:				
Shallow (<800 m)	63,000	43.2	310	>200
Deep (>800 m)	63,000	43.2	6	>10,000
Lakes:				
Freshwater	19,000	13.0	190	100
Salt	58	<0.1	6	>10
Soil moisture (root zone – 1 m)	630	0.4	3,100	<0.1
Stream channels[a]	50	<0.1	1,900	
Atmospheric water vapor	190	0.1	6,200	>0.1
Ice, glaciers	67	<0.1	2	>40

From van der Leeden et al. 1990.

[a] *Average.*

In arid regions of the United States and regions with periodic drought conditions, allocation of water for irrigation of turfgrass often has been a controversial policy issue (Rossillon 1985; Watson 1985). Although consumption of water for turfgrass is relatively small compared to agricultural and industrial use (van der Leeden et al. 1990), highly visible irrigation of recreational and leisure turfgrass facilities has been considered as "luxury" use by certain sectors of the public. Development of cultural practices to conserve water used in management of golf courses and turfgrass, even in humid regions, remains a critical research need for turfgrass scientists (Carrow et al. 1990; Shearman 1985).

The relationship of turfgrass management and water resources on a national scale has been reviewed by Watson (1985) and Walker et al. (1990). Water for irrigation and removal of excess water is critical for maintenance of high-quality turfgrass on golf courses. Technical aspects of these water management practices have been reviewed by Beard (1982, 1985), Carrow et al. (1990), Marsh (1969), Meyer and Camenga (1985), and Olson (1985). Surface water, including small streams, ponds, lakes, and drainage channels, are often incorporated into the overall design and irrigation scheme of golf courses. Groundwater also is a common source of water for golf course, recreational, and residential turfgrass. In arid and semiarid regions, use of effluent water for irrigation of turfgrass, especially on golf courses, has become relatively common (Butler et al. 1985; Harivandi 1982). In humid regions, turfgrass areas have been used for disposal and treatment of municipal waste effluent.

Table 2. Use of Water Resources in the United States

Purpose	Description of activities
Domestic and industrial water supply	Provides water for domestic, industrial, municipal, and other uses
Irrigation	Used for agricultural production, management of golf course turfgrass, management of parks and other turfgrass areas
Watershed management	Conservation and improvement of soil, sediment control, reduction of runoff, improvement of forests and grasslands, and protection of water supplies
Pollution control	Protection and improvement of water supplies for municipal, domestic, industrial, agricultural, turfgrass, and recreational uses; also used for protection of aquatic life
Erosion control	Reduction of sediment loading in surface waters from agricultural production and urban development
Flood control	Flood damage abatement and reduction, protection of rural and urban development, conservation storage, river regulation, groundwater recharge, protection of water supplies, and protection of human life and wildlife
Salinity control	Reduction or prevention of salt water contamination of municipal, agricultural, and industrial water supplies
Drainage	Agricultural production, turfgrass management, urban development, and insect control for protection of public health
Fish and wildlife habitat	Improvement and protection of habitat for wildlife, fish, and other aquatic organisms; enhancement of recreational activities and commercial fisheries
Hydroelectricity	Supply power for development and maintenance of economic and domestic activities
Navigation	Transportation of goods and passengers
Recreation	Increased enjoyment of life through leisure activities

From van der Leeden et al. 1990.

Increased use of water resources for intensified exploitation by agriculture, industry, and municipalities has been associated with problems of deterioration of groundwater quantity and quality. Significant depletion of regional aquifers has occurred in

California, Arizona, and the Central Plains region (Pye and Patrick 1983; Watson 1985; van der Leeden et al. 1990). Contamination of groundwater, as well as surface water, with substances making them unsuitable for human use has become a matter for intense public concern. Protection of groundwater as a source of potable water is currently a national priority issue.

Groundwater is an important component of available water resources in the United States (Table 1) (van der Leeden et al. 1990). Half of the population of the United States uses groundwater as their primary source of drinking water (Table 3); 36% of municipal drinking water is supplied from groundwater; 75% of major U.S. cities depend on groundwater for much of their total supply; and 97% of all rural domestic drinking water in the United States is supplied by groundwater wells (EPA 1977, 1987; Lee and Nielsen 1987; Water Resources Council 1978). Of the 88.5 billion gallons of fresh groundwater used per day in the United States in 1980, 65% was used for irrigated agriculture (Pye and Patrick 1983). Groundwater withdrawals increased 162% from 1950 to 1980, compared to a 107% increase in surface water withdrawals (EPA 1986; Lee and Nielsen 1987). Approximately 30% of surface water is supplied by groundwater emerging as springs or in other seepage areas (Pye and Patrick 1983). During dry periods, groundwater may form all of the base flow of streams and rivers.

2.2 COMPONENTS OF THE WATER CYCLE, SOIL, AND TURFGRASS SYSTEMS

Water is the dominant factor that controls the structure and functions of managed and natural ecosystems. Moisture flux within the atmosphere-plant-soil continuum is the dynamic linkage between terrestrial, aquatic, and atmospheric systems (Figure 1). Movement of water within the hydrologic cycle determines the availability of water for turfgrass growth and development, surface runoff from urban and rural watersheds, recharge and quality of groundwater, and potential transport of applied pesticides and fertilizers (Balogh and Gordon 1987; Beard 1985; Carrow 1985). An understanding of the hydrologic cycle, its impacts on management practices, and its interaction with applied chemicals is essential for improved evaluation of environmental effects resulting from golf course and turfgrass management (Figure 1). Irrigation and drainage practices have a direct relationship with cultural, nutrient, and pest management practices on golf courses, lawns, and other turfgrass systems (Biran et al. 1981; Colbaugh and Elmore 1985; Nus and Hodges 1985; Shearman 1985; Youngner 1985).

Precipitation, irrigation, interception, infiltration, surface runoff, subsurface flow, soil water balance, evapotranspiration, drainage, and deep leaching to groundwater

Table 3. Reliance of the Population of the United States on Groundwater as a Source of Drinking Water

State	Percent of population using groundwater as a source of drinking water
Alabama	44
Alaska	79
Arizona	65
Arkansas	63
California	70
Colorado	21
Connecticut	32
Delaware	61
Florida	91
Georgia	47
Hawaii	93
Idaho	91
Illinois	48
Indiana	66
Iowa	81
Kansas	59
Kentucky	34
Louisiana	57
Maine	47
Maryland	33
Massachusetts	36
Michigan	34
Minnesota	80
Mississippi	93
Missouri	47
Montana	53
Nebraska	90
Nevada	40
New Hampshire	57
New Jersey	49
New Mexico	90
New York	34
North Carolina	58
North Dakota	63
Ohio	44
Oklahoma	35
Oregon	43
Pennsylvania	42
Puerto Rico	30
Rhode Island	24
South Carolina	43
South Dakota	85

Table 3. Reliance of the Population of the United States on Groundwater as a Source of Drinking Water (Continued)

State	Percent of population using groundwater as a source of drinking water
Tennessee	51
Texas	48
Utah	66
Vermont	56
Virgin Islands	58
Virginia	37
Washington	54
West Virginia	51
Wisconsin	70
Wyoming	64
Total United States	56

From van der Leeden et al. 1990.

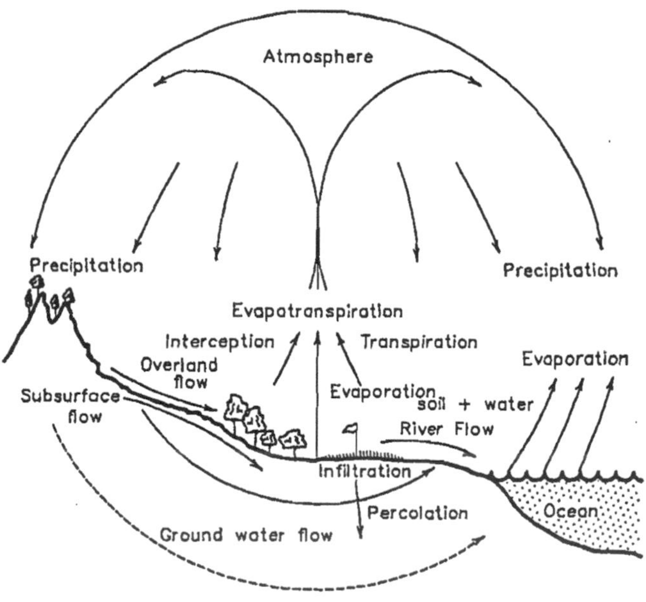

Fig. 1. The water cycle.

are essential components of the hydrologic cycle (Figure 1). Excellent general reviews of hydrologic processes in relation to management of biological systems, water quality issues, and environmental effects have been presented by Beard (1985), Carrow (1985), Schwab et al. (1981), and Woolhiser (1976). Processes within the hydrologic cycle will be briefly summarized in this chapter. In-depth discussions of water transport components of the water cycle have been presented by Blad (1983), Ghildyal and Tripathi (1987), Hanks and Ashcroft (1976), and Thien (1983).

The principles of mass and energy conservation govern the amount and distribution of water in the hydrologic cycle (Beard 1985; Woolhiser 1976). Mass balance of soil water within the hydrologic cycle is represented by Equation 1.

$$P + Ir = Q_s + QB + D + S + U + E \tag{1}$$

where P = precipitation received on area, A
Ir = imported water (irrigation)
Q_s = net surface runoff (runoff leaving A less runoff entering A)
QB = net subsurface lateral flow (includes saturated and unsaturated flow; and subsurface drainage effluent)
D = change in surface depression or detention storage
U = net vertical outflow through soil or bedrock (transport to groundwater)
E = evaporation from soil and plants (evapotranspiration)

Water yield, surface (Q_s) and subsurface (QB + U), from an area (watershed) is the difference between the total water input (P + Ir) and evaporation, assuming soil and surface storage changes are insignificant. Water available for transport to groundwater is the difference between the total water input and evaporation, assuming all storage terms are satisfied (Beard 1985; Rosenberg 1974).

Evaporation is controlled by the amount of energy available at the soil-air interface. The energy balance equation at the soil-plant-air interface is expressed as net rate of incoming energy per unit area = net rate of outgoing energy,

$$R_s(1 - p) = R_l + G + H + LE \tag{2}$$

where R_s = flux density of total short-wave (solar) radiation at the soil surface
p = albedo (reflected fraction of R_s at soil surface)
R_l = net flux density of long-wave (terrestrial) radiation at the soil surface
G = heat flux density into the soil surface
H = heat flux density into the atmosphere
L = latent heat of evaporation of water
E = evaporation rate

Heat storage in vegetation and photosynthetic energy constitute less than 1% of R_s and are ignored in Equation 2 for watershed consideration. The magnitude of terms in the energy balance equation vary substantially. Under conditions with wet soil and transpiring vegetation, most of the solar energy is partitioned to evapotranspiration. Dry soil surface conditions will result in R_s being partitioned to H, heating the air. Equations 1 and 2 are linked by the evaporation term, E. Evaporation and transpiration are limited by the energy available at the turfgrass surface.

Precipitation is the water equivalent of all forms of moisture which falls as rain, drizzle, snow, fog, hail, ice pellets, and glaze ice (Shaw 1982). A sequence of precipitation events in a single area forms a time series. In a deterministic sense, a time series of precipitation is unpredictable. This stochastic or probabilistic element within the water cycle is a controlling feature of many hydrological dependent processes (Balogh and Gordon 1987).

The timing of precipitation events affects many cultural and management practices on golf courses. The timing of precipitation and soil moisture conditions at that time have considerable impact on surface and subsurface moisture, moisture available for turfgrass maintenance and growth, trafficability, quality, and chemical transport processes (Beard 1985; Carrow 1985). The probabilistic nature of atmospheric moisture and energy gives the entire hydrologic cycle a stochastic or unpredictable character. As a result of the probabilistic processes in the hydrologic cycle, water-related transport of pollutants and environmental impacts also will have a stochastic element. Spatial variability of soil properties is an additional complication to assessing or predicting hydrological dependent events (Jury 1986a; Peck 1983; Starr 1990; Starr et al. 1978).

During and following precipitation or irrigation, a portion of the water is intercepted by turfgrass and some infiltrates the soil surface. Water not infiltrating into the soil or not evaporating will run off of the turfgrass as runoff water to local drainage systems, streams, lakes, and eventually oceans. Most water infiltrating the turfgrass canopy, thatch layer, and soil surface is subject to evapotranspiration. Evapotranspiration is a combination of (1) evaporation directly from the grass surface (interception loss), (2) transpiration through plant leaves, and (3) evaporation from thatch or soil surfaces if a turfgrass canopy is thin or absent. The remaining portion of the precipitation will percolate to subsurface drainage systems, groundwater, or eventually seep slowly to streams and to the ocean (Carrow 1985; Schwab et al. 1981). Because these are dynamic and complex processes, the amounts of water and associated chemicals have both seasonal and regional variation. Climate, chemical characteristics, soil physical and chemical properties, and management practices also influence partitioning of water and chemicals in the water cycle. Interception, infiltration, depression storage, runoff, and soil water transport are primary components of the hydrologic cycle influencing nonpoint pollutant transport.

Interception is the process in which precipitation is retained by a turfgrass canopy and redistributed to the ground (throughfall and stemflow), to the atmosphere (evaporation), or into the plant. Interception loss is the portion of the precipitation returned to the atmosphere by evaporation from plant surfaces or absorbed into the plant. In the first comprehensive review of interception, Horton (1919) described interception loss as the sum of water stored on plant surfaces at the end of a precipitation event and the evaporation of moisture from wet plant surfaces during a storm.

Several attempts have been made to develop models based on physical processes to describe interception; however, most of these equations are crude approximations (Schwab et al. 1981; Woolhiser 1976). Many hydrologic transport models use an explicit lumped storage term with depletion by evaporation to incorporate interception into the hydrologic portion of the model (e.g., Carsel et al. 1984). From several data surveys, interception storage and loss from grass canopies ranges from 0.10 to 0.90 cm (Woolhiser 1976; Zinke 1967). Interception capacity of dense grass and crop canopies generally ranges from 0.10 to 0.30 cm (Knisel 1980).

Infiltration describes the entry of water into the thatch layer and soil through pores at the surface. Water entering the soil originates as melting snow, direct precipitation and irrigation, and throughfall from the plant canopy (Carrow 1985; Carrow et al. 1990). High rates of infiltration reduces the hazard of surface runoff and detention storage on turfgrass surfaces, but enhances leaching losses of water through the seed bed and soil. Maintaining open surface pores in turfgrass maximizes initial rates of infiltration. Many soil properties and turfgrass management practices affect soil porosity. The major properties and practices including soil texture and structure, aggregate stability, organic matter, thatch layer, initial water content, irrigation scheduling, drainage control, cultivation, and traffic control all influence infiltration (Beard 1982; Carrow 1985; Woolhiser 1976). Management practices that reduce the volume of surface runoff by increasing infiltration have been found to be the most effective in reducing runoff losses of soluble and sediment-bound chemicals (Caro 1976; Frere 1976; Koehler et al. 1982a; Watschke and Mumma 1989).

The rate of infiltration declines rapidly after the start of precipitation or irrigation. Within 1–2 hr, infiltration reaches a constant value as the surface layers of soil are saturated (Carrow 1985). Sandy soils have higher rates of infiltration than silts and clays, although well-aggregated clay soils may have very high initial rates of infiltration (Thien 1983). Surface runoff of water occurs when the rate of rainfall or irrigation exceeds the infiltration rate. Soil compaction by rain drop impact, machinery, or excessive traffic reduces infiltration capacity, increases surface runoff, and increases ponding (Beard 1982; Carrow 1985). Nutrient management, traffic control, thatch control, and periodic cultivation mitigate surface compaction by increasing soil organic matter content, increasing aggregate stability, and reducing soil compaction (Carrow 1985; O'Neil and Carrow 1982, 1983; Woolhiser 1976). Turfgrass protects the soil surface from compaction, as well as reducing initial surface water content. These conditions enhance infiltration rates. Maintenance of high infiltration rates is critical for reduction of runoff losses of water and maintenance of quality playing surfaces on golf courses (Beard 1982).

Development of thatch, black, and hydrophobic layers under turfgrass will influence rates of infiltration (Carrow 1985; Hurto et al. 1980; Schmidt and Blaser 1969; Taylor and Blake 1982; Templeton et al. 1989). These layers may delay infiltration by changing the hydraulic properties of the surface soil (Harris 1978; Hurto et al.

1980). Taylor and Blake (1982) investigated the effect of a thatch layer over sand in field and column studies. Once a constant rate of infiltration was attained, the rate of infiltration was not different between treatments with or without thatch. The initial infiltration rate for thatch-covered sand was much lower than the sustained infiltration rate. The initial movement of water into sand without thatch was much higher than the sustained rate of infiltration. The initial rate of infiltration was very low on thatch-covered sand (0.3 cm min^{-1}), although within 12 min infiltration reached a constant rate. The effect of thatch on initial rates of infiltration should be taken into account in irrigation scheduling. Use of wetting agents and surfactants has been found to increase infiltration rates and irrigation use efficiency on turfgrass systems with hydrophobic thatch, sand, or black layers (Carrow 1985; Moore 1979; Templeton and Rodriquez 1988; Templeton et al. 1989). Use of a short prewetting cycle prior to irrigation may enhance initial rates of infiltration (Carrow 1985; Taylor and Blake 1982).

Evapotranspiration (ET) is the process of water transfer into the atmosphere from vegetated land surfaces. ET is a composite process of evaporation from soil surfaces and transpiration from plants (Beard 1985; Rosenberg 1974). ET is a critical component of the hydrologic cycle linking transport of water between turfgrass, soil, and the atmosphere. Rates of ET have been used as an approximation of water use by turfgrass (Beard 1985, 1989; Carrow et al. 1990). ET and soil texture are the dominant factors in determining (1) the amount of water available for plant growth and stored in the soil at any given time, (2) the amount of excess water available for subsurface leaching and groundwater recharge (drainage), and (3) indirect control of surface runoff by the influence on antecedent soil moisture conditions.

ET transports about 70% of the water falling as precipitation on the continental United States back to the atmosphere (Woolhiser 1976). In turfgrass systems, up to 90% of irrigation water and precipitation may be lost by ET. Water consumption of most turfgrasses ranges between 2.5–9 mm d^{-1} (Beard 1973, 1985). Water availability and cost of irrigation is one of the major factors in growing turfgrass on golf courses and other lawns (Butler et al. 1985; Watson 1985). Relatively few studies on turfgrass, ET, and water conservation were published prior to the 1970s (Beard 1985). Research on water conservation, water use and irrigation practices, turfgrass survival, and evaporative demand has expanded considerably in the 1980s (e.g., Aronson et al. 1987a, 1987b; Baker and van Bavel 1986; Biran et al. 1981; Kneebone and Pepper 1982, 1984; Krans and Johnson 1974; Nus and Hodges 1986; Throssell et al. 1987).

The specific rate and amount of ET is influenced by soil water content, soil physical properties, vapor pressure gradient, soil-air temperatures, wind, pattern of precipitation and irrigation, turfgrass cultivar, and cultural practices (Beard 1985; Blad 1983; Ritchie and Johnson 1990). The physics of ET and concepts of potential ET have been extensively reviewed in several texts and literature reviews (Blad 1983; Ghildyal and Tripathi 1987; Howell 1990; Sharma 1984; Sinclair 1990).

Basic methods of predicting ET have been grouped into three categories by Hatfield (1990) and Schwab et al. (1981). Mass transfer equations use the concept that water moves away from evaporating and transpiring surfaces in response to vapor pressure gradients combined with turbulent mixing of air. Energy balance approaches have provided the most practical methods for evaluation of ET. Empirical methods have been developed from experience and field research. These methods generally rely on the assumption that energy available for evaporation is proportional to temperature. Excellent reviews of the prolific literature on methods to estimate ET have been presented by Beard (1985), Blad (1983), Hatfield (1990), and Schwab et al. (1981).

Surface runoff, or overland flow, and drainage effluent is an important link between golf courses and other turfgrass systems with surface waters. Surface runoff transports both dissolved chemicals and entrained sediment (Daniel et al. 1979; Morton et al. 1988; Watson 1985; Woolhiser 1976). Surface runoff occurs after the demands of evaporation, interception, infiltration, and detention storage have been satisfied (Schwab et al. 1981). Although the volume of surface runoff is relatively low from developed turfgrass systems (EPA 1983; Gross et al. 1990; Hayes et al. 1978; Tollner et al. 1977; Welterlen et al. 1989), the volume of runoff from bare soil on turfgrass construction sites could be very high (Daniel et al. 1979; Schuler 1987).

The rate of surface runoff is influenced by the specific hydraulic and resistance characteristics of individual slopes (Stewart et al. 1975; Woolhiser 1976). Factors affecting runoff rates and volumes include (1) precipitation duration, intensity, and spatial extent; (2) size, shape, orientation, topography, and geology of the golf course watershed; (3) soil physical and chemical properties, infiltration capacity, and antecedent soil moisture conditions; (4) type and extent of grass cover (sod vs seed); and (5) cultural practices (Watson 1985; Welterlen et al. 1989). Runoff is a problem on sloped sites, soils with fine texture (e.g., silt loam and clayey soils), thatched turfgrass, compacted soils, and precipitation or application of irrigation at rates exceeding the rate of infiltration. Reduction of runoff requires site remeditation with cultivation, control of thatch, and control of irrigation rates (Carrow 1985).

Following infiltration of water into the soil, gravity, moisture and temperature gradients, evaporation, plant uptake, tile drainage systems, and soil storage are all forces acting on soil water. These forces control the movement and redistribution of water throughout the soil profile (Carrow 1985; Hillel 1971; Thien 1983). Quantitative description of this redistribution of soil water is difficult as conditions vary from the high tension gradients at the wetting front to saturated conditions just below the infiltration zone (Thien 1983). Description of soil water movement under turfgrass is further complicated by soil layering, the presence of thatch, spatial variability of soil physical properties, and creation of nonuniform hydraulic conditions as water moves through the soil (Hanks and Ashcroft 1976; Starr et al. 1978).

The movement, or percolation, of water through soil is determined by the hydraulic conductivity of the soil and the hydraulic gradient. The hydraulic conductivity refers to the ability of a soil to transmit water through its pore-channel system. The hydraulic gradient is the driving force of water movement, which reflects the steepness of the gradient between the forces that moves water through the soil. The three types of soil water movement are saturated flow, unsaturated flow, and vapor transport (Table 4). A description of saturated and unsaturated flow processes is discussed in Section 2.3, "Transport Processes."

2.3 TRANSPORT PROCESSES

2.3.1 Surface Water

Turfgrass nutrients and pesticides are transported to surface water in runoff water and on eroded sediment. Erosion and surface runoff processes in relation to water quality and environmental effects have been examined in detail by Anderson et al. (1989), Leonard (1988, 1990), and Stewart et al. (1975). During construction and development of turfgrass systems, when the mineral soil is exposed to the action of wind and water, considerable erosion and runoff will occur (Daniel et al. 1979). After turfgrass has been established, the level of runoff water, sediment, and chemical transport by surface processes is significantly reduced (Bennett 1979; Gross et al. 1990, 1991; Schuler 1987; Watson 1985).

Energy provided by the action of wind and water is the factor responsible for most cases of accelerated soil erosion. In general, soil erosion involves (1) soil particle detachment, (2) soil particle transport, and (3) mitigating forces retarding erosion. Both mineral and organic soil fractions are lost with erosion. The major variables affecting soil erosion are climate, soil, vegetation, and topography (Schwab et al. 1981).

Climatic factors influencing erosion are precipitation, temperature, wind, humidity, and solar radiation. Turfgrass significantly reduces erosive forces by (1) interception of rainfall, (2) reduction of the magnitude and velocity of surface runoff, (3) physical binding of soil particles, (4) improvement of soil physical properties involved with structural stability and particle cohesiveness, and (5) reduction of soil moisture by transpiration resulting in increased water holding capacity. The efficacy of turfgrass as a soil and water conservation practice varies with season, cover type, soil, and climate (Stewart et al. 1975).

Soil physical properties influence water infiltration capacity and the extent to

Table 4. Conditions for Water Movement in Soils

Conditions/ factors	Type of movement Saturated	Unsaturated	Vapor
Conditions necessary for flow	Soil water is at zero potential	Soil water is between –0.033 MPa and –3.1 MPa, continuous water films	Soil water potential greater than –3.1 MPa, discontinuous water films
Factors determining:			
Hydraulic conductivity	Pore size	Degree of unsaturation	Water-free pore space
Driving force	Hydrostatic head	Potential gradient	Vapor pressure gradient
Comparative flow rates	Sand>loam>clay	Clay>loam>sand	Warm, moist > cool, dry

From Carrow 1985; Thien 1983.

which the soil can be dispersed and transported. Soil properties directly influencing potential soil erosion include soil texture, soil structure, organic matter content, moisture content, bulk density, and biological activity. Length and gradient of slope, as well as the size and shape of the runoff area, are the important topographic features controlling the potential extent of water erosion (Wischmeier 1976). Different categories of water erosion have been classified by mechanisms of soil particle detachment and transport. These erosive processes include raindrop, sheet, rill, gully, and stream channel.

Erosion is a selective process with respect to particle size and density. Runoff has greater potential to carry smaller and lighter particles to surface waters. The finer particles have higher capacity per unit surface area to adsorb organic nitrogen, absorbed or adsorbed phosphorus, and adsorbed pesticides. These sorbed nutrients and pesticides will be concentrated by water transport in relation to bulk soil (Frere 1976; Sharpley 1980). Enrichment ratios (the ratio of sediment to soil chemical concentration) of 2 to 6 have been reported for phosphorus (Sharpley 1980). Barrows and Kilmer (1963) reported an average sediment enrichment ratio for nitrogen as 2.7. Pesticide enrichment in sediment depends on the specific chemical properties of the pesticide (Smolen et al. 1984; Wauchope 1978).

Bailey et al. (1974) described the basic modes of chemical transport within the moving water. The mechanisms are (1) diffusion and turbulent transport of the dissolved compound from soil solution into overland flow, (2) desorption of the compound from soil particles into the moving water, (3) dissolution of stationary particulate matter trapped at the soil/runoff boundary, (4) scouring of particulate matter with adsorbed compounds into overland flow, and (5) dissolution of adsorbed compounds from the entrained sediments. Partitioning of chemicals between particulate and solution phase depends on the chemical characteristics of specific compounds, water solubility of specific compounds, charge of ionic species, nature of the adsorbing surfaces of the sediment, and duration of runoff event.

The rate, timing, and volume of runoff water have considerable impact on the form and quantity of sediment and chemicals transported in surface runoff (Knisel 1980; Welterlen et al. 1989). Location of surface water, runoff patterns, golf course design and drainage systems, and location of areas used for repeated chemical handling are significant contributing factors to potential surface and subsurface movement of applied chemicals (Erickson and Kuhlman 1987).

Attempts to predict rates and volume of runoff still require considerable research and refinement. Several computer models of chemical surface and subsurface transport use the simplistic Soil Conservation Service (SCS) runoff curve technique to evaluate the timing of overland flow (Carsel et al. 1984; Haith 1980; SCS 1972, 1973). However, several more sophisticated models using one-dimensional flow equations to predict runoff hydrographs are being developed (Hernandez et al. 1989).

Predictive methods have been developed to evaluate the potential for loss of surface soils by accelerated erosion. The universal soil loss equation (USLE) is an empirical method for evaluation of soil loss related to raindrop, sheet, and rill erosion (Peterson and Swan 1979; Schwab et al. 1981). The empirical USLE formula is based on (1) estimated average annual loss of soil per unit area; (2) a rainfall factor; (3) a soil erodibility index; (4) a topographic factor; and (5) a cover and management factor to adjust the soil erodibility factor for soil surface cover, soil surface mulch, coarse fragments, and land management practices. The USLE equation does not predict soil losses for a particular year. The USLE estimates long-term (10-year intervals) soil erosion as an average of annual soil losses. Several computer models based on the USLE concept have been used to estimate relative erosion, sediment transport, and sediment entrained transport of chemicals from agricultural watersheds (Haith 1980; Knisel 1980; Young et al. 1986).

Recently, a cooperative USDA SCS, Forest Service, and U.S Department of the Interior-Bureau of Land Management (BLM) project developed a mechanistic model of soil erosion to replace the traditional index approach for prediction of surface runoff and erosion (Lane and Nearing 1989; Nearing et al. 1990). The USDA Water Erosion Prediction Project (WEPP) hillslope model is based on the fundamental processes of infiltration, surface and soil hydrology, surface runoff, ET, residue decomposition, erosion mechanics, plant-soil-atmosphere interactions, climate, and irrigation practices (Lane and Nearing 1989). This new technology for simulation of water erosion processes will accommodate spatial and temporal variability in topography, surface roughness, soil properties, and various land use conditions (Nearing et al. 1990). Although the WEPP model has not been specifically designed for turfgrass systems, the simulation capabilities of the computer model easily could be modified for predictive purposes during golf course design and construction.

Soil erosion by wind occurs when the wind exerts sufficient force on the soil surface to dislodge easily detachable soil aggregates, soil particles, and sand grains (Skidmore 1982; Zobeck 1991). These detached particles are then transported away by the wind. Wind erosion is a major problem on about 30 million hectares of land in the western United States, but generally has not been considered a major component of environmental issues on well-developed turfgrass or grassland systems (Bennett 1979; Hagen 1991; Skidmore 1982; Stewart et al. 1975, 1976). However, on golf course and turfgrass construction sites, bare soil surfaces are temporarily exposed to the potential erosive action of wind. Under these conditions, wettable powder formulations of several pesticides applied to undeveloped turfgrass surfaces may be susceptible to loss by wind erosion (Richards et al. 1987; Taylor and Glotfelty 1988). These wind-blown pesticide particles may impose additional risk for nonpoint source contamination of unintended nearby areas and surface waters.

Conditions conducive to aeolian transport of soils are (1) loose, dry, and finely subdivided soil; (2) smooth surfaces and absent or sparse vegetative cover; (3) extensive open areas; and (4) wind with sufficient force to move soil particles (Skidmore 1982; Schwab et al. 1981; Zobeck 1991). These conditions are most prevalent in semiarid regions when precipitation is inadequate, but also occur in subhumid and humid regions, particularly in areas with noncohesive soils (Schwab et al. 1981). Suspension, saltation, and surface creep are the dominant mechanisms involved in aerial transport of soil particles. The process of wind erosion and soil-conserving practices of grasses has been described in greater detail in several texts and literature reviews (Schwab et al. 1981; Skidmore 1982; Skidmore and Woodruff 1968; Skidmore et al. 1970).

The mechanics and major factors influencing wind erosion are incorporated into an empirical relationship similar to the USLE. The empirical relationship for the wind erosion equation (WEQ) is based on (1) estimated average annual loss of soil per unit area; (2) a climatic factor; (3) a soil erodibility index; (4) a microrelief or soil roughness factor; (5) field length along the direction of the prevailing wind; and (6) equivalence factor for vegetative cover (Skidmore 1982). The relationships among these variables are complex and cannot be expressed in simple mathematical terms as derived for the USLE (Zobeck 1991). Skidmore and Woodruff (1968) developed graphs, tables, maps, and nomographs to estimate the factors for wind erosion. Recently, the SCS developed a comprehensive field guide for field application of the WEQ (SCS 1988). A process-oriented model, the Wind Erosion Prediction System (WEPS), is being developed by the USDA Agricultural Research Service Wind Erosion Unit (Hagen 1991; Zobeck 1991). This user-oriented model is in development and should replace the traditional empirical approach to estimation of soil wind erosion. The USDA WEPS model is based on the fundamental climatic, soil, microrelief, vegetative, and cultural processes that influence airborne transport of soil particles. This new technology for simulation of wind erosion will accommodate spatial and temporal variability in microrelief, soil particle and aggregate properties, crust formation, and various land use practices (Zobeck 1991).

2.3.2 Groundwater

Groundwater is water which occurs in fully saturated soils and geological formations. The top of this saturated zone is the water table, the level to which water rises at atmospheric pressure in a well or hole bored into the earth. Above the water table in the unsaturated or vadose zone, all of the pore spaces are not filled with water (Figure 2). Groundwater moves through permeable saturated strata of rock or unconsolidated materials called aquifers (Pye 1983; Watson 1985). Productive or high water-yielding aquifers consist of (1) unconsolidated materials such as sands and gravels, (2) permeable rock such as sandstone, or (3) rocks with large fractures or cavities including limestone (EPA 1986; Pye et al. 1983).

There are two main types of aquifers, unconfined and confined (Figure 2). Aquifers have been characterized by their ability to yield relatively large amounts of water. An unconfined aquifer is not protected by a layer of impermeable material. Unconfined aquifers are recharged through the soil surface and unsaturated zone. Deep leaching of precipitation or irrigation water through soil and the unsaturated zone percolates directly to unconfined aquifers. Unconfined aquifers are directly exposed to nonpoint source contamination by chemicals transported by water infiltrating from the surface (EPA 1986).

Water in unconfined aquifers is at atmospheric pressure. The depth of the water table is dependent on the amount of water in storage, which in turn is dependent upon seasonal recharge. Perched water tables occur above true water tables when a limiting layer, a lens of lower or higher permeable materials, restricts the downward movement of water. A thin zone of saturation will form above the impermeable lens, forming a temporary aquifer. Perched water tables also form due to disruption of internal soil drainage resulting from textural discontinuities (Carrow 1985). The capillary rise of water above textural discontinuities will provide additional water for turfgrass systems. The United States Golf Association (USGA), Green Section golf green mix has an intentional perched water table at the pea gravel-coarse sand interface (Bengeyfield 1989). This perched water table increases the amount of water in the zone of saturation and capillary fringe of the green root mix (Carrow 1985). The perched water table exists in the finer seed bed soil, or root mix, until sufficient hydraulic head causes the saturated fine-textured interface to release the perched water into the coarse sand layer.

Aquifers overlain and separated by layers of impermeable geologic formations are referred to as confined or artesian aquifers (Pye 1983). Confined aquifers generally have restricted recharge areas and are at saturated pressure greater than atmospheric pressure. The impermeable layers, which are soil and rock formations of low hydraulic

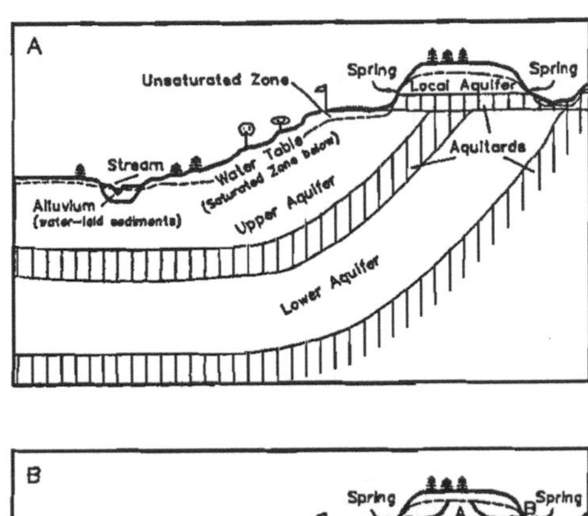

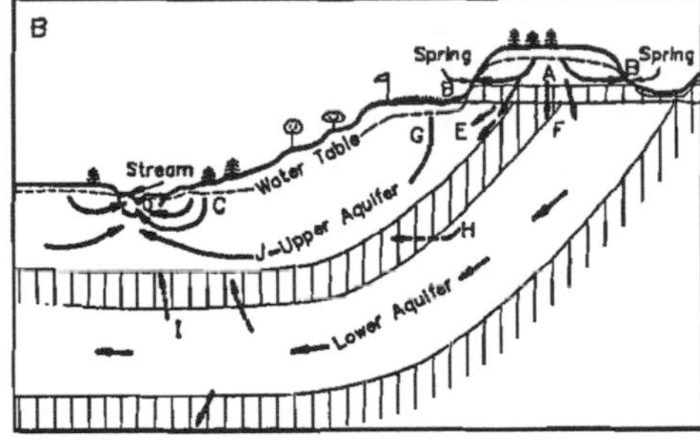

Fig. 2. Cross-section (A) and generalized direction of water flow (B) of aquifers, aquitards, and the unsaturated zone.

conductivity, such as shale, are called aquitards (Figure 2). Aquitards are not entirely impermeable. They will permit some recharge from the surface or discharge to lower aquifers. This allows for potential movement of contaminants into confined aquifers (Pye et al. 1983).

Groundwater moves in response to gravity, pressure, diffusion, and friction. Groundwater flows from areas of high pressure to areas of low pressure (Pye et al. 1983). The rates and direction of flow is dependent on hydraulic gradients, distance between flow points, thickness of the aquifer, and hydraulic conductivity of intervening materials (Pye 1983). The velocity of water flow is the primary difference between surface water flow and groundwater flow. The movement of groundwater is slow and usually occurs under laminar conditions. Groundwater flow rates vary from millimeters to meters per day (Pye and Patrick 1983).

Recharge of groundwater is a complex and variable process (Freeze and Cherry 1979; Pettyjohn 1982). The quantity and quality of water which enters the soil and eventually percolates to groundwater are dependent on movement of water within the hydrologic cycle and land management practices. Variability of regional groundwater recharge and depth to the water table depends on climate variability, soils, topography, and subsurface geology. The volume and timing of precipitation and ET demand largely control groundwater recharge. Subsurface geology affects the rate of water movement within the aquifer. Considering the variability of recharge and groundwater flux, the residence time of groundwater is also highly variable. Residence time in sand/gravel and carbonate aquifers may be days or weeks. In regional aquifers, groundwater movement from the point of entry to discharge area may require hundreds to thousands of years (Freeze and Cherry 1979).

Water and chemical transport in the saturated and unsaturated zones are also seasonably variable. With high rates of ET in dry or hot seasons, shallow water tables will be lowered, and aquifers will not be subject to recharge. During wet and cool seasons, water will be available to recharge groundwater, as well as transport surface applied chemicals toward the water table (Anderson et al. 1989; Pratt 1985). Nutrients and pesticides are transported to groundwater primarily in solution with water percolating through the soil rooting zone. These transport processes are subject to the same seasonal variation as water flow. Application of chemicals, rates of degradation or transformation, and uptake by plants are seasonably variable as well. Potential movement of chemicals into groundwater will vary with time and rate of application and the amount of water passing beyond the vadose zone (Anderson et al. 1989; Balogh and Gordon 1987; Gold et al. 1989). These subsurface transport processes in relation to water quality and environmental impacts have been extensively reviewed by Anderson et al. (1989), Bouchard et al. (1989), Enfield and Yates (1990), Fairchild (1987a), Jury et al. (1987), Keeney (1986), Petrovic (1990), and Pratt (1985).

2.3.3 Water and Chemical Movement in the Unsaturated Zone

The unsaturated zone includes the soil and underlying materials above the water table (Figure 2). Soil consists of minerals, organic matter, water, and air (Figure 3). The relative distribution of each component determines the capacity of a given soil to attenuate or transmit water and dissolved nutrients and pesticides. The surface layers of the soil are differentiated by horizons due to pedogenic processes and management practices (Buol et al. 1989). Soil also is characterized by organic matter content, which usually decreases with depth. The proportion of soil solids has a tendency to increase with depth. The soil pore space, occupied by water and air, tends to decrease with depth. The proportion of the total pore space containing air and water varies with depth, proximity to the water table, soil texture, degree of compaction, cultural operations, season, and precipitation and temperature relations (Carrow 1985; Hanks and Ashcroft 1976).

Turfgrass chemicals that penetrate the dense turfgrass canopy enter the thatch and soil surface layers with percolating water. The turfgrass nutrients and pesticides are distributed between soil water, soil air, on the surface of soil solids, or as undissolved particles and precipitates. Chemicals are generally transported in solution phase to groundwater and not as gases. Therefore, the focus of chemical movement in the unsaturated zone is on solute transport. Water and solute transport processes in soil

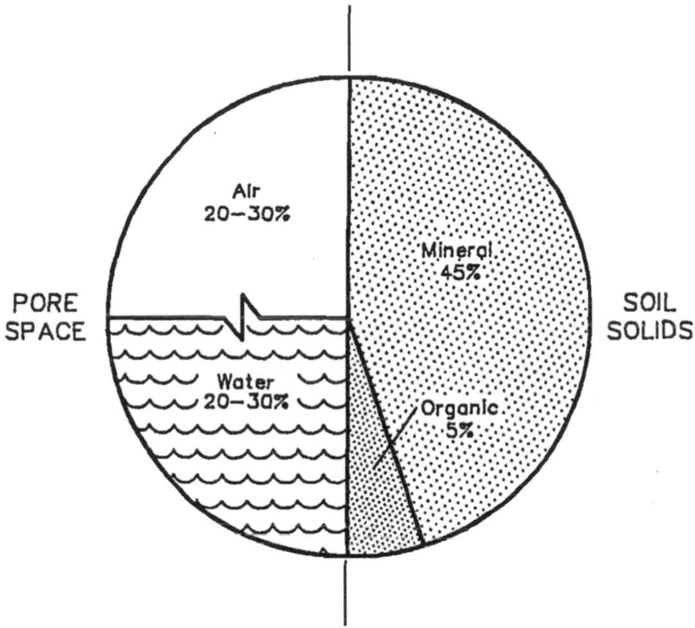

Fig. 3. Distribution of mineral, organic, water, and air fractions of a hypothetical soil.

and the unsaturated zone have been extensively reviewed and discussed by Anderson et al. (1989), Bouchard et al. (1989), Davidson et al. (1983), Hutzler et al. (1989), Jury (1986b), Jury et al. (1987), Rao et al. (1988), and Weber and Miller (1989).

Movement of water and chemicals occurs in a complex, three-dimensional network of pores of varying size, shape, and configuration. A variety of conceptual models have been devised to describe water and solute transport (Davidson et al. 1983; Jury and Ghodrati 1989). The movement of water occurs at differing rates through soil pores and channels depending on their size and shape. As a result, incoming water simultaneously mixes with and displaces existing soil water. This overall process is called hydrodynamic dispersion (Bigger and Nielsen 1976). The rate of water flow is dependent on the hydraulic gradient and hydraulic conductivity (Carrow 1985; Hanks and Ashcroft 1976; Rao et al. 1988). The hydraulic gradient is the energy gradient or driving force required for flow of water. The hydraulic gradient is formed as a combination of gravitational, matrix, pressure, and osmotic potentials. Hydraulic conductivity describes the soil's ability to transmit water.

Under saturated soil conditions, water flows primarily under gravitational potential through macropores and soil channels. Soil macropores and channels are created by earthworm activity, roots, formation of soil cracks, and formation of interfaces between adjacent soil peds. Hydraulic conductivity is at a maximum during saturated flow with the rate of flow limited only by pore size and pore continuity. Saturated conductivity increases with pore size. Sandy soils will transmit water much faster than fine-textured soils under saturated conditions. Significant soil water drainage and solute leaching occurs only when soils are wetter than field capacity (10–33 kPa). As soils dry, unsaturated conditions prevail. The rate of water transport and conductivity decreases rapidly with decreasing soil water content. With decreasing soil water potential, water is transmitted through micropores at exponentially reduced rates. Matric potential is the driving force for water movement under unsaturated conditions. Saturated and unsaturated flow of water are dependent on soil texture, bulk density, organic matter, and structure.

Chemicals dissolved in soil water move through soil and the unsaturated zone by two processes: (1) mass transport, convective movement in water flow, and (2) diffusion, random motion of ions and molecules (Davidson et al. 1983; Rao et al. 1988). Mass flow is the passive transport of dissolved solute within moving soil water. Mass transport is approximated by the product of the volume of water flow and the concentration of dissolved solute. Liquid diffusion is the transport of solutes by molecular collisions which move ions and molecules down concentration gradients (Jury and Valentine 1986).

Only convective/dispersive transport is effective in moving solutes from the soil surface to groundwater (Jury et al. 1987). Many single and multidimensional convective/dispersive models of water and solute flux have been proposed to simulate water and chemical transport (Balogh and Gordon 1987; Hern and Melancon 1986; Javandel et al. 1984; Oster 1982; Wagenet and Hutson 1986). Mathematical description and derivation of water and solute transport equations are provided by Davidson et al. (1983) and Jury (1986b).

Few models of water and chemical transport have been modified for use on turfgrass systems. Several surface and subsurface simulation models could be modified to meet the unique circumstances and soil conditions occurring on golf courses and heavily used turfgrass systems (e.g., Balogh and Gordon 1987; Carsel et al. 1984, 1985; Lane and Nearing 1989; Wagenet and Hutson 1986). Use of program modules simulating artificial drainage and perched water tables could be incorporated into existing transport models for simulation of water, sediment, and chemical movement in turfgrass systems (Perry et al. 1988; Stuff and Dale 1978; Ward et al. 1988).

In layered soils, especially soils with thin layers and many textural discontinuities, water movement may be unstable at textural interfaces and move rapidly in fingers of flow through coarser subsoil (Baker and Hillel 1990; Starr et al. 1978). The phenomena of fingering through layered soils has been theorized to result from accelerated flow in constricted flow regions (Hillel and Baker 1988). Tamai et al. (1987) described fingering as a response to differential air entrapment within the soil matrix. Fingering and preferential flow complicate the simplified assumptions in many simulation models.

Rapid movement of water through soil macropores and restricted regions is described as preferential flow. Clothier and White (1981) defined macropores as voids and channels with diameters exceeding 750 μm (10^{-6} meters, micrometer). Luxmoore (1981) designated three classes of soil pores based on diameter: macropores (>1000 μm), mesopores (10–1000 μm), and micropores (<10 μm). The percent of the total soil volume occupied by pores that actually transmit water and solutes may be quite small. Bouma et al. (1979) reported that water-conducting macropores often occupy less than 1% of the total volume of soil. Dunn and Phillips (1991) also observed concentration of water flow through soil macropores in both conventional tillage and no-tillage plots of corn grown on silt loam soils. For the no-tillage plots with receiving minimal physical disturbance, 83% of the total flux of water was conducted through an effective porosity of 0.07% of the total volume of the bulk soil. In the conventional tillage plots, 73% of the total flux moved through 0.03% of the total volume of the bulk soil. In studies of water infiltration on undisturbed forest soils, Watson and Luxmoore (1986) and Wilson and Luxmoore (1988) quantified the distribution of water flow. Watson and Luxmoore (1986) observed that 96% of the water flux moved through soil macropores which constituted only 0.32% of the soil volume. They also observed a direct relationship between larger flow volumes and the degree macropores contributed to total moisture flux. Wilson and Luxmoore (1988) reported that 85% of ponded flow was transmitted through soil macropores and large mesopores. Considering that most turfgrass soils are relatively undisturbed, preferential flow through soil macropores also occurs through turfgrass soil. This important water and solute transport process required further research under actual golf course field conditions.

Preferential movement of water and solutes in macropores or restricted regions in the bulk soil bypasses most of the soil matrix that reduces the downward flux of chemicals applied to the soil surface or turfgrass canopy. The assumption of one-dimensional movement of water, piston flow, through heavily layered soils will lead to large errors in estimating the movement of water and dissolved solutes. Accurate prediction of this type of preferential flow pattern requires further research under actual turfgrass management conditions. Preferential flow patterns and fingering also have been observed in uniform sandy soil (Ghodrati and Jury 1990). Preferential flow and finger patterns which occur on sandy golf greens are a result of specific construction requirements, followed by excessive topdressing, and formation of thatch layers.

Most transport equations and models are deterministic and do not account for spatial variability of flux, spatial variability of soil hydraulic properties, preferential flow patterns, drainage systems, and perched water tables in the field and turfgrass systems (Jury 1986a; Nielsen et al. 1983; Peck 1983; Starr et al. 1978). A stochastic approach to modeling spatially variable transport of solutes and to modeling preferential flow has been proposed (Jury 1982). Reviews of scaling and geostatistical methods for analysis of spatially variable transport processes have been presented by Bigger and Nielsen (1976), Nielsen et al. (1983), Vieira et al. (1983), and Wagenet and Rao (1990).

2.3.4 Soil Properties Affecting Movement of Turfgrass Chemicals

The factors influencing both surface and subsurface transport of nutrients and pesticides include soil texture and layering, soil porosity, soil organic matter content, adsorption processes, chemical characteristics, adsorption, degradation, plant uptake, volatile losses, and turfgrass management practices (Table 5). Soil-specific factors will be briefly discussed in this section, and management-specific issues will be discussed in the chapters related to pest and nutrient management-practices (Balogh and Anderson 1992; Walker and Branham 1992).

The distribution of soil particle sizes is defined as soil texture (Hillel 1971). One of the traditional methods of classifying soil texture is to divide soil particles into three size classes: sand (0.05–2.00 mm), silt (0.002–0.05 mm), and clay (<0.002 mm). Soil texture is a primary soil physical property influencing bulk density, soil structure, infiltration, porosity, water retention characteristics, water available for turfgrass and other vegetation, and hydraulic conductivity. The degree of play, cultural, and management considerations, especially those related to heavy traffic, dictate that golf greens have a sandy to sandy loam texture (Beard 1982). However, other turfgrass components of golf courses have a much wider range of texture.

Sandy soils are dominated by macropores which allow rapid infiltration and drainage of excess water (Carrow 1985). With few small pores to hold available moisture for turfgrass, sandy soils are susceptible to drought conditions. Rapid infiltration and leaching in sandy soils result in reduced risk of surface runoff, but increased potential for subsurface transport (Carrow 1985). Clay particles have large surface areas compared to sand, thus are capable of greater absorption potential of pollutants (Freeze and Cherry 1979). Clay soils are dominated by micropores with reduced rates of infiltration and conductivity. With the exception of soils with shrink-swell cracks, clay soils have reduced potential for subsurface transport of applied chemicals, but much higher potential for runoff losses. Intermediate-textured soils, loams and silt loams, have intermediate transport characteristics (Hanks and Ashcroft 1976).

Soil structure is determined by the aggregation of individual soil particles into secondary shapes. Soil structure directly influences bulk density, macropore space and configuration, infiltration rates, hydraulic conductivity, and erosive potential. Soil structure is influenced by degree of compaction, turfgrass management practices, cultivation, texture, and organic matter-clay interactions.

Table 5. Major Factors Influencing Transport of Applied Nutrients and Pesticides

Soil parameters	Environmental parameters	Management parameters
Water content	Precipitation	Chemical type, rate, formulation, concentration
Bulk density or porosity	Temperature	
Structure		Irrigation
Hydraulic conductivity, water retention characteristics	Evapotranspiration	Drainage of greens
	Parent material	
Texture and clay content		Cultivation and traffic control
Organic matter content		Top dressing and development of soil layers and discontinuities
		Development of thatch layers
pH and additions of lime or gypsum		
Depth to water table		Additions of fill or drainage (if water table is too close to the soil surface)

Soil texture and stability of soil structure influences the soil porosity, the volume of air and water (total pore space) that occurs in a soil. Porosity in the soil rooting zone varies with changes in soil texture and structure. The total pore space in a soil ranges from 30 to 60%, with an overall average of 50% (Figure 3) (Hillel 1971). Although the coarser sandy texture soils (approximately 30–40% porosity) tend to be less porous than finer-textured soils, the median size of the pores is greater in sandy soils (Brady 1974; Hillel 1971; Carrow 1985). Texture, structure, and porosity all affect the capacity of soil to retain water and entrained compounds. Sands with larger pore size distribution, rapid conductivity, and low water holding capacity are at greater risk of leaching than silt loams with greater water holding capacity and lower conductivity (Anderson et al. 1989). Puddled or compacted surface soils, having lost surface structure, are at a greater risk of surface runoff losses with reduced infiltration (Carrow 1985).

Clay minerals and soil organic matter provide absorptive surfaces for retention of inorganic and organic chemicals. These sorption processes are produced by physical and chemical interactions of nutrients and pesticides with clay and organic matter surfaces. Adsorption reduces transport of chemicals to surface water and groundwater by reduction of the amount of an applied compound in solution (Rao and Davidson 1982). Adsorption may enhance loss of phosphorus and pesticides with accelerated erosion from bare soil surfaces during construction and renovation (Daniel et al. 1979; Koehler et al. 1982a; Wauchope 1978). Although pesticides are primarily adsorbed on organic matter surfaces, a few charged pesticides (e.g., glyphosate) are strongly adsorbed to both the organic and inorganic cation exchange complex (CEC) (Hassett and Banwart 1989; Zielke et al. 1989).

Positively charged nutrient ions (e.g., calcium, magnesium, potassium, sodium, and ammonium) are also strongly adsorbed by the clay-organic CEC. Negatively charged ions, nitrate and chloride, are not readily adsorbed by soil organic and clay surfaces. These soluble ions will move readily with surface and subsurface water. The phosphate anion is adsorbed or precipitated from soil solution under both acidic and basic conditions. Phosphate does not move readily in solution that is in contact with soil (de Camargo et al. 1979; Enfield and Ellis 1983; Ryden and Pratt 1980). However, in sandy soils receiving high rates and frequent applications of available nitrogen and phosphorus fertilizer, there is some elevated risk of subsurface movement of ammonium and phosphate (Anderson et al. 1989).

2.3.5 Additional Concerns for Susceptibility of Groundwater to Contamination

One of the primary concerns regarding contamination of groundwater is that aquifers are out of sight (Pye 1983). Groundwater processes are complex, monitoring is difficult, and remedial action often is prohibitively expensive (Cohen et al. 1990; Sharefkin et al. 1983). Another concern is the potential that once groundwater contamination has been discovered, remedial action could be impossible. The extent of groundwater contamination is unknown, but reports of groundwater pollution have occurred in every region of the United States (Bell et al. 1991; Canter 1987; Daugherty 1991; Fairchild 1987b; Water Resources Council 1981). Research on groundwater contamination has focused on industrial point sources, agricultural systems, and hydrocarbon transport from leaking storage tanks (Anderson et al. 1989; Cheng 1990; Kostecki and Calabrese 1991). However, recent concern has been expressed regarding nonpoint sources from urban and turfgrass systems (Grant 1987; Leslie and Metcalf 1989; Morton et al. 1988).

Some of the general principles and concepts for mitigation of adverse environmental impacts on turfgrass systems have been identified in agricultural systems. Contamination of groundwater by nitrates and pesticides has been limited primarily to upper aquifers (EPA 1988; Fairchild 1987a). The shallow distribution of management compounds may be a consequence of slow movement and a short period of intense application of chemicals in turfgrass systems. However, regional nonpoint pollution of deep aquifers has been observed with increasing frequency (Anderson et al. 1989; Freeze and Cherry 1979).

Fractured aquifers, shallow aquifers, and intensified irrigation practices have additional unique features related to nonpoint impacts on groundwater quality (Keeney 1986; Koehler et al. 1982b; Pratt 1985; Smolen et al. 1984). Large channels and underground caves in limestone rock and tubes in lava flows allow rapid underground transport of groundwater. If these openings are exposed to the surface, both soil water flux and surface runoff sources of chemicals will be transported rapidly with negligible dilution. Karst topography is characterized by sinkholes, disappearing streams, and disappearing valleys. These features act as direct conduits for the entry of surface water, entrained sediments, and dissolved chemicals into underground channels and caverns in limestone aquifers. Karst aquifers are highly susceptible to contamination not only by surface runoff, but also by sediment-adsorbed chemicals, microorganisms, and low solubility pesticides usually not transmitted by drainage water (Hallberg et al. 1983; Libra et al. 1987). Golf courses associated with karst topography are especially sensitive to groundwater impacts and public concern.

Shallow aquifers and unconfined sand and gravel aquifers overlain by sandy soil are also at high risk of nonpoint contamination from surface-applied chemicals (EPA 1986, 1988; Pratt 1985). Rapid convective transport, low absorptive capacity, and short water retention time are all factors in enhanced risk of contamination in this situation. Shallow depth to bedrock will decrease the residence time in soil for pollutant load reduction by degradation, adsorption, and plant uptake. Unconfined aquifers beneath shallow soils are also at high risk to nonpoint source contamination. Golf courses built on sandy soils over shallow aquifers are considered at risk for subsurface transport of frequently and heavily applied nutrients and pesticides (Erickson and Kuhlman 1987).

Underground storage tanks (USTs) are frequently used by golf courses and many other industries as a convenient storage facility for gasoline and other petroleum products (Beard 1982; Marian and Brown 1987). Gasoline and other related petroleum products are usually stored in underground tanks for prevention of fires and explosions. Unfortunately, gasoline and other potentially toxic chemicals leaking from USTs are one of the most common causes of groundwater contamination. One of the primary concerns with USTs on golf courses is plumes of soil contamination with potential movement of certain components of gasoline to groundwater (Mackay et al. 1985; Marian and Brown 1987).

The immediate concern at the site of a leaking tank is public safety, fire, and explosion hazards. However, the potential health effects, especially long-term effects, are not as well understood. Most of the public concern and uncertainty associated with hydrocarbon-contaminated soil and groundwater is related to potential exposure to benzene and ethyl dibromide. These compounds are suspected cancer-causing agents (ICF 1989; Daugherty 1991). Calabrese et al. (1988) has reviewed the public health implications of soils and water contaminated with petroleum products. In addition to the immediate human health and groundwater problems, leaking USTs can contaminate surface waters, cause fires and explosions, and generate toxic fumes that may seep into homes and businesses.

The main causes of releases from USTs are piping failure, spills and overfills, and tank corrosion. The U.S. Environmental Protection Agency (EPA) estimates that 80% of all releases of hydrocarbons from USTs occur as a result of failures and fatigue in piping systems (ICF 1989). Many of these failures result from improper installation and maintenance. Corrosion of tank walls, especially unprotected older steel tanks, and structural failure of fiberglass reinforced tanks are another common source of site contamination with petroleum products. Cleanup and remediation of soil and groundwater contaminated with chemicals from leaking USTs can be prohibitively expensive. The New England Interstate Water Pollution Control Commission (NEIWPCC) has suggested that prevention of UST failures, leaks, and spills is the best strategy for environmental and economic protection. Installation of double-walled tanks, early detection of leaks, inventory control, monitoring, and tightness testing are recommended by the NEIWPCC to prevent costly environmental cleanup projects (NEIWPCC 1985).

Although the extent of leaking USTs on golf course facilities has not been documented, the EPA estimates that at least 20% of the approximately 2 million USTs in the United States may be leaking (ICF 1989). In 1990, California alone reported over 13,000 cases of leaking USTs (Daugherty 1991). Other cases of extensive contamination of soil and groundwater from leaking USTs have been reported in Florida (Rudy and Bedosky 1988) and Long Island, New York (Peterec 1988). Given the potential magnitude of the problem, environmental regulators at the federal, state, and local level are rapidly implementing and refining the regulations regarding installation, maintenance, and site remediation associated with leaking USTs (Bell et al. 1991).

From a regulatory perspective, the basic concerns regarding petroleum-contaminated soil and water are (1) uncertain exposure pathways and effects, (2) determination of acceptable cleanup levels, (3) the publics expectations of protection, (4) the need for documentation, (5) the large number of contaminated sites, and (6) the need for timely decisions. The complex array of regulatory concerns and site remediation requirements have been reviewed by Daugherty (1991) and Bell et al. (1991). The rules regarding USTs and petroleum-contaminated soils change frequently, and turfgrass managers should be aware of new stringent tank and environmental standards (Marian and Brown 1987). Additional information on the complex subject of hydrocarbon-contaminated soils and groundwater, site cleanup, and environmental regulation has been compiled in reviews by Kostecki and Calabrese (1991), Noonan and Curtis (1990), and Testa and Winegardner (1990).

The processes that control the environmental fate and movement of leaking

Table 6. Processes Influencing the Environmental Fate of Petroleum Compounds Released from Underground Storage Tanks

Process	Description	General comments
Mass movement of nonaqueous phase liquid (NAPL) compounds	Saturated flow source of contamination; residual compounds trapped in pore spaces where gravity and hydraulic forces are balanced	Compounds penetrate unsaturated zone and spread as a complex plume toward groundwater. Complex distribution of residual contaminants could remain as a long-term source of contamination.
Partitioning in the root and unsaturated zone	Selective solubility, volatilization, and adsorption of individual chemical components of petroleum mixture	Individual chemicals are partitioned in soil air, soil pore water, or adsorption surfaces based on individual chemical characteristics. Partition coefficients suggest aromatic hydrocarbons (e.g., benzene) move into soil water and air; long chain aliphatics are partitioned into soil air.
Movement in soil pore water through the unsaturated zone	Aqueous diffusion to groundwater	Slow process, but may be significant for shallow aquifers.
Leaching through the unsaturated zone in percolating preciptation, irrigation, or leaking water lines	Convective-dispersive movement of dissolved contaminants in periodic flux of water through the soil	Potential source for movement of hydrocarbons, especially aromatics, to groundwater. Movement is retarded by textural and lithic barriers. Concentration of contaminants is reduced by dispersion.
Movement through soil air	Movement of contaminant vapors to the atmosphere or indoor air	Rapid process of hydrocarbon losses to the atmosphere. Losses are reduced if plume of leakage occurs under pavement. Atmospheric losses constitute an immediate public health concern.

Table 6. Processes Influencing the Environmental Fate of Petroleum Compounds Released from Underground Storage Tanks (continued)

Process	Description	General comments
Adsorption-desorption	Temporary retention on organic matter and possibly mineral colloid surfaces	Adsorption is a significant process in surface soils high in organic matter content. Adsorptive processes may not be as important in subsoil and aquifers with low organic matter and mineral colloid surfaces.
Biological transformation	Biological degradation of petroleum products ultimately to carbon dioxide and water; rate and effectiveness is difficult to predict and depends on interaction of many factors	Biodegradation permanently reduces environmental load of spilled contaminants. Occurs to some extent under all conditions, but is enhanced at aerobic margins of contaminant plumes. Use of bacteria has been suggested as a bioremediation technique for contaminated soils and aquifers. However, as analytical techniques improve, bacteria may no longer be able to reduce residual contaminant levels below regulatory detection limits.
Chemical transformation	Oxidation, reduction, and hydrolysis	Generally not considered significant for petroleum products in subsurface environment, although processes are not well understood.

Groundwater flow	Saturated flow with convection, dispersion, and dissolution	Rate of flow is generally slow except in karst topography or fractured bedrock. Public health and water resources are affected if direct human exposure to contaminated water supplies occur.
Immobilization by restriction of flow by impervious layers or aquitards	Flow of water reduced in fine-textured layers or fine-grained formations	Extent of contaminant retardation by restricting layers is difficult to predict. Most sedimentary formations are not completely impervious. The structure of clay in layers and lens is changed by petroleum hydrocarbons which increases permeability. Undetected macropore and/or preferential flow will increase rate of contaminant movement to groundwater.

From Daugherty 1991; Mackay et al. 1985.

gasoline and petroleum products from USTs are similar to the processes for other organic compounds in the soil (Tables 5 and 6). These processes have been discussed in Section 2.3.3 of this chapter and by Balogh and Anderson (1992) in Chapter 4 of this book. Critical components of the processes that affect the fate of hydrocarbons in contaminated soils are (1) multiphase partitioning, (2) transport and attenuation processes, and (3) transformation processes (Table 6). Hydrocarbons released from underground storage tanks exist in different phases including (1) a nonaqueous phase liquid (NAPL) or bulk phase; (2) aqueous phase in soil water or groundwater; (3) vapor phase in soil gas; and (4) adsorbed phase on soil particles and other solid surfaces (Mackay et al. 1985). The liquid phase may exist as a free moving or adsorbed nonaqueous fluid or emulsion (Bonazountas 1991). Organic liquids less dense than water will spread across the top of the water table. The nonaqueous liquids that have a higher density than water will sink to the bottom of a contaminated aquifer. Residual NAPLs become trapped in soil pore spaces and act as a long-term source of contamination. Even after NAPLs are removed from the soil by various physical and biological remediation processes, up to as much as 30% of the soil pore space may contain residual organic liquids (Daugherty 1991).

Other organic compounds in petroleum are relatively water soluble, such as benzene (1780 mg L^{-1}), and dissolve into the aqueous phase of soil and groundwater. Many of the water soluble hydrocarbons from leaking petroleum products are partitioned into unsaturated zone water. These compounds are subject to movement to groundwater in water leaching through the plume of petroleum contamination (Mackay et al. 1985). Subsurface contaminants are subject to convective, dispersive, and diffusive processes (Bonazountas 1991). The role of adsorption in the fate and movement of petroleum contamination in soils is not as well-known. The factors controlling adsorption are dependent on the individual chemical characteristics of the contaminants, soil and aquifer properties, and flow characteristics. As with pesticides, adsorption of petroleum compounds is generally related to the adsorptive capacity of the solid phase media, especially organic matter substrate, and the relative mobility of the individual compound (Mackay et al. 1985). Roy and Griffen (1985) classified benzene as highly mobile (low adsorption), o-xylene as moderately mobile (medium adsorption), and m-xylene and p-xylene are considered as low mobility petroleum compounds (high adsorption).

Biological degradation by bacteria is the most important transformation process that permanently reduces the environmental load of organic compounds in petroleum-contaminated soils. Use of bacteria has been suggested and tested as a method to bioremediate contaminated sites (Piotrowski 1991). Although the rates of microbial degradation of petroleum products tend to be unpredictable, interaction between the oxygen and the organic contaminants is considered to be the rate-limiting step (Borden and Bedient 1986).

Several attempts have been made to combine the complex fate and transport processes of petroleum-contaminated soils into computer simulation models

(Bonazountas 1991). Although several of these models are mathematically elegant, the series of equations often require unavailable input parameters and immense computational effort in order to simulate a single site (Daugherty 1991). Lack of field verification, variable transport parameters, macropore and preferential flow, layering, and lack of necessary input variables are additional sources of uncertainty for efforts to predict movement of organic compounds released from USTs (Hillel 1988). Additional information on the fate, transport, and modeling of organic compounds in soils are provided in Chapters 4 and 5 of this book (Balogh and Anderson 1992; Balogh et al. 1992).

2.4 SOIL AND WATER CONSERVATION IN TURFGRASS MANAGEMENT

Conservation of soil resources, reduction of sediment loss to streams, and conservation of water resources remain important national environmental priorities (Larson et al. 1983; Matheson 1991; Pierce et al. 1983). Development and implementation of soil and water conservation practices have reduced loss of valuable topsoil, conserved soil moisture during drought conditions, and decreased the erosion and sedimentation in rural and urban environments (Daniel et al. 1982; Koehler et al. 1982a; Schuler 1987; Stewart et al. 1975, 1976). Conservation of water resources, reduction in runoff losses from golf courses, reduction in erosion, and stabilization of aquatic systems are important components of integrated systems for management of turfgrass.

Conservation turfgrass systems in agricultural and urban environments have long been recognized as a significant component of programs to reduce surface losses of soil and enhance the quality of water resources (Anderson et al. 1989; Bennett 1979; Hughes 1946; Schuler 1987). Sod-based rotations and high-quality sod strips, grassed waterways, and turfgrass settling basins are all considered excellent practices for erosion control and soil conservation (Moldenhauer 1979). It has been amply demonstrated that erosive losses from pastures and turfgrass systems are far less than losses from agricultural fields and similar to losses from undisturbed forested watersheds (Table 7). Research on the development of soil and water conservation practices using turfgrass and forage grasses has been reviewed by Bennett (1979), Gibeault and Cockerham (1985), Koehler et al. (1982a), Richards and Middleton (1978), Schuler (1987), and Stewart et al. (1975, 1976).

2.4.1 Sediment and Runoff Control with Turfgrass

2.4.1.1 Processes and Control for Established Turfgrass. Use of turfgrass and forage grasses in pastures and crop rotations, reduced tillage systems, construction sites, highway rights-of-way, stabilization of mine spoils, and agricultural buffer

Table 7. Average Annual Sediment Losses from Selected Soils Under Different Management Systems

Soil texture/ percent slope	Management system: Fallow	Continuous cropping	Rotation	Grass	Forest
	Annual sediment loss (tons of soil ac^{-1})				
Loam/4	41.6	19.7	2.7	0.3	—
Loam/8	—	—	—	—	<0.1–0.4
Silt loam/8	112.8	85.5	11.4	0.3	—
Silt loam/16	151.9	84.1	25.3	<0.1	—
Fine sandy loam/8	20.3	28.1	5.5	<0.1	—
Sandy loam/8	—	—	—	<0.01	—
Sandy loam/4	—	—	—	—	<0.1
Sandy clay loam/10	64.7	25.8	10.8	<0.1	—

From Balogh et al. 1990; Bennett 1939; Gross et al. 1990.

strips have mitigated serious erosion and sediment loss in both rural and urban environments (Barfield and Albrecht 1982; Barnett 1965; Barnhisel et al. 1990; Bennett 1979; Dillaha et al. 1988; Hafenrichter et al. 1968; Koehler et al. 1982a, 1982b; Moldenhauer 1979; Richards and Middleton 1978; Schwab et al. 1981; Stewart et al. 1975, 1976). Use of pasture and forage grasses also reduces sediment and nutrient loading of surface water through reduction of sediment transport and movement of soluble chemicals in runoff water (Khaleel et al. 1980; Sweeten and Reddel 1978). However, under certain conditions, overloading of conservation turfgrass systems with nutrients (e.g., vegetative filter strips) potentially may increase the content of soluble nutrients in surface runoff (Balogh and Madison 1985; Reddy et al. 1980).

The process of surface runoff and channeled drainage links turfgrass systems with streams, rivers, and lakes. Runoff water is the medium of transport for eroded soil and dissolved chemicals (Daniel and Schneider 1979; Watson 1985). Improvement of infiltration, soil stability, and reduction of runoff losses of water are factors associated with improved soil and water conservation under turfgrass and forage grasses (Bennett 1979; Watson 1985). Numerous studies have demonstrated that loss of water, sediment, and entrained chemicals in surface runoff is effectively reduced by established turfgrass plots and grass buffer strips (Barfield et al. 1979; EPA 1983; Gross et al. 1990, 1991; Hayes et al. 1978; Schuler 1987; Tollner et al. 1977; Welterlen et al. 1989; Young et al. 1980). Reduction in the amount of water lost as runoff increases the amount of water available for plant uptake and infiltration through the soil.

Turfgrass is especially effective in reducing runoff and sediment losses in comparison to bare soil and conventional agricultural systems (Table 7) (Gross et al.

1987, 1990, 1991; Koehler et al. 1982a). Factors controlling the extent of erosion and runoff from turfgrass are (1) rainfall intensity and timing of initiation of runoff, (2) antecedent soil moisture conditions, (3) sod vs seeded turfgrass, (4) soil texture and structure, and (5) site topography and landscape design. Tall fescue, Kentucky bluegrass, bermudagrass, orchardgrass, bromegrass, perennial ryegrass, and bahiagrass all have been used as conservation turfgrasses (e.g., Barnhisel et al. 1990; Bennett 1979; Hafenrichter et al. 1968; Jean and Juang 1979).

Gross et al. (1990) conducted a study of sediment and runoff losses from fertilized and unfertilized plots of sodded tall fescue and Kentucky bluegrass. Event and annual losses of runoff water and sediment were extremely low for the rainfall events that produced runoff. Total annual runoff from the sodded treatment plots ranged from 0.97 to 14.1 mm ha^{-1} (Table 8). Runoff was significantly greater from the plots receiving a granular vs a liquid urea treatment during the first year of the experiment. However, runoff from the fertilized plots was not significantly different from the sodded control plots (Table 8). Total annual loss of suspended sediment from the turfgrass plots ranged from 2.01 to 24.87 kg ha^{-1} (0.001 to 0.01 tons ac^{-1}). Fertilizer treatment did not affect the level of suspended sediment in the surface runoff (Table 8). This experimental site was previously used to examine runoff losses related to different tobacco management practices (Angle 1985). During this agronomic study, Angle (1985) estimated that annual runoff averaged 33.4 mm ha^{-1}. Reported runoff and sediment losses from other row crops also are much higher than from turfgrass plots (see Alberts et al. 1978; Andraski et al. 1985a; Klausner et al. 1974; McDowell and McGregor 1980; Romkens et al. 1973; Sharpley 1985).

In a simulated rainfall study of tall fescue turfgrass plots with variable seeding and tiller density, Gross et al. (1991) also observed low rates of runoff and sediment losses (Table 9). The mean density of the turfgrass plots ranged from 0 to 5692 tillers m^{-2}. Increasing intensity of simulated rainfall did increase the runoff and sediment load from all of the turfgrass plots. The plots with turfgrass cover (867–5692 tillers m^{-2}) significantly reduced runoff and sediment losses compared to the plots with a bare soil surface for the high-intensity rainfall treatment (Table 9). There was no significant difference in runoff or sediment transport between the 867, 2056, 3080, and 5692 tillers m^{-1} density plots at the medium- and low-intensity rainfall treatments. Even at relatively low density, turfgrass plots significantly reduce runoff and sediment losses compared to bare soil surfaces. Gross et al. (1991) attributed the reduction in runoff and sediment movement to an increase in infiltration, hydraulic resistance, and surface storage capacity. Runoff and sediment losses reported for row crops using simulated rainfall techniques are significantly greater than those reported for this turfgrass study (see Andraski et al. 1985b; Baker and Laflen 1982; Baker et al. 1978; Sauer and Daniel 1987).

The favorable results of studies demonstrating reduced surface runoff, sediment, and chemical losses from turfgrass plots (e.g., Gross et al. 1990, 1991; Harrison

Table 8. Mean Runoff and Sediment Losses from Sodded and Fertilized Tall Fescue/Kentucky Bluegrass Plots[a]

Date of runoff producing rainfall event	Total rainfall (mm)	Runoff			Suspended sediment		
		Fertilizer treatment			Fertilizer treatment		
		Liquid	Granular (mm ha^{-1})	Control	Liquid	Granular (kg ha^{-1})	Control
Year 1							
Nov. 29, 1985	21	0.30	0.16	0.40	0.42	1.75	0.41
May 20, 1986	137	0.12	0.13	0.09	0.73	2.32	0.19
July 3, 1986	28	0.08	0.33	0.08	0.03	0.21	0.11
July 14, 1986	25	0.06	0.12	0.08	0.10	0.33	0.07
July 20, 1986	26	0.17	0.30	0.35	0.25	0.36	0.75
July 29, 1986	7	0.13	0.21	0.17	0.39	0.32	0.24
Aug. 6, 1986	18	0.11	0.20	0.12	0.09	0.17	0.22
Total	262	0.97[a,b]	1.45[b]	1.29[a,b]	2.01[a]	5.46[a]	1.99[a]
Year 2							
Nov. 13, 1986	21	0.19	0.57	0.16	0.16	0.11	0.03
Dec. 5, 1986	35	2.75	0.78	0.32	0.89	0.27	0.35
Dec. 27, 1986	61	7.57	0.77	0.45	17.11	1.87	0.07
Jan. 19, 1987	22	0.09	0.26	0.51	0.09	0.95	0.67
March 2, 1987	16	1.05	0.22	0.12	0.40	0.11	0.19
April 25, 1987	46	0.18	0.72	0.29	0.86	1.43	0.50
May 21, 1987	26	1.46	1.48	0.60	3.65	5.73	0.56

June 4, 1987	19	0.08	0.51	0.17	0.20	0.79	2.45
June 5, 1987	35	0.09	0.87	1.21	0.18	0.44	0.96
Aug. 17, 1987	13	0.40	0.87	0.58	1.02	2.64	1.50
Sept. 18, 1987	34	0.20	1.31	0.20	0.31	0.05	2.03
Total	328	14.06[a]	3.36[a]	4.61[a]	24.87[a]	14.39[a]	9.31[a]

From Gross et al. 1990.

[a] *Based on a randomized block design on fine sandy loam soils with a slope of 5–7%. Plots were located at the Chesapeake Bay Foundation Clagget Farm in Upper Marlboro, MD. Fertilizer treatments included an unfertilized control, liquid urea spray, and granular application of urea. Details of fertilizer rates and scheduling are available in Gross et al. (1990).*

[b] *Variable totals for each row that are followed by the same letter are not significantly different.*

Table 9. Mean Rates of Runoff and Sediment Losses from Variable Density Tall Fescue Plots Irrigated with Variable Intensity of Simulated Rainfall[a]

Seeding rate (kg ha^{-1})	Mean tiller density (tillers m^{-2})	Runoff rate			Sediment loss rate		
		Low	Medium (mm h^{-1})	High	Low	Medium (kg m^{-2} h^{-1})	High
0	0	36.8	53.0	62.2	0.044	0.068	0.104
98	867	2.9	8.3	18.1	0.002	0.007	0.016
244	2056	30.6	41.3	44.9	0.012	0.015	0.021
390	3080	14.8	18.8	31.1	0.005	0.007	0.021
488	5692	17.7	24.6	29.1	0.005	0.007	0.011

From Gross et al. 1991.

[a] *Based on a randomized block design on sandy loam soils with a slope of 8%. Plots were located at the University of Maryland Turfgrass Research Facility in Silver Spring, MD. Intensity of simulated rainfall at the low, medium, and high treatment levels was 76, 94, and 120 mm hr^{-1}, respectively. Details of turfgrass density and simulated rainfall treatments are available in Gross et al. (1991).*

1989) should be extrapolated to the field, golf course, or watershed level with care. Many of the hydraulic and transport processes operating at the field scale are difficult to predict compared to relatively small (20–150 m^2) and well-tended research plots (Anderson et al. 1989; Leonard 1988, 1990). Although runoff and sediment losses from turfgrass plot studies indicate surface losses are in general very low, none the less, runoff and some sediment losses still can occur from turfgrass systems (Tables 8 and 9). The reduced volume and velocity of runoff from turfgrass and conservation systems will selectively erode smaller soil particles. Compared to the bulk soil, the smaller eroded soil particles have a higher capacity per unit area to absorb nutrients and organic chemicals (Caro 1976; Frere 1976; Sharpley 1985; Wauchope 1978). Also in well-drained turfgrass soils, rapid infiltration and leaching reduces surface runoff, but increases the potential for subsurface transport (Carrow 1985). This drainage water will eventually (1) migrate to tile drains and emerge as translocated surface runoff or (2) percolate past the root zone. Displacement of drainage water with entrained chemicals or sediment is not necessarily the intention of conservation and environmental quality practices. During intense rainfall events or antecedent soil saturation with irrigation water, sufficient runoff could occur to transport sediment or relatively mobile chemicals (Balogh and Anderson 1992; Hall et al. 1987). Movement of water and chemical partitioning at the field scale should be considered when evaluating the potential edge of golf course losses of water, sediment, and applied chemicals (Balogh and Anderson 1992; Walker and Branham 1992).

2.4.1.2 Runoff and Sediment Control During Site Construction and Turfgrass Establishment. Land disturbance during residential construction or construction of new golf course facilities exposes bare soil to water and wind erosion, surface crusting, and loss of soil physical structure (Richards and Middleton 1978). These conditions may result in significant surface loss of water, sediment, and nutrients (Beard 1973, 1982; Daniel et al. 1982).

Daniel et al. (1979) observed that annual sediment losses from residential construction sites ranged from 13.4 to 26.9 Mg ha^{-1}(6 to 12 tons ac^{-1}). On average, sediment yield from the construction sites was 20 times higher compared to adjacent agricultural watersheds. Wolman and Schick (1967) determined that annual sediment yield from construction sites varies from 38 to 251 Mg ha^{-1} (17 to 112 tons ac^{-1}) of sediment annually. The amount of sediment lost from disturbed construction sites depends primarily on (1) the duration and extent of disturbance, (2) the volume and rate of surface runoff, (3) the configuration and topography of the exposed site, and (4) the use of mitigating soil conservation practices. After turfgrass has been established, the level of runoff water, sediment, and chemical transport by surface processes is significantly reduced. Reduction in sediment losses during construction, sediment losses from altered drainage systems, and thermal pollution of waterways is needed for protection of aquatic environments (Klein 1990).

2.4.1.3 Design Considerations to Reduce Runoff and Erosion. Management practices designed to reduce erosion and loss of runoff water during and after construction projects have been reviewed by Forrest (1988), Klein (1990), Richards and Middleton (1978), Schwab et al. (1981), and Schuler (1987). Guidelines for soil and runoff control measures and water conservation practices are summarized from these sources. Immediate coverage of bare soil surfaces by seeding with turfgrass or placement of sod reduces the risk of erosion. Surface coverage provides important protection against wind erosion. Restoration of turfgrass, wooded areas along streambanks, and adapted natural vegetation are some of the best means for long-term control of erosion.

Potential adverse environmental impacts, including surface runoff, flooding, and erosion, should be an *essential consideration in golf course design. Design criteria should minimize the need for site disturbance and reshaping of natural landscape elements.* The intensity of site disturbance and reshaping with heavy equipment is directly related to serious short- and long-term erosion and runoff problems. Loss of topsoil and compaction, resulting from construction and equipment traffic, will significantly increase sediment transport from the construction site. Avoiding design elements that encourage development of gullies, reroute streams, or change the natural surface and subsurface drainage is critical for long-term site stability. Long steep slopes with smooth surfaces and compaction of surface soils are conditions especially conducive to erosion problems.

During construction, temporary erosion control devices will mitigate offsite transport of eroded sediment. These practices include (1) construction of temporary silt fences to stop particle transport; (2) construction of small check dams or weirs to flatten upstream slopes and decrease the velocity of runoff; and (3) use of temporary mulches, matting, or blankets to reduce erosive forces until vegetation or long-term measures are in place.

Long-term erosion and runoff control techniques should be used on sites with highly erosive soils, steep banks, or design elements conducive to rapid runoff and sediment transport. A wide range of techniques include the following:

1. Planting of native vegetation with soil stabilizing canopies and root systems
2. Placement of short (6 in.) silt fences on steep slopes to trap sediment, reduce runoff velocity, and allow development of vegetation
3. Use wattling on sites with highly erosive soils
4. Construction of terraces on steep slopes with drainage swales that collect and divert runoff water
5. Construction of detention ponds within drainage channels which reduce runoff velocity and provides temporary storage for eroded sediment

6. Avoid irrigation rates or duration which may cause runoff of water, resulting from irrigation of turfgrass at rates greater than soil infiltration rates and soil storage capacity
7. Repair leaking irrigation systems to reduce potential erosion problems on steep or unstable slopes

Golf courses also should be constructed to minimize disturbance of vegetation in the vicinity of drainage ditches or stream banks. Disturbed embankments will be susceptible to erosion from stream scour and slumping. On areas with concentrated flow velocities, permanent stream bank and shoreline stabilization should be practiced. Long-term stabilization structures include interlocking block walls, riprap, gabion walls, and planting trees with stabilizing root systems.

2.4.2 Turfgrass and Water Conservation

In addition to water quality issues, public concern again has focused on issues regarding allocation of water resources (Matheson 1991; Rossillon 1985). In the United States, historical policies for the management of water have treated it as an unlimited resource (Freeze and Cherry 1979; Watson 1985). Traditionally, many turfgrass managers have used water on golf courses as if inexhaustible in supply (Youngner 1970; Shearman 1985). This attitude has been changing in the last decade. On a national and regional level, deterioration of drinking water supplies have resulted in changing public attitudes on the distribution of limited water resources (Gilliom 1985; Kircher et al. 1985; Mann 1985; Meade and Parker 1985; Rossillon 1985). Water use issues in regions with continually limited or periodic drought conditions require the attention of both turfgrass researchers and managers. Water stress, consumptive use of water, and water conservation are important considerations in integrated management of turfgrass.

2.4.2.1 Water Stress and Mechanisms to Resist Drought. Turfgrasses use water for growth and transpiration. Water stress in turfgrass can be characterized as both chronic and acute (Beard 1989). Acute water stress is associated with short-term water demand which typically occurs during peak evaporative demand in midsummer on closely mowed turfgrasses with restricted root systems. Despite sufficient soil moisture, acute moisture stress occurs if the rate of transpiration exceeds the ability of grass root systems to supply sufficient moisture. Chronic moisture stress results from drought conditions with progressively greater decreases in soil moisture. Heat stress often occurs concurrently with acute and chronic moisture stress (Beard 1989). Although long-term moisture stress usually occurs in midsummer, winter desiccation of dormant turfgrass is another form of chronic moisture stress. The severity and timing of drought conditions are difficult to predict for specific locations. However, on a regional basis, chronic moisture stress is more prevalent in the arid and semiarid climatic zones of the United States (Beard 1985).

The United States Golf Association (USGA), Green Section has an active commitment to research for support of water conservation in turfgrass systems (USGA 1989). Research on water conservation practices in turfgrass management has focused on (1) selecting species and breeding of cultivars capable of efficient water use or drought resistance, (2) establishment of evapotranspiration rates for turfgrass, and (3) development of cultural practices to reduce consumptive use of water by turfgrasses (Beard 1985, 1989; Shearman 1985). Research has demonstrated that total water consumption resulting from turfgrass management is related to water supply and evaporation demand, length of growing season, turfgrass species and cultivars, amount and rate of vegetative growth, and turfgrass cultural practices.

In recent reviews of research on water use and irrigation of turfgrass, Beard (1985, 1989) and Carrow et al. (1990) summarized several important factors regarding water stress and consumptive use in turfgrass systems. Drought resistance is a combination of mechanisms in plants to withstand the stress of dry weather (Beard 1973, 1989). It is important to distinguish the following terms related to water stress in discussions of water conservation and turfgrass management. Drought resistance includes (1) *dehydration tolerance:* the ability of the plant to endure high water potentials caused by drought by minimizing tissue damage despite negative internal water deficits (low tissue water); (2) *dehydration avoidance:* the ability of the plant to avoid tissue damage during water stress periods by maintaining an internal positive water balance; and (3) *escape:* the ability of the plant to complete its life cycle before damaging internal water stress is initiated. Tolerance and avoidance are the two resistance mechanisms considered in selection of drought resistant turfgrasses (Beard 1989).

Consumptive rates of water use for turfgrass species (Table 10) are a distinctly different process from drought resistance (Table 11). Each condition is regulated by different physiological processes (Beard 1989; Youngner 1985). For example, tall fescue is a drought-tolerant cool-season turfgrass, but has a high rate of water use (e.g., Kneebone and Pepper 1982). Compared to other plants and landscape options, turfgrasses have relatively high rates of water consumption. In turfgrass systems, up to 90% of irrigation water and precipitation may be lost by evapotranspiration.

Reducing overall water use is a water conservation strategy associated with irrigated turfgrass (Carrow et al. 1990). Selection of grasses resistant to sustained drought is another water conservation strategy. If it is acceptable to temporarily lower turfgrass quality, growing drought-resistant grasses species is an acceptable strategy to reduce annual water consumption (Table 11). However, species with good drought resistance may use water at a high consumptive rate under conditions of unlimited water supply (Biran et al. 1981). Data from several studies suggest that when water use is not limited, selection of species with low consumption rates has the best

Table 10. Summary of Mean Rates of Turfgrass Evapotranspiration

Turfgrass species[a,b]		Mean summer ET rate	Relative
Cool season	Warm season	(mm d^{-1})	ranking[c]
	Buffalograss	5–7	Very low
	Bermudagrass hybrids[d]	6–7	Low
	Centipedegrass	6–9	
	Bermudagrass[d]	6–9	
	Zoysiagrass[d]	5–8	
Hard fescue		7–8.5	Medium
Chewings fescue		7–8.5	
Red fescue		7–8.5	
	Bahiagrass	6–8.5	
	Seashore paspalum	6–8.5	
	St. Augustinegrass	6–9	
Perennial ryegrass		6.6–11.2	High
	Carpetgrass	8.8–10	
	Kikuyugrass	8.5–10	
Tall fescue		7.2–12.6	Very high
Creeping bentgrass		5–10	
Annual bluegrass		>10	
Kentucky bluegrass[d]		4–>10	
Italian ryegrass		>10	

[a] *Mean rates of water use based on research by Aronson et al. (1987a, 1987b), Beard (1985), Biran et al. (1981), Gibeault et al. (1985), Johns et al. (1983), Kim and Beard (1988), Kim et al. (1987), Kneebone and Pepper (1982, 1984), Kopec et al. (1988), Krans and Johnson (1974), Meyer et al. (1985), O'Neil and Carrow (1982), Pruitt (1964), Shearman and Beard (1972, 1973), Sifers et al. (1987a, 1987b), Tovey et al. (1969), van Bavel (1966), and Youngner et al. (1981).*

[b] *Based on the most widely used cultivars of each species.*

[c] *Based on ranking by Beard (1989).*

[d] *Variable among cultivars within species.*

potential as a water conservation strategy (Table 10) (Carrow et al. 1990). Depending on the management criteria, selection of turfgrasses requiring the least amount of irrigation has been a significant water conservation strategy.

Until recently, major research efforts have focused on water use rates in warm-season grasses (Carrow et al. 1990). Zoysiagrass, bermudagrass, buffalograss, and centipedegrass have shown the lowest overall use rates. The limited research on cool-season grasses indicates that Kentucky bluegrass has somewhat lower rates of water consumption compared to tall fescue, perennial ryegrass, and creeping bentgrass (Table 10) (Beard 1989). Turfgrass cultivars have a fairly wide range of water use rates making generalizations difficult.

Growth and morphological factors associated with low water use rates are slow vertical leaf extension and a high canopy resistance to evaporative demand. Critical

Table 11. Comparative Resistance of Turfgrasses Grown in Region of Climatic Adaption and Preferred Cultural Regime

Turfgrass species[a]		
Cool season	Warm season	Relative ranking
	Bermudagrass[b] Bermudagrass hybrids[b]	Superior
	Buffalograss Seashore paspalum[b] Zoyiagrass Bahiagrass	Excellent
Fairway wheatgrass	St. Augustinegrass[b] Centipedegrass Carpetgrass	Good
Tall fescue Perennial ryegrass[b] Kentucky bluegrass[b] Creeping bentgrass[b] Hard fescue Chewings fescue Red fescue		Moderate Fair
Colonial bentgrass Annual bluegrass		Poor
Rough bluegrass		Very poor

From Beard 1989.

[a] *Based on the most widely used cultivars of each species.*
[b] *Variable among cultivars within species.*

components of these water use limiting factors are management selection for high shoot density, narrow leaves, and horizontal leaf orientation. Factors that stimulate leaf growth, leaf area, reduce canopy resistance, and ET demand (turbulence, low humidity, high temperature) increase water consumption (Shearman 1985). Water use efficiency by turfgrass also should be based on the amount of water required to meet turfgrass quality objectives. Water use levels for golf greens will be higher than for low-maintenance conservation turfgrass (Shearman 1985).

Appropriate use of cultural practices including mowing, fertilization scheduling, timing and duration of irrigation, soil cultivation, antitranspirants, and growth regulators will increase water use efficiency and conservation (Carrow et al. 1990). Cultural practices and considerations used to enhance water conservation include (1) increasing mowing frequency and optimizing height within economic and physical constraints, (2) overfertilizing with nitrogen decreases water use efficiency, and (3) adjusting irrigation frequency based on moisture sensing methods increases water use efficiency and conservation.

2.4.2.2 Water Use and Drought Resistance. Comparison of consumptive use rates demonstrates a relatively wide range of water use efficiencies for cool-season and warm-season turfgrasses (Table 10). Cool-season grasses, with an optimum soil temperature range of 15–24°C, have an ET range of 7–12 mm d^{-1} under conditions of high evaporative demand. The fine-leafed fescues have moderate water use rates compared to very high rates of water consumption for Kentucky bluegrass, annual bluegrass, and creeping bentgrass. Additional research is still needed to characterize maximum water use rates of cool-season grass species and cultivars (Beard 1985, 1989). Warm-season grasses, with an optimum soil temperature range from 27 to 35°C, have a range in ET of 4.5–8.5 mm d^{-1} under conditions of high evaporative demand. Buffalograss, bermudagrass, and centipedegrass have relatively low rates of water consumption compared to the intermediate rate of St. Augustinegrass and bahiagrass. In addition to interspecies variation in water use, there is a large diversity in rates of ET on an interspecific level (Beard 1985; Kim and Beard 1988; Kopec et al. 1988; Shearman 1986). Selection of both species and cultivars is an important water conservation consideration.

Research on mechanisms of drought tolerance and avoidance by turfgrasses has been reviewed by Beard (1989). Avoidance of dehydration is potentially a preferred resistance mechanism as a green vegetative canopy is maintained during water stress periods. Depending on the degree of water stress, avoidance mechanisms achieve reduction in water loss by (1) increased canopy, stomatal, or cuticular resistance to ET; (2) reduced leaf area; and (3) decreased absorption of radiant energy (Beard 1989; Johns et al. 1981, 1983; Kim and Beard 1988). Morphological factors contributing to canopy resistance are high shoot density, high verdure, high leaf number, and horizontal leaf orientation (Kim and Beard 1988; Shearman 1986; Sifers et al. 1987a). Another component of avoidance of water stress is the ability of the turfgrass to enhance water uptake. Deep rooting, high root density, and extensive development of root hairs contribute to water stress avoidance (Beard 1989).

The two major components of drought tolerance are the maintenance of turgor pressure and tolerance of desiccation (Beard 1989). Adjustment of osmotic pressure by changing solute concentration, increased elasticity, and decreased cell size are physiological mechanisms used to maintain turgor pressure under water stress conditions. Physiological research on tolerance to desiccation is limited. Storage of proline has been associated with the ability of warm-season grasses to recover from chronic water stress. The capacity to become dormant during chronic moisture stress is a drought tolerance mechanism of bahiagrass and buffalograss (Beard 1989). Changes in development of secondary lateral stems and rhizome development are morphological traits associated with increased drought resistance in several perennial turfgrasses (Beard 1989; Nus and Hodges 1986).

Individual species may differ in their drought resistance mechanisms (Beard 1985, 1989). Texas common St. Augustinegrass has high dehydration tolerance, but very low drought resistance. In contrast, bermudagrass has lower tolerance to dehydration, but has superior drought avoidance. Additional research on other warm-season and cool-season grasses is needed to determine mechanisms and status of drought resistance.

Compared to the cool-season grasses, the warm-season grasses have better drought resistance capacity (Table 11). Species and cultivars with low ET rates and deep, extensive root systems have a good capability to avoid dehydration. There is considerable diversity in ability to resist moisture stress among cultivars of St. Augustinegrass, bermudagrass, Kentucky bluegrass, and perennial ryegrass (Dernoeden and Butler 1978; Kim et al. 1987; Minner and Butler 1985). Turfgrass managers should note that water stress relations established for grass seedlings may not necessarily apply to mature stands of turfgrass (Wood and Buckland 1966; Wood and Kingsbury 1971).

2.4.2.3 Water Conservation and Cultural Practices: Mowing, Fertilization, and Irrigation. Mowing, fertilization, and irrigation are the primary cultural practices used to minimize water losses and increase water conservation on golf courses (Carrow et al. 1990; Shearman 1985). *Mowing* height, frequency of cutting, and mowing equipment affect water use on golf courses. Increased water use has been observed for both warm-season and cool-season grasses with increased mowing height (Biran et al. 1981; Feldhake et al. 1983; Madison and Hagan 1962; Mitchell and Kerr 1966; Parr et al. 1984; Shearman and Beard 1973; Sprague and Graber 1938). High cut turfgrass with an open canopy, lower shoot density, and greater leaf area has lower canopy resistance to ET and increased water consumption. Although the effects of mowing frequency are not always consistent, reducing the leaf area, shoot size, and rooting depth by more frequent mowing decreases water consumption (Shearman 1985). Several other studies indicate that increased mowing frequency will increase shoot density and leaf succulence (Hart and Burton 1966; Madison 1960; Madison and Hagan 1962; Prine and Burton 1956). Manipulation of mowing height and frequency for specific turfgrass species and cultivars to enhance depth of rooting, decrease leaf area, and increase canopy resistance can be used to enhance water use efficiency.

Sharpness of the mowing blade also influences water use efficiency of turfgrass (Beard 1973; Shearman et al. 1980a). It has been speculated that water loss will increase, resulting from mutilation of grass leaves with a dull blade. However, in field studies comparing sharp and dull blades, mowing with a dull blade decreased long-term water use by decreasing leaf area, turfgrass quality, and verdure (Shearman et al. 1980a; Steinegger et al. 1983).

Fertilization influences water use indirectly through changes in growth processes. Research has demonstrated that turfgrass growth and water use rates increase with increased mineral, especially nitrogen, nutrition (Carroll 1943; Feldhake et al. 1983;

Krogman 1967; Mantell 1966; Power 1985; Shearman and Beard 1973). Feldhake et al. (1983) observed a 13% increase in water use when 0.4 kg 100 m^2 of nitrogen was applied monthly to Kentucky bluegrass during the spring and summer compared to a single spring application. Shearman and Beard (1973) observed the direct relationship of increased water use of "Penncross" creeping bentgrass and nitrogen fertility levels in a controlled environment study. They observed an increase in the water use rate as nitrogen application increased from 0 to 2 lb. of nitrogen per 1000 ft^2. Water use declined as fertilization rates increased from 2 to 6 lb. of nitrogen per 1000 ft^2. Changes in leaf width, shoot growth, and shoot density, resulting from nitrogen fertility levels, were positively correlated with changes in the water use pattern. Increased root and shoot growth occurs as nitrogen nutrition is raised from the deficiency level, but continues after the growth optimum is achieved (Beard 1973). Root growth is suppressed or ceases entirely for turfgrasses fertilized at rates beyond growth requirements. With diminished root capacity, turfgrasses grown under excessive nitrogen fertilization regimes are more susceptible to wilt and drought injury when exposed to high evaporative demand (Juska and Hanson 1967).

Kneebone and Pepper (1982) compared the water use efficiency of several warm-season and cool-season turfgrasses grown under low- and high-maintenance conditions. Low-maintenance grasses received 5 g of nitrogen m^2 bimonthly and high-maintenance grasses received 5 g of nitrogen m^2 monthly. Although the quality of the high-maintenance turfgrass was much better than the low-maintenance turfgrass, consumptive water use was approximately 334 mm less with a low-maintenance system. This research indicates there is considerable potential for water conservation by manipulating fertility levels within specified quality criteria (Shearman 1985).

Several studies on timing of application and interaction of nitrogen, phosphorus, and potassium suggest that fall application of nitrogen and phosphorus compared to spring application results in enhanced ability of turfgrass to recover from moisture stress (Powell et al. 1967; Schmidt and Breuninger 1981; Snyder and Schmidt 1974). Sufficient application of potassium benefited turfgrass recovery from moisture stress regardless of nitrogen and phosphorus practices (Schmidt and Breuninger 1981). However, application of nitrogen after the onset of dormancy may result in excess leaching losses of nitrogen.

Research on water use in relation to potassium and phosphorus fertilization is limited (Carrow et al. 1990; Shearman 1985). Increased rooting density and depth, resulting from potassium fertilization (Monroe et al. 1969), may be associated with increased drought resistance in turfgrass (Schmidt and Breuninger 1981; Shearman 1985).

The relationship of mineral nutrition and water use is complicated, yet very important. Turfgrass managers should consider fertility programs that meet the needs of turfgrasses, but also avoid excessive growth and unnecessary water stress. Water use efficiency of turfgrass should be considered in relation to turfgrass quality, the ability of turfgrass to recuperate from stress, and reduced water use. Production of dry matter should not be a water use issue for turfgrass on golf courses, residential lawns, and conservation turfgrass systems. Additional research on the relationship of (1) rate of potassium application, (2) nitrogen carrier, (3) timing of application, and (4) drought resistance is needed.

Irrigation, in many parts of the United States, is required to supplement precipitation for maintenance of high-quality turfgrass (Carrow et al. 1990; Watson 1985). Irrigation practices alone and their interaction with other practices affect water use and drought resistance of turfgrass (Power 1971; Shearman 1985; Tovey et al. 1969). Although an accepted practice for turfgrass irrigation is deep and infrequent watering (Shearman 1985), this recommendation does not provide specific information on the amount needed or the timing of application.

Adjustment in the frequency of irrigation will affect water use requirements. Several studies have shown that irrigation on the basis of observed need rather than calendar scheduling results in adequate maintenance of turfgrass quality and reduction of water use. Delays in irrigation scheduling on the basis of visual need reduced water use in several studies (Biran et al. 1981; Peacock and Dudeck 1984; Shearman and Beard 1973; Tovey et al. 1969). Use of visual observation resulted in a 33% reduction of water use on "Penncross" creeping bentgrass (Shearman and Beard 1973). However, use of visual indicators of wilt alone to reduce irrigation frequency may result in a decline of vegetative cover and turfgrass quality (Shearman and Beard 1973; Tovey et al. 1969).

Using visual indicators as a control and with tensiometers and pan evaporation as the guide to irrigation scheduling, Marsh et al. (1980) and Youngner et al. (1981) used 55% less water on warm-season grasses without loss of turfgrass quality. However, on the cool-season grasses, reduced irrigation frequency resulted in decreased quality of tall fescue and Kentucky bluegrass. In the year after full root development, irrigation scheduling by soil moisture deficit did not reduce the quality of the tall fescue grown as turfgrass. Using neutron probe techniques, Gerst and Wendt (1983) assessed the interaction of irrigation frequency and rate on a mature bermudagrass lawn. Irrigation was scheduled when total soil water content was 41–42 cm in the upper 1.83 m of the soil profile. The best sustained turfgrass growth, with lower frequency and poor quality, was observed at the most frequent and lowest rate of application (2.3 cm). Intermediate (6.1 cm) and high (11.9 cm) rates of irrigation with less frequent application resulted in deeper movement of water in the profile, and good turfgrass quality. The high application rate also required greater frequency of mowing and resulted in water losses by leaching. The intermediate irrigation treatment resulted in the highest water use rate. In addition to water use considerations, irrigation strategies should be designed to reduce the potential for the negative impacts of moisture leaching losses.

Depth of soil wetting affects water use and drought avoidance by turfgrass. Light applications of water encourage shallow rooting, while deep wetting encourages deeper root growth (Madison and Hagan 1962). With deeper and more extensive root systems, turfgrass are able to use moisture from a greater portion of the soil profile. Deep rooting enhances the ability of turfgrass to avoid stress, wilt, and the need for increased supplemental irrigation as compared to shallow rooted turfgrass (Shearman 1985).

Several studies using pan evaporation, electronic soil moisture monitoring systems, tensiometers, canopy temperature indices, and soil moisture-controlled irrigation systems have all resulted in significant reductions in water use without reduction of Kentucky bluegrass quality (Carrow et al. 1990; Kneebone and Pepper 1982; Marsh et al. 1980; O'Neil and Carrow 1982; Throssell et al. 1987). Adjustment of irrigation frequency using electronic soil moisture sensing devices, tensiometers, canopy temperature indices, and pan evaporation reduces unnecessary applications of water. Turfgrass managers utilizing this approach to water conservation should consult local data on acceptable levels of soil moisture deficits in relation to maintenance of turfgrass quality. Onsite experimentation may be required to program the equipment for microclimate variations (Shearman 1985). Manipulation of irrigation rates and schedules has high potential as a water conservation practice.

Irrigation systems and design methods which are state-of-the-art irrigation systems that when properly *designed*, programmed correctly, and installed within design criteria, conserve water and apply it in accordance with the actual needs of the turfgrass plant (Carrow et al. 1990). Further, systems equipped with single-head controls and a recycling capability apply water commensurate with infiltration and percolation capacity of the soil. Appropriate integration of sprinkler heads, valves, and controllers are all important criteria for conserving irrigation water (Meyer and Camenga 1985).

Controllers are the key to water conservation in irrigation systems (Carrow et al. 1990). Controllers are electrical timing devices used to open or close valves that regulate the flow of water to the sprinkler heads. They usually are of two types: (1) solid state and (2) electromechanical. Electronic (solid state) controllers are usually accurate within 1 min, an obvious saving of water over the 1–4 min accuracy range of electromechanical types. Computer-controlled systems may be operated from ET rates calculated from on-site weather stations or other programs based on regional ET values. The California Irrigation and Management Information System (CIMIS) is an example of a program that supplies regional ET values for computer-controlled irrigation systems. These generalized ET values are based on generalized average solar radiation and historical data (Anonymous 1987; Sasso 1988). Annual savings of up to 40% water use and 40% energy cost have been speculated when using low pressure sprinkler heads with a computerized controller (Anonymous 1987).

Use of soil moisture sensing devices to determine the frequency of operation for controllers is another means of effective conservation of water for irrigation of turfgrass (Carrow et al. 1990; Meyer and Camenga 1985). These devices may be used to cancel preselected start and stop times of irrigation systems when soil moisture levels reach a preselected level. Soil moisture sensors are subject to soil variability. However, when properly programmed, the combination of sensors and controllers have been shown to conserve 25–40% of the water recommended for home lawns (Carrow et al. 1990).

Soil manipulation and other cultural practices also influence the need for irrigation of turfgrass. In turfgrass systems, soil compaction reduces bulk density, aeration, and water retention capacity, as well as turfgrass shoot/root growth and development (O'Neil and Carrow 1982; Thurman and Pokorny 1969). Changes in soil physical properties will affect soil hydraulic processes and irrigation requirements (Carrow 1985; Carrow et al. 1990). In two field studies, soil compaction reduced the rate of evaporation by 20% in old Kentucky bluegrass turfgrass (O'Neil and Carrow 1982) and 21–49% in perennial ryegrass turfgrass (O'Neil and Carrow 1983). Similar results were obtained in a greenhouse study using perennial ryegrass (Stills and Carrow 1983). Cultivation and aeration practices are designed to reduce the negative effects of soil compaction (Carrow 1985). Although limited data is available regarding the effects of turfgrass cultivation on water use, increased hydraulic capacity, water retention, and turfgrass growth may increase water use on cultivated soils (Shearman 1985).

Other practices and soil conditions will influence irrigation scheduling and water use patterns. Syringing *may* influence the water use efficiency of turfgrass. However, no data is currently available to document speculation on enhanced water use efficiency (Dipaola 1982; Shearman 1985) associated with syringing. Maintaining irrigation equipment to insure proper rates of application and minimizing leakage also reduces unnecessary loss of water during turfgrass irrigation (Marsh 1969).

Use of antitranspirants, wetting agents, and plant growth regulators have all been suggested as potential chemical strategies for increased water use efficiency and water conservation. Antitranspirants have been used to decrease evapotranspiration in several nongrass species (Shearman 1985). Testing of five potential antitranspirants on "Penncross" creeping bentgrass and "Tifway" bermudagrass demonstrated ephemeral reduction of ET (Stahnke and Beard 1982). Application rates that produced significant reductions in ET using a mixture of phenylmecuric acetate, Aqua-Gro (a commercial wetting agent), and beta-naphthoxy-acetic acid caused browning of the grass shoots and eventual death of the turfgrass. Several pesticides, including siduron and benefin, also have indirect antitranspirant effects (Shearman 1985; Shearman et al. 1980b). The observed decline in water use was associated with other detrimental effects of the pesticides on shoot and root growth, turfgrass density, and reduction in drought tolerance.

Use of wetting agents and surfactants has been found to increase infiltration rates and irrigation use efficiency on turfgrass systems with hydrophobic thatch, sand, or black layers (Carrow 1985; Letey et al. 1963; Moore 1979; Pelishek et al. 1962;

Templeton and Rodriquez 1988; Templeton et al. 1989; Wilkenson and Miller 1978). Improved soil water retention and reduced ET also have been attributed to the application of soil wetting agents (Law 1964; Madison 1966). Shearman (1985) suggested that use of wetting agents has only a limited and indirect role in water conservation on turfgrass systems.

Use of plant growth regulators (PGRs) or turfgrass growth retardants for suppression of growth and reduction of water use has also been suggested by several investigators. Research has shown that PGRs will reduce transpiration rates in warm-season and cool-season turfgrasses (Johns and Beard 1982; Shearman 1985). Use of PGRs has shown promise for reduction of turfgrass growth rates and enhancement of water conservation. However, development of chemical management strategies for water conservation has limited potential considering their current social and environmental disadvantages.

Use of alternate sources of water is another water conservation strategy practiced on golf courses (Butler et al. 1985; Watson 1985). Use of nonpotable water and wastewater reduces the pressure on potable drinking water sources, reduces depletion of groundwater, and potentially enhances the quality of water emerging from turfgrass systems. However, alternate sources of water, especially saline water, must be used with care. High salt content, contamination with toxic levels of micronutrients, reduction in available oxygen, silting of water storage areas, and increased damage to irrigation equipment caused by heavy loads of suspended solids are all potential problems associated with use of nonpotable water resources (Butler et al. 1985). Use of pH amendments, appropriate irrigation and drainage line spacing, periodic leaching of accumulated salts, and selection of salt-tolerant grasses are strategies to avoid problems associated with saline water and wastewater sources (Butler et al. 1985; Devitt and Miller 1988; Harivandi 1982; Lunt et al. 1961, 1964; Youngner et al. 1967).

2.5 RESEARCH NEEDS

Water and soil conservation practices are an integral component of integrated systems for management of turfgrass. Research and development for implementation of water conservation on turfgrass continues to be an important research priority. This section briefly outlines critical research needs in relation to water conservation on golf course turfgrass.

Investigation of evapotranspiration and drought resistance in cool-season turfgrass cultivars is needed to understand water use in areas with periodic drought conditions. The interaction of evapotranspiration, precipitation, drainage, soil properties, and chemical characteristics ultimately control the potential for water and chemical movement. Basic research on the impact of both drought and unlimited water supply on many turfgrass cultivars is still needed. Development of cultural practices to reduce water use or enhance water efficacy will reduce demands for irrigation water and decrease potential for excessive drainage. Research on cultural practices that potentially reduce consumptive and luxury water use in turfgrass should include (1) climate-specific selection or breeding of turfgrass cultivars that are resistant or tolerant to continual or periodic drought stress; (2) mowing height and frequency; (3) irrigation scheduling, application techniques, and computerized control; and (4) use of growth retardants and antitranspirants.

Characterization of hydraulic processes that occur in turfgrass soil environments is essential for development of realistic environmental fate and chemical transport models. Research should be conducted to incorporate current knowledge regarding hydrology of turfgrass systems into water and chemical transport models. Realistic simulation of water flux is essential for accurate prediction of both water availability for irrigation scheduling and potential movement of applied chemicals. Government agencies regulating the allocation of water resources and environmental impacts of turfgrass systems currently are using simulation software that is not specifically modified for turfgrass conditions.

Attempts to predict rates and volume of drainage and runoff from turfgrass watersheds still require considerable research and refinement. Several essential research areas include (1) evaluation of runoff and sediment losses from turfgrass watersheds (golf courses) rather than small plots; (2) characterization of preferential flow paths; (3) identification and quantitative research on leaching patterns through layered soils; (4) characterization of the effects of cultural practices, thatch layers, soil structure, and compaction on soil hydraulic properties under turfgrass; (5) determination of the effects of perched water tables and capillary fringe on water and chemicals transport processes; (6) characterization of the effects of topographic, spatial, and temporal variability on water and solute transport processes; and (7) assessment of the effects of construction and golf course maintenance on sedimentation and deterioration of surface water quality. Control of erosion and runoff by turfgrass is well-documented. Erosion control practices during construction of golf courses and protection of streams from sedimentation and thermal pollution also should be included in comprehensive guidelines on development of integrated management systems for turfgrass.

There are certainly regional concerns regarding the impacts of consumptive use of water, irrigation, water conservation, soil erosion and sediment transport during construction, and water transport of chemicals in turfgrass systems. Understanding the hydrologic cycle, irrigation practices, and drainage systems in relation to other management practices should be an essential component of research on the environmental impacts of turfgrass systems. Regionally specific information on water management guidelines for water and soil conservation practices should be developed at a regional level. This information would be useful for developing consistent guidelines and recommended practices at the state and local levels. A comprehensive source of information on environmentally sound management of turfgrass must include information on water use and conservation, soil management, pest control, fertilization, and other cultural practices specific to turfgrass.

REFERENCES

Alberts, E. E., Schuman, G. E. and Burwell, R. E. 1978. Seasonal runoff losses of nitrogen and phosphorus from Missouri Valley loess watersheds. *J. Environ. Qual.* 7(2):203–208.

Anderson, J. L., Balogh, J. C. and Waggoner, M. 1989. Soil Conservation Service Procedure Manual: Development of Standards and Specifications for Nutrient and Pesticide Management. Section I & II. USDA Soil Conservation Service. State Nutrient and Pest Management Standards and Specification Workshop, St. Paul, MN, July 1989, 453 pp.

Andraski, B. J., Daniel, T. C., Lowery, B. and Mueller, D. H. 1985a. Runoff results from natural and simulated rainfall for four tillage systems. *Trans. ASAE* 28(4):1219–1225.

Andraski, B. J., Mueller, D. H. and Daniel, T. C. 1985b. Effects of tillage and rainfall simulation date on water and soil losses. *Soil Sci. Soc. Am. J.* 49(6):1512–1517.

Angle, J. S. 1985. Effect of cropping practices on sediment and nutrient runoff losses from tobacco. *Tob. Sci.* 29:107–110.

Anonymous. 1987. Irrigation revolutions. *Landscape Manage.* 26(7):34–36.

Aronson, L. J., Gold, A. J. and Hull, R. J. 1987a. Cool-season turfgrass responses to drought stress. *Crop Sci.* 27(6):1261–1266.

Aronson, L. J., Gold, A. J., Hull, R. J. and Cisar, J. L. 1987b. Evapotranspiration of cool-season turfgrasses in the humid northeast. *Agron. J.* 79(5):901–905.

Bailey, G. W., Swank, R. R. and Nicholson, H. P. 1974. Predicting pesticide runoff from agricultural land: A conceptual model. *J. Environ. Qual.* 3(2):95–102.

Baker, J. L. and Laflen, J. M. 1982. Effects of corn residue and fertilizer management on soluble nutrient runoff. *Trans ASAE* 25(2):344–348.

Baker, J. L., Laflen, J. M. and Johnson, H. P. 1978. Effect of tillage systems on runoff losses of pesticides, A rainfall simulation study. *Trans. ASAE* 21(5):886–892.

Baker, J. M. and van Bavel, C. H. M. 1986. Resistance of plant roots to water loss. *Agron. J.* 78(4):641–644.

Baker, R. S. and Hillel, D. 1990. Laboratory tests of a theory of fingering during infiltration into layered soils. *Soil Sci. Soc. Am. J.* 54:20–30.

Balogh, J. C. and Anderson, J. L. 1992. Environmental impacts of turfgrass pesticides. in Balogh, J. C. and Walker, W. J. (Eds.). *Golf Course Management and Construction: Environmental Issues.* Lewis Publishers, Chelsea, MI, Chapter 4.

Balogh, J. C. and Gordon, G. A. 1987. Climate Variability and Soil Hydrology in Water Resource Evaluation. Phase I Project Report. National Science Foundation, ISI-87053. NTIS #PB90-203183/A05, 92 pp.

Balogh, J. C., Gordon, G. A., Murphy, S. R. and Tietge, R. M. 1990. The Use and Potential Impacts of Forestry Herbicides in Maine. 89-12070101. Final Report Maine Dept. of Cons., 237 pp.

Balogh, J. C., Leslie, A. R., Walker, W. J. and Kenna, M. P. 1992. Development of integrated management systems for turfgrass. in Balogh, J. C. and Walker, W. J. (Eds.). *Golf Course Management and Construction: Environmental Issues*. Lewis Publishers, Chelsea, MI, Chapter 5.

Balogh, J. C. and Madison, F. W. 1985. Runoff treatment from a turkey production facility. *Trans. ASAE* 28(5):1476–1481.

Barfield, B. J. and Albrecht, S. C. 1982. Use of vegetative filter zone to control fine-grained sediments from surface mines. in Symp. Surface Mining Hydrol., Sedimentol., Reclam. Univ. Kentucky. Lexington, KY, pp 481–490.

Barfield, B. J., Tollner, E. W. and Hayes, J. C. 1979. Filtration of sediment by simulated vegetation. I. Steady-state with homogeneous sediment. *Trans. ASAE* 22(3):540–545,548.

Barnett, A. P. 1965. Using perennial grasses and legumes to control runoff and erosion. *J. Soil Water Cons.* 20:212–215.

Barnhisel, R. I., Ebelhar, M. W., Powell, J. L. and Akin, G. W. 1990. Nitrogen response of common bermudagrass grown on surface-mined coal spoils. *J. Prod. Agric.* 3(1):56–60.

Barrows, H. L. and Kilmer, V. J. 1963. Plant nutrient losses from soil by water erosion. *Adv. Agron.* 15:303–316.

Beard, J. B. 1973. *Turfgrass Science and Culture*. Prentice-Hall, Inc. Englewood Cliffs, NJ, 658 pp.

Beard, J. B. 1982. *Turf Management for Golf Courses*. Macmillan Publishing Co., New York, 642 pp.

Beard, J. B. 1985. An assessment of water use by turfgrass. in Gibeault, V. A. and Cockerham, S. T. (Eds.). *Turfgrass Water Conservation*. University of California, Riverside, Division of Agriculture and Natural Resources, pp. 45–60.

Beard, J. B. 1989. Turfgrass Water Stress: Drought Resistance Components, Physiological Mechanisms, and Species Genotype Diversity. in Sixth Int. Turf. Res. Conf. Tokyo, pp. 23–28.

Bell, C. E., Kostecki, P. T. and Calabrese, E. J. 1991. Review of state cleanup levels for hydrocarbon contaminated soils. in Kostecki, P. T. and Calabrese, E. J. (Eds.). *Hydrocarbon Contaminated Soils and Groundwater*. Lewis Publishers, Chelsea, MI, pp. 23–63.

Bengeyfield, W. H. (Ed.). 1989. Specifications for a Method of Putting Green Construction. United States Golf Association, Green Section, Far Hills, NJ, 24 pp.

Bennett, O. L. Conservation. 1979. in Buckner, R. C. and Bush, L. P. (Eds.). *Tall Fescue*. Agron. Monogr. 20:319–340. Am. Soc. Agron., Crop Sci. Soc. Am., Soil Sci. Soc. Am., Madison,WI.

Bigger, J. W. and Nielsen, D. R. 1976. Spatial variability of the leaching characteristics of a field soil. *Water Resour. Res.* 12(1):78–84.

Biran, I., Bravdo, B., Bushkin-Harav, I. and Rawitz, E. 1981. Water consumption and growth rate of 11 turfgrasses as affected by mowing height, irrigation frequency, and soil moisture. *Agron. J.* 73(1):85–90.

Blad, B. L. 1983. Atmospheric demand for water. in Teare, I. D. and Peet, M. M. (Eds.). *Crop-Water Relations*. John Wiley & Sons. New York, pp 1–44.

Bonazountas, M. 1991. Fate of hydrocarbons in soils: Review of modeling practices. in Kostecki, P. T. and Calabrese, E. J. (Eds.). *Hydrocarbon Contaminated Soils and Groundwater*. Lewis Publishers, Chelsea, MI, pp. 167–185.

Borden, R. C. and Bedient, P. B. 1986. Transport of dissolved hydrocarbon influenced by oxygen-limited biodegradation. *Water Resour. Res.* 22(13):1973–1982.

Bouchard, D. C., Enfield, C. G. and Piwoni, M. D. 1989. Transport processes involving organic chemicals. in Sawhney, B. L. and Brown, K. (Eds.). *Reactions and Movement of Organic Chemicals in Soils*. SSSA Special Publication No. 22. Soil Sci. Soc. Am., Am. Soc. Agron., Madison, WI, pp. 349–371.

Bouma,, J. A., Jongerius, A. and Schoonderbeek, D. 1979. Calculations of saturated hydraulic conductivity of some pedal clay soils using micromorphometric data. *Soil Sci. Soc. Am. J.* 43(2):261–264.

Brady, N. C. 1974. *The Nature and Properties of Soils*. MacMillan Publishing Co., New York, 639 pp.

Buol, S. W., Hole, F. D. and McCracken, R. J. 1989. *Soil Genesis and Classification*. 3rd ed. Iowa State University Press, Ames, IA, 446 pp.

Butler, J. D., Rieke, P. E. and Minner, D. D. 1985. Influence of water quality on turfgrass. in Gibeault, V. A. and Cockerham, S. T. (Eds.). *Turfgrass Water Conservation*. University of California, Riverside, Division of Agriculture and Natural Resources, pp. 71–84.

Calabrese, E. J., Kostecki, P. T. and Leonard, D. A. 1988. Public health implications of soils contaminated with petroleum products. in Calabrese, E. J. and Kostecki, P. T. (Eds.). *Soils Contaminated by Petroleum: Environmental and Public Health Effects*. John Wiley & Sons, New York, pp. 191–229.

Canter, L. W. 1987. Nitrates and pesticides in ground water: An analysis of a computer-based literature search. in Fairchild, D. M. (Ed.). *Ground Water Quality and Agricultural Practices*. Lewis Publishers, Chelsea, MI, pp. 153–174.

Caro, J. H. 1976. Pesticides in agricultural runoff. in Stewart, B. A., Woolhiser, D. A., Wischmeier, W. H., Caro, J. H. and Frere, M. H. (Eds.) Control of Water Pollution from Cropland: An Overview. Vol. 2. EPA 600/2-75-026b. U.S. Environmental Protection Agency, Athens, GA and USDA Agricultural Research Service, Washington D.C., pp. 91–119.

Carroll, J. C. 1943. Effects of drought, temperature, and nitrogen on turfgrasses. *Plant Phys.* 18:19–36.

Carrow, R. N. 1985. Soil/water relationships in turfgrass. in Gibeault, V. A. and Cockerham, S. T. (Eds.). *Turfgrass Water Conservation*. University of California, Riverside, Division of Agriculture and Natural Resources, pp. 85–102.

Carrow, R. N., Shearman, R. C. and Watson, J. R. 1990. Turfgrass. in Stewart, B. A. and Nielsen, D. R. (Eds.). *Irrigation of Agricultural Crops*. Agron. Monogr. 30:889–919. Am. Soc. Agron., Crop Sci. Soc. Am., Soil Sci. Soc. Am., Madison, WI.

Carsel, R. F., Mulkey, L. A., Lorber, M. N. and Baskin, L. B. 1985. The pesticide root zone model (PRZM): A procedure for evaluating pesticide leaching threats to groundwater. *Ecol. Model.* 30:49–69.

Carsel, R. F., Smith, C. N., Mulkey, J. T., Dean, J. D. and Jowise, P. 1984. Users Manual for the Pesticide Root Zone Model (PRZM). Release 1. EPA-600/3-84-109. U.S. Environmental Protection Agency, Athens, GA, 216 pp.

Cheng, H. H. (Ed.). 1990. *Pesticides in the Soil Environment: Processes, Impacts, and Modeling*. SSSA Book Series No. 2. Soil Science Society of America, Inc., Madison, WI, 530 pp.

Clothier, B. E. and White, I. 1981. Measurement of sorptivity and soil water diffusivity in the field. *Soil Sci. Soc. Am. J.* 45:241–245.

Cohen, S. Z., Nickerson, S., Maxey, R., Dupuy, A., Jr. and Senita, J. A. 1990. Ground water monitoring study for pesticides and nitrates associated with golf courses on Cape Cod. *Ground Water Monit. Rev.* 10(1):160–173.

Colbaugh, P. F. and Elmore, C. L. 1985. Influence of water on pest activity. in Gibeault, V. A. and Cockerham, S. T. (Eds.). *Turfgrass Water Conservation*. University of California, Riverside, Division of Agriculture and Natural Resources, pp. 113–129.

Daniel, T. C. and Schneider, R. R. 1979. Nonpoint pollution: Problem assessment and remedial measures; economic and planning considerations for designing control methods. in *Planning the Uses and Management of Land*. American Society of Agronomy, Madison, WI, pp. 829–851.

Daniel, T. C., McGuire, P. E., Stoffel, D. and Miller, B. 1979. Sediment and nutrient yield from residential construction sites. *J. Environ. Qual.* 8(3):304–308.

Daniel, T. C., Wendt, R. C., McGuire, P. E. and Stoffel, D. 1982. Nonpoint source loading rates from selected land uses. *Water Res. Bull.* 18(1):117–120.

Daugherty, S. J. 1991. Regulatory approaches to hydrocarbon contamination from underground storage tanks. in Kostecki, P. T. and Calabrese, E. J. (Eds.). *Hydrocarbon Contaminated Soils and Groundwater*. Lewis Publishers, Chelsea, MI, pp. 23–63.

Davidson, J. M., Rao, P. S. C. and Nkedi-Kizza, P. 1983. Physical processes influencing water and solute transport in soils. in Nelsen, D. W., Elrick, D. E. and Tanji, K. K. (Eds.). *Chemical Mobility and Reactivity in Soil Systems*. Soil Sci. Soc. Am. Special Publ. No. 11, pp. 35–47.

de Camargo, O. A., Bigger, J. W. and Nielsen, D. R. 1979. Transport of inorganic phosphorus in an alfisol. *Soil Sci. Soc. Am. J.* 43:884–890.

Dernoeden, P. H. and Butler, J. D. 1978. Drought resistance of Kentucky bluegrass cultivars. *Hort. Sci.* 13:667–668.

Devitt, D. A. and Miller, W. W. 1988. Subsurface drip irrigation of Bermuda grass with saline water. *Appl. Agric. Res.* 3(3):133–143.

Dillaha, T. A., Sherrard, J. H., Mostaghimi, L. S. and Shanholtz, V. O. 1988. Evaluation of vegetative filter strips as a best management practice for feed lots. *J. Water Pollut. Control. Fed.* 60(7):1231–1238.

Dipaola, J. M. 1982. The influence of syringing on the canopy temperatures of bentgrass greens. *Agron. Abstr.* p. 141.

Dunn, G. H. and Phillips, R. E. 1991. Macroporosity of a well-drained soil under no-till and conventional tillage. *Soil Sci. Soc. Am. J.* 55(3):817–823.

Enfield, C. G. and Ellis, R. Jr.,1983. The movement of phosphorus in soil. in Nelsen, D. W., Elrick, D. E. and Tanji, K. K. (Eds.). *Chemical Mobility and Reactivity in Soil Systems*. Soil Sci. Soc. Am. Special Publ. No. 11. pp. 93–107.

Enfield, C. G. and Yates, S. R. 1990. Organic chemical transport to groundwater. in Cheng, H. H. (Ed.). *Pesticides in the Soil Environment: Processes, Impacts, and Modeling*. SSSA Book Series No. 2. Soil Science Society of America, Inc., Madison, WI, pp. 271–302.

Environmental Protection Agency. 1977. The Report to Congress. Waste Disposal Practices and Their Effects on Ground Water. U.S. Environmental Protection Agency, Office of Water Supply and Office of Solid Waste Management Programs, Washington, D.C., 512 pp.

Environmental Protection Agency. 1983. Reducing Runoff Pollution Using Vegetated Borderland for Manure Application Sites. EPA-600/52-83-022. U.S. Environmental Protection Agency, Washington, D.C.

Environmental Protection Agency. 1986. Pesticides in Ground Water: Background Document EPA-WH550G, U.S. Environmental Protection Agency, Office of Ground-Water, Washington, D.C., 72 pp.

Environmental Protection Agency. 1987. Agricultural Chemicals in Groundwater: Proposed Pesticide Strategy. Proposed Strategy Document, U.S. Environmental Protection Agency, Office of Pesticides and Toxic Substances, Washington, D.C., 149 pp.

Environmental Protection Agency. 1988. Protecting Ground Water: Pesticides and Agricultural Practices. EPA-440/6-88-001. U.S. Environmental Protection Agency, Office of Ground-Water Protection, Washington, D.C., 53 pp.

Erickson, L. E. and Kuhlman, D. 1987. What you need to know about groundwater concerns. *Grounds Maintenance* 22:74–75,128.

Fairchild, D. M. (Ed.). 1987a. *Ground Water Quality and Agricultural Practices.* Lewis Publishers, Chelsea, MI, 402 pp.

Fairchild, D. M. 1987b. A national assessment of groundwater contamination from pesticides and fertilizers. in Fairchild, D. M. (Ed.). *Ground Water Quality and Agricultural Practices.* Lewis Publishers, Chelsea, MI, pp. 273–294.

Feldhake, C. M., Danielson, R. E. and Butler, J. D. 1983. Turfgrass evapotranspiration. I. Factors influencing rate in urban environments. *Agron. J.* 75:(5):824–830.

Feldhake, C. M., Danielson, R. E. and Butler, J. D. 1984. Turfgrass evapotranspiration. II. Responses to deficit irrigation. *Agron. J.* 76(1):85–89.

Forrest, C. 1988. New treatments for soil erosion. *Grounds Maintenance* 23(8):60–61,108.

Freeze, R. A. and Cherry, J. A. 1979. *Groundwater.* Prentice-Hall, Inc., Englewood Cliffs, NJ, 604 pp.

Frere, M. H. 1976. Nutrient aspects of pollution from cropland. in Stewart, B. A., Woolhiser, D. A., Wischmeier, W. H., Caro, J. H. and Frere, M. H. (Eds.) Control of Water Pollution from Cropland: An Overview. Vol. 2. EPA 600/2-75-026b. U.S. Environmental Protection Agency, Athens, GA and USDA Agricultural Research Service, Washington, D.C., pp. 59–90.

Gerst, M. D. and Wendt, C. W. 1983. Effect of irrigation frequency on turf water requirements. in Texas Turfgrass Research — 1983. PR-4157. The Texas Agric. Exp. Stn., The Texas A&M Univ. Sys., College Station, TX, pp. 41–53.

Ghildyal, B. P. and Tripathi, R. P. 1987. *Soil Physics: Theory and Practice.* John Wiley & Sons, New York, 656 pp.

Ghodrati, M. and Jury, W. A. 1990. A field study using dyes to characterize preferential flow of water. *Soil Sci. Soc. Am. J.* 54(6):1558–1563.

Gibeault, V. A. and Cockerham, S. T. (Eds.). 1985. *Turfgrass Water Conservation.* University of California, Riverside, Division of Agriculture and Natural Resources, 155 pp.

Gibeault, V. A., Meyer, J. L., Youngner, V. B. and Cockerham, S. T. 1985. Irrigation of Turfgrass Below Replacement of Evapotranspiration as a Means of Water Conservation: Performance of Commonly Used Turfgrasses. in Lemaire, F. L. Proc. Fifth Int. Turf. Res. Conf., Angers, France, pp. 347–356.

Gilliom, R. J. 1985. Pesticides in rivers of the United States. in U.S.G.S. National Water Summary 1984: Hydrologic Events, Selected Water-Quality Trends, and Ground-Water Resources. USGS Water-Supply Paper 2275, pp. 85–92.

Gold, A. J., Sullivan, W. M. and Hull, R. J. 1989. Influence of fertilization and irrigation practices on waterborne nitrogen losses from turfgrass. in Leslie, A. R. and Metcalf, R. L. (Eds.). Integrated Pest Management for Turfgrass and Ornamentals. Office of Pesticide Programs. 1989-625-030. U.S. Environmental Protection Agency, Washington, D.C., pp. 143–150.

Grant, Z. 1987. One aspect of government relations: Groundwater issues in pesticide regulation. *Golf Course Manage.* 55(6):14–20.

Gross, C. M., Angle, J. S., Hill, R. L. and Welterlen, M. S. 1987. Natural and simulated runoff from turfgrass. *Agron. Abstr.* p. 135.

Gross, C. M., Angle, J. S. and Welterlen, M. S. 1990. Nutrient and sediment losses from turfgrass. *J. Environ. Qual.* 19(4):663–668.

Gross, C. M., Angle, J. S., Hill, R. L. and Welterlen, M. S. 1991. Runoff and sediment losses from tall fescue under simulated rainfall. *J. Environ. Qual.* 20(3):604–607.

Hafenrichter, A. L., Schwendiman, J. L., Harris, H. L., MacLauchlan, R. S. and Miller, H. W. 1968. Grasses and Legumes for Soil Conservation in the Pacific Northwest and Great Basin States. Agricultural Handbook 339. USDA Soil Cons. Serv., Washington, D.C., 68 pp.

Hagen, L. J. 1991. A wind erosion prediction system to meet user needs. *J. Soil Water Cons.* 46(2):106–111.

Haith, D. A. 1980. A mathematical model for estimating pesticide losses in runoff. *J. Environ. Qual.* 9(3):428–433.

Hall, C. J., Bowhey, C. S. and Stephenson, G. R. 1987. Lateral Movement of 2,4-D from Grassy Inclines. *Proc. British Crop Protect. Conf.* 2(3):593–599.

Hallberg, H. R., Hoyer, B. E., Bettis, E. A., III and Libra, R. D. 1983. Hydrogeology, Water Quality, and Land Management in the Big Spring Basin, Clayton County, Iowa. Iowa Geol. Surv., Open-File Rept. 83-3, 191 pp.

Hanks, R. J. and Ashcroft, G. L. 1976. Physical Properties of Soils. UMC 48. Dept. Soil Sci. Biometeor., Utah State University, Logan, UT, 127 pp.

Harivandi, M. A. 1982. The use of effluent water for turfgrass irrigation. *Calif. Turf. Cult.* 32(3–4):1–7.

Harris, J. R. 1978. Maintenance of soil structure under playing turf. in Emerson, W. W. (Ed.). *Modification of Soil Structure.* John Wiley & Sons, New York, pp. 289–296.

Harrison, S. A. 1989. Effects of Turfgrass Establishment Method and Management on the Quantity and Nutrient and Pesticide Content of Runoff and Leachate. M.S. thesis. The Pennsylvania State University, University Park, PA, 125 pp.

Hart, R. H. and Burton, G. W. 1966. Prostrate vs common dallisgrass under different clipping frequencies and fertility levels. *Agron. J.* 58:521–522.

Hassett, J. J. and Banwart, W. L. 1989. Sorption dynamics of organic compounds in soils and sediments. in Sawhney, B. L. and Brown, K. (Eds.). *Reactions and Movement of Organic Chemicals in Soils.* SSSA Special Publication No. 22. Soil Sci. Soc. Am., Am. Soc. Agron., Madison, WI, pp. 31–80.

Hatfield, J. L. 1990. Methods of estimating evapotranspiration. in Stewart, B. A. and Nielsen, D. R. (Eds.). *Irrigation of Agricultural Crops.* Agron. Monogr. 30:435–474. Am. Soc. Agron., Crop Sci. Soc. Am., Soil Sci. Soc. Am., Madison, WI.

Hayes, J. C., Barfield, B. J. and Barnhisel, R. I. 1978. Filtration of sediment by simulated vegetation. II. Unsteady flow with non-homogeneous sediment. *Trans. ASAE* 21(10): 1063–1067.

Hern, S. C. and Melancon, S. M. (Eds.) 1986. *Vadose Zone Modeling of Organic Pollutants.* Lewis Publishers. Chelsea, MI, 295 pp.

Hernandez, M., Land, L. J. and Stone, J. J. 1989. Surface runoff. in Lane, L. J. and Nearing, M. A. (Eds.) USDA-Water Erosion Prediction Project: Hillslope Profile Model Documentation. NSERL Report No. 2. USDA-ARS Nat. Soil Erosion Res. Lab., West Lafayette, IN, pp. 5.1–5.18.

Hillel, D. 1971. *Soil and Water: Physical Principles and Processes.* Academic Press, New York, 288 pp.

Hillel, D. 1988. Movement and retention of organics in soil: A review and a critque of modeling. in Calabrese, E. J. and Kostecki, P. T. (Eds.). *Soils Contaminated by Petroleum: Environmental and Public Health Effects.* John Wiley & Sons, New York, pp. 81–86.

Hillel, D. and Baker, R. S. 1988. A descriptive theory of fingering during infiltration into layered soils. *Soil Sci.* 146(1):51–56.

Horton, R. E. 1919. Rainfall interception. *Mon. Weather Rev.* 47(9):603–623.

Howell, T. A. 1990. Relationships between crop production and transpiration, evapotranspiration, and irrigation. in Stewart, B. A. and Nielsen, D. R. (Eds.). *Irrigation of Agricultural Crops*. Agron. Monogr. 30:391–434. Am. Soc. Agron., Crop Sci. Soc. Am., Soil Sci. Soc. Am., Madison, WI.

Hughes, H. D. 1946. The role of sod crops in production and conservation programs. *J. Am. Soc. Agron.* 38(12):1035–1048.

Hurto, K. A., Turgeon, A. J. and Spomer, L. A. 1980. Physical characteristics of thatch as a turfgrass growing medium. *Agron. J.* 72:165–167.

Hutzler, N. J., Gierke, J. S. and Krause, L. C. 1989. Movement of volatile organic chemicals in soils. in Sawhney, B. L. and Brown, K. (Eds.). *Reactions and Movement of Organic Chemicals in Soils*. SSSA Special Publication No. 22. Soil Sci. Soc. Am., Am. Soc. Agron. Madison, WI, pp. 373–403.

ICF Inc. 1989. Overview of leaking underground storage tanks. in Tank Tour: Your Guide to Federal Underground Storage Tank Program. Contract No. 68-01-7385. U.S. Environmental Protection Agency, Office of Underground Storage Tanks, pp. 1–8.

Javandel, I., Doughty, C., and Baker, J. L., 1984. *Groundwater Transport: Handbook of Mathematical Models*. Water Resources Monograph 10, American Geophysical Union, Washington, D.C., 228 pp.

Jean, S. and Juang, T. 1979. Effect of bahia grass mulching and covering on soil physical properties and losses of water and soil of slopeland (first report). *J. Agric. Assoc. China* 105:57–66.

Johns, D. and Beard, J. B. 1982. Water conservation —a potentially new dimension in the use of growth regulators. in Texas Turfgrass Research — 1982. PR-4040. The Texas Agric. Exp. Stn., The Texas A&M Univ. Sys., College Station, TX, p. 35.

Johns, D., Beard, J. B. and van Bavel, C. H. M. 1983. Resistances to evapotranspiration from a St. Augustinegrass turf canopy. *Agron. J.* 75(3):419–422.

Johns, D., van Bavel, C. H. M. and Beard, J. B. 1981. Determination of the resistance to sensible heat flux density from turfgrass for estimation of its evaporation rate. *Agric. Meteorol.* 25(1):15–25.

Jury, W. A. 1982. Simulation of solute transport with a transfer function model. *Water Resour. Res.* 18(2):363–368.

Jury, W. A. 1986a. Spatial variability of soil properties. in Hern, S. C. and Melancon, S. M. (Eds.). *Vadose Zone Modeling of Organic Pollutants*. Lewis Publishers, Chelsea, MI, pp. 245–269.

Jury, W. A. 1986b. Mathematical derivation of chemical transport equations. in Hern, S. C. and Melancon, S. M. (Eds.). *Vadose Zone Modeling of Organic Pollutants*. Lewis Publishers, Chelsea, MI, pp. 271–288.

Jury, W. A. 1986c. Chemical movement through soil. in Hern, S. C. and Melancon, S. M. (Eds.). *Vadose Zone Modeling of Organic Pollutants*. Lewis Publishers, Chelsea, MI, pp. 135–158.

Jury, W. A. and Ghodrati, M. 1989. Overview of organic chemical environmental fate and transport modeling approaches. in Sawhney, B. L. and Brown, K. (Eds.). *Reactions and Movement of Organic Chemicals in Soils*. SSSA Special Publication No. 22. Soil Sci. Soc. Am., Am. Soc. Agron., Madison, WI, pp. 271–304.

Jury, W. A. and Valentine, R. L. 1986. Transport mechanisms and loss pathways for chemicals in soil. in Hern, S. C. and Melancon, S. M. (Eds.). *Vadose Zone Modeling of Organic Pollutants*. Lewis Publishers, Chelsea, MI, pp. 37–60.

Jury, W. A., Winer, A. M., Spencer, W. F. and Focht, D. D. 1987. Transport and transformations of organic chemicals in the soil-air-water ecosystem. *Rev. Environ. Contam. Toxicol.* 99:119–164.

Juska, F. V. and Hanson, A. A. 1967. Effect of nitrogen sources, rates and time of application on performance of Kentucky bluegrass turf. *Proc. Am. Soc. Hort. Sci.* 90:413–419.

Keeney, D. R. 1986. Sources of nitrate to groundwater. *CRC Crit. Rev. Environ. Control* 16:257–304.

Khaleel, R., Reddy, K. R. and Overcash, M. R. 1980. Transport of potential pollutants in runoff water from land areas receiving animal wastes: A Review. *Water Res.* 14:421–436.

Kim, K. S. and Beard, J. B. 1988. Comparative turfgrass evapotranspiration rates and associated plant morphological characteristics. *Crop Sci.* 28(2):328–331.

Kim, K. S., Sifers, S. I. and Beard J. B. 1987. Comparative drought resistances among the major warm-season turfgrass species and cultivars. in Texas Turfgrass Research — 1986. PR-4319. The Texas Agric. Exp. Stn., The Texas A&M Univ. Sys., College Station, TX, pp. 28–30.

Kircher, J. E., Gilliom, R. J. and Hickman, R. E. 1985. Loads and concentrations of dissolved solids, phosphorus, and inorganic nitrogen at U.S. geological survey national stream quality accounting network stations. in USGS National Water Summary 1984: Hydrologic Events, Selected Water-Quality Trends, and Ground-Water Resources. USGS Water-Supply Paper 2275, pp. 61–73.

Klausner, S. D., Zwerman, P. J. and Ellis, D. F. 1974. Surface runoff losses of soluble nitrogen and phosphorus under two systems of soil management. *J. Environ. Qual.* 3(1):42–46.

Klein, R. D. 1990. *Protecting the Aquatic Environment from the Effects of Golf Courses.* Community & Environmental Defense Associates, Maryland Line, MD, 59 pp.

Kneebone, W. R. and Pepper, I. A. 1982. Consumptive water use by sub-irrigated turfgrasses under desert conditions. *Agron. J.* 74(3):419–423.

Kneebone, W. R. and Pepper, I. L. 1984. Luxury water use by bermudagrass turf. *Agron. J.* 76(6):999–1002.

Knisel, W. G. (Ed.). 1980. CREAMS: A Field Scale Model for Chemicals, Runoff and Erosion from Agricultural Management Systems. USDA Conservation Research Report No. 26, Washington, D.C., 640 pp.

Koehler, F. A., Humenik, F. J., Johnson, D. D., Kreglow, J. M., Dressing, R. P. and Maas, R. P. 1982a. Best Management Practices for Agricultural Nonpoint Source Control. III. Sediment. USDA Coop. Agree. 12-05-300-472, EPA Interagency Agree. AD-12-F-0-037-0. North Carolina Agricult. Ext. Serv., 49 pp.

Koehler, F. A., Humenik, F. J., Johnson, D. D., Kreglow, J. M., Dressing, R. P. and Maas, R. P. 1982b. Best Management Practices for Agricultural Nonpoint Source Control. II. Commercial Fertilizer. USDA Coop. Agree. 12-05-300-472, EPA Interagency Agree. AD-12-F-0-037-0. North Carolina Agricult. Ext. Serv., 55 pp.

Kopec, D. M., Shearman, R. C. and Riordan, T. P. 1988. Evapotranspiration of tall fescue turf. *HortScience.* 23:300–301.

Kostecki, P. T. and Calabrese, E. J. (Eds.). 1991. *Hydrocarbon Contaminated Soils and Groundwater.* Lewis Publishers, Chelsea, MI, 354 pp.

Krans, J. V. and Johnson, G. V. 1974. Some effects of subirrigation on bentgrass during heat stress in the field. *Agron. J.* 66:(4):526–530.

Krogman, K. K. 1967. Evapotranspiration by irrigated grass as related to fertilizer. *Can. J. Plant Sci.* 47:281–287.

Lane, L. J. and Nearing, M. A. (Eds.). 1989. USDA-Water Erosion Prediction Project: Hillslope Profile Model Documentation. NSERL Report No. 2. USDA-ARS Nat. Soil Erosion Res. Lab., West Lafayette, IN.

Larson, W. E., Pierce, F. J. and Dowdy, R. H. 1983. The threat of soil erosion to long-term crop production. *Science* 219:458–465.

Law, J. P. 1964. The effect of fatty alcohol and nonionic surfactant on soil moisture evaporation in controlled environment. *Soil Sci. Soc. Am. Proc.* 28:695–699.

Lee, L. K. and Nielsen, E. G. 1987. The extent and costs of groundwater contamination by agriculture. *J. Soil Water Cons.* 42(4):243–248.

Leonard, R. A. 1988. Herbicides in surface waters. in Grover, R. (Ed.). *Environmental Chemistry of Herbicides.* Vol. 1. CRC Press, Boca Raton, FL, pp. 45–87.

Leonard, R. A. 1990. Movement of pesticides into surface waters. in Cheng, H. H. (Ed.). *Pesticides in the Soil Environment: Processes, Impacts, and Modeling.* SSSA Book Series No. 2. Soil Science Society of America, Inc., Madison, WI, pp. 303–349.

Leslie, A. R. and Metcalf, R. L. (Eds.). 1989. Integrated pest management for turfgrass and ornamentals. 1989-625-030. U.S. Environmental Protection Agency, Office of Pesticide Programs, Washington, D.C., 337 pp.

Letey, J., Welch, N., Pelishek, R. E. and Osborn, J. 1963. Effect of wetting agents on irrigation of water repellent soils. *Turf. Cult.* 13(1):1–2.

Libra, R. D., Hallberg, G. R. and Hoyer, B. E. 1987. Impacts of agricultural chemicals on ground water in Iowa. in Fairchild, D. M. (Ed.). *Ground Water Quality and Agricultural Practices.* Lewis Publishers, Chelsea, MI, pp. 185–215.

Lunt, O. R., Youngner, V. B. and Oertli, J. J. 1961. Salinity tolerance of five turfgrass species. *Agron. J.* 53:247–249.

Lunt, O. R., Kaempffe, C. and Youngner, V. B. 1964. Tolerance of five turfgrass species to soil alkali. *Agron. J.* 56:481–483.

Luxmoore, R. J. 1981. Micro-, meso-, and macroporosity of soil. *Soil Sci. Soc. Am. J.* 45(3):671.

Mackay, D. M., Roberts, P. V. and Cherry, J. A. 1985. Transport of organic contaminants in groundwater. *Environ. Sci. Tech.* 19(5):384–392.

Madison, J. H. 1960. Mowing of turfgrass. I. The effect os season, interval and height of mowing on the growth of Seaside bentgrass turf. *Agron. J.* 52:449–452.

Madison, J. H. 1966. Effects of wetting agents on water movement in the soil. *Agron. Abstr.* p. 35.

Madison, J. H. and Hagan, R. M. 1962. Extraction of soil moisture by Merion bluegrass (*Poa pratensis* L. 'Merion') turf, as affected by irrigation frequency, mowing height and other cultural operations. *Agron. J.* 54(2):157–160.

Mann, L. J. 1985. Ground-water-level changes in five areas of the United States. in U.S.G.S. National Water Summary 1984: Hydrologic Events, Selected Water-Quality Trends, and Ground-Water Resources. USGS Water-Supply Paper 2275, pp. 106–113.

Mantell, A. 1966. Effect of irrigation frequency and nitrogen fertilization on growth and water use of Kikuyugrass lawn (*Pennisetum clandestinum* Hochst.). *Agron. J.* 58:559–561.

Marian, D. F. and Brown, J. K. 1987. Planning for compliance with environmental regulations. *Golf Course Manage.* 55(11):38,42,44,46,48.

Marsh, A. W. 1969. Soil water — Irrigation and drainage. in Hanson, A. A. and Juska, F. V. (Eds.). *Turfgrass Science.* Agron. Monogr. 14:151–186. Am. Soc. Agron. Madison, WI.

Marsh, A. W., Strohman, R. A., Spaulding, S., Youngner, V. and Gibeault, V. 1980. Turfgrass research at the University of California. *Irrig. J.* 30:20–21,32–33.

Matheson, S. M. 1991. Future water issues: Confrontation or compromise. *J. Soil Water Cons.* 46(2):96–99.

McDowell, L. L. and McGregor, K. C. 1980. Nitrogen and phophorus losses in runoff from no-till soybeans. *Trans. ASAE* 23:643–648.

Meade, R. H. and Parker, R. S. 1985. Sediment in rivers of the United States. in USGS National Water Summary 1984: Hydrologic Events, Selected Water-Quality Trends, and Ground-Water Resources. U.S.G.S. Water-Supply Paper 2275, pp. 49–60.

Meyer, J. L. and Camenga, B. C. 1985. Irrigation systems for water conservation. in Gibeault, V. A. and Cockerham, S. T. (Eds.). *Turfgrass Water Conservation*. University of California, Riverside, Division of Agriculture and Natural Resources, pp. 103–114.

Meyer, J. L., Gibeault, V. A. and Youngner, V. B. 1985. Irrigation of Turfgrass Below Replacement of Evapotranspiration as a Means of Water Conservation: Determining Crop Coefficients of Turfgrasses. in Lemaire, F. L. Proc. Fifth Int. Turf. Res. Conf., Angers, France, pp. 357–364.

Minner, D. D. and Butler, J. D. 1985. Drought Tolerance of Cool Season Turfgrasses. in Lemaire, F. L. (Eds.) Proc. Fifth Int. Turf. Res. Conf., Angers, France, pp. 199–212.

Mitchell, K. J. and Kerr, J. R. 1966. Differences in rate and use of soil moisture by stands of perennial ryegrass and white clover. *Agron. J.* 58(1):5–8.

Moldenhauer, W. C. 1979. Erosion control obtainable under conservation practices. in Peterson, A. E. and Swan, J. B. (Eds.). *Universal Soil Loss Equation: Past, Present, and Future*. SSSA Spec. Publ. No. 8. Soil Science Society of America, Madison, WI, pp. 33–43.

Monroe, C. A., Coorts, G. D. and Skogley, C. R. 1969. Effects of nitrogen-potassium levels on the growth and chemical composition of Kentucky bluegrass. *Agron. J.* 61:294–296.

Moore, R. A. 1979. Wetting agents and their role in water conservation today. *Weeds, Trees Turf* 18(1):30–31.

Morton, T. G., Gold, A. J. and Sullivan, W. M. 1988. Influence of overwatering and fertilization on nitrogen losses from home lawns. *J. Environ. Qual.* 17(1):124–130.

Nearing, M. A., Lane, L. J., Alberts, E. E. and Laflen, J. M. 1990. Prediction technology for soil erosion by water: Status and research needs. *Soil Sci. Soc. Am. J.* 54(6):1702–1711.

NEIWPCC 1985. Here Lies the Problem: Leaking Underground Storage Systems. New England Interstate Water Pollution Control Commission, Boston, MA, 4 pp.

Nielsen, D. R., Wierenga, P. J. and Bigger, J. W. 1983. Spatial soil variability and mass transfers from agricultural soils. in Nelsen, D. W., Tanji, K. K. and Elrick, D. E. (Eds.). *Chemical Mobility and Reactivity in Soil Systems*. Soil Sci Soc. Am. Special Publ. No. 11, Madison, WI, pp. 65–78.

Noonan, D. C. and Curtis, J. T. 1990. *Groundwater Remediation and Petroleum: A Guide for Underground Storage Tanks*. Lewis Publishers, Chelsea, MI, 250 pp.

Nus, J. L. and Hodges, C. F. 1985. Effect of water stress and infection by *Ustilago striiformis* or *Urocystis agropyri* on leaf turgor and water potentials of Kentucky blue grass. *Crop Sci.* 25(2):322–326.

Nus, J. L. and Hodges, C. F. 1986. Differential sensitivity of turfgrass organs to water stress. *HortScience*. 21(4):1014–1015.

Olson, C. O. 1985. Site design for water conservation. in Gibeault, V. A. and Cockerham, S. T. (Eds.). *Turfgrass Water Conservation*. University of California, Riverside, Division of Agriculture and Natural Resources, pp. 131–149.

O'Neil, K. J. and Carrow, R. N. 1982. Kentucky bluegrass growth and water use under different soil compaction and irrigation regimes. *Agron. J.* 74(6):933–936.

O'Neil, K. J. and Carrow, R. N. 1983. Perennial ryegrass growth, water use and soil aeration status under soil compaction. *Agron. J.* 75(3):177–180.

Oster, C. A. 1982. Review of Ground-Water Flow and Transport Models in the Unsaturated Zone. PNL-4427, NUREG/CR-2917. Battelle, Pacif. Northwest Lab., Richland, WA.

Parr, T. W., Cox, R. and Plant, R. A. 1984. The effects of cutting height on root distribution and water use of ryegrass (*Lolium perenne* L. S23) turf. *J. Sports Turf. Res. Inst.* 60:45–53.

Peacock, C. H. and Dudeck, A. E. 1984. Physiological response of St. Augustinegrass to irrigation scheduling. *Agron. J.* 76(2):275–279.

Peck, A. J. 1983. Field variability of soil physical properties. *Adv. Irrig.* 2:189–221.

Pelishek, R. E., Osborn, J. and Letey, J. 1962. The effect of wetting agents on infiltration. *Soil Sci. Soc. Am. Proc.* 26:529–530.

Perry, C. D., Thomas, D. L., and Smith, M. C. 1988. Expert Systems Based Coupling of SOYGRO and DRAINMOD. ASAE Winter Meeting, Chicago, IL.

Peterec, L. J. 1988. A case study in petroleum contamination: The North Babylon, Long Island experience. in Calabrese, E. J. and Kostecki, P. T. (Eds.). *Soils Contaminated by Petroleum: Environmental and Public Health Effects.* John Wiley & Sons, New York, pp. 231–255.

Peterson, A. E. and Swan, J. B. (Eds.). 1979. *Universal Soil Loss Equation: Past, Present, and Future.* Soil Science Society America, Madison, WI. Special Publ. No. 8. Proc. of Symp. Div. S-6 of SSSA, 53 pp.

Petrovic, A. M. 1990. The fate of nitrogenous fertilizers applied to turfgrass. *J. Environ. Qual.* 19(1):1–14.

Pettyjohn, W. A. 1982. Cause and effect of cyclic changes in groundwater quality. *Ground Water Mon. Rev.* 2:43–49.

Pierce, F. J., Larson, W. E., Dowdy, R. H. and Graham, W. A. P. 1983. Productivity of soils: Assessing long-term changes due to erosion. *J. Soil Water Cons.* 38(1):39–44.

Piotrowski, M. R. 1991. Bioremediation of hydrocarbon contaminated surface water, groundwater, and soils: The microbial ecology approach. in Kostecki, P. T. and Calabrese, E. J. (Eds.). *Hydrocarbon Contaminated Soils and Groundwater.* Lewis Publishers, Chelsea, MI, pp. 203–238.

Powell, A. J., Blaser, R. E. and Schmidt, R. E. 1967. Effect of nitrogen on winter root growth of bentgrass. *Agron. J.* 59:529–530.

Power, J. F. 1971. Evaluation of water and nitrogen stress on bromegrass growth. *Agron. J.* 63(5):726–728.

Power, J. F. 1985. Nitrogen- and water-use efficiency of several cool season grasses receiving ammonium nitrate for 9 years. *Agron. J.* 77(2):189–192.

Pratt, P. F. (Chairman). 1985. Agriculture and Groundwater Quality. Council for Agricultural Science and Technology. CAST Report No. 103, 62 pp.

Prine, G. M. and Burton, G. W. 1956. The effect of nitrogen rate and clipping frequency upon yield, protein content, and certain morphoplogical characteristics of coastal bermudagrass (*Cyndon dactylon* (L.) Pers.). *Agron. J.* 48:296–301.

Pruitt, W. O. 1964. Evapotranspiration — A guide to irrigation. *Calif. Turf. Cult.* 14(4): 27–32.

Pye, V. I. 1983. Groundwater contamination in the United States. in Goundwater Resources and Contamination in the United States. PRA Report No. 83-12. National Science Foundation, Washington, D.C., pp. 17–63.

Pye, V. I. and Patrick. R. 1983. Ground water contamination in the United States. *Science* 221:713–221.

Pye, V. I., Patrick, R. and Quarles, J. 1983. *Groundwater Contamination in the United States.* University of Pennsylvania Press, Philadelphia, 315 pp.

Rao, P. S. C., and Davidson, J. M. 1982. Retention and transformation of selected pesticides and phosphorus in soil-water systems: A critical review. EPA-600/3-82-060. Environmental Research Laboratory, U.S. Environmental Protection Agency, Athens, GA.

Rao, P. S. C., Jessup, R. E. and Davidson, J. M. 1988. Mass flow and dispersion. in Grover, R. (Ed.). *Environmental Chemistry of Herbicides.* Vol. 1. CRC Press, Boca Raton, FL, pp. 21–43.

Reddy, K. R., Overcash, M. R., Khaleel, R. and Westermann, P. W. 1980. Phosphorus adsorption-desorption characteristics of two soils utilized for disposal of animal wastes. *J. Environ. Qual.* 9(1):86–92.

Richards, D. L. and Middleton, L. M. 1978. Best Management Practices for Erosion and Sediment Control. FHWA-HD-15-1. Department of Transportation, Federal Highway Administration, Region 15, Arlington, VA, 90 pp.

Richards, R. P., Kramer, J. W., Baker, D. B. and Krieger, K. A. 1987. Pesticides in rainwater in the northeastern United States. *Nature* 327(1560):129–131.

Ritchie, J. T. and Johnson, B. S. 1990. Soil and plant factors affecting evaporation. in Stewart, B. A. and Nielsen, D. R. (Eds.). *Irrigation of Agricultural Crops.* Agron. Monogr. 30: 363–390. Am. Soc. Agron., Crop Sci. Soc. Am., Soil Sci. Soc. Am., Madison, WI.

Romkens, M. J. M., Nelson, D. W. and Mannering, J. V. 1973. Nitrogen and phosphorus composition of surface runoff as affected by tillage method. *J. Environ. Qual.* 2(2):292–295.

Rosenberg, N. J. 1974. *Microclimate: The Biological Environment.* John Wiley & Sons, New York, 315 pp.

Rossillon, J. P. 1985. Water: Whose is it and who gets it? in Gibeault, V. A. and Cockerham, S. T. (Eds.). *Turfgrass Water Conservation.* University of California, Riverside, Division of Agriculture and Natural Resources, pp. 13–17.

Roy, W. R. and Griffen, R. A. 1985. Mobility of organic solvents in water-saturated soils. *Environ. Geol. Water Sci.* 7(4):241–247.

Rudy, R. J. and Bedosky, S. J. 1988. Segregation and distribution of gasoline compounds in the Floridan aquifer at Wacissa, Florida. in Proc. of Petroleum Hydrocarbons and Organic Chemicals in Ground Water: Prevention, Detection and Restoration. National Well Water Association, Dublin, OH, pp. 897–913.

Ryden, J. C. and Pratt, P. F. 1980. Phosphorus removal from wastewater applied to land. *Hilgardia* 48:1–36.

Sasso, C. M. 1988. Centralized irrigation control. *Southwest. Lawn Landscape.* 3(8):4–15.

Sauer, T. J. and Daniel, T. C. 1987. Effect of tillage system on runoff losses of surface-applied pesticides. *Soil Sci. Soc. Am. J.* 51(2):410–415.

Schmidt, R. E. and Blaser, R. E. 1969. Ecology and turf management. in Hanson, A. A. and Juska, F. V. (Eds.). *Turfgrass Science.* Agron. Monogr. 14:217–239. Am. Soc. Agron., Madison, WI.

Schmidt, R. E. and Breuninger, J. M. 1981. The effects of fertilization on recovery of Kenntucky bluegrass turf from summer drought, in Sheard, R. W. (Ed.). Proc. Fourth Int. Turf. Res. Conf., Guelph, Ontario, Canada, pp. 333–341.

Schuler, T. R. 1987. Controlling Urban Runoff: A Practical Manual for Planning and Designing Urban BMPs. Dept. of Environ. Programs. Washington Metropolitan Water Resources Planning Board, Washington, D.C.

Schwab, G. O., Frevert, R. K., Edminster, T. W. and Barnes K. K. 1981. *Soil and Water Conservation Engineering.* John Wiley & Sons, New York, 525 pp.

Sharefkin, M. F., Schechter, M. and Kneese, A. V. 1983. Impacts, costs, and techniques for mitigation of contaminated groundwater. in Goundwater Resources and Contamination in the United States. PRA Report No. 83-12. National Science Foundation, Washington, D.C., pp. 93–126.

Sharma, M. L. (Ed.). 1984. *Evapotranpiration from Plant Communities*. Devel. Agricult. Water Manag., Elsevier Science Publishing Co., Inc., New York, 344 pp.

Sharpley, A. N. 1980. The enrichment of soil phosphorus in runoff sediments. *J. Environ. Qual.* 9(3):521-526.

Sharpley, A. N. 1985. The selective erosion of plant nutrients in runoff. *Soil Sci. Soc. Am. J.* 49(6):1527–1534.

Shaw, R. H. 1982. Climate of the United States. in Kilmer, V. J. (Ed.). *Handbook of Soils and Climate in Agriculture*. CRC Press, Boca Raton, FL, pp. 1–101.

Shearman, R. C. 1985. Turfgrass culture and water use. in Gibeault, V. A. and Cockerham, S. T. (Eds.). *Turfgrass Water Conservation*. University of California, Riverside, Division of Agriculture and Natural Resources, pp. 61–70.

Shearman, R. C. 1986. Kentucky bluegrass cultivar evapotranspiration rates. *HortScience*, 21(3):455–457.

Shearman, R. C. and Beard, J. B. 1972. Stomatal density and distribution of *Agrostis* as influenced by species, cultivar, and leaf blade surface and position. *Crop Sci.* 12: 822–823.

Shearman, R. C. and Beard, J. B. 1973. Environmental and cultural preconditioning effects on the water-use rate of *Agrostis palustris* Huds., cultivar Penncross. *Crop Sci.* 13: 424–427.

Shearman, R. C., Kinbacher, E. J. and Reierson, K. A. 1980b. Siduron effects on tall fescue (*Festuca arundinacea*) emergence, growth and high temperature injury. *Weed Sci.* 28(2): 194–196.

Shearman, R. C., Steinegger, D. H., Riorden, T. P. and Kinbacher, E. J. 1980a. Mowing height, mowing frequency, and mower sharpness influence on Kentucky bluegrass (*Poa pratensis* L.) turfs. in Turfgrass research summary — 1980. Univ. Nebr. Dept. Hort. Prog. Rept. No. 81-1. pp. 57–64.

Sifers, S. I., Beard, J. B. and Kim, K. S. 1987a. Criteria for visual prediction of low water use rates of bermudagrass cultivars. in Texas Turfgrass Research — 1986. PR-4319. The Texas Agric. Exp. Stn., The Texas A&M Univ. Sys., College Station, TX, pp. 22–23.

Sifers, S. I., Beard, J. B., Kim, K. S. and Walker, J. R. 1987b. 1985 bermudagrass cultivar characterizations: College Station, TX. in Texas Turfgrass Research — 1986. PR-4319. The Texas Agric. Exp. Stn., The Texas A&M Univ. Sys., College Station, TX, pp. 24–27.

Sinclair, T. R. 1990. Theoretical considerations in the description of evaporation and transpiration. in Stewart, B. A. and Nielsen, D. R. (Eds.) *Irrigation of Agricultural Crops*. Agron. Monogr. 30:343–361. Am. Soc. Agron., Crop Sci. Soc. Am., Soil Sci. Soc. Am., Madison, WI.

Skidmore, E. L. 1982. Soil and water management and conservation: Wind erosion. in Kilmer, V. J. (Ed.). *Handbook of Soils and Climate in Agriculture*. CRC Press, Boca Raton, FL, pp. 371–399.

Skidmore, E. L. and Woodruff, N. P. 1968. Wind Erosion Forces in the U.S. and Their Use in Predicting Soil Loss. U.S. Dept. Agricult. Agron. Handbook No. 346. Washington, D.C., 42 pp.

Skidmore, E. L., Fischer, P. S. and Woodruff, N. P. 1970. Wind erosion equation: Computer solution and application. *Soil Sci. Soc. Am. Proc.* 34:931–935.

Smolen, M. D., Humenik, F. J., Spooner, J., Dressing, S. A. and Maas, R. P. 1984. Best Management Practices for Agricultural Nonpoint Source Control. IV. Pesticides. USDA Coop. Agree. 12-05-300- 472, EPA Interagency Agree. AD-12-F-0-037-0. North Carolina Agricult. Ext. Serv., 87 pp.

Snyder, V. and Schmidt, R. E. 1974. Nitrogen and Iron Fertilization of Bentgrass. in Roberts, E. C. (Ed.). Proc. Second Turf, Res. Conf. Blacksburg, VA, 1973. Am. Soc. Agron., Madison, WI, pp. 176–185.

Soil Conservation Service. 1972. Hydrology. in National Engineering Handbook, Section 4. USDA, Soil Cons. Serv., Washington, D.C.

Soil Conservation Service. 1973. A Method for Estimating Volume and Rate of Runoff in Small Watersheds. SCS-TP-149. USDA, Soil Cons. Serv., Washington, D.C.

Soil Conservation Service. 1988. National Agronomy Manual. Wind Erosion. Part 502. Title 190. Government Printing Office, Washington, D.C.

Sprague, V. G. and Graber, L. F. 1938. The utilization of water by alfalfa (*Medicago sativa*) and by blue grass (*Poa pratensis*) in relation to managerial treatments. *J. Am. Soc. Agron.* 30:986–997.

Stahnke, G. K. and Beard, J. B. 1982. An assessment of antitranspirants on creeping bentgrass and bermudagrass turfs. in Texas Turfgrass Research — 1982. PR-4014. The Texas Agric. Exp. Stn., The Texas A&M Univ. Sys., College Station, TX, pp. 36–37.

Starr, J. L. 1990. Spatial and temporal variation of ponded infiltration. *Soil Sci. Soc. Am. J.* 54(3):629–636.

Starr, J. L., DeRoo, H. C., Frink, C. R. and Parlange, J. Y. 1978. Leaching characteristics of a layered field soil. *Soil Sci. Soc. Am. J.* 42(3):386–391.

Steinegger, D. H., Shearman, R. C., Riordan, T. P. and Kinbacher, E. J. 1983. Mower blade sharpness effects on turf. *Agron. J.* 75(3):479–480.

Stewart, B. A., Woolhiser, D. A., Wischmeier, W. H., Caro, J. H. and Frere, M. H. 1975. Control of Water Pollution from Cropland: A Manual for Guideline Development. Vol. 1. EPA 600/2-75-026a. U.S. Environmental Protection Agency, Athens, GA and USDA Agricultural Research Service, Washington, D.C., 111 pp.

Stewart, B. A., Woolhiser, D. A., Wischmeier, W. H., Caro, J. H. and Frere, M. H. 1976. Control of Water Pollution from Cropland: An Overview. Vol. 2. EPA 600/2-75-026b. U.S. Environmental Protection Agency, Athens, GA and USDA Agricultural Research Service, Washington, D.C., 187 pp.

Stills, M. J. and Carrow, R. N. 1983. Turfgrass growth, N use, and water use under soil conpaction and N fertilization. *Agron. J.* 75(3):488–492.

Stuff, R. G. and Dale, R. F. 1978. A soil moisture budget model accounting for shallow water table influences. *Soil Sci. Soc. Am. J.* 42:637–643.

Sweeten, J. M. and Reddel, D. L. 1978. Nonpoint sources: state-of-the-art overview. *Trans. ASAE* 21(3):474–483.

Tamai, N., Asaeda, T. and Jeevaraj, C. G. 1987. Fingering in two-dimensional, homogeneous, unsaturated porus media. *Soil Sci.* 144(2):107–112.

Taylor, A. W. and Glotfelty, D. E. 1988. Evaporation from soils and crops. in Grover, R. (Ed.). *Environmental Chemistry of Herbicides*. Vol. 1. CRC Press, Boca Raton, FL, pp. 89–129.

Taylor, D. H. and Blake, G. R. 1982. The effect of turfgrass thatch on water infiltration rates. *Soil Sci. Soc. Am. J.* 46:616–619.

Templeton, A. R. and Rodriquez, D. A. 1988. Water infiltration using soil surfactants. *ALA* 9:1–4 .

Templeton, A. R., Helbing, P. G., Powell, D. M. and Moore, R. A. 1989. Effect of a Soil Wetting Agent Program on Hydrophobicity, Root Depth and Black Layer. 6th Internat. Turf. Res. Conf., Tokyo, Japan, pp. 197–199.

Testa, S. M. and Winegardner, D. L. 1990. *Restoration of Petroleum-Contaminated Aquifers*. Lewis Publishers, Chelsea, MI, 242 pp.

Thien, S. J. 1983. The soil as a water reservoir. in Teare, I. D. and Peet, M. M. (Eds). *Crop-Water Relations*. John Wiley & Sons, New York, pp. 45–72.

Throssell, C. S., Carrow, R. N. and Milliken, G. A. 1987. Canopy temperature based irrigation scheduling indices for Kentucky bluegrass turf. *Crop Sci.* 27(1):126–131.

Thurman, P. C. and Pokorny, F. A. 1969. The relationship of several amended soils and compaction rates on vegetative growth, root development, and cold resistance of Tifgreen bermudagrass. *J. Am. Soc. Hort. Sci.* 94:463–465.

Tollner, E. W., Barfield, B. J., Vachirakornwatana, C. and Haan, C. T. 1977. Sediment deposition patterns in simulated grass filters. *Trans. ASAE* 20(4):940–944.

Tovey, R., Spencer, J. S. and Muckel, D. C. 1969. Turfgrass evapotranspiration. *Agron. J.* 61:863–867.

USGA, Green Section. 1989. *1989 Turfgrass Research Summary*. United States Golf Association, Green Section, Golf Course Super. Assoc. Am., Far Hills, NJ, 37 pp.

van Bavel, C. H. M. 1966. Potential evapotranspiration: The combination concept and its experimental verification. *Water Resour. Res.* 2(3):455–468.

van der Leeden, F., Troise, F. L. and Todd, D. K. 1990. *The Water Encyclopedia*. Lewis Publishers, Chelsea, MI, 808 pp.

Vieira, S. R., Hatfield, J. L., Nielsen, D. R. and Biggar, J. W. 1983. Geostatistical theory and application to variability of some agronomical properties. *Hilgardia* 51:1–75.

Wagenet, R. J. and Hutson, J. L. 1986. Predicting the fate of non-volatile pesticides in the unsaturated zone. *J. Environ. Qual.* 15(4):315–322.

Wagenet, R. J. and Rao, P. S. C. 1990. Modeling pesticide fate in soils. in Cheng, H. H. (Ed.). *Pesticides in the Soil Environment: Processes, Impacts, and Modeling*. SSSA Book Series No. 2. Soil Science Society of America, Inc., Madison, WI, pp. 351–399.

Walker, W. J. and Branham, B. 1992. Environmental impacts of turfgrass fertilization. in Balogh, J. C. and Walker, W. J. (Eds.). *Golf Course Management and Construction: Environmental Issues*. Lewis Publishers, Chelsea, MI, Chapter 4.

Walker, W. J., Balogh, J. C., Tietge, R. M. and Murphy, S. R. 1990. *Environmental Issues Related to Golf Course Construction and Management*. United States Golf Association, Green Section, Far Hills, NJ, 378 pp.

Ward, A. D., Alexander, C. A., Fausey, N. R. and Dorsey, J. D. 1988. The ADAPT Agricultural Drainage and Pesticide Transport Model. Proc. ASAE Internat. Symp. Modeling Agricult., Forest and Range, Chicago, IL.

Water Resources Council. 1978. The Nation's Water Resources. 1975–2000. Second National Water Assessment. Vol. 1: Summary. Superintendents of Documents, Washington, D.C., 86 pp.

Water Resources Council. 1981. A Summary of Groundwater Problems. Washington, D.C.

Watschke, T. L. and Mumma, R.O. 1989. The Effect of Nutrients and Pesticides Applied to Turf on the Quality of Runoff and Percolating Water. A Final Report to the U.S Dept. Int., U.S. Geol. Sur. ER 8904. Env. Res. Res. Inst., The Pennsylvania State University, University Park, PA, 64 pp.

Watson, J. R., Jr. 1985. Water resources in the United States. in Gibeault, V. A. and Cockerham, S. T. (Eds.). *Turfgrass Water Conservation*. University of California, Riverside, Division of Agriculture and Natural Resources, pp. 19–36.

Watson, K. W. and Luxmoore, R. J. 1986. Estimating macroporosity in a forest watershed by use of a tension infiltrometer. *Soil Sci. Soc. Am. J.* 50(3):578–582.

Wauchope, R. D. 1978. The pesticide content of surface water draining from agricultural fields — A review. *J. Environ. Qual.* 7:459–472.

Weber, J. B. and Miller, C. T. 1989. Organic chemical movement over and through soil. in Sawhney, B. L. and Brown, K. (Eds.). *Reactions and Movement of Organic Chemicals in Soils*. SSSA Special Publication No. 22. Soil Sci. Soc. Am., Am. Soc. Agron., Madison WI, pp. 305–334.

Welterlen, M. S., Gross, C. M., Angle, J. S. and Hill, R. L. 1989. Surface runoff from turf. in Leslie, A. R. and Metcalf, R. L. (Eds.). Integrated Pest Management for Turfgrass and Ornamentals. Office of Pesticide Programs. 1989-625-030. U.S. Environmental Protection Agency, Washington, D.C., pp. 153–160.

Wilkenson, J. F. and Miller, R. H. 1978. Investigation and treatment of localized dry spots on sand golf greens, *Agron. J.* 70(2):299–304.

Wischmeier, W. H. 1976. Cropland erosion and sedimentation. in Stewart, B. A., Woolhiser, D. A., Wischmeier, W. H., Caro, J. H. and Frere, M. H. (Eds.). Control of Water Pollution From Cropland: An Overview. Vol. 2. EPA 600/2-75-026b. U.S. Environmental Protection Agency, Athens, GA and USDA Agricultural Research Service, Washington, D.C., pp. 31–57.

Wilson, G. V. and Luxmoore, R. J. 1988. Infiltration, macroporosity, and mesoporosity distributions on two forested watersheds. *Soil Sci. Soc. Am. J.* 52(2):329–335.

Wolman, M. G. and Schick, A. P. 1967. Effects of construction on fluvial sediment in urban and suburban areas of Maryland. *Water Resour. Res.* 3:451–464.

Wood, G. M. and Buckland, H. E. 1966. Survival of turfgrass seedlings subjected to induced drought stress. *Agron. J.* 58(1):19–23.

Wood, G. M. and Kingsbury, P. A. 1971. Emergence and survival of cool season grasses under drought stress. *Agron. J.* 63(6):949–951.

Woolhiser, D. A. 1976. Hydrologic aspects of nonpoint pollution. in Stewart, B. A., Woolhiser, D. A., Wischmeier, W. H., Caro, J. H. and Frere, M. H. (Eds.). Control of Water Pollution From Cropland: An Overview. Vol. 2. EPA 600/2-75-026b. U.S. Environmental Protection Agency, Athens, GA and USDA Agricultural Research Service, Washington, D.C., pp. 7–29.

Young, R. A., Huntrods, T. and Anderson, W. 1980. Effectiveness of vegetated buffer strips in controlling pollution from feedlot runoff. *J. Environ. Qual.* 9(3):483–487.

Young, R. A., Onstad, C. A., Bosch, D. D. and Anderson, W. P. 1986. Agricultural Nonpoint Source Pollution Model: A Watershed Analysis Tool; A Guide to Model Users. USDA ARS, Morris, MN and Minn. Poll. Cont. Agency, St. Paul, MN, 87 pp.

Youngner, V. B. 1970. Turfgrass varieties and irrigation practices. *Golf Superind.* 38:66,68.

Youngner, V. B. 1985. Physiology of water use and water stress. in Gibeault, V. A. and Cockerham, S. T. (Eds.). *Turfgrass Water Conservation*. University of California, Riverside. Division of Agriculture and Natural Resources, pp. 37–43.

Youngner, V. B., Lunt, O. R. and Nudge, F. 1967. Salinity tolerance of seven varietes of creeping bentgrass, *Agrostis palustris* Hud. *Agron. J.* 59:335–336.

Youngner, V. B., Marsh, A. W., Strohman, R. A., Gibeault, V. A. and Spaulding, S. 1981. Water Use and Turf Quality of Warm-Season and Cool-Season Turfgrasses. in Sheard, R. W. (Ed.). Proc. Fourth Int. Turf. Res. Conf., Guelph, Ontario, Canada, pp. 251–257.

Zielke, R. C., Pinnavaia, T. J. and Mortland, M. M. 1989. Adsorption and reactions of selected organic molecules on clay mineral surfaces. in Sawhney, B. L. and Brown, K. (Eds.). *Reactions and Movement of Organic Chemicals in Soils*. SSSA Special Publication No. 22. Soil Sci. Soc. Am., Am. Soc. Agron., Madison WI, pp. 81–110.

Zinke, P. J. 1967. Forest inteception studies in the United States. in Sopper, W. E. and Lull, H. R. (Eds.). *Forest Hydrology*. Pergamon Press, New York, pp. 137–161.

Zobeck, T. M. 1991. Soil properties affecting wind erosion. *J. Soil Water Cons.* 46(2):112–118.

CHAPTER 3

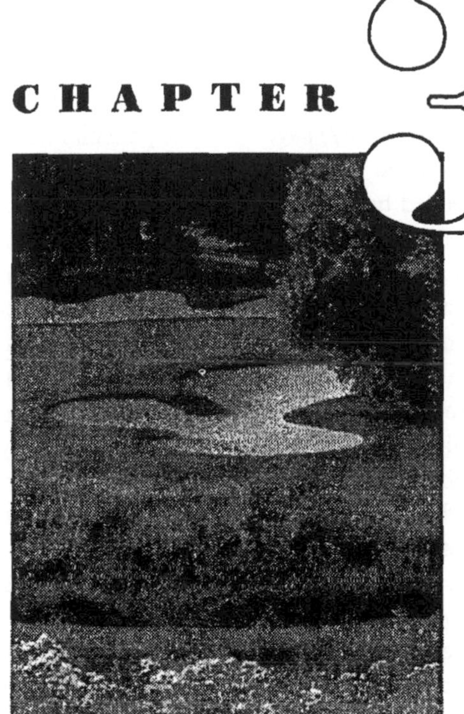

Environmental Impacts of Turfgrass Fertilization

William J. Walker and Bruce Branham

3.1 USE OF TURFGRASS FERTILIZERS AND ENVIRONMENTAL IMPACTS

The use of fertilizer for maintaining both an acceptable turfgrass growth rate and aesthetic quality is an important component of turfgrass management. However, recent concern over the potential for contamination of water supplies associated with turfgrass systems, such as golf courses, has prompted scientists, health officials, and others to evaluate the effect of turfgrass maintenance practices on the environment. This chapter, then, is a review of the current scientific literature concerning the effect of fertilizer addition and management practices on the potential for contamination of associated water supplies.

3.1.1 Use of Fertilizers in Golf Course Construction and Maintenance

Nitrogen (N), phosphorus (P), and potassium (K) are generally the most widely applied fertilizers for turfgrass and golf course construction and maintenance. The use of fertilizers and associated nutrient requirements of turfgrass systems have been

Photo courtesy of Michael French.

reviewed by Beard (1973, 1982), Davis (1969), Murray and Powell (1979), Petrovic (1990), and Wilkinson and Mays (1979). In addition to the fertilizer requirements for turfgrass systems, Petrovic (1990) and Schuler (1987) have noted the potential water quality effects posed by the use of nitrogen and phosphorus in turfgrass systems.

The specific fertilizer requirements needed both during golf course construction and maintenance have been reviewed by Beard (1982). In general, turfgrass is most responsive to nitrogen fertilization (Beard 1982). Since nitrogen is essential for maintaining both growth and color, it is usually added to achieve a range with respect to both properties. Due to its dynamic nature in the soil, nitrogen levels tend to decrease over time and, therefore, require regular additions to maintain a sufficient, preestablished level. Beard (1982) has reviewed the various symptoms associated with nutrient deficiencies in turfgrasses. Nitrogen deficiency tends to occur frequently and is characterized by the stunting of shoot growth in the initial stages of deficiency, followed by a yellowish chlorosis across the entire leaf blade in intermediate stages. In the advance stages of nitrogen deficiency, necrosis of the leaves occurs (Beard 1982). This condition is exacerbated on coarse, sandy soils and on soils subject to leaching from intensive rainfall or irrigation (Beard 1982).

Turfgrasses need potassium in relatively large amounts as well (Beard 1982). Recent research has demonstrated that increasing potassium levels results in improved root growth; an enhancement of heat, cold, drought, and wear tolerance; and reduced incidence of disease (Street 1988). Potassium deficiency in turfgrass systems occurs less frequently than nitrogen (Beard 1982). It is characterized by excessive tillering in initial stages and by leaf scorching in advanced stages. Potassium deficiency is also exacerbated by conditions favoring losses of nitrogen, such as extensive leaching occurring in coarse-textured soils (Beard 1982).

Phosphorus usually enhances the rate of turfgrass establishment from seed or vegetative plantings and enhances root growth. Phosphorus deficiency does not occur as commonly as nitrogen or potassium deficiencies. Reduced levels of phosphorus are usually related to low soil levels or to soil pHs that are either too low (acid) or too high (alkaline) (Beard 1982). Deficiency symptoms include a darkening of leaves, followed by the appearance of blue-green or purplish coloration, followed by leaf tip withering and necrosis (Beard 1982).

Other essential elements of importance to turfgrass systems include sulfur (S), calcium (Ca), magnesium (Mg), and micronutrients. Sulfur deficiency occurs at a frequency similar to potassium. Deficiency of sulfur is characterized by the loss of green color from older leaves in the initial stages (Beard, 1982). Sulfur deficiency has been observed on coarse, sandy soil low in organic matter and subject to intensive leaching. Calcium and magnesium deficiencies are extremely rare, but are more likely to occur in coarse, acid soils. Micronutrient levels are usually adequate in most soils. In addition, micronutrients are needed in very small quantities and are often supplied as impurities in commonly used fertilizers, liming materials, top-dressing mixes, certain pesticides, and irrigation water. Sandiness increases the possibility for micronutrient deficiency. However, most sands used for soil modification are not pure and are usually modified to some extent with soil or organic matter. In general, micronutrient deficiencies are most likely to occur in alkaline soils. They are further aggravated by high soil phosphorus and high soil levels of other micronutrients (Beard 1982).

The most common approach to the prevention of nutrient deficiency is through soil testing. Evaluation of soil nutrient levels ensures the best possible efficiency and economy of fertilization (Turner and Waddington 1983). Addition of soil nutrients far in excess of plant growth and uptake requirements has been associated with nonpoint source pollution in intensely managed pasture and forage systems (Anderson et al. 1989). Use of soil testing is considered a best management practice (BMP) for reduction of potential movement of applied nutrients in surface runoff or leaching (Anderson et al. 1989; Koehler et al. 1982a).

Beard (1982) notes that soil testing is perhaps the best means of assessing phosphorus and potassium needs, but is generally ineffective for determining the nitrogen status of soils. He further recommends that soil tests be made prior to the establishment of turfgrass and then conducted annually or until soil levels show signs of stabilizing. After this, a soil test is needed at 1–3 year intervals with longer intervals for fine-textured soils and more frequent intervals for greens and tees constructed of sandy root zone mixes.

The following is a summary of typical fertilizer management practices used for different parts of a golf course, largely recommended by Beard (1982).

The Putting Green — The establishment phase of a golf course requires different fertilization practices than would be done on a routine basis with established turfgrass. Preplant incorporation of phosphorus fertilizer into the putting green root zone is suggested in most situations because phosphorus is strongly sorbed to the soil and is considered immobile unless phosphorus levels exceed the adsorption capacity of the soil. Phosphorus will benefit from incorporation and the application rate should be based on soil test results. Nitrogen and potassium generally do not need to be incorporated because surface applications provide adequate levels of these nutrients. Incorporation of nitrogen into a putting green root zone is not necessary and should be considered a poor management practice since much of the applied nitrogen would be leached from the root zone before the turfgrass root system would develop sufficiently to take up the nutrients. A typical program for establishing a green would require the application of a starter fertilizer at seeding. Most starter fertilizers have 1:1:1 ratio of N/P/K, although frequently 1:2:1 ratio sources are used. One pound of actual nitrogen per 1000 ft^2 (M) is commonly applied, although less Nitrogen could be used particularly if using the higher phosphorus analysis materials. Since fertilizer will be applied frequently during the grow-in phase, a good management practice would be to use the 1:2:1 analysis starter fertilizer and apply at a rate to give 0.5 lb actual N/M. Following germination, it is very important to fertilize new seedlings adequately in order to ensure rapid closure of the turfgrass canopy which will prevent weed invasion and allow the turfgrass to be put into use more quickly. Typically, a soluble N source such as urea is used to fertilize new seedlings. In recent years, golf course builders have used high rates of fertilizer to push the turfgrass to fill in so that the course can be opened for play more quickly. Rates of 1 lb N/M week^{-1} have been used for this purpose. This cannot be considered an acceptable management practice, although the area of application, generally just the greens, would be small and the potential for nitrate leaching would be quite high. A more reasonable program would be to apply 0.5 lbs N/M at 10–14 d intervals.

The nutrient requirements of established putting greens vary with the amount of water applied, nutrient holding capacity, climate, turfgrass species/cultivar, degree of traffic, and certain other cultural practices such as verticutting and mowing height. No one fertilizer program will cover all situations. Specific application procedures for turfgrass situations will be described in subsequent sections. Some of the common characteristics can, however, be summarized.

Nitrogen must be applied to maintain turfgrass shoot density, adequate recuperative potential, moderate shoot growth rate, and, to a lesser extent, color. Fertilization rates typically range from 3 to 6 lb 1000 ft^{-2} year^{-1} for bentgrass and annual bluegrass greens and from 6 to 18 lb year^{-1} on bermudagrass greens. Application is usually at 2–4 week intervals during periods of normal shoot growth. Specific intervals, amounts applied, and placement techniques will depend on type of nitrogen carrier used and the turfgrass species fertilized (Hummel and Waddington 1984; Impithuska et al. 1979; Jackson and Burton 1962; Lamond and Moyer 1983; Landschoot and Waddington 1987).

Potassium is prone to leaching through the soil, especially in sand root zones. Potassium fertilization is often determined by the results of a soil test. Potassium is frequently applied at 75% the level of applied nitrogen, although higher levels are sometimes applied. The nutritional requirements of different turfgrass species have been determined (Adams et al. 1967; Belesky and Wilkinson 1983; Juska and Hanson 1969; Woodhouse 1968). Recent research by Shearman (1985, 1986) and Shearman and Beard (1973) has shown dramatic benefits of potassium on stress tolerance. Turfgrasses under high potassium fertilization levels had a greatly reduced wilting tendency and an increased wear tolerance. Based upon this research, many superintendents currently use fertilizers with 1:1 ratio of N/K and some go as high 1:3 ratio. On athletic field turf, some potassium is added on a routine basis regardless of soil test levels. Spring and late summer-early fall are the times when potassium applications are most commonly made. The effect of source, rate, and frequency of potassium application on turfgrass growth and response has been noted by Adams et al. (1967), Laudua et al. (1973), and Nelson et al. (1983).

Compared to nitrogen and potassium, phosphorus is required in much smaller amounts on greens. The rate of application should be based on soil test results. Research by Goss (1974) and Waddington et al. (1978) has indicated that phosphorus fertilization may encourage invasion by *Poa annua*. Therefore, some superintendents have used fertilizers with a zero phosphorus analysis. Varco and Sartain (1986) also examined the influence of phosphorus fertilization, sulfur addition, and pH on *Poa annua* invasion and found that infestation of turfgrass may be reduced by limiting phosphorus fertilization and by maintaining a soil pH near 5.0. On high sand content greens, this practice combined with the removal of clippings can result, over a period of years, in a phosphorus deficiency. When phosphorus is used, it is commonly applied one or two times per year as one component of a complete fertility treatment. King and Skogley (1969) have determined the phosphorus requirements for a bluegrass-red fescue turfgrass establishment,while Jackson et al. (1959) have determined the phosphorus requirements for coastal bermudagrass. Phosphorus is usually applied during spring and late summer or early fall. Application of phosphorus fertilizers is preferable just after coring to achieve deep soil penetration of this relatively immobile nutrient. It is sometimes questionable whether application of phosphorus is required, particularly since visible growth responses are seldom observed. High phosphate levels are quite common on older greens, although research by Juska et al. (1965) suggests that excessive soil phosphorus may not pose a nutritional problem for turfgrass. Again, the BMP is to base phosphorus fertilization on soil test results and to withhold phosphorus fertilization when soil test levels approach 100 lb P_2O_5 ac^{-1}. In fact, phosphorus soil test levels in the range of 50–75 lb P_2O_5 ac^{-1} are sufficient for most turfgrass situations, and phosphorus fertilization could be discontinued when soil tests fall in this range.

The Tee — The philosophy of tee fertilization is slightly different from that of greens. Extensive damage caused by divots dictates the need for rapid shoot growth rate to enhance turfgrass recovery (Beard 1982). Sufficient nitrogen must be applied to maintain adequate turfgrass shoot density and recuperative rate in terms of lateral shoot growth and tillering. The nitrogen fertilization rate ranges from 3 to 8 lb 16 NM^{-1} $year^{-1}$ on creeping bentgrass, perennial ryegrass, annual bluegrass, and Kentucky bluegrass tees and from 5 to 10 lb NM^{-1} on bermudagrass (Beard 1982). Higher rates are also required on small and heavily trafficked tees in many areas. The frequency of application typically ranges from 15 to 30 growing days. The rate of fertilization utilized on individual tees also may vary even though root zone mix and irrigation practices are similar for all tees. Tees that receive exhaustive divots, such as the par-3s and the first tee, may require up to twice the amount of nitrogen applied to larger tees (Beard, 1982). Larger tees are generally subject to less stress and concentrated traffic. Seasonal timing of application is basically the same as that described for putting greens.

The Fairway — The nutrient requirements for fairways vary with soil type, soil nutrient holding capacity, amount of water applied, climate, turfgrass species, and amount of play and traffic. Sufficient nitrogen must be applied to fairways to maintain proper turfgrass density, recuperative potential, moderate shoot growth rate, and to a lesser extent, color. Among the cool-season grasses, the perennial ryegrass and Kentucky bluegrasses usually require 2–4 lb N/M^{-1} $year^{-1}$ In contrast, bentgrass fairways frequently require only 1–3 lb N/M^{-1} $year^{-1}$. Perennial ryegrass and annual bluegrass usually require higher levels of nitrogen fertilization, in the range of 2–6 lb N/M^{-1} $year^{-1}$. Little or no nitrogen should be applied during heat stress periods for cool-season turfgrasses (Beard 1982). Among the warm-season grasses, bermudagrass requires 0.2–0.8 lb N/M^{-1} per growing month. The higher level of nitrogen fertilizer is commonly used on coarse-textured soils, where leaching is a greater problem, or on common bermudagrass which requires higher nitrogen levels to provide acceptable turfgrass density and quality (Beard 1982). This practice may be associated with undesirable loss of nitrogen, constituting a potential nonpoint pollution problem in natural or designed subsurface drainage systems.

Phosphorus application is based, as before, on the results of soil tests and applied one or two times per year as part of a complete fertilizer. Potassium applications may be made at supplemental rates above the level indicated by soil tests. The same guidelines for supplemental phosphorus and potassium application to greens generally are followed on fairways.

The Rough — Optimum fertilizer treatment should be applied during the establishment of roughs (Beard 1982). A relatively high fertility level is maintained through the first growing season or until a mature sod is established. At this point, fertilization is reduced to a lower level. The fertility level required for maintenance of the rough depends on species, soil conditions, tree root competition, cart traffic, and climate. Roughs on fertile soils not prone to nutrient loss usually require less fertilization than roughs on sandy soils with intense nutrient leaching losses. A typical fertilization program on high-quality rough turfgrass is one application per year of a complete fertilizer. This single application is usually applied in the fall in the case of cool-season species and in the spring in the case of warm-season turfgrasses (Beard 1982).

3.1.2 Environmental Issues Related to Fertilizer Use in Turfgrass Systems

There have been several concerns, in recent years, regarding the application of pesticides or fertilizer nutrients to turfgrass systems. These have been noted by

Watschke and Mumma (1989) and reviewed by Daniel et al. (1982) and Schuler (1987). The primary concerns include

1. Turfgrass is frequently maintained in highly populated areas and, therefore, requires judicious management practices. Improper use of pesticides or fertilizers in these systems may decrease water quality (Watschke and Mumma 1989).
2. Surface water quality also can be affected by land use and management practices. As noted by Watschke and Mumma (1989), urban and suburban environments contain a high percentage of runoff surfaces which can contribute significantly to the recharge and hence quality of surface water. Thus, it is important to implement practices designed to prevent the transport of sediment from urban runoff surfaces, as well as prevent excessive fertilization which might lead to high concentrations of fertilizer chemicals in storm runoff.
3. Cultivation and fertilizer application techniques in turfgrass systems can be quite different from those employed in agricultural systems. Because turfgrass chemicals are frequently added to the surface with less mixing than in agricultural systems, there exists a potential for enhanced surface runoff or transport from turfgrass systems if BMPs are not employed (Anderson et al. 1989; Koehler et al. 1982a).

3.1.3 Health Issues of Inorganic Fertilizers and Nitrate in Drinking Water

Potential adverse effects of nitrogen on health and the environment have been identified (Table 1). Health issues and environmental impacts of nitrate-nitrogen have been extensively reviewed by Brezonik (1978), Cantor et al. (1988), Keeney (1982), and Pratt (1985). Environmental effects of nitrogen include changes in productivity of natural and managed ecosystems, potential eutrophication of surface waters, contamination of groundwater with nitrates, a role in acid deposition, and partial depletion of stratospheric ozone by nitrous oxides.

The potential adverse human health effects of nitrates in drinking water include birth defects, cancer, nervous system impairment, and methemoglobinemia. The inorganic forms of phosphorus are not toxic. The environmental problems associated with phosphorus are concerned with the control of undesirable fertility levels and eutrophication of surface waters (Taylor and Kilmer 1980).

Acute nitrate poisoning of humans is rare (Keeney 1983b), requiring a single oral ingestion of 1–2 g of nitrate by an adult. Chronic health effects of nitrates have been attributed to the reduction of nitrate to nitrite by saliva and intestinal flora of some animals and of infants during the first 3–4 months of life. Newborn infants are susceptible to this condition, although clinical reports of infant mortality from methemoglobinemia currently are rare (Keeney 1986). Physicians usually recommend

Table 1. Potential Adverse Effects of Agricultural Nutrients on Health and the Environment

Effect	Causative agents
Human health:	
Methemoglobinemia in infants	Excess NO_3^- and NO_2^- in water and food.
Cancer	Nitrosamines from NO_2^-, secondary amines.
Respiratory illness	Peroxyacyl nitrates, alkyl nitrates, NO_3^- aerosols, NO_2^-, HNO_3 vapor in urban atmospheres.
Animal health:	
Environment	Excess NO_3^- in feed and water.
Eutrophication	Inorganic and organic N and P in surface waters.
Material and ecosystem damage	HNO_3 aerosols in rainfall.
Plant toxicity	High levels of NO_2^- in soils.
Excessive vegetative growth	Excess available nitrogen.
Stratospheric ozone depletion	Nitrous oxide from nitrification, denitrification, and stack emissions.

From Keeney 1982.

use of bottled water in areas with excess nitrate-nitrogen in drinking water supplies (Pratt 1985).

The relationship of nitrates in drinking water and methemoglobinemia is the only human health effect that has been verified (Cantor et al. 1988). The degree of risk to human health from moderate to high levels of nitrate concentration in drinking water is still being debated (Keeney 1983b). The current U.S. Public Health Service drinking water standard and U.S. EPA Maximum Contaminant Level (MCL) for nitrate-nitrogen is 10 mg L^{-1} (45 mg L^{-1} as nitrate). This standard has withstood several critical examinations, and current evidence suggests the nitrate standard provides reasonable protection to newborns against methemoglobinemia (National Research Council 1978).

Evidence linking nitrate or nitrite to subclinical effects in humans has not been definitely established. However, there are an increasing number of studies that indicate an association between consumption of nitrate-nitrite with incidents of cancer and other subclinical effects (Cantor et al. 1988). Exposure to high levels of nitrate in drinking water has been linked by some studies to gastric cancer; however, other investigations have not confirmed this association (Fraser et al. 1980). Most positive evidence of nitrate-nitrite caused stomach cancer comes from circumstantial evidence and geographic correlations (Cantor et al. 1988). In general, strong epidemiologic association of nitrates in drinking water and health risks has been inconclusive.

Even at excess concentrations, there is limited evidence that nitrate-nitrite is toxic to animals with the exception of ruminants. Moderate levels of nitrate poisoning of livestock from drinking water have been associated with poor growth, infertility, abortions, and general unhealthiness. Further research is needed to substantiate the response of livestock to moderate and low levels of nitrate contamination (Hansen et al. 1987).

3.1.4. Environmental Effects of Inorganic Fertilizers

Eutrophication is the overenrichment of lakes, bays, and slow-moving streams with nutrients causing subsequent proliferation of aquatic plants (Frere 1976). In addition to nutrient content, lake productivity is the result of basin, water, and limnological properties (Vollenweider and Kerekes 1980). The natural progression of lakes from oligotrophic to eutrophic status may be greatly accelerated by human activities and nonpoint source nutrient loading (Daniel et al. 1982; Keeney 1983; Schuler 1987). The concepts, processes, and nutrient sources responsible for eutrophication have been extensively reviewed (Loehr 1974; National Academy of Sciences 1969; Vollenweider and Kerekes 1980). The undesirable symptoms of eutrophication are algal blooms, algal mats, luxuriant development of certain aquatic macrophytes, depletion of oxygen on lake bottoms, and a decrease in water clarity (Brezonik 1969). Depletion of oxygen on lake bottoms and the subsequent release of toxins have been associated with fish kills (Fry 1969).

Nutrients available for algal growth and eutrophication are of greater concern for water quality than total nutrient loading. Phosphorus and, to a lesser extent, nitrogen are associated with eutrophication of surface water. The role of soluble and sediment-bound phosphorus in algal growth and eutrophication has been reviewed by Kramer et al. (1972) and Lee (1973). Soluble polyphosphates are regarded as completely available for algal growth, but are rapidly converted to orthophosphorus (Frere 1976). Phosphorus has been consistently found to be almost totally sediment bound rather than dissolved in surface runoff (Reddy et al. 1978). The bulk of phosphorus is transported to surface water by sediment transport during runoff. However, depending on timing of application and phosphorus content of surface vegetation and organic residues, soluble phosphorus losses in surface runoff may be relatively high (Anderson et al. 1989).

Depending on sediment composition, recent studies suggest that 10–20% of sediment-bound phosphorus is available for algal growth (Taylor and Kilmer 1980). Sediment low in phosphorus will usually remove phosphorus from solution. When sediments have high concentrations of adsorbed phosphorus, some phosphorus will be released to solution. Water quality criteria for phosphate-phosphorus (PO_4^{3-}) have not been established. Average PO_4^{3-} concentrations of 25 $\mu g\ L^{-1}$ and entry values of 50 $\mu g\ L^{-1}$ in lakes and reservoirs are considered the upper limits for protection against biological nuisances (Koehler et al. 1982a).

The complex role of nitrogen in aquatic systems is difficult to quantify (Keeney 1982). In addition to nonpoint sources of nitrogen, natural fixation of nitrogen by algae and current inputs of atmospheric nitrogen may provide sufficient nitrogen to offset management reductions of nitrogen inputs. Total nitrogen concentrations as low as 1–2 ppm will support prolific algal growth when other conditions are satisfied. As with phosphorus, establishing water quality criteria to control nitrogen-induced eutrophication of surface waters has been difficult. The maximum permissible nitrate-nitrogen concentration in domestic water supply is 10 ppm.

3.2 NITROGEN CYCLE AND CHEMICAL PROPERTIES

Nitrogen is one of the essential plant nutrients required in substantial amounts to maintain adequate turfgrass and plant growth. Nitrogen is contained in turfgrass protein, nucleic acids, and protoplasm. It is absorbed by turfgrass primarily through the roots as ammonium (NH_4^+) or as nitrate (NO_3^-). Nitrogen promotes above-ground vegetative growth, regulates phosphorus and potassium uptake, and turfgrass color. Oversupply of nitrogen predisposes plants to disease and insect pests, reduces root growth, delays maturation, causes stem weakness, causes possible accumulation of thatch, and increases nitrogen available for offsite transport to surface water and groundwater (Aldrich 1984; Keeney 1983b; White and Dickens 1984).

One of the guiding principles of sound nutrient management is to provide sufficient nutrient to meet turfgrass needs, while minimizing surface and subsurface losses. In order to meet this criteria a thorough knowledge of the nitrogen cycle is important (Figure 1).

Organic nitrogen, nitrate-nitrogen, and ammonium nitrogen are the primary forms of nitrogen in the soil environment (Stevenson 1982a). The dominant form of nitrogen in the soil is the organic nitrogen, which is slowly made available for plant uptake by biologically mediated transformations (Frere 1976). Woodmansee et al. (1978) determined that greater than 95% of nitrogen in shortgrass prairie systems was present as organic nitrogen. The complex and interrelated succession of chemical and biological transformations of nitrogen in turfgrass, agricultural, forest, and other natural systems is summarized as the nitrogen cycle (Figure 1 and Table 2).

The nitrogen cycle is a simplified, but convenient, description of the potential sources, sinks pathways, and transformations of nitrogen in terrestrial systems (Figure 1) (Stevenson 1982a, 1982b). The critical biological transformations of nitrogen include (1) immobilization, the assimilation of inorganic nitrogen by plants and microorganisms to form organic compounds of nitrogen; (2) mineralization, the microbial decomposition of organic nitrogen to inorganic forms; (2a) ammonification, the microbial transformation of organic nitrogen to ammonium; (2b) nitrification, the

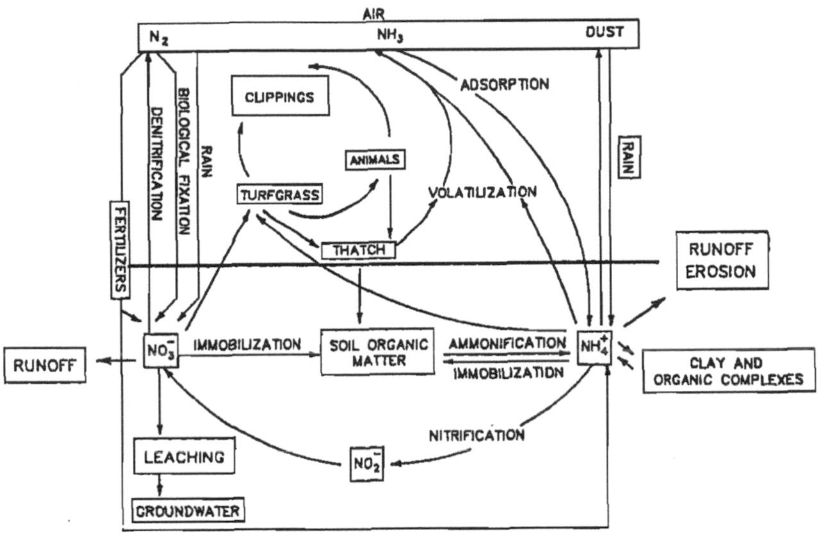

Fig. 1. The nitrogen cycle.

microbial oxidation of ammonium to nitrate and nitrite; (3) denitrification, reduction of nitrate to nitrite, N_2O and N_2; and (4) nitrogen fixation, the reduction of N_2 to ammonium by biological and nonbiological transformations (Alexander 1977; Keeney 1983b; Stevenson 1982a). Important soil chemical reactions involving soil nitrogen are (1) ammonia volatilization or sorption, (2) ammonium exchange with the cation exchange complex of soils, (3) ammonium fixation in clay mineral interlayers, and (4) chemical denitrification and chemodenitrification (Keeney 1983b, 1986).

The main natural sources of nitrogen in soils include fixation of N_2 by microorganisms, symbiotic and nonsymbiotic, and nitrogen deposition in precipitation (Stevenson 1982a). Supplemental inputs include fertilizer, manure, mowing residues, and waste materials. Losses of nitrogen from the cycle mainly occur through volatilization, leaching from the root zone, and surface runoff (Keeney 1982; Legg and Meisinger 1982; Petrovic 1990).

The chemical form of the different nitrogen species determines the mode of nitrogen transport from the field to surface water or groundwater. The inorganic forms in soil and water include nitrate, nitrite, ammonium, N_2O, and N_2. Nitrate is chemically unreactive in dilute solutions and has a relatively low tendency to form complexes with dissolved metals or to sorb on the surface of soil colloids.

The high solubility of nitrate in soil solutions makes it particularly susceptible to subsurface leaching and surface transport in the solution phase of runoff (Nielsen et al. 1982). Once past the zone of biological activity at the soil surface, there are few mechanisms that will attenuate the transport of nitrates to groundwater. Limited levels of denitrification have been observed in the lower portion of the vadose zone

Table 2. Brief Description of Critical Components of the Nitrogen Cycle

Processes	Description	General ref.
Sources	Major nitrogen sources include commercial fertilizers; nonsymbiotic microbial fixation, fixation by symbiotic microbes (e.g., rhizobium-soybeans); atmospheric deposition (electrical fixation, industrial pollution); application of animal waste, crop residue, or industrial/municipal waste.	Bouldin et al. (1984) Havelka et al. (1982) Munson (1982), Phillips and DeJong (1984), Sommers and Giordano (1984), Russel (1984), Stevenson (1982a)
Transformations		
Immobilization	Biological assimilation of inorganic forms of nitrogen by plants and microbes to form amino acids, amino sugars, proteins, purines, pyrimidines, and nucleic acids. Net immobilization of nitrogen continues until C:N ratio reaches 22 or less.	Jansson and Personn (1982), Alexander (1977)
Mineralization	Release of inorganic nitrogen as microbial waste produced during decomposition of soil organic matter. Mineralization and immobilization are opposing processes that occur continuously. Mineralization is subdivided into ammonification and nitrification. The net result of mineralization is the release of both inorganic nitrogen and carbon. In temperate zones, net mineralization is about 2–4% of the total soil nitrogen per year. Factors influencing mineralization processes include nature of organic substrate, soil temperature, and water and aeration status.	Jansson and Personn (1982) Alexander (1977)
Ammonification	Formation of ammonium as a result of biological degradation of soil organic matter. Ammonium ions (cations) are held by the soil exchange complex and some are trapped in interlayers of nonexpanding 2:1 clays, where it is temporarily protected from soluble losses in runoff and leaching.	Ladd and Jackson (1982), Alexander (1977)

Nitrification	Nitrification is the microbial oxidation of ammonia to nitrite and then to nitrate. With the exception of a few atmospheric reactions, this process is the sole natural source of nitrate in the biosphere. Nitrification is the key process in relation to potential nitrogen losses in the N cycle. Nitrification transforms the relatively immobile ammonium ion into nitrate, which is readily leached or denitrified. The specific organisms mediating this process derive energy from the oxidation of ammonium to nitrite (*Nitrobacter* sp.) and oxidation of nitrite to nitrate (*Nitrosomonas* sp.). The nitrifying bacteria have an optimal environmental range near neutral soil pH (pH 6–8), warm temperatures (20–40°C), and moist soil conditions. A plentiful supply of ammonium is also necessary for nitrification to proceed. Nitrification has received considerable attention as a component of biological control for nitrogen management. Various compounds have been studied as nitrification inhibitors; only nitrapyrin and etradiazol have shown any effectiveness.	Schmidt (1982), Alexander (1977)
Denitrification	Although still not well understood, denitrification is the use of nitrate by anaerobic microbes in place of oxygen and results in the production of nitrogen and nitrogen oxide gases. This process constitutes a major volatile loss pathway for nitrogen. Conditions required for denitrification are waterlogging or high soil water content, sufficient source of carbon, and temperatures within the range of microbial activity. This process is difficult to quantify and makes nitrogen budgeting difficult. Overestimation of denitrification losses may cause subsequent overfertilization with corresponding surface water and groundwater quality problems.	Firestone (1982), Alexander (1977)
Off-site transport		
Volatilization	Volatile losses of ammonia (NH_3) to the atmosphere occur primarily in cases of surface-applied urea fertilizers, animal wastes, and municipal wastes. This is not a major source of nitrogen to surface water or groundwater. However, this process also makes prediction of nitrogen availability and nitrogen credits from surface-applied fertilizers more difficult.	Nelson (1982)

Table 2. Brief Description of Critical Components of the Nitrogen Cycle (continued)

Processes	Description	General ref.
Off-site transport (continued)		
Surface runoff	Nitrogen is transported to surface water in runoff as organic nitrogen and ammonium associated with eroded sediment and soluble ammonium- and nitrate-nitrogen. The bulk of surface losses is in the form of organic nitrogen. The enrichment ratio for sediment-bound nitrogen relative to soil nitrogen ranges from 2.5 to 7.5 for cropland. In many cases inorganic nitrogen added in precipitation will exceed the soluble nitrate content of runoff. The amount of precipitation, timing of precipitation in relation to fertilizer application, form and method of nitrogen application, soil chemical and physical properties all influence nitrogen losses in runoff.	Keeney (1982, 1983), Petrovic (1990)
Leaching	Nitrate-nitrogen is the principal form of nitrogen transported in subsurface water. Ammonium is retained by the soil exchange complex, and organic nitrogen is associated with soil particle surfaces. Nitrate leached below the root zone may be transported in interflow, transported to groundwater, or reappear in surface flow in tile drains or base flow. Subsurface losses of nitrate may be as important as volatilization in determining nitrogen budgets. The main transport processes are bulk flow (convective flow) of nitrates in soil solution and to a lesser extent molecular diffusion.	Keeney (1986), Nielsen et al. (1982), Pratt (1985), Starr (1983), Petrovic (1990)

and groundwater. Although initial infiltration during storm events generally leaches nitrate beyond the surface zone, changing form and application techniques, amount of thatch, root uptake activity, and incorporation practices will affect the levels of nitrate lost in surface runoff (Koehler et al. 1982a; Petrovic 1990). Over 90% of nitrogen occurring in the surface layer of soils is in the form of organic nitrogen (Volk and Loeppert 1982). Organic nitrogen is directly available for release into soil solution, but has a microbially mediated exchange with inorganic soil nitrogen through the processes of mineralization and immobilization. Organic nitrogen is a component of all phases of soil organic matter including humin, humic, and fulvic acids and is often the primary form of nitrogen in surface runoff (Frere 1976).

3.3 THE PHOSPHORUS CYCLE AND CHEMICAL PROPERTIES

Phosphorus and nitrogen combined are the two most critical nutrient elements required for turfgrass development. Phosphorus is a key element in plant RNA and DNA molecules; it is an integral component of plant energy reactions, photosynthesis, and respiration mediated through phosphorylation and conversion of adenosine triphosphate (Munson 1982). Low phosphorus concentrations generally result in delayed maturity, reduced yields, and stunted leaf growth (Munson 1982). The addition of phosphorus fertilizer to deficient soils generally enhances plant productivity. The importance of phosphorus in agriculture has been reviewed by Fox (1981) and Khasawneh (1980). Use of phosphorus in turfgrass has been reviewed by Beard (1973, 1982) and Wilkinson and Mays (1979).

The phosphorus cycle is similar to the nitrogen cycle, but considerably less complicated (Figure 2). Within the biological portion of the cycle, phosphorus exists at a higher free-energy level than the forms present in soil, water, and geological systems (Taylor and Kilmer 1980). Without biotical energy inputs, the high-energy forms of phosphorus are reduced in the soil system. Phosphorus is removed from active cycling by chemical precipitation, adsorption on mineral surfaces, plant and microbial uptake (immobilization), surface runoff, and limited subsurface transport (Anderson 1980; Sample et al. 1980; Taylor and Kilmer 1980).

Sources of phosphorus include geological and soil minerals, commercial fertilizer, animal manures, plant residues, and waste materials. Phosphate concentrations in soil solution are usually quite low, ranging from 0.01 to 0.1 ppm, although total soil levels may range from 100 to 1000 ppm (Anderson et al. 1989). During soil development, phosphorus is initially derived from parent material in inorganic form. As organisms become established, requiring phosphorus as a nutrient, it becomes immobilized in biological tissue. Phosphorus is taken up by plant roots as the orthophosphate form ($H_2PO_4^-$ or HPO_4^{2-}). When organisms die, phosphorus is released. The conversion of organic forms to inorganic is the mineralization process.

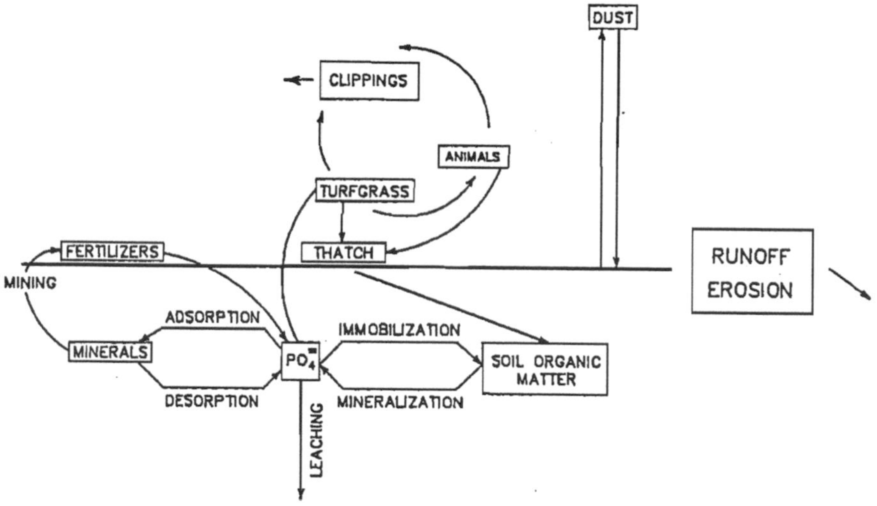

Fig. 2. The phosphorus cycle.

In solution, inorganic orthophosphate is subject to several chemical reactions that reduce solubility and availability (Sample et al. 1980). Although soils vary considerably in their ability to retain or adsorb phosphates, phosphate availability in soil decreases exponentially over time (Anderson et al. 1989). Precipitation reactions, adsorption on mineral surfaces, and retention by soil constituents will immobilize more than 50% of soluble phosphate within a few hours of application (Sample et al. 1980). Rapid retention of initially soluble phosphates accounts for observation of low concentrations in soil solution (Frere 1976; Sample et al. 1980).

Given the uncertainty of phosphate chemistry in soil and the dynamics of the phosphorus cycle, few generalities are possible in regard to the state of phosphorus in soil (Anderson et al. 1989). Phosphate (PO_4^{3-}) is the only stable oxidation state of inorganic phosphorus under a wide range of soil conditions. In the pH ranges commonly encountered in soils, PO_4^{3-}, HPO_4^{2-}, and H_2PO^{4-} are the predominant forms of inorganic phosphorus. In acid soils, the concentrations of these species are governed by solid phases of iron (Fe) and aluminum (Al), while in alkaline soils, solubility is governed by calcium phosphate solid phases.

Except for very coarse-textured soils, phosphate fixation is significant. Both Langmuir and Freundlich adsorption isotherms have been used to describe phosphate adsorption reactions by soil (Sample et al. 1980; Stuanes 1984). The Langmuir model implies that all increments of adsorbed phosphorus have the same surface bonding energy and that the surface reaches a maximum coverage with respect to adsorbed phosphorus. The Freundlich model implies that the energy of adsorption decreases with increasing levels of added phosphate and that no maximum is reached, a condition commonly observed in soils (Olsen and Khasawneh 1980). The Freundlich model is often incorporated into transport models for describing phosphate adsorption (Mansell et al. 1985).

Although successful for site-specific analysis and use in assessment of localized waste treatment, isotherms tend to oversimplify phosphate dynamics. Equilibrium techniques tend to neglect the influence of changing solution pH, oxidation-reduction potential, and speciation on adsorption maxima of phosphates. Adsorption isotherms assume that phosphate adsorption is achieved under equilibrium conditions and is completely reversible, conditions rarely met in natural systems (Barrow and Shaw 1975). Combinations of kinetic and transport models may eventually have better success in describing the fate of phosphate in soil (Enfield and Ellis 1983).

Biological immobilization of phosphorus fertilizers in soil organic matter and organic forms of soil phosphorus have been reviewed by Alexander (1977) and Anderson (1980). Naturally occurring organic forms of phosphorus include a wide range of esters, inositol phosphates, nucleic acids, and phospholipids. Net mineralization of organic phosphorus does not occur until phosphorus constitutes about 0.2–1% of organic carbon content and 5–20% of the organic nitrogen content (Alexander, 1977). Mineralization is favored by conditions that encourage microbial activity. Phosphorus is rapidly retained as insoluble inorganic compounds and sorbed to soil surfaces; therefore, soluble losses of phosphorus in subsurface flow and runoff tend to be quite low. Offsite transport of phosphorus tends to be associated with sediment erosion (Koehler et al. 1982b; Taylor and Kilmer 1980).

3.4 NUTRIENT LOSSES FROM CULTIVATED SYSTEMS

Limited scientific information has been published in the peer reviewed literature regarding nutrient losses from turfgrass and golf courses. *Losses of nutrients from other cultivated systems provide a valuable perspective on potential losses and possible options to mitigate pollution of water resources associated with golf courses* (e.g., Anderson et al. 1989). The research discussed in this section is presented as a source of information and conceptual guide for potential problems in turfgrass systems. Section 3.5 of this chapter summarizes research on nutrient losses from turfgrass systems.

3.4.1 Nutrient Loss from Surface Runoff

Edge-of-field losses of nitrogen and phosphorus transport of nitrogen have been associated with intensive use of fertilizers in row crop systems (Tables 3 and 4) (Burwell et al. 1977; Keeney 1982, 1983; Koehler et al. 1982a, 1982b). Total

Table 3. Factors Associated with Nitrogen Transport in Surface Runoff for Both Agricultural and Turfgrass Systems

Factors	General comments
Climatic	
Rainfall/runoff related to fertilizer application	*Highest concentration of nitrogen in runoff occurs in the first significant runoff after fertilizer application.* Nitrogen concentration and availability at soil surface dissipates with time.
Rainfall intensity	Runoff occurs when precipitation exceeds infiltration. Increasing intensity increases runoff rate and energy available for nitrogen extraction and transport, but time to runoff within a storm is reduced. Intensity also may affect the depth of surface interaction.
Rainfall duration/ amount	Affects runoff volume and subsurface soil leaching of nitrogen past zone of surface runoff extraction and transport.
Time to runoff after fertilizer application	Runoff concentrations of nitrogen increase as time to runoff is decreased. Period between application and the first runoff event critical in determining total nitrogen transported in surface runoff. As the period between the first runoff event and application is extended, a greater proportion of nitrogen will be immobilized by plants or soil or leached past the active mixing zone. Peak sediment nitrogen losses occur at the onset of runoff; peak soluble-nitrogen losses are often delayed as part of surface losses occur as interflow.
Soil	
Soil texture and organic matter content	Soil texture affects infiltration rates, soil erodibility, particle transport potential, and chemical enrichment factors. Soil texture and organic matter content affects adsorption and mobility of organic nitrogen and ammonium. *Runoff is usually higher on finer-textured soils. Time to runoff is greater on sandy soils, possibly reducing initial runoff losses of soluble nitrogen.*
Surface crusting and compaction	Decreases infiltration rates, reduces time to runoff, and increases initial concentrations of soluble nitrogen.
Water content	Initial soil water content may increase runoff potential, reduce time to runoff, and reduce subsurface leaching of soluble nitrogen prior to start of runoff.

Table 3. Factors Associated with Nitrogen Transport in Surface Runoff for Both Agricultural and Turfgrass Systems (continued)

Factors	General comments
Soil (continued)	
Slope	Increasing slope may increase runoff rate, soil detachment and transport, and effective surface depth for chemical extraction.
Degree of aggregation and stability	Affects infiltration rates, crusting potential, effective depth for entrainment, sediment transport potential, and adsorbed nitrogen enrichment in sediments.
Nitrogen Management Practices	
Application rate	*Runoff potential and concentrations are proportional to amounts of nitrogen present. Application rates commensurate with realistic growth and quality objectives significantly reduces potential surface runoff losses of nitrogen.*
Placement	*Exposure of applied nitrogen losses in surface runoff is reduced by fertilizer or organic residue incorporation below the soil surface.*
Timing of application	*Application of nitrogen fertilizer as close as possible to the time required for plant uptake reduces exposure to surface runoff losses. Split applications increase nitrogen efficiency. Fall application is an economically and environmentally unsound management practice. Late-fall and winter application of organic wastes have been associated with increased loss of nitrogen.*
Formulation	The major source of nitrogen in surface runoff is sediment sorbed-nitrogen (which is minimal in high quality turf). Initial leaching of soluble forms beneath the effective surface depth of interaction reduces soluble losses. Organic nitrogen and ammonium are sorbed on soil surfaces and sequestered from transport in runoff water. Urea is water soluble, but rapidly transformed to ammonium. Losses of nitrate in runoff water are associated with heavy application of organic residues or nitrate based commercial fertilizers. When time to runoff is short, increased amounts of soluble nitrogen at the soil surface may enhance runoff concentration. Slow release formulations will reduce surface losses if time of release is synchronized with plant uptake and does not coincide with runoff producing events.

Table 3. Factors Associated with Nitrogen Transport in Surface Runoff for Both Agricultural and Turfgrass Systems (continued)

Factors	General comments
Agricultural and soil management practices	
Soil and water conservation practices	Reduces transport of organic and sediment nitrogen. Reduces transport of soluble nitrogen compounds if, after nitrogen application, runoff volumes are reduced sufficiently during critical periods to offset increases in runoff concentrations. Reductions in surface runoff losses of nitrogen are often at the expense of increased leaching losses of nitrate.
Vegetative buffer strips	Transport of sediment and soluble nitrogen may be reduced, if grass buffer strips are located around fields. Reduction occurs due to secondary infiltration, sediment deposition, and uptake by buffer strip vegetation. Shock loading of surface waters with soluble nitrogen is possible from leaching of nitrogen-rich vegetation in grassed buffer strips and waterways.
Irrigation	Chemigation application of nitrogen may move soluble nitrogen into soil surface and reduce runoff potential. Plant uptake of soluble nitrogen applied at the appropriate rate also reduces nitrogen in return flows. Effect on antecedent soil moisture is critical. By keeping soil moisture levels higher than necessary for turfgrass growth, turfgrass managers increase the liklihood of nitrogen losses by both surface and subsurface transport.

Table 4. Factors Associated with Phosphorus Transport in Surface Runoff for Both Agricultural and Turfgrass Systems

Factors	General comments
Climatic	
Rainfall/runoff related to fertilizer application	*Highest concentration and loss of phosphorus in runoff occurs in first significant runoff after fertilizer application.* Phosphorus concentration and availability at soil surface dissipates rapidly with time. Mass loss of phosphorus is related to sediment transport, therefore peak runoff loading of phosphorus coincides with peak sediment loads. Again, this is most important for the establishment phase of turfgrass and of less importance after establishment.
Rainfall intensity	Runoff occurs when precipitation exceeds infiltration. Increasing intensity increases runoff rate and energy available for sediment transport, but time to runoff within a storm is reduced. Intensity also may affect depth of surface interaction.
Rainfall duration/ amount	Affects runoff volume and depth of surface interaction with soil adsorbed phosphorus. Relationship of runoff volume and phosphorus transport is not as strong as relationship with surface transport of nitrogen.
Time to runoff after fertilizer application	*Runoff concentrations of phosphorus increases as time to runoff decreased. Period between application and first runoff event important in determining soluble phosphorus transported in surface runoff.* As period between first runoff event and application is extended, a greater proportion of applied phosphorus is adsorbed/precipitated on soil surfaces or immobilized by plants. Peak sediment and total phosphorus losses occur at the onset of runoff.
Soil	
Soil texture	Soil texture affects infiltration rates, soil erodibility, particle transport potential, phosphorus adsorption sites, and chemical enrichment factors. Soil texture affects adsorption and mobility of phosphorus. Runoff is usually higher on finer-textured soils. *Time to runoff is greater on sandy soils, possibly reducing initial runoff losses of soluble phosphorus.*
Surface crusting and compaction	Decreases infiltration rates, reduces time to runoff, and increases initial concentrations of soluble phosphorus.
Water content	Initial soil water content may increase runoff potential and reduce time to runoff.
Slope	Increasing slope may increase runoff rate, soil detachment and transport, and effective surface depth for chemical extraction.
Degree of aggregation and stability	Affects infiltration rates, crusting potential, effective depth for entrainment, sediment transport potential, and adsorbed phosphorus enrichment in sediments.

Table 4. Factors Associated with Phosphorus Transport in Surface Runoff for Both Agricultural and Turfgrass Systems (continued)

Factors	General comments
Phosphorus management practices	
Application rate	*Runoff potential and concentrations are proportional to amounts of phosphorus present.* Application rates commensurate with realistic yield objectives significantly reduces potential surface runoff losses. Heavy applications of phosphorus create exceptionally increased risk of surface runoff losses of sediment and soluble phosphorus.
Placement	Phosphorus exposure to runoff losses is reduced by fertilizer incorporation beneath the soil surface. Incorporation of commercial fertilizer by injection reduces soluble losses of phosphorus.
Timing of application	Application of phosphorus fertilizer as close as the time required for plant uptake may reduce exposure to surface runoff losses.
Formulation	Major surface losses of phosphorus are associated with sediment sorbed phosphorus. Both organic and inorganic phosphorus are rapidly sorbed on soil surfaces sequestering soluble phosphorus from transport in runoff water. Heavy losses of soluble phosphorus may result from leaching of frozen or dead vegetation, leaching of unincorporated crop residues, and from heavy applications of animal waste.
Agricultural and soil management practices	
Soil and water conservation practices	Reduces transport of total and sediment phosphorus. Reduces transport of soluble phosphorus if, after phosphorus application, runoff volumes are reduced sufficiently during critical periods to offset increases in runoff concentrations. Reduced tillage systems decrease total phosphorus lost in surface runoff, but increase both concentration of algal available and soluble phosphorus lost in runoff water.
Irrigation	*Incorporation by light irrigation after application is routinely done with all fertilizer applications in turfgrass and should reduce both nitrogen and phosphorus runoff potential.*

nitrogen concentration in agricultural runoff ranges from less than 0.5 ppm to 92 ppm (Koehler et al. 1982a). Total phosphorus concentration in agricultural runoff ranges from less than 0.25 ppm to 1.25 ppm (Koehler et al. 1982a). Baker (1985) cites illustrative examples of nitrogen and phosphorus pollution of the Lake Erie Basin. Contamination of both surface and groundwater in this region is linked to intensive row crop production systems and edge-of-field losses of nutrients.

Anderson et al. (1989), Frere (1976), Keeney (1982, 1983), Koehler et al. (1982a, 1982b), Legg and Meisinger (1982), and Taylor and Kilmer (1980) describe numerous examples of runoff losses of nitrogen and phosphorus from agricultural fields (Tables 5 and 6). Runoff losses of nutrients are greatly reduced if (1) occurrence of runoff is delayed after application; (2) rainfall or irrigation leaches the chemical beneath the soil surface prior to occurrence of runoff; (3) nutrient sources are incorporated into the soil; (4) the rate of application is limited by the amount required for crop production, as determined by soil testing and nutrient budgets; (5) nutrient sources are applied at the time required for crop growth and at the time nutrients are least likely to be exposed to interaction with runoff; and (6) residue and cultivation practices are implemented that reduce the volume of both runoff sediment and water (Table 3 and 4) (Aldrich 1984; Anderson et al. 1989; Koehler 1982a, 1982b).

The rate of fertilizer application and the period between nutrient application and occurrence of runoff are critical factors in determining storm event losses of nutrients (Anderson et al. 1989; Koehler et al. 1982a). Total nitrogen in runoff from fertilized agricultural plots was 60% greater when runoff occurred 3 d after application than in runoff occurring 3 months later (Bradford 1974). Dunigan et al. (1976) observed greater nutrient losses from surface applications from ryegrass and millet plots in 1 year with greater total precipitation. The first few rain events of the year did not produce runoff and allowed movement of the applied fertilizer into the soil. The highest surface losses of nutrients were from plots receiving the highest fertilizer rate. A greater proportion of applied nutrients will be absorbed by plants, adsorbed by the soil, or leached into the soil beyond the mixing zone by light rain as the period between application and runoff is extended (Bradford 1974; Frere et al. 1980). Klausner et al. (1974) reported that the greatest surface losses of nitrate occurred after heavy fall fertilization of wheat with nitrogen just prior to intense rainfall. However, neither ammonium nor inorganic phosphorus loading of surface runoff was related to time of application.

Chichester (1977) observed the greatest surface losses of nitrogen from corn and meadow plots when intense rainfall events occurred shortly after fertilizer application. Surface transport of total nitrogen was reduced as effective surface cover was increased. The results of this study emphasized the importance of overall good

Table 5. Selected Examples of Runoff Losses of Nitrogen from Treated Plots and Watersheds; Studies Represent Range of Soil and Management Conditions Associated with Nitrogen Fertilization

Catchment type/ moisture input	Duration of input/ variables investigated	General comments and conclusions	Ref.
		Georgia	
0.34 ha watershed/ rainfall	3-year study of surface and subsurface transport of nitrate; continuous corn, 3.2% slope, sandy loam	Average concentration of nitrate in surface runoff ranged from less than 1 to 3 ppm. Annual nitrate-N losses in surface runoff (in kg ha^{-1}) from 1969 to 1971 were 0.12, 0.43, and 0.34. Surface runoff losses of nitrate-N were relatively small compared to losses in subsurface flow. On the nearly level watershed off-site transport of solution was primarily by subsurface flow. Subsurface transport accounted for 80% of water transport and for 99.1% of nitrate-N losses.	Jackson et al. (1973)
Upland agricultural watersheds/rainfall	5-year study of sediment and nutrient transport, conservation practices	Annual losses of ammonium runoff and sediment N were 35–40% less on watersheds with terraces and double cropping of corn compared to watersheds with no conservation practices. There were no real annual differences in surface transport of nitrates. Runoff water and sediment transport accounted for 82–93% of the variation in nutrient losses on the conventional watersheds, but for as little as 61% of the variation on watersheds with conservation treatments. Total seasonal losses were affected by the quantity of sediment transported during the highly erosive period from May through July.	Langdale et al. (1979)

		Indiana	
38.8 m^2 plots/ simulated rain	Runoff loss of nitrogen, sod vs fallow, initial soil moisture, surface sealing	Soluble and sediment nitrogen losses occur from NH_4-NO_3 pellets applied to the surface of fallow or sod covered fields. Total N and soluble N losses were generally greater for the fallow treatment compared to the sod treatment. *Nitrogen losses in runoff were greater under conditions of higher initial soil moisture content for both surface treatments.* Surface losses of nitrogen were also elevated with soil sealing of the fallow treatment. Measured losses of fertilizer nitrogen were low for both treatments. *The greatest loss of fertilizer nitrogen (15% of amount applied) occurred after 5 in. of heavy rainfall.*	Moe et al. (1967)
38.8 m^2 plots/ simulated rain	Runoff loss of nitrogen, NH_4–NO_3 vs urea, sod vs fallow	*Soluble and sediment nitrogen losses were greater for broadcast pellets of NH_4–NO_3 compared to broadcast pellets of urea. Ionized NH_4–NO_3 was held at the soil surface and available for runoff extraction, while the nonionized urea was initially leached further into the soil.* Soluble N losses were greater from the sod-covered plots than the fallow plots.	Moe et al. (1968)
84.5 m^2 plots/ simulated rain	Runoff loss of nutrients, tillage method: conventional, till-plant, chisel, double disk, and coulter	Coulter and chisel tillage systems controlled soil and total nutrient losses, but soluble N losses from surface-applied nitrogen were higher than the conventional tillage treatment. Disk and till-plant systems were less effective in sediment control, but had lower soluble N losses in runoff water. Conventional tillage, which allowed the highest degree of fertilizer incorporation, had the highest soil and total N losses, but the smallest loss of soluble N in runoff water.	Romkens et al. (1973)

Table 5. Selected Examples of Runoff Losses of Nitrogen from Treated Plots and Watersheds; Studies Represent Range of Soil and Management Conditions Associated with Nitrogen Fertilization (continued)

Catchment type/ moisture input	Duration of input/ variables investigated	General comments and conclusions	Ref.
		Iowa	
6–61 ha watersheds/ rainfall	Surface and subsurface loss of nutrients, sediment, and runoff water; conservation practices	Terraced watersheds reduced sediment N and soluble N losses in surface runoff by controlling sediment and runoff losses compared to contoured watersheds. Increased infiltration of water resulted in increased subsurface transport of soluble N on the terraced watersheds compared to the contoured watersheds.	Burwell et al. (1976, 1977), Schuman et al. (1973a)
30–60 ha watersheds/ rainfall	Surface loss of nutrients, season, conservation practices, 7-year study	Most of the average annual total loss of nitrogen occurred during the period of seedbed preparation, fertilizer application, and crop establishment from April through June. Seasonal discharges of runoff, sediment, sediment nitrogen, and total nitrogen were effectively reduced by a level terraced watershed compared to contour-cropped watersheds. Average annual soluble nitrogen losses never exceeded 1% of the annual application.	Alberts et al. (1978)
32 m^2 plots/ simulated rain	88.9 mm rain for 1.4 hr, 63 mm rain for 1 hr, 63.5 mm rain for 0.5 hr, runoff losses compared for variable conservation tillage practices	Loss of total N was inversely related to percentage of soil covered with residue. Losses of soluble N were small compared to sediment N. Soluble N concentrations were correlated to percentage of soil covered with crop residue. Conservation tillage practices were ineffective in controlling soluble N losses; total N losses were reduced by control of erosion.	Barisas et al. (1978)

0.55–1.75 ha watersheds/ rainfall	3-year study of nutrients transported in runoff, tillage treatments: conventional, ridge-till, and no-till	Conservation tillage systems reduce both runoff and erosion compared to conventional tillage. With the reduction of both sediment and runoff water, total N, sediment N, and soluble N losses in no-till and ridge-till systems were also reduced compared to conventional tillage systems. Movement of runoff water was reduced sufficiently by the conservation tillage treatments to offset increased concentration of soluble N in runoff.	Johnson et al. (1979)
13.7 m^2 plots/ simulated rain	63.5 mm rain for 2 hr; compared levels of residue and fertilization, placement of fertilizer above or below residue; fertilizer incorporated on 0 residue plots	Crop residue increased soluble N concentration in runoff insignificantly. Surface fertilization increased concentrations of N, especially NH_4, in runoff. Placement of fertilizer above or below residue did not affect concentrations in runoff. Increases in runoff losses of nitrate on fertilized plots were less than 1% of that applied; however, losses of ammonium increased by as much as 5% on surface-fertilized plots. *Increases in surface runoff concentrations of soluble N with reduced tillage results from lack of fertilizer incorporation.*	Baker and Laflen (1982)
13.7 m^2 plots/ simulated rain	111 mm rain applied for 1.75 hr followed by 72 mm rain applied for 1.13 hr 11 d later; fall fertilization after bean harvest; differential application: surface applied vs point injection; tillage treatments were no-till.	Chemical concentrations and losses of soluble N in surface runoff water were highest for fall fertilized, surface applied, nonincorporated treatments. Point injection of liquid fertilizer significantly reduced losses of nutrients in surface runoff from the no-till and disk till treatments. No-till reduced total N losses, but soluble N losses were equal to or greater than the alternate tillage treatments.	Baker and Laflen (1983a)

Table 5. Selected Examples of Runoff Losses of Nitrogen from Treated Plots and Watersheds; Studies Represent Range of Soil and Management Conditions Associated with Nitrogen Fertilization (continued)

Catchment type/ moisture input	Duration of input/ variables investigated	General comments and conclusions	Ref.
		Iowa (continued)	
32 m^2 plots/ simulated rain	Rain applied at 6.3 cm hr^{-1}, comparison of conservation tillage and crop rotation	Concentration and mass loss of soluble N increased for the reduced tillage systems compared to conventional tillage practice. The reduction of sediment loss in conservation tillage systems significantly reduced total N losses despite enrichment of sediment with adsorbed nitrogen. *When fertilizer was not incorporated in the no-till treatment, concentrations and mass loss of soluble N increased 5 to 9 times the amount lost in the conventional practice.*	Laflen and Tabatabai (1984)
		Louisiana	
4.5 m^2 plots/ rainfall	2-year study comparing rate, placement, and type of nutrient element on millet and ryegrass	Only small amounts of soluble N were lost in surface runoff when fertilizer was soil incorporated. Loss of soluble N did not exceed 1% of the total amount applied when soil incorporated regardless of application rate. Soluble N losses in surface runoff were directly related to application rate. Top-dressed fertilizer N losses ranged from 1.82 to 2.68% of N applied. NH_4 losses exceeded NO_3 losses in surface water in year with greater rainfall. Low intensity rain may have leached NO_3 beyond the surface mixing zone. Surface loss of soluble N and urea N were greater for uncoated urea than sulfur-coated urea. *Large losses of nitrogen occurred (9.5%) from uncoated urea when heavy precipitation occurred soon after application.*	Dunigan et al. (1976)

		Minnesota	
89.2 m^2 plots/ simulated rain	62.75 mm rain applied on consecutive days, comparison of runoff losses with different fertilizer placement	Deep incorporation of nitrogen fertilizer reduced nitrogen losses to the level of losses from unfertilized plots. Highest losses in runoff occurred when fertilizer was broadcast on a disked surface. Disking-in of nitrogen fertilizer was not effective in reducing surface losses of sediment or soluble N.	Timmons et al. (1973)
90 m^2 plots/ rainfall	10-year study of nutrient losses; crop cover, seasonal periods	Much of the annual losses of sediment and nutrients occurred during the critical erosion period from the time of corn planting to 2 months later. Snowmelt runoff accounted for much of the annual water and soluble nutrient losses. Average quantities of nitrate-N and ammonium N in precipitation exceeded annual losses in runoff. *Study emphasized importance of soil cover and nutrients lost in runoff. The amounts of soluble N and sediment N, respectively, for five cover conditions (in kg ha^{-1}) were continuous clean fallow, 3.0 and 51.3; continuous corn, 2.3 and 21.5; corn in rotation, 1.0 and 13.2; oats in rotation 2.6 and 1.9; and hay in rotation, 3.1 and 0.2.*	Burwell et al. (1975)
		Mississippi	
0.01 ha plots/ rainfall	2-year study comparing tillage treatment and soybean rotations	Total N lost from no-till soybeans was 4.7 kg ha^{-1} compared to 46.4 kg ha^{-1} lost in surface runoff from conventional tillage treatments. Concentration of soluble N increased in runoff from all no-till treatments. However, runoff volume in the bean-bean rotation was reduced sufficiently to offset the increased concentration, and soluble N losses were reduced. In the no-till bean-wheat and bean-corn rotations, both concentration and mass loss of soluble N exceeded the conventional treatment. No-till reduces total nutrient losses, but soluble nutrient losses in surface runoff remains a water quality concern.	McDowell and McGregor (1980)

Table 5. Selected Examples of Runoff Losses of Nitrogen from Treated Plots and Watersheds; Studies Represent Range of Soil and Management Conditions Associated with Nitrogen Fertilization (continued)

Catchment type/ moisture input	Duration of input/ variables investigated	General comments and conclusions	Ref.
		New York	
0.32 ha fields/ rainfall	1-year study on plots with 15 years of crop rotations and soil management; compare fertility levels on corn and wheat plots with good and poor soil management; poor soil management doubled volume of runoff water	Ammonium losses were not significantly related to crop rotation, fertilization levels, and soil management practices. NH_4 losses ranged from 0.14 to 1.3 kg ha^{-1} $year^{-1}$. NO_3 losses were related to crop rotation, fertility level, and soil management. Nitrate losses ranged from 0.39 to 29.23 kg ha^{-1} $year^{-1}$. *Heavy fall fertilization resulted in soluble N losses greater than soluble N delivered in rainfall. The quantity of nitrogen lost in runoff was directly related to time of application and occurrence of rain.* Concentration of inorganic N in runoff is a function of the time of year and timing of application in relation to runoff events.	Klausner et al. (1974)
0.25 ha plot/ irrigation	Runoff loss of nitrate, placement and antecedent moisture conditions, 8% slope	Nitrate is lost exponentially as a function of surface runoff. Nitrate in macropore space is lost more rapidly than from micropore space. *Surface-applied nitrate on saturated soil was completely lost in runoff. Between 5 and 30% of soil incorporated nitrate was lost in runoff under antecedent moisture between field capacity to near saturation. Five percent of applied nitrate was lost in runoff when it was soil incorporated on soil below field capacity.* Application of fertilizer to drier soil and incorporation resulted in reduction of loss in surface runoff.	Steenhuis and Muck (1988)

		Nevada	
Field size watersheds	3 year study on pollutant load in irrigation return flow from fields with continuous vegetation, pastured and unpastured	No net loss of nitrate and total N from irrigated field of alfalfa hay and grass pastures in return flow was observed. Nitrate and total N concentrations were consistently lower in surface return flow than in irrigation water. The presence or absence of grazing animals did not affect soluble nitrogen concentrations in return flows. Statistical variation was attributed to variation in volume of return flow. Continual vegetation cover did not contribute to nitrogen loading via surface runoff and return flow. Nitrogen infiltrated into the soil rather than transported in surface runoff.	Miller et al. (1984)
		North Carolina	
85 m^2 plots/ rainfall and irrigation with swine effluent	5.5-year study, swine effluent applied at three rates: low = 335 kg N ha^{-1} $year^{-1}$, moderate = 670 kg N ha^{-1} $year^{-1}$, and high = 1340 kg N ha^{-1} $year^{-1}$	The combination of low rainfall runoff losses and moderate concentration of soluble N losses from swine effluent irrigated plots resulted in generally low annual mass loss of nitrogen in surface runoff (<1% of applied nitrogen). *Concentration and annual mass loss of soluble N from the high application level (4.6 kg ha^{-1}, 7.7 ppm) was higher than either the moderate (2.8 kg ha^{-1}, 5.2 ppm) or low (2.2 kg ha^{-1} 4.6 ppm) effluent application level.* Application levels are limited by subsurface loss of nitrates rather than by surface loss of soluble N.	Westerman et al. (1985)

Table 5. Selected Examples of Runoff Losses of Nitrogen from Treated Plots and Watersheds; Studies Represent Range of Soil and Management Conditions Associated with Nitrogen Fertilization (continued)

Catchment type/ moisture input	Duration of input/ variables investigated	General comments and conclusions	Ref.
		North Carolina (continued)	
49 m^2 plots/ rainfall and irrigation with swine effluent	3-year study, waste treatments compared to conventional fertilization of tall fescue on sandy clay loam soil; fertilizer (201 kg N ha^{-1} $year^{-1}$), manure (670 kg N ha^{-1} $year^{-1}$), swine effluent 600 kg N ha^{-1} $year^{-1}$), swine effluent (1200 kg N ha^{-1} $year^{-1}$)	Although some nutrient concentrations were high, rainfall runoff volume from all treatments was low; mass transport of nutrients did not exceed 1.5% of nutrients applied. Nitrate concentrations were higher in the liquid manure and high N lagoon effluent treatments than the commercial fertilizer and moderate N lagoon treatments. Total N losses after manure application were three times higher than average when runoff occurred 5 d after spring application. *Concentrations of most pollutant elements are much higher, if runoff occurs within 1 or 2 d of application.* The manure treatment (670 N kg ha^{-1}), and both swine effluent treatments (600 and 1200 N kg ha^{-1}) resulted in fairly high nitrate concentrations in runoff water. Average nitrate concentrations were 4.0, 7.3, 4.6, and 13.2 mg L^{-1} for the fertilizer, manure, moderate N effluent, and high N effluent treatments.	Westerman et al. (1987)
		Oklahoma	
7.2–11.6 ha watersheds/ rainfall	1-year monitoring of nutrients from cropland and grazed rangeland	Total N in runoff ranged from 2 to 15 kg ha^{-1} $year^{-1}$; flow weighted mean concentration of total N ranged from 1 to 6 ppm and nitrate-N ranged from 0.2 to 1.9 ppm. Runoff losses of soluble inorganic nitrogen were less than input by precipitation. Loss of fertilizer nitrogen did not exceed 5% of most recent application	Olness et al. (1975)

Six 0.6–17.9 ha watersheds/ rainfall	5-year study monitoring sediment and nutrients in runoff from cropland and rangeland	Annual variations in nutrient content of runoff is large. Variations may be greater than the effects of treatments designed to control both sediment and nutrient discharges. Long historical records are required to evaluate the impact of efforts to mitigate nonpoint source pollution from agricultural sources. Maximum annual nutrient losses were 13 kg ha^{-1} of total N and 4 kg ha^{-1} of nitrate-N. Nitrate accounted for 10–30% of total N discharge in runoff. Average annual nutrient losses were about half of the maximum value for each land use.	Menzel et al. (1978)
Four 7.8–11.1 ha watersheds/ rainfall	1-year monitoring study of nutrient losses in runoff, fertilization, and grazing management	Initially, fertilization of grazed native grassland, continuous and rotation, increases soluble N losses in runoff. However, over time increased forage production may reduce soil and water losses and reduce both sediment and soluble N losses. *Broadcast fertilization on grazed land is vulnerable to surface transport, but these losses will not exceed 5% of the amount applied up to rates of 75 kg N ha^{-1}. Fertilizer N losses were detected for 1 month after application.*	Olness et al. (1980)
Twelve 1.6–5.6 ha watersheds in OK, eight 1.1–122 ha watersheds in TX/rainfall	8-year monitoring of nutrients in runoff and groundwater wells on cultivated and native grassland	22–93% of surface transported N was in the particulate form, sorbed NH_4 or organic N. The range of mean annual flow weighted concentration of nitrogen in surface water was 1.77–14.20 ppm for total N, 0.24–3.56 ppm for nitrate-N, and 0.02–1.72 for ammonium-N. Change from grassland to cultivation increased nitrate and total N in runoff; NH^4 losses decreased with cultivation due to increased conversion to nitrate. Nitrogen losses were greatest for fertilized wheat and rotational cropland, although total mass trasport was small. Total N loss represented 3 and 9% of the amount applied in wheat and rotational cropland, respectively.	Sharpley et al. (1987)

Table 5. Selected Examples of Runoff Losses of Nitrogen from Treated Plots and Watersheds; Studies Represent Range of Soil and Management Conditions Associated with Nitrogen Fertilization (continued)

Catchment type/ moisture input	Duration of input/ variables investigated	General comments and conclusions	Ref.
		Saskatchewan, Canada	
4–5 ha agricultural fields/snowmelt and rainfall	6-year monitoring study of nutrients in runoff, summerfallow vs stubble mulch, effect of fall fertilization	The amount of nutrients lost from agricultural fields in spring snowmelt exceeded algal growth and water quality limits. Average total sediment N in snowmelt runoff was 0.4 kg, 2.5, and 4.1 kg ha^{-1} for stubble mulch, summerfallow, and fall fertilized summerfallow, respectively. Average total nitrate-N in snowmelt runoff was 0.1, 0.3, and 0.3 kg ha^{-1} for stubble mulch, summerfallow, and fall fertilized summerfallow, respectively.	Nicholaichuk and Read (1978)
		South Dakota	
0.7–1.0 ha prairie basins, 92.4 m^2 plots/ rainfall	Nutrient losses in runoff from native prairie and fertilized corn, oats, alfalfa, and fallow plots	Soluble N losses in runoff were similar for both native prairie grass and cultivated cropland. *Variations in runoff losses from prairie basins were attributed to differences in prarie vegetation and surface mulches of cropland; differences in overland flow distances that affect time to dissolve nutrient elements; and dilution of runoff in basin sites compared to erosion plots.* NH_4 concentrations in cropland and prairie runoff ranged from 0.3 to 2.6 ppm and 0.16 to 3.3 ppm, respectively. NO_3 concentrations in cropland and prairie runoff ranged from 0.1 to 1.1 ppm and 0.07 to 1.68 ppm, respectively. Soluble N concentrations in runoff from each crop cover type were similar.	White and Williamson (1973)

		Texas	
Duplicate 4 ha watersheds/rainfall	5-year study of nitrate in runoff on swelling clay soil, sorghum, cotton, and oats rotation	At the recommended application rates, annual average concentration of nitrate-N in runoff ranged from 2.3 to 2.9 ppm. Mean total loss of nitrate-N was 3.2 kg ha^{-1} $year^{-1}$. Losses of sediment nitrogen were about 5 kg N ha^{-1} $year^{-1}$. Storm event concentrations of nitrate-N depended on timing of runoff event in relation to fertilizer application. *Concentrations of nitrate-N in runoff were highest just after fertilizer application when soil was near field capacity. Concentrations of nitrate-N were lowest when large amounts of nutrients infiltrated dry soil prior to initiation of runoff. During runoff producing storms, immediately after fertilization, nitrate concentrations were lowest in initial runoff and highest near the end of the runoff event.*	Kissel et al. (1976)

Table 6. Selected Examples of Runoff Losses of Phosphorus from Treated Plots and Watersheds; Studies Represent Range of Soil and Management Conditions Associated with Phosphorus Fertilization

Catchment type/ moisture input	Duration of input/ variables investigated	General comments and conclusions	Ref.
		Indiana	
84.5 m^2 plots/ simulated rain	Runoff loss of nutrients, tillage methods: conventional, till-plant, chisel, double disk, and coulter	Coulter and chisel tillage systems controlled soil and total nutrient losses, but soluble P losses from surface-applied phosphorus were higher than the conventional tillage treatment. Disk and till-plant systems were less effective in sediment control, but had lower soluble P losses in runoff water. Conventional tillage, which allowed the highest degree of fertilizer incorporation, had the highest soil and total P losses, but the smallest loss of soluble P in runoff water.	Romkens et al. (1973)
20 m^2 plots/ simulated rain	Five applications of 1 hr duration for a total of 41 cm over a 5-month period; fertilizer broadcast and incorporated; three rates of application	Rate of P fertilization is directly related to levels of soluble *ortho*-PO_4 and sediment extractable P levels in runoff. Soluble and sediment extractable P in runoff are not related to total P and organic P levels in runoff. Average *ortho*-P content in runoff ranged from 0.07 (no treatment) to 0.44 (113 P kg ha^{-1}); average available P on sediment ranged from 15 (no treatment) to 58 ppm of sediment (113 P kg ha^{-1}); organic P concentration in runoff ranged from 99 to 152 ppm; and average total P ranged from 461 to 558 ppm of sediment.	Romkens and Nelson (1974)

		Iowa	
6–61 ha watersheds/ rainfall	Surface and subsurface loss of nutrients, sediment and runoff water; conservation practices	Terraced watersheds reduced sediment P and soluble p losses in surface runoff by controlling sediment and runoff losses compared to contoured watersheds. Increased infiltration of water resulted in increased subsurface transport of soluble P on the terraced watershed compared to the contoured watersheds.	Burwell et al. (1976, 1977), Schuman et al. (1973b)
30–60 ha watersheds/rainfall	Surface loss of nutrients, season, conservation practices, 7-year study	Most of the average annual total loss of phosphorus occured during the period of seedbed preparation, fertilizer application, and crop establishment from April through June. Seasonal discharges of runoff, sediment, sediment P, and total P was effectively reduced by a level terraced watershed compared to contour-cropped watersheds. Average annual soluble P losses never exceeded 1% of the annual application.	Alberts et al. (1978)
32 m^2 plots/ simulated rain	88.9 mm rain for 1.4 hr, 63 mm rain for 1 hr, 63.5 mm rain for 0.5 hr, runoff losses compared to variable conservation tillage practices	Available P content of eroded soil was directly related to percentage of soil covered with residue. Losses of soluble P were small compared to sediment P. Soluble P concentrations were correlated to percentage of soil covered with crop residue. Conservation tillage practices were ineffective in controlling soluble P losses; total-nutrient losses were reduced by control of erosion.	Barisas et al. (1978)
0.55–1.75 ha watersheds/rainfall	3-year study of nutrients transported in runoff, tillage treatments: conventional, ridge-till, and no-till	Conservation tillage systems reduced both runoff and erosion compared to conventional tillage. With the reduction of sediment transport, total P losses were reduced by conservation tillage methods. Soluble P and available P losses in surface runoff were increased in the reduced tillage systems compared to the conventional tillage treatments. *Decreased nutrient incorporation, leaching of crop residues and sediment enrichment were factors contributing to increases in runoff concentrations and total losses.*	Johnson et al. (1979)

Table 6. Selected Examples of Runoff Losses of Phosphorus from Treated Plots and Watersheds; Studies Represent Range of Soil and Management Conditions Associated with Phosphorus Fertilization (continued)

Catchment type/ moisture input	Duration of input/ variables investigated	General comments and conclusions	Ref.
		Iowa (continued)	
13.7 m^2 plots/ simulated rain	63.5 mm rain for 2 hr compared levels of residue and fertilization, placement of fertilizer above or below residue; fertilizer incorporated on 0 residue plot	Crop residue increased soluble P concentration in runoff insignificantly. Surface fertilization increased concentrations of soluble P in runoff. Placement of P did not affect concentrations in runoff. Losses of soluble P increased by as much as 5% on surface-fertilized plots. *Increases in surface runoff concentrations of soluble P with reduced tillage resulted from lack of fertilizer incorporation.* Point injection of liquid fertilizer P resulted in soluble losses of phosphorus at levels similar to unfertilized plots.	Baker and Laflen (1982)
13.7 m^2 plots/ simulated rain	111 mm rain applied for 1.75 hr followed by 72 mm rain applied for 1.13 hr 11 d later; fall fertilization after bean harvest; differential application: surface applied vs point injection; tillage treatments were no-till, disk, and chisel	*Chemical concentrations and losses of soluble P in surface runoff water were highest for fall fertilized surface-applied, nonincorporated treatments.* Point injection of liquid fertilizer significantly reduces losses of nutrients in surface runoff from the no-till and disk till treatments. No-till with surface-applied fertilizer phosphorus reduces total P losses, but soluble P concentration and losses were much greater than the alternate tillage treatments. Fall fertilization of soybeans was not recommended due to water quality concerns.	Baker and Laflen (1983a)

$32\ m^2$ plots/ simulated rain	Rain applied at 6.3 cm hr^{-1}, comparison of conservation tillage and crop rotation	Concentration and mass loss of soluble P increased for the reduced tillage systems compared to conventional tillage practice. The reduction of sediment loss in conservation tillage systems significantly reduced total P losses despite enrichment of sediment with adsorbed phosphorus. *When fertilizer was not incorporated in the no-till treatment, concentrations and mass loss of soluble P increased five to nine times the amount lost in the conventional practice.*	Laflen and Tabatabai (1984)
		Louisiana	
$4.5\ m^2$ plots/ rainfall	2-year study comparing rate, placement, and type of nutrient element on millet and ryegrass	*Only small amounts of soluble P were lost in surface runoff when fertilizer was soil incorporated. Loss of soluble P did not exceed 0.35% of the total amount applied when soil incorporated regardless of application rate.*	Dunigan et al. (1976)
		Minnesota	
$89.2\ m^2$ plots/ simulated rain	62.75 mm rain applied on consecutive days, comparison of runoff losses with different fertilizer placement	Deep incorporation of phosphorus fertilizer reduced phosphorus losses to the level of losses from unfertilized plots. Highest losses in runoff occurred when fertilizer was broadcast on a disked surface. Disking-in of phosphorus fertilizer was not effective in reducing surface losses of sediment or soluble P.	Timmons et al. (1973)
$90\ m^2$ plots/ rainfall	10-year study of nutrient losses; crop cover, seasonal periods	Much of the annual losses of sediment and nutrients occurred during the critical erosion period from the time of corn planting to 2 months later. Snowmelt runoff accounted for much of the annual water and soluble nutrient losses. Annual losses of *ortho*-P in surface runoff exceeded	Burwell et al. (1975)

Table 6. Selected Examples of Runoff Losses of Phosphorus from Treated Plots and Watersheds; Studies Represent Range of Soil and Management Conditions Associated with Phosphorus Fertilization (continued)

Catchment type/ moisture input	Duration of input/ variables investigated	General comments and conclusions	Ref.
		Minnesota (continued)	
		contributions in precipitation. Study emphasized importance of soil cover and nutrients lost in runoff. The amounts of soluble P, sediment available P, and total P, respectively, for five cover conditions (in kg ha^{-1}) were continuous clean fallow, 0.07, 0.27, and 11.67; continuous corn, 0.18, 0.29, and 5.22; corn in rotation, 0.11, 0.13, and 2.97; oats in rotation 0.08, 0.02, and 0.43; and hay in rotation, 0.17, 0.05, and 1.15. No difference in soluble P losses was related to cover treatment. Sediment P losses were greatly reduced by cover and associated reduction of erosion.	
		Mississippi	
0.01 ha plots/ rainfall	2 year study comparing tillage treatments and soybean rotation	Total P lost in surface runoff from no-till soybeans was 2.8 kg ha^{-1} compared to 17.6 kg ha^{-1} lost from conventional tillage treatments. Concentration and mass loss of soluble P was related to tillage and crop management practices. Solube P losses in runoff increased in the following order: conventional till soybeans (continuous) < no-till corn-soybeans (rotation) < no-till soybeans (continuous) < no-till soybeans, wheat (double cropped). Increase of soluble P concentration in runoff was attributed to insufficient phosphorus sorption by sediment, addition of phosphorus fertilizer in the double-cropped	McDowell and McGregor (1980)

		treatment, limited phosphorus sorption without incorporation, and release of phosphorus from crop residues. No-till reduces total nutrient losses, but soluble nutrient losses in surface runoff remains a water quality concern.	
		New York	
0.32 ha fields/ rainfall	1-year study on plots with 15 years of crop rotations and soil management; compare fertility levels on corn and wheat plots with good and poor soil management; poor soil management doubled volume of runoff water	Losses of inorganic P were related to crop rotation, fertilization levels, and soil management. PO_4^{-3} losses ranged from 0.04 to 0.49 kg ha^{-1} $year^{-1}$. Greatest losses of inorganic P occurred under conditions of high phosphorus fertility and poor soil management, which provided high volume of runoff water and high levels of inorganic P available for increased runoff concentration. Although the time of year influenced soluble P losses in runoff, no relationship between inorganic P concentration and runoff volume was detected.	Klausner et al. (1974)
		Nevada	
Field size watersheds	3-year study on pollutant load in irrigation return flow from fields with continuous vegetation, pastured and unpastured	*A net loss of ortho-P and total P from irrigated fields of alfalfa hay and grass pastures. Ortho-P losses were associated with fertilized alfalfa hay.* Presence or absence of grazing animals and cropping pattern were not as important to water quality as site characteristics, water management, and runoff variations. *Ortho-P concentrations were consistently higher in surface return flow than in irrigation water. Continual vegetation did not contribute to phosphorus loading via surface runoff and return flow.*	Miller et al. (1984)

Table 6. Selected Examples of Runoff Losses of Phosphorus from Treated Plots and Watersheds; Studies Represent Range of Soil and Management Conditions Associated with Phosphorus Fertilization (continued)

Catchment type/ moisture input	Duration of input/ variables investigated	General comments and conclusions	Ref.
		North Carolina	
85 m^2 plots/ rainfall and irrigation with swine effluent	5.5-year study, swine effluent applied at three rates: low = 90 kg P ha^{-1} $year^{-1}$, moderate = 180 kg P ha^{-1} $year^{-1}$, and high = 360 kg P ha^{-1} $year^{-1}$	The combination of low rainfall losses and moderate concentration of soluble P losses from swine effluent irrigated plots resulted in generally low annual mass loss of phosphorus in surface runoff (<1% of applied phosphorus). Concentration and annual mass loss of soluble P from the high application level (3.8 kg ha^{-1}, 5.3 ppm) was not higher than either the moderate (1.2 kg ha^{-1}, 2.6 ppm) or low (0.9 kg ha^{-1}, 1.9 ppm) effluent application level. All application levels have phosphorus concentration levels in runoff in excess of eutrophication levels. Maximum concentrations of P in runoff occurred during the irrigation season. *Increases in P accumulation in soil increase potential for increased surface and subsurface transport of phosphorus to receiving waters.*	Westerman et al. (1985)
49 m^2 plots/ rainfall and irrigation with swine effluent	3-year study on fertilizer, compared to conventional fertilization of tall fescue on sandy clay loam soil; fertilizer (34 kg P ha^{-1} $year^{-1}$), manure (200 kg P ha^{-1} $year^{-1}$), swine effluent (120 kg P ha^{-1} $year^{-1}$), swine effluent (240 kg P ha^{-1} $year^{-1}$)	Although some nutrient concentrations were high, rainfall runoff volume from all treatments was low; mass transport of nutrients did not exceed 2% of nutrients applied. Phosphorus concentrations were higher in the liquid manure (6 ppm) and high P lagoon effluent (9 ppm) treatments than the commercial fertilizer and moderate P lagoon treatments.	Westerman et al. (1987)

		Ohio	
0.37 m^2 greenhouse microplots, simulated rain	Five combination of rainfall intensity/duration and slope to increase runoff potential; compared soil, cover crop, nutrient source, rate, and form	Most of surface transported P moved as a component of sediment. Cover crops decreased sediment losses of P, but increased total soluble P lost in runoff water. Inorganic concentration of P in surface runoff was 3 times greater than organic P in solution; however soluble organic P was the dominant leachate form of P in subsurface transport except when inorganic P did not react with the soil during percolation. More solution and sediment P was lost from manure and manure treated plots. *Effect of soil on phosphorus losses is reflected in differences in runoff potential between different soils. Total P lost was directly related to sediment transport; soluble and extractable P losses related to runoff volume, source of P, and cover conditions.*	Reddy et al. (1978)
		Oklahoma	
Six 0.6–17.9 ha watersheds, rainfall	5-year study monitoring of sediment and nutrients in runoff from cropland and rangeland	Annual variations in nutrient content of runoff is large. Variations may be greater than the effects of treatments designed to control both sediment and nutrient discharges. Long historical records are required to evaluate the impact of efforts to mitigate nonpoint source pollution from agricultural sources. Maximum annual nutrient losses were 11 kg ha^{-1} of total P and 2 kg ha^{-1} of soluble P. Soluble P accounts for 20% of the total P discharged from cropland and <10% of discharge from rangeland. Average annual nutrient losses were about half of the maximum value for each land use.	Menzel et al. (1978)

Table 6. Selected Examples of Runoff Losses of Phosphorus from Treated Plots and Watersheds; Studies Represent Range of Soil and Management Conditions Associated with Phosphorus Fertilization (continued)

Catchment type/ moisture input	Duration of input/ variables investigated	General comments and conclusions	Ref.
		Oklahoma (continued)	
Small boxes, simulated rain	Effects of soil roughness on runoff losses of dissolved reactive P; rainfall intensity was 6.8 cm hr^{-1} for 1 hr	Concentration of dissolved reactive P increased 15% in soil treatments with surface ridges compared to controls. Soils with ridges exposed 12% greater surface area to runoff losses compared to controls. Increased soil roughness decreased runoff volume, but increases in dissolved reactive P concentration offset any reduction in total load of soluble P in runoff water.	Ahuja et al. (1983)
Four 7.8–11.1 ha watersheds, rainfall	1-year monitoring study of nutrient losses in runoff, fertilization and grazing management	Initially, fertilization of grazed native grassland, continuous and rotation, increases soluble P losses in runoff. However, over time increased forage production may reduce soil and water losses and reduce both sediment and soluble P losses. *Broadcast fertilization on grazed land is vulnerable to surface transport, but these losses will not exceed 5% of the amount applied up to rates of 75 kg P ha^{-1} $year^{-1}$.* Fertilizer P losses decreased with time, but were detected for one year after application. Continuous grazing management will also increase total P and soluble P losses in comparison to rotation grazing.	Olness et al. (1980)

Twelve 1.6–5.6 ha watersheds in OK, 11 1.1–125 ha watersheds TX, rainfall; small boxes, simulated rain	4-year watershed monitoring, simulated runoff studies, watershed management practices, soil type, cover type	Measured and predicted concentrations of soluble and sediment P were compared for a range of cropland and rangeland watersheds in the Southern Plains. Soluble P was predicted from the water-extractable phosphorus content of the surface soil (0–1 cm) from samples collected in March of each year. Particulate P concentration in runoff was predicted from total P content of the surface soil from samples collected in March of each year. Predictions of mean annual soluble and sediment P concentrations in runoff were good over a range of concentrations, fertilizer and watershed management practices, soil types, and vegetative covers.	Sharpley et al. (1982)
Twelve 1.6–5.6 ha watersheds in OK, 8 1.1–122 ha watersheds in TX rainfall	8-year monitoring of nutrients in runoff and groundwater wells on cultivated and native grassland	40–93% of surface transported P was in the particulate form, sorbed P or organic P. The range of mean annual flow weighted concentration of phosphorus in surface water was 180–3070 ppb for total P; 20–430 ppb for soluble P. Soluble and total P concentrations in cultivated and uncultivated watersheds consistently exceeded critical levels for eutrophication (10 and 20 ppb, respectively). *Dramatic increases in total P concentrations and loads were associated with cultivation of native grassland.* Phosphorus losses were greatest for fertilized wheat and rotational cropland, although total mass transport was small. Total P loss represented 8 and 4% of the amount applied in wheat and rotational cropland, respectively.	Sharpley et al. (1987)

Table 6. Selected Examples of Runoff Losses of Phosphorus from Treated Plots and Watersheds; Studies Represent Range of Soil and Management Conditions Associated with Phosphorus Fertilization (continued)

Catchment type/ moisture input	Duration of input/ variables investigated	General comments and conclusions	Ref.
		Oklahoma (continued)	
7.2–11.6 ha watersheds/rainfall	1-year monitoring of nutrients from cropland and grazed rangeland	Total P in runoff ranged from 1 to 11.5 kg ha^{-1} $year^{-1}$; flow weighted mean concentration of total P ranged from 0.5 to 4.8 ppm and of soluble P ranged from 0.04 to 0.9 ppm. Concentration of soluble P was greater from cropland watersheds than from rangeland watersheds. Loss of fertilizer phosphorus did not exceed 5% of most recent application.	Olness et al. (1975)
		Ontario, Canada	
930 m^2 plots, Rainfall and snowmelt	Runoff concentrations and and loss from nearly level plots on clay soils, 12 plots monitored for two seasons, p applied at two rates (0 and 30 kg ha^{-1}), subsurface loss of sediment and dissolved P also monitored	Low levels of runoff were observed in continuous corn and bluegrass sod plots grown on nearly level clay soils. Dissolved and sediment P concentrations in runoff were highly variable. Average snowmelt concentrations of total dissolved P were 0.10 and 0.13 mg L^{-1} for corn and bluegrass, respectively. Average rainfall concentrations were 0.08 and 0.10 mg L^{-1} for corn and bluegrass, respectively. *Average snowmelt concentrations of sediment P were 0.14 and 0.13 mg L^{-1} for corn and bluegrass, respectively. Average rainfall concentrations were 0.35 and 0.06 mg L^{-1} for corn and bluegrass, respectively. Tile drainage contributes significantly to dissolved and sediment P losses from plots. Drainage losses are directly affected by rate of fertilization, crop practices, soil type, antecedent moisture conditions, and tile depth.*	Culley et al. (1983)

50.8 km^2, agricultural watershed, rainfall and snowmelt	2-year study on contribution of surface runoff and tile drainage on runoff losses of total and dissolved P from clay soils	On a watershed basis total P and *ortho*-P concentrations in runoff range from 0.015–3.30 and 0.001–0.450. Total P concentrations decreased in the order stormflow > snowmelt > baseflow; *ortho*-P concentrations decreased in the order snowmelt = stormflow > baseflow. Stormflow contributed to 73% of annual total P losses and to 56% of annual ortho P losses. 35% of the total P load lost to surface waters results from dissolved and sediment P lost in tile drainage and in streambank erosion. *Up to 50% of ortho-P losses are associated with surface flow from tile drains.* With low sediment delivery, conservation practices will have marginal impact on surface water.	Culley and Bolton (1983)
		Saskatchewan, Canada	
4–5 ha agricultural fields, snowmelt and rainfall	6-year monitoring study of nutrients in runoff, summer fallow vs stubble mulch, effect of fall fertilization	The amount of nutrients lost from agricultural fields in spring snowmelt exceeded algal growth and water quality limits. Average total P in snowmelt runoff was 0.4 kg ha^{-1}, 1.4 kg ha^{-1}, and 2.9 kg ha^{-1} for stubble mulch, summerfallow, and fall-fertilized summerfallow, respectively. Average total soluble P in snowmelt runoff was 0.1 kg ha^{-1}, 0.2 kg ha^{-1}, and 1.2 kg ha^{-1} for stubble mulch, summerfallow, and fall-fertilized summer fallow, respectively.	Nicholaichuk and Read (1978)
		South Dakota	
0.7–1.0 ha prairie basins, 92.4 m^2 plots, rainfall	Nutrient losses in runoff from native prairie and fertilized corn, oats, alfalfa, and fallow plots	*Total P and soluble P losses in runoff were similar for both relic prairie grass and cultivated cropland. Variations in runoff losses from prairie basins was attributed to differences in prairie vegetation and surface mulches of cropland; differences in overland*	White and Williamson (1973)

Table 6. Selected Examples of Runoff Losses of Phosphorus from Treated Plots and Watersheds; Studies Represent Range of Soil and Management Conditions Associated with Phosphorus Fertilization (continued)

Catchment type/ moisture input	Duration of input/ variables investigated	General comments and conclusions	Ref.
		South Dakota (continued)	
		flow distances that affects time to dissolve nutrient elements; and dilution of runoff in basin sites compared to erosion plots. Soluble P concentrations in cropland and prairie runoff ranged from 0.1–0.2 ppm and 0.05–19.5 ppm, respectively. Sediment P losses in surface runoff from cropland increased in the following order, alfalfa > oats > corn > fallow.	
		Wisconsin	
1.35 m^2 plots, simulated rain	Total and algal available-P in runoff from cropland and forestland, effect of manure application	Phosphorus in surface runoff from agricultural lands receiving manure applications is a potential nonpoint pollution problem. Land use, time of year, and manure application affects the level of surface transport of phosphorus. Losses of algal available-P were highest from newly established alfalfa in the fall after the foliage had been killed by frost. The fraction of total P lost in the algal available form was higher for alfalfa than for corn or a prepared seed bed. *Greatest total P losses were associated with sediment transport from corn or seedbed prepared treatments. Surface applied manure on corn and alfalfa increased the concentration of phosphorus in runoff, but increases in infiltration reduced mass losses of phosphorus. The effect of surface applied manure on frozen ground may increase*	Wendt and Corey (1980)

		phosphorus loads in runoff. Phosphorus losses in runoff from forestland was less than from cropland.	
1.35 m^2 plots/ simulated rain	Total, soluble, and algal compared in conventional (CN), chisel plow (CH), and no-till (NT) tillage systems for manured and unmanured plots for corn produced for dairy operations; 2-year study, rain applied May–July and Sept. at 14.5 cm hr^{-1}; manure was surface applied at 8 metric tons ha^{-1}; 10–26–26 was band applied to all plots	*Differences in runoff volumes influenced P losses.* Runoff losses were high for NT, and losses of soluble and algal available P were very high when manure was surface applied. In all plots, unincorporated manure increased soluble P lost in runoff. Manure application increased algal available P losses for NT relative to CH and CN. Runoff and total P losses were significantly reduced on the manured CH plots, relative to either the manured CN and NT plots. CH is the tillage method of choice for reducing phosphorus losses in surface runoff when manure is applied. Differences in soluble P losses among all tillage systems were insignificant for unmanured plots.	Mueller et al. (1984a, 1984b)
234 m^2 plots/ simulated rain	Runoff losses of phosphorus compared in conventional (CN), no-till (NT), till-plant (TP), and chisel plow (CH) tillage systems; simulated runoff trials conducted over 4 years; trials conducted in June–July and Sept.–Oct.; at planting fertilizer was subsurface banded at 8–10 cm	NT, CH, and TP treatments reduced total P losses on the average by 81, 70, and 59%, respectively, relative to CN. Concentrations of molybdate-reactive P (DMRP) were lowest for CN, but differences among all tillage treatments were generally small and insignificant. Unincorporated crop residue was not a major source of dissolved P. Although conservation tillage treatments increased algal available P concentrations in runoff water, reduction in runoff volume did reduce total runoff loads of algal available P. NT, CH, and TP treatments reduced algal available P by an average of 63, 58, and 27%, respectively, compared to CN. *Subsurface banding of fertilizer phosphorus at planting reduced dissolved and sediment associated runoff losses of phosphorus.*	Andraski et al. (1985a, 1985b)

management, including nitrogen application in balance with realistic yield objectives, soil incorporation of fertilizers, and adequate surface cover to improve nutrient use efficiency and minimize erosion and runoff water losses of nutrients (Legg and Meisinger 1982).

Application of nitrogen or phosphorus nutrient sources at rates higher than those recommended by soil tests or estimates based on realistic growth requirements increases the risk of offsite movement of nutrients either by surface runoff or leaching (Aldrich 1984; Anderson et al. 1989; Koehler et al. 1982a). Field studies of nitrogen and phosphorus losses in surface runoff demonstrated direct correlation with fertilization rates (Zwerman et al. 1971). Although soluble ammonium loading of surface runoff was not related to fertilization rate in the same region, soluble inorganic phosphorus and nitrate movement in surface runoff was directly related to application rates (Klausner et al. 1974). Research in the Corn Belt on fertilization rates and crop yield has demonstrated conclusive relationships with the quality of surface runoff and application rates (Koehler et al. 1982a). While having an insignificant effect on crop yields, excessive nitrogen and phosphorus fertilization in the Midwest has resulted in potentially greater runoff losses of nutrients (Burwell et al. 1977; Romkens and Nelson 1974; Whitaker et al. 1978). Total Kjeldahl-nitrogen in surface runoff from corn-cropped watersheds in Georgia was directly related to the rate of nitrogen application (Langdale et al. 1979). In most of the cited research, *excessive applications of nitrogen or phosphorus resulted in increased offsite transport of nutrients* (Tables 3–6).

Recent reviews on nutrient applications and yield suggest that overfertilization does not result in increased crop yields or improved crop quantity (Anderson et al. 1989; Bock 1984; Hanway and Olson 1980; Lorenz and Vittum 1980; Olson and Kurtz 1982; Ozanne 1980; Stanford and Legg 1984). Optimizing yield in relation to the economic inputs and water quality issues requires nutrient inputs at levels designed to achieve realistic yield goals. The point of greatest economic return from fertilization is usually below the "maximum" yield level. Setting realistic yield goals in advance should allow adjustment of rate, timing, and method of application in order to maximize economic return and minimize adverse environmental impacts. *Improved efficiency of nitrogen use through soil tests, timing of application to periods of maximum plant uptake, use of alternate types of nutrient sources, and method of application are effective practices to reduce the rate of application and potential for offsite transport of nutrients* (Anderson et al. 1989; Koehler et al. 1982a).

The concentration of nutrients in runoff is highly variable between fields and storm events (Dunigan et al. 1976; Frere 1976). The unreliability of predictions of nutrients in runoff is related to unexplained variations in both sediment yield and runoff volume (Daniel et al. 1982; Wendt et al. 1986). Spatial and management variability in the effective depth of interaction (EDI) or surface mixing zone further complicates estimates of chemicals transported in runoff (Sharpley 1985a).

Although quantitative prediction of nutrient losses in runoff is extremely difficult (Frere et al. 1980), qualitative storm event patterns have been identified (Baker 1985). Concentrations of sediment-bound nutrients, including organic nitrogen, organic phosphorus, and adsorbed ammonium and phosphate, increase rapidly at the beginning of runoff events and peak well in advance of peak runoff flow. Soluble nutrient concentrations, especially nitrate-nitrogen, are highest either during peak flow or during the falling portion of runoff events. *In regions with substantial tile drainage systems, edge-of-field losses of tile drainage and interflow occur well after the peak of surface runoff flow* (Baker 1985). Therefore, soluble nutrients in tile effluent or emerging interflow will appear in surface runoff after peak flow has occurred.

Reported losses of nitrogen in runoff are extremely variable (Anderson et al. 1989). Nitrogen losses are controlled by the amount and technique of nitrogen application, timing and type of postapplication rain events, antecedent soil moisture and infiltration conditions, site topography, and soil and water conservation practices (Tables 3 and 5). Surface losses of nitrogen are generally low, except when nitrogen rates are high and are surface applied prior to rainfall events (Legg and Meisinger 1982). Incorporation of fertilizer nitrogen is one of the most important measures in reduction of surface losses of fertilizer nitrogen (Legg and Meisinger 1982; Timmons et al. 1973). Other critical factors in the amount of nitrogen available for loss in surface runoff are (1) the amount of nitrogen applied in relation to the amount and placement of nitrogen utilized by the crop, (2) the timing of nitrogen application in relation to the first significant runoff event, and (3) site-specific cultivation practices (Tables 3 and 5) (Anderson et al. 1989).

Generally, the bulk of nitrogen lost in surface runoff is in the form of organic nitrogen associated with eroded soil (Armstrong et al. 1974; Keeney et al. 1983). The finer particle size fraction of eroded soil has higher organic matter content than the source material. The ratio of nitrogen in the eroded sediment to the original soil is the nitrogen enrichment ratio. Menzel (1980) reviewed the literature values of nitrogen enrichment ratios and observed that the enrichment ratios declined as the erosion increased. The nitrogen enrichment ratios ranged from 2.5:7.5 for cropland and were similar in value to phosphorus. Sharpley (1985b) also observed that nitrogen and phosphorus enrichment ratios are inversely related to surface soil losses. Sharpley (1980) observed that enrichment ratios for phosphorus were inversely related to rainfall intensity and soil slope reflecting slope and precipitation effects on particle size distribution of sediment. An increase of the phosphorus enrichment ratio from 2.43:6.29 was also reported at phosphorus applications of 0 and 100 kg ha^{-1}, respectively. Although not directly related to turfgrass systems, this information can be an important consideration during the establishment phase of a turfgrass.

In addition to organic nitrogen, runoff water also contains soluble nitrogen, primarily in the forms of nitrate and ammonium (Keeney 1983; Legg and Meisinger 1982). As precipitation or irrigation starts, inorganic nitrogen applied to the soil surface is leached into the soil. As previously described, ammonium is (1) retained by the soil cation exchange complex, (2) subject to nitrification, (3) erodes with soil particles, or (4) reenters solution through equilibrium processes. Nitrate is subject to (1) surface transport in runoff water, (2) entry into the soil nitrogen cycle, (3) reappearance in downslope surface water, or (4) leaching past the active rooting zone. When runoff occurs during a storm event, inorganic nitrogen enters runoff water from leaching of vegetation and surface crop residues, as well as from the active runoff-soil mixing zone (EDI). Timmons and Holt (1980) demonstrated in a review of soluble nutrient concentrations in runoff that leaching of vegetation and crop residue are a significant source of inorganic nitrogen and phosphorus. Leaching of nitrogen and phosphorus from crop and vegetation residues is particularly significant during spring snowmelt and precipitation (Timmons and Holt 1980; White and Williamson 1973). Timing and placement of fertilizer and manure in relation to the occurrence of runoff also affect the quantity of inorganic nitrogen in runoff (Keeney 1983; Timmons and Holt 1980).

Phosphorus losses are associated with sediment and soluble surface runoff (Koehler et al. 1982a, 1982b; Stewart et al. 1975; Taylor and Kilmer 1980). The bulk of phosphorus lost in surface runoff is sediment-bound and organic phosphorus (Frere 1976; Koehler et al. 1982b; Stewart et al. 1975; Taylor and Kilmer 1980). The ratio of total phosphorus to soluble phosphorus in runoff ranges from 50:400 (Burwell et al. 1975; Hanway and Laflen 1974; White and Williamson 1973). Johnson et al. (1976) observed that 78% of the total phosphorus transported in the Fall Creek, New York area was in the form of suspended particles, organic and adsorbed to sediment. Approximately 4% of the phosphorus associated with suspended particles was readily exchangeable with dissolved phosphorus. As with nitrogen, it is likely that sediment losses of phosphorus will occur only during the establishment phase of the turfgrass.

Assessing the impact of sediment-bound phosphorus in the aquatic environment is difficult (Taylor and Kilmer 1980). Approximately 10–20% of the total sediment-adsorbed phosphorus is potentially available for eutrophication processes. The amount of phosphorus available for algae uptake is dependent on the amount of sediment delivered to the aquatic system, duration of suspension in the active algae zone, temperature, pH, algal species and population size, and amount of light and availability of other essential nutrients. Various chemical and resin extraction procedures have been proposed to estimate algal available phosphorus in runoff suspensions (Huettl et al. 1979). Sediment with high levels of adsorbed and organic phosphorus and nitrogen will act as long-term sinks for nutrient input into aquatic systems. Phosphorus-deficient sediments, such as stream bank and subsoil sediments, will act as long-term phosphorus adsorption sinks (Kunishi et al. 1972; Taylor and Kunishi 1971).

Under certain conditions, considerable amounts of soluble phosphates may be released in runoff water. Timmons et al. (1970) observed that a single leaching of frozen alfalfa could release 0.65 kg ha^{-1} of total soluble phosphorus. Eighty percent of the total loss was in the form of inorganic phosphates. Under similar conditions, bluegrass released 0.3 kg ha^{-1}; however, losses of soluble phosphates from oat stubble were small. Burwell et al. (1975) observed losses of up to 0.3 kg ha^{-1} of soluble phosphorus in snowmelt runoff from alfalfa hay. Leaching of surface residues, mulches, and standing dead or dormant vegetation can release significant levels of soluble phosphorus to surface runoff (Anderson et al. 1989; Koehler et al. 1982a).

Rate of application, potential for sediment transport, runoff volume, cultivation, and residue and tillage management practices determine edge-of-field losses of both soluble and sediment-bound phosphorus (Tables 4 and 6). Frere (1976), Koehler et al. (1982a), Romkens and Nelson (1974), and Taylor and Kilmer (1980) describe several studies that demonstrate increasing losses of soluble and sediment-bound phosphorus as the rate of application increases. The level of both nitrogen and phosphorus losses in runoff are increased by surface application without incorporation, fertilization beyond crop uptake requirements, and runoff leaching of surface crop residues (Frere 1976; Keeney 1982; Koehler et al. 1982a; Taylor and Kilmer 1980).

3.4.2. Nutrient Loss from Subsurface Transport

Infiltration of soil water beyond the rooting zone has the potential to transport solution phase nutrients to groundwater (Keeney 1986; Pratt 1985). Leaching is the process of mass and diffusive transport of solutes in water percolating through the soil. *The nonadsorbed anion, nitrate, is the principal nutrient from agricultural sources detected in appreciable quantities in subsurface water* (Frere 1976; Keeney 1986; Pratt 1985).

Organic forms of nitrogen and phosphorus, ammonium and orthophosphate, are readily adsorbed by surface soils and are generally not detected to any extent in subsurface soil water or groundwater (Keeney 1982, 1986; Taylor and Kilmer 1980). Phosphorus seldom contaminates groundwater, with concentrations usually lower than 0.05 mg L^{-1} in subsurface flow (Reddy et al. 1979). Baker et al. (1975) found soluble orthophosphate concentrations in subsurface flow in Iowa to be less than 0.038 ppm without seasonal trends. Nitrate-nitrogen is the primary nutrient associated with potential contamination of groundwater resulting from subsurface transport. The factors controlling nitrogen movement and case studies in agricultural systems have been reviewed by Anderson et al. (1989) and Keeney (1986) (Tables 7 and 8).

In New York, Zwerman et al. (1972) found concentrations of orthophosphate consistently below 0.01 ppm. However, higher levels of orthophosphate have been detected in subsurface water under conditions of heavy surface applications of waste

Table 7. Factors Associated with Potential Nitrate Transport to Groundwater in Agricultural and Turfgrass Systems

Factor	Impact on transport of nitrate to groundwater
Climate	
Precipitation	*Precipitation and/or irrigation has a dominant effect on mass transport of nitrate. Degree of nitrate leaching depends directly on the amount of water infiltrating the soil surface.* Timing, duration, and amount of rainfall is determined by season and geographic location. In general, nitrate is more likely to leach below the root zone when soil is at or near saturation (maximum hydraulic conductivity) after heavy nutrient application. Leaching of nitrate is potentially high when extremely intense precipitation occurs immediately after application on highly permeable soils. Leaching losses can be minimized by proper timing of application in relation to crop uptake requirements, seasonal trends, and by observation of current forecasts. Reduction of solution runoff by management or natural level of precipitation intensity, timing, and duration will increase the amount of water available for subsurface transport of nitrate.
Evapo-transpiration	The extent of evapotranspiration (ET) in relation to the amount of precipitation or irrigation affects the mass flux of water and dissolved nitrate past the root zone. If ET does not exceed precipitation or irrigation, there will generally be a long-term downward flux of water and dissolved solutes past the root zone. When ET exceeds infiltration, there is usually insufficient water to transport nitrates past the root zone. Unlike pesticides, nitrate is not subject to mass reduction by decomposition. Even in regions where precipitation rarely exceeds ET, when sufficient moisture flux does occur, accumulated nitrates in the soil profile will be subject to deeper leaching. ET is regulated by meteorological conditions including solar radiation, wind, air temperature, and moisture gradients. Soil and plant resistances also play a major role in regulation of ET under dry soil conditions.
Temperature	Temperature affects all nitrogen transformation processes including immobilization, mineralization, nitrification, and denitrification. Temperature has an important impact on the amount of nitrate in the soil available for subsurface transport. Higher temperatures generally increase microbial-mediated processes. Depending on soil moisture, soil aeration, and plant growth conditions, increased temperatures may or may not increase nitrate formation and subsequent exposure to subsurface transport processes. Temperature impacts on mass transport of water in soils is poorly understood and is likely to have minor influence on mass transport of solutes.

Table 7. Factors Associated with Potential Nitrate Transport to Groundwater in Agricultural and Turfgrass Systems (continued)

Factor	Impact on transport of nitrate to groundwater
Soil properties	
Water content	Soil water ultimately transports applied chemicals not adsorbed on soil surfaces. Soil water content influences both diffusive and mass transport. Increased soil water levels will increase both water and mass flux of nitrate within and beyond the root zone. *Mass transport is the primary means of nitrate transport to groundwater. Under dry soil conditions, nitrate may accumulate in and below the soil root zone. When sufficient moisture flux occurs, residual nitrate will continue to be leached toward groundwater.* Soil water and temperature control soil aeration and potential conditions for denitrification. Denitrification may reduce soil nitrate levels and decrease potential for leaching past the root zone. However, determining the appropriate rate of nitrogen fertilizer may be difficult when relatively unpredictable volatile losses of nitrogen occur.
Bulk density	Porosity is related to bulk density. Porosity is a function of pore size and distribution as determined by soil texture, structure and particle shape. Decreasing porosity or increasing bulk density decreases mass transport of nitrogen by decreasing the cross-sectional area available for mass flow and increasing path lengths of water flow.
Hydraulic conductivity	Hydraulic conductivity is the proportionality coefficient between water flux and moisture gradient. Hydraulic conductivity is a measure of a soil's ability to transmit water and dissolved solutes. *Soils with high hydraulic conductivity in relation to initial infiltration of water (e.g., sands) have a greater potential for mass transport of water and dissolved solutes past the root zone.*
Texture	The particle size distribution will determine in part water retention, porosity, hydraulic conductivity, and adsorption site density of a soil. In general, *sandy soils have a greater capacity for mass transport and the least capacity for adsorption of nitrogen species (e. g., ammonium) as compared to finer-textured soils including silt loams and clays.* When adsorbed on soil particle surfaces, organic-nitrogen and ammonium are temporarily protected from conversion to nitrate.
Soil structure	Aggregation of soil particles, soil structure, will affect potential nitrate transport. Structure is influenced by both texture, organic matter content, and soil management. Both texture and structure determine the amount of infiltration and subsurface movement of water. Highly structured soils will have preferential pathways allowing mass transport of water and solutes past the root zone.

Table 7. Factors Associated with Potential Nitrate Transport to Groundwater in Agricultural and Turfgrass Systems (continued)

Factor	Impact on transport of nitrate to groundwater
Soil properties (continued)	
Organic matter and clay content	The amount of adsorption of ammonium is directly related to the soil cation exchange complex: the negatively charged sites on mineral and organic colloid surfaces. Higher adsorption of both organic-nitrogen and ammonium will temporarily attenuate mass transport of nitrate to groundwater, allowing for slower release of nitrate to soil solution, and increasing the amount of time available for nitrogen uptake by crop growth. Increased adsorption also decreases diffusion and volatile nitrogen losses. *Increased adsorption of nitrogen and low hydraulic conductivity will reduce mass transport of nitrates in clay soils. However, potential formation of preferential pathways coupled with precipitation immediately after nutrient application may result in deep leaching of nitrates.*
Depth to groundwater	Depth to groundwater affects the travel time of a chemical leached from the root zone to groundwater. *Shallow groundwater has been found to have a greater potential for contamination with nitrates.* Depth to groundwater also influences the potential for upward transport of water from groundwater to the surface under high evaporation conditions. This will also influence upward transport of leached nitrates.
Nutrient management practices	
Application rate	*The amount of nitrogen applied is directly proportional to the risk of subsurface transport.* Many documented cases of groundwater contamination are associated with heavy and long-term use of nitrogen fertilizers. Use of banded and split application techniques may reduce potential for leaching by increasing nitrogen uptake efficiency and reducing total site loading. *The most effective method to reduce the loss of fertilizer-derived nitrate to groundwater is to reduce the quantity of nitrogen fertilizer applied.*
Timing	*Timing of nitrogen application relative to precipitation and plant uptake events is a major factor in determining potential leaching of nitrate beyond the root zone.* Leaching is minimized when the applied nitrogen is fully utilized during crop development. Application of nitrogen fertilizer prior to irrigation or precipitation will increase potential losses of nitrate from the root zone to groundwater.

Table 7. Factors Associated with Potential Nitrate Transport to Groundwater in Agricultural and Turfgrass Systems (continued)

Factor	Impact on transport of nitrate to groundwater
Nutrient management practices (continued)	
Formulation	The form of applied nutrients will affect the potential for leaching of nitrate to groundwater. Soluble nitrogen formulations (e.g., nitrate and urea) are more readily leached into the soil than ammonium or organic nitrogen from crop residues or animal wastes. *Use of slow release technology or nitrification inhibitors may reduce release of nitrate during critical periods of water flux beneath the root zone prior to development of root systems.*
Placement	The potential for nitrate leaching to groundwater is higher for broadcast applied nitrogen fertilizer than banded, knifed, or split application. These incorporation techniques reduce the total amount of nitrogen required for a given crop yield by increasing fertilizer efficiency. Soil incorporated and knifed applications increase nitrogen contact with adsorptive soil surfaces. Leaching potential is particularly high, if nitrogen fertilizer is applied to a site conducive to leaching (low organic matter content and sandy texture). Improper calibration and maintenance of application equipment will result in uneven distribution of nitrogen fertilizer and possible over application. Nonuniform application may increase leaching potential from concentrated bands of underutilized fertilizer.
Volatility	Volatilization of nitrogen through atmospheric losses of ammonia or through denitrification is a major loss pathway for applied nitrogen. Although volatile losses represent an effective economic reduction of applied fertilizer, nitrogen lost to the atmosphere is not at risk of transport to groundwater. Volatility is primarily a function of nutrient form, placement, environmental conditions, and vapor concentration gradients.
Fertilizer handling	*Proper handling of fertilizers during equipment loading and mixing is essential to avoid both subsurface and surface contamination.* Spills of fertilizers near well heads may result in localized surface and subsurface contamination.

Table 7. Factors Associated with Potential Nitrate Transport to Groundwater in Agricultural and Turfgrass Systems (continued)

Factor	Impact on transport of nitrate to groundwater
Water and crop management practices	
Irrigation	*Intensive fertilization and irrigation practices have been linked to transport of nitrate to groundwater in many regions of the United States. Irrigation has an effect similar to precipitation on transport processes, although there is a greater degree of management potential with its application.* Salinity control requires application of sufficient water to exceed evapotranspiration and maintain drainage conditions. It is rarely possible to manage irrigation with less than a leaching fraction of 0.2. The primary effect of irrigation is on soil water content of the root zone. Irrigation at higher intensity will increase soil water content, mass transport, and diffusive transport. Irrigation practices that influence the potential for nitrate leaching include method of irrigation, timing of application, volume, and frequency of application. Over-watering promotes mass flux and desorption processes, especially for permeable soils. Trickle or drip irrigation has the least potential for over-irrigation and potential for subsurface transport. *Irrigation of intensively fertilized and shallow rooted crops has a high potential for moving nitrate past the effective root uptake zone. Subsequent irrigation will gradually move accumulated nitrates to groundwater.*
Chemigation	Chemigation involves application of dissolved nutrients with irrigation water prior to field application. Research is limited on the potential for leaching of nutrients (and pesticides). Concern has been expressed that leaching may be increased due to continuous application or pulse application of dissolved chemicals. Another potential source of groundwater contamination is faulty, leaking, or nonexisting antibacksiphoning devices.
Vegetation characteristcs	Rooting depth will influence water uptake and chemical characteristics transport. Shallow rooted turfs (e.g., annual bluegrass) have greater potential for leaching of nitrates by reduction of water uptake within the soil profile, limiting adsorption after heavy nitrogen applications and requiring intensive water management. Uptake of deeply leached nitrates helps eliminate movement of residual nitrogen to groundwater. Deeper rooted turfgrass species (creeping bentgrass, tall fescue, and most warm-season grasses) have a more extensive root zone for nitrogen uptake.

From Anderson et al. 1989.

Table 8. Selected Examples of Subsurface Transport of Nitrogen from Treated Plots and Watersheds; Studies Represent Range of Soil and Management Conditions Associated with Nitrogen Fertilization

Catchment type/ moisture input	Duration and method of analysis/ variables investigated	General comments and conclusions	Ref.
		California	
96 m^2 plots/ irrigation and rainfall	6-year study of subsurface movement and fate of applied nitrate, rate of application, vegetable production, irrigation rate, sandy loam, and soil cores	Heavy applications of nitrogen fertilizer in vegetable production results in concentrations of nitrate of 25 ppm in soil water beneath the root zone. Unless lost by denitrification, accumulated soil nitrates are available for deep leaching losses. The concentration of nitrate in the unsaturated zone is directly related to the rate of fertilizer application, but inversely related to drainage volumes.	Adriano et al. (1972a)
12 citrus plots/ irrigation and rainfall	Nitrate leaching beneath root zone, drainage rate, application rate, citrus production, sandy loam, and soil cores	Important factors in determining the amount of nitrate leaching from the unsaturated to saturated zones are (1) the volume of drainage water, (2) annual excess of nitrate available for leaching, and (3) estimate of denitrification. Transit time of excess nitrate in surface soil to reach groundwater ranges from 12 to 49 years. *Adjusting fertilizer application rates, irrigation rates, and improved plant growth reduces excess nitrate available for subsurface transport.*	Pratt et al. (1972)
		Florida	
1 ha plots/ rainfall and irrigation	Nitrate transport to submerged drains in a sandy spodosol; citrus production; tillage treatment: surface	Subsurface losses of nitrate in tile effluent from the surface tilled (15 cm), deep-tilled (107 cm), and deep-tilled with deep incorporation of limestone were equivalent to 32%, 8%, and 15%, respectively,	Calvert (1975)

Table 8. Selected Examples of Subsurface Transport of Nitrogen from Treated Plots and Watersheds; Studies Represent Range of Soil and Management Conditions Associated with Nitrogen Fertilization (continued)

Catchment type/ moisture input	Duration and method of analysis/ variables investigated	General comments and conclusions	Ref.
		Florida (continued)	
	(15 cm), deep-till (107 cm), deep-till and incorporation of limestone; also measured K and PO_4^{-3} concentration in drainage water	of the total nitrogen applied. Total losses were affected by both volume of discharge water and concentration of nitrate. Deep-tillage and incorporation of nitrogen fertilizer significantly reduced nitrate concentration in drainage water compared to surface-tillage or deep-tillage with limestone. Addition of limestone increased subsurface losses of nitrate-nitrogen, but decreased subsurface losses of soluble P. Peak concentrations of nutrients in drainage water was a function of rainfall, irrigation, and timing of nutrient application.	
0.5 ha plots/ irrigation and rainfall	Nitrate losses in submerged drainage system in acid sandy spodosol with shallow water table; citrus production; tillage treatment: shallow (15 cm), deep-till (105 cm), deep-till with incorporation of limestone	Nitrate losses in the shallow tilled plot were less than 2% of the applied fertilizer nitrogen; subsurface losses of nitrate-N from the deeply tilled plots were much less than the shallow tilled treatment. Increased root penetration in the deeply tilled citrus plots, especially with incorporation of limestone, enhanced subsurface root development. Increased root uptake efficiencies reduced residual soil nitrate available for loss in the drainage system.	Mansell et al. (1986)

		Georgia	
0.34 ha watershed/ rainfall	3-year study of surface and subsurface transport of nitrate; continuous corn, 3.2% slope, sandy loam	Average concentration of nitrate in subsurface flow ranged from 7 to 10 ppm. Annual nitrate-N losses in subsurface flow (in kg ha^{-1}) from 1969 to 1971 were 34.1, 49.9, and 22.0, respectively. Subsurface runoff losses of nitrate-N were relatively large compared to losses in surface runoff. On the nearly level watershed, offsite transport of solution was primarily by subsurface flow. Subsurface transport accounted for 80% of water transport and for 99.1% of nitrate-N losses.	Jackson et al. (1973)
72 m^2 plots/ irrigation and rainfall	2-year study of nitrogen fate with intensive vegetable production, irrigated sandy soil, soil cores, and method of fertilization	Nitrogen uptake efficiency for total aboveground biomass averaged 59% of N applied to intensively grown vegetable crops on sandy soil. Only 28% of applied nitrogen was removed in vegetable harvest. With low rates of denitrification (low soil organic matter and water content) and rapid mineralization most of the remaining nitrogen was available for leaching to groundwater. Method of fertilizer application (broadcast vs chemigation) had no effect on denitrification rates or crop uptake of nitrogen. *Continued use of irrigated sandy soil for intensive vegetable production requires increased removal of applied fertilizer nitrogen in crop biomass.* Increased biomass harvesting would reduce residual nitrogen lost to groundwater as nitrate.	Lowrance and Smittle (1988)

Table 8. Selected Examples of Subsurface Transport of Nitrogen from Treated Plots and Watersheds; Studies Represent Range of Soil and Management Conditions Associated with Nitrogen Fertilization (continued)

Catchment type/ moisture input	Duration and method of analysis/ variables investigated	General comments and conclusions	Ref.
		Indiana	
17 ha field/ rainfall	3-year monitoring study of sediment, nutrient, and pesticide losses from tile drainage system; flat silty clay soil; wheat-bean/corn/ wheat-bean rotation	Tile drainage systems substantially reduced surface runoff losses of water. Average annual nitrate-N concentrations in tile effluent ranged from 5 to 12 ppm. Soluble organic N in tile effluent ranged from 0.3 to 3 ppm. A storm event occurring soon after urea application in 1977 resulted in initially heavy subsurface losses of soluble organic N. In a storm event occurring 10 d later, significant losses of nitrate occurred in subsurface drainage. Within 2 weeks after fertilization with soluble urea, the organic-N was almost totally immobilized or transformed to inorganic forms. Low yields of ammonium in subsurface drainage resulted from transformation to nitrate or attachment to the soil cation exchange complex. Over 70% of annual nitrogen lost in subsurface flow was in the form of nitrate. Approximately 70% of phosphorus lost in subsurface flow was in the form of sediment-bound phosphorus. Sediment-bound phosphorus and pesticides were able to move through the soil profile into the subsurface drainage system indicating the presence of macropore flow channels. *Heavy rainfall after fertilizer application results in heavy soluble losses of both nitrogen and phosphorus.*	Bottcher et al. (1981)

		Iowa	
Four 0.41-0.46 ha field/ rainfall	4-year monitoring study of nutrients in tile effluent, level silt loam	Annual subsurface losses of nitrate ranged from 0 to 93 kg ha^{-1}. Tile drainage water had consistently high concentrations of nitrate (>10 ppm) despite relatively low fertility management (112 kg N ha^{-1} $year^{-1}$). Concentrations of nitrate in tile drainage water were extremely variable. Trends in nitrate content of unsaturated and saturated zones of the soil indicated that pulses of water, with variable nitrate concentration, moving through the soil profile, were responsible for observed differences in nitrate concentrations in tile drainage effluent. *Large variations in water quality data make inferences from short-term studies difficult to interpret.*	Baker et al. (1975)
Two 30 ha contoured watersheds/rainfall	4-year study of subsurface transport of nitrate, soil profile (core) samples, and application rate	Accumulated soil nitrate moved from the 1.0 to 3.1 m depth over a 4-year period in an excessively fertilized (448 kg N ha^{-1} $year^{-1}$) watershed. Nitrate concentration in baseflow from the watershed also increased, indicating part of the leached nitrate had reached groundwater. Seventy-five percent of the nitrate leached past the root zone between June 1972 and April 1973, when 80 cm of precipitation resulted in 21 cm of subsurface flow. Fertilization at the recommended fertilization rate (116 kg N ha^{-1} $year^{-1}$) did not result in accumulation of nitrate in the soil profile. Average concentration of nitrate at the water table under the excessively fertilized treatment increased from 3.7 to 12.9 ppm between April 1973 and April 1974. Under the normally fertilized watershed, the average concentration of nitrate at the water table increased from 2.0 to 4.5 ppm during the same sample period.	Schuman et al. (1975)

Table 8. Selected Examples of Subsurface Transport of Nitrogen from Treated Plots and Watersheds; Studies Represent Range of Soil and Management Conditions Associated with Nitrogen Fertilization (continued)

Catchment type/ moisture input	Duration and method of analysis/ variables investigated	General comments and conclusions	Ref.
		Iowa (continued)	
30–61 ha watersheds/rainfall	3 to 5-year monitoring study of nutrients in surface and subsurface flow, soil and water conservation practices, and rates of application	Subsurface nitrate losses accounted for 84–95% of soluble N discharged to streamflow. Although level terraces reduces surface losses of nitrogen, subsurface losses of nitrate were greater in the terrace watersheds than the contoured watersheds. Application of nitrogen fertilizer at the recommended rate (174 kg ha^{-1} $year^{-1}$) did not increase nitrate movement past the root zone under corn production. Excessive nitrogen applications (605 kg ha^{-1} $year^{-1}$) increased movement of nitrate past the root zone and increased nitrate concentration at the groundwater surface from 3.7 to 12.9 ppm after 3 years. Nitrate accumulation in the soil profile did not occur at recommended rates, but increased by 719 kg ha^{-1} at the excessive rate of fertilization. Residual soil nitrates will continue to be leached toward groundwater resources.	Burwell et al. (1977)
2 plots/rainfall	4-year monitoring study of nitrate in tile effluent; corn/oats and corn/bean rotation, application rate	Concentrations of nitrates in tile effluent from higher fertility plots (240–250 kg ha^{-1} $year^{-1}$) were significantly higher than from lower fertility plots (90–100 kg ha^{-1} $year^{-1}$). A 2-month delay was observed before the differential effect of application rate	Baker and Johnson (1981)

		was observed in tile effluent. Although, concentration of nitrate in effluent from a higher fertility plot decreased after termination of differential treatments, the influence of application rate was still 3 years after completion of the treatments. Annual subsurface loss of nitrate from the high fertility treatment was 48 kg ha^{-1} and from the low fertility treatment was 27 kg ha^{-1}.	
4 agricultural drainage wells	2-year monitoring study of nitrate and pesticide concentrations	Drainage wells remove subsurface drainage from fields and may directly inject contaminants to aquifers beneath tile drained fields. 85% of monitored samples had concentrations in excess of 10 mg L^{-1}, the drinking water health standard. A survey of area drinking water wells indicated that nitrate in drainage injection wells increased nitrate levels in the local aquifer. The overall average of nitrate concentrations in samples from farm drinking water wells ranged from 3.0 to 10.9 mg L^{-1}.	Baker et al. (1985)
0.42 ha plot/ rainfall, simulated data	Comparison of 6 years of tile effluent to simulated data, management level capacity model, and corn-soybean management	Concentrations of nitrate in tile effluent from field data were compared to computer simulation of hydrologic and nitrate transport. Predicted values from the management level model deviated from measured daily values, but variability was small when considered over the entire crop growth period. Measured and predicted data indicated that an equivalent of nearly 50% of applied nitrogen is discharged in tile effluent. Larger nitrogen applications would increase subsurface losses of nitrogen, which would have considerable economic and environmental impacts.	Kanwar et al. (1983)

Table 8. Selected Examples of Subsurface Transport of Nitrogen from Treated Plots and Watersheds; Studies Represent Range of Soil and Management Conditions Associated with Nitrogen Fertilization (continued)

Catchment type/ moisture input	Duration and method of analysis/ variables investigated	General comments and conclusions	Ref.
		Iowa (continued)	
25 m^2 plots/ simulated rain	Nitrate movement of surface applied nitrate, soil cores, tillage systems, and surface application before or after tillage	Plots under no-till management retained significantly higher amounts of nitrate in the upper 30 cm of the soil profile than did plots receiving surface application of nitrate before and after conventional tillage. The amount of nitrate leached from the root zone (150 cm) was greater for the moldboard-plow treatments (122 kg ha^{-1}) than for the no-till treatment (29 kg ha^{-1}). The effects of methods of fertilizer applications were inconsistent. More nitrate was retained in the upper 30 cm of soil when nitrate fertilizer was not incorporated, but more nitrate was missing from the entire root zone (150 cm) after precipitation treatments.	Kanwar et al. (1985)
		Kansas	
16 microplots/ rainfall	5-year monitoring of soil profile using 15N, rate and timing of application	Approximately 50% of applied nitrogen fertilizer remained in soil profile after 5 years of treatment. Most of the residual soil N was in an organic form in the upper 0.1 m of soil. *Spring application of fertilizer resulted in greater uptake efficiency. Fall applied nitrogen was subject to greater losses by nitrification and immobilization. Subsurface losses were greater for the higher rate of nutrient application.*	Olson and Swallow (1984)

		Kentucky	
59 m^2 plots/ rainfall	Nitrate movement related to tillage treatment, soil cores, and silt loam soil	Increased nitrate leaching from the soil surface of a no-till, killed sod occurred in comparison to a conventionally plowed plot. Loss of nitrate in the no-till treatment was attributed to increased infiltration, decreased ET from the mulched surface, and deeper subsurface penetration of percolating soil water.	Thomas et al. (1973)
13.3 m^2 plots/ rainfall	Nitrate movement related to tillage treatment, Ebermayer lysimeters, and silt loam soil	Subsurface nitrate losses were greater under a killed sod no-till treatment compared to conventional tillage treatments. Subsurface losses of applied nitrogen occurred for 1–2 months after application. *Concentrations of nitrate in subsurface flow indicate that surface-applied anions are susceptible to preferential flow in natural soil cracks and channels. Nonreactive chemicals may be transported deeper in the soil profile than predicted by piston flow models.*	Tyler and Thomas (1977)
35 m^2 plots/ rainfall	3-year study of nitrate accumulation in soil, tillage treatment, rate of application, soil cores, and silt loam	Nitrogen recovery in corn grain was less under no-till treatment than the conventional tillage treatment at a low level of fertilization (84 kg ha^{-1}); however, nitrogen recovery was not different for the tillage methods at the higher rate of fertilization (168 kg N ha^{-1}). Nitrogen was surface applied as a solution of NH_4NO_3. Neither tillage treatment nor application rate had a significant effect on denitrification or leaching losses of nitrate. Increased immobilization of nitrogen in the surface soil of the no-till system accounted for reduced crop uptake efficiency.	Kitur et al. (1984)

Table 8. Selected Examples of Subsurface Transport of Nitrogen from Treated Plots and Watersheds; Studies Represent Range of Soil and Management Conditions Associated with Nitrogen Fertilization (continued)

Catchment type/ moisture input	Duration and method of analysis/ variables investigated	General comments and conclusions	Ref.
		Minnesota	
2 agricultural fields/rainfall	Nitrate movement in tile drained fields, soil cores, and loam and silty clay loam soils	When fertilizer nitrogen was applied at recommended rates for corn production, nitrates did not accumulate in the soil profile and leaching losses were minimal. Denitrification, uptake and removal by crop production, and to a lesser extent incorporation of nitrogen in soil organic matter were the major loss pathways of applied fertilizer nitrogen.	Gast et al. (1974)
210 m^2 plots/ rainfall	Nitrate movement in tile drained fields, 3-year monitoring study of tile effluent, and rate of application	Loss of nitrate through tile lines (25 kg ha^{-1}) and accumulation in the soil profile (100 kg ha^{-1}) was small for nitrogen application at the recommended rates (112 kg ha^{-1}). As fertilizer application increased to 224 and 448 kg ha^{-1}, soil nitrate accumulation (426 and 770 kg ha^{-1}) and loss in tile effluent (59 and 120 kg ha^{-1}) increased significantly. Maximum depth of nitrate accumulation occurred at 1 m in the soil profile. There was little evidence of nitrate movement below 2.2 m.	Gast et al. (1978)
198 m^2 plots/ rainfall and irrigation	1-year study of nitrate movement using groundwater wells, soil sampling, and porous ceramic cups; rate and method of application, irrigated corn, and sandy loam	Split applications of nitrogen at 179 and 269 kg ha^{-1} had minimal impact on nitrate concentration of a 4.5 m deep aquifer. Split application significantly increased plant uptake efficiency and reduced nitrate leaching. Single applications of nitrogen increased aquifer nitrate concentration by 7–10 ppm.	Gerwing et al. (1979)

84 m^2 plots/ rainfall and irrigation	5-year study of nitrate transport beneath to root zone using nonweighing lysimeters, two application methods, and two irrigation levels	Average annual leaching losses ranged from 29 kg (nonirrigated, nonfertilized corn) to 112 kg ha^{-1}. Low moisture holding capacity, variable rainfall, and variable soil nitrate content resulted in highly variable nitrate leaching losses between fertilization, irrigation, and annual treatments. *Irrigation increased leaching losses of nitrate for the fertilized treatments. Periodic chemigation of nitrogen reduced nitrate leaching in the full replenishment treatment, but no difference was observed in the partial replenishment treatment.*	Timmons and Dylla (1981)
		Nebraska	
93 agricultural sites/rainfall and irrigation	2-year monitoring of irrigation wells	Statewide nitrate concentration increased 24% between 1961–62 and 1971–72. During this period, nitrogen fertilization quadrupled and irrigated acreage doubled. *Only on sites with intensive irrigation, sandy soils, and shallow water tables has there been a significant decrease in groundwater quality.*	Muir et al. (1973)
18 agricultural sites/rainfall and irrigation	Deep soil profile sampling, variable crop, and water management systems	Nitrate leaching is limited under nonirrigated crop management systems. Leaching of nitrate to groundwater does occur on irrigated sandy soils. Production of deep rooted alfalfa in rotation with irrigated row crops is an effective method of recovering nitrogen leached past the root zone of the row crops.	Muir et al. (1976)

Table 8. Selected Examples of Subsurface Transport of Nitrogen from Treated Plots and Watersheds; Studies Represent Range of Soil and Management Conditions Associated with Nitrogen Fertilization (continued)

Catchment type/ moisture input	Duration and method of analysis/ variables investigated	General comments and conclusions	Ref.
		Nebraska (continued)	
Irrigated corn field	Vacuum extractors sample nitrate in soil water percolate, rate and method of fertilizer application, irrigation scheduling, and sandy soil	Irrigation of sandy soils with a shallow water table increases potential for nitrate leaching to groundwater. Percolation of nitrate was affected more by total deep percolation of nitrogen than by method of application. *More frequent irrigations of smaller amounts will reduce total amount of deep percolation and leaching loss of nitrate.*	Linderman et al. (1976)
1 ha plots/ rainfall and irrigation	One growing season, soil cores, rate and method of application, irrigation management, and sandy soil	Nitrogen fertilizer application method, rate, and water management all affected nitrate movement and amount of nitrate remaining in a sandy soil at the end of the growing season. Very small amounts of nitrate remained in the soil profile, when NH_4NO_3 was applied in a single broadcast application at planting, regardless of the amount of irrigation. Water application rate and total amount of nitrogen applied determined the amount of leaching and residual soil nitrate when nitrogen was applied through the irrigation system or injected during the growing season. *Proper nitrogen application rates and water management will reduce the potential of nitrate leaching past the root zone on irrigated sandy soils.*	Smika and Watts (1978)

Simulation/ irrigation	3-year simulation of nitrate leaching, irrigation, corn, sandy soil, rate and source of nitrogen application	A field calibrated mechanistic model of nitrogen uptake and leaching was used to evaluate management practices and nitrate leaching losses from the root zone of irrigated corn on sandy soil. *Leaching losses of nitrate are reduced by control of irrigation practices to reduce mass flow of water and proper selection of nitrogen application rates and sources.* Model results indicate that reduction of nitrate leaching past the root zone to zero amounts is impossible.	Watts and Martin (1981)
Tillage plots/ rainfall	11-year study of soil nitrate accumulation in soil profile, soil cores, tillage system, and rate of application	Increased accumulation of nitrate content of the soil profile was not observed among the no-till, stubble mulch, and conventional fallow tillage treatments. Increased rate of nitrogen fertilization did increase accumulation of soil nitrates, irrespective of tillage system. *Timing of soil sampling is critical to evaluate the correct rate for fertilizer application.*	Lamb et al. (1985)
		New York	
24 32 ha plots/ rainfall	Nitrate losses in tile effluent, rate of application, crop management, use of cover crops, and residue management	Nitrate concentration in tile effluent ranged from 3 to 51 ppm. Concentration of nitrate in subsurface flow was directly related to the rate of application. Ammonium concentration in tile effluent ranged from 0.02 to 0.03 ppm and was only slightly related to rate of application. Nitrate concentration in drainage water was not related to crop removal of fertilizer nitrogen. The influence of crop on subsurface losses of nitrate is mediated through timing of fertilizer application. The presence of a cover crop and incorporation of crop residue did not significantly influence nitrate concentration in tile effluent.	Zwerman et al. (1972)

Table 8. Selected Examples of Subsurface Transport of Nitrogen from Treated Plots and Watersheds; Studies Represent Range of Soil and Management Conditions Associated with Nitrogen Fertilization (continued)

Catchment type/ moisture input	Duration and method of analysis/ variables investigated	General comments and conclusions	Ref.
		North Dakota	
10.2 cm dia. subsurface soil cylinders, rainfall	5-year study of residual fertilizer nitrogen, N-15_N labeled fertilizer in soil cores, grassland management, and silt loam soil	*Under semiarid grassland management conditions, residual fertilizer nitrogen is rapidly incorporated into the soil organic fraction.* At the fertilizer application rate of 84 kg ha^{-1} $year^{-1}$, little fertilizer derived nitrogen is available for subsurface leaching. However, the residual soil nitrogen is incorporated into newly formed and more accessible organomineral complexes. Therefore, residual fertilizer nitrogen is more accessible to mineralization to inorganic forms than the original soil N.	Smith and Power (1985)
		Ohio	
123 ha water-shed/rainfall	2-year monitoring study of nitrate transport to groundwater and returned to surface flow via springs and baseflow, crop management, and rate of application	Nitrate leaching to groundwater, measured as nitrate concentration in springflow and stream baseflow, is directly related to fertilizer management, timing of application, and volume of subsurface flow. Nitrogen concentration in springs draining highly fertile cropland was greater than concentrations in nonagricultural areas. *Subsurface transport of nitrates was directly related to the rate of fertilizer application. Subsurface flow decreased throughout the cropping season until a minimum was reached in the fall.* Concentration of nitrate in subsurface flow increased from a fall minimum to a maximum in the l early spring months.	Chichester (1976)

Eleven 0.0008 ha monolith lysimeters/ rainfall and irrigation	5-year monitoring study of runoff and leaching losses of nitrogen, application rate, cropping practices, seasonal losses, and silt loam	Nitrogen leaching losses were greatest in the winter when ET demand was low and mass flow of water was high. The effect of different management practices on the subsurface transport of nitrate was a function of soluble N remaining in the soil at the end of the growing season. Nitrogen transport under corn production exceeded 250 kg ha^{-1} $year^{-1}$ with concentrations of nitrate-N in percolate exceeding 10 ppm and at times reaching 70 ppm. Nitrogen transport under meadow was less than 10 kg ha^{-1} $year^{-1}$ with concentration of nitrate-N not exceeding 10 ppm. Fertilizer applications under corn production (336 kg ha^{-1}) were in excess of uptake requirements in comparison to meadow production (179 kg ha^{-1}). *Balancing nitrogen inputs with crop requirements is needed to decrease leaching losses of nitrate and maintain crop yield goals.*	Chichester (1977)
0.22 ha plots/ irrigation and rainfall	3-year monitoring study of surface and subsurface transport of nutrients, tile effluent, and silty clay soil	At the recommended levels of fertilizer application, nitrate concentrations and annual mass losses in tile effluent were slightly greater under conventional tillage (average 18 kg ha^{-1}) than no-till management (14.3 kg ha^{-1}).	Schwab et al. (1973)
0.22 ha plots/ rainfall	10-year monitoring study of subsurface transport of nutrients in tile effluent, different drainage systems (swallow and deep), and silty clay soil	At the recommended rates of fertilizer application, nitrate losses in deep tile drainage (18.7 kg ha^{-1} $year^{-1}$) were greater than losses in surface tile drainage (11.2 kg ha^{-1} $year^{-1}$). Rainwater falling on the study site had considerably more nitrate content (21.1 kg ha^{-1} $year^{-1}$) than that removed in drainage effluent.	Schwab et al. (1980)

Table 8. Selected Examples of Subsurface Transport of Nitrogen from Treated Plots and Watersheds; Studies Represent Range of Soil and Management Conditions Associated with Nitrogen Fertilization (continued)

Catchment type/ moisture input	Duration and method of analysis/ variables investigated	General comments and conclusions	Ref.
		Ohio (continued)	
Two 1.1 ha pasture watersheds, three 8 m^2 monolith lysimeters/rainfall	2-year monitoring study of long-term impact of single application of a nonreactive ion (Br^-), silt loam soil	Leaching of a one-time application of a soluble, nondegradable chemical constituent had a multiyear impact on groundwater quality. Peak bromide concentration in percolate under the grassed lysimeters (24 ppm) occurred 52–78 weeks after application. Bromide concentration in groundwater had lower peak concentrations (9.2 ppm) that occurred 84–104 weeks after application. Watersheds with longer flow paths illustrated the long-term impacts of surface applications of nonreactive anions on groundwater quality better than lysimeters. Bromide uptake and movement is similar to nitrate behavior. This study emphasizes the need to apply nitrogen at appropriate rates. Nitrogen application in excess of plant requirements, either in single or multiple applications, impacts groundwater quality for several years.	Owens et al. (1985)
Three 8-m^2 monolith lysimeters/rainfall	6-year monitoring study of nitrate leaching, nitrification inhibitor, and silt loam soil	Nitrate concentrations in leachate were higher under no-till corn production than during a meadow period. Leachate from lysimeters receiving treatment with a nitrification inhibitor, nitrapyrin, and urea fertilizer (336 kg ha^{-1} $year^{-1}$) had flow weighted average concentrations of nitrate ranging from 6 to 40 ppm. Leachate from lysimeters receiving untreated urea had	Owens (1987)

		flow weighted nitrate concentrations ranging from 20 to 54 ppm. The average annual loss of nitrate was 117 and 160 kg ha^{-1} from the treated urea and untreated urea lysimeters, respectively. Nitrification inhibitors have the potential for reducing nitrate leaching when applied with ammonia based fertilizers.	
		Oklahoma	
Microplots/ rainfall	3-year study of fate of nitrate in soil, soil cores, and sorghum-sudangrass	Use of nitrate fertilizer at the prescribed rate does not result in nitrate leaching beyond the root zone. Profile sampling showed little evidence of nitrate leaching past the root zone. On a seasonal average, less than 5% of applied nitrogen remained in the profile as inorganic nitrogen.	Smith et al. (1982)
Twelve 1.6–5.6 ha watersheds, OK, eight 1.1–122 ha watersheds, TX/ rainfall	8-year monitoring study of well water beneath nonirrigated grass and small grain watersheds; level of fertilization, surface runoff	In most cases nitrate-N (range of 0.3–16.1 ppm) and ammonium-N (range of 0.03–0.5 ppm) concentrations in groundwater beneath agricultural watersheds were within drinking water standards. Land use in nonirrigated watersheds had little effect on N and P concentrations in groundwater. However, there was increased potential for nitrate movement to groundwater under minimum tillage and intensive grass production.	Sharpley et al. (1987)

Table 8. Selected Examples of Subsurface Transport of Nitrogen from Treated Plots and Watersheds; Studies Represent Range of Soil and Management Conditions Associated with Nitrogen Fertilization (continued)

Catchment type/ moisture input	Duration and method of analysis/ variables investigated	General comments and conclusions	Ref.
		Texas	
260 m^2 plots/ rainfall and irrigation	Movement of residual fertilizer-nitrate in irrigation return flow, soil cores, and irrigation treatment	Fertilizer uptake and plant nitrate content was not significantly different between irrigation treatments. Soil nitrate concentrations were affected by the interaction of irrigation system, fertilizer timing, and placement. Nitrogen immobilization and denitrification also affected the fate of applied nitrogen under different irrigation treatments. *Subirrigation methods with fertilizer placement above the subirrigation lateral reduces movement of nitrate beneath the root zone compared to furrow or sprinkler irrigation with banded application.*	Onken et al. (1979)
		Wisconsin	
49 m^2 plots/ rainfall and irrigation with swine effluent	3-year study, waste treatments compared to conventional fertilization of tall fescue on sandy clay loam soil; Fertilizer (201 kg N ha^{-1} $year^{-1}$), manure (670 kg N ha^{-1} $year^{-1}$), swine effluent (600 kg N ha^{-1} $year^{-1}$), swine effluent (1200 kg N ha^{-1} $year^{-1}$)	Heavy application rates of swine manure and swine effluent resulted in increased accumulation of soil nitrates ranging from 1120 to 1720 kg ha^{-1}. *At the application rates in this study, the grass vegetation was unable to utilize sufficient nitrogen to reduce subsurface leaching losses. Disposal rates of animal waste application pose a definite groundwater pollution hazard by nitrate contamination.*	Westerman et al. (1987)

From Anderson et al. 1989.

materials with a high concentration of soluble phosphorus (Hook 1983; Koehler et al. 1982c). Soils have a large, yet finite, capacity to fix phosphorus. Heavy applications of soluble orthophosphate fertilizer, manure, or other waste materials may exceed the capacity of a soil to adsorb additional phosphorus. Under these conditions, there is an elevated risk of increased concentrations of phosphorus in both runoff and soil leachate (Dickey and Vanderholm 1980; Khaleel et al. 1980; Koehler et al. 1982c).

Latterell et al. (1982) observed increased levels of organic phosphorus in soils after 5 years of irrigation with moderate and high levels of municipal wastewater effluent. Increases in available phosphorus were observed to a depth of 60 cm with the water table at a depth of 140–150 cm. At the high rate of effluent application (237 cm $year^{-1}$), the phosphorus adsorption capacity of the soil was reduced. Higher phosphorus concentrations (0.07 ppm) in soil water on the high effluent treatment sites were observed at 60 cm, indicating some phosphorus leaching was occurring. Transport of phosphorus to groundwater is also possible if excessive loading of fertilizer or manure phosphorus is applied to sandy soils with limited phosphorus adsorption or buffering capacity (Koehler et al. 1982a; Peaslee and Phillips 1981). Duxbury and Peverly (1978) and Miller (1979) have observed that subsurface transport of organic matter will enhance the downward movement of phosphorus in soils high in organic matter content.

Leaching loss of nitrate is one of the most important avenues of nitrogen loss from agricultural fields other than losses from volatilization and plant uptake (Legg and Meisinger 1982). Considerable subsurface losses of nitrogen occur when (1) soil nitrate levels are high, and (2) water movement is large. Nitrate leached below the root zone will eventually be transported to subsurface interflow, groundwater, or reappear in surface flow in tile drainage or stream base flow (Keeney 1982). The dominant processes in the subsurface movement of nitrate, or other nutrient solutes, is (1) convective or mass flow of dissolved nutrients in bulk flow of soil solution, (2) diffusion resulting from concentration gradients, and (3) hydrodynamic dispersion due to local turbulent mixing of soil solution (Keeney 1986).

Factors associated with nitrate leaching and contamination of groundwater include rates, form and placement of fertilizer nitrogen, timing of application in relation to plant uptake and leaching events, amount of percolating water associated with rainfall or irrigation, soil and climate conditions, and cultivation and waste management practices (Table 7). Intensity of nitrogen use and soil leaching potential are critical factors in potential for groundwater contamination (Keeney 1986; Pratt 1985). Depth to groundwater, concentration of nitrate in geologic substrates, irrigation practices, occurrence of preferential flow paths, and well location also have a profound influence on potential transport of nitrate beyond the root zone (Canter 1987; Davidson et al. 1983; Jury 1983; Keeney 1986; Pratt 1985). Fractured aquifers, karst topography, shallow aquifers, and intensified irrigation practices are additional concerns for nonpoint contamination of groundwater with nitrates (Keeney 1986; Pratt 1984, 1985).

Use of tile drainage effluent may not provide reliable quantitative estimates of subsurface losses of nitrate-nitrogen (Thomas and Barfield 1974). Nitrate concentrations tend to be higher in tile effluent than in nontile flow. Caution must also be observed in interpreting the results of tile drainage studies, as tile drains do not always intercept all subsurface flow. Tile drainage changes the flow pattern within agricultural fields, reduces the time of saturation after precipitation, and lowers the local water table.

Row crop agriculture has been identified as a significant nonpoint source of nitrates to groundwater. The large extent of heavily fertilized cropland, current dependence on external sources of nitrogen in commercial fertilizers and waste, and inefficient uptake of applied nitrogen are all nonpoint source pollution factors. *The objective of efficient nutrient management is to minimize subsurface losses of nitrates. The basic concept relies on preventing the accumulation of excess nitrate in the root zone at times when climate and soil conditions are conducive to high rates of leaching* (Aldrich 1984; Keeney 1982). However, lack of knowledge about site-specific effects on nitrogen availability, the imprecise nature of the relative availability of nitrogen from soil organic matter, crop residues and animal wastes, and the difficulty in predicting annual climatic patterns makes implementation of the simple concept of nutrient management difficult to achieve (Keeney 1986). Despite these limitations, many studies on subsurface transport of nitrate in the midwest and corn belt states demonstrate some of the relationships between subsurface losses of nitrogen and certain nutrient management practices (Table 8).

In the humid areas of the corn belt, nitrate is completely removed from the soil profile by a combination of leaching and denitrification; or nitrates may accumulate in the soil and move to groundwater depending on soil hydraulic characteristics, climate, fertilization practices, and crop management practices (Legg and Meisinger 1982). Considerable discharge of nitrate-nitrogen has been observed from tile-drained agricultural fields when nitrogen is applied at rates higher than recommended for optimum or realistic crop yields (Baker et al. 1975; Miller 1979). Nitrates did not accumulate beneath the rooting zone of poorly drained soils or were not observed at excessive levels in tile drainage effluent when nitrogen fertilizer was applied at recommended rates (Gast et al. 1974; Nelson and MacGregor 1973). Considerable denitrification in the water-saturated soils near the tile drains accounted for appreciable reduction of nitrate concentration in these soils (Gast et al. 1974). Nelson and MacGregor (1973) observed that after 12 years of nitrogen fertilization on tile-drained soils, 23% of the nitrogen applied at 112 kg ha^{-1} was lost by subsurface transport. At the rate of 448 kg ha^{-1}, annual subsurface losses of nitrates were as high as 41%. Leaching of nitrates below the rooting zone will occur with over-fertilization. Once nitrate is beyond the zone of biological activity and plant uptake, the potential for transport to groundwater is significant (Anderson et al. 1989; Pratt 1985).

MacGregor et al. (1974) observed soil profile distribution of nitrates on a well-drained loam in western Minnesota and poorly drained silty clay loam in southern Minnesota after 15 and 11 years of treatment, respectively. Increased potential for subsurface loss of nitrates to groundwater under these conditions was observed at the highest rates of nitrogen fertilization. In general, there was less movement of nitrates through the fine-textured soil with reduced rates of percolation, despite higher rainfall levels compared to the well-drained soil with less rainfall.

In Ohio, Chichester (1977) observed the highest flux of nitrogen in lysimeters during winter months when evapotranspiration was low. Regardless of crop management practice, during the winter months water flux through the monitored soils was similar. *The critical factor determining the mass of nitrate-nitrogen leached was the amount of soluble nitrogen remaining in the soil after plant uptake and crop harvest.* Subsurface losses of nitrogen did not exceed 5% of the amount applied at the application rate of 179 kg N ha^{-1} $year^{-1}$. However, at high rates of application, 336 kg N ha^{-1} $year^{-1}$, subsurface losses of nitrogen often equaled or exceeded the amount of nitrogen utilized for crop growth.

In the southeastern states, the highest rates of leaching occur in the winter when soils are wet, but not frozen. High rates of evapotranspiration during the summer months limits movement of water and solutes past the root zone (Keeney 1986; Thomas 1970). In the coastal plain of North Carolina nitrates accumulate in shallow groundwater as a result of agricultural activity (Daniels et al. 1975; Gilliam et al. 1974). However, impermeable and reduced zones within this aquifer prevent deeper migration of nitrates into groundwater. Development of center pivot irrigation systems in the coastal plain of Georgia has increased the potential for nitrate transport to shallow aquifers in the area (Hubbard et al. 1984). Nitrate-nitrogen concentrations in groundwater beneath intensive multiple cropping systems with center pivot irrigation averaged 20 ppm and concentrations under a nearby forest were 1 ppm. In Florida citrus producing areas, large losses of nitrates from soils may occur on poorly drained sandy soils during heavy rains (Calvert et al. 1981). Messer and Brezonik (1983) reported that widespread contamination of Florida groundwater has not occurred. Relatively heavy precipitation on sandy surface soils dilutes the nitrate concentration of groundwater.

In the nonirrigated sections of the Great Plains, nitrate transport to groundwater is usually low because annual precipitation is low and evapotranspiration is high (Pratt 1985). Several studies involving nonirrigated crops grown on the Great Plains (e.g., winter wheat, sorghum-sudangrass) demonstrate that with good management fertilizer nitrogen does not contribute to groundwater pollution (Olson and Swallow 1984; Smith et al. 1982; Sommerfeldt and Smith 1973). However, periodic heavy summer rains will result in the deep movement of nitrate past the rooting zone. Deep transport of nitrates was observed after a wet winter in Kansas (Olson 1982). Significant amounts of nitrate were leached from wheat fields in Saskatchewan in years with above-average precipitation or during heavy spring rains (Campbell et al. 1983).

Timing and placement of fertilizer are additional factors in determining nutrient uptake efficiency, crop growth, and potential subsurface transport of nitrates to groundwater (Keeney 1982; Koehler et al. 1982a; Randall 1988). Timing and placement techniques are available to increase nutrient uptake efficiency at the time it is required for crop growth and development. These techniques, such as split applications, deep placement, delayed application, and banding, reduce the total quantity of nitrogen needed to produce a given yield (Gerwing et al. 1979; Randall 1984; Bock 1984; Whitaker et al. 1978). Decreasing total rate of application reduces residual nitrogen concentrations in the soil and the potential for subsequent transport to groundwater (Pratt 1985; Keeney 1982). Vegetation type, soil type, climate variability, and date of planting all influence critical demand periods for nutrients (Olson and Kurtz 1982). Split application techniques and timing of application to periods of peak uptake have reduced nonpoint source losses of nutrients without decreasing yields in several other studies throughout the United States (Bandel 1981; Kamprath et al. 1973; Langdale et al. 1979). Split and side-dressed application of fertilizer nitrogen has the advantage of adjustment of fertilizer rates for changes in climate and anticipated yield. However, incorporation is required and root damage is a potential disadvantage of later applications of nitrogen (Keeney 1982).

In cool, moist climates, fall application of fertilizer nitrogen for the next growing season remains controversial (Beauchamp 1977; Pratt 1985; Stevenson and Baldwin 1969). Fall application has the advantage of distributing labor and equipment demands throughout the year. Variation in response is attributed to site-specific factors and climate variations. Considerable leaching and runoff losses of fall- and winter-applied nutrients have been observed (Anderson et al. 1989; Koehler et al. 1982a). Fall fertilization is not recommended (1) on coarse or poorly drained soils or (2) at soil temperatures greater than 10°C. Although nitrification is reduced at low temperature, it commences with spring warming. Production and accumulation of mobile nitrates in a warm, wet spring could result in extensive leaching and/or denitrification of fertilizer nitrogen (Keeney 1982; Pratt 1985), although turfgrass systems generally have a fully developed root system in the spring. Fall application of nitrogen has been discouraged as both an economically and environmentally unsound practice (National Research Council 1978).

Subsurface transport of nitrate-nitrogen and irrigation management on many soils is a potentially serious concern in regard to groundwater quality (Anderson et al. 1989). Residual nitrate in the soil profile is subject to leaching with addition of irrigation water. Irrigation is energy and capital intensive. Therefore, irrigated crops usually have high value and are fertilized heavily (Pratt 1984). High potential rates of leaching in the permeable soils, which typically receive irrigation and high rates of nitrogen, result in high subsurface losses of nitrate (Branson et al. 1975; Keeney 1982, 1986; Pratt 1984).

The impacts of irrigated agriculture on nitrate concentrations in both surface water and groundwater have been assessed most extensively in the southeast and in the arid west, especially California (Keeney 1986; Pratt 1978, 1984). In a review of many California studies, Pratt (1978, 1984) observed that average nitrate-nitrogen concentrations in groundwater under irrigated croplands are 25 to 30 ppm. Significant loss of fertilizer nitrates occurred only when (1) crop roots were unable to access the entire soil profile, (2) excessive amounts of nitrogen were applied as fertilizer or for animal waste disposal, and (3) soils were overirrigated. Anderson et al. (1989) presented a concise review of the relationship of nitrogen fertilization, irrigation, and contamination of groundwater.

Management of nitrogen fertilizer and irrigation water is particularly difficult when shallow-rooted crops are grown on coarse-textured soils (Keeney 1986; Pratt 1985). Coarse-textured soils have limited water and nutrient retention capacity which exaggerates nitrogen leaching potential. Potato production requires limited periods of moisture stress and fairly high levels of nitrogen fertility to produce a commercially marketable crop (Saffigna et al. 1977; Saffigna and Keeney 1977b). The roots of potatoes do not intercept the entire volume of the surface soil. With the limited spatial uptake capacity, nitrates leached below 15–20 cm are not recovered by the crop (Keeney 1986). Saffigna and Keeney (1977a) observed significantly greater nitrate concentrations in groundwater under irrigated potatoes than uncropped areas in the sand plain region of central Wisconsin. Elevated levels of nitrate in groundwater, ranging from 4 to 23 ppm, were attributed to fertilizer nitrogen sources. Potato production has also contributed to nitrate contamination of aquifers in Long Island, NY (Meisinger 1976).

3.5 NUTRIENT LOSSES FROM TURFGRASS SYSTEMS

Nitrogen is the nutrient used in turfgrass systems of most importance from a water quality perspective. This is in part due to the (1) production of highly mobile nitrate from fertilization and (2) establishment of turfgrass systems on well-drained soils or soil mixtures, such as putting greens and sports fields, that lend themselves to deep leaching patterns. As noted by Flipse et al. (1984), nitrate loss from turfgrass systems has been identified as a potential major contributor to groundwater contamination in areas where turfgrass is a major component of the landscape.

Until recently, no comprehensive review of nitrogen losses from turfgrass systems existed. Therefore, no means of assessing the magnitude or importance of nitrogen losses from turfgrass systems were available. Recently, Petrovic (1990) provided an excellent review on this important subject in order to summarize the current state of scientific information and to provide information for the development of BMPs for the prevention of groundwater deterioration. We have included much of this work in the summary of nitrogen losses that follows and have, in some cases, expanded upon the original work. The reader is advised to consult the work by Petrovic (1990) for specific information regarding nitrogen removal or loss through plant uptake and nitrogen storage in soil.

As noted by Petrovic (1990) and others, nitrogen will escape from local nutrient cycling systems in turfgrass through (1) gaseous processes including volatilization of applied fertilizer and denitrification, (2) leaching from the root zone, and (3) surface runoff. Subsurface and surface runoff transports of soluble and sediment-bound phosphorus are the primary mechanisms of phosphorus loss from turfgrass systems. Removal and offsite disposal of grass clippings is potentially another significant source of nutrient loss from the turfgrass nutrient cycle (Petrovic 1990). The nitrogen cycle (N cycle) for turfgrass, and the processes by which nitrogen may escape the system, have been presented schematically by Petrovic (1990) and are shown in Figure 1.

3.5.1 Nitrogen Loss Through Volatilization

Nitrogen applied to turfgrass can be lost to the atmosphere as ammonia or as nitrous oxide compounds. The primary factors associated with volatilization and denitrification of nitrogen from turfgrass have been reviewed by Petrovic (1990) (Table 9).

Ammonia volatilization is known to occur rapidly following application of urea (Bowman et al. 1987; Torello et al. 1983; Volk 1959). Factors influencing the rate and amount of ammonia volatilization are nitrogen source, nitrogen form, rate of application, soil pH, antecedent and postapplication soil water content, and thatch development (Petrovic, 1990). Volk (1959) observed that ammonia volatilization was greater from a turfgrass system than from a bare soil. Under similar conditions, turfgrass will accelerate the process of ammonia volatilization.

The results of studies from closed system experiments have shown that the absence or presence of thatch dramatically affects ammonia volatilization. Nelson et al. (1980) found that within 8 d after application of urea-nitrogen, 39% of the applied nitrogen volatilized from Kentucky bluegrass containing about 5 cm of thatch. Only 5% volatilized in the cores with no thatch below the sod. Bowman et al. (1987) observed substantial urease activity in thatch layers. Urease is the enzyme necessary to convert urea to ammonia. The activity of this enzyme explains the role of thatch in enhancing the rate and quantity of ammonia volatilization in turfgrass systems. The relation of thatch and thatch development to nitrogen application have been noted by Carrow it al. (1987), Meinhold et al. (1973), Nelson et al. (1980), and Smith (1979).

The source, rate, and form of nitrogen influences the pool of ammonia available for volatilization (Petrovic, 1990). The volatile losses of ammonia from agricultural

Table 9. Atmospheric Loss of Fertilizer Nitrogen Applied to Turfgrass

Species	Experimental conditions	Duration of sampling	Source	Single/total N application rate (kg N ha^{-1})	Precipitation/ irrigation (cm)	Temp. (°C)	Soil texture	NH_3 Volatilization	Denitrification	Ref.
								% of applied N		
Poa pratensis	Field	3 d	Urea	58	0	27–39	Loam	3–36	—	Bowman et al. (1987)
					0.5			2–21		
					1.0			1–8		
					2.0			1–5		
					4.0			0–3		
Poa pratensis	Growth chamber	10 d	KNO_3	52	—	22[a]	Silt	—	0.02	Mancino et al. (1988)
						>30[a]	Silt	—	0.11	
						22[a]	Silt loam	—	0.4	
						>30[a]	Silt loam	—	—	
						22[b]	Silt	—	5.4	
						>30[b]	Silt	—	94	
						22[b]	Silt loam	—	2.2	
						>30[b]	Silt loam	—	46	
Poa pratensis	Growth chamber	8 d	Urea	253	2.27 d^{-1}	—	Silt loam	5	—	Nelson et al. (1980)
							Thatch	39	—	
			IBDU			—	Silt loam	2	—	
							Thatch	4		

Table 9. Atmospheric Loss of Fertilizer Nitrogen Applied to Turfgrass (continued)

Species	Experimental conditions	Duration of sampling	Source	Single/total N application rate (kg N ha^{-1})	Precipitation/ irrigation (cm)	Temp. (°C)	Soil texture	NH_3 Volatilization	Denitrification	Ref.
								% of applied N		
Poa pratensis/ *Festuca rubra*	Field	8 d (July)	Urea	100	0	—	—	15.1	—	Sheard and Beauchamp (1985)
		5 d (August)			0.19	—	—	6.7	—	
Poa pratensis/ *Festuca rubra*	Field	Growing season	$(^{15}NH_4)_2SO_4$	90/180	—	—	Sandy loam	—	24[c] 36[d]	Starr and DeRoo (1981)
Poa pratensis	Growth chamber	84 h	Urea (granular)	73	—	10	Silt loam	18	—	Titko et al. (1987)
					—	22		43	—	
					—	32		61	—	
					—	(31)[e]		39	—	
					—	(68)[e]		61	—	
					0	—		51	—	
					2.5	—		2	—	

Poa pratensis	Growth chamber	21 d	Urea	293	—	24	Silt loam	10	—	Torello et al. (1983)
			SCU		—			2	—	
			Urea (granular)	49	—	24		2	—	
			Urea (dissolved)		—			5	—	
			Ureaformal-dehyde	49	—	24		3	—	
			Methyol urea		—			5	—	

From Petrovic 1990.

[a] *Soil moisture at 75% saturation.*
[b] *Soil moisture at 100% saturation.*
[c] *Combination of ammonia volatilization and denitrification for plots with clippings returned.*
[d] *Combination of ammonia volatilization and denitrification for plots with clippings removed.*
[e] *Relative humidity.*

systems, as influenced by the form of nitrogen added, has been reviewed by Terman (1979) and from turfgrass systems by Petrovic (1990). Within 21 d after a single application of urea at a rate of 293 kg ha^{-1}, Torello et al. (1983) noted that 10% of the applied urea volatilized as ammonia, whereas, only 1–2% of sulfur-coated urea (SCU) was volatilized as ammonia. At a lower rate of urea application, 49 kg ha^{-1}, only about 2% volatilized as ammonia. Titko et al. (1987) found more ammonia volatilization with granular than dissolved urea, while Torello et al. (1983) observed a reverse relationship to formulation of urea and volatile losses.

Hargrove and Kissel (1979) attempted to directly measure ammonia losses of surface-applied urea on coastal bermudagrass and then to compare these results to losses determined under similar conditions in the laboratory. They noted, in general, that ammonia losses in the field were small compared to those determined in the laboratory. Nine percent of applied nitrogen in the field volatilized compared to 13–31% of applied nitrogen in the laboratory. The results suggested that previous estimates of ammonia losses, as determined in the laboratory, might not be as severe as thought. Perhaps the most important aspect of the study was the suggestion that ammonia losses should be directly measured in the field rather than extrapolated from laboratory studies.

Sheard and Beauchamp (1985) estimated ammonia volatilization from turfgrass under field conditions using an aerodynamic procedure. They noted that 15% of urea was lost by ammonia volatilization from a bluegrass/red fescue turfgrass fertilized at 100 kg N ha^{-1}. Lightner et al. (1990) determined ammonia volatilization losses from several different urea-based fertilizers broadcast onto orchardgrass. The formulations included solid urea, urea solution, cogranulated urea-urea phosphate, ammonium nitrate, urea-KCL (solution), and urea-$CaCL_2$ (solution). Fertilizer nitrogen was applied at rates of 200 kg N ha^{-1} in the spring and 100 kg N ha^{-1} in late summer. They noted that measurable ammonia losses occurred within 24 hr of application and that the highest losses were from the solid urea, urea solution, and urea-KCl solution. Losses ranged from 27 to 41% of the total applied. They noticed no significant loss of ammonia from the ammonium nitrate treated plots.

Volatilization of ammonia is also influenced by original fertilizer placement and postapplication position as determined by precipitation and irrigation (Petrovic 1990). Bowman et al. (1987) found that 36% of applied ammonia volatilized when no irrigation followed the surface application, whereas 1–4 cm of water within 5 min of application reduced ammonia volatilization from urea to 8 or 1%, respectively. Titko et al. (1987) also found reduced ammonia volatilization from either dry or dissolved urea applied to turfgrass with less than 2.5 cm of irrigation. Irrigation after application also affects the urea position in turfgrass. For example, Bowman et al. (1987) found that in a turfgrass system without irrigation, 68% of the urea was located in the shoots and thatch, while irrigation with 0.5 and 1.0 cm of water reduced shoot and thatch borne urea to 31 and 26%, respectively. Sheard and Beauchamp (1985) noted similar reductions of ammonia volatilization when 1.2 cm of rainfall occurred within 72 hr of a urea application. Movement of urea into the soil and conversion to ammonium and nitrate reduces initial volatile losses of nitrogen.

The rate at which liquid urea dries also influences ammonia volatilization (Petrovic 1990). Petrovic (1990) used data from Bowman et al. (1987) to show that ammonia volatilization from urea on nonirrigated sites appears to be independent of the maximum air temperature within the first 24 hr after application. Volatilization of ammonia was inversely related to daily evaporation rate. Titko et al. (1987) also noted more volatilization at 68% relative humidity than at 31% with both granular and dissolved urea.

As noted by Petrovic (1990), there are few studies designed to measure the magnitude of denitrification under turfgrass conditions. Mancino et al. (1988) measured denitrification from Kentucky bluegrass fertilized with KNO_3. They found that denitrification was less than 1% of the applied nitrogen and was denitrified up to soil moisture levels of 75% of saturation. However, at soil saturation, denitrification became significant. Temperature became important with more denitrification occurring when the saturated soils were 30°C or above. Starr and DeRoo (1981) used ^{15}N-labeled $(NH_4)_2SO_4$ to calculate a mass balance for nitrogen in a turfgrass system. They found that about 30% of fertilizer nitrogen applied to Kentucky bluegrass was lost to the atmosphere by volatilization or denitrification.

3.5.2 Nitrogen Loss Through Subsurface Drainage

The general process of subsurface transport of soil nutrients and solutes has been extensively reviewed by Davidson et al. (1983), Ellis et al. (1983), Enfield and Ellis (1983), Keeney (1982, 1983, 1986), Nielsen et al. (1982, 1983), and Starr (1983). General environmental and soil factors associated with potential runoff and leaching losses of applied nutrients were discussed in Section 3.4 (Tables 3, 4, and 7). The methods used to study leaching of fertilizer nitrogen in turfgrass and the assumptions associated with these methods have been noted by Petrovic (1990).

The degree of nitrate leaching from turfgrass systems is variable (Petrovic 1990) (Table 10). Dramatic differences in nitrate leaching have been determined by different researchers (Table 10). Some researchers report little or no leaching, while others suggest leaching of 80% or more of applied nitrogen. Factors that influence the degree of nitrate leaching include soil type, irrigation, nitrogen source and rate, and season of application (Petrovic 1990; Anderson et al. 1989).

Soil texture has a direct effect on leaching losses of nitrate because of its influence on the rate and amount of water moving through the turfgrass (Table 8). On an irrigated site, Rieke and Ellis (1974) monitored the movement of nitrate in a sandy loam soil to a depth of 60 cm by periodic sampling. During the 2-year study, application of nearly 300 kg N ha^{-1} elevated nitrate levels only in 2 of the 20 samples

Table 10. Subsurface Loss of Nitrate Resulting from Application of Fertilizer Nitrogen to Turfgrass

Species	Source	Nitrogen: Single N application rate	Nitrogen: Total annual N rate	Season applied	Irrigation (mm d^{-1})	Soil texture	Percent of applied N leached	Concentration of nitrate in water (mg L^{-1})	Ref.
		(kg N ha^{-1})							
Cynodon dacytylon	Ureaformal–dehyde	224	224	June	6–8	Sand/peat	—	0	Brown et al. (1977)
					8–10		—	<1	
					10–12		—	<1	
	NH_4NO_3	163	163	February	6–8[a]		—	<1	
					8–10[a]		—	>10 for 20 d	
				10–12		—	>10 for 28 d		
Milorganite	146	146	October	6–8		—	<3		
				8–10		—	<6		
				10–12		—	<5		
$(NH_4)SO_4$	24	24	Summer	12		37	<10		
	49	49		12		25	<10		
	73	73		12		22	>10 on 3 d		
	98	98		12		16	>10 on 3 d		
Cynodon dacytylon	IBDU	146	146	June	12	Sand/peat	0.9	0	Brown et al. (1982)
						Sand/soil/peat	0.7	<2	
						Sandy loam	0.1	<1	
	Milorganite	146	146	October	12	Sand/peat	7.7	0	
						Sand/soil/peat	2.4	<2.2	
						Sandy loam	0.5	0	

Cynodon dacytylon (continued)	Ureaformal–dehyde	224	224	June	12	Sand/peat	0.2	0	
						Sand/soil/peat	0.3	0	
						Sandy loam	0.1	0	
	NH_4NO_3	163	163	February	12[a]	Sand/peat	22	>10 for 25 d	
						Sand/soil/peat	22	>10 for 25 d	
						Sandy loam	8.6	>10 for 25 d	
Poa pratensis/ *Festuca rubra*	Urea + fluf	49	98	June/Nov.	1.8	Sandy loam	—	0.87	Morton et al. (1988)
					5.4		—	1.77	
		49	245	June/July Aug./Nov.	1.8		—	1.24	
		49	245	June/July Aug./Nov.	5.4		—	4.02	
		0	0		1.8		—	0.51	
		0	0		5.4		—	0.36	
Poa pratensis	NH_4NO_3	74	74	Cool	3.6	Silt loam	0	—	Mosdell and Schmidt (1985)
					7.2		0	—	
				Warm	3.6		1.2	—	
					7.2		2.6	—	
	IBDU	74	74	Cool	3.6		2.7	—	Mosdell and Schmidt (1985)
					7.2		0	—	
				Warm	3.6		0	—	
					7.2		0	—	
Poa pratensis	IBDU	245	245		2.3	Silt loam	26	—	Nelson et al. (1980)
					—	Thatch	7	—	
	Urea			Warm	—	Silt loam	32	—	
					—	Thatch	84	—	

Table 10. Subsurface Loss of Nitrate Resulting from Application of Fertilizer Nitrogen to Turfgrass (continued)

Species	Source	Nitrogen: Single N application rate (kg N ha^{-1})	Nitrogen: Total annual N rate (kg N ha^{-1})	Season applied	Irrigation (mm d^{-1})	Soil texture	Percent of applied N leached	Concentration of nitrate in water (mg L^{-1})	Ref.
Poa pratensis	Ureaformaldehyde	98	98	November	None	Sandy loam	0–4	—	Perovic et al. (1986)
	PCU (150D)						0	—	
	Milorganite						0–3	—	
	Urea						29–47	—	
	SCU						11–12	—	
Agrostis palustris	Urea	24	294	All year	—	Sand	2.0	<1.3	Sheard et al. (1985)
	SCU				—		1.2	<1.3	
Poa pratensis/ *Festuca rubra*	$(NH_4)_2SO_4$	88	176	May/Sept.	None	Sandy loam	0	0	Starr and DeRoo (1981)
							0	—	
Cynodon X magneissii	Control	0	0	All year	As needed	Sand	0	0	Snyder et al. (1981)
	Methylene Urea	39	245				<1	<1	
	Ureaformaldehyde						<1	<1	
	SCU						0	<1	
	IBDU						0.5	<1	
	Urea						0	<1	

Cynodon X magneissii (continued)	Methylene Urea	78	490				2.0	<1	
	Ureaformaldehyde						0.1	1	
	SCU						0.8	<1	
	IBDU						5.5	1.4	
	Urea						0.9	<1	
	$Ca(NO_3)_2$						9.3	2.4	
Cynodon X magneissii	NH_4NO_3			Feb./March	6 (Daily)	Sand	54.6	9.4	Snyder et al. (1984)
	SCU						33.1	6.5	
	Fertigation						7.0	1.2	
	NH_4NO_3				1.5 (Sensor)		40.5	14.4	
	SCU						11.2	4.0	
	Fertigation						6.3	2.2	
	NH_4NO_3			April/May	8 (Daily)		56.1	18.9	
	SCU						14.4	4.8	
	Fertigation						3.5	1.2	
	NH_4NO_3				3 (Sensor)		1.9	6.2	
	SCU						0.3	1.0	
	Fertigation						0.3	1.0	
	NH_4NO_3			June/July	12 (Daily)		22.2	3.2	
	SCU						10.1	3.2	
	Fertigation						15.3	2.1	
	NH_4NO_3				3 (Sensor)		8.3	3.2	
	SCU						1.6	0.8	
	Fertigation						0.8	0.1	

Table 10. Subsurface Loss of Nitrate Resulting from Application of Fertilizer Nitrogen to Turrgrass (continued)

Species	Source	Nitrogen: Single N application rate (kg N ha^{-1})	Nitrogen: Total annual N rate (kg N ha^{-1})	Season applied	Irrigation (mm d^{-1})	Soil texture	Percent of applied N leached	Concentration of nitrate in water (mg L^{-1})	Ref.
Festuca	Control	0	0	March/May	—	Sandy loam	—	0.2–0.5	Gross et al.
arundinacea	Urea (liquid)	49	49		—		—	0.1–3.2	(1990)
Poa pratensis	Urea (granular)	49	49		—		—	0.1–0.7	
	Control	0	0	June/Aug.	—		—	0.1–1.0	
	Urea (liquid)	49	49		—		—	0.1–1.2	
	Urea (granular)	49	49		—		—	0.2–0.7	
	Control	0	0	Sept./Oct.	—		—	0.2–0.3	
	Urea (liquid)	49	49		—		—	0.1–2.2	
	Urea (granular)	49	49		—		—	0.2–0.7	
	Control	0	0	Nov./Feb.	—		—	0.1–0.4	
	Urea (liquid)	49	49		—		—	0.5–2.2	
	Urea (granular)	49	49		—		—	0.1–0.5	
—	—	—	—	—	—	Sand/gravel	—	5.6–12.0[b]	Cohen et al.
			100					0.1–7.0[c]	(1990)
			95					1.3–10.0[d]	
			214					0.1–9.0[e]	
—	—	—	—	—	—	Sand/gravel	—	<0.1–0.1[b]	
			139			Clay lens		<0.1–5.0[c]	
			126					0.1–20.0[d]	
			164					1.4–30.0[e]	

—	—	—	—	—	—	Sand	—	<0.1–0.1[b]
			122					0.4–2.4[c]
			98					—
			249					0.4–6.5[e]
—	—	—	—	—	—	Sand	—	<0.1–0.2[b]
			149					0.8–4.8[c]
			105					0.6–6.5[d]
			183					1.4–10.2[e]

Adapted from Petrovic 1990.

[a] *Irrigation applied every other day.*
[b] *Background monitoring well.*
[c] *Tee monitoring well.*
[d] *Fairway monitoring well.*
[e] *Green monitoring well.*

at the 60 cm depth. Elevated levels of residual nitrate were primarily observed in the top 30 cm of the plots. The lack of downward movement of the applied nitrogen may have been due to the presence of sandy clay loam subsoil capable of promoting denitrification (Exner et al. 1991). Applying the same amount of fertilizer in three applications showed similar results.

Sheard et al. (1985) observed that creeping bentgrass sand greens only lost a maximum of 2% of applied nitrogen in the drainage water after application of nitrogen at levels similar to Rieke and Ellis (1974). The results of leaching from a U.S. Golf Association specification putting green were somewhat higher. These greens have a minimum of 93% sand, a maximum of 3% silt and 5% clay with an infiltration rate of at least 5 cm hr^{-1} (Petrovic 1990). Brown et al. (1982) found that 22% of applied ammonium nitrate leached as nitrate when the nitrogen source was applied in February at a rate of 163 kg ha^{-1}. However, Snyder et al. (1981) found that only about 1% of applied nitrogen as urea was lost as nitrate on bermudagrass sand greens.

Petrovic (1990) noted that information on nitrate leaching from cool- and warm-season grasses is more extensive on sandy loam soils than on finer and coarser textured soils. Brown et al. (1982) found that 9% of the ammonium nitrate leached as nitrate from a single application of 163 kg N ha^{-1} to bermudagrass greens over a sandy loam soil. Rieke and Ellis (1974) observed limited nitrate leaching in a sandy loam soil even at high application rates (290 kg ha^{-1}). Elevated nitrate concentrations were observed in the upper soil, but deeper leaching did not occur during the experimental period. Starr and DeRoo (1981) also found little or no nitrate leaching into groundwater using ^{15}N techniques. They concluded that the annual total application of ammonium nitrate at 180 kg ha^{-1} for fertilization of Kentucky bluegrass-red fescue turfgrass would not result in contamination of groundwater.

Exner et al. (1991) conducted a study to determine the amount of nitrate released from turfgrass plots fertilized with ammonium nitrate at rates ranging from 0 to 2.4 kg 100 m^{-2}. It was noted that 34 d after application all the plots showed significant nitrate concentrations below the rooting zone (municipal sports turfgrass, Kentucky bluegrass/creeping red fescue over a fine sandy loam). Nitrate concentrations were highest at about 3 m depth for all treatments and were in excess of 50 mg L^{-1} in all treatments except the control. A second pulse of high nitrate was also observed at a depth of about 4 m, although much less than in the first. The high nitrate concentrations observed in this study were attributed to the fact that large amounts of irrigation water were used frequently (51 mm every third day) which effectively leached the nitrate out of the rooting zone before the turfgrass was able to utilize it. Another factor was that the irrigation water itself was nearly 10 mg L^{-1} nitrate and may have exceeded the crop nitrogen requirement. Clearly excessive watering regimes coupled with higher than needed nitrogen application rates can pose water quality problems.

In a monitoring study of nitrate leaching beneath golf courses on the sandy soils of Cape Cod, MA, Cohen et al. (1990) observed a wide range of nitrate concentrations in groundwater from less than 0.1 to 30 mg L^{-1}. Nitrate concentrations did not exceed the drinking water standard of 45 mg L^{-1} (10 mg L^{-1} nitrate-nitrogen). In general, decreased subsurface losses of nitrogen were directly related to decreased fates of application and use of slow-release formulations.

Information on nitrate leaching from application of nitrogen fertilizer to fine-textured turfgrass systems is limited (Petrovic 1990). In general, the studies were short-term growth experiments that lack information regarding the long-term fate of field applied nitrogen (Petrovic 1990). For example, Nelson et al. (1980) studied nitrate leaching from urea and isobutylidene diurea (IBDU) applied to Kentucky bluegrass over 5 cm of silt loam or thatch. Application of 253 kg urea ha^{-1} caused 32–81% of the applied urea to leach from the soil or thatch, respectively. Only 5–23% of the IBDU was leached. Nitrogen leaching losses from the thatch were lower than from the soil. Decreased leaching in thatch was probably due to its lower moisture retention capacity and subsequent inability to efficiently hydrolyze the applied IBDU (Petrovic 1990).

The impact of the source and rate of nitrogen on the leaching of nitrogen was also noted by Petrovic (1990). In worst case scenario studies (sandy soils, high application rates, heavy irrigation), it has been shown that as the rate of nitrogen applied increases, the percent of total nitrogen leached decreases. However, the total amount of nitrate leaching on an area basis increases with increasing rates of application (Petrovic 1990). Brown et al. (1977) noted that on putting greens containing root zone mixtures of about 80% sand, 5–10% clay, and 10% peat, the percent of nitrogen from ammonium sulfate that leached decreased from 38 to 16% as the rate of nitrogen applied was increased from 24 to 98 kg ha^{-1}. However, the total amount of nitrate that leached actually increased from 9 to 15 kg ha^{-1}. The volume of water and mass of nitrate lost by subsurface movement are the important factors related to the nitrate concentration in drainage water. Brown et al. (1977) also noted that when a fine sandy loam was used, the percent of nitrogen lost by leaching was reduced from 15 to 5% as the rate of applied nitrogen increased. With a finer-textured root zone mix, it is important to note the amount of nitrate lost by subsurface movement on an area basis was basically unchanged with increasing nitrogen application rates (Petrovic 1990).

Snyder et al. (1981) studied the effect nitrogen rate and source had on nitrogen leaching from sand. With low application rates such as 39 kg ha^{-1} applied bimonthly, very little leaching of nitrate was observed regardless of source. Most leaching occurred with calcium nitrate as the source, and only 2.9% of the total leached over a 2-year period. At higher rates (78 kg ha^{-1}), applied bimonthly, leaching loss of nitrate was less than 10% of the total applied from either calcium nitrate or IBDU. Less than 1% of the nitrogen from ureaformaldehyde, sulfur-coated urea (SCU), and urea was lost by subsurface movement. The mean concentration of nitrate found in the leachate was 2.4 and 1.4 mg L^{-1} for the calcium nitrate and IBDU, respectively.

Sheard et al. (1985) found only 1.2 and 2.0% of the applied nitrogen (293 kg ha^{-1} $year^{-1}$) leached to the drainage water from creeping bentgrass sand greens fertilized with SCU or urea, respectively. The investigators also observed insignificant differences between nitrogen leaching from acid and alkaline greens when urea was applied. They attributed the lower leaching from urea to the greater ammonia volatilization on the slightly alkaline sands. Gross et al. (1990) observed limited leaching losses of nitrogen leached from turfgrass grown on sandy loam soil. Average nitrate-nitrogen concentrations in subsurface leachate ranged from 0 to 3.2 mg L^{-1} on plots fertilized with granular and liquid formulations of urea applied at 220 kg ha^{-1} $year^{-1}$. However, with an infrequent and fixed sampling protocol, the investigators may have missed significant leaching events associated with heavy precipitation.

Rieke and Ellis (1974) also assessed the subsurface movement of nitrate in relation to fertilizer rate and source. They noted that more nitrate leaching occurred where ammonium nitrate, ureaformaldehyde, or IBDU were applied than activated sewage sludge. The nitrogen sources in this study were applied in the spring at eight times the normal rate (378 kg ha^{-1}) to a sandy soil that received over 80 cm of rainfall per year. Nitrate concentrations were elevated in the surface 30 cm of soil, but did not exceed the control plots at depths of 60 cm, except on a single sampling date.

Brown et al. (1982) studied the interaction of nitrogen source and soil texture on nitrate leaching from U.S. Golf Association specification bermudagrass greens with some irrigation. They found that leaching losses were greatest for applications of ammonium nitrate and activated sewage sludge compared to comparable rates of either IBDU or ureaformaldehyde when applied to root zone mixtures containing more than 80% sand or greens constructed with sandy loam soils. In another study, Bredakis and Steckel (1963) observed that in the first 3 weeks after fertilization of turfgrass, inorganic sources of nitrogen resulted in higher amounts of residual nitrate in soil as compared to organic and slow-release sources of nitrogen. Elevated levels of leachable nitrogen in soil, especially nitrate, increase the risk of subsurface losses of nitrogen from turfgrass.

Several reports discuss the effect of irrigation practices and the potential for leaching of fertilizer nitrogen (Petrovic, 1990). Morton et al. (1988) studied the effect of nitrogen rate and irrigation on the potential leaching of nitrate from a Kentucky bluegrass-red fescue lawn. The nitrogen rate was typical of a moderate to high lawn fertility program of 50% urea and 50% flowable ureaformaldehyde applied at 98 and 244 kg ha^{-1} $year^{-1}$ (Petrovic 1990). Two irrigation regimes were used. The first regime applied 1.2 cm of water when the tensiometer readings reached –0.05 MPa. The second regime applied 3.75 cm of water per week. The first treatment, irrigation on the basis of soil moisture status, did not result in water drainage out of the root zone. The second treatment, irrigation on a fixed schedule, did result in subsurface drainage of water. The first irrigation regime did not cause a significant increase in nitrogen leaching for either nitrogen source. Irrigating at the higher rate did result in measurable leaching losses of nitrogen, although the observed nitrate levels were still well below the drinking water standard.

In another study, Snyder et al. (1984) studied the effect of irrigation, fixed schedule vs tensiometer, and nitrogen source, ammonium nitrate vs SCU applied at 98 kg ha^{-1}, on nitrogen leaching from sand under bermudagrass. Nitrogen was also supplied in the irrigation water (chemigation). The percent of applied nitrogen leached ranged from 0.3 to 56% with leaching loss greatest for plots treated with ammonium nitrate rather than SCU. In addition, the researchers noted that the greatest leaching occurred in February and March and the least observed in June and July. The decline in leaching was attributed to increased plant growth and evapotranspiration (Petrovic 1990). In addition, leaching of nitrogen was greater from the fixed schedule irrigation plots compared to the tensiometer sensored plots.

Brown et al. (1977) also evaluated the effect of nitrogen source and rate of irrigation on leaching. Irrigation had little effect on nitrate leaching from plots treated with very high rates of nitrogen from sludge or ureaformaldehyde. When ammonium nitrate was applied, however, substantial increases in nitrate movement occurred.

In a study comparing surface and subsurface irrigation, chemigation practices, and sources of nitrogen, Mitchell et al. (1978) monitored nitrogen concentrations in drainage water from creeping bentgrass greens. The observed pattern of nitrogen movement was similar for the surface and subsurface irrigation systems, but concentration of nitrate and ammonium in drainage water was generally higher from greens fertilized through the subsurface irrigation system. The low initial cost of nitrogen applied through a subsurface irrigation system should be balanced against the high rate of subsurface nitrogen losses. Solubility of nitrogen source did influence the pattern and total losses of nitrogen. Compared to slow-release forms of nitrogen, soluble sources of nitrogen had greater and more rapid losses of nitrate in drainage water. Activated sewage sludge and anhydrous ammonia with nitrapyrin were the slow-release formulations of nitrogen. A solution of urea (15%) and ammonium nitrate (15%) was the soluble formulation. Immediately after application of the soluble nitrogen formulation, the concentration of nitrate in drainage effluent (96–109 mg L^{-1}) temporarily increased to a level well above the drinking water standard for nitrate (45 mg L^{-1}). Peak nitrate concentration from greens fertilized with anhydrous ammonia and anhydrous ammonia with nitrapyrin was delayed by two weeks. The delay in release of nitrogen resulted in greater concentrations of nitrate (126–148 mg L^{-1}) in the effluent water. The onset of dormancy may have decreased the ability of the bentgrass to utilize the slowly released nitrate. Overall, the level of nitrate in drainage water was more closely related to environmental conditions favoring nitrification of residual and added nitrogen than to the sources of nitrogen, green mix, or timing of application.

Petrovic (1990) also noted that the season in which nitrogen is applied has an effect on the nitrate leaching potential. He observed that leaching can be significant during periods of low temperature and high precipitation. Cool temperatures reduce denitrification and ammonia volatilization, microbial immobilization, and plant uptake. Since evapotranspiration is also low with relatively high precipitation, more subsurface drainage occurs. Recently, late fall fertilization of turfgrass with nitrogen has become an important nutrient management practice for cool-season grasses (Street 1988). If nitrogen uptake by turfgrass does not continue through this period, there will be a high potential for nitrate leaching. When applied in the fall, Petrovic et al. (1986) found that soluble nitrogen sources such as urea can lead to significant nitrate leaching. Depending on site characteristics, estimates of nitrogen leaching ranged from 21 to 47% of applied soluble nitrogen. On gravelly sand there was more nitrate leaching from urea than from sludge, ureaformaldehyde or resin-coated urea, although nitrate leaching was less than 2% of applied fertilizer.

A generalized description of subsurface solute and water leaching processes was discussed in Chapter 2 (Balogh and Watson 1992). Reviews of the quantitative descriptions of nitrogen and phosphorus transport have been presented by Anderson et al. (1989), Enfield and Ellis (1983), Nielsen et al. (1982), Starr (1983), and Wagenet (1983). These reviews contain mathematical derivations of solute transport equations. Many mechanistic/management level models of nutrient transport (e.g., Wagenet and Hutson 1987) utilize the instantaneous and reversible linear model to describe adsorption-desorption in these transport equations. Attempts have been made to use nonlinear models, kinetic approaches, and ion exchange equations to evaluate the fate and movement of nitrogen (Wagenet 1983); however, many of these adsorption constants must be evaluated for each individual model system. Accuracy of these models also requires a description of irreversible sources and sinks of nitrogen including (1) biological and nonbiological fixation, (2) microbial transformations, and (3) adsorption or exudation by plant roots.

The complex relationship between microbial transformations, plant uptake, and nutrient concentrations can only be approximated (Cushman 1983; Nielsen et al. 1982). Computer simulation of nitrogen transport using finite differencing techniques allows solution of the complex transport model for use in multilayered soils under field management conditions, including simplified plant uptake of nutrients (e.g., Tillotson and Wagenet 1982; Wagenet and Hutson 1986). The potential use of several models for simulation of the fate and transport of nitrogen in turfgrass systems is discussed in Chapter 5 (Balogh et al. 1992).

3.5.3 Nutrient Loss through Surface Runoff

The generalized process of surface transport of applied turfgrass chemicals was described in Chapter 2 (Balogh and Watson 1992). Nutrients lost in runoff include dissolved, suspended particles and those adsorbed on sediment. These nutrient phases all potentially move in water from fertilized land surfaces (Anderson et al. 1989; Daniel et al. 1979; Koehler et al. 1982a, 1982b). Runoff occurs when precipitation, less interception, exceeds the rate of infiltration. Surface-applied nutrients are dissolved in runoff water and transported on the surface of entrained sediment (Keeney 1983; Taylor and Kilmer 1980). Initial extraction and movement of nutrients from the soil surface layers is controlled by (1) diffusion and turbulent movement from soil pores and surfaces into runoff water, (2) desorption of nutrients on soil exchange sites, (3) dissolution of fertilizer granules, (4) entrainment of undissolved fertilizer elements into runoff, and (5) entrainment of organic and inorganic nutrients adsorbed to soil particles moving on eroded sediment in runoff water. The amount of fertilizer lost in runoff is determined by a combination of factors including (1) the volume of runoff water; (2) the timing, amount, and placement of fertilizer in relation to subsequent runoff events; and (3) the magnitude of immobilization, adsorption, volatilization, and cycling processes.

The EDI, effective depth of interaction, is the thickness of the surface soil in which the degree of chemical and physical interaction between soil and runoff is equal to the interaction at the soil surface. The EDI is the depth in the soil that is subject to extraction of nutrients by runoff processes. Sharpley (1985a) observed a range of 1.3–37.4 mm for the effective depth of interaction between runoff water and adsorbed chemicals. Both Ahuja et al. (1982) and Sharpley (1985a) demonstrated a direct relationship between rainfall intensity, slope, soil structure, and runoff energy with the depth of the active mixing zone. The amount of surface cover or crop residue was inversely related to the EDI. Rainfall intensity, slope, soil structure, and surface residue affect the energy of runoff. Runoff energy is directly related to the amount of sediment transport and depth of interaction for solution transport of nutrients. Accurate EDIs have not been estimated for turfgrass systems. Considering the limited sediment and soluble losses of nutrients from natural grasslands and unfertilized pastures (Anderson et al. 1989; Gross et al. 1990, 1991) (Tables 5 and 6), the depth of interaction for developed turfgrass should be quite small.

Application of fertilizer nitrogen or phosphorus poses the potential for some runoff into surface waters. Few studies have investigated this process in turfgrass systems. In a 2-year field study in Rhode Island, Morton et al. (1988) studied the influence of overwatering and fertilization on nitrogen losses from home lawns (Kentucky bluegrass). In general, soil water percolate accounted for greater than 93% of total water and inorganic nitrogen from all treatments. They observed only two natural events that resulted in runoff. This runoff was the result of unusual climatic conditions: (1) frozen ground and (2) from plots that received 12.5 cm of water in 1 week. The runoff drainage was noted to contain a concentration of nitrate from 1.1 to 4.2 mg L^{-1} for the two events. In another study investigating the impact of source, rate, and soil texture, Brown et al. (1977) observed only one case of nitrate concentration in runoff water from turfgrass exceeding the 10 mg L^{-1} drinking water standard.

In an unpublished thesis from the Pennsylvania State University, Harrison (1989) observed low runoff volumes and nutrient concentrations from irrigated lysimeters of Kentucky bluegrass grown on clay soil. Concentrations of soluble nitrogen and soluble phosphorus in runoff rarely exceeded 5 and 2 mg L^{-1}, respectively.

Gross et al. (1990) also observed limited soluble and sediment losses of phosphorus and nitrate in runoff from Kentucky bluegrass plots fertilized at a rate of 220 kg N ha^{-1} $year^{-1}$. The maximum loss of nitrogen in runoff was observed when a runoff event occurred the day after fertilization. However, the loss of nitrogen in the surface runoff was less than 0.1% of the applied granular formulation of urea. A large portion of soluble nitrogen losses in runoff water occurred when a significant precipitation event occurred shortly after fertilizer application. Total and soluble losses of phosphorus were not significantly different between fertilized and control plots. Significant losses of phosphate for a liquid treatment were associated with high amounts of runoff from the turfgrass plots in December. This loss accounted for 73% of total runoff losses of phosphorus. Timing of fertilizer application to avoid precipitation and runoff events will reduce potential surface losses of nutrients.

3.6 NUTRIENT LOSS FROM TURFGRASS AND MANAGEMENT GUIDELINES

On the basis of the review of the scientific literature, nitrogen, and in particular nitrate, is the nutrient posing the most serious threat from an environmental and water quality perspective. From the scientific literature on potential transport of nitrogen from turfgrass systems, several conclusions can be inferred.

If nitrogen fertilizer residues are present in a soluble form above a concentration that can be used by turfgrass, and if sufficient water is available to percolate through the soil water, then leaching can occur. Use of insoluble or less available nitrogen sources reduces nitrate leaching (Petrovic 1990).

Increasing the rate of nitrogen application to sandy greens will lead to a deterioration in drainage water quality. On sandy loam greens, increased nitrogen fertilization commensurate with uptake capacity of specific turfgrass species should not further reduce water quality.

Late fall nitrogen fertilization of cool-season grasses is an attractive turfgrass practice. However, the turfgrass quality benefits derived from this practice may be reduced by the adverse environmental impacts especially in areas with shallow or sensitive groundwater. As noted by Petrovic (1990), nitrate losses are greater on warm-season grasses fertilized in the cooler part of the year compared with warmer seasons (Brown et al. 1977; Snyder et al. 1984).

Turfgrass ecosystems generally result in soils with a high infiltration capacity, thereby reducing the potential for nutrient loss and subsequent water quality deterioration by runoff. Several studies demonstrate that runoff of water from turfgrass is limited (Gross et al. 1990, 1991; Welterlen et al. 1989).

The leaching of fertilizer nitrogen applied to turfgrass is related to soil texture and degree of conductivity of soil water; the amount of subsurface movement of water; nitrogen source, formulation, rate, and timing; and irrigation and rainfall. Significant leaching of nitrate will occur when soluble nitrogen is applied at a rate higher than normal on turfgrass grown on sandy soil. The potential for subsurface loss of nitrogen is further increased when the turfgrass is irrigated at a rate in excess of plant use, evapotranspiration, and soil storage.

If fertilization of turfgrass does pose a threat to surface water and groundwater quality, several management options are available to minimize or eliminate the problem (Petrovic 1990). Limiting irrigation to replacing soil moisture used by turfgrass, use of slow-release sources of nitrogen, and use of less sandy soils would significantly reduce or eliminate nitrate leaching from turfgrass. Timing of fertilizer application in relation to active uptake and potential runoff events, as well as the use of realistic nitrogen application rates, are primary methods to reduce nonpoint source losses of nitrogen from turfgrass systems.

REFERENCES

Adams, W. E., White, A. W., McCreery, R. A. and Dawson, R. N. 1967. Coastal bermudagrass forage production and chemical composition as influenced by potassium source, rate, and frequency of application. *Agron. J.* 59:247–250.

Adriano, D. C., Pratt, P. F. and Takatori, F. H. 1972a. Nitrate in unsaturated zone of an alluvial soil in relation to fertilizer nitrogen rate and irrigation level. *J. Environ. Qual.* 1(4):418–422.

Ahuja, L. R., Lehman, O. R. and Sharpley, A. N. 1983. Bromide and phosphate in runoff water from shaped and cloddy surfaces. *Soil Sci. Soc. Am. J.* 47(4):746–748.

Ahuja, L. R., Sharpley, A. N. and Lehman, O. R. 1982. Effect of soil slope and rainfall characteristics on phosphorus in runoff. *J. Environ. Qual.* 11(1):9–13.

Alberts, E. E., Schuman, G. E. and Burwell, R. E. 1978. Seasonal runoff losses of nitrogen and phosphorus from Missouri Valley loess watersheds. *J. Environ. Qual.* 7(2):203–208.

Aldrich, S. R. 1984. Nitrogen management to minimize adverse effects on the environment in Hauck, R. D. (Ed.). *Nitrogen in Crop Production*. Am. Soc. Agron., Crop Sci. Soc. Am. Soil Sci. Soc. Am., Madison, WI, pp. 663–673.

Alexander, M. 1977. *Introduction to Soil Microbiology*. John Wiley & Sons, New York 467 pp.

Anderson, G. 1980. Assessing organic phosphorus in soils. in Khasawneh, F. E. (Ed. chair.) *The Role of Phosphorus in Agriculture*. Am. Soc. Agron., Crop Sci. Soc. Am., Soil Sci. Soc Am., Madison, WI, pp. 411–431.

Anderson, J. L., Balogh, J. C. and Waggoner, M. 1989. Soil Conservation Service Procedur Manual: Development of Standards and Specifications for Nutrient and Pesticid Management. Section I & II. USDA Soil Conservation Service. State Nutrient and Pes Management Standards and Specification Workshop, St. Paul, MN, July 1989, 453 pp.

Andraski, B. J., Mueller, D. H. and Daniel, T. C. 1985a. Phosphorus losses in runoff as affecte by tillage. *Soil Sci. Soc. Am. J.* 49(6):1523–1527.

Andraski, B. J., Mueller, D. H. and Daniel, T. C. 1985b. Effects of tillage and rainfal simulation date on water and soil losses. *Soil Sci. Soc. Am. J.* 49(6):1512–1517.

Armstrong, D. E., Lee, K. W., Uttormark, P. D., Keeney, D. R. and Harris, R. F. 1974 *Pollution of the Great Lakes by Nutrients from Agricultural Land. Vol I.* Internationa Reference Group on Great Lakes Pollution from Land Use Activities. Category 6, An Arbor, MI.

Baker, D. B. 1985. Regional water quality impacts of intensive row-crop agriculture: A Lak Erie Basin case study. *J. Soil Water Cons.* 40:125–132.

Baker, J. L. and Johnson, H. P. 1981. Nitrate-nitrogen in tile drainage as affected by fertilization *J. Environ. Qual.* 10(4):519–522.

Baker, J. L. and Laflen, J. M. 1982. Effects of corn residue and fertilizer management o soluble nutrient runoff. *Trans. ASAE* 25(2):344–348.

Baker, J. L. and Laflen, J. M. 1983a. Runoff losses of nutrients and soil from ground fall fertilized after soybean harvest. *Trans. ASAE* 26(4):1122–1127.

Baker, J. L. and Laflen, J. M. 1983b. Water quality consequences of conservation tillage. *J Soil Water Cons.* 38:186–193.

Baker, J. L., Campbell, L., Johnson, H. P. and Hanway, J. J. 1975. Nitrate, phosphorus an sulfate in subsurface drainage water. *J. Environ. Qual.* 4(3):406–412.

Baker, J. L., Kanwar, R. S., and Austin, T. A. 1985. Impact of agricultural drainage wells o groundwater quality. *J. Soil Water Cons.* 40(6):516–620.

Balogh, J. C. and Watson, J. R., Jr. 1992. Role and conservation of water resources. in Balogh J. C. and Walker, W. J. (Eds.). *Golf Course Management and Construction: Environmenta Issues*. Lewis Publishers, Chelsea, MI, Chapter 2.

Balogh, J. C., Leslie, A. R., Walker, J. C. and Kenna, M. P. 1992. Development of integrate management systems for turfgrass. in Balogh, J. C. and Walker, W. J. (Eds.). *Golf Cours Management and Construction: Environmental Issues*. Lewis Publishers, Chelsea, MI Chapter 5.

Bandel, V. A. 1981. Fertilization techniques for no-tillage corn. *Agrichem. Age* 25(7):14–15

Barisas, S. G., Baker, J. L., Johnson, H. P. and Laflen, J. M. 1978. Effect of tillage system on runoff losses of nutrients: A rainfall simulation study. *Trans. ASAE* 21(5):893–897.

Barrow, N. J. and Shaw, T. C. 1975. The slow reactions between soil and anions: 3. The effec of time and temperature on the decrease in isotopically exchangeable phosphate. *Soil Sci* 119:190–197.

Beard, J. B. 1973. *Turfgrass Science and Culture*. Prentice-Hall, Englewood Cliffs, NJ 658 pp.

Beard, J. B. 1982. *Turf Management for Golf Courses*. Macmillan Publishing Co., New York, 642 pp.

Beauchamp, E. G. 1977. Slow release N fertilizers applied in fall for corn. *Can. J. Soil Sci.* 57:487–496.

Belesky, D. P. and Wilkinson, S. R. 1983. Response of Tifton 44 and coastal bermudagrass to soil pH, K, and N source. *Agron. J.* 75(1):1-4.

Bock, B. R. 1984. Efficient use of nitrogen in cropping systems. in Hauck, R. D. (Ed.). *Nitrogen in Crop Production*. Am. Soc. Agron., Crop Sci. Soc. Am., Soil Sci. Soc. Am., Madison, WI, pp. 273–294.

Bottcher, A. B., Monke, E. J. and Huggins, L. F. 1981. Nutrient and sediment loading from a subsurface drainage system. *Trans. ASAE* 24(5):1221–1226.

Bouldin, D. R., Klausner, S. D. and Reid, S. W. 1984. Use of nitrogen from manure. in Hauck, R. D. (Ed.). *Nitrogen in Crop Production*. Am. Soc. Agron., Crop Sci. Soc. Am., Soil Sci. Soc. Am., Madison, WI, pp. 221–245.

Bowman, D. C., Paul, J. L., Davis, W. B. and Nelson, S. H. 1987. Reducing ammonia volatilization from Kentucky bluegrass turf by irrigation. *HortScience* 22(1):84–87.

Bradford, R. R. 1974. Nitrogen and Phosphorus Losses from Agronomy Plots in North Alabama. EPA-660/2-74-033. U.S. Environmental Protection Agency, Washington, D.C.

Branson, R. L., Pratt, P. F., Rhoades, J. D. and Oster, J. D. 1975. Water quality in irrigated watersheds. *J. Environ. Qual.* 4(1):33–40.

Bredakis, E. J. and Steckel, J. E. 1963. Leachable nitrogen from soils incubated with turfgrass fertilizers. *Agron. J.* 53:145–147.

Brezonik, P. L. 1969. Eutrophication: The process and its modeling potential. in Putnam, H. D. (Ed.). Modeling the Eutrophication Process. Proc. Wrkshop., Dept. Environ. Eng., University of Florida, Gainesville, pp. 68–116.

Brezonik, P. L. (Chairman). 1978. Nitrates: An environmental assessment. National Academy of Sciences, Washington, D.C., Chapters 1, 8, 9.

Brown, K. W., Duble, R. W. and Thomas, J. C. 1977. Influence of management and season on fate of N applied to golf greens. *Agron. J.* 69(4):667–671.

Brown, K. W., Thomas, J. C. and Duble, R. L. 1982. Nitrogen source effect on nitrate and ammonia leaching and runoff losses from greens. *Agron. J.* 74(6):947–950.

Burwell, R. E., Schuman, G. E., Heinemann, H. G. and Spomer, R. G. 1977. Nitrogen and phosphorus movement from agricultural watersheds. *J. Soil Water Cons.* 32(5): 226–230.

Burwell, R. E., Schuman, G. E., Saxton, K. E. and Heinemann, H. E. 1976. Nitrogen in subsurface discharge from agricultural watersheds. *J. Environ. Qual.* 5(3):325–329.

Burwell, R. E., Timmons, D. R. and Holt, R. F. 1975. Nutrient transport in surface runoff as influenced by seasonal cover and seasonal periods. *Soil Sci. Soc. Am. Proc.* 39: 523–528.

Calvert, D. V., Stewart, E. H., Mansell, R. S., Fiskell, J. S. A., Rogers, J. S., Allen, L. H., Jr. and Graetz, D. A. 1981. Leaching losses of nitrate and phosphate from a spodosol as influenced by tillage and irrigation level. *Soil Crop Sci. Soc. Fla. Proc.* 40:62–71.

Campbell, C. A., Read, D. W. L., Biederbeck, V. O. and Winkleman, G. E. 1983. The first 12 years of long-term crop rotation study in southwestern Saskatchewan — Nitrate-N distribution in soil and N uptake by the plant. *Can. J. Soil Sci.* 63:563– 578.

Canter, L. W. 1987. Nitrates and pesticides in ground water: An analysis of a computer-based literature search. in Fairchild, D. M. (Ed.). *Ground Water Quality and Agricultural Practices*. Lewis Publishers, Chelsea, MI, pp. 153–174.

Cantor, K. P., Blair, A. and Zahn, S. H. 1988. Health Effects of Agrichemicals in Groundwater: What Do We Know? in Agricultural Chemicals and Groundwater Protection: Emerging Management and Policy. Proceedings of a Conference, Freshwater Foundation, Oct. 22–23, 1987, St. Paul, MN, pp. 27–42.

Carrow, R. N., Johnson, B. J. and Burns, R. E. 1987. Thatch and quality of Tifway Bermudagrass in relation to fertility and cultivation. *Agron. J.* 79:525–530.

Chichester, F. W. 1976. The impact of fertilizer use and crop management on nitrogen content of subsurface water draining from upland agricultural watersheds. *J. Environ. Qual.* 5(4): 413–416.

Chichester, F. W. 1977. Effects of increased fertilizer rates on nitrogen content of runoff and percolate from monolith lysimeters. *J. Environ. Qual.* 6(2):221–216.

Cohen, S. Z., Nickerson, S., Maxey, R., Dupuy, A., Jr. and Senita, J. A. 1990. Ground water monitoring study for pesticides and nitrates associated with golf courses on Cape Cod. *Ground Water Monit. Rev.* 10(1):160–173.

Culley, J. L. B. and Bolton, E. F. 1983. Suspended solids and phosphorus loads from a clay soil: II Watershed study. *J. Environ. Qual.* 12(4):498–503.

Culley, J. L. B., Bolton, E. F. and Bernyk, V. 1983. Suspended solids and phosphorus loads from a clay soil: I Plot studies. *J. Environ. Qual.* 12(4):493–498.

Cushman, J. H. 1983. Incorporation of rhizosphere into plant root models. in Nelsen, D. W., Elrick, D. E. and Tanji, K. K. (Eds.). *Chemical Mobility and Reactivity in Soil Systems.* Special Publ. No. 11. Soil Sci Soc. Am., Madison, WI, pp. 165–181.

Daniel, T. C., McGuire, P. E., Stoffel, D. and Miller, B. 1979. Sediment and nutrient yield from residential construction sites. *J. Environ. Qual.* 8(3):304–308.

Daniel, T. C., Wendt, R. C., McGuire, P. E. and Stoffel, D. 1982. Nonpoint source loading rates from selected land uses. *Water Res. Bull.* 18(1):117–120.

Daniels, R. B., Gilliam, J. W., Gamble, E. E. amd Skaggs, R. W. 1975. Nitrogen movement in a shallow aquifer system of the North Carolina coastal plain. *Water Resour. Res.* 11:1121–1130.

Davidson, J. M., Rao, P. S. C. and Nkedi-Kizza, P. 1983. Physical processes influencing water and solute transport in soils. in Nelsen, D. W., Elrick, D. E. and Tanji, K. K. (Eds.). *Chemical Mobility and Reactivity in Soil Systems.* Special Publ. No. 11. Soil Sci Soc. Am., Madison, WI, pp. 35–47.

Davis, R. R. 1969. Nutrition and fertilizers. in Hanson, A. A. and Juska, F. V. (Eds.). *Turfgrass Science.* Agron. Mongr. 14:130–150. Am. Soc. Agron., Madison, WI.

Dickey, E. C. and Vanderholm, D. H. 1980. Performance and design of vegetative filters for feedlot runoff treatment. in Livestock Wastes: A Renewable Resource. Proc. of the 4th Internat. Symp. on Livestock Wastes. ASAE, St. Joseph, MI, pp. 257–260.

Dunigan, E. P., Phelan, R. A. and Mondart, C. L., Jr. 1976. Surface runoff losses of fertilizer elements. *J. Environ. Qual.* 5(3):339–342.

Duxbury, J. M. and Peverly, J. H. 1978. Nitrogen and phosphorus losses from organic soils. *J. Environ. Qual.* 7:566–570.

Ellis, B. G., Knezek, B. D. and Jacobs, L. W. 1983. The movement of micronutrients in soils. in Nelsen, D. W., Elrick, D. E. and Tanji, K. K. (Eds.). *Chemical Mobility and Reactivity in Soil Systems.* Special Publ. No. 11. Soil Sci Soc. Am., Madison, WI, pp. 109–122.

Enfield, C. G. and Ellis, R., Jr. 1983. The movement of phosphorus in soil. in Nelsen, D. W., Elrick, D. E. and Tanji, K. K. (Eds.). *Chemical Mobility and Reactivity in Soil Systems.* Special Publ. No. 11. Soil Sci Soc. Am., Madison, WI, pp. 93–107.

Exner, M. E., Burbach, M. E., Watts, D. G., Shearman, R. C. and Spalding, R. F. 1991. Deep nitrate movement in the unsaturated zone of a simulated urban lawn. *J. Environ. Qual.* 20(3):658–662.

Firestone, M. K. 1982. Biological denitrification. in Stevenson, F. J. (Ed.). *Nitrogen in Agricultural Soils*. Agron. Monogr. 22:289–326 Am. Soc. Agron., Crop Sci. Soc. Am., Soil Sci. Soc. Am., Madison, WI.

Flipse, W. J., Katz, B. G., Lindner, J. B. and Markel, R. 1984. Sources of nitrate in groundwater in a sewered housing development, central Long Island, New York. *Ground Water*. 32: 418–426.

Fox, R. L. 1981. External phosphate requirements of crops. in Stelly, M. (Ed. in chief). *Chemistry in the Soil Environment*. ASA Spec. Publ. No. 40. Am. Soc. Agron. and Soil Sci. Soc. Am., Madison, WI, pp. 223–239.

Fraser, P., Chilvers, C., Beral, V. and Hill, M. J. 1980. Nitrate and human cancer: A review of the evidence. *Int. J. Epidemiol.* 9:3–11.

Frere, M. H. 1976. Nutrient aspects of pollution from cropland. in Stewart, B. A., Woolhiser, D. A., Wischmeier, W. H., Caro, J. H. and Frere, M. H. (Eds.). Control of Water Pollution from Cropland: An Overview. Vol. 2. EPA 600/2-75-026b. U.S. Environmental Protection Agency, Athens, GA and USDA Agricultural Research Service, Washington, D.C., pp. 59–90.

Frere, M. H., Ross, J. D. and Lane, L. J. 1980. The nutrient submodel. in Knisel, W. G. (Ed.). CREAMS: A Field Scale Model for Chemicals, Runoff and Erosion from Agricultural Management Systems. USDA Conservation Research Report No. 26, Washington, D.C., pp. 65–87.

Fry, F. E. J. 1969. Some possible physiological stresses induced by eutrophication. in *Eutrophication: Causes, Consequences, Correctives*. National Academy of Sciences, Washington, D.C. pp. 531–536.

Gast, R. G., Nelson, W. W. and MacGregor, J. M. 1974. Nitrate and chloride accumulation and distribution in fertilized tile-drained soils. *J. Environ. Qual.* 3(3):209–213.

Gast, R. G., Nelson, W. W. and Randall, G. W. 1978. Nitrate accumulation in soils and loss in tile drainage following nitrogen application to continuous corn. *J. Environ. Qual.* 7: 258–262.

Gerwing, J. R., Caldwell, A. C. and Goodroad, L. L. 1979. Fertilizer nitrogen distribution under irrigation between soil, plant, and aquifer. *J. Environ. Qual.* 8(3):281–284.

Gilliam, J. W., Daniels, R. B. and Lutz, J. F. 1974. Nitrogen content of shallow ground water in the North Carolina Coastal Plain. *J. Environ. Qual.* 3(2):147–151.

Goss, R. L. 1974. Effects of variable rates of sulfur on the quality of putting green bentgrass. in Roberts, E. C. (Ed.). Proc. of Second Int. Turfgrass Res. Conf. Blacksburg, VA, July 1969, ASA, Madison, WI, p. 172–175.

Gross, C. M., Angle, J. S. and Welterlen, M. S. 1990. Nutrient and sediment losses from turfgrass. *J. Environ. Qual.* 19(4):663–668.

Gross, C. M., Angle, J. S., Hill, R. L. and Welterlen, M. S. 1991. Runoff and sediment losses from tall fescue under simulated rainfall. *J. Environ. Qual.* 20(3):604–607.

Halevy, J. 1987. Efficiency of isobutylidene diurea, sulphur-coated urea and urea plus nitrapyrin, compared with divided dressings of urea, for dry matter production and nitrogen uptake of ryegrass. *Expl. Agric.* 23:167–179.

Hansen, S. A., Linn, J. G. and Plegge, S. D. 1987. Water quality for beef and dairy cattle. in Understanding Nitrogen and Agricultural Chemicals in the Environment. AG-BU-3166. University of Minnesota Agric. Ext. Serv. St. Paul, MN, pp. 7–17.

Hanway, J. J. and Laflen, J. M. 1974. Plant nutrient losses from tile-outlet terraces. *J. Environ. Qual.* 3:351–356.

Hanway, J. J. and Olson, R. A. 1980. Phosphate nutrition of corn, sorghum, soybeans and small grains. in Khasawneh, F. E. (Ed. chair.). *The Role of Phosphorus in Agriculture.* Am. Soc. Agron., Crop Sci. Soc. Am., Soil Sci. Soc. Am., Madison, WI, pp. 681–692.

Hargrove, W. L. and Kissel, D. E. 1979. Ammonia volatilization from surface applications of urea in the field and laboratory. *Soil Sci. Soc. Am. J.* 43:359–363.

Harrison, S. A. 1989. Effects of Turgrass Establishment Method and Management on the Quantity and Nutrient and Pesticide Content of Runoff and Leachate. Unpubl. Thesis. The Pennsylvania State University, University Park, PA, 125 pp.

Havelka, U. D., Boyle, M. G. and Hardy, R. W. F. 1982. Biological nitrogen fixation. in Stevenson, F. J. (Ed.). Nitrogen in agricultural soils. *Agronomy* 22:365–422.

Hook, J. E. 1983. Movement of phosphorus and nitrogen in soil following application of municipal wastewater. in Nelsen, D. W., Elrick, D. E. and Tanji, K. K. (Eds.). *Chemical Mobility and Reactivity in Soil Systems.* Special Publ. No. 11. Soil Sci. Soc. Am., Madison, WI, pp. 241–262.

Hubbard, R. K., Asmussen, L. E. and Allison, H. D. 1984. Shallow ground water quality beneath an intensive multiple-cropping system using center pivot irrigation. *J. Environ. Qual.* 13(1):156–161.

Huettl, P. J., Wendt, R. C. and Corey, R. B. 1979. Prediction of algal-available phosphorus in runoff suspensions. *J. Environ. Qual.* 8(1):130–132.

Hummel, N. W. and Waddington, D. V. 1981. Evaluation of slow-release nitrogen sources on Baron Kentucky bluegrass. *Soil Sci. Soc. Am. J.* 45:966–970.

Hummel, N. W. and Waddington, D. V. 1984. Sulfur coated urea for turfgrass fertilization. *Soil Sci. Soc. Am. J.* 48(1):191–195.

Impithuksa, V., Dantzman, C. L. and Blue, W. G. 1979. Fertilizer nitrogen utilization by three warm-season grasses on an Alfic haplaquod as indicated by nitrogen-15. *Soil Crop Sci. Soc. Fla. Proc.* 38:93–97.

Jackson, J. E. and Burton, G. W. 1962. Influence of sod treatment and nitrogen placement on the utilization of urea nitrogen by Coastal Bermudagrass. *Agron. J.* 54(1):47–49.

Jackson, J. E., Walker, M. E. and Carter, R. L. 1959. Nitrogen, phosphorus, and potassium requirements of coastal bermudagrass on a Tifton loamy sand. *Agron. J.* 51(1):129–131.

Jackson, W. A., Asmussen, L. E., Hauser, E. W. and White, A. W. 1973. Nitrate in surface and subsurface flow from a small agricultural watershed. *J. Environ. Qual.* 2(4):480–482.

Jansson, S. L. and Personn, J. 1982. Mineralization and immobilization of soil nitrogen. in Stevenson, F. J. (Ed.). *Nitrogen in Agricultural Soils.* Agron. Monogr. 22:229–252. Am. Soc. Agron., Crop Sci. Soc. Am., Soil Sci. Soc. Am., Madison, WI.

Johnson, A. H., Bouldrin, D. R., Goyette, E. A. and Hedges, A. H. 1976. Phosphorus loss by stream transport from a rural watershed: Quantities, processes, and sources. *J. Environ. Qual.* 5(1):148–157.

Johnson, H. P., Baker, J. L., Shrader, W. D. and Laflen, J. M. 1979. Tillage system effects on sediment and nutrients in runoff from small watersheds. *Trans. ASAE* 22:1110–1114.

Jury, W. A. 1983. Chemical transport modeling: Current approaches and unresolved problems. in Nelsen, D. W., Elrick, D. E. and Tanji, K. K. (Eds.). *Chemical Mobility and Reactivity in Soil Systems.* Soil Sci. Soc. Am. Special Publ. No. 11. Madison, WI, pp. 49–64.

Juska, F. V. and Hanson, A. A. 1969. Effect of nitrogen sources, rates and time of application on performance of Kentucky bluegrass turf. *Proc. Am. Soc. Hort. Sci.* 90:413–419.

Juska, F. V., Hanson, A. A. and Erickson, C. J. 1965. Effects of phosphorus, and other treatments on the development of red fescue, merion, and common Kentucky bluegrass. *Agron. J.* 57(1):75–78.

Kamprath, E. J., Broome, S. W., Raja, M. E., Tonapa, S., Baird, J. V. and Rice, J. C. 1973. Nitrogen Management, Plant Populations and Row Width Studies with Corn. North Carolina Agric. Exp. Stn. Tech. Bull. No. 217. Raleigh, NC, 16 pp.

Kanwar, R. S., Johnson, H. P. and Baker, J. L. 1983. Comparison of simulated and measured nitrate losses in tile effluent. *Trans. ASAE* 26(5):1451–1457.

Kanwar, R. S., Baker, J. L. and Laflen, J. M. 1985. Nitrate movement through the soil profile in relation to tillage system and fertilization application method. *Trans. ASAE* 28(6):1802–1807.

Keeney, D. R. 1982. Nitrogen management for maximum efficiency and minimum pollution. in Stevenson, F. J. (Ed.). *Nitrogen in Agricultural Soils.* Agron. Monogr. 22:605–649. Am. Soc. of Agron., Crop Sci. Soc. Am., Soil Sci. Soc. Am., Madison, WI.

Keeney, D. R. 1983. Transformations and transport of nitrogen. in Schaller, F. (Ed.). Proc. Natl. Conf. Agric. Mgmt. and Water Quality, Ames, IA, 26–29 May, 1981, pp. 48–63.

Keeney, D. R. 1986. Sources of nitrate to groundwater. *CRC Crit. Rev. Environ. Control* 16:257–304.

Khaleel, R., Reddy, K. R. and Overcash, M. R. 1980. Transport of potential pollutants in runoff water from land areas receiving animal wastes: A review. *Water Res.* 14:421–436.

Khasawneh, F. E. (Ed. chair.). 1980. *The Role of Phosphorus in Agriculture.* Am. Soc. Agron., Crop Sci. Soc. Am., Soil Sci. Soc. Am., Madison, WI, 910 pp.

King, J. W. and Skogley, C. R. 1969. Effect of nitrogen and phosphorus placements on turfgrass establishments. *Agron. J.* 61(1):4–6.

Kissel, D. E., Richardson, C. W. and Burnett, E. 1976. Losses of nitrogen in surface runoff in the Blackland Prairie of Texas. *J. Environ. Qual.* 5(3):288–293.

Kitur, B. K., Smith, M. S., Blevins, R. L. and Frye, W. W. 1984. Fate of N-depleted ammonium nitrate applied to no-tillage and conventional tillage corn. *Agron. J.* 76:240–242.

Klausner, S. D., Zwerman, P. J. and Ellis, D. F. 1974. Surface runoff losses of soluble nitrogen and phosphorus under two systems of soil management. *J. Environ. Qual.* 3(1):42–46.

Koehler, F. A., Humenik, F. J., Johnson, D. D., Kreglow, J. M., Dressing, S. A. and Maas, R. P. 1982a. Best Management Practices for Agricultural Nonpoint Surce Control. II. Commercial fertilizer. USDA Coop. Agree. 12-05-300-472. EPA Interagency Agree. AD-12-F-0-037-0. North Carolina Agric. Ext. Serv., Raleigh, NC, 55 pp.

Koehler, F. A., Humenik, F. J., Johnson, D. D., Kreglow, J. M., Dressing, S. A. and Maas, R. P. 1982b. Best Management Practices for Agricultural Nonpoint Source Control. III. Sediment. USDA Coop. Agree. 12-05-300-472. EPA Interagency Agree. AD-12-F-0-037-0. North Carolina Agric. Ext. Serv., Raleigh, NC, 49 pp.

Koehler, F. A., Humenik, F. J., Johnson, D. D., Kreglow, J. M., Dressing, S. A. and Maas, R. P. 1982c. Best Management Practices for Agricultural Nonpoint Source Control. I. Animal waste. USDA Coop. Agree. 12-05-300-472. EPA Interagency Agree. AD-12-F-0-037-0. North Carolina Agric. Ext. Serv., Raleigh, NC, 67 pp.

Kramer, J. R., Herbes, S. E. and Allen, H. E. 1972. Phosphorus: analysis of water, biomass, and sediment. in Allen, H. E. and Kramer, J. R. (Eds.). *Nutrients in Waters.* John Wiley & Sons, New York, pp. 51–100.

Kunishi, H. M., Taylor, A. W., Heald, W. R., Gbrrek, W. J. and Weaver, R. N. 1972. Phosphate movement from an agricultural watershed during two rainfall periods. *J. Agric. Food Chem.* 20(4):900–905.

Ladd, J. N. and Jackson, R. B. 1982. Biochemistry of ammonification. in Stevenson, F. J. (Ed.). *Nitrogen in Agricultural Soils*. Agron. Monogr. 22:173–228. Am. Soc. Agron., Crop Sci. Soc. Am., Soil Sci. Soc. Am., Madison, WI.

Laflen, J. M. and Tabatabai, M. A. 1984. Nitrogen and phosphorus losses from corn-soybean rotations as affected by tillage practices. *Trans. ASAE* 27(1):58–63.

Lamb, J. A., Peterson, G. A. and Fenster, C. R. 1985. Fallow nitrate accumulation in a wheat-fallow rotation as affected by tillage system. *Soil Sci. Soc. Am. J.* 49(6):1441–1446.

Lamond, R. E. and Moyer, J. L. 1983. Effects of knifed vs. broadcast fertilizer placement on yield and nutrient uptake by tall fescue. *Soil Sci. Soc. Am. J.* 47(1):145–149.

Landschoot, P. J. and Waddington, D. V. 1987. Response of turfgrass to various nitrogen sources. *Soil Sci. Soc. Am. J.* 51(1):225–230.

Landua, D. P., Swoboda, A. R. and Thomas, G. W. 1973. Response of Coastal bermudagrass to soil applied sulfur, magnesium, and potassium. *Agron. J.* 65:541–544.

Langdale, G. W., Leonard, R. A., Fleming, W. G. and Jackson, W. A. 1979. Nitrogen and chloride movement in small upland piedmont watersheds: II. Nitrogen and chloride transport in runoff. *J. Environ. Qual.* 8(1):57–63.

Latterell, J. J., Dowdy, R. H., Clapp, C. E., Larson, W. E. and Linden, D. R. 1982. Distribution of phosphorus in soils irrigated with municipal waste-water effluent: A 5-year study. *J. Environ. Qual.* 11(1):124–128.

Lee, G. F. 1973. Role of phosphorus in eutrophication and diffuse source control. in Jenkins, S. H. and Ives, K. J. (Eds.). *Phosphorus in Fresh Water and the Marine Environment*. Pergamon Press, Oxford, pp. 111–128.

Legg, J. O. and Meisinger, J. J. 1982. Soil nitrogen budgets. in Stevenson, F. J. (Ed.). *Nitrogen in Agricultural Soils*. Agron. Monogr. 22:503–566. Am. Soc. Agron., Crop Sci. Soc. Am., Soil Sci. Soc. Am., Madison, WI.

Lightner, J. W., Mengel, D. B. and Rhykerd, C. L. 1990. Ammonia volatilization from nitrogen fertilizer surface applied to orchardgrass sod. *Soil Sci. Soc. Am. J.* 54(6):1478–1482.

Linderman, C. L., Mielke, L. N. and Schuman, G. E. 1976. Deep percolation in a furrow-irrigated sandy soil. *Trans. ASAE* 19:250–258.

Loehr, R. C. 1974. Characteristics and comparative magnitude of nonpoint sources. *J. Water Pollut. Control Fed.* 46:1849–1970.

Lorenz, O. A. and Vittum, M. T. 1980. Phosphorus nutrition of vegetable crops and sugar beets. in Khasawneh, F. E. (Ed. chair.). *The Role of Phosphorus in Agriculture*. Am. Soc. Agron., Crop Sci. Soc. Am., Soil Sci. Soc. Am., Madison, WI, pp. 737–762.

Lowrance, R. and Smittle, D. 1988. Nitrogen cycling in a multiple-crop vegetable production system. *J. Environ. Qual.* 17(1):158–162.

MacGregor, J. M., Blake, R. G. and Evans, S. D. 1974. Mineral nitrogen and pH of tiled and untiled soils following continued annual nitrogen fertilization for corn. *Soil Sci. Soc. Am. Proc.* 38(1):110–113.

Mancino, C. F., Torello, W. A. and Wehner, D. J. 1988. Denitrification losses from Kentucky bluegrass sod. *Agron. J.* 80(1):148–153.

Mansell, R. S., Fiskell, J. G. A., Calvert, D. V. and Rogers, J. S. 1986. Distributions of labeled nitrogen in the profile of a fertilized sandy soil. *Soil Sci.* 141(2):120–126.

Mansell, R. S., McKenna, P. J., Flaig, E. and Hall, M. 1985. Phosphate movement in columns of sandy soil from a wastewater-irrigated site. *Soil Sci.* 140(1):59–68.

McDowell, L. L. and McGregor, K. C. 1980. Nitrogen and phosphorus losses in runoff from no-till soybeans. *Trans. ASAE* 23:643– 648.

Meinhold, V. H., Duble, R. L., Weaver, R. W., and Holt, E. C. 1973. Thatch accumualtion in bermudagrass turf in relation to management. *Agron. J.* 65:833–835.

Meisinger, J. J. 1976. Nitrogen application rates consistent with environmental constraints for potatoes on Long Island. *Search Agric.* 6(7):1–19.

Menzel, R. G. 1980. Enrichment ratios for water quality modeling. in Knisel, W. G. (Ed.). CREAMS: A Field Scale Model for Chemicals, Runoff and Erosion from Agricultural Management Systems. USDA Conservation Research Report No. 26. Washington, D.C.

Menzel, R. G., Rhoades, E. D., Olness, A. E. and Smith, S. J. 1978. Variability of annual nutrient and sediment discharges in runoff from Oklahoma cropland and rangeland. *J. Environ. Qual.* 7(3):401–406.

Messer, J. and Brezonik, P. L. 1983. Agricultural nitrogen model: A tool for regional environmental management. *Environ. Manage.* 7:177–187.

Miller, M. H. 1979. Contribution of nitrogen and phosphorus to subsurface drainage water from intensively cropped mineral and organic soils in Ontario. *J. Environ. Qual.* 8(1): 42–48.

Miller, W. W., Guitjens, J. C. and Mahannah, C. N. 1984. Water quality irrigation and surface return flows from flood-irrigated pasture and alfalfa hay. *J. Environ. Qual.* 13(4):543–548.

Mitchell, W. H., Morehart, A. L., Cotnoir, L. J., Hesseltine, B. B. and Langston, D. N., III 1978. Effect of soil mixtures and irrigation methods on leaching of N in golf greens. *Agron. J.* 70(1):29–35.

Moe, P. G., Mannering, J. V. and Johnson, C. B. 1967. Loss of fertilizer nitrogen in surface runoff water. *Soil Sci.* 104(6):389–394.

Moe, P. G., Mannering, J. V. and Johnson, C. B. 1968. A comparison of nitrogen losses from urea and ammonium nitrate in surface runoff water. *Soil Sci.* 105(6):428–433.

Morton, T. G., Gold, A. J. and Sullivan, W. M. 1988. Influence of overwatering and fertilization on nitrogen losses from home lawns. *J. Environ. Qual.* 17(1):124–130.

Mosdell, D. K. and Schmidt, R. E. 1985. Temperature and Irrigation Influences on Nitrate Losses of *Poa pratensis* L. Turf. in Proc. Fifth Int. Turf. Res. Conf. Angers, France, pp. 487–494.

Mueller, D. H., Wendt, R. C. and Daniel, T. C. 1984a. Phosphorus losses as affected by tillage and manure applications. *Soil Sci. Soc. Am. J.* 48(4):901–905.

Mueller, D. H., Wendt, R. C. and Daniel, T. C. 1984b. Soil and water losses as affected by tillage and manure application. *Soil Sci. Soc. Am. J.* 48(4):896–900.

Muir, J., Boyce, J. S., Seim, E. C., Mosher, P. N., Deibert, E. J. and Olson, R. A. 1976. Influence of crop management practices on nutrient movement below the root zone in Nebraska soils. *J. Environ. Qual.* 5(3):255–259.

Muir, J., Seim, E. C. and Olson, R. A. 1973. A study of factors influencing the nitrogen and phosphorus contents of Nebraska waters. *J. Environ. Qual.* 2(4):466–470.

Munson, R. D. 1982. Soil Fertility, fertilizers, and plant nutrition. in Kilmer, V. J. (Ed.). *Handbook of Soils and Climate in Agriculture.* CRC Press, Boca Raton, FL, pp. 269–293.

Murray, J. J. and Powell, J. B. 1979. Turf. in Buckner, R. C. and Bush, L. P. (Eds.). *Tall Fescue.* Agron. Monogr. 20:293–306. Am. Soc. Agron., Crop Sci. Soc. Am., Soil Sci. Soc. Am., Madison, WI.

National Academy of Sciences. 1969. *Eutrophication: Causes, Consequences, and Correctives.* National Academy Sciences, Washington, D.C.

National Research Council. 1978. *Nitrates: An Environmental Assessment.* National Academy of Sciences, Washington, D.C.

Nelson, D. W. 1982. Gaseous losses of nitrogen other than through dentrification. in Stevenson, F. J. (Ed.). *Nitrogen in Agricultural Soils*. Agron. Monogr. 22:327–363. Am. Soc. Agron., Crop Sci. Soc. Am., Soil Sci. Soc. Am., Madison, WI.

Nelson, K. E., Turgeon, A. J., and Street, J. R. 1980. Thatch influence on mobility and transformation of nitrogen carriers applied to turf. *Agron. J.* 72(3):487–492.

Nelson, L. R., Keisling, T. C. and Rouquette, F. M. 1983. Potassium rates and sources for coastal bermudagrass. *Soil Sci. Am. J.* 47(5):963–966.

Nelson, W. W. and MacGregor, J. M. 1973. Twelve years of continuous corn fertilization with ammonium nitrate or urea nitrogen. *Soil Sci. Soc. Am. Proc.* 37:583–586.

Nicholaichuk, W. and Read, D. W. L. 1978. Nutrient runoff from fertilized and unfertilized fields in western Canada. *J. Environ. Qual.* 7(4):542–544.

Nielsen, D. R., Bigger, J. W. and Wierenga, P. J. 1982. Nitrogen transport processes in soil. in Stevenson, F. J. (Ed.). *Nitrogen in Agricultural Soils*. Agron. Monogr. 22:423–448. Am. Soc. Agron., Crop Sci. Soc. Am., Soil Sci. Soc. Am., Madison, WI.

Nielsen, D. R., Wierenga, P. J. and Bigger, J. W. 1983. Spatial soil variability and mass transfers from agricultural soils. in Nelsen, D. W., Tanji, K. K. and Elrick, D. E. (Eds.). *Chemical Mobility and Reactivity in Soil Systems*. Soil Sci Soc. Am. Special Publ. No. 11. Madison, WI, pp. 65–78.

Olness, A., Rhoades, E. D., Smith, S. J. and Menzel. R. G. 1980. Fertilizer nutrient losses from rangeland in central Oklahoma. *J. Environ. Qual.* 9(1):81–86.

Olness, A., Smith, S. J., Rhoades, E. D. and Menzel, R. G. 1975. Nutrient and sediment discharge from agricultural watersheds in Oklahoma. *J. Environ. Qual.* 4(3):331–336.

Olsen, S. R. and Khasawneh, F. E. 1980. Use and limitations of physical-chemical criteria for assessing the status of phosphorus in soils. in Khasawneh, F. E. (Ed. chair.). *The Role of Phosphorus in Agriculture*. Am. Soc. Agron., Crop Sci. Soc. Am., Soil Sci. Soc. Am., Madison, WI, pp. 361–410.

Olson, R. A. and Kurtz, L. T. 1982. Crop nitrogen requirements, utilization, and fertilization. in Stevenson, F. J. (Ed.). *Nitrogen in Agricultural Soils*. Agron. Monogr. 22:567–604. Am. Soc. Agron., Crop Sci. Soc. Am., Soil Sci. Soc. Am., Madison, WI.

Olson, R. V. 1982. Immobilization, nitrification, and losses of fall-applied labeled ammonium-nitrogen during growth of winter wheat. *Agron. J.* 74:991–995.

Olson, R. V. and Swallow, C. W. 1984. Fate of labeled nitrogen fertilizer applied to winter wheat for five years. *Soil Sci. Soc. Am. J.* 48(3):583–586.

Onken, A. B., Wendt. C. W., Wilke, O. C., Hargrove, R. S., Bausch, W. and Barnes, L. 1979. Irrigation system effects on applied fertilizer nitrogen movement in soil. *Soil Sci. Soc. Am. J.* 43:367–372.

Owens, L. B. 1987. Nitrogen leaching losses from monolith lysimeters as influenced by nitapyrin. *J. Environ. Qual.* 16(1):34–38.

Owens, L. B., Van Keuren, R. W. and Edwards, W. W. 1985. Groundwater quality changes resulting from a bromide application to a pasture. *J. Environ. Qual.* 14(4): 543–548.

Ozanne, P. G. 1980. Phosphate nutrition of plants — A general treatise. in Khasawneh, F. E. (Ed. chair.). *The Role of Phosphorus in Agriculture*. Am Soc. Agron., Crop Sci. Soc. Am., Soil Sci. Soc. Am., Madison, WI, pp. 559–589.

Peaslee, D. E. and Phillips, R. E. 1981. Phosphorus dissolution-desorption in relation to bioavailability and environmental pollution. in Stelly, M. (Ed. in chief). *Chemistry in the Soil Environment*. ASA Spec. Publ. #40. Am. Soc. Agron. and Soil Sci. Soc. Am., Madison, WI, pp. 241–259.

Petrovic, A. M., Hummel, N. W. and Carroll, M. J. 1986. Nitrogen source effects on nitrate leaching from late fall nitrogen applied to turfgrass. *Agron. Abstr.* p. 137.

Petrovic, A. M. 1990. The fate of nitrogenous fertilizers applied to turfgrass. *J. Environ. Qual.* 19(1):1–14.

Phillips, D. A. and DeJong T. M. 1984. Dinitrogen fixation in leguminous crop plants. in Hauck, R. D. (Ed.). *Nitrogen in Crop Production.* Am. Soc. Agron., Crop Sci. Soc. Am., Soil Sci. Soc. Am., Madison, WI, pp. 121–132.

Pratt, P. F. (Ed.). 1978. Proceedings, National Conference on Management of Nitrogen in Irrigated Agriculture. Department of Soil and Environmental Sciences, University of California, Riverside, 442 pp.

Pratt, P. F. 1984. Nitrogen use and nitrate leaching in irrigated agriculture. in Hauck, R. D. (Ed.). *Nitrogen in Crop Production.* Am. Soc. Agron., Crop Sci. Soc. Am., Soil Sci. Soc. Am., Madison, WI, pp. 319–333.

Pratt, P. F. (Chairman). 1985. Agriculture and Groundwater Quality. Council for Agricultural Science and Technology. CAST Report No. 103. 62 pp.

Pratt, P. F., Jones, W. W. and Hunsaker, V. E. 1972. Nitrate in deep soil profiles in relation to fertilizer rates and leaching volume. *J. Environ. Qual.* 1(1):97–102.

Randall, G. W. 1984. Efficiency of fertilizer nitrogen use as related to application methods. in Hauck, R. D. (Ed.). *Nitrogen in Crop Production.* Am. Soc. Agron., Crop Sci. Soc. Am., Soil Sci. Soc. Am., Madison, WI, pp. 521–533.

Randall, G. W. 1988. Effective nitrogen management: Considerations for growers. in Agricultural chemicals and groundwater protection: Emerging management and policy. Proceedings of a Conference, Oct. 22–23, 1987. Freshwater Foundation, St. Paul, MN, pp. 117–123.

Reddy, G. Y., McLean, E. O., Hoyt, G. D. and Logan, T. J. 1978. Effects of soil, cover crop, and nutrient source on amounts and forms of phosphorus movement under simulated rainfall conditions. *J. Environ. Qual.* 7(1):50–54.

Reddy, K. R., Khaleel, R., Overcash, M. R. and Westerman, P. W. 1979. Phosphorus — A potential nonpoint source pollution problem in the land areas receiving long-term animal wastes. in *Best Management Practices for Agriculture and Silviculture.* Ann Arbor Science, Ann Arbor, MI, pp. 193–211.

Rieke, R. E. and Ellis, B. G. 1974. Effects of Nitrogen Fertilization on Nitrate Movements Under Turfgrass. in Proc. 2nd Int. Turfgrass Res. Conf., pp. 121–130.

Romkens, M. J. M. and Nelson, D. W. 1974. Phosphorus relationships in runoff from fertilized soils. *J. Environ. Qual.* 3(1):10–13.

Romkens, M. J. M., Nelson, D. W. and Mannering, J. V. 1973. Nitrogen and phosphorus composition of surface runoff as affected by tillage method. *J. Environ. Qual.* 2(2):292–295.

Russel, D. A. 1984. Conventional nitrogen fertilizers. in Hauck, R. D. (Ed.). *Nitrogen in Crop Production.* Am. Soc. Agron., Crop Sci. Soc. Am., Soil Sci. Soc. Am., Madison, WI, pp. 183–194.

Saffigna, P. G. and Keeney, D. R. 1977a. Nitrate and chlorine in ground water under irrigated agriculture in central Wisconsin. *Ground Water* 15:170–177.

Saffigna, P. G. and Keeney, D. R. 1977b. Nitrogen and chloride uptake by irrigated Russet Burbank potatoes. *Agron. J.* 69:258–264.

Saffigna, P. G., Keeney, D. R. and Tanner, C. B. 1977. Nitrogen, chlorine, and water balance with irrigated Russet Burbank potatoes in a sandy soil. *Agron. J.* 69:251–257.

Sample, E. C., Soper, R. J. and Racz, G. J. 1980. Reactions of phosphate fertilizer in soils. in Khasawneh, F. E. (Ed. chair.). *The Role of Phosphorus in Agriculture.* Am. Soc. Agron., Crop Sci. Soc. Am., Soil Sci. Soc. Am., Madison, WI, pp. 263–310.

Schmidt, E. L. 1982. Nitrification in soil. in Stevenson, F. J. (Ed.). *Nitrogen in Agricultural Soils*. Agron. Monogr. 22:253–288. Am. Soc. Agron., Crop Sci. Soc. Am., Soil Sci. Soc. Am., Madison, WI.

Schuler, T. R. 1987. Controlling Urban Runoff: A Practical Manual for Planning and Designing Urban BMPs. Dept. of Environ. Programs. Washington Metropolitan Water Resources Planning Board, Washington, D.C.

Schuman, G. E., Burwell, R. E., Piest, R. F. and Spomer, R. G. 1973a. Nitrogen losses in surface runoff from agricultural watersheds on Missouri Valley loess. *J. Environ. Qual.* 2(2):229–302.

Schuman, G. E., McCalla, T. M., Saxton, K. E. and Knox, H. T. 1975. Nitrate movement and its distribution in the soil profile of differentially fertilized corn watersheds. *Soil Sci. Soc. Am. Proc.* 39(6):1192–1197.

Schuman, G. E., Spomer, R. G. and Piest, R. F. 1973b. Phosphorus losses from four agricultural watersheds on Missouri Valley loess. *Soil Sci. Soc. Am. Proc.* 37:424–427.

Schwab, G. O., Frevert, R. K., Edminster, T. W. and Barnes K. K. 1981. *Soil and Water Conservation Engineering*. John Wiley & Sons, New York, 525 pp.

Schwab, G. O., McLean, E. O., Waldron, A. C., White, R. K. amd Michener, D. W. 1973. Quality of drainage water from a heavy-textured soil. *Trans. ASAE* 16:1104–1107.

Schwab, G. O., Fausey, N. R. and Kopcak, D. E. 1980. Sediment and chemical content of agricultural drainage water. *Trans. ASAE* 23(6):1446–1449.

Sharpley, A. N. 1980. The enrichment of soil phosphorus in runoff sediments. *J. Environ. Qual.* 9(3):521–526.

Sharpley, A. N. 1985a. Depth of surface soil-runoff interaction as affected by rainfall, soil slope and management. *Soil Sci. Soc. Am. J.* 49(4):1010–1015.

Sharpley, A. N., 1985b. The selective erosion of plant nutrients in runoff. *Soil Sci. Soc. Am. J.* 49(6):1527–1534.

Sharpley, A. N., Smith, S. J. and Menzel, R. G. 1982. Prediction of phosphorus losses in runoff from Southern Plains watersheds. *J. Environ. Qual.* 11(2):247–251.

Sharpley, A. N., Smith, S. J. and Noney, J. W. 1987. Environmental impact of agricultural nitrogen and phosphorus use. *J. Agric. Food Chem.* 35(5):812–817.

Sheard, R. W., Haw, M. A., Johnson, G. B. and Ferguson, J. A. 1985. Mineral Nutrition of Bentgrass on Sand Rooting Systems. in Proc. Fifth Int. Turf. Res. Conf. Angers, France, pp. 469–485.

Sheard, R. W. and Beauchamp, E. G. 1985. Aerodynamic Measurement of Ammonium Volatilization from Urea Applied to Bluegrass-Fescue Turf. in Lemaire, F. L. Proc. Fifth Int. Turf. Res. Conf. Angers, France, pp. 549–556.

Shearman, R. C. 1985. Turfgrass culture and water use. in Gibeault, V. A. and Cockerham, S. T. (Eds.). *Turfgrass Water Conservation*. University of California, Riverside, Division of Agriculture and Natural Resources, pp. 61–70.

Shearman, R. C. 1986. Kentucky bluegrass cultivar evapotranspiration rates. *HortScience* 21(3):455–457.

Shearman, R. C. and Beard, J. B. 1973. Environmental and cultural preconditioning effects on the water-use rate of *Agrostis palustris* Huds., cultivar Penncross. *Crop Sci.* 13: 424–427.

Smika, D. E. and Watts, D. G. 1978. Residual nitrate-N in fine sand as influenced by N fertilizer and water management practices. *Soil Sci. Soc. Am. J.* 42(6):923–926.

Smith, G. S. 1979. Nitrogen and aerification influence on putting green thatch and soil. *Agron. J.* 71:681–684.

Smith, S. J. and Power, J. F. 1985. Residual forms of fertilizer nitrogen in a grassland soil. *Soil Sci.* 140(5):362–367.

Smith, S. J., Dillow, D. W. and Young, L. B. 1982. Disposition of fertilizer nitrate applied to sorghum-sudangrass in the Southern Plains. *J. Environ. Qual.* 11(3):341–344.

Smolen, M. D., Humenik, F. J., Spooner, J., Dressing, S. A. and Maas, R. P. 1984. Best Management Practices for Agricultural Nonpoint Source Control. IV. Pesticides. USDA Coop. Agree. 12-05-300- 472. EPA Interagency Agree. AD-12-F-0-037-0. North Carolina Agricult. Ext. Serv., Raleigh, NC, 87 pp.

Snyder, G. H., Burt, E. O. and Davidson, J. M. 1981. Nitrogen leaching in Bermudagrass Turf: Effect of Nitrogen Sources and Rates. in Proc. Fourth Int. Turfgrass Res. Conf. Guelph, Ontario, Canada, pp. 185–193.

Snyder, G. H., Augustin, B. J. and Davidson, J. M. 1984. Moisture sensor-controlled irrigation for reducing N leaching in bermudagrass turf. *Agron. J.* 76(6):964–969.

Sommerfeldt, T. G. and Smith, A. D. 1973. Movement of nitrate nitrogen in some grassland soils of southern Alberta. *J. Environ. Qual.* 2:112–115.

Sommers, L. E. and Giordano, P. M. 1984. Use of nitrogen from agricultural, industrial, and municipal wastes. in Hauck, R. D. (Ed.). *Nitrogen in Crop Production.* Am. Soc. Agron., Crop Sci. Soc. Am., Soil Sci. Soc. Am., Madison, WI, pp. 208–220.

Stanford, G. and Legg, J. O. 1984. Nitrogen and yield potential. in Hauck, R. D. (Ed.). *Nitrogen in Crop Production.* Am. Soc. Agron., Crop Sci. Soc. Am., Soil Sci. Soc. Am., Madison, WI, pp. 263–272.

Starr, J. L. 1983. Assessing nitrogen movement in the field. in Nelsen, D. W., Elrick, D. E. and Tanji, K. K. (Eds.). *Chemical Mobility and Reactivity in Soil Systems.* Special Publ. No. 11. Soil Sci. Soc. Am., Madison, WI, pp. 79–92.

Starr, J. L. and DeRoo, H. C. 1981. The fate of nitrogen fertilizer applied to turfgrass. *Crop Sci.* 21(4):531–535.

Steenhuis, T. S. and Muck, R. E. 1988. Preferred movement of nonadsorbed chemicals on wet, shallow, sloping soils. *J. Environ. Qual.* 17(3):376–384.

Stevenson, C. K. and Baldwin, C. S. 1969. Effect of time and method of nitrogen application and source of nitrogen on the yield and nitrogen content of corn (*Zea mays* L.). *Agron. J.* 61:381–384.

Stevenson, F. J. 1982a. Origin and distribution of nitrogen in soil. in Stevenson, F. J. (Ed.). *Nitrogen in Agricultural Soils.* Agron. Monogr. 22:1–42. Amer. Soc. Agron., Crop Sci. Soc. Amer., Soil Sci. Soc. Amer., Madison, WI.

Stevenson, F. J. (Ed.). 1982b. *Nitrogen in Agricultural Soils.* Agron. Monogr. 22:1–940. Amer. Soc. Agron., Crop Sci. Soc. Amer., Soil Sci. Soc. Amer., Madison, WI.

Stewart, B. A., Woolhiser, D. A., Wischmeier, W. H., Caro, J. H. and Frere, M. H. 1975. Control of Water Pollution from Cropland: A Manual for Guideline Development. Vol. 1. EPA 600/2-75-026a. U.S. Environmental Protection Agency, Athens, GA and USDA Agricultural Research Service, Washington, D.C., 111 pp.

Street, J.R. 1988. New concepts in turf fertilization. *Landscape Manage.* 2:38–46.

Stuanes, A. O. 1984. Phosphorus sorption of soils to be used in wastewater renovation. *J. Environ. Qual.* 13(2):220–224.

Taylor, A. W. and Kilmer, V. J. 1980. Agricultural phosphorus in the environment. in Khasawneh, F. E. (Ed. chair.). *The Role of Phosphorus in Agriculture.* Am. Soc. Agron., Crop Sci. Soc. Am., Soil Sci. Soc. Am., Madison, WI, pp. 545–557.

Taylor, A. W. and Kunishi, H. M. 1971. Phosphate equilibria on stream sediment and soil in streams draining an agricultural region. *J. Agric. Food Chem.* 19:827–831.

Terman, G. L. 1987. Volatilization losses of nitrogen as ammonia from surface applied fertilizers, organic amendments, and crop residues. *Adv. Agron.* 31:189–223.

Thomas, G. W. 1970. Soil and climatic factors which affect nutrient mobility. in Englestad, O. P. (Ed.). *Nutrient Mobility in Soils: Accumulation and Losses.* Special Publ. No. 4. Soil Sci. Soc. Am., Madison, WI, pp. 1–20.

Thomas, G. W. and Barfield, B. J. 1974. The unreliability of tile effluent for monitoring subsurface nitrate-nitrogen losses from soils. *J. Environ. Qual.* 3(2):183–185.

Thomas, G. W., Blevins, R. L., Phillips, R. E. and McMahon, M. A. 1973. Effect of killed sod mulch on nitrate movement and corn yield. *Agron. J.* 65:736–739.

Tillotson, W. R. and Wagenet, R. J. 1982. Simulation of fertilizer nitrogen under cropped situations. *Soil Sci.* 133(3):133–143.

Timmons, D. R. and Dylla, A. S. 1981. Nitrogen leaching as influenced by nitrogen management and supplemental irrigation level. *J. Environ. Qual.* 10(3):421–426.

Timmons, D. R. and Holt, R. F. 1980. Soluble N and P concentrations in surface runoff water. in Knisel, W. G. (Ed.). CREAMS: A Field Scale Model for Chemicals, Runoff and Erosion from Agricultural Management Systems. USDA Conservation Research Report No. 26. Washington, D.C.

Timmons, D. R., Burwell, R. E. and Holt, R. F. 1973. Nitrogen and phosphorus losses in surface runoff from agricultural land as influenced by placement of broadcast fertilizer. *Water Resour. Res.* 9(3):658–667.

Timmons, D. R., Holt, R. F. and Latterwell, J. J. 1970. Leaching of crop residues as a source of nutrients in surface runoff water. *Water Resour. Res.* 6(5):1367–1375.

Titko, S., Street, J. R., and Logan, T. J. 1987. Volatiliztion of ammonia from granular and dissolved urea applied to turfgrass. *Agron. J.* 79(3):535–540.

Torello, W. A., Wehner, D. J., and Turgeon, A. J. 1983. Ammonia volatilization from fertilized turfgrass stands. *Agron. J.* 75(3):454–456.

Turner, T. R. and Waddington, D. V. 1983. Soil test calibration for establishment of turfgrass monostands. *Soil Sci. Soc. Am. J.* 47(6):1161–1166.

Tyler, D. D. and Thomas, G. W. 1977. Lysimeter measurements of nitrate and chloride losses from soil under conventional and no-tillage corn. *J. Enviran. Qual.* 6(1):63–66.

Varco, J. J. and Sartain, J. B. 1986. Effects of phosphorus, sulfur, calcium hydroxide, and pH on growth of annual bluegrass. *Soil Sci. Soc. Am. J.* 50(1):128–132.

Volk, G. M. 1959. Volatile loss of ammonia following surface applications of urea to turf or bare soil. *Agron. J.* 75:454–456.

Volk, R. G. and Loeppert, R. H. 1982. Soil organic matter. in Kilmer, V. J. *Handbook of Soils and Climate in Agriculture.* CRC Press, Boca Raton, FL, pp. 211–268.

Vollenweider, R. A. and Kerekes, J. 1980. The loading concept as basis for controlling eutrophication philosophy and preliminary results of the OECD programme on eutrophication. *Prog. Water. Tech.* 12:5–38.

Waddington, D. V., Turner, T. R., Duich, J. M. and Moberg, E. L. 1978. Effect of fertilization on penncross creeping bentgrass. *Soil Sci. Soc. Am. Proc.* 70:713–718.

Wagenet, R. J. 1983. Principles of salt movement in soils. in Nelsen, D. W., Elrick, D. E. and Tanji, K. K. (Eds.). *Chemical Mobility and Reactivity in Soil Systems.* Special Publ. No. 11. Soil Sci. Soc. Am., Madison, WI, pp. 123–140.

Wagenet, R. J. and Hutson, J. L. 1986. Predicting the fate of nonvolatile pesticides in the unsaturated zone. *J. Environ. Qual.* 15(4):315–322.

Wagenet, R. J. and Hutson, J. L. 1987. *LEACHM Leaching Estimation and Chemistry Model.* Continuum, Water Resources Institute, Vol 2. Center for Environ. Res. Cornell University, Ithaca, NY, 80 pp.

Watschke, T. L. and Mumma, R. O. 1989. The Effect of Nutrients and Pesticides Applied to Turf on the Quality of Runoff and Percolating Water. A Final Report to the U.S. Dept. Int., U.S. Geol. Sur. ER 8904. Env. Res. Res. Inst. Penn State University, University Park, PA, 64 pp.

Watts, D. G. and Martin, D. L. 1981. Effects of water and nitrogen management on nitrate leaching loss from sands. *Trans. ASAE* 24(4):911–916.

Welterlen, M. S., Gross, C. M., Angle, J. S. and Hill, R. L. 1989. Surface runoff from turf. in Leslie, A. R. and Metcalf, R. L. Integrated Pest Management for Turfgrass and Ornamentals. Office of Pesticide Programs, 1989-625-030. U.S. Environmental Protection Agency, Washington, D.C. pp. 153–160.

Wendt, R. C. and Corey, R. B. 1980. Phosphorus variations in surface runoff from agricultural lands as a function of land use. *J. Environ. Qual.* 9(1):130–136.

Wendt, R. C., Alberts, E. E. and Hjelmfelt, A. T., Jr. 1986. Variability of runoff and soil loss from fallow experimental plots. *Soil Sci. Soc. Am. J.* 50(3):730–736.

Westerman, P. W., King, L. D., Burns, J. C., Cummings, G. A. and Overcash, M. R. 1987. Swine manure and lagoon effluent applied to a temperate forage mixture: II. Rainfall runoff and soil chemical properties. *J. Environ. Qual.* 16(2):106–112.

Westerman, P. W., Overcash, M. R., Evans, R. O., King, L. D., Burns, J. C. and Cummings, G. A. 1985. Swine lagoon effluent applied to "Coastal" Bermudagrass: III. Irrigation and rainfall runoff. *J. Environ. Qual.* 14(1):22–25.

Whitaker, F. D., Heinemann, H. G. and Burwell, R. E. 1978. Fertilizing corn adequately with less nitrogen. *J. Soil Water Cons.* 33:28–32.

White, E. M. and Williamson, E. J. 1973. Plant nutrient concentrations in runoff from fertilized cultivated erosion plots and prairie in eastern South Dakota. *J. Environ. Qual* 2(4):453–455.

White, R. H. and Dickens, R. 1984. Thatch accumulation in bermudagrass as influenced by cultural practices. *Agron. J.* 76:19–22.

Wilkinson, S. R. and Mays, D. A. 1979. Mineral nutrition. in Buckner, R. C. and Bush, L. P. (Eds.). *Tall Fescue*. Agron. Monogr. 20:41–73. Am. Soc. Agron., Crop Sci. Soc. Am., Soil Sci. Soc. Am., Madison, WI.

Woodhouse, W. W. 1968. Long term fertility requirements for coastal bermudagrass, I. Potassium. *Agron. J.* 60:508–511.

Woodmansee, R. G., Dodd, J. L., Bowman, R. A., Clark, F. E. and Dickinson, C. E. 1978. Nitrogen budget of a shortgrass prairie ecosystem. *Oecologia* 34:363–376.

Zwerman, P. J., Bouldin, D. R., Greweling, T., Klausner, S. D., Lathwell, D. J. and Wilson, D. O. 1971. Management of Nutrients on Agricultural Land for Improved Water Quality. EPA-13020-DPB-08/71. U.S. Environmental Protection Agency and Cornell University, Ithaca, New York.

Zwerman, P. J., Greweling, T., Klausner, S. D. and Lathwell, D. J. 1972. Nitrogen and phosphorus content of water from tile drains at two levels of management and fertilization. *Soil Sci. Soc. Am. Proc.* 36:134–137.

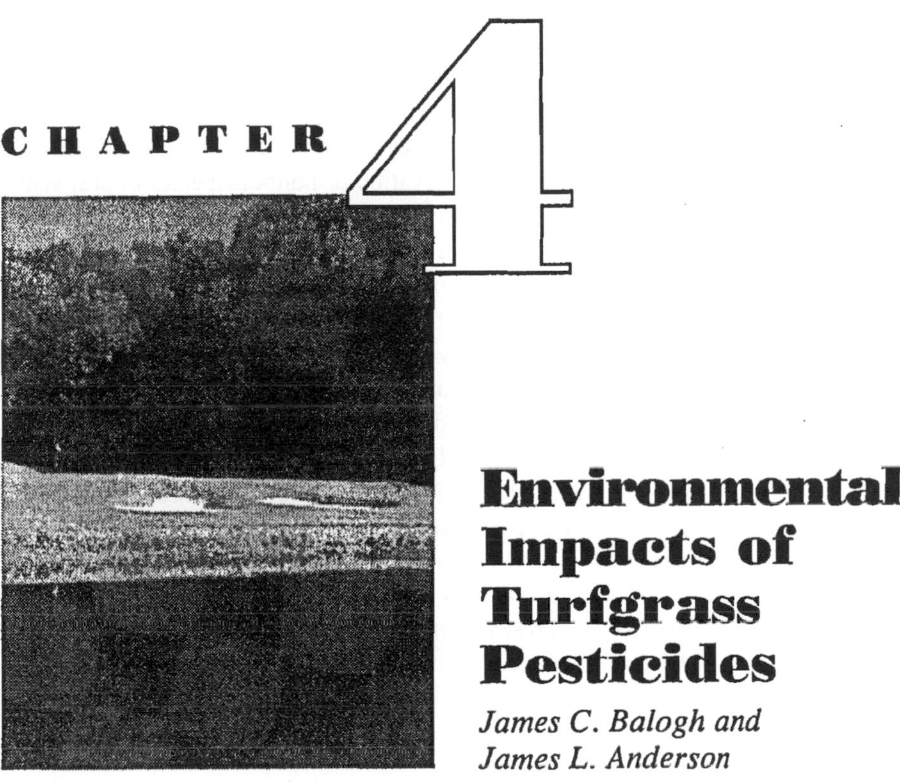

CHAPTER 4

Environmental Impacts of Turfgrass Pesticides

James C. Balogh and James L. Anderson

4.1 INTRODUCTION

Synthetic organic pesticides are potent agents used for the control of undesirable organisms. In the last 40 years, these chemicals have become a major component of agricultural, forestry, and turfgrass management systems (Balogh et al. 1990, 1992a; Cheng 1990a; Walker et al. 1990). Use of chemical control of turfgrass pests in conjunction with nutrient and cultural practices has had a tremendous effect on the function and quality of turfgrass grown for golf courses and lawns (Beard 1982). The public demand for turfgrass of high quality for home lawns and as a uniform playing surface on golf courses often requires intensive management to control pests. Chemical control of pests is only one of the techniques used on golf courses to promote sustained turfgrass quality and reduce labor and energy costs.

Despite the obvious cultural and economic benefits, conflicts have developed over pesticide use in relation to environmental quality issues (Balogh et al. 1992a). Pesticide residues have been associated with adverse environmental and potential human health effects including (1) reduction of certain predator bird populations; (2) appearance of detectable residues in aquatic ecosystems on a global scale; (3) implication of many pesticides as potential human carcinogens; (4) long-term

Photo courtesy of Michael French.

contamination of soils with persistent pesticides; (5) contamination of drinking water, surface water, and groundwater; (6) destruction of nontarget organisms (fish kills and beneficial soil organisms); (7) elevation of nonpest species to pest status; and (8) evolution of resistant insect strains (Madhun and Freed 1990; Metcalf 1980, 1989; Potter et al. 1989). Also, growing evidence indicates that small quantities of pesticide residues can migrate from agricultural and turfgrass areas to impact other lands, the atmosphere, and water (Canter 1987; EPA 1990; Kurtz 1990; Pratt 1985).

The purpose of this chapter is to review scientific and technical literature on the environmental effects of pesticides used on turfgrass. Understanding the objectives and potential effects of chemicals used for control of turfgrass pests is critical for formulation of rational strategies by turfgrass managers and for development of regulatory policy by local, state, and federal agencies. Subjects to be discussed in this review include (1) use of turfgrass pesticides, (2) insect resistance to insecticides, (3) environmental and health issues, (4) fate and persistence of turfgrass pesticides, (5) potential effects of turfgrass pesticides on water quality, (6) perspectives on management guidelines, and (7) research needs. The focus of the review will center on turfgrass pesticides. However, general information on the environmental impacts of pesticides from appropriate agricultural systems is used to supplement areas lacking for turfgrass systems.

4.2 USES OF PESTICIDES IN TURFGRASS MANAGEMENT

The importance of pesticides for protection of high-quality turfgrass on golf courses and lawns has been well-documented (Beard 1973, 1982; Britton 1969; Engel and Ilnicki 1969; Madison 1982). The factors controlling the quality of turfgrass, such as color, uniformity, and density, often are affected adversely by invasions of weeds, disease, and insects. Properly maintained and healthy turfgrass will tolerate the presence of low levels of pest populations without suffering permanent damage. Healthy turfgrass also will recover more rapidly from major pest or disease infestations (Beard 1982; Murray and Powell 1979). However, under conditions of massive insect, disease, or weed infestations of turfgrass, chemical pest control is often the only effective control option. This is especially true after cultural and alternate control options have not proven effective (Potter and Braman 1991).

In this chapter, and throughout this book, pesticide refers to manufactured organic compounds or biochemical control agents for use in turfgrass management. This includes all insecticides, fungicides, herbicides, nematicides, miticides, rodenticides, fumigants, and other organic chemicals used for pest control practices (Table 1). These compounds are used to control or reduce the adverse effects of pests including harmful insects, unwanted plants, and pathogenic microorganisms (Tables 2–7). The

Table 1. Summary of Major and Minor Use Pesticides Registered for Use for Turfgrass Management; Pesticides Assembled from Databases Compiled by the U. S. Environmental Protection Agency (EPA), Golf Course Superintendents Association of America (GCSAA), and Professional Lawn Care Association of America (PLCAA)

Chemical or common name	Typical[a] trade name(s)	CAS number	Pesticide[b] Category	Family	Major or minor use[c] Golf course	Lawn
Acephate	Orthene	30560-19-1	I	Organophosphate	No	Major
Bacillus thuringiensis	Dipel	68038-71-1	I	Selective microbial insecticide	—	—
Bacillus popillae	Doom, Grub Attack	—	I	Selective microbial insecticide	—	—
Bendiocarb	Dycarb, Turcam	22781-23-3	I	Carbamate	Major	Major
Carbaryl	Sevin	63-25-2	I	Carbamate	Major	Major
Chlorpyrifos	Dursban	2921-88-2	I	Organophosphate	Major	Major
DDVP, Dichlorvos		62-73-7	I	Organophosphate	No	Minor
Diazinon[d]	D.z.n diazinon	333-41-5	I, N	Organophosphate	No[d]	Major
Ethoprop	Mocap	13194-48-4	I, N	Organophosphate	Major	Minor
Fenamiphos	Nemacur	22224-92-6	N	Organophosphate	Major	No
Fluvalinate	Mavrik	69409-94-5 42509-80-8	I		No	Minor
Isazofos	Triumph	67329-04-8	I	Organophosphate	Major	Major
Isofenphos	Oftanol	25311-71-1	I	Organophosphate	Major	Major
Malathion		121-75-5	I	Organophosphate	No	Minor
Methoxychlor		72-43-5	I	Biphenyl chlorine	No	Minor
Propoxur	Baygon	114-26-1	I	Carbamate	—	Major
Trichlorfon	Proxol	52-68-6	I	Organophosphate	Major	Major

Table 1. Summary of Major and Minor Use Pesticides Registered for Use for Turfgrass Management; Pesticides Assembled from Databases Compiled by the U. S. Environmental Protection Agency (EPA), Golf Course Superintendents Association of America (GCSAA), and Professional Lawn Care Association of America (PLCAA) (continued)

Chemical or common name	Typical[a] trade name(s)	CAS number	Pesticide[b]		Major or minor use[c]	
			Category	Family	Golf course	Lawn
Anilazine	Dyrene	101-05-3	F	Triazine	Major	No
Benomyl	Tersan 1991 DF	17804-35-2	F	Carbamate	Major	Minor
Chloroneb	Terraneb SP	2675-77-6	F	Halobenzene	Major	No
Chlorothalonil	Daconil 2787	1897-45-6	F	Halobenzene	Major	Minor
Etridiozole (ethazole)	Terrazole 35%	2593-15-9	F		Major	No
Fenarimol	Rubigan AS	60168-88-9	F		Major	Minor
Fosetyl Al	Alliette	39148-24-8	F		Major	No
Iprodione	Chipco 26019	36734-19-7	F	Nitrogen heterocyclic	Major	Minor
Mancozeb	Dithane Fore	8018-01-7	F	Dithiocarbamate	Major	No
Maneb	Manzate 200 DF	12427-38-2	F	Dithiocarbamate	Major	Minor
Metalaxyl	Apron 25W Subdue 2E	57837-19-1	F		Major	No
PCNB (quintozene)	Terraclor 75% WP Turfcide 2 lb	82-68-8	F	Nitrobenzene, fumigant	Major	Minor
Propiconazole	Banner	60207-90-1	F	Triazole	Major	No
Propamocarb HCl	Banol Prevex	25606-41-1	F	Carbamate	Major	No
Sulfur		7704-34-9	F	Sulfur	No	Minor

Thiophanate-methyl	Fungo 50 Topsin M Turf Topsin M 4.5F	23564-05-8	F	Carbamate	Major	No
Thiram	Spotrete	137-26-8	F		Major	No
Triadimefon	Bayleton	43121-43-3	F	Fumigant	Major	Minor
Vinclozolin	Vorlan	50471-44-8	F		Major	No
Ziram		137-30-4	F	Dithiocarbamate	No	Minor
Atrazine	AAtrex	1912249	H	Triazine	Major	Major
Benefin (Benfluralin)	Balan	1861-40-1	H	Dinitroaniline	Major	Major
Bensulide	Betasan	741-58-2	H	Organophosphate	Major	Minor
Bentazon	Basagran	50723-80-3 25057-89-0	H		Minor	Minor
2,4-D 2,4-D Isoctylester 2,4-D Monobutylester 2,4-D Butoxyethanol ester Sodium salt of 2,4-D 2,4-D Dimethylamine salt	Many formulations	94-75-7 25168267 1320-18-9 1929733 2702729 2008-39-1	H	Chlorophenoxy	Major	Major
2,4-D + MCPP + Dicamba	Trimec many formulations	94-75-7 7085-19-0 2300-66-5	H	Chlorophenoxy	No	Major
DCPA	Dacthal	1861-32-1 1918-00-9 62610393	H	Aryl aliph acid	Major	Major
Dicamba	Banvel	2300-66-5	H	Aryl aliph acid	Major	Major

Table 1. Summary of Major and Minor Use Pesticides Registered for Use for Turfgrass Management; Pesticides Assembled from Databases Compiled by the U. S. Environmental Protection Agency (EPA), Golf Course Superintendents Association of America (GCSAA), and Professional Lawn Care Association of America (PLCAA) (continued)

Chemical or common name	Typical[a] trade name(s)	CAS number	Pesticide[b]		Major or minor use[c]	
			Category	Family	Golf course	Lawn
DSMA	Numerous names	144-21-8 129-67-9	H	Organic arsenical	Major	Minor
Endothall	Endothal	145-73-3	H	Aryl aliph acid	Major	Minor
Ethofumesate	Prograss	26225-79-6 66441-23-4 1330-20-7	H		Major	No
Fenoxaprop-ethyl	Acclaim	68-12-2 1071-83-6	H		No	Minor
Glyphosate	Roundup	38641940	H	Glyphosate	Major	Major
Isoxaben	Gallery	82558-50-7	H		Major	Minor
MCPA	Rhonox	94-74-6	H	Chlorophenoxy	Major	Major
MCPP (Mecoprop)		7085-19-0	H	Chlorophenoxy	Major	Major
Metribuzin	Sencor	21087-64-9	H	Triazine	No	Minor
MSMA	Daconate 6	2163-80-6 6484522	H	Organic arsenical	Major	Major
Oxadiazon	Ronstar	19666-30-9	H		Major	Major
Pendimethalin	Prowl	40487-42-1	H	Dinitroaniline	Major	Major
Pronamide, propyzamide	Kerb	23950-58-5	H	Amide herbicide	Major	Minor

Siduron	Tupersan	1982-49-6	H	Urea derivative	Major	Minor
Simazine	Princep	122-34-9	H	Triazine	Major	Major
Triclopyr	Turflon Amine	5721369	H		Major	Major
	Turflon Ester	69633-04-1				
Trifluralin	Treflan	1582-09-8	H	Dinitroaniline	Major	No

From Leslie 1991.

[a] *Use of trade names is for the purpose of illustration and should not be construed as a product endorsement or condemnation. All chemicals used for pest control should be applied in strict compliance with local, state, and federal regulations. This is not a comprehensive list of all pesticides registered for use on turfgrass. Registration and restriction of pesticide use changes constantly, and indication or no indication of restricted use status is intended as a guideline. Prior to the use of any pesticide, the applicator must comply with all current regulations.*

[b] *F = fungicide, G = growth regulator-retardant, H = herbicide, I = insecticide, N = nematicide.*

[c] *Major and minor use based on U.S. EPA, GCSAA, and PLCAA databases (Leslie 1991).*

[d] *Not registered for use on golf courses.*

Table 2. Projected Total Use of Pesticides in 1982 for Golf Courses Compared to Agricultural Usage

Pesticide category	Estimated pounds of active ingredient (a.i.) (millions of pounds)	
	Golf courses	Farm use
Herbicides	4.57	485.3
Fungicides	4.57	6.6
Insecticides	3.18	62.3
Total	12.32	554.2

From Kriner 1985 and Smolen et al. 1984.

Table 3. Twenty of the Most Heavily Used Pesticides on Golf Courses in the United States in 1982

Chemical name	Estimated total pounds of active ingredients
Chlorothalonil	1,298,581
MCPP (mecoprop)	1,096,157
MSMA	834,830
Iprodione	815,694
Thiram	635,185
Diazinon	512,112
Benomyl	500,912
Paraffin oil	487,427
Dimethylamine 2,4-D	462,006
Pentachlorophenol	456,909
EDB (ethylene dibromide)	402,067
DCPA	400,016
Maneb	376,763
Isofenphos	374,718
Dicamba	297,262
Mancozeb	218,083
Trichlorfon	215,975
Bensulide	210,510
Aromatic petroleum derivative/solvent	190,516
Xylene/Solvent	152,703

Table 4. Estimated Regional Use of Pesticides in 1982 for Golf Courses

States	Estimated pounds of active ingredient (in millions)			
	Herbicides	Fungicides	Insecticides	Total
CT, MA, ME, NH, RI, VT	0.284	0.122	0.049	0.455
NJ, NY, PR, VI	0.439	0.643	0.241	1.323
DE, MD, PA, VA, WV, DC	1.249	0.462	0.229	1.940
AL, FL, GA, KY, MS, NC, SC, TN	1.035	0.395	1.976	3.406
IL, IN, MI, MN, OH, WI	0.899	2.640	0.288	3.827
AR, LA, NM, OK, TX	0.247	0.070	0.129	0.446
IA, KS, MO, NE	0.261	0.134	0.080	0.475
CO, MT, ND, SD, UT, WY	0.053	0.023	0.020	0.096
AZ, CA, HI, NV, AS, GU, TT	0.067	0.052	0.165	0.284
AK, ID, OR, WA	0.036	0.034	0.005	0.075

From Kriner 1985.

Note: PR = Puerto Rico; DC = Washington, D.C.; AS = American Samoa; GU = Guam; TT = Federated States of Micronesia Marshall Islands Palau.

Table 5. Common Herbicides, Mode of Action, and Common Target Weeds

Herbicide	Mode of action	Typical target weed species
Atrazine	Systemic, selective, pre- and postemergent	Selective control of several broadleaf and grassy weeds
Benefin	Selective, preemergent	Annual bluegrass, bamyardgrass, crabgrass, goosegrass, and foxtail
Bensulide	Selective, preemergent	Annual bluegrass, bamyardgrass, crabgrass, fall panicum, and foxtail
2,4-D	Systemic, selective, postemergent	Selective control of many broadleaf weeds
DCPA	Selective, preemergent	Annual bluegrass, bamyardgrass, crabgrass, foxtail, goosegrass, carpetweed, purslane, spurge, common chickweed, and field sandbur
Dicamba	Systemic, selective, postemergent	Selective control of many broadleaf weeds
DSMA	Systemic, selective, postemergent	Bahiagrass, barnyardgrass, crabgrass, dallisgrass, and field sandbur
Endothall	Systemic, selective, postemergent	Annual bluegrass, rescuegrass and selective control of several broadleaf weeds

Table 5. Common Herbicides, Mode of Action, and Common Target Weeds (continued)

Herbicide	Mode of Action	Typical target weed species
Glyphosate	Systemic, nonselective, postemergent	Nonselective control of perennial grassy weeds: bentgrass, Bermuda grass, kikuyugrass, nimblewill, orchardgrass, quackgrass, tall fescue, torpedograss, velvetgrass
MCPA	Systemic, selective, postemergent	Selective control of many broadleaf weeds
MCPP (mecoprop)	Systemic, selective, postemergent	Selective control of many broadleaf weeds
MSMA	Systemic, selective, postemergent	Bahiagrass, barnyardgrass, crabgrass, dallisgrass, and field sandbur
Oxadiazon	Selective preemergent	Annual bluegrass, barnyardgrass, crabgrass, fall panicum, green foxtail, and goosegrass
Pendimethalin	Selective, preemergent	Many annual grassy weeds
Pronamide	Systemic, selective, pre- and postemergent	Annual bluegrass and certain broadleaf weeds
Siduron	Selective, preemergent	Crabgrass, foxtail, and barnyardgrass
Simazine	Systemic, selective, preemergent	Selective control of several broadleaf weeds
Triclopyr	Systemic, selective, postemergent	Selective control of many broadleaf and woody weeds
Trifluralin	Selective, preemergent	Annual bluegrass, barnyardgrass, signalgrass, bromegrass, cheat, crabgrasses, fall panicum, goosegrass, johnsongrass (seedlings and rhizomes), and selective control of several broadleaf weeds

From Beard 1982; CPCR 1991; Farm Chemicals Handbook 1991; T&OCR 1991.

Table 6. Common Insecticides Used to Control Insects, Mites, and Nematodes on Golf Courses

Insecticide	Controlled pest insect
Bendiocarb	Chinch bug, mole cricket, and selected species of white grubs
Carbaryl	Ants, billbug, chinch bug, cutworm, leaf-hoppers, sod webworm, mole cricket[a]; and marginal control of fall armyworm and white grubs
Chlorpyrifos	Ants, armyworm, billbug, chinch bug, fall armyworm, cutworm, mole cricket, sod web-worm, turfgrass weevil, and selected species of white grubs
Diazinon[b]	Ants, armyworm, Bermuda grass mite and scale, billbug, chinch bug, cutworm, fall armyworm, frit fly, leafhoppers, mole cricket, rhodegrass mealybug, sod webworm, turfgrass weevil,[c] and white grubs, nematodes
Ethoprop	Chinch bug, mole cricket, sod webworm, selected species of white grubs, and nematodes
Isazofos[c]	Sod webworms, white grubs, and mole crickets
Isofenphos[d]	Chinch bugs, sod webworms, and white grubs
Trichlorfon	Armyworm, cutworm, fall armyworm, mole cricket, sod webworm, and white grubs

From Beard 1982; CPCR 1991; Farm Chemicals Handbook 1991; T&OCR 1991.

[a] *Bait.*
[b] *Not labeled for use on golf courses, but is labeled for control of pests on home lawns.*
[c] *Not for use on golf course fairways.*
[d] *Labeled for use on turfgrass only in selected states (CPCR 1991).*

Table 7. Common Fungicides Used to Control Diseases of Turfgrass on Golf Courses

Fungicide	Controlled disease
Anilazine	Brown patch,[a] copper spot, dollar spot, *Drechslera* and *Bipolaris* diseases, rusts, and typhula blight
Benomyl[b]	Brown patch, dollar spot, *Fusarium* blight and patch, and smuts
Chloroneb	Brown patch, dollar spot, *Pythium* blight, and typhula blight
Chlorothalonil	Brown patch, copper spot, dollar spot, *Fusarium* patch, gray leaf spot, *Drechslera* and *Bipolaris* diseases, red thread, and rusts
Etridiozole	*Pythium* and *Phytophthora* blight and damping-off
Iprodione[b]	Brown patch, dollar spot, *Fusarium* blight and patch,[c] *Drechslera* and *Bipolaris* diseases, and typhula blight
Mancozeb	Brown patch, copper spot, dollar spot,[a] *Fusarium* patch and blight,[a] *Drechslera* and *Bipolaris* diseases, Pythium blight,[a] red thread,[a] rusts, slime mold, and *Fusarium* snow mold
Maneb	Brown patch, dollar spot,[a] *Drechslera* and *Bipolaris* diseases, and rusts
Metalaxyl	*Pythium* blight
PCNB	Brown patch, dollar spot, *Fusarium* patch, *Drechslera* and *Bipolaris* diseases, rusts, smuts, and typhula blight
Propamocarb[b]	*Pythium* blight[c]
Thiophanatemethyl[b]	Brown patch, dollar spot, *Fusarium* blight, and smut
Thiram	Brown patch,[a] dollar spot, *Fusarium* patch, gray leafspot, and snow molds
Thiophanate-methyl + mancozeb	Anthracnose, brown patch, copper spot, dollar spot, *Fusarium* patch, *Drechslera* and *Bipolaris* diseases, red thread, and rusts
Triadimefon[b]	Brown patch, copper spot, dollar spot, *Fusarium* blight and patch, powdery mildew, rusts, and typhula blight
Vinclozolin	Dollar spot, *Drechslera* and *Bipolaris* diseases, red thread, pink patch, and *Fusarium* patch

From Beard 1982; CPCR 1991; Farm Chemicals Handbook 1991; T&OCR 1991.

[a] *Use is labeled, but control is marginal.*

[b] *Systemic fungicide: disease tolerance to the fungicide may develop. Other fungicides are nonsystemic or contact fungicides.*

[c] *Labeled for use on turfgrass in selected states.*

efficacy of these techniques has been extensively reviewed (App and Kerr 1969; Beard 1973, 1982; Britton 1969; Chapman 1979; Engel and Ilnicki 1969; Heald and Perry 1969; Leistra and Green 1990; Madison 1982; Murray and Powell 1979; Shearman 1987). Biological control practices, cultural practices, and management of environmental stress in relation to pest and disease control are discussed in Chapter 5, Section 5.4, "Alternate Methods of Pest and Disease Control" (Balogh et al. 1992b).

Several attempts have been made to document general pesticide use, chemical properties, formulations, application techniques, target organisms, trade chemical names, regional and federal regulation, toxicity, and labeling requirements. Sources include the Turf & Ornamental Chemicals Reference (T&OCR 1991), *Crop Protection Chemicals Reference* (CPCR 1991), *Farm Chemicals Handbook* (1991), and *Herbicide Handbook of the Weed Science Society of America* (Humburg et al. 1989). The mode of action and environmental fate of pesticides have been reviewed by Anderson et al. (1989), Ashton and Crafts (1981), Bovey and Young (1980), Cheng (1990b), Grossbard and Atkinson (1985), Grover (1988), Grover and Cessna (1991), Kearney and Kaufman (1988), Pimentel (1981), and Sawhney and Brown (1989).

Understanding the quantity, type, and pattern of pesticide use is important in order to establish integrated practices for efficient pest control and maintenance of environmental quality (Anderson et al. 1989; Leistra and Green 1990). The use of pesticides in general increased significantly between 1970 through 1986 (EPA 1987). Although overall herbicide and insecticide use has reached a plateau in the United States, use of specific chemicals is still increasing (Anderson et al. 1989). *However, quantitative data on current national, regional, or even local pesticide use on golf courses and residential lawns is not generally available.*

The most recent survey of pesticide use on golf courses was conducted in 1982 by the American Association of Retired Persons (AARP) (Kriner 1985). Although the information in the AARP report is now out of date, the report gives a general indication of the extent of pesticide use on golf courses in the United States. The AARP reported that 126 different active ingredients were used for turfgrass maintenance on golf course facilities. Total estimated use in 1982 for herbicides, fungicides, and insecticides was 4.6, 4.6, and 3.2 million lb of active ingredients, respectively (Table 2). Total use of pesticides for turfgrass management is significantly less than pesticide use for crop production (Table 2). The 12 most heavily used pesticides on golf courses according to the AARP report are provided in Table 3. However, information on when, how, and where these pesticides were used is insufficient to evaluate their potential environmental effects. On a regional basis, the heaviest use of pesticides on golf courses is in the north central, south Atlantic, and east south central states (Table 4). These states also have the highest proportion of golf facilities (see also Table 5, Balogh et al. 1992a). The heaviest use of fungicides in 1982 occurred in the north central states (Table 4). Insecticide use was heaviest in the south eastern states (Table 4). Herbicide use was more evenly distributed among the states and appears to be proportional to the number of golf course facilities within a region (Table 4) (see also Table 5, Balogh et al. 1992a).

A wide range of chemicals are used to provide control of pest and disease infestations on golf courses (Table 1). In qualitative reviews of pesticides used for weed control on golf courses, atrazine, benefin, bensulide, bentazon, 2,4-D, DCPA, dicamba, DSMA, endothall, glyphosate, MCPA, MCPP, siduron, and triclopyr are common turfgrass herbicides (Table 5) (Beard 1982; Dernoeden 1984; Freeborg 1984; Leslie 1991; Shearman 1987). Bendiocarb, carbaryl, chlorpyrifos, ethoprop, fenamiphos, isazofos, isofenphos, and trichlorfon are commonly used insecticides on golf course turfgrass (Table 6) (Beard 1982; *Farm Chemicals Handbook* 1991; Leslie 1991). Selection and application rate of a herbicide or insecticide is dependent on the specific pest, turfgrass type and condition, climate and temperature, soil conditions, management practices, and pesticide chemical/physical properties. Appropriate selection of the chemical formulation and timing of chemical application in relation to temperature, precipitation, irrigation, and pest lifestage is critical for effective suppression of turfgrass pests (Niemczyk 1987; Reinert 1976; Tashiro and Kuhr 1978; Villani et al. 1988; Vittum 1985).

Chemical options for insect control are limited compared to weed control with herbicides (Table 6) (CPCR 1991). Most labeled insecticides have relatively short persistence, which may require frequent application for chronic insect problems (Beard 1982). Most insecticides are more toxic to humans, animals, and nontarget organisms, when compared to herbicides. Proper application rates and techniques are necessary to avoid unnecessary risk of exposure. A few of the typical insect pests of golf course turfgrass include white grubs, sod webworms, chinch bugs, billbugs, cutworms, turfgrass weevils, and mole crickets (Table 6).

Although many nematodes are beneficial for turfgrass and soil (Potter et al. 1989), several nematodes feed on turfgrass and are considered parasites (Dernoeden 1989). Turfgrass nematodes are parasitic microscopic animals (Heald and Perry 1969). They inhabit the soil as eggs, larvae, and adults. Adult nematodes are transparent, eel shaped, and range in length from 0.3 to 2.5 mm. In the later stages of life, some adult nematodes, such as root-knot nematodes, become swollen to pear or kidney shape (Heald and Perry 1969). With proper turfgrass management, the turfgrass will tolerate fairly high populations of nematodes without reducing turfgrass quality (Beard 1982).

Typically parasitic nematodes are associated with a complex of factors resulting from disease, pest, environmental, and nutrient stress. Consistent and clearly defined relationships between nematode activity and turfgrass damage have been difficult to document (Dernoeden 1989; Vargas 1981). In the past, many cases of assumed nematode problems were a result of misdiagnosis, and damage was due to some other cause. Stress conditions in turfgrass can be exacerbated by the presence of parasitic nematodes. Nematicides are used on golf course turfgrass when visual confirmation of a very high nematode population has been made. Currently, insecticides, such as ethoprop, are used as nematicides (Table 1). Selection of specific nematicides is dependent on specific label approval. Appropriate handling and careful application of nematicides, insecticides, and fungicides is extremely important, as many of these compounds are highly toxic in comparison to turfgrass herbicides (*Farm Chemicals Handbook* 1991; Murphy 1992).

Turfgrass diseases are a significant problem on golf courses, even under good management conditions (Smiley 1983). Close mowing, high rates of nutrient application, intense irrigation and the moist condition of the turfgrass canopy, and constant injury to the vegetation from traffic and divoting are conditions that favor the occurrence of infectious diseases (Britton 1969). Disease problems result from infestations of turfgrass with fungi, viruses, and bacteria. Fungal infections are the most serious pathogens in turfgrass (Beard 1982; Chapman 1979; Smiley 1983). Disease problems are usually most severe on putting greens composed of bentgrass or annual bluegrass in contrast to bermudagrass (Beard 1982). Tees are also subject to similar levels of disease infestation (Beard 1982). Fairway turfgrass has intermediate disease problems. Diseases tend to be more severe on fairways composed of bentgrass and/or annual bluegrass compared to fairways composed of Kentucky bluegrass, zoysiagrass, and bermudagrass (Beard 1982; Smiley 1983). The intensity of diseases is greater on cool-season grasses than on warm-season grasses, especially in the humid climatic portions of the United States (Smiley 1983).

Several fungicides commonly used on golf courses include benomyl, chloroneb, chlorothalonil, etridiozole, iprodione, mancozeb, maneb, metalaxyl, PCNB (quintozene), thiophanate-methyl, and thiram (Table 7). Selection and use of specific fungicides are dependent on the specific pathogenic fungi and intensity of infection, climatic conditions, soil and thatch condition, turfgrass species and type, and intensity of management practices. Fungicides are considered to be most effective and can be applied at lower rates if diagnosis and treatment occurs prior to a virulent fungal infection (Beard 1982).

In recent years, preventative disease control has become a pervasive turfgrass management practice in certain parts of the United States (Smiley 1983). Preventative programs are based on the application of broad spectrum fungicides. Specific deterrence programs are tailored to season, climate conditions, and irrigation practices (Beard 1982). However, use of this type of management strategy for disease prevention is no longer considered an environmentally acceptable practice (Leslie 1989a). Integrated pest management techniques should be employed to provide early warning of turfgrass disease problems (Balogh et al. 1992b). With these techniques, the proper fungicide

Table 8. The Number of Documented Cases of Insect Pests with Multiple Resistance to Insecticide Groups

		Chemical Groups			
Year	Total number resistant	DDT cyclodienes	DDT cyclodienes organophosphorus	DDT cyclodienes organophosphorus carbamates	DDT cyclodienes organophosphorus carbamates pyrethroids
1938[a]	7	0	0	0	0
1948[a]	14	1	0	0	0
1954[b]	25	18	3	0	0
1970[c]	224	42	23	4	0
1975[d]	364	70	44	22	7
1980[e]	428	105	53	25	14
1984[f]	447	119	54	25	17

From Metcalf 1986.

[a] *Brown and Pal 1971.*
[b] *Metcalf 1955.*
[c] *Brown 1971.*
[d] *Georghiou and Taylor 1986.*
[e] *Georghiou 1981.*
[f] *Georghiou 1986.*

can be selected for control and the application can be limited to the affected area. This approach helps control the total environmental load of fungicides (Leslie 1989b; Miller et al. 1989).

4.3 DEVELOPMENT OF INSECT RESISTANCE

Acquiring resistance or tolerance to pesticides by insect pests, and to a lesser extent diseases and weeds, has been identified as a serious problem for agricultural, turfgrass, and urban pest management (Metcalf 1980, 1989; Reinert 1982a). Since the introduction of DDT as an effective broad spectrum insecticide, the total number of documented cases of insecticide resistance has increased exponentially (Table 8). On a worldwide basis, pesticide resistance has been identified in over 440 insect and acarine pests (Georghiou 1980; Metcalf 1989; Tashiro 1982a). The phenomenon and history of increasing pest resistance to chemical control strategies has been reviewed by Forgash (1984), Georghiou (1980, 1984, 1986), and Metcalf (1989). Tashiro (1982a), Reinert (1982a), and Reinert and Niemczyk (1982) have reviewed the incidence of insect resistance to DDT, chlordane, dieldrin, and organophosphate and carbamate insecticides in turfgrass.

Development of pesticide resistance is an example of accelerated microevolution. The step-wise evolutionary process leading to chemical tolerance depends on the presence of resistant genes within the pest population. During initial exposure to chemicals, the latent period, resistant genes are segregated into surviving members of treated pest populations. Initial tolerance to a single chemical may be associated with other environmental factors that are favorable to pest survival (Forgash 1984). Following the initial development of pesticide tolerance, resistance to chemical control strategies develops as populations become resistant to both related and unrelated compounds. Although cross-resistance does occur between different classes of pesticides such as organophosphates and carbamates (Table 8), it does *not* necessarily occur in all cases nor is the development of cross-resistance always rapid.

Multiple and cross-resistance to pesticides often develops in populations after repeated failure of the initial compound used for pest control (Metcalf 1989). Cross-resistance of insects occurs within a family of chemically related compounds. Multiple resistance is defined as resistance of insects to several chemically unrelated compounds. Reinert (1982b) observed development of cross-resistance of southern chinch bugs in turfgrass to two organophosphate insecticides, chlorpyrifos and isofenphos. In a series of laboratory and field tests, Reinert and Niemczyk (1982) observed cross-resistance of several epigeal insect pests to organophosphates in Florida. Reinert and Portier (1983) reported development of multiple resistance to pesticides by southern chinch bugs. In addition to developing resistance to an organophosphate, chlorpyrifos, these insects were also tolerant to propoxur, a carbamate. Ng and Ahmad (1979) observed multiple resistance to both chlopyrifos, an organophosphate, and bendiocarb, a carbamate, in a population of dieldrin-resistant Japanese beetles. Reinert (1982a) and Tashiro (1982a) reviewed several additional studies of turfgrass epigeal and soil inhabiting insects with multiple and cross-resistance to pesticides.

The time frame for development of resistance depends on many genetic, biological, and operational factors (Table 9) (Forgash 1984; Metcalf 1989). Factors that determine the onset and extent of resistance to pesticides are (1) the frequency and dominance of resistance genes in the target pest; (2) the level of previous chemical selection pressures; (3) the frequency of generation turnover within a pest population; (4) chemical properties such as persistence in the environment, chemical relationship to previously applied compounds, and the mode of activity; (5) the extent and frequency of application; and (6) timing of application in relation to pest life stage (Forgash 1984; Metcalf 1989).

The effect of exponentially increasing costs for development of insecticides, the rapidly decreasing rate of commercialization of new chemicals, and the development of cross- and multiple resistance to new classes of chemicals have made insect resistance to insecticides one of the most demanding problems facing entomologists and turfgrass managers (Metcalf 1980, 1989). The usual response to chemical tolerance is to increase the frequency, rate, and number of compounds used to control the pest

Table 9. Factors Influencing the Natural Selection of Insects Resistant to Insecticides

Genetic	Biological	Management/operational
	Biotic	**Chemical properties**
Frequency of R alleles	Generation turnover	Structure/activity
Number of R alleles	Offspring per generation	Relationship to earlier used chemicals
Dominance of R alleles	Monogamy/polygomy/ parthenogenesis	Persistence of residues
Penetrance; interaction of R alleles		
Previous chemical selection	**Behavioral**	**Application**
Extent of integration of R genome with other fitness factors	Isolation/mobility/ migration	Application threshold
	Monophagy/polyphagy	Selection threshold
	Fortuitous survival/ refugia	Life stage(s) selected
		Method of application
		Space-limited selection
		Alternating selection

From Forgash 1984.

infestation (Metcalf 1980; Tashiro 1982a). It is precisely this response that increases selective pressures and accelerates development of resistant populations. Increased extent and frequency of insecticide applications used to combat pest resistance also increases the potential for adverse environmental impacts (Metcalf 1989). The research effort currently dedicated to the problem of insect resistance and alternate control strategies is surprisingly limited given the extent of insect resistance and potential for adverse economic impacts (Forgash 1984; Metcalf 1980, 1989).

There are fewer documented cases of resistance to insecticides by turfgrass pests compared to agricultural situations. However, turfgrass insects are no less susceptible to accelerated evolution and development of cross- and multiple resistant populations (Reinert 1982a; Tashiro 1982a). The European chafer, Japanese beetle, other scarabaeids, and chinch bugs all have demonstrated increasing resistance to commonly used turfgrass compounds, such as diazinon, chlorpyrifos, and bendiocarb (Dunbar and Beard 1975; Kuhr et al. 1972; Ng and Ahmad 1979; Niemczyk and Lawrence 1973; Reinert 1982b; Reinert and Niemczyk 1982; Reinert and Portier 1983; Tashiro et al. 1971). Carbamates and synthetic pyrethroids have been used for control of resistant insect populations in turfgrass (Reinert 1982b, 1983; Reinert and Niemczyk 1982; Reinert and Portier 1983); however, under current management strategies, the emergence of insect populations with tolerance and resistance to these newer compounds is inevitable (Metcalf 1989; Reinert 1982a; Tashiro 1982a).

Development of cross- and multiple resistance has greatly complicated the chemical control of individual insect pests. Insect resistance has jeopardized the success of highly structured insect and disease control programs in both agricultural and turfgrass management systems (Metcalf 1989; Tashiro 1982a). Many of the traditional insect control techniques used in agriculture, public health, and turfgrass management have attempted to suppress insect species to unrealistically low levels. Pest species exposed to insecticide levels designed for eradication will ultimately breed new populations capable of surviving periodic environment exposure to insecticides.

Strategies to avoid breeding resistant *insect* populations in turfgrass systems have been developed. Decreasing selection pressures and using nonchemical control methods may be the best strategies for reducing development of resistant *insect* populations (Georghiou and Taylor 1977; Metcalf 1980). Methods to mitigate selection pressures include (1) reducing the frequency of insecticide treatment by applying chemicals only when necessary; (2) reducing the extent of treatment; (3) avoiding use of insecticides with prolonged persistence, slow-release formulations, and certain tank mixtures; (4) reducing the use of residual treatment strategies; (5) avoiding treatments that apply selection pressure to larval and adult stages; (6) discontinuing use of specific compounds before their effectiveness fails completely; and (7) incorporating nonchemical methods into control programs (Leeper et al. 1986; Metcalf 1980, 1989). These are all well-documented components of integrated management strategies (Balogh et al. 1992b). Many of these guidelines also are components of systems designed to mitigate potential adverse effects of pest management on the environment and water quality.

Recently, use of nonchemical biocontrol practices in integrated pest management (IPM) programs has been advocated as a systematic replacement for chemical methods (e.g., Leslie and Metcalf 1989). Turfgrass and agricultural managers should recognize that the evolutionary processes involved in development of chemically resistant pest populations also operate in relation to use of biological control strategies. Introduction of competitive, predatory, or parasitic organisms to control turfgrass pests will accelerate microevolution of populations resistant to any new selection and breeding pressure (Waldrop 1990). An integrated system for turfgrass management should include judicious use of chemical and nonchemical control methods combined with a selection of pest and disease resistant turfgrass cultivars. This concept of management ultimately results in effective and environmentally sound control of pests and diseases in golf course turfgrass.

4.4 PESTICIDES, ENVIRONMENT, AND HEALTH ISSUES

Often emotional issues surrounding pest management are directly linked to the larger issue of pesticide use in both urban and rural environments and the potential health impacts. Since pesticides are toxic or biologically active by design, there is a natural concern regarding their effects on human health and environmental quality. Exposure of natural populations to agricultural, silvicultural, and turfgrass pesticides has been associated with nonpoint pollution and reported incidences of acute toxicity (Balogh et al. 1990; Madhun and Freed 1990; Stewart et al. 1975; Stone 1979; Stone and Knoch 1982). Since the 1960s there has been slowly accumulating evidence of adverse effects of certain pesticides on the environment and human health (Canter 1987; Madhun and Freed 1990). Toxicology of pesticides in laboratory animals and natural populations has been the subject of considerable research (see Tietge 1992; Murphy 1992). Pesticide toxicity information is available in the FDA Surveillance Index (FDA 1984), *Farm Chemicals Handbook* (1991), and various federal databases such as ACQUIRE (Murphy 1990) and the U.S. EPA's Integrated Risk Information System (IRIS).

From an environmental perspective, pesticide-induced effects on aquatic organisms have been well-documented (Brown 1978; Madhun and Freed 1990; Smolen et al. 1984). Several prominent cases involved acute aquatic exposure and subsequent fish kills during the 1960s in which insecticides were implicated as contributing factors (e.g., Caro 1976; Cottam 1965; Hunt and Linn 1970; Mount and Putnicki 1966). A fish kill on the Mississippi River was linked to endrin contamination and a second occurred on the Rhine River in Germany associated with endosulfan contamination (Brown 1978; Mount and Putnicki 1966). These and several other kills were attributed to high pesticide concentrations from point sources rather than nonpoint sources. However, in 1969 agricultural runoff of DDT was partially responsible for contamination of commercial salmon catches in Lake Michigan (Johnson and Ball 1972).

Herbicides are often considered less toxic to aquatic organisms than insecticides. However, several commonly used herbicides are acutely toxic to fish including 2,4-D (ester formulation) and trifluralin (Murphy 1992). Algal photosynthesis is reduced by the presence of many herbicides at levels well below acute or sublethal levels (Caro 1976). Herbicidal action on aquatic plants may cause excessive algae growth with the release of nutrients from decaying vegetation (Boyle 1980). Total impact on the aquatic environment due to low levels of pesticides is just now being addressed by researchers.

Unless pesticide concentrations are very high, the response of aquatic organisms is difficult to predict. In general, nonpoint source concentration of pesticides in surface waters and groundwater are well below acute toxicity levels for aquatic and terrestrial species (Anderson et al. 1989; Murphy 1992). However, the issue of chronic exposure to low doses of pesticides is less clear. Sublethal doses, chronic exposure, and indirect effects on aquatic organisms resulting from pesticide contamination has been recognized, despite difficulties in direct measurement. Smolen et al. (1984) identified over 130 studies of sublethal effects of pesticides on aquatic organisms. Impacts varied depending on the type of pesticide and organism. The issue of acute and chronic toxicity for aquatic organisms and wildlife is discussed by Madhun and Freed (1990) and Tietge (1992).

Understanding and evaluating the human health effects of pesticides in drinking water is a critical goal for both state and federal regulatory agencies (EPA 1986, 1987). Human health effects of pesticides are classified as acute and chronic. Acute effects are the result of exposure to high levels of a chemical over short periods of time. Chronic effects occur as a result of long-term exposure to low concentrations of chemicals. Acute effects, including nausea and skin rashes, occur rapidly and are usually easier to diagnose. Chronic effects are difficult to identify and document due to long intervals between exposure and symptoms.

Human health concerns, especially for groundwater contamination, have focused on acute effects, since these are easier to document. Health effects from chronic pesticide exposure are currently considered equivocal (Pratt 1985). Despite difficulties in evaluating studies with a low level of statistical validation, some survey and epidemiological studies have linked long-term pesticide exposure to several potential chronic effects including cancer, mutations, birth defects, and immunological changes (Albert 1987; Cantor et al. 1988; Fiore 1987). *Although conclusive evidence of health effects of long-term exposure to pesticides has yet to be established, there is intense public perception of risk concerning pesticides in drinking water.*

Generally, pesticides in surface water and groundwater used for drinking water have been found at low levels (e.g., Anderson et al. 1989). The U.S. Environmental Protection Agency (EPA) recently completed a 5-year national survey of chemical contamination of drinking water wells (EPA 1990). The EPA estimates that 10.4% of community water systems and 4.2% of rural domestic wells are contaminated with at least 1 of the 126 pesticides included in the analysis. Although DCPA metabolites, atrazine, or simazine were detected in fewer than 7% of the water system wells and fewer than 3% of rural domestic wells, these compounds are associated with both agricultural and turfgrass management (Tables 10 and 11) and are cause for concern.

The EPA is currently working to establish drinking water standards or reference doses for surface water and groundwater (EPA 1986, 1987; Kimm and Barles 1988). Regulatory decisions used by the EPA are based on a combination of separate assessments for both hazard and risk of exposure (Severn and Ballard 1990). Hazard assessments are based on the inherent toxicity of the pesticide. Risk of exposure is based on an evaluation of the potential effects of pesticides on humans or wildlife based on likely levels of exposure. Standards will be based on the same toxicological research used to establish reference doses (formerly called Acceptable Daily Intake, or ADI) for food. The reference dose (RfD) is a benchmark dose derived from studies that determined no observed adverse effect level (NOAEL). These standards will be the maximum contaminant levels (MCLs) allowed for pesticide concentrations in drinking waters and food. RfD levels will be used as benchmarks for evaluation of relative risk of exposure to pesticides. Recommended MCLs have been established

Table 10. Estimated Number and Percent of Community Water System Wells Containing Nitrates or Pesticides Based on U. S. EPA National Pesticide Survey

Chemical	Estimated[a] number	Estimated[a] percent	Minimum reporting limit
Nitrate	49,300	52.1	0.15 mg L^{-1}
DCPA acid metabolites	6,010	6.4	0.10 µg L^{-1}
Atrazine	1,570	1.7	0.12 µg L^{-1}
Simazine	1,080	1.1	0.38 µg L^{-1}
Prometon	520	0.5	0.15 µg L^{-1}
Hexachlorobenzene	470	0.5	0.06 µg L^{-1}
DBCP	370	0.4	0.01 µg L^{-1}
Dinoseb	25	<0.1	1.3 µg L^{-1}

From EPA 1990.

[a] *Estimates based on a nationwide sample of 1300 community water wells and rural domestic wells representing a total of 94,600 drinking water wells at 38,800 community water systems.*

for 2,4-D at 70 ppb, atrazine at 3 ppb, alachlor at 2 ppb, and aldicarb at 9 ppb (van der Leeden et al. 1990). Health advisory levels (HALs) have been proposed by the National Research Council (1977). HALs are calculated from a no observed effect level (NOEL) in pesticide feeding trials, average amount of water consumed, and a 100- to 1000-fold safety factor. Proposed MCLs and HALs are currently not enforceable. Establishment of water and environmental quality standards will balance risk and benefits associated with pesticide use and public perception. Further information on the EPA risk/benefit assessment for pesticides is presented by Severn and Ballard (1990).

In addition to federal efforts to alleviate environmental quality concerns, state governments are also developing water quality regulations (Fairchild 1987; Morandi 1988). State governments recognize the need to protect valuable surface and groundwater resources through both education and enforcement. Some states, such as California, New York, Nebraska, and Wisconsin, have selected a regulatory approach to water quality issues (Morandi 1988). Others are legislating a combined approach of education, research, and demonstration (Johnson 1988).

A basic understanding of the factors influencing the fate and transport of pesticides in the soil environment is necessary to understanding the potential impacts on target and nontarget organisms (Cheng 1990b). Many of the concerns regarding pesticide use are associated with the persistence and movement of pesticides in soil. The behavior of pesticides in the soil environment ultimately controls the potential for

Table 11. Estimated Number and Percent of Rural Domestic Wells Containing Nitrates or Pesticides Based on U. S. EPA National Pesticide Survey

Chemical	Estimated[a] number	Estimated[a] percent	Minimum reporting limit
Nitrate	5,990,000	57.0	0.15 mg L^{-1}
DCPA acid metabolites	264,000	2.5	0.10 μg L^{-1}
Atrazine	70,800	0.7	0.12 μg L^{-1}
DBCP	38,400	0.4	0.01 μg L^{-1}
Prometon	25,600	0.2	0.15 μg L^{-1}
Simazine	25,100	0.2	0.38 μg L^{-1}
Ethylene dibromide	19,200	0.2	0.01 μg L^{-1}
Lindane	13,100	0.1	0.043 μg L^{-1}
Ethylene thiourea	8,470	0.1	4.5 μg L^{-1}
Bentazon	7,160	0.1	0.25 μg L^{-1}
Alachlor	3,140	<0.1	0.50 μg L^{-1}

From EPA 1990.

[a] *Estimates based on a nationwide sample of 1300 community water wells and rural domestic wells representing a total of 10.5 million rural domestic wells in the United States.*

chemical exposure to both target and nontarget organisms. Many of the specific factors and effects of pesticides on the environmental, nontarget organisms, soil, surface water, and groundwater are not fully understood. However, research is beginning to elucidate the general relationship between pesticide management, environmental impacts, and effects on water quality. Movement of pesticides out of the intended target zone has been a critical environmental concern. The primary mechanisms for movement of pesticides out of the intended target zone are direct application by dumping, improper storage, and spills; drift, volatilization, and deposition on surface water; overland flow and erosion losses; and subsurface flow to both surface water and groundwater. The relative importance of each process depends on site-specific conditions and the characteristics of the pesticide itself. Pesticide volatility, mobility, and persistence are the primary factors that determine potential exposure of nontarget and beneficial organisms to pesticide residues. Combining the risk of exposure with pesticide toxicity determines the hazard of the presence of a chemical in the environment. The following sections will emphasize the processes and factors affecting the environmental fate of pesticides in the turfgrass environment.

4.5 FATE AND PERSISTENCE OF PESTICIDES IN TURFGRASS

Pesticides begin to disperse from the target area immediately after application by the action of air, water, and degradation. Entry of pesticides into the environment and partitioning of pesticides to the atmosphere, soil, groundwater, and surface water are determined by innumerable interacting factors and conditions (Figures 1 and 2) (Himel et al. 1990; Jury et al. 1987a). Some of the critical factors may be classified by their impact on four dominant processes: (1) volatilization, (2) adsorption, (3) decomposition, and (4) water transport (Table 12) (Jury et al. 1987a).

Several excellent general reviews on the fate of pesticide residues have been presented by Anderson et al. (1989), Cheng (1990b), Grover (1988), Grover and Cessna (1991), Howard (1991), and Jury et al. (1987a). Although the focus of this and the following sections is on turfgrass systems, examples of fate and persistence of pesticides in other systems are used to highlight principles and potential trends in turfgrass systems. Although turfgrass and turfgrass soils have unique characteristics, many of the processes and factors affecting the fate of pesticides in agricultural systems are similar to the turfgrass environment. Published research on the fate and transport of agrichemicals in the turfgrass environment is limited compared to agricultural or silvicultural systems (Leonard 1990; Walker et al. 1990). Examples from systems with limited surface soil disturbance including grassland soils, forage production systems, and reduced and no-till systems may supply the best surrogate information on the environmental fate of pesticides when direct information on turfgrass soils is restricted.

4.5.1 Initial Distribution

The initial distribution of pesticide applied to turfgrass is extremely important (Figures 1 and 2). The initial distribution of the applied chemical ultimately determines the amount of pesticide reaching the intended target and the amount of pesticide that will be lost from the turfgrass ecosystem before reaching the intended target. A pesticide that is primarily distributed in the turfgrass canopy may undergo significant volatilization or photolysis. These initial losses may result in insufficient pesticide concentrations to affect the intended target pest. A pesticide that is initially distributed in the thatch layer may be strongly adsorbed. Sequestering a pesticide in the thatch may reduce its efficacy, possibly delay degradation, or enter complexes with water-soluble organic compounds which are susceptible to leaching losses.

Pesticides are introduced directly into the turfgrass environment in a liquid phase; as a spray solution or dispersion; or in a solid phase such as a powder dust,

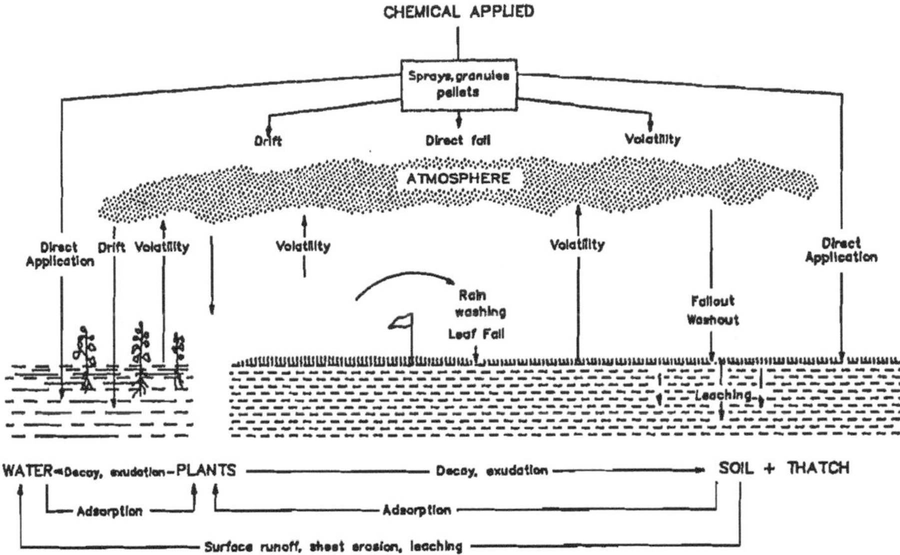

Fig. 1. The initial distribution and disposition of chemicals applied to turfgrass.

microcapsule, or granule (Beard 1982; Himel et al. 1990). The specific pesticide formulation may be applied directly to the established canopy or to the soil prior to turfgrass establishment. Pesticides may be incorporated into the soil by watering-in on established turfgrass or soil incorporated prior to establishment. Initial pesticide distribution and potential pathways for transport into or out of a turfgrass system is a function of the type and rate of pesticide applied; the method used to apply the pesticide; and the rate and timing of subsequent precipitation, watering, or irrigation. Several studies have been conducted on thc initial distribution and fate of pesticides applied to turfgrass.

Tashiro (1982b) studied the initial distribution of the insecticides chlorpyrifos and diazinon applied to Kentucky bluegrass. Approximately 50% of the applied insecticides were initially distributed in the turfgrass layer and 50% in the soil layer. Although slight effects were noted for different formulations of the insecticides, watering immediately after application had little effect on the initial distribution of pesticide.

Sears and Chapman (1979, 1982) studied the initial distribution of chlordane, chlorpyrifos, diazinon, and CGA 12223 (isazofos) when applied to annual bluegrass and watered immediately afterwards. Their results indicated that less than 10% of the applied insecticide reached the soil root zone with primary distribution occurring in the turfgrass-thatch zone. Movement of the insecticide from the thatch layer to the root zone was limited. Insecticide movement from the grass-thatch and primary root zones into the underlying 2.5 cm of soil was minimal. Less than 1% of the initial

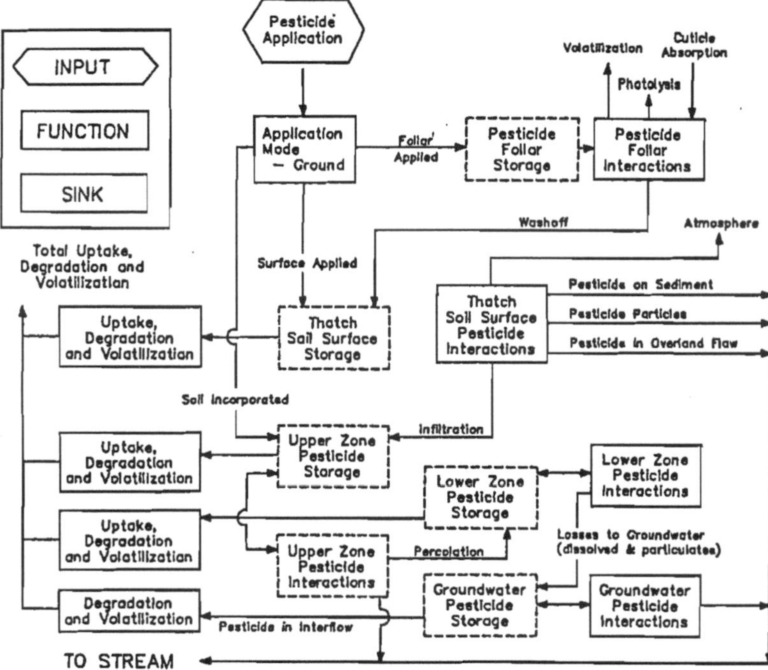

Fig. 2. Transformations, transport, and dissipation of pesticides in the turfgrass canopy, soil environment, and vadose zone. (Adapted from Himel et al. 1990).

Table 12. Behavior of Water and Pesticides in the Atmospheric and Soil Environments

Environmental zone	Water behavior	Pesticide behavior
Atmosphere	**Addition** by evaporation from land and water surfaces, transported by air movement, and sprinkler irrigation. **Loss** by precipitation, transport by air movement, condensation, and sprinkler irrigation.	**Addition** by application, volatilization from plants and soils, and transport by drift. **Loss** by deposition on plants and soils, transport by air movement, and decomposition.
Aboveground plant canopy	**Addition** by precipitation, sprinkler irrigation, and condensation. **Loss** by throughfall, evaporation, and transpiration.	**Addition** by application to foliage, uptake from soil, and by air movement. **Loss** by volatilization, throughfall, washoff, and decomposition by sunlight and plant metabolism.

Table 12. Behavior of Water and Pesticides in the Atmospheric and Soil Environments (continued)

Environmental zone	Water behavior	Pesticide behavior
Soil surface and thatch	**Addition** by precipitation, irrigation, infiltration, lateral surface movement, condensation, and upward movement from the root zone. **Loss** by evaporation, runoff, and leaching.	**Addition** by application to the soil and infiltration, transport by air movement, throughfall and washoff from plants, transport by lateral movement on the surface, and upward transport from the root zone. **Loss** by volatilization, runoff, leaching to zones beneath, and chemical and biological decomposition. **Retention** by the soil and thatch.
Root zone (unsaturated soil)	**Addition** by transport from the surface and from the root zone. **Loss** by root uptake and movement to the soil surface and leaching to zones beneath.	**Addition** by subsurface application, incorporation, and by transport from the soil surface and zones beneath. **Loss** by chemical and biological decomposition, by root uptake, and by transport to the soil surface and zones beneath. **Retention** by the soil.
Unsaturated zone below root zone	**Addition** by movement from the zones above and below. **Loss** by movement into the zones above and below.	**Addition** by transport from the zones above and beneath. **Loss** by chemical and biological decomposition and by transport into the zones above and beneath. **Retention** by the soil.
Saturated zone (groundwater)	**Addition** by movement from the zones above and by lateral and vertical movement within the saturated zone. **Loss** by movement into the zones above and by lateral and vertical movement within the saturated zone.	**Addition** by transport from the zones above and by lateral and vertical transport in groundwater. **Loss** by transport into the zones above, by lateral and vertical transport within the unsaturated zone, by discharge to bodies of surface water, and by chemical and biological decomposition. **Retention** by the soil.

concentration of chlordane and chlorpyrifos was detected in the soil after 56 d and less than 1% of the CGA 12223 (isazofos) and diazinon was present in the soil layer 14 d after treatment. Discrepancies between these two studies (Sears and Chapman 1979; Tashiro 1982b) may be attributed to differences in method of application, formulation, or amount and duration of watering.

Kuhr and Tashiro (1978) conducted a study to determine the distribution of chlorpyrifos and diazinon to established turfgrass in watered and unwatered systems when applied to bluegrass in a fine sandy loam. They found diazinon residues higher than chlorpyrifos in both soil and turfgrass. Watering-in had a minor effect on the distribution of chlorpyrifos between foliage and soil. Diazinon appeared to move more rapidly from grass to soil when subjected to water treatment. However, residue from either insecticide was found only at extremely low concentrations below the 1.3 cm level.

Goh et al. (1986) applied a formulated mixture of chlorpyrifos and dichlorvos to irrigated and nonirrigated plots of Kentucky bluegrass. They found that residual chlorpyrifos on turfgrass foliage was initially low. The initial residue concentrations decreased significantly 48 hr after application, especially in the irrigated plots. Immediately after application, residues of dichlorvos on foliage exceeded the safe level of 0.06 mg cm^{-2}. The level dropped below this level in the next 23 hr. No significant difference was noted between irrigated and nonirrigated plots.

Nash and Beall (1980) reported on the fate of 2,4-D in four compartments of microagroecosystems (air, grass, soil, and water) containing Kentucky bluegrass. The concentration of the herbicide in the surface 2 cm of soil was much greater than in the subsurface soil or foliage immediately after application.

After surface application, Sears et al. (1987) observed fairly high sorption of diazinon, chlorpyrifos, and isofenphos by the vegetative canopy of Kentucky bluegrass. Dislodgeable residues declined from 10% of the total immediately after application to 0.3% within 1 d. A considerable portion of the applied insecticides was absorbed to the waxy surface of the grass blades. Liquid formulations were lost more rapidly from the vegetative surfaces than granular formulations. Within 24 hr of application, dislodgeable residues on the surface of the turfgrass leaves were well below mammalian toxicity levels.

In addition to chemical properties and application methods, several other factors influence the final initial concentration of pesticides available for intended targets in turfgrass canopy and soils. These factors include volatilization and drift, photochemical degradation, chemical and biological transformations, foliar washoff and drip, sorption, plant uptake, runoff, and leaching (Tables 12 and 13, Figure 2) (Leistra and Green 1990; Pimentel and Levitan 1986). After initial application, the major sources of pesticides for offsite movement include the turfgrass canopy, thatch, clippings, the soil surface, foliar washoff, dissolved in runoff, and leachate (Table 13). During the short-term establishment phase of turfgrass and during site construction, there may be some losses of pesticides bound to eroded soil, granules, dusts, powders, and microcapsules (Table 13). In a review of initial pesticide losses after application, Pimentel and Levitan (1986) observed that often less than 0.1% of applied pesticides reach the intended target pest. The rest of the applied pesticide is then available for movement and transformation throughout the soil, water, and atmospheric environment. Management strategies need to be used which maximize the amount of pesticide reaching the intended pest, while reducing the overall application. Many of these processes also account for the long term environmental dissipation or loss of applied pesticides (Jury et al. 1987a). These processes are discussed in the following sections.

4.5.2 Drift and Volatilization

Volatilization and air transport are the primary means of regional and global dispersion of pesticides in the environment (Bidleman and Olney 1975; Glotfelty et al. 1990; Levy 1990; Muir et al. 1990; Winer and Atkinson 1990). In general, atmospheric losses of applied pesticides result from (1) vaporization and particulate drift during application; (2) dust from the site of application, especially during turfgrass establishment after construction or renovation; and (3) volatilization from soil, turfgrass leaf surfaces, and water (Table 14). These processes have been extensively reviewed by Grover (1991), Himel et al. (1990), Jury (1986a), Spencer and Cliath (1990), Taylor and Glotfelty (1988), and Taylor and Spencer (1990). The following is a brief discussion of atmospheric losses due to drift and volatilization.

Drift is the airborne loss of pesticide from the intended foliar or soil target. Pesticide applications, until recently, have relied primarily on spray techniques. The active pesticide ingredient in the form of an emulsion, suspension, or solution is injected through the narrow orifice of a nozzle (Grover 1991). This fluid sheet disperses toward the intended target as a large number of droplets of varying sizes (Himel et al. 1990). The objective of spray operations in turfgrass management is to deliver the pesticide to the intended foliar or soil target. However, not all of the pesticide spray necessarily reaches the intended target (Table 15). There are two undesirable effects of spray application techniques, short- and long-distance airborne transport. A fraction of the pesticide spray is deposited on surfaces a short distance downwind of the target as a result of spray drift. Drift losses of pesticides constitutes a potential economic, legal, and environmental hazard (Grover 1991). Long-distance transport of small droplets, usually less than 50 μm in diameter, can result in rapid atmospheric dispersal of pesticide sprays. These small droplets are redeposited over long and variable periods of time. Airborne transport and subsequent deposition of

Table 13. Initial Sources and Potential Environmental Exposure Pathways for Pesticides Applied to Turfgrass

Chemical source	Transport process	Environmental or exposure pathway	Nontarget populations at risk to exposure
Canopy	Dislodgeable and volatilization	Inhalation/skin contact	Humans, animals
Thatch	Volatilization	Inhalation/skin contact	Humans, animals
Soil surface	Volatilization	Inhalation/skin contact	Humans, animals
Root zone and vadose zone	Volatilization	Inhalation/skin contact	Humans, animals
Clippings	Dislodgeable, feeding, and volatilization	Inhalation/skin contact feeding	Humans, animals
Thatch, soil and runoff	Overland flow	Surface water/ingestion	Humans, aquatic organisms
Eroded soil and bound formulation	Overland flow	Ingestion surface water/ingestion	Humans, aquatic organisms
Leachate, dissolved or bound to organic colloids	Leaching	Groundwater surface water/ingestion	Humans, aquatic organisms, animals

Table 14. Sources and Forms of Pesticide Released to the Atmosphere

Site activity	Source	Form of release
Turfgrass/ urban gardens/ agriculture/ forestry	Loss during application Postapplication loss from targets	Droplet, vapor Vapor, dust
Industrial	Effluent, exhaust systems	Vapor, dust
Waste disposal	Improper disposal/dump sites	Vapor, dust
Accidents	Spills, improper handling in storage, spills	Vapor, dust

From Grover 1991.

Table 15. Types of Pesticide Losses During Spray Operations

Loss component	Percent of spray
Delivery losses:	60–80
Directly to soil and peripheral vegetation	
Volatilization during spray operation	3–10[a]
Particulate or droplet losses	3–5

From Himel et al. 1990.

[a] *Depends on pesticide volatility.*

small pesticide droplets can occur at considerable distances from the spray target (Bidleman and Olney 1975; Bidleman et al. 1990; Richards et al. 1987; Weber and Montone 1990).

Economic and legal considerations have focused research on the short-distance fraction of pesticide drift (Grover 1991). In a review of early work on phenoxy herbicides, Bovey and Young (1980) determined that potentially high drift losses (20%) of 2,4-D could be controlled by enlarging spray droplet size, use of higher viscosity formulations, and use of low-volatility carriers. Volatility of the phenoxy herbicides is directly related to formulation. The acid, sodium salt, triethanolamine, and amide forms of 2,4-D have low-volatility hazard, but the various ester formulations had increased risk of volatile losses (Grover 1976; Grover et al.1972).

Drift of applied pesticides is a complex scientific, environmental, social, and at times legal problem. Efficacy of spray operations is limited by the amount of spray reaching its intended target. Drift and secondary volatilization of applied pesticides constitutes a direct reduction in application efficacy. In general, the early field studies found that the magnitude of downwind drift is primarily dependent on nozzle type, boom or spray height, spray pressure, and wind speed (Grover 1991; Himel et al. 1990). Drift losses range can range from <0.5 to 50% of the original spray, depending on the type of equipment and the environmental conditions (e.g., Grover et al. 1985, 1988; Smith et al. 1986).

Early assumptions on drift losses were based on sedimentation calculations (Akesson and Yates 1984). However, Himel et al. (1990) observed that the literature basis for the sedimentation approach to drift ignores the underlying principles of spray physics. Determining both pesticide drift characteristics and environmental risk factors are important considerations when planning and conducting pesticide spray operations (Himel et al. 1990). Factors influencing drift characteristics of specific pesticide formulations and climate conditions are (1) wind speed and direction, (2) droplet density and size distribution, (3) evaporation rate, (4) height and swath pattern of spray delivery, (5) volume and amount of applied pesticide, and (6) turbulent or stable atmospheric conditions. Additional factors that affect risk of impacts include (1) proximity of sensitive areas or species of nontarget organisms, (2) pesticide toxicity and safe exposure levels, and (3) the persistence and degradation rate of applied pesticide.

Environmental, legal, and economic considerations make it critical to manage pesticide spray operations for maximum pest control and minimum offsite movement. Areas on a golf course may be next to and commingled with sensitive nontarget areas such as ornamental vegetation, wetlands, lakes and streams, and residential property. Establishment of buffer zones and prespray planning are frequently used to protect sensitive nontarget areas.

Himel et al. (1990) made several observations regarding the current state of knowledge on pesticide drift and their environmental implications. These observations are

1. Efficacy of pesticide delivery is limited by droplet size.
2. Current spray technology does not conform to the fundamental principles of spray physics. Methods and equipment for application need to be changed, but this process will be slow as current technology appears to be sensible and is deeply rooted. Efficient pesticide spray systems must be developed.
3. Most current spray technologies are inefficient and require enhancement of efficacy. New spray technologies should be based on efficient target delivery.
4. Spray droplets >100 to 150 μm are filtered out in the first few millimeters of foliage. The outer (e.g., turfgrass) canopy acts as a major barrier to current pesticide sprays.
5. Insects and disease occupy protected microhabitats in the turfgrass and soil. Effective control requires the delivery of all control agents into the pest microhabitat.

6. Mass and droplet transport has never really been quantified for herbicides. Spray drift is often an overriding factor in use of herbicides. Control of drift using ultralarge droplet or polydisperse sprays may not be the ultimate solution.

7. Most sprayed pesticides are subject to volatile losses from the spray, foliage, and the ground. This is a significant environmental management problem that requires further investigation.

Volatilization is the loss of chemicals in vapor phase from soil and plant surfaces to the atmosphere. Atmospheric dispersion of pesticides involves two processes: (1) evaporation from the plant, soil, or water surface into the air; and (2) dispersion of the vapor into the atmosphere by diffusion and turbulent mixing. The process of volatilization is the balance of several rate processes including flux from plant canopy surfaces, the flux of chemical through the soil to the soil surface, and atmospheric transport away from the soil surface (Jury 1986a). Volatilization of many pesticides controls the environmental dispersal and length of effectiveness in the target zone (Taylor and Spencer 1990). Potential volatility of a chemical is related to its inherent vapor pressure (Table 16), while actual rates of volatilization depend on the environmental factors and chemical processes occurring at the soil-air-water interface (Table 17). Accurate evaluation of environmental and water quality impacts resulting from use of pesticides requires assessment of volatile pesticide losses (Jury et al. 1983, 1984a, 1987b; Spencer et al. 1982a, 1988).

Diffusion and convection processes control the movement of pesticides within the soil profile and away from the soil surface. In the stagnant boundary layer at the soil surface, mass transfer is proportional to the vapor density or vapor pressure of the chemical at the evaporating surface. Convective air movement also affects the rate of volatilization by continuous mixing and replacement of the air above the evaporative surface (Taylor and Spencer 1990). Once the pesticide has evaporated from the soil, plant, or water surface into the atmosphere, convective transport occurs very rapidly except when the atmosphere is exceptionally stable.

Distribution of pesticide residues between various environmental phases, including soil solids, soil solution, plant surfaces, and the vapor phase is a diffusive process. Quantifying, or partitioning, of residues between these phases is accomplished by use of adsorption isotherms (Taylor and Glotfelty 1988; Taylor and Spencer 1990). Henry's Law is used to describe the nearly linear equilibrium relationship of pesticide vapor pressure (or density) and the concentration in solution (Spencer et al. 1982a, 1988). Henry's Law, as well as solution to soil partitioning, uses the concept of fugacity or "escaping tendency" to describe partitioning of a chemical into different

Table 16. Range of Saturation Vapor Pressures and Henry's Law Constants Reported for Pesticides Used on Turfgrass

Compound	Molecular wt (g)	Temperature (°C)	Vapor pressure (Pa)	Henry's Law Constant, K_h[a] (Pa m^3 mol^{-1})
		Insecticides and Nematicides		
Bendiocarb	223.2	25	6.9E-4	—
Carbaryl	201.2	20	2.0E-4	1.3E-3
		24	1.6E-4	—
		25	1.8E-4	—
		30	1.9E-4–1.7E-2	—
Chlorpyrifos	350.6	20	1.2E-3–2.3E-3	—
		25	2.5E-3	1.8–7.9
		35	1.2E-2	—
Diazinon	304.3	20	2.4E-3–1.9E-2	1.1E-2–6.7E-2
Ethoprop	242.3	26	4.8E-2–5.1E-2	—
Fenamiphos[b]	303.4	25	1.3E-2	—
		30	1.3E-4	—
Isazofos	313.7	20	4.3E-3	—
		25	1.2E-2	—
Isofenphos	345.4	20	4.0E-4–5.3E-4	—
Trichlorfon	293.8	20	9.5E-4–1.1E-3	1.7E-6
		Fungicides		
Benomyl	290.3	20	<1.3E-3	—
		25	<1.3E-8	—
Chloroneb	207.1	25	4.0E-1	—
Chlorothalonil	265.9	40	1.3	—
Etridiozole	247.5	20	1.3E-2	—
Fenarimol	331.2	25	1.3E-5–2.9E-5	—
Fosetyl Al	354.1	20	<1.3E-3–<1.3E-5	—
Iprodione	330.2	20	<1.3E-5–<2.7E-5	—
Mancozeb	—	20	<1.3E-2	—
Maneb	(265.3)x	20	<1.3E-4	—
Metalaxyl	279.3	20	2.9E-4	—
		25	6.4E-4–7.5E-4	—
PCNB	295.3	10	2.1E-3	—
		20	6.7E-3	—
		25	1.5E-2–3.2E-1	—
Propamocarb HCl	224.7	25	0–8E-1	—
Propiconazole	342.2	20	1.3E-4	—
		25	5.6E-5	—

Table 16. Range of Saturation Vapor Pressures and Henry's Law Constants Reported for Pesticides Used on Turfgrass (continued)

Compound	Molecular wt (g)	Temperature (°C)	Vapor pressure (Pa)	Henry's Law Constant, K_h[a] (Pa m^3 mol^{-1})
		Fungicides (continued)		
Thiophanate -methyl	342.4	20	<1.3E-5	—
Thiram	240.4	20	<1.3E-3	—
		25	<1.0E-3	<8.0E-3
Triadimefon	293.8	20	1.1E-4	—
		40	2.0E-3	—
		Herbicides		
Atrazine	215.7	20	4.0E-5	2.5E-4–2.9E-4
		25	3.9E-5–8.8E-5	—
		30	1.9E-4	
Benefin	335.3	20	4.0E-3	—
		25	3.7E-3–1.0E-2	—
		30	5.2E-3	1.3
Bensulide	397.5	25	1.1E-4–1.3E-4	—
2,4-D acid	221.0	20	<1.3E-5–1.1E-3	—
		25	8.0E-5–1.0E-3	1.4E-5–5.5E-1
2,4-D dimethyl-amine salt	264.1	38	1.1E-7	—
2,4-D butoxyethyl ester	321.2	25	6.0E-4–2.3E-1	1.0E–2-1.0
2,4-D Isooctyl ester	333.3	25	2.7E-4	4.4
DCPA	332.0	25	3.3E-4	—
Dicamba	221.1	25	2.7E-3–4.9E-1	1.2E-4–9.1E-2
Endothall	186.2	25	<1.0E-3	—
Ethofumesate	286.3	25	8.7E-5–6.5E-4	—
Glyphosate	169.1	25	Negligible	—
Glyphosate, amine salt	228.0	25	Negligible	—
MCPA, sodium salt	243.7	20	Negligible	—
MCPA ester	200.6	20	2.0E-4	—
MCPP acid	214.6	20	<1.3E-5	—
MCPP, amine salt (Mecoprop)	260.0	20	Negligible	—

Table 16. Range of Saturation Vapor Pressures and Henry's Law Constants Reported for Pesticides Used on Turfgrass (continued)

Compound	Molecular wt (g)	Temperature (°C)	Vapor pressure (Pa)	Henry's Law Constant, K_h[a] (Pa m^3 mol^{-1})
		Herbicides (continued)		
MSMA	183.9	20	Negligible	—
Oxadiazon	345.2	20	<1.3E-4	—
Pendimethalin	281.3	25	1.3E-3–4.0E-3	—
Siduron	232.3	25	5.3E-7–<8.0E-4	—
Simazine	201.7	20	7.2E-8–8.1E-7	3.4E-4
Triclopyr, amine	357.5	25	1.6E-4	—
Triclopyr, ester	357.7	25	1.6E-4–9.5E-3	—
Trifluralin	335.3	20	6.5E-3–1.5E-2	4.1–6.6
		25	1.5E-2	—
		30	2.7E-2–3.2E-2	—

From Ashton and Crafts 1981; Helling 1976; Howard 1991; Taylor and Spencer 1990; Wauchope et al. 1991.

[a] $K_h = d_o C^{-1}$, *where* d_o = *vapor density and C = solution concentration.*
[b] *Bilkert and Rao 1985.*

Table 17. Principal Soil, Environmental, and Management Factors Influencing Chemical Volatilization

Parameters	Influence on volatilization
Soil	
Water content	Vapor diffusion in soil decreases as water content increases depending on the effective vapor-liquid diffusion coefficient. In very dry soils vaporization is suppressed by increased adsorption of compounds on soil surface.
Bulk density, or porosity	Decreasing porosity or increasing bulk density generally decreases volatilization because diffusive transport to the soil surface is reduced by decreasing cross-sectional area for flow and increasing flow length. Increased bulk density also increases the density of adsorptive sites on soil surface increasing surface partitioning of pesticides.
Adsorption site density	The quantity of clay minerals and soil organic matter determines the partitioning of pesticides on adsorption sites. Adsorption increases the concentration of immobilized pesticide at the expense of liquid and vapor phases. Increased adsorption always implies decreased volatilization as adsorption decreases the transport of chemicals from the soil to the surface to replace chemicals lost by volatilization. Adsorption decreases the concentration of chemical vapor at volatilization sites.

Table 17. Principal Soil, Environmental, and Management Factors Influencing Chemical Volatilization (continued)

Parameters	Influence on volatilization
Environmental	
Temperature	The influence of temperature on volatilization is undoubtedly important, but has received little research attention. Inferences are generally based on speculation. At ambient pesticide concentrations in soil, temperature effects on volatilization are complicated by soil moisture relations, changes in saturation vapor pressure, and adsorption relations. As a general rule, Henry's Law constant increases, as temperature increases, and it is expected that volatilization will also increase. Solution transport by mass flow and diffusion will decrease compared to vapor transport as temperature increases.
Wind	Wind increases boundary layer mixing, or equivalently decreases the stagnant boundary layer above the soil surface that limits volatilization. Depending on compound vapor pressure and Henry's Law coefficient, increased wind will increase volatile losses of pesticides at the surface of the soil and plant canopy. Wind will increase application (drift) losses of pesticides, although turbulent and thermal upwelling also may increase airborne losses on days with low wind conditions.
Evaporation	The influence of water evaporation on pesticide volatilization is complex. For all chemicals, water evaporation will increase mass flow of chemicals in solution to the surface sites of volatilization. The degree of a compound's vapor pressure and upward flux in solution will determine the extent of evaporation on volatilization. Volatilization of strongly sorbed pesticides (e.g., glyphosate) will be insignificantly influenced by evaporation.
Precipitation	The influence of rainfall on volatilization is indirect. Precipitation will transport pesticides away from the soil surface and volatilization sites. Volatilization of soluble pesticides (e.g., atrazine) is limited by the degree of chemical leaching into the soil by precipitation.
Management	
Chemical concentration	Volatilization rate of a chemical is proportional to its vapor concentration at the soil surface, increasing vapor concentration creates higher vapor density and higher volatilization. If the pesticide partitions linearly, volatilization is proportional to chemical concentration.
Depth of incorporation	Volatilization rates are reduced by incorporation of surface-applied chemicals. Incorporation reduces the concentration of pesticides at the surface, thus reducing volatilization losses.

Table 17. Principal Soil, Environmental, and Management Factors Influencing Chemical Volatilization (continued)

Parameters	Influence on volatilization
Management (continued)	
Chemical formulation	Different formulations of the same pesticide will have different vapor pressures (e.g., 2,4-D). Microencapsulated formulations of certain pesticides reduces volatile losses. Wettable powder formulations of normally low volatile chemicals are susceptible to airborne transport. Under dry soil and plant canopy conditions, the wettable powder leaves a thin deposit of dry powder on the soil surface. The small particles of pesticide are susceptible to wind erosion under dry soil conditions.
Irrigation	The effect of irrigation is similar to rainfall as the amount of water infiltrating the soil leaches applied chemicals below the volatilization sites. The intensity and spacing between irrigation will affect the duration of the evaporation cycle which also affects the upward transport and deposition of pesticides at the soil surface. High-frequency irrigation will most likely significantly decrease volatilization of incorporated pesticides.
Soil and cultivation	The same factors that increase diffusion of oxygen downward into the plant root zone also will increase the volatilization of pesticides applied to the soil surface. Such practices that promote good structure and soil aeration will promote volatilization.

phases of environmental systems (Mackay and Paterson 1981, 1982; Mackay and Stiver 1991). According to Henry's Law,

$$d = K_h \, C_l \tag{1}$$

where d = the vapor density of a pesticide residue
K_h = the proportionality constant, Henry's Law coefficient, for a given residue (Table 16)
C_l = the residue concentration in solution

Vapor density is the ratio of the mass of pesticide vapor per unit volume and is converted from vapor pressure by the equation,

$$P = d\,(RT\,M^{-1}) \tag{2}$$

where P = the vapor pressure
R = the molecular gas constant
T = the absolute temperature (°K)
M = the molecular weight of the chemical

Although the range of pesticide vapor pressures is quite large, the generally low values suggest that volatilization of most compounds applied on golf courses is quite low (Table 16). However, the background levels of most pesticides in the atmosphere also is quite low. Therefore, pesticides will evaporate from soil and plant surfaces into the atmosphere as if the atmosphere were a vacuum (Taylor and Spencer 1990). The pressure gradient for pesticide evaporation from soil and turfgrass surfaces is quite high, despite the low vapor pressure of most pesticides.

Adsorption of pesticides on soil and thatch will influence volatile losses of pesticides. The relationship of solution concentration of a pesticide and adsorption is described in the next section. Increased adsorption of residues on soils decreases volatilization by sequestering residues on colloid surfaces. Pesticide residue concentration in solution and ultimately its vapor density is related to pesticide solubility, amount of pesticide applied, adsorption on soil colloid surfaces, and temperature (Hance 1988). Additional factors influencing volatile losses of pesticides applied in turfgrass systems are soil moisture, soil organic matter content and colloid type, bulk density, and management practices (Table 17).

Volatilization is one of the most important pathways for environmental dissipation of applied pesticides from land surfaces (Jury et al. 1987a). Volatile loss and air transport are the principal means of widespread dispersion of pesticides in the environment. Glotfelty et al. (1984a), Grover (1991), Spencer et al. (1973), Spencer and Cliath (1990), Taylor (1978), Taylor and Glotfelty (1988), and Taylor and Spencer (1990) have reviewed field studies of volatile losses of applied pesticides. Volatile losses from plant and moist soil surfaces can be large with losses approaching 90% for volatile compounds (Table 18). Glotfelty et al. (1984a) found that measured volatile losses were dependent on both the pesticide properties and management practices. Shallow incorporation during turfgrass establishment or watering-in, presence of organic colloids, and dry soil surfaces reduced volatile losses. Volatile pesticides with low water solubility tend to be lost from soil to the atmosphere and are unlikely to be transported into or over soil surfaces.

Although the importance of volatile pesticide losses from agricultural systems has been well-documented, research describing pesticide volatilization from turfgrass systems is limited. Branham and Wehner (1985) determined that less than 4% of diazinon applied to a thatched turfgrass was lost through volatilization when application was followed by immediate watering.

Using a field volatility chamber over Kentucky bluegrass, Cooper et al. (1990) and Jenkins et al. (1990) measured the loss of pendimethalin applied as a liquid suspension. Within 24 hr, 4.8% of the applied pendimethalin was lost by volatilization. During the first 5-d period following application, approximately 13% of the pendimethalin was lost to the atmosphere. During the first 2 d of the experiment, the investigators observed a direct correlation between vapor flux and the temperature of the turfgrass canopy. These investigators hypothesized that dislodgeable foliar residues were an

Table 18. Summary of Field Measurements of Pesticide Volatilization

Compound/ usage	Surface	Season, location	Fraction volatilized in period
Alachlor/ surface	Fallow, silt loam	May, Maryland	26% in 24 d
Atrazine/ surface	Fallow, silt	May, Maryland	2.4% in 24 d
Chlorpropham/ surface	Soil under soybeans	May–July, Maryland	49% in 50 d
Microencapsulated	Soybeans	May–July, Maryland	20% in 50 d
DCPA/ surface	Moist fallow, silt loam	August, Maryland	2% in 34 hr
Surface[a]	Onions and parsley, fine silt loam	April, California	10% in 21 d
Diazinon/ surface[b]	Kentucky bluegrass turfgrass, silt loam	Microecosystem, Illinois	5% in 21 d
EPTC/ surface in irrigation water	Soil under 25 cm alfalfa	May, California	74% in 52 hr
Heptachlor/ incorporated to 7.5 cm	Soil under corn	May–October, Ohio	7% in 170 d
Surface	Moist fallow, silt loam	August, Maryland	50% in 6 hr 90% in 6 d
Surface	Dry fallow, sandy loam	June, Maryland	40% in 50 hr
Vegetation	Short orchard grass	July, Maryland	90% in 7 d
Lindane/ surface	Moist fallow, silt loam	August, Maryland	50% in 6 hr 90% in 6 d
Surface	Dry fallow, sandy loam	June, Maryland	12% in 50 hr
Pendimethalin/ surface[c]	Kentucky bluegrass turfgrass, silt loam	May, Massachusetts	4.8% in 24 hr 6.1% in 48 hr 13% in 5 d
Simazine/ surface	Fallow, silt sandy loam	May, Maryland	1.1% in 24 hr
Triallate/ incorporated[d] to 5 cm	Wheat, heavy clay	April, Saskatchewan	9.8% in 7 d 17.6% in 67 d

Table 18. Summary of Field Measurements of Pesticide Volatilization (continued)

Compound/ usage	Surface	Season, location	Fraction volatilized in period
Trifluralin/ incorporated to 2.5 cm	Soybeans, sandy loam, 0.6% o.m.[e]	June–October, Georgia	22% in 24 d
Incorporated to 7.5 cm	Soybeans, loam, 4.0% o.m.[e]	May–September, New York	3.4% in 90 d
Surface	Dry fallow, sandy loam	June, Maryland	25% in 50 hr
Surface	Moist fallow, silt loam	August, Maryland	50% in 7.5 hr 90% in 7 d
Incorporated[c] to 5 cm	Wheat, heavy clay	April, Saskatchewan	12.2% in 7 d 23.7% in 67 d
2,4-D (Octyl ester)/ surface	20 cm high spring wheat	June, Saskatchewan	21% in 5 d

From Taylor and Glotfelty 1988, and Taylor and Spencer 1990.

[a] *Ross et al. 1990.*
[b] *Branham and Wehner 1985, Branham et al. 1985.*
[c] *Cooper et al. 1990, Jenkins et al. 1990.*
[d] *Grover et al. 1988.*
[e] *Organic matter.*

important source of pendimethalin volatility. The decline in foliar residues matched the observed decline of volatile losses during the entire 15-d sample period.

Loss of pesticides to the atmosphere by drift or volatilization reduces the potential for direct transport to groundwater or surface water by leaching and runoff processes (Jury and Valentine 1986). However, airborne pesticides have the potential to pollute terrestrial and aquatic systems through drift and deposition in precipitation (Glotfelty et al. 1987; Kurtz 1990; Richards et al. 1987). Volatile losses of pesticides, especially drift losses, constitute a concern to turfgrass managers from a perspective of pest control efficacy, potential legal issues, potential exposure of sensitive areas to pesticides, and a potential source of pesticides for long-range transport.

4.5.3 Adsorption and Retention

The process of adsorption, desorption, and bound residue in plants and microorganisms directly or indirectly influences all the major factors contributing to the fate, degradation, and transport of pesticide residues (Hance 1988; Kahn 1991). Adsorption of pesticides on the surface of soil colloids and thatch layers controls the initial partitioning of residues in the soil-air-water continuum. Adsorption is the equilibrium process of retention or bonding of a solute to adsorptive sites on soil solids, either mineral or organic matter surfaces (Jury 1986b). Retention of pesticides on soil surfaces or in plant tissues temporarily immobilizes the molecule from transport in either solution or vapor phases. The degree of immobilization determines, in part, the susceptibility of pesticide transport to groundwater and surface water in solution or attached to sediment (Hance 1988; Koskinen and Harper 1990).

Adsorption provides only temporary storage of pesticides in the environment. The amount of pesticide adsorbed on soil and organic matter surfaces is in equilibrium with the amount of pesticide in soil water. As soil solution moves through the soil, the amount of pesticide decreases in solution and more pesticide is released from the soil. The rate of retention and release processes is a function of pesticide chemistry and concentration, adsorptivity of the organic and mineral soil, the length or residence time of the pesticide on the soil-organic matter complex, the degree of soil heterogeneity, and the amount of water leaching through the soil (Koskinen and Harper 1990; Rao et al. 1988).

The removal of pesticides from soil solution by adsorption and the return of adsorbed molecules to solution by desorption are major factors controlling pesticide mobility and persistence in the soil environment. Given the importance of adsorption processes, there are many excellent reviews available in the literature (Hance 1988; Jury 1986b; Koskinen and Harper 1990; Mingelgrin and Gerstl 1983; Rao and Davidson 1979, 1980; Zielke et al. 1989). Theoretical, kinetic, and nonequilibrium aspects of the adsorption-desorption processes have been discussed by Brusseau and Rao (1989), Green and Karickhoff (1990), Koskinen and Harper (1990), and Rao et al. (1979). Despite the voluminous literature on adsorption of organic chemicals on soil surfaces, knowledge of adsorption and desorption mechanisms remains largely empirical (Jury 1986b).

The principle bonding mechanisms between pesticides and soil surfaces involve high- and low-energy bonds (Table 19). The most important component for adsorption of pesticides on soil surfaces are the soil colloids, clay minerals and organic matter. Soil organic matter content is the most significant soil factor for partitioning of most pesticides between the soil surface and soil solution (Hance 1988). The concept of hydrophobic partitioning of nonpolar organic compounds has been used extensively to describe sorption processes between soil and soil solution (Chiou et al. 1979; Mackay et al. 1985). However, the overall extent and importance of this mechanism has not been definitively established (Mingelgrin and Gerstl 1983).

Sorption of pesticides usually involves a continuum of possible mechanisms (Koskinen and Harper 1990; Senesi and Testini 1980). Adsorption by ionic bonds is due to ion exchange on both clay mineral and organic matter surfaces. Ionic bonds occur between organic cations or anions and negative or positive charged surfaces on

Table 19. Intermolecular Bonding Mechanisms Involved in Adsorption of Pesticides

High energy bonds	Low energy bonds
Ionic bonds	Charge-dipole and dipole-dipole bonds
Ligand bonds	Hydrogen bonds
	Charge transfer bonds
	Cation and water bridging
	van der Waal's-London bonds
	Entropy generation
	Magnetic bonds

From Jury 1986b.

adsorbents. Most pesticides are uncharged, but some are weak acids (phenoxyacetic acids) or weak bases (s-triazines). At soil pH values below the zero point of charge of iron and aluminum oxides, anionic herbicides, such as 2,4-D, or zwitterionic herbicides, such as glyphosate, are adsorbed as anions. At pH values below the pK_a of the s-triazine herbicides, these weak bases become cationic and are adsorbed on negatively charged soil colloid surfaces. Adsorption of weakly acidic or basic herbicides increases as soil pH decreases (Paulson 1981). Ligand exchange of partially chelated transition metals on clays and humic acids is another possible binding mechanism for some triazines and substituted urea compounds (Jury 1986b; Koskinen and Harper 1990).

Hydrogen bonding involves an electrostatic attraction of covalently bonded H, usually to N or O, to another electronegative atom. Virtually all pesticides contain groups that have potential to form hydrogen bonds. Dipole type bonds are involved between polar organic molecules and electrically charged or polar adsorbing surfaces. Charge transfer bonds involve electron transfer across the surface of the adsorbent or organic molecule. Entropy generation, van der Waal's bonds, and magnetic interactions are additional weak attractive forces postulated to be involved in organic molecule adsorption on colloid surfaces (Hance 1988).

Soil properties, pesticide characteristics, and environmental conditions are the important factors determining the degree of pesticide adsorptivity (Table 20) (Hance 1988; Jury 1986b; Koskinen and Harper 1990; Zielke et al. 1989). Soil properties influencing adsorption of pesticides include organic matter content, clay composition and exchange capacity, soil water content, bulk density, and pH. Water solubility, electronic structure, and solution concentration are important chemical characteristics influencing pesticide adsorption. Temperature is the primary environmental property controlling the degree of chemical adsorption.

Adsorption isotherms are equilibrium models used to describe equilibrium sorption or partitioning of pesticides between the soil-water interface. The linear and Freundlich

Table 20. Soil, Chemical, and Environmental Properties Influencing Pesticide Adsorption on Soils

Property	Influence on adsorption
Soil	
Organic matter content	A positive linear relationship observed soil organic matter content and adsorption of pesticides. Highest correlations were observed in soils with relatively high organic matter contents and for nonionic/weakly polar compounds.
Clay content and exchange capacity	Clay minerals have a net negative charge. This charge will influence the adsorption for virtually all chemicals. Inorganic cations (e.g., Ca^{+2}, K^{+}, trace metals) and organic chemicals with permanent positive charge (e.g., paraquat, diquat) are strongly adsorbed on the exchange complex of clay minerals. Depending on solution pH, polar compounds and ionizable compounds (e.g., s-triazine and phenoxyacetic herbicides) are also adsorbed on clay mineral surfaces. However, considering the nonpolar nature of most pesticides and the large capacity of soil organic matter surfaces, usually no direct correlation is observed between most pesticides and clay content.
Soil water content	Soil water content influences adsorption by (1) modifying the solution pathway leading to adsorptive surfaces and (2) modifying the solution concentration in comparison to the concentration on the adsorptive surfaces. Usually, the influence of soil water content on adsorption is not strong until the soil is very dry. Pesticide adsorption increases dramatically when soils are very dry.
Bulk density	Bulk density increases the density of adsorptive sites per unit volume soil. A direct correlation between bulk density and adsorption between groups of soils may be small as soils high in organic matter and clay content tend to have lower bulk density than coarse-textured soil with low organic matter content.
Pesticide Chemical Properties	
Electronic structure	Electronic structure of pesticide is extremely important in adsorption processes. Structure of a compound determines the strength and type of adsorptive forces. Molecules with permanent charge are strongly adsorbed on inorganic and organic exchange sites. Neutral, nonpolar compounds (e.g., ureas) are only weakly adsorbed by van der Waal's interaction and other minor bonding mechanisms. The electronic configuration of a compound will determine, in a large part, the adsorption partition coefficient, K_{oc}, for a given compound.
Water solubility	There is an inverse, linear relationship between water solubility and adsorption coefficients within classes of similar organic compounds. Regression equations have been developed within pesticide classes to estimate K_{oc} from pesticide water solubility. In general, as the water solubility of a nonionic pesticide increases,

Table 20. Soil, Chemical, and Environmental Properties Influencing Pesticide Adsorption on Soils (continued)

Property	Influence on adsorption
Pesticide Chemical Properties (continued)	
	adsorption on organic surfaces decreases. This relationship is not valid for ionic (e.g., paraquat) and zwitterionic (e.g., glysophate) pesticides, which are very water soluble, but also strongly adsorbed on colloid surfaces.
Solution composition and concentration	Few quantitative relationships have been developed between adsorption and macroscopic chemical properties of the soil solution. Inorganic cations will compete with positively charged organic compounds for adsorption on negative exchange sites. However, for nonionic pesticides, only the organic matter content of the soil has any measurable impact on adsorption. As the solution concentration of a pesticide increases, the amount of chemical adsorbed on soil surfaces also increases. Organic solvents, used as chemical carriers for some pesticides, also influence the degree of pesticide adsorption. Pesticide solubility in the carrier will often exceed the solubility of the pesticide in water.
Environmental	
pH	The pH of the soil system will have a strong influence on the adsorption of pesticides which are weakly acidic or basic. In general, as soil acidity increases, adsorption of weak acids and weak bases increases. Weak acids in the undisassociated form at lower pH values are sorbed more strongly than the disassociated anionic form. Weak bases are converted to cations at lower pH values and are also more highly sorbed than as free bases. Polar compounds actually may not form cationic or anionic forms, but are capable of forming hydrogen bonds which are somewhat pH dependent. In general, the correlation between pH and adsorption of pesticides is much lower than between organic matter and adsorption. Adsorption of pesticides subject to disassociation in solution is likely to be strongly affected by pH, while other pesticides are relatively independent of pH.
Soil temperature	Although the adsorption of some compounds increases with temperature, the adsorption of the majority of pesticides decreases with increases in temperature. Temperature affects both solute-surface interactions and water solute interactions. The effect of temperature on the adsorption equilibrium is a direct indication of the strength of adsorption. Therefore, the stronger a compound is adsorbed, the larger is the influence of soil temperature on adsorption.

From Hance 1988 and Jury 1986b.

isotherm models have been most commonly used to describe pesticide sorption on soils (Green and Karickhoff 1990; Rao and Jessup 1983). These equations are

$$S = K_d(C_l);\ \text{Linear isotherm} \tag{3}$$

and

$$S = K_f(C_l)^N,\ N < 1;\ \text{Freundlich isotherm} \tag{4}$$

where S = the adsorbed-phase concentration
C_l = the solution-phase concentration
K_d = the linear partition coefficient
K_f = the Freundlich partition coefficient
N = an empirical constant

The partition coefficients and the N constant are usually obtained using curve fitting techniques for specific soil-pesticide combinations. N values range from 0.7 to 1 for many pesticides. When N is approximately equal to 1, Equation 4 is reduced to the linear form of Equation 3. At low aqueous concentrations and for neutral and weakly polar compounds, the linear adsorption isotherm adequately describes adsorption partitioning of pesticides (Karickhoff 1981; Karickhoff et al. 1979). The linear adsorption model has been used successfully in several computer models assessing chemical transport for many nonionic and weakly polar pesticides (Boesten and van der Linden 1991; Carsel et al. 1984; Wagenet and Hutson 1986).

The most important soil property determining adsorption or soil-water partitioning is the soil organic matter content (Hassett and Banwart 1989; Jury 1986b; Pignatello 1989). The adsorption partition coefficient, K_d, of most nonionic and polar pesticides is highly correlated with soil organic matter content (% OC) (Jury 1986b; Karickhoff 1981; Kenaga and Goring 1980; Rao and Jessup 1983). For ionic or ionizable compounds, the effects of pH and pK_a should be included in statistical estimations of K values (Rao and Jessup 1983). Several studies on the fate and persistence of pesticides in turfgrass systems have shown a high degree of affinity of organic and inorganic pesticides with turfgrass thatch (e.g., Branham and Wehner 1985; Niemczyk and Chapman 1987). The high organic matter content of the thatch layer makes it a potentially highly absorptive media.

Most research strongly suggests that for nonionic or weakly polar pesticides, the sorption coefficient (K_d and K_f) is independent of soil type when normalized for soil organic matter content (Hance 1988; Jury 1986b). Reduction of variability in the soil-based adsorption coefficient, K_d, is achieved by defining this partition coefficient per unit of soil organic carbon. The soil independent adsorption coefficient has been defined as K_{oc} according to the following equation:

$$K_{oc} = K_d\,(f_{oc})^{-1} \tag{5}$$

where f_{oc} = the organic matter fraction
K_d = the distribution coefficient
K_{oc} = the organic carbon adsorption coefficient

Recent research suggests that within certain limitations K_{oc} is independent of particle size fractions within and between different soils (Green and Karickhoff 1990; Hance 1988; Karickhoff et al. 1979). The K_{oc} value is consistent for a given chemical, having been normalized for differences in soil organic carbon content (Table 21) (Hance 1988). Despite claims of consistency, there is considerable variability in the K_{oc} values reported in the literature (Anderson et al. 1989; Wauchope et al. 1991). A study of soil adsorption of triazine and thiocarbamate herbicides by Singh et al. (1990) suggests that the current range of many K_{oc} values may be a result of inaccurate determination of the original K_d partition constants.

Values of K_{oc} for a broad range of pesticides are available in reviews by Anderson et al. (1989), Hanmaker and Thompson (1972), Hassett et al. (1983), Karickhoff (1981), Kenaga and Goring (1980), Rao and Davidson (1980), and Wauchope et al. (1991). These investigators have all concluded that soil organic matter is the principal sorbent of pesticides in soils, despite some variability due to interactions with inorganic adsorption sites. In soils with extremely low or high organic matter contents, the use of K_{oc} to estimate the soil-pesticide distribution coefficient (K_d or K_f) is subject to error (Green and Karickhoff 1990; Hanmaker and Thompson 1972; Rao and Davidson 1979).

With a high organic matter content, turfgrass thatch has a high sorption affinity for many pesticides. As a temporary sink for many applied pesticides, thatch is a potential long-term source of pesticides for both degradation, leaching, and volatilization. Niemczyk and Krueger (1982) determined the binding characteristics of six different insecticides to turfgrass thatch in a laboratory experiment and found that chlorpyrifos had the highest binding affinity for thatch followed by diazinon, fesulfothion, CGA 12223 (isazofos), bendiocarb, and trichlorfon. Unfortunately, no comparisons were made to the affinity of the same pesticides to mineral soil or green mixes.

Hurto and Turgeon (1979) applied paraquat and glyphosate to Kentucky bluegrass turfgrass at thatchy and thatch-free sites. Establishment of perennial ryegrass on paraquat treated thatchy sites was inhibited due to high herbicide residues in the thatch. Modification of the thatch by soil incorporation following core cultivation or removal of the thatch layer by vertical mowing significantly improved the percent cover of ryegrass on the treated sites. Sites treated with glyphosate had ryegrass cover similar to the modified paraquat sites suggesting that glyphosate adsorption on thatch and residual activity was minimal.

Table 21. Pesticide Solubility, Soil Adsorption Coefficients, and Octanol-Water Partition Coefficients

Pesticide	Water solubility (ppm)	Soil adsorption coefficient, K_{oc}	Octanol-water partition coefficient K_{ow}	Ref.[a]
		Insecticides and Nematicides		
Bendiocarb	40	570	—	10,11
Carbaryl	40	230	230	1
	40	230	646	4
	—	—	176	6
	—	229	—	8
	—	—	363	9
	40	104–423	—	11
	32	79–316	229	12
Chlorpyrifos	0.3	13,600	97,700	1
	1.2	6,100	—	2
	—	14,000	—	3
	2.0	13,490	2,042–128,825	4,6
	—	6,070	—	8
	2	—	93,325	9
	2.0–4.8	2,500–14,800	—	11
	1.1	4,381–13,600	48,980	12
Diazinon	40	—	—	1
	—	—	1,052	6
	—	85	—	8
	—	—	1,047	9
	40	85–570	—	11
	69	40–432	6,457	12
Ethoprop	750	120	—	7,8
	700–750	26–120	—	11
Fenamiphos	—	171	—	8
	400	—	—	9
	400–700	26–120	—	11
	700	148–249	—	13
Isazofos	69	44–143	—	11
Isofenphos	20–24	17–536	—	11
Trichlorfon	154,000	—	3	1
	120,000	—	—	9
	154,000	6	—	10
	120,000–154,000	2–6	—	11

Table 21. Pesticide Solubility, Soil Adsorption Coefficients, and Octanol-Water Partition Coefficients (continued)

Pesticide	Water solubility (ppm)	Soil adsorption coefficient, K_{oc}	Octanol-water partition coefficient K_{ow}	Ref.[a]
		Fungicides		
Anilazine	8	1070–3000[b]	—	11
Benomyl	—	—	264	6,9
	2–4	200–2100	—	11,12
Chloroneb	8	1,159	—	1
	—	1,653	—	6
Chlorothalonil	0.6	1,380–5,800	—	11
Etridiozole	50–<200	1,000–4,400[b]	—	11
Fenarimol	14	600–1,030	—	11
Fosetyl Al	120,000	20	—	11
Iprodione	13–14	500–1300	—	11
Mancozeb	0.5	2,000	—	11
Maneb	0.5[b]	2,000[b]	—	11
Metalaxyl	7,100–8,400	29–287	—	11
PCNB	0.44	350–10,000	—	11
	0.03	7,965[b]	43,650	12
Propamocarb HCl	700,000–1,000,000	1,000,000[b]	—	11
Propiconazole	100–110	387–1147	—	11
Thiophanate-methyl	3.5	1,830[b]	—	11
Thiram	30	670	—	11
	30	672[b]	—	12
Triadimefon	70	73[b]	—	9,11
		Herbicides		
Atrazine	33	149	476	1
	—	170	—	3
	33	148–214	214–512	4
	—	216	513	5
	—	163	226	6
	—	160	—	8
	—	—	282	9
	33–70	38–170	—	11
	30	39–216	—	12

Table 21. Pesticide Solubility, Soil Adsorption Coefficients, and Octanol-Water Partition Coefficients (continued)

Pesticide	Water solubility (ppm)	Soil adsorption coefficient, K_{oc}	Octanol-water partition coefficient K_{ow}	Ref.[a]
		Herbicides (continued)		
Benefin	<1	10,700	—	1
	0.1	781–10,700	—	11
Bensulide	5.6–25	740–10,000[b]	—	11
2,4-D acid	900	20	37	1
	900	60	—	2
	—	20	416–646	6,8,9
	703–1,072	20–57	—	11
	682	20–109	—	12
2,4-D dimethyl-amine salt	300,000–790,000	18–20	—	11
	3,000,000	0.1–136	—	12
2,4-D butoxyethyl ester	12	1,100–6,900[c]	3,390	12
2,4-D Isooctyl ester	0.02	25,000–68,000	2,089,296[b]	12
DCPA	0.5	4,000–6,400	—	11
Dicamba, acid	4,500	0.4	—	1
	5,000–8,000	2–4.4	162	11,12
Dicamba, salt	80,000	2.2	3	6,8,9
Dicamba, salt, sodium dimethylamine	400,000	—	—	11
	850,000	—	—	11
DSMA	254,000	770	—	1
Endothall	100,000	8–138	—	1,11,12
Ethofumesate	50–110	340[b]	—	11
Glyphosate, acid	12,000	2,640	—	1,11
Glyphosate, amine salt	900,000	24,000[b]	—	11
MCPA ester	5	1,000	—	11
MCPA, salt dimethylamine	866,000	20	—	11
MCPA, sodium salt	270,000	20	—	11
MCPP amine (Mecoprop)	660,000	20[b]	—	11

Table 21. Pesticide Solubility, Soil Adsorption Coefficients, and Octanol-Water Partition Coefficients (continued)

Pesticide	Water solubility (ppm)	Soil adsorption coefficient, K_{oc}	Octanol-water partition coefficient K_{ow}	Ref.[a]
		Herbicides (continued)		
MSMA, sodium	—	—	0.0008	6,9
salt	570,000–1,400,000	100,000–300,000	—	11
Oxadiazon	0.7	3,241	—	1
	0.7	3,421–5,300	—	11
Pendimethalin	0.5	—	—	9
	0.275–0.5	5,000	—	
11				
Siduron	18	420–890	—	11
Simazine	3.5	135	155	1
	5	135–214	145	4,5
	—	140	106	6,8
	—	—	87	9
Triclopyr (triethylamine salt)	2,100,000	1.5–27	3–27	1,11
Triclopyr (butoxyethyl ester)	23	780	12,300	1,11
Trifluralin	0.6	13,700	220,000	1
	—	3,900	—	3
	—	30,550	208,000	5
	—	—	1,150	6
	—	7,300	—	8
	24	—	56,234	9

[a] *1 = Kenaga and Goring (1980), 2 = McCall et al. (1983), 3 = McCall et al. (1981), 4 = Karickhoff (1981), 5 = Brown and Flagg (1981), 6 = Rao and Davidson (1980), 7 = Jury et al. (1984b), 8 = Jury et al. (1987b), 9 = Carsel et al. (1984), 10 = Kenaga (1980), 11 = Wauchope et al. 1991, and 12 = Howard (1991).*

[b] *Estimated values.*

[c] *Degrades rapidly to 2,4-D acid.*

The K_{oc} value is a quantitative indicator of pesticide adsorptivity (Table 21). The K_{oc} values for different pesticides range over several orders of magnitude (Table 21). Pesticides with high K_{oc} values, such as glyphosate, are strongly adsorbed to soil particles and are not subject to significant movement in runoff water or subsurface leachate. Pesticides with low to moderate K_{oc} values, such as dicamba (low) and atrazine (moderate), are not strongly adsorbed to soil surfaces and are subject to movement in runoff water or leaching. Molecular properties of pesticides influencing K_{oc} include aqueous solubility, disassociation constants, charge status, structure and conformation, polarity and polarization potential, and molecular size (Table 20). For nonionic pesticides, it is generally true that K_{oc} is inversely related to water solubility and directly related to the octanol-water partition coefficient (Chiou 1989). Log-log regression equations relating molecular properties of pesticides and K_{oc} values have been developed (Anderson et al. 1989; Chiou 1989; Hassett and Banwart 1989). These relationships are useful in estimating potential adsorptivity of new pesticide formulations.

Nonpolar organic contaminants in soil and porous media do not always exhibit ideal equilibrium and reversible adsorption-desorption behavior (Brusseau and Rao 1989; Pignatello 1990; Pignatello and Huang 1991; Weber et al. 1986). Conventional 24-hr batch studies produce sorption coefficients that represent macroscale flow conditions. These short-term extraction studies of pesticides and nonorganic soil contaminants usually demonstrate completely reversible sorption (Boesten and van der Pas 1988). However, longer term extraction studies demonstrate the importance of nonequilibrium and slowly reversible sorption of pesticides in soils and sediments. Slowly reversible sorption is defined as sorptive processes that are slow compared to immediate transport and degradation conditions (Pignatello and Huang 1991).

Nonideal or nonequilibrium phenomena have been observed as hysteresis in pesticide adsorption and desorption isotherms (Felsot and Dahm 1979). Generally, less chemical desorbs from colloid in a given period of time than is sorbed. Also, investigators have observed decreased extractability of organic compounds with increased incubation time (Brusseau and Rao 1989). Decreased extractability with increased residence time may be due to physical trapping in the soil matrix and micropores, colloid interlayers, and matrix of soil organic matter (Brusseau and Rao 1989; Pignatello 1990a; Steinberg et al. 1987; Weber et al. 1965, 1986). Kahn (1982) has suggested that pesticides become chemically bound and physically trapped within the interior structure of humic organic structures. Nonequilibrium adsorption also has been attributed to diffusion of compounds to sites not readily accessible to bulk flow of water or laboratory extractants (Pignatello 1989). These sites occur within the microstructure of soil particles, interstitial layers of clays, and within the matrix of soil organic matter. Degradation of pesticides by microorganisms and uptake by plants and soil microorganisms will also contribute to apparent hysteresis sorption behavior of pesticides (Hance 1988; Kahn 1991).

There is evidence that long-term sorption and entrapment of organic contaminants in soil and sediments contributes to increased environmental persistence (Harmon et al. 1989; Karickhoff and Morris 1985; Pignatello 1989; Steinberg et al. 1987). Persistence of residues bound at restricted sites may be increased by reducing transport in soil solution and reducing degradation by microorganisms. A pesticide may be adsorbed initially by low-energy bonding mechanisms. These compounds are readily accessible for degradation and equilibrium partitioning into soil solution and subsequent loss in surface runoff or leaching. However, over time pesticide residues and metabolites may be converted to more stable high-energy bond sites (Koskinen and Harper 1990). This process decreases the concentration of a compound in solution, but increases the long-term persistence and eventual loss of low levels of pesticide residues in surface runoff and subsurface flow. The highly bound pesticide residues are protected from immediate degradation and desorption into soil solution and runoff water. However, these bound compounds are slowly released into soil solution over longer periods of time than used under equilibrium experimental conditions (Hance 1988; Koskinen and Harper 1990). All of the nonequilibrium sorption processes affect surface and subsurface movement of pesticides. Incorporation of these concepts into pesticide transport models and field testing is critical for accurate prediction of the potential fate and effect of pesticide application (van Genuchten and Wagenet 1989).

Several investigators have attempted to use a "two-site" or time dependent processes to describe nonequilibrium sorption (e.g., Lee et al. 1988; Pignatello 1990b; Rao et al. 1979; Wu and Gschwend 1986, 1988). The two-site sorption concept assumes that sorption sites on soils are classified into two types: (1) sorption is instantaneous and reversible on the labile sites (Type 1); and (2) the time-dependent sorption results in slow release from the restricted sites (Type II) (Koskinen and Harper 1990; van Genuchten and Wagenet 1989). The two-site chemical process approach assumes that sorption is not at equilibrium and has a time-dependent or kinetic component. Pignatello and Huang (1991) describe a method to determine both the labile sorption coefficient (K_d) and the apparent long-term field sorption coefficient (K_{app}). In general, the labile coefficient was much smaller than the apparent field sorption coefficient. A considerable portion of applied atrazine and metoachlor residues in the field were observed as a slowly reversible form. Pignatello and Huang (1991) have hypothesized an increase in the restricted fraction of residues in field soils over time.

Another nonequilibrium approach to explain the apparent hysteresis observed in adsorption-desorption studies is the "two-region" model (van Genuchten and Wagenet 1989; Gamerdinger et al. 1990). This model of adsorption processes partitions soil water into mobile (flowing) and immobile (stagnant) regions. In the two-region model the rate of adsorption is limited by the rate at which compounds are transported to the adsorption sites within each region. The two-region model assumes equilibrium adsorption, but transport to the adsorption sites is controlled by differential rates for diffusive processes (Gamerdinger et al. 1990; Saleh et al. 1990).

4.5.4 Decomposition and Persistence

Degradation of pesticides is the only process that permanently reduces the total environmental load of pesticides resulting from application in turfgrass systems (Alexander 1981; Alexander and Scow 1989; Bollag and Liu 1990; Nash 1988; Wolfe et al. 1990). All pesticides applied to turfgrass are degraded, although the rates of degradation and hence persistence are extremely variable (Table 22). Degradation of pesticides into simpler compounds over time is the result of physical, chemical, and biological action of soil, air, water, and sunlight. Rates of pesticide degradation should not be confused with rates of general pesticide dissipation or disappearance from the soil. Permanent degradation, especially biological decomposition, is the primary mechanism reducing the environmental load of applied pesticides (Alexander and Scow 1989; Nash 1988).

Nonbiological degradation of pesticides is mediated by both physical and chemical processes (Wolfe et al. 1990). Photochemical decomposition of pesticides occurs in the atmosphere, on foliage, and on soil surfaces (Valentine 1986). 2,4-D, bromoxynil, and simazine are examples of pesticides effectively degraded by sunlight at the soil surface (Pratt 1985). Many pesticides are decomposed at measurable rates by sunlight (Haque and Freed 1974; Helling et al. 1971). However, once the pesticide has penetrated the soil surface or the zone of light penetration, photochemical reactions are no longer significant (Valentine 1986). Other nonbiological transformation reactions include chemical hydrolysis, aqueous hydrolysis, and oxidation-reduction reactions (Nash 1988; Wolfe et al. 1990).

Biodegradation of pesticides is mediated by soil microorganisms and, to a smaller extent, by plant uptake (Alexander 1981; Bollag and Liu 1990). The speed of pesticide degradation is dependent on the factors promoting the population size and metabolism of the soil microorganisms (Alexander and Scow 1989; Nash 1988; Ou 1984). These factors include soil moisture, temperature, aeration, energy and nutritional status, and pH (Table 23). The most rapid biodegradation of pesticides occurs in the warm, moist, fertile surface layers of the soil where microorganisms use pesticides as an energy source. Choi et al. (1988) observed that the effect of soil temperature, moisture, and texture on degradation of DCPA applied to turfgrass was mediated by the effects on soil microbial populations. The fastest rate of DCPA degradation, half-life of 11 d, occurred under warm and moist conditions that are favorable to soil microorganisms. Cold and dry conditions unfavorable for soil microbes resulted in a slow rate of DCPA degradation, with a half-life of 105 d.

Soil bacteria, actinomycetes, and fungi are the microorganisms primarily responsible for biodegradation of pesticides. Specific degradation mechanisms and reactions have been extensively reviewed by Alexander and Scow (1989), Bollag and Liu (1990), Kuhn and Suflita (1989), Smith (1988), and Valentine and Schnoor (1986). Microbial degradation of pesticides involves five basic processes. These transformation processes include the following:

1. Biodegradation — Pesticide is a substrate for growth.
2. Cometabolism — Pesticide is transformed by metabolic processes, but does not function as an energy source.

Table 22. Range of Reported Values for Rates of Pesticide Degradation in the Soil Root Zone

Pesticide	Degradation rate, k_s (days^{-1})	Half-life, $t_{0.5}$ (days)	Persistence[a] classification	Ref.[b]
	Insecticides and Nematicides			
Bendiocarb	0.033–0.231	3–21	3–5	8
Carbaryl	0.077–0.116	6–9	4	1
Lab	0.006–0.037	19–110	1–3	2
Field	0.10	7	4	2
Chlorpyrifos	0.005–0.116	6–139	1–4	4
Typical	0.023	30	3	4
	0.008–0.099	7–84	2–4	8
Field	0.017–0.046	15–42	2–4	8
Diazinon	0.007–0.033	21–103	1–3	1,4
Lab	0.022–0.023	30–32	2	2
	0.022	32	2	3
Turfgrass	0.012–0.099	7–56	2–4	6
	0.015–0.099	<7–45	2–4	8
Ethoprop	0.014	50	2	3
	0.011–0.049	14–63	2–4	4
Fenamiphos	0.023–0.231	3–30	3–5	4
Fenamiphos, sulfoxide	0.008	87	2	4
Isazofos	0.020	34	2	4
Isofenphos	0.002–0.023	30–365	1–3	4
Trichlorfon	0.026–0.231	3–27	3–5	4
	Fungicides			
Anilazine	0.693–1.386	0.5–1	5	4
Benomyl	0.002–0.005	119–360	1	4
Turfgrass	0.004–0.008	90–180	1–2	8
Chloroneb[c]	0.004–0.008	90–180	1–2	4
Chlorothalonil	0.008–0.050	14–90	2–4	4
Etridiozole[c]	0.035	20	3	4
Fenarimol	0.007	360	1	4
Fosetyl Al	1.386	1	5	4
Iprodione	0.023–0.099	7–30	3–4	4
Mancozeb	0.005–0.020	35–139	1–2	4
Maneb	0.012–0.058	12–56	2–4	4
Metalaxyl	0.004–0.099	7–160	1–4	4
PCNB	0.033	21	3	4
	0.002–0.005	141–434	1	8

Table 22. Range of Reported Values for Rates of Pesticide Degradation in the Soil Root Zone (continued)

Pesticide	Degradation rate, k_s ($days^{-1}$)	Half-life, $t_{0.5}$ (days)	Persistence[a] classification	Ref.[b]
	Fungicides (continued)			
Propamocarb HCl	0.023	30	3	4
Propiconazole	0.006	109–123	1	4
Thiophanate-methyl[c]	0.069	10	4	4
Thiram	0.046	15	4	4
Triadimefon	0.025–0.116	6–28	3–4	4
	Herbicides			
Atrazine	0.006–0.015	47–110	1–2	1
Lab	<0.001–0.019	36–6930	1–2	2
Field	0.042	17	3	2
	0.009	71	2	3
	0.006–0.039	18–119	1–3	4
Benefin	0.335	2	5	1
	0.005–0.029	24–130	1–3	4
Bensulide	0.004–0.023	30–180	1–3	4
2,4-D acid	0.023–0.069	10–30	3–4	1
Lab	0.051–0.063	11–14	4	2
Field	0.139	5	4	2
	0.046	15	3	3
	0.050–0.347	2–14	4–5	4,8
2,4-D dimethyl-amine salt	0.043–0.347	2–16	3–5	4
	0.030–0.173	4–23	3–5	8
DCPA	0.006–0.050	14–100	1–3	4
Turfgrass	0.002–0.053	13–295	1–3	7
Dicamba	0.020–0.214	3–35	2–5	1
Lab	0.002–0.022	32–315	1–2	2
Field	0.093	7	4	2
	0.020–0.087	8–35	2–4	4
	0.022–0.050	14–31	2–4	8
Endothall	0.099–0.347	2–7	4–5	4
	0.080–0.173	4–9	4–5	8

Table 22. Range of Reported Values for Rates of Pesticide Degradation in the Soil Root Zone (continued)

Pesticide	Degradation rate, k_s ($days^{-1}$)	Half-life, $t_{0.5}$ (days)	Persistence[a] classification	Ref.[b]
	Herbicides (continued)			
Ethofumesate	0.023–0.035	20–30	3–4	4
Glyphosate (Lab)	0.009–0.099	7–81	2–4	2
Glyphosate, amine salt	0.014–0.023	30–50	2–3	4
MCPA	0.010–0.087	8–69	2–4	10
MCPA salt	0.050	14	3	5,9
MCPA, salt (acclimated)	0.087–0.173	4–8	4–5	9
MCPA, dimethylamine salt	0.033	21	3	4
MCPP, amine salt	0.033	21	3	4
MSMA[c]	0.001	1000	1	4
Oxadiazon	0.004–0.023	30–180	1–3	4
Pendimethalin	0.001–0.087	8–480	1–4	4
Siduron	0.008	90	2	4
Simazine	0.007–0.054	13–94	2–4	1
Lab	0.014	50	2	2
Field	0.022	32	2	2
	0.009	75	2	3
Triclopyr, amine	0.008–0.023	30–90	2–3	4
Triclopyr, ester	0.008–0.023	30–90	2–3	4
	0.015	46	2	5
Trifluralin	0.003–0.096	7–267	1–4	1,4
Lab	0.001–0.025	28–533	1–3	2
Field	0.02	35	2	2
	0.005	132	1	3

[a] *Persistence classes: 1 = highly persistent, 2 = moderately persistent, 3 = moderately short-lived, 4 = short-lived, and 5 = very short-lived.*

[b] *1 = Carsel et al. (1984), 2 = Rao and Davidson (1980), 3 = Jury et al. (1984b), 4 = Wauchope et al. (1991), 5 = Balogh et al. (1990), 6 = Branham and Wehner (1985), 7 = Choi et al. 1988, 8 = Howard (1991), 9 = Kirkland and Fryer (1972), and 10 = Nash (1988).*

[c] *Estimated value.*

Table 23. Major Soil and Environmental Factors Affecting Biological Degradation of Pesticides

Property	Influence on biological degradation
Soil	
Organic matter content	Amount of organic carbon and associated nutrients are the primary food source for microbial populations degrading organic chemicals in soil. The size of the degradation community is directly related to the amount of the carbon substrate available for energy. However, increases in readily available carbon substrate have been associated with decreased pesticide degradation rates. Decreased degradation may result from avoidance of refractory pesticide substrate due to greater ease of utilization of added carbon sources (e.g., clipping residues). Organic matter also is strongly associated with pesticide adsorption and persistence. Adsorption processes and organic polymer stabilization of pesticide residues will sequester pesticides from microbial degradation. Potential decreases in degradation due to adsorption may be countered by increased microbial population size with additions of organic carbon sources. Thatch often increases the rate or pesticide degradation by providing a media conducive to high levels of microbial activity.
Clay content	Clay contributes to exchange surfaces for polarized (e.g., triazines) and cationic pesticides (e.g., paraquat). Sorbed pesticides may be sequestered from microbial attack. Clay content also will affect oxygen and moisture content of soil that will influence the microbial environment.
Oxygen content	Oxygen is the terminal electron acceptor in energy systems of many microorganisms. Aerobic or anaerobic conditions will determine the specific species available to degrade pesticide residues. Filamentous fungi and actinomycetes are strict aerobes. Soil bacteria include aerobes, strict anaerobes, and facultative aerobes. Whether aerobic or anaerobic conditions favor biodegradation is compound specific. Soil oxygen conditions will also affect the end products of degradation processes. Degradation rates are usually lower and produce intermediate metabolites under anaerobic conditions.
Soil water content	Water is required for the growth of microorganisms responsible for degradation of pesticide residues. An over supply of soil moisture will reduce gas exchange, limit soil oxygen content by microbial depletion, and thus reduce aerobic microbial activity and create anaerobic soil conditions. Optimum moisture levels for aerobic organisms range from 50 to 75% volumetric water content. Soil water content also influences the concentration of soluble pesticide in solution. Decreasing levels of soil moisture increase adsorption and sequestering of residues from degradation. Increased moisture levels will decrease pesticide concentration in solution that may

Table 23. Major Soil and Environmental Factors Affecting Biological Degradation of Pesticides (continued)

Property	Influence on biological degradation
Soil (continued)	
	reduce the rate of biotransformation. However, decreased solution concentration will dilute the concentration of toxic levels of pesticide residues, potentially increasing the rate of degradation.
pH	The level of microbial activity is affected by soil pH. Microorganisms grow in a limited pH range. Growth of soil bacteria is generally favored near neutral (6–8) pH. Actinomycetes are less tolerant of acidic (pH 5) conditions, but are better competitors with bacteria under alkaline conditions. Fungi tolerate a wide range of pH conditions, although certain species may favor a limited pH range. Soil pH will also affect the availability of pesticide residues as a carbon substrate by influencing the amount of adsorption. Increases or decreases in sorption of polar compounds are dependent on whether protonated or nonprotonated pesticide residues are preferentially adsorbed.
Nutrients	Soil nutrients, in addition to carbon, are required for microbial growth. Nutrient limitations will inhibit population development and reduce rates of biodegradation. The availability of nitrogen and phosphorus is of particular importance for proliferation of soil microorganisms.
Environmental	
Soil temperature	Microbial degradation of pesticides is generally stimulated by increases in temperature within the range tolerated by specific microorganisms. Optimum temperature ranges for microbial metabolism include mesophilic optimum activity range, 25–35°C; psychropilic optimum range, below 20°C; and thermophilic optimum range, 45–65°C. Increasing rates of degradation within temperature ranges is a result of increased rates of chemical reactions. Increases in temperature are generally expected to decrease the amount of sorbed residue and result in increasing solution concentrations. Temperature-dependent increases in solution concentration and increases in microbial activity generally will increase degradation rates.
Microbial Population	
Nature of microbial population	Degradation of specific chemicals is characterized by microbial specificity. Both chemical types and metabolic pathways are determined by the specific suite of microorganisms present in the soil. Degradation of certain pesticides may require a sequence of decomposer organisms acting simultaneously or on a series of metabolites (e.g., DDT). Population size, level of activity, and spatial distribution all affect degradation processes. In general, any increase in population size and level of activity will increase the rate of pesticide degradation.

Table 23. Major Soil and Environmental Factors Affecting Biological Degradation of Pesticides (continued)

Property	Influence on biological degradation
Microbial Population (continued)	
Acclimation	Chemical degradation frequently does not occur at an appreciable rate until well after introduction of a compound into the soil. The lag period has been attributed to the period required to allow for development of required enzyme systems or preferential development of specific degrading microorganisms. After initial acclimation of a population to a specific pesticide, the lag period for the second pesticide application usually is reduced.
Pesticide Properties	
Concentration	Solution concentration affects the availability of pesticides to microbes. High concentrations may be bactericidal or bacteriostatic, reducing the rate of degradation. Low concentrations will reduce substrate levels and possibly not induce the enzyme systems or support microbial growth required for residue degradation.
Chemical properties	Electronic structure, water solubility, polarity, and steric configuration all determine the solution concentrations of pesticide residues and their availability to microbial degradation. Use of pesticide residues as organic substrate in microbial metabolism is also dependent on the structure and composition of individual compounds. Halogenated hydrocarbon insecticides (e.g., DDT and Toxaphene) are not readily decomposed by soil microorganisms as compared to phosphorus-based insecticides and many herbicides. The degree of halogenation, number of aromatic constituents, and steric configuration will partially determine the ease of residue decomposition by soil microorganisms.

From Nash 1988 and Valentine and Schnoor 1986.

3. Conjugation — Pesticide molecules are linked with other organic compounds including other pesticides or naturally occurring chemicals.
4. Accumulation — Pesticide is incorporated into the microbial population.
5. Secondary effects — Microbial-mediated changes in soil conditions such as pH, redox potential, and addition of other reactive products may result in pesticide transformations.

Acclimation of microbial population to pesticides as an energy source, pesticide accessibility or degree of adsorption, and nature of microbial population also influences the rate and degree of pesticide degradation (Figure 3) (Moyer et al. 1972; Racke and Coats 1987; Valentine and Schnoor 1986). Acclimation of microbial population to applied pesticides and the nature of soil microorganisms may substantially enhance

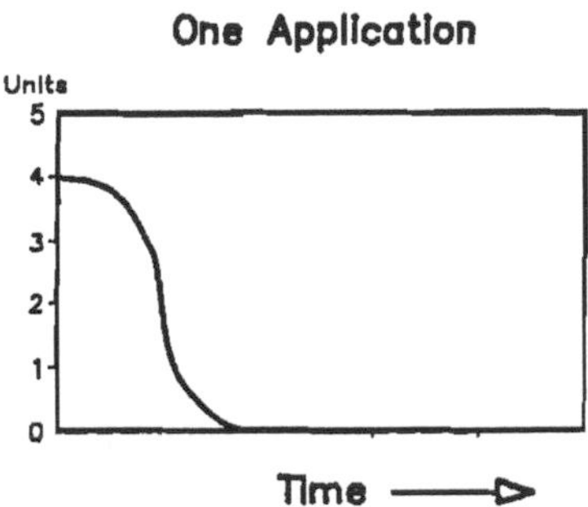

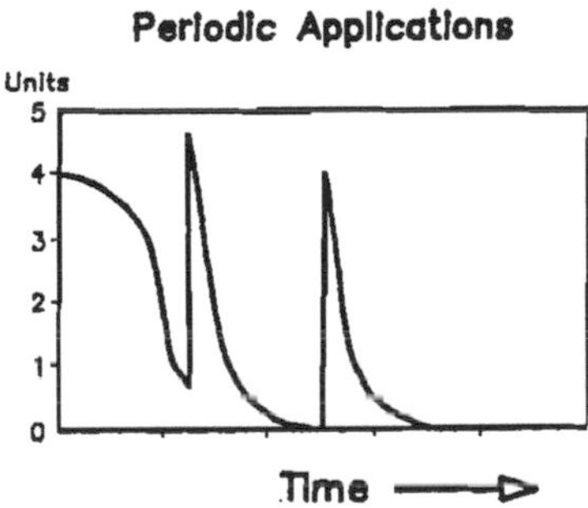

Fig. 3. Enhanced biological degradation of pesticides resulting from multiple applications.

the rate and degree of pesticide degradation. A substantial body of research, reviewed in depth by Felsot (1989), demonstrates that some labile pesticides degrade at accelerated rates after repeated application of the same or similar compounds. The enrichment of soil microbial populations and associated increased rates of degradation after repeated chemical application were first observed and later confirmed for 2,4-D, MCPA, and MCPP (Kirkland and Fryer 1972; Newman et al. 1952; Smith and Aubin 1991; Spain and van Veld 1983). Enhanced degradation has been most thoroughly studied for carbamate insecticides on agricultural soils (Felsot 1989; Read 1986). Several investigations have shown that soil microorganisms adapted to a certain pesticide enhance the degradation of compounds with closely related chemical structure (Kearney and Kellog 1985; Leistra and Green 1990). Preconditioning of soil with carbofuran resulted in enhanced degradation of other carbamate compounds such as carbosulfan and aldicarb (e.g., Harris et al. 1984; Suett 1987). Cross-conditioning of soils is the term used to describe accelerated degradation of a chemical after the soil has been originally treated with another chemical. Crossconditioned compounds are usually structurally related (Felsot 1989). Not all soils preconditioned by prior treatment with the same or similar compounds exhibit accelerated degradation (Racke and Coats 1987; Spain and van Veld 1984). Enhanced degradation has been associated with ineffective pest control in both agricultural (Felsot 1989; Leistra and Green 1990; Walker and Suett 1986) and turfgrass systems (Niemczyk and Chapman 1987).

Although accelerated degradation has not been as well-documented for organophosphate insecticides, cases of enhanced degradation and pest control failure have been reported for diazinon, isofenphos, and ethoprop (e.g., Chapman et al. 1986; Forrest et al. 1981; Racke and Coats 1987; Rhode et al. 1980; Sethunathan 1971). In a study of the fate, persistence, and accelerated degradation of isofenphos on golf course fairways, Niemczyk and Chapman (1987) speculated that the thatch layer contains a "degradative system" capable of rapidly degrading isofenphos. Isofenphos disappeared from the thatch zone within 3 d of application in thatch areas compared to almost zero degradation in nonthatch areas over the same time interval. The degradative system capable of rapid decomposition probably formed as a result of preconditioning of the site with previous isofenphos treatments. No accelerated degradation was observed in control broths of inocula from previously untreated turfgrass soil and thatch samples.

Enhanced degradation does have a positive effect by reducing the environmental load of pesticides at a rate faster than anticipated. Use of microbial enrichment of soils has been suggested as a means to increase the detoxification of soils contaminated with high levels of organic contaminants and pesticide residues (Alexander 1981; Kilbane et al. 1983; Piotrowski 1991). However, increased frequency of applications required to control pest populations as a result of shorter persistence may negate the environmental benefit of enhanced degradation.

Resistance of pests to chemical control has resulted in the use of insecticide combinations. In addition to potential for enhanced degradation as a result of preconditioning, decreased rates of degradation for some pesticide combinations may result from pesticide interactions. Smith and Willis (1987) observed that when applied as tank mixes, flucthrinate retarded the dissipation of methyl parathion, although the presence of methyl parathion increased the rate of degradation of permethrin. Similar results were observed by Brown et al. (1982). The results of these studies indicate that when two or more pesticides are applied in combination, interactions may occur that alter the persistence and environmental fate of each chemical. Increased persistence may enhance pesticide efficacy, but also may increase the potential for offsite movement in surface runoff or leaching losses.

Theoretical and observed rates of pesticide dissipation are described by the following simplified rate equation:

$$k_s = kc^n \tag{6}$$

where k_s = the rate of loss
k = the rate constant
c = the concentration (Alexander and Scow 1989; Nash 1988)

This power decay-rate model has been used to estimate persistence of pesticides in the environment. Pesticide dissipation is usually described by first-order kinetics, even when dissipation is primarily microbial (Nash 1988). When pesticide degradation results from microbes unaccustomed to the pesticide as a food source, initial dissipation is slow. As microbial populations increase in size or adaptive enzyme systems develop in the population, the rate of pesticide degradation increases with subsequent applications (Figure 3) (Anderson et al. 1989; Kirkland and Fryer 1972; Smith et al. 1989).

Use of first-order rates simplifies development of equations describing pesticide dissipation and degradation. The pseudo first-order equation for dissipation is developed by integrating Equation 6 (n = 1) such that:

$$C_d = C_o e^{-k_s d} \tag{7}$$

where C_d = the pesticide concentration on day = d
C_0 = the initial pesticide concentration or concentration when d = 0
k_s = the pesticide dissipation rate

This equation is used often to describe both general dissipation and degradation of pesticides (Nash 1988). Many management level and screening models combine biological and nonbiological degradation rates into one lumped degradation rate (Carsel et al. 1984; Jury et al. 1983; Wagenet and Hutson 1986; Wagenet and Rao 1990). The first-order description of pesticide degradation allows development of the half-life concept. Half-life of a compound is the time required for half of the chemical to degrade. To calculate the half-life of an applied pesticide, Equation 7 is rearranged into the form:

$$d_{0.5} = 0.693\,(k_s)^{-1} \tag{8}$$

where $d_{0.5}$ = the number of days required for one half of the pesticide to dissipate or degrade
k_s = the rate of loss

A list of pesticide degradation rates and general dissipation rates has been compiled from Anderson et al. (1989), Carsel et al. (1984), Nash (1988), Rao and Davidson (1980), and Wauchope et al. (1991) (Table 22). Jury et al. (1984b) categorized pesticides into five persistence classes based on rates of dissipation. The classes and ranges of K_s for each class are as follows: Class 1, highly persistent, $k_s < 0.007$; Class 2, moderately persistent, $0.007 < k_s < 0.023$; Class 3, moderately short-lived, $0.023 < k_s < 0.046$; Class 4, short-lived, $0.046 < k_s < 0.139$; and Class 5, very short-lived, $k_s > 0.139$. Field and laboratory degradation and dissipation rates are valid for the conditions of their derivation. Specific rates of pesticide loss vary by the method of determination (e.g., laboratory vs field), pesticide formulation, application history, climatic and soil conditions, and depth in the soil profile (Table 23) (Nash 1988; Valentine and Schoor 1986). Differences in observed rates of degradation result from different field and climatic conditions. Several regression models have been developed in an attempt to incorporate these environmental factors into equations for estimation of pesticide dissipation (Nash 1988).

Chemical persistence of applied pesticides has a critical influence on their fate and movement in the environment. Degradation and sorption are two of the most critical processes controlling potential pesticide leaching and runoff losses (Leonard 1990; Wagenet and Rao 1990). Chemical half-life serves one of the benchmarks for assessing potential environmental impacts of applied compounds (Table 22) (Mackay and Stiver 1991). Short-lived compounds (Classes 4–5) are less likely to be transported in surface runoff and groundwater than highly persistent compounds (Classes 1–2). Persistent compounds (Classes 1–2), even highly adsorbed compounds such as chlordane, may eventually be leached to groundwater under certain soil and hydraulic conditions (Cohen et al. 1990). The half-life of an applied compound also can be used as an indicator of potential chemical exposure to nontarget organisms (see Table 11 in Balogh et al. 1992b).

Rates of biological degradation of pesticides decrease as the chemical moves deeper in the soil profile. Reduced rates of biological degradation are a result of reduced nutrient levels, organic matter content, temperature, and size of microbial populations (Nash 1988). There is limited research available on pesticide degradation beneath the soil root zone (Nash 1988). Degradation of pesticides in the subsoil is hypothesized to be dominated by slower nonbiological reactions and reduced levels of biological degradation (Carsel et al. 1985; Nash 1988). Considerable spatial variability of pesticide degradation rates at the soil surface has been observed. Spatial variation in the degradation and recovery rates for simazine and metribuzin was attributed to initial unevenness of the initial application (Walker and Brown 1983).

Several studies have been conducted to evaluate the fate and persistence of pesticides applied to turfgrass systems. Kuhr and Tashiro (1978) studied the dissipation rates of liquid and granular formulations of diazinon and chlorpyrifos applied to bluegrass. They reported a half-life of about one week for total dissipation of the applied insecticides from the grass. The investigators were unable to distinguish the amount of pesticide dissipation due to degradation and movement into the soil. In general, chlorpyrifos persisted longer than diazinon. Concentrations of chlorpyrifos residues in soil and grass above 1 ppm were found even after 6 weeks. Very little of either insecticide was found below the 2 cm soil depth.

Sears and Chapman (1979, 1982) determined the rates of degradation of four common insecticides applied under optimum field conditions on a golf course fairway. Chlordane was found to be the most persistent when compared to chlorpyrifos, diazinon, and CGA 12223 (isazofos). Sixty percent of the initial application was found in the grass-thatch layer 56 d after the application. Chlorpyrifos was the most persistent of the nonorganophosphorus insecticides. Only 9% of the initial application was found in the grass-thatch layer 56 d after application. Both diazinon and CGA 12223 (isazofos) were completely degraded within 14 d of application. The authors also noted that only 2–8% of the initial application of the insecticide from any source was found in the upper 2 cm of soil. Residue levels in soil quickly decreased to less than 1% within 14 d of application.

Nash and Beall (1980) reported residue half-concentration time of 10 d for granular 2,4-D applied to Kentucky bluegrass in microecosystem chambers. The granular formulation was concentrated in the upper 2 cm of soil. The effect of the grass-thatch layer was difficult to assess due to failure in establishment of a healthy stand in the chambers.

Bingham and Schmidt (1967) investigated the persistence of bensulide applied in four annual applications of 15 lb ac^{-1} to Kentucky bluegrass. The results of this study showed that large amounts of bensulide were found in soil 11 months after the last annual application. Detectable quantities were found at depths approaching 13 cm into the soil. This corresponded to a dissipation rate of about 80% of the total bensulide applied over the 4-year period. Callahan et al. (1983) observed similar rates of dissipation over a 9-year period for application rates of 60 lb ac^{-1} of carbyne and 10 and 20 lb ac^{-1} of bensulide. DCPA, terbutol, benefin, simazine, and siduron were found to be much less persistent at similar application rates and frequency. Actual degradation rates were not determined in these investigations.

Murray et al. (1983) investigated the long-term persistence of six common turfgrass herbicides applied to Kentucky bluegrass. Eight annual applications of benefin, 2,4-D, DCPA, dicamba, and DSMA resulted in concentrations of 87, 5, 19, 0, and 17,800 ppb, respectively, in the upper 2 cm of soil. All levels were lower than the accepted phytotoxic level. They concluded that annual application of the herbicides caused no observable injury to the grass. The investigators also believed that despite detection of residues in the soil 1 year after application, no significant accumulation was occurring. A possible exception to this observation was the DSMA-treated bluegrass. A slight discoloration of the turfgrass occurred immediately after treatment. Residue analysis of arsenic (As) during the 8-year application period suggested an annual loss of about 20–30% of the applied DSMA with soil residue levels approaching 18 ppm. The loss of As from the system was thought to be occurring primarily through volatilization of alkylarsines. Depending on the formulation and speciation,

As is a heavy metal with demonstrated toxicity to humans, wildlife, and aquatic organisms (Eisler 1988). The proposed EPA drinking water MCL for As is 30 μg L^{-1} (ppb) (van der Leeden et al. 1990). Considering the toxicity of As, the long residual half-life of 100 d, comparatively high adsorption, and soil residuals at 18 ppm suggests that continual annual applications of DSMA may not always be an environmentally sound turfgrass management practice.

Past use of mercurial (Hg) compounds for the control of turfgrass diseases is well-documented (Britton 1969; Madison 1971; Monteith and Dahl 1932). Gilmour and Miller (1973) investigated the short-term distribution and persistence of the combination of mercurous chloride and mercuric chloride applied at a rate of 0.6 g m^{-2} to golf greens. They found that most of the applied Hg resided in the upper thatch zone and decreased by two orders of magnitude at the 2 cm soil depth (100 ppm thatch to <1 ppm soil) for both the older, previously treated green and the young, untreated green. A comparison between Hg in the thatch zone of both greens showed that Hg was accumulating in the thatch layer where downward movement under irrigation was restricted. The primary mechanism of dissipation appeared to be through volatilization of Hg(0). MacLean et al. (1973) found similar rates of accumulation in three Canadian golf courses in operation for at least 50 years. Mercury in the upper 5 cm of green showed a mean value of 57 ppm for turfgrass and 53 ppm for soil. However, on one golf course, significant movement of Hg to deeper soil layers was occurring and could pose a threat to nearby waterways. Fushtey and Frank (1981) examined Hg distribution in two other Canadian golf course greens. They also observed accumulation of Hg in the upper thatch and soil zone and noted at one site significant movement of Hg to lower soil horizons. Surface and subsurface accumulation of potentially toxic heavy metals, such as Hg, suggests that sustained use of mercurial compounds for control of fungal infestations is not a good environmental practice. Soil residue of Hg is a long-term source for low level environmental contamination.

Several studies demonstrate the significance of thatch layers on the dissipation and persistence of pesticides in turfgrass systems. The presence of thatch has been shown to affect the behavior of pesticides applied to turfgrass stands and specifically has been implicated in changing the efficacy of diazinon in field studies (Beard 1973). Several studies have been conducted to evaluate these potential effects. Branham and Wehner (1985) and Branham et al. (1985) monitored the weekly rate of loss of radio-labeled diazinon from turfgrass with and without a thatch layer in microecosystems. They found that after two weeks, more diazinon was lost from turfgrass systems with a thatch layer than without one. When a thatch layer was present, variable and poor control of soil-borne insects was attributed to the interaction of thatch and the applied diazinon. Increased adsorption of diazinon to the thatch layer reduced downward movement of the insecticide into the target soil zone and increased the rate of degradation. Well-aerated and moist thatch with high organic carbon content provides an excellent habitat for microorganisms and pesticide adsorption sites. Both of these conditions will limit the potential efficacy of diazinon for control of target insects.

Hurto et al. (1979) observed that benefin and DCPA degraded significantly faster in thatch than in soil. The authors of this study speculated that preemergence application of insecticides and herbicides to turfgrass will persist less in thatch than in soil. This condition may result in more frequent or higher rates of application to achieve the desired level of pest control when a significant thatch layer exists.

Herbicides are a valuable aid in maintaining the quality of turfgrass systems. However, as shown previously, herbicide residues may accumulate and persist in turfgrass systems. These residues can become problems during attempts to establish new turfgrass either by seeding or by sodding. After application of herbicides such as 2,4-D, dicamba, oxadiazon, and MCPP, persistence of soil and thatch residues can adversely affect seedlings for several weeks (Berry and Buchanan 1974; Bingham and Hall 1985). Residues from herbicides such as benefin, bensulide, DCPA, and pronamide, used for preemergent crabgrass or annual bluegrass control, may persist and affect seedlings for several months (Jagschitz 1971).

Several studies have been conducted to determine if activated charcoal could adsorb residual chemicals and thereby nullify their harmful effects (Jagschitz 1971, 1974). Experimental work with cabbage, beets, and strawberries show that the harmful effects of herbicide residues can be reduced with activated charcoal (Ahrens 1965; Schubert, 1967). Jagschitz (1971, 1974) found that activated charcoal improved turfgrass stands when seedbeds contained 2,4-D, A-820,* benefin, bensulide, DCPA, dicamba, MBR-8521,* mecoprop, nitralin, picloram, oxadiazon, simazine, and terbutol. Twelve other chemicals either did not inhibit seedlings or were not deactivated by the charcoal. Addition of charcoal also alleviated damage to established turfgrass from 2,4-D, bromoxynil, dicamba, endothall, linuron, and simazine. Injury from nine other chemicals was not reduced by charcoal. The herbicidal activity of bensulide was not reduced where charcoal was used in the seedbeds the year before herbicide application. The activity of bensulide, and possibly DCPA and siduron, was reduced where charcoal was used five months earlier on the turfgrass. Jagschitz (1974) noted that higher rates of charcoal caused greater reduction in herbicide efficacy due to the increase in adsorptive surface area in the soil.

4.6 PESTICIDES AND WATER QUALITY

The effects of turfgrass, farm, and forest management on environmental quality have received considerable attention over the past several years (Anderson et al.

* Numbered compounds are reported for the purpose of illustration and may not be currently used on turfgrass. However, these compounds have been included as an important component of the turfgrass research literature.

1989; Balogh et al. 1990; Leonard 1990; Walker et al. 1990). The movement of pesticide residues into surface water and groundwater has been documented as a result of pesticide use on agricultural lands (Anderson et al. 1989; Canter 1987; Leonard 1990; Smolen et al. 1984; Pratt 1985). However, the extent of surface water and groundwater contamination from individual agricultural, forested, and urban watersheds is still difficult to predict (Leonard 1990; Wagenet and Rao 1990). Transport distance, dilution, sorption processes, sedimentation, degradation, and the ability of surface water to recover from pulses of contamination are all factors confounding the establishment of relationships between field losses in water and contamination of surface water and groundwater.

From studies surveying the presence of pesticides in surface water samples, the major compounds found are heavily used, persistent, and relatively soluble pesticides (e.g., Baker 1985; Baker et al. 1981; Frank et al. 1982; Gilliom 1985; Glotfelty et al. 1984b; Muir and Grift 1987; Wu et al. 1983). These compounds include atrazine, simazine, alachlor, metolachlor, linuron, 2,4-D, and dicamba. Although the use of many organochlorine pesticides has been discontinued, residues of very persistent compounds such as DDT, DDE, DDD, toxaphene, dieldrin, endrin, and heptachlor are still regularly detected in stream and lake sediment samples (e.g., Cooper et al. 1987; Gilliom 1985; Ricci et al. 1983).

Pesticides detected in groundwater are usually either herbicides or soluble insecticides associated with agriculture (Anderson et al. 1989; EPA 1988, 1990). Several recent reviews on water quality report regional contamination of groundwater at low concentrations throughout the United States (EPA 1986, 1990; Fairchild 1987; Pratt 1985) (Tables 10, 24, and 25). Persistent and relatively mobile compounds such as atrazine, aldicarb, alachlor, DCPA metabolites, and simazine are those most often found. Atrazine, DCPA, and simazine are all major use golf course herbicides (Table 1). Detection of pesticide residues in water at any level, whether from rural or urban sources, leads to public concerns due to the unknown long-term environmental and health effects of pesticides.

The effects on water quality from pesticide use in turfgrass systems is not well-documented (Leonard 1990; Watschke and Mumma 1989). Research information on pesticide transport in agricultural systems is a reliable indicator of potential problems due to use in turfgrass systems. Many of the critical factors identified by research on pesticide loss in agricultural systems also can be identified in turfgrass systems. Research on pesticide losses from pastures, grassed waterways, and reduced tillage systems provide the best surrogate information to indicate *potential losses from golf course watersheds.*

The potential for pesticide transport to surface water and groundwater is determined by the interaction of factors including partitioning in the soil-air-water continuum, environmental persistence, climate, and management practices (Jury et al. 1987a;

Table 24. Typical Positive Pesticide Values Observed in Drinking Water Wells from Agricultural Land

Pesticide	Use[a]	States	Typical positive value (ppb)	Suggested health advisory concentration (ppb)
Alachlor	H	MD, IA, NE, PA, KS, MN	0.1–10	700
Aldicarb (sulfoxide and sulfone)	I, N	AR, AZ, CA, FL, MA, ME, NC, NJ, NY, OR, RI, TX, VA, WA, WI, NM	1–50	10–50
Atrazine	H	PA, IA, NE, WI, MD, KS, DE, MN	0.3–3	150
Bromacil	H	FL	300	7.5
Carbofuran	I, N	NY, WI, MD	1–50	50
Cyanazine	H	IA, PA, MN	0.1–1	
DCPA (and metabolites)	H	NY	50–700	5000
DBCP	N	AZ, CA, HI, MD, SC	0.02–20	0.05
1,2-Dichloropropane	N	CA, MD, NY, WA	1–50	5–10
Dinoseb	H	NY	1–5	12.5
Dyfonate	I	IA	0.1	
EDB	N	CA, FL, GA, SC, WA, AZ, MA, CT	0.05–20	0
Metolachlor	H	IA, PA	0.1–0.4	
Metribuzin	H	IA	1.0–4.3	
Oxamyl	I, N	NY, RI	5–65	250
Simazine	H	CA, PA, MD, MN	0.2–3.0	1500

From EPA 1986; Fairchild 1987; Pratt 1985.

[a] *H = Herbicide, I = Insecticide, N = Nematicide.*

Table 25. Examples of Case Studies of Pesticides in Groundwater

Source	General comments
Cohen et al. (1990)	Chlordane and nitrates were detected in groundwater monitoring wells beneath golf courses with sandy soils on Cape Cod, MA. Although chlordane is no longer used on the golf courses, the authors attributed heavy past use coupled with preferential flow through coarse soils as the factors responsible for groundwater contamination at levels that at times exceed health advisory levels. Trace levels of chlorothalonil, chlorpyrifos, 2,4-D, DCPA, dicamba, and isofenphos were detected sporadically in monitoring wells beneath heavy use areas such as greens and tees.
Frank et al. (1987)	Atrazine, cyanazine, simazine, 2,4-D, and dichlorprop were detected in well water samples associated with nonpoint runoff processes. Safe handling of pesticides near well heads, use of antibackflow devices, prompt cleanup of spills, and protection of well heads from contaminated runoff will reduce groundwater contamination.
Hallberg et al. (1983, 1984)	Pesticides have been detected in karst groundwater of the Big Spring Basin, IA. Maximum concentrations for several pesticides were 16.6 ppb, alachlor; 10.0 ppb, atrazine; 1.2 ppb, cyanazine; 0.6 ppb, metolachlor; 4.4 ppb, metribuzin; and 0.1 ppb, fonofos.
Harkin et al. (1984)	There is a potential for natural degradation of aldicarb in Wisconsin groundwater. Low levels of aldicarb are transported to high water tables primarily in the spring and fall from potato fields on sandy soils.
Hindall (1978)	There was DDT in groundwater beneath the sand plains of central Wisconsin.
Isensee et al. (1990)	Contamination of shallow groundwater beneath conventional and no-till plots with silt loam soil was monitored in Maryland. Atrazine was detected in groundwater throughout the year. Less persistent pesticides were detected in groundwater early in the growing season, but were not present in groundwater later in the season. These pesticides included alachlor, cyanazine, and carbofuran. Rainfall timing in relation to pesticide application and preferential flow processes were critical in relation to the magnitude of leaching events. No statistical difference was detected in pesticide concentration in groundwater beneath the no-till and conventional tillage plots.
Junk et al. (1980)	Atrazine at concentrations up to 3 ppb was identified in groundwater beneath irrigated sandy soil in Nebraska. Atrazine is significantly correlated with nitrates in groundwater in wells down gradient from irrigated fields in the fall. Low levels of atrazine in well water in the spring suggest atrazine may be sorbed on aquifer sediments. Dieldren was detected at levels well below 0.1 ppb.

Table 25. Examples of Case Studies of Pesticides in Groundwater (continued)

Source	General comments
Klaseus et al. (1988)	Pesticides have been detected in groundwater in agricultural regions of Minnesota. Atrazine was detected in 31% of the sampled wells (range of 0.01–42.4 ppb); alachlor was detected in 3% of the samples (0.07–4.03 ppb); 2,4-D was detected in 1.4% of the samples (0.07–5.70 ppb); dicamba, cyanazine, picloram, pentachlorophenol, metoachlor, propachlor, MCPA, aldicarb, simazine, 2,4,5-T, and EPTC were detected in fewer than 1% of the sampled wells.
Marlin and Droste (1986)	EDB, a soil fumigant, is detected in wells sampled in Connecticut. There were 321 wells that had concentrations of EDB in excess of the state water standard of 0.1 ppb.
Neil et al. (1987)	Aldicarb is detected in water samples from 50% of 274 sites sampled in the vicinity of Maine potato production. In a state survey, low levels of dinoseb, metribuzin, methamidophos, endosulfan, and degradation products of maneb and mancozeb were detected in groundwater. Pesticides were detected in sand, gravel, and bedrock aquifers. Twenty percent of samples taken from all aquifer types under potato production had low levels of pesticide contamination.
Oki and Giambelluca (1987)	The nematicides DBCP and EDB have been detected in groundwater wells in Hawaii. Maximum DBCP concentrations ranged from 2 to 115 ng L^{-1}. Maximum EDB concentrations ranged from 58 to 190 ng L^{-1}. Location of contaminated wells was correlated with past and present use of fumigants when groundwater flow patterns were considered.
Pignatello (1987)	Despite discontinued use in 1967, EDB is still detectable in shallow groundwater in Connecticut. Slower degradation rates and lower abundance of bacteria result in reduced degradation rates. EDB continues to percolate to groundwater from surface soils, where it is protected from microbial degradation in intraaggregate micropores.
Pionke et al. (1988)	Widespread contamination of groundwater with low concentrations of herbicides in the Mahantango Creek Watershed, Pennsylvania was reported. Atrazine was detected in 70% of the sampled wells with concentrations ranging from 13 ppt to 1.1 ppb. Simazine was detected in 35% of the wells within the range of 10–170 ppt, despite a single application. Degree and distribution of atrazine contamination depended on frequency of application and distribution of corn production.
Rothschild et al. (1982)	There was occurrence and movement of aldicarb in groundwater in the central sand plain of Wisconsin. Highest concentrations were found in shallow wells. Concentrations of aldicarb ranged from <1 to 210 ppb for well samples with detectable levels.

Table 25. Examples of Case Studies of Pesticides in Groundwater (continued)

Source	General comments
Scarano (1986)	Aldicarb was detected in 25% of wells sampled in a potato growing region in Massachusetts.
Spalding et al. (1980)	There was occurrence of atrazine, alachlor, and possibly 2,4-D in groundwater beneath irrigated Nebraska farmland. Levels of various organochlorine insecticides, silvex, and EPTC were below detection limits. Atrazine levels ranged from 0.06 to 3.12 ppb and were correlated to nitrate levels in groundwater. Alachlor levels ranged from <0.01 to 0.71 ppb.
Wehtje et al. (1983)	Concentrations of atrazine in groundwater from irrigated cropland in the central Platte Valley, Nebraska ranged from 0.01 to 8.29 ppb.
Wehtje et al. (1984)	Leaching is the primary mechanism of atrazine contamination of groundwater beneath sandy irrigated cropland in Nebraska. Despite limited solubility, 33 mg L^{-1} of atrazine, soils lacking appreciable levels of soil colloids are subject to considerable leaching losses of pesticides.
Zaki et al. (1982)	Occurrence and countermeasures for aldicarb in groundwater used for water supplies in Suffolk County, New York. Thirteen and a half percent of all samples had concentrations of aldicarb exceeding the recommended state limit of 7 ppb.

Jury and Valentine 1986; McCall et al. 1983). Adsorption, mobility and persistence of pesticides are the primary factors that determine their effectiveness as pest control agents, as well as their potential for surface water and groundwater contamination (Rao et al. 1988). Although research on surface water quality is often separated from groundwater quality, surface water and groundwater are often intimately linked in certain regions. Depending on hydraulic gradients and regional geology, surface water may directly recharge groundwater (Leonard 1990).

Loss of pesticides in surface runoff, subsurface drainage to surface water, and subsurface leaching directly affect the quality of water associated with golf course turfgrass. The process of surface runoff, erosion, and subsurface movement of water in turfgrass was discussed in detail by Balogh and Watson (1992). The following sections discuss the movement of pesticides from turfgrass by surface and subsurface transport.

4.6.1 Runoff and Surface Water Quality

Loss of agricultural pesticide residues into the environment by surface runoff has been a major research and water quality concern for the last 25 years (Anderson et al. 1989). Pesticide runoff includes dissolved, suspended particulate, and sediment-adsorbed pesticides that are transported by water from a treated land surface (Leonard 1988, 1990; Wauchope 1978). Several extensive literature reviews on the mechanisms and occurrence of surface runoff of pesticides are available (Anderson et al. 1989; Caro 1976; Leonard 1988, 1990; Stewart et al. 1975). These reviews provide insight into the factors that influence surface losses of pesticides from agricultural and turfgrass systems. Although sediment and runoff losses from turfgrass systems are small compared to agricultural systems (Balogh and Watson 1992), surface transport processes must be understood for sound management of construction, renovation, and turfgrass establishment projects.

Pesticide runoff includes dissolved, suspended particulate, and sediment-bound pesticides that moves with surface runoff water (Leonard 1990). The principal factors determining runoff (soluble and sediment-bound) losses of pesticides include rainfall characteristics, interval between pesticide application and onset of runoff, pesticide chemical properties, rate and method of pesticide application, rate of field degradation, soil properties and antecedent moisture conditions, ground cover, and transport distance (Table 26) (Leonard 1988, 1990; Wauchope 1978). Partitioning of pesticides between surface transport and subsurface transport depends on chemical characteristics and soil properties (EPA 1988; Leonard 1988, 1990; Smolen et al. 1984; Stewart et al. 1975). The high variability of pesticide losses observed in runoff result from (1) the complexity of precipitation and erosion processes, (2) foliar washoff patterns,(3) spatial heterogeneity of soil physical and chemical properties, (4) differences in experimental and monitoring techniques, and (5) differences in management objectives all introduce variations between observed and predicted pesticide losses in runoff.

Entrainment and transport of pesticides in runoff from turfgrass systems is a micro- and macroscale process (Leonard 1990). At the microscale entrainment of pesticides in runoff is described by (1) diffusion and turbulent transport of dissolved pesticide from the soil pores to the stream of runoff, (2) desorption of pesticides from organic and soil surfaces into the moving transport stream, (3) dissolution of pesticide particulates, (4) scouring of pesticide particulates into the runoff stream, and (5) entrainment of pesticides adsorbed to soil particles moving in the sediment stream by erosive forces. Macroscale processes affecting runoff losses include surface topography, soil erosivity, and cultural practices. The factors and specific examples of research on these processes are summarized in Tables 26 and 27.

Concentration of pesticides in runoff generally decreases exponentially with time (Leonard et al. 1979; Wauchope and Leonard 1980). The decrease in runoff concentrations for pesticides, other than strongly adsorbed compounds (e.g., glyphosate), reflects the decrease in pesticide present in the runoff-active zone at the soil surface (Leonard et al. 1979). Runoff potential of pesticides is greatly reduced if (1) runoff is delayed by 30 d or more after application, (2) rainfall or irrigation leaches the chemical beneath the soil surface prior to occurrence of runoff, or (3) the pesticide is incorporated into the soil (Leonard 1990; Stewart et al. 1975; Wauchope 1978).

Table 26. Factors Affecting Pesticide Entrainment and Transport in Runoff

Factors	General comments
Climatic	
Rainfall/runoff related to pesticide application	The highest concentration of pesticide in runoff occurs during the first major runoff event after pesticide application. After initial losses, pesticide concentration and availability at the soil and turfgrass surfaces dissipate, often exponentially with time.
Rainfall intensity	Runoff occurs when rainfall rate exceeds the infiltration rate. Increasing intensity of rain increases the rate of runoff water and energy available for pesticide extraction and transport and decreases the time before the onset of runoff within a storm. Rainfall intensity also may affect depth of surface interaction.
Rainfall/duration/ amount	Affects runoff volume and subsurface soil leaching. Pesticide washoff from foliage is related to amount of rainfall.
Time to runoff after start of rainfall	Runoff concentrations increased as time to runoff decreased. Pesticide concentrations and availability are greater in the first part of rainfall before reduction occurs due to leaching and incorporation by raindrop impact. Greatest surface loss of pesticides occurs when storm runoff occurs immediately after application.
Water temperature	Little information is available on temperature effects. Increasing temperature usually increases pesticide solubility and decreases adsorption.
Soil	
Soil texture and organic matter content	Infiltration rates are affected. Soil texture affects soil erodibility, particle transport potential, and chemical enrichment factors. Organic matter content affects pesticide adsorption and mobility. Runoff is usually higher on finer-textured soils. Time to runoff is greater on sandy soils, reducing initial runoff concentrations of soluble pesticides.
Surface crusting and compaction	This decreases infiltration rates, reduces time to runoff, and increases initial concentrations of soluble pesticides in runoff water.
Water content	Increasing initial soil water content may increase runoff potential, reduce time to onset of runoff, and reduce subsurface leaching of soluble chemicals before start of runoff.
Slope	Increasing slope may increase runoff rate, soil detachment and transport, and effective surface depth for chemical extraction.

Table 26. Factors Affecting Pesticide Entrainment and Transport in Runoff (continued)

Factors	General comments
Soil (continued)	
Degree of aggregation and stability	This affects infiltration rates, crusting, effective depth for chemical entrainment, sediment transport potential, and chemical enrichment in sediments.
Sediment concentration	Sediment concentration also controls pesticide partitioning in runoff. Solution transport dominates when sediment concentration is low. When sediment concentrations are high, surface movement of pesticides on eroded particles is high for pesticides with either high or intermediate adsorption.
Pesticide	
Solubility	Soluble pesticides may be more readily leached into the soil or removed from surface residue and foliage during the initial rainfall. Therefore, very soluble chemicals are less likely to be lost in surface runoff compared to pesticides with intermediate and low solubility. When time to runoff is short, increased solubility may enhance the initial concentration of pesticides in runoff water.
Sorption properties	Pesticides strongly adsorbed on soil or plant surfaces are retained near the soil surface and possibly are more susceptible to runoff if not incorporated into the soil. Runoff amounts are then dependent on soil erosion and sediment transport. Highly adsorbed pesticides are transported almost exclusively on sediment; highly water-soluble pesticides are lost primarily in leachate; pesticides with intermediate solubility are lost in surface runoff water and on eroded soils.
Polarity/ionic characteristics	Adsorption of nonpolar compounds is determined by soil organic matter; adsorption of ionized compounds and weak acids/bases is affected more by mineral surfaces and soil pH. Lipophilic compounds (nonpolar) are retained on foliage by leaf surfaces and waxes. Polar compounds are more easily removed from foliage by rainfall.
Persistence	Higher probability of runoff occurs in pesticides which remain at the soil surface for longer periods of time because of their resistance to volatilization, chemical, photochemical, and biological degradation.
Formulation	Wettable powders are susceptible to entrainment and transport. Liquid forms may be more readily transported than granular. Esters less soluble than salts produce higher runoff concentrations where initial leaching into soil is important.

Table 26. Factors Affecting Pesticide Entrainment and Transport in Runoff (continued)

Factors	General comments
Pesticide (continued)	
Application rate	Runoff potential and concentrations are proportional to the amounts of pesticide present. At the usual rates of application for pest control, sorption and degradation are not affected by initial residue levels of pesticides.
Placement	Concentrations exposed to runoff are reduced by pesticide incorporation or placement below the soil surface.
Management and Cultural Practices	
Erosion control practices	Reduces transport of adsorbed/insoluble compounds. Reduces transport of soluble compounds if, after pesticide application, runoff volumes are reduced during critical periods.
Residue management	Turfgrass canopies can reduce pesticide runoff by increasing the time to the onset of runoff, decreasing runoff volumes, and decreasing erosion and sediment transport. Pesticide runoff may be increased where pesticides are washed from clippings, thatch, or turfgrass surfaces directly into runoff water under conditions with high initial soil water, clay soil, and intense rainfall immediately after pesticide application.
Vegetative buffer strips	Transport of some pesticides may be reduced if buffer strips are around treated areas. Reduction can occur due to secondary infiltration, sediment deposition, and sorption on plant surfaces and detritus.
Irrigation practices	Pesticide application by sprinkler irrigation may move soluble pesticides into soil surface and reduce runoff potential. Watering in pesticides on turfgrass will increase movement into the thatch layer and soil, where the chemical is less susceptible to surface runoff losses in subsequent precipitation events.
Cultivation and aeration	Cultivation and aeration of turfgrass soils will maintain high rates of infiltration and soil drainage. Although high rates of infiltration and drainage reduces pesticide losses in surface flow, movement of chemicals to tile drain systems may reemerge as surface runoff in drainage effluent.
Tile drainage	Tile drainage reduces surface runoff by increasing subsurface movement of excess water. Pesticides in tile drainage effluent may become a component of surface runoff, if the effluent drains directly into surface water or back onto the golf course watershed.

From Caro 1976; Leonard 1988, 1990; Wauchope 1978.

Table 27. Runoff Losses of Selected Pesticides from Treated Plots and Watersheds. Pesticides Represent the Extremes of Types and Formulations

Pesticide/ catchment type	Precipitation/ irrigation input	Duration of input/ variables investigated	General comments and conclusions	Ref.
		Alabama		
Atrazine/ 20 m^2 plots	Simulated rain	Single severe storms/effects of soil type	Soil type affected runoff loss; greatest portion of loss occurred in water phase in first 50 min of runoff.	Bailey et al. (1974)
		Arizona		
Picloram/ 146 ha Pinyon-Juniper watershed	Rainfall	Persistence and movement of picloram observed for several years	Total of 1.1% of application lost from target area in runoff; maximum concentration in runoff was 320 $\mu g\ L^{-1}$; picloram not detected in streamflow further than 5.6 km downstream.	Johnsen (1980)
Picloram/ 0.85 ha brush watershed	Rainfall	2-month study on brush treated at high application rate (10.4 kg ha^{-1}) and 20% slope	Total of 4.5% of application lost in runoff; maximum concentration was 370 $\mu g\ L^{-1}$; concentration was diluted by water from untreated areas.	Davis and Ingebo (1973)

Table 27. Runoff Losses of Selected Pesticides from Treated Plots and Watersheds. Pesticides Represent the Extremes of Types and Formulations (continued)

Pesticide/ catchment type	Precipitation/ irrigation input	Duration of input/ variables investigated	General comments and conclusions	Ref.
		California		
Trifuluralin, DCPA, chlorpyrifos, diazinon/ 28 ha agricultural fields	Flood/ irrigation	Studies conducted 3 years/data-base of several compounds developed	Seasonal losses of total application were 0.29 and 0.14% for trifluralin, 1.22 to 1.40% for DCPA, 0.02 to 0.24% for chlorpyrifos, and 0.04 to 0.07% for diazinon of total application; maximum concentrations in runoff were 19 µg L^{-1} for trifluralin; 189 µg L^{-1} for DCPA, 480 µg L^{-1} for chlorpyrifos, and 22 µg L^{-1} for diazinon.	Spencer et al. (1982b, 1985)
		Georgia		
Atrazine/ 20 m^2 plots	Simulated rain	Single storms/rainfall timing and amount	Runoff concentrations highest immediately after application and decreased rapidly during storm.	White et al. (1967)
Atrazine/ 0.02 ha watershed	Rainfall	Two growing seasons/ten plots, two rates	Atrazine detected in runoff only during first 26 d after application; seasonal losses ranged from 0.22 to 2.24%.	Rohde et al. (1981)

Atrazine, trifluralin, cyanazine/ 1.3–1.4 ha watersheds	Rainfall	2–3 year duration/conservation practices	Seasonal losses ranged from 0.2 to 2% of application and primarily were lost in water phase of runoff.	Leonard et al. (1979)
Cyanazine, sulfometuron-methyl/ 3 m² plots of Bermuda grass and bahiagrass	Simulated rain	Runoff from storm events/ effects of rates, formulation, and grass cover	Losses of both herbicides in different formulations ranged from 1 to 2% of total application. Total losses over time were sensitive to length of time between application, initiation of rainfall, and the initiation of runoff. Leaching of herbicide into the soil made them unavailable for runoff losses. For a given runoff volume, herbicide losses were a constant fraction of the total amount applied with or without grass cover. Surface losses of chemicals from grassed plots was one half of the losses from bare soil because of reduced runoff from the turfgrass plots.	Wauchope et al. (1990)
2,4-D acid/ 30 m² plots	Simulated rain	Transport through adjacent grassed buffer strips	Losses during simulated storm 1 d after application were 10.3 and 2.5% of application from initially wet and dry soils, respectively.	Asmussen et al. (1977)
2,4-D salt/ 20 m² plots	Simulated rain	Runoff from extreme storm events/effects of initial soil moisture and salt formulation	Runoff losses were 3–8% of application after 127 mm 2 hr^{-1} storm from soils initially dry and wet, respectively.	Barnett et al. (1967)

Table 27. Runoff Losses of Selected Pesticides from Treated Plots and Watersheds. Pesticides Represent the Extremes of Types and Formulations (continued)

Pesticide/ catchment type	Precipitation/ irrigation input	Duration of input/ variables investigated	General comments and conclusions	Ref.
		Georgia (continued)		
2,4-D ester/ 20 m^2 plots	Simulated rain	Runoff from intense storms/ effect of soil water content and ester formulations	After 127 mm 2 hr^{-1} storm losses were 27 and 26% of that applied from initially dry and wet soils, respectively.	Barnett et al. (1967)
2,4-D, trifluralin/ 1.3 and 1.4 ha watersheds	Rainfall	Monitor runoff 2 months after application for 2 years	Seasonal losses were 1.0 and 0.007% of application for 2,4-D and 0.1 to 0.3% of application for trifluralin; Maximum concentration for 2,4-D was 309 and 1.2 μg L^{-1} with the highest losses from watershed with runoff occurring soon after application. Maximum concentration of trifluralin in runoff was 400 μm L^{-1} occurring soon after application.	Smith et al. (1978)
2,4-D salt/ 0.34 ha watershed and 30 m^2 subplots	Rainfall + simulated rain	Watershed study conducted for three growing seasons with natural rain/simulated rain on subplots; salt formulation	Seasonal losses in runoff were 0.33, 0.27, and 0.04% under natural conditions; intense simulated rain 1 d after application resulted in 1.5% loss in runoff.	White et al. (1976)

Ethoprop/ small watershed	Rainfall	Runoff losses for different formulations	Losses in runoff for liquid formulation (0.1% of application) were much greater than runoff losses of granular formulation (0.01% of application).	Rohde et al. (1979)
Trifluralin/ 0.34 ha watershed	Simulated rain	2-year growing season/subplots subjected to intense rain in addition to natural rain	Seasonal losses 0.03 to 0.17% under natural conditions; simulated rain produced losses up to 1.3%; immediate off-site transport limited by grassed buffer strips.	Rohde et al. (1980)
		Iowa		
Atrazine/ 0.5–1.2 ha watersheds	Rainfall	Growing seasons/total herbicide losses and mode of transport	60% of losses in water phase; seasonal losses from 2.5 to 15.9%, highest transport value may be overestimated.	Ritter et al. (1974) Wauchope (1978)
Atrazine, alachlor, cyanazine, fonofos/ small watershed	Rainfall	Growing seasons/4 years, tillage variables	2.1, 1, 2.1, and 0.4% average seasonal losses atrazine, alachlor, cyanazine, and fonofos, respectively. Losses depend on time between application and runoff. Treatments that decrease runoff and erosion decreased herbicide losses. Decreased losses were not proportional as concentrations were higher in reduced tillage systems.	Baker and Johnson (1979)

Table 27. Runoff Losses of Selected Pesticides from Treated Plots and Watersheds. Pesticides Represent the Extremes of Types and Formulations (continued)

Pesticide/ catchment type	Precipitation/ irrigation input	Duration of input/ variables investigated	General comments and conclusions	Ref.
		Iowa (continued)		
Atrazine, alachlor/ small plots	Simulated rain	122 mm rain applied in 1.75 hr/ surface application vs incorporation; effects of wheel-track compaction	Total losses were 4.7, 1.6, and 18.3% for surface, incorporated, and surface + wheel-track compaction, respectively, for atrazine. Total losses of alachlor were similar to or greater than losses of atrazine.	Baker and Laflen (1979)
Atrazine/ small boxes	Simulated rain	70 mm rain applied/limed vs unlimed soil compared	Runoff losses very low from lime soil due to little runoff water; when runoff occurred, losses were 3.7% of application.	Gaynor and Volk (1981)
Atrazine, alachlor/ 14 m^2 plots	Simulated rain	127 mm rain applied in 2 hr; simulation of 100 year storm/ effect of herbicide placement with respect to residue	Residue increased time to runoff and reduced losses; placement had no effect on runoff concentration, but concentration negatively correlated with time to runoff.	Baker et al. (1982)

Alachlor, cyanazine, fonofos/ 33 m^2 plots in three locations	Simulated rain	216 mm rain applied in three events/different amounts of residue on plots	Losses of alachlor, cyanazine, and fonofos were 7.9, 11.0, and 1.8%, respectively; reduced herbicide losses due to reductions in runoff water and sediment loss was offset by higher herbicide concentrations in runoff water from conservation systems.	Baker et al. (1978)
		Louisiana		
Trifluralin 445 m^2 plots	Rainfall	3 years	Total seasonal losses were 0 to 0.05% of application; concentrations did not exceed 0.5 $\mu g\ L^{-1}$.	Willis et al. (1975)
		Maryland		
Atrazine, alachlor/ field-catchment	Rainfall	Growing season; quantify seasonal losses/mode of transport	Seasonal losses of atrazine in runoff were 1% of total applied; seasonal losses of alachlor were 0.2% of total applied.	Wu (1980)
Atrazine/ estuary of Chesapeake Bay	Rainfall	Seasonally; 3 years	2–3% of applied atrazine moved to estuary in 1 year with significant runoff within 2 weeks of application; 15 $\mu g\ L^{-1}$ was maximum loading of estuary.	Glotfelty et al. (1984b)

Table 27. Runoff Losses of Selected Pesticides from Treated Plots and Watersheds. Pesticides Represent the Extremes of Types and Formulations (continued)

Pesticide/ catchment type	Precipitation/ irrigation input	Duration of input/ variables investigated	General comments and conclusions	Ref.
		Maryland (continued)		
Atrazine, alachlor/ estuary of Chesapeake Bay	Rainfall	3 years/eight watersheds with varying land use; corn as major crop	About 1% of applied atrazine lost in runoff seasonally; conservation tillage reduced runoff volume, atrazine losses were not reduced in proportion because concentration was higher under reduced tillage; alachlor concentrations in runoff were significantly lower than observed for atrazine.	Wu et al. (1983)
		Mississippi		
Atrazine/ 2.7 m^2 tilt-bed trays	Simulated rain	23–33 mm rain applied in 5–7 min/effect of formulation; effect of application rate	Total losses ranged from 4 to 8% for emulsions and dispersible liquid formulation; total losses ranged from 9 to 12% for wettable powder and dispersible granule formulation. First flush of runoff carries the highest concentration (23 ppm) of atrazine in water phase of runoff.	Wauchope (1987a)

Trifluralin/ 15.6 ha watershed	Rainfall	2-year observations/other pesticides also monitored	Seasonal losses were about 0.2%; trifluralin yields in runoff were linearly related to sediment yield in runoff.	Willis et al. (1976)
Trifluralin/ 15.6 ha watershed	Rainfall	6-year observation/sediment yield compared to pesticide yield	Mean discharge (weighted) concentration was 0.4 μg L^{-1}; trifluralin yield was poorly correlated with annual sediment yield, but was directly proportional to clay and organic matter concentration of sediment.	Willis et al. (1983)
		Nebraska		
Atrazine/ 2025 ha agricultural watershed	Rainfall	3 years	Maximum runoff concentrations were 24 μg L^{-1} for atrazine and 1.4 μg L^{-1} for alachlor.	Schepers et al. (1980)
		North Carolina		
Trifluralin/ 17 m^2 plots	Rainfall	6–8 month study/fate of several applied pesticides	Total seasonal runoff losses ranged from 0.3 to 0.5% of application; maximum concentrations in runoff were 8–24 μg L^{-1}.	Sheets et al. (1972)

Table 27. Runoff Losses of Selected Pesticides from Treated Plots and Watersheds. Pesticides Represent the Extremes of Types and Formulations (continued)

Pesticide/ catchment type	Precipitation/ irrigation input	Duration of input/ variables investigated	General comments and conclusions	Ref.
		North Dakota		
Sulfometuron/ 21.8 m^2 plots of bluegrass and wheatgrass	Rainfall	2-year study/fate and movement of herbicide applied to grassed slopes of 2, 8, and 16%.	The highest concentration of herbicide was 1 $\mu g\ kg^{-1}$, regardless of slope. Downslope movement of herbicide attributed to movement of both plant litter and soil particles.	Lym and Swenson (1991)
		Ohio		
Atrazine/ 2.7 ha watershed	Rainfall	Growing season/tillage effects	Runoff losses were 4.7% of application; runoff losses dependent on runoff timing and volume.	Edwards (1972)
Atrazine, simazine/ 0.4–3.5 ha watersheds	Rainfall	Growing season; 3 years/ compare conventional till and no-till	Average seasonal losses ranged from 0.2 to 5.7% of total applied and were primarily lost in water phase. Time of runoff relative to application was one of the most important factors in runoff losses of herbicides.	Triplett et al. (1978)

Atrazine alachlor, simazine, metribuzin, metoachlor, butylate, carbofuran, fonofos/, complex agri-cultural basin	Rainfall	Growing season; 1981	7.5% of applied atrazine in streamflow; rainfall was two to three times the normal during study period; maximum streamflow concentration was 87 $\mu g\ L^{-1}$ for atrazine, 105 $\mu g\ L^{-1}$ for alachlor, 7.4 $\mu g\ L^{-1}$ for simazine, 2.3 $\mu g\ L^{-1}$ for metribuzin, 0.49 $\mu g\ L^{-1}$ for butylate, 104 $\mu g\ L^{-1}$ for metoachlor, 45 $\mu g\ L^{-1}$ for carbofuran, and 1 $\mu g\ L^{-1}$ for fonofos.	Baker et al. (1981)
Glyphosate/ tall fescue pasture	Rainfall	Three growing seasons/preseeding treatment for establishment of a tall fescue pasture	Highest concentration of glyphosate occurred in runoff (5.2 ppm) 1 d after application. At low application rates (1.1–3.3 kg ha^{-1}) concentration in runoff rarely exceeded 100 ppb.	Edwards et al. (1980)
Picloram/ 0.0008 ha	Rainfall	11 months	0.007% of application lost in runoff from grass sod.	Glass and Edwards (1974)
		Oregon		
2,4-D/ hill pastures	Rainfall	Brush control and fate of applied pesticides	Losses in stream flow were 0.014% of application; source of streamflow contamination was from channel banks; no overland flow was observed.	Norris et al. (1982)
Atrazine/ 40 m^2 plots, 14% slope	Rainfall	Application rates/seasonal relationships	Seasonal losses range from 1.7 to 4.7% of application.	Hall et al. (1972)

Table 27. Runoff Losses of Selected Pesticides from Treated Plots and Watersheds. Pesticides Represent the Extremes of Types and Formulations (continued)

Pesticide/ catchment type	Precipitation/ irrigation input	Duration of input/ variables investigated	General comments and conclusions	Ref.
		Oregon (continued)		
Atrazine/ 40 m^2 plots, 14% slope	Rainfall	Growing season/mode of transport in runoff	Four to eight times greater losses in solution phase compared to sediment.	Hall (1974)
Atrazine/ 40 m^2 plots 14% slope	Rainfall	Growing season/application rates and method; cropping patterns	Cropping patterns had the highest effect on transport of atrazine in a growing season with a 100-year storm event. Losses of 3.5 and 0.33% without and with an oat strip were observed. Preplant incorporation also reduced surface losses. Amount of nonrunoff producing precipitation between runoff events and length of period between runoff and application was related to concentrations of atrazine in runoff water and total atrazine delivery in runoff, respectively.	Hall et al. (1983)

		Texas		
Picloram/ semiarid rangelands	Rainfall irrigation	Persistence and runoff potential	Concentrations in runoff were 1 μg L–1 when runoff occurred 30 d or more after application.	Scrifes et al. (1971)
Picloram/ rangeland	Rainfall	Picloram in reservoirs after application	Concentrations in reservoirs adjacent to treated areas ranged from 55 to 180 μg L^{-1} when runoff occurred within 2 weeks of application.	Haas et al. (1971)
Picloram/ 9 m^2 plots	Irrigation simulated rain	2 years/runoff timing, application rates, and application patterns	Runoff losses were 3.2% of application when upper half of plots were treated; runoff losses were 5.5% when entire plot was treated.	Trichell et al. (1968)
Picloram/ 3.7–7.2 ha rangeland watersheds	Rainfall irrigation	Persistence and transport in runoff from grass/180 d after application	Concentrations in runoff were 17 μg L^{-1} 10 d after application; 20 to 30 d after application concentrations in runoff were 1 μg L^{-1}. 0.002% of application lost in runoff.	Scrifes et al. (1977)
Picloram/ 1.2 ha grassed watershed	Rainfall	Concentrations in runoff monitored during 6 month spray period/5 applications	Concentrations ranged from 400 to 800 μg L^{-1} in runoff if heavy rainfall occurred immediately after application; runoff concentrations were 5 μg L^{-1}, if rain occurred 1 month after application; plant washoff was main source of picloram in runoff.	Bovey et al. (1974)

Table 27. Runoff Losses of Selected Pesticides from Treated Plots and Watersheds. Pesticides Represent the Extremes of Types and Formulations (continued)

Pesticide/ catchment type	Precipitation/ irrigation input	Duration of input/ variables investigated	General comments and conclusions	Ref.
		Texas (continued)		
Picloram/ 8 ha grassed watershed	Rainfall	Monitor pesticide dissipation dilution from application zone	Maximum concentrations were 48 and watershed 250 μg L^{-1} in initial runoff in 1978 and 1979; concentrations decreased with distance and time from application; 6% of application lost in runoff during 1 month period when conditions were conducive to runoff events.	Mayeux et al. (1984)
		Wisconsin		
Atrazine, alachlor/ 234 m^2 plots	Simulated rain	Growing season/compare tillage treatments	Runoff volumes were reduced by conservation tillage, but higher concentration in runoff offset reductions in losses. Maximum average loss of applied atrazine and alachlor was 5.8 and 4.0%, respectively, when high intensity rain event occurred within 7 d of application.	Sauer and Daniel (1987)

Loss of sediment and surface runoff from established turfgrass is limited compared to agricultural systems (Balogh and Watson 1992; Gross et al. 1990, 1991; Koehler et al. 1982; Welterlen et al. 1989). The primary mode of surface movement of pesticides from established turfgrass is in runoff water. However, under heavy or sustained runoff conditions, small amounts of sediment are still lost from turfgrass plots (Balogh and Watson 1992; Gross et al. 1991). During golf courses construction, renovation of existing facilities, or turfgrass establishment, bare soil surfaces are exposed to considerable loss of runoff water and eroded soils. Research has shown that considerable sediment and runoff water is lost from construction sites or sites with bare soil compared to sites with established turfgrass (Daniel et al. 1979, 1982; Wauchope et al. 1990; Wolman and Schick 1967). Use of preemergence herbicides and preventative application of insecticides or fungicides on areas under construction on golf courses may result in pesticide losses in runoff water and entrained as a sediment-bound phase. Understanding the processes and relationships of runoff, sediment transport, and surface movement of pesticides is critical for environmentally sound management of golf courses and turfgrass.

Several studies indicate the greatest loss of pesticides results from storms that occur soon after initial applications (Tables 26 and 27). Since turfgrass systems often receive several applications per year, the *potential* for loss of pesticides in runoff water from turfgrass may actually be greater than from agricultural systems (Watschke et al. 1988). Hall et al. (1987) conducted experiments under controlled indoor growth chambers and outdoor conditions to determine the amount of 2,4-D recovered in runoff from Kentucky bluegrass sod. When runoff was collected immediately after application of 1 kg 2,4-D ha^{-1} under controlled and outdoor conditions, 91 and 71% of the 2,4-D was found in the effluent water. After 10 d, the amount of pesticide recovered in the runoff water decreased to 32%. Within 14 d after herbicide application, approximately 10% of the applied 2,4-D was still present in the runoff water from sod grown outdoors.

Edwards et al. (1980) applied glyphosate to several watersheds as a preseeding herbicide for establishment of a tall fescue pasture in a no-tillage system. Glyphosate was applied at three rates of 1.1, 3.3, and 8.96 kg ha^{-1}. The highest concentration of glyphosate, 5.2 ppm, was observed in runoff occurring 1 d after application of the herbicide at the highest application rate. Glyphosate at levels up to 2 ppb were detected in runoff at this site for up to 4 months after application. At sites receiving the lower application rates, maximum concentrations rarely exceeded 100 ppb in runoff for events close to the time of application. Concentration in runoff decreased to <2 ppb within 2 months of treatment. In each of the three study years, herbicide transport in the first runoff event following treatment accounted for 99% of the total runoff transport. The maximum amount of glyphosate lost in runoff was 1.85% of the total applied.

Brejda et al. (1988) studied the offsite transport of atrazine from a dense grass sod in a subirrigated meadow in the sandhills of Nebraska. Following application of atrazine to this semiarid site, they observed insignificant movement of this herbicide which has intermediate mobility. They did note that late season application may have greater potential for contamination of surface water than spring application due to slower degradation in cooler soils.

Wauchope et al. (1990) applied 4.5 kg ha^{-1} of cyanazine and 0.4 kg ha^{-1} sulfometuron-methyl in different forms to mixed stands of bahiagrass and bermudagrass. Although losses of sediment and runoff water were reduced significantly on the turfgrass plots compared to the plots with bare soil, surface losses of all formulations were 1–2% of the total applied regardless of grass cover. The first sample in each runoff event always had the highest herbicide concentration. Total losses of all formulations were sensitive to the length of time between herbicide application, initiation of rainfall, and the start of runoff from the plots. Leaching of the herbicide below the active mixing zone made it unavailable for movement in runoff. The investigators observed rapid infiltration of water into the soil prior to initiation of runoff. Initial high leaching of water and pesticides into the soil increased the time to the start of runoff and lowered the concentration of pesticides in the runoff. Using computer simulation, the investigators also compared herbicide losses from the grassed and bare plots. The total amount of herbicide lost from grassed plots was less than one half of the losses for the bare plots.

Rhodes and Long (1974) studied the runoff and leaching characteristics of benomyl in soils and turfgrass and found that less than 0.1% of the amount applied was found in runoff from the turfgrass site. They concluded that benomyl is immobile in soil and turfgrass and does not leach or move significantly from the site of application.

Duble et al. (1978) evaluated the drainage characteristics of calcium arsenate applied to golf greens for control of grubs. Arsenic was found in significant concentrations in both drainage and runoff from lysimeters. Concentrations of As in the runoff from the sites were as high as 14 ppm when a heavy rain occurred shortly after application. Concentrations exceeding 0.8 ppm were found in runoff even 115 d after application. Concentrations of As in the runoff and drainage from all plots exceeded acceptable limits for irrigation water established by the U.S. Public Health Service and water quality MCLs proposed by the EPA (van der Leeden et al. 1990).

In reviews of agricultural losses of pesticides, Anderson et al. (1989), Nash (1988), Smolen et al. 1984, and Wauchope (1978) reached the following conclusions regarding pesticide losses in surface runoff (see also Table 27). Surface losses of water-insoluble pesticides are less than 1% of the total amount applied. Depending on slope and hydrologic conditions, wettable powders produce the highest surface losses ranging from 2 to 5% of the total applied. Water-soluble and soil-incorporated pesticides show losses of 0.5% or less. Catastrophic surface losses of pesticides are associated with runoff events occurring immediately after surface application of pesticides. Catastrophic losses are defined as surface transport of 2% of the applied pesticide in a single event. If runoff events occur within 2 weeks of application, surface losses of wettable powders and soluble compounds will increase threefold. In general, annual runoff losses of pesticides are quite low (2–3% of total applied) when compared to dissipation by volatilization and degradation. However, under intense precipitation conditions, the upper limit of pesticide loss may be 18% for a single event.

A wide variety of cultural practices have been suggested to reduce nonpoint contamination of surface water and groundwater by agricultural pesticides (Anderson et al. 1989; EPA 1988; Leonard 1988, 1990; Smolen et al. 1984; Stewart et al. 1975). Practices which are useful in turfgrass management include alternating use of pesticides, integrated pest management techniques, control and timing of pesticide application, and soil and water conservation practices. The amount and exposure of a pesticide at the soil surface during a runoff event is a key variable in determining the amount of pesticide lost in runoff (Wauchope 1978). Changes in the rate and method of application, formulation, and type of pesticide will influence the concentration and mass of pesticides lost in surface runoff (Caro 1976; Leonard 1988). Management practices that influence infiltration, runoff timing, runoff volume, soil water content, and sediment yield will determine changes in pesticide transport in runoff.

In several extensive literature reviews, estimated edge-of-field seasonal losses have been established (Table 27). Observed runoff losses of pesticides on an event basis have been characterized as nearly random (Smith et al. 1978). The cited reviews and other studies from pasture, vegetative buffer areas, and reduced tillage systems illustrate the complex interaction of chemical properties and site-specific conditions influencing pesticide transport in runoff (Table 27). Several of these studies indicate that similar pesticide losses from turfgrass systems are to be expected.

4.6.2 Subsurface Movement of Pesticides

Research on leaching of pesticides from the soil surface to groundwater is limited in comparison to studies on surface transport of pesticides (Anderson et al. 1989; Fairchild 1987). The lack of published research on pesticide management practices for turfgrass and their impact on groundwater quality reflects the recent attention these issues have received (Walker et al. 1990). Evaluation of pesticide contamination of groundwater resources has relied heavily on extrapolation of transport processes in the root zone, limited field soil- and water-monitoring studies, soil column studies, well water surveys, and computer simulation (Cohen et al. 1990; EPA 1986, 1988, 1990; Fairchild 1987; Fermanich et al. 1991; Kladivko et al. 1991; Leonard and Knisel 1988; Shoemaker et al. 1990).

The results of studies identifying the occurrence of pesticides in groundwater beneath agricultural systems has been extensively reviewed by Anderson et al. (1989), Canter (1987), and Fairchild (1987). Studies on potential contamination of groundwater with agricultural chemicals has focused on relatively persistent chemicals such as aldicarb, alachlor, and atrazine (Table 25). Based on chemical characteristics, aldicarb is considered a mobile compound, while atrazine and alachlor have intermediate mobility (Table 21).

The potential for transport of pesticides from the root zone to groundwater is determined by a complex interaction of chemical, soil, climatic, and management factors (Tables 12 and 28) (EPA 1988; Pratt 1985; Smolen et al. 1984). The chemical properties that influence leaching potential include the rate of application, application technique and formulation of the pesticide, water solubility, soil adsorption, volatility, degradation, and persistence. Critical soil properties related to subsurface movement of pesticides include texture, organic matter content, cation exchange capacity, structure, porosity, and moisture conditions. Other site-specific factors related to subsurface losses of pesticides include depth to groundwater, geologic substrate, specific type of turfgrass cover, and cultivation practices. Climate factors such as the precipitation amount, timing, intensity and frequency, and antecedents soil moisture content all affect leaching (Table 28).

The variation and timing of precipitation or irrigation in relation to pesticide application, occurrence of preferential flow paths, pesticide handling and disposal practices, and well location have profound effects on pesticide transport (Bailey and White 1970; EPA 1988). These factors are associated with increased leaching potential despite chemical and soil properties that attenuate pesticides within the root zone. Fractured aquifers, shallow aquifers, and intensified irrigation practices are additional unique concerns for nonpoint impacts on groundwater quality (Pratt 1985; Smolen et al. 1984). Development of golf courses on sites with shallow sandy soils, shallow aquifers, or karst topography have greater potential for subsurface losses of applied chemicals.

The critical influence of precipitation and irrigation events on surface transport in relation to timing of pesticide application is well recognized (Table 27) (Leonard 1988, 1990; Wauchope 1978). Climate also has considerable impact on pesticide transport through the root zone to groundwater (Anderson et al. 1989). However, the impact of precipitation and temperature interactions on partitioning of pesticides between surface runoff and subsurface leaching has not been well-characterized. The total amount of rainfall or irrigation water, the intensity, and frequency of precipitation all affect the movement of pesticides in soils (e.g., Hall et al. 1989). Based on computer simulation of pesticide movement by Balogh and Gordon (1987), the amount and depth of pesticide leaching within surface soils is directly related to the amount and frequency of precipitation or irrigation.

Table 28. Factors Affecting Potential Pesticide Transport to Groundwater in Agricultural and Turfgrass Systems

Factor	Impact on transport of pesticides to groundwater
Climate	
Precipitation	Precipitation and/or irrigation has a dominant effect on mass transport of pesticides. Degree of pesticide leaching depends directly on the amount of water infiltrating the soil surface. Timing, duration, and amount of rainfall is determined by the season and geographic location. In general, pesticides are more likely to leach below the root zone when soil is at or near saturation (maximum hydraulic conductivity) after heavy chemical application. Leaching of pesticides is potentially high when extremely intense precipitation occurs immediately after application on highly permeable soils. Leaching losses can be minimized by proper timing of application in relation to seasonal trends and current meteorological forecasts. Reduction of solution runoff by management practices or increases in the intensity and duration, frequency, and timing of precipitation or irrigation events will increase the amount of water available for subsurface transport of pesticides.
Evapotranspiration	Evapotranspiration (ET) is the amount of applied water transpired by plants and evaporation from the soil surface. ET represents the volume of water lost to the atmosphere and not available for subsurface drainage. The extent of ET in relation to the amount of precipitation or irrigation affects the mass flux of water and pesticides beyond the root zone. If ET does not exceed precipitation or irrigation, there will generally be a long-term downward flux of water and dissolved chemicals past the root zone. When ET exceeds infiltration, there is usually insufficient water to transport pesticides past the root zone. ET is regulated by meteorological conditions including solar radiation, wind, air temperature, and moisture gradients. Soil and plant resistances also play a major role in regulation of ET under dry soil conditions.
Temperature	Temperature affects all rate processes including pesticide volatilization, adsorption, and degradation. Temperature has an important impact on the amount of pesticide in solution and available for subsurface transport. Higher temperatures generally increase pesticide degradation and reduce residues available for transport to groundwater. Increased temperatures may decrease pesticide adsorption and increase the amount of residues in soil solution available for transport beyond the root zone. Temperature impacts on mass transport of water in soils is poorly understood and is likely to have minor influence on mass transport of solutes.

Table 28. Factors Affecting Potential Pesticide Transport to Groundwater in Agricultural and Turfgrass Systems (continued)

Factor	Impact on transport of pesticides to groundwater
Soil Properties	
Water content	Soil water ultimately transports pesticides not adsorbed on soil surfaces. Soil water content influences both diffusive and mass transport. Soil water content also influences partitioning of pesticides between solution, vapor, and adsorbed phases in soil. Higher soil water content decreases vapor, diffusion, and mass transport path lengths. Increased soil water levels will increase both water and chemical mass flux within and beyond the root zone. Mass transport is the primary means of pesticide transport to groundwater. Under dry soil conditions, adsorption of pesticides is significantly increased and volatile losses are decreased. Under dry conditions pesticides are immobilized due to both increased adsorption and lack of water movement.
Bulk density	Porosity is related to bulk density. Porosity is a function of pore size and distribution as determined by soil texture, structure, and particle shape. Decreasing porosity or increasing bulk density decreases mass transport of pesticides by increasing adsorption surfaces, decreasing the cross-sectional area available for mass flow, and increasing path lengths of water flow.
Hydraulic conductivity	Hydraulic conductivity is the proportionality coefficient between water flux and moisture gradient. Hydraulic conductivity is a measure of a soil's ability to transmit water and entrained solutes. Soils with high hydraulic conductivity in relation to initial infiltration of water (e.g., sands) have a greater potential for mass transport of both water and dissolved solutes past the root zone.
Texture	The particle size distribution will determine, in part, water retention, porosity, hydraulic conductivity, and adsorption site density of a soil. In general, sandy soils have greater capacity for mass transport and the least capacity for adsorption of pesticides as compared to heavier-textured soils including silt loams and clays.
Soil structure	Aggregation of soil particles into soil structure will affect potential pesticide transport. Structure is influenced by both texture, organic matter content, and soil management. Both texture and structure determine the amount of infiltration and subsurface movement of water. Highly structured soils will have preferential pathways allowing mass transport of water and chemicals beyond the root zone.

Table 28. Factors Affecting Potential Pesticide Transport to Groundwater in Agricultural and Turfgrass Systems (continued)

Factor	Impact on transport of pesticides to groundwater
Soil Properties (continued)	
Organic matter content	The amount of adsorption of most nonionic pesticides is directly related to the soil organic matter content. Higher adsorption will temporarily attenuate mass transport of pesticides to groundwater, allows for slower release of pesticides to soil solution, and increases the amount of time available for residue reduction by degradation. Increased adsorption also decreases diffusion and volatile pesticide losses.
Thatch layer	The thatch is a continuous organic medium which may form between the turfgrass canopy and mineral soil. This organic layer intercepts applied pesticides before they enter the mineral soil. The rate and extent of pesticide movement through the thatch layer is reduced in many cases by increased adsorption and decomposition. Thatch may act as a barrier to direct access to preferential pathways in well-structured soils.
Clay content	Polar and ionic pesticides will be attracted to the charged surfaces of the soil exchange complex which includes both mineral (clay) and organic surfaces. Increased adsorption for polar and ionic residues decreases potential transport to groundwater as previously described for organic matter content. Adsorption processes and low hydraulic conductivity will reduce mass transport of applied pesticides in clay soils. However, potential formation of preferential pathways (e.g., surface cracks and root channels) coupled with precipitation immediately after application may result in deep leaching of pesticides even in soils high in clay or organic matter content.
Depth to groundwater	Depth to groundwater affects the travel time of a chemical leached from the root zone to groundwater. Shallow groundwater has been found to have a greater potential for contamination with pesticides. Depth to groundwater also influences the potential for upward transport of water from groundwater to the surface under high evaporation conditions. This also will influence the upward transport of leached pesticide residues.
Pesticide Properties	
Water solubility	The ability of a pesticide to dissolve in water is directly related to adsorption characteristics. The solubility of nonionic pesticides is proportional to the mass carried by soil solution to both surface water and groundwater.

Table 28. Factors Affecting Potential Pesticide Transport to Groundwater in Agricultural and Turfgrass Systems (continued)

Factor	Impact on transport of pesticides to groundwater
Pesticide Properties (continued)	
Sorption properties	The degree of adsorption of a pesticide to soil particles is defined as the ratio of pesticide concentration in soil to pesticide concentration in water (K_d). This adsorption partition coefficient of nonionic pesticides is highly correlated with the organic carbon content. The partition coefficient, K_d, is standardized for organic carbon content as the organic carbon partition coefficient (K_{oc}). The lower the K_d and K_{oc} values, the less likely a pesticide will be adsorbed to soil particles and the more likely they will be transported to groundwater. Most pesticides currently found in groundwater have had site-specific K_d values less than 1. Generally these compounds have standardized adsorption partition coefficients, K_{oc}, less than 300.
Volatility	Volatilization of pesticides from soils and plant surfaces is a major loss pathway for applied pesticides. Although volatile losses represent an effective economic reduction of applied pesticides, pesticides lost to the atmosphere are generally not at risk for transport to groundwater. Volatility is primarily a function of vapor pressure of the chemical, environmental conditions, and vapor concentration gradients. Certain compounds have high vapor pressure, but are also water soluble. These compounds may be leached to significant depth in soil when applied just prior to irrigation or rainfall.
Persistence/ dissipation	Degradation is the primary mechanism of environmental reduction of pesticide residues. In general, pesticides with greater resistance to degradation have a greater potential for subsurface transport when other dissipation processes (e.g., runoff and volatilization) are equal. Persistent and pesticides with high or intermediate solubility are particularly susceptible to mass transport to groundwater.
Formulation	The form of a pesticide will affect the potential for leaching to groundwater. Wettable powders, concentrated emulsions, liquid concentrates, and aqueous solutions all have different impacts on transport characteristics. Use of surfactants increases penetration and translocation of pesticides applied to foliage, but these additives will also increase potential for leaching in soil by decreasing adsorption to soil particles. Wettable powders, dusts, and microgranuales are highly soluble and generally susceptible to leaching losses. These solid forms do not volatilize as rapidly as liquid and aqueous forms and may persist in soil, affording more time in which leaching may occur. Ester formulations are less soluble than salts and are less susceptible to subsurface leaching.

Table 28. Factors Affecting Potential Pesticide Transport to Groundwater in Agricultural and Turfgrass Systems (continued)

Factor	Impact on transport of pesticides to groundwater
Pesticide Properties (continued)	
Application rate	The amount of pesticide is directly proportional to the risk of subsurface transport. Many cases of documented contamination are associated with heavy and/or long-term use of relatively soluble and persistent pesticides. Use of band application or application to areas only requiring treatment will reduce potential for leaching by use of less compound per unit area of application.
Placement	The potential for pesticide leaching to groundwater is higher for surface-applied and soil incorporated pesticides than for foliar-applied pesticides. The leaching potential is particularly high if pesticides are applied to a site conducive to leaching (low organic matter content and sandy texture). Nonuniform application of pesticides will result in pesticide leaching from areas of overconcentration. Proper equipment calibration and uniform application will reduce the quantity of pesticide required and reduce pesticide leaching. Improper calibration, nozzle height, and nozzle spacing will result in nonuniform application which will increase the potential for leaching from concentrated bands.
Timing	Timing of application relative to precipitation events is a major factor in determining potential leaching of pesticides beyond the root zone. Leaching is minimized when the applied pesticide is fully utilized or when soil conditions promote degradation. Application of pesticides prior to irrigation or precipitation will increase the potential for leaching losses of pesticides.
Management	
Irrigation	Irrigation has an effect similar to precipitation on transport processes, although there is a greater degree of management potential with its application. Salinity control requires application of sufficient water to exceed ET and maintain drainage conditions. Irrigation has a complex interaction with mass transport of water and dissolved pesticides. Chemical concentration tends to increase as water passes through the root zone, potentially increasing adsorption and decreasing mass chemical transport. The primary effect of irrigation is on soil water content of the root zone. Irrigation at a higher intensity will increase soil water content, mass transport, and diffusive transport. Irrigation practices that influence the potential for pesticide leaching include method of irrigation, timing of application, volume, and frequency of application. Overwatering promotes mass flux and desorption processes, especially for permeable soils. Sprinkle irrigation will promote pesticide leaching when it washes off residues from foliar applications prior to full utilization.

Table 28. Factors Affecting Potential Pesticide Transport to Groundwater in Agricultural and Turfgrass Systems (continued)

Factor	Impact on transport of pesticides to groundwater
Management (continued)	
Chemigation	Chemigation involves application of pesticides mixed with irrigation water prior to field application. Research is limited on the potential for leaching of pesticides and nutrients. Concern has been expressed that leaching may be increased due to continuous application or pulse of dissolved chemicals. Another potential source of contamination is faulty, leaking, or nonexisting antibacksiphoning devices.
Chemical handling	Proper handling and disposal of containers is essential to avoid both subsurface and surface contamination. Spills of concentrated pesticides near well heads may result in well contamination and plumes of contamination. Backsiphoning during tank preparation also will result in point source well and groundwater contamination. Improper disposal of unused pesticides also may result in subsurface transport of chemicals past the root zone.
Residue and clipping management	Surface residues may reduce runoff and surface transport by increasing the amount of infiltration. Increased infiltration of water will increase the potential for subsurface mass transport by increasing soil water content, conductivity, and decreasing adsorption of applied chemicals. These effects may be offset by increased levels of organic carbon and degradation rates. Where runoff volumes are increased by residue management, leaching potential is decreased at the expense of increased losses to surface water. Leaching of turfgrass clippings in compost piles may be a source of subsurface movement of pesticides if (1) clippings have a high concentration of recently applied chemicals and (2) precipitation occurs before degradation processes are increased by elevated temperatures and microbial activity in the compost.
Erosion control and cultivation	Cultivation and erosion control practices both influence the quantity of pesticides used and site leaching potential. Potential losses of preemergent herbicides and pesticides applied during establishment of turfgrass may result from sediment transport, surface runoff, and leaching. The overall impact of cultivation and aeration on subsurface transport is unclear and probably site and climate specific. Water conservation, cultivation, and soil aeration practices that reduce runoff by enhancing infiltration will increase the potential for mass transport of dissolved pesticides to groundwater.

Table 28. Factors Affecting Potential Pesticide Transport to Groundwater in Agricultural and Turfgrass Systems (continued)

Factor	Impact on transport of pesticides to groundwater
Management (continued)	
Plant characteristics	Plant rooting depth influences water uptake and chemical transport. Shallow rooted plants have greater potential for leaching of pesticides by reducing water uptake within the soil profile. Roots left by decaying plants may be involved in forming macropores and preferential flow phenomena. Selection of pest and disease resistant turfgrass cultivars will reduce the quantities of pesticides applied on a given site and reduce overall application rates. Maintaining healthy turfgrass using both chemical and alternate control strategies reduces the need for periodic and heavy pesticide application during significant pest or disease infestations. Improved growth conditions and periodic rotation of host varieties will help eliminate pest species and the need for chemical control methods. Adjusting planting and sequences of turfgrass establishment may reduce potential groundwater contamination by reducing the need for application of pesticides. These are all components of integrated systems for turfgrass management.

From EPA 1988; Jury 1986C; Walker et al. 1990.

There is a wide range of field and chemical conditions that influence potential transport of pesticides to groundwater. When persistent compounds that have high or intermediate mobility are applied to soils with minimal adsorptive capacity and shallow water tables, the leaching process may result in groundwater contamination. On the opposite end of the risk spectrum is the use of pesticides that are rapidly decomposed and/or strongly adsorbed. When these chemicals are applied to soils with high adsorptive capacity, deep water tables, and appropriate water management, the probability of pesticide leaching is significantly reduced. However, under certain hydraulic conditions, even very persistent and immobile compounds, such as chlordane, will be leached to groundwater (Cohen et al. 1990). Therefore, the complex suite of conditions controlling the movement of applied chemicals must be evaluated on a site-specific basis.

Research on aquifer contamination in Wisconsin, Nebraska, and New York has demonstrated that agricultural pesticides with high or intermediate solubility (e.g., aldicarb, alachlor, atrazine) applied for sustained periods will result in pesticide leaching and potential groundwater contamination (Tables 10, 11, and 25) (EPA 1986; Hall et al. 1989; Rothschild et al. 1982; Spalding et al. 1980; Wehtje et al. 1983, 1984). Many of these studies were located in areas with sandy soils with low organic matter content. These soil conditions are conducive to rapid movement of organic chemicals. Many of the reported cases of pesticide contamination of groundwater are associated with "normal and approved uses." There has been speculation that increased use of approved pesticides and inadvertent overuse may be associated with increased leaching potential (Anderson et al. 1989). Therefore, it is to be expected that in sensitive soil and geological areas, similar low level losses will result from pesticide use on turfgrass.

Mass transport of pesticides is the generalized process of water and chemical movement in the unsaturated zone as was discussed in Chapter 2 (Balogh and Watson 1992). The total movement of pesticides in soil solution is the sum of both mass flow and hydrodynamic dispersion. Excellent reviews of the quantitative aspects of pesticide transport have been presented by Enfield and Yates (1990), Jury et al. (1987a), and Rao et al. (1988). A complete derivation of the coupled water and solute transport equation using a convective/dispersive rationale (Bear 1972) has been presented by Enfield and Yates (1990), Jury et al. (1987a), and Rao et al. (1988). Combining the dispersive/convective equations for water movement and the equation for conservation of mass, the vertical transient flux of nonvolatile pesticides is described by the general coupled solute-water transport equation,

$$\frac{\partial}{\partial t}(\theta C+\rho S)=\frac{\partial}{\partial z}\left[D_h\theta\left(\frac{\partial C}{\partial z}\right)-qC\right]+U(z,t) \qquad (9)$$

where, C = the pesticide concentration in solution
θ = the volumetric water content
S = the pesticide concentration on the soil-organic matter matrix
ρ = the soil bulk density
D_h = the combined hydrodynamic dispersion coefficient
q = the soil water flux determined by coupled solution of the Richard's equation
t = time
z = depth

U(z,t) represents pesticide losses and gains due to processes including biological and abiotic degradation, plant uptake, and solubilization/precipitation. The sorbed phase pesticide concentration, S, is determined using the previously defined adsorption partition equation

$$S = K_dC = (K_{oc}f_{oc})C \qquad (10)$$

Simple first-order kinetics are often used to describe pesticide transformations which lead to the following expression for U:

$$U = -(k_c\theta C + k_s\rho S) \qquad (11)$$

where k_c and k_s are the first-order rate coefficients for pesticide degradation in solution and sorbed phases, respectively. Substituting Equation 10 into Equation 11 and assuming $k_c = k_s = k$ then

$$U = -[k(\theta + \rho K_d)C] \tag{12}$$

where k is the pooled degradation coefficient for all mechanisms and pathways (Table 22). Substitution Equations 10 and 12 into Equation 9 gives a general equation for convective/dispersive pesticide transport defined in Equation 13.

$$\left(\theta+\rho K_d\right)\frac{\partial C}{\partial t} = \theta D_h \frac{\partial^2 C}{\partial z^2} - q\frac{\partial C}{\partial z} - \left[k\left(\theta+\rho K_d\right)C\right] \tag{13}$$

By defining additional variables such as the retardation factor,

$$R = [1 + (\rho K_d\, \theta^{-1})] \tag{14}$$

the average water velocity,

$$v = q\, \theta^{-1} \tag{15}$$

$$D_h{}^* = D_h\, R^{-1} \text{ and } v^* = v\, R^{-1} \tag{16}$$

and using these variables in Equation 13, the general pesticide transport equation is expressed as

$$\frac{\partial C}{\partial t} = D_h^* \frac{\partial^2 C}{\partial z^2} - v*\frac{\partial C}{\partial z} - kC \tag{17}$$

Equation 17 describes nonvolatile pesticide transport, sorption, and degradation in soils during one-dimensional, steady, vertical flow of water in homogeneous soil profiles. A variety of analytical solutions and numerical modeling techniques have been developed to solve this equation under different boundary conditions (Oster 1982). Computer simulation of pesticide transport using finite differencing techniques allows solution of this equation for use in multilayered soils under field management conditions. Use of simulation techniques are discussed in the following chapter (Balogh et al. 1992b).

Limitations in use of the convective transport model for field assessment of pesticide transport to groundwater has been discussed by Jury (1983), Nielsen et al. (1983), and Rao et al. (1988). Problems encountered in evaluation of transport theory include validity of model assumptions, failings in solution of the Richard's equation, problems in describing preferential and macropore flow phenomena, inability to determine model parameters, lack of field testing, and spatial variation of physical and chemical processes.

A direct consequence of preferential movement of water through soil macropores is that dissolved pesticides can bypass the soil matrix that attenuate leaching processes (Everts et al. 1989; Starr and Glotfelty 1990; Thomas and Phillips 1979). This enables surface-applied chemicals to move faster and further than would be expected based on equilibrium adsorption and dispersion/convective transport theory. Computer simulation models based on convective transport and equilibrium adsorption process will not account for these preferential flow patterns (Balogh et al. 1992b)

Starr and Glotfelty (1990) observed a significant potential for accelerated movement of agricultural chemicals through preferential flow channels when major leaching events occur immediately after application. Preferential flow and the attendant potential for nonequilibrium adsorption suggests that preferential flow may have a significant affect on the leaching potential of chemicals usually adsorbed by soils. Ghodrati and Jury (1990) demonstrated with dye patterns that water is preferentially channeled through the bulk soil even in structureless sandy soils. Differences in surface preparation, irrigation patterns, and fingering patterns of infiltration into layered sandy soil will result in preferential movement of water and solutes through the soil matrix (Baker and Hillel 1990; Ghodrati and Jury 1990; Hillel and Baker 1988). This type of preferential flow and potential for accelerated transport of dissolved pesticides will affect the movement of pesticides through layered sandy soils beneath golf course greens. In a monitoring study by Cohen et al. (1990), leaching of the immobile compound chlordane to groundwater beneath golf courses on sandy soils was attributed to preferential flow processes and perhaps poor well construction techniques.

Kladivko et al. (1991) studied the chemical content of tile drain effluent from agricultural fields. They detected small amounts of atrazine, cyanazine, carbofuran, and alachlor in subsurface flow within 3 weeks of application and after less than 2 cm of tile flow. The early arrival or rapid flux of pesticides in tile drainage effluent is consistent with the concept of preferential movement through soil macropores and channels. Kladivko et al. (1991) also found greater subsurface losses of pesticides with increasing density of tile drainage lines. Rapid movement of pesticides and chemical tracers has been observed in several other studies of tile-drained soils (Bottcher et al. 1981; Everts et al. 1989; Richard and Steenhuis 1988).

Tile drainage systems also are used on golf course greens and fairways (Beard 1982). Rapid removal of excess water is essential for turfgrass management on greens and wet areas on fairways. However, golf course developers and turfgrass managers should be aware of the potential relationship of rapid flux of pesticides in tile-drained systems. In addition to promoting the subsurface movement of water and including any entrained chemicals, effluent from tile drainage systems may become a component of surface runoff from drained turfgrass areas. The coarse aggregate surrounding the tile drains also is directly exposed or brought to the surface. Therefore, these drains accept both surface runoff and subsurface flow.

Under different conditions, soluble pesticides are applied to the soil and precipitation or irrigation does not occur immediately or at a level insufficient to leach chemicals out of the root zone. In this situation, the soluble pesticide will have sufficient time to diffuse into micropores. Subsequent movement of water through macropores will bypass micropores and will not result in movement of pesticides to significant depths in the soil (Wagenet 1987; Wauchope 1987b). Assessment of the impact of macropore flow on subsurface movement of nitrate and pesticides has become a research priority of the USDA (1989a, 1989b).

In addition to pesticide transport in moving water, research has suggested that under certain conditions pesticide will be translocated through the soil profile bound to the surface of suspended particles and water-soluble organic matter. Vinten et al. (1983) demonstrated that pesticides with high K_d values, usually associated with surface transport on sediment, also may be transported vertically through the soil profile on mobile colloid surfaces. Under conditions causing dispersion and downward transport of soil colloids, strongly adsorbed pesticides including glyphosate and the organochlorine insecticides could be transported through the unsaturated zone. Binding of pesticides by water-soluble organic constituents such as fulvic and humic acid has been associated with subsurface movement of pesticides not normally associated with leaching losses (Lee and Farmer 1989; Madhun et al. 1986). Under these conditions, use of equilibrium adsorption isotherms requires modification.

There are several case studies of pesticide leaching from turfgrass that specifically address the mobility of pesticides in turfgrass systems. Niemczyk et al. (1988) traced the downward movement of the insecticides, isazofos, isofenphos, and ethoprop, after application to an Ohio golf course fairway. They found that even at times as long as 91 d after application, as much as 90% of the residues could still be found in the thatch layer. Despite the high water solubility of the formulations, regular rainfall, and irrigation, very little movement of the insecticides into the underlying soil was observed. The thatch layer acted as an efficient "trap" which prevented the downward movement of the insecticides. Branham and Wehner (1985) found similar results for the vertical movement of diazinon applied to Kentucky bluegrass in a microecosystem chamber. These investigators observed that 96% of the diazinon residue remained in the thatch layer and upper soil zone (0 – 10 mm) 22 d after application.

In similar research, Niemczyk and Krueger (1987) and Niemczyk (1987) showed that application of isazofos and isofenphos insecticides to turfgrass with a definite thatch layer resulted in little leaching of the active ingredient into the first 2.5 cm of soil. They noted that movement of the insecticides into the soil beneath thatch is not accelerated by irrigation applied soon after pesticide treatment. However, differences in the levels of pesticide residues in thatch between the immediately irrigated and nonirrigated sites were observed. Irrigation washed the insecticides off the turfgrass blades into the target zone for the intended pests. Irrigation after pesticide treatment reduces the potential for exposure to golfers, domestic animals, and wildlife.

Brejda et al. (1988) followed the mobility and dissipation for 1 year after application of atrazine to a subirrigated meadow in the semiarid sandhills of Nebraska. They noted that over 90% of the residue was located in the upper 5 cm of soil. Low concentrations were observed in the 5–15 cm depth, but decreased to less than detectable levels in about 100 d. The leaching potential under these conditions appeared to be minimal. Rhodes and Long (1974) determined in laboratory and greenhouse studies that benomyl in turfgrass and soil plots was relatively immobile and did not move or leach significantly from the site of application.

Gold et al. (1988) monitored the leaching characteristics of 2,4-D and dicamba applied at different rates to home lawns overlying a sandy loam soil. 2,4-D has intermediate mobility characteristics, while dicamba is considered to be relatively mobile in the environment (Balogh et al. 1990). Leaching of these herbicides was monitored by Gold et al. (1988) with ceramic extraction plates located 0.2 m below the surface. Regular irrigation was applied to the lawn. The geometric mean concentrations in the extraction plates ranged from 0.55 to 0.87 $\mu g\ L^{-1}$ for 2,4-D and from 0.26 to 0.55 $\mu g\ L^{-1}$ for dicamba. These concentrations are much lower than the recommended MCLs for the two chemicals. Low concentrations in soil solution were attributed to the existence of excellent degradation conditions in the turfgrass root zone.

Watschke and Mumma (1989) examined the leaching and runoff characteristics of pendimethalin, chlorpyrifos, 2,4-D, and 2,4-DP applied to turfgrass plots. Pendimethalin and chlorpyrifos were not present in detectable amounts either in leachate or runoff due to their high binding affinity for soil and thatch (Niemczyk et al. 1977). For 2,4-D, dicamba, and 2,4-DP, the highest transport rates were noted for the postapplication interval of 2 d after application. Total losses in runoff and subsurface flow still represented less than 2% of the total applied pesticide. Sufficient dissipation of the herbicides occurred between 25 and 46 d since no chemical was detectable in leachate after that period. Slight differences were noted between dicamba and the phenoxy herbicides in their partitioning between runoff water and leachate. These differences were probably related to the solubility and adsorption characteristics of the different chemicals. The relatively soluble dicamba was more effectively leached than moved in overland flow. The less soluble phenoxy compounds were more evenly distributed between surface and subsurface flow.

The results of these studies on turfgrass soils suggest that with few exceptions, low percentages of applied pesticides are available for movement beneath the turfgrass root zone. Losses of 1 or 2% of an applied pesticide may translate to concentrations in the range of 1–3 ppb. Since this level would be at the current MCL for atrazine, these leaching levels could be considered by some to be highly significant. In addition, processes occurring on small plots, soil columns, and microecosystems must be extrapolated with care to the larger watershed or field conditions of a golf course facility.

Cohen et al. (1990) monitored 19 wells located on four golf courses on Cape Cod, MA. They sampled both soil and well water sporadically over a 1.5-year period and determined the concentrations of 17 different pesticides commonly used in golf course maintenance. The soils of the area are sandy and have limited subsurface organic matter content. These conditions are considered highly conducive to pesticide leaching (EPA 1986) (see Table 13 in Balogh et al. 1992b). Seven of the 17 pesticides were never detected in the water samples on any of the sampling dates. Chlordane, a highly adsorbed, immobile, yet persistent compound, was detected at concentrations exceeding the current HALs and suggested MCLs. Leaching of this immobile chemical could be attributed to repeated heavy applications coupled with preferential flow of bound particulate phase or cross-contamination during well installation. Heptachlor residues in groundwater were also detected at levels exceeding current HALs. Chlorothalonil, chlorpyrifos, 2,4-D, DCPA, DCBA, dicamba, isofenphos, and 3,5,6 trichloro-2-pyridinol were periodically detected in the monitoring wells.

Based on these monitoring results and considering that the other pesticides were not detected in the water samples, Cohen et al. (1990) concluded there was no cause for concern regarding the use of currently *registered* pesticides. However, detection of a highly immobile compound in the groundwater suggests other interpretations of this study are possible. In this monitoring study, sampling did not necessarily coincide with major precipitation events nor pulses of water moving through the unsaturated zone. Preferential pulses of relatively mobile compounds, such as dicamba, could easily move past the monitoring wells and escape detection. Surrogate monitoring data does not always provide definitive answers on pesticide fate and transport processes. Sole reliance on periodic monitoring of groundwater wells, especially in sandy soils underlying many golf courses, may not detect transient surges of chemicals passing through permeable vadose zones and aquifers. Well-designed, process-oriented research with appropriate sampling schedules should be drawn from a wide range of climatic, soil, and management conditions. Results obtained under these conditions substantiates claims of movement or nonleaching of pesticides to groundwater.

4.7 MANAGEMENT GUIDELINES RELATED TO PESTICIDES AND WATER QUALITY

A wide range of management guidelines have been suggested to reduce nonpoint contamination of surface water and groundwater with pesticides (Anderson et al. 1989; Caro 1976; EPA 1988; Felsot et al. 1990; Leonard 1988, 1990; Smolen et al. 1984; Stewart et al. 1975). On the basis of these reviews of the current scientific literature, pesticide persistence and transport of pesticides in surface water and groundwater are the critical concerns from an environmental and water quality perspective. General guidelines for environmentally sound management of pesticides on golf courses have been summarized from these sources:

1. Only chemicals specifically labeled for application should be applied. All pesticide use should comply strictly with local, state, and federal regulations. Following Cooperative Extension Service recommendations, training in use of pesticides, and training in the appropriate pesticide application techniques are essential components for continued use of pesticides and maintenance of water quality.
2. Equipment maintenance and proper calibration is essential for even applications at the volumes intended by the user. All label instructions, storage requirements, and regulations must be followed to insure safe handling of pesticides. Proper mixing, handling, and loading prior to application will reduce fill-site contamination. Closed systems for loading and mixing pesticides are especially useful in reducing water source contamination for products where these systems are available.
3. Proper disposal of unused chemicals and containers will insure safety of the user, water resources, and nontarget organisms. Pesticide applicators should avoid chemical exposure by safe handling practices including use of protective clothing, respirators, gloves, and shoes.
4. Use of antibacksiphoning devices in chemigation equipment will reduce potential groundwater contamination with pesticides applied during irrigation. Pesticides should be applied through irrigation equipment only when appropriate and specific label instructions are available. Management practices for environmentally safe chemigation include the (a) flushing of injection equipment to prevent pesticide accumulation; (b) flushing the irrigation system after pesticide injection; (c) use of properly calibrated equipment; (d) prevention of runoff of water-pesticide mixture; avoid application to permanent or semipermanent standing water on or near fields; and (e) periodic monitoring of equipment to insure proper application to the intended target.

5. Selection of pesticides should be based on efficacy of treatment and criteria that reduce offsite movement and potential adverse environmental impacts. Selection of less toxic, less mobile, and less persistent chemicals with greater selective control of pests is an important management consideration.

6. The rate and timing of pesticide application relative to precipitation or irrigation that produces runoff or leaching episodes is a critical management consideration. The amount of pesticide movement may be partially controlled by judicious use of pesticides. Restriction of application prior to anticipated storm events is essential for reduction of runoff and leaching losses. Light irrigation after surface application will reduce exposure to losses in runoff water. Incorporation of pesticides or any placement below the soil surface reduces exposure to runoff process and enhances adsorption by turfgrass thatch and soil.

7. Unless compacted, turfgrass soils have high rates of water infiltration and conductivity. Numerous studies demonstrate that runoff of water from turfgrass plots is limited. However, studies on runoff patterns from entire golf course facilities are limited. Although, the threat of pesticide loss in surface runoff from turfgrass plots is limited, these results have not been extrapolated to a systems level. High rates of infiltration reduce runoff losses of water, soluble pesticides, and sediment-sorbed pesticides. However, in exchange for reduced surface losses, the amount of leaching is increased if irrigation significantly exceeds water use demand of the turfgrass.

8. Maintaining a buffer between surface water and areas treated with pesticides increases the transport distance and reduces the potential for offsite movement.

9. Application and handling of pesticides in regions of karst topography should be conducted with care to avoid direct runoff of chemicals into fractures and sinkholes.

10. Movement of applied pesticides is related to (a) soil texture and degree of hydraulic conductivity, (b) thatch development, (c) the amount of subsurface movement of water, (d) pesticide adsorption and degradation characteristics, and (e) cultural management practices. Leaching of pesticides will occur when persistent and soluble pesticides are frequently applied at high rates on sandy soils. The potential for subsurface loss of pesticides is further increased when the turfgrass is irrigated at a rate in excess of plant use, evapotranspiration, and soil storage.

11. Although extremely limited, currently published research has demonstrated low levels of water-related losses of pesticides from turfgrass systems. Low levels of pesticide losses from turfgrass could potentially be equivalent to low level losses found under some agricultural systems, especially for herbicides. Understanding the site-specific relationship of pesticide fate, persistence, and transport is an essential component of understanding the environmental impacts of pesticide management.

4.8 RESEARCH AND TECHNOLOGY TRANSFER NEEDS

A suggested list of research and development priorities has been developed based on gaps in the scientific literature, current regulatory objectives, and availability or applicability of relevant scientific information from other biological systems. Understanding and quantifying the pathways and rates of basic processes is required for accurate assessment of environmental impacts of managed turfgrass systems. Given the current status of research on environmental impacts of turfgrass, a higher priority should be assigned to basic and applied research rather than directed toward modeling efforts. Once water transport and pesticide processes have been fully characterized, development or modification of existing computer models can proceed. Research topics related to processes, fate, persistence, degradation, and reduction in pesticide use should receive immediate attention.

Another high priority is to develop regional management guidelines for turfgrass and golf course managers. Providing consistent information on guidelines to mitigate adverse environmental impacts is essential for environmentally sound management.

Process Research on the Fate of Pesticides — Understanding the mass balance, fate, and persistence of pesticides in the environment requires research to quantify (1) pesticide adsorption, volatility, degradation, and mobility; (2) turfgrass uptake and metabolism of pesticides; (3) the influence of thatch; (4) the fate of pesticides in clippings used as mulch or as a nitrogen source; (5) solute and water transport considering the hydrology of tiled drained greens and fairways; and (6) the influence of temperature, freeze/thaw cycles, and seasonal effects on pesticide transport. Resolution of questions involving the fate of applied pesticides can only be completely addressed using a mass balance approach.

Investigation of the mobility and persistence of parent compounds and transformation products of commonly applied pesticides is needed to assess the fate and persistence of pesticides applied to turfgrass. Since microbial degradation is the primary mechanism for reduction of the environmental load of applied pesticides, quantitative studies should be initiated to determine the mechanisms of microbial adaptation to turfgrass pesticides. These investigations will determine realistic rates of pesticide degradation after long-term use of pesticides. Investigations integrating adsorption, degradation, and environmental partitioning should be given preferential consideration. Development and improvement in the understanding of microbial degradation of turfgrass pesticides for a wide range of soils, turfgrass type, turfgrass cultivars, and climate conditions is needed. Past studies on the fate and persistence of pesticides applied to turfgrass have had inadequate design regarding the impact of climate, soil type, turfgrass cultivars, and thatch thickness. Many of the studies involving microbial dissipation and environmental fate lack quality control and adequate experimental design. The ability to extrapolate research results from greenhouse or plot conditions to the watershed level of a golf course has not been proven.

Studies to determine pesticide adsorption processes in turfgrass systems are needed. Specific studies should address the following specific topics:

1. Nonequilibrium adsorption of pesticides as a function of residence time, soil type and layering, and thatch conditions.
2. Quantify sorption coefficients and mobility processes as a function of soil properties, climatic factors, and turfgrass management practices. Adsorption and degradation are the primary factors controlling pesticide mobility in the environment. Past studies of pesticide adsorption, mobility, and fate have not incorporated realistic golf course conditions and turfgrass management practices.
3. Understanding the influence of turfgrass type, thatch thickness, and soil mix on pesticide transport is another important research topic.
4. Potential attenuation on solid organic surfaces and transport of pesticides in soluble organic matter complexes depends on thatch thickness, soil mix, and occurrence of soil organic matter.

Research on volatile losses of pesticides from turfgrass systems is very limited. Volatilization and drift of pesticides is a major pathway for offsite movement of pesticides. Prediction of pesticide residues available for leaching or runoff losses requires knowledge of pesticide losses to the atmosphere.

Monitoring studies of runoff or leaching losses of applied chemicals are often necessary for site-specific projects. Monitoring is often required to satisfy environmental regulation or litigation involving construction of new golf courses. However, at best, past monitoring studies provide surrogate information and lack a process orientation. Results of monitoring studies are often difficult to extrapolate beyond the specific testing site. Monitoring of runoff or leaching losses of turfgrass chemicals should be developed in conjunction with research designed to quantitatively evaluate improved management strategies, evaluate a wide range of climate and site conditions, or test alternate methods of disease and pest control. Runoff and leaching studies should be conducted under realistic turfgrass practices appropriate to golf course conditions. Methods to evaluate spatial variability, preferential flow, nonequilibrium rate processes, and the relationship of tile drainage with surface and subsurface flow need to be incorporated into research plans.

Models of Pesticide Fate and Transport — Several pesticide transport models could be used to evaluate pesticide movement on turfgrass systems. Several of these models are discussed in Chapter 5 (Balogh et al. 1992b). These models need to be calibrated and modified for simulation of water and chemical movement in turfgrass systems. Additional process-oriented research is required on water transport, adsorption, and degradation before pesticide transport models will be capable of realistic predictions. Current simulation technology for turfgrass systems is insufficient to produce appropriate results for site-specific management, research, or regulatory decisions. However, regulatory agencies will use current simulation strategies to establish management guidelines, unless efforts are made to provide more suitable predictive models.

Effects on Nontarget Organisms — Evaluation of the effects of pest management strategies on nontarget organisms would establish whether adverse impacts occur under normal golf course management conditions. Development of consistent management guidelines should suggest use of chemicals having the least potential for adverse effects on humans, wildlife, and other beneficial organisms. A rating of pesticide toxicity, potential for surface and subsurface transport of specific chemicals, conditions conducive to adverse impacts, and suggestions for alternate chemicals or methods of control will assist all turfgrass managers in the selection of appropriate management strategies.

Scope of Pesticide Use — The scope and extent of nutrient and pesticide use in golf course, lawn, sports turfgrass, and other turfgrass management systems has not been documented. Surveys designed to examine criteria for pesticide selection, rates and methods of application, and total volume of specific chemical use is needed to document the scope of pesticide use. Assessment of the potential environmental risk requires knowledge of the intrinsic danger of a practice (e.g., pesticide toxicity) and the probability of exposure to a specific risk (potential for exposure to applied chemicals).

Education and Technology Transfer — Development of a comprehensive source of information on pest management practices to reduce adverse environmental effects and water quality impacts is needed. Transfer current information and the results of the current U.S. Golf Association research program on environmental quality issues is essential to meet turfgrass and environmental quality objectives.

Specific issues that should be addressed in any source of information on environmental quality and turfgrass management are methods to (1) determine timing, quantity, and form of pesticides needed to achieve acceptable economic control of target pests with minimum use; (2) determine the threshold of pest infestations needed to adversely affect turfgrass quality and net operating profits; (3) determine the appropriate application technology for different chemical and biocontrol methods; and (4) develop principles for proper handling of chemicals to avoid concentrated spills or soil contamination. *Methods to establish realistic turfgrass quality goals in relation to other environmental quality objectives should be the first priority of any education and research program.* Specific practices can be developed within individual states; however, consistent approaches to establish these methods should be outlined in a series of environmental guidelines or objectives.

Several areas of research may be required to produce information needed to summarize information on environmental effects. Additional research would include development of improved application technology for chemical and biological control agents, selection or development of insect and disease resistant cultivars, and expansion of current research techniques to a wide range of climatic conditions and turfgrass management conditions. Establishing timing, method, and rate of pesticide application in relation to climatic and soil conditions will ultimately control the efficacy and environmental impacts of traditional chemical pesticides and biocontrol agents. Development or selection of disease- and pest-resistant turfgrass cultivars could have a significant impact on reducing pesticide use on turfgrass systems. However, the cost of introducing new germplasm into a perennial management system must be considered for this approach.

Currently, there is no single source of summarized information available to golf course managers regarding management guidelines and options for turfgrass management and environmental protection. These guidelines should be developed for a wide range of climatic regimes and turfgrass conditions. Specific guidelines should be field tested under realistic turfgrass and golf course conditions. Education on environmental quality issues and integrated systems management should be directed toward golf course developers, golf course superintendents, environmental regulators, and golfers.

REFERENCES

Ahrens, J. F. 1965. Detoxification of simazine and atrazine treated soil with activated carbon. *Proc. Northeast. Weed Control Conf.* 19:364–365.

Akesson, N. B. and Yates, W. E. 1984. Physical parameters affecting aircraft spray applications. in Garner, W. Y. and Harvey, J., Jr. (Eds.). *Chemical and Biological Controls in Forestry.* ACS Symp. Ser. 238. American Chemical Society, Washington, D.C., pp. 95–115.

Albert, R. E. 1987. Health concerns and chronic exposures to pesticides assessing the risk from carcinogens. in Pesticides and Groundwater: A Health Concern for the Midwest. Proc. of a Conf. Oct. 16–17, 1986, Freshwater Foundation, St. Paul, MN, pp. 163–168.

Alexander, M. 1981. Biodegradation of chemicals of environmental concern. *Science* 211(4478):132–138.

Alexander, M. and Scow, K. M. 1989. Kinetics of biodegradation in soil. in Sawhney, B. L. and Brown, K. (Eds.). *Reactions and Movement of Organic Chemicals in Soils.* SSSA Special Publication No. 22. Soil Sci. Soc. Am., Am. Soc. Agron., Madison, WI, pp. 243–269.

Anderson, J. L., Balogh, J. C. and Waggoner, M. 1989. Soil Conservation Service Procedure Manual: Development of Standards and Specifications for Nutrient and Pesticide Management. Section I & II. USDA Soil Conservation Service. State Nutrient and Pest Management Standards and Specification Workshop, St. Paul, MN, July 1989, 453 pp.

App, B. A. and Kerr, S. H. 1969. Harmful insects. in Hanson, A. A. and Juska, F. V. (Eds.). *Turfgrass Science*. Agron. Mongr. 14:336–357. Am. Soc. Agron., Madison, WI.

Ashton, F. M. and Crafts, A. S. 1981. *Mode of Action of Herbicides*. John Wiley & Sons, New York, 525 pp.

Asmussen, L. E., White, A. W., Hauser, E. W. and Sheridan, J. M. 1977 Movement of 2,4-D in vegetated waterway. *J. Environ. Qual.* 6:159–162.

Bailey, G. W. and White, J. L. 1970. Factors influencing adsorption, desorption, and movement of pesticides in soil. *Residue Rev.* 32:29–92.

Bailey, G. W., Mulkey, L. A. and Swank, R. R. 1985. Environmental implication of conservation tillage: A systems approach. in D'Itri, F. M. (Ed.). *A Systems Approach to Conservation Tillage*. Lewis Publishers, Chelsea, MI, pp. 240–265.

Bailey, G. W., Swank, R. R. and Nicholson, H. P. 1974. Predicting pesticide runoff from agricultural land: A conceptual model. *J. Environ. Qual.* 3(2):95–102.

Baker, D. B. 1985. Regional water quality impacts of intensive row-crop agriculture: A Lake Erie Basin case study. *J. Soil. Water Cons.* 40(2):125–132.

Baker, D. B., Kruger, K. A. and Stezler, J. V. 1981. The Concentrations and Transport of Pesticides in Northwestern Ohio Rivers — 1981. Gov. Rep. Announce. (U.S.) U.S. Army Corps Engineers, Buffalo, NY, pp. 1700–1733.

Baker, J. L. and Johnson, H. P. 1979. The effect of tillage systems on pesticides in runoff from small watersheds. *Trans. ASAE* 22(3):554–559.

Baker, J. L. and Laflen, J. M. 1979. Runoff losses of surface-applied herbicides as affected by wheel tracks and incorporation. *J. Environ. Qual.* 8(4):602–607.

Baker, J. L., Laflen, J. M. and Hartwig, R. O. 1982. Effects of corn residue and herbicide placement on herbicide runoff losses. *Trans. ASAE* 25(2):340–343.

Baker, J. L., Laflen, J. M. and Johnson, H. P. 1978. Effect of tillage systems on runoff losses of pesticides, A rainfall simulation study. *Trans. ASAE* 21(5):886–892.

Baker, R. S. and Hillel, D. 1990. Laboratory tests of a theory of fingering during infiltration into layered soils. *Soil Sci. Soc. Am. J.* 54(1):20–30.

Balogh, J. C. and Gordon, G. A. 1987. Climate Variability and Soil Hydrology in Water Resource Evaluation. Phase I Project Report. National Science Foundation, ISI-87053. NTIS #PB90-203183/A05. 92 pp.

Balogh, J. C., Gordon, G. A., Murphy, S. R. and Tietge, R. M. 1990. The Use and Potential Impacts of Forestry Herbicides in Maine. 89-12070101. Final Report. Maine Dept. Cons. 237 pp.

Balogh, J. C. and Watson, J. R., Jr. 1992. Role and conservation of water resources. in Balogh, J. C. and Walker, W. J. (Eds.). *Golf Course Management and Construction: Environmental Issues*. Lewis Publishers, Chelsea, MI, Chapter 2.

Balogh, J. C., Gibeault, V. A., Walker, W. J., Kenna, M. P., and Snow, J. T. 1992a. Background and overview of turfgrass and environmental issues. in Balogh, J. C. and Walker, W. J. (Eds.). *Golf Course Management and Construction: Environmental Issues*. Lewis Publishers, Chelsea, MI, Chapter 1.

Balogh, J. C., Leslie, A. R., Walker, J. C. and Kenna, M. P. 1992b. Development of integrated management systems for turfgrass. in Balogh, J. C. and Walker, W. J. (Eds.). *Golf Course Management and Construction: Environmental Issues*. Lewis Publishers, Chelsea, MI, Chapter 5.

Barnett, A. P., Hauser, E. W., White, A. W. and Holladay, J. H. 1967. Loss of 2,4-D in washoff from cultivated fallow land. *Weeds* 15:133–137.

Bear, J. 1972. *Dynamics of Fluids in Porus Media*. American Elsevier, New York.

Beard, J. B. 1982. *Turf Management for Golf Courses*. Macmillan Publishing Co., New York, 642 pp.

Beard, J. B. 1973. *Turfgrass Science and Culture*. Prentice-Hall, Englewood Cliffs, NJ, 658 pp.

Berry, C. D. and Buchanan, G. A. 1974. Tolerance of *Phalaris tuberosa* L. and *Fetuca arundinacea* Schreb. to preemergence herbicide treatment. *Crop Sci.* 14(1):96–99.

Bidleman, T. F. and Olney, C. E. 1975. Long range transport of toxaphene insecticide in the atmosphere of the western north Atlantic. *Nature* 257:475–477.

Bidleman, T. F., Patton, G. W., Hinckley, D. A., Walla, M. D., Cotham, W. E. and Hargrave, B. T. 1990. Chlorinated pesticides and polychlorinated biphenyls in the atmosphere of the Canadian Arctic. in Kurtz, D. A. (Ed.). *Long Range Transport of Pesticides*. Lewis Publishers, Chelsea, MI, pp. 347–372.

Bilkert, J. N. and Rao, P. S. C. 1985. Sorption and leaching of three nonfumigant nematicides in soils. *J. Environ. Sci. Health* B20(1):1–26.

Bingham, S. W. and Hall, J. R., III. 1985. Effects of herbicides on bermudagrass (*Cynodon* spp.) spring establishment. *Weed Sci.* 33(2):253–257.

Bingham, S. W. and Schmidt, R. E. 1967. Residue of bensulide in turfgrass soil following annual treatments for crabgrass control. *Agron. J.* 59(4):327–329.

Boesten, J. J. T. I. and van der Linden, A. M. A. 1991. Modeling of sorption and transformation on pesticide leaching and persistence. *J. Environ. Qual.* 20(2):425–435.

Boesten, J. J. T. I. and van der Pas, L. J. T. 1988. Modeling adsorption/desorption kinetics of pesticides in a soil suspension. *Soil Sci.* 146(4):221–231.

Bollag, J.-M. and Liu, S.-Y. 1990. Biological transformation processes of pesticides. in Cheng, H. H. (Ed.). *Pesticides in the Soil Environment: Processes, Impacts, and Modeling*. SSSA Book Series No. 2. Soil Science Society of America, Madison, WI, pp. 169–211.

Bottcher, A. B., Monke, E. J. and Huggins, L. F. 1981. Nutrient and sediment loading from a subsurface drainage system. *Trans. ASAE* 24(5):1221–1226.

Bovey, R. W. and Young, A. L. 1980. *The Science of 2,4,5-T and Associated Phenoxy Herbicides*. John Wiley & Sons, New York, 462 pp.

Bovey, R. W., Burnett, E., Richardson, C., Merkle, M. G., Baur, J. R. and Knisel, W. G. 1974. Occurrence of 2,4,5-T and picloram in surface water in the Blacklands of Texas. *J. Environ. Qual.* 3(1):61–64.

Boyle, T. P. 1980. Effects of the aquatic herbicide 2,4-D DMA on the ecology of experimental ponds. *Environ. Pollut. (Ser. A)* 21(1):35–49.

Branham, B. E. and Wehner, D. J. 1985. The fate of diazinon applied to thatched turf. *Agron. J.* 77(1):101–104.

Branham, B. E., Wehner, D. J., Torello, W. A. and Turgeon, A. J. 1985. A microecosystem for fertilizer and pesticide fate research. *Agron. J.* 77(1):176–180.

Brejda, J. J., Shea, P. J., Moser, L. E. and Waller, S. S. 1988. Atrazine dissipation and off-plot movement in a Nebraska sandhills subirrigated meadow. *J. Range Manage.* 41(5):416–420.

Britton, M. P. 1969. Turfgrass diseases. In Hanson, A. A. and Juska, F. V. (Eds.). *Turfgrass Science*. Agron. Mongr. 14:288–335. Amer. Soc. Agron., Madison, WI.

Brown, A. W. A. 1971. Pest resistance to insecticides. in White-Stevens, R. (Ed.). *Pesticides in the Environment*. Vol. 1. Part II. Marcel Dekker, Inc., New York, pp. 457–552.

Brown, A. W. A. 1978. *Ecology of Pesticides*. John Wiley & Sons, New York.

Brown, A. W. A. and Pal. R. 1971. *Insecticide Resistence in Arthropods*. World Health Organization, Geneva, Switzerland, 491 pp.

Brown, D. S. and Flagg, E. W. 1981. Empirical prediction of organic pollutant sorption in natural sediments. *J. Environ. Qual.* 10(3):382–386.

Brown, T. M., Johnson, D. R., Hopkins, A. R., Durant, J. A. and Montefiori, D. C. 1982. Interactions of pyrethroid insecticides and toxaphene in cotton. *J. Agric. Food Chem.* 30:542–545.

Brusseau, M. L. and Rao, P. S. C. 1989. Sorption nonideality during organic contaminant transport in porous media. *CRC Crit. Rev. Environ. Control* 19(1):33–99.

Callahan, L. M., Overton, J. R. and Sanders, W. L. 1983. Initial and residual herbicide control of crabgrass in bermudagrass turf. *Weed Sci.* 31:619–622.

Canter, L. W. 1987. Nitrates and pesticides in ground water: An analysis of a computer-based literature search. in Fairchild, D. M. (Ed.). *Ground Water Quality and Agricultural Practices*. Lewis Publishers, Chelsea, MI, pp. 153–174.

Cantor, K. P, Blair, A. and Zahn, S. H. 1988. Health effects of Agrichemicals in groundwater: What do we know? in Agricultural Chemicals and Groundwater Protection: Emerging Management and Policy. Proceedings of a Conference, Oct. 22–23, 1987, Freshwater Foundation, St. Paul, MN, pp. 27–42.

Caro, J. H. 1976. Pesticides in agricultural runoff. in Stewart, B. A., Woolhiser, D. A., Wischmeier, W. H., Caro, J. H. and Frere, M. H. (Eds.) Control of Water Pollution from Cropland: An Overview. Vol. 2. EPA 600/2-75-026b. U.S. Environmental Protection Agency, Athens, GA and USDA Agricultural Research Service, Washington, D.C., pp. 91–119.

Carsel, R. F., Mulkey, L. A., Lorber, M. N. and Baskin, L. B. 1985. The pesticide root zone model (PRZM): A procedure for evaluating pesticide leaching threats to groundwater. *Ecol. Modelling* 30:49–69.

Carsel, R. F., Smith, C. N., Mulkey, J. T., Dean, J. D. and Jowise, P. 1984. User's Manual for the Pesticide Root Zone Model (PRZM). Release 1. EPA-600/3-84-109. U.S. Environmental Protection Agency, Athens, GA, 216 pp.

Chapman, R. A. 1979. Diseases and nematodes. in Buckner, R. C. and Bush, L. P. (Eds.). *Tall Fescue*. Agron. Monogr. 20:307–318. Am. Soc. Agron., Crop Sci. Soc. Am., Soil Sci. Soc. Am., Madison, WI.

Chapman, R. A., Harris, C. R., Moy, P. and Henning, K. 1986. Biodegradation of pesticides in soil: Rapid degradation of isofenphos in a clay loam after a previous treament. *J. Environ. Sci. Health* B21:269–276.

Cheng, H. H. 1990a. Pesticides in the soil environment — An overview. in Cheng, H. H. (Ed.). *Pesticides in the Soil Environment: Processes, Impacts, and Modeling*. SSSA Book Series No. 2. Soil Science Society of America, Madison, WI, pp. 1–5.

Cheng, H. H. (Ed.). 1990b. *Pesticides in the Soil Environment: Processes, Impacts, and Modeling*. SSSA Book Series No. 2. Soil Science Society of America, Madison, WI, 530 pp.

Chiou, C. T., Peters, L. J. and Freed, V. H. 1979. A physical concept of soil-water equilibria for nonionic organic compounds. *Science* 206:831–832.

Chiou, C. T. 1989. Theoretical considerations of the partition uptake of nonionic organic compounds by soil organic matter. in Sawhney, B. L. and Brown, K. (Eds.). *Reactions and Movement of Organic Chemicals in Soil*. SSSA Special Publication No. 22. Soil Sci. Soc. Am., Am. Soc. Agron., Madison, WI, pp. 1–44.

Choi, J. S., Fermanian, T. W., Wehner, D. J. and Spomer, L. A. 1988. Effect of temperature, moisture, and soil texture on DCPA degradation. *Agron. J.* 80(1):108–113.

Cohen, S. Z., Nickerson, S., Maxey, R., Dupuy, A., Jr. and Senita, J. A. 1990. Ground water monitoring study for pesticides and nitrates associated with golf courses on Cape Cod. *Ground Water Monit. Rev.* 10(1):160–173.

Cooper, C. M., Dendy, F. E., McHenry, J. R. and Ritchie, J. C. 1987. Residual pesticide concentrations in Bear Creek, Mississippi, 1976 to 1979. *J. Environ. Qual.* 16(1):69–72.

Cooper, R. J., Jenkins, J. J. and Curtis, A. S. 1990. Pendimethalin volatility following application to turfgrass. *J. Environ. Qual.* 19(3):508–513.

Cottam, C. 1965. The ecologists' role in problems of pesticide pollution. *BioScience* 15: 457–463.

CPCR. 1991. *Crop Protection Chemicals Reference.* Chemical and Pharmaceutical Press, John Wiley & Sons, New York, 2170 pp.

Daniel, T. C., McGuire, P. E., Stoffel, D. and Miller, B. 1979. Sediment and nutrient yield from residential construction sites. *J. Environ. Qual.* 8(3):304–308.

Daniel, T. C., Wendt, R. C., McGuire, P. E. and Stoffel, D. 1982. Nonpoint source loading rates from selected land uses. *Water Res. Bull.* 18(1):117–120.

Davis, E. A. and Ingebo, P. A. 1973. Picloram movement from a chapparral watershed. *Water Resour. Res.* 9:1304–1313.

Dernoeden, P. H. 1984. Management of preemergence herbicides for crabgrass control in transition-zone turf. *HortScience* 19(3):443–445.

Dernoeden, P. H. 1989. Symptomatology and management of common turfgrass diseases in transition zone and northern regions. in Leslie, A. R. and Metcalf, R. L. (Eds). Integrated Pest Management for Turfgrass and Ornamentals. Office of Pesticide Programs, 1989-625-030. U.S. Environmental Protection Agency, Washington, D.C., pp. 273–296.

Duble, R. L., Thomas, J. C., and Brown, K. W. 1978. Arsenic pollution from underdrainage and runoff from golf greens. *Agron. J.* 70(1):71–74.

Dunbar, D. M. and Beard, R. L. 1975. Status of control of Japanese beetle and oriental beetles in Connecticut. *Conn. Agric. Exp. Stn. Bull.* No. 757. 5 pp.

Edwards, W. M. 1972. Agricultural chemical pollution as affected by reduced tillage systems. in Proc. No-tillage Systems Symp., Feb. 21, 1972, Ohio State University, Columbus, OH, pp. 30–40.

Edwards, W. M., Triplett, G. B. and Kramer, R. M. 1980. A watershed study of glysophate transport in runoff. *J. Environ. Qual.* 9(4):661–665.

Eisler, R. 1988. Arsenic Hazards to Fish, Wildlife, and Invertegrates: A Synoptic Review. Contam. Haz Rev. Rept. No. 12. U.S. Fish and Wildlife Serv., U.S. Dept. Int. Biol. Report 85(1.2), 92 pp.

Enfield, C. G. and Yates, S. R. 1990. Organic chemical transport to groundwater. in Cheng, H. H. (Ed.). *Pesticides in the Soil Environment: Processes, Impacts, and Modeling.* SSSA Book Series No. 2. Soil Science Society of America, Madison, WI, pp. 271–302.

Engel, R. E. and Ilnicki, R. D. 1969. Turfweeds and their control. in Hanson, A. A. and Juska, F. V. (Eds.). *Turfgrass Science.* Agron. Mongr. 14:240–287. Am. Soc. Agron., Madison, WI.

Environmental Protection Agency. 1986. Pesticides in Ground Water: Background Document. EPA-WH550G. Office of Ground-Water, U.S. Environmental Protection Agency, Washington, D.C., 72 pp.

Environmental Protection Agency. 1987. Agricultural Chemicals in Groundwater: Proposed Pesticide Strategy. Proposed Strategy Document, U.S. Environmental Protection Agency, Office of Pesticides and Toxic Substances, Washington, D.C., 149 pp.

Environmental Protection Agency. 1988. Protecting Ground Water: Pesticides and Agricultural Practices. EPA-440/6-88-001. U.S. Environmental Protection Agency, Office of Ground-Water Protection, Washington, D.C., 53 pp.

Environmental Protection Agency. 1990. National Pesticide Survey: Summary Results of EPA's National Survey of Drinking Water Wells. U.S. Environmental Protection Agency. Office of Pesticides and Toxic Substances, Washington, D.C., 16 pp.

Everts, C. J., Kanwar, R. S., Alexander, E. C., Jr. and Alexander, S. C. 1989. Comparison of tracer mobilities under laboratory and field conditions. *J. Environ. Qual.* 18(4):491–498.

Fairchild, D. M. 1987. A national assessment of groundwater contamination from pesticides and fertilizers. in Fairchild, D. M. (Ed.). *Ground Water Quality and Agricultural Practices.* Lewis Publishers, Chelsea, MI, pp. 273–294.

Farm Chemicals Handbook. 1991. *Pesticide and Fertilizer Dictionary.* Meister Publishing Company, Willoughby, OH, 770 pp.

FDA. 1984. Surveillance Index for Pesticides Update. NTIS PB84-913200. Food and Drug Administration, Washington, D.C.

Felsot, A. S. 1989. Enhanced biodegradation of insecticides in soil: Implications for agroecosystems. *Ann. Rev. Entomol.* 34:453–476.

Felsot, A. and Dahm, P. A. 1979. Sorption of organophosphorus and carbamate insecticides by soil. *J. Agric. Food Chem.* 27(3):557–563.

Felsot, A. S., Mitchell, J. K. and Kenimer, A. L. 1990. Assessment of management practice for reducing pesticide runoff from sloping cropland in Illinois. *J. Environ. Qual.* 19(3):539–545.

Fermanich, K. J., Daniel, T. C. and Lowry, B. 1991. Microlysimeter soil columns for evaluating pesticide movement through the root zone. *J. Environ. Qual.* 20(1):189–195.

Fiore, M. 1987. Chronic exposure to aldicarb contaminated groundwater and human immune function. in Pesticides and Groundwater: A Health Concern for the Midwest. Proc. of a Conf. Oct. 16–17, 1986, Freshwater Foundation, St. Paul, MN, pp. 199–203.

Forgash, A. J. 1984. History, evolution, and consequences of insecticide resistance. *Pest. Biochem. Physiol.* 22:178–186.

Forrest, M., Lord, K. A., Walker, N. and Woodville, H. C. 1981. The influence of soil treatments in the bacterial degradation of diazinon and other organophosphorus insecticides. *Environ. Pollut.* A24:93–104.

Frank, R., Braun, H. E., Holdrinet, M. V. and Sirons, G. J. 1982. Agriculture and water quality in the Canadian Great Lakes Basin: V. Pesticide use in 11 agricultural water sheds and presence in stream water, 1975–1977. *J. Environ. Qual.* 11(3):497–505.

Frank, R., Clegg, B. S. Ripley, B. D. and Braun, H. E. 1987. Investigation of pesticide contamination in rural wells, 1979–1984, Ontario, Canada. *Arch. Contamin. Toxicol.* 16(1):9–22.

Freeborg, R. P. 1984. Preemergent weed control. *ALA* 5(1):7–11.

Fushtey, S. G. and Frank, R. 1981. Distribution of mercury residues from the use of mercurial fungicides on golf course greens. *Can. J. Soil Sci.* 61:525–527.

Gamerdinger, A. P., Wagenet, R. J. and van Genuchten, M. Th. 1990. Application of two site/two region models for studying simultaneous nonequilibrium transport and degradation of pesticides. *Soil Sci. Soc. Am. J.* 54(4):957–963.

Gaynor, J. D. and Volk, V. V. 1981. Runoff losses of atrazine and terbutryn from unlimed and limed soil. *Environ. Sci. Tech.* 15:440–443.

Georghiou, G. P. 1980. Insectcide resistance and prospects for its management. *Res. Rev.* 76:131–143.

Georghiou, G. P. 1981. The Occurrence of Resistance to Pesticides in Arthropods. Plant Protect. Ser., FAO, UN, Rome.

Georghiou, G. P. 1984. Management of resistance in arthropods. in Georghiou, G. P. and Saito, T. (Eds). *Pest Resistance to Pesticides.* Plenum Press, New York, pp. 769–792.

Georghiou, G. P. 1986. The magnitude of the resistance problem. in *Pesticide Resistance: Strategies and Tactics for Management.* National Academy Press, Washington, D.C., pp. 14–43.

Georghiou, G. P. and Taylor, C. E. 1977. Operational influences in the evolution of insecticide resistance. *J. Econ. Entomol.* 70:653–658.

Georghiou, G. P. and Taylor, C. E. 1986. Factors influencing the evolution of resistance. in *Pesticide Resistance: Strategies and Tactics for Management.* National Academy Press, Washington, D.C., pp. 157–169.

Ghodrati, M. and Jury, W. A. 1990. A field study using dyes to characterize preferential flow of water. *Soil Sci. Soc. Am. J.* 54(6):1558–1563.

Gilliom, R. J. 1985. Pesticides in rivers of the United States. in USGS National Water Summary 1984: Hydrologic Events, Selected Water-Quality Trends, and Ground-Water Resources. USGS Water-Supply Paper 2275, pp. 85–92.

Gilmour, J. T. and Miller, M. S. 1973. Fate of mercuric-mercurous chloride fungicide added to turfgrass. *J. Environ. Qual.* 2(1):145–148.

Glass, B. L. and Edwards, W. M. 1974. Picloram in lysimeter runoff and percolation water. *Bull. Environ. Contam. Toxicol.* 11:109–112.

Glotfelty, D. E., Seiber, J. N. and Liljedahl, L. A. 1987. Pesticides in fog. *Nature* 325:602–605.

Glotfelty, D. E., Taylor, A. W., Isensee, A. R., Jersey, J. and Glenn, S. 1984b. Atrazine and simazine movement to Wye River estuary. *J. Environ. Qual.* 13(1):115–121.

Glotfelty, D. E., Taylor, A. W., Turner, B. C. and Zoller, W. H. 1984a. Volatilization of surface-applied pesticides from fallow soil. *J. Agric. Food Chem.* 32:638–643.

Glotfelty, D. E., Williams, G. H., Freeman, H. P. and Leech, M. M. 1990. Regional atmospheric transport and deposition of pesticides in Maryland. in Kurtz, D. A. (Ed.). *Long Range Transport of Pesticides.* Lewis Publishers, Chelsea, MI, pp. 199–221.

Goh, K. S., Edmiston, S., Maddy, K. T., Meinders, D. D. and Margetich, S. 1986. Dissipation of dislodgeable foliar residue of chlopyrifos and dichlorvos on turf. *Bull. Environ. Contam. Toxicol.* 37:27–32.

Gold, A. J., Morton, T. G., Sullivan, W. M. and McClory, J. 1988. Leaching of 2,4-D and dicamba from home lawns. *Water Air Soil Poll.* 37:121–129.

Green, R. E. and Karickhoff, S. W. 1990. Sorption estimates for modeling. in Cheng, H. H. (Ed.). *Pesticides in the Soil Environment: Processes, Impacts, and Modeling.* SSSA Book Series No. 2. Soil Science Society of America, Madison, WI, pp. 79–101.

Gross, C. M., Angle, J. S. and Welterlen, M. S. 1990. Nutrient and sediment losses from turfgrass. *J. Environ. Qual.* 19(4):663–668.

Gross, C. M., Angle, J. S., Hill, R. L. and Welterlen, M. S. 1991. Runoff and sediment losses form tall fescue under simulated rainfall. *J. Environ. Qual.* 20(3):604–607.

Grossbard, E. and Atkinson, D. (Eds.). 1985. *The Herbicide Glyphosate.* Butterworths & Co., Boston, MA, 490 pp.

Grover, R. 1976. Relative volatilities of ester and amine forms of 2,4-D. *Weed Sci.* 24:26–28.

Grover, R. 1991. Nature, transport, and fate of airborne residues. in Grover, R., and Cessna, A. J. (Eds.). *Enviromental Chemistry of Herbicides.* Vol 2. CRC Press, Boca Raton, FL, pp. 89–117.

Grover, R. (Ed.). 1988. *Environmental Chemistry of Herbicides.* Vol. 1. CRC Press, Boca Raton, FL, 207 pp.

Grover, R. and Cessna, A. J. (Eds.). 1991. *Environmental Chemistry of Herbicides.* Vol 2. CRC Press, Boca Raton, FL, 302 pp.

Grover, R., Maybank, J. and Yoshida, K. 1972. Droplet and vapor drift from butyl ester and dimethylamine salt of 2,4-D. *Weed Sci.* 20:320–324.

Grover, R., Shewchuk, S. R., Cessna, A. J., Smith, A. E. and Hunter, J. H. 1985. Fate of 2,4-D *iso*-octyl ester after application to a wheat field. *J. Environ. Qual.* 14(2):203–210.

Grover, R., Smith, A. E., Shewchuk, S. R., Cessna, A. J. and Hunter, J. H. 1988. Fate of triflualin and triallate applied as a mixture to a wheat field. *J. Environ. Qual.* 17(4):543–550.

Haas, R. H., Scifres, C. J., Merkle, M. G., Hahn, R. R. and Hoffman, G. O. 1971. Occurrence and persistence of picloram in grassland water sources. *Weed Res.* 11:54–62.

Hall, C. J., Bowhey, C. S. and Stephenson, G. R. 1987. Lateral movement of 2,4-D from grassy inclines. *Proc. British Crop Protect. Conf.* 2(3):593–599.

Hall, J. K. 1974. Erosional losses of s-triazine herbicides. *J. Environ. Qual.* 3(1):174–180.

Hall, J. K., Hartwig, N. L. and Hoffman, L. D. 1983. Atrazine mode and alternate cropping effects on atrazine losses from a hillside. *J. Environ. Qual.* 12(3):336–340.

Hall, J. K., Murray, M. R. and Hartwig, N. L. 1989. Herbicide leaching and distribution in tilled and untilled soil. *J. Environ. Qual.* 18(4):439–445.

Hall, J. K., Pawlus, M. and Higgins, E. R. 1972. Losses of atrazine in runoff water and soil sediment. *J. Environ. Qual.* 1:172–176.

Hallberg, H. R., Hoyer, B. E., Bettis, E. A., III, and Libra, R. D. 1983. Hydrogeology, Water Quality, and Land Management in the Big Spring Basin, Clayton County, Iowa. Iowa Geol. Surv., Open-File Rept. 83–3. 191 pp.

Hallberg, H. R., Libra, R. D., Bettis, E. A., III, and Hoyer, B. E. 1984. Hydrogeologic Investigations in the Big Spring Basin, Clayton County, Iowa: 1983 Water-Year. Iowa Geol. Surv., Open-File Rept. 84–4.

Hance, R. J. 1988. Adsorption and bioavailability. 1988. in Grover, R. (Ed.). *Environmental Chemistry of Herbicides*. Vol. 1. CRC Press, Boca Raton, FL, pp. 1–19.

Hanmaker, J. W. and Thompson, J. M. 1972. Adsorption. in Goring, C. A. I. and Hamaker, J. W. (Eds.). *Organic Chemicals in the Environment*. Marcel-Dekkar, New York, pp. 49–143.

Haque, R. and Freed, V. H. 1974. Behavior of pesticides in the environment: Environmental chemodynamics. *Residue Rev.* 52:89– 116.

Harkin, J. M., Jones, F. A., Fathulla, R., Dzantor, E. K., O'Neill, E. J., Kroll, D. G. and Chesters, G. 1984. Pesticides in Groundwater Beneath the Central Sand Plain of Wisconsin. OWRT-A-094-WIS(1). Office of Water Research and Technology, U.S. Department of Interior, Washington, D.C., 46 pp.

Harmon, T. C., Ball, W. P. and Roberts, P. V. 1989. Nonequilibrium transport of organic contaminants in groundwater. in Sawhney, B. L. and Brown, K. (Eds.). *Reactions and Movement of Organic Chemicals in Soil.* SSSA Special Publication No. 22. Soil Sci. Soc. Am., Am. Soc. Agron., Madison, WI, pp. 405–438.

Harris, C. R., Chapman, R. A., Harris, C. and Tu, C. M. 1984. Biodegradation of pesticides in soil: rapid induction of carbamate degrading factors after carbofuran treatment. *J. Environ. Sci. Health* B19:1–11.

Hassett, J. J. and Banwart, W. L. 1989. Sorption dynamics of organic compounds in soils and sediments. in Sawhney, B. L. and Brown, K. (Eds.). *Reactions and Movement of Organic Chemicals in Soils.* SSSA Special Publication No. 22. Soil Sci. Soc. Am., Am. Soc. Agron., Madison, WI, pp. 31–80.

Heald, C. M. and Perry, V. G. 1969. Nematodes and other pests. in Hanson, A. A. and Juska, F. V. (Eds.). *Turfgrass Science*. Agron. Mongr. 14:358–369. Am. Soc. Agron., Madison, WI.

Helling, C. S., 1976. Dinitroaniline herbicides in soils. *J. Environ. Qual.* 5(1):1–15.

Helling, C. S., Kearney, P. C. and Alexander, M. 1971. Behavior of pesticides in soils. *Adv. Agron.* 23:147–240.

Hillel, D. and Baker, R. S. 1988. A descriptive theory of fingering during infiltration into layered soils. *Soil Sci.* 146(1):51–56.

Himel, C. M., Loats, H. and Bailey, G. W. 1990. Pesticide sources to the soil and principles of spray physics. in Cheng, H. H. (Ed.). *Pesticides in the Soil Environment: Processes, Impacts, and Modeling*. SSSA Book Series No. 2. Soil Science Society of America, Madison, WI, pp. 7–50.

Hindall, S. M. 1978. Effects of Irrigation on Water Quality in the Sand Plain of Central Wisconsin. Information Circ. No. 36. Wisc. Geol. Survey and Nat. Hist. Survey, Madison, WI.

Howard, P. H. (Ed.). 1991. *Handbook of Environmental Fate and Exposure Data for Organic Chemicals. Vol. 3. Pesticides*. Lewis Publishers, Chelsea, MI, 684 pp.

Humburg, N. E., Colby, S. R., Hill, E. R., Kitchen, L. M., Lym, R. G., McAvoy, W. J. and Prasak, R. 1989. *Herbicide Handbook of the Weed Science Society of America*. 6th ed. Weed Science Society of America, Champaign, IL, 301 pp.

Hunt, E. G. and Linn, J. D. 1970. Fish kills by pesticides. in Gillett, J. W. (Ed.). *The Biological Impact of Pesticides in the Environment*. Oregon State University Press, Corvallis, OR, pp. 97–103.

Hurto, K. A., Turgeon, A. J. and Cole, M. A. 1979. Degradation of benefin and DCPA in thatch and soil from a Kentucky bluegrass (*Poa pratensis*) turf. *Weed Sci.* 27(2): 154–157.

Isensee, A. R., Nash, R. G. and Helling, C. S. 1990. Effect of conventional vs. no-tillage on pesticide leaching to shallow groundwater. *J. Environ. Qual.* 19(3):434–440.

Jagschitz, J. A. 1971. Effect of Activated Charcoal on the Future Performance of Chemicals for Pre-emergence Crabgrass Control in Turfgrass. *Proc. Ann. Northeast. Weed Sci. Soc.* 25:109–114.

Jagschitz, J. A. 1974. Use of activated charcoal to nullify the harmful effects of chemicals in turfgrass. Proc. Int. Turfgrass Research Conf. Proc., 2nd Int. Turf. Res. Conf. ASA-CSSA Special Publication. Madison, WI, pp. 399–409.

Jenkins, J. J., Cooper, R. J. and Curtis, A. S. 1990. Comparison of pendimethalin airobrne and dislodgeable residues following application to turfgrass. in Kurtz, D. A. (Ed.). *Long Range Transport of Pesticides*. Lewis Publishers, Chelsea, MI, pp. 29–46.

Johnsen, T. N., Jr. 1980. Picloram in water and soil from a semiarid Pinyon-Juniper watershed. *J. Environ. Qual.* 9(4):601–605.

Johnson, H. E. and Ball, R. C. 1972. Organic pesticide pollution in an aquatic environment. in *Fate of Organic Pesticides in the Aquatic Envirnoment*. Advan. Chem. Ser. 111, Am. Chem. Soc., Washington, D.C., pp. 1–10.

Johnson, P. W. 1988. Iowa's 1987 groundwater protection act: Purpose and process. in Agricultural Chemicals and Groundwater Protection: Emerging Management and Policy. Proceedings of a Conference, Oct. 22–23, 1987, Freshwater Foundation, St. Paul, MN, pp. 167–170.

Junk, G. A., Spalding, R. F. and Richard, J. J. 1980. Areal, vertical, and temporal differences in groundwater chemistry. II. *Org. Constituents. J. Environ. Qual.* 9(3):479–483.

Jury, W. A. 1983. Chemical transport modeling: Current approaches and unresolved problems. in Nelsen, D. W., Elrick, D. E. and Tanji, K. K. (Eds.). *Chemical Mobility and Reactivity in Soil Systems*. Soil Sci Soc. Am. Special Publ. No. 11. pp. 49–64.

Jury, W. A. 1986a. Volatilization from soil. in Hern, S. C. and Melancon, S. M. (Eds.). *Vadose Zone Modeling of Organic Pollutants*. Lewis Publishers, Chelsea, MI, pp. 159–176.

Jury, W. A. 1986b. Adsorption of Organic chemicals onto soil. In Hern, S. C. and Melancon, S. M. (Eds.). *Vadose Zone Modeling of Organic Pollutants*. Lewis Publishers, Chelsea, MI, pp. 177–189.

Jury, W. A. 1986c. Chemical movement through soil. in Hern, S. C. and Melancon, S. M. (Eds.). *Vadose Zone Modeling of Organic Pollutants.* Lewis Publishers, Chelsea, MI, pp. 135–158.

Jury, W. A. and Valentine, R. L. 1986. Transport mechanisms and loss pathways for chemicals in soil. in Hern, S. C. and Melancon, S. M. (Eds.). *Vadose Zone Modeling of Organic Pollutants.* Lewis Publishers, Chelsea, MI, pp. 37–60.

Jury, W. A., Focht, D. D. and Farmer, W. J. 1987b. Evaluation of pesticide groundwater pollution potential from standard indices of soil-chemical adsorption and biodegradation. *J. Environ. Qual.* 16:422–428.

Jury, W. A., Spencer, W. F. and Farmer, W. J. 1983. Behavior assessment model for trace organics in soil. I. Model description. *J. Environ. Qual.* 12(4):558–564.

Jury, W. A., Spencer, W. F. and Farmer, W. J. 1984a. Behavior assessment model for trace organics in soil. II. Chemical classification and parameter sensitivity. *J. Environ. Qual.* 13(4):567–572.

Jury, W. A., Spencer, W. F. and Farmer, W. J. 1984b. Behavior assessment model for trace organics in soil: III. Application of screening model. *J. Environ. Qual.* 13(4):573–579.

Jury, W. A., Winer, A. M, Spencer, W. F. and Focht, D. D. 1987a. Transport and transformations of organic chemicals in the soil-air-water ecosystem. *Rev. Environ. Contam. Toxicol.* 99:119–164.

Kahn, S. U. 1982. Bound pesticde residues in soil and plants. *Residue Rev.* 84:1–25.

Kahn, S. U. 1991. Bound residues. in Grover, R. and Cessna, A. J. (Eds.). *Environmental Chemistry of Herbicides.* Vol 2. CRC Press, Boca Raton, FL, pp. 265–279.

Karickhoff, S. W. 1981. Semi-empirical estimation of sorption of hydrophobic pollutants on natural sediments and soils. *Chemosphere* 10:833–846.

Karickhoff, S. W. and Morris, K. R. 1985. Sorption dynamics of hydrophobic pollutants in sediment suspensions. *Environ. Toxicol. Chem.* 4(4):469–479.

Karickhoff, S. W., Brown, D. S. and Scott, T. A. 1979. Sorption of hydrophobic pollutants on natural sediment. *Water Res.* 13:241–284.

Kearney, P. C. and Kaufman, D. D. (Eds.). 1988. *Herbicides: Chemistry, Degradation, and Mode of Action.* Vol. 1 (1975), Vol. 2 (1976), Vol. 3 (1988). Marcel Dekker, Inc., New York.

Kearney, P. C. and Kellog, S. T. 1985. Microbial adaption to pesticides. *Pure Appl. Chem.* 57:390–403.

Kenaga, E. E. 1980. Predicted bioconcentration factors and soil sorption coefficients of pesticides and other chemicals. *Ectotoxicol. Environ. Safety* 4:26–38.

Kenaga, E. E. and Goring, C. A. I. 1980. Relationship between water solubility, soil sorption, octanol-water partitioning, and concentration of chemicals in biota. in Eaton, J. G., Parrish, P. R. and Hendricks, A. C. (Eds.). *Aquatic Toxicology,* ASTM STP 707. American Society for Testing and Materials, Philadelphia, PA, pp. 78–115.

Kilbane, J. J., Chatterjee, D. K. and Chakrabarty, A. M. 1983. Detoxification of 2,4,5-Trichlorophenoxyacetic acid from contaminated soil by *Pseudomonas cepacia. Appl. Environ. Microbiol.* 45:1697–1700.

Kimm, V. J. and Barles, R. 1988. EPA's pesticides in groundwater strategy. in Agricultural Chemicals and Groundwater Protection: Emerging Management and Policy. Proceedings of a Conference, Oct. 22–23, 1987, Freshwater Foundation, St. Paul, MN, pp. 135–145.

Kirkland, K. and Fryer, J. D. 1972. Degradation of several herbicdes in a soil previously treated with MCPA. *Weed Res.* 12:90–95.

Kladivko, E. J., Van Scoyoc, G. E., Monke, E. J., Oates, K. M. and Pask, W. 1991. Pesticide and nutrient movement into subsurface tile drains on a silt loam soil in Indiana. *J. Environ. Qual.* 20(1):264–270.

Klaseus, T. G., Buzicky, G. C. and Schneider, E. C. 1988. Pesticides and Groundwater: Surveys of Selected Minnesota Wells. Minn. Dept. of Health and Minn. Dept. of Agric., Legistative Comm. on Minn. Res., St. Paul, MN, 95 pp.

Koehler, F. A., Humenik, F. J., Johnson, D. D., Kreglow, J. M., Dressing, R. P. and Maas, R. P. 1982b. Best Management Practices for Agricultural Nonpoint Source Control. III. Sediment. USDA Coop. Agree. 12-05-300-472. EPA Interagency Agree. AD-12-F-0-037-0. North Carolina Agricult. Ext. Serv., Raleigh, NC, 49 pp.

Koskinen, W. C. and Harper, S. S. 1990. The retention process: Mechanisms. in Cheng, H. H. (Ed.). *Pesticides in the Soil Environment: Processes, Impacts, and Modeling*. SSSA Book Series No. 2. Soil Science Society of America, Madison, WI, pp. 51–77.

Kriner, R. E. 1985. Final Report on the Results of a National Survey of Pesticide Usage on Golf Courses in the U.S. conducted in July–September 1982. EPA-H-7503W. Coop. Agree. Amer. Assoc. Retired Persons and U.S. Environmental Protection Agency, Washington, D.C., 31 pp. and appendices.

Kuhn, E. P. and Suflita, J. M. 1989. Dehalogenation of pesticides by anaerobic microorganisms in soils and groundwater — A review. in Sawhney, B. L. and Brown, K. (Eds.). *Reactions and Movement of Organic Chemicals in Soil*. SSSA Special Publication No. 22. Soil Sci. Soc. Am., Am. Soc. Agron., Madison, WI, pp. 81–180.

Kuhr, R. J. and Tashiro, H. 1978. Distribution and persistence of chlorpyriphos and diazinon applied to turf. *Bull. Environ. Contam. Toxicol.* 20:652–656.

Kuhr, R. J., Schohon, J. L., Tashiro, H. and Fiori, B. J. 1972. Dieldrin resistance in the European chafer grub. *J. Econ. Entomol.* 7:167–173.

Kurtz, D. A. (Ed.). 1990. *Long Range Transport of Pesticides*. Lewis Publishers, Chelsea, MI, 462 pp.

Lee, D.-Y. and Farmer, W. J. 1989. Dissolved organic matter interaction with napropamide and four other nonionic pesticides. *J. Environ. Qual.* 18(4):468–474.

Lee, L. S., Rao, P. S. C., Brusseau, M. L. and Ogwada, R. A. 1988. Nonequilibrium sorption of organic contaminants during flow through columns of aquifer materials. *Environ. Toxicol. Chem.* 7:779–793.

Leeper, J. R., Roush, R. T. and Reynolds, H. T. 1986. Preventing or managing resistance in arthropods. in *Pesticide Resistance: Strategies and Tactics for Management*. National Academy Press, Washington, D.C., pp. 335–346.

Leistra, M. and Green, R. E. 1990. Efficacy of soil-applied pesticides. in Cheng, H. H. (Ed.). *Pesticides in the Soil Environment: Processes, Impacts, and Modeling*. SSSA Book Series No. 2. Soil Science Society of America, Madison, WI, pp. 401–428.

Leonard, R. A. 1988. Herbicides in surface waters. in Grover, R. (Ed.). *Environmental Chemistry of Herbicides*. Vol. 1. CRC Press, Boca Raton, FL, pp. 45–87.

Leonard, R. A. 1990. Movement of pesticides into surface waters. in Cheng, H. H. (Ed.). *Pesticides in the Soil Environment: Processes, Impacts, and Modeling*. SSSA Book Series No. 2. Soil Science Society of America, Madison, WI, pp. 303–349.

Leonard, R. A. and Knisel, W. G. 1988. Evaluating groundwater contamination potential from herbicide use. *Weed. Technol.* 2(2):207–216.

Leonard, R. A., Langdale, G. W. and Fleming, W. G. 1979. Herbicide runoff from Upland Piedmont watersheds — Data and implications for modeling pesticide transport. *J. Environ. Qual.* 8(2):223–229.

Leslie, A. R. 1989a. Societal problems associated with pesticide use in the urban sector. in Leslie, A. R. and Metcalf, R. L. (Eds.). Integrated Pest Management for Turfgrass and Ornamentals. Office of Pesticide Programs, 1989-625-030. U.S. Environmental Protection Agency, Washington, D.C., pp. 93–96.

Leslie, A. R. 1989b. Development of an IPM program for turfgrass. in Leslie, A. R. and Metcalf, R. L. (Eds.). Integrated Pest Management for Turfgrass and Ornamentals. Office of Pesticide Programs, 1989-625-030. U.S. Environmental Protection Agency, Washington, D.C., pp. 315–318.

Leslie, A. R. 1991. Database of Pesticides Registered for Use on Turfgrass from Compilations by the U.S. Environmental Protection Agency, Golf Course Superintendents Association of America, Professional Lawn Care Association of America, and the University of Maryland. Personal Communication. U.S. Environmental Protection Agency, Office of Pesticide Programs, Washington, D.C.

Leslie, A. R. and Metcalf, R. L. (Eds.). 1989. Integrated Pest Management for Turfgrass and Ornamentals. Office of Pesticide Programs, 1989-625-030. U.S. Environmental Protection Agency, Washington, D.C., 337 pp.

Levy, H., II. 1990. Regional and global transport and distribution of trace species released at the earth's surface. in Kurtz, D. A. (Ed.). *Long Range Transport of Pesticides.* Lewis Publishers, Chelsea, MI, pp. 83–95.

Lym, R. G. and Swenson, O. R. 1991. Sulfometuron persistence and movement in soil and water in North Dakota. *J. Environ. Qual.* 20(1):209–215.

Mackay, D. and Paterson, S. 1981. Calculating fugacity. *Environ. Sci. Technol.* 15(9): 1006–1014.

Mackay, D. and Paterson, S. 1982. Fugacity revisited. *Environ. Sci. Technol.* 16(12): 654A–660A.

Mackay, D. and Stiver, W. 1991. Predictability and environmental chemistry. in Grover, R. and Cessna, A. J. (Eds.). *Environmental Chemistry of Herbicides.* Vol 2. CRC Press, Boca Raton, FL, pp. 281–297.

Mackay, D., Paterson, S., Cheung, B. and Neely, W. B. 1985. Evaluating the environmental behavior of chemicals with a level III fugacity model. *Chemosphere* 14(3/4):335–374.

MacLean, A. J., Stone, B. and Cordukes, W. E. 1973. Amounts of mercury in soil of some golf course sites. *Can. J. Soil Sci.* 53:130–132.

Madhun, Y. A. and Freed, V. H. 1990. Impact of pesticides on the environment. in Cheng, H. H. (Ed.). *Pesticides in the Soil Environment: Processes, Impacts, and Modeling.* SSSA Book Series No. 2. Soil Science Society of America, Madison, WI, pp. 429–466.

Madhun, Y. A., Young, J. L. and Freed, V. H. 1986. Binding of herbicides by water-soluble organic materials from soil. *J. Environ. Qual.* 15(1):64–68.

Madison, J. H. 1971. *Principles of Turfgrass Culture.* Van Nostrand Reinhold Co., New York, 420 pp.

Marlin, P. A. and Droste, E. X. 1986. Contamination of groundwater as a result of agricultural use of ethylene dibromide. in *Proc. Third Annual Eastern Regional Ground Water Conf.*, Springfield, MA, 28–30 July. National Water Well Association, Dublin OH, pp. 277–305.

Mayeux, H. S., Jr., Richardson, C. W., Bovey, R. W., Burnett, M., Merkle, G. and Meyer, R. E. 1984. Dissipation of picloram in storm runoff. *J. Environ. Qual.* 13(1):44–49.

McCall, P. J., Laskowski, D. A., Swann, R. L. and Dishburger, H. J. 1981. Measurement of sorption coefficients of organic chemicals and their use in environmental fate analysis, in *Test Protocols for Environmental Fate and Movement of Toxicants.* Association of Official Analytical Chemists, 94th Annual Meeting, Arlington, VA, pp. 89–109.

McCall, P. J., Laskowski, D. A., Swann, R. L., and Dishburger, H. J. 1983. Estimation of environmental partitioning of organic chemicals in model ecosystems. *Residue Rev.* 85: 231–244.

Metcalf, R. L. 1955. Physiological basis for insecticide resistance in insects. *Physiol. Rev.* 35:197–232.

Metcalf, R. L. 1980. Changing role of insecticides in crop protection. *Ann. Rev. Entomol.* 25:219–256.

Metcalf, R. L. 1986. The ecology of insecticides and the chemical control of insects. in Kogan, M. (Ed.). *Ecological Theory and Integrated Pest Management Practice.* John Wiley & Sons, New York, pp. 251–297.

Metcalf, R. L. 1989. Insect resistance to insecticides. in Leslie, A. R. and Metcalf, R. L. (Eds.). Integrated Pest Management for Turfgrass and Ornamentals. Office of Pesticide Programs, 1989-625-030. U.S. Environmental Protection Agency, Washington, D.C., pp. 3–44.

Miller, S. A., Grothaus, G. D., Peterson, F. P., Rittenburg, J. H., Plumley, K. A. and Lankow, R. K. 1989. Detection and monitoring of turfgrass pathogens by immunoassay. in Leslie, A. R. and Metcalf, R. L. (Eds.). Integrated Pest Management for Turfgrass and Ornamentals. Office of Pesticide Programs, 1989-625-030. U.S. Environmental Protection Agency, Washington, D.C., pp. 109–120.

Mingelgrin, U. and Gerstl, Z. 1983. Reevaluation of partitioning as a mechanism of nonionic chemical adsorption in soil. *J. Environ. Qual.* 12(1):1–11.

Monteith, J. M. and Dahl, D. S. 1932. Turf diseases and their control. *Bull. USGA Green Comm.* 12:85–187.

Morandi, L. 1988. Overview of innovative State policy initiatives. in Agricultural Chemicals and Groundwater Protection: Emerging Management and Policy. Proceedings of a Conference, Oct. 22–23, 1987, Freshwater Foundation, St. Paul, MN, pp. 163–166.

Mount, D. I. and Putnicki, G. J. 1966. Summary report of the 1963 Mississippi fish kill. *Trans. N. Am. Wildl. Conf.* 31:177–184.

Moyer, J. R., Hance, R. J. and McKone, C. E. 1972. The effect of adsorbents on the rate of degradation of herbicides incubated with soil. *Soil Biol. Biochem.* 4:307–311.

Muir, D. C. G. and Grift, N. P. 1987. Herbicide levels in rivers draining two prairie agricultural watersheds (1984). *J. Environ. Sci. Health* 22(3):259–284.

Muir, D. C. G., Grift, N. P., Ford, C. A., Reiger, A. W., Hendzel, M. R. and Lockhart, W. L. 1990. Evidence for long-range transport of toxaphene to remote arctic and subarctic waters from monitoring of fish tissues. in Kurtz, D. A. (Ed.). *Long Range Transport of Pesticides.* Lewis Publishers, Chelsea, MI, pp. 329–346.

Murphy, S. R. 1990. Aquatic Toxicity Information Retrieval (AQUIRE): A Microcomputer Database. System Documentation. Spectrum Research, Inc., Duluth, MN, 37 pp.

Murphy, R. A. 1992. Aquatic and terrestrial toxicity tables. in Balogh, J. C. and Walker, W. J. (Eds.). *Golf Course Management and Construction: Environmental Issues.* Lewis Publishers, Chelsea, MI, Chapter 8.

Murray, J. J. and Powell, J. B. 1979. Turf. in Buckner, R. C. and Bush, L. P. (Eds.). *Tall Fescue.* Agron. Monogr. 20:293–306. Am. Soc. Agron., Crop Sci. Soc. Am., Soil Sci. Soc. Am., Madison, WI.

Murray, J. J., Klingman, D. L., Nash, R. G. and Woolson, E. A. 1983. Eight years of herbicide and nitrogen fertilizer treatments on Kentucky bluegrass turf. *Weed Sci.* 31:825–831.

Nash, R. G. 1988. Dissipation from soil. in Grover, R. (Ed.). *Environmental Chemistry of Herbicides.* Vol. 1. CRC Press, Boca Raton, FL, pp. 131–169.

Nash, R. G and Beall, M. L., Jr. 1980. Distribution of silvex, 2,4-D, and TCDD applied to turf in chambers and field. *J. Agric. Food Chem.* 28(3):614–623.

Neil, G. D., Williams, J. S. and Weddle, T. K. 1987. Second Annual Report — Pesticides in Gound Water Study. Maine Geol. Survey. Open-File No. 87–20. 27 pp.

Newman, A. S., Thomas, J. R. and Walker, R. L. 1952. Disapperance of 2,4-dichlorophenoxyacetic acid and 2,4,5-trichlorophenoxeacetic acid from soil. *Soil Sci. Soc. Am. Proc.* 21(1):21–24.

Ng, Y.-S. and Ahmad, S. 1979. Resistance to dieldrin and tolerance to chloryrifos and bendiocarb in northern New Jersey population of Japanese beetle. *J. Econ. Entomol.* 72:167–173.

Nielsen, D. R., Wierenga, P. J. and Bigger, J. W. 1983. Spatial soil variability and mass transfers from agricultural soils. in Nelsen, D. W., Tanji, K. K. and Elrick, D. E. (Eds.). *Chemical Mobility and Reactivity in Soil Systems*. SSSA Special Publ. No. 11. Soil Science Society of America, Madison, WI, pp. 65–78.

Niemczyk, H. D. 1987. The influence of application timing and posttreatment irrigation on the fate and effectiveness of isofenphos for control of Japanese beetle (*Coleoptera: Scarabaeidae*) larvae in turfgrass. *J. Econ. Entomol.* 80(2):465–470.

Niemczyk, H. D. and Chapman, R. A. 1987. Evidence of enhanced degradation of isofenphos in turfgrass thatch and soil. *J. Econ. Entomol.* 80(4):880–882.

Niemczyk, H. D. and Krueger, H. R. 1987. Persistence and mobility of isazofos in turfgrass thatch and soil. *J. Econ. Entomol.* 80(4):950–952.

Niemczyk, H. D. and Lawrence, K. O. 1973. Japanese beetles: Evidence of resistance to cyclodiene insecticides in larvae and adults in Ohio. *J. Econ. Entomol.* 66:520–521.

Niemczyk, H. D., Filary, A. and Krueger, H. R. 1988. Movement of insecticide residues in turfgrass thatch and soil. *West. Views* 1:7.

Niemczyk, H. D., Krueger, H. R. and Lawrence, K. O. 1977. Thatch influences movement of soil insecticides. *Ohio Rep.* 62:26–28.

Norris, L. A., Montgomery, M. L., Warren, L. E., and Mosher, W. D. 1982. Brush control with herbicides on hill pasture sites in southern Oregon. *J. Range Manage.* 35(1):75–80.

Oki, D. S. and Giambelluca, T. W. 1987. DBCP, EDB, and TCP contamination of groundwater in Hawaii. *Ground Water* 25(6):693–702.

Oster, C. A. 1982. Review of ground-water flow and transport models in the unsaturated zone. PNL-4427, NUREG/CR-2917. Battelle, Pacif. Northwest Lab., Richland, WA.

Ou, L.-T. 1984. 2,4-D degradation and 2,4-D degrading microorganisms in soils. *Soil Sci.* 137(2):100–107.

Paulson, D., Jr. 1981. Effect of pH, organic matter and soil texture on herbicides. *Solutions* 81(1):40–54.

Pignatello, J. J. 1987. Microbial degradation of 1,2-dibromoethane in shallow aquifer materials. *J. Environ. Qual.* 16(4):307–312.

Pignatello, J. J. 1989. Sorption dynamics of organic compounds in soils and sediment. in Sawhney, B. L. and Brown, K. (Eds.). *Reactions and Movement of Organic Chemicals in Soil.* SSSA Special Publication No. 22. Soil Sci. Soc. Am., Am. Soc. Agron., Madison, WI, pp. 45–80.

Pignatello, J. J. 1990a. Slowly reversible sorption of aliphatic halocarbons in soils. I. Formation of residual fractions. *Environ. Toxicol.. Chem.* 9:1107–1115.

Pignatello, J. J. 1990b. Slowly reversible sorption of aliphatic halocarbons in soils. II. Mechanistic aspects. *Environ. Toxicol. Chem.* 9:1117–1126.

Pignatello, J. J. and Huang, L. Q. 1991. Sorptive reversibility of atrazine and metoachlor residues in field soil samples. *J. Environ. Qual.* 20(1):222–228.

Pignatello, J. J., Frink, C. R., Marin, P. A. and Droste, E. X. 1990. Field-observed ethylene dibromide in an aquifer after two decades. *J. Contam. Hydrol.* 5:195–214.

Pimentel, D. (Ed.). 1981. *Handbook of Pest Management in Agriculture.* Vols. 1–3. CRC Press, Boca Raton, FL.

Pimentel, D. and Levitan, L. 1986. Pesticides: Amounts applied and amounts reaching pests. *Bioscience* 36(2):86–90.

Pionke, H. B., Glotfelty, D. E., Lucas, A. D. and Urban, J. B. 1988. Pesticide contamination of groundwaters in the Mahantango Creek watershed. *J. Environ. Qual.* 17(1):76–84.

Piotrowski, M. R. 1991. Bioremediation of hydrocarbon contaminated surface water, groundwater, and soils: The microbial ecology approach. in Kostecki, P. T. and Calabrese, E. J. (Eds.). *Hydrocarbon Contaminated Soils and Groundwater*. Lewis Publishers, Chelsea, MI, pp. 203–238.

Potter, D. A. and Braman, S. K. 1991. Ecology and management of turfgrass insects. *Annu. Rev. Entomol.* 36:383–406.

Potter, D. A., Cockfield, S. D. and Morris, T. A. 1989. Ecological side effects of pesticide and fertilizer use on turfgrass. in Leslie, A. R. and Metcalf, R. L. (Eds.). Integrated Pest Management for Turfgrass and Ornamentals. Office of Pesticide Programs, 1989-625-030. U.S. Environmental Protection Agency, Washington, D.C., pp. 33–44.

Pratt, P. F. (Chairman). 1985. Agriculture and Groundwater Quality. Council for Agricultural Science and Technology. CAST Report No. 103. 62 pp.

Racke, K. D. and Coats, J. R. 1987. Enhanced degradation of isofenphos by soil microorganisms. *J. Agric. Food Chem.* 35(1):94–99.

Rao, P. S. C. and Davidson, J. M. 1979. Adsorption and movement of selected pesticides at high concentrations in soils. *Water Res.* 13:375–380.

Rao, P. S. C. and Davidson, J. M. 1980. Estimation of pesticide retention and transformation parameters required in nonpoint source pollution models. in Overcash, M. R. and Davidson, J. M. (Eds.). *Environmental Impact of Nonpoint Source Pollution.* Ann Arbor Science Publishing, Inc., Ann Arbor, MI, pp. 23–67.

Rao, P. S. C. and Jessup, R. E. 1983. Sorption and movement of pesticides and other toxic organic substances in soils. in Nelsen, D. W., Elrick, D. E. and Tanji, K. K. (Eds.). *Chemical Mobility and Reactivity in Soil Systems.* Soil Sci. Soc. Am. Special Publ. No. 11. Soil Science Society of America, Madison, WI, pp. 183–201.

Rao, P. S. C., Davidson, J. M., Jessup, R. E. and Selim, H. M. 1979. Evaluation of conceptual models for describing nonequilibrium adsorption-desorption of pesticides. *Soil Sci. Soc. Am. J.* 43(1):22–28.

Rao, P. S. C., Jessup, R. E. and Davidson, J. M. 1988. Mass flow and dispersion. in Grover, R. (Ed.). *Environmental Chemistry of Herbicides.* Vol. 1. CRC Press, Boca Raton, FL. pp. 21–43.

Read, D. C. 1986. Accelerated microbial breadkdown of carbofuran in soils from previously treated fields. *Agric. Ecosys. Environ.* 15:51–61

Reinert, J. A. 1976. Control of sod webworms (*Herpetogramma* spp. and *Crambus* spp.) on bermudagrass. *J. Econ. Entomol.* 69(5):669–672.

Reinert, J. A. 1982a. Insecticide resistance in epigeal insect pests of turfgrass: I. A review. in Niemczyk, H. D. and Joyner, B. G. (Eds.). *Advances in Turfgrass Entomology.* Hammer Graphics, Inc., Piqua, OH, pp. 71–76.

Reinert, J. A. 1982b. Carbamate and synthetic pyrethroid insecticides for control of organophosphate-resistant southern chinch bugs (*Heteroptera: Lygaeidae*). *J. Econ. Entomol.* 75(4):716–718.

Reinert, J. A. 1983. Field experiments for insecticidal control of sod webworms (*Lepidoptera: Pyralidae*) in Florida Turfgrass. *J. Econ. Entomol.* 76(1):150–153.

Reinert, J. A. and Niemczyk, H. D. 1982. Insecticide resistance in epigeal insect pests of turfgrass. II. A review. in Niemczyk, H. D. and Joyner, B. G. (Eds.). *Advances in Turfgrass Entomology.* Hammer Graphics, Inc., Piqua, OH, pp. 77–80.

Reinert, J. A. and Portier, K. M. 1983. Distribution and characterization of organophosphate-resistant southern chinch bugs (*Heteroptera: Lygaeidae*) in Florida. *J. Econ. Entomol.* 76(5):1187–1190.

Rhode, W. A., Johnson, A. W., Dowler, C. C. and Glaze, N. C. 1980. Influence of climate and cropping patterns on the efficacy of ethoprop, methyl bromide, and DD-MENCS for control of root-knot nematodes. *J. Nematol.* 12:33–39.

Rhodes, R. C. and Long, J. D. 1974. Run-off and mobility studies on benomyl in soils and turf. *Bull. Environ. Contam. Toxic.* 12(4):385–393.

Ricci, E. D., Hubert, W. A. and Richard, J. J. 1983. Organochlorine residues in sediment cores of a midwestern reservoir. *J. Environ. Qual.* 12(3):418-421.

Richard, T. L. and Steenhuis, T. S. 1988. Tile drain sampling of preferential flow on a field scale. *J. Contam. Hydrol.* 3:307–325.

Richard, T. L. and Steenhuis, T. S. 1988. Tile drain sampling of preferential flow on a field scale. *J. Contam. Hydrol.* 3:307–325.

Richards, R. P., Kramer, J. W., Baker, D. B. and Krieger, K. A. 1987. Pesticides in rainwater in the northeastern United States. *Nature* 327(1560):129–131.

Ritter, W. F., Johnson, H. P., Lovely, W. G. and Molnau, M. 1974. Atrazine, propachlor, and diazonon residues on small agricultural watersheds. *Environ. Sci. Technol.* 8(1):38–42.

Rohde, W. A., Asmussen, L. E., Hauser, E. W., Hester, M. L. and Allison, H. D. 1981. Atrazine persistence in soil and transport in surface and subsurface runoff from plots in the Coastal Plain of the southern United States. *Agro-Ecosystems* 7:225–238.

Rohde, W. A., Asmussen, L. E., Hauser, E. W. and Johnson, A. W. 1979. Concentrations of Ethoprop in the Soil and Runoff Water of a Small Agricultural Watershed. USDA-SEA Agric. Res. Results ARR-S-2. U.S. Government Printing Office, Washington, D.C.

Rohde, W. A., Asmussen, L. E., Hauser, E. W., Wauchope, R. D. and Allison, H. D. 1980. Trifluralin movement in runoff from a small agricultural watershed. *J. Environ. Qual.* 9(1):37–42.

Ross, L. J., Nicosia, S., McChesney, M. M., Hefner, K. L., Gonzalez, D. A. and Seiber, J. N. 1990. Volatilization, off-site deposition, and dissipation of DCPA in the field. *J. Environ. Qual.* 19(4):715–722.

Rothschild, E. R., Manser, R. J. and Anderson, M. P. 1982. Investigation of aldicarb in ground water in selected areas of the central sand plain of Wisconsin. *Ground Water* 20:437–445.

Saleh, F. M. A., Bishop, D. J., Dietrich, S. F., Knezovich, J. P. and Harrison, F. L. 1990. Transport of nonsorbed chemicals in the subsurface environment: Proposed model with experimental verification. *Soil Sci.* 149(1):23–34.

Sauer, T. J. and Daniel, T. C. 1987. Effect of tillage system on runoff losses of surface-applied pesticides. *Soil Sci. Soc. Am. J.* 51(2):410–415.

Sawhney, B. L. and Brown, K. (Eds.). 1989. *Reactions and Movement of Organic Chemicals in Soils*. SSSA Special Publication No. 22. Soil Sci. Soc. Am., Am. Soc. Agron., Madison, WI, 474 pp.

Scarano, L. J. 1986. The Massachusetts Aldicarb Well Water Survey 1983–1985. in Proc. Third Annual Eastern Regional Ground Water Conf., Springfield, MA, 28–30 July. National Water Well Association, Dublin, OH, pp. 261–276.

Schepers, J. S., Vavricka, E. J., Anderson, D. R., Wittmuss, H. D. and Schuman, G. E. 1980. Agricultural runoff during a drought period. *J. Water Pollut. Control Fed.* 52:711–719.

Schubert, O. E. 1967. Can activated charcoal protect crops from herbicide injury. *Crop Soils.* 19:10–11.

Scrifes, C. J., Hahn, R. R., Diaz-Colon, J. and Merkle, M. G. 1971. Picloram persistence in semiarid rangeland soils and water. *Weed Sci.* 19:381–384.

Scrifes, C. J., McCall, H. G., Maxey, R. and Tai, H. 1977. Residual properties of 2,4,5-T and picloram in sandy rangeland soils. *J. Environ. Qual.* 6(1):36–42.

Sears, M. K. and Chapman, R. A. 1979. Persistence and movement of four insecticides applied to turfgrass. *J. Econ. Entomol.* 72(2):272–274.

Sears, M. K. and Chapman, R. A. 1982. Persistence and movement of four insecticides applied to turfgrass. in Niemczyk, H. D. and Joyner, B. G. (Eds.). *Advances in Turfgrass Entomology.* Hammer Graphics, Inc., Piqua, OH, pp. 57–59.

Senesi, N. and Testini, C. 1980. Adsorption of some nitrogenated herbicides by soil humic acids. *Soil Sci.* 130(6):314–320.

Sethunathan, N. 1971. Degradation of diazinon in paddy fields as a cause of its inefficiency for controlling brown planthoppers in rice fields. *PANS* 17:18–19.

Severn, D. J. and Ballard, G. 1990. Risk/benefit and regulations. in Cheng, H. H. (Ed.). *Pesticides in the Soil Environment: Processes, Impacts, and Modeling.* SSSA Book Series No. 2. Soil Science Society of America, Madison, WI, pp. 467–491.

Shearman, R. 1987. Pre-emergence weed control of cool-season turf. *Weeds, Trees, & Turf* 26(2):70–71.

Sheets, T. J., Bradley, J. R., Jr. and Jackson, M. D. 1972. Contamination of Surface and Ground Water with Pesticides Applied to Cotton. Univ. North Carolina Water Res., Res. Rept. No. 60. Chapel Hill, NC, 63 pp.

Shoemaker, L. L., Magette, W. L. and Shirmohammadi, A. 1990. Modeling management practice effects on pesticide movement to ground water. *Ground Water Monit. Rev.* 10(1):109–115.

Singh, G., Spencer, W. F., Cliath, M. M. and van Genuchten, M. Th. 1990. Sorption behavior of s-triazine and thiocarbamate herbicides on soils. *J. Environ. Qual.* 19(3):520–525.

Smiley, R. W. 1983. *Compendium of Turfgrass Diseases.* The Am. Phytopath. Soc., APS Press, St. Paul, MN, 101 pp.

Smith, A. E. 1988. Transformations in soil. in Grover, R. (Ed.). *Environmental Chemistry of Herbicides.* Vol. 1. CRC Press, Boca Raton, FL, pp. 172–200.

Smith, A. E. and Aubin, A. J. 1991. Effects of long-term 2,4-D, and MCPA field application on the soil breakdown of 2,4-D, MCPA, Mecoprop, and 2,4,5-T. *J. Environ. Qual.* 20(2):436–438.

Smith, A. E., Grover, R., Cessna, A. J., Shewchuk, S. R. and Hunter, J. H. 1986. Fate of diclofop-methyl after application to a wheat field. *J. Environ. Qual.* 15(3):234–238.

Smith, A. T., Aubin, A. J. and Biederbeck, V. O., 1989. Effects of long-term 2,4-D and MCPA field applications on soil residues and their rates of breakdown. *J. Environ. Qual.* 18(3):299–302.

Smith, C. N., Leonard R. A., Langdale, G. W. and Bailey, G. W. 1978. Transport of Agricultural Chemicals from Small Upland Piedmont Watersheds. EPA-600/3-78-056. U.S. Environmental Protection Agency, Washington, D.C.

Smith, S. and Willis, G. H. 1987. Interaction of methyl parathion with permethrin and flucythrinate as related to their mutual persistence in soil. *Soil Sci.* 144(1):67–71.

Smolen, M. D., Humenik, F. J., Spooner, J., Dressing, S. A. and Maas, R. P. 1984. Best Management Practices for Agricultural Nonpoint Source Control. IV. Pesticides. USDA Coop. Agree. 12-05-300-472. EPA Interagency Agree. AD-12-F-0-037-0. North Carolina Agricult. Ext. Serv., Raleigh, NC, 87 pp.

Spain, J. C. and van Veld, P. A. 1983. Adaption of nautral microbial communities to degradation of xenobiotic compounds: Effects of concentration, exposure time, inoculum, and chemical structure. *Appl. Environ. Microbiol.* 45(2):428–435.

Spalding, R. F., Junk, G. A. and Richards, J. J. 1980. Pesticides in groundwater beneath irrigated farmland in Nebraska, August, 1978. *Pestic. Monit. J.* 14:70–73.

Spencer, W. F. and Cliath, M. M. 1990. Movement of pesticides from soil to the atmosphere. in Kurtz, D. A. (Ed.). *Long Range Transport of Pesticides*. Lewis Publishers, Chelsea, MI, pp. 1–16.

Spencer, W. F., Cliath, M. M., Blair, J. and LeMest, R. A. 1982b. Transport of Pesticides from Fields in Surface Runoff and Tile Drain Waters. EPA-78-D-X0117. U.S. Environmental Protection Agency, Washington, D.C.

Spencer, W. F., Cliath, M. M., Blair, J. and LeMest, R. A. 1985. Transport of Pesticides from Irrigated Fields in Surface Runoff and Tile Drain Waters, USDA-ARS Conserv. Res. Rept. 31. U.S. Government Printing Office, Washington, D.C.

Spencer, W. F., Cliath, M. M. Jury, W. A. and Zhang, L. 1988. Volatilization of organic chemicals from soil as related to Henry's Law constants. *J. Environ. Qual.* 17(3):504–509.

Spencer, W. F., Farmer, W. J. and Cliath, M. M. 1973. Pesticide volatilization. *Residue Rev.* 49:1–45.

Spencer, W. F., Farmer, W. J. and Jury, W. A. 1982a. Review: Behavior of organic chemicals at soil, air, water interfaces as related to predicting the transport and volatilization of organic pollutants. *Environ. Toxicol. Chem.* 1:17–26.

Starr, J. L. and Glotfelty, D. E. 1990. Atrazine and bromide movement through a silt loam soil. *J. Environ. Qual.* 19(3):552–558.

Steinberg, S. M., Pignatello, J. J. and Sawhney, B. L. 1987. Persistence of 1,2-dibromoethane in soils: Entrapment in intraparticle micropores. *Environ. Sci. Technol.* 21(12):1201–1208.

Stewart, B. A., Woolhiser, D. A., Wischmeier, W. H., Caro, J. H. and Frere, M. H. 1975. Control of Water Pollution from Cropland: A Manual for Guideline Development. Vol. 1. EPA 600/2-75-026a. U.S. Environmental Protection Agency, Athens, GA and USDA Agricultural Research Service, Washington, D.C., 111 pp.

Stone, W. B. 1979. Poisoning of wild birds by organophosphate and carbamate pesticides. *N.Y. Fish Game J.* 26(1):37–47.

Stone, W. B. and Knoch H. 1982. American brant killed on golf courses by diazinon. *N.Y. Fish Game J.* 29(2):95–96.

Suett, D. L. 1987. Influence of treatment of soil with carbofuran on the subsequent performance of insecticides against cabbage root fly (*Delia radicum*) and carrot fly (*Psila rosae*). *Crop Prot.* 6:371–378.

Suntio, L. R., Shiu, W. Y., Mackay, D., Seiber, J. N. and Glotfelty, D. 1988. Critical review of Henry's Law constants for pesticides. *Rev. Environ. Contam. Toxicol.* 103:1–59.

Tashiro, H. 1982a. The incidence of insecticide resistance in soil inhabiting turfgrass insects. in Niemczyk, H. D. and Joyner, B. G. (Eds.). *Advances in Turfgrass Entomology*. Hammer Graphics, Inc., Piqua, OH, pp. 81–84.

Tashiro, H. 1982b. Distribution and persistence of chlorpyrifos and diazinon in soil when applied to turf. in Niemczyk, H. D. and Joyner, B. G. (Eds.). *Advances in Turfgrass Entomology*. Hammer Graphics, Inc., Piqua, OH, pp. 53–56.

Tashiro, H. and Kuhr, R. J. 1978. Some factors influencing the toxicity of soil applications of chlorpyrifos and diazinon to European chafer grubs. *J. Econ. Entomol.* 71(6):904–907.

Tashiro, H., Personius, K. E., Zinter, D. and Zinter, M. 1971. Resistance of the European chafer to cyclodiene insecticides. *J. Econ. Entomol.* 64:242–245.

Taylor, A. W. and Glotfelty, D. E. 1988. Evaporation from soils and crops. in Grover, R. (Ed.). *Environmental Chemistry of Herbicides.* Vol. 1 CRC Press, Boca Raton, FL, pp. 89–129.

Taylor, A. W. and Spencer, W. F. 1990. Volatilization and vapor transport processes. in Cheng, H. H. (Ed.). *Pesticides in the Soil Environment: Processes, Impacts, and Modeling.* SSSA Book Series No. 2. Soil Science Society of America, Madison, WI, pp. 213–269.

Tietge, R. M. 1992. Wildlife and golf courses. in Balogh, J. C. and Walker, W. J. (Eds.). *Golf Course Management and Construction: Environmental Issues.* Lewis Publishers, Chelsea, MI, Chapter 6.

Thomas, G. W. and Phillips, R. E. 1979. Consequences of water movement in macropores. *J. Environ. Qual.* 8(2):149–152.

Trichell, D. W., Morton, H. L. and Merkle, M. G. 1968. Loss of herbicides in runoff water. *Weed Sci.* 16:447–449.

Triplett, G. B., Jr.,Conner, B. J. and Edwards, W. M. 1978. Transport of atrazine and simazine in runoff from conventional and no-tillage corn. *J. Environ. Qual.* 7(1):77–84.

T & OCR. 1991. *Turf & Ornamental Chemicals Reference.* Chemical and Pharmaceutical Press, John Wiley & Sons, New York, 929 pp.

USDA Agricultural Research Service. 1989a. USDA Research Plan for Water Quality. United States Dept. of Agric., Coop. State Res. Serv., State Agric. Exp. Stns., Washington, D.C., 14 pp.

USDA Agricultural Research Service. 1989b. Midwest Water Quality Research Initiative Management Systems Evaluation Areas. FY 1990 Guidelines for Proposal Preparation and Submission. Agricultural Chemical Management Research: A Multiagency Focus on Ground Water Quality. United States Dept. of Agric., Coop. State Res. Serv., State Agric. Exp. Stns., Washington, D.C., 17 pp.

Valentine, R. L. 1986. Nonbiological transformation. in Hern, S. C. and Melancon, S. M. (Eds.). *Vadose Zone Modeling of Organic Pollutants.* Lewis Publishers, Chelsea, MI, pp. 223–224.

Valentine, R. L. and Schnoor, J. L. 1986. Biotransformation. in Hern, S. C. and Melancon, S. M. (Eds.). *Vadose Zone Modeling of Organic Pollutants.* Lewis Publishers, Chelsea, MI, pp. 191–222.

van der Leeden, F., Troise, F. L. and Todd, D. K. 1990. *The Water Encyclopedia.* Lewis Publishers, Chelsea, MI, 808 pp.

van Genuchten, M. Th. and Wagenet, R. J. 1989. Two site/Two region models for pesticide transport and degradation: Theoretical development and analytical solutions. *Soil Sci. Soc. Am. J.* 53(5):1303–1310.

Vargas, J. M., Jr. 1981. *Management of Turfgrass Diseases.* Burgess Co. Minneapolis, MN, 204 pp.

Villani, M. G., Wright, R. J. and Baker, P. B. 1988. Differential susceptibility of Japanese beetle, oriental beetle, and European chafer (*Coleoptera: Scarabaeidae*) larvae to five soil insecticides. *J. Econ. Entomol.* 81(3):785–788.

Vinten, A. J., Yaron, B. and Nye, P. H. 1983. Vertical transport of pesticide into soil when adsorbed on suspended particles. *J. Agric. Food. Chem.* 31:662–664.

Vittum, P. J. 1985. Effect of timing of application on effectiveness of isofenphos, isazophos, and diazinon of Japanese beetle (*Coleoptera: Scarabaeidae*) grubs in turf. *J. Econ. Entomol.* 78(1):172–180.

Wagenet, R. J. 1987. Processes influencing pesticide loss with water under conservation tillage. in Logan, T. J., Davidson, J. M., Baker, J. L. and Overcash, M. R. (Eds). *Effects of Conservation Tillage on Groundwater Quality.* Lewis Publishers, Chelsea, MI, pp. 189–204.

Wagenet, R. J. and Hutson, J. L. 1986. Predicting the fate of nonvolatile pesticides in the unsaturated zone. *J Environ. Qual.* 15(4):315–322.

Wagenet, R. J. and Rao, P. S. C. 1990. Modeling pesticide fate in soils. in Cheng, H. H. (Ed.). *Pesticides in the Soil Environment: Processes, Impacts, and Modeling*. SSSA Book Series No. 2. Soil Science Society of America, Madison, WI, pp. 351–399.

Waldrop, M. M. 1990. Spontaneous order, evolution, and life. *Science* 247(4950):1543–1545.

Walker, A. and Brown, P. A. 1983. Spatial variability in herbicide degradation rates and residues in soil. *Crop Prot.* 2(1):17–25.

Walker, A. and Suett, D. L. 1986. Enhanced degradation of pesticides in soil: A potential problem for continued pest, disease and weed control. *Aspects Appl. Biol.* 12:95–103.

Walker, W. J., Balogh, J. C., Tietge, R. M. and Murphy, S. R. 1990. *Environmental Issues Related to Golf Course Construction and Management*. United States Golf Association, Green Section. Far Hills, NJ, 378 pp.

Watschke, T. L. and Mumma, R.O. 1989. The Effect of Nutrients and Pesticides Applied to Turf on the Quality of Runoff and Percolating Water. A Final Report to the U.S Dept. Int., U.S. Geol. Sur. ER 8904. Env. Res. Inst., Penn State University, College Park, PA, 64 pp.

Watschke, T. L., Hamilton, G. and Harrison, S. 1988. Is Pesticide runoff from turf increasing. *ALA* 9(7):43–44.

Wauchope, R. D. 1978. The pesticide content of surface water draining from agricultural fields — A review. *J. Environ. Qual.* 7(4):459-472.

Wauchope, R. D. 1987a. Tilted-bed simulation of erosion and chemical runoff from agricultural fields: II. Effects of formulation on atrazine runoff. *J. Environ. Qual.* 16(3):212–216.

Wauchope, R. D. 1987b. Effects of conservation tillage on pesticide loss with water. in Logan, T. J., Davidson, J. M., Baker, J. L. and Overcash, M. R. (Eds). *Effects of Conservation Tillage on Groundwater Quality*. Lewis Publishers, Chelsea, MI, pp. 205–215.

Wauchope, R. D. and Leonard, R. A. 1980. Maximum pesticide concentrations in agricultural runoff: A semi-empirical prediction formula. *J. Environ. Qual.* 9(4):665–672.

Wauchope, R. D., Buttler, T. M., Hornsby, A. G., Augustijn-Beckers, P. W. M. and Burt, J. P. 1991. The SCS/ARS/CES pesticide database for environmental decision-making. *Rev. Environ. Contam. Toxicol.* 123:1–155.

Wauchope. R. D., Williams, R. G. and Marti, L. R. 1990. Runoff of sulfometuron-methyl and cyanazine from small plots: effects of formulation and grass cover. *J. Environ. Qual.* 19(1):119–125.

Weber, J. B. and Weed, S. B. 1968. Adsorption and desorption of diquat, paraquat, and prometone by montmorillonite and kaolinite clay minerals. *Soil Sci. Soc. Am. Proc.* 32: 485–487.

Weber, J. B., Perry, P. W. and Upchurch, R. P. 1965. The influence of temperature and time on the adsorption of paraquat, 2,4-D and prometone by clays, charcoal, and an ion-exchange resin. *Soil Sci. Soc. Am. Proc.* 29:678–688.

Weber, J. B., Shea, P. H. and Weed, S. B. 1986. Fluidone retention and release in soils. *Soil Sci. Soc. Am. J.* 50(3):582–588.

Weber, R. R. and Montone, R. C. 1990. Distribution of organochlorines in the atmosphere of the south Atlantic and Antarctic oceans. in Kurtz, D. A. (Ed.). *Long Range Transport of Pesticides*. Lewis Publishers, Chelsea, MI, pp. 185–197.

Wehtje, G., Mielke, L. N., Leavitt, J. R. C. and Schepers, J. S. 1984. Leaching of atrazine in the root zone of an alluvial soil in Nebraska. *J. Environ. Qual.* 13(4):507–513.

Wehtje, G. R., Spalding, R. F., Burnside, O. C., Lowry, S. R. and Leavitt, J. R. 1983. Biological significance and fate of atrazine under aquifer conditions. *Weed Sci.* 31:610–618.

Welterlen, M. S., Gross, C. M., Angle, J. S. and Hill, R. L. 1989. Surface runoff from turf. in Leslie, A. R. and Metcalf, R. L. Integrated Pest Management for Turfgrass and Ornamentals. Office of Pesticide Programs, 1989-625-030. U.S. Environmental Protection Agency, Washington, D.C., pp. 153–160.

White, A. W., Asmussen, L. E., Hauser, E. W. and Turnbull, J. W. 1976. Loss of 2,4-D in runoff from plots receiving simulated rainfall and from a small agricultural watershed. *J. Environ. Qual.* 5:487–490.

White, A. W., Barnett, A. D., Wright, B. G. and Holladay, J. H. 1967. Atrazine losses from fallow land caused by runoff and erosion. *Environ. Sci. Technol.* 1:740–744.

Willis, G. H. and McDowell, L. L. 1982. Review: Pesticides in agricultural runoff and their effects on downstream water quality. *Environ. Toxicol. Chem.* 1:267–279.

Willis, G. H., McDowell, L. L., Murphree, C. E., Southwick, L. M. and Smith, S. 1983. Pesticide concentrations and yields in runoff from silty soils in the lower Mississippi Valley. *J. Agric. Food Chem.* 31:1171–1177.

Willis, G. H., McDowell, L. L., Parr, J. F. and Murphree, C. E. 1976. Pesticide Concentrations and Yields in Runoff and Sediment from a Mississippi Delta Watershed. In Proc. 3rd Interagency Sediment Conf. March 22–23, 1976, Denver, CO, pp. 53–64.

Willis, G. H., Rogers, R. L. and Southwick. L. M. 1975. Losses of diuron, linuron, fenac, and trifluralin in surface drainage water. *J. Environ. Qual.* 4:399–402.

Winer, A. M. and Atkinson, R. 1990. Atmospheric reaction pathways and lifetimes for organophosporous compounds. in Kurtz, D. A. (Ed.). *Long Range Transport of Pesticides*. Lewis Publishers, Chelsea, MI, pp. 115–126.

Wolfe, N. L., Mingelgrin, U. and Miller, G. C. 1990. Abiotic transformations in water, sediments, and soil. in Cheng, H. H. (Ed.). *Pesticides in the Soil Environment: Processes, Impacts, and Modeling*. SSSA Book Series No. 2. Soil Science Society of America, Madison, WI, pp. 103–168.

Wolman, M. G. and Schick, A. P. 1967. Effects of construction on fluvial sediment in urban and suburban areas of Maryland. *Water Resour. Res.* 3:451–464.

Wu, S.-C. and Gschwend, P. M. 1986. Sorption kinetics of hydrophobic organic compounds to natural sediments and soils. *Environ. Sci. Technol.* 20(7):717–725.

Wu, S.-C. and Gschwend, P. M. 1988. Numerical modeling of sorption kinetics of organic componds to soil and sediment particles. *Water Resour. Res.* 24(8):1373–1383.

Wu, T. L. 1980. Dissipation of the herbicides atrazine and alachlor in a Maryland corn field. *J. Environ. Qual.* 9:459–465.

Wu, T. L., Correll, D. L. and Remenapp, H. E. H. 1983. Herbicide runoff from experimental watersheds. *J. Environ. Qual.* 12(3):330–336.

Zaki, M. H., Moran, D. and Harris, D. 1982. Pesticides in groundwater: The aldicarb story in Suffolk County, *N.Y. Am. J. Public Health* 72(12):1391–1395.

Zielke, R. C., Pinnavaia, T. J. and Mortland, M. M. 1989. Adsorption and reactions of selected organic molecules on clay mineral surfaces. in Sawhney, B. L. and Brown, K. (Eds.). *Reactions and Movement of Organic Chemicals in Soils*. SSSA Special Publication No. 22. Soil Sci. Soc. Am., Am. Soc. Agron., Madison, WI, pp. 81–110.

CHAPTER 5

Development of Integrated Management Systems for Turfgrass

James C. Balogh, Anne R. Leslie, William J. Walker, and Michael P. Kenna

5.1 CONCEPT OF INTEGRATED MANAGEMENT SYSTEMS FOR TURFGRASS

Management of water, nutrients, pests, and diseases is essential for production and maintenance of high-quality turfgrass on golf courses. Depending on the intensity of golf course management, several potentially adverse environmental impacts have been identified in other chapters in this book. These effects include (1) contamination of surface water and groundwater resources with nitrates and residual pesticides; (2) contamination of surface water with sediment during construction and interference with natural drainage systems; (3) actual and perceived health effects on humans, wildlife, and other nontarget organisms; (4) development of insect and disease populations resistant to currently used pesticides.

Reduction in the use of pesticides, fertilizers, and irrigation water is not always a direct economic issue to many golf course managers. However, the perception that these materials, as well as construction and cultural activities, have adverse effects on the environment has led to concerns about their use and to intense opposition to

---

Photo courtesy of Michael French.

the construction of new golf courses. This has made it essential that golf course architects, developers, and managers consider programs designed to reduce water use and adverse environmental effects. These actions will reduce potential financial losses resulting from litigation, detrimental publicity, construction delays, and loss of permits for development of new golf courses, in addition to the positive environmental quality benefits (Cohen at al. 1990; Grant 1989; Leslie 1989a).

It is relatively easy to identify environmental issues. It is much more difficult to develop economically feasible systems to mitigate or resolve environmental problems related to turfgrass management. The systems approach to turfgrass management and the reduction of potentially adverse environmental effects requires a significant basic and applied research effort. Current initiatives in agricultural research are directed toward reduction of nonpoint source pollution problems in production agriculture. This research has focused on development of *integrated crop management* (ICM) systems (USDA 1989a, 1989b). Integrated crop management is an important conceptual expansion of *integrated pest management* (IPM) practices. ICM integrates all production factors in order to sustain long-term productivity, optimize yields and profitability, and maintain the integrity of the local ecosystem. Critical ICM factors include crop and cultivar selection; crop rotations; soil building practices; tillage systems; nutrient management; chemical, biological, and cultural pest management; livestock management; and soil, water, energy, and natural resource conservation.

The concepts originally developed for agricultural IPM programs have been suggested as a format for ecologically sound management of turfgrass pests (Potter and Braman 1991; Raupp et al. 1989). The implementation of ICM concepts for integrated management of turfgrass pests, nutrients, and water could be just as effective for reducing the perceived adverse environmental impacts associated with management of golf courses and other lawns. This integrated systems approach is defined as turfgrass management systems (TMS®). TMS combines all cultural management factors for sustained productivity of an acceptable level of quality for turfgrass, course profitability, and the integrity of ecosystems on and in the vicinity of the golf course.

IPM programs developed for turfgrass are a significant component of turfgrass management. However, traditionally defined IPM protocols *are only a part of the expanded functional concept of TMS.* Critical components of TMS include selection of (1) turfgrass species and cultivars; (2) soil management practices; (3) clipping and cultivation practices; (4) nutrient management; (5) irrigation and drainage management; and (6) chemical, biological, and cultural pest management. Conservation of soil, water, energy, and other natural resources during construction and maintenance of golf courses is one of the primary goals of TMS. The following discussion of TMS concepts are derived from research on turfgrass IPM programs and agricultural ICM systems. These programs may serve as a model for the development of profitable and environmentally sound integrated systems for golf course management. Ultimately, the goal of TMS is to balance costs, benefits, public health, and environmental quality with acceptable levels of playability.

The principles of IPM are not new and were originally developed by Forbes (1880) over 100 years ago. Many of the IPM concepts were practiced prior to introduction of wide-spectrum pesticides in the 1940s (Metcalf 1980). The IPM concepts of economic thresholds, chemical management, and biological control originally were systematized in the late 1950s by Stern et al. (1959). By the mid 1970s, IPM was recognized as an economically and ecologically sound system for insect control in agriculture (Wearing 1988). Most agricultural scientists and many producers recognize that total reliance on broad spectrum insecticides, and even herbicides, is not necessarily the only effective pest management strategy (Allen and Bath 1980; Leslie 1989b; Metcalf 1980).

Intensive use and overuse of pesticides, fertilizers, and irrigation water in agriculture and turfgrass management are major factors in the rapid growth of interest in IPM and, more recently, ICM (Leslie 1989a). IPM, ICM, and TMS seek to minimize the disadvantages associated with intensive nutrient, pesticide, and water inputs and to maximize the advantages of their use (Allen and Bath 1980; Leslie 1989c). Development of site-specific TMS programs offers one of the few comprehensive solutions for systematic control of environmental problems related to management of ecosystems. These integrated approaches offer systematic options for selection of alternate control strategies and to maximize chemical efficiency. TMS does not preclude the use of pesticides and inorganic fertilizers when needed. The basic components of TMS are

1. A system using multiple control methods
2. A decision process based on intensive use of information
3. A risk reduction system
4. A cost effective and site specific management strategy (Table 1)

The operational strategies used for ICM programs in agriculture could be used as a model for expansion of turfgrass IPM concepts for development of TMS (Allen and Bath 1980; Metcalf 1980; Wearing 1988). Development of TMS programs should progress through three iterative phases: (1) basic and applied research, (2) development of field programs, and (3) implementation and economic assessment of field programs (Table 1). Utilizing a systems analysis approach will help avoid significant lag periods between initial research and final field implementation of TMS programs (Allen and Bath 1980). TMS combines pest and nutrient control, irrigation scheduling, other cultural practices, and golfer use patterns. When properly implemented, the systems approach will produce turfgrass programs that are economically feasible, profitable, and acceptable to turfgrass managers. Compatibility with the game of golf is another long-term objective in development of TMS golf courses. All integrated

Table 1. Stages in Development of TMS Programs

Basic and applied research	Components of field programs	Implementation, education, and assessment
· Generate or compile basic knowledge required to understand interrelationship of (a) ecology of turfgrass, weeds, insects, and diseases; (b) nutrition; (c) water relations; and (d) cultural practices.	· Define the roles of all people in the integrated management process to assure communication and cooperation.	· Develop effective education programs and sources of information on site-specific TMS programs.
· Conduct applied research to develop specific control techniques and related methods including (a) breeding resistant grasses, (b) development of alternate pest and nutrient control plans, (c) modify existing cultural practices, and (d) incorporate weather factors in planning.	· Determine the pest, nutrient, and cultural objectives for specific areas on each site. Expand the concept of action thresholds to include pests, nutrients, and water.	· Develop a one-on-one communication system to initially transfer integrated management concepts.
· Conduct interdisciplinary systems research to integrate water, pest, disease, nutrient,	· Establish action and economic thresholds based on monitoring environmental conditions that require use of control practices for irrigation, fertilization, pest control, and clipping.	· Disseminate articles, bulletins, and newsletters to assist in the development and maintenance of integrated systems. Computer-assisted information systems provide additional support.
	· Monitor for nutrient, water, pests, and diseases on a periodic and consistent basis to determine when action thresholds are reached.	· Incentives for adoption include reduction in pest resistance and potential adverse environmental problems; increase in long-term profitability; and reduction in legal, financial, social, and environmental risk.

and cultural requirements for quality turfgrass systems. Action thresholds are established.	· Modify the turfgrass environment to reduce the need for chemical management practices.	· Obstacles to implementation include complexity, high cost of initial training and site conversion programs, training of golfers on integrated systems management, and lack of consistent and comprehensive sources of information.
· Research and development for effective training and education programs are established.	· Use appropriate alternate cultural practices to reduce infestations or deficiencies.	
· Incorporate information from field programs into research.	· Evaluate the results of integrated practices and maintain written records.	

management approaches should be compatible with long-term water quality, ecological, and societal goals (Dalton 1975).

Basic and applied research at an interdisciplinary level is currently required for development of practical IPM and TMS programs (Potter and Braman 1991). Limited initial research and pilot programs have shown that IPM systems in urban settings reduce overall use of insecticides without compromising the quality of turfgrass or landscape plants (Brown et al. 1989; Grant 1989; Raupp et al. 1989; Smith and Raupp 1986). However, broad scale effectiveness and acceptance of IPM and alternate control methods have yet to be demonstrated (Potter and Braman 1991). Interdisciplinary research, extension, and education programs are required to expand the current single or multiple pest control strategies on golf courses.

Within 30 years, over 85% of the North American population will be concentrated in urban areas (Potter and Braman 1991). This urban population is likely to have a strong desire for quality turfgrass with minimum adverse effects on the environment. The growing concern for environmental quality and expanding urban populations should strongly motivate turfgrass researchers and managers to develop IPM and TMS approaches for golf courses and lawns. Greater research on the ecology of turfgrass and pests in relationship to water use, cultural practices, and fertility is needed to meet the objectives of integrated management programs (Potter and Braman 1991). Understanding the interrelationship of turfgrass management, pests, water, and nutrients requirements will reduce dependence on chemical control methods. Also, the enhanced efficacy of chemical treatments in a systems approach will reduce the frequency of chemical application and environmental loading of pesticides and fertilizers (Leistra and Green 1990).

5.2 COMPONENTS OF INTEGRATED MANAGEMENT SYSTEMS FOR TURFGRASS

5.2.1 System Components

Development of economically feasible programs compatible with long-term environmental and societal goals is an essential ingredient for the success of IPM and TMS. The basic steps in development of site-specific IPM programs for turfgrass management have been outlined by Leslie (1989b, 1989c) (Table 1). These IPM components should be expanded to a systems management level for TMS. Systems type research for ICM has been defined by recent water quality and environmental research programs (USDA 1989a, 1989b). Using ICM as a model, TMS has eight basic components.

System Component 1 — *Definition* of the roles of all people involved in the management of turfgrass assures understanding of goals and promotes communication. Important individuals in the development of TMS on golf courses are the superintendent, members of the green and golf committees, the golf professional, the golfers, and possibly adjacent landowners. During golf course design or improvements, the golf course architect should be informed of site-specific conditions required for systematic development of TMS programs.

System Component 2 — *Establish objectives* for realistic cultural, water, nutrient, and pest management of specific areas on the course. These objectives will serve as the basis for establishing control methods and action thresholds. Management of tees and greens will require different control strategies than for fairways and rough. Integration of irrigation, fertilization, pest and other cultural control methods is essential even for the preliminary development of IPM and TMS programs.

System Component 3 — *Establish action thresholds* based on regional research and prevailing economic conditions. In TMS programs, action thresholds are expanded beyond the traditional definition used for IPM practices. They are based on pest populations, turfgrass/soil nutrient tests, soil water conditions, soil and thatch physical properties, turfgrass playing conditions, and environmental conditions. All of these conditions indicate whether action must be taken to maintain turfgrass quality. No action, whether chemical, physical, or cultural, is taken until the predetermined action point is reached. Actions are taken when the impact of pest populations or nutrient/water deficiency affect turfgrass quality sufficiently to threaten the biological and economic viability of the turfgrass. These effect levels are defined by TMS objectives.

System Component 4 — *Monitoring* the golf course for climatic conditions, soil condition, pest populations, and turfgrass quality on a periodic and consistent basis determines when the action thresholds are reached. Monitoring also may be used to determine whether a specific set of practices has been successful. Monitoring and identification of pests and deficiencies is the cornerstone of integrated management technologies. Use of automatic monitoring equipment, innovative identification techniques, climatic phenological indicators, and use of computer simulation models provide assistance in determining when action thresholds have been reached (Fermanian and Michalski 1989; Fermanian et al. 1989; Miller et al. 1989; Nyrop et al. 1989; Snyder et al. 1984; Vargas et al. 1989; Villani and Wright 1988a).

Potter and Braman (1991) have suggested that simple, reliable, and cost-effective monitoring techniques, other than visual inspection, are lacking for pest control. Survey techniques based on research include direct soil sampling; flotation; irritant drenches; pit-fall traps; sound, sweep, suction, food, and pheromone sampling traps; and heat extraction (Potter and Braman 1991). Direct insect population survey techniques are suitable for research, but are too time consuming and destructive for practical use on golf course and lawn turf. The utility of current monitoring techniques is further compromised by the lack of established damage or action thresholds (Potter

1986). Additional research on the ecology of turfgrass and turf pests is needed to establish the relationship of the distribution of pest populations, turfgrass quality, and sample strategies.

System Component 5 — *Specific management practices* to suppress pest populations, reduce nutrient and water deficiencies, or maintain turfgrass quality for playability should be selected from a range of options. Depending on site-specific conditions, management options include physical, biological, or chemical treatment. Habitat modification and understanding the relationship of nutrient, water, and climatic conditions is especially important for control and prevention of conditions conducive to pest and disease infestations. Use of alternate nutrient and pest control options should always be considered. Alternate biological and cultural options are discussed in the following section.

System Component 6 — In *chemical usage,* when necessary, appropriate chemical control, alternate control, irrigation, or cultural action should be taken. Preferred chemical practices would reduce movement of the applied chemical off the target site and provide maximum contact with the intended pest (pesticide) or root system (fertilizer), while presenting the least possible hazard to nontarget organisms. Chemicals should be applied on the basis of need. Fertilizers should be applied during periods of active uptake (growth), and pesticides should be applied when the pest or disease is in its most vulnerable life stage. Calendar, global broadcast, and preventative applications are not always consistent with integrated systems, economic, environmental quality, or societal goals.

System Component 7 — *Maintaining written records* of course management objectives, monitoring methods, data collection, management actions, and the results of management practices is essential for evaluation of TMS and development of future plans.

System Component 8 — *Evaluation* of the results of pest habitat alteration, pesticide application, use of alternate control options, fertilization, and water management (irrigation/drainage) should be conducted periodically to assess the success of the TMS program. The results of the evaluation are used to modify the initial program to meet changing environmental, cultural, and pest conditions. Flexibility and economic feasibility ultimately determine the long-term success of TMS (Smith and Raupp 1986; Wearing 1988).

5.2.2 Management Guidelines in TMS

Selection of appropriate chemical practices for nutrient and pest management is an essential component of TMS (Leslie 1989b; Metcalf 1980). Continued availability of both multicomponent chemical and nonchemical control strategies are necessary to avoid future pest resistance, pest resurgence, and to maintain turfgrass health and quality. The aesthetic standards for turfgrass in North America are so high that even limited insect and disease damage is considered intolerable (Beard 1982; Potter and Braman 1991). Despite advances in alternate control strategies, application of insecticides and fungicides may still be the only effective response to heavy infestations of insects and disease (Leslie and Metcalf 1989; Potter and Braman 1991). Pesticides and fertilizers can be applied using management practices to reduce adverse water quality and environmental effects. Many of these practices have been proposed for use in agricultural, urban, and turfgrass systems (Anderson et al. 1989; Leslie 1989c; Metcalf 1980).

The following discussion of management practices is intended as an outline of general principles. Specific management of pesticides and fertilizers must be designed based on site-specific conditions, as well as federal, state, and local regulations. Managing the type, amount, formulation, placement, and timing of pesticide and fertilizer applications will help to accomplish pest control, nutrient, water quality, and environmental goals. The following methods have been considered useful in meeting agronomic and environmental goals (Anderson et al. 1989; EPA 1988; Koehler et al. 1982; Leslie and Metcalf 1989; Metcalf 1980, 1989; Pratt 1985; Smolen et al. 1984; Stewart et al. 1975; Walker et al. 1990). The critical principles of *pest management* consistent with TMS concepts are

1. Only pesticides specifically *labeled* for application should be applied, and then only by properly registered, certified, and trained personnel. All pesticide use should comply strictly with local, state, and federal regulations. Following label recommendations, obtaining certification to apply pesticides, and training in the appropriate pesticide application techniques are essential components for continued use of pesticides in TMS.

2. *Selection criteria* for the type of pesticide should include consideration of the target species or disease, pesticide characteristics, and site characteristics. Important pesticide properties include efficacy; solubility; formulation; degradation rate; volatility; adsorption; potential toxicity to natural pest enemies; and toxicity to wildlife, nontarget or beneficial organisms. Critical site characteristics and soil properties related to pesticide efficacy and fate are soil texture and organic matter content, local and regional geology, depth to groundwater, proximity to well heads and surface water, topography, and climate. Pesticides that minimize the pollution potential should be selected.

3. *Reducing the frequency* of pesticides applied to turfgrass may be the single most effective practice to reduce potential adverse environmental effects.

4. *Selection of less toxic, less mobile, and less persistent pesticides, or use of alternate control strategies* will help reduce potentially adverse environmental effects. Depending on site characteristics, consideration of the potential mode of chemical loss could reduce environmental loading, potential for contamination of water resources, and adverse effects on nontarget organisms.

5. *Controlling the timing and amount* of a pesticide application in relation to local environmental conditions, especially rainfall, determines the potential for offsite movement and degradation characteristics. Restricting application prior to anticipated storm events is effective in reducing surface and subsurface losses of pesticides. Loss of wildlife can be avoided, in part, by restricting application of chemicals with high toxicity during critical migratory or lifestage periods. Selection of less toxic compounds during the period wildlife are using golf courses is another consideration in timing of applications.

6. *Action thresholds* permit control of pests within economic constraints and reduced levels of pesticides. Other factors to consider are the use of resistant turfgrass species and cultivars and those concepts of TMS as described in Section 5.1 (Table 1). Applying pesticides only when and where necessary will significantly decrease chemical loading and adverse effects on the environment.

7. Selection of pesticide *formulations* also influences pesticide fate and losses. Wettable powders, dusts, and microgranules are generally most susceptible to surface and leaching losses.

8. *Application methods* influence the partitioning and potential effectiveness of pesticides. Proper application rates, equipment selection and calibration, and careful application to the target site will insure effective use of the applied pesticide. Spot applications will reduce the amount of chemical applied to turfgrass and limit total environmental loading.

9. *Incorporation* of pesticides, placement below the soil/thatch surface, and "watering-in" reduces exposure to runoff process and enhances soil adsorption.

10. *Proper equipment maintenance and calibration* is essential for even applications at the volumes intended by the user. All label instructions, storage requirements, and regulations must be followed to insure safe handling of pesticides. Proper mixing, handling, and loading prior to application will reduce fill-site contamination. Closed systems for loading and mixing pesticides are especially useful in reducing contamination of the site and nearby waters.

11. *Proper disposal* of unused chemicals and containers will ensure safety of the user, water resources, and nontarget organisms. Pesticide applicators should avoid chemical exposure by safe handling practices including use of protective clothing, gloves, and shoes. Respirators should only be used after proper training has been conducted.

12. Use of antibacksiphoning devices in *chemigation* equipment will reduce the potential for pesticide contamination of groundwater or other water supplies during irrigation. Pesticides should be applied through irrigation equipment only when appropriate and when specific label instructions are available. Environmentally safe chemigation practices include (1) flushing of injection equipment to prevent pesticide accumulation, (2) flushing the irrigation system after pesticide injection, (3) use of properly calibrated equipment, (4) preventing

runoff of water-pesticide mixture, (5) avoiding application to permanent or semipermanent standing water on or near fields, and (6) periodic monitoring of equipment to ensure proper application to the intended target.

13. *Assessment* of potential offsite transport of chemicals by runoff or leaching losses prior to application will provide essential information on selection of pesticides appropriate for a specific site. Computer models and qualitative indexes could be adapted for this approach (see Section 5.5) (e.g., Vargas et al. 1989).
14. Application and handling of pesticides in regions of *karst topography* should be conducted with care to avoid movement of chemicals into fractures and sinkholes via surface runoff.
15. Development and implementation of *quality control* (QC) and *quality assurance* (QA) guidelines for pest management will ensure that TMS practices are used with reasonable accuracy by field personnel.

The critical principles of *nutrient management* consistent with TMS programs are

1. *Using minimal rates* of nitrogen and phosphorus to maintain nutrient levels needed to sustain turfgrass quality is one of the primary management practices used to minimize both surface and subsurface losses of nutrients.
2. *Improved uptake efficacy* decreases environmental loading of nutrients. This is achieved through (a) selection of realistic goals for turfgrass quality, (b) selection of application rates to meet these quality goals, (c) use of soil and tissue tests to establish proper application rates, and (d) use of nutrient application history or credits. Records of all forms and sources of nutrients applied to turfgrass is essential to determine rates of fertilizer application. Records of nutrient applications should include the types and amounts commercial fertilizers, clippings returned, effluent water used, as well as other organic sources such as compost and top-dressing.
3. Application of nutrients at the times and amounts commensurate with turfgrass growth requirements is one of the single most important management practices used for reduction of offsite transport of nutrients and mitigation of adverse environmental effects. The optimum time of application depends on turfgrass species and cultivar, climate, soil conditions, and chemical formulation of the fertilizer. Application of nitrogen after turfgrass uptake of nitrogen has ceased may lead to possible surface and subsurface losses of nitrogen. In agricultural systems, fall application of nitrogen for the next growing season remains controversial in cold moist climates (Beauchamp 1977; Keeney 1982, 1986; Pratt 1985). Considerable leaching and runoff losses of fall- and winter-applied fertilizer has been observed in pasture, forage, and row crop systems (Anderson et al. 1989; Gilbertson et al. 1979; Kamprath 1973; Klausner et al. 1974; Koehler et al. 1982; Zwerman et al. 1971). Although, fall application of

nutrients has become an important turfgrass management practice (Street 1988), Petrovic et al. (1986) have shown that fall application of soluble forms of nitrogen (urea) can lead to significant nitrogen leaching. Fall-applied inorganic nitrogen and residual soil nitrates are at risk of leaching past the root zone during the fall and early spring, especially on coarse texture or shallow soils (Keeney 1982; Pratt 1985). Fall application of nitrogen has been discouraged as an environmentally unsound practice by the National Research Council (1978). Application of nutrients, especially organic wastes in the late fall or winter, increases the risk of loss in snowmelt and spring runoff (Anderson et al. 1989; Gilbertson et al. 1979; Timmons et al. 1970). Although turfgrass is very efficient in the uptake of nutrients, leaching of nutrients from living and dead grass and other vegetation during snowmelt and storm flow may be a significant source of both soluble nitrogen and phosphorus (Balogh and Madison 1985; Koehler et al. 1982; Timmons and Holt 1980; Timmons et al. 1970; White and Wiliamson 1973). Fate and potential losses of nutrients applied to turfgrass is discussed in detail by Walker and Branham (1992).

4. *Patterns and intensity of traffic* on golf courses can affect turfgrass density, soil compaction, and the rates and timing required for fertilization. These factors could affect pollution potential, especially for surface runoff. Management of traffic on golf courses to minimize surface runoff, soil compaction, pest infestations, and the need for frequent fertilizer and pesticide applications will reduce potential losses of water and applied chemicals.

5. *Application techniques* that will reduce surface and leaching losses include nutrient incorporation into soil when possible and frequent applications of reduced amounts of fertilizer. These techniques reduce movement of nutrients in solution and increase application efficiency.

6. The *source and formulation* of fertilizer used also influences the potential for offsite transport. Where leaching is a problem, slowly available sources of nitrogen should be used in place of readily available sources. Urea, though chemically an organic compound, is readily available and subject to movement in water. Many organic sources of nutrients offer slower release and may delay nutrient availability until required for turfgrass growth. The organic forms of nitrogen are released at rates commensurate with the rate of turfgrass growth. However, appropriate rates and placement are critical even for organic sources of nutrients. *Overapplication* of slow release and organic forms of nitrogen must be avoided to reduce the long-term potential for groundwater contamination.

7. *Slow-release nitrogen fertilizers and nitrification inhibitors* have the potential for reducing the environmental impacts resulting from losses of nitrates. However, some managers may have problems in matching the timing of slow-

release fertilizers with critical turfgrass growth periods. Some nitrification inhibitors (e.g., nitrapyrin) have proven effective in slowing conversion of ammonium to nitrate. However, the effectiveness of the inhibitor for reducing nitrogen losses in runoff, leaching, and by volatilization is dependent on climatic conditions, soil conditions, and water management practices.

8. *Proper calibration of equipment* will ensure proper placement and rate of nutrient delivery. Improper calibration and equipment maintenance will result in under or overapplication and uneven distribution of nutrients. Appropriate handling and loading procedures will prevent localized spills and concentrations of fertilizers.

9. *Irrigation, drainage, water management, and traffic effects* are critical factors for the potential leaching and surface runoff of nitrogen. Leaching losses are increased with irrigation of shallow rooted crops or shallow rooted turfgrass on sandy soils. Accumulated nitrates in irrigated soils will be leached past the rooting zone when excess irrigation water is applied to reduce salt accumulation in the surface soil. On sandy soils, excess nitrates will accumulate in the soil during years with normal fertilizer application, but with low moisture input and poor turfgrass development. In subsequent years with normal levels of precipitation or irrigation, excess nitrates accumulated in the soil may be available for leaching. Subsurface losses are reduced if the turfgrass or microbial population recovers the nitrogen from the soil or adequate nutrient credits are applied in determining rates of additional nitrogen application.

10. Application and handling of organic and commercial fertilizers in regions of *karst topography* should be conducted with care to avoid movement of chemicals into fractures and sinkholes via surface runoff.

11. *Maintaining good turfgrass growing conditions* will reduce both surface runoff losses and subsurface losses of plant nutrients. Preventing pest damage to turfgrass, adjusting soil pH for optimum growth, providing good soil tilth for root development, planting suitable turfgrass species and cultivars, and improving water management practices will increase turfgrass efficiency of nutrient uptake.

12. *Assessment of potential offsite transport* of chemicals by runoff or leaching losses prior to application will provide essential information on selection of nutrient management practices appropriate for a specific site. Site characteristics such as soil texture and organic matter content, geology, depth to groundwater, proximity of loading areas to well heads, proximity to surface water, topography, climate, and the effect of traffic on turfgrass density and soil compaction provide qualitative indications of site potential for runoff and leaching losses of nutrients. Computer models and qualitative indexes currently could be adapted for this approach (see Section 5.5) (e.g., Vargas et al. 1989).

13. Again, development and implementation of *quality control* (QC) and *quality assurance* (QA) guidelines for nutrient management will insure that integrated practices are used with reasonable accuracy by field personnel.

5.3 ADOPTION OF INTEGRATED SYSTEMS FOR TURFGRASS MANAGEMENT

IPM and recently ICM have received acclaim as part of systems required for cost-effective solutions to pest management and environmental problems (e.g., Anderson et al. 1989; Leslie and Metcalf 1989). However, as early as 1965, proponents of integrated management of biological systems observed a slow rate of adoption (Wearing 1988). Systematic studies on the process of IPM implementation has been limited (Leslie and Metcalf 1989; Smith and Raupp 1986; Wearing 1988). Most of the literature on adoption of IPM in urban settings is anecdotal (e.g., Grant 1989; Raupp et al. 1989). Information on the development of TMS for management of pests, nutrients, irrigation, and protection of wetlands and wildlife on golf courses is very limited (Walker et al. 1990).

Problems associated with implementation, education, and transfer of integrated management technology have been identified as the principal obstacles limiting progress of integrated programs in agricultural and turfgrass systems (Corbet 1981; Lincoln and Blair 1977; Raupp et al. 1989; Wearing 1988). Difficulty with technology transfer, slow acceptance by superintendents, and conceptual complexity has limited implementation despite rising chemical costs, pest resistance, and societal concerns related to environmental impacts.

Successful implementation of TMS requires education of both turfgrass managers and individuals using golf courses (Table 1). Education and provision of comprehensive source material on development of IPM, ICM, and TMS are the key components to successful implementation of working programs (Lincoln and Blair 1977; Poe 1981; Wearing 1988). Written and verbal communication on the concepts and integrated control tactics are essential components of educational programs. Success in previous IPM programs involved a high level of technical coordination and a strong cooperative attitude by producers (Grant 1989; Raupp et al. 1989; Reardon et al. 1987; Wearing 1988). Fragmented technical and scientific reviews developed for academics, scientists, and regulatory agencies do not provide information in a format usable by turfgrass managers. The lack of organized educational and technical assistance on monitoring, selection of chemicals, and selection of cost-effective alternate control options can be a formidable obstacle to the acceptance of IPM and TMS by golf course and turfgrass managers.

Demonstrating the cost-effectiveness of TMS strategies for maintaining turfgrass quality is necessary for its general acceptance. Despite the potential of IPM and TMS for reduction of adverse environmental impacts (Leslie and Metcalf 1989), these approaches will not be used to maintain the quality of golf course turfgrass without appropriate training and marketing. Cost advantage, long-term maintenance of turfgrass quality, reduction of pesticide resistance, and reduction of risk associated with environmental and human health issues are the primary motivation for implementation of integrated programs (Norton 1982; Wearing 1988). Onsite assistance by technical experts has been cited as essential for adoption of system management approaches (Wearing 1988).

A large number of technical, financial, educational, institutional, and social/marketing constraints to implementation of TMS have been identified (Table 1) (Corbet 1981; Wearing 1988). The primary technical obstacle is the lack of simple monitoring techniques to establish action thresholds (Potter 1982; Potter and Braman 1991; Wearing 1988). There is a critical need to simplify current TMS monitoring methodology. In order to implement effective monitoring programs, there is a need for considerable enhancement of (1) immunoassay methods (Miller 1988; Miller and Martin 1988; Miller et al. 1989); (2) effective trapping devices (Raupp 1985); (3) scouting and counting services (Cooper et al. 1987); (4) diagnosis of deficiencies and pest symptoms (Dernoeden 1989; Metcalf 1980); (5) documentation for climate and pest infestation relationships (Davidson et al. 1972; Nus and Hodges 1985; Potter and Braman 1991; Regniere et al. 1981); (6) identification and application of appropriate alternate control methods (e.g., Klein 1989; Poinar 1979; Siegel et al. 1989); and (7) development of effective computer assisted technology to monitor climate conditions, soil moisture, and soil nutrient status (Carrow et al. 1990; Nyrop et al. 1989; Welch 1984).

Establishment of economically based action thresholds for pest control, water management, and nutrient management is a complex issue for turfgrass managers. The complexity and expense of developing integrated systems on golf courses is confounded by the financial risk associated with maintaining high-quality turfgrass. The complexity of integrated systems for pest, crop, and turfgrass management is widely acknowledged as a major obstacle to implementation (Potter and Braman 1991; Wearing 1988). The organization and operation of the multicomponent innovations required for TMS to succeed on golf courses requires a high level of participation among golf course superintendents; USGA Green Section agronomists; local, state, and federal regulators; cooperative extension specialists; turfgrass researchers; and golf course users.

Current lack of financing at the state and federal level limits education, IPM and TMS training, and the necessary basic and applied research. The current strong recommendations by regulatory agencies for implementation of IPM as a component of overall TMS programs should be backed by financial assistance, at least for the necessary educational and applied research programs. Information on specific criteria for development of IPM and TMS programs is fragmented in detailed technical publications and reviews (e.g., Beard 1982; Gibeault and Cockerham 1985; Hanson and Juska 1969; Leslie and Metcalf 1989). Although current sources, including this book, contain the necessary information to develop regional TMS guidelines, it is necessary to tailor the information for local turfgrass requirements. Without development of detailed training programs, instructional manuals, and onsite technical assistance, TMS and IPM will not be successfully marketed to turfgrass and urban landscape managers (Holmes and Davidson 1984; Koehler et al. 1985; Raupp and Noland 1984; Raupp et al. 1989).

Development of educational and marketing strategies should be based on interagency cooperation of the USGA, Golf Course Superintendents Association (GCSAA), state and U.S. Environmental Protection Agencies, and scientists in academia and the private sector. Unless TMS programs are tailored to the needs of local turfgrass managers, current chemical and cultural management practices will continue to be used despite societal and environmental problems.

5.4 ALTERNATE METHODS OF PEST AND DISEASE CONTROL

The choice of the best method to suppress turfgrass pests and diseases in TMS programs depends on careful monitoring, knowledge of cultural practices and ecological relationships, and access to information on both chemical and nonchemical control options. Although chemical pesticides are currently the primary defense against pests and disease, fewer compounds are being developed for use on turfgrasses. Public concern over the potential health effects and relatively small market for recovery of massive development and registration costs limit incentives for development of new turfgrass pesticides (Potter and Braman 1991). These conditions provide incentives for development and use of alternate control strategies for turfgrass management. Biological control options are being stressed as an important component of the IPM systems in literature published by the U.S. Environmental Protection Agency (Leslie and Metcalf 1989). The role of alternatives and biological control options has been reviewed by Baker (1986) and Potter and Braman (1991).

Some of the possible nonchemical options for management of turfgrass pests and diseases include (1) use of biological control agents such as microorganisms, endophytic fungi, nematodes, and parasitoids (e.g., Funk et al. 1989; Klein 1988, 1989; Poinar 1979); (2) use of traps and attractants (Klein 1982, 1989); (3) encouragement of natural pest predators (Cockfield and Potter 1985; Reinert 1978); (4) selection and breeding of pest- and disease-resistant turfgrass species and cultivars (Busey 1989; Meyer and Funk 1989); (5) use of growth regulators (Kageyama and Widell 1989); (6) modification of pest and disease habitat using cultural treatments

(Colbaugh and Elmore 1985); and (7) incorporation of turfgrass ecological and allelopathic relations into TMS planning (Potter et al. 1989; Schmidt and Blaser 1969). The following is a brief summary of several of these alternatives and biological control options.

Microorganisms — Biological control of insect and disease pests using microorganisms has been extensively reviewed (Baker 1986; Cook and Baker 1983; Klein et al. 1976, 1981a, 1982, 1989). Bacterial pathogens used in commercial formulations for successful control of scarab grubs, sod webworms, and other insect pests include the milky disease bacteria (*Bacillus popillae* Dutky) and B.t. (*Bacillus thuringiensis* Berliner) (Klein 1988, 1989) (Table 2). Numerous studies have shown successful control of scarab grubs with *B. popillae*. Factors influencing the efficacy of commercial formulations of *B. popillae* and other milky disease bacteria are (1) rate and extent of application, (2) method of application, (3) timing of treatment in relation to soil incorporation and life stage of grubs, (4) soil temperature, and (5) long-term attenuation of spores (Hanula and Andreadis 1988; Klein 1988; Ladd and Klein 1982a; Tashiro 1987; Tashiro and Steinkraus 1966; Warren and Potter 1983). Unfortunately, the cost of formulating and registering strains of *B. popillae* for control of turfgrass grubs and other organisms may not be supported by the turfgrass market (Klein 1989). Although B.t. has proven successful for control of sod webworms (Reinert 1976), it has not been effective against other soil-inhabiting insect pests (Klein 1989). Other potential microorganisms being considered as biocontrol organisms include the bacteria *Serratia* spp. (Jackson et al. 1986), fungal species *Beauveria* and *Metarhizium* (Klein 1988), and the protozoan *Ovavesicula popilliae* (Hanula and Andreadis 1988).

Microbial-based inoculants have considerable potential for reducing fungicide usage. This alternate control approach uses introduced microbial antagonists that interfere with pathogen populations and reduce disease development through several modes of action. Commonly studied antagonists include fungi in the genera *Gliocladium, Laetisaria, Penicillium, Sporidesmium, Talaromyces, Trichoderma,* and *Verticillium* and bacteria in the genera *Bacillus, Enterobacter, Erwinia,* and *Pseudomonas* (Cook and Baker 1983). Currently, several types of biological control products are commercially available for control of plant pathogens on agricultural crops.

Although the use of microbial antagonists for biological control of turfgrass diseases is still in the early stages of development, preliminary research has demonstrated some initial success. Laboratory and greenhouse studies have described microbial antagonists suppressive to pythium blight caused by *Pythium aphanidermatum* (Nelson and Craft 1989a; O'Leary et al. 1988; Wilkinson and Avenius 1984), brown patch caused by *Rhizoctonia solani* (Burpee and Goulty 1984; Nelson 1988; O'Leary et al. 1988), gray snow mold caused by *Typhula incanata* (Harder and Troll 1973), and take-all patch caused by *Gaeumannomyces graminis*

Table 2. Summary of Alternate Control Options for Control of Turfgrass Pests and Diseases

Alternate control agent or turfgrass species	Typical target pest species	General ref.
Microorganisms		
Bacillus popillae, *Bacillus thuringiensis*	Scarab grubs, sod webworm	Klein (1982), Klein et al. (1976), Tashiro (1987)
Antagonistic microbial based inoculants and composts	Pythium blight, brown patch, gray snow mold, dollar spot, pythium root rot, summer patch, red thread, necrotic ringspot	Cook and Baker (1983), Nelson (1988), Nelson and Craft (1989a, 1989b, 1990a, 1990b)
Entomogenous nematodes		
Neoaplectana spp., *Heterorhabditis* spp.	Japanese beetles, masked and European chafer, mole crickets, cutworms, armyworms, billbugs, sod webworms	Georgis and Poinar (1989), Klein (1990), Poinar (1979)
Other parasitoids		
Parasitic wasps, *Neodusmetia sangwani* (Rao)	White grubs, southern chinch bugs, mole crickets, Rhodesgrass mealy bugs	Klein (1982, 1989)
Endophytic Fungi		
Acremonium spp.	Argentine stem weevil, sod webworms, aphids, chinch bugs, and billbugs; possible tolerance to pathogenic fungi, summer stress, and weeds; imparts increased vigor, persistence, and	Siegel et al. (1989), Funk et al. (1989), Bacon et al. (1986),

	tolerance to wear, summer, and winter stress	
Natural predators		
Birds, mammals, amphibians	Broad range of insect pests	Klein (1982)
Invertebrates: ants, mites, and other predatory arthropods	Broad range of insect pests: e.g., chinch bugs, sod webworms, mole crickets	Reinert (1978), Potter et al. (1989), Crocker and Whitcomb (1980)
Traps and attractants	Japanese beetle	Klein (1981b)
Breeding and cultivar selection		
Kentucky bluegrass	*Bipolaris* and *Drechslera* diseases, smut, rust, powdery mildew, dollar spot; chinch bug, billbugs, sod webworms	Meyer (1982), Meyer and Funk (1989)
Perennial ryegrass	Brown patch, stem rust, snow mold, net blotch; greenbug aphids, Argentine stem weevil, billbugs, sod webworm	Meyer (1982), Meyer and Funk (1989)
Turf-type Fescues	Brown patch, dollar spot, net blotch, rusts; resistant to a wide range of insect pests	Meyer (1982), Meyer and Funk (1989)
Bermudagrass	Mole crickets, ectoparasitic nematodes, tropical sod webworm, Bermuda grass mite, spittlebugs	Busey (1989)
St. Augustinegrass	Gray leaf spot, downy mildew,	Busey (1989)

Table 2. Summary of Alternate Control Options for Control of Turfgrass Pests and Diseases (continued)

Alternate control agent or turfgrass species	Typical target pest species	General ref.
	Panicum mosaic virus; Southern chinch bugs	
Zoysiagrass	Sod webworms, nematodes	Busey (1989)
Allelopathy		
Grass root exudates	Tall fescue exudates suppress a wide range of weeds; seed leachates of ryegrass, annual bluegrass, and Kentucky bluegrass may suppress some broadleaf plants; root exudates of various grasses suppress growth of trees and woody ornamentals	Rice (1976, 1979)
Ecological and Competitive Relationships		
Manipulation of species mixes or cultural practices	Manipulation of competitive relationships used to control annual ryegrass	Schmidt and Blaser (1969)
Nontoxic compounds		
Lime and suffocants	Japanese beetle grubs	Klein (1982)

var. *avenae* (Wong and Baker 1984; Wong and Siviour 1979). In field studies, applying preparations of *Typhula phacorrhiza* to creeping bentgrass swards provided up to 74% control of gray snow mold caused by *T. incarnata* and *T. ishikariensis* (Burpee et al. 1987). Isolates of binucleate *Rhizoctonia* spp. and *Laetisaria arvalis* provided up to 90% control of brown patch on creeping bentgrass greens and tall fescue turfgrass (Burpee and Goulty 1984; Sutker and Lucas 1987). Isolates of *Gliocladium virens* have been effective in suppressing dollar spot on bermudagrass (Haygood and Mazur 1990). Strains of *Enterobacter cloacae* have been effective in suppressing dollar spot on creeping bentgrass in the field (Nelson and Craft 1991) and pythium blight and brown patch in greenhouse trials (Nelson 1988; Nelson and Craft 1989a). *Fusarium* spp. isolates provided up to 87% control of dollar spot in turfgrass field trials in Ontario, Canada (Goodman and Burpee 1987, 1988).

Composts have proven to be the best sources of complex mixtures of antagonistic microorganisms used for disease control. Monthly applications to greens of composted top-dressing are effective in suppressing diseases such as dollar spot, brown patch, gray snow mold, and red thread (Nelson and Craft 1989a, 1989b, 1990a, 1990b). The reason that composts are such a rich source of microbial antagonists is that composting relies on a diverse assemblage of microorganisms to carry out the process of decomposition. The composting process involves successions of both moderate (mesophilic) and high (thermophilic) temperature microorganisms during various stages of organic decomposition (Alexander 1977). The mesophilic microflora, predominant during the later stabilization phase of decomposition, are the most important in suppressing turfgrass diseases (Hoitink and Fahy 1986; Nelson and Craft 1990a, 1990b). Microbial activity is critical for the expression of disease-suppressive properties in composts (Hoitink and Fahy 1986). However, the specific disease-suppressing microorganisms have not been identified in turfgrass composts. Identification of specific organisms in composts with biological activity is an important research objective in order to understand the disease-suppressive properties of compost applied to turfgrass.

Entomogenous Nematodes — Parasitic *entomogenous nematodes* are recognized as excellent candidates for agents of biological control for turfgrass pests (Table 2) (Gaugler 1981; Georgis and Poinar 1989; Klein 1990; Shetlar 1989). Nematodes are soil microfauna living freely in the soil or as parasites in plants, insects, and other animals. Entomophilic nematodes of the genera *Neoaplectana* and *Heterorhabditis* are regarded as good candidates for biological control agents (Georgis and Poinar 1989; Kaya 1985). These nematodes have a mutualistic association with bacteria in the genus *Xenorhabdus* (Poinar 1979). The infective and ensheathed juvenile nematodes enter the hosts via natural body openings. Some *Heterorhabditis* will directly penetrate the insect cuticle. Once the infective nematode enters the host insect, the bacteria are liberated, causing septic death of the infected insect within 24–48 hr. The nematodes feed on the bacteria and the dead insect tissue. The nematodes pass through several generations in which their abundance increases enormously. Eventually, ensheathed juveniles emerge from the depleted cadaver, carrying the bacteria in their guts (Poinar 1979).

Attempts to control foliar insects with entomophilic nematodes has not been successful (Kaya 1985). Poor control was attributed to drying of spray formulations and surface exposure to sunlight and extreme temperatures (Gaugler and Boush 1978; Georgis and Poinar 1989; Shetlar 1989). Nematodes occur naturally in a moist, dark soil environment (Alexander 1977). Therefore, soil-inhabiting pests are most likely to be effectively controlled by entomophilic nematodes. Entomogenous nematodes have been reported to control European chafer and Japanese beetle larvae in turfgrass as effectively as chemical pesticides (Villani and Wright 1988b). Under controlled conditions, nematodes have been demonstrated as an effective alternative to turfgrass insecticides for broad spectrum control of Japanese beetles, masked and European chafer, mole crickets, black cutworms, armyworms, billbugs, and sod webworms. Improved nematode formulations and application technologies are needed before nematodes will be effective for control of social insects such as ants, wasps, and yellow jackets (Poinar and Georgis 1989). Problems of storage, cost, availability, handling, compatibility with other turf pesticides, and reliability of nematode formulations must be resolved before being widely accepted by the turfgrass industry (Potter and Braman 1991; Zimmerman and Cranshaw 1990).

Efficacy of field applied nematodes is dependent on the following:

1. Nematode strain and specific target pest (e.g., Georgis and Poinar 1989; Shetlar 1987)
2. Sufficient spray volume, concentration of organisms, thatch conditions, and irrigation practices (Georgis and Poinar 1989; Shetlar et al. 1988)
3. Persistence of formulation, soil texture, and soil moisture conditions (Georgis and Poinar 1983; Molyneux and Bedding 1984; Schroeder and Beavers 1987)
4. Temperature (Molyneux 1984; Schmiege 1963)
5. Effects of solar radiation (Gaugler and Boush 1978)
6. Compatibility with other pesticides (Zimmerman and Cranshaw 1990)
7. Timing, placement, and method of application (Georgis and Poinar 1989; Morris 1987)

Surface spraying of nematodes is the most widely used application technique (Georgis and Poinar 1989; Shetlar et al. 1988). A high concentration of nematodes in the spray is required to ensure that a sufficient number of surviving nematodes come in contact with pest targets. Other methods of application have been devised to assist in nematode survival. These techniques include use of injection techniques, baits, encapsulated formulations, and application through irrigation systems (Georgis et al. 1989; Glaser and Farrell 1935; Kaya 1985; Kaya and Nelson 1985; Reed et al. 1986). Additional research is needed to improve production of commercial nematode formulations and application techniques under different habitat conditions.

Parasitoids — Other parasitic organisms with potential for control of selected turfgrass pests have been identified by Fleming (1968) and Klein (1982). Parasitic wasps have been observed and used to suppress white grub populations (Fleming 1968; Jarvis 1966), southern chinch bugs (Reinert 1972), and mole crickets (Wolcott 1941). A well-known successful use of biological control involved introduction of a parasitoid, *Neodusmetia sangwani* (Rao), from India for control of rhodesgrass mealybugs in Texas. This organism has been described as having completely controlled the mealybug (Dean et al. 1979; Schuster et al. 1971). Approximately $17 million year^{-1} was saved as a result of this biological control agent.

Endophytic Fungi — The relationship of nonsporulating *endophytic fungi* with forage and turfgrasses has been known for many years. However, only recently has research on the association of endophytic fungi with animal toxicoses, stimulation of growth, and insect resistance received attention (Bacon et al. 1986; Bush et al. 1979; Johnson et al. 1985; Latch et al. 1985; Newton et al. 1987; Read and Camp 1986; Saha et al. 1987; Siegel et al. 1989). Endophytic fungi are contained or grow entirely within the infected grass species and usually produce no external symptoms of infection. Currently, no measurable detrimental effects of the fungi on grass hosts have been observed (Siegel et al. 1987, 1989). Several endophytic fungi form mutualistic relationships with grass species. The fungi receive nutrition, enhanced protection and survival, and dissemination through seed from the host grass. The beneficial endophytes impart improved growth and persistence to the host plants, tolerance to herbivore feeding, tolerance to winter and summer stress, and improved resistance to diseases and pests (Funk et al. 1989; Latch et al. 1985; Siegel et al. 1985, 1987, 1989).

Five species of *Acremonium* are known to infect turfgrasses as endophytes. The association of these fungal endophytes is considered a desirable attribute in conservation and turfgrasses (Funk et al. 1989; Siegel et al. 1987). Increased resistance of fescues and perennial ryegrasses to the Argentine stem weevil (Prestidge et al. 1982; Stewart 1985), sod webworms (Funk et al. 1983), aphids (Johnson et al. 1985), chinch bugs (Funk et al. 1985; Saha et al. 1987), and billbugs (Ahmad et al. 1986) has been attributed to their association with endophytic fungi.

The endophytes produce a wide range of alkaloid compounds within the host grasses (Siegel et al. 1985, 1987). The clavine ergot and pyrolizidine alkaloids have insecticidal properties and have been isolated from endophytes in association with tall fescue (Siegel et al. 1985). The technology currently exists for production of synthetic grass-endophyte complexes capable of producing these natural insecticides. The synthesis involves transfer of endophytic fungi to grasses by (1) inoculation of

1-week-old seedlings (Latch and Christensen 1985), (2) inoculation of callus tissue (Johnson et al. 1986), and (3) selection of infected maternal lines (Funk et al. 1985).

Currently, breeding research is being conducted to select infected varieties and improve methods to incorporate endophytes into existing turfgrass cultivars (Funk and Clarke 1989; Funk et al. 1989). This research has focused on improvement of turfgrass germplasm and enhancement by endophytic infection. Promising turfgrass species are perennial ryegrass and hard, chewings, fine, and tall fescues. White (1987) observed infection of other important conservation grasses and turfgrasses including *Agrostis, Bromus, Cinna, Digitaria, Elymus, Festuca, Lolium, Melica, Poa, Sitanion,* and *Stipa.* Endophytic infection of *Cynodon* and *Zoysia* grasses has not been observed (Newton et al. 1987).

In addition to the intrinsic resistance to insects imparted by the endophytic fungi, increased tolerance to insects by infected turfgrass results from (1) enhancement of genetically mediated resistance, (2) growth enhancement of healthy dense turfgrass, and (3) increased tolerance to weed and moisture stress (Funk et al. 1985, 1989; Read and Camp 1986). Preliminary results from laboratory trials suggest that infection with endophytic fungi may increase tolerance of turfgrass to disease-inducing fungi (Bayaa et al. 1987; White and Cole 1985).

Natural predators — Natural predator-prey relationships are an important factor in suppression of insect pests in turfgrass (Arnold and Potter 1987; Crocker and Whitcomb 1980; Klein 1982; Reinert 1978). Although at times problematic, birds and mammals are the most visible predator of turfgrass insects (Klein 1982; Reinert and Short 1980). The giant toad, *Bufo marinis*, was successfully introduced into Puerto Rico for suppression of June beetles and mole crickets (Klein 1982). Invertebrate predators are also critical for control of insect pests in turfgrass. Several researchers have demonstrated the importance of predatory arthropods and mites for active control of chinch bugs, sod webworms, and other soft-bodied turfgrass pests (Arnold and Potter 1987; Cockfield and Potter 1984a; Crocker and Whitcomb 1980; Reinert 1978).

Maintenance of balanced turfgrass ecosystems is an important strategy for control of undesirable insect pests. Pesticides applied for control of turfgrass insect pests are also toxic to beneficial insect and spider predators and parasites. Removal of effective predator populations with pesticides may result in long term increases in prey/pest populations. Cockfield and Potter (1984b) observed that a single surface application of chlorpyrifos reduced the insect predator-induced mortality of sod webworm eggs for at least 3 weeks after treatment. In another study, Cockfield and Potter (1983) observed that populations of predatory mites, spiders, and insects in Kentucky bluegrass were reduced by 60% after a single application of chlorpyrifos or isofenphos. Predatory arthropods were found to be less abundant and diverse on Kentucky bluegrass sites subject to high maintenance levels of pesticides and fertilizers (Arnold and Potter 1987; Cockfield and Potter 1985).

Direct evidence does exist, although limited, suggesting that repeated pesticide treatments could reduce the stability of turfgrass communities resulting in increased pest infestations (Cockfield and Potter 1984b; Reinert 1978; Streu 1969, 1973; Streu and Cruz 1972; Streu and Gingrich 1972). Reinert (1978) observed low levels of southern chinch bugs in untreated lawns in Florida where insect predators and parasite populations were high. However, lawns receiving repeated insecticide treatments had chinch bug densities reaching outbreak levels. Resurgence of hairy chinch bug populations following years of chlordane treatments was attributed to reduction in predatory pressure from mites and predatory hemipterans (Streu 1969, 1973). Timing of insecticide applications should be adjusted to minimize adverse effects on natural insect predators. Use of selective insecticides or introduction of insecticide-resistant predators are alternate methods to avoid disruption of beneficial prey populations (Graves et al. 1978; Klein 1982).

Traps and Attractants — Traps and attractants have been used for suppression of Japanese beetles (Hamilton, et al. 1971; Klein 1981b, 1982; Ladd and Klein 1982b; Langford et al. 1940; Tumlinson et al. 1977). Commercial beetle traps are used in combination with a food-type (floral) lure alone or in combination with a sex attractant. Food attractants are floral lures and are effective for capture of both male and female beetles (Ladd et al. 1981). The sex attractant, or pheromone, has been isolated from female Japanese beetles. The pheromone has been synthesized and made commercially available for traps (Tumlinson et al. 1977). When the sex attractant is used in combination with a food attractant, both females and males are lured to traps at rates two to three times higher than either attractant used alone (Ladd et al. 1981).

The utility of mass trapping of Japanese beetles for suppression of grub populations was demonstrated in Massachusetts by Petraitis (1981). Using a dual bait system in 47 traps, the need for insecticidal suppression of grubs was eliminated. However, a recent study of a heavy infestation of Japanese beetles in Kentucky showed increased damage to plants in the vicinity of the beetle traps (Gordon and Potter 1985). It was suggested that despite trapping thousands of beetles per day, the traps attracted more beetles to an area than could be caught. Although turfgrass damage in the vicinity of the traps did not increase, damage to foliage of other desirable plants did increase. Research is currently being conducted to develop a commercial trapping/attractant system for sod webworms, masked chafers, and green June bugs (Banerjee 1969; Klein 1989; Domek and Johnson 1987, 1988; Potter 1980, 1989). Pheromone attractants also have been identified for fall armyworms, armyworms, black cutworms, and bluegrass webworms (Clark and Haynes 1990; Hill et al. 1979; Kamm 1982). Use of pheromone attractants and food-baited traps may be more significant for insect monitoring technology, than as a means of direct insect control (Clark and Haynes 1990; Potter and Braman 1991).

Breeding and Selection — Breeding and selection of turfgrass to be free or tolerant of pests and diseases are integral components of most breeding programs

(Table 2) (Busey 1989; Meyer 1982; Meyer and Funk 1989). The potential benefits of selecting pest and disease-resistant turfgrasses were not addressed in recent literature regarding alternate strategies in turfgrass IPM programs (Leslie and Metcalf 1989). Research is currently needed to improve methods for screening turfgrass cultivars with resistance to diseases and pests (Johnson-Cicalese et al. 1989). In a review of turfgrass insect management, Potter and Braman (1991) suggest that greater research and development should be placed on combining genetic insect resistance with other desirable turfgrass traits.

Substantial success has been achieved in breeding Kentucky bluegrass for resistance to many important turfgrass diseases (Table 2). Bluegrass cultivars have been developed to resist or tolerate leafspot and melting out disease; stripe smut; leaf, stem, and stripe rusts; powdery mildew; and dollar spot (Meyer 1982; Meyer and Funk 1989). Several Kentucky bluegrasses have been identified with significant levels of tolerance to damage by billbugs, chinch bugs, green bug aphids, and sod webworms (Johnson-Cicalese et al. 1989; Meyer and Funk 1989). Although relatively stable genetic resistance has been developed for the cool-season turfgrasses, additional research is needed to blend resistant strains for increased stability (Meyer and Funk 1989; Smiley and Craven-Fowler 1984). Natural mutations in pest populations often overcome resistance in turfgrass species with "single gene" resistance. Breeding for multigene resistance in turfgrass would slow down the natural process of pest acclimation to resistant turfgrass varieties.

Perennial ryegrass is another important cool-season turfgrass species. However, susceptibility to disease and poor performance under heat and drought conditions in warmer climatic regions has been a problem (Funk et al. 1985). Clark et al. (1989) has reported development of a perennial ryegrass with improved resistance to brown patch. Several turf-type perennial ryegrasses also have been developed with improved genetic resistance to winter diseases and stem rust (Rose-Fricker et al. 1986; Meyer and Funk 1989). Improved resistance of perennial ryegrass cv. Derby to damage by greenbug aphids was demonstrated by Jackson et al. (1981). Enhancement of ryegrass performance by selection of cultivars infected with endophytic fungi has increased ryegrass resistance to stem weevils, sod webworms, and billbugs (Ahmad et al. 1986; Funk et al. 1983). Reduced invasion by weeds and improved summer and winter recovery has been reported in perennial ryegrass and several fescues selected for endophyte-enhanced performance (Funk et al. 1985).

Progress has been made to improve the disease and insect resistance of the lower maintenance fine, hard, chewings, and tall fescue turfgrasses (Meyer 1982; Meyer and Funk 1989; Murray and Powell 1979). Improved cultivars have shown resistance to brown patch, net blotch, powdery mildew, red thread, dollar spot, and several rusts (Berry and Gudauskas 1972; Clarke et al. 1989; Meyer 1982). Additional research is needed to develop resistance to a number of *Pythium* spp., stem and crown rusts, and brown patch diseases (Meyer and Funk 1989). Improved resistance to drought stress could reduce damage of tall fescues from brown patch. Many of the tall fescue species have good overall resistance to insect damage (Jackson et al. 1981; Johnson-Cicalese et al. 1989; Meyer and Funk 1989).

Published research on the breeding of disease-resistant bentgrasses and many warm-season turfgrasses is limited compared to the cool-season turfgrasses (Busey 1989; Meyer and Funk 1989). Creeping, colonial, and velvet bentgrasses are very susceptible to fungal diseases (Beard 1982). Research is needed to develop disease and insect-resistant varieties of these important golf course turfgrasses. Economical methods to incorporate new resistant varieties into perennial turfgrass systems also require research and development.

Several species of warm-season turfgrasses are considered resistant to insect infestation (Busey 1989). Additional research is needed to demonstrate that host resistance to insects observed in the laboratory is associated with field tolerance to insect damage. There is considerable undocumented field information on insect resistance of bermudagrass cultivars (Busey 1989). This information is not useful to turfgrass managers until it is made available. Nonpreference screening trials have limited utility considering the monoculture conditions on many golf courses (e.g., Jackson et al. 1981; Leuck et al. 1968; Reinert and Busey 1983). Maintenance of genetic purity of clonal selections is also a problem encountered by turfgrass managers (Busey 1989).

Despite these technical difficulties, several cultivars of bermudagrass have shown potential resistance to mole crickets (Reinert and Busey 1984), ectoparasitic nematodes (Tarjan and Busey 1985), tropical sod webworm (Reinert and Busey 1983), bermudagrass mite (Reinert et al. 1978), and spittlebugs (Stimmann and Taliaferro 1969). A genotype of Bell rhodesgrass resistant to rhodesgrass mealybug has been released (Reinert 1982). Different genetic resistance to sod webworms and sting nematodes was reported for zoysiagrass. The level of resistance was related to the quality of zoysiagrass performance in selection trials conducted by Busey et al. (1982).

Cultivars of St. Augustinegrass have shown resistance to gray leafspot, downy mildew, and *Panicum* mosaic virus (Atilano and Busey 1983; Bruton et al. 1983; Reinert et al. 1981; Toler et al. 1983, 1985). Cultivars of Floratam and Floralawn St. Augustinegrass with genetic resistance to damage by southern chinch bugs have been released (Dudeck et al. 1986; Reinert and Dudeck 1974). Recently, extensive southern chinch bug damage to previously resistant Floratam and Floralawn turfgrass has occurred in Florida (Busey and Center 1987). However, several St. Augustinegrass cultivars have retained their resistance to damage by the southern chinch bug (Crocker et al. 1989; Reinert and Dudeck 1974; Reinert et al. 1981, 1986). Overcoming the genetic resistance to insect damage demonstrates the ability of genetically variable pest populations to overcome biological control options. Resistance to biocontrol options would be similar to development of pest resistance to chemical control strategies as described by Balogh and Anderson (1992).

Allelopathy — Allelopathy is the direct or indirect effect of one plant on another through production and release of natural compounds (Rice 1974, 1979). Allelopathy and competition for light, water, and nutrients are both components of plant interference interactions. The effects of allelopathy have been recognized as a growth factor in both manipulated and natural ecosystems (Rice 1979). Mechanisms of action responsible for allelopathy include (1) effects on cell elongation and root tip structure, (2) effects on hormone production, (3) effects on membrane permeability, (4) impact on mineral uptake and nutrient availability, (5) detrimental effects on physiological processes, and (6) clogging of the xylem. Rice (1974, 1976, 1979) has reviewed allelopathy in agriculture, forestry, and natural grassland ecosystems.

Allelopathic relationships are often difficult to separate from the effects of competition. However, researchers have demonstrated that various crops, barley in particular, successfully suppress weeds through inhibition by root exudates (Glass 1973; Overland 1966; Peters 1968). Detrimental allelopathic effects also have been observed in crop rotations (e.g., Kimber 1973; Tamura et al. 1969). Suppression of weeds in thin and dense fields of Kentucky-31 tall fescue (*Festuca arundinacea* Schreb.) has been attributed to root exudate of toxic materials (Peters 1968).

The potential of allelopathy for biological control of weeds has been recognized in agriculture and aquatic systems (Fay and Duke 1977; Rice 1974, 1979; Robson 1977; Schweizer 1988). Research is relatively limited on development of allelopathy as an alternate control technique for control of weeds in agriculture and turfgrass management (Rice 1976, 1979). Recent reviews of biocontrol options for turfgrass management have not included allelopathy as a management option (Leslie and Metcalf 1989). However, under natural conditions and forage management, several grass species have exhibited allelopathic relationships (Cope 1982; Newman and Miller 1977; Newman and Rovira 1975; Rice 1976, 1979; Tang and Young 1982). For example, in laboratory assays, seed leachates of tall fescue, perennial ryegrass, Kentucky bluegrass, and annual bluegrass inhibited growth of lettuce seedlings (Buta et al. 1987). Organic fractions of grass seed leachate were inhibitory to lettuce growth. Inorganic fractions of grass seed leachates inhibited lettuce growth at high concentrations, but stimulated growth at low concentrations.

Growth of trees and ornamental shrubs may be adversely affected by turfgrass and sod. This effect may result from competition for growth resources or exudates of allelopathic compounds by grass roots (Fales and Wakefield 1981; Nielsen and Wakefield 1978; Whitcomb and Roberts 1973). It is difficult to separate the effect of competition for growth resources and root exudates in research on suppression of trees by grass. Allelopathic suppression of trees and shrubs grown in sod is usually identified by unsubstantiated inference (Fales and Wakefield 1981).

Competitive Relationships of Turfgrasses and Associated Cultural Practices — Understanding and manipulating the competitive ecology of turfgrass and weed species is an important component for maintenance of high-quality turfgrass. Ecological factors in the development and maintenance of single species or complex mixtures of turfgrass change progressively as microclimate, nutrient, and cultural conditions change (Potter and Braman 1991; Schmidt and Blaser 1969). Grasses that become dominant in turfgrass mixtures are characterized by (1) wide adaption to natural and imposed environmental conditions, (2) excellent persistence under stress, and (3) a rhizomatous or stoloniferous growth habit.

Initial seeding ratios, clipping schedules and techniques, soil fertility, irrigation practices, seasonal variation in light and temperature, and use of growth regulators will affect the competitive relationship of turfgrasses. Mowing practices altered the competitive relationship of seedling mixtures of Kentucky bluegrass and perennial ryegrass (Brede and Duich 1984a). Close mowing after planting favored the bluegrass and aided in its establishment. However, early emergence of the ryegrass did stabilize the seedbed, allowing mowing to begin sooner than would have been possible with other turfgrasses. Brede and Brede (1988) observed early soil stabilization after rapid establishment of annual ryegrass in a tall fescue-annual ryegrass mixture. A close clipping shortly after emergence of the ryegrass allowed the slower developing tall fescue to achieve at least 50% ground coverage within 60 d.

Brede and Duich (1986) also observed the competitive interaction of Kentucky bluegrass and perennial ryegrass with the grass weed species, annual bluegrass. The investigators observed that dominance of a turfgrass species in this mixture results from their interaction aboveground, belowground, or both. The competition for light, moisture, and nutrients tends to induce cyclic declines among the competitors. Kentucky bluegrass dominated the ryegrass when competition was confined to aboveground factors. When interaction was belowground, ryegrass outcompeted Kentucky bluegrass for moisture and nutrients. The ability of ryegrass species to aggressively compete for moisture and nutrients has been well documented (King 1971; Milthrope 1961; Rhodes 1968). Under good growth conditions, Kentucky bluegrass dominates both annual bluegrass and ryegrass. However, rapid tiller growth of both annual bluegrass and ryegrass gives these species a competitive advantage during initial stand establishment or under resource stress conditions (Brede and Duich 1984b, 1986).

Gaussoin and Branham (1989) observed that removal of clippings and overseeding without a growth regulator significantly reduced the density of annual bluegrass in creeping bentgrass plots. Use of high nitrogen fertility levels with a growth regulator (mefluidide) significantly increased the competitive ability of annual bluegrass compared to creeping bentgrass.

Seasonal variation in temperature will influence the ability of grasses to compete for dominance. Cultivars of Kentucky bluegrass have differential tolerance to heat stress, while several cultivars of perennial ryegrass had little significant response to changing heat stress (Minner et al. 1983). Tolerance to heat stress may change seasonally within a single cultivar (Wehner et al. 1985). Selection of cultivars and species mixtures with stress tolerance to regional and local climatic patterns will minimize the decline of turfgrass quality (Holt 1969; Juska et al. 1969; Keen 1969; Minner et al. 1985; Youngner 1985; Wehner et al. 1985). Additional research to establish competitive patterns and the ecology of turfgrass species and cultivars is needed. Ecological and environmental information should be used to establish turfgrass mixes and cultural practices to maintain high-quality turfgrass without implementation of chemical management strategies.

Use of turfgrass growth retardants has been suggested as a management option for manipulation of competitive turfgrass growth (Devitt and Morris 1989; Haley and Fermanian 1989; Kageyama and Widell 1989). Growth regulators or retardants are a diverse group of compounds used to block the plant hormone activity such as gibberellic acid synthesis (Kane and Smiley 1983; Shive and Sisler 1976). Use of growth retardants results in shortened stem internodes and reduced elongation of leaves and rhizomes (Kageyama and Widell 1989). Examples of growth retardant compounds include ethephon, flurprimidol, mefluidide, malic hydrazide, metsulfuron methyl, paclobutrazol, and sulfometuron methyl (*Farm Chemicals Handbook* 1991).

Growth retardants have been used in an attempt to (1) control the invasion of annual bluegrass in creeping bentgrass (Haley and Fermanian 1989; Kageyama and Widell 1989), (2) retard growth of warm-season turfgrass to reduce frequency of mowing (Devitt and Morris 1989; Rogers et al. 1987), (3) increase seed yields (Buettner et al. 1976), (4) reduce turfgrass growth and demand for water during drought periods (Mathias et al. 1971; Nielsen and Wakefield 1975; Watschke 1981), and (5) suppress the growth of overseeded cool-season species during conversion of greens to warm-season summer species (Fry and Dernoeden 1986; Mazur 1988). Suppression of annual bluegrass and eliminating the potential of weed invasion during seasonal species conversion will reduce the need for herbicide treatments.

Long-term quality of turfgrass has not been consistently improved with the use of growth retardants (Fry and Dernoeden 1986; Mazur 1988). Often additional nitrogen fertilization is required to offset damage to the turfgrass caused by use of growth retardants (Devitt and Morris 1989; Gaussoin and Branham 1989; Kageyama and Widell 1989). Use of organic compounds for growth suppression has the potential for generating the same environmental and societal problems associated with the use of chemical pesticides. Assessment of the environmental impacts from the use of growth retardants should be included in the category of other chemical control options. Risk factors associated with chemical management options were discussed in Chapter 4 (Balogh and Anderson 1992). Although growth retardants allow manipulation of competitive interactions among turfgrass species, use of these chemical compounds does not necessarily constitute an alternate control option.

Alternate Nontoxic Compounds — The use of nontoxic compounds has recently received renewed attention for alternate control of turfgrass pests and diseases (Klein 1982). Van Leeuwen and van der Muelen (1928) observed that china clay, chalk, and slaked lime effectively repelled Japanese beetles. A relationship between higher soil pH levels and lower Japanese beetle populations was observed in Ohio (Polinka 1960a, 1960b; Wessel and Polinka 1952). This relationship was established on sites with naturally occurring high pH values and sites treated with lime. However, recent laboratory studies failed to demonstrate reduced hatching and survival of grubs in soil at pH ranging from 5.5 to 7.5, which is the general range for growth of turfgrass (Vittum and Tashiro 1980). Additional research is needed to clarify the potential relationship of soil pH and disease infestations.

Environmental Stress, Water and Nutrient Management, and Cultural Control — Environmental stress, especially the extremes of temperature and water, have a direct and indirect impact on turfgrass health and resistance to pest infestation and disease (Beard 1973, 1982; Chapman 1979; Colbaugh and Elmore 1985; Nus and Hodges 1985; Smiley 1983). Many of the routine cultural practices on golf course turfgrass, including irrigation, fertilization, and mowing, have a direct effect on the incidence of pest and disease damage (Beard 1973, 1982). Understanding the relationship of environmental, soil moisture and nutrient conditions, and cultural impacts to pest and disease control is essential in development of TMS (Potter and Braman 1991; Villani and Wright 1981, 1989). The following are examples of research regarding the relationship of nutrient and water management to pest and disease control.

Winter and low temperature-related stress results in increased susceptibility of turfgrass to snow molds, dead spot, and snow scald (Smiley 1983). Low light conditions, especially under shade trees, combined with high humidity and moderate temperatures make succulent grass shoot tissue especially susceptible to fungal infection (Beard 1982; Britton 1969; Nilsen and Hodges 1980). Seasonal changes in photoperiod, quality of light, and age of turfgrass affects the ability of turfgrass to tolerate pathogenesis of leafspot disease (*Drechslera sorokiniana*) (Nilsen et al. 1979a, 1979b). Longer photoperiods (14 hr) and balanced light conditions during the summer inhibits infection of Kentucky bluegrass by leaf spot disease. Depending on leaf age, shorter photoperiods (10 hr) with changed color-biased light quality (blue-biased) increases expression of the fungal disease.

Climatic conditions and water management also influence weed infestations on golf courses (Beard 1982; Colbaugh and Elmore 1985). Climate, in part, dictates the particular set of weed species in a specific region (e.g., Holt 1969; Juska et al. 1969; Keen 1969). However, annual bluegrass, crabgrass, and similar species are exceptions and have become widespread across North America (Beard 1982). Seasonal fluctuations in temperature and moisture conditions will influence germination, survival, and competitive adaptation to the turfgrass environment. Crabgrass and goosegrass are favored by high temperature conditions, while shepherds purse is favored by cool temperatures in the spring and fall. Many weed species are favored by frequent irrigation and overwatering (Beard 1982). Overwatering of golf courses is a major factor encouraging invasion of turfgrass by annual bluegrass (Colbaugh and Elmore 1985). A combination of wet soil and excessive irrigation is also associated with invasions of crabgrass, chickweed, and nutsedge. Quackgrass, goosegrass, dandelions, plantain, yarrow, and white clover are very tolerant of low soil moisture conditions.

The frequency and duration of precipitation and irrigation often determine the suitability of turfgrass as a habitat for fungal pathogens (Colbaugh and Elmore 1985; Dernoeden 1989; Smiley 1983). Continuously wet conditions favor development of gray leaf spot, *Pythium* blight, brown patch, and various seedling diseases. Diseases favored by wetting-drying cycles or low moisture conditions include dollar spot, fairy ring diseases, *Fusarium* diseases, *Bipolaris* and *Drechslera* diseases, and rusts. Frequent light irrigation reduces the susceptibility of turfgrass to damage by these diseases. Reduction in excessive accumulation of thatch decreases habitat for survival of pathogenic saprophytic fungi (Beard 1973; Smiley 1983).

Water management practices and temperature influence insect and nematode infestations in golf course turfgrass and lawns (Colbaugh and Elmore 1985). Insect life cycles, patterns of dispersal, and survival are closely connected with seasonal variation in precipitation and temperature (Davidson and Roberts 1968; Davidson et al. 1972; Gaylor and Frankie 1979; Potter 1981, 1983; Potter and Gordon 1984; Regniere et al. 1981). Egg deposition and survival of white grubs are reduced in very wet or dry soils (Ladd and Buriff 1979). Dry conditions and low soil fertility were associated with increased damage of pasture Kentucky bluegrass by white grubs (Graber et al. 1931). Drought and heat stress also reduces the viability of grubs of the southern masked chafer (Potter 1981, 1983; Potter and Gordon 1984). Continually moist soils are more conducive to grub feeding and mole cricket activity than drier soils (Colbaugh and Elmore 1985). Frequent application of water increases the survival of sod webworm and fall army worms in turfgrass. However, drier turfgrass conditions and high temperatures are conducive to damage by scale insects, chinch bugs, leaf hoppers, and eriophyid mites (Colbaugh and Elmore 1985).

Water management influences nematode-related diseases on golf courses. Soil nematodes are dependent on free-water in soil for mobility, parasitic activity, and reproduction. Moist and porous sandy soils receiving heavy irrigation are susceptible to nematode infestation and damage. In the absence of free-water, nematode damage is limited. Water conservation practices will restrict the movement and activity of nematodes (Colbaugh and Elmore 1985).

Effectiveness of soil pesticides also is influenced by soil moisture conditions, water management practices, and the lifestage of the pest population (Ahmadi et al. 1980; Colbaugh and Elmore 1985, Harris 1972, 1982). Infiltration of insecticides is required for effective control of several insect and nematode species (e.g., Colbaugh and Elmore 1985; Reinert 1974, 1979; Shetlar et al. 1988; Vittum 1985). However, infiltration of insecticides past the thatch layer or surface soil layers may result in increased and undesirable leaching losses of pesticides. In addition to turfgrass growth and maintenance requirements, irrigation schedules should be designed to reduce site-specific infestations of weeds, insects, and disease. Consideration of site-specific soil, climate, irrigation, and cultural practices need to be integrated into pest management programs (Engel and Ilnicki 1969; Schmidt and Blaser 1969; Short et al. 1982). In general, the duration and timing of irrigation cycles should be scheduled to limit current pest pressures. Frequent and light irrigation is often necessary on severely compacted soils to ensure infiltration; however, this practice (1) favors germination and survival of weed seedlings, (2) ensures a humid environment that favors many insect pests, and (3) enhances production of fungal spores and infection of drought-stressed plants.

Fertilization and other cultural practices also affect pest and disease resistance of turfgrass. Mowing practices have been shown to affect disease resistance. The potential for the occurrence of disease in turfgrass is enhanced when (1) the mowing height is shorter than the optimum for a species, and (2) increased mowing frequency causes excessive damage to leaf tips (Beard 1982). Fertilization practices also influence disease control in turfgrass culture (Beard 1982; Dernoeden 1987; Smiley 1983). High potassium levels usually decrease the severity of fungal diseases. Except for dollar spot, red thread, and rust, most turfgrass diseases are promoted by excessive nitrogen fertilization (Beard 1982; Chapman 1979). Adequate nitrogen fertilization of turfgrass, however, is necessary to maintain turfgrass vigor and density and to reduce the invasion by weeds. Using appropriate nitrogen fertilization practices will reduce turfgrass diseases and their symptoms. Application of nitrogen to chewings fescue sod significantly reduced the intensity of yellow tuft disease (*Scleropphthora macrospora*) (Dernoeden and Jackson 1980). Despite masking of the disease symptoms, the investigators observed that fertilization did increase the sporulation capacity of this downy mildew fungus. Use of ammonium chloride alone and phenyl mercury acetate plus ammonium sulfate has been used to control take-all patch disease (*Gaeumannomyces graminis* var. *avenae*) in creeping bentgrass (Dernoeden 1987). Granular sulfur and pH control were not effective for control of this disease in bentgrass.

The relationship of soil fertility and pH to insect and weed infestation is complex. Reduced fertility levels will slow development of succulent tissue preferred as insect and fungal food sources. Lower fertilization rates will favor low nutrient weeds and also decrease the nutrient levels available for high nutrient weeds. However, nutrient deficiencies will weaken turfgrass resistance to disease and limit the ability of turfgrasses to compete with weeds for limited nutrient and water resources (Beard 1982; Murray and Powell 1979).

High levels of soil phosphorus have been found to be conducive for encroachment of annual bluegrass to turfgrass, increasing annual bluegrass density, and increasing the survival and growth of bluegrass in monoculture (Dest and Allinson 1981; Varco and Sartain 1986; Waddington et al. 1978). Waddington et al. (1978) also observed that increased levels of potassium increased the growth of annual bluegrass and its invasion of turfgrass. Application of sulfur and reduction of soil pH have been used to reduce invasion of turfgrass by annual bluegrass and to reduce its growth in monoculture (Goss 1974; Varco and Sartain 1986). Varco and Sartain (1986) observed an interaction between added phosphorus, calcium hydroxide, and sulfur on the growth of annual bluegrass. Establishment and growth of annual bluegrass were enhanced overall by additions of phosphorus and calcium. However, this positive influence on growth was modified by the acidifying effect of sulfur in the root zone. Annual bluegrass infestations of turfgrass could be reduced by limiting additions of phosphorus and acidifying the soil pH to the range of 5. However, turfgrass infested with annual bluegrass must be able to tolerate these conditions (Varco and Sartain 1986).

The effect of pH on white grub survival in turfgrass has been considered as a control option (Klein 1982, 1989). As previously mentioned, Polinka (1960a, 1960b) observed that Japanese beetle oviposition and subsequent grub survival was adversely affected by high soil pH levels. However, Vittum (1984) and Vittum and Tashiro (1980) found that changes in soil pH levels had little effect on either Japanese beetles or European chafer grubs. The preferred pH range for growth of turfgrass is 5.5–6.5 for bentgrasses and fine-leafed fescues; 6.0–6.5 for ryegrasses; 6.0–7.0 for bermudagrass, Kentucky bluegrass, and zoyiagrass; and 6.2–7.2 for annual bluegrass (Beard 1973, 1982; Madison 1982). Changing soil pH outside of the optimum range for growth of good quality turfgrass is not a practical approach for control of pests.

Effective long-term control of disease and pests on golf courses requires cultural practices that encourage vigorous growth of turfgrass. Use of pesticides without appropriate cultural practices provides only short-term control of disease and pests in turfgrass (Murray and Powell 1979; Short et al. 1982). Manipulation of cultural practices to promote high-quality turfgrass, control undesirable growth conditions, and reduce pest and disease damage is a significant component of TMS.

5.5 QUALITATIVE AND QUANTITATIVE TECHNIQUES TO ASSESS ENVIRONMENTAL IMPACTS

Assessment of water, pesticide, and nutrient practices prior to their use provides an excellent opportunity for cost-effective water and soil management (Jury et al. 1987). Use of simulation models is an important educational and monitoring component of TMS. Qualitative assessment techniques, empirical methods, and computer simulation models provide a logical mechanism to integrate the complex factors influencing water quality, fate and persistence of chemicals, and the environmental effects of different management practices (Donigian and Rao 1986a; Wagenet and Rao 1990).

Models of water resources and environmental impacts are most useful in analyses of current and anticipated conditions. Substantial progress has been made in the last 10 years in the development of quantitative and qualitative software technology used for analysis of water resources and fate of chemicals in the environment (Anderson et al. 1989; Bonazountas 1991; Shoemaker et al. 1990; Wagenet and Rao 1990). Computer models are currently being used by the U.S. Environmental Protection Agency, state regulatory agencies, and private consulting companies for assessment of environmental impacts and regulation of pesticides, fertilizers, and hydrocarbon contamination of soil (e.g., Daugherty 1991; Dean et al. 1984; Horsely and Moser 1990; Jones et al. 1991; Wagenet and Rao 1990). Simulation models also are used to predict or screen the mobility and persistence of pesticides currently under development or considered for use on a particular site (e.g., Cohen et al. 1990; Jury et al. 1984, 1987). Another major use of simulation models is by agricultural, forestry, and turfgrass managers in designing effective and environmentally sound plant, soil, water, and chemical management plans (e.g., Crowder et al. 1985; Donigian and Carsel 1987; Leonard et al. 1987; Sauer et al. 1990; Summer et al. 1990; Shoemaker et al. 1990; Wagenet et al. 1989). Simulation models can be used to identify agricultural or turfgrass management practices with potentially favorable or adverse effects on the environment. Computer simulation is used to (1) evaluate alternate management practices; (2) select alternate compounds for pest and nutrient control; and (3) design optimum water, cultural, and chemical management strategies to meet turfgrass and environmental quality goals.

The most effective end use of simulation and screening models is in policy and management decisions (Donigian and Rao 1986a; Mackay and Stiver 1991). Process-oriented computer simulation models provide an opportunity to ask "what if" questions concerning nutrient and pesticide management without the high cost of experimentation or field monitoring. Screening models rank chemicals and management practices for preliminary assessment of potential partitioning and fate in the environment. Appropriate simulation strategies will identify chemicals and cultural practices with potential adverse impacts (e.g., leaching to groundwater). Limited funding for field research and monitoring programs can then be allocated to those chemicals and practices identified in the simulation programs that have the greatest risk or benefit.

Periodic collection of monitoring data without process-oriented guidance has relatively little value other than supplying site- and point-specific responses to nutrient and pesticide management. A fixed schedule for collection of field data in monitoring programs is rarely sufficient to determine the long-term impacts of water and chemical management on ecosystems. A combination of process-oriented research combined with long-term simulation is more likely to yield results consistent with environmentally sound systems management practices. Computer models, with varying degrees of success, attempt to combine the primary determinants of water and chemical behavior in the soil, vadose zone, surface water, and groundwater (Jury and

Valentine 1986; Leonard 1988, 1990; Wagenet and Rao 1990). The primary processes controlling the fate of pesticides, nutrients, and other contaminants in the environment are

1. Biological and chemical transformations in the soil
2. Sorption and desorption by the organic and mineral phases of soil, which controls chemical mobility in soil
3. Uptake of the compounds and water by plants
4. Volatilization or loss of compounds by evaporation from the soil
5. Dilution of compounds in surface and subsurface water flow during transport to surface water or groundwater

Computer simulation integrates these processes with chemical properties, site characteristics, management options, and climatic conditions over space and time (Figure 1) (Jury and Ghodrati 1989; Wagenet and Rao 1990). The factors controlling water and environmental fate of chemicals in managed ecosystems are reviewed extensively in Chapters 2, 3, and 4 of this book.

There is a wide range of computer models and assessment techniques that address water, sediment, and chemical pollutant transport on agricultural and forest land (Table 3). Available computer models and simulation strategies for assessment of water, pesticide, and nutrient fate in the environment have been reviewed by Addiscott and Wagenet (1985), Balogh and Gordon (1987), Donigian and Rao (1986a), Jury and Ghodrati (1989), Leonard (1988, 1990), Shoemaker et al. (1990), and Wagenet and Rao (1990). Leaking underground petroleum storage tanks also have been identified as a potential problem for golf course managers (Balogh and Watson 1992). Fate and chemical transport in soils contaminated by hydrocarbons from leaking underground storage tanks have been reviewed by Bonazountas (1988, 1991). Many of these research and management level models range from simplistic or demonstration models (e.g., Nofziger and Hornsby 1986) to more theoretically rigorous research models (e.g., van Genuchten 1978; Wagenet and Hutson 1986).

Criteria for selection of chemical transport models have been discussed by Balogh and Gordon (1987), Leonard (1988), and Rao et al. (1982). Model selection criteria include (1) availability of computational facilities, (2) spatial and temporal scales of application, (3) intended use and interpretation of model output, (4) availability of model software, (5) availability and resolution of model input data, (6) appropriate calibration of model input parameters; and (7) confidence in the output. Selecting a model compatible with the goals of a simulation or screening project, availability of software or ability to modify the program to meet current needs, and the availability of input data are three crucial criteria (Wagenet and Rao 1990; Balogh and Gordon 1987).

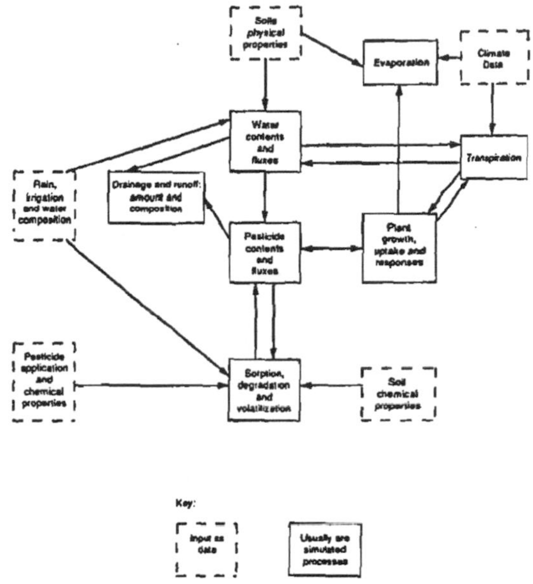

Fig. 1. Processes influencing chemical fate in soils.

Several studies have recommended the use of computer simulation models for assessment of environmental impacts on golf courses (e.g., Horsley and Moser 1990; Klein 1990; Cohen et al. 1990). However, modification of computer models to adequately simulate the hydrology, transport processes, and environmental conditions for golf course turfgrass has not been published. Several simulation models currently available for agricultural assessment could be adapted for use in turfgrass systems (Table 4). Specific modifications are required for use of these models. Improvement of the hydrologic and chemical transport components are necessary to simulate the effects of thatch, perched water tables, and artificial drainage conditions. Availability of daily climatic data, physical and chemical parameters for chemical fate and persistence under turfgrass conditions, and management strategies have compromised recent attempts to use simulation models in many ecosystems (Balogh and Gordon 1987). Recently released databases on soil and chemical properties were developed specifically for use with simulation models (e.g., Carsel et al. 1991; Wauchope et al. 1991). Despite many of the computation, validation, and data limitations, the following is a discussion of currently available software directly related to potential use for turfgrass systems.

Matching the objectives of simulation projects with the many types of available models is a critical step in understanding and deciding on specific simulation software. Addiscott and Wagenet (1985), and more recently Wagenet and Rao (1990),

Table 3. Classification of Water and Chemical Transport Models

Deterministic models

Mechanistic — Based on rate parameters

1. Analytical — (e.g., Nielsen and Bigger 1962; van Genuchten and Wierenga 1976)
2. Numerical — (e.g., Childs and Hanks 1975; Robbins et al. 1980)

Functional — Based on simplified rate or capacity factors

1. Analytical — (e.g., Enfield et al. 1982; van Genuchten and Alves 1982)
2. Partially Analytical — (e.g., De Smedt and Wierenga 1978; Rose et al. 1982a, 1982b)
3. Numerical — (e.g., Wagenet and Hutson 1986)
4. Layer and other capacity approaches — (e.g., Addiscott 1977; Bresler 1967; Burns 1974; Carsel et al. 1984; Nofziger and Hornsby 1986; Tanji et al. 1972)

Stochastic models

1. Mechanistic — (Amoozegar-Fard et al. 1982; Dagan and Bresler 1979)
2. Nonmechanistic — transfer function approach (e.g., Jury 1982; Jury et al. 1982)

Screening models

1. Simplified capacity and partitioning approaches — (e.g., Jury et al. 1983, 1987; Mackay and Stiver 1991)
2. Algebraic combination of partitioning parameters — (e.g., Aller et al. 1985; Gustafson 1989)

From Addiscott and Wagenet 1985; Balogh and Gordon 1987; Wagenet and Rao 1990.

categorized computer models of water and chemical fate into groups based on their computational and functional approaches to simulation of chemical movement (Table 3). There is a critical distinction in simulation approaches between deterministic and stochastic approaches. Deterministic models simulated behavior based on the best knowledge of the operational processes within that system. A given set of events will always lead to a unique outcome. Stochastic models operate on the assumption that events are uncertain and there is a range of probable outcomes based on initially defined events.

A second distinction exists for mechanistic and functional modeling approaches. Mechanistic models incorporate the currently understood processes of a system into a series of algorithms and solutions to rate expressions. Many mechanistic models are used for research purposes and provide quantitative results (e.g., Wagenet and Hutson 1986). Mechanistic models require specific rate parameters for such terms as water flow, solute diffusion-dispersion, and chemical transformations. On an operational level, this type of data often is not readily available to land use managers and environmental regulators. Functional models incorporate certain simplifications into the treatment of water flow and chemical processes. Functional models use simplified terms for rates of change or substitute capacity terms to define changes in water content and solute concentration (e.g., Carsel et al. 1984, 1985; Knisel 1980).

Although functional models are not as accurate for prediction of water and chemical movement under field conditions, they do provide a management level approach to assessing movement of water, sediment, and chemicals (Table 4). Functional or management level models usually require input data generally available to resource managers and require less extensive computational facilities (Shoemaker et al. 1990; Wagenet and Rao 1990). Reviews of both surface and subsurface computer models for assessment of water, sediment, nutrients, and pesticide fate and movement have been published by Balogh and Gordon (1987), Donigian and Rao (1986a), Jury and Ghodrati (1989), Keeney (1986), Leonard (1988, 1990), Melancon et al. (1986), Oster (1982), Tanji (1982), and Wagenet and Rao (1990).

Several management level models incorporate both subsurface leaching and surface transport of nutrients and pesticides used in agricultural and turfgrass systems. PRZM, developed by the U.S. Environmental Protection Agency (Carsel et al. 1984, 1985), and GLEAMS/CREAMS, developed by the U.S. Department of Agriculture (USDA) (Knisel 1980; Leonard et al. 1987), are examples of management level models for evaluation of surface and subsurface transport of pesticides (Table 4). PRZM and GLEAMS/CREAMS are continuous daily simulation models developed by the U.S. EPA and USDA, respectively.

PRZM and GLEAMS/CREAMS simulate surface runoff, erosion, leaching, and related movement of agricultural chemicals. Data inputs and model outputs are relatively easy to understand. Both models utilize a simplified advective-dispersive water balance method for water and nonvolatile solute movement within a multiple layered profile (Carsel et al. 1985; Leonard et al. 1987). PRZM bases surface runoff and erosion on the USDA Soil Conservation Service (SCS) runoff curve number and the Universal Soil Loss Equation (USLE). PRZM uses a simplified approach to snowmelt and plant uptake of pesticides, linear sorption partitioning strategy, and lumped first-order biological and chemical transformations. The surface runoff and erosion component of CREAMS/GLEAMS is based on the USDA SCS runoff curve number and a physically based erosion component developed by Foster et al. (1980, 1985).

CREAMS, a surface sediment and chemical transport model, has been extensively tested (Knisel et al. 1983; Leonard 1988, 1990). GLEAMS, a subsurface expansion of the computationally efficient CREAMS, retains the ability to simulate sophisticated management practices in agriculture and forestry (Leonard et al. 1987; Shoemaker et al. 1990). PRZM has undergone limited validation compared to field data (Carsel et al. 1986; Donigian and Rao 1986b; Hedden et al. 1986) and has been used to evaluate

Table 4. Selected Examples of Currently Available Management Level Models for Simulation of Water, Sediment, and Chemical Transport

	Transport process		Chemical species		
Model[a]	Surface	Subsurface	Pesticide	Nutrients	Ref.
ADAPT	Yes	Yes	Yes	N & P	Ward et al. (1988)
CREAMS/ GLEAMS	Yes	Yes	Yes	N & P	Knisel (1980); Leonard et al. (1987)
PRZM	Yes	Yes	Yes	No	Carsel et al. (1984, 1985)
CPM/CPS	Yes	Yes	Yes	No	Steenhuis and Walter (1980); Steenhuis (1979); Lorber and Mulkey (1982)
LEACHM	No	Yes	Yes	N & salts	Wagenet and Hutson (1986, 1989)
PESTAN	No	Yes	Yes	No	Enfield et al. (1982); Donigian and Rao (1986a)
CMIS	No	Yes	Yes	No	Nofziger and Hornsby (1984, 1986)
HSPF[b]	Yes	**[c]	Yes	N & P	Johanson et al. (1981)
AGNPS[b,d]	Yes	No	No	N & P	Young et al. (1986)
BASIN	Yes	No	No	N & P	Heatwole et al. (1986)
NTRM	Yes	Yes	No	N	Shaffer and Pierce (1982); Shaffer (1985)
NITROSIM	No	Yes	No	N	Rao et al. (1984); Rolston et al. (1984)
—	No	Yes	No	N	Tillotson and Wagenet (1982)

—	Yes	No	No	P	Persson et al. (1983); Moore and Madison (1985)
SWNP	No	Yes	No	N	Martin and Watts (1982)
"NAPL"	No	***[e]	No	No	Abriola and Pinder (1985a, 1985b); Corapcioglu and Baehr (1987a, 1987b)

[a] *Based in part on a review of 150 transport models by Balogh and Gordon (1987).*
[b] *Basin and field scale capabilities.*
[d] *Event basis simulation.*
[c] ***Nitrogen, phosphorus, and groundwater contribution to streamflow.*
[e] ****Multiphase models for groundwater contamination by nonaqueous phase organic liquids (e.g.,petroleum products from leaking underground storage tanks).*

the nonpoint source pollution effects of agricultural and turfgrass practices (e.g., Donigian and Carsel 1987; Horsely and Moser 1990; Sauer et al. 1990). These models could be modified to simulate runoff, leaching, and sediment transport from turfgrass systems.

Many research and management level models of vadose zone processes are not linked to groundwater transport models. The need to link surface models with subsurface transport models has been identified as a research priority for evaluation of nonpoint source pollution potential (Shoemaker et al. 1990; Wagenet and Rao 1990). At the Ohio State University, the Agricultural Drainage and Pesticide Transport Model (ADAPT) is being developed by combining GLEAMS and DRAINMOD (Ward et al. 1988) (Table 4). GLEAMS, as previously described, is a deterministic model developed to evaluate the effects of agricultural management systems on the movement of water and chemicals in the plant root zone (Leonard et al. 1987). DRAINMOD is a field-scale mathematical model developed to describe the performance of a drainage system. When these models have been integrated, ADAPT will be capable, on an hourly or daily basis, of simulating evapotranspiration, surface runoff, infiltration, deep seepage to groundwater, surface and subsurface drainage, sediment yields, surface and subsurface nutrient and chemical discharges, and plant growth. Incorporation of subsurface drainage and irrigation components into simulation models is critical for use in turfgrass systems.

LEACHM, developed at Cornell University, is a theoretically and processed-based research model used for deterministic evaluation of both nutrient and pesticide leaching on agricultural soils (Table 4) (Wagenet and Hutson 1986, 1989). Using a finite differencing technique to simulate water and nonvolatile solute flux, LEACHM is intended for use in a single growing season with surface-applied chemicals. LEACHM currently has sophisticated components to simulate nitrogen and pesticide transformations in the soil. Another feature of LEACHM is its ability to simulate transport of the parent compound and up to two daughter products. This is essential for modeling the behavior of compounds such as aldicarb, atrazine, diazinon, and fenamiphos. Many transport models do not have these capabilities.

LEACHM does provide accurate predictions of pesticide movement at the expense of large mainframe or microcomputer time and extensive field data requirements. This model also has several other conditions that limits its use as a management level model. Although LEACHM incorporates a variable depth multilayer soil profile under transient moisture conditions, it does not incorporate variations in surface management techniques such as pesticide incorporation, tillage, slope, or surface roughness. Also, LEACHM currently lacks a surface runoff and sediment transport component which compromises its practical utility. Despite its utility as a research and academic tool (Wagenet and Rao 1990; Wagenet et al. 1989), several authors suggest that LEACHM does not fulfill the need for evaluation of management activities on surface water and groundwater quality (e.g., Leonard et al. 1987; Shoemaker et al. 1990).

Many of the deterministic models are incapable of evaluating the inherent spatial variation in water transport processes and site characteristics. The Agricultural Nonpoint Source Pollution (AGNPS) model, a single event surface model (Young et al. 1986) is an example of a spatial model developed for comparison of long-term averages of surface transport of nutrients and pesticides (Table 4). AGNPS is a well-documented watershed level model that could be specifically adapted for assessment of sediment and chemical losses from golf courses during construction. Using a cellular grid-based system, AGNPS incorporates both the spatial and temporal distribution of surface transport processes to evaluate the impacts of different management practices. This model can also be linked to other process-based models to evaluate the effects resulting from edge-of-field losses of chemicals and sediments on aquatic systems (e.g., Summer et al. 1990). Other watershed level models capable of modeling surface transport of sediment and associated chemicals in agricultural or urban environments include the Hydrologic Simulation Program — Fortran (HSPF) (Donigian et al. 1983), Small Watershed Agricultural Model (SWAM) (DeCoursey 1982), and the Pesticide Runoff Simulator (PRS/SWRRB) (Computer Sciences Corp. 1980; Williams et al. 1985). All of these models could be modified to simulate potential runoff from golf courses during construction or from established turfgrass.

The USDA-ARS has expended considerable resources to develop the Water Erosion Prediction Project (WEPP) model (Lane and Nearing 1989). In its current form, it is a very effective tool for prediction of hydrological response and erosion rates. Using a limited set of data, the model does have the ability to estimate the aggregate size distribution of eroded sediment. This mechanistic model is a considerable improvement over the use of the Universal Soil Loss Equation used in older models (e.g., AGNPS and CREAMS). Incorporation of the WEPP technology into a water transport system would enhance the ability of golf course designers and managers to simulate the effects of golf course construction and maintenance on soil erosion. Currently, researchers at Ohio State University are attempting to integrate WEPP into ADAPT to simulate the effects of surface hydrology, irrigation management, drainage practices, and chemical transport.

Several compartmental mass balance models have been developed for assessment of nitrogen transport (Burns 1980a, 1980b; Fried et al. 1976; Mehran et al. 1981; Tanji et al. 1975, 1977). Although the mass balance approach will predict site-specific nitrogen transport (Messer and Brezonik 1983), this approach is best suited for regional evaluation of nutrient management strategies. These models are not suitable for site-specific assessments. A series of models based on the Soil-Water-Nitrogen-Plant (SWNP) model have been used to simulate the effects of nitrogen management and irrigation on sandy soils (Table 4) (Martin and Watts 1982; Watts and Hanks 1978; Watts and Martin 1981). Although this model has several simplifying assumptions, it could be modified for site-specific use in irrigated/drained turfgrass systems. As previously discussed, LEACHM (Wagenet and Hutson 1986), has a component to simulate nitrogen fate and subsurface movement in the soil profile. Several other research-based models have been developed to evaluate the movement, transformation, and chemical reactions of nitrogen (Table 4). These models would require extensive modification and testing for use in conjunction with management of turfgrass.

Highly simplified deterministic models of water and chemical movement also have been developed (e.g., CMIS, Nofziger and Hornsby 1986; MOUSE, Pacenka and Steenhuis 1984). These models use very simplified portrayals of surface and subsurface transport processes. However, these models are computationally efficient and can be implemented with generally available input data. Simplified models are primarily intended to illustrate basic principles and the generalized effect of management practices on major transport processes (Wagenet and Rao 1990). By providing strictly qualitative estimates of water, sediment, and solute movement, simplified deterministic models are closely related to screening models for establishing initial priorities for further investigation.

Quantitative and qualitative methods evaluating the movement of water and chemicals using empirical or simulation approaches have several limitations. These limitations have been reviewed by Jones et al. (1991), Jury (1983, 1986a, 1986b), Rao et al. (1988), and Rao and Wagenet (1985). Critical problems encountered in extrapolation of model results for turfgrass management decisions include the following:

1. Validity of model assumptions concerning interaction of sorption, microbial transformations, and plant uptake
2. Oversimplification of processes in construction of management level models
3. Failings in numerical solution of the solute/solvent transport functions
4. Problems in describing stochastic, preferential, and macropore flow phenomena
5. Inability to determine model parameters and lack of readily available spatially derived input data
6. Lack of software that is sufficiently free of program errors and developed for use on computer hardware generally available to natural resource and turfgrass managers
7. Lack of field testing
8. Failure to incorporate spatial variation of physical, chemical, and biological processes in assessment strategies
9. Need to incorporate a thatch layer for turfgrass soils and a perched water table for putting greens into deterministic and screening model approaches

Three alternate approaches to the exclusive use of deterministic models have been proposed to resolve the problems and limitations of deterministic models using convective dispersive equations (CDE) (Jury and Ghodarti 1989; Wagenet and Rao 1990). Recent research on the spatial and temporal variability of water movement, solute transport, climatic conditions, and soil properties has stimulated development of stochastic and statistical approaches (Balogh and Gordon 1987; Brusseau and Rao 1989; Jury 1983, 1986a, 1986b). The transport equations and partitioning components of deterministic models do not account for (1) spatially variable flux in the field; (2) preferential flow patterns; and (3) spatially and temporally variable transport, sorption, and degradation processes. Agricultural research on tile-drained soils has demonstrated the need to account for preferential flow on the fate of applied pesticides and nutrients (Kladivko et al. 1991). Early arrival of small amounts of pesticides and nutrients in tile drainflow is related to initial rapid water flow through the soil profile. This phenomenon is not described by deterministic convective-dispersive and many screening model approaches (Utermann et al. 1990). Modeling approaches to evaluate chemical transport from tile-drained turfgrass and agricultural soils need to be developed.

Several stochastic, scaling, and geostatistical approaches are available based on the assumption that soil-water processes are basically uncertain and only can be defined in statistical terms (Jury 1983; Wagenet and Rao 1990). A stochastic approach to modeling spatially variable transport of solutes and modeling preferential flow has been proposed in a series of papers (e.g., Jury 1982, 1983; Jury and Gruber 1989; Jury et al. 1982; Knighton and Wagenet 1987a, 1987b; Small and Mular 1987; Utermann et al. 1990). Reviews of scaling methods and geostatistical techniques for analysis of spatially variable transport processes have been presented by Nielsen et al. (1983), Russo and Bresler (1980), Tillotson and Nielsen (1984), and Vieira et al. (1983).

Stochastic and statistical models are derived from descriptions of variable transport processes. Most stochastic and statistical models focus on mean values and their deviations for a particular heterogeneous field condition. These stochastic, scaling, and statistical models describe the condition of inherent field variability, but validation and field testing has proven to be extremely difficult and limited (Jury et al. 1982; Nielsen et al. 1983; Wagenet and Rao 1990). Although spatially derived simulation of uncertain transport processes is conceptually sound, currently available models have not been developed or implemented as software in a form suitable for making management decisions.

Deterministic models can be used to represent spatially and temporally variable soil and water systems by repeated execution of the model with the spatial range of input parameters. Running deterministic models using a variable range of soil and climatic conditions can accurately describe the potentially variable fate of water and chemicals applied in the field (Balogh and Gordon 1987; Bresler et al. 1979; Carsel et al. 1988; Jones et al. 1987; Wagenet and Hutson 1989; Wagenet and Rao 1983). Using process-oriented models in a stochastic manner offers the option to use research and management level models to describe the effects of TMS on water and environmental quality. This stochastic approach takes advantage of the strengths and availability of management models, while incorporating the uncertainty of inherently variable field processes. Before being used for site assessments, most models need to incorporate components to simulate the thatch layer, the influence of tile drainage and irrigation, and the effect of enhanced degradation with multiple applications.

Several modeling approaches are available for qualitative assessment of the environmental impacts of water and chemical management on soil and water resources. Many of these approaches are amenable to modification for use on turfgrass systems. Use of qualitative and analytical solutions of water and solute transport is intended for evaluation of chemical behavior under limited conditions, but with relatively few data input requirements (Jury et al. 1983). Screening models are very useful for categorization of chemicals, especially pesticides, into broad behavioral classes.

DRASTIC (Aller et al. 1985) and other screening models developed by Jury et al. (1983, 1987) and Steenhuis and Naylor (1987) provide rapid methods for qualitative assessment of chemical leaching potential. These approaches focus on chemical parameters and/or site characteristics as indexes of potential chemical mobility. DRASTIC, created by the National Water Well Association, is a systematic index of site factors associated with chemical leachability (Aller et al. 1985). The mappable hydrologic factors include depth to groundwater, net aquifer recharge, soil media, topography, impact of vadose zone, and hydraulic conductivity of the aquifer.

LEACH, developed by the U.S. Environmental Protection Agency, is based on chemical and site characteristics linked to leachability (Dean et al. 1984). These characteristics include the retardation factor (R), the rate of chemical decay (K_s), and the U.S. Soil Conservation Service Curve Number. Both the DRASTIC and LEACH create an index of pollution potential based on the algebraically weighted combination of leachability factors. These indexes of pollution potential have been successful in delineating regional areas at risk to groundwater contamination. However, these pollution potential indexes have not been successful in predicting site-specific cases of groundwater contamination (Curry 1987).

Assessment of relative pesticide mobility and differential potential for transport to groundwater requires consistent testing under uniform conditions in the field and laboratory. Helling et al. (1971) reviewed and classified pesticides into five general mobility groups (Table 5). This classification was based on empirical and experimental evidence of pesticide transport through soils with moving water. The mobility classes are in general agreement with values for pesticide adsorption (see Table 21 in Chapter 4). The disadvantage of mobility classifications based on adsorption partition coefficients, column studies, and soil thin layer chromatography (Helling et al. 1971;

Table 5. Relative Mobility of Pesticides in Soils Ranked According to Estimated Order of Decreasing Mobility Within Each Class

Very mobile	Moderately mobile	Slightly mobile	Nearly immobile	Immobile
Dicamba	MCPA	Endothall	Siduron	DCPA
Dalapon	2,4-D	Atrazine	Bensulide	Benomyl
Chloramben		Simazine	Diazinon	Trifluralin
		Alachlor		Benefin

From Helling et al. 1971.

Helling 1971; Rao and Davidson 1980) is that adsorption and solubility indices and short-term experiments neglect the major pesticide loss pathways including volatilization and decomposition.

Several investigators have attempted to include many of the environmental pathways of pesticides into simplified screening methods. Steenhuis and Naylor (1987) and Jury et al. (1983, 1987) developed simple mechanistic approaches for screening the relative risk of chemical leaching based on chemical and site characteristics. The leachability index properties in the Steenhuis and Naylor (1987) model include amount of chemical applied, half-life of the chemical, depth to groundwater, depth or zone of soil organic matter, saturated hydraulic conductivity, and soil adsorption. The screening model developed by Jury et al. (1983) is based on simplified convective-diffusive transport, first-order degradation rates, sorption, and pesticide volatilization. Assuming constant water flow and uniform soil properties, an analytical solution was derived to describe pesticide movement and concentrations. Using this model solution, pesticides were classified into specific mobility classes (Table 6) (Jury et al. 1984).

A screening model described by Jury et al. (1987), behavior assessment model (BAM), uses average soil drainage rates, uniform soil properties, uniform application rates, adsorption partition coefficients, first-order degradation rates, and depth to groundwater as priority screening parameters. Two screening scenarios were developed for two hypothetical soils with low groundwater pollution potential and high groundwater leaching potential (Tables 7 and 8). Residual mass from a single application and time required for the compound to leach 3 m in the soil was simulated for 50 pesticides using each screening scenario. The high groundwater leaching scenario could be used for assessment of turfgrass pesticides (Table 8). These semiquantitative approaches have proven fairly accurate in evaluating site-specific potential for chemical transport to groundwater (Jury et al. 1984, 1987; Steenhuis and Naylor 1987). These approaches could be readily adapted to turfgrass conditions.

A second approach to development of screening models focuses on chemical properties related to surface and subsurface transport processes (Mackay and Stiver 1991; Wagenet and Rao 1990). These models combine sorption, degradation, and

Table 6. Mobility Classification for Several Pesticides and Other Compounds Found in Golf Course Soils Based on Screening Model Results

Chemical[a]	Convection[b] time (days)	Convection[c] mobility class	Diffusion[b] time (days)	Diffusion[d] mobility class
Atrazine	31	3	—[e]	1
2,4-D	6	5	—	1
DDT	41000	1	—	1
Diazinon	146	2	—	1
Mercury	6930	1	—	1
Methyl bromide	7	5	1	3
Simazine	26	4	—	1
Trifluralin	1242	1	—	1

From Jury et al. 1984.

[a] *Not all chemicals are necessarily registered for current use on golf course or lawn turfgrass.*
[b] *Time to travel 10 cm in soil with moisture content = 0.3 m^3 m^{-3}, organic matter fraction = 0.0125, and bulk density = 1350 kg m^{-1}.*
[c] *1 = very immobile; 2 = moderately immobile; 3 = slightly mobile; 4 = moderately mobile; and 5 = very mobile.*
[d] *1 = insignificant diffusive mobility; 2 = moderate diffusive mobility; and 3 = high diffusive mobility.*
[e] *Time to travel 10 cm by diffusion exceeds 1000 d.*

Table 7. Residual Mass of Pesticide and Travel Time Required to Reach 3 m Depth in the Soil for Pesticides Leached Under *Low Leaching Potentials*

Pesticide	Residual of applied compound reaching 3 m depth in soil (percent)	Time required to reach 3 m depth in soil (years)
Carbaryl	0.00	52.5
Chlorpyrifos	0.00	1314.0
DDT	0.00	51837.8
Diazinon	0.00	21.4
Fenamiphos	0.00	39.9
Malathion	0.00	390.9
Trifluralin	0.00	1579.6
2,4-D	0.00	7.3
Chlorothalonil	0.00	301.0
Atrazine	0.00	37.6
Simazine	0.00	33.2
Dicamba	0.00	3.5
Metribuzin	0.00	8.2

From Jury et al. 1987.

Table 8. Residual Mass of Pesticide and Travel Time Required to Reach 3 m Depth in the Soil for Pesticides Leached Under *High Leaching Potentials*

Pesticide	Residual of applied compound reaching 3 m depth in soil (percent)	Time required to reach 3 m depth in soil (years)
Chlorpyrifos	0.00	137.2
DDT	0.00	5400.1
Malathion	0.00	41.0
Trifluralin	0.00	164.8
Chlorothalonil	0.00	31.6
Fenamiphos	0.00	4.4
Carbaryl	0.00	5.8
2,4-D	0.00	1.0
Diazinon	0.40	2.5
Atrazine	0.99	4.2
Simazine	2.98	3.7
Dicamba	3.84	0.6
Metribuzin	11.47	1.1

From Jury et al. 1987.

volatilization into relative mobility classes based on index scores or environmental partitioning potential.

Mackay and Stiver (1991) developed an environmental partitioning or fugacity model similar to the analytical model of Jury et al. (1983) as a pesticide screening tool. The fugacity or partitioning approach uses linear sorption and volatilization functions, first-order lumped degradation, and linear moisture flux in a surface soil horizon. Using simplified boundary conditions for volatilization, water transport, degradation, and sorption processes (Table 9), this pesticide partitioning model focuses on potential distribution and flux of pesticides in the environment (Table 10). This model uses molecular weight, solubility, vapor pressure, organic carbon partition coefficient, degradation half-life, and soil water content as input data. The model computes the distribution of a pesticide between environmental compartments, volatile losses, leaching losses, and degradation losses (Table 10). Overall persistence of an applied chemical is also calculated based on pesticide flux from the surface soil (Table 10). The rate of water movement, rates of application, and organic matter content can be modified to reflect site-specific conditions.

The pesticide partitioning approach is amenable to preliminary assessment of rates and pathways of pesticide loss. As a screening approach, the pesticide partitioning model provides an overall profile of potential chemical behavior in the environment. Use of the overall pesticide persistence calculated in the model (Table 10) could be

Table 9. Initial Boundary Conditions for the Surface of a Field Moist Sandy Soil for Qualitative Partitioning of Selected Pesticides Used on Turfgrass

Properties	Dimension
Soil and Air	
Area	1 m^2
Depth	0.02 m
Volume	0.02 m^3
Volumetric water content	10%
Porosity of dry soil	50%
Fraction organic matter content	0.025 kg kg^{-1} soil
Water leaching rate	0.0022 m^3 m^{-2} h^{-1}
Water diffusivity	4.3E-5 m^2 d^{-1}
Air diffusivity	0.43 m^2 d^{-1}
Equivalent air mass transfer coefficient	90.5 m d^{-1}
Air boundary layer thickness	0.00475 m
Pesticide properties	
Rate of pesticide application	1 kg a.i. ha^{-1}
Vapor density (Pa)	See Table 16, Chapter 4
Organic carbon partition coefficient (K_{oc})	
Water solubility (g m^3)	See Table 21, Chapter 4
Degradation half-life (d)	See Table 22, Chapter 4

used for initial assessment of the potential for exposure of nontarget organisms (e.g., Table 11).

Use of basic pesticide properties as an index of leaching potential has also been suggested as a screening approach for management and regulatory assessments. The basic premise in this approach is to combine essential parameters influencing pesticide transport into a single rating score. Any method of assessing leachability must account for the mobility *and* persistence of a chemical. Sorption and degradation are two of the most critical processes controlling potential pesticide leaching (Wagenet and Rao 1990). Gustafson (1989) combined the organic carbon partition coefficient (K_{oc}) and degradation half-life ($t_{0.5}$) into a Groundwater Ubiquity Score (GUS). This score is based on a weighted logarithm function such that

$$GUS = \log_{10}(t_{0.5}) * (4 - \log_{10}(K_{oc})) \qquad (1)$$

where $t_{0.5}$ = the soil degradation half-life of a pesticide in days
K_{oc} = organic matter partition coefficient

The GUS provides a consistent index of leachability based on pesticide mobility and persistence (Table 12). Gustafson (1989) classified chemicals with a GUS greater

Table 10. Results of Pesticide Partitioning Ranking Procedure[a]

Chemical name	Percent distribution of chemicals			Percent loss by			Field dissipation[c]
	Air	Water	Soil	Volatilization	Leaching	Degradation[b]	half-life (d)
Insecticides and Nematicides							
Bendiocarb	<0.0	0.6	99.4	<0.1	4.3	95.7	5
Carbaryl	<0.0	1.1	98.9	<0.1	14.6	85.4	9
Chlorpyrifos	<0.0	0.1	99.9	0.7	2.5	96.8	29
Diazinon	<0.0	1.5	98.5	2.6	46.9	50.5	20
Ethoprop	<0.0	2.7	97.3	0.7	50.9	48.4	12
Fenamiphos	<0.0	1.9	98.1	0.1	59.9	40.0	20
Isazofos	<0.0	3.2	96.8	2.4	61.6	36.0	12
Isofenphos	<0.0	0.5	99.5	0.3	56.2	43.5	65
Trichlorfon	<0.0	24.6	75.4	<0.0	79.6	20.3	2
Fungicides							
Anilazine	<0.0	0.3	99.7	<0.0	0.5	99.5	1
Benomyl	<0.0	0.2	99.8	<0.0	38.3	61.7	148
Chloroneb	<0.0	0.3	99.7	73.0	9.9	17.1	22
Chlorothalonil	<0.1	0.2	99.8	81.0	1.9	17.1	5
Etridiozole	<0.0	0.3	99.7	0.4	9.3	90.2	18
Fenarimol	<0.0	0.3	99.7	0.1	65.0	34.9	126
Fosetyl Al	<0.0	14.1	85.9	<0.0	18.2	81.8	1
Iprodione	<0.0	0.5	99.5	<0.1	9.4	90.6	13
Mancozeb	<0.0	0.3	99.7	0.3	15.3	84.4	30
Maneb	<0.0	0.3	99.7	0.1	15.3	84.6	30
Metalaxyl	<0.0	8.5	91.5	<0.1	77.2	22.8	6

Table 10. Results of Pesticide Partitioning Ranking Procedure[a] (continued)

Chemical	Percent distribution of chemicals			Percent loss by			Field dissipation[c]
name	Air	Water	Soil	Volatilization	Leaching	Degradation[b]	half-life (d)
				Fungicides (continued)			
PCNB	<0.0	0.1	99.9	13.0	1.9	85.1	18
Propamocarb HCl	<0.0	<0.0	100.0	<0.0	0.2	99.8	30
Propiconazole	<0.0	0.3	99.7	<0.1	34.1	65.9	66
Thiophanate-methyl	<0.0	0.2	99.8	<0.1	2.8	97.2	10
Thiram	<0.0	0.5	99.5	0.1	10.4	89.5	13
Triadimefon	<0.0	1.1	98.9	<0.0	30.8	69.2	18
				Herbicides			
Atrazine	<0.0	2.1	97.9	0.1	66.9	33.0	20
Benefin	<0.0	<0.1	100.0	1.8	1.9	96.3	39
Bensulide	<0.0	0.3	99.7	0.2	38.2	61.6	74
2,4-D, acid	<0.0	14.1	85.9	0.1	69.0	30.9	3
2,4-D, amine	<0.0	3.2	96.8	0.9	33.1	66.0	7
DCPA	<0.0	0.1	99.9	1.8	11.3	86.9	87
Dicamba	<0.0	62.0	38.0	<0.0	93.2	6.8	1
DSMA	<0.0	0.4	99.6	<0.0	40.2	59.8	60
Endothall	<0.0	14.1	85.9	<0.0	60.9	39.1	3
Ethofumesate	<0.0	1.0	99.0	0.1	31.2	65.7	21
Glyphosate	<0.0	<0.1	100.0	<0.0	2.4	97.6	46
MCPA Amine salt	<0.0	14.1	85.9	<0.0	84.8	15.2	4
MCPA Ester	<0.0	0.3	99.7	0.1	11.4	88.5	22
MCPP	<0.0	14.0	86.0	<0.0	82.4	17.6	4

MSMA	<0.0	<0.1	100.0	<0.0	34.2	65.8	658
Oxadiazon	<0.0	0.1	99.9	0.4	8.8	90.8	54
Pendimethalin	<0.0	0.1	99.9	19.9	6.8	73.2	66
Siduron	<0.0	0.8	99.2	<0.1	52.5	47.5	43
Simazine	<0.0	2.4	97.6	0.1	70.0	30.0	18
Triclopyr							
Amine	<0.0	14.0	86.0	<0.0	91.1	8.9	4
Ester	<0.0	0.4	99.6	0.1	23.3	76.6	35
Trifluralin	<0.0	0.4	99.6	18.3	3.1	78.6	47

From Mackay and Stiver 1991.

[a] *Initial boundary conditions for soil, air, water movement, and pesticide application are defined in Table 9.*

[b] *Degradation includes biological and chemical reactions.*

[c] *Overall half-life indicates persistence of the chemical in the upper 2 cm of soil with loss mechanisms, including degradation, volatilization, and leaching losses.*

Table 11. Relative Ranking for Risk of Pesticide Exposure to Terrestrial Organisms on Turfgrass Based on Estimates of Field Dissipation[a]

Level of potential exposure	Range of field dissipation (d)				
Low			Half-life	<	5
Moderate	5	<	half-life	<	20
Moderately high	20	<	half-life	<	45
High			Half-life	>	45

[a] *Ranking is for purposes of preliminary assessment and comparison only. Field dissipation includes all forms of loss from the soil surface as calculated in Table 10.*

than 2.8 as "leachers" and chemicals with a GUS less than 1.8 as "nonleachers." Transition compounds have intermediate scores between 1.8 and 2.8.

Results of the GUS classification are consistent with the leaching potential criteria developed by Cohen et al. (1984) and the U.S. Environmental Protection Agency (1988). The EPA has developed a suggested list of pesticide characteristics and soil conditions conducive to chemical leaching conditions (Table 13). Based on established soil and pesticide characteristics, these EPA guideline characteristics can also be used to establish preliminary estimates of pesticide leaching potential.

Wauchope (1978) and Wauchope and Leonard (1980a) developed a simple empirical approach for predicting the maximum concentration of pesticides in runoff. This approach also creates a simple index or comparative score for potential losses of pesticides in surface runoff. The predicted maximum or "worst case" concentration of pesticide in runoff from agricultural land at any time after pesticide application is

$$C_t = AR(1 + 0.44t)^{-1.6} \tag{2}$$

where C_t = the runoff concentration in an event after time, t (ppb)
R = the rate of pesticide application (kg ha^{-1})
A = an "availability index" (ppb ha kg^{-1})

The prediction formula was based on data compiled by Wauchope (1978) and Wauchope and Leonard (1980a, 1980b). This formula could be used as a qualitative estimate of both sediment-bound and dissolved pesticides lost in runoff from a turfgrass site under initial construction (Table 12).

Use and development of CREAMS has been extensively documented by Knisel (1980). Examples of the utility of using this functional modeling approach for evaluation of chemical transport have been presented by Leonard (1988, 1990), Leonard and Knisel (1988), and Crowder and Young (1987). Using GLEAMS and CREAMS, Goss (1991) has developed a tabular hazard rating for both surface and

Table 12. Qualitative Assessment of Potential Surface and Subsurface Losses of Pesticides

Chemical name	Potential surface losses: Maximum concentration in runoff[a] ($g\ m^{-3}$)	SCS runoff rating[b]: sediment	SCS runoff rating[b]: soluble	Potential subsurface losses: GUS[c]	GUS ranking	SCS leaching[b] rating
Insecticides and Nematicides						
Bendiocarb	5.6	Small	Large	0.87	Nonleacher	Small
Carbaryl	1.7	Small	Medium	1.52	Nonleacher	Small
Chlorpyrifos	0.6	Medium	Small	0.32	Nonleacher	Small
Diazinon	1.7	Large	Large	2.65	Intermediate	Small
Ethoprop	1.7	Small	Medium	2.68	Intermediate	Large
Fenamiphos	1.7	Medium	Large	3.01	Leacher	Large
Isazofos	1.7	Small	Large	3.06	Leacher	Large
Isofenphos	1.7	Medium	Large	2.65	Intermediate	Medium
Trichlorfon	1.7	Small	Medium	3.00	Leacher	Large
Fungicides						
Anilazine	0.6	Small	Small	0.00	Nonleacher	Small
Benomyl	5.6	Large	Large	1.66	Nonleacher	Small
Chloroneb	5.6	Large	Large	1.98	Intermediate	Small
Chlorothalonil	1.7	Medium	Medium	1.27	Nonleacher	Small
Etridiozole	0.6	Medium	Medium	1.30	Nonleacher	Small
Fenarimol	0.6	Medium	Large	2.55	Intermediate	Large
Fosetyl Al	5.6	Small	Medium	0.00	Nonleacher	Small
Iprodione	1.7	Small	Large	1.32	Nonleacher	Small
Mancozeb	5.6	Large	Large	1.54	Nonleacher	Small

Table 12. Qualitative Assessment of Potential Surface and Subsurface Losses of Pesticides (continued)

Chemical	Potential surface losses			Potential subsurface losses		
	Maximum concentration	SCS runoff rating[b]			GUS	SCS leaching[b]
name	in runoff[a] ($g\ m^{-3}$)	sediment	soluble	GUS[c]	ranking	rating
		Fungicides (continued)				
Maneb	5.6	Large	Large	1.54	Nonleacher	Small
Metalaxyl	5.6	Medium	Large	3.43	Leacher	Large
PCNB	0.6	Medium	Small	0.39	Nonleacher	Small
Propamocarb HCl	0.6	Medium	Small	−1.48	Nonleacher	Extra small
Propiconazole	0.6	Large	Large	2.00	Intermediate	Medium
Thiophanate-methyl	5.6	Medium	Medium	0.74	Nonleacher	Small
Thiram	5.6	Small	Large	1.38	Nonleacher	Small
Triadimefon	5.6	Small	Large	2.15	Intermediate	Medium
		Herbicides				
Atrazine	0.2–5.6	Medium	Large	3.24	Leacher	Large
Benefin	0.6	Large	Medium	−0.05	Nonleacher	Small
Bensulide	0.6	Large	Large	2.08	Intermediate	Medium
2,4-D, acid	1.7	Small	Medium	2.69	Intermediate	Medium
2,4-D, amine	1.7	Small	Medium	2.00	Intermediate	Medium
DCPA	5.6	Large	Medium	0.80	Nonleacher	Small
Dicamba	1.7	Small	Medium	4.24	Leacher	Large
DSMA	5.6	Large	Small	2.31	Intermediate	Extra small
Endothall	0.6	Small	Medium	2.28	Intermediate	Medium
Ethofumesate	1.7	Small	Medium	2.17	Intermediate	Medium

Glyphosate	5.6	Large	Large	0.00	Nonleacher	Extra small
MCPA Amine salt	1.7	Small	Medium	3.77	Leacher	Large
MCPA Ester	0.6	Medium	Medium	1.39	Nonleacher	Small
MCPP	1.7	Small	Medium	3.51	Leacher	Large
MSMA	5.6	Large	Small	0.00	Nonleacher	Extra small
Oxadiazon	0.6	Large	Medium	0.88	Nonleacher	Small
Pendimethalin	0.6	Large	Medium	0.59	Nonleacher	Small
Pronamide	5.6	Medium	Large	3.02	Leacher	Large
Siduron	5.6	Medium	Large	2.69	Intermediate	Medium
Simazine	5.6	Medium	Large	3.35	Leacher	Large
Triclopyr amine	1.7	Medium	Large	4.49	Leacher	Large
Triclopyr ester	1.7	Medium	Large	1.84	Intermediate	Medium
Trifluralin	0.6	Large	Medium	0.17	Nonleacher	Small

[a] *Based on worst case model developed by Wauchope and Leonard (1980) and assumes rain occurs 1 d after application of a pesticide at a rate of 1 kg a.i. ha^{-1}. Concentration is edge-of-field runoff loss of sediment-bound pesticide and pesticide in solution. Edge-of-field losses of pesticides are not the concentrations of pesticides found in lakes and streams. In general, estimated losses represent a "worse case" situation, concentrations may be overestimated by one order of magnitude, and estimates vary depending on the formulation of the applied pesticide.*

[b] *USDA Soil Conservation Service pesticide and water quality screening ratings developed by Goss (1991) and reported in USDA SCS pesticide database (Wauchope et al. 1991).*

[c] *Groundwater Ubiquity Score and leaching potential rating based on pesticide degradation and organic matter partitioning was developed by Gustafson (1989). Potential difference in GUS rating and SCS rating is due to selection of different partition coefficients for development of ratings.*

Table 13. Chemical/Physical Properties of Pesticides: Values Which Indicate Potential for Groundwater Contamination

Pesticide characteristic	Parameter value or range indicating potential for groundwater contamination
Water solubility	Greater than 30 ppm
K_d	Less than 5, usually less than 1
K_{oc}	Less than 300–500
Henry's Law Constant	Less than 10^{-2} atm m^{-3} mol
Speciation	Negatively charged, fully or partially at ambient pH
Hydrolysis half-life	Greater than 175 d
Photolysis half-life	Greater than 7 d
Field dissipation half-life	Greater than 21 d

From EPA 1988.

subsurface losses of pesticides for the USDA SCS (Table 12). This type of hazard rating could be modified for turfgrass systems within different climatic zones and for soils with variable properties (Goss 1991). Pesticides are ranked in the SCS system with extra small, small, medium, or large probability for losses in surface runoff or subsurface leaching. The surface runoff losses are partitioned into soluble and sediment-bound phases (Table 12). The rating for potential leaching losses of pesticides in the approach developed by Goss (1991) is similar to the hazard score system developed by Gustafson (1989).

As a screening approach, the SCS pesticide ranking procedure and the GUS index provides another screening profile of potential chemical movement to surface water and groundwater. Use of the ranking scores and probability ratings (Table 12) could be used for preliminary assessment of the potential for exposure of aquatic organisms (e.g., Table 14).

Growing healthy turfgrass in an environmentally sound manner is an important goal of TMS. Healthy turfgrass is resistant to loss of quality due to disease and pest problems. Use of expert systems to assist in the proactive management of water and pests in turfgrass has been proposed as a component of TMS (e.g., Nyrop et al. 1989; Vargas et al. 1989). A computer program, WEEDER, has been developed by Fermanian and Michalski (1989) to assist in identification of grass weeds and turfgrass species. Use of this technology-based expert system allows the user to select variables that are available for unknown species. Operating on multiple levels of uncertainty, the system provides an efficient means for weed identification. This type of technology is essential for simplifying pest monitoring programs. The WEEDER system is based on a nonspecialist artificial intelligence system, AGASSISTANT (Fermanian et al. 1989). This rule-based technology could be used to develop identification systems

Table 14. Relative Ranking for Risk of Pesticide Exposure to Aquatic Organisms in the Vicinity of Turfgrass Based on Potential Soluble Losses in Runoff and Subsurface Drainage[a]

Level of potential exposure	Potential for loss in runoff		Potential for loss in[b] subsurface drainage		
Low	Small	and	GUS	<	1.8
Moderate	Small	and	GUS	>	1.8
	Medium	and	GUS	<	1.8
Moderately high	Medium	and	GUS	>	1.8
	Large	and	GUS	<	1.8
High	Large	and	GUS	>	1.8

[a] *Ranking is for purposes of preliminary assessment and comparison only.*

[b] *Loss in runoff is based on SCS Soluble Runoff Ranking (Table 12). Sediment losses from established turfgrass are very small; therefore, erosion-related pesticide losses are not included in this ranking. Risk of subsurface drainage losses are based on GUS (Table 12).*

for nutrient, pest, and disease management in turfgrass systems (Nyrop et al. 1989). Vargas et al. (1989) is currently developing computer prediction models for management strategies to control diseases and encourage development of pest predator populations.

Development of expert system technology is an expensive and time-consuming effort (Nyrop et al. 1989). Development of rule-based, prototype systems may require several months to develop thousands of rules. The cost and effort to develop a finished system is likely to triple that of development of the initial prototype. Nyrop et al. (1989) summarized the criteria necessary to justify development of knowledge-based or expert systems. The principal criteria are (1) solution of the management problem with an expert system must have a high payoff; (2) the expertise required to solve the management problem must exist, but is relatively scarce; (3) the management problem must be sufficiently well understood to formalize the decision information; and (4) the management process must be sufficiently complex to warrant development of an expert system. Many of these criteria are satisfied regarding development of expert systems for integrated management of turfgrass (e.g., Wearing 1988).

Appropriate interpretation of model results in relation to the intended scale and use of the model requires a sophisticated understanding of the model on the part of the user. Use of research, deterministic, stochastic, or screening models all have specific advantages and limitations. Despite many limitations in modeling of chemical and environmental impacts, computer simulation remains a viable management strategy for assessment of the impacts of agricultural, forestry, and turfgrass management practices. Environmental regulators, turfgrass scientists, and turfgrass managers have the option to use a wide range of simulation models. Many of these models, mainframe to microcomputer, can be used as one of the tools to assess the effects of integrated management practices on the environment prior to implementing the practice for a specific turfgrass system.

5.6 CONCLUSIONS

The concepts developed for ICM in agriculture are certainly suitable for development of ecologically sound management of turfgrass. The use of TMS would be very effective for growing of high-quality turfgrass and for reducing potentially adverse environmental quality effects. TMS practices would coordinate all factors required for long-term sustained productivity and quality of turfgrass, golf course profitability, and ecological soundness of selected management options. The critical components for planning and implementing TMS must include

1. Proper design and construction of golf courses
2. Selection of appropriate turfgrass species and cultivars
3. Selection of timely soil and cultural practices
4. Integrated planning of water, nutrient, and pest management in relation to other management goals.
5. Consideration of alternate chemical, biological, and cultural pest, nutrient, and water management practices when possible
6. Conservation of soil, water, energy, and natural resources during construction and maintenance of golf courses

Although IPM programs are critical components of an entire integrated management system, pest management alone is insufficient for overall planning of an economically and environmentally sound turfgrass program. The goal of the TMS approach is to balance turfgrass quality, costs, benefits, public health, and environmental quality. Use of integrated water, pest, and nutrient management strategies will ultimately resolve the issue of maintaining high-quality turfgrass with a minimum of environmental disturbance.

Providing consistent information on specifications and practices to mitigate adverse environmental impacts is essential for implementation of TMS. There is no single source of summarized information available to golf course managers regarding management options for TMS and environmental protection. A systematic approach and formulation of guidelines for environmentally sound turfgrass management would provide regulatory and environmental protection agencies with an understanding that turfgrass managers are using consistent practices to mitigate potential environmental hazards. The series of papers on IPM developed by the U.S. EPA

(Leslie and Metcalf 1989) could serve as an initial model for documenting a systems perspective. However, an integrated water, nutrient, pest, and water resources perspective should be maintained rather than an alternate pest control perspective.

Compilation of information on a systems approach to turfgrass management requires a direct link to environmental and water quality protection in addition to issues on maintenance of high-quality turfgrass. This type of information should be modified for different climatic regions, soils, irrigation and drainage systems, and golf course management requirements. A single site-specific management handbook may not be possible. However, the combined resources of the turfgrass industry, environmental regulatory agencies, university extension services, and private sector information services would be capable of producing a guidance system of principles on TMS.

The general principles for TMS need to be included in the development of any environmentally sound guidance system. These principles should include consideration of nutrient and pesticide sources, rates and timing of application to meet turfgrass and environmental goals, chemical and alternate control formulations, use of alternate practices for nutrient and pest management, irrigation scheduling, proximity of surface water and groundwater sources, proper equipment calibration, impacts of other cultural practices, and assessment for offsite movement of applied chemicals.

Prior to implementing local TMS programs, regional IPM and TMS approaches will require additional information from applied research projects. Determining economic thresholds and scouting programs for implementing nutrient and pest management strategies should be a high research priority. Development of commercially viable alternate control technologies requires additional research on pest and disease monitoring, improved application technology for chemical and biological control agents, selection or development of insect and disease resistant cultivars, and expansion of current research techniques to a wide range of climatic conditions and turfgrass management conditions. Research on timing, method, and rate of nutrients and pest control agents in relation to climatic and soil conditions ultimately controls their efficacy and environmental impacts. Development or selection of disease- and pest-resistant turfgrass cultivars also will have a significant impact on reducing pesticide use on turfgrass systems. Education of turfgrass managers in appropriate integrated systems technology is required before changes can be made in current management strategies.

Several transport and environmental fate models could be used to evaluate pesticide and nutrient movement on turfgrass systems. Simulation approaches provide an opportunity to evaluate the environmental effects of different management plans prior to their implementation. These models need to be calibrated and modified for simulation of water and chemical movement in turfgrass systems. Research is required to improve simulation of water transport, volatilization, adsorption, and degradation. Accurate projections of these processes is needed before chemical fate models will be capable of realistic predictions. Current simulation technology for turfgrass is not adequate to produce appropriate results for site-specific management, research, or regulatory decisions. However, regulatory agencies will use current simulation strategies to establish management guidelines, unless efforts are made to provide more suitable predictive approaches. Development of modeling and risk assessment approaches is an essential component of TMS.

REFERENCES

Abriola, L. M. and Pinder, G. F. 1985a. A multiphase approach to modeling of porus media contamination by organic compounds. I. Equation development. *Water Resour. Res.* 21(1):11–18.

Abriola, L. M. and Pinder, G. F. 1985b. A multiphase approach to modeling of porus media contamination by organic compounds. I. Numerical Simulation. *Water Resour. Res.* 21(1):19–26.

Addiscott, T. M. 1977. A simple computer model for leaching in structured soils. *J. Soil Sci.* 28:554–563.

Addiscott, T. M. and Wagenet, R. J. 1985. Concepts of solute leaching in soils: A review of modeling approaches. *J. Soil Sci.* 36:411–424.

Ahmad, S., Johnson-Cicalese, J. M., Dickson, W. K. and Funk, C. R. 1986. Endophyte-enhanced resistance in perennial ryegrass to the bluegrass billbug, *Sphenophorus parvulus. Entomol. Exp. Appl.* 41:3–10.

Ahmadi, M. S., Haderlie, L. C. and Wicks, G. A. 1980. Effect of growth stage and water stress on Barnyardgrass (*Echinochloa crus-galli*) control and on glyphosate absorption and translocation. *Weed Sci.* 28(3):277–282.

Alexander, M. 1977. *Introduction to Soil Microbiology.* John Wiley & Sons, New York, 467 pp.

Allen, G. E. and Bath, J. E. 1980. The conceptual and institutional aspects of integrated pest management. *BioScience* 30(10):658–664.

Aller, L., Bennett,T., Lehr, J. H. and Petty, R. H. 1985. DRASTIC: A Standardized System for Evaluating Ground Water Pollution Potential Using Hydrogeologic Setting. EPA-600/S2-85-018. Robert S. Kerr Environmental Research Laboratory, U.S. Environmental Protection Agency, Ada, OK.

Amoozegar-Fard, A., Nielsen, D. R. and Warrick, A. W. 1982. Soil solute concentration distributions for spatially-varying pore water velocities and apparent diffusion coefficients. *Soil Sci. Soc. Am. J.* 46(1):3–9.

Anderson, J. L., Balogh, J. C. and Waggoner, M. 1989. Soil Conservation Service Procedure Manual: Development of Standards and Specifications for Nutrient and Pesticide Management. Section I & II. USDA Soil Conservation Service. State Nutrient and Pest Management Standards and Specification Workshop, St. Paul, MN, July 1989, 453 pp.

Arnold, T. B. and Potter, D. A. 1987. Impact of a high-maintenance lawn-care program on nontarget invertebrates in Kentucky bluegrass turf. *Environ. Entomol.* 16(1):100–105.

Atilano, R. A. and Busey, P. 1983. Susceptibility of St. Augustinegrass germ plasm to *Pyricuaria grisea*. *Plant Dis*. 67(7):782–783.

Bacon, C. W., Lyons, P. C., Porter, J. K. and Robbins, J. D. 1986. Ergot toxicity from endophyte-infected grasses: A review. *Agron. J*. 78:106–116.

Baker, R. 1986. Biological control: an overview. *Can. J. Plant Pathol*. 8:218–221.

Balogh, J. C. and Anderson, J. L. 1992. Environmental impacts of turfgrass pesticides. in Balogh, J. C. and Walker, W. J. (Eds.). *Golf Course Management and Construction: Environmental Issues*. Lewis Publishers, Chelsea, MI, Chapter 4.

Balogh, J. C. and Gordon, G. A. 1987. Climate Variability and Soil Hydrology in Water Resource Evaluation. Phase I Project Report. National Science Foundation, ISI-87053. NTIS #PB90-203183/A05. 92 pp.

Balogh, J. C. and Watson, J. R., Jr. 1992. Role and conservation of water resources. in Balogh, J. C. and Walker, W. J. (Eds.). *Golf Course Management and Construction: Environmental Issues*. Lewis Publishers, Chelsea, MI, Chapter 2.

Balogh, J. C. and Madison, F. W. 1985. Runoff treatment from a turkey production facility. *Trans. ASAE* 28(5):1476–1481.

Banerjee, A. C. 1969. Sex attractants in sod webworms. *J. Econ. Entomol*. 62(3):705–708.

Bayaa, B. O., Halisky, P. M. and White, J. F., Jr. 1987. Inhibitory interactions between *Acermonium* spp. and the mycoflora from seeds of *Festuca* and *Lolium*. *Phytopathol*. 77:115.

Beard, J. B. 1973. *Turfgrass Science and Culture*. Prentice Hall, Englewood Cliffs, NJ, 658 pp.

Beard, J. B. 1982. *Turf Management for Golf Courses*. Macmillan Publishing Co., New York, 642 pp.

Beauchamp, E. G. 1977. Slow release N fertilizers applied in fall for corn. *Can. J. Soil Sci*. 57:487–496.

Berry, C. D. and Gudauskas, R. T. 1972. Susceptibility of tall fescue, *Fetuca arundinacea* Schreb., to crown rust. *Crop Sci*. 12:101–102.

Bonazountas, M. 1988. Mathematical pollutant fate modeling of petroleum products in soil systems. in Calabrese, E. J. and Kostecki, P. T. (Eds.). *Soils Contaminated by Petroleum: Environmental & Public Health Concerns*. John Wiley & Sons, New York, pp. 31–111.

Bonazountas, M. 1991. Fate of hydrocarbons in soils: Review of modeling practices. in Kostecki, P. T. and Calabrese, E. J. (Eds.). *Hydrocarbon Contaminated Soils and Groundwater*. Lewis Publishers, Chelsea, MI, pp. 167–185.

Brede, A. D. and Brede, J. L. 1988. Establishment clipping of tall fescue and companion annual ryegrass. *Agron. J*. 80(1):27–30.

Brede, A. D. and Duich, J. M. 1984a. Initial mowing of Kentucky bluegrass-perennial ryegrass seedling turf mixtures. *Agron. J*. 76(5):711–714.

Brede, A. D. and Duich, J. M. 1984b. Establishment characteristics of Kentucky bluegrass-perennial ryegrass turf mixtures as affected by seeding rate and ratio. *Agron. J*. 76(6):875–879.

Brede, A. D. and Duich, J. M. 1986. Plant interaction among *Poa annua*, *Poa pratensis*, and *Lolium perenne* turfgrasses. *Agron. J*. 78(1):179–184.

Bresler, E. 1967. A model for tracing salt distribution in the soil profile and estimating efficient combination of water quality and quantity under varying field conditions. *Soil Sci*. 104:227–233.

Bresler, E. H., Bieloria, H. and Laufer, A. 1979. Field test of solution flow models in a heterogeneous irrigated cropped soil. *Water Resour. Res*. 15(3):645–652.

Britton, M. P. 1969. Turfgrass diseases. in Hanson, A. A. and Juska, F. V. (Eds.) *Turfgrass Science*. Agron. Monogr. 14:288–335. Am. Soc. Agron., Madison, WI.

Brown, W., Jr., Cranshaw, W. and Rasmussen-Dykes, C. 1989. Educational, environmental and economic impacts of integrated pest management programs for landscape plants. in Leslie, A. R. and Metcalf, R. L. (Eds.). Integrated Pest Management for Turfgrass and Ornamentals. Office of Pesticide Programs, 1989-625-030. U.S. Environmental Protection Agency, Washington, D.C., pp. 57–75.

Brusseau, M. L. and Rao, P. S. C. 1989. Sorption nonidality during organic contaminant transport in porous media. *Crit. Rev. Environ. Chem.* 19(1):33–99.

Bruton, B. D., Toler, R. W. and Reinert, J. A. 1983. Combined resistance in St. Augustinegrass to the southern chinch bug and the St. Augustine decline strain of panicum mosaic virus. *Plant Dis.* 67:171–172.

Buettner, M. R., Ensign, R. D. and Boe, A. A. 1976. Plant growth regulator effects on flowering of *Poa pratensis* L. under field conditions. *Agron. J.* 68(2):410–413.

Burns, I. G. 1974. A model for predicting the redistribution of salts applied to fallow soils after excess rainfall or evaporation. *J. Soil Sci.* 25:165–178.

Burns, I. G. 1980a. A simple model for predicting the effects of leaching of fertilizer nitrate during the growing season of the nitrogen fertilizer needs of crops. *J. Soil Sci.* 31:175–187.

Burns, I. G. 1980b. A simple model for predicting the effects of winter leaching of residual nitrate on the nitrogen fertilizer need of spring crops. *J. Soil Sci.* 31:187–202.

Burpee, L. L. and Goulty, L. G. 1984. Suppression of brown patch disease of creeping bentgrass by isolates of nonpathogenic *Rhizoctonia* spp. *Phytopathol.* 74:692–694.

Burpee, L. L., Kaye, L. M., Goulty, L. G. and Lawton, M. B. 1987. Suppression of gray snow mold on creeping bentgrass by an isolate of *Typhula pharcorrhiza*. *Plant Dis.* 71:97–100.

Busey, P. 1989. Progress and Benefits to Humanity from Breeding Warm-Season Grasses for Turf. in Sleper, D. A., Asay, K. H. and Pedersen, J. F. (Eds.). *Contributions from Breeding Forage and Turf Grasses*. CSSA Spec. Publ. No. 15. Crop Sci. Soc. Am., Madison, WI, pp. 49–70.

Busey, P. and Center, B. J. 1987. Southern chinch bug (*Hemiptera: Heteroptera: Lygaeidae*) overcomes resistance in St. Augustinegrass. *J. Econ. Entomol.* 80:(3):608–611.

Busey, P., Reinert, J. A. and Atilano, R. A. 1982. Genetic and environmental determinants of zoysiagrass adaption in a subtropical region. *J. Am. Soc. Hort. Sci.* 107(1):79–82.

Bush, L., Boling, J. and Yates, S. 1979. Animal disorders. in Buckner, R. C. and Bush, L. P. (Eds.). *Tall Fescue*. Agron. Monogr. 20:247–306. Am. Soc. Agron., Crop Sci. Soc. Am., Soil Sci. Soc. Am., Madison, WI.

Buta, J. G., Spaulding, D. W. and Reed, A. N. 1987. Differential growth responses of fractionated turfgrass seed leachates. *HortScience* 22(6):1317–1319.

Carrow, R. N., Shearman, R. C. and Watson, J. R. 1990. Turfgrass. in Stewart, B. A. and Nielsen, D. R. (Eds.). *Irrigation of Agricultural Crops*. Agron. Monogr. 30:889–919. Am. Soc. Agron., Crop Sci. Soc. Am., Soil Sci. Soc. Am., Madison, WI.

Carsel, R. F., Imhoff, J. C., Little, J. L., Jr. and Hummel, P. R. 1991. Development of a database and model parameter analysis system for agricultural soils. *J. Environ. Qual.* 20(3):642–647.

Carsel, R. F., Jones, R. L., Hansen, J. H., Lamb, R. L. and Anderson, M. P. 1988. A simulation procedure for groundwater quality assessment of pesticides. *J. Contam. Hydrol.* 2:125–138.

Carsel, R. F., Mulkey, L. A., Lorber, M. N. and Baskin, L. B. 1985. The pesticide root zone model (PRZM): A procedure for evaluating pesticide leaching threats to groundwater. *Ecol. Modelling* 30:49–69.

Carsel, R. F., Nixon, W. B. and Balentine, L. G. 1986. Comparison of pesticide root zone model predictions with observed concentrations for the tobacco pesticide metalaxyl in unsaturated zone soils. *Environ. Toxicol. Chem.* 5:345–353.

Carsel, R. F., Smith, C. N., Mulkey, J. T., Dean, J. D. and Jowise, P. 1984. Users Manual for the Pesticide Root Zone Model (PRZM). Release 1. EPA-600/3- 84-109. U.S. Environmental Protection Agency, Athens, GA, 216 pp.

Chapman, R. A. 1979. Diseases and nematodes. in Buckner, R. C. and Bush, L. P. (Eds.). *Tall Fescue.* Agron. Monogr. 20:307–318. Am. Soc. Agron., Crop Sci. Soc. Am., Soil Sci. Soc. Am., Madison, WI.

Childs, S. W. and Hanks, R. J. 1975. Model for soil salinity effects on crop growth. *Soil Sci. Soc. Am. Proc.* 39:617–622.

Clark, J. D. and Haynes, K. F. 1990. Sex attractant for the bluegrass webworm (*Lepidotera: Pyralidae*). *J. Econ. Entomol.* 83(3):856–859.

Clarke, B. B., Johnson-Cicalese, J. M. and Funk, C. R. 1989. Development of Perennial Ryegrass Cultivars with Improved Resistance to *Rhizoctonia* Brown Patch. in Sixth Int. Turf. Res. Conf., Tokyo, Japan, pp. 337–340.

Cockfield, S. D. and Potter, D. A. 1983. Short-term effects of insecticidal applications on predaceous arthropods and oribatid mites in Kentucky bluegrass turf. *Environ. Entomol.* 12(4):1260–1264.

Cockfield, S. D. and Potter, D. A. 1984a. Predatory insects and spiders from suburban lawns in Lexington, Kentucky. *Great Lakes Entomol.* 17(3):179–184.

Cockfield, S. D. and Potter, D. A. 1984b. Predation of sod webworm (*Lepidoptera: Pyralidae*) eggs as affected by chlorpyrifos application to Kentucky bluegrass. *J. Econ. Entomol.* 77(6):1542–1544.

Cockfield, S. D. and Potter, D. A. 1985. Predatory arthropods in high- and low-maintenance turfgrass. *Can. Entomol.* 117:423–429.

Cohen, S. Z., Creeger, S. M., Carsel, R. F. and Enfield, C. G. 1984. Potential for pesticide contamination of groundwater resulting from agricultural uses. in Kruger, R. F. and Seiber, J. N. (Eds.). *Treatment and Disposal of Pesticide Wastes.* ACS Symposium Series No. 259. American Chemical Society, Washington, D.C., pp. 297–325.

Cohen, S. Z., Nickerson, S., Maxey, R., Dupuy, A., Jr. and Senita, J. A. 1990. Ground water monitoring study for pesticides and nitrates associated with golf courses on Cape Cod. *Ground Water Monit. Rev.* 10(1):160–173.

Colbaugh, P. F. and Elmore, C. L. 1985. Influence of water on pest activity. in Gibeault, V. A. and Cockerham, S. T. (Eds.). *Turfgrass Water Conservation.* University of California, Riverside, Division of Agriculture and Natural Resources, pp. 113–129.

Computer Sciences Corporation. 1980. Pesticide Runoff Simulator. User's Manual. U.S. Environmental Protection Agency, Falls Church, VA.

Cook, R. J. and Baker, K. F. 1983. *The Nature and Practice of Biological Control of Plant Pathogens.* American Phytopathological Society, St. Paul, MN, 539 pp.

Cooper, R. J., Vittum, P. J. and Bhowmik, P. C. 1987. Urban lawn integrated pest management (IPM): Results of a pilot program. in Elzerman, A. W. (Ed.). *Proc. Symp. New Directions in Urban Int. Pest. Manag.: An Environ. Mandate.* Am. Chem. Soc. Symp. Series, pp. 543–544.

Cope, W. A. 1982. Inhibition of germination and seedling growth of eight forage species by leachates from seeds. *Crop Sci.* 22(6):1109–1111.

Corapcioglu, M. Y. and Baehr, A. L. 1987a. A compositional multiphase model for groundwater contamination by petroleum products. I. Theoretical considerations. *Water Resour. Res.* 23(1):191–200.

Corapcioglu, M. Y. and Baehr, A. L. 1987b. A compositional multiphase model for groundwater contamination by petroleum products. I. A numerical solution. *Water Resour. Res.* 23(1):201–213.

Corbet, P. S. 1981. Non-entomological impediments to the adoption of integrated pest management. *Prot. Ecol.* 3:183–202.

Crocker, R. L. and Whitcomb, W. H. 1980. Feeding niches of the big-eyed bugs *Geocoris bullatus, G. punctipes,* and *G. uliginosus (Hemiptera: Lygaeidae: Geocorinae). Environ. Entomol.* 9(5):508–513.

Crocker, R. L., Toler, R. W., Beard, J. B., Engelke, M. C. and Kubica-Breier, J. S. 1989. St. Augustinegrass antibiosis to southern chinch bug (*Hemiptera: Lygaeidae*) and to St. Augustine decline strain of panicum Mosaic virus. *J. Econ. Entomol.* 82(6):1729–1732.

Crowder, B. M. and Young, C. E. 1987. Soil Conservation practices and water quality, is erosion control the answer. *Water Res. Bull.* 23(5):879–902.

Crowder, B. M., Pionke, H. B., Epp, D. J. and Young, C. E. 1985. Using CREAMS and economic modeling to evaluate conservation tillage practices: An application. *J. Environ. Qual.* 14(3):428– 434.

Curry, D. S. 1987. Assessment of empirical methodologies for predicting ground water pollution from agricultural chemicals. in Fairchild, D. M. (Ed.). *Ground Water Quality and Agricultural Practices.* Lewis Publishers, Chelsea, MI, pp. 227–245.

Dagan, G. and Bresler, E. 1979. Solute dispersion in unsaturated heterogeneous soil at field scale. I. Theory. *Soil Sci. Soc. Am. J.* 43:461–466.

Dalton, G. E. 1975. *Study of Agricultural Systems.* Applied Science Publishers, London, England.

Daugherty, S. J. 1991. Regulatory approaches to hydrocarbon contamination from underground storage tanks. in Kostecki, P. T. and Calabrese, E. J. (Eds.). *Hydrocarbon Contaminated Soils and Groundwater.* Lewis Publishers, Chelsea, MI, pp. 23–63.

Davidson, R. L. and Roberts, R. J. 1968. Influence of plants, manure, and soil moisture on survival and liveweight gain of two scarbaeid larvae. *Entoml. Exp. Appl.* 11:305–314.

Davidson, R. L., Wiseman, J. R. and Wolf, V. J. 1972. Environmental stress in the pasture scarab *Sericesthis nigrolineata* Boisd. II. Effects of soil moisture and temperature on survival of first-instar larvae. *J. Appl. Ecol.* 9:799–806.

Dean, H. A., Schuster, M. F., Boling, J. C. and Riherd, P. T. 1979. Complete biological control of *Antonina graminis* in Texas with *Neodusmetia sangwani* (a classic example). *Bull. Entomol. Soc. Am.* 25:262–267.

Dean, J. D., Jowise, P. P., and Donigian, A. S., Jr. 1984. Leaching Evaluation of Agricultural Chemicals (LEACH) Handbook. EPA–600/3–84–068. Environmental Research Laboratory, U.S. Environmental Protection Agency, Athens, GA.

DeCoursey, D. G. 1982. ARS Small Watershed Model. in Proc. Summer Meet. Madison, WI. ASAE Paper No. 82–2094. St. Joseph, MI, pp. 1–33.

Dernoeden, P. H. 1987. Management of take-all patch of creeping bentgrass with nitrogen, sulfur, and phenyl acetate. *Plant Dis.* 71(3):226–229.

Dernoeden, P. H. 1989. Symptomatology and management of common turfgrass diseases in transition zone and northern regions. in Leslie, A. R. and Metcalf, R. L. (Eds.). Integrated Pest Management for Turfgrass and Ornamentals. Office of Pesticide Programs, 1989-625-030. U.S. Environmental Protection Agency, Washington, D.C., pp. 273–296.

Dernoeden, P. H. and Jackson, N. 1980. Managing yellow tuft disease. *J. Sports Turf Res.* 56:9–17.

De Smedt, F. and Wierenga, P. J. 1978. Approximate analytical solution for solute flow during infiltration and redistribution. *Soil Sci. Soc. Am. J.* 42:407–412.

Dest, W. M. and Allinson, D. W. 1981. Influence of Nitrogen and Phosphorus Fertilization on the Growth and Development of *Poa annua* L. in Sheard, R. W. (Ed.). Proc. 4th Int. Turfgrass Res. Conf. Guelph, Ontario, pp. 325–332.

Devitt, D. A. and Morris, R. L. 1989. Growth of common bermudagrass as influenced by plant growth regulators, soil type and nitrogen fertility. *J. Environ. Hort.* 7(1):1–8.

Domek, J. M. and Johnson, D. T. 1987. Evidence of a sex pheromone in the green June beetle, *Cotinus nitida* (L.) (*Coleoptera: Scarabaeidae*). *J. Entomol. Sci.* 22(3):264–267.

Domek, J. M. and Johnson, D. T. 1988. Demonstration of semiochemically induced aggregation in the green June beetle, *Cotinus nitida* (L.) (*Coleoptera: Scarabaeidae*). *Environ. Entomol.* 17(2):147–149.

Donigian, A. S., Jr. and Carsel, R. F. 1987. Modeling impact of conservation tillage practices on pesticide concentrations in ground and surface waters. *J. Environ. Toxicol. Chem.* 6(4):241–250.

Donigian, A. S., Jr. and Rao, P. S. C. 1986a. Overview of terrestrial processes and modeling. in Hern, S. C. and Melancon, S. M. (Eds.). *Vadose Zone Modeling of Organic Pollutants.* Lewis Publisers, Chelsea, MI, pp. 3–35.

Donigian, A. S., Jr. and Rao, P. S. C. 1986b. Example model testing studies. in Hern, S. C. and Melancon, S. M. (Eds.). *Vadose Zone Modeling of Organic Pollutants.* Lewis Publishers, Chelsea, MI, pp. 103–131.

Donigian, A. S., Jr., Imhoff, J. C. and Bichnell, B. R. 1983. Predicting water quality resulting from agricultural nonpoint source pollution via simulation — HSPF. in Schaller, F. W. and Bailey, G. W. (Eds.). *Agricultural Management and Water Quality.* Iowa University Press, Ames, IA, pp. 200–249.

Dudeck, A. E., Reinert, A. E. and Busey, P. 1986. Florlawn St. Augustinegrass. Fla. Agric. Exp. Stat. Circ. No. S-327.

Enfield, C. G., Carsel, R. F., Cohen, S. Z., Phan, T. and Walters, D. M. 1982. Approximating pollutant transport to ground water. *Ground Water* 20:711–722.

Engel, R. E. and Ilnicki, R. D. 1969. Turfweeds and their control. in Hanson, A. A. and Juska, F. V. (Eds.). *Turfgrass Science.* Agron. Monogr. 14:240–287. Am. Soc. Agron., Madison, WI.

Environmental Protection Agency. 1988. Protecting Ground Water: Pesticides and Agricultural Practices. EPA-440/6-88-001. U.S. Environmental Protection Agency, Office of Ground-Water Protection, Washington, D.C., 53 pp.

Fales, S. L. and Wakefield, R. C. 1981. Effects of turfgrass on the establishment of woody plants. *Agron. J.* 73(4):605–610.

Farm Chemicals Handbook. 1991. *Pesticide and Fertilizer Dictionary.* Meister Publishing Company, Willoughby, OH, 770 pp.

Fay, P. K. and Duke, W. P. 1977. An assessment of allelopathic potential in *Avena* germplasm. *Weed Sci.* 25:224–228.

Fermanian, T. W. and Michalski, R. S. 1989. Weeder: an advisory system for the identification of grasses in turf. *Agron. J.* 81(2):312–316.

Fermanian, T. W., Michalski, R. S., Katz, B. and Kelly, J. 1989. Agassistant: and artificial intelligence system for discovering patterns in agricultural knowledge and creating diagnostic advisory systems. *Agron. J.* 81(2):306–312.

Fleming, W. E. 1968. Biological control of the Japanese beetle. *U.S. Expt. Agricult. Tech. Bull.* 1383:1–78.

Forbes, S. A. 1880. Some interaction of organisms. III. *Nat. Hist. Surv. Bull.* 1:3–17.

Foster, G. R., Young, R. A. and Neibling, W. H. 1985. Sediment composition for nonpoint source pollution analysis. *Trans. ASAE* 28(1):133–139, 146.

Foster, G. R., Lane, L. J., Nowlin, J. D. Laflen, J. M. and Young, R. A. 1980. A model to estimate sediment yield from field-size areas: Development of model. in Knisel, W. G. (Ed.). CREAMS: A Field-Scale Model for Chemicals, Runoff, and Erosion from Agricultural Management Systems. USDA Conservation Research Report No. 26. pp. 36–64.

Fried, M., Tanji, K. K. and Van De Pol, R. M. 1976. Simplified long-term concept for evaluating leaching of nitrogen from agricultural land. *J. Environ. Qual.* 5(2):197–200.

Fry, J. D. and Dernoeden, P. H. 1986. Zoysiagrass competition in two cool-season turfgrasses treated with plant growth regulators. *HortScience* 21(3):464–466.

Funk, C. R. and Clarke, B. B. 1989. Turfgrass Breeding — with Special Reference to Turf-Type Perennial Ryegrass, Tall Fescue and Endophytes. in Sixth Int. Turf. Res. Conf., Tokyo, pp. 3–10.

Funk, C. R., Clarke, B. B. and Johnson-Cicalese, J. M. 1989. Role of endophytes in enhancing the performance of grasses used for conservation and turf. in Leslie, A. R. and Metcalf, R. L. (Eds.). Integrated Pest Management for Turfgrass and Ornamentals. Office of Pesticide Programs, 1989-625-030. U.S. Environmental Protection Agency, Washington, D.C., pp. 203–210.

Funk, C. R., Halisky, P. M., Ahmad, S. and Hurley, R. H. 1985. How Endophytes Modify Turfgrass Performance and Response to Insect Pests in Turfgrass Breeding and Evaluation Trials. in Lemaire, F. (Ed.). Proc. Fith Int. Turf. Res. Conf., Avignon, France, pp. 137–145.

Funk, C. R., Halisky, P. M., Johnson, M. C., Siegel, M. R., Stewart, A. V., Hurley, R. H. and Harvey, I. C. 1983. An endophytic fungus and resistance to sod webworms: Association in *Lolium perenne* L. *Bio/Technology* 1:189–191.

Gaylor, M. J. and Frankie, G. W. 1979. The relationship of rainfall to adult flight activity and of soil moisture to oviposition behavior and egg and first instar survival in *Phyllophaga crinita. Environ. Entomol.* 8:591–594.

Gaugler, R. 1981. Biological control potential of neoaplectanid nematodes. *J. Nematol.* 13(3):241–249.

Gaugler, R. and Boush, G. M. 1978. Effects of ultraviolet radiation and sunlight on the entomogenous nematode, *Neoaplectana carpocapsae. J. Invertebr. Pathol.* 32:291–296.

Gaussoin, R. E. and Branham, B. E. 1989. Influence of cultural factors on species dominance in a mixed stand of annual bluegrass/creeping bentgrass. *Crop Sci.* 29(2):480–484.

Georgis, R. and Poinar, G. O., Jr. 1983. Effect of soil texture on the distribution and infectivity of *Neoaplectana carpocapsae* (*Nematoda: Steinernematidae*). *J. Nematol.* 15(2):308–311.

Georgis, R. and Poinar, G. O., Jr. 1989. Field effectiveness of entomophilic nematodes Neoaplectana and Heterohabditis. in Leslie, A. R. and Metcalf, R. L. (Eds.). Integrated Pest Management for Turfgrass and Ornamentals. Office of Pesticide Programs. 1989-625-030. U.S. Environmental Protection Agency, Washington, D.C., pp. 213–224.

Georgis, R., Wojcik, W. F. amd Shetlar, D. J. 1989. Use of *Steinernema feltiae* for the control of black cutworms (*Agrotis ipsilon*) and tawny mole crickets (*Scapteriscus vicinus*). *Fla. Entomol.* 72:203–205.

Gibeault, V. A. and Cockerham, S. T. (Eds.) 1985. *Turfgrass Water Conservation.* University of California, Riverside, Division of Agriculture and Natural Resources, 155 pp.

Gilbertson, C. B., Norstadt, F. A., Mathers, A. C., Holt, R. F., Barnett, A. P., McCalla, T. M., Onstad, C. A. and Young, R. A. 1979. Animal Waste Utilization on Cropland and Pastureland: A Manual for Evaluating Agronomic and Environmental Effects. USDA Util. Rept. 6. EPA-600/2-79-059. Washington, D.C., 135 pp.

Glaser, R. W. and Farrell, C. C. 1935. Field experiments with the Japanese beetle and its nematode parasite. *J. N.Y. Entomol. Soc.* 43:345–371.

Glass, A. D. M. 1973. Influence of phenolic acids on ion uptake. I. Inhibition of phosphate uptake. *Plant Physol.* 51(6):1037–1041.

Goodman, D. M. and Burpee, L. L. 1987. Biological control of dollar spot disease of creeping bentgrass. in *1987 Annual Research Report.* The Guelph Turfgrass Institute, Guelph, Ontario, pp. 76–80.

Goodman, D. M. and Burpee, L. L. 1988. Isolate PBA7 of *Fusarium* as an antagonist of *Sclerotinia homoeocarpa.* in *1988 Annual Research Report.* The Guelph Turfgrass Institute, Guelph, Ontario, pp. 71–76.

Gordon, F. C. and Potter, D. A. 1985. Efficiency of Japanese beetle (*Coleoptera: Scarabaeidae*) traps in reducing defoliation of plants in the urban landscape and effect on larval density in turf. *J. Econ. Entomol.* 78(4):774–778.

Goss, R. L. 1974. Effects of Variable Rates of Sulfur on the Quality of Putting Green Bentgrass. in Roberts, E. C. (Ed.). Proc. 2nd Annual Int. Turfgrass Res. Conf., Blacksburg, VA, pp. 172–175.

Goss, D. 1991. Screening procedure for soils and pesticides relative to potential water quality impacts. in *Using Computer Simulation Models in Pesticide Registration Decision Making.* A Symposium/Workshop, Weed Sci. Soc. Am., Louisville, KY, (in press).

Graber, L. F., Fluke, C. L. and Dexter, S. T. 1931. Insect injury of bluegrass in relation to the environment. *Ecology* 12(3):547–566.

Grant, Z. 1989. Integrated pest management in the golf course industry: A case study and some general considerations. in Leslie, A. R. and Metcalf, R. L. (Eds.). Integrated Pest Management for Turfgrass and Ornamentals. Office of Pesticide Programs. 1989-625-030. U.S. Environmental Protection Agency, Washington, D.C., pp. 85–91.

Graves, J. B., Mohamad, R. B. and Clower, D. F. 1978. Beneficial insects also developing resistance. *La. Agric.* 22(1):10–11.

Gustafson, D. I. 1989. Groundwater ubiquity score: A simple method for assessing pesticide leachability. *Environ. Toxicol. Chem.* 8:339–357.

Haley, J. E. and Fermanina, T. W. 1989. Flurprimidol effect on the emergence and growth of annual bluegrass and creeping bentgrass. *Agron. J.* 81(2):198–202.

Hamilton, D. W., Schwartz, P. H., Townshend, B. G. and Jester, C. W. 1971. Traps reduce an isolated infestation of Japanese beetle. *J. Econ. Entomol.* 64(1):150–153.

Hanson, A. A. and Juska, F. V. (Eds.). 1969. *Turfgrass Science.* Agron. Monogr. 14. Am. Soc. Agron., Madison, WI, 715 pp.

Hanula, J. L. and Andreadis, T. G. 1988. Parasitic microorganisms of the Japanese beetle (*Coleoptera: Scarabaeidae*) and associated Scarabaeidae larvae in Connecticut soils. *Environ. Entomol.* 17:709–714.

Harder, P. R. and Troll, J. 1973. Antagonism of *Trichoderma* spp. to sclerotia of *Typhula incarnata. Plant Dis. Rep.* 57:924–926.

Harris, C. R. 1972. Factors influencing the effectiveness of insecticides in soil. *Ann. Rev. Entomol.* 17:177–198.

Harris, C. R. 1982. Factors influencing the toxicity of insecticides in soil. in Niemczyk, H. D. and Joyner, B. G. (Eds.). *Advances in Turfgrass Entomology*. Hammer Graphics, Inc., Piqua, OH, pp. 47–52.

Haygood, R. A. and Mazur, A. R. 1990. Evaluation of *Gliocladium virens* as a biocontrol agent of dollar spot on bermudagrass. *Phytopathol.* 80:435.

Heatwole, C. D., Bottcher, A. B. and Baldwin, L. B. 1986. Basin scale model for evaluating best management practice implementation programs. *Trans. ASAE* 29(2):439–444.

Hedden, K. F. 1986. Example field testing of soil fate and transport model, RRZM, Dougherty Plain, Georgia. in Hern, S. C. and Melancon, S. M. (Eds.). *Vadose Zone Modeling of Organic Pollutants*. Lewis Publishers, Chelsea, MI, pp. 81–101.

Helling, C. S. 1971. Pesticide mobility in soils: I. Parameters of thin-layer chromatography. II. Applications of soil thin-layer chromatography. *Soil Sci. Soc. Am. Proc.* 35:732–743.

Helling, C. S., Kearney, P. C. and Alexander, M. 1971. Behavior of pesticides in soils. *Adv. Agron.* 23:147–240.

Hill, A. S., Rings, R. W., Swier, S. R. and Roelofs, W. L. 1979. Sex pheromone of the black cutworm moth, *Agrotis ipsilos*. *J. Chem. Ecol.* 5:439–457.

Hoitink, H. A. J. and Fahy, P. C. 1986. Basis for control of soilborne plant pathogens with composts. *Ann. Rev. Phytopathol.* 24:93–114.

Holmes, J. J. and Davidson, J. A. 1984. Integrated pest management for arborist: Implementation of a pilot program. *J. Arbor.* 10(3):65–70.

Holt, E. C. 1969. Turfgrasses under warm, humid conditions. in Hanson, A. A. and Juska, F. V. (Eds.). *Turfgrass Science*. Agron. Monogr. 14:513–528. Am. Soc. Agron., Madison, WI.

Horsley, S. W. and Moser, J. A. 1990. Monitoring ground water for pesticides at a golf course — A case study on Cape Cod, Massachusetts. *Ground Water Monit. Rev.* 10(1):101–108.

Jackson, D. W., Vessels, K. J. and Potter, D. A. 1981. Resistance of selected cool and warm season turfgrasses to the greenbug (*Schizaphis graminum*). *HortScience* 16(4):558–559.

Jackson, T. A., Pearson, J. F. and Stucki, G. 1986. Control of grass grub, *Costelytra zealandica* (White) (*Coleoptera: Scarabaeidae*) by application of the bacteria *Serratia* spp. causing honey disease. *Bull. Entomol. Res.* 76:69–79.

Jarvis, J. L. 1966. Studies of *Phyllophaga anxia* (*Coleoptera: Scarabaeidae*) in the sandhills area of Nebraska. *J. Kans. Entomol. Soc.* 39:410–419.

Johanson, R. C., Imhoff, J. C., Davis, H. H., Kittle, J. L. and Donigian, A. S., Jr. 1981. User's Manual for the Hydrologic Simulation Program — Fortran (HSPF), Release 7.0. U.S. Environmental Protection Agency, Environmental Research Laboratory, Athens, GA.

Johnson, M. C., Bush, L. P. and Siegel, M. R. 1986. Infection of tall fescue with *Acremonium coenophialum* by means of callus culture. *Plant Dis.* 70:380–382.

Johnson, M. C., Dahlman, D. L., Siegel, M. R., Bush, L. P., Latch, G. C. M., Potter, D. A. and Varney, D. R. 1985. Insect feeding deterents in endophyte-infected tall fescue. *Appl. Environ. Microbiol.* 49(3):568–571.

Johnson-Cicalese, J. M., Hurley, R. H., Wolf, G. W. and Funk, C. R. 1989. Developing Turfgrasses with Improved Resistance to Billbugs. in Sixth Int. Turf. Res. Conf., Tokyo, pp. 107–111.

Jones, K. C., Keating, T., Diage, P. and Chang, A. C. 1991. Transport and food chain modeling and its role in assessing human exposure to organic chemicals. *J. Environ. Qual.* 20(2):317–329.

Jones, R. L. and Back, R. C. 1984. Monitoring aldicarb residues in Florida soil and water. *Environ. Toxicol. Chem.* 3(1):9–20.

Jones, R. L., Hornsby, A. G., Rao, P. S. C. and Anderson, M. P. 1987. Movement and degradation of aldicarb residues in the saturated zone under citrus groves on the Florida ridge. *J. Contam. Hydrol.* 1:265–285.

Jury, W. A. 1982. Simulation of solute transport with a transfer function model. *Water Resour. Res.* 18(2):363–368.

Jury, W. A. 1983. Chemical transport modeling: Current approaches and unresolved problems. in Nelsen, D. W., Elrick, D. E. and Tanji, K. K. (Eds.). *Chemical Mobility and Reactivity in Soil Systems.* Special Publ. No. 11. Soil Sci. Soc. Am., Madison, WI, pp. 49–64.

Jury, W. A. 1986a. Chemical movement through soil. in Hern, S. C. and Melancon, S. M. (Eds.). *Vadose Zone Modeling of Organic Pollutants.* Lewis Publishers, Chelsea, MI, pp. 271–288.

Jury, W. A. 1986b. Spatial variability of soil properties. in Hern, S. C. and Melancon, S. M. (Eds.). *Vadose Zone Modeling of Organic Pollutants.* Lewis Publishers, Chelsea, MI, pp. 245–269.

Jury, W. A. and Ghodrati, M. 1989. Overview of organic chemical environmental fate amd transport modeling approaches. in Sawhney, B. L. and Brown, K. (Eds.). *Reactions and Movement of Organic Chemicals in Soils.* SSSA Special Publication No. 22. Soil Sci. Soc. Am., Am. Soc. Agron., Madison, WI, pp. 271–304.

Jury, W. A. and Gruber, J. 1989. A stochastic analysis of the influence of soil and climatic variability on the estimate of pesticide groundwater pollution potential. *Water Resour. Res.* 25(12):2465–2474.

Jury, W. A. and Valentine, R. L. 1986. Transport mechanisms and loss pathways for chemicals in soil. in Hern, S. C. and Melancon, S. M. (Eds.). *Vadose Zone Modeling of Organic Pollutants.* Lewis Publishers, Chelsea, MI, pp. 37–60.

Jury, W. A., Focht, D. D. and Farmer, W. J. 1987. Evaluation of pesticide groundwater pollution potential from standard indicies of soil-chemical adsorption and biodegradation. *J. Environ. Qual.* 16(4):422–428.

Jury, W. A., Spencer, W. F. and Farmer, W. J. 1983. Behavior assessment model for trace organics in soil. I. Model description. *J. Environ. Qual.* 12(4):558–564.

Jury, W. A., Spencer, W. F. and Farmer, W. J. 1984. Behavior assessment model for trace organics in soil. II. Chemical classification and parameter sensitivity. *J. Environ. Qual.* 13(4):567–572.

Juska, F. V., Cornman, J. F. and Hovin, A. W. 1969. Turfgrasses under cool, humid conditions. in Hanson, A. A. and Juska, F. V. (Eds.). *Turfgrass Science.* Agron. Monogr. 14:513–528. Am. Soc. Agron., Madison, WI.

Kageyama, M. E. and Widell, L. R. 1989. Annual bluegrass to bentgrass conversion with turf growth retardents. in Leslie, A. R. and Metcalf, R. L. (Eds.). Integrated Pest Management for Turfgrass and Ornamentals. Office of Pesticide Programs. 1989–625–030. U.S. Environmental Protection Agency, Washington, D.C., pp. 307–311.

Kamm, J. A. 1982. Use of insect sex pheromones in turfgrass management. in Niemczyk, H. D. and Joyner, B. G. (Eds.). *Advances in Turfgrass Entomology.* Hammer Graphics, Piqua, OH, pp. 39–41.

Kamprath, E. J., Broome, S. W., Raja, M. E., Tonapa, S., Baird, J. V. and Rice, J. C. 1973. Nitrogen Management, Plant Populations and Row Width Studies with Corn. North Carolina Agricult. Exp. Stn. Tech. Bull. No. 217. Raleigh, NC, 16 pp.

Kane, R. T. and Smiley, R. W. 1983. Plant growth-regulating effects of systemic fungicides applied to Kentucky blue grass. *Agron. J.* 75:469–473.

Kaya, H. K. 1985. Entomogenous nematodes for insect control in IPM systems. in Hoy, M. A. and Herzog, D. C. (Eds.). *Biological Control in Agricultural IPM Systems*. Academic Press, New York, pp. 283–302.

Kaya, H. K. and Nelsen, C. E. 1985. Encapsulation of Steinernematid and Heterorhabditid nematodes with calcium alginate: A new approach for insect control and other applications. *Environ. Entomol.* 14(5):572–574.

Keen, R. A. 1969. Turfgrasses under semi-arid and arid conditions. in Hanson, A. A. and Juska, F. V. (Eds.). *Turfgrass Science*. Agron. Monogr. 14:513–528. Am. Soc. Agron., Madison, WI.

Keeney, D. R. 1982. Nitrogen management for maximum efficiency and minimum pollution. in Stevenson, F. J. (Ed.). *Nitrogen in Agricultural Soils*. Agron. Monogr. 22:605–649. Am Soc. of Agron., Crop Sci. Soc. Am., Soil Sci. Soc. Am., Madison, WI.

Keeney, D. R. 1986. Sources of nitrate to groundwater. *CRC Crit. Rev. in Environ. Control* 16:257–304.

Kimber, R. W. L. 1973. Phytotoxicity from plant residues. III. The relative effect of toxins and nitrogen immobilization on the germination and growth of wheat. *Plant Soil* 38:543–555.

King, J. 1971. Competition between established and newly sown grass species. *J. Br. Grassl. Soc.* 26:221–229.

Kladivko, E. J., Van Scoyoc, G. E., Monke, E. J., Oates, K. M. and Pask, W. 1991. Pesticide and nutrient movement into subsurface tile drains on a silt loam soil in Indiana. *J. Environ. Qual.* 20(1):264–270.

Klausner, S. D., Zwerman, P. J. and Ellis, D. F. 1974. Surface runoff losses of soluble nitrogen and phosphorus under two systems of soil management. *J. Environ. Qual.* 3(1):42–46.

Klein, M. G. 1981a. Advances in the use of *Bacillus popilliae* for pest control. in Burges, H. D. (Ed.). *Microbial Control of Pests and Plant Diseases 1970–1980*. Academic Press, London, pp. 183–192.

Klein, M. G. 1981b. Mass trapping for supression of Japanese beetles. in Mitchell, E. R. (Ed.). *Management of Insect Pests with Semiochemicals*. Plenum Publishers, New York, pp. 183–190.

Klein, M. G. 1982. Biological suppression of turf insects. in Niemczyk, H. D. and Joyner, B. G. (Eds.). *Advances in Turfgrass Entomology*. Hammer Graphics, Inc., Piqua, OH, pp. 91–97.

Klein, M. G. 1988. Pest management of soil-inhabiting insects with microorganisms. *Agric. Ecosys. Environ.* 24:337–349.

Klein, M. 1989. Suppression of white grubs with microorganisms and attractants. in Leslie, A. R. and Metcalf, R. L. (Eds.). Integrated Pest Management for Turfgrass and Ornamentals. Office of Pesticide Programs, 1989-625-030. U.S. Environmental Protection Agency, Washington, D.C., pp. 297–305.

Klein, M. G. 1990. Efficacy against soil-inhabiting insect pests. in Gaulger, R. and Kaya, H. K. (Eds.). *Entomopathogenic Nematodes in Biological Control*. CRC Press, Boca Raton, FL.

Klein, M. G., Johnson, C. H. and Ladd, T. L., Jr. 1976. A bibliography of the milky disease bacteria (*Bacillus* spp.) associated with the Japanese beetle, *Popillia japonica* and closely related Scarabaeidae. *Bull. Entomol.* 22(3):305–310.

Klein, R. D. 1990. *Protecting the Aquatic Environment from the Effects of Golf Courses*. Community & Environmental Defense Associates, Maryland Line, MD, 59 pp.

Knighton, R. E. and Wagenet, R. J. 1987a. Simulation of solute transport using a continuous time Markov process. I. Theory and steady-state application. *Water Resour. Res.* 23:1911–1916.

Knighton, R. E. and Wagenet, R. J. 1987b. Simulation of solute transport using a continuous time Markov process. II. Application to transient field conditions. *Water Resour. Res.* 23:1917–1925.

Knisel, W. G. (Ed.). 1980. CREAMS: A Field Scale Model for Chemicals, Runoff and Erosion from Agricultural Management Systems. USDA Conservation Research Report No. 26. Washington, D.C., 640 pp.

Knisel, W. G., Foster, G. R. and Leonard, R. A. 1983. CREAMS: A system for evaluating management practice. in Schaller, F. W. and Bailey, G. W. (Eds.). *Agricultural Management and Water Quality*. Iowa University Press, Ames, IA, pp. 178–199.

Koehler, F. A., Humenik, F. J., Johnson, D. D., Kreglow, J. M., Dressing, R. P. and Màas, R. P. 1982. Best Management Practices for Agricultural Nonpoint Source Control. II. Commercial Fertilizer. USDA Coop. Agree. 12-05-300-472. EPA Interagency Agree. AD-12-F-0-037-0. North Carolina Agricult. Ext. Serv., Raleigh, NC, 55 pp.

Ladd, T. L., Jr. and Buriff, C. R. 1979. Japanese beetle: Influence of larval feeding on bluegrass yields at two levels of soil moisture. *J. Econ. Entomol.* 72(3):311–314.

Ladd, T. L. and Klein, M. G. 1982a. Controlling the Japanese beetle. Home Gard. Bull. No. 159. U.S. Dept. Ag. Govt. Printing Off., 14 pp.

Ladd, T. L. and Klein, M. G. 1982b. Trapping Japanese beetles with synthetic female sex pheromone and food-type lures. in Kydonieus, A. F. and Beroza, M. (Eds.). *Insect Suppression with Controlled Release Pheromone Systems*. CRC Press, Boca Raton, FL.

Ladd, T. L., Klein, M. G. and Tumlinson, J. H. 1981. Phenethyl propionate + eugenol + geramol-(3:7:3) and Japonilure: a highly effective joint lure for Japanese beetles. *J. Econ. Entomol.* 74(6):665–667.

Lane, L. J. and Nearing, M. A. (Eds.). 1989. USDA-Water Erosion Prediction Project: Hillslope Profile Model Documentation. NSERL Report No. 2. USDA-ARS Nat. Soil Erosion Res. Lab., West Lafayette, IN.

Langford, G. S., Crosthwait, S. L. and Whittington, F. B. 1940. The value of traps in Japanese beetle control. *J. Econ. Entomol.* 33(2):318–320.

Latch, G. C. M. and Christensen, M. J. 1985. Artificial infection of grasses with endophytes. *Ann. Appl. Biol.* 107:17–24.

Latch, G. C. M., Hunt, W. F. and Musgrave, D. R. 1985. Endophytic fungi affect growth of perennial ryegrass. *N. Zeal. J. Agric. Res.* 28:165–168.

Leistra, M. and Green, R. E. 1990. Efficacy of soil-applied pesticides. in Cheng, H. H. (Ed.). *Pesticides in the Soil Environment: Processes, Impacts, and Modeling*. SSSA Book Series No. 2. Soil Science Society of America, Madison, WI, pp. 401–428.

Leonard, R. A. 1988. Herbicides in surface waters. in Grover, R. (Ed.). *Environmental Chemistry of Herbicides*. Vol. 1. CRC Press, Boca Raton, FL, pp. 45–87.

Leonard, R. A. 1990. Movement of pesticides into surface waters. in Cheng, H. H. (Ed.). *Pesticides in the Soil Environment: Processes, Impacts, and Modeling*. SSSA Book Series No. 2. Soil Science Society of America, Madison, WI, pp. 303–349.

Leonard, R. A. and Knisel, W. G. 1988. Evaluating groundwater contamination potential from herbicide use. *Weed. Technol.* 2(2):207–216.

Leonard, R. A., Knisel, W. G. and Still, D. A. 1987. GLEAMS: Groundwater loading effects of agricultural management systems. *Trans. ASAE* 30(5):1403–1418.

Leslie, A. R. 1989a. Societal problems associated with pesticide use in the urban sector. in Leslie, A. R. and Metcalf, R. L. (Eds.). Integrated Pest Management for Turfgrass and Ornamentals. Office of Pesticide Programs. 1989-625-030. U.S. Environmental Protection Agency, Washington, D.C., pp. 93–96.

Leslie, A. R. 1989b. Development of an IPM program for turfgrass. in Leslie, A. R. and Metcalf, R. L. (Eds.). Integrated Pest Management for Turfgrass and Ornamentals. Office of Pesticide Programs. 1989-625-030. U.S. Environmental Protection Agency, Washington, D.C., pp. 315–318.

Leslie, A. R. 1989c. Pesticide management for local government. *ICMA MIS Report* 21(8):1–14.

Leslie, A. R. and Metcalf, R. L. (Eds.). 1989. Integrated Pest Management for Turfgrass and Ornamentals. Office of Pesticide Programs.1989-625-030. U.S. Environmental Protection Agency, Washington, D.C., 337 pp.

Leuck, D. B., Taliaferro, C. M., Burton, G. W., Burton, R. L. and Bowman, M. C. 1968. Resistance in bermudagrass to the fall armyworm. *J. Econ. Entomol.* 61(5):1321–1322.

Lincoln, C. and Blair, B. D. 1977. Extension entomology: A critique. *Ann. Rev. Entomol.* 22:139–155.

Lorber, M. N. and Mulkey, L. E. 1982. An evaluation of three pesticide runoff loading models. *J. Environ. Qual.* 11(3):519–529.

Mackay, D. and Stiver, W. 1991. Predictability and environmental chemistry. in Grover, R. and Cessna, A. J. (Eds.). *Environmental Chemistry of Herbicides.* Vol 2. CRC Press, Boca Raton, FL, pp. 281–297.

Madison, J. H. 1982. *Principles of Turfgrass Culture.* Robert E. Krieger Publishing Co., Melbourne, FL, 431 pp.

Martin, D. L. and Watts, D. G. 1982. Potential purification of high nitrate groundwater through irrigation management. *Trans. ASAE* 25(6):1662–1667.

Mathias, E. L., Bennet, O. L., Jung, G. A. and Lundberg, P. E. 1971. Effect of two growth-regulating chemicals on yield and water use of three perennial grasses. *Agron. J.* 63:480–483.

Mazur, A. R. 1988. Influence of plant growth regulators on transition of bermudagrass putting green overseeded with perennial ryegrass. *J. Am. Soc. Hort. Sci.* 113(3):367–373.

Mehran, M., Tanji, K. K. and Iskandar, I. K. 1981. Compartmental modeling for prediction of nitrate leaching. in Iskandar, I. K. (Ed.). *Modeling Wastewater Renovation, Land Treatment.* John Wiley & Sons, New York, Chapter 16.

Melancon, S. M., Pollard, J. E. and Hern, S. C. 1986. Evaluation of SESOIL, PRZM and PESTAN in a column leachning experiment. *Environ. Toxicol. Chem.* 5:865–878.

Messer, J. and Brezonik, P. L. 1983. Agricultural nitrogen model: A tool for regional environmental management. *Environ. Manage.* 7:177–187.

Metcalf, R. L. 1980. Changing role of insecticides in crop protection. *Ann. Rev. Entomol.* 25:219–256.

Metcalf, R. L. 1989. Insect resistance to insecticides. in Leslie, A. R. and Metcalf, R. L. (Eds.). Integrated Pest Management for Turfgrass and Ornamentals. Office of Pesticide Programs. 1989-625-030. U.S. Environmental Protection Agency, Washington, D.C., pp. 3–44.

Meyer, W. A. 1982. Breeding disease-resistant cool-season turfgrass cultivars for the United States. *Plant Dis.* 66(4):341–344.

Meyer, W. A. and Funk, C. R. 1989. Progress and benefits to humanity from breeding cool-season grasses for turf. in Sleper, D. A., Asay, K. H. and Pedersen, J. F. (Eds.). *Contributions from Breeding Forage and Turf Grasses.* CSSA Spec. Publ. No. 15. Crop Sci. Soc. Am., Madison, WI, pp. 31–48.

Miller, S. A. 1988. Biotechnology-based disease diagnostics. *Plant Dis.* 72:188.

Miller, S. A. and Martin, R. R. 1988. Molecular diagnosis of plant disease. *Ann Rev. Phytopathol.* 26:409–432.

Miller, S. A., Grothaus, G. D., Peterson, F. P., Rittenburg, J. H., Plumley, K. A. and Lankow, R. K. 1989. Detection and monitoring of turfgrass pathogens by immunoassay. in Leslie, A. R. and Metcalf, R. L. (Eds.). Integrated Pest Management for Turfgrass and Ornamentals. Office of Pesticide Programs. 1989-625-030. U.S. Environmental Protection Agency, Washington, D.C., pp. 109–120.

Milthrope, F. L. 1961. The nature and analysis of competition between plants of different species. in *Mechanisms in Biological Competition*. Symp. Soc. Exp. Biol. No. 15. Academic Press, New York, pp. 330–355.

Minner, D. D. and Butler, J. D. 1985. Drought Tolerance of Cool Season Turfgrasses. in Lemaire, F. L. (Ed.). Proc. Fifth Int. Turf. Res. Conf., Angers, France, pp. 199–212.

Minner, D. D., Demoeden, P. H., Wehner, D. J. and McIntosh, M. S. 1983. Heat tolerance screening of field-grown cultivars of Kentucky bluegrass and perennial ryegrass. *Agron. J.* 75(5):772–775.

Molyneux, A. S. 1984. The Influence of Temperature on the Infectivity of Hererorhabditid and Steinernematid Nematodes for Larvae of the Sheep Blowfly, *Lucilia cuprina*. Proc. 4th Austral. Appl. Entomol. Res. Conf., pp. 344–351.

Molyneux, A. S. and Bedding, R. A. 1984. Influence of soil texture and moisture on the infectivity of *Heterorhabditis* sp. D1 and *Steinernema glaseri* for larvae of the sheep blowfly. *Nematologica* 30:358–365.

Moore, I. C. and Madsion, F. W. 1985. Description and application of an animal waste phosphorus loading model. *J. Environ. Qual.* 14(3):364–369.

Morris, O. N. 1987. Evaluation of the nematode, *Steinernema feltiae* Filipjev, for the control of the flea beetle, *Phyllotreta curciferae* (Goeze) (*Coleoptera Chrysomelidae*). *Can. Entomol.* 119:95–102.

Murray, J. J. and Powell, J. B. 1979. Turf. in Buckner, R. C. and Bush, L. P. (Eds.). *Tall Fescue*. Agron. Monogr. 20:293–306 Am. Soc. Agron., Crop Sci. Soc. Am., Soil Sci. Soc. Am., Madison, WI.

National Research Council. 1978. *Nitrates: An Environmental Assessment*. National Academy of Sciences, Washington, D.C.

Nelson, E. B. 1988. Biological control of Pythium Blight and Brown Patch of creeping bentgrass. *1988 Cornell Res. Rep.* 2:44–47.

Nelson, E. B. and Craft, C. M. 1989a. Evaluation of bacterial antagonists for the biological control of Pythium Blight of turfgrass. *1989 Cornell Res. Rep.* 3:73–77.

Nelson, E. B. and Craft, C. M. 1989b. Biological control of Dollar Spot on a creeping bentgrass/annual bluegrass putting green. *1989 Cornell Res. Rep.* 3:68–72.

Nelson, E. B. and Craft, C. M. 1990a. Use of disease-suppressive top-dressings for the control of Dollar Spot (*Sclerotinia homoeocarpa*) on a creeping bentgrass putting green. *Phytopathol.* 80:122.

Nelson, E. B. and Craft, C. M. 1990b. Application of top-dressings amended with composts and organic fertilizers for the suppression of Brown Patch on a creeping bentgrass putting green. *Phytopathol.* 80:122.

Nelson, E. B. and Craft, C. M. 1991. Introduction and establishment of strains of *Enterobacter cloacae* in golf course turf for the biological control of dollar spot. *Plant Dis.* (in press).

Newman, E. I. and Miller, M. H. 1977. Allelopathy among some British grassland species. II. Influence of root exudates on phosphorus uptake. *J. Ecol.* 65(2):399–411.

Newman, E. I. and Rovira, A. D. 1975. Allelopathy among some British grassland species. *J. Ecol.* 63(3):727–737.

Newton, P. J., Johnson-Cicalese, J. M. and Halinsky, P. M. 1987. The good & bad of endophytes. *Greenworld* 17(2):1–6.

Nielsen, A. P. and Wakefield, R. C. 1975. Effects of growth retardants on the top growth of turfgrasses. *Proc. Northeast. Weed Sci. Soc.* 29:403–408.

Nielsen, A. P. and Wakefield, R. C. 1978. Competitive effects of turfgrass on the growth of ornamental shrubs. *Agron J.* 70(1):39–42.

Nielsen, D. R. and Bigger, J. W. 1962. Miscible displacement. III. Theoretical considerations. *Soil Sci. Soc. Am. Proc.* 26:216–221.

Nielsen, D. R., Wierenga, P. J. and Bigger, J. W. 1983. Spatial soil variability and mass transfers from agricultural soils. in Nelsen, D. W., Tanji, K. K. and Elrick, D. E. (Eds.). *Chemical Mobility and Reactivity in Soil Systems.* Special Publ. No. 11. Soil Sci Soc. Am., Madison, WI, pp. 65–78.

Nilsen, K. N. and Hodges, C. F. 1980. Photomorphogenically defined light and resistance of *Poa pratensis* to *Dreshslera sorokiniana. Plant Physiol.* 65(5):569–573.

Nilsen, K. N., Hodges, C. F. and Madsen, J. P. 1979a. Enhanced *Drechslera sorokiniana* leaf spot expression on *Poa pratensis* in response to photoperiod and blue-biased light. *Physiol. Plant Pathol.* 14:57–69.

Nilsen, K. N., Hodges, C. F. and Madsen, J. P. 1979b. Pathogenesis of *Drechslera sorokiniana* leaf spot on progressively older leaves of *Poa pratensis* as influenced by photoperiod and light quality. *Physiol. Plant Pathol.* 15:171–176.

Nofziger, D. L. and Hornsby, A. G. 1984. *Chemical Movement in Soil: User's Guide.* University of Florida, Gainesville, FL.

Nofziger, D. L. and Hornsby, A. G. 1986. A microcomputer-based management tool for chemical movement in soil. *Appl. Agric. Res.* 1:50–56.

Norton, G. A. 1982. Crop protection decision making — An overview. in Austin, R. B. (Ed.). *Decision Making in the Practice of Crop Protection.* Monogr. No. 25. Br. Crop Prot. Counc., London, pp. 3–11.

Nus, J. L. and Hodges, C. F. 1985. Effect of water stress and infection by *Ustilago striiformis* or *Urocystis agropyri* on leaf turgor and water potentials of Kentucky blue grass. *Crop Sci.* 25(2):322–326.

Nyrop, J. P., Huber, B. and Wolf, W. 1989. Knowledge based systems for use in integrated pest management: Requirements, pitfalls, and opportunities. in Leslie, A. R. and Metcalf, R. L. (Eds.). Integrated Pest Management for Turfgrass and Ornamentals. Office of Pesticide Programs. 1989-625-030. U.S. Environmental Protection Agency, Washington, D.C., pp. 319–330.

O'Leary, A. L., O'Leary, D. J. and Woodhead, S. H. 1988. Screening potential bioantagonists against turf pathogens. *Phytopathol.* 78:1593.

Oster, C. A. 1982. Review of Ground-Water Flow and Transport Models in the Unsaturated Zone. PNL-4427, NUREG/CR-2917. Battelle, Pacif. Northwest Lab., Richland, WA.

Overland, L.. 1966. The role of allelopathic substances in the "smother crop" barley. *Am. J. Bot.* 53:423–432.

Pacenka, S. and Steenhuis, T. 1984. *User's Guide for the MOUSE Computer Program.* Cornell University, Ithaca, NY.

Persson, L. A., Peterson, J. O. and Madison, F. W. 1983. Evaluation of sediment and phosphorus management practices in the White Clay Lake Watershed. *AWRA Water Resour. Bull.* 19:753–762.

Peters, E. J. 1968. Toxicity of tall fescue to rape and birdsfoot trefoil seeds and seedlings. *Crop Sci.* 8:650–653.

Petraitis, J. 1981. Controlling the Japanese beetle without spraying. *Golf Course Manage.* 49(6):48.

Petrovic, A. M., Hummel, N. W. and Carroll, M. J. 1986. Nitrogen source effects on nitrate leaching from late fall nitrogen applied to turfgrass. *Agron. Abstr.* p. 137.

Poe, S. L. 1981. An overview of integrated pest managment. *HortScience.* 16:501–506.

Poinar, G. O., Jr. 1979. *Nematodes for Biological Control of Insects.* CRC Press, Boca Raton, FL, 277 pp.

Poinar, G. O., Jr. and Georgis, R. 1989. Biological control of social insects with nematodes. in Leslie, A. R. and Metcalf, R. L. (Eds.). Integrated Pest Management for Turfgrass and Ornamentals. Office of Pesticide Programs. 1989-625-030. U.S. Environmental Protection Agency, Washington, D.C., pp. 255–269.

Polinka, J. B. 1960a. Effect of lime applications to soil on Japanese beetle larval populations. *J. Econ. Entomol.* 53:476–477.

Polinka, J. B. 1960b. Grub populations in turf varies with pH levels in Ohio soils. *J. Econ. Entomol.* 53:860–863.

Potter, D. A. 1980. Flight activity and sex attraction of northern and southern masked chafers in Kentucky turfgrass. *Ann. Entomol. Soc. Am.* 73(4):414–417.

Potter, D. A. 1981. Seasonal emergence and flight of northern and southern masked chafers in relation to air and soil temperature and rainfall patterns. *Environ. Entomol.* 10(5):793–797.

Potter, D. A. 1982. Influence of feeding by grubs of the southern masked chafer on quality and yield of Kentucky bluegrass. *J. Econ. Entomol.* 75(1):21–24.

Potter, D. A. 1983. Effect of soil moisture on oviposition, water absorption, and survival of southern masked chafer (*Coleoptera: Scarabaeidae*) eggs. *Environ. Entomol.* 12(4):1223–1227.

Potter, D. A. 1986. Urban landscape pest management. in Bennet, G. W. and Owens, J. M. (Eds.). *Advances in Urban Pest Management.* Van Nostrand Reinhold, New York, pp. 219–252.

Potter, D. A. 1989. Flight activity and sex attraction of northern and southern masked chafers in Kentucky turfgrass. *Ann. Entomol. Soc. Am.* 73:414–417.

Potter, D. A. and Braman, S. K. 1991. Ecology and management of turfgrass insects. *Annu. Rev. Entomol.* 36:383–406.

Potter, D. A. and Gordon, F. C. 1984. Susceptibility of *Cyclocephala immaculata* (*Coleopter: Scarabaeidae*) eggs and immatures to heat and drought in turf grass. *Environ. Entomol.* 13(3):794–799.

Potter, D. A., Cockfield, S. D. and Morris, T. A. 1989. Ecological side effects of pesticide and fertilizer use on turfgrass. in Leslie, A. R. and Metcalf, R. L. (Eds.). Integrated Pest Management for Turfgrass and Ornamentals. Office of Pesticide Programs. 1989-625-030. U.S. Environmental Protection Agency, Washington, D.C., pp. 33–44.

Pratt, P. F. (Chairman). 1985. Agriculture and Groundwater Quality. Council for Agricultural Science and Technology. CAST Report No. 103. 62 pp.

Prestidge, R. A., Pottinger, R. P. and Barker, G. M. 1982. An Association of Lolium Endophyte with Rye Grass Resistance to Argentine Stem Weevil. in Proc. 35th N.Z. Weed and Pest Control Conf., pp. 119–122.

Rao, P. S. C. and Davidson, J. M. 1980. Estimation of pesticide retention and transformation parameters required in nonpoint source pollution models. in Overcash, M. R. and Davidson, J. M. (Eds.). *Environmental Impact of Nonpoint Source Pollution.* Ann Arbor Science Publishing Inc., Ann Arbor, MI, pp. 23–67.

Rao, P. S. C. and Wagenet, R. J. 1985. Spatial variablity of pesticides in field soils: Methods for data analysis and consequences. *Weed Sci.* 33(Suppl. 2):18–24.

Rao, P. S. C., Davidson, J. M., Jessup, R. E. and Reddy, K. R. 1984. Nitrogen Behavior in Croplands Receiving Organic Wastes: Development and Verification of a Simulation Model (NITROSIM). Inst. Food Agricult. Sci. Monogr. University of Florida, Gainesville, FL.

Rao, P. S. C., Jessup, R. E. and Davidson, J. M. 1988. Mass flow and dispersion. in Grover, R. (Ed.). *Environmental Chemistry of Herbicides.* Vol. I. CRC Press, Inc., Boca Raton, FL, pp. 21–43.

Rao, P. S. C., Jessup, R. E. and Hornsby, A. G. 1982. Simulation of nitrogen in agro-ecosystems: Criteria for model selection and use. *Plant Soil* 67(1):35–43.

Raupp, M. J. 1985. Monitoring: An essential factor to managing pests of landscape trees and shrubs. *J. Arbor.* 11(12):349–355.

Raupp, M. J. and Noland, R. M. 1984. Implementing landscape plant management programs in institutional and residential settings. *J. Arbor.* 10(6):161–169.

Raupp, M. J., Smith, M. F. and Davidson, J. A. 1989. Educational, environmental and economic impacts of integrated pest management programs for landscape plants. in Leslie, A. R. and Metcalf, R. L. (Eds.). Integrated Pest Management for Turfgrass and Ornamentals. Office of Pesticide Programs. 1989-625-030. U.S. Environmental Protection Agency, Washington, D.C., pp. 77–83.

Read, J. C. and Camp. B. J. 1986. The effect of the fungal endophyte *Acremonium coenophialum* in tall fescue on animal performance, toxicity, and stand maintenance. *Agron. J.* 78:848–850.

Reardon, R., McManus, M., Kolodny-Hirsh, D. Tichenor, R., Raupp, M., Schwalbe, C., Webb, R. and Meckley, P. 1987. Development and implementation of a gypsy moth integrated pest management program. *J. Arbor.* 13(9):209–216.

Reed, D. K., Reed, G. L. and Creighton, C. S. 1986. Introduction of entomogenous nematodes into trickle irrigation systems to control striped cucumber beetle, *Acalymma vittatum* (*Coleoptera: Chrysomelidae*). *J. Econ. Entomol.* 79:1330–1333.

Regniere, J., Rabb, R. L. and Stinner, R. E. 1981. *Popillia japonica:* Effect of soil moisture and texture on survival and development of eggs and first instar grubs. *Environ. Entomol.* 10(5):654–660.

Reinert, J. A. 1972. New distribution and host record for the parasitoid *Eumicrosoma benefica. Flor. Entomol.* 55(3):143–144.

Reinert, J. A. 1974. Tropical sod webworm and southern chinch bug control in Florida. *Flor. Entomol.* 57(3):275–279.

Reinert, J. A. 1976. Control of sod webworms (*Herpetogramma* spp. and *Crambus* spp.) on bermudagrass. *J. Econ. Entomol.* 69(5):669–672.

Reinert, J. A. 1978. Natural enemy complex of the southern chinch bug in Florida. *Ann. Entomol. Soc. Am.* 71(5):728–731.

Reinert, J. A. 1979. Response of white grubs infesting bermudagrass to insecticides. *J. Econ. Entomol.* 72(4):546–548.

Reinert, J. A. 1982. A review of host resistance to insects and acarines with emphasis on the southern chinch bug. in Niemczyk, H. D. and Joyner, B. G. (Eds.). *Advances in Turfgrass Entomology*. Hammer Graphics, Piqua, OH, pp. 3–12.

Reinert, J. A. and Busey, P. 1983. Resistance of bermudagrass selection to the tropical sod webworm (*Lepidoptera: Pyralidae*). *Environ. Entomol.* 12(6):1844–1845.

Reinert, J. A. and Busey, P. 1984. Resistant varieties. in Walker, T. J. (Ed.). *Mole Crickets in Florida*. Fl. Agricult. Exp. Bull. No. 846. University of Florida, Gainesville, FL, pp. 35–40.

Reinert, J. A. and Dudeck, A. E. 1974. Southern chinch bug resistance in St. Augustine-grass. *J. Econ. Entomol.* 67(2):275–277.

Reinert, J. A. and Short, D. E. 1980. Southern Turfgrass Insect Pests with Emphasis on Mole Cricket Biology and Management. in Proc. 28th Ann. Flor. Turf. Manage. Conf., pp. 33–43.

Reinert, J. A., Busey, P. and Bilz, F. G. 1986. Old world St, Augustinegrasses resistant to the southern chinch bug (*Heteroptera: Lygaeidae*). *J. Econ. Entomol.* 79(4):1073–1075.

Reinert, J. A., Dudbeck, A. E. and Snyder, G. H. 1978. Resistance in bermudagrass to the bermudagrass mite. *Environ. Entomol.* 7(6):885–888.

Reinert, J. A., Toler, R. W., Burton, B. D. and Busey, P. 1981. Retention of resistance by mutants of "Floratam" St. Augustinegrass to the southern chinch bug and St. Augustine decline. *Crop Sci.* 21:464–466.

Rhodes, I. 1968. The growth and development of some grass species under competitive stress. I. Competition between seedlings, and between seedlings and established plants. *J. Br. Grassl. Soc.* 23:129–136.

Rice, E. L. 1974. *Allelopathy.* Academic Press, New York.

Rice, E. L. 1976. Allelopathy and grassland improvement. in Estes, J. R. and Tyrl, R. J. (Ed.). *The Grasses and Grasslands of Oklahoma.* Noble Foundation, Ardmore, OK, pp. 90–111.

Rice, E. L. 1979. Allelopathy — An update. *Bot. Rev.* 45(1):15–109.

Robson, T. O. 1977. Perspectives of biological control of aquatic weeds in temperate climatic zones. *Aquat. Bot.* 3:125–132.

Robbins, C. W., Wagenet, R. J. and Jurinak, J. J. 1980. A combined salt transport-chemcial equilibrium model for calcareous and gypsiferous soils. *Soil Sci. Soc. Am. J.* 44:1191–1194.

Rogers, J. N., III, Miller, E. M. and King, J. W. 1987. Growth retardation of bermudagrass with metsulfuron methyl and sulfometuron methyl. *Agron. J.* 79(2):225–229.

Rolston, D. E., Rao, P. S. C., Davidson, J. M. and Jessup, R. E. 1984. Simulation of denitrification losses of nitrate fertilizer applied to uncropped, cropped, and manure-amended field plots. *Soil Sci.* 137(4):270–279.

Rose, C. W., Chichester, F. W., Williams, J. R. and Ritchie, J. T. 1982a. A contribution to simplified models of field solute transport. *J. Environ. Qual.* 11(1):146–150.

Rose, C. W., Chichester, F. W., Williams, J. R. and Ritchie, J. T. 1982b. Application of an approximate analytic method of computing solute profiles with dispersion in soils. *J. Environ. Qual.* 11(1):151–155.

Rose-Fricker, C. A., Meyer, W. A. and Kronstad, W. E. 1986. Inheritance of resistance to stem rust (*Puccinia graminis* subsp. *graminicola*) in six perennial ryegrass (*Lolium perenne*) crosses. *Plant Dis.* 70(7):678–681.

Russo, D. and Bresler, E. 1980. Scaling soil hydraulic properties of a heterogeneous field. *Soil Sci. Soc. Am. J.* 44:681–683.

Saha, D. C., Johnson-Cicalese, J. M., Halisky, P. M., van Heemstra, M. I. and Funk, C. R. 1987. Occurrence and significance of endophytic fungi in fine fescues. *Plant Dis.* 71(11):1021–1024.

Sauer, T. J., Fermanich, K. J. and Daniel, T. C. 1990. Comparison of the pesticide root zone model simulated and measured pesticide mobility under two tillage systems. *J. Environ. Qual.* 19(4):727–734.

Schmidt, R. E. and Blaser, R. E. 1969. Ecology and turf management. in Hanson, A. A. and Juska, F. V. (Eds.). *Turfgrass Science.* Agron. Monogr. 14:217–239. Am. Soc. Agron., Madison, WI.

Schmiege, D. C. 1963. The feasibility of using a neoaplectanid nematode for control of some forest insect pests. *J. Econ. Entomol.* 56:427–431.

Schroeder, W. J. and Beavers, J. B. 1987. Movement of the entomongenous nematodes of the families Heterorhabditidae and Steinernematidae in soil. *J. Nematol.* 19(2):257–259.

Schuster, M. F., Boling, J. C. and Marony, J. J. 1971. Biological control of Rhodesgrass scale by airplane release of an introduced parasite of limited dispersing ability. in Huffaker, C. B. (Ed.). *Biological Control.* Plenum Press, New York, pp. 227–250.

Schweizer, E. E. 1988. New technological developments to reduce groundwater contamination by herbicides. *Weed Tech.* 2(2):223–227.

Shaffer, M. J. 1985. Simulation model for soil erosion- productivity relationships. *J. Environ. Qual.* 14(1):144–150.

Shaffer, M. J. and Pierce, F. J. 1982. Nitrogen-Tillage-Residue Management (NTRM) Model: User's Manual. Res. Rep. USDA-ARS, St. Paul, MN.

Shetlar, D. J. 1987. Use of Entomogenous Nematodes for Control of Turfgrass Infesting Insects. Div. Environ. Chem. ACS Meeting, New Orleans, LA, 27(2):822–825.

Shetlar, D. J. 1989. Entomogenous nematodes for control of turfgrass insects with notes on other biological control agents. in Leslie, A. R. and Metcalf, R. L. (Eds.). Integrated Pest Management for Turfgrass and Ornamentals. Office of Pesticide Programs. 1989-625-030. U.S. Environmental Protection Agency, Washington, D.C., pp. 225–253.

Shetlar, D. J., Suleman, P. E. and Georgis, R. 1988. Irrigation and use of entomogenous nematodes, *Neoaplectana* spp. and *Heterorhabditis heliothidis* (Rhabditida: Steinernematidae and Heterorhabditidae), for control of Japanese bettle (Coleoptera: Scarabaeidae) grubs in turfgrass. *J. Econ. Entomol.* 81(5):1318–1322.

Shive, J. B., Jr. and Sisler, H. S. 1976. Effects of ancymidol (a growth retardant) and triarimol (a fungicide) on the growth, sterols and gibberellins of *Phaseolus vulgaris* (L.). *Plant Physiol.* 57:640–644.

Shoemaker, L. L., Magette, W. L. and Shirmohammadi, A. 1990. Modeling management practice effects on pesticide movement to ground water. *Ground Water Monit. Rev.* 10(1):109–115.

Short, D. E., Reinert, J. A. and Atilano, R. A. 1982. Integrated pest management for urban turfgrass culture in Florida. in Niemczyk, H. D. and Joyner, B. G. (Eds.). *Advances in Turfgrass Entomology.* Hammer Graphics, Inc., Piqua, OH, 150 pp.

Siegel, M. R., Dahlman, D. L. and Bush, L. P. 1989. The role of endophytic fungi in grasses: new approaches to biological control of pests. in Leslie, A. R. and Metcalf, R. L. (Eds.). Integrated Pest Management for Turfgrass and Ornamentals. Office of Pesticide Programs. 1989-625-030 U.S. Environmental Protection Agency, Washington D.C., pp. 169–179.

Siegel, M. R., Latch, C. G. M. and Johnson, M. C. 1985. *Acremonium* fungal endophytes of tall fescue and perennial rygrass: significance and control. *Plant Dis.* 69(2):179–183.

Siegel, M. R., Latch, C. G. M. and Johnson, M. C. 1987. Fungal endophytes of grasses. *Ann. Rev. Phytopathol.* 25:293–315.

Small, M. J. and Mular, J. R. 1987. Long-term pollutant degradation in the unsaturated zone with stochastic rainfall infiltration. *Water Resour. Res.* 23(12):2246–2256.

Smiley, R. W. 1983. *Compendium of Turfgrass Diseases.* The Am. Phytopath. Soc. APS Press, St. Paul, MN, 101 pp.

Smiley, R. W. and Craven-Fowler, M. 1984. *Leptosphaeria korrae* and *Phialophora graminicola* associated with Fusarium blight syndrome of *Poa pratensis* in New York. *Plant Dis.* 68:440–442.

Smith, D. C. and Raupp, M. J. 1986. Economic and environmental assessment of an integrated pest management program for community-owned landscape plants. *J. Econ. Entomol.* 79(1):162–165.

Smolen, M. D., Humenik, F. J. Spooner, J., Dressing, S. A. and Maas, R. P. 1984. Best Management Practices for Agricultural Nonpoint Source Control. IV. Pesticides. USDA Coop. Agree. 12-05-300-472. EPA Interagency Agree. AD-12-F-0-037-0. North Carolina Agricult. Ext. Serv., Raleigh, NC, 87 pp.

Snyder, G. H., Augustin, B. J. and Davidson, J. M. 1984. Moisture sensor-controlled irrigation for reducing N leaching in bermudagrass turf. *Agron. J.* 76(6):964–969.

Steenhuis, T. S. 1979. Simulation of the action of soil and water conservation practices in controlling pesticides. in Haith, D. A. and Loehr, R. C. (Eds.). Effectiveness of Soil and Water Conservation Practices for Pollution Control. EPA-600/3-79-106. U.S. Environmental Protection Agency, Environmental Research Laboratory, Athens, GA.

Steenhuis, T. S. and Naylor, L. M. 1987. A screening method for preliminary assessment of risk to groundwater from land-applied chemicals. *J. Contam. Hydrol.* 1:395–406.

Steenhuis, T. S. and Walter, M. F. 1980. Closed form solution for pesticide loss in runoff water. *Trans. ASAE* 23:615–620, 628.

Stern, V. M., Smith, R. F., Van den Bosch, R. and Hagen, K. S. 1959. The integration of chemical and biological conrol of the spotted alfalfa aphid. *Hilgardia* 29:1–154.

Stewart, B. A., Woolhiser, D. A., Wischmeier, W. H., Caro, J. H. and Frere, M. H. 1975. Control of Water Pollution from Cropland: A Manual for Guideline Development. Vol. 1. EPA 600/2-75-026a. U.S. Environmental Protection Agency, Athens, GA and USDA Agricultural Research Service, Washington, D.C. 111 pp.

Stewart, A. V. 1985. Perennial ryegrass seedling resistance to Argentine stem weevil. N.Z. *J. Agric. Res.* 28:403–407.

Stimmann, M. W. and Taliaferro, C. M. 1969. Resistance of selected accession of bermudagrass to phytotoxemia caused by adult tow lined spittlebugs. *J. Econ. Entomol.* 62:(5)1189–1190.

Street. J.R. 1988. New concepts in turf fertilization. *Landscape Manage.* 2:38–46.

Streu, H. T. 1969. Some Cumulative Effects of Pesticides in the Turfgrass Ecosystem. in *Proc. Scotts Turfgrass Res. Conf.* I. Entomology. Marysville, OH, pp. 53–59.

Streu, H. T. 1973. The turfgrass ecosystem: Impact of pesticides. *Bull. Entomol. Soc. Am.* 19:89–90.

Streu, H. T. and Cruz, C. 1972. Control of the chinch bug in turfgrass in the northeast with Dursban insecticide. *Down Earth* 28(1):1–4.

Streu, H. T. and Gingrich, J. B. 1972. Seasonal activity of the winter grain mite in turfgrass in New Jersey. *J. Econ. Entomol.* 65:427–430.

Summer, R. M., Alonso, C. V. and Young, R. A. 1990. Model linked watershed and lake processes for water quality management decisions. *J. Environ. Qual.* 19(3):421–427.

Sutker, E. M. and Lucas, L. T. 1987. Biocontrol of *Rhizoctonia solani* in tall fescue turfgrass. *Phytopathol.* 77:1721.

Tamura, S., Chang, C., Suzuki, A. and Kumai, S. 1969. Chemical studies on "clover sickness." Part I. Isolation and structural elucidation of two new isoflavonoids in red clover. *Agric. Biol. Chem.* 33:391–397.

Tang, C. and Young, C. 1982. Collection and identification of allelopathic compounds from the undisturbed root system of bigalta limpograss (*Hemarthria altissima*). *Plant Physiol.* 69(1):155–160.

Tanji, K. K. 1982. Modeling of the soil nitrogen cycle. in Stevenson, F. J. (Ed.). *Nitrogen in Agricultural Soils.* Agron. Monogr. 22:721–772. Am. Soc. of Agron., Crop Sci. Soc. Am., Soil Sci. Soc. Am., Madison, WI.

Tanji, K. K., Broadbent, F. E., Mehran, M. and Fried, M. 1977. An extended version of a conceptual model for evaluating annual nitrogen leaching losses from croplands. *J. Environ. Qual.* 8(1):114–120.

Tanji, K. K., Doneen, J. D., Ferry, G. V. and Ayers, R. S. 1972. Computer simulation analysis on reclamation of salt-affected soils in San Joaquin Valley, California. *Soil Sci. Soc. Am. Proc.* 36:127–133.

Tanji, K. K., Fried, M. and Van De Pol, R. M. 1975. A steady-state conceptual nitrogen model for estimating nitrogen emissions from cropped lands. *J. Environ. Qual.* 6(1):155–159.

Tarjan, A. C. and Busey, P. 1985. Genotypic variability in bermudagrass damage by ectoparasitic nematodes. *HortScience* 20(4):675–676.

Tashiro, H. 1987. *Turfgrass Insects of the United States and Canada.* Cornell University Press, Ithaca, NY, 391 pp.

Tashiro, H. and Steinkraus, K. H. 1966. Virulence of species and strains of milky disease bacteria in the European chafer, *Amphimallon majalis* (Razoumowsky). *J. Invert. Pathol.* 8:382–389.

Tillotson, P. and Nielsen, D. R. 1984. Scale factors in soil science. *Soil Sci Soc. Am. J.* 48(5):953–959.

Tillotson, W. R. and Wagenet, R. J. 1982. Simulation of fertilizer nitrogen under cropped situations. *Soil Sci.* 133(3):133–143.

Timmons, D. R. and Holt, R. F. 1980. Soluble N and P concentrations in surface runoff water. in Knisel, W. G. (Ed.). CREAMS: A Field Scale Model for Chemicals, Runoff and Erosion from Agricultural Management Systems. USDA Conservation Research Report No. 26. Washington, D.C.

Timmons, D. R., Holt, R. F. and Latterwell, J. J. 1970. Leaching of crop residues as a source of nutrients in surface runoff water. *Water Resour. Res.* 6(5):1367–1375.

Toler, R. W., Burton, B. D. and Grisham, M. P. 1983. Evaluation of St. Augustinegrass assessions and cultivars for resistance to *Sclerophthora macrospora. Plant Dis.* 67:1008–1010.

Toler, R. W., Beard, J. B., Grisham, M. P. and Crocker, R. L. 1985. Registration of TXSA 8202 and TXSA 8218 St. Augustinegrass germplasm resistant to panicum mosaic virus St. Augustine decline strain. *Crop Sci.* 25:371.

Tumlinson, J. H., Klein, M. G., Doolittle, R. E., Ladd, T. L. and Proveaux, A. T. 1977. Identification of the female Japanese beetle sex pheromone: Inhibition of male response by an enantiomer. *Science* 197:789–792.

USDA Agricultural Research Service. 1989a. USDA Research Plan for Water Quality. United States Dept. of Agric., Coop. State Res. Serv., State Agric. Exp. Stns., Washington, D.C., 14 pp.

USDA Agricultural Research Service. 1989b. Midwest Water Quality Research Initiative Management Systems Evaluation Areas. FY 1990 Guidelines for Proposal Preparation and Submission. Agricultural Chemical Management Research: A Multiagency Focus on Ground Water Quality. United States Dept. of Agric., Coop. State Res. Serv., State Agric. Exp. Stns., Washington, D.C., 17 pp.

Utermann, J., Kladivko, E. J. and Jury, W. A. 1990. Evaluating pesticide migration in tile-drained soils with a transfer function model. *J. Environ. Qual.* 20(4):707–714.

van Genuchten, M. Th. 1978. Mass Transport in Saturated-Unsaturated Media: One-Dimensional Solutions. Res. Dept. 78-WR-11. Water Resour. Prog. Dept. Civil Eng., Princeton University, Princeton, NJ.

van Genuchten, M. Th. and Alves, W. J. 1982. Analytical Solutions of the One-Dimensional Convective Dispersive Solute Transport Equation. USDA Bulletin 1661, U.S. Dept. of Agriculture, Washington, D.C.

van Genuchten, M. Th. and Wierenga, P. J. 1976. Mass transfer studies in porous sorbing media. I. Analytical solutions. *Soil Sci. Soc. Am. J.* 40:473–480.

Van Leeuwen, E. R. and Van der Meulen, P. A. 1928. Some phases of the Japanese beetle insecticide investigations. *J. Econ. Entomol.* 21(6):805–813.

Varco, J. J. and Sartain, J. B. 1986. Effects of phosphorus, sulfur, calcium hydroxide and pH on growth of annual bluegrass. *Soil Sci. Soc. Am. J.* 50(1):128–132.

Vargas, J. M., Roberts, D., Danneburger, T. K., Otto, M. and Detweller, R. 1989. Biological management of turfgrass pests and the use of prediction models for more accurate pesticide applications. in Leslie, A. R. and Metcalf, R. L. (Eds.). Integrated Pest Management for Turfgrass and Ornamentals. Office of Pesticide Programs. 1989-625-030. U.S. Environmental Protection Agency, Washington D.C., pp. 121–141.

Vieira, S. R., Hatfield, J. L., Nielsen, D. R. and Biggar, J. W. 1983. Geostatistical theory and application to variability of some agronomical properties. *Hilgardia* 51:1–75.

Villani, M. G. and Wright, R. J. 1981. Environmental considerations in soil insect pest management. in Pimentel, D. (Ed.). *Handbook of Pest Management in Agriculture.* Vol. 1. CRC Press, Inc., Boca Raton, FL, 26 pp.

Villani, M. G. and Wright, R. J. 1988a. Use of radiography in behavioral studies of turfgrass-infesting scarab grub species (*Coleoptera: Scarabaeidae*). *Bull. Ent. Soc. Am.* 13:132–144.

Villani, M. G. and Wright, R. J. 1988b. Entomogenous nematodes as biological control agents of european chafer and Japanese beetle (*Coleoptera: Scarabaeidae*) larvae infesting turfgrass. *J. Econ. Entomol.* 81(2):484–487.

Villani, M. and Wright, R. J. 1989. Managing the scarab grub pest complex in turfgrass: Some ecological considerations. in Leslie, A. R. and Metcalf, R. L. (Eds.). Integrated Pest Management for Turfgrass and Ornamentals. Office of Pesticide Programs. 1989-625-030. U.S. Environmental Protection Agency, Washington D.C., pp. 127–141.

Vittum, P. J. 1984. Effect of lime applications on Japanese beetle (*Coleoptera: Scarabaeidae*) grub populations in Massachusetts soils. *J. Econ. Entomol.* 77:687–90.

Vittum, P. J. 1985. Effect of timing of application on effectiveness of isofenphos, isazophos, and diazinon of Japanese beetle (*Coleoptera: Scarabaeidae*) grubs in turf. *J. Econ. Entomol.* 78(1):172–180.

Vittum, P. J. and Tashiro, H. 1980. Effect of soil pH on survival of Japancese beetle and European chafer larvae. *J. Econ. Entomol.* 73:577–579.

Waddington, D. V., Turner, T. R. Duich, J. M. and Moberg, E. L. 1978. Effect of fertilization on penncross creeping bentgrass. *Soil Sci. Soc. Am. Proc.* 70(5):713–718.

Wagenet, R. J. and Hutson, J. L. 1986. Predicting the fate of nonvolatile pesticides in the unsaturated zone. *J Environ. Qual.* 15(4):315–322.

Wagenet, R. J. and Hutson, J. L. 1989. LEACHM: A Finite Difference Model for Simulating Water, Salt and Pesticide Movement in the Plant Root Zone. Continuum. Vol. 2. Ver. 2.0. New York State Water Resour. Inst., Cornell University, Ithaca, NY.

Wagenet, R. J. and Rao, P. S. C. 1983. Description of nitrogen movement in the presence of spatially variable soil hydraulic properties. *Agric. Water Manage.* 6:227–242.

Wagenet, R. J. and Rao, P. S. C. 1990. Modeling pesticide fate in soils. in Cheng, H. H. (Ed.). *Pesticides in the Soil Environment: Processes, Impacts, and Modeling.* SSSA Book Series No. 2. Soil Science Society of America, Madison, WI, pp. 351–399.

Wagenet, R. J., Hutson, J. L. and Biggar, J. W. 1989. Simulating the fate of a volatile pesticide in unsaturated soil: A case study with DBCP. *J. Environ. Qual.* 18(1):78–84.

Walker, W. J. and Branham, B. 1992. Environmental impacts of turfgrass fertilization. in Balogh, J. C. and Walker, W. J. (Eds.). *Golf Course Management and Construction: Environmental Issues.* Lewis Publishers, Chelsea, MI, Chapter 3.

Walker, W. J., Balogh, J. C., Tietge, R. M. and Murphy, S. R. 1990. *Environmental Issues Related to Golf Course Construction and Management*. United States Golf Association, Green Section. Far Hills, NJ, 378 pp.

Ward, A. D., Alexander, C. A., Fausey, N. R. and Dorsey, J. D. 1988. The ADAPT Agricultural Drainage and Pesticide Transport Model. Proc. ASAE Internat. Symp. Modeling Agricult., Forest and Range. Chicago, IL.

Warren, G. W. and Potter, D. A. 1983. Pathogenicity of *Bacillus popilliae* (*Cyclociphala* strain) and other milky disease bacteria in grubs of the southern masked chafer (*Coleoptera: Scarabaeidae*). *J. Econ. Entomol.* 76(1):69–73.

Watschke, T. L. 1981. Effect of four growth retardants on two Kentucky bluegrasses. *Proc. Northeast. Weed Sci. Soc.* 35:322–330.

Watts, D. G. and Hanks, D. L. 1978. A soil-water-nitrogen model for irrigated corn on sandy soils. *Soil Sci. Soc. Am. J.* 42(3):492–499.

Watts, D. G. and Martin, D. L. 1981. Effects of water and nitrogen management on nitrate leaching loss from sands. *Trans. ASAE* 24(4):911–916.

Wauchope, R. D. 1978. The pesticide content of surface water draining from agricultural fields — A review. *J. Environ. Qual.* 7(4):459–472.

Wauchope, R. D. and Leonard, R. A. 1980a. Maximum pesticide concentrations in agricultural runoff: A semi-empirical prediction formula. *J. Environ. Qual.* 9(4):665–672.

Wauchope, R. D. and Leonard, R. A. 1980b. Pesticide concentrations in agricultural runoff: Available data and an approximation formula. in Knisel, W. G. (Ed.). CREAMS: A Field Scale Model for Chemicals, Runoff and Erosion from Agricultural Management Systems. USDA Conservation Research Report No. 26. pp. 544–559.

Wauchope, R. D., Buttler, T. M., Hornsby, A. G., Augustijn-Beckers, P. W. M. and Burt, J. P. 1991. The SCS/ARS/CES pesticide database for environmental decision-making. *Rev. Environ. Contam. Toxicol.* 123:1–155.

Wearing, C. H. 1988. Evaluating the IPM implementation process. *Ann. Rev. Entomol.* 33:17–38.

Wehner, D. J., Minner, D. D., Dernoeden, P. H. and McIntosh, M. S. 1985. Heat tolerance of Kentucky bluegrass as influenced by pre- and post-stress environment. *Agron. J.* 77(3):376–378.

Welch, S. M. 1984. Developments in computer-based IPM extension delivery systems. *Ann. Rev. Entomol.* 29:359–381.

Wessel, R. D. and Polinka, J. B. 1952. Soil pH in relation to Japanese beetle populations. *J. Econ. Entomol.* 45:733–735.

Whitcomb, C. E. and Roberts, E. C. 1973. Competition between established tree roots and newly seeded Kentucky bluegrass. *Agron. J.* 65:126–129.

White, E. M. and Williamson, E. J. 1973. Plant nutrient concentrations in runoff from fertilized cultivated erosion plots and prairie in eastern South Dakota. *J. Environ. Qual.* 2(4):453–455.

White, J. F., Jr. 1987. The widespread distribution of endophytes in the Poaceae. *Plant Dis.* 71:340–342.

White, J. F., Jr. and Cole, G. T. 1985. Endophyte-host associations in forage grasses. III. *In vitro* inhibition of fungi by *Acremonium coenophialum*. *Mycologia* 77(3):487–489.

Williams, J. R., Nicks, A. D. and Arnold, J. G. 1985. Simulation for water resources in rural basins. *ASCE J. Hydraul.* 111:970–986.

Wilkinson, H. T. and Avenius, R. 1984. The selection of bacteria antagonistic to *Pythium* spp. pathogenic to turfgrass. *Phytopathol.* 74:812.

Wolcott, G. N. 1941. The establishment into Puerto Rico of *Larra americana* Saussure. *J. Econ. Entomol.* 34(1):53–56.

Wong, P. T. W. and Baker, R. 1984. Suppression of wheat take-all and Ophiobolus patch by fluorescent pseudomonads from a Fusarium-suppressive soil. *Soil Biol. Biochem.* 16:397–403.

Wong, P. T. W. and Siviour, T. R. 1979. Control of Ophiobolus patch in *Agronstis* turf using a virulent fungi and take-all suppressive soils in pot experiments. *Ann. Appl. Biol.* 92:191–197.

Young, R. A., Onstad, C. A., Bosch, D. D. and Anderson, W. P. 1986. Agricultural Nonpoint Source Pollution Model: A Watershed Analysis Tool; A Guide to Model Users. USDA ARS and Minn. Poll. Cont. Agency, St. Paul, MN, 87 pp.

Youngner, V. B. 1985. Physiology of water use and water stress. in Gibeault, V. A. and Cockerham, S. T. (Eds.). *Turfgrass Water Conservation.* University of California, Riverside, Division of Agriculture and Natural Resources, pp. 37–43.

Zimmerman, R. J. and Cranshaw, W. S. 1990. Compatibility of three entomogenous nematodes (*Rhabditida*) in aqueous solutions of pesticides used in turfgrass maintenance. *J. Econ. Entomol.* 83(1):97–100.

Zwerman, P. J., Bouldin, D. R., Greweling, T., Klausner, S. D., Lathwell, D. J. and Wilson, D. O. 1971. Management of Nutrients on Agricultural Land for Improved Water Quality. EPA-13020-DPB-08/71. Environmental Protection Agency and Cornell University, Ithaca, New York.

CHAPTER 6

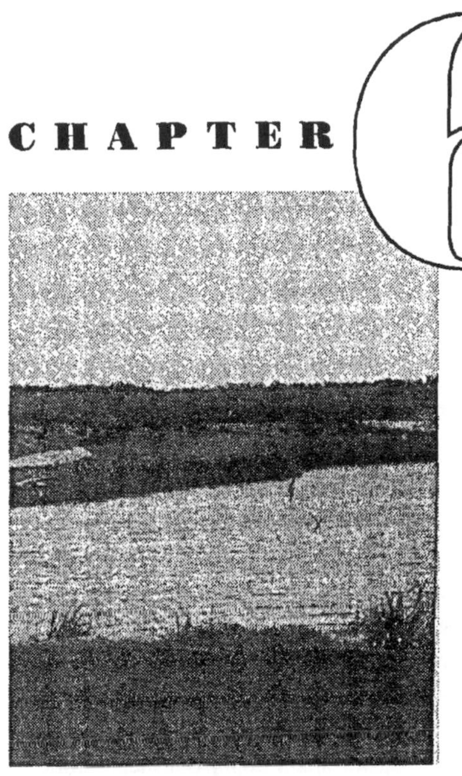

Wildlife and Golf Courses

Roberta M. Tietge

"Your profession and the (golf) industry are just scratching the surface. The growth that you are going to see — both nationally and internationally — is going to be absolutely staggering."

—Mark McCormack, entrepreneur and keynote speaker of 62nd GCSAA International Golf Course Conference and Show, Las Vegas, Nevada, February 8, 1991.

6.1 INTRODUCTION: VALUES AND AESTHETICS OF WILDLIFE

Golf courses are wildlife habitats. Wetlands, roughs, trees, and wildflowers, all of these support wildlife. In fact, historically, animals helped to make the game what it is; rabbit scrapes formed targets and sand-pit burrows created by sheep later became sand bunkers (Beard 1982). Somehow, with time, technology, and "Americanization," golf courses evolved into a much different creature which could no longer tolerate native animals and thus they became "pests."

In the evolutionary process leading to modern methods of course design and construction, many of the natural features were plowed and leveled, allowing for a uniform surface upon which a golf ball could travel virtually unhindered. Natural

Photo courtesy of *Golf Course Management*, GCSAA.

grass species were replaced with less tolerant, domesticated cultivars. Chemical and mechanical management allowed inferior turfgrass species to thrive. Pest resistance to chemical control; potential groundwater contamination; die offs of animals, birds, and fish; and occasional human health effects have changed the regulatory and management perspective for both agricultural and turfgrass systems. Many of these circumstances necessitate strict permitting and licensing regulations for many turfgrass and agricultural chemicals. In all of these related management and environmental decision processes, wildlife and wildlife habitat is often virtually forgotten. As natural and agricultural lands disappear under parking lots and urban industrial development, golf courses assume an increasingly important role for wildlife.

6.1.1 Wildlife Values

A diversity of habitat is found on a golf course. Smooth, green tees and gently rolling fairways are soothing to the eye and the spirit. It is cooler and much quieter on the course than in the city or suburban business areas. A small flock of ducks float on a pond nearby. Wooded areas and other roughs add particular interest to the eye and the game. A colorful songbird darts among the tree leaves, gleaning insects. Songs of many different types of birds can be heard. Several butterflies taste the flowers blooming at the edge of the rough. A hawk soars lazily overhead, coming down to seek out a rodent for its lunch. At night a different world awakens: bats fly silently, gobbling up thousands of insects; a chorus of frogs sing from the edge of a wetland; an owl pounces on a mouse. The golf course is a complex ecosystem which is capable of supporting a great diversity of natural wildlife.

A surprising number of wildlife species inhabit golf courses (Maffei 1978). Golf courses possess several features which attract wildlife. Food sources include insects, nuts, and berries. Cover is provided by trees and shrubs. Water exists in the form of ponds and wetlands, and open areas surrounded by a variety of vegetation provide habitat diversity and edge effect for many species. Peripheral regions of low- or nonmaintained areas, such as roughs, provide critical habitat for wildlife (Maffei 1978). Golf course managers now realize that environmental consciousness on their courses is crucial for the well-being of water resources, humans, and wildlife alike. Golf courses have become important multiple use environments in urban and suburban landscapes.

6.1.2 Golf Course Aesthetics and Wildlife

Aesthetics of the golf course are at least as important as the game itself. In addition to vegetation, people expect to see wildlife. This is perceived in a positive way by golfers and nongolfers alike. Seeing a bluebird, for example, adds to the golfing experience and the importance of both their roles in the urban landscape.

Golf courses create valuable green space in urban areas, reducing temperature and noise away from the asphalt and traffic of neighboring city suburbs (Kaplan and Kaplan 1989; Roberts and Roberts 1989). Golf courses are, in part, responsible for greatly enhancing real estate value in urban developments. In some urban areas, home values may increase by 60% in as little as 2 years. At least half the real estate projects for year-round or resort living in the U.S. include golf site lots for potential buyers (Anonymous 1988). Golf advertisements now boast of the beauty of wildlife and natural areas on their courses, attracting recreation seekers (GCSAA 1991). As urban development continues, the green space available on golf courses will become increasingly vital as a wildlife habitat. Given this condition, the use of practices consistent with maintaining the value of wildlife is paramount.

Golf courses must be seen as diversified ecosystems for wildlife. In every aspect of golf course management, from planning and construction to pesticide and fertilizer application, aquatic and terrestrial species should be integral factors in the plan. In previous years, texts published on the topic of golf courses did not consider wildlife as a legitimate component. It is now widely recognized that wildlife provides aesthetic value and acts as a measure of general environmental health (NCC 1989). Its presence and diversity is essential for the quality of life we enjoy on this planet.

Public awareness of the effects of pesticides in our environment is increasing, and the public is now demanding a verification of the risks versus the benefits of pesticide use. They want to be assured that the course they play on or live next to is not contaminated with potentially harmful chemicals. Course managers are certainly interested in making their course safe for wildlife and players alike. Policy planners are beginning to acknowledge the value of wildlife species as indicators of environmental health. Therefore, with the increasing number of chemicals available each year on the market, the need is perhaps greater than ever to put stock into pesticide and wildlife research, particularly in field studies. The welfare of wildlife is also the welfare of human beings.

6.2 IMPACTS OF TURFGRASS PESTICIDES ON WILDLIFE

6.2.1 What is the Toxicity of Pesticides to Wildlife? An Overview of the Regulatory Perspective

The U.S. Environmental Protection Agency (EPA) has been mandated by law to reregister all existing pesticides containing any active ingredient which was registered before November 1, 1984 (*Farm Chemicals Handbook* 1991). After this date, Congress determined that most pesticides would have up-to-date databases and will not require reregistration. There are about 600 pesticide active ingredient cases requiring reevaluation (EPA 1990).

The pesticides registrant, such as a chemical company, must pay reregistration fees. Annual costs of reregistration were $697 million in 1990. A government relations briefing recently reported this cost will soar to $1.3 billion by 1995 (Anonymous 1991). In 1988, the Federal Insecticide, Fungicide and Rodenticide Act (FIFRA) was amended to give the EPA regulatory authority in this area. Due in part to mounting public concern regarding the timeliness of the reevaluation with current use, the EPA has been mandated to accelerate the reregistration process.

The EPA has grouped "chemically-related" active ingredients, thus reducing the 600 pesticides to 194. These 194 cases represent 85–90% of the total volume of pesticides used in the United States. The evaluation is to be done in five phases and will theoretically provide a complete scientific database for each pesticide product before evaluation and reregistration by the EPA (EPA 1990).

It becomes evident when reviewing information on pesticide products that there is a dynamic concatenation of events from chemical production to sale. Given the number and total volume of chemicals in use each year, there is a tremendous backlog of toxicity information to be assessed. This assessment of toxicity and potential for exposure is required to determine the short-term and long-term effects of pesticides used on golf courses and the environment in general.

6.2.2 Direct Effects: Acute and Chronic Toxicity

All species, including humans, are ultimately subject to potential exposure as the result of pesticides applied on golf course turfgrass. Consideration of the potential for exposure and toxic effects of applied chemicals should be an important component of golf course policy decisions regarding chemical management of turfgrass pests and diseases. Wildlife can be exposed to pesticides in a variety of ways. Grue et al. (1983) lists several routes of exposure: direct consumption of granular formulations or solutions from standing water, ingestion of poisoned insects and other contaminated prey, residues on treated vegetation or seeds, and dermal absorption and inhalation.

One of the hazards of chemical management is the risk of adverse effects on nontarget animals and plants through onsite and offsite exposure. The level of risk is determined by the *intrinsic toxicity* of the chemical and the *potential for exposure* to toxic doses of the chemical. Effective chemical action of applied chemicals requires (1) exposure to a sufficient quantity of chemical, (2) presence of the chemical in an active form, and (3) exposure for sufficient time to produce a toxic effect. Toxicity of a chemical alone is insufficient to produce chronic or acute effects without sufficient exposure. One of the ultimate goals of a golf course manager should be to eliminate or avoid the potential of pesticide exposure to nontarget wildlife. Careful planning, timing and rate of application, supervision of personnel, and using appropriate chemicals are all components of safe handling practices for wildlife, golfers, and golf course workers.

6.2.2.1 Definitions of Toxicity Tests and Effects. There are two types of direct effects of pesticides on wildlife: acute and chronic. Specific test results of acute and chronic effects of pesticides used on golf courses are available in Chapter 8 of this book (Murphy 1992). The type of response, acute or chronic, depends on both the *magnitude of the dose* and the *duration of exposure*. The acute effects are dramatic or rapid responses of organisms exposed to large quantities of pesticides over a short period of time. When they are exposed to sufficient quantities of applied pesticides, acute effects in wildlife may occur from chemical applications, immediately followed by a heavy rainfall. Runoff or surface pooling of drainage water may result in potentially toxic concentrations of pesticides in temporary surface detention ponds. Toxic concentrations of pesticides also will occur as a result of a spill or overapplication or if they have not been properly watered-in (Stone and Gradoni 1985; Edmondson 1987).

Different laboratory and field tests have been developed to test for the toxic effects of pesticides. The definitions of various toxicity tests are

Acute test	Large dose administered over a short period of time producing a lethal result
Subacute toxicity test	A somewhat smaller dose administered over a longer period of time, but in sufficient quantity to be lethal
Chronic test	Small doses administered over a long or extended period of time which may produce lethal or other deleterious physiological effects
Sublethal effects	Dose applied is not lethal, but may cause a variety of aberrations in behavior, reproduction, health, and physiology

Standard tests for acute toxicity include LD_{50}, LC_{50}, and EC_{50}. These terms are defined as the following:

1. LD_{50} is the lethal dose at which 50% of the test organisms die from a single dose.
2. LC_{50} is the concentration of a substance in water that is lethal to 50% of a population of test organisms within a specific period of time (24–96 h).

3. EC_{50} is the concentration of a chemical that has a specific effect on 50% of a population of test organisms. A few of these specific acute effects include avoidance, immobilization, behavior, respiration, food consumption, growth, physiological effects, reproductive effects, regeneration, and teratogenesis.

Chronic effects are less evident than dramatic episodes of acute toxicity. Chronic effects result from long-term exposure to small doses of chemical. Chronic physiological effects may not be seen by casual field observation and thus are much more difficult to determine (Grue et al. 1983). Reproductive effects, carcinogenicity, bioconcentration, changes in population density, and teratogenesis are a few of the typical adverse responses to chronic exposure to chemicals (see Chapter 8; Murphy 1990, 1992). Unfortunately, scientific research is very limited on the process of acute and chronic exposure of nontarget organisms to pesticides used on golf courses. The sublethal effects of pesticides should be considered as important as the lethal effects.

Extensive data have been collected on chemicals using a variety of terrestrial and aquatic laboratory test organisms (Murphy 1990, 1992; Sheedy et al. 1991). This information provides a point of reference when considering the overall toxicity of a particular chemical. The overall toxicity of a chemical can be very difficult to assess due to the complex interaction of environmental and chemical variables that affect potential for exposure. These factors may include concentration, formulation, timing and duration of precipitation, runoff to aquatic systems, and soil transport (see Chapter 4; Balogh and Anderson 1992). Toxicity also depends on the level of exposure of bird, mammal, reptile, amphibian, insect, or aquatic species, as well as the age of the organism (Eisler 1986). Different species respond uniquely to different pesticide formulations. Some pesticides act as a synergist when mixed with another chemical, enhancing or increasing the effect of the pesticide.

The relative ratings for the acute toxicity of pesticides have been established according to ranges of LC_{50} and LD_{50} values (Table 1). Specific acute toxicity tests are presented in Chapter 8 (Murphy 1992) and are summarized in Table 2.

While pesticides come with application instructions, the fact remains that application conditions can be highly variable. In addition, some pesticides are highly toxic to a variety of nontarget organisms including aquatic and terrestrial groups (Table 2). Highly toxic chemicals typically fall into the group of restricted use pesticides (RUPs). Examples of RUPs include anilazine, chloroneb, chlorothalonil, etridiazole, mancozeb, thiophanate-methyl, pendimethalin, ethoprop, fenamiphos, and isofenphos (see Table 1 in Chapter 4; Balogh and Anderson 1992). Due to their extreme toxicity, special care must be taken during their application by a licensed individual. Some concern exists regarding the variation in licensing requirements from state to state. In addition, the risk of licensed applicators handing over the task of applying RUPs to nonlicensed personnel may be an existing problem in some areas, having potentially catastrophic results.

Table 1. Relative Pesticide Toxicity Ratings

Toxicity[a] rating	LC_{50} (mg L^{-1})	Class	LD_{50} (mg kg^{-1})	Class
1	<0.1	Very highly toxic	<5	Extremely toxic
2	0.1–0.99	Highly toxic	5–49	Very toxic
3	1–9.99	Moderately toxic	50–499	Moderately toxic
4	10–99.99	Slightly toxic	500–4999	Slightly toxic
5	>100	Practically nontoxic	>4999	Almost nontoxic

[a] *Adapted from EPA (1985a) and USDA (1969).*

With many pesticides, following manufacturers' directions during application should minimize the chance of toxic contamination to wildlife or nontarget organisms (Anderson et al. 1989; Balogh et al. 1990). However, the wide variety of circumstances and nontarget species involved in chemical use makes it impossible to predict the toxic effects of pesticides. Many reported cases of wildlife fatalities due to pesticide exposure on golf courses and other managed ecosystems occur every year (Madhun and Freed 1990; Stone 1979; Stone and Gradoni 1985; Stone and Knoch 1982; White et al. 1979; Zinkl et al. 1978).

The focus of scientific literature appears to be on larger, high-profile animals actually found dead on golf courses. Toxic effects, however, are being experienced by low-profile species as well. Animals including amphibians, reptiles, and nocturnal species are also at risk. As part of the food chain, they are quite essential to the overall health of wildlife in general.

Toxic effects may occur despite mild intrinsic toxicity or rapid dissipation of the pesticide. For example, birds may ingest the affected insects following chemical application, or feed on the grass which has been treated. Waterfowl and other birds often will walk into and drink from puddles, pools, or ponds which form as a result of rain or watering-in after pesticide treatment (Stone 1979; Stone and Knoch 1982).

Wildlife may be exposed to pesticides in a variety of ways. These exposure pathways include residues on treated turfgrass, in standing water or wetland areas, ingestion of poisoned prey such as insects or small birds and mammals, and through inhalation and dermal exposure. Granular formulations of pesticides pose a specific risk to many species. These formulations can be picked up and eaten, thus becoming very detrimental to some species, especially seed-eating birds (Grue et al. 1983).

Most laboratory species used for pesticide toxicity evaluation, such as rats, mice, and aquatic organisms, are chosen due to their ability to live and reproduce under laboratory conditions. An important distinction exists between animals in a laboratory environment and wildlife on a golf course. Certain pesticides may be only slightly toxic to some laboratory animals, yet for certain wildlife species the same chemical may be highly toxic (Table 2). In fact, although laboratory studies demonstrate low

Table 2. Summary and Range of Turfgrass Pesticide Toxicity Tests

Chemical	CAS number	Chemical type	Species class	Toxicity test	Range of concentration ($\mu g\ L^{-1}$)	Range of toxicity classes[a]
Atrazine	1912249	Herbicide	Fish	LC_{50}	220–>100,000	2–5
	1912249	Herbicide	Insect	LC_{50}	720	2
	1912249	Herbicide	Annelida	LC_{50}	6,300–9,900	3
	1912249	Herbicide	Mullusca	LC_{50}	>100,000	5
	1912249	Herbicide	Crustacea	LC_{50}	94–54,000	1–4
	1912249	Herbicide	Protozoa	LC_{50}	6–38	1
	1912249	Herbicide	Tracheophy	LC_{50}	100–540	2
Benefin (Benfluralin)	1861401	Herbicide	Mammal	LD_{50}	790,000	4
	1861401	Herbicide	Fish	LC_{50}	800–<1,000	2
	1861401	Herbicide	Amphibia	LC_{50}	11,000	4
	1861401	Herbicide	Crustacea	LC_{50}	1,100–8,200	3
Bensulide	741582	Herbicide	Fish	LC_{50}	379–970	2
	741582	Herbicide	Mammal	LD_{50}	241,000–1,470,000	3–4
	741582	Herbicide	Crustacea	LC_{50}	1,400–7,600	3
2,4-D Acid	94757	Herbicide	Bird	LD_{50}	472,000–668,000	3–4
	94757	Herbicide	Fish	LC_{50}	208–7,000,000	2–5
	94757	Herbicide	Fish	LD_{50}	1,000,000–2,000,000	4
	94757	Herbicide	Insect	LC_{50}	1,600–91,800	3–4
	94757	Herbicide	Annelida	LC_{50}	122,200	5
	94757	Herbicide	Amphibia	LC_{50}	8,050–13,770	3–4
	94757	Herbicide	Mullusca	LC_{50}	259,000	5
	94757	Herbicide	Crustacea	LC_{50}	1,000–1,389,000	2–5
	94757	Herbicide	Protozoa	LC_{50}	104,000–485,000	5
2,4-D Amine	2008391	Herbicide	Bird	LD_{50}	2,000–>2,025	1

2,4-D Butyl ester	1929733	Herbicide	Bird	LC_{50}	>5,000	3
	1929733	Herbicide	Mammal	LD_{50}	380–848	1
	1929733	Herbicide	Fish	LC_{50}	303–43,400	2–4
	1929733	Herbicide	Insect	LC_{50}	1,600–8,500	3
	1929733	Herbicide	Mullusca	LC_{50}	740–740	2
	1929733	Herbicide	Crustacea	LC_{50}	440–>100,000	2–5
2,4-D PGBE	1320189	Herbicide	Fish	LC_{50}	170–9,300	2–3
	1320189	Herbicide	Insect	LC_{50}	2,400–2,600	3
	1320189	Herbicide	Crustacea	LC_{50}	400–>100,000	2–5
2,4-D Sodium salt	2702729	Herbicide	Fish	LC_{50}	129,00–2,971,000	4–5
	2702729	Herbicide	Insect	LC_{50}	>40,000	4
	2702729	Herbicide	Mullusca	LC_{50}	64,290–>100,000	4–5
	2702729	Herbicide	Amphibia	LC_{50}	>40,000	4
	2702729	Herbicide	Crustacea	LC_{50}	2,224–2,644,000	3–5
2,4-DP Dichlorprop	120365	Herbicide	Mammal	LD_{50}	532	1
DCPA	1861321	Herbicide	Mammal	LD_{50}	>10,000,000	5
	1861321	Herbicide	Fish	LC_{50}	>10,000–1,000,000	4–5
	1861321	Herbicide	Annelida	LC_{50}	286,000	5
	1861321	Herbicide	Crustacea	LC_{50}	>40,000	4
Dicamba	1918009	Herbicide	Bird	LC_{50}	>4,640–>10,000	3–4
	1918009	Herbicide	Bird	LD_{50}	>2,510	1
	1918009	Herbicide	Mammal	LD_{50}	757	1
Dicamba, Acid	1918009	Herbicide	Fish	LC_{50}	28,000–516,000	4–5
	1918009	Herbicide	Amphibia	LC_{50}	106,000–220,000	5
	1918009	Herbicide	Crustacea	LC_{50}	3,900–>100,000	3–5
Dicamba, Sodium salt	62610393	Herbicide	Fish	LC_{50}	20,000–600,000	4–5
DSMA	144218	Herbicide	Mammal	LC_{50}	22,100	4
	144218	Herbicide	Mammal	LD_{50}	1,800,000	4

Table 2. Summary and Range of Turfgrass Pesticide Toxicity Tests (continued)

Chemical	CAS number	Chemical type	Species class	Toxicity test	Range of concentration ($\mu g\ L^{-1}$)	Range of toxicity classes[a]
DSMA	144218	Herbicide	Fish	LC_{50}	>10,000–1,500,000	4–5
(continued)	144218	Herbicide	Amphibia	LC_{50}	271,000–600,000	5
	144218	Herbicide	Crustacea	LC_{50}	>40,000	4
Endothall, disodium salt	129679	Herbicide	Fish	LC_{50}	310–>560,000	2–5
	129679	Herbicide	Mullusca	LC_{50}	12,500–48,080	4
Endothall	145733	Herbicide	Bird	LD_{50}	229,000	3
	145733	Herbicide	Fish	LC_{50}	180–1,740,000	2–5
	145733	Herbicide	Insect	LC_{50}	3,250	3
	145733	Herbicide	Crustacea	LC_{50}	50–510	1–2
Fenoxaprop-ethyl	1330207	Herbicide	Fish	LC_{50}	8,200–1,080,000	3–5
	66441234	Herbicide	Mammal	LD_{50}	3,310,000–3,400,000	4
	1330207	Herbicide	Crustacea	LC_{50}	350–1,000,000	2–5
Glyphosate	1071836	Herbicide	Bird	LC_{50}	163–>4,640	2–3
	1071836	Herbicide	Bird	LD_{50}	>2,000–>3,800	1
	1071836	Herbicide	Mammal	LD_{50}	3,800–>7,900	1–2
	1071836	Herbicide	Fish	LC_{50}	12,800–240,000	4–5
	1071836	Herbicide	Insect	LC_{50}	673,430	5
	1071836	Herbicide	Crustacea	LC_{50}	22,000–47,310	4
Glyphosate, Rodeo	1071836	Herbicide	Mammal	LD_{50}	>5,000–5,400	2
Glyphosate, Roundup	38641940	Herbicide	Fish	LC_{50}	1,300–54,800	3–4
	38641940	Herbicide	Insect	LC_{50}	30,000–43,000	4
	38641940	Herbicide	Crustacea	LC_{50}	43,000–62,000	4

MCPA	94746	Herbicide	Bird	LC_{50}	>5,000	3
	94746	Herbicide	Fish	LC_{50}	>10,000–163,500	4–5
	94746	Herbicide	Insect	LC_{50}	335,000	5
	94746	Herbicide	Crustacea	LC_{50}	11,000–>40,000	4
MSMA	6484522	Herbicide	Fish	LC_{50}	74,000–79,000	4
MSMA + Surfactant	2163806	Herbicide	Mammal	LC_{50}	>20,000	4
	2163806	Herbicide	Mammal	LD_{50}	300,000–1,738,000	3–4
	2163806	Herbicide	Fish	LC_{50}	1,200–4,700,000	3–5
	2163806	Herbicide	Crustacea	LC_{50}	37,600–5,100,000	4–5
Oxadiazon	19666309	Herbicide	Fish	LC_{50}	1,460–3,200	3
	19666309	Herbicide	Amphibia	LC_{50}	1,300–2,800	3
Pronamide	23950585	Herbicide	Mammal	LD_{50}	5,620,000–10,000,000	5
	23950585	Herbicide	Amphibia	LC_{50}	>40,000	4
Simazine	122349	Herbicide	Bird	LC_{50}	8,800–51,200	3–4
	122349	Herbicide	Mammal	LD_{50}	>4,000	1
	122349	Herbicide	Fish	LC_{50}	90–8,100,000	1–5
	122349	Herbicide	Insect	LC_{50}	1,900–>40,000	3–4
	122349	Herbicide	Mullusca	LC_{50}	>100,000	5
	122349	Herbicide	Amphibia	LC_{50}	>100,000	5
	122349	Herbicide	Crustacea	LC_{50}	13,000–424,000	4–5
Triclopyr, Ester	55335063	Herbicide	Bird	LC_{50}	9,026–>10,000	3–4
	55335063	Herbicide	Bird	LD_{50}	>4,640	1
	55335063	Herbicide	Mammal	LD_{50}	2,140–2,460	1
Triclopyr, Acid	55335063	Herbicide	Bird	LC_{50}	2,935–>5,640	3
	55335063	Herbicide	Bird	LD_{50}	1,698	1
	55335063	Herbicide	Mammal	LD_{50}	310–729	1
	55335063	Herbicide	Fish	LC_{50}	1,200–>100,000	3–5

Table 2. Summary and Range of Turfgrass Pesticide Toxicity Tests (continued)

Chemical	CAS number	Chemical type	Species class	Toxicity test	Range of concentration ($\mu g\ L^{-1}$)	Range of toxicity classes[a]
Triclopyr, Amine	57213691	Herbicide	Bird	LC_{50}	>10,000–11,622	4
	57213691	Herbicide	Mammal	LD_{50}	2,140–2,830	1
	57213691	Herbicide	Fish	LC_{50}	101,000–498,000	5
	57213691	Herbicide	Crustacea	LC_{50}	1,140,000–1,170,000	5
Trifluralin	1582098	Herbicide	Fish	LC_{50}	10–28,000	1–4
	1582098	Herbicide	Amphibia	LC_{50}	100–14,000	2–4
	1582098	Herbicide	Insect	LC_{50}	2,800–>40,000	3–4
	1582098	Herbicide	Crustacea	LC_{50}	50–>40,000	1–4
	1582098	Herbicide	Annelida	LC_{50}	>300	2
	1582098	Herbicide	Mullusca	LC_{50}	8,000–35,000	3–4
Bendiocarb	22781233	Insecticide	Mammal	LD_{50}	40,000–179,000	2–3
	22781233	Insecticide	Insect	LC_{50}	57–1,700	1–3
	22781233	Insecticide	Crustacea	LC_{50}	5,550	3
Carbaryl	63252	Insecticide	Mammal	LD_{50}	112,000–588,000	3–4
	73252	Insecticide	Mammal	LD_{50}	1,800,000	4
	63252	Insecticide	Fish	LC_{50}	140–204,000	2–5
	63252	Insecticide	Fish	LD_{50}	>10	1
	63252	Insecticide	Insect	LC_{50}	2–100,000	1–5
	63252	Insecticide	Annelida	LC_{50}	7,200–13,000	3–4
	63252	Insecticide	Annelida	LD_{50}	>50,000–>100,000	3
	63252	Insecticide	Mullusca	LC_{50}	1,200–2,000,000	3–5
	63252	Insecticide	Amphibia	LC_{50}	4,700–150,000	3–5
	63252	Insecticide	Crustacea	LC_{50}	1–22,900	1–4
	63252	Insecticide	Protozoa	LC_{50}	7,900–105,000	3–5

Chlorpyrifos	2921882	Insecticide	Fish	LC_{50}	0.8–8,950	1–3
	2921882	Insecticide	Insect	LC_{50}	0.16–71	1
	2921882	Insecticide	Crustacea	LC_{50}	0.035–6	1
	2921882	Insecticide	Mullusca	LC_{50}	>806	2
Diazinon	333415	Insecticide	Bird	LC_{50}	47,000–245,000	4–5
	333415	Insecticide	Bird	LD_{50}	3,540–4,330	1
	333415	Insecticide	Fish	LC_{50}	1–23,400	1–4
	333415	Insecticide	Fish	LD_{50}	310	1
	333415	Insecticide	Insect	LC_{50}	0–6,700	1–3
	333415	Insecticide	Annelida	LC_{50}	3,160	3
	333415	Insecticide	Mullusca	LC_{50}	48–20,000	1–4
	333415	Insecticide	Amphibia	LC_{50}	14,000	4
	333415	Insecticide	Crustacea	LC_{50}	0–>20,000	1–4
Ethoprop	13194484	Insecticide	Bird	LD_{50}	4,210–12,600	1–2
Trichlorfon	52686	Insecticide	Mammal	LD_{50}	866–726,970	1–4
Anilazine	101053	Fungicide	Fish	LC_{50}	8.5–675	1–2
	101053	Fungicide	Crustacea	LC_{50}	0.27–5,550	1–3
	101053	Fungicide	Mullusca	LC_{50}	580–1,800	2–3
Benomyl	17804352	Fungicide	Fish	LC_{50}	6–2,200	1–3
Chlorothalonil	1897456	Fungicide	Mammal	LC_{50}	310	2
	1897456	Fungicide	Mammal	LD_{50}	>10,000,000	5
	1897456	Fungicide	Fish	LC_{50}	8–170	1–2
	1897456	Fungicide	Insect	LC_{50}	1,800	3
	1897456	Fungicide	Mullusca	LC_{50}	9,000–37,000	3–4
	1897456	Fungicide	Amphibia	LC_{50}	160	2
	1897456	Fungicide	Crustacea	LC_{50}	140–>10,000	2–4
Etridiozole	2593159	Fungicide	Mammal	LD_{50}	1,077,000–4,000,000	4
	2593159	Fungicide	Amphibia	LC_{50}	12,000	4

Table 2. Summary and Range of Turfgrass Pesticide Toxicity Tests (continued)

Chemical	CAS number	Chemical type	Species class	Toxicity test	Range of concentration ($\mu g\ L^{-1}$)	Range of toxicity classes[a]
Mancozeb	8018017	Fungicide	Mammal	LD_{50}	11,200,000	5
	8018017	Fungicide	Fish	LC_{50}	400–2,600	2–3
	8018017	Fungicide	Crustacea	LC_{50}	1,300	3
Maneb	12427382	Fungicide	Mammal	LD_{50}	7,990,000	5
	12427382	Fungicide	Fish	LC_{50}	165–6,000	2–3
	12427382	Fungicide	Mullusca	LC_{50}	100,000–330,000	5
	12427382	Fungicide	Amphibia	LC_{50}	25,000–125,000	4–5
	12427382	Fungicide	Crustacea	LC_{50}	110–10,000	2–4
PCNB	82688	Fungicide	Mammal	LD_{50}	1,700,000	4
	82688	Fungicide	Fish	LC_{50}	1,000–40,000	3–4
Thiophanate-methyl	23564058	Fungicide	Mammal	LD_{50}	790,000–7,500,000	4–5
	23564058	Fungicide	Fish	LC_{50}	7,800–130,000	3–5
	23564058	Fungicide	Crustacea	LC_{50}	16,000	4
Thiram	137268	Fungicide	Fish	LC_{50}	0.3–430	1–2
	137268	Fungicide	Amphibia	LC_{50}	13–25	1
	137268	Fungicide	Annelida	LC_{50}	670	2
	137268	Fungicide	Platyhelmi	LC_{50}	48–530	1–2
	137268	Fungicide	Insect	LC_{50}	390–1,920	2–3
	137268	Fungicide	Crustacea	LC_{50}	8–1,882,000	1–5

From Murphy 1990, 1992.

[a] *Range of toxicity classes are defined in Table 1.*

toxicity of certain chemicals to laboratory species, the same compounds have proven to be highly toxic on nontarget species which inhabit golf courses (Murphy 1992; Zinkl et al. 1978; also see Chapter 8). Additionally, test data shows the delayed lethal effects of some pesticides. The C-12 trichlorfon LD_{50} data have a dual toxicity pattern. Initially, bird mortality occurs rapidly (between 15 min and 3 hr). There also is a delayed toxic response to exposure. This delayed response occurs from several hours to as long as eight days after treatment (Hudson 1984). Use of localized or short-term field observations definitely obscures the complete pattern of toxic effects of such compounds.

A few studies have been conducted on reptiles and amphibians (Hudson et al. 1984; Pacces Zaffaroni et al. 1986; Powell et al. 1982). It has generally been found that these species metabolize pesticides rapidly and thus pass them out of their system in a short period of time. This suggests these species may not be as vulnerable to the toxic effects of certain chemicals as birds, for example.

6.2.2.2 Impacts of Turfgrass Pesticides on Wildlife: A Risk Assessment. About 50 different pesticides are currently in major use on golf courses. These include insecticides, herbicides, fungicides, and nematicides (see Table 1 in Chapter 4; Balogh and Anderson 1992). Several different groups of chemically related pesticides exist. These groups generally possess a particular toxic action. The toxicity of each group and perhaps even a specific chemical may depend on the species of wildlife involved. Some of these chemical groups are the organophosphates, phenoxy herbicides, carbamates, and heavy metals.

The *organophosphorus* (OP) *pesticides* are one of the most widely used groups on golf courses. This group includes ethoprop, trichlorfon, chlorpyrifos, fenamiphos, isofenphos, and isazophos. Another commonly used OP insecticide, diazinon, is no longer registered for use on golf course turfgrass, but is still registered for use on residential lawns. The majority of wildlife related research exists for these compounds (Table 3), which are one of the more toxic groups to wildlife, especially to birds (Grue et al. 1982; Hill and Fleming 1982; Stone and Gradoni 1985; Stone and Knoch 1982; White et al. 1979; Zinkl et al. 1978).

The OP compounds depress cholinesterase (ChE) enzyme levels in the brain and act as a neurotoxin in nontarget species. In addition, other enzymes may be inhibited which normally metabolize and detoxify the organophosphates (Grue et al. 1983). Young birds and mammals are more sensitive to OPs than adults. It is believed that wildlife affected at sublethal levels die of exposure, predation, starvation, or dehydration (Eisler 1986). Behavioral modifications resulting from exposure to OP pesticides may include learning impairment and reproductive effects. Grue et al. (1982) reports behavioral aberrations such as reduced nestling feeding and nest attentiveness from OP intoxication. Effects on reproduction include reduced egg production, reduced litter sizes, low fledgling weights, and decreased postfledgling

Table 3. Effects of Turfgrass and Related Pesticides on Wildlife and Test Animals

Cover type	Pesticide	Wildlife type	Wildlife response and effects
Organophosphate pesticides			
Hay meadows	Fenthion	Frog breeding ground	No organophosphate (OP) residues were detected in any frog samples. After 24 hr no residues were detected in water samples. The frogs were not a probable source of secondary toxicity for predatory species in the food chain (Powell et al. 1982).
Fairway	Triumph 4E Isazofos 5 kg a.i.ha^{-1}	Bird breeding range	Greatest routes of exposure were in water, turfgrass, and invertebrates. Thousands of dead or intoxicated mole crickets in treated area were food items for several different bird species. Necropsy evaluation was done. No brain cholinesterase activity analyses were done to determine organophosphate exposure. Mortality seemed correlated to application. Subsequent increased mortality from golf balls, carts, mowing equipment, and predation probably due to sublethal effects of pesticide (Brewer et al. 1988).
Golf course	Diazinon	Canada geese	One living and 14 dead geese were found. Bird could fly weakly, but salivated profusely and could not stand. Brain ChE activity was depressed by 69% compared to controls. Diazinon concentration of grass taken from the mouth of one goose was 20 ppm (about four times the LD_{50} for pheasant, five times that of mallard). Diazinon was applied at least 3 months prior to die off. Cover application, spill, pooling, or runoff is possible (Zinkl et al. 1978).
Golf course (case 1)	Diazinon AG500 48% 95 ml 100 m^{-2} irrigated	Brant feed area	Approximately 700 brant died on three fairways. Periods of rain over the next 3 days occurred after application. Analysis of grass residue from the esophagus of one bird showed 79 ppm diazinon. Heavy metal residues were low enough to be eliminated as cause of death. No parasitic infestation or disease was evident. Death was attributed to diazinon poisoning based on history, grass pathology, and chemistry. Country club paid a maximum civil

			penalty for commercial application of pesticide without a valid applicator's certificate (Stone and Gradoni 1985).
Golf course (case 2)	Diazinon AG500 127 ml 100 m^{-2}	Canada goose feeding area	Actual application rate was probably closer to 95 ml/100 m^2; however, treated areas were not watered-in, as required. About 10 geese were poisoned. Necropsies showed birds were in otherwise good flesh. Typical signs of OP poisoning were evident. Golf course owner donated $750.00 to state wildlife program in exchange for violation being dropped (Stone and Gradoni 1985).
Home lawn (case 3)	Diazinon 5% granular	Robin feeding	Accidental application was made by a homeowner. Pesticide was not watered-in as directed. Light rain followed 1 day later, with heavy rain 2 days later. One adult robin was found dead in the yard, one was found immobile and several others were weakened. Fourteen ppm diazinon were found in the gizzard and 0.5 ppm in the intestine. Unwashed earthworms from treated area contained 13 ppm diazinon. Label directions for average homeowner may be too complex to follow (Stone and Gradoni 1985).
Golf course	Diazinon 2.2 kg a.i. ha^{-1}	Waterfowl feeding area	Eighty five American widgeon died after feeding on a golf course study site. Post-irrigation residues of diazinon were higher than expected. A surge of diazinon residue was reported in runoff from a 48-hr heavy rain that occurred 12 days after the application of diazinon. Acetylcholinesterase (AChE) activity in the dead widgeon was depressed by 44 to 87%. Necropsy results indicated that some sublethally affected widgeon were probably killed by predators. Canada geese also were affected sublethally, but appeared to recover. One American robin was found dead on the site. Migration studies strongly suggest there are definite time periods in the year when widgeon and other waterfowl are present or absent. Avoiding pesticide application during migratory periods reduces the potential for pesticide exposure (Kendall et al. 1987).

Table 3. Effects of Turfgrass and Related Pesticides on Wildlife and Test Animals (continued)

Cover type	Pesticide	Wildlife type	Wildlife response and effects
—	Dicrotophos 2.5 mg kg^{-1} body wt	Starling nest area	Affected females made fewer trips to feed young and were away from nests for longer period than controls. Significant weight loss was found in affected young compared to controls. Care of nestlings by adult was significantly reduced due to severe but sublethal exposure to OP pesticide (Grue et al. 1982).
Laboratory	OP pesticide bioaccumulation	Amphibians	OP compounds may bioaccumulate in amphibians to levels which may be detrimental to predatory wildlife species which feed on them. Bioconcentration factors up to 335[a] for parathion in model studies have been reported. Fish have been documented to have residues of 26.4 ppm of parathion when exposed to a water concentration of 0.5 ppm. Frogs are resistant to ChE inhibition, and tadpoles can concentrate the pesticides up to 60 times the concentration in water. Tadpoles exposed to 1 ppm parathion and 5 ppm fenthion killed mallard ducklings. Dicrotofos, malathion, and acephate did not bioaccumulate to lethal levels when eaten in a single meal by ducks. Mortality was attributed to the accumulation of the parent compound in the tadpoles, rather than effect of a metabolite (Hall and Kolbe 1980).
Carbamates, chlorophenoxy herbicides, and fungicides			
Soil	Multiple	Earthworm, food chain item	Ninety chemicals were tested on earthworms and used as an indicator of relative chemical toxicity to soil invertebrates. Only carbofuran and eserine salicylate, both carbamates, were supertoxic. Metabolic products of parathion, carbaryl, 2,4-D (and 2,4,5-T) were as toxic, or more toxic, than the parent compound, all being extremely toxic or very toxic. Chemicals moderately to relatively nontoxic to mammals, but extremely or very toxic to earthworms included carbaryl, malathion, cypermethrin, and benomyl. This showed how unpredictable chemical toxicity can be on various animal

			species. Thus environmental risk to one species based on results from another can be difficult to assess (Roberts and Dorough 1984).
Golf course	Multiple fungicides	Earthworm, food chain item	Fungicides affect populations as seen by the number of casts produced by earthworms. The fungicide applied and number of earthworm casts observed per plot after application are as follows: control, 100 casts; methyl thiophanate, 56 casts; and chlorothalonil, 38 casts; dithiocarbamate and mancozeb, 32 casts; anilazine, 29 casts; benomyl, 20 casts; and triadimefon 18 and 10 casts. If repeated applications are needed, use of a low toxicity fungicide is especially important (King and Dale 1977).
Laboratory	2,4-D, 2,4,5-T, MCPA, MCPP, dichlorprop	Hen's eggs	Phenoxy herbicides were injected into eggs or they were submerged. Injection of 2 mg of herbicide into a 60 mg egg decreased hatching percentage and chick viability. Immersion into a 1% solution had no effect, and 5% solution had moderate effects. Dioxin was 100 times more toxic than any of the herbicides (Gyrd-Hansen and Dalgaard-Mikkelsen 1974).
Laboratory	2,4-D and picloram 2.8 kg ha^{-1} 11.2 kg ha^{-1}	Pheasant eggs	Fertile eggs were sprayed with two doses of a 2,4-D and picloram mixture. No adverse effects were observed on hatching success, occurrence of deformed embryos, or chick death (Somers et al. 1974).
Laboratory	MCPA	Amphibian	A correlation existed between herbicide concentrations and mortality in both sexes. Crested newts seem to be less sensitive than other aquatic species to MCPA. Severe neuromuscular disorders were evident in individuals exposed to 1600 and 3200 ppm MCPA. The sodium salt of MCPA appears to be of low toxicity to the crested newt (Pacces Zaffaroni et al. 1986).
Heavy metals			
Laboratory	Cadmium	Wood ducks	Cadmium (Cd) is probably incorporated into feather structure. Cd concentration increases in feather tissue with increased Cd ingestion. Chronic toxicity results show liver concentrations highest with increasing doses. Liver:kidney ratios higher than 1 show renal damage, which will aid in identifying Cd intoxication in waterfowl (Mayack et al. 1981).

Table 3. Effects of Turfgrass and Related Pesticides on Wildlife and Test Animals (continued)

Cover type	Pesticide	Wildlife type	Wildlife response and effects
Laboratory	Cadmium	Mallard ducks	About 97% of the Cd found in tissues was located in the liver and kidney. Cd levels of 75 ppm significantly decreased body weight of young chicks after 3 weeks. Sixty ppm of Cd in the diet of chickens was found to suppress egg production, and 600 ppm had the same effect in ducks, as well as reducing hatchability in eggs before laying stopped. Duckling growth was also reduced. Cd transfer to eggs apparently is low. There seems to be a correlation between levels of Cd in the body and environmental contamination (White and Finley 1978).
Golf course	Cadmium	Japanese quail	Earthworms may ingest and concentrate amounts of Cd from the soil. Earthworms were collected from a golf course treated with fungicides and municipal sewage sludge containing Cd. Quail were fed 47.8 ppm dry weight of Cd for 27 days. Cd levels were highest in the kidney and liver tissue. Concentrations of Cd were 13.70, 4.71, 0.14, 17.70, 5.29, and 0.08 ppm, dry weight. The pH of avian (bird) digestive fluids helps determine Cd availability from the soil and food items such as earthworms, as Cd solubility increases as soil pH decreases. The birds appeared normal at the measured treatment levels. (Pimental et al. 1984).

[a] *The bioconcentration factor (BCF) is a unitless proportionality constant relating the concentration of a chemical in water to its concentration in the animal at steady state equilibrium. The BCF is an estimate of a chemicals predisposition to accumulate in aquatic organisms.*

survival. Reduced visual acuity, hypothermia, and reduced immune response resistance also have been observed (Grue et al. 1982, 1983). Loss of appetite is a common factor in pesticide exposure, which may result in weight loss. This is particularly significant for birds and mammals with high metabolic rates.

In a review of OP effects, Grue et al. (1983) reported that OPs are best detected in brain tissue, as they are rapidly metabolized in body tissues. In birds and reptiles, brain ChE, which is inhibited by 50% or more in dead specimens, indicates OP poisoning. A criteria for detecting anti-ChE compounds has not been found for amphibians and mammals. However, brain ChE activity has been used to determine sublethal exposure of OPs in amphibians, reptiles, birds, and mammals. Brain ChE depression of 20% or more indicates OP exposure in birds, and by 40% for reptiles.

Even at recommended rates of application, diazinon has been responsible for the deaths of honey bees, and particularly waterfowl which consume treated grass (Brewer et al. 1988; Eisler 1986; Stone and Gradoni 1985; Zinkl et al. 1978). It has also been known to result in the death of songbirds feeding directly on treated insects (Maffei 1978). Diazinon became enough of a risk to the welfare of some bird populations that in 1985 the EPA discontinued the use of diazinon on golf courses and sod farms (EPA 1985b).

Different formulations of pesticides have variable toxicity for many wildlife and aquatic species. An experimental insecticide, C-12 trichlorfon, differs from trichlorfon by the addition of a long carbon chain. This added carbon chain has substantially decreased the toxicity of the derivative to mallards, pheasants, bobwhites, and possibly to California quail (Hudson 1984). Another OP pesticide, ethoprop, is highly toxic to aquatic organisms, fish and invertebrates, and birds (CPCR 1991; Hudson et al. 1984). Many of its formulations are designated as RUP. The use of this particular chemical has been discouraged by Stone (1980) due to its potential toxicity to local wildlife species. The deaths of brant (geese) have been diagnosed and attributed to the application of isofenphos, an OP pesticide, on a golf course (Stone and Gradoni 1987).

Chlorpyrifos demonstrates a high degree of cumulative action for an OP pesticide. At recommended treatment levels, chlorpyrifos has documented harmful effects on nontarget resident species, including fish, aquatic invertebrates, honeybees, birds (including ducklings), and humans (Odenkirchen and Eisler 1988). In general, mammals appear to be relatively tolerant to lethal toxic effects of chlorpyrifos. For most test species, decreasing the dosage levels did not produce proportional decreases in response. Mortalities occur at levels much lower than the calculated LD_{50}. There is a function of effect of age on sensitivity of mallards to this insecticide as well (Hudson et al. 1984). The effectiveness of chlorpyrifos, and other OP pesticides, depends on several variables. These include formulation, route of administration, pond substrate, dose, and water temperature (Odenkirchen and Eisler 1988).

Carbamates are another chemically related pesticide group. Examples of carbamates include the insecticides bendiocarb and carbaryl and the fungicide benomyl. The mode of action of these compounds are similar to the organophosphates by inhibiting cholinesterase. The toxicity of the different carbamate formulations is highly variable (see Chapter 8) (Murphy 1992). Some forms are considered acutely toxic for several wildlife species, particularly bees, and wildlife that consume fish or treated water (see Chapter 8) (*Farm Chemicals Handbook* 1991; Murphy 1992). Several carbamate pesticides used in golf course management exhibit extreme and very high toxicities to terrestrial test species (toxicity rating = 1 and 2) (Tables 1 and 2). Symptoms of carbamate or OP poisoning in birds and mammals may include muscular incoordination, wing spasms, wing droop, hunched back, labored breathing, tear production, salivation, prostration, diarrhea, sweating, vomiting, convulsions, loss of reflex, and coma (Hudson et al. 1984).

Less wildlife research exists for the currently registered *phenoxy herbicides* than for the organophosphates. The phenoxy herbicides used for turfgrass management include 2,4-D, MCPA, and MCPP. Several studies were conducted using eggs of game birds and domestic fowl. Somers et al. (1974) reported one study that found extensive embryo mortality and malformed game bird chicks when 2,4-D was sprayed at recommended rates onto the eggs. Consecutive studies failed to produce the same deleterious effects using hen and ring-neck pheasant eggs (Gyrd-Hansen and Dalgaard-Mikkelsen 1974; Somers et al. 1974). Studies with amphibians have shown MCPA to be low to moderately toxic (Pacces Zaffaroni et al. 1986) (Table 3). However, dogs possess a poor ability to excrete organic acids and are thus particularly sensitive to the phenoxy group (Schwetz 1977).

Many pesticides exhibit a wide range of toxicities across one group of animals and within a single species. Toxicity of atrazine, a selective preemergence *triazine herbicide*, ranges from highly toxic to practically nontoxic to fish (Table 2). This range occurs even within species such as in bullhead, guppies, and bluegill (see Chapter 8) (Murphy 1992). In this situation, managers should use discretion with directions for standard application. Pesticide application should be conducted to meet specific needs, while avoiding potential problems for local wildlife. Careful consideration of circumstances on a particular golf course can help mitigate or even eliminate wildlife casualties and environmental contamination with regard to pesticide use. Some chemicals are considered safe due to their high adsorption to soil with low potential for leaching. These compounds, however, can be made available to bottom feeders including fish, diving water fowl, and benthic organisms, which are all important parts of the food chain (Isensee 1991). Highly adsorbed compounds are transported to aquatic systems by sediment transport and become part of bottom sediments (Muir 1991).

Another group of potentially toxic compounds contain the *heavy metals* such as mercurial compounds (mercurous and mercuric chloride), organic arsenicals, and cadmium. Although many of these compounds are no longer used to any extent on golf courses, turfgrass fungicides containing heavy metals, such as the organic arsenical MSMA and DSMA, are still major use herbicides on golf course turfgrass (see Table 1 in Chapter 4; Balogh and Anderson 1992). Although a variety of aquatic organisms are known to accumulate arsenic, no magnification has been known to occur along the aquatic food chain (Odenkirchen and Eisler 1988).

Cadmium is no longer a major use chemical on golf courses, but it may still persist in the soil from previous use. Current sources of cadmium on golf courses are fertilizers derived from municipal sewage sludges. The liming of golf courses helps to reduce mobility of cadmium in the soil. Research has clearly shown that earthworms do ingest and concentrate cadmium from the soil. It is a highly toxic heavy metal which accumulates in living tissue when ingested. The uptake and bioconcentration of cadmium by worm-eating songbirds is a concern (Pimentel et al. 1984).

Cadmium causes a wide variety of physiological damage and alterations, including intestinal mucosa and testicular damage in Japanese quail; intestinal lesions and hypertension in rabbits, rats, and possibly humans; and permanent sterility in rats (Richardson et al. 1974; Spivey Fox et al. 1971). Since heavy metals are very persistent in the environment and highly toxic to all forms of life, concern should be exercised regarding long- and short-term effects on the golf course as much as anywhere else.

6.2.2.3 Additional Concerns and General Guidelines on Toxicity of Pesticides to Wildlife. Several guidelines and techniques have been made for assessing pesticide exposure and effects on wildlife. Using radio telemetry techniques to monitor wildlife following pesticide exposure may be particularly helpful. Colonial species may provide a more adequate sample size for study. Census methods which rely on visual or auditory data alone are not advised due to difficulty in interpreting results (Grue et al. 1983).

For potential OP poisoning, measurements of brain ChE activity of the animal from immediately frozen brain tissue and residue analysis on food sources must be included. Creation of a reliable database from field cases using *species-specific* ChE activity and associated effects can be valuable. This can assist in quantifying direct and indirect effects of OP application on reproduction and survival (Grue et al. 1983; Hill and Fleming 1982). Data from field research needs to be acquired and analyzed particularly for sublethal and synergistic (combined enhancing) effects on nontarget species.

Pesticide application methods should be planned to minimize potential exposure and limit conflicts with the activity of wildlife species. Most species are very active in the morning hours when most pesticide application occurs. Late evening treatments in areas frequented by wildlife will help minimize exposure by most species (Grue et al. 1983). In the past, golf course maintenance programs, as well as agricultural systems, often followed a rigid schedule of nutrient, water, and pesticide application with little regard for climatic events, appropriate rates of application based on plant uptake requirements or economic thresholds, pest life cycles, and proper handling and response to spills (Leslie and Metcalf 1989). The concept of integrated pest and turfgrass management (IPM) is a significant concept regarding all environmental impacts and maintenance of quality turfgrass.

The use of IPM practices to reduce chemical loading on golf courses also will mitigate impacts on wildlife. Consideration of wildlife should be a factor in developing integrated management systems for golf courses. One facet of IPM, the use of biological controls and cultivars resistant to disease and pests, can help reduce direct or indirect toxic effects on wildlife. Another component of IPM is the development of economic thresholds and scouting for pests and diseases prior to chemical treatment. Reduced chemical loading of golf courses resulting from prescription treatment wherever possible rather than preventative control will also benefit wildlife (Stone 1979). Less rigid adherence to single management strategies not only benefits the quality of turfgrass for golf, but enhances the other positive attributes of golf courses.

Alternate strategies may incorporate wildlife species into pest control plans. Sugarloaf Country Club in Kingfield, ME has been plagued with thriving populations of black flies and mosquitoes. Responding to a legislative mandate which prevented the course's former method of insect control, they undertook a more natural approach (Kahler 1990). After consulting with a local bat expert, they constructed and placed bat houses on the course. In addition, bird boxes were added to the golf course to accommodate swallows, another insect-eating species. The local bat biologist has helped course managers learn that an average of only ten bats can consume more than one million mosquitoes during a summer. Each bat house is small, but can house from 30 to 60 individuals, depending on the species of bat. These small structures may be placed on a course in inconspicuous places. Although bats are a migratory species, they return each year to raise their young and usually bring more females along with them. Managers should be aware that it will take more than 1 year to establish a population of bats in an area where bat boxes are put up. However, once the females are attracted to an area, they can have between 25 and 30 offspring in 1 year.

There are many other advantages to this natural insect control program. Bats are nocturnal animals. While golfers are out playing the game, the bats are in their boxes. At dusk bats come out to begin their work of eating insects. Construction of the bat houses is a one-time investment, virtually eliminating the large annual expense of repeated chemical applications. Bats do their job quietly and neatly, without creating any concerns to the greenskeeper. Bats exist in nearly every part of the country, which makes them a likely candidate for insect control on almost any golf course. They have proven to be a valuable mammalian species. The concept of bats for insect control should be included in an environmentally minded pest program.

6.3 GOLF COURSES AS WILDLIFE HABITAT

6.3.1 Wildlife Habitat Selection

As urban areas continue to expand into places once used as wildlife habitats, golf courses are becoming increasingly critical as natural habitats in urban and suburban environments. There are a surprising number of wildlife species which inhabit golf courses (Table 4). Not only do golf courses provide a habitat for wildlife reproduction, they serve as critical overwintering areas for migratory and nonmigratory species. Andrew et al. (1987) reported on the diversity of species which exist throughout the year on a midwestern U.S. golf course (Table 4). A list of species in the southwestern states would be completely different. In addition, the variety of species using a particular golf course may change on an annual or seasonal basis. Factors influencing the selection of golf courses as wildlife habitats include climatic patterns, migratory ranges, food availability, habitat changes, and successional changes in vegetation.

Golf courses provide food, cover, water, space, and nesting habitats for many wildlife species (Maffei 1978). Food and water are essential for growth and development. Food sources include insects, seeds, nuts, and berries. Cover is provided by trees and shrubs. Water that exists in the form of ponds and wetlands, and open areas surrounded by a variety of vegetation provide habitat diversity for many species. Vegetative cover provides critical shelter for wildlife (Maffei 1978). Low maintenance areas, especially the rough, trees, and shrubs, are used for reproductive activities, escape, and shelter for resting. All of these factors are fundamental components of a wildlife habitat. Intensive turfgrass management practices modify the structure of wildlife communities on golf courses. Many of these management practices have an indirect effect on the structure and dimensions of a wildlife habitat.

6.3.2 Indirect Effects of Pest Management on Wildlife Habitat

One type of indirect effect of pesticide use on turfgrass involves changes in insect food populations. Wildlife survival and reproduction corresponds directly to the abundance of food. Many species rely on peaks in insect populations during their reproductive season. Insecticides and other pesticides affect the reproduction of wildlife by decreasing their food supply. Birds and mammals emigrate from pesticide-treated areas due to the reduction in the abundance of insects (Grue et al. 1983). This can have serious consequences for breeding birds. They are affected by changes in territory size and structure, increased risk of predation, increased difficulty in locating new nest sites and building new nests, and increased competition for food. For those species remaining in a treated area, the reduction in nontarget food items could mean reduced survival of young offspring (Grue et al. 1983).

Table 4. Wildlife Species Using a Golf Course Area in Massachusetts

Butterflies	**Insects**
Browns/Satyrs: 3 species	Beetles: 5 species
Brush-footed: 6 species	Bugs: 3 species
Hairstreaks/Coppers/Blues: 6 species	Dragon Fly: 1 species
Milkweed butterfly (Monarch): 1 species	Katydids: 1 species
Skippers: 8 species	Long-horned Grasshopper: 1 species
Swallowtails: 5 species	Moths: 4 species
Whites/Sulphurs: 4 species	Northern Masked Chafer: 1 species
	Spiders: 3 species
	Wasps: 3 species
	Water Skipper: 1 species
Mammals	**Reptiles**
Cottontail Rabbit	Frogs:
Eastern Chipmunk	Bull
Ground Hog	Leopard
Meadow Vole	Snakes:
Muskrat	Copperhead
Opossum	Garter
Raccoon	Juvenile
Skunk	Toads:
Squirrels:	American
Fox	Fowlers
Gray	Turtles:
White-Footed Mouse	Box
White Tailed Deer	Snapping
	Softshell

Birds

American Goldfinch
American Redstart
Belted Kingfisher
Blackbirds:
  Rusty
  Red-winged
Blue Gray Gnatcatcher
Blue Jay
Brown Creeper
Brown Thrasher
Brown Headed Cowbird
Carolina Chickadee
Cedar Waxwing
Chimney Swift
Common Crow
Common Flicker
Common Grackle
Common Redpoll
Common Yellowthroat
Cuckoos:
  Black-Billed
  Yellow-Billed
  Yellow-Tailed
Doves:
  Mourning
  Rock
Eastern Bluebird
Eastern Kingbird
Eastern Meadowlark
Eastern Peewee
Eastern Phoebe
European Starling
Finches:
  House
  Purple
Flycatchers:
  Acadian
  Great Crested
  Least
  Willow
Gray Catbird
Grosbeaks:
  Blue
  Evening
  Rose-Breasted
Horned Lark
Indigo Bunting
Juncos:
  Northern
  Oregon
Louisiana Waterthrush
Northern Cardinal
Northern Mockingbird
Northern Parula
Night Hawk
Nuthatches:
  Red-Breasted
  White-Breasted
Orioles:
  Northern
  Orchard
Kinglets:
  Golden Crowned
  Ruby Crowned
Ovenbird
Pine Siskin
Purple Martin
Robin
Ruby-Throated Hummingbird
Rufous-Sided Towhee
Sparrows:
  Chipping
  Field
  Fox
  Grasshopper
  Henslow
  House
  Savannah
  Song
  Swamp
  Vesper
  White-Crowned
  White-Throated
Snow Bunting
Swallows:
  Bank

Table 4. Wildlife Species Using a Golf Course Area in Massachusetts (continued)

Birds (continued)	Prothonotary	Merlin
Barn	Worm-Eating	Northern Harrier
Rough-Winged	Yellow	Owls:
Tree	Yellow-Throated	Barn
Tanagers:	Yellow-Rumped	Barred
Scarlet	Water Pipit	Great Horned
Summer	Wrens:	Long-Eared
Thrushes:	Bewick's	Saw-Whet
Hermit	Carolina	Screech
Wood	House	Short-Eared
Townsend's Solitaire	Winter	Red-Tailed Hawk
Tufted Titmouse	Whip-Poor-Will	Red-Shouldered Hawk
Vireos:	Woodpeckers:	Rough-Legged Hawk
Red-Eyed	Downy	Sharp Shinned Hawk
Warbling	Hairy	Turkey Vulture
Yellow-Eyed	Pileated	
Yellow-Throated	Red-Bellied	**Quail, Herons, and Sandpipers**
Warblers:	Red-Headed	Bobwhite Quail
Blackpoll	Yellow-Bellied Sapsucker	Herons:
Black and White	Yellow-Breasted Chat	Black Crowned
Blue Winged		Great Blue
Cerulean	**Birds of Prey**	Green
Hooded	American Kestrel	American Woodcock
Kentucky	Broad-Winged Hawk	Common Snipe
Magnolia	Coopers Hawk	Killdeer
Prairie	Goshawk	Spotted Sandpiper

Waterfowl	Canvasback	Hooded
American Coot	Gadwall	Red-Breasted
American Wigeon	Grebes:	Pintail
Black Duck	Horned	Redhead
Blue-Winged Teal	Pied-Billed	Ring-Billed Gull
Bufflehead	Lesser Scaup	Ring-Necked Duck
Canada Goose	Mallard	Ruddy Duck
Common Goldeneye	Merganzers:	Shoveler
Common Loon	Common	Wood Duck

From Andrew et al. 1987.

Perhaps the greatest indirect impact of pesticide use and cultural practices on wildlife is that created by a change in habitat characteristics. This can occur through the use of herbicides and other chemicals which change the vegetation structure. Insecticides eliminate food sources. Herbicides also affect prey populations by eliminating their vegetative food source and habitat (Eisler 1989). Wildlife habitat loss also occurs during construction and redevelopment of golf courses (Edmondson 1987; Holing 1987).

6.3.3 Protection of Wildlife Habitat During Golf Course Development

As with any other type of urban development, golf course construction impacts wildlife. Initial site preparation, with heavy machinery and change of vegetation to a turfgrass monoculture, profoundly alters the area, causing an exodus of wildlife. Depending on the season and region of the country, nesting sites and breeding grounds are destroyed. Feeding areas are altered. With proper reclamation and consultation with local wildlife experts, these may be replaced and enhanced, attracting preexisting and possibly new species upon completion. It is important to remember that *ongoing management* of these resources is critical to maintaining wildlife.

Careful selection of golf course sites can sometimes preempt certain problems toward and with wildlife, and consequently, potential legal battles. The value of cooperation with local biologists, conservation experts, and local organizations cannot be understated in each case. For example, courses have been placed in the middle of elk migration routes, heavy waterfowl use areas, and habitats of endangered and threatened wildlife species (Edmondson 1987; Holing 1987). Elk, with their historic and well-established migration areas, are not able to easily "re-route." This results in effects on the elk and troubles for the greenskeeper as well.

The dynamic state of environmental, public, and regulatory concerns has increased the difficulty for permission to develop new golf courses using existing construction practices. In some coastal areas where wetlands are incorporated into the course, buffer strips must be incorporated to protect bogs and other wetland areas from fertilizer and pesticide contamination (Jones 1989). The question remains whether current criteria safeguard a wetland area and the waterfowl which use it. For example, buffer zone size must be sufficient to trap persistent pesticides with potential for surface losses. Factors regarding sand substrate, slope, and drainage systems should be incorporated into the design of buffer strips.

Studies by the Audubon Society of New York State Inc. indicate that 64% of the habitat of endangered bird species is lost to wetland drainage and habitat destruction (Linder 1990). Therefore, a component of their current program caters to golf courses and proposed golf course sites to help them qualify as a wildlife sanctuary under the Cooperative Sanctuary System (CSS). To receive information on how to become a participant in the CSS program contact Audobon Cooperative Sanctuary System; c/o The Audubon

Society of New York State, Hollyhock Hollow Sanctuary, Route 2, Box 131, Selkirk, NY 12158.

In Great Britain, some courses are recognized as important wildlife habitats due to the diversity of species, including rare species. In fact, at least half of Britain's top courses are designated Sites of Special Scientific Interest (SSSI). They receive management assistance from the Nature Conservancy Council (NCC) who helps protect habitat and the wildlife it supports on golf courses (NCC 1989).

In a recent report on mitigating the negative effects of golf courses on the aquatic habitat, Klein (1990) identified several problems associated with golf course construction and maintenance. These factors include the following:

1. Stream channelization
2. Loss or permanent destruction of wetlands
3. Loss of wooded buffer zones on waterways
4. Increasing the temperatures of streams and lakes as a result of shade reduction, reduction of interflow recharge, release of heated water from shallow ponds, and release of heated storm flow from impervious surfaces
5. Reduction of stream base flow resulting from irrigation withdrawal
6. Release of toxic substances and oxygen deficient water from ponds
7. Pollution of surface water and groundwater from periodic spills of pesticides, fertilizer, and fuels
8. Pollution of surface water and groundwater resulting from nonpoint movement of pesticides and fertilizers
9. Movement of pollutants in storm flow from impervious surfaces
10. Acceleration of channel scouring due to increased duration and/or velocity of storm water runoff
11. Altering the frequency and magnitude of flooding in surrounding terrestrial and aquatic systems
12. Poor erosion control during design and construction
13. Inadequate treatment of sewage and other wastewaters generated on golf courses

Klein (1990) outlined a series of guidelines for (1) screening potentially sensitive sites, (2) design criteria to reduce erosion and alteration of natural runoff features of the landscape, (3) plans for planting or maintenance of wooded buffer zones, and (4) irrigation and water management practices to reduce withdrawals.

Many new courses are constructed on urban fringes and agricultural land. Often, however, the uniqueness of natural features is what draws the developer. At this time, questions must be asked and balanced decisions made. Will the environmental quality of this area be degraded, despite the utmost care to prevent such impacts? Are the needs of public recreation greater than the need to preserve or conserve the wildlife for future generations to enjoy? Planning authorities or commissions will ultimately decide these issues. It is important for developers, local organizations, and the public to meet and discuss the issues early in the planning stages (NCC 1989).

Conservation easements provide a means of preserving wildlife habitat and wetlands while allowing other development projects to continue in environmentally sensitive areas. The concept of conservation easements was reviewed by Linder (1990). Conservation easements are parcels of land which are deeded to a charitable trust or environmental organization to maintain natural land or water areas. This in no way gives the nonprofit organization title to the property. Rather, it gives individual rights to an organization to maintain the land in an undeveloped, natural state. For example, a part of a proposed golf course may contain a wetland or environmentally sensitive area. This property may be deeded to a land trust or environmental group to maintain in its natural state without any loss to the course. In addition, the course would gain public recognition. This would not require conversion of private golf courses to public use.

The only restrictions on the land in conservation easements would be those stated by the actual land owner. It is vital to have these restrictions documented in a written agreement. If a golf course should violate the agreement, such as filling in a wetland included in the conservation easement area, the grantee has legal power to stop the violation. A conservation easement is a good example of *cooperative development strategies*.

Another classic example of productive cooperation between groups involves a lizard and a multimillion dollar golf resort located in the Coachella Valley (Holing 1987). This valley near Palm Springs, CA, is one of the fastest growing urban areas and largest golfing hotspots in the world. It is also the home of the Coachella Valley fringe-toed lizard, an endangered species. This lizard has evolved an impressive list of adaptations for surviving in the baking heat of blow-sand deposits where surface temperatures can reach 180°F. It possesses hinged scales on its long toes, enabling it to "swim" underneath the sand. An overlapping upper jaw and adjustable nostrils with a U-shaped trap keep the sand out, and a third parietal eye on its head apparently monitors solar radiation.

Incredibly, several state and federal agencies, a county, nine cities, a Native American tribe, the real estate industry, construction companies, environmental groups, resource experts, and wildlife biologists cooperated to find a solution that satisfied all groups involved, including the Coachella Valley lizard. This cooperation resulted in the creation of a 20 mi^2 preserve set aside for the fringe-toed lizard, which exists in the midst of the golf course resort area. Realizing that all interested parties were at risk of losing what mattered to them, they decided to cooperate. After much time and effort, they found a solution through a clause in the Endangered Species Act. This was a particular set of conservation steps, called the habitat conservation plan (HCP). They decided to create a preserve for the lizard consisting of one of the more beautiful and undeveloped natural areas left in the valley. Funds for the project came from all of the involved interest groups. These included congressional appropriations, the Nature Conservancy, the Bureau of Land Management, California Fish & Game, and the land developers and builders. The methods utilized by this group serve as a blueprint for other groups across the country.

The long-term result of this cooperative effort is that the Coachella Valley Preserve is quickly becoming the only genuine oasis left in the middle of urban sprawl. Its intrinsic beauty will further enhance real estate values in surrounding areas. It harbors 25 nesting bird species, as well as many others that overwinter there; at least 130 plant species; and 180 mammalian species including bobcats, and other desert wildlife (Holing 1987).

It is critical to be aware of potential golf course lands which are biologically productive areas rich in wildlife and unique vegetation. Identify plant materials that must not be driven over during the construction phase. Cordon off these areas, or if this is impossible, move them to temporary holding areas until completion of construction. This results in a course pleasing to golfers and useful to wildlife.

6.4 GENERAL GUIDELINES FOR WILDLIFE HABITAT ENHANCEMENT

Grigg (1990) gives four basic phases of new golf course development: conception and planning, construction, establishment and growth of turfgrass and other habitats via sound agronomic practices, and ongoing management. Selecting a golf course construction company with a good environmental track record is important. Superintendents must be concerned about the long-term future of the golf facility in relationship to preserving the well-being of water resources, wildlife, and surrounding areas. Construction machinery traffic can be detrimental to natural or sensitive areas. Proper selection of turfgrass types can help minimize amounts of fertilizer, pesticides and water needed for maintenance. Care must be taken to prevent contamination of roughs when fertilizers or pesticides are used on tees, fairways, and greens. Selecting and placing native vegetation such as trees, bushes, flowers, herbs, grasses, and aquatic plants is crucial to how wildlife will use an area. While the turfgrass is being established, the staff can be educated on project and environmental goals. Conservation of all the resources on a golf course is a continuous process.

Irrigation practices often make golf courses into an ecosystem possessing constant moisture compared to the surrounding area. This factor is particularly critical in the desert Southwest, where there has been a huge amount of activity in the way of golf course construction. The desert ecosystem, with its cactus species, sand and rock, and other unique features, is profoundly altered by irrigation. Reptiles such as Gila monsters and other lizards, snakes, tortoises, insects, toads, frogs, owls and other birds, and mammals who rely on a desert environment to dwell, feed, and reproduce must emigrate (Edmondson 1987). The resulting habitat is such that preexisting species cannot return. Roughs may be preserved in their existing desert state, allowing some species to find it habitable. Ecosystem size may be an important consideration for many species that require larger territories to find food and reproduce (Gavareski 1976).

Considering the information presented in this chapter, some guidelines may be made concerning enhancement of wildlife habitat on golf courses.

1. Roughs can be made into wildlife sanctuaries. By preserving or maintaining native local vegetation (trees, shrubs, herbaceous plants, and flowers), a rough will become an oasis for endemic wildlife species (Gavareski 1976). This is especially true in desert areas such as the Southwest (Edmondson 1987).
2. The introduction and culture of exotic vegetation is generally discouraged. Not only will it be less suitable for wildlife, but it may be poorly adapted to the area and require more labor and chemical-intense management for its success (NCC 1989).
3. Native trees, such as oak for example, are capable of supporting several hundred different wildlife species, including invertebrates, birds, and mammals. The opposite is often true of exotic tree species (NCC 1989).
4. An inventory, with the help of local experts, should be conducted before construction to identify any special or unique habitat types, rare or endangered species, or special geological or ecological situations. The extra care taken at this time may very well enhance the course and make it perhaps more attractive in the end (Kahler 1990; NCC 1989).
5. Golf courses, including roughs, need to integrate an *ongoing management scheme* to insure continued success of a habitat and wildlife conservation plan. Besides not creating valuable wildlife habitat, an unmanaged woodland can be an eyesore on a course (NCC).
6. Even dead trees and snags provide vital wildlife habitats in safe areas where they are left standing (NCC).
7. The use of integrated management strategies will enable the manager to understand the natural processes which operate within a course (Leslie and Metcalf 1989).

6.5 CONCLUSION

Public pressure creates the need for course managers to produce impeccable turfgrass on golf courses. The same public pressure requires the golf course to be managed in a way which benefits players and wildlife alike. Perhaps American golfers should reassess some of their high expectations for the course in light of environmental quality and wildlife goals. Some aspects of the game and course may be available for compromise which would suit golfers, natural resources, and the environment alike. Events requiring change are already taking place and will most likely continue. The replacement of high-quality drinking water with treated sewage effluent for irrigation is one example. Integrative management strategies for turfgrass may be tailored for individual courses, proving beneficial and quite successful.

Golf course developers and managers now realize that the costs involved in maintaining natural areas are usually much lower than the potential cost of litigation. Development of natural areas on golf courses could reduce long-term maintenance expenses. Choosing an alternate site to an environmentally unique area or preserving that area within a golf course allows conservation and golf to coexist, often to the benefit of the other. Urban areas continue to expand as natural wildlife habitat dwindles in the United States. Wildlife is given the ever-burgeoning problem of finding natural areas in which to survive. Golf courses can provide a critical natural habitat. Enjoying the game of golf and preserving our wildlife heritage are mutually compatible goals.

REFERENCES

Anderson, J. L., Balogh, J. C. and Waggoner, M. 1989. Soil Conservation Service procedure manual: Development of standards and specifications for nutrient and pesticide management. Section I & II. USDA Soil Conservation Service. State Nutrient and Pest Management Standards and Specificaton Workshop, St. Paul, MN, July 1989. 453 pp.

Andrew, N. J., Stockdale, T. M., Kelley, D. L. and Holtzman, J. 1987. Wildlife Related Values of Park Golf Course Ecosystems. Hamilton County Park District and The Ohio State University, Cincinnati, OH, 103 pp.

Anonymous. 1991. *Pesticide Regulation Costs Rising*. Vol. 3. Number 3. Golf Course Superint. Assoc. Am., Lawrence, KS, p. 1.

Anonymous. 1988. Palm Springs turf benefits from reuse project. *SportsTURF* 5(2):28.

Balogh, J. C. and Anderson, J. L. 1992. Environmental impacts of turfgrass pesticides. in Balogh, J. C. and Walker, W. J. (Eds.). *Golf Course Management and Construction: Environmental Issues*. Lewis Publishers, Chelsea, MI, Chapter 4.

Balogh, J. C., Gordon, G. A., Murphy, S. R. and Tietge, R. M. 1990. The use and potential impacts of forestry herbicides in Maine. 89-12070101. Final Report Maine Dept. of Cons. Duluth, MN and Orono, ME. 237 pp.

Beard, J. B. 1982. *Turf Management for Golf Courses*. Macmillan Publishing Co., New York, 642 pp.

Brewer, L. W., Driver, C. J., Kendall, R. J., Lacher, T. E., Jr. and Galindo, J. C. 1988. Avian response to a turf application of Triumph 4E. *Environ Toxicol Chem*. 7:391–401.

CPCR. 1991. *Crop Protection Chemicals Reference*. Chemical and Pharmaceutical Press, John Wiley & Sons, New York, 2170 pp.

Cain, B. W., Sileo, L., Franson, J. C. and Moore, J. 1983. Effects of dietary cadmium on mallard ducklings. *Environ. Res*. 32:286–297.

Edmondson, J. 1987. Hazards of the game. *Audubon* 89(11):25–37.

Eisler, R. 1985. Cadmium hazards to fish, wildlife, and invertebrates: a synoptic review. U.S. Fish Wildl. Serv. Biol. Rep. 85(1.2). 46 pp.

Eisler, R. 1986. Diazinon Hazards to Fish, Wildlife, and Invertebrates: A Synoptic Review. U.S. Fish Wildl. Serv. Biol. Rep. 85(1.9). 37 pp.

Eisler, R. 1989. Atrazine Hazards to Fish, Wildlife, and Invertebrates: A Synoptic Review. U.S. Fish Wildl. Serv. Biol. Rep. 85(1.18). 53 pp.

Environmental Protection Agency. 1985a. Hazard Evaluation Division, Standard Evaluation Procedure, Acute Toxicity Test for Freshwater Fish. EPA-540/9-85-006. U.S. Environmental Protection Agency, Washington D.C.

Environmental Protection Agency. 1985b. Diazinon Support Document. Office of Pesticide Programs and Toxic Substances, U.S. Environmental Protection Agency, 401 M Street, S. W., Washington D.C. 20460.

Environmental Protection Agency. 1990. Environmental Fact Sheet. Office of Pesticides and Toxic Substances, U.S. Environmental Protection Agency, Washington D.C.

Farm Chemicals Handbook. 1991. *Pesticide and Fertilizer Dictionary*. Meister Publishing Company, Willoughby, OH, 770 pp.

Gavareski, C. A. 1976. Relation of park size and vegetation to urban bird populations in Seattle, Washington. *The Condor* 78:375–382.

GCSAA. 1991. We keep golf green. *Golf Course Manage*. 59(4):42.

Grigg, G. T. 1990. Seeking a fresh vision of environmental responsibility. *Golf Course Manage*. 58(9):38–46.

Grue, C. E., Fleming, W. J., Busby, D. G. and Hill, E. F. 1983. Assessing Hazards of Organophosphate Pesticides to Wildlife. in Trans. 48th No. Am. Wildlife and Nat. Res. Conf., pp. 200–220.

Grue, C. E., Powell, G. V. N. and McChesney, M. J. 1982. Care of nestlings by wild female starlings exposed to an organophosphate pesticide. *J. Appl. Ecol*. 19:327–335.

Gyrd-Hansen, N. and Dalgaard-Mikkelsen, Sv. 1974. The effect of phenoxy herbicides on the hatchability of eggs and the viability of the chicks. *Acta Pharmacol. Toxicol*. 35:300–308.

Hall, R. J. and Kolbe, E. 1980. Bioconcentration of organophosphorus pesticides to hazardous levels by amphibians. *J. Toxicol. Environ. Health* 6:853–860.

Hill, E. F. and Fleming, W. J. 1982. Anticholinesterase poisoning of birds: field monitoring and diagnosis of acute poisoning. *Environ. Toxicol. Chem*. 1:27–38.

Holing, D. 1987. Lizard and the links. *Audubon* 89(11):39–49.

Hudson, R. H., Tucker, R. K. and Haegele, M. A. 1984. Handbook of Toxicity of Pesticides to Wildlife. U.S. Dept. of Interior, Fish Wildl. Serv. Res. Pub. 153. 90 pp.

Isensee, A. R. 1991. Bioaccumulation and food chain accumulation. in Grover, R. and Cessna, A. J. (Eds.). *Environmental Chemistry of Herbicides.* Vol 2. CRC Press, Boca Raton, FL, pp. 187–197.

Jones, R. T., Jr. 1989. Use of wetlands in golf course design. *Golf Course Manage.* 57(7):6–16.

Kahler, K. E. 1990. Bats! *Golf Course Manage.* 58(11):36–42.

Kaplan, R. and Kaplan, S. 1989. *The Experience of Nature — A Psychological Perspective.* Cambridge University Press, New York, 340 pp.

Kendall, R. J., Brewer, L. W. and Hitchcock, R. R. 1987. Avian Response to and Environmental Chemistry of Turf Applications of Diazinon AG500 in Western Washington. Final Report. Institute of Wildlife Toxicology, Huxley College of Environmental Studies, Western Washington University, Bellingham, WA (unpublished).

King, J. W. and Dale, J. L. 1977. Reduction of earthworm activity by fungicides applied to putting green turf. *Arkansas Farm Res.* 26(5):12.

Klein, R. D. 1990. *Protecting the Aquatic Environment from the Effects of Golf Courses.* Community & Environmental Defense Associates, Maryland Line, MD, 59 pp.

Leslie, A. R. and Metcalf, R. L. (Eds.). 1989. Integrated Pest Management for Turfgrass and Ornamentals. Office of Pesticide Programs. 1989-625-030. U.S. Environmental Protection Agency, Washington, D.C., 337 pp.

Linder, T. 1990. Conservation easements. *Golf Course Manage.* 58(11):20–26.

Madhun, Y. A. and Freed, V. H. 1990. Impact of pesticides on the environment. in Cheng, H. H. (Ed.). *Pesticides in the Soil Environment: Processes, Impacts, and Modeling.* SSSA Book Series No. 2. Soil Science Society of America, Madison, WI, pp. 429–466.

Maffei, E. J. 1978. Golf courses as wildlife habitat. *Trans. Northeast Sect. Wildl. Soc.* 35: 120–129.

Mayack, L. A., Bush, P. B., Fletcher, O. J., Page, R. K. and Fendley, T. T. 1981. Tissue residues of dietary cadmium in wood ducks. *Arch. Environ. Contam. Toxicol.* 10:637–645.

Muir, D. C. G. 1991. Dissipation and transformations in water and sediment. in Grover, R., and Cessna, A. J. (Eds.). *Environmental Chemistry of Herbicides.* Vol 2. CRC Press, Boca Raton, FL, pp. 1–87.

Murphy, S. R. 1990. *Aquatic Toxicity Information Retrieval (AQUIRE): A Microcomputer Database.* System documentation. Specturm Research, Inc., Duluth, MN, 37 pp.

Murphy, R. A. 1992. Aquatic and terrestrial toxicity tables. in Balogh, J. C. and Walker, W. J. (Eds.). *Golf Course Management and Construction: Environmental Issues.* Lewis Publishers, Chelsea, MI, Chapter 8.

NCC. 1989. On Course Conservation: Managing Golf's Natural Heritage. Dept. OCC, Nature Conservancy Council, Northminster House, Peterborough PE1 1UA. 46 pp.

Odenkirchen, E. W. and Eisler, R. 1988. Chlorpyrifos Hazards to Fish, Wildlife, and Invertebrates: A Synoptic Review. U.S. Fish Wildl. Serv. Biol. Rep. 85(1.13). 34 pp.

Pacces Zaffaroni, N., Zavanella, T., Ferrari, M. L. and Arias, E. 1986. Toxicity of 2-methyl-4-chlorophenoxyacetic acid to the adult crested newt. *Environ. Res.* 41:201–206.

Pimentel, D., Culliney, T., Burgess, M. N., Stoewsand, G. S., Anderson, C. L., Bache, C. A., Gutenmann, W. H. and Lisk, D. J. 1984. Cadmium in Japanese quail fed earthworms inhabiting a golf course. *Nut. Rep. Int.* 30(2):475–481.

Powell, G. V. N., DeWeese, L. R., and Lamont, T. G. 1982. A field evalutaion of frogs as a potential source of secondary organophosphorus insecticide poisoning. *Can. J. Zool.* 60: 2233–2235.

Richardson, M. E., Spivey Fox, M. R. and Fry, B. E., Jr. 1974. Pathological changes produced in Japanese quail by ingestion of cadmium. *J. Nutr.* 104:323–338.

Roberts, B. L. and Dorough, H. W. 1984. Relative toxicities of chemicals to the earthworm *Eisenia foetida. Environ. Toxicol. Chem.* 3:67–78.

Roberts, E. C. and Roberts, B. C. 1989. *Lawn and Sports Turf Benefits.* The Lawn Institute, Pleasant Hill, TN, 31 pp.

Schwetz, B. A. 1977. Toxicology of Phenoxy Acid Herbicides. Proc. Soc. Am. For. Nat. Conv. pp. 256–260.

Sheedy, B. A., Lazorchak, J. M., Grunwald, D. J., Pickering, Q. H., Pilli, A., Hall, D. and Webb, R. 1991. Effects of pollution on freshwater organisms. *J. Water Pollut. Control Fed.* 63(4):619–696.

Somers, J., Moran, E. T., Jr. and Reinhart, B. S. 1974. Effect of external application of pesticides to the fertile egg on hatching success and early chick performance 2. Commercial-herbicide mixtures of 2,4-D with picloram or 2,4,5-T using the pheasant. *Bull. Environ. Contam. Toxicol.* 11(4):339–342.

Spivey Fox, M. R., Fry, B. E., Jr., Harland, B. F., Schertel, M. E. and Weeks, C. E. 1971. Effect of ascorbic acid on cadmium toxicity in the young *Coturnix. J. Nutr.* 101:1295–1306.

Spivey Fox, M. R., Jacobs, R. M., Lee Jones, A. O. and Fry, B. E., Jr. 1979. Effects of nutritional factors on metabolism of dietary cadmium at levels similar to those of man. *Environ. Health Perspect.* 28:107–114.

Stone, W. B. 1979. Poisoning of wild birds by organophosphate and carbamate pesticides. *N.Y. Fish Game J.* 26(1):37–47.

Stone, W. B. 1980. Bird deaths caused by pesticides used on turfgrass. *Proc. N.Y. State Turfgrass Conf.* 4:58–64.

Stone, W. B. and Gradoni, P. B. 1985. Recent poisonings of wild birds by diazinon and carbofuran. *Northeast. Environ. Sci.* 4(3/4):160–164.

Stone, W. B. and Gradoni, P. B. 1987. Poisoning of Birds by Cholinesterase Inhibitor Pesticides. N.Y. Dept. Environ. Cons. Report. March 1987. 15 pp.

Stone, W. B. and Knoch, H. 1982. American brant killed on golf courses by diazinon. *N.Y. Fish Game J.* 29(2):95–96.

USDA. 1969. *Agricultural Handbook No. 332.* U.S. Government Printing Office, Washington, D.C.

White, D. H. and Finley, M. T. 1978. Uptake and retention of dietary cadmium in mallard ducks. *Environ. Res.* 17:53–59.

White, D. H., King, K. A., Mitchell, C. A., Hill, E. F. and Lamont, T. G. 1979. Parathion causes secondary poisoning in a laughing gull breeding colony. *Bull. Environ. Contam. Tox.* 23: 281–284.

Zinkl, J. G., Rathert, J. and Hudson, R. R. 1978. Diazinon poisoning in wild Canada geese. *J. Wildl. Manage.* 43(2):406–408.

CHAPTER 7

Wetlands and Golf Courses

Patricia A. Kosian, Mary E. Balogh, and Roberta M. Tietge

7.1 INTRODUCTION

In the past, wetlands were considered areas of little environmental or economic value. As a result, many of the original wetlands in the United States were destroyed for development of agricultural and urban land area. However, with an increased understanding of ecological processes and environmental value, wetlands are now recognized as a valuable natural resource. Wetlands provide essential habitat for fish and wildlife, aesthetic and recreation areas, flood control, improvements in water quality, and shoreline erosion control.

The current national trend regarding wetlands and land development is for no net loss of wetland areas. Environmental legislation both at the state and federal level has been enacted to establish a comprehensive program to conserve and manage wetlands in the United States. As land use developers and managers, it is essential for the golf industry to understand the environmental issues regarding wetland conservation. The purpose of this chapter is to provide an introduction to wetlands and examine some of the critical issues regarding wetland conservation relative to golf course development and turfgrass management practices.

Photo courtesy of Michael French.

7.2 DEFINITIONS OF WETLANDS

"Wetlands" is a general term used to describe swamps, bogs, marshes, mires, ponds, potholes, and similar areas. These wet areas develop in transition zones between dry uplands and deepwater aquatic systems. They tend to be located in depressions or low-lying areas along inland lakes and rivers and along coastal regions. Three major characteristics used to define and identify wetlands are hydrology, vegetation, and soil. Wetlands are primarily distinguished by the presence of water. They are either inundated or saturated by water for varying periods of time. The presence of water creates conditions that support vegetation adapted to wet conditions, hydrophytes, and produce characteristic soils, hydric soils, that develop under anaerobic conditions.

Wetlands are difficult to define since they usually have no clear boundary with upland systems. Rather, wetlands evolve continuously from aquatic to terrestrial systems. Many definitions of wetlands have been attempted; however, there is no universally accepted one. The selection of a wetland definition depends on the objective and the field of interest of the user. For example, a wetland scientist is interested in a flexible definition that facilitates wetland research, classification, and inventory. A wetland manager or prospective developer is more concerned with regulations designed to control wetland modifications and needs a clear legally defensible definition. Understanding these different definitions is important when interacting with different managers and regulatory agencies during the permitting process.

One of the earliest wetland definitions was developed by the U.S. Fish and Wildlife Service (FWS) in 1956 (Shaw and Fredine 1956). This is frequently referred to as the "Circular 39" definition and is still used by wetland scientists and managers. The Circular 39 definition is as follows:

> "The term 'wetlands'... refers to lowlands covered with shallow and sometimes temporary or intermittent waters. They are referred to by such names as marshes, swamps, bogs, wet meadows, potholes, sloughs and river-overflow lands. Shallow lakes and ponds usually with emergent vegetation as a conspicuous feature, are included in the definition, but the permanent waters of streams, reservoirs, and deep lakes are not included. Neither are water areas that are so temporary as to have little or no effect on the development of moist-soil vegetation."

In 1979, the FWS adopted a new definition (Cowardin et al. 1979).

> "Wetlands are lands transitional between terrestrial and aquatic systems where the water table is usually at or near the surface or the land is covered by shallow water. Wetlands must have one or more of the following three attributes: (1) at least periodically the land supports predominantly hydrophytes, (2) the substrate is predominantly undrained hydric soil, and (3) the substrate is nonsoil and is saturated with water or covered by shallow water at some time during the growing season of each year."

The U.S. Environmental Protection Agency (EPA) and the U.S. Army Corps of Engineers (Corps) produced a definition that has been incorporated into Section 404 of the 1979 Clean Water Act (CWA) Amendments. The latest Section 404 version of a wetland definition is as follows:

"The term 'wetlands' means those areas that are inundated or saturated by surface or ground water at a frequency and duration sufficient to support, and that under normal circumstances do support, a prevalence of vegetation typically adapted for life in saturated soil conditions. Wetlands generally include swamps, marshes, bogs and similar areas." (33 CFR323.2(c); 1984)

7.3 WETLAND TYPES AND CLASSIFICATION

Wetlands are ubiquitous throughout the world and are present on every continent except Antarctica. In the United States they can be found in every state. A wide variety of wetlands have formed in the United States due to regional and local differences in vegetation, flooding patterns, water chemistry, topography, climate, soils, and other factors (Mitsch and Gosselink 1986). With an abundance and diversity of wetlands, it is important to consistently categorize wetlands. Several classification schemes have been developed over the last century (Cowardin et al. 1979; Mitsch and Gosselink 1986; Shaw and Fredine 1956). Two of these classification schemes will be briefly outlined.

A plan developed by Mitsch and Gosselink (1986) is a simple wetland classification scheme that uses common terminology. Their classification also includes generally recognizable wetland types currently found in the United States. In this scheme, seven major types of wetlands found in the United States are divided into two main groups, coastal and inland wetland ecosystems (Table 1). Coastal wetland ecosystems include tidal salt marshes, tidal freshwater marshes, and mangrove wetlands. Inland wetland ecosystems include inland freshwater marshes, northern peatlands, southern deepwater swamps, and riparian wetlands.

Coastal wetlands are found along the Atlantic, Pacific, Gulf, and Alaskan coasts. They are influenced by tidal activity and their link with estuarine systems form an environment of varying salinity. Near the coastline, the salinity of the water approaches that of the ocean, while further inland the salinity can approach that of freshwater. Many of the coastal wetland areas are contiguous with regions experiencing considerable golf course development.

Tidal salt marshes occur primarily in the United States along the Atlantic and Gulf Coasts. Narrow belts of these wetlands can also be found along the Pacific coast and along much of the Alaskan coast. Salt marshes are dominated by halophytic (salt tolerant) grasses such as cord grass, salt meadow grass, black grass, and spike grass. This vegetation can tolerate fluctuations in salinity, water levels, and temperature.

Table 1. Types and Regional Distribution of Wetlands in the United States

Wetland type	U.S. location
Coastal wetlands	
Tidal salt marshes	East and Gulf coasts, Alaska
Tidal freshwater marshes	Middle and southern Atlantic Gulf coasts
Mangrove wetlands	Southern Florida and Hawaii
Inland wetlands	
Inland freshwater marshes	Many inland regions of United States
Northern peatlands	WI, MI, MN, and north central United States
Southern deepwater swamps	Southeastern United States
Riparian wetlands	Inland rivers and streams, arid and semiarid regions

From Mitsch and Gosselink 1986.

Tidal freshwater marshes are located inland from tidal salt marshes. They experience tidal effects, but are not influenced by salinity stress. These wetlands can be found primarily along the middle and south Atlantic coasts and along the Gulf coast. Tidal freshwater marshes are dominated by a variety of grasses and broadleaf aquatic plants. Examples include pickerelweed, smartweed, arrow-arum, cattails, giant cutgrass, and big cordgrass.

Mangrove wetlands are located in tropical and subtropical regions such as the south and Gulf coasts of Florida and Hawaii. These coastal forested wetlands are dominated by halophytic trees such as red and black mangroves. Mangrove wetlands are influenced by a wide range of salinity and tidal conditions.

Inland wetlands are found in the interior regions of the United States. They account for approximately 95% of the total wetlands in the lower 48 states. Inland freshwater marshes are generally characterized by the presence of (1) soft stemmed aquatic plants (emergents) that grow partly under and above water, (2) shallow water, and (3) shallow peat deposits. Inland freshwater marshes are ubiquitous throughout the United States. Some of the major regions include Wisconsin, Minnesota, the prairie pothole region of the Dakotas, and the Everglades of Florida. Typical freshwater marsh plants include cattails and various species of grasses and sedges.

Northern peatlands are used to describe deep peat deposits of the cold northern forested regions of North America. In the United States, they are located primarily in the north central region. Similar deposits called pocosins which lie in elevated depressions are located in the coastal plains of the Carolinas. Peatlands are characterized by acidic water and poorly drained soils of low fertility. Common vegetation include sphagnum moss, cranberry, tamarack, and black spruce.

Southern deepwater swamps are located in the southeastern United States. They are freshwater forested wetlands that have standing water for most of the growing season. The dominant vegetation is cypress and gum/tupelo trees.

Riparian wetlands occur along rivers and streams. These wetlands are periodically influenced by flooding. In the southeastern United States, they are referred to as bottomland hardwood forests. Riparian wetlands also occur in arid and semiarid regions of the United States. The vegetation in most bottomland hardwood forests is dominated by trees that are adapted to a wide variety of environmental conditions that can occur on floodplains.

The United States National Wetland Inventory Program (NWI) currently utilizes a classification scheme developed by the FWS. It was published in 1979 and entitled "Classification of Wetlands and Deepwater Habitats of the United States" (Cowardin et al. 1979). This classification is based on a hierarchical approach consisting of systems, subsystems, and classes (Figure 1). A system refers to "a complex of wetlands and deepwater habitat that share the influence of similar hydrologic, geomorphologic, chemical, or biological factors." A subsystem subdivides the system into specific categories. For example, subtidal and intertidal are subsystems of marine system. A class describes the general appearance of the habitat in terms of either dominant vegetation or substrate type.

Definitions of the systems described in the FWS publication include the following:

1. A *marine system* "consists of the open ocean overlying the continental shelf and its associated high-energy coastline." An example of a marine system is coastal wetlands.
2. An *estuarine system* "consists of deepwater tidal habitats and adjacent tidal wetlands that are usually semienclosed by land, but have open, partly obstructed, or sporadic access to the open ocean, and in which ocean water is at least occasionally diluted by freshwater runoff from the land." Examples include tidal marshes and mangrove swamps.
3. A *riverine system* "includes all wetlands and deepwater habitats contained within a channel with two exceptions: (1) wetlands dominated by tress, shrubs, persistent emergents, emergent mosses, or lichens, and (2) habitats with water containing ocean-derived salts in excess of 0.5‰." Examples include wetlands associated with rivers and streams.
4. A *lacustrine system* "includes wetlands and deepwater habitats with all of the following characteristics:(1) situated in a topographic depression or a dammed river channel: (2) lacking trees, shrubs, persistent emergents, emergent mosses or lichens with greater than 30% areal coverage: and (3) total area exceeds 8 ha (20 acres)." Examples include wetlands associated with lakes.
5. A *palustrine system* "includes all nontidal wetlands dominated by trees, shrubs, persistent emergents, emergent mosses or lichens, and all such wetlands that occur in tidal areas where salinity due to ocean-derived salts is below 0.5‰. It also includes wetlands lacking such vegetation, but with all of the following four characteristics: (1) area less than 8 ha (20 ac); (2) active wave-formed or bedrock shoreline features lacking; (3) water depth in the deepest part of the basin less than 2 m at low water; and (4) salinity due to ocean-derived salts less than 0.5‰." Examples include marshes, swamps, and bogs.

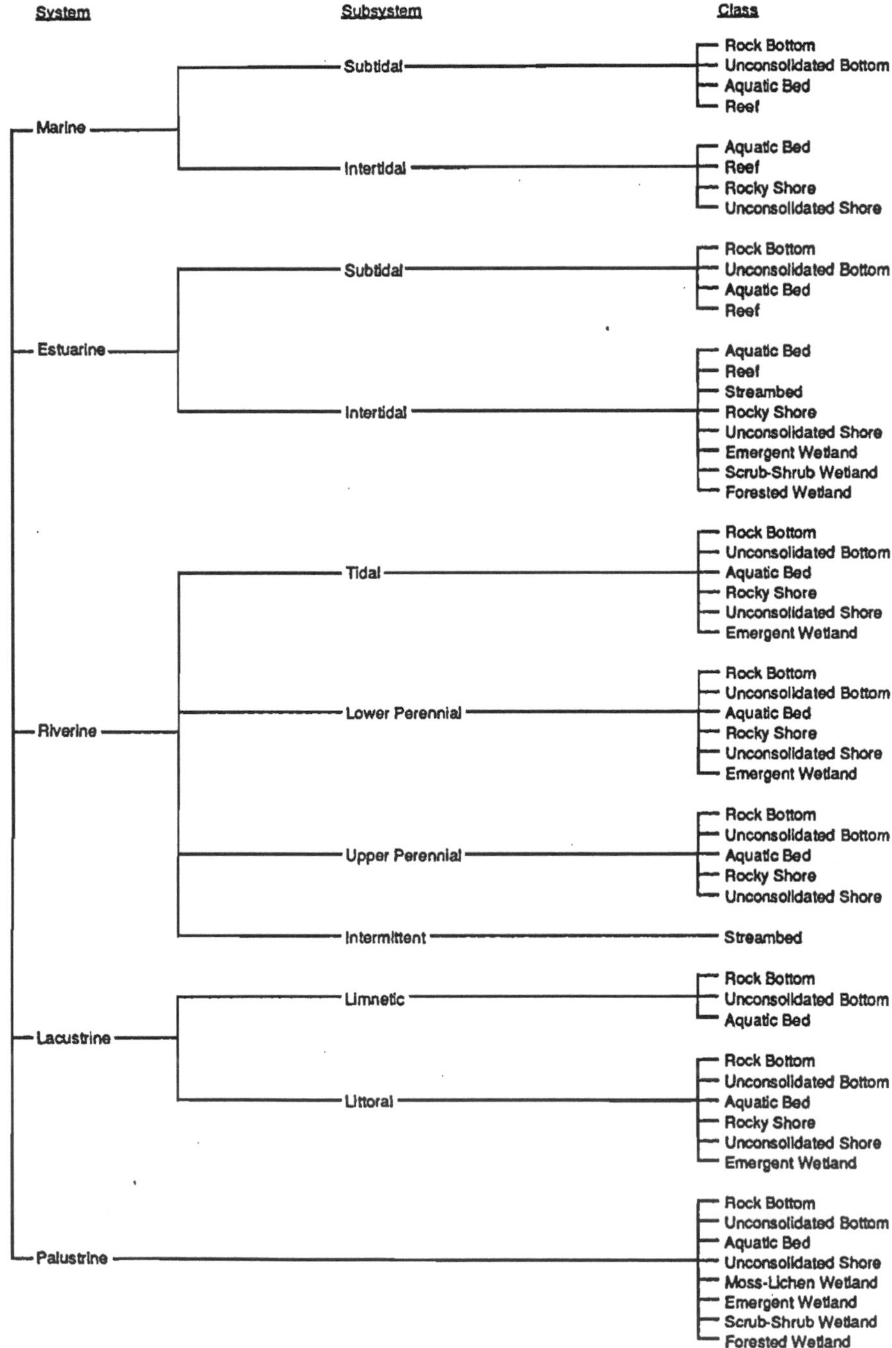

Fig. 1. Classification of wetlands (Cowardin et al. 1979).

7.4 EXTENT OF WETLANDS AND GOLF COURSES

A recent survey by the American Society of Golf Course Architects (ASGCA) indicates that environmental concerns over wetland loss and degradation was the primary reason for difficulties obtaining permits for golf course projects (Anonymous 1990a, 1990b). However, published research on wetlands in relation to the golf course industry is very limited. The authors could find no specific published information concerning the extent of wetlands located on or adjacent to golf courses or proposed golf course sites. This lack of information should not preclude use of wetlands as important landscape features on golf courses or for consideration by the golf course developer and manager. There is, however, extensive information regarding the extent and distribution of wetlands in the United States (Dahl and Pywell 1989; Tiner 1984). Coupling this information with the distribution of golf courses in the United States allows one to assume that (1) wetlands are associated with golf courses in certain regions and (2) wetland areas are likely to be encountered on sites proposed for golf course construction.

7.4.1 Status and Trends of Wetlands: 1954–1974

In 1982, the FWS completed a study of the status and trends of wetlands for the conterminous United States (Tiner 1984). The study included wetland data from the mid 1950s to the mid 1970s. The two main objectives of the study were to (1) produce comprehensive information on the status and trends of wetlands and (2) produce detailed wetland maps for the United States. As a result, the study provided statistically valid estimates of the total acreage of wetlands by wetland category and by certain deepwater habitats. Also, the average rate of wetland loss between 1954 and 1974 was determined. The information generated from this study resulted in the enactment of various federal policies concerning wetland conservation. In 1986, Congress mandated through the Emergency Wetland Resources Act the periodic updating of wetland status and trends in the United States (Dahl and Pywell 1989).

The results of the initial wetland status and trends assessment indicated that roughly half of the 81 million ha (200 million ac) of wetlands that originally existed in the United States had been destroyed by the mid 1970s. It was estimated that in 1974 only 40 million ha (99 million ac) of wetlands remained in the lower 48 states. In the 20-year span from 1954 to 1974, an estimated 185,350 ha (458,000 ac) of wetlands were lost annually. It was estimated that losses were unevenly distributed between inland and coastal wetlands. Approximately 178,066 ha (440,000 ac) of inland wetlands and 7,284 ha (18,000 ac) of coastal wetlands were lost prior to 1974.

Table 2. Wetlands Lost in the United States from 1953 to 1973

State or region	Original wetlands (ha)	Current wetlands (ha)	Percent of wetlands lost
Iowa's natural marshes	944,200	10,700	99
California	2,023,500	182,100	91
Nebraska's rainwater basin	38,000	3,400	91
Mississippi alluvial plain	9,712,700	2,104,400	78
Michigan	4,532,600	1,295,000	71
North Dakota	2,023,500	809,400	60
Minnesota	7,446,400	3,520,800	53
Louisiana's forested wetlands	4,573,000	2,280,500	50
Connecticut's coastal marshes	12,140	6,100	50
North Carolina's pocosins	1,011,700	608,300	40
South Dakota	809,400	526,100	35
Wisconsin	4,046,900	7,731,700	32

From van der Leeden et al. 1990.

There is considerable range in the extent of wetland losses in various states and regions of the United States (Table 2).

Wetland loss and degradation have resulted from a combination of human activities and natural causes (Table 3). Impact on wetlands resulting from human activities include (1) drainage for agricultural conversion, (2) mining for peat and coal, (3) dredging and channelization for navigation, (4) filling for urban development, and (5) use for disposal of urban and industrial waste. Natural losses of wetlands have resulted from erosion, subsidence of sea level rise, droughts, storms, and overgrazing by wildlife.

The initial study of wetland trends indicated that 87% of wetlands lost prior to 1974 was due to draining and clearing of wetlands for agricultural conversion (Tiner 1984). Agricultural activities had the greatest impact on forested wetlands, inland marshes, and wet meadows (EPA 1988). Urban development including residential and industrial development accounted for 8% of the losses and the remaining 5% were lost to other causes. In coastal states such as New Jersey, Texas, New York, California, and Florida, urbanization was responsible for over 90% of coastal wetland losses. Also, the study showed that the states with the greatest losses in overall wetland areas were Louisiana, Mississippi, Arkansas, North Carolina, North Dakota, South Dakota, Nebraska, Florida, and Texas.

The second objective of the original status and trends study was to examine the extent of wetlands in the United States (Tiner 1984). In the mid 1970s, it was estimated that the United States had approximately 40 million ha (99 million ac) of

Table 3. Major Causes of Wetland Loss and Degradation in the United States

Type of effect	Activity
Direct human impacts	
	1. Drainage for agriculture, timber production, and mosquito control
	2. Dredging and stream channelization for navigation, flood protection, coastal housing development, and reservoir maintenance
	3. Filling for dredge spoil and other waste disposal, roads and highways, commercial and industrial development, and residential development
	4. Construction of dikes, dams, levees, and seawalls for flood control, water supply, irrigation, and storm protection
	5. Discharge of materials into surface waters and wetlands: Discharged materials include: pesticides and other organic pollutants; nutrient loading from domestic sewage and agricultural and urban runoff; and sediments from dredging and filling, agricultural sources, and rural and urban land development.
	6. Mining of wetland soils for peat, coal, sand, gravel, phosphate, and other materials
Indirect human impacts	
	1. Sediment diversion by dams, deep channels, and other structures
	2. Hydrologic alterations by canals, spoil banks, roads, and other structures
	3. Subsidence due to extraction of groundwater, oil, gas, coal, and other minerals
Natural impacts	
	1. Subsidence, including natural rise of sea level
	2. Droughts
	3. Hurricanes and other storms
	4. Erosion and geological weathering
	5. Biological effects such as overgrazing by wildlife

Adapted from van der Leeden et al. 1990.

wetlands in the lower 48 states, Alaska contained an additional 90 million ha (200 million ac), and Hawaii had 40,469 ha (100,000 ac). The regions that contained the majority of wetlands (excluding Alaska) are located along the eastern United States (Maine, New Jersey, Virginia, North Carolina, South Carolina, and Georgia), the southern Gulf region (Florida, Alabama, Mississippi, Louisiana, and Arkansas), and the north central United States (North Dakota, South Dakota, Minnesota, Wisconsin, and Michigan). Louisiana and Florida contain the largest number of wetlands in the conterminous United States.

The extent of wetlands by type in the mid 1970s also has been reported (EPA 1988) and can be seen in Figures 2 and 3. Approximately half of the 40 million ha (99

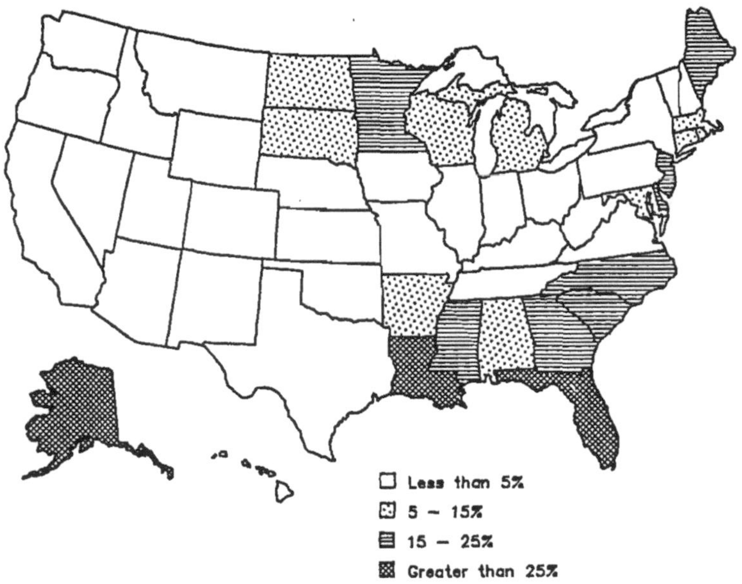

Fig. 2. Relative abundance of wetlands in the United States (EPA 1988).

million ac) were inland forested wetlands. There were 11 million ha (28 million ac) of inland marshes and wet meadows, 4 million ha (10 million ac) of inland shrubs and swamps, 2 million ha (5 million ac) of coastal wetlands, and 2 million ha (5 million ac) of inland wetland.

Complete results of the status and trends report can be found in "Wetlands of the United States: Current Status and Recent Trends" (Tiner 1984) and "Status and Trends of Wetlands and Deepwater Habitats in the Conterminous United States, 1950's to 1970's" (Frayer et al. 1983).

7.4.2 Status and Trends: 1979–1991

In 1979, the FWS initiated the NWI program to update the status and trends of wetlands in the United States. The study is currently in progress and contains wetland inventory data from 1979 to present. To date, wetlands in 65% of the conterminous United States have been mapped. A cooperative agreement exists between the FWS and the U.S. Geological Survey (USGS) for the sale and distribution of the NWI maps. To obtain further information on wetlands distribution in this study, contact the USGS's Earth Science Information Center at 1-800-USA-MAPS.

In addition to the wetland maps, the NWI is constructing a georeferenced wetland database using geographic information systems (GIS) technologies. Currently, more than 7,421 NWI maps representing 13.7% of the conterminous United States are in

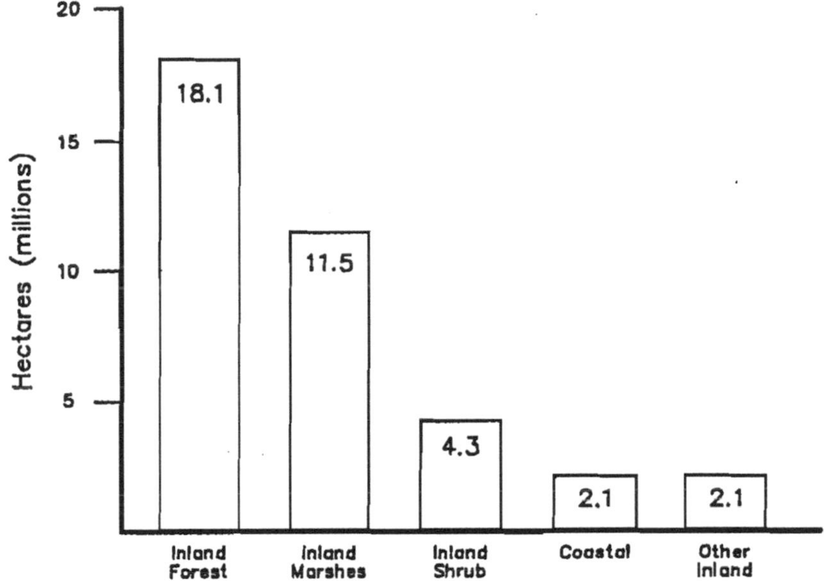

Fig. 3. Extent of wetlands by type in the lower 48 states (EPA 1988).

digital format. Statewide databases have been developed for New Jersey, Delaware, Illinois, and Maryland and are in progress for Indiana and Washington. NWI digital data is also available for portions of 14 other states (U.S. FWS 1991). Intended applications of the digital data include resource management planning, impact assessment, wetland trend analysis, and information retrieval. Information on the NWI digital database can be obtained by calling 1-800-USA-MAPS.

7.4.3 Estimated Distribution of Golf Courses

The distribution of golf courses and golfers in the United States on a state and regional level has been discussed previously (Balogh et al. 1992). The authors identified three regions in the lower United States that contain the largest proportion of golfers and golf courses. These areas are the middle Atlantic (NJ, NY, PA), the south Atlantic (DC, DE, FL, GA, MD, NC, SC, VA, WA), and east north central (IL, IN, MI, OH, WI) regions. The relative abundance of golf courses on a statewide basis is illustrated in Figure 4. Comparing the regional distribution of golf courses (Figure 4) to the regional distribution of wetlands (Figure 2) in the lower 48 states, it appears an important spatial relationship exists. In many cases, the regions with the greatest distribution of golfers and golf courses are also the regions with the greatest abundance of wetlands. It should be noted that this comparison is based on wetland data from the mid 1970s status and trends report since the current NWI has not been completed.

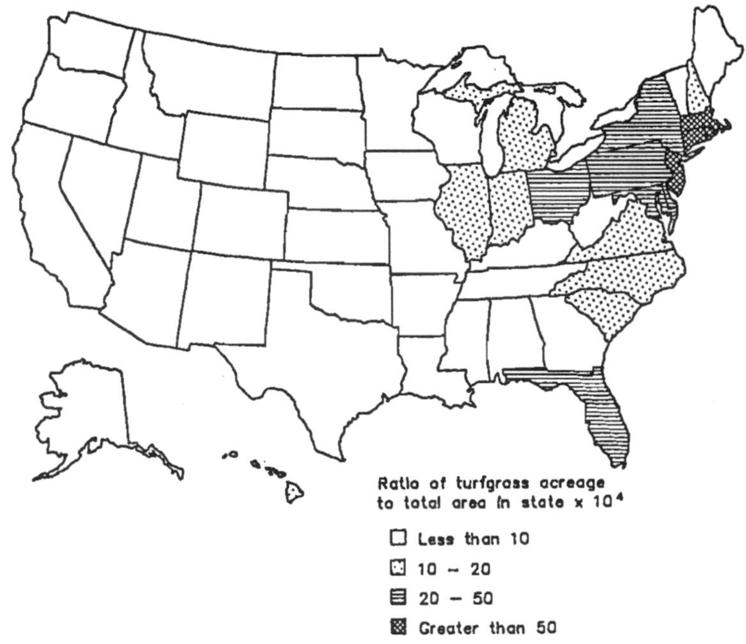

Fig. 4. Relative abundance of golf courses in the United States (Balogh et al. 1992).

Golf course developers should obtain current NWI information from the previously listed sources to identify specific trends and distribution of wetlands in specific areas. Although the NWI maps do not identify wetlands regulated by the Section 404 permit program, they can be helpful in locating wetlands during the initial planning stages of any golf course development project.

7.5 WETLAND REGULATIONS AND PERMITTING PROCESS

The National Golf Foundation predicts that one new golf course must be built per day for the next 11 years to meet projected growth of the golf industry (Pinch 1989). This potential for golf course construction coupled with environmental concerns and government permitting processes presents many challenges to the golf course developer.

In a recent survey by the American Society of Golf Course Architects (ASGCA), 39 out of 40 golf course architecture firms report difficulties obtaining permits for golf projects because of environmental concerns by local, state, or federal agencies (Anonymous 1990a, 1990b). Of the firms surveyed, 56% cited wetlands as the primary problem they encountered during the permit process. Other areas of concern, ranked in order by the ASGCA, are habitat, nitrates and chemical contamination, groundwater protection, and pesticide usage. The survey also addresses issues of project time delays and additional costs. Approximately 50% of the survey respondents indicated that some of their new projects had been delayed by 8–12 months. The cost of delays ranged from $10,000 to $1,000,000.

The survey clearly demonstrates that environmental concerns have a significant impact on the golf industry. Difficulties obtaining permits have become a serious problem for the golf course developer. To understand the problems associated with permit approval, a historical perspective is needed for the complex federal and state laws that exist for the regulation of wetland development.

7.5.1 Federal Wetland Regulations

In recent years, the government has addressed the issue of wetland preservation. Numerous laws and regulations for management and protection of wetlands have been enacted (Table 4). The most significant legislation for wetland regulation is Section 404 of the Federal Water Pollution Control Act (FWPCA) of 1972. In 1977, this act was renamed the Clean Water Act (CWA) and was later amended in 1987.

Section 404 of the CWA has become the federal government's primary tool for protecting wetlands (Salvesen 1990). It gave the Corps authority to establish a permit system designed to regulate the dredging and filling of materials into waters of the United States. Initially, the Corps applied this act only to navigable waters. However, in 1975, the definition of waters of the United States was expanded to include wetlands.

Section 404 of the CWA consists of the following three main components: Section 404 (a), (b), and (c). Additional components include Section 404 (e), (f), and (g). Section 404 (a) authorizes the Corps to issue permits for filling navigable waters (including wetlands) after public notification and hearings. Section 404 (b) requires the Corps to issue permits in accordance with the (b)(1) guidelines (Guidelines) developed by the EPA. The Guidelines are written hierarchically and provide environmental criteria used to evaluate permit applications. In general, the Guidelines prohibit the Corps from granting a permit if (1) there is a practicable alternative that would have less adverse impact on the aquatic ecosystem, (2) the project contributes to a significant degradation of the waters of the United States, and (3) appropriate and practicable steps have not been taken that will minimize potential adverse impacts on the aquatic ecosystem. For illustrative purposes, a summary of the specific guidance that EPA Region IX provides to applicants is presented by Yocom et al. (1989).

Section 404 (c) grants EPA the authority to veto the Corps decision to issue a permit. EPA will veto the permit if they find that the proposed project will have an adverse impact on the aquatic ecosystem.

Table 4. Major Federal Laws Affecting Wetlands

Executive order or act	Date	Agency
National Environmental Policy Act (NEPA)	1969	All agencies
Executive Order 11990 Protection of Wetlands	1977	All agencies
Executive Order 11988 Floodplain Management	1977	All agencies
Federal Water Pollution Control Act (FWPCA): Amended and renamed — Clean Water Act (CWA): Section 404: Dredge and Fill Permit Program Section 401: Water Quality Standards	1972 1977 1987	Army Corps of Engineers, EPA, and FWS
Coastal Zone Management Act	1972	Office of Coastal Zone Management
Endangered Species Act	1973	Federal agencies with FWS and NMFS
Fish and Wildlife Coordination Act	1934 1946 1958 1977	Army Corps of Engineers and FWS

From Mitsch and Gosselink 1986; Salvesen 1990.

Section 404 (e) authorizes the Corps to issue general permits on a state, regional, or national level. A general permit grants authorization for certain types of fill in certain size wetlands. General permits are granted for activities that will cause only minimal individual and cumulative adverse impacts. General permits issued nationwide are called nationwide permits.

Section 404 (f) exempts certain activities from the permit requirement. These activities include normal farming, silviculture, and ranching activities. However, these activities must be part of an established farming, silviculture, or ranching operation. Section 404 (g) authorizes the states to assume the permit program from the Corps provided the program has been approved by the EPA.

In the *Section 404 Permit Process*, the general procedure for obtaining a 404 permit for development of areas with wetlands consists of several steps (Figure 5). The basic steps in this process are discussed in detail by Salvesen (1990). Initially, a preapplication consultation may be arranged. This is an optional meeting between the applicant and the district office of the Corps. During this meeting the Corps discusses permit requirements, identifies potential problem areas, and advises the applicant about the public interest review and the 404 (b)(1) Guidelines.

The next step is to submit the completed application to the Corps district office. The type of information included in the application is the project purpose, a detailed description of the proposed project (including maps or sketches), a construction schedule, and information on the type of discharge (sand, gravel). The Corps office reviews the application and performs a preliminary assessment. Not all activities in wetlands require a 404 permit. Some require only a general permit or a nationwide permit as discussed in Section 404 (e). If the proposed project does not meet the requirements for a general or nationwide permit, the Corps district engineer issues a public notice. Generally, the public notice is issued within 15 d upon receipt of the completed application.

The public notice is sent to various federal, state, and local agencies. Federal agencies include the EPA, FWS, and the National Marine Fisheries Service (NMFS). The notice is also sent to special interest groups and individuals. After the public notice has been issued, the Corps generally allow 15–30 d for comments. A public hearing may be held at this time.

During the review of the 404 permit application, the Corps is required by the National Environmental Policy Act (NEPA) to determine whether an environmental impact statement (EIS) or if only an environmental assessment is needed. This decision may involve additional consultations with state and federal agencies. The proposed project must be consistent with the state coastal zone management program, the Endangered Species Act, and the state water quality standards (Section 401 of CWA) before a permit can be issued. States may veto a Corps permit if they determine that the proposed project is not consistent with its coastal zone management program or if it violates water quality standards.

The Corps district engineer will decide whether a permit should be issued to the land developer. The decision is based on the public interest review and evaluation of the 404 (b)(1) Guidelines. The decision to issue a permit also is based on a number of considerations such as the projects potential impact on conservation, aesthetics, navigation, environmental values, economics, water quality, and fish and wildlife values. If the Corps determines that adverse impacts will result, then the permit may be denied or the applicant must modify the design or mitigation may be required. If the proposed project is judged not to have adverse impacts, then the Corps may grant a permit providing that the project complies with EPA's 404 (b)(1) Guidelines.

7.5.2 State Wetland Regulations

Many states already have or are actively developing inland and coastal wetland laws and policies. Some states may require state permits in addition to the Corps permits for wetland development. As a result, golf developers face the increasing challenge to design their projects to meet both state and federal requirements to protect wetlands.

A database called WETLANDS may prove useful to the land developer. It was recently developed by the EPA's Office of Wetlands Protection and the Council of State Governments' Center for the Environment and Natural Resources. The database

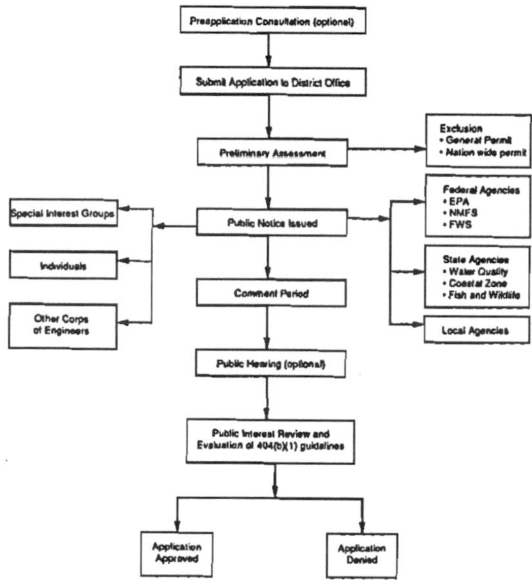

Fig. 5. U.S. Army Corps of Engineers' permit process (adapted from Salveson 1990).

contains a directory of 475 state wetland officials and provides detailed information on 350 state programs affecting wetlands. Available information includes acquisition, drainage regulation, coastal and inland laws, 401 permits, executive orders, flood plain regulations, pending legislation, state environmental policy acts, taxation programs, research, and other programs. Organizations or individuals interested in using WETLANDS should contact R. Steven Brown at the Center for the Environment and Natural Resources, Council of State Governments, Iron Works Pike, P.O. Box 11910, Lexington, Kentucky 40578, (606)-252-2291 (Anonymous 1989).

7.5.3 Summary of the Regulatory Process

The complexity and problems associated with the Section 404 permit process should not be underestimated. The golf course developer must recognize that jurisdiction of wetlands is distributed between several federal and state agencies. The Corps issues the 404 permit, but it must follow the 404 (b) (1) Guidelines established and monitored by EPA. Also, the Corps must give full consideration to comments of the FWS, NMFS, and state agencies before it reviews the permit.

Recently, the Comprehensive Wetlands Conservation and Management Act of 1991 (H.R. 1330) was introduced in the House of Representatives (Anonymous 1991). The bill was introduced to revamp Section 404 of the CWA and to establish a comprehensive program for conserving and managing U.S. wetlands. The measure would categorize wetlands based on their ecological functions and values. Also, the term wetlands would be redefined to ensure that only properties with true wetland functions and values would be protected under federal regulation. Introduction of this bill serves as an example to the land developer of the ever-changing legislation regarding wetland conservation.

The golf course developer must recognize the importance of proper guidance when applying for a Section 404 permit. The developer should select an individual or organization that will (1) provide reliable guidance when filing the application, (2) assist with the public hearing, and (3) be available to provide additional information that may be requested by the governmental agencies. Additional consultants such as civil engineers and environmental consultants may be needed. It is also important that the golf industry initiate uniform development strategies that recognize wetland protection guidelines established by federal and state agencies. This will facilitate and expedite the permitting process.

7.6 WETLAND MITIGATION

Development of mitigation strategies is often an important component of land development on areas with wetlands. Wetlands should be delineated before the initial design phase of any golf course construction project to avoid project delay, premature close-out, and payment of severe penalties due to regulatory restrictions. Developers must meet with both federal and state agencies to discuss proposed plans early in the planning stage to avoid delays and design modifications. Golf course developers should expect to redesign a project several times. Design revisions may be necessary to meet state requirements for wetland protection and again to meet requirements of the Corps or EPA. Part of this permitting process may require the developer to develop a plan to mitigate any loss of wetlands on the proposed site. It is difficult for developers to be apprised of the complex federal and state regulations concerning wetlands, as well as having the technical expertise to prepare a mitigation plan. Therefore, it is recommended that wetland mitigation activities be performed with the aid of an environmental consulting firm qualified in wetland science.

The Council of Environmental Quality (CEQ) defines mitigation as (1) avoiding impact by not taking a certain action; (2) minimizing impacts by limiting the degree of action; (3) rectifying an impact by repairing, rehabilitating, or restoring the affected environment; (4) reducing or eliminating the impact over time by preservation and maintenance operations; and (5) compensation for adverse environmental effects by replacement or providing substitute resources (40 C.F.R. Section 1508.20). The term wetland mitigation is broadly defined to include activities taken to avoid or minimize damage to wetlands, and to rectify, reduce, eliminate or compensate for adverse environmental effects (40 C.F.R. Section 1508.20). Depending on the situation, wetland mitigation is required as a condition of (1) federal permits issued under Section 404 and other laws; (2) federal projects subject to the National Environmental Policy Act (NEPA), the Fish and Wildlife Coordination Act, or the Executive Order on Wetlands; (3) licenses issued by the Federal Energy Regulatory Commission (FERC); or (4) state or local regulation of private activities (The Conservation Foundation 1988). A few states also have established formal mitigation policies. Compliance with federal and state wetland regulations is the responsibility of the land owner and developer. It is common practice for local governing bodies, reviewing land subdivision and site development plans, to request verification of wetland regulatory compliance.

The following sections briefly outline the wetland mitigation policies established by the FWS and the Corps in cooperation with the EPA. The emphasis, however, is placed on EPA and Corps regulations concerning wetlands development. Additional topics discussed are the role of mitigation in the Section 404 Permit Process, options to consider in preparation of mitigation plans, and types of available mitigation techniques. State wetland mitigation regulations are not discussed. It should be noted that some state and local wetland regulations are more restrictive than federal regulations. Additional information on specific state regulations is discussed in greater detail by Want (1989).

7.6.1 Mitigation Policy of the U.S. Fish and Wildlife Service

The mitigation policy instituted by the FWS in 1981 include several guideline (Clean Water Act Section 404 Guidelines) precepts or rules (Federal Record 456,15:7644–7663). FWS allows mitigation for proposals that do the following:

1. Are ecologically sound
2. Select the least environmentally damaging alternative
3. Avoid or minimize loss of fish and wildlife resources
4. Adopt all measures to compensate for unavoidable loss
5. Demonstrate a public need and are clearly water dependent

The Fish and Wildlife Coordination Act requires that the Corps give the same consideration to fish and wildlife resources as to all other factors considered while evaluating Section 404 permits. The Corps must consult with field offices of the FWS to identify and attempt to mitigate a project's impacts on fish and wildlife.

7.6.2 Mitigation Policy of the EPA and Corps

In the past, the Corps had opinions that often differed with the objectives of EPA and FWS on wetland mitigation. These differences existed despite mutual adoption of the CEQ mitigation definition. Recently, in a 1990 Memorandum of Agreement (MOA), the Corps and EPA developed a joint mitigation policy that identified procedures to determine the type and level of mitigation required to demonstrate compliance with the CWA Section 404(b)(1) Guidelines. In the MOA, the Corps agrees to strive to meet a goal of no overall net loss of the nation's remaining wetlands base. Under the MOA, the Corps follows the sequential approach to mitigation as stated by the EPA in Section 404 (b)(1) Guidelines when reviewing 404 permit applications.

The approach to mitigation planning as defined in the Guidelines states that wetland mitigation occurs in the following sequence: (1) avoidance, (2) minimize wetland impacts, and (3) compensation of unavoidable impacts. In evaluating Section 404 permits, the Corps, except for explicit cases stated in the MOA, must make the following determination before a permit is granted.

1. Potential impacts have been avoided to the maximum extent practicable.
2. Remaining unavoidable impacts will then be mitigated to the extent appropriate and practicable by taking steps to minimize environmental impacts.
3. Unavoidable impacts to aquatic resource values are compensated.

7.6.3 Mitigation Strategies

7.6.3.1 Wetland Avoidance. If a development plan for a golf course results in impacts to wetlands, the developer must make an initial analysis of alternative sites or designs, construction methods, or other logistical consideration to avoid impact to wetlands. Avoidance as used in Section 404 (b)(1) Guidelines and the MOA does not include compensatory mitigation. In a summary of the Guidelines, Yocom et al. (1989) states that the Guidelines establish a regulatory presumption that a less environmentally damaging practicable alternative exists if a project is not water dependent, but proposes to impact wetlands. The permit applicant must clearly demonstrate that no other practicable alternative exists before applying for a Section 404 permit [40 CFR 230.10 (a) (3)]. Examples of water-dependent activities include development of port facilities, boat landings, and docks. Golf courses may not be considered a water-dependent activity by the Corps.

Since golf courses are not considered a surface water dependent activity, the Corps does not allow certain mitigation plans as an option in initial Section 404 permit applications (Kruczynski 1990a). In other words, when the golf course developer applies for the Section 404 permit, the permit cannot contain wetland mitigation plans that would result in net "zero impacts" alone. Other options for mitigation can only be considered when all practicable options for avoidance or minimization have been considered (Kruczynski 1990a). The best mitigation strategy for a developer is to avoid building in wetlands. The developer should attempt to conduct all nonwater-dependent development activity in upland areas and keep wetlands open.

7.6.3.2 Minimize Wetland Impacts. If a developer clearly demonstrates that impacts to wetlands are unavoidable, options for less environmentally damaging alternatives can be considered. Once efforts have been made to avoid impacts to wetlands, the golf course developer still must take appropriate and practicable steps to minimize potential adverse impacts to wetlands. During permitted project development, wetland scientists should be employed to design wetland mitigation strategies. These strategies could incorporate existing wetlands into the proposed project design to avoid or minimize wetland impacts.

7.6.4 Wetland Compensation

Compensatory mitigation of wetland impacts is an attempt to replace the functions and values of the affected wetlands. Compensatory mitigation is achieved by restoration of impacted wetlands, creation of new wetlands or enhancement of existing wetlands. Considering the sequence established by the Guidelines however, compensatory mitigation cannot be considered as a mitigation option if alternatives that avoid or minimize impacts to wetlands exist.

The goals of compensatory mitigation should be consistent with the CWA. These objectives are "to restore and maintain the chemical, physical, and biological integrity of our Nation's waters." Compensatory mitigation of wetlands should be designed to replace the ecological functions supplied by the impacted wetlands. These functions include wildlife habitat, water quality, and flood storage. The preferred option is onsite, same locale, and in-kind, same wetland community type, mitigation. This ensures the least disruption to surrounding ecosystems that depend on the impacted wetland.

7.6.4.1 Compensatory Mitigation Strategies. Permitted activities that (1) meet state and EPA Section 404(b)(1) Guidelines and (2) require wetland mitigation must be designed and judged on an individual basis. Kruczynski (1990b) states that successful mitigation plans are very complex. These complexities are due to the site-specific characteristics of the many types of wetlands. Additional confusion is introduced by the many options available for manipulating the biotic and abiotic factors at mitigation sites. These options for manipulation include (1) availability of plant materials; (2) genetic compatibility of stock material with local populations and environmental conditions; and (3) handling of plant material, planting schemes, slopes, water depth and periodicity, soils, and fertilization rates. Seasonal timing of planting, flooding, and fertilization are critical to the success of mitigation projects.

It is beyond the scope of this chapter to outline a step-by-step approach to wetland mitigation design since site-specific characteristics of wetlands impacted and available mitigation sites cannot be anticipated. It is suggested that compensatory mitigation be conducted by qualified environmental scientists and engineers with technical background necessary to produce a successful mitigation project.

7.6.4.2 Types of Compensatory Mitigation. Three basic types of compensatory mitigation are used to replace impacted or lost wetlands. In the recommended order of implementation these basic types are restoration, creation, and enhancement. The choice of one of these three mitigation options depends upon site-specific characteristics of available locations. The choice is based on an analysis of factors that limit ecological functioning of the watershed, ecosystem, or region. Restoration and enhancement are part of a continuum which can be extended to include a fourth but least optimal mitigation type called wetland exchange. Definitions of these types of mitigation follow.

Wetland Restoration — Wetland restoration involves two base functions. Restoration requires (1) improvement of the condition of existing degraded wetlands so the functions they provide are improved and (2) reestablishment of wetlands in areas they once occupied. Historic wetland functions may have been disturbed through human activities such as draining, filling, channelization, or eutrophication. Natural events such as lake level rise, shoreline erosion, sediment deposition, beavers, or decreased flooding are other causes of wetland loss of function. Wetland soils often remain at disturbed sites, but are altered from drainage, oxidation, or burial (Conservation Foundation 1988). Restoration involves removing the perturbations and reestablishing the soils, plants, and hydrology at the site where a wetland originally existed.

Wetland Creation — The most controversial form of wetland mitigation is the creation of a new wetland to replace one lost to development. Wetland creation involves the development of a wetland in an area where wetlands do not or have not recently occurred. Most often, wetlands are created by removal of upland soils to levels that support wetland species. Removal of soil alone may alter the hydrology to a point causing inundation or seasonal flooding necessary to support hydrophytic vegetation (Conservation Foundation 1988). Construction of the correct elevation and establishment of the proper hydroperiod is the critical factor in the success of created wetlands (Kruczynski 1990b).

Wetland Enhancement — Enhancement refers to changing one or more functions of an existing wetland type with another. A functional enhancement may include increasing the productivity or habitat value by modifying environmental parameters such as elevation, subsidence rate, or wind fetch. Enhancement may imply a net benefit, but a positive change in one wetland function may have adverse effects on other wetland functions. For example, the habitat value for wood ducks in a forested wetland may be increased by increasing the amount of open water. Although flooding may provide more food and nesting areas for ducks, the value of the wetland for other species such as deer and marsh rabbit may be significantly reduced (Kruczynski 1990b).

Wetland Exchange — On a continuum from restoration to enhancement mitigation, the extreme end of enhancement is referred to as exchange of wetland types. An example involves enhancing the habitat value for some species by establishing an emergent or forested wetland on fill material deposited in an open water wetland. This type of enhancement is an exchange since the submerged habitat is replaced with an emerged habitat. Another illustration is the replacement of a complex food chain of a natural forested wetland with a monoculture, mass producing tree species. The monoculture forest provides food for one mammal at the expense of eliminating a complex food chain.

It is suggested that wetland exchange be allowed only under unusual circumstances. Gains in wetland type cannot be equated with losses of another type. Each type of wetland performs different functions and are unique assemblages of physical, chemical, and biological variables. A justification of a wetland exchange as compensatory mitigation is cited by Kruczynski (1990b), in which a particularly abundant type wetland is exchanged for a particularly rare wetland.

Wetlands Preservation — An uncommon form of compensatory mitigation is preservation of existing wetlands through acquisition. This option is not normally considered since most wetlands, even those proposed for preservation, are already regulated by Section 404 programs. If preservation of existing wetlands is to be considered as compensation for loss of another wetland, a net loss of wetland functions and acreage would occur. However, some situations may justify wetland preservation as a mitigation option. For example, if the environmental effects of proposed wetland impacts are minimal and the benefits of placing large areas of wetlands into public ownership are great, preservation may be consistent with the goals of wetland protection despite some loss of wetlands (Kruczynski 1990b). This is true if an area proposed for preservation is a unique habitat subject to general or nationwide permitting or otherwise vulnerable to development. Agreements for preservation of existing wetlands as compensatory mitigation will require that (1) preservation be in perpetuity and (2) must be assured through an appropriate method. An example of an appropriate method of assurance would be a fee title conveyance to a well-established, responsible conservation organization. If the ecological benefits of preservation outweigh environmental losses resulting from permitted activity, then preservation of unique or valuable wetlands vulnerable to development should be considered (Kruczynski 1990b).

Suggested Choice of Options — The choice of mitigation options for any golf course project depends upon the site-specific characteristics of available locations. The decision should be based on an analysis of factors that limit the ecological functioning of the watershed, ecosystem, or region. Very little information exists to date on the success of various wetland mitigation strategies since monitoring wetland mitigation projects is still being developed. The following guidelines are the authors' opinions on the best mitigation planning strategy available as suggested by the currently available literature. The following recommendations have not been endorsed in the form of any formal federal, state, or local policy or guidance recommendation.

Ecologically, the preferred choice of mitigation options is restoration of historic functions of a degraded system if that system exists on-site or nearby. Restoration reestablishes the natural order and ratio of community composition in the regional ecosystem. Often, the efforts required to restore a wetland are minimal. Kruczynski (1990b) explains that the success rate of restoration is greater than in other options. For example, a pasture that has been diked and drained through ditching may be restored to its original state and function if the dike is removed. If the soils are still intact, the chances of restoring this wetland to its original functions are good if the hydrology is reestablished. However, if the organic portion of the soils are oxidized and the wetland area subsided, reflooding of the area may create an open water lake, rather than an emergent wetland.

In summary, golf course developers should consider restorations of degraded wetlands as the preferred practice for wetland mitigation. The least preferred option, wetlands exchange, should only be used when scientifically justified. Enhancement and creation are intermediate options. Since wetland creation will add to total wetland acreage and enhancement may only provide one or more additional functions to an existing wetland, creation has priority over enhancement (Conservation Foundation 1988; Kruczynski 1990b). Each wetland mitigation project has site-specific characteristics often requiring the assistance of ecologists, soils and wetland specialists, and engineers. These individuals should be an important component of golf course development, design, and construction efforts.

7.6.4.3 Credit for Compensatory Mitigation. Hectare-for-hectare ratios of wetlands lost to wetlands replaced by mitigation have not been established. One strategy the Corps may use to determine acreage ratio of replacement mitigation wetlands to lost wetlands is based on the replacement of ecological functions of wetlands proposed for destruction. There are several problems, however, with this approach. First, the acreage of wetlands proposed to replace functions of lost wetlands may result in a net loss of acreage. Second, the combined effects of the ecological components contributed by a wetland such as habitat type and primary productivity are difficult to assess in the time allowed for the permitting process. It is the responsibility of the permit applicant to supply information on the acreage of wetlands necessary to return wetland functions if the lowest ratio (0.405:0.405 ha or 1:1 ac) replacement is expected. The following ratios of wetland credit have been suggested by Kruczynski (1990b).

7.6.4.4 Ratios for Wetland Restoration. Ratios of wetlands restored to wetlands lost for compensation projects are difficult to determine. This results from the highly variable success rates of restoration projects performed on a wide range of wetland types. However, if a developer demonstrates that a certain technique is successful for restoration of a particular type of wetland, a 1.5:1 mitigation is suggested on a hectare-for-hectare or acre-for-acre basis. Higher rates are suggested for projects where success of restoration is questionable. If restoration success for a particular wetland type cannot be demonstrated, the mitigation plan may be rejected. A 1:1 ratio of wetland acreage impacted to wetland acreage restored is acceptable if the replacement wetland has been constructed, monitored, and proven to be fully functional according to an approved plan.

Ratios for Wetland Creation — Generally, ratios of 1.5:1 or 2:1 are granted for projects permitted for mitigation through wetland creation. Ratios can be adjusted for several reasons. First, the ratio may be increased to 2:1 if it has not been established that a particular wetland type can be successfully created. If a successful creation has been performed before the time of permit application, the ratio may be reduced to 1:1.

Ratios for Wetland Enhancement — In a project involving wetland enhancement, it is difficult to prove that changing one component of the wetland ecosystem does not degrade other components. Therefore, a 3:1 mitigation may be required on a hectare-for-hectare or acre-for-acre basis. A 2:1 ratio may be considered if mitigation is performed before commencement of permitted activities.

Ratios For Wetland Preservation — Hectare-for-hectare ratios for mitigation through wetland preservation are considered only in unusual circumstances since preservation may result in a net overall loss in wetland function and acreage. Establishing an exchange ratio for mitigation through preservation may be considered ethically and legally dubious.

7.6.4.5 Timing of Compensatory Mitigation. Regulators prefer that mitigation projects occur before a Section 404 permit is issued, especially for projects with considerable ecological risk. If mitigation cannot occur in advance, it should proceed simultaneously with project construction and become an integral part of the development project. Mitigation activities can be planned for implementation during the season when the maximum success rate is anticipated. For instance, forested wetlands should be transplanted in winter when plant material is dormant. Mitigation is discouraged after the development phase of a project, since incentives to complete the mitigation phase on time is lost.

7.6.4.6 Location of Compensatory Mitigation: On-site and Off-site Location. Ecological values of a wetland extend past the immediate wetland area into the surrounding watershed and ecosystem. Therefore, mitigation should be performed on-site (same location) and off-site (different location) in the same watershed and ecosystem of the impacted wetland. The majority of restoration, creation, and enhancement mitigation projects should occur on-site. On-site mitigation reestablishes the natural order and ratio of community composition in the regional ecosystem. On-site mitigation ensures that the ecosystem will remain unchanged. The chance of success of the mitigation is maximized if it is close to an area that already supports the vegetative community that is being replaced.

The location of off-site mitigation should be chosen within the same embayment, stream reach, or watershed (ecosystem), if potential on-site mitigation sites are not available. A proposed project may not receive a permit if mitigation sites within the ecosystem do not exist.

Wetland *mitigation banking* is an off-site compensatory mitigation technique. Theoretically, this technique involves (1) wetland restoration and creation to form a wetland "bank"; (2) quantification of wetland values in the bank; and (3) use of wetland bank values as "credits" that are withdrawn, at a price, to compensate for unavoidable wetland impacts. The mitigation bank concept is based on the concept of wetland aggregation. Mitigation banks aggregate smaller wetland impacts into projects that restore and create larger contiguous wetlands (Salvesen 1990).

Mitigation banking has the potential for serious misuse. A developer should not expect to apply credits from a wetland mitigation bank created in one watershed and ecosystem to another ecosystem. Golf course developers should realize that before the expense of construction of wetland bank are incurred, there is no guarantee that any credits will ever be used.

Mitigation banking credits may not be acceptable in the permit process if the ecological issues of compensatory mitigation are not recognized. Most likely permits will not be issued if mitigation sites cannot be located within the ecosystem. Without on-site and properly located off-site mitigation, the equilibrium and ratio of community composition in the ecosystem will not be reestablished. The wetland banks should be located in the same geographical area and consist of wetland types similar to wetlands where the impacts will eventually occur.

It is difficult to meet regulatory processes with wetland mitigation banking. Individual projects must still comply with all regulatory requirements such as demonstration of unavoidability of wetland impact or loss. Mitigation banking requires considerable legal, scientific, and administrative complexity. A mitigation plan involving banking must involve a bank that is operational before a permit application is based on that bank's values. The mitigation plan must also include long-term operational, maintenance, and monitoring plans. In addition, the mitigation plan involving banking should include legal guarantees to assure that these tasks are feasible and will be undertaken by the appropriate parties under the force of law (Kruczynski 1990b).

In the *selection of community type for mitigation projects,* wetlands that are unavoidably lost in development projects should be replaced by the same type of wetland community. This type of in-kind mitigation is preferred since it restores the natural order of community types. Out-of-kind mitigation, replacement of wetlands by a different wetland community, is approved in specific cases where it is proven that it clearly benefits the ecosystem under evaluation. Examples of benefits include (1) providing a more diverse habitat and ecological function, (2) improving ecological functions of the ecosystem, and (3) achieving stabilization of an area under the assumption that the original community will eventually invade the site (Kruczynski 1990b).

7.6.5 Monitoring Mitigation Projects

Little quantitative information currently exists on the success of replacement of wetland communities, especially forested communities. Therefore, it is often recommended that all mitigation plans include an approved monitoring program. Criteria to evaluate a project's success should be established and agreed upon before the permit is issued. This plan should ensure that all objectives are met and the site is maintained to keep the wetland functioning properly. The plan should specify who will monitor, how often, and who will report and review the results. If the criteria for success are not met, the applicant must take corrective actions until criteria are met. The Corps may set the criteria to define the plan and determine the success and corrective actions. These criteria may then be included as special conditions of the Section 404 permit.

7.6.6 Conclusions on Mitigation Strategies

Golf course developers should consider avoidance as the best and most effective mitigation option. Otherwise, the golf course developer must demonstrate that impacts to wetlands are unavoidable before applying for Section 404 permits and initiating mitigation planning. Federal and numerous state agencies require compensatory mitigation in the form of wetland restoration, creation, and enhancement for activities that result in unavoidable wetland impacts. Regulatory agencies prefer on-site, in-kind, and up front mitigation. This results in the least disturbance to surrounding ecosystems that depend on the impacted wetland. Additional information regarding wetland mitigation can be found in Kusler and Kentula (1990), Want (1989), Wolf et al. (1986), and U.S. EPA (1990).

7.7 WETLAND VALUES AND FUNCTIONS

Wetlands, once called America's wastelands, are now recognized for their environmental and economic values. Wetlands represent many values, functions, and utilities for our society and environment. However, only recently have wetland values and functions been widely recognized (Salvesen 1990). This broadened awareness of wetland values has resulted from increased activities in wetland scientific research; enacted federal, state, and local legislation; and public education. In general, wetland values and functions include wildlife and fish habitat, shoreline and erosion control, flood water control, aquifer recharge; water quality and stormwater management, and recreation and aesthetics. Many of these values are important in both rural and urban wetland environments. Wetland values have been reviewed extensively by Mitsch and Gosselink (1986) and Salvesen (1990).

7.7.1 Wildlife and Fish Habitat

Wetlands are one of the most diverse and productive ecosystems in the world (Mitsch and Gosselink 1986). They are critical for the survival of zooplankton, worms, insects, crustaceans, reptiles, amphibians, fish, birds, mammals, and plants. Their productivity has been estimated to exceed even the most fertile farm areas in the United States. As a result, wetlands sustain an abundance of vegetation that provide excellent habitats for a variety of terrestrial and aquatic species. Organisms also come from upland areas to utilize wetland resources and help form a complex, but integrative system.

The diversity, productivity, and economic importance of wetlands is demonstrated by the diversity of species they contain. It has been estimated that wetlands provide habitats for 190 species of amphibians, 270 species of birds, 11 orders of insects, and 5,000 species of plants (Rucker 1988; Hammer 1989). Wetland ecosystems comprise a large part of critical habitats listed under the Endangered Species Act. In fact, 26% of the plants and 45% of the animals listed as threatened or endangered depend directly or indirectly on wetlands for their survival (Rucker 1988). Many of these wetlands can be found on or in close proximity to golf courses or proposed course sites.

Invertebrates are an infrequently mentioned type of wetland wildlife which provide a critical function as part of the food chain. Mollusks belong to this group and include snails, clams, and mussels. These are largely benthic or bottom dwelling or may be associated with aquatic vegetation. Mollusks are often abundant and are consumed by ducks, fish, mink, otter, muskrat, raccoons, turtles, and salamanders. Populations of these invertebrates influence the abundance and diversity of fish, amphibians, reptiles, birds, and mammals. Aquatic insect larvae occur largely along wetland edges, on water surfaces, and in bottom soils and organic matter. Insects and their larvae are important food sources for fish, frogs, ducks, and shorebirds.

Amphibians including frogs, toads, and salamanders exist in virtually every wetland. These adult animals and their egg masses and tadpoles provide a critical food supply for ibises, egrets, herons, pelicans, snakes, fish, and several small mammals. Virtually every avian group utilizes wetland areas during one or more of their life cycle stages, including endangered species such as the peregrine falcon. Birds use wetlands as nesting and feeding areas. Also they may use wetlands as a forage and shelter stopover during migration in the spring and fall. Knowledge about requirements of most nongame birds using wetlands is sparse and needs further study.

Wetlands are perhaps best known for their abundance of waterfowl such as ducks and geese. Due to the migratory nature of waterfowl, they encounter a wide diversity of habitats. Habitat diversity is important for growth and survival of waterfowl. For example, freshwater prairie potholes of the north central United States are the primary breeding place for 60–70% of the 10–12 million waterfowl in the lower 48 states (Chafee 1983). Successful duck reproduction depends on the periodic drying and flooding of wetlands. Periodic drying releases nutrients to the ecosystem. During flooding, there is a rapid increase of plant and insect life which provides a maximum protein source for egg production (Harmon 1971). However, if these ducks find poor unsuitable water conditions, they will leave their historic nesting area and move north. As a result, these pairs usually do not nest that year (Harmon 1971).

Other waterfowl show the following preferences. Forested wetlands are preferred by wood ducks. During the winter, diving ducks are found in brackish marshes that are adjacent to deep ponds and lakes. Dabbling ducks prefer freshwater marshes. Gadwalls prefer shallow ponds (Mitsch and Gosselink 1986).

Abundant wetland waterfowl provide substantial economic and recreational benefits. Hunters spend over \$300 million year^{-1} to hunt waterfowl (Salvesen 1990). Fifty percent of mallards, which comprise about one third of the total of harvested ducks, are shot in wetlands (Mitsch and Gosselink 1986).

Wetlands are important habitats for aquatic species such as fish and shellfish. The aquatic species found in wetlands are diverse and their dependence on wetlands varies greatly. Several species exist that utilize wetlands for their entire life cycle and could not adapt if this habitat type were destroyed, lost, or vastly changed. Coastal wetlands are important habitats for estuarine and marine fish and shellfish. Most commercial and game fish use coastal marshes and estuaries as nursery and/or spawning grounds (Sather and Smith 1984). The most common fish and shellfish that depend on coastal wetlands include menhaden, spot salmon, striped bass, sea trout, blue fish, Atlantic croaker, flounder, shrimp, blue crab, oysters, and clams. It is estimated that two thirds of the fish and shellfish species harvested commercially are associated with wetlands (Salvesen 1990). In 1983, wetlands supported a \$12 billion year^{-1} commercial and recreational fishing industry (Chafee 1983). Inland wetlands are also important habitat areas for fish. These wetlands provide food and nursery grounds. Common freshwater fish include perch, pickerel, sunfish, trout, and catfish (Mitsch and Gosselink 1986; Salvesen 1990).

Mammals use wetlands as primary and secondary habitats. Wetlands provide primary habitats for muskrat, nutria, otter, and beaver. Muskrat are the most abundant and can be found in wetlands throughout the United States. They prefer freshwater inland marshes, but can be found along the Gulf coast in brackish marshes. Additional mammals such as deer, moose, raccoon, mink, meadow mice, bog lemmings, bobcat, fox, coyote, and black bear use wetlands as secondary habitats (Rucker 1988; Sather and Smith 1984). Wetlands provide important sources of food, water, and cover for these mammals.

Since the value of natural and constructed wetlands to wildlife is well-known, it is critical for the golf course industry to examine those practices which affect wetland areas. This includes construction and maintenance of wetlands, as well as chemical management practices used on the upland portion of the course. For example, chemicals used on turfgrass have some potential to runoff into the lower wetland areas on and adjacent to the course. It is important to exercise proper management practices to maintain both healthy turfgrass and healthy wildlife populations.

7.7.2 Shoreline and Erosion Control

Wetlands located along coastlines, riverbanks, and lakeshores play a valuable role in erosion control (Mitsch and Gosselink 1986). Wetland vegetation has been identified as a key component aiding in shoreline erosion control. They protect the shorelines against the erosive action of waves, tides, storms, and wind. The vegetation helps anchor the sediment in place, dampen wave action, and reduce current velocity.

Wetlands are so effective in erosion control that in some coastal regions planting wetland vegetation is recommended. For example, along the Chesapeake Bay, private land owners rely on manmade wetlands to help reduce shoreline erosion. The relative cost of planting marsh grasses for erosion control is low compared to constructing manmade bulkheads or stone jetties. In 1983, it was estimated that contractors in Chesapeake charged \$40–\$75 ft^{-1} to build stone jetties. In contrast, the fee for planting a shoreline with marsh grasses was \$8–\$12 (Temple 1983).

7.7.3 Flood Protection

One of the greatest benefits of inland wetlands is their ability to provide natural flood control. Wetlands act as buffer zones where the flow of flood waters from tributaries and rivers can be slowed. Wetlands act as overflow areas in which flood waters can be (1) temporarily stored, (2) distributed over larger areas, and (3) slowed by vegetation. In addition, wetlands help prevent the suspended sediments in flood water from being carried downstream. This upstream detention of sediment helps prevent clogging of rivers, reservoirs, and flood control basins.

In urban areas, development projects can alter the rate and volume of surface water runoff. Wetlands have proven so effective in flood control that in some urban areas wetlands have been used rather than building dams. For example, in the Boston area, the Corps purchased 3,440 ha (8,500 ac) of wetlands for flood control. This is part of the Charles River Natural Valley Storage Project. The wetlands provide an effective and inexpensive alternative to dam construction. It is estimated that the Charles River Project prevents \$3–\$17 million year^{-1} in flood damages (Chafee 1983; Mitsch and Gosselink 1986).

7.7.4 Water Quality and Storm Water Management

Numerous studies demonstrate the effectiveness of wetlands to remove pollutants from point sources such as effluents from wastewater treatment plants and from nonpoint sources such as agricultural and urban runoff. The ability of wetlands to alter water quality is a result of complex chemical transport and transformation processes that occur in wetlands. Due to the great diversity of wetlands and the complexity of these chemical processes the reader is directed to additional sources on this subject (Hammer 1989; Mitsch and Gosselink 1986).

In general terms, wetlands have several attributes that can influence water quality. Wetlands reduce the velocity of water as it enters wetland areas which causes transported chemicals sorbed to suspended solids to be removed through sedimentation. The shallow water increases the contact time of water containing contaminants and the adsorptive surfaces of deposited sediments. This process may lead to significant chemical partitioning between water and sediment. A variety of anaerobic and aerobic processes such as denitrification and chemical precipitation can remove certain chemicals from the water column in wetlands. Finally, the high rate of productivity in some wetlands can result in high rates of mineral uptake by vegetation and subsequent burial in sediments when plants die.

An increasing number of developers recognize the advantage of using wetlands for storage of storm water runoff. For example, in Addision, IL, a 2.6 ha (6.5 ac) wetland retention basin was built to compensate for 2.3 ha (5.8 ac) of wetlands that had been destroyed for an industrial park (Salvesen 1990). Prior to entering the wetland basin, the storm water runoff was passed through a sediment trap and a biological filter. The sediment trap reduced the amount of suspended solids entering the wetlands. The biological filter consisted of a series of shallow wetland areas. The emergent vegetation in these areas helped to reduce the concentration levels of storm water pollutants such as oil, grease, heavy metals, trace organics, salt, and fertilizers. The biological filter was periodically burned to rid the area of weeds and some of the contaminates.

Despite the growing number of developers using wetlands for storm water management, the impacts of suspended solids and contaminants have on wetlands should not be ignored. A survey conducted in King County, WA, assessed the impact of urban storm water runoff on wetlands (Salvesen 1990). Forty-six wetlands receiving storm water runoff from urban areas were compared to 27 control sites. An important finding of the study indicated that natural plant composition was impacted in the wetlands receiving urban storm water runoff. In many of the impacted areas, reed canary grass was the dominant vegetation. This decrease in vegetative diversity can result in a decrease in wildlife diversity.

In addition to using wetlands to manage storm water runoff, a growing number of communities are relying on natural and constructed wetlands to treat municipal wastewater. These wetlands are used to remove solids, biological oxygen demand (BOD), nutrients, toxic material, and bacteria from wastewater. Many of these wetland treatment systems have been summarized by Hammer (1989). Golf course developers also have constructed wetlands to treat recycled water from golf course drainage systems (Kahler 1991). Ocean Course on Kiawah Island, SC, has incorporated 8.9 ha (22 ac) of constructed freshwater wetlands into the course's extensive drainage system. Irrigation water pumped into the wetland areas is filtered through native vegetation before it is pumped back on the course.

When using wetlands to improve the water quality of municipal wastewater, storm water runoff, or runoff from golf courses, a number of management issues need to be addressed. In recent years, the city of Bellevue, WA, raised several management issues (Bissonnette 1987). The Bellevue Office of the City Manager posed the following questions regarding wetlands and water quality treatment.

1. What are the loading rates of sediment entering wetlands, and what are their impacts?
2. Do wetlands used for runoff treatment need periodic dredging, and how should dredge spoils be treated?
3. What are the expected nutrient uptake rates?
4. To what extent can wetlands absorb heavy metals and toxic organics before their adsorptive capacity is exceeded?

5. What is the fate of sediment, nutrients, and toxic chemicals in the food web?
6. Are BOD, nutrients, persistent toxic organics, and heavy metals rereleased during winter die back of vegetation and to what extent?
7. Is harvesting necessary to prevent release of adsorbed contaminants?
8. How are wetlands reestablished once they have been disturbed?

Additional considerations regarding the impact on wetlands for the use of treating municipal wastewater and storm water runoff can be found in Godfrey et al. (1985) and Tilton (1981).

7.7.5 Aquifer Recharge

The role of wetlands in groundwater recharge is unclear and not well-documented. Some hydrologists believe only a limited number of wetlands are involved in groundwater recharge (Sather and Smith 1984). Temporary or seasonal wetlands are more likely candidates for groundwater recharge than wetlands that are permanent or semipermanent.

It has been suggested that the following wetland features may be closely associated with groundwater recharge: nature of the substrate, water permanence, nature of surface outlets, edge to volume ratio, and the type and amount of vegetation. However, further research is needed to examine the interaction of these wetland features with groundwater recharge (Sather and Smith 1984).

7.7.6 Recreation and Aesthetics

It is widely recognized that rural and urban wetlands have recreational and aesthetic value. They can provide recreational areas for fishing, hunting, canoeing, and bird watching. It is estimated that annually 50 million people spend nearly $10 billion to observe and photograph wetland-dependent birds (U.S. EPA 1988).

In urban areas, wetlands are used to preserve precious green space. They provide urban separation and visual relief in often dense and artificial environments. Residents have been known to use urban wetlands as retreats to experience solitude and view wilderness without leaving the city (Bissonnette 1987).

7.7.7 Summary of Wetland Values

Wetlands provide many functions to society. In the past, these benefits were largely unrecognized or not appreciated. Wetland values have become strong arguments for the preservation and, in some cases, for the construction of wetlands.

7.8 WETLANDS, GOLF COURSE DEVELOPMENT, AND MANAGEMENT PRACTICES

The preservation and protection of wetlands have become important concerns in the golf industry (Dye 1989). Increased awareness of these environmentally sensitive areas have presented many new challenges for the golf course developer, superintendent, and architect. As a result, golf course management practices, site selections, and course designs are being reexamined and modified.

Water features which include wetlands are important areas on golf courses. They not only provide beauty and challenges to the game, but are now recognized as valuable areas for conservation. The relative size and position of wetlands on golf courses can be quite variable. Typically, wetlands can be incorporated into the main playing area where they act as a water hazard or they can be located in the rough. The location and size of the wetland on or adjacent to golf courses will influence how they are maintained. Routine golf course management practices may need to be altered to avoid physical and chemical impact to the wetland.

7.8.1 Wetlands and Golf Course Management Practices

Published literature examining alternative golf course management practices to avoid impacting wetlands is extremely limited and brief. In chapter 3 and 4 of this book, Walker and Branham (1992) and Balogh and Anderson (1992) conducted a review of the scientific literature on the transport of pesticides and nutrients from turfgrass systems. In general, chemical transport and persistence were identified as critical concerns from an environmental quality and water quality perspective. Since water features, such as wetlands, are an important area on golf courses, the management guidelines certainly apply when wetlands are present. The reader is directed to Chapters 3 and 4 of this book for a complete summary of management practices related to pesticide and nutrient loss.

Peacock et al. (1990) recommended the following golf course management practices when treating areas bordering wetlands.

1. Mow turfgrass located adjacent to wetlands with small maneuverable equipment.
2. Limit the distribution of fertilizers in areas adjacent to wetlands.
3. Avoid nutrient overloading in wetlands by using slow-release fertilizers.
4. Design and operate irrigation and drainage systems to prevent the movement of water into wetlands. This practice will reduce chemical transport and disruption of the natural hydrologic cycle of wetlands.

5. Practice prudent application of pesticides on land adjacent to wetlands to avoid detrimental nontarget impacts to wetland populations.
6. Consider chemical properties such as solubility, leaching potential, half-life, and degradation products when selecting chemicals for application.
7. Establish Integrated Pest Management (IPM) as a key component of golf course maintenance in areas adjacent to wetlands.

In addition to these management practices, Klein (1990) recommended the following practices to avoid impacting wetlands associated with golf courses.

1. Establish buffer strips along the perimeter of wetlands.
2. Chemicals such as pesticides, fertilizers and fuel should be stored in a location where a spill or leak would not result in their transport into wetlands.
3. Avoid storm water runoff into wetlands from impervious surfaces such as parking lots and paths.

A growing practice in golf course management is irrigation of turfgrass using domestic wastewater effluent or reclaimed water. In California and Florida, where freshwater has rapidly become a limited natural resource, irrigation of turfgrass with secondary effluent has become an attractive concept. Reclaimed water has been applied to parks, roadsides, landscapes, golf courses, cemeteries, and athletic fields (Anonymous 1985; Harivandi 1982; Kahler 1991). Secondary effluent water can contain a variety of chemicals and biological material such as soluble salts, nutrients, heavy metals, bacteria, and virus. As a result, runoff and drift from irrigation systems should be prevented from entering wetlands.

7.8.2 Wetlands and Development of New Golf Courses

In some regions of the United States, the demand for new golf course construction often exceeds the amount of available land for development. As a result, developers have proposed new golf course sites on or near government protected areas such as wetlands. The government policy to preserve and protect wetlands may seem to exclude these areas from development. However, several case studies will be presented to demonstrate that developers, architects, and superintendents have worked in cooperation to ensure that wetlands and golf course development projects can coexist.

The exclusive 182 ha (450 ac) Old Marsh Golf Course in North Palm Beach, FL, is a well-documented example of course development on an environmentally sensitive area (Anonymous 1988b; Jones 1988; Mackay 1987). The course site for Old Marsh was a series of wetland prairies that were protected under the jurisdiction of several local, state, and federal agencies. The main concerns of the regulatory groups were related to alteration of the wetland hydrology and runoff from the course that would impact the wetlands.

One of the most challenging issues for the Old Marsh project was to design irrigation and drainage systems to protect the 30 ha (75 ac) of natural wetlands and the 12 ha (30 ac) of created wetlands. Regulatory agencies required that 70% of golf course runoff be prevented from entering the wetlands. The course architect designed a system intended to prevent 100% of the runoff from entering the marsh areas. An extensive tile draining system was laid beneath the entire course. Drain tiles were networked with strategically placed 6 in. catch basins. The water from the catch basins was pumped to artificial lakes where vegetation was to provide initial improvements in water quality. The water from the lakes was then force pumped to the South Florida Management District for further treatment. To further protect the adjacent marshes, a moat was constructed around the entire course.

Additional protection measures were taken during construction of the network of paths that linked the holes across the marshes on the Old Marsh site. In an effort to preserve the natural vegetation, paths that transected the marshes were elevated to allow sunlight to penetrate the grasses. Also during path construction, pilings were driven in rather then being set into predug holes. Due to the proximity of the Old Marsh course to other wetland areas, the site required some special maintenance techniques. These included (1) prevention of irrigation spray from crossing the moat into the marsh, (2) hand mowing of the sharp drop-offs around the tees and green, (3) avoiding transport of fertilizers and pesticides into the marshes, and (4) planting low maintenance turfgrass on the banks of the moat to reduce mowing and fertilizer application.

Construction of Old Marsh in an area where wetlands exist has provided a prototype for future course development projects. Advocates of Old Marsh attribute its success to cooperation between the regulatory agencies and the golf course developer, architect, and superintendent.

Other case studies where golf course development projects have been incorporated into environmentally sensitive areas have been reviewed by Anonymous (1988a, 1988c), Dye (1989), and Jones (1989). Three examples from these reviews were selected for illustration purposes.

The Links at Spanish Bay in Monterey, CA, is an interesting illustration of a land reclamation project. Prior to the 1930s, Spanish Bay consisted of native sand dunes and wetlands which provided valuable wildlife habitat. However, during an extended period from the 1930s through 1973, these areas were destroyed by extensive sand quarrying. As a result, the land selected for course development required considerable restoration. Permit approval was obtained after developers agreed to reconstruct 40.5 ha (100 ac) of silicon sand dunes and enhance two wetland areas. The wetlands serve as areas for flood control and natural drainage.

Willowbend Club in Cotuit, MA, is an example of a golf course that was constructed on a site containing a mixture of salt marshes, freshwater streams, seasonal wetlands, and cranberry bogs. The presence of cranberry bogs was a key issue in the permit process. Sand and vegetation buffer strips were incorporated into course design to prevent the movement of golf course fertilizers and pesticides into wetland areas.

The last example is the development of Bonita Bay near Naples, FL. The course, which was constructed on a vast expanse of Florida marshlands, illustrates some interesting design features and golf course management practices. Wetland areas were incorporated into the course design so they require no irrigation and little maintenance. One of the most unique aspects of the course is that only 29 ha (65 ac) of the 81 ha (200 ac) course is irrigated, maintained turfgrass. This practice avoids altering native vegetation by artificial irrigation in natural areas. Also, in areas where native vegetation meets the highly maintained turfgrass, undesirable plant growth is removed by hand grooming.

7.8.3 Conservation Easements

Conservation easements offer the golf industry another approach to preserve and protect environmentally sensitive areas from development (Linder 1990). This practice allows a parcel of property to be deeded to an environmental organization or charitable trust for the purpose of retaining the area in a natural condition. The nonprofit organization does not hold the title to the property, but instead has the development right to the deeded property. Conservation easements on golf courses can include wetlands.

State governments provide economic incentives to land owners for establishing conservation easements. Economic benefits to the golf course owner can be realized through property tax and income tax savings. For tax purposes, golf courses are generally classified as agricultural, recreational, or commercial property. When the property is deeded to an environmental organization, the land is set aside for environmental purposes so the general property classification does not apply. In some cases, property owners can significantly reduce their tax base. Donation of conservation easements to qualified conservation organizations or public agencies can also reduce the land owners income tax. The donated easement must be perpetual and be used for conservation purposes. Further information regarding the benefits and state regulations on conservation easement can be obtained by contacting the state departments of natural resources or the Golf Course Superintendents Association of America (GCSAA) office of government relations (Linder 1990).

7.9 CONCLUSIONS AND FUTURE RESEARCH

Given the national trend to conserve wetlands, it is necessary for the golf industry to examine the potential environmental impacts associated with current development and management practices. The intent of this chapter was to identify some of the critical issues regarding wetland conservation in relation to the golf industry.

The golf industry identified that environmental concerns over wetland loss and degradation was one of the primary reasons for difficulties in obtaining permits for golf course projects. It is the authors' opinion that three steps are necessary to facilitate the permit process. First, it is prudent for the developer to select qualified scientists and organizations that are experts in the field of wetland mitigation, knowledgeable of current wetland regulations, and have experience in the permit process. Second, the golf industry should develop a comprehensive public relation program with environmental groups and the general public. Fostering a good relationship with the public is critical to the success of the proposed project. Finally, due to the complexities of wetland legislation, the golf industry must develop uniform strategies that recognize wetland protection guidelines established by federal and state agencies. Developing uniform interpretation of guidelines between the golf industry and regulatory agencies will expedite the permit process.

Currently, there is limited published scientific literature that examines the impacts of golf course management on associated wetlands. Due to the intensive turfgrass management practices on golf courses, the potential for impact exists. Therefore, scientific research is needed to address this issue. Current management practices and their alternatives must be examined to avoid physical and chemical impacts to wetlands. Given the prohibitive expense of mitigating environmental problems, it should be recognized that prevention strategies are preferred to mitigation of adverse effects.

REFERENCES

Anonymous. 1985. Effluent reuse in Southeast Florida. *Biocycle* 26(2):39–40.

Anonymous. 1988a. Building the links at Spanish Bay. *Landscape Manage.* 27(1):58–60.

Anonymous. 1988b. The making of a great course. *Landscape Manage.* 27(1):44–48.

Anonymous. 1988c. A peaceful co-existence: Bonita Bay golf course. *Grounds Maintenance* 23(2):110–111.

Anonymous. 1989. Database of state wetland programs and contacts. *Natl. Wetlands Newslett.* 11(1):7–8.

Anonymous. 1990a. Permit problems getting serious, say architects. *Fla. Green* Summer:16–18.

Anonymous. 1990b. Survey indicates environmental standardization needed. *Golf & Sportsturf* 6(7):7.

Anonymous. 1991. Bill seeks overhaul of section 404. *J. Soil Water Conserv.* 46(2):125.

Aungst, H. 1986. A home for the prairie. *Weeds Trees Turf* 25(10):58.

Balogh, J. C. and Anderson, J. L. 1992. Environmental impacts of turfgrass pesticides. in Balogh, J. C. and Walker, W. J. (Eds.). *Golf Course Management & Construction: Environmental Issues*. Lewis Publishers, Chelsea, MI, Chapter 4.

Balogh, J. C., Gibeault, V. A., Walker, W. J., Kenna, M. P. and Snow, J. T. 1992. Background and overview of turfgrass and environmental issues. in Balogh, J. C. and Walker, W. J. (Eds.). *Golf Course Management & Construction: Environmental Issues*. Lewis Publishers, Chelsea, MI, Chapter 1.

Bissonnette, P. 1987. Wetlands in an urban environment. *Natl. Wetlands Newslett.* 9(4):7–9.

Chafee, J. H. 1983. Saving our nation's wetlands. *EPA J.* 9(2):3–4.

The Conservation Foundation. 1988. Protecting America's Wetlands: An Action Agenda. The Final Report of the National Wetlands Policy Forum. Washington, D.C., 69 pp.

Cowardin, L. M., Carter, V., Golet, F. G. and LaRoe, E. T. 1979. Classification of Wetlands and Deepwater Habitats of the United States. U.S. Fish and Wildlife Service Pub. FWS/OBS-79/31. Washington, D.C., 103 pp.

Dahl, T. E. and Pywell, H. R. 1989. National status and trends study: Estimating wetland resources in the 1980's. in Fisk, D. W. (Ed.). *Wetlands: Concerns and Successes*. American Water Resources Association, Bethesda, MD, pp. 25–31.

Dye, P. 1989. Golf course architects adapt to environmental challenges. *SportsTURF* 5(5):29.

Frayer, W. E., Monahan, T. J., Bowden D. C. and Graybill F. A. 1983. Status and Trends of Wetlands and Deepwater Habitats in the Conterminous United States, 1950's to 1970's. Dept. of Forest and Wood Sciences, Colorado State University, Ft. Collins, 32 pp.

Godfrey, P. J., Kaynor, E. R., Pelczarski, S. and Benforado, J. (Eds.). 1985. *Ecological Considerations in Wetlands Treatment of Municipal Wastewaters*. Van Nostrand Reinhold Co., New York, 473 pp.

Hammer, D. A. (Ed.). 1989. *Constructed Wetlands for Wastewater Treatment; Municipal, Industrial and Agricultural*. Lewis Publishers, Chelsa, MI, 831 pp.

Harivandi, M. A. 1982. The use of effluent water for turfgrass irrigation. *Calif. Turf. Cult.* 32(3/4):1–7.

Harmon, K. W. 1971. Prairie potholes. *Natl. Parks Cons. Mag.* 43(3):25–28.

Jones, P. 1988. Innovative construction: A model of cooperation. *Golf Course Manage.* 56(11):6–16.

Jones, R. T., Jr. 1989. A challenging environmental issue: Use of wetlands in golf course design. *Golf Course Manage.* 57(7):6–16.

Kahler, K. E. 1991. Kiawah Island's Ocean Course: Vision in the sand. *Golf Course Manage.* 59(8):6–20.

Klein, R. D. 1990. *Protecting the Aquatic Environment from the Effects of Golf Courses*. Community and Environmental Defense Associates, Maryland Line, MD, 59 pp.

Kruczynski, W. L. 1990a. Mitigation and the section 404 program: A perspective. in Kusler, J. H. and Kentula, M. E. (Eds.). *Wetland Creation and Restoration: The Status of the Science*. Island Press, Washington, D.C., pp. 549–554.

Kruczynski, W. L. 1990b. Options to be considered in preparation and evaluation of mitigation plans. in Kusler, J. A. and Kentula, M. E. (Eds.). *Wetland Creation and Restoration: The Status of the Science*. Island Press, Washington, D.C., pp. 555–570.

Kusler, J. A. and Kentula, M. E. (Eds.). 1990. *Wetland Creation and Restoration: The Status of the Science*. Island Press, Covelo, CA, 591 pp.

Linder, T. 1990. Conservation easements. *Golf Course Manage.* 58(11):20–26.

Mackey, D. 1987. Golf course design maximizes grass potential. *Seed World.* 125(12): 71–74.

Mayack, L. A., Bush, P. B., Fletcher, O. J., Page, R. K. and Fendley, T. T. 1981. Tissue residue of dietary cadmium in wood ducks. *Arch. Environ. Contam. Toxicol.* 10:637–645.

Mitsch, W. J. and Gosselink, J. G. 1986. *Wetlands*. Van Nostrand Reinhold, New York, 539 pp.

Peacock, C. H., Bruneau, A. H. and Spak, S. P. 1990. Protecting a valuable resource: Preservation of wetlands from a technical perspective. *Golf Course Manage.* 58(11): 6–16.

Pinch, N. O. 1989. Things to do before contacting a golf course architect. *USGA Green Section Rec.*27(2):20–22.

Rucker, D. (Ed.). 1988. Turnabout on wetlands: From destruction to preservation and construction. *Wetlands* 11(1/2):1–20.

Salvesen, D. 1990. *Wetlands: Mitigating and Regulating Development Impacts*. The Urban Land Institute. Washington, D.C., 117 pp.

Sather, J. H. and Smith, R. D. 1984. An Overview of Major Wetland Functions and Values. U.S. Dept. Int., Fish Wildl. Serv. FWS/OBS-84/18. 69 pp.

Shaw, S. P. and Fredine, C. G. 1956. Wetlands of the United States, Their Extent and Their Value for Waterfowl and Other Wildlife. U.S. Dept. Int., Fish Wildl. Serv., Circular 39. Washington, D.C., 67 pp.

Temple, T. 1983. The marsh makers of St. Michaels. *EPA J.* 9(2):9–12.

Tilton, D. L. 1981. The environmental impact associated with development in wetlands. in Richardson, B. (Ed.). *Selected Proceedings of the Midwest Conference on Wetland Values and Management*. Freshwater Inst., Navarre, MN, pp. 357–362.

Tiner, R. W., Jr. 1984. Wetlands of the United States: Current Status and Recent Trends. U.S. Dept. Int. Fish Wildl. Serv., Washington, D.C., 59 pp.

U.S. Environmental Protection Agency. 1988. America's Wetlands: Our Vital Link Between Land and Water. OPA-87-016. U.S. Environmental Protection Agency, Washington, D.C., 8 pp.

U.S. Environmental Protection Agency and the Department of the Army. 1990. Memorandum of Agreement Concerning the Determination of Mitigation Under the Clean Water Act Section 404(b)(1) Guidelines. February 6, 1990.

U.S. Fish and Wildlife Service. 1991. National Wetlands Inventory Handout Fact Sheet. U. S. Dept. Int., Fish Wildl. Serv., Monroe Building, Suite 101, 9720 Executive Center Drive, St. Petersburg, FL, 2 pp.

van der Leeden, F., Troise, F. L. and Todd, D. K. 1990. *The Water Encyclopedia*. Lewis Publishers, Inc. Chelsea, MI, 808 pp.

Walker, W. J. and Branham, B. 1992. Environmental impacts of turfgrass fertilization. in Balogh, J. C. and Walker, W. J. (Eds.). *Golf Course Management & Construction: Environmental Issues*. Lewis Publishers, Chelsea, MI, Chapter 3.

Want, W. L. 1989. *Law of Wetland Regulation*. Clark Boardman Company, New York.

Wolf, R. B., Lee, L. C. and Sharitz, R. R. 1986. Special issue: Wetland creation and restoration in the United States from 1970 to 1985: An annotated bibliography. *Wetlands* 6(1):1–88.

Yocom, T. G., Leidy, R. A. and Morris, C. A. 1989. Wetlands protection through impact avoidance: a discussion of the 404 (b)(1) alternative analysis. *Wetlands* 9(2):283–297.

CHAPTER 8

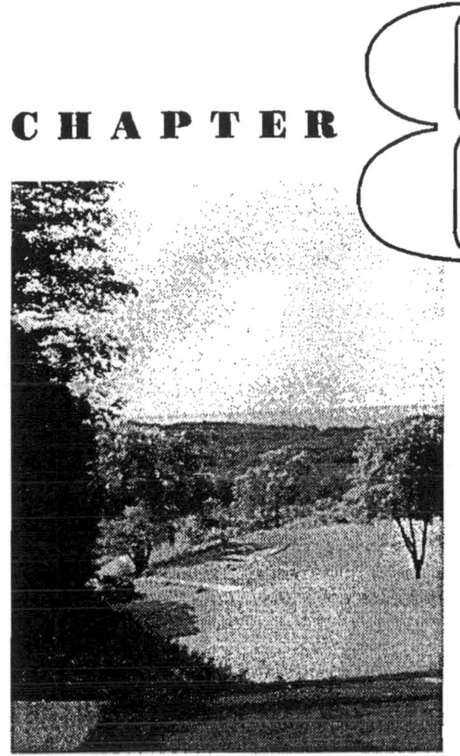

Aquatic and Terrestrial Toxicity Tables

Sheila R. Murphy

8.1 ACUTE AND CHRONIC TOXICITY

Golf course and turfgrass managers control turfgrass pests and disease with a wide range of cultural and chemical practices. Given the current public concern for the environmental impacts of chemical management strategies, the potential for exposure and toxic effects of applied chemicals should be an important component of golf course policy decisions. One of the primary hazards of chemical management is the risk of adverse effects to humans, nontarget animals, and plants through onsite and offsite exposure. The level of risk is determined by the *intrinsic toxicity* of the chemical and the *potential for exposure* to toxic doses of the chemical. These concepts were discussed by Balogh and Anderson (1992) and Tietge (1992) in earlier chapters of this book.

Effective chemical action of applied chemicals requires (1) exposure to a sufficient quantity of chemical, (2) presence of the chemical in an active form, and (3) exposure for sufficient time to produce a toxic effect. Toxicity alone is insufficient to produce chronic or acute effects without sufficient exposure to the toxic agent. There are two types of direct effects of pesticides on nontarget plants and animals: acute and chronic. The type of response (acute or chronic) depends on both the *magnitude*

Photo courtesy of Michael P. Kenna, USGA.

Table 1. Relative Pesticide Toxicity Ratings

Toxicity rating	LC_{50}[a] (mg L^{-1})	Class[a]	LD_{50}[b] (mg kg^{-1})	Class[b]
1	<0.1	Very highly toxic	<5	Extremely toxic
2	0.1–1	Highly toxic	5–49	Very toxic
3	1–9.99	Moderately toxic	50–499	Moderately toxic
4	10–99.99	Slightly toxic	500–4999	Slightly toxic
5	>100	Practically nontoxic	>4999	Almost nontoxic

[a] *EPA 1985a.*
[b] *USDA 1969.*

of the dose and the *duration of exposure.* Acute effects are dramatic or rapid responses of organisms to large quantities of pesticides over a short period of time. Exposure to sufficient quantities of applied pesticides to produce acute effects in wildlife or other nontarget organisms may occur with chemical application immediately followed by a heavy rainfall. Runoff or surface pooling of drainage effluent may result in potentially toxic concentrations of pesticides in temporary surface detention ponds. Toxic concentrations of pesticides also will occur as a result of spills during tank filling or overapplication.

8.2 SUMMARY OF ACUTE AND CHRONIC TOXICITY TESTS

Extensive data have been collected on chemicals using a variety of terrestrial and aquatic test organisms. These data are available for the assessment of chemical toxicity on "representative" laboratory animals. The relative acute toxicities of pesticides have been established according to ranges of LC_{50} and LD_{50} values (Table 1). A summary of toxicity tests for several turfgrass pesticides are presented in the following section. The types of toxicity tests are defined in Table 2. Toxicity information on nontarget organisms was obtained from literature reviews (Atkinson 1985; Balogh et al. 1990; Bovey and Young 1980; Heath et al. 1972; Kenaga 1975; Kline et al. 1987; Murty 1986; Pickering et al. 1989; Pilli et al. 1988; Sheedy et al. 1991; Stephan et al. 1986) and computerized toxicity databases (AQUIRE: Murphy 1990; Pilli et al. 1989; TERRETOX: Meyers and Shiller 1985). These databases were originally developed by the U.S. Environmental Protection Agency and The University of Montana, Missoula. The results of acute and chronic tests for turfgrass pesticides are summarized in Tables 3 through 57.

Table 2. Description of Acute and Chronic Toxicity Test Codes

Effect code[a]	Test description
	Acute toxicity tests
EC_{50}	Median effective concentration — used when an effect other than death is the observed endpoint
EC_{xx}	Effective concentration to xx% of tested organisms
	EC_{50}AB - Abnormalities
	EC_{50}BA - Byssal Attachment
	EC_{50}BH - Behavior
	EC_{50}BM - Biomass
	EC_{50}CH - Chlorophyll
	EC_{50}DT - Detachment
	EC_{50}DV - Development
	EC_{50}EQ - Equilibrium
	EC_{50}EZ - Enzyme Activity
	EC_{50}FC - Food Consumption
	EC_{50}FR - Filtration Rate
	EC_{50}GR - Growth
	EC_{50}HA - Hatchability
	EC_{50}IM - Immobilization
	EC_{50}IN - Inhibition
	EC_{50}IR - Irritation
	EC_{50}NF - Nitrogen Fixation
	EC_{50}OX - Oxygen Production
	EC_{50}PG - Pigment Change
	EC_{50}PH - Physiological Effect
	EC_{50}PP - Population Size Reduction
	EC_{50}PS - Photosynthesis Effect
	EC_{50}RE - Reproduction
	EC_{50}SW - Swimming
	EC_{50}TE - Teratogenesis
ET_{50}	Enzyme effect: change in enzyme activity
LC_{50}	Median lethal concentration is the amount of a substance necessary to kill 50% of a population; used only when death is the observed endpoint
LC_{xx}	Lethal concentration to xx% of tested organisms
LD_{50}	Median lethal dose or concentration is the amount of a substance necessary to kill 50% of a population when administered through injection or diet
LD_{xx}	Median lethal dose to xx% of tested organisms
LET	Lethal: 100% mortality or 0% survival
LT_{50}	Lethal Threshold Concentration
MOR	Mortality: effects expressed as % death or % survival

Table 2. Description of Acute and Chronic Toxicity Test Codes (continued)

Effect code[a]	Test description
	Chronic toxicity tests
ABD	Abundance: number of organisms within the same species changes
ABN	Abnormalities: physical malformation during developmental stages
AVO	Avoidance: avoidance or attraction to a chemical gradient
BCF	Bioconcentration factor: accumulation of a toxicant in the tissues of the test organism compared to the measured toxicant concentration in the water in which the organism was exposed, e.g., $$BCF = \frac{\text{g/kg chemical in fish tissue (wet weight)}}{\text{g/L chemical in } H_2O}$$
BEH	Behavior: quantifiable change in activity
BIO	Biochemical effect: change in physiological processes
BMS	Biomass: change in the amount of living matter
CEL	Cellular effect: change in organelle structure
CLN	Rate of colonization: change in ability to colonize an uninhibited substrate under toxicant stress
CLR	Chlorophyll: measurable change in chlorophyll content
CYT	Cytogenetic effect: changes in the RNA and DNA in the cell
DRF	Drift: change in number of larval aquatic insects to travel a given distance in a stream
DVP	Development: change in the ability to grow to a more mature lifestage in time between lifestages
ENZ	Enzyme effect: change in enzyme activity
EQU	Equilibrium: change in ability to maintain balance
FCR	Change in food consumption rate
FLT	Filtration rate: change in rate of filtration by molluscs and crustaceans
GRO	Measurable change in length and/or weight
HAT	Change in % hatch or time to hatch
HEM	Change in hematological or various blood parameters
HIS	Histological effect: presence of physical damage to tissues
HRM	Hormone effect: change in hormone concentration
LET	Lethal: 100% mortality or 0% survival including algicidal and herbicidal effects
LOC	Locomotor behavior: quantifiable change in direct movement or activity
MOR	Mortality: effect expressed as % death or % survival
MOT	Motility: change in the ability to move
NFX	Nitrogen fixation: change in ability to fix nitrogen
NOEL	No effect level for a given test
OC	Oxygen consumption: change in oxygen uptake of test organism or its tissues
PATH	Pathological
PRG	Change in population growth rate
PHY	Physiological effects: changes in the organic processes or functions of organs
POP	Population: change in number of species groups in a given community; species diversity

Table 2. Description of Acute and Chronic Toxicity Test Codes (continued)

Effect code[a]	Test description
PSE	Photosynthesis effect: change in plant productivity
REP	Change in male and/or female ability to reproduce
RES	Respiration effect: change in rate of ventilation or breathing activity
RGN	Regeneration: change in ability to regenerate a body part
RSD	Residue: used for tissue concentration when bioconcentration factor is not reported
STR	Stress: observed physiological tension in test organism
TER	Teratogenesis: quantifiable occurrence of abnormal offspring
THL	Thermal effect: change in tolerance to temperature change
TMR	Tumor occurrence: presence of a mass of abnormal tissue
VTE	Vertebral effect: physical change in vertebral structure and/or composition leading to scoliosis, lordosis, and bone composition

[a] *Effect code used in toxicity table in Section 8.2. D = Delayed effect endpoint, indicated in the effect code field.*

Table 3. Toxicology of Aquatic and Terrestrial Species: Chemical: 2,4-D Acid, CAS Number: 94757

Class	Species	Effect	Duration (days)	Concentration ($\mu g\ L^{-1}$)	Lifestage	Ref.
Algae	Algae, phytoplankton, algal mat	ABD	3	0–68	NR	Inabinet (1979)
		CLR	90	0–68	NR	Inabinet (1979)
		POP	90	0–68	NR	Inabinet (1979)
		PSE	4 hr	1,000	NR	Butler (1963)
	Blue-green algae (*Agmenellum* sp)	MOR	7	220,000	NR	Elder et al. (1970)
	Blue-green algae (*Anabaena variabilis*)	EC_{50}PP	2	>2,200	NR	Hawxby et al. (1977)
		EC_{50}PS	NR	>2,200	NR	Hawxby et al. (1977)
		CEL	NR	>2,200	NR	Hawxby et al. (1977)
	Blue-green algae (*Anabaena flos-aquae*)	CLR	NR	10,000	NR	Chinnaswamy and Patel (1983)
	Blue-green algae (*Anabaena cylindrica*)	MOR	15	100,000	NR	Venkataraman and Rajyalakshmi (1972)
	Blue-green algae (*Anabaena* sp)	MOR	15	100,000	NR	Venkataraman and Rajyalakshmi (1972)
	Blue-green algae (*Anabaena variabilis*)	MOR	15	500,000	NR	Venkataraman and Rajyalakshmi (1972)
	Blue-green algae (*Anabaena* sp)	MOR	15	500,000	NR	Venkataraman and Rajyalakshmi (1972)
		PGR	15		NR	Venkataraman and Rajyalakshmi (1971)
	Blue-green algae (*Anacystis flos-aquae*)	CEL	7	10,000	NR	El-Ayouty et al. (1978)
		CEL	7	18,000	NR	El-Ayouty et al. (1978)
	Blue-green algae (*Anacystis* sp)	MOR	7	220,000	NR	Elder et al. (1970)

Blue-green algae (*Anacystis nidulans*)	MOR	15	100,000	NR	Venkataraman and Rajyalakshmi (1972)
Blue-green algae (*Anacystis aeruginosa*)	PGR	1	250,000	5 D LOG PHASE, (1–2)E+6 CELLS ML^{-1}	Fitzgerald et al. (1952)
Blue-green algae (*Anacystis nidulans*)	PGR	7	50,000	NR	Voight and Lynch (1974)
	PGR	7	90,000	NR	Voight and Lynch (1974)
	PGR	15	200	NR	Venkataraman and Rajyalakshmi (1971)
Blue-green algae (*Anacystis aeruginosa*)	PGR	21	2,000	INITIAL CONC. 125,000 CELLS ML^{-1}	Palmer and Maloney (1955)
Blue-green algae (*Anacystis nidulans*)	PGR	NR	80,000	NR	Voight and Lynch (1974)
Blue-green algae (*Aulosira fertilissima*)	MOR	15	100,000	NR	Venkataraman and Rajyalakshmi (1972)
	PGR	15	200	NR	Venkataraman and Rajyalakshmi (1971)
Blue-green algae (*Cylindrospermum* sp)	MOR	7	220,000	NR	Elder et al. (1970)
Blue-green algae (*Cylindrospermum licheniforme*)	PGR	3	2,000	INITIAL CONC. 125,000 CELLS ML^{-1}	Palmer and Maloney (1955)
Blue-green algae (*Lyngbya* sp)	EC_{50}PP	1	>2,200	EXPONENTIAL PHASE	Hawxby et al. (1977)
	EC_{50}PS	NR	>2,200	EXPONENTIAL PHASE	Hawxby et al. (1977)
	CEL	NR	>2,200	EXPONENTIAL PHASE	Hawxby et al. (1977)
Blue-green algae (*Microcoleus* sp)	MOR	7	220,000	NR	Elder et al. (1970)
Blue-green algae (*Nostoc* sp)	MOR	7	220,000	NR	Elder et al. (1970)
	MOR	15	5,000	NR	Venkataraman and Rajyalakshmi (1972)
	MOR	15	50,000	NR	Venkataraman and Rajyalakshmi (1972)

Table 3. Toxicology of Aquatic and Terrestrial Species: Chemical: 2,4-D Acid, CAS Number: 94757 (continued)

Class	Species	Effect	Duration (days)	Concentration ($\mu g\ L^{-1}$)	Lifestage	Ref.
Algae (continued)	Blue-green algae (*Nostoc calcicola*)	MOR	15	100,000	NR	Venkataraman and Rajyalakshmi (1972)
	Blue-green algae (*Nostoc* sp)	MOR	15	100,000	NR	Venkataraman and Rajyalakshmi (1972)
		PGR	15		NR	Venkataraman and Rajyalakshmi (1971)
	Blue-green algae (*Oscillatoria amphibia*)	CEL	7	10,000	NR	El-Ayouty et al. (1978)
		CEL	7	18,000	NR	El-Ayouty et al. (1978)
	Blue-green algae (*Oscillatoria* sp)	MOR	7	220,000	NR	Elder et al. (1970)
	Blue-green algae (*Tolypothrix tenuis*)	MOR	15	100,000	NR	Venkataraman and Rajyalakshmi (1972)
		PGR	15	200	NR	Venkataraman and Rajyalakshmi (1971)
	Coccolithophorid (*Coccolithus* sp)	MOR	7	220,000	NR	Elder et al. (1970)
	Cuvie, tangleweed	GRO	NR	100,000	GAMETOPHYTE	Hopkin and Kain (1978)
		GRO	NR	100,000	SPOROPHYTE	Hopkin and Kain (1978)
		PGR	28	1,000	ZOOSPORES	Hopkins and Kain (1971)
		PSE	NR	1,000,000	MATURE FROND TISSUE	Hopkin and Kain (1978)
	Diatom (*Gomphonema parvulum*)	PGR	21	2,000	INITIAL CONC. 125,000 CELLS ML^{-1}	Palmer and Maloney (1955)
	Diatom (*Navicula* sp)	MOR	7	220,000	NR	Elder et al. (1970)
	Diatom (*Nitzschia palea*)	PGR	3	2,000	INITIAL CONC. 125,000 CELLS ML^{-1}	Palmer and Maloney (1955)

Diatom (*Phaeodactylum tricornutum*)	EC_{50}OX	1.5 hr	60,000	LOG PHASE	Walsh (1972)
	PGR	10	50,000	LOG PHASE	Walsh (1972)
Diatom (*Phaeodactylum* sp)	PSE	NR	10,000	NR	Walsh et al. (1970)
Flagellate euglenoid (*Euglena* sp)	MOR	7	220,000	NR	Elder et al. (1970)
Flagellate euglenoid (*Euglena gracilis*)	MOT	3	220,000	NR	Elder et al. (1970)
	PGR	1	100,000	1 WEEK	Poorman (1973)
	PGR	7	100,000	1 WEEK	Poorman (1973)
	RSD	6 hr	10–1,000	NR	Valentine and Bingham (1974)
Green algae (*Ankistrodesmus* sp)	PGR	14	110,000	NR	Lembi and Coleridge (1975)
Green algae (*Bracteacoccus cinnabarinus*)	PGR	2–10	40,000–70,000	NR	Voight and Lynch (1974)
	PGR	NR	100,000	NR	Voight and Lynch (1974)
Green algae (*Chlamydomonas reinhardtii*)	EC_{50}GR	5	320,000–560,000	500,000 CELLS ML^{-1}	Benijts-Claus and Persoone (1975)
Green algae (*Chlamydomonas eugametos*)	PGR	4	200,000	2ND D OF LOG PHASE	Vance and Smith (1969)
Green algae (*Chlamydomonas reinhardtii*)	PGR	5	300,000	500,000 CELLS ML^{-1}	Benijts-Claus and Persoone) (1975)
	RSD	6 hr	10–1,000	NR	Valentine and Bingham (1974)
Green algae (*Chlorococcum* sp)	EC_{50}OX	1.5 hr	60,000	LOG PHASE	Walsh (1972)
Green algae (*Chlorella pyrenoidosa*)	EC_{50}PP	1	>2,200	EXPONENTIAL PHASE	Hawxby et al. (1977)
Green algae (*Chlorococcum* sp)	EC_{50}PP	1	>2,200	EXPONENTIAL PHASE	Hawxby et al. (1977)
Green algae (*Chlorella pyrenoidosa*)	EC_{50}PS	NR	>2,200	EXPONENTIAL PHASE	Hawxby et al. (1977)

Table 3. Toxicology of Aquatic and Terrestrial Species: Chemical: 2,4-D Acid, CAS Number: 94757 (continued)

Class	Species	Effect	Duration (days)	Concentration ($\mu g\ L^{-1}$)	Lifestage	Ref.
Algae (continued)	Green algae (*Chlorococcum* sp)	EC_{50}PS	NR	>2,200	EXPONENTIAL PHASE	Hawxby et al. (1977)
	Green algae (*Chlorella pyrenoidosa*)	CEL	NR	>2,200	EXPONENTIAL PHASE	Hawxby et al. (1977)
	Green algae (*Chlorococcum* sp)	CEL	NR	>2,200	EXPONENTIAL PHASE	Hawxby et al. (1977)
	Green algae (*Chlorella pyrenoidosa*)	CLR	3	50,000	NR	Huang and Gloyna (1968)
		PGR	1.67	1,000,000	NR	Tomisek et al. (1957)
		PGR	4	200,000	2ND D OF LOG PHASE	Vance and Smith (1969)
	Green algae (*Chlorococcum* sp)	PGR	10	50,000	LOG PHASE	Walsh (1972)
	Green algae (*Chlorella variegata*)	PGR	21	2,000	INITIAL CONC. 125,000 CELLS ML^{-1}	Palmer and Maloney (1955)
	Green algae (*Chlorella pyrenoidosa*)	PSE	1 hr	442,000	2 WEEK CULTURE	Wedding et al. (1954)
		PSE	>2 hr	0.00302	3 D CULTURE, 20 MG DRY WT	Erickson et al. (1955)
	Green algae (*Chlorella* sp)	PSE	NR	10,000	NR	Walsh et al. (1970)
	Green algae (*Chlorococcum* sp)	PSE	NR	10,000	NR	Walsh et al. (1970)
	Green algae (*Chlorella pyrenoidosa*)	RSD	6 hr	10–1,000	NR	Valentine and Bingham (1974)

Green algae (*Coelastrum microporum*)	PGR	2–10	5,000–50,000	NR	Voight and Lynch (1974)
	PGR	2–10	55,000–65,000	NR	Voight and Lynch (1974)
	PGR	7	1,800	NR	El-Ayouty et al. (1978)
	PGR	7	3,200	NR	El-Ayouty et al. (1978)
	PGR	NR	100,000	NR	Voight and Lynch (1974)
Green algae (*Dunaliella tertiolecta*)	EC_{50}OX	1.5 hr	50,000	LOG PHASE	Walsh (1972)
Green algae (*Dunaliella bioculata*)	PGR	2	100	LOG PHASE	Felix et al. (1988)
Green algae (*Dunaliella tertiolecta*)	PGR	10	75,000	LOG PHASE	Walsh (1972)
Green algae (*Dunaliella* sp)	PSE	NR	10,000	NR	Walsh et al. (1970)
Green algae (*Enteromorpha linza*)	BCF	1	25	GERMLINGS	Sikka et al. (1976)
	PSE	21	220,000	GERMLINGS	Sikka et al. (1976)
Green algae (*Hydrodictyon* sp)	MOR	7	220,000	NR	Elder et al. (1970)
Green algae (*Nannochloris* sp)	PSE	NR	10,000	NR	Walsh et al. (1970)
Green algae (*Pandorina* sp)	MOR	7	220,000	NR	Elder et al. (1970)
Green algae (*Pediastrum* sp)	ABD	14	22,000	NR	Elder et al. (1970)
	PGR	14	110,000	NR	Lembi and Coleridge (1975)
Green algae (*Pithophora* sp)	MOR	7	220,000	NR	Elder et al. (1970)
Green algae (*Platymonas* sp)	PSE	NR	10,000	NR	Walsh et al. (1970)
Green algae (*Scenedesmus* sp)	EC_{50}GR	5	870,000–1,000,000	500,000 CELLS ML^{-1}	Benijts-Claus and Persoone (1975)
Green algae (*Scenedesmus obliquus*)	BCF	NR	11.05	NR	Bohm and Muller (1976)
Green algae (*Scenedesmus opoliensis*)	CEL	7	5,600	NR	El-Ayouty et al. (1978)
	CEL	7	10,000	NR	El-Ayouty et al. (1978)

Table 3. Toxicology of Aquatic and Terrestrial Species: Chemical: 2,4-D Acid, CAS Number: 94757 (continued)

Class	Species	Effect	Duration (days)	Concen-tration ($\mu g\ L^{-1}$)	Lifestage	Ref.
Algae (continued)	Green algae (*Scenedesmus quadricauda*)	PGR	4	200,000	2ND D OF LOG PHASE	Vance and Smith (1969)
	Green algae (*Scenedesmus* sp)	PGR	5	300,000	500,000 CELLS ML^{-1}	Benijts-Claus and Persoone (1975)
		PGR	14	110,000	NR	Lembi and Coleridge (1975)
	Green algae (*Scenedesmus obliquus*)	PGR	21	2,000	INITIAL CONC. 125,000 CELLS ML^{-1}	Palmer and Maloney (1955)
	Green algae (*Scenedesmus quadricauda*)	PSE	2	1,000	NR	Stadnyk et al. (1971)
		PSE	4	1,000	NR	Stadnyk et al. (1971)
		PSE	6	100	NR	Stadnyk et al. (1971)
		PSE	8	100	NR	Stadnyk et al. (1971)
		PSE	10	1,000	NR	Stadnyk et al. (1971)
	Green algae (*Scenedesmus obliquus*)	RSD	4.5 hr	100–1,000	1,000,000 CELLS ML^{-1}	Ellgehausen et al. (1980)
	Green algae (*Scenedesmus quadricauda*)	RSD	6 hr	10–1,000	NR	Valentine and Bingham (1974)
	Green algae (*Selenastrum capricornutum*)	EC_{50}	NR	12,900	NR	Miller et al. (1985)
		EC_{50}	NR	95,800	NR	Miller et al. (1985)
	Green algae (*Spirogyra* sp)	MOR	7	220,000	NR	Elder et al. (1970)
	Green algae (*Ulva lactuca*)	BCF	1	25	FIELD THALLI	Sikka et al. (1976)
		PSE	7	220,000	FIELD THALLI	Sikka et al. (1976)
	Green algae (*Volvox* sp)	MOR	7	220,000	NR	Elder et al. (1970)
	Green algae (*Zygnema* sp)	MOR	7	220,000	NR	Elder et al. (1970)
	Haptophyte (*Isochrysis galbana*)	$EC_{50}OX$	1.5 hr	60,000	LOG PHASE	Walsh (1972)
		PGR	10	50,000	LOG PHASE	Walsh (1972)

	Haptophyte (*Isochrysis* sp)	PSE	NR	10,000	NR	Walsh et al. (1970)
	Haptophyte (*Pavlova* sp)	PSE	NR	10,000	NR	Walsh et al. (1970)
	Haptophyte (*Pleurochrysis* sp)	MOR	7	220,000	NR	Elder et al. (1970)
	Red algae (*Champia parvula*)	GRO	11–14	21,600	TETRASPOROPHYTES	Thursby et al. (1985)
		REP	11–14	100,000	TETRASPOROPHYTES	Thursby et al. (1985)
	Red algae (*Rhodymenia pseudopalmata*)	BCF	1	25	LAB THALLI	Sikka et al. (1976)
		PSE	21	220,000	LAB THALLI	Sikka et al. (1976)
Amphibia	Common indian toad	LC_{50}	1	13,770	LARVAE	Vardia et al. (1984)
		LC_{50}	2	9,030	LARVAE	Vardia et al. (1984)
		LC_{50}	4	8,050	LARVAE	Vardia et al. (1984)
	Frog (*Rana temporaria*)	STR-D	2	50,000	TADPOLE	Cooke (1972)
Annelida	Annelid worm class	ABD	21	44	NR	Marshall and Rutschsky (1974)
	Leech class	ABD	7	44	NR	Marshall and Rutschsky (1974)
	Oligochaete (*Lumbriculus variegatus*)	LC_{50}	7	122,200	NR	Bailey and Liu (1980)
		LC_{50}	8	122,200	NR	Bailey and Liu (1980)
Aquatic intertebrate	Invertebrates	ABD	3	22.4	NR	Hooper (1953)
Arachnida	Water mite (*Hydracarina*)	ABD	35	44	NR	Marshall and Rutschsky (1974)
Chondricht	Spiny dogfish	MOR	3	20	NR	Guarino et al. (1976)
Cnidaria	Hydra	ABN	3.75	4,000	ADULT	Kudla (1984)

Table 3. Toxicology of Aquatic and Terrestrial Species: Chemical: 2,4-D Acid, CAS Number: 94757 (continued)

Class	Species	Effect	Duration (days)	Concentration (μg L^{-1})	Lifestage	Ref.
Crustacea	Blue crab	EC_{50}EQ	1	2,900	JUVENILE	Butler (1963)
	Blue crab	EC_{50}EQ	2	2,900	JUVENILE	Butler (1963)
	Brown shrimp	EC_{50}EQ	1	550	ADULT	Butler (1963)
	Brown shrimp	EC_{50}EQ	2	550	ADULT	Butler (1963)
	Calanoid copepod (*Eudiaptomus gracilis*)	LC_{50}	4	144,100	NR	Presing and Ponyi (1986)
	Calanoid copepod	LC_{50}	1	2,400	ADULT, 2.2–2.8 MM	Kader et al. (1976)
	(*Spicodiaptomus chilospinus*)	LC_{50}	2	1,850	ADULT, 2.2–2.8 MM	Kader et al. (1976)
	Crayfish family	ABD	21	44	NR	Marshall and Rutschsky (1974)
	Cyclopoid copepod (*Acanthocyclops vernalis*)	EC_{50}IM	2	37,420	0–4 HR, NAUPLII	Robertson and Bunting (1976)
		EC_{50}IM	4	8,720	0–4 HR, NAUPLII	Robertson and Bunting (1976)
	Cyclopoid copepod (*Mesocyclops leuckarti*)	MOR	30	50,000	NR	George et al. (1982)
	Dungeness or edible crab	DVP	1	100,000	PRE ZOEAE	Caldwell et al. (1979)
		DVP	50	1,000	1ST STAGE ZOEAE	Caldwell et al. (1979)
		MOR	80	590	JUVENILES, 3RD INSTAR	Caldwell et al. (1979)
		MOR	85	10,000	ADULT, 80–100 MM	Caldwell et al. (1979)
	Grass shrimp (*Palaemonetes pugio*)	AVO	0.5 hr	1,000	10–40 MM	Hansen et al. (1973)
	Red swamp crayfish	LC_{50}	4	1,389,000	JUVENILE, 0.4 G, 25–35 MM	Cheah et al. (1980)
	Scud (*Gammarus fasciatus*)	LC_{50}	2	3,200	EARLY INSTAR	Sanders (1970a)

Water flea (*Daphnia magna*)	EC_{50}	NR	13,100	NR	Miller et al. (1985)
	EC_{50}	NR	>240,000	NR	Miller et al. (1985)
	EC_{50}IM	1.08	>100,000	1ST INSTAR	Crosby and Tucker (1966)
Water flea (*Daphnia pulex*)	EC_{50}IM	2	3,200	1ST INSTAR	Sanders and Cope (1966)
Water flea (*Daphnia magna*)	EC_{50}IM	2	>100,000	EARLY INSTAR	Sanders (1970a)
	LC_{50}	1	180,000	NR	Benijts-Claus and Persoone (1975)
	LC_{50}	1	225,000	NR	Benijts-Claus and Persoone (1975)
	LC_{50}	1	235,000	NR	Benijts-Claus and Persoone (1975)
	LC_{50}	1	240,000	NR	Benijts-Claus and Persoone (1975)
	LC_{50}	1	>100,000	NEONATE	Alexander et al. (1985)
	LC_{50}	2	25,000	NEONATE	Alexander et al. (1985)
	LC_{50}	2	36,400	NEONATE	Alexander et al. (1985)
	LC_{50}	2	135,000	NR	Benijts-Claus and Persoone (1975)
	LC_{50}	2	90,000–120,000	NR	Benijts-Claus and Persoone (1975)
	LC_{50}	2	195,000–220,000	NR	Benijts-Claus and Persoone (1975)
	LC_{50}	21	1,000	NR	Benijts-Claus and Persoone (1975)
	LC_{50}	21	100,000	NR	Benijts-Claus and Persoone (1975)
Water flea (*Daphnia lumholzi*)	LT_{50}	0.88	20,000	NR	George et al. (1982)
	LT_{50}	1.58	10,000	NR	George et al. (1982)
	LET	1.29	20,000	NR	George et al. (1982)
	LET	2.96	10,000	NR	George et al. (1982)
Water flea (*Daphnia magna*)	RSD	2	100–1,000	1 MG ML^{-1}	Ellgehausen et al. (1980)

Table 3. Toxicology of Aquatic and Terrestrial Species: Chemical: 2,4-D Acid, CAS Number: 94757 (continued)

Class	Species	Effect	Duration (days)	Concentration ($\mu g\ L^{-1}$)	Lifestage	Ref.
Crustacea (continued)	Water flea (*Simocephalus serrulatus*)	$EC_{50}IM$	2	4,900	1ST INSTAR	Sanders and Cope (1966)
Fish	American eel	LC_{50}	1	427,200	YOUNG OF THE YEAR	Rehwoldt et al. (1977)
		LC_{50}	2	390,200	YOUNG OF THE YEAR	Rehwoldt et al. (1977)
		LC_{50}	4	300,600	YOUNG OF THE YEAR	Rehwoldt et al. (1977)
	Banded killifish	LC_{50}	1	306,200	YOUNG OF THE YEAR	Rehwoldt et al. (1977)
		LC_{50}	2	261,100	YOUNG OF THE YEAR	Rehwoldt et al. (1977)
		LC_{50}	4	26,700	YOUNG OF THE YEAR	Rehwoldt et al. (1977)
	Banded tetra	MOR	2	600	NR	Rodrigues et al. (1980)
		MOR	2	900	NR	Rodrigues et al. (1980)
		MOR	2	1,350	NR	Rodrigues et al. (1980)
		MOR	2	2,020	NR	Rodrigues et al. (1980)
		MOR	2	3,030	NR	Rodrigues et al. (1980)
	Bluegill	LC_{50}	1	8,000	FINGERLING, 2.5–7.6 CM	Hughes and Davis (1963)
		LC_{50}	1	305,000	19.5 MM, 0.15 G	Alexander et al. (1985)
		LC_{50}	2	8,000	FINGERLING, 2.5–7.6 CM	Hughes and Davis (1963)
		LC_{50}	2	290,000	19.5 MM, 0.15 G	Alexander et al. (1985)
		LC_{50}	3	263,000	19.5 MM, 0.15 G	Alexander et al. (1985)
		LC_{50}	4	263,000	19.5 MM, 0.15 G	Alexander et al. (1985)
		LD_{50}	7	1,000,000	6.4–10.2 CM, 5 G	Harrisson and Rees (1946)
		LET	1	5,000	FRY	Hiltibran (1967)
		MOR	7	100,000	FINGERLING, 6.2 CM	King and Penfound (1946)

	RSD	1	125–312	NR	Schultz and Harman (1974)
	RSD	14	25–617	NR	Schultz and Harman (1974)
	STR	1	5,000	FINGERLING, <=10 CM	Applegate et al. (1957)
Brown bullhead	LD_{50}	7	2,000,000	12.7–15.2 CM, 8.5 G	Harrisson and Rees (1946)
	LET	7	3,000,000	12.7–15.2 CM, 85 G	Harrisson and Rees (1946)
	MOR	7	1,000,000	12.7–15.2 CM, 85 G	Harrisson and Rees (1946)
	MOR	7	2,000,000	12.7–15.2 CM, 85 G	Harrisson and Rees (1946)
	MOR	7	2,500,000	12.7–15.2 CM, 85 G	Harrisson and Rees (1946)
Channel catfish	MOR	2	10,000	1 Y, 14 G, 12 CM, FINGERLINGS	Mccorkle et al. (1977)
	RSD	1	125–312	NR	Schultz and Harman (1974)
	RSD	4	100–1,000	1–2 G	Ellgehausen et al. (1980)
	RSD	5	1,000	3–4 G, 50–75 MM	Rodgers and Stalling (1972)
	RSD	14	25–617	NR	Schultz and Harman (1974)
Chum salmon	MOR	4	10,000	FRY	Meehan et al. (1974)
	MOR	4	50,000	FRY	Meehan et al. (1974)
Class — bony fishes	RSD	60	3.9	YOUNG — ADULT FROM 4 FAMILIES	Frank et al. (1987)
Coho salmon, silver salmon	BCF	3	200	FRY	Sears and Meehan (1971)
	BCF	NR	430,000	6–18 MONTHS, 8.9–25.5 G	Walsh and Ribelin (1973)
	MOR	4	10,000	FRY	Meehan et al. (1974)
	MOR	4	10,000	FINGERLING	Meehan et al. (1974)
	MOR	4	50,000	FRY	Meehan et al. (1974)
	MOR	4	50,000	FINGERLING	Meehan et al. (1974)
	MOR	6	200,000	YEARLING	Lorz et al. (1979)

Table 3. Toxicology of Aquatic and Terrestrial Species: Chemical: 2,4-D Acid, CAS Number: 94757 (continued)

Class	Species	Effect	Duration (days)	Concentration ($\mu g\ L^{-1}$)	Lifestage	Ref.
Fish (continued)	Common, mirror, colored, carp	LC_{100}	0.67	3,200,000	EGGS, 27 HR	Kamler et al. (1974)
		LC_{100}	2 hr	3,200,000	LARVAE, 2 HR	Kamler et al. (1974)
		LC_{50}	1	5,930	6.52(5.97–6.97)CM, 3.52(3.43–3.67)G	Vardia and Durve (1981b)
		LC_{50}	1	20,000	6.52(5.97–6.97)CM, 3.52(3.43–3.67)G	Vardia and Durve (1981b)
		LC_{50}	1	35,720	6.52(5.97–6.97)CM, 3.52(3.43–3.67)G	Vardia and Durve (1981b)
		LC_{50}	1	35,720	1.01–3.63 G	Vardia and Durve (1981a)
		LC_{50}	1	38,720	6.52(5.97–6.97)CM, 3.52(3.43–3.67)G	Vardia and Durve (1981b)
		LC_{50}	1	45,000	6.52(5.97–6.97)CM, 3.52(3.43–3.67)G	Vardia and Durve (1981b)
		LC_{50}	1	175,200	YOUNG OF THE YEAR	Rehwoldt et al. (1977)
		LC_{50}	2	5,800	6.52(5.97–6.97)CM, 3.52(3.43–3.67)G	Vardia and Durve (1981b)
		LC_{50}	2	18,400	6.52(5.97–6.97)CM, 3.52(3.43–3.67)G	Vardia and Durve (1981b)
		LC_{50}	2	21,300	6.52(5.97–6.97)CM, 3.52(3.43–3.67)G	Vardia and Durve (1981b)
		LC_{50}	2	21,300	1.01–3.63 G	Vardia and Durve (1981a)
		LC_{50}	2	35,380	6.52(5.97–6.97)CM, 3.52(3.43–3.67)G	Vardia and Durve (1981b)
		LC_{50}	2	35,720	6.52(5.97–6.97)CM, 3.52(3.43–3.67)G	Vardia and Durve (1981b)

	LC_{50}	2	42,500	6.52(5.97–6.97)CM, 3.52(3.43–3.67)G	Vardia and Durve (1981b)
	LC_{50}	2	100,200	YOUNG OF THE YEAR	Rehwoldt et al. (1977)
	LC_{50}	4	5,100	6.52(5.97–6.97)CM, 3.52(3.43–3.67)G	Vardia and Durve (1981b)
	LC_{50}	4	15,300	6.52(5.97–6.97)CM, 3.52(3.43–3.67)G	Vardia and Durve (1981b)
	LC_{50}	4	20,000	6.52(5.97–6.97)CM, 3.52(3.43–3.67)G	Vardia and Durve (1981b)
	LC_{50}	4	20,000	1.01–3.63 G	Vardia and Durve (1981a)
	LC_{50}	4	24,150	6.52(5.97–6.97)CM, 3.52(3.43-3.67)G	Vardia and Durve (1981b)
	LC_{50}	4	31,250	6.52(5.97–6.97)CM, 3.52(3.43–3.67)G	Vardia and Durve (1981b)
	LC_{50}	4	35,000	6.52(5.97–6.97)CM, 3.52(3.43–3.67)G	Vardia and Durve (1981b)
	LC_{50}	4	96,500	YOUNG OF THE YEAR	Rehwoldt et al. (1977)
	DVP	34	50,000	EGG	Kamler et al. (1974)
	MOR	1	100,000	EGGS, 27 HR	Kamler et al. (1974)
	MOR	2	800,000	LARVAE, 3 HR	Kamler et al. (1974)
	MOR	2	1,600,000	LARVAE, 3 HR	Kamler et al. (1974)
	MOR	8.3	50,000	EMBRYO	Kamler (1973)
	MOT	1	200,000	EGGS, 27 HR	Kamler et al. (1974)
	OC	9	50,000	EGG	Kamler et al. (1974)
	PHY	10	5,000	FERTILIZED EGGS	Kamler (1972)
Cutthroat trout	LC_{50}	4	64,000	0.3 G	Johnson and Finley (1980)
Cyprinid fish	LC_{50}	1	7,800	1.01–3.63 G	Vardia and Durve (1981a)
(*Rasbora neilgherriensis*)	LC_{50}	2	6,800	1.01–3.63 G	Vardia and Durve (1981a)
	LC_{50}	4	5,600	1.01–3.63 G	Vardia and Durve (1981a)
Dolly varden	MOR	4	50,000	FINGERLING	Meehan et al. (1974)

Table 3. Toxicology of Aquatic and Terrestrial Species: Chemical: 2,4-D Acid, CAS Number: 94757 (continued)

Class	Species	Effect	Duration (days)	Concen-tration ($\mu g\ L^{-1}$)	Lifestage	Ref.
Fish (continued)	Fathead minnow	LC_{50}	1	344,000	20.4 MM, 0.14 G	Alexander et al. (1985)
		LC_{50}	2	325,000	20.4 MM, 0.14 G	Alexander et al. (1985)
		LC_{50}	3	325,000	20.4 MM, 0.14 G	Alexander et al. (1985)
		LC_{50}	4	263,000	20.4 MM, 0.14 G	Alexander et al. (1985)
	Fish (*Labeo boga*)	LC_{50}	1	6,700	1.01–3.63 G	Vardia and Durve (1981a)
		LC_{50}	2	3,800	1.01–3.63 G	Vardia and Durve (1981a)
		LC_{50}	4	3,800	1.01–3.63 G	Vardia and Durve (1981a)
	Goldfish	LC_{50}	4	>187,000	EGGS, 4 D POSTHATCH	Birge et al. (1979)
		LC_{50}	4	>201,000	EGGS, 4 D POSTHATCH	Birge et al. (1979)
		LC_{50}	8	119,100	EGGS, 4 D POSTHATCH	Birge et al. (1979)
		LC_{50}	8	133,100	EGGS, 4 D POSTHATCH	Birge et al. (1979)
	Green sunfish	HAT	8	5,000	FERTILIZED EGGS	Hiltibran (1967)
		MOR	1.71	110,000	50–150 G	Sergeant et al. (1970)
		MOR	1.71	110,000	50–150 G	Sergeant et al. (1971)
	Guppy	LC_{50}	1	10,210	1.7 CM, 0.10 G	Vardia and Durve (1984)
		LC_{50}	1	76,700	NR	Rehwoldt et al. (1977)
		LC_{50}	2	9,616	1.7 CM, 0.10 G	Vardia and Durve (1984)
		LC_{50}	2	81,200	NR	Rehwoldt et al. (1977)
		LC_{50}	4	8,356	1.7 CM, 0.10 G	Vardia and Durve (1984)
		LC_{50}	4	70,700	NR	Rehwoldt et al. (1977)
	Harlequinfish, red rasbora	LC_{50}	1	3,950,000	1.3–3 CM	Alabaster (1969)
		LC_{50}	1	7,000,000	1.3–3 CM	Alabaster (1969)
		LC_{50}	2	3,100,000	1.3–3 CM	Alabaster (1969)

Killifish, topminnow family	LD_{50}	7	2,000,000	3.8–6.4 CM, 1.5 G	Harrisson and Rees (1946)
	LET	7	2,500,000	3.8–6.4 CM, 1.5 G	Harrisson and Rees (1946)
	MOR	7	100,000	3.8–6.4 CM, 1.5 G	Harrisson and Rees (1946)
	MOR	7	1,000,000	3.8–6.4 CM, 1.5 G	Harrisson and Rees (1946)
	MOR	7	1,500,000	3.8–6.4 CM, 1.5 G	Harrisson and Rees (1946)
	MOR	7	2,000,000	3.8–6.4 CM, 1.5 G	Harrisson and Rees (1946)
Lake trout, siscowet	LC_{50}	4	45,000	0.3 G	Johnson and Finley (1980)
	BCF	NR	430,000	6–18 MONTHS, 8.9–25.5 G	Walsh and Ribelin (1973)
Largemouth bass	LC_{50}	3.5	160,700	EGGS, 4 D POSTHATCH	Birge et al. (1979)
	LC_{50}	3.5	165,400	EGGS, 4 D POSTHATCH	Birge et al. (1979)
	LC_{50}	7.5	81,600	EGGS, 4 D POSTHATCH	Birge et al. (1979)
	LC_{50}	7.5	108,600	EGGS, 4 D POSTHATCH	Birge et al. (1979)
	MOR	7	100,000	FINGERLING, 7.8 CM	King and Penfound (1946)
	RSD	1	125–[illegible]12	NR	Schultz and Harman (1974)
	RSD	14	25–617	NR	Schultz and Harman (1974)
Longnose killifish	LC_{50}	1	3,500	JUVENILE	Butler (1963)
	LC_{50}	2	3,000	JUVENILE	Butler (1963)
Medaka, high-eyes	LC_{50}	1	110,000	NR	Shim and Self (1973)
	MOR	3	14	NR	Shim and Self (1973)
	MOR	3	28	NR	Shim and Self (1973)
Northern squawfish	MOR	1	10,000	5–10 CM	MacPhee and Ruelle (1969)
Oikawa	LC_{50}	1	140,000	NR	Shim and Self (1973)
	MOR	3	14	NR	Shim and Self (1973)
	MOR	3	28	NR	Shim and Self (1973)
Pink salmon	LET	4	50,000	FRY	Meehan et al. (1974)
	MOR	4	1,000	FRY	Meehan et al. (1974)
	MOR	4	5,000	FRY	Meehan et al. (1974)
	MOR	4	10,000	FRY	Meehan et al. (1974)

Table 3. Toxicology of Aquatic and Terrestrial Species: Chemical: 2,4-D Acid, CAS Number: 94757 (continued)

Class	Species	Effect	Duration (days)	Concentration ($\mu g\ L^{-1}$)	Lifestage	Ref.
Fish (continued)	Pumpkinseed	LC_{50}	1	120,000	YOUNG OF THE YEAR	Rehwoldt et al. (1977)
		LC_{50}	2	118,300	YOUNG OF THE YEAR	Rehwoldt et al. (1977)
		LC_{50}	4	94,600	YOUNG OF THE YEAR	Rehwoldt et al. (1977)
		LET	7	1,500,000	6.4–10.2 CM, 5 G	Harrisson and Rees (1946)
		MOR	7	10,000	6.4–10.2 CM, 5 G	Harrisson and Rees (1946)
		MOR	7	100,000	6.4–10.2 CM, 5 G	Harrisson and Rees (1946)
		MOR	7	500,000	6.4–10.2 CM, 5 G	Harrisson and Rees (1946)
		MOR	7	1,000,000	6.4–10.2 CM, 5 G	Harrisson and Rees (1946)
	Rainbow trout, donaldson trout	LC_{50}	1	3,000	UNDER-YEARLING	Alabaster (1969)
		LC_{50}	1	358,000	27.7 MM, 0.34 G	Alexander et al. (1985)
		LC_{50}	1	3,400,000	UNDER-YEARLING	Alabaster (1969)
		LC_{50}	1	7,000,000	1.3–3 CM	Alabaster (1969)
		LC_{50}	2	2,200	UNDER-YEARLING	Alabaster (1969)
		LC_{50}	2	358,000	27.7 MM, 0.34 G	Alexander et al. (1985)
		LC_{50}	2	2,400,000	UNDER-YEARLING	Alabaster (1969)
		LC_{50}	2	4,800,000	1.3–3 CM	Alabaster (1969)
		LC_{50}	3	358,000	27.7 MM, 0.34 G	Alexander et al. (1985)
		LC_{50}	4	358,000	27.7 MM, 0.34 G	Alexander et al. (1985)
		LC_{50}	23	4,200	EGGS, 4 D POSTHATCH	Birge et al. (1979)
		LC_{50}	23	11,000	EGGS, 4 D POSTHATCH	Birge et al. (1979)
		LC_{50}	27	4,200	EGGS, 4 D POSTHATCH	Birge et al. (1979)
		LC_{50}	27	11,000	EGGS, 4 D POSTHATCH	Birge et al. (1979)
		ENZ	NR	20,000	YEARLING	Davis et al. (1972)
		MOR	4	50,000	FINGERLING	Meehan et al. (1974)
		RSD	1–24 hr	0.1–10	125–175 G, 20–28 CM	Carpenter and Eaton (1983)

	RSD	5	1,000	3–4 G, 50–75 MM	Rodgers and Stalling (1972)
	STR	1	5,000	FINGERLING, ≤ 10 CM	Applegate et al. (1957)
	STR	6	1,000,000	FINGERLING, 3.5–9.1 G	Doe et al. (1988)
Red breasted bream	LET	2	8,000	NR	Rodrigues et al. (1980)
	MOR	2	4,000	NR	Rodrigues et al. (1980)
Sea lamprey	STR	1	5,000	LARVAE, 8–13 CM	Applegate et al. (1957)
Sheepshead minnow	AVO	NR	100	20–40 MM	Hansen (1970)
Sockeye salmon	MOR	4	10,000	SMOLT	Meehan et al. (1974)
	MOR	4	50,000	SMOLT	Meehan et al. (1974)
Striped bass	LC_{50}	1	8[illegible],600	YOUNG OF THE YEAR	Rehwoldt et al. (1977)
	LC_{50}	2	70,200	YOUNG OF THE YEAR	Rehwoldt et al. (1977)
	LC_{50}	4	70,100	YOUNG OF THE YEAR	Rehwoldt et al. (1977)
Tooth carp	LC_{50}	2	207.80	NR	Boumaiza et al. (1979)
	LC_{50}	4	2[illegible]2.22	NR	Boumaiza et al. (1979)
Two spotted, tic tac toe barb	LC_{50}	1	1,600	MATURE, 2.85 CM, 614 MG	Verma et al. (1984b)
	HIS	1	1,000	MATURE, 2.85 CM, 614 MG	Verma et al. (1984b)
White mullet	LC_{50}	1	1,500	JUVENILE	Butler (1963)
	LC_{50}	2	1,500	JUVENILE	Butler (1963)
	GRO	2	5[illegible],000	JUVENILE	Butler (1963)
White perch	LC_{50}	1	5[illegible],500	YOUNG OF THE YEAR	Rehwoldt et al. (1977)
	LC_{50}	2	4[illegible],200	YOUNG OF THE YEAR	Rehwoldt et al. (1977)
	LC_{50}	4	4[illegible],000	YOUNG OF THE YEAR	Rehwoldt et al. (1977)
Winter flounder	BCF	1 hr	2,000	NR	Pritchard and James (1979)
Zebra danio, zebrafish	LC_{50}	1	18[illegible],000	NR	Benijts-Claus and Persoone (1975)
	LC_{50}	2	16[illegible],000	NR	Benijts-Claus and Persoone (1975)
	LC_{50}	4	16[illegible],000	NR	Benijts-Claus and Persoone (1975)

Table 3. Toxicology of Aquatic and Terrestrial Species: Chemical: 2,4-D Acid, CAS Number: 94757 (continued)

Class	Species	Effect	Duration (days)	Concentration ($\mu g\ L^{-1}$)	Lifestage	Ref.
Insect	Biting midge family	ABD	7	44	NR	Marshall and Rutschsky (1974)
	Damselfly family	ABD	21	44	NR	Marshall and Rutschsky (1974)
	Dragonfly family (*Gomphidae*)	ABD	35	44	NR	Marshall and Rutschsky (1974)
	Dragonfly family (*Libellulidae*)	ABD	7	44	NR	Marshall and Rutschsky (1974)
	Mayfly (*Hexagenia* sp)	ABD	21	44	NR	Marshall and Rutschsky (1974)
	Mayfly family	ABD	35	44	NR	Marshall and Rutschsky (1974)
	Midge family	ABD	7	44	NR	Marshall and Rutschsky (1974)
	Mosquito (*Aedes aegypti*)	MOR	1	10,000	LARVAE, 3RD INSTAR	Lichtenstein et al. (1973)
	Mosquito (*Culex tritaeniorhynchus*)	LC_{50}	1	91,800	LARVAE	Shim and Self (1973)
	Springtail beetle	ABD	7	44	NR	Marshall and Rutschsky (1974)

	Stonefly	LC_{50}	1	8,500	NYMPH	Cope (1965)
	(*Pteronarcys californica*)	LC_{50}	1	56,000	30–35 MM	Sanders and Cope (1968)
		LC_{50}	2	1,800	NYMPH	Cope (1965)
		LC_{50}	2	44,000	30–35 MM	Sanders and Cope (1968)
		LC_{50}	4	1,600	NYMPH	Cope (1965)
		LC_{50}	4	15,000	30–35 MM	Sanders and Cope (1968)
	Water scavenger beetle family	ABD	35	44	NR	Marshall and Rutschsky (1974)
	Water strider family	ABD	7	44	NR	Marshall and Rutschsky (1974)
Mullusca	American or virginia oyster	GRO	4	2,000	2.5–5.1 CM	Butler (1963)
	Bivalve, clam, mussel class	ABD	35	44	NR	Marshall and Rutschsky (1974)
	Common bay mussel, blue mussel	EC_{50}BA	4	262,000	ADULT, 3–4 CM	Liu and Lee (1975)
		EC_{50}DV	2	211,700	EGGS	Liu and Lee (1975)
		LC_{50}	4	259,000	ADULT, 3–4 CM	Liu and Lee (1975)
		DVP	40	176,000	29–30 D, LARVAE	Liu and Lee (1975)
		GRO	20	45,700	LARVAE	Liu and Lee (1975)
		GRO	20	91,400	LARVAE	Liu and Lee (1975)
	Pouch snail family	ABD	21	44	NR	Marshall and Rutschsky (1974)
Platyhelmi	Planarian, flatworm	ABD	7	44	NR	Marshall and Rutschsky (1974)
Protozoa	Ciliate (*Colpidium campylum*)	PGR	1.79	>10,000	>96 HR	Dive et al. (1980)
	Ciliate (*Paramecium aurelia*)	MOR	1.5 hr	1,000	10,000 CELLS L^{-1}	Joshi and Misra (1986)
		MOR	1.5 hr	5,000	10,000 CELLS L^{-1}	Joshi and Misra (1986)

Table 3. Toxicology of Aquatic and Terrestrial Species: Chemical: 2,4-D Acid, CAS Number: 94757 (continued)

Class	Species	Effect	Duration (days)	Concentration ($\mu g\ L^{-1}$)	Lifestage	Ref.
Protozoa (continued)	Ciliate (*Stylonychia mytilus*)	LC_{50}	3	104,000	NR	Benijts-Claus and Persoone (1975)
		LC_{50}	3	485,000	NR	Benijts-Claus and Persoone (1975)
Rotifera	Rotifer	LT_{50}	1	5,000	NR	George et al. (1982)
	(*Brachionus calyciflorus*)	LET	1.29	5,000	NR	George et al. (1982)
Tracheophy	Duckweed (*Lemna perpusilla*)	GRO	11	10	NR	Schott and Worthley (1974)
		GRO	11	100	NR	Schott and Worthley (1974)
	Duckweed (*Lemna minor*)	PHY	2	200	FROND	O'Brien and Prendeville (1979)
		PHY	4	200	FROND	O'Brien and Prendeville (1979)
	Parrot's-feather	GRO	14	200	1 WEEK	Sutton and Bingham (1970)
		GRO	14	220	2 WEEK	Sutton and Bingham (1970)
		GRO	14	550	2 WEEK	Sutton and Bingham (1970)
		GRO	14	2,000	1 WEEK	Sutton and Bingham (1970)
		PHY	14	20	1 WEEK	Sutton and Bingham (1970)
		PHY	14	200	1 WEEK	Sutton and Bingham (1970)
		PHY	14	220	2 WEEK	Sutton and Bingham (1970)
		PHY	14	550	2 WEEK	Sutton and Bingham (1970)
	Water-meal	LET	12–30	1,000,000	NR	Worthley and Schott (1972)

Bird	Japanese quail	LD_{50}	14	66[illegible] mg/kg	60 D	Hudson et al. (1984)
		LO REP	6	25[illegible] mg/kg	ADULT	Haegele and Tucker (1974)
		LO SIGNS	6	25[illegible] mg/kg	ADULT	Haegele and Tucker (1974)
	Mallard duck	LO REP	6	1,500 mg/kg	ADULT	Haegele and Tucker (1974)
	Pheasant	NO REP	NR	150 m	ADULT	Solomon et al. (1973)
	Pigeon	LD_{50}	14	668 mg/kg		Dahlgren and Linder (1971)
	Ring-necked pheasant	LD_{50}	14	472 mg/kg	90–120 D	Hudson et al. (1984)
Mammal	Deer mouse	NO POP	NR	3 lb/ac		Johnson and Hansen (1969)
		NO REP	NR	3 lb/ac		Johnson and Hansen (1969)
	Least chipmunk	LO POP	NR	2 lb/ac		Johnson and Hansen (1969)
		NO REP	NR	2 lb/ac		Johnson and Hansen (1969)
	Montane vole	LO POP	NR	2 lb/ac		Johnson and Hansen (1969)
		LO REP	NR	2 lb/ac		Johnson and Hansen (1969)
	Mouse (*Mus* sp.)	LO PHY	7	100 mg/kg	56–84 D	Jenssen and Renberg (1976)
		LO PHY	7	100 mg/kg	84 D	Jenssen and Renberg (1976)
	Northern pocket gopher	LO POP	NR	2 lb/ac		Johnson and Hansen (1969)
	Rat	LO GRO	NR	50 mg/kg	ADULT	Khera and Mckinley (1972)
		LO GRO	NR	150 mg/kg	ADULT	Khera and Mckinley (1972)
		LO REP	NR	100 mg/kg	ADULT	Khera and Mckinley (1972)
		NO GRO	NR	25 mg/kg	ADULT	Khera and Mckinley (1972)
		NO GRO	NR	100 mg/kg	ADULT	Khera and Mckinley (1972)
		NO REP	NR	50 mg/kg	ADULT	Khera and Mckinley (1972)

Table 4. Toxicology of Aquatic and Terrestrial Species: Chemical: 2,4-D Butyl Ester, CAS Number: 1929733

Class	Species	Effect	Duration (days)	Concentration ($\mu g\ L^{-1}$)	Lifestage	Ref.
Algae	Algae, phytoplankton, algal mat	PGR	4 hr	1,000	NR	Butler (1964)
	Diatom	$EC_{50}OX$	1.5 hr	200,000	LOG PHASE	Walsh (1972)
	(*Phaeodactylum tricornutum*)	PGR	10	150,000	LOG PHASE	Walsh (1972)
	Green algae	$EC_{50}OX$	1.5 hr	100,000	LOG PHASE	Walsh (1972)
	(*Chlorococcum* sp)	PGR	10	75,000	LOG PHASE	Walsh (1972)
	Green algae	$EC_{50}OX$	1.5 hr	100,000	LOG PHASE	Walsh (1972)
	(*Dunaliella tertiolecta*)	PGR	10	75,000	LOG PHASE	Walsh (1972)
	Green algae division	PGR	14	4,000	NR	Butler et al. (1975)
		PGR	14	<1,000	NR	Butler et al. (1975)
	Haptophyte	$EC_{50}OX$	1.5 hr	100,000	LOG PHASE	Walsh (1972)
	(*Isochrysis galbana*)	PGR	10	75,000	LOG PHASE	Walsh (1972)
Aquatic	Invertebrates	POP	338	1,000	BENTHIC	Stephenson and Mackie (1986)
Crustacea	Aquatic sowbug	LC_{50}	2	3,200	EARLY INSTAR	Sanders (1970a)
	(*Asellus brevicaudus*)	LC_{50}	4	2,600	MATURE	Johnson and Finley (1980)
	Blue crab	MOR	21		SMALL	Rawls (1977)
	Crayfish (*Orconectes nais*)	LC_{50}	2	>100,000	EARLY INSTAR	Sanders (1970a)
	Grass shrimp (*Palaemonetes kadiakensis*)	LC_{50}	2	1,400	EARLY INSTAR	Sanders (1970a)
	Grass shrimp (*Palaemonetes pugio*)	MOR	21		SMALL	Rawls (1977)
	Harpacticoid copepod	LC_{50}	4	3,100	0.6–0.8 MM	Linden et al. (1979)

	Ostracod (*Cypridopsis vidua*)	EC_{50}IM	2	1,800	EARLY INSTAR	Sanders (1970a)
		EC_{50}IM	2	2,200	MATURE	Johnson and Finley (1980)
	Pink shrimp	MOR	2	1,000	ADULT	Butler (1964)
	Scud (*Gammarus lacustris*)	LC_{50}	1	1,400	2 MONTHS OLD	Sanders (1969)
	Scud (*Gammarus fasciatus*)	LC_{50}	1	6,500	EARLY INSTAR	Sanders (1970a)
	Scud (*Gammarus lacustris*)	LC_{50}	2	760	2 MONTHS OLD	Sanders (1969)
	Scud (*Gammarus fasciatus*)	LC_{50}	2	5,900	EARLY INSTAR	Sanders (1970a)
		LC_{50}	2	5,900	EARLY INSTAR	Sanders (1970a)
	Scud (*Gammarus lacustris*)	LC_{50}	4	440	2 MONTHS OLD	Sanders (1969)
	Scud (*Gammarus fasciatus*)	LC_{50}	4	5,900	EARLY INSTAR	Sanders (1970a)
		LC_{50}	4	6,100	MATURE	Johnson and Finley (1980)
	Water flea (*Daphnia magna*)	EC_{50}IM	2	5,600	EARLY INSTAR	Sanders (1970a)
		EC_{50}IM	2	6,400	1ST INSTAR	Johnson and Finley (1980)
	Water flea (*Daphnia pulex*)	OC	0.25–0	3,300	MATURE	Sigmon (1979b)
Fish	Black bullhead	LC_{50}	1	8,700	125 MM, 15 G	Inglis and Davis (1972)
		LC_{50}	1	8,300	125 MM, 15 G	Inglis and Davis (1972)
		LC_{50}	2	7,100	125 MM, 15 GM	Inglis and Davis (1972)
		LC_{50}	2	7,700	125 MM, 15 GM	Inglis and Davis (1972)
		LC_{50}	6 hr	8,900	125 MM, 15 G	Inglis and Davis (1972)
		LC_{50}	6 hr	10,600	125 MM, 15 G	Inglis and Davis (1972)
		LC_{50}	6 hr	11,200	125 MM, 15 G	Inglis and Davis (1972)
	Bleak	LC_{50}	4	3,200–3,700	8 CM	Linden et al. (1979)
	Bluegill	LC_{50}	0.5	2,000	40 MM, 0.75 GM	Inglis and Davis (1972)
		LC_{50}	0.5	2,020	40 MM, 0.75 GM	Inglis and Davis (1972)
		LC_{50}	0.5	2,120	40 MM, 0.75 GM	Inglis and Davis (1972)
		LC_{50}	1	1,460	40 MM, 0.75 GM	Inglis and Davis (1972)
		LC_{50}	1	1,540	40 MM, 0.75 GM	Inglis and Davis (1972)
		LC_{50}	1	1,680	40 MM, 0.75 GM	Inglis and Davis (1972)
		LC_{50}	1	1,710	40 MM, 0.75 GM	Inglis and Davis (1972)
		LC_{50}	1	1,740	40 MM, 0.75 GM	Inglis and Davis (1972)

Table 4. Toxicology of Aquatic and Terrestrial Species: Chemical: 2,4-D Butyl Ester, CAS Number: 1929733 (continued)

Class	Species	Effect	Duration (days)	Concen-tration ($\mu g\ L^{-1}$)	Lifestage	Ref.
Fish (continued)	Bluegill	LC_{50}	1	1,980	40 MM, 0.75 GM	Inglis and Davis (1972)
		LC_{50}	1	2,100	FINGERLING, 2.5–7.6 CM	Hughes and Davis (1963)
		LC_{50}	1	2,100	NR	Hughes and Davis (1962)
		LC_{50}	1	36,500	NR	Hughes and Davis (1962)
		LC_{50}	1	43,400	NR	Hughes and Davis (1962)
		LC_{50}	2	1,380	40 MM, 0.75 GM	Inglis and Davis (1972)
		LC_{50}	2	1,470	40 MM, 0.75 GM	Inglis and Davis (1972)
		LC_{50}	2	1,510	40 MM, 0.75 GM	Inglis and Davis (1972)
		LC_{50}	2	2,100	FINGERLING, 2.5–7.6 CM	Hughes and Davis (1963)
		LC_{50}	2	2,100	NR	Hughes and Davis (1962)
		LC_{50}	2	34,500	NR	Hughes and Davis (1962)
		LC_{50}	2	41,400	NR	Hughes and Davis (1962)
		LC_{50}	4	1,200	1.4 G	Johnson and Finley (1980)
		LC_{50}	6 hr	2,030	40 MM, 0.75 GM	Inglis and Davis (1972)
		LC_{50}	6 hr	2,090	40 MM, 0.75 GM	Inglis and Davis (1972)
		LC_{50}	6 hr	2,160	40 MM, 0.75 GM	Inglis and Davis (1972)
		LC_{50}	6 hr	2,500	40 MM, 0.75 GM	Inglis and Davis (1972)
		LC_{50}	6 hr	2,500	40 MM, 0.75 GM	Inglis and Davis (1972)
		LC_{50}	6 hr	2,650	40 MM, 0.75 GM	Inglis and Davis (1972)
		BCF	1	770	3–4 G, 50–75 MM	Rodgers and Stalling (1972)
		BCF	1	770	3–4 G, 50–75 MM	Rodgers and Stalling (1972)
		BCF	1 hr	530	3–4 G, 50–75 MM	Rodgers and Stalling (1972)
		BCF	1 hr	530	3–4 G, 50–75 MM	Rodgers and Stalling (1972)
		BCF	2 hr	680	3–4 G, 50–75 MM	Rodgers and Stalling (1972)
		BCF	2 hr	680	3–4 G, 50–75 MM	Rodgers and Stalling (1972)

	BCF	3 hr	650	3–4 G, 50–75 MM	Rodgers and Stalling (1972)
	BCF	3 hr	650	3–4 G, 50–75 MM	Rodgers and Stalling (1972)
	BCF	4 hr	620	3–4 G, 50–75 MM	Rodgers and Stalling (1972)
	BCF	4 hr	620	3–4 G, 50–75 MM	Rodgers and Stalling (1972)
	BCF	6 hr	760	3–4 G, 50–75 MM	Rodgers and Stalling (1972)
	BCF	6 hr	760	3–4 G, 50–75 MM	Rodgers and Stalling (1972)
	MOR	NR	6,000	SMALL	Hiltibran (1969)
	OC	0.5–1 hr	3,000	0.28–23.30 G	Sigmon (1979a)
	OC	0.5–1 hr	3,000	0.28–23.30 G	Sigmon (1979a)
	OC	0.5–1 hr	3,000	0.28–23.30 G	Sigmon (1979a)
	OC	NR	54	SMALL	Hiltibran (1969)
	OC	NR	540,000	SMALL	Hiltibran (1969)
	RSD	1	<1–37	NR	Smith and Isom (1967)
Channel catfish	BCF	1	770	3–4 G, 50–75 MM	Rodgers and Stalling (1972)
	BCF	1	770	3–4 G, 50–75 MM	Rodgers and Stalling (1972)
	BCF	1 hr	530	3–4 G, 50–75 MM	Rodgers and Stalling (1972)
	BCF	1 hr	530	3–4 G, 50–75 MM	Rodgers and Stalling (1972)
	BCF	2 hr	680	3–4 G, 50–75 MM	Rodgers and Stalling (1972)
	BCF	2 hr	680	3–4 G, 50–75 MM	Rodgers and Stalling (1972)
	BCF	3 hr	650	3–4 G, 50–75 MM	Rodgers and Stalling (1972)
	BCF	3 hr	650	3–4 G, 50–75 MM	Rodgers and Stalling (1972)
	BCF	4 hr	620	3–4 G, 50–75 MM	Rodgers and Stalling (1972)
	BCF	4 hr	620	3–4 G, 50–75 MM	Rodgers and Stalling (1972)
	BCF	6 hr	760	3–4 G, 50–75 MM	Rodgers and Stalling (1972)
	BCF	6 hr	760	3–4 G, 50–75 MM	Rodgers and Stalling (1972)

Table 4. Toxicology of Aquatic and Terrestrial Species: Chemical: 2,4-D Butyl Ester, CAS Number: 1929733 (continued)

Class	Species	Effect	Duration (days)	Concentration ($\mu g\ L^{-1}$)	Lifestage	Ref.
Fish (continued)	Chinook salmon	LC_{50}	4	303	FRY, 37 MM, 0.44 G	Finlayson and Verrue (1985)
		LC_{50}	4	327	FRY, 3.8 MM, 0.38 G	Finlayson and Verrue (1985)
		LC_{50}	4	332	SMOLT, 111 MM, 15.4 G	Finlayson and Verrue (1985)
		LC_{50}	4	418	SMOLT, 109 MM, 14.6 G	Finlayson and Verrue (1985)
		GRO	86	40	FRY, 36 D POSTHATCH	Finlayson and Verrue (1985)
		GRO	86	60	FRY, 36 D POSTHATCH	Finlayson and Verrue (1985)
		MOR	86	40	ALEVIN TO FRY	Finlayson and Verrue (1985)
		MOR	86	60	ALEVIN TO FRY	Finlayson and Verrue (1985)
		MOR	86	118	EMBRYO	Finlayson and Verrue (1985)
	Fathead minnow	LC_{50}	4	3,300	0.9 G	Johnson and Finley (1980)
		MOR	14	310	EGG TO FRY	Mount and Stephan (1967)
		MOR	300	310	2.5 CM	Mount and Stephan (1967)
	Goldfish	LC_{50}	0.5	3,950	41 MM, 1.25 G	Inglis and Davis (1972)
		LC_{50}	0.5	4,500	41 MM, 1.25 GM	Inglis and Davis (1972)
		LC_{50}	0.5	4,550	41 MM, 1.25 G	Inglis and Davis (1972)
		LC_{50}	1	3,630	41 MM, 1.25 GM	Inglis and Davis (1972)
		LC_{50}	1	3,980	41 MM, 1.25 GM	Inglis and Davis (1972)
		LC_{50}	1	4,100	41 MM, 1.25 GM	Inglis and Davis (1972)
		LC_{50}	2	2,650	41 MM, 1.25 GM	Inglis and Davis (1972)
		LC_{50}	2	2,720	41 MM, 1.25 GM	Inglis and Davis (1972)
		LC_{50}	2	3,620	41 MM, 1.25 GM	Inglis and Davis (1972)
		LC_{50}	6 hr	5,000	41 MM, 1.25 GM	Inglis and Davis (1972)
		LC_{50}	6 hr	5,170	41 MM, 1.25 GM	Inglis and Davis (1972)
		LC_{50}	6 hr	5,780	41 MM, 1.25 GM	Inglis and Davis (1972)

Green sunfish	LET	1 hr	110,000	50–150 G	Sergeant et al. (1971)
	MOR	1 hr	110,000	50–150 G	Sergeant et al. (1970)
Harlequinfish, red rasbora	LC_{50}	1	1,000	1.3–3 CM	Alabaster (1969)
	LC_{50}	2	1,000	1.3–3 CM	Alabaster (1969)
Longnose killifish	LC_{50}	1	5,000	JUVENILE	Butler (1963)
	LC_{50}	2	5,000	JUVENILE	Butler (1963)
Mosquitofish	LC_{50}	1	7,000	NR	Hansen et al. (1972)
(*Gambusia affinis*)	AVO	1 hr	1,000	20–45 MM	Hansen et al. (1972)
Mummichog	MOR	21		NR	Rawls (1977)
Rainbow trout,	LC_{50}	0.5	1,940	46 MM, 0.67 G	Inglis and Davis (1972)
donaldson trout	LC_{50}	0.5	1,960	46 MM, 0.67 GM	Inglis and Davis (1972)
	LC_{50}	0.5	1,980	46 MM, 0.67 GM	Inglis and Davis (1972)
	LC_{50}	1	1,520	46 MM, 0.67 G	Inglis and Davis (1972)
	LC_{50}	1	1,560	46 MM, 0.67 GM	Inglis and Davis (1972)
	LC_{50}	1	1,560	46 MM, 0.67 GM	Inglis and Davis (1972)
	LC_{50}	2	1,460	46 MM, 0.67 G	Inglis and Davis (1972)
	LC_{50}	2	1,500	46 MM, 0.67 G	Inglis and Davis (1972)
	LC_{50}	2	1,550	46 MM, 0.67 G	Inglis and Davis (1972)
	LC_{50}	4	452	SMOLT, 81 MM, 7.8 G	Finlayson and Verrue (1985)
	LC_{50}	4	484	SMOLT, 85 MM, 6.4 G	Finlayson and Verrue (1985)
	LC_{50}	4	512	FRY, 24 MM, 0.11 G	Finlayson and Verrue (1985)
	LC_{50}	4	525	FRY, 23 MM, 0.10 G	Finlayson and Verrue (1985)
	LC_{50}	4	1,206	SMOLT, 82 MM, 7.8 G	Finlayson and Verrue (1985)
	LC_{50}	4	3,689	SMOLT, 80 MM, 7.6 G	Finlayson and Verrue (1985)
	LC_{50}	6 hr	2,810	46 MM, 0.67 GM	Inglis and Davis (1972)
	LC_{50}	6 hr	3,200	46 MM, 0.67 GM	Inglis and Davis (1972)
	LC_{50}	6 hr	3,440	46 MM, 0.67 GM	Inglis and Davis (1972)
	MOR	1	14,000	YEARLING, 15–27 CM	Dodson and Mayfield (1979a)
Sheepshead minnow	AVO	0.5 hr	100	20–40 MM	Hansen (1969)
Two spotted, tic tac toe barb	HEM	1	1,000	2.85 CM, 614.06 MG	Chouhan et al. (1983)

Table 4. Toxicology of Aquatic and Terrestrial Species: Chemical: 2,4-D Butyl Ester, CAS Number: 1929733 (continued)

Class	Species	Effect	Duration (days)	Concentration ($\mu g\ L^{-1}$)	Lifestage	Ref.
Insect	Alderfly family	ABD	21	44	NR	Marshall and Rutschsky (1974)
	Caddisfly family	ABD	360	<1–37	NR	Smith and Isom (1967)
	Cranefly family	ABD	35	44	NR	Marshall and Rutschsky (1974)
	Crawling water beetle family	ABD	21	44	NR	Marshall and Rutschsky (1974)
	Dragonfly suborder	ABD	360	<1–37	NR	Smith and Isom (1967)
	Horsefly family	ABD	35	44	NR	Marshall and Rutschsky (1974)
	Mayfly (*Caenis* sp)	ABD	360	<1–37	NR	Smith and Isom (1967)
	Mayfly (*Hexagenia* sp)	ABD	360	<1–157	NR	Smith and Isom (1967)
		ABD	360	<1–37	NR	Smith and Isom (1967)
	Midge (*Chironomus* sp)	DVP	11	3,000	LARVAE	Sigmon (1979c)
		DVP	13	3,000	LARVAE	Sigmon (1979c)
		DVP	NR	3,000	LARVAE	Sigmon (1979c)
		DVP	NR	3,000	LARVAE	Sigmon (1979c)
		DVP	NR	3,000	LARVAE	Sigmon (1979c)
		DVP	NR	3,000	LARVAE	Sigmon (1979c)
		DVP	NR	3,000	LARVAE	Sigmon (1979c)
		DVP	NR	1,000–3,000	LARVAE	Sigmon (1979c)
		DVP	NR	1,000–3,000	LARVAE	Sigmon (1979c)
		MOR	NR	3,000	LARVAE	Sigmon (1979c)
		MOR	NR	3,000	LARVAE	Sigmon (1979c)
		MOR	NR	1,000–3,000	LARVAE	Sigmon (1979c)

	Mosquito (*Anopheles quadrimaculatus*)	MOR	NR	100	3RD and 4TH INSTAR LARVAE	Smith and Isom (1967)
	Riffle beetle family	ABD	360	<1–37	NR	Smith and Isom (1967)
	Stonefly (*Pteronarcys californica*)	LC_{50}	1	8,500	30–35 MM	Sanders and Cope (1968)
		LC_{50}	2	1,800	30–35 MM	Sanders and Cope (1968)
		LC_{50}	4	1,600	30–35 MM	Sanders and Cope (1968)
Mullusca	American or Virginia oyster	EC_{50}DV	2	8,000	EGGS, 2 CELL STAGE	Davis and Hidu (1979)
		EC_{50}GR	4	3,750	2.5–5.1 CM	Butler (1963)
		LC_{50}	12	740	2 D LARVAE	Davis and Hidu (1979)
		GRO	12	25	2 D LARVAE	Davis and Hidu (1979)
		GRO	21		6.25 CM	Rawls (1977)
		GRO	66		6.25 CM	Rawls (1977)
	Hooked mussel	MOR	21		SMALL	Rawls (1977)
	Mollusk phylum	ABD	30	<1–37	NR	Smith and Isom (1967)
	Mussel (*Elliptio crassidens*)	RSD	1	<1–37	NR	Smith and Isom (1967)
Tracheophy	Creeping bentgrass	MOR	NR	2,000	NR	Hiltibran and Turgeon (1977)
	Pondweed (*Potamogeton perfoliatus*)	STR	21		NR	Rawls (1977)
	Vascular plants	RSD	0–88	<115.5	AQUATIC MACROPHYTES	Carpentier et al. (1988)
	Water-celery, tapegrass	MOR	39	10	NR	Quinn et al. (1977)
		MOR	46	10	NR	Quinn et al. (1977)

Table 4. Toxicology of Aquatic and Terrestrial Species: Chemical: 2,4-D Butyl Ester, CAS Number: 1929733 (continued)

Class	Species	Effect	Duration (days)	Concentration ($\mu g\ L^{-1}$)	Lifestage	Ref.
Tracheophy (continued)	Water-milfoil (*Myriophyllum spicatum*)	CLR	7		NR	Rawls (1977)
		MOR	14		NR	Haven (1963)
		MOR	14	10	NR	Quinn et al. (1977)
		MOR	21		NR	Rawls (1977)
		MOR	22	10	NR	Quinn et al. (1977)
		MOR	30	44	NR	Marshall and Rutschsky (1974)
		MOR	114		NR	Rawls (1977)
		RSD	1	<1–37	NR	Smith and Isom (1967)
	Waterweed (*Elodea canadensis*)	MOR	13	10	NR	Quinn et al. (1977)
		MOR	35	10	NR	Quinn et al. (1977)
	Widgeon-grass	STR	21		NR	Rawls (1977)
Bird	Bobwhite quail	LC_{50}	8	>5,000		Heath et al. (1972)
	Japanese quail	LC_{50}	8	>5,000		Heath et al. (1972)
	Mallard duck	LC_{50}	8	>5,000		Heath et al. (1972)
	Ring-necked pheasant	LC_{50}	8	>5,000		Heath et al. (1972)
Invertebrate	Honey bee	LO SIGNS	NR	1 lb/ac	ADULT	Moffett and Morton (1972)
		NO MOR	NR	1 lb/ac	ADULT	Moffett and Morton (1972)
		NO MOR	NR	1 lb/ac	ADULT	Moffett and Morton (1972)
Mammal	Guinea pig	LD_{50}	NR	848		USDA Forest Service (1984)
	Mouse	LD_{50}	NR	380		USDA Forest Service (1984)
	Rabbit	LD_{50}	NR	424		USDA Forest Service (1984)
	Rat	LD_{50}	NR	620		USDA Forest Service (1984)

Table 5. Toxicology of Aquatic and Terrestrial Species: Chemical: 2,4-D Ester, CAS Number: 25168267

Class	Species	Effect	Duration (days)	Concentration (μg L^{-1})	Lifestage	Ref.
Invertebrate	Honey bee	LO MOR	NR	4 lb ac^{-1}		Moffett and Morton (1972)
		NO MOR	NR	4 lb ac^{-1}		Moffett and Morton (1972)
Mammal	Rat	LO REP	NR	150 mg kg^{-1}	ADULT	Khera and Mckinley (1972)
		NO REP	NR	50 mg kg^{-1}	ADULT	Khera and Mckinley (1972)

Table 6. Toxicology of Aquatic and Terrestrial Species: Chemical: 2,4-D Ester, CAS Number: 64047354

Class	Species	Effect	Duration (days)	Concentration (μg L^{-1})	Lifestage	Ref.
Mammal	Rat	LO REP	NR	150 mg kg^{-1}	ADULT	Khera and Mckinley (1972)
		NO REP	NR	50 mg kg^{-1}	ADULT	Khera and Mckinley (1972)

Table 7. Toxicology of Aquatic and Terrestrial Species: Chemical: 2,4-D PGBE, CAS Number: 1320189

Class	Species	Effect	Duration (days)	Concen-tration (µg L^{-1})	Lifestage	Ref.
Algae	Algae, phytoplankton, algal mat	PGR	4 hr	1,000	NR	Butler (1964)
	Stonewort (*Chara* sp)	MOR	196	0–8,000	NR	Cope et al. (1970)
Crustacea	Aquatic sowbug (*Asellus brevicaudus*)	LC_{50}	2	2,200	EARLY INSTAR	Sanders (1970a)
	Crayfish (*Orconectes nais*)	LC_{50}	2	>100,000	EARLY INSTAR	Sanders (1970a)
	Grass shrimp	LC_{50}	2	2,700	EARLY INSTAR	Sanders (1970a)
	(*Palaemonetes kadiakensis*)	LC_{50}	4	400	MATURE	Johnson and Finley (1980)
	Ostracod (*Cypridopsis vidua*)	EC_{50}IM	2	320	EARLY INSTAR	Sanders (1970a)
		EC_{50}IM	2	400	MATURE	Johnson and Finley (1980)
	Pink shrimp	MOR	2	1,000	ADULT	Butler (1964)
	Scud (*Gammarus lacustris*)	LC_{50}	1	2,100	2 MONTHS	Sanders (1969)
	Scud (*Gammarus fasciatus*)	LC_{50}	1	4,100	EARLY INSTAR	Sanders (1970a)
	Scud (*Gammarus lacustris*)	LC_{50}	2	1,800	2 MONTHS	Sanders (1969)
	Scud (*Gammarus fasciatus*)	LC_{50}	2	2,600	EARLY INSTAR	Sanders (1970a)
		LC_{50}	2	2,600	EARLY INSTAR	Sanders (1970a)
	Scud (*Gammarus lacustris*)	LC_{50}	4	1,600	2 MONTHS	Sanders (1969)
	Scud (*Gammarus fasciatus*)	LC_{50}	4	2,500	EARLY INSTAR	Sanders (1970a)
		LC_{50}	4	2,900	MATURE	Johnson and Finley (1980)
	Water flea (*Daphnia magna*)	EC_{50}IM	2	100	EARLY INSTAR	Sanders (1970a)
		EC_{50}IM	2	1,200	1ST INSTAR	Johnson and Finley (1980)
	Water flea (*Simocephalus serrulatus*)	EC_{50}IM	2	4,900	1ST INSTAR	Johnson and Finley (1980)

Fish	Bluegill	LC_{50}	0.5	1,280	40 MM, 0.75 G	Inglis and Davis (1972)
		LC_{50}	0.5	1,340	40 MM, 0.75 G	Inglis and Davis (1972)
		LC_{50}	0.5	1,430	40 MM, 0.75 G	Inglis and Davis (1972)
		LC_{50}	1	1,020	40 MM, 0.75 G	Inglis and Davis (1972)
		LC_{50}	1	1,150	40 MM, 0.75 G	Inglis and Davis (1972)
		LC_{50}	1	1,200	40 MM, 0.75 G	Inglis and Davis (1972)
		LC_{50}	1	2,100	FINGERLING, 2.5–7.6 CM	Hughes and Davis (1963)
		LC_{50}	1	2,100	NR	Hughes and Davis (1962)
		LC_{50}	1	9,300	NR	Hughes and Davis (1962)
		LC_{50}	2	990	40 MM, 0.75 G	Inglis and Davis (1972)
		LC_{50}	2	1,030	40 MM, 0.75 G	Inglis and Davis (1972)
		LC_{50}	2	1,180	40 MM, 0.75 G	Inglis and Davis (1972)
		LC_{50}	2	2,100	FINGERLING, 2.5–7.6 CM	Hughes and Davis (1963)
		LC_{50}	2	2,100	NR	Hughes and Davis (1962)
		LC_{50}	2	9,300	NR	Hughes and Davis (1962)
		LC_{50}	4	600	1.0 G	Johnson and Finley (1980)
		LC_{50}	6 hr	1,960	40 MM, 0.75 G	Inglis and Davis (1972)
		LC_{50}	6 hr	2,080	40 MM, 0.75 G	Inglis and Davis (1972)
		LC_{50}	6 hr	2,140	40 MM, 0.75 G	Inglis and Davis (1972)
		LET	2	1,000	FRY	Hiltibran (1967)
		MOR	12	2,000	SMALL	Hiltibran (1967)
		MOR	NR	2,000	SMALL	Hiltibran (1969)
		OC	NR	64	SMALL	Hiltibran (1969)
		REP	196	100–4,000	80–110 MM	Cope et al. (1970)
	Central stoneroller	HAT	8	25,000	FERTILIZED EGGS	Hiltibran (1967)
	Chinook salmon	LC_{50}	4	170	SMOLT, 109 MM, 14.6 G	Finlayson and Verrue (1985)
		LC_{50}	4	311	SMOLT, 111 MM, 15.4 G	Finlayson and Verrue (1985)
	Coho salmon, silver salmon	LET	4	5,000	FINGERLING	Meehan et al. (1974)
		MOR	4	1,000	FINGERLING	Meehan et al. (1974)

Table 7. Toxicology of Aquatic and Terrestrial Species: Chemical: 2,4-D PGBE, CAS Number: 1320189 (continued)

Class	Species	Effect	Duration (days)	Concentration ($\mu g\ L^{-1}$)	Lifestage	Ref.
Fish (continued)	Cutthroat trout	LC_{50}	4	490	FINGERLING, 0.34–1.1 G	Woodward and Mayer (1978)
		LC_{50}	4	770	FINGERLING, 0.4–0.8 G	Woodward (1982)
		LC_{50}	4	780	FINGERLING, 0.34–1.1 G	Woodward and Mayer (1978)
		LC_{50}	4	860	FINGERLING, 0.34–1.1 G	Woodward and Mayer (1978)
		LC_{50}	4	930	FINGERLING, 0.34–1.1 G	Woodward and Mayer (1978)
		LC_{50}	4	930	FINGERLING, 0.34–1.1 G	Woodward and Mayer (1978)
		LC_{50}	4	1,000	1.0 G	Johnson and Finley (1980)
		LC_{50}	4	1,000	FINGERLING, 0.34–1.1 G	Woodward and Mayer (1978)
		LC_{50}	4	1,000	FINGERLING, 0.34–1.1 G	Woodward and Mayer (1978)
		LC_{50}	4	1,030	FINGERLING, 0.34–1.1 G	Woodward and Mayer (1978)
		LC_{50}	4	1,220	FINGERLING, 0.34–1.1 G	Woodward and Mayer (1978)
		GRO	60	124	FRY	Woodward and Mayer (1978)
		LET	4	500	ALEVIN	Woodward and Mayer (1978)
		MOR	60	31	ALEVIN	Woodward and Mayer (1978)
		MOR	60	60	ALEVIN	Woodward and Mayer (1978)
	Green sunfish	HAT	8	1,000	FERTILIZED EGGS	Hiltibran (1967)
		LET	5	1,000	FRY	Hiltibran (1967)
	Lake chubsucker	LET	5	1,000	FRY	Hiltibran (1967)

	Lake trout, siscowet	LC_{50}	4	630	FINGERLING, 0.34–1.1 G	Woodward and Mayer (1978)
		LC_{50}	4	700	FINGERLING, 0.34–1.1 G	Woodward and Mayer (1978)
		LC_{50}	4	840	FINGERLING, 0.34–1.1 G	Woodward and Mayer (1978)
		LC_{50}	4	1,000	FINGERLING, 0.34–1.1 G	Woodward and Mayer (1978)
		LC_{50}	4	1,050	FINGERLING, 0.34–1.1 G	Woodward and Mayer (1978)
		LC_{50}	4	1,075	FINGERLING, 0.34–1.1 G	Woodward and Mayer (1978)
		LC_{50}	4	1,100	0.6 G	Johnson and Finley (1980)
		LC_{50}	4	1,125	FINGERLING, 0.34–1.1 G	Woodward and Mayer (1978)
		LC_{50}	4	1,150	FINGERLING, 0.34–1.1 G	Woodward and Mayer (1978)
		LC_{50}	4	1,200	FINGERLING, 0.34–1.1 G	Woodward and Mayer (1978)
		DVP	60	100	ALEVIN	Woodward and Mayer (1978)
		GRO	60	100	FRY	Woodward and Mayer (1978)
		MOR	60	52	FRY, ALEVIN	Woodward and Mayer (1978)
		MOR	60	100	FRY, ALEVIN	Woodward and Mayer (1978)
	Longnose killifish	LC_{50}	1	5,000	JUVENILE	Butler (1963)
		LC_{50}	2	4,500	JUVENILE	Butler (1963)
	Rainbow trout, donaldson trout	LC_{50}	4	304	FRY, 24 MM, 0.11 G	Finlayson and Verrue (1985)
		LC_{50}	4	329	SMOLT, 81 MM, 7.8 G	Finlayson and Verrue (1985)
		LC_{50}	4	354	FRY, 23 MM, 0.10 G	Finlayson and Verrue (1985)
		LC_{50}	4	355	SMOLT, 85 MM, 6.4 G	Finlayson and Verrue (1985)
		LC_{50}	4	1,000	1.5 G	Johnson and Finley (1980)
	Smallmouth bass	LET	5	1,000	FRY	Hiltibran (1967)
Insect	Stonefly (*Pteronarcella badia*)	LC_{50}	4	2,400	NAIAD	Johnson and Finley (1980)
	Stonefly (*Pteronarcys californica*)	LC_{50}	4	2,600	2ND YEAR CLASS	Johnson and Finley (1980)
Tracheophy	Pondweed (*Potamogeton* sp)	MOR	196	0–8,000	NR	Cope et al. (1970)
	Water nymph	MOR	196	0–8,000	NR	Cope et al. (1970)

Table 8. Toxicology of Aquatic and Terrestrial Species: Chemical: 2,4-D Sodium Salt, CAS Number: 2702729

Class	Species	Effect	Duration (days)	Concentration ($\mu g\ L^{-1}$)	Lifestage	Ref.
Algae	Blue-green algae (*Anabaenopsis raciborskii*)	CYT	0.33	1,000,000	NR	Das and Singh (1977)
	Blue-green algae (*Anabaena aphanizomendoides*)	CYT	0.33	1,000,000	NR	Das and Singh (1977)
	Blue-green algae (*Anabaena spiroides*)	CYT	0.33	1,000,000	NR	Das and Singh (1977)
	Blue-green algae (*Anacystis flos-aquae*)	CYT	0.33	1,000,000	NR	Das and Singh (1977)
	Blue-green algae (*Cylindrospermum* sp)	MOR	8	900,000	INITIAL CONC. 1,500 FILAMENTS ML^{-1}	Singh (1974)
		PGR	1	1,000,000	INITIAL CONC. 4.5E+5 CELLS ML^{-1}	Singh (1974)
		PGR	8	100,000	INITIAL CONC. 1,500 FILAMENTS ML^{-1}	Singh (1974)
		PGR	8	150,000	INITIAL CONC. 4.5E+5 CELLS ML^{-1}	Singh (1974)
		PGR	8	800,000	INITIAL CONC. 4.5E+5 CELLS ML^{-1}	Singh (1974)
		PGR	8	800,000	INITIAL CONC. 1,500 FILAMENTS ML^{-1}	Singh (1974)

	Green algae (*Chlorella pyrenoidosa*)	BIO	0.75	110,000	NR	Bertagnolli and Nadakavukaren (1974)
		CEL	6 hr	1,215,000	5 D CULTURES	Bertagnolli and Nadakavukaren (1970)
		CLR	3	5,000,000	NR	Huang and Gloyna (1968)
		ENZ	0.75	1,100,000	NR	Bertagnolli and Nadakavukaren (1974)
		OC	0.75	110,000	NR	Bertagnolli and Nadakavukaren (1974)
		PGR	0.75	1,100,000	NR	Bertagnolli and Nadakavukaren (1974)
		PSE	0.75	110,000	NR	Bertagnolli and Nadakavukaren (1974)
Amphibia	Frog (*Bufo bufo japonicus*)	LC_{50}	1	>40,000	NR	Nishiuchi (1980a)
		LC_{50}	1	>40,000	NR	Nishiuchi (1980a)
		LC_{50}	1	>40,000	NR	Nishiuchi (1980a)
		LC_{50}	1	>40,000	NR	Nishiuchi (1980a)
		LC_{50}	1	>40,000	NR	Nishiuchi (1980a)
		LC_{50}	1	>40,000	NR	Nishiuchi (1980a)
		LC_{50}	2	>40,000	TADPOLE	Hashimoto and Nishiuchi (1981)
	Frog (*Rana temporaria*)	DVP	1	1,000	TADPOLE	Buslovich and Borushko (1976)
Bryophyta	Floating moss	MOR	6	5,000	NR	Titova (1978)

Table 8. Toxicology of Aquatic and Terrestrial Species: Chemical: 2,4-D Sodium Salt, CAS Number: 2702729 (continued)

Class	Species	Effect	Duration (days)	Concentration ($\mu g\ L^{-1}$)	Lifestage	Ref.
Crustacea	Aquatic sowbug	EC_{50}IM	1	1,070,000	NR	Zimakowska-Gnoinska (1977)
	(*Asellus aquaticus*)	EC_{50}IM	2	530,000	NR	Zimakowska-Gnoinska (1977)
	Calanoid copepod	LC_{50}	4	68,600	NR	Presing and Ponyi (1986)
	(*Eudiaptomus gracilis*)	LC_{50}	4	75,200	NR	Presing and Ponyi (1986)
		LC_{50}	4	81,900	NR	Presing and Ponyi (1986)
		LC_{50}	4	90,900	NR	Presing and Ponyi (1986)
		LC_{50}	4	73,400	NR	Presing and Ponyi (1986)
	Calanoid copepod (*Eudiaptomus graciloides*)	MOR	1	2,430,000	NAUPLIUS, COPEPODIDS I–V, ADULT	Wierzbicka (1974)
	Calanoid copepod (*Eudiaptomus gracilis*)	REP	21	2,500	NR	Presing and Ponyi (1986)
	Cyclopoid copepod (*Cyclops strenuus*)	MOR	1	2,430,000	NAUPLIUS, COPEPODIDS I–V, ADULT	Wierzbicka (1974)
	Cyclopoid copepod (*Cyclops bohater*)	MOR	4	2,430,000	COPEPODID V	Wierzbicka (1974)
	Cyclopoid copepod (*Cyclops vicinus vicinus*)	MOR	4	3,650,000	COPEPODID IV	Wierzbicka (1974)
	Cyclopoid copepod (*Cyclops vicinus kikuchii*)	MOR	4	3,650,000	COPEPODID IV	Wierzbicka (1974)
	Cyclopoid copepod (*Cyclops strenuus*)	MOR	4	3,650,000	COPEPODID IV	Wierzbicka (1974)
	Freshwater prawn (*Macrobrachium dayanum*)	LC_{50}	1	2,474,000	46–55 MM	Omkar and Shukla (1984)
	Freshwater prawn (*Macrobrachium naso*)	LC_{50}	1	2,644,000	65–75 MM	Omkar and Shukla (1984)

Freshwater prawn (*Macrobrachium dayanum*)	LC_{50}	2	2,331,000	46–55 MM	Omkar and Shukla (1984)
Freshwater prawn (*Macrobrachium naso*)	LC_{50}	2	2,536,000	65–75 MM	Omkar and Shukla (1984)
Freshwater prawn (*Macrobrachium dayanum*)	LC_{50}	3	2,333,000	46–55 MM	Omkar and Shukla (1984)
Freshwater prawn (*Macrobrachium naso*)	LC_{50}	3	2,435,000	65–75 MM	Omkar and Shukla (1984)
Freshwater prawn (*Macrobrachium dayanum*)	LC_{50}	4	2,275,000	46–55 MM	Omkar and Shukla (1984)
Freshwater prawn (*Macrobrachium naso*)	LC_{50}	4	2,397,000	65–75 MM	Omkar and Shukla (1984)
Prawn	LC_{50}	1	2,224	NR	Shukla and Omkar (1983a)
	LC_{50}	1	2,267	NR	Shukla and Omkar (1983a)
	LC_{50}	1	2,309	NR	Shukla and Omkar (1983a)
	LC_{50}	1	2,342	NR	Shukla and Omkar (1983a)
Water flea (*Daphnia pulex*)	LC_{50}	3 hr	>40,000	NR	Hashimoto and Nishiuchi (1981)
Water flea (*Daphnia magna*)	MOR	6	50,000	NR	Matlak (1972)
Water flea (*Moina macrocopa*)	LC_{50}	3 hr	>40,000	NR	Hashimoto and Nishiuchi (1981)
Water flea (*Simocephalus vetulus*)	LT_{50}	4	1,700,000	1.5 MM	Kaniewska-Prus (1975)
	LET	4	564,000	1.5 MM	Kaniewska-Prus (1975)
	MOR	0.5	610,000–1,200,000	MATURE FEMALE, 50–70 µG DRY WEIGHT	Klekowski and Zvirgzds (1971)
	MOR	6 hr	3,600,000	MATURE FEMALE, 50–70 µG DRY WEIGHT	Klekowski and Zvirgzds (1971)
	OC	4	282,000	1.5 MM	Kaniewska-Prus (1975)
	RES	3 hr	1,800,000	MATURE FEMALE, 50–70 µG DRY WEIGHT	Klekowski and Zvirgzds (1971)

Table 8. Toxicology of Aquatic and Terrestrial Species: Chemical: 2,4-D Sodium Salt, CAS Number: 2702729 (continued)

Class	Species	Effect	Duration (days)	Concentration ($\mu g\ L^{-1}$)	Lifestage	Ref.
Fish	Bleak	LC_{50}	0.5	111,200	LARVAE, >55 HR, 2 MG	Biro (1979)
		LC_{50}	0.5	159,400	EMBRYO 16–30 HR, <4 MG	Biro (1979)
		LC_{50}	1	70,600	LARVAE, >55 HR, 2 MG	Biro (1979)
		LC_{50}	1	129,000	EMBRYO 16–30 HR, <4 MG	Biro (1979)
		LC_{50}	1.5	62,100	LARVAE, >55 HR, 2 MG	Biro (1979)
		LC_{50}	1.5	63,900	EMBRYO 16–30 HR, <4 MG	Biro (1979)
		LC_{50}	2	12,900	EMBRYO 16–30 HR, <4 MG	Biro (1979)
		LC_{50}	2	51,600	LARVAE, >55 HR, 2 MG	Biro (1979)
	Bluegill	BIO	NR	37	SMALL	Hiltibran (1969)
		MOR	12	100,000	SMALL	Hiltibran (1967)
		MOR	NR	100,000	SMALL	Hiltibran (1969)
	Common, mirror, colored, carp	LC_{50}	2	>40,000	NR	Hashimoto and Nishiuchi (1981)
		GRO	10	5,000	EGGS	Matlak (1972)
		GRO	10	50,000	EGGS	Matlak (1972)
		LET	2	3,200,000	LARVAE, 3 HR	Kamler et al. (1974)
		LET	2 hr	3,200,000	EGGS, 27 HR	Kamler et al. (1974)
		MOR	1	200,000	EGGS, 27 HR	Kamler et al. (1974)
		MOR	2	1,600,000	LARVAE, 3 HR	Kamler et al. (1974)
		MOT	1	400,000	EGGS, 27 HR	Kamler et al. (1974)

	Goldfish	LC_{50}	2	>40,000	NR	Hashimoto and Nishiuchi (1981)
	Guppy	LC_{50}	1	2,971,000	1.7 CM, 0.10 G	Vardia and Durve (1984)
		LC_{50}	2	2,348,000	1.7 CM, 0.10 G	Vardia and Durve (1984)
		LC_{50}	4	1,418,000	1.7 CM, 0.10 G	Vardia and Durve (1984)
	Harlequinfish, red rasbora	LC_{50}	1	1,160,000	1.3–3 CM	Alabaster (1969)
	Loach (*Misgurnus fossilis*)	LC_{90}	0.42	3,645,000	EGG	Klekowski et al. (1977)
		OC	0.42	3,645,000	EGG	Klekowski et al. (1977)
		OC	6 hr	3,645,000	EGG	Klekowski et al. (1977)
	Medaka, high-eyes	LC_{50}	2	>40,000	NR	Hashimoto and Nishiuchi (1981)
		AVO	NR	177	NR	Hidaka et al. (1984)
	Oriental weatherfish	LC_{50}	2	>40,000	NR	Hashimoto and Nishiuchi (1981)
Insect	Mayfly (*Cloeon dipterum*)	LC_{50}	2	>40,000	LARVAE	Hashimoto and Nishiuchi (1981)
Mullusca	American or Virginia oyster	$EC_{50}DV$	2	20,440	EGGS, 2 CELL STAGE	Davis and Hidu (1979)
		LC_{50}	12	6,290	2 D, LARVAE	Davis and Hidu (1979)
		GRO	12	25	2 D, LARVAE	Davis and Hidu (1979)
	Bladder snail	LC_{50}	2	>100,000	NR	Hashimoto and Nishiuchi (1981)
	Marsh snail	LC_{50}	2	>100,000	NR	Hashimoto and Nishiuchi (1981)
	Snail (*Indoplanorbis exustus*)	LC_{50}	2	>100,000	NR	Hashimoto and Nishiuchi (1981)
	Swan mussel	$EC_{50}PH$	0.5 hr	0.25 g L^{-1}	LARVAE	Varanka (1979)
		$EC_{50}PH$	0.5 hr	7 g L^{-1}	LARVAE	Varanka (1979)

Table 8. Toxicology of Aquatic and Terrestrial Species: Chemical: 2,4-D Sodium Salt, CAS Number: 2702729 (continued)

Class	Species	Effect	Duration (days)	Concentration ($\mu g\ L^{-1}$)	Lifestage	Ref.
Tracheophy	Coon-tail	MOR	>6	5,000	UPPER LEAFY PARTS, 10 CM	Titova (1978)
	Duckweed (*Lemna minor*)	MOR	>6	5,000	NR	Titova (1978)
	Water-milfoil (*Myriophyllum spicatum*)	MOR	>6	5,000	UPPER LEAFY PARTS, 10 CM	Titova (1978)
	Waterweed (*Elodea canadensis*)	MOR	6	5,000	UPPER LEAFY PARTS, 10 CM	Titova (1978)

Table 9. Toxicology of Aquatic and Terrestrial Species: Chemical: 2,4-D Amine, CAS Number: 2008391

Class	Species	Effect	Duration (days)	Concentration ($\mu g\ L^{-1}$)	Lifestage	Ref.
Bird	Mallard duck	LD_{50}	NR	2000		Tucker and Crabtree (1970)
		LD_{50}	NR	>2025		Hudson et al. (1984)
Invertebrate	Honey bee	NO MOR	NR	1 lb ac^{-1}	ADULT	Moffett and Morton (1972)
		NO MOR	NR	4 lb ac^{-1}	MULT	Moffett and Morton (1972)

Table 10. Toxicology of Aquatic and Terrestrial Species: Chemical: 2,4-DP Dichlorprop, CAS Number: 120365

Class	Species	Effect	Duration (days)	Concentration ($\mu g\ L^{-1}$)	Lifestage	Ref.
Algae	Green algae (*Ankistrodesmus falcatus*)	EC_{50}PP	NR	<500 $\mu L\ L^{-1}$	NR	Tscheu-Schluter (1972)
		PGR	NR	200 $\mu L\ L^{-1}$	NR	Tscheu-Schluter (1972)
		PGR	NR	300 $\mu L\ L^{-1}$	NR	Tscheu-Schluter (1972)
Fish	Minnow, carp family	LC_{50}	2	774 $\mu L\ L^{-1}$	NR	Tscheu-Schluter (1972)
Mammal	Rat	LD_{50}	NR	532		USDA Forest Service (1989)

Table 11. Toxicology of Aquatic and Terrestrial Species: Chemical: Acephate, CAS Number: 30560191

Class	Species	Effect	Duration (days)	Concentration ($\mu g\ L^{-1}$)	Lifestage	Ref.
Amphibia	Bullfrog	MOR	4	5,000	TADPOLE	Hall and Kolbe (1980)
	Frog (*Bufo bufo japonicus*)	LC_{50}	1	>40,000	25–35 D, 0.34 G, TADPOLE	Nishiuchi and Yoshida (1974)
	Frog (*Bufo vulgaris formosus*)	LC_{50}	1	>40,000	NR	Nishiuchi (1980b)
	Frog (*Bufo bufo japonicus*)	LC_{50}	2	>40,000	25–35 D, 0.34 G, TADPOLE	Nishiuchi and Yoshida (1974)
		LC_{50}	3	>40,000	25–35 D, 0.34 G, TADPOLE	Nishiuchi and Yoshida (1974)
		LC_{50}	4	>40,000	25–35 D, 0.34 G, TADPOLE	Nishiuchi and Yoshida (1974)
	Salamander	LC_{50}	4	8,816,000	LARVAE, 68 D POSTHATCH	Geen (1984b)
		GRO	49	798,000	LARVAE	Geen (1984b)
		HAT	NR	798,000	EGG	Geen (1984b)
Aquatic invertebrate	Invertebrates	ABD	8	113–140	NR	Rabeni and Stanley (1979)
		BCF	2.5 hr	1,199	NYMPH AND LARVAE	Geen et al. (1981)
		BCF	3 hr	974	NYMPH AND LARVAE	Geen et al. (1981)
		BCF	5 hr	1,141	NYMPH AND LARVAE	Geen et al. (1981)
		BCF	5.75 hr	75	NYMPH AND LARVAE	Geen et al. (1981)
		DRF	4	113–140	NR	Rabeni and Stanley (1979)
		MOR-D	5 hr	1,100–1,200	NYMPH AND LARVAE	Geen et al. (1981)

Crustacea	Scud	LC_{50}	4	>25,000	NR	Woodward and Mauck (1980)
	(*Gammarus pseudolimnaeus*)	LC_{50}	4	>25,000	NR	Woodward and Mauck (1980)
		LC_{50}	4	>25,000	NR	Woodward and Mauck (1980)
		LC_{50}	4	>50,000	MATURE	Sanders et al. (1983)
		LC_{50}	4	>50,000	MATURE	Sanders et al. (1983)
		LC_{50}	4	>50,000	MATURE	Johnson and Finley (1980)
		LC_{50}	4	>100,000	MATURE	Schoettger and Mauck (1978)
	Water flea (*Daphnia magna*)	EC_{50}IM	2	>50,000	1ST INSTAR	Sanders et al. (1983)
		EC_{50}IM	2	>50,000	1ST INSTAR	Sanders et al. (1983)
Fish	Atlantic salmon	ENZ	8	113	NR	Rabeni and Stanley (1979)
		GRO	120	113–140	NR	Rabeni and Stanley (1979)
	Bluegill	LC_{50}	4	>50,000	NR	Sanders et al. (1983)
		LC_{50}	4	>50,000	NR	Sanders et al. (1983)
		LC_{50}	4	1,000,000	0.4 G	Johnson and Finley (1980)
		LC_{50}	4	1,000,000	0.4 G	Johnson and Finley (1980)
	Brook trout	LC_{50}	4	>100,000	NR	Schoettger and Mauck (1978)
		LC_{50}	4	>100,000	NR	Schoettger and Mauck (1978)
		LC_{50}	4	>100,000	0.2 G	Johnson and Finley (1980)
		ENZ	8	113–140	NR	Rabeni and Stanley (1979)
		FCR	2	113	NR	Rabeni and Stanley (1979)
		FCR	2	140	NR	Rabeni and Stanley (1979)
		GRO	120	113–140	NR	Rabeni and Stanley (1979)
	Channel catfish	LC_{50}	4	>50,000	NR	Sanders et al. (1983)
		LC_{50}	4	>50,000	NR	Sanders et al. (1983)
		LC_{50}	4	1,000,000	2.0 G	Johnson and Finley (1980)
		LC_{50}	4	560,000–1,000,000	0.5 G	Johnson and Finley (1980)

Table 11. Toxicology of Aquatic and Terrestrial Species: Chemical: Acephate, CAS Number: 30560191 (continued)

Class	Species	Effect	Duration (days)	Concentration ($\mu g\ L^{-1}$)	Lifestage	Ref.
Fish (continued)	Cutthroat trout	LC_{50}	4	>5,000	NR	Woodward and Mauck (1980)
		LC_{50}	4	>50,000	NR	Woodward and Mauck (1980)
		LC_{50}	4	>60,000	NR	Woodward and Mauck (1980)
		LC_{50}	4	800,000–900,000	10–25 G	Geen et al. (1984a)
		LC_{50}	4	>100,000	NR	Woodward and Mauck (1980)
		LC_{50}	4	>100,000	NR	Woodward and Mauck (1980)
		LC_{50}	4	>100,000	NR	Woodward and Mauck (1980)
		LC_{50}	4	>100,000	0.7 G	Johnson and Finley (1980)
		LC_{50}	4	>100,000	0.9 G	Johnson and Finley (1980)
	Fathead minnow	LC_{50}	4	>50,000	NR	Sanders et al. (1983)
		LC_{50}	4	>50,000	NR	Sanders et al. (1983)
		LC_{50}	4	1,000,000	1.0 G	Johnson and Finley (1980)
		LC_{50}	4	1,000,000	1.0 G	Johnson and Finley (1980)
	Motsuga, stone moroko	ENZ	0.2 hr	0.0091 M	4–8 CM, 1–5 G	Kanazawa (1983)
	Rainbow trout, donaldson trout	LC_{50}	1	1,940	FINGERLING, 8.71 G, 8.63 CM	Duangsawasdi (1977)
		LC_{50}	1	2,740	FINGERLING, 7.63 G, 8.18 CM	Duangsawasdi (1977)
		LC_{50}	1	4,320	FINGERLING, 10.56 G, 9.15 CM	Duangsawasdi (1977)
		LC_{50}	1	895,000	6.9 CM, 8.2 G	Klaverkamp et al. (1975)
		LC_{50}	1	1,050,000	8.1 CM, 6.7 G	Klaverkamp et al. (1975)
		LC_{50}	1	1,880,000	FINGERLING, 8.85 G, 8.6 CM	Duangsawasdi and Klaverkamp (1979)

LC_{50}	1	2,890,000	FINGERLING, 8.85 G, 8.6 CM	Duangsawasdi and Klaverkamp (1979)
LC_{50}	1	3,160,000	FINGERLING, 8.85 G, 8.6 CM	Duangsawasdi and Klaverkamp (1979)
LC_{50}	2	1,670	FINGERLING, 7.63 G, 8.18 CM	Duangsawasdi (1977)
LC_{50}	2	1,720	FINGERLING, 8.71 G, 8.63 CM	Duangsawasdi (1977)
LC_{50}	2	2,370	FINGERLING, 10.56 G, 9.15 CM	Duangsawasdi (1977)
LC_{50}	4	1,340	FINGERLING, 10.56 G, 9.15 CM	Duangsawasdi (1977)
LC_{50}	4	1,390	FINGERLING, 8.71 G, 8.63 CM	Duangsawasdi (1977)
LC_{50}	4	1,440	FINGERLING, 7.63 G, 8.18 CM	Duangsawasdi (1977)
LC_{50}	4	2,660	4–7 G	Geen et al. (1984a)
LC_{50}	4	2,820	3–4 G	Geen et al. (1984a)
LC_{50}	4	730,000	1.2 G	Johnson and Finley (1980)
LC_{50}	4	>50,000	NR	Sanders et al. (1983)
LC_{50}	4	>50,000	NR	Sanders et al. (1983)
LC_{50}	4	1,100,000	1.5 G	Johnson and Finley (1980)
BCF	0.33	24	YEARLING	Geen et al. (1981)
BCF	0.33	24	YEARLING	Geen et al. (1981)
BCF	0.33	37	YEARLING	Geen et al. (1981)
BCF	0.33	37	YEARLING	Geen et al. (1981)
BCF	1–24	880	6–10 G	Geen et al. (1984a)
BCF	1–24	4,290	6–10 G	Geen et al. (1984a)
BCF	1–24	8,460	6–10 G	Geen et al. (1984a)
BCF	1–24	20,100	6–10 G	Geen et al. (1984a)
BCF	1–60	880	4–10 G	Geen et al. (1984a)

Table 11. Toxicology of Aquatic and Terrestrial Species: Chemical: Acephate, CAS Number: 30560191 (continued)

Class	Species	Effect	Duration (days)	Concentration ($\mu g\ L^{-1}$)	Lifestage	Ref.
Fish (continued)	Rainbow trout, donaldson trout	BCF	1–60	4,750	4–10 G	Geen et al. (1984a)
		BCF	1–60	9,670	4–10 G	Geen et al. (1984a)
		BCF	1–60	23,100	4–10 G	Geen et al. (1984a)
		BCF	2.5 hr	974	YEARLING	Geen et al. (1981)
		BCF	2.5 hr	1,199	YEARLING	Geen et al. (1981)
		BCF	5 hr	987	YEARLING	Geen et al. (1981)
		BCF	5 hr	987	YEARLING	Geen et al. (1981)
		BCF	5 hr	1,141	YEARLING	Geen et al. (1981)
		BCF	5 hr	1,141	YEARLING	Geen et al. (1981)
		ENZ	0.66 hr	3,000,000	ADULT, 29 CM, 400 G	Klaverkamp and Hobden (1980)
		ENZ	1	100,000	30–80 G, 14–21 CM	Zinkl et al. (1987a)
		ENZ	1	400,000	30–80 G, 14–21 CM	Zinkl et al. (1987a)
		ENZ	1	2,000,000	ADULT, 514 G, 34.7 CM	Duangsawasdi (1977)
		ENZ	4	300,000	FINGERLING, 8.71 G, 8.63 CM	Duangsawasdi (1977)
		ENZ	4	300,000	FINGERLING, 10.56 G, 9.15 CM	Duangsawasdi (1977)
		ENZ	4	300,000	FINGERLING, 7.63 G, 8.18 CM	Duangsawasdi (1977)
		HEM	1	2,000,000	ADULT, 514 G, 34.7 CM	Duangsawasdi (1977)
		MOR-D	5 hr	1,141	YEARLING	Geen et al. (1981)
		MOR-D	5 hr	1,100–1,200	YEARLING	Geen et al. (1981)
		PHY	1	2,000,000	ADULT, 514 G, 34.7 CM	Duangsawasdi (1977)

		PHY	2	2,000,000	ADULT, 514 G, 34.7 CM	Duangsawasdi and Klaverkamp (1979)
		RES	1	2,000,000	ADULT, 514 G, 34.7 CM	Duangsawasdi (1977)
	Sucker family	ENZ	8	113–140	NR	Rabeni and Stanley (1979)
	Yellow perch	LC_{50}	4	>50,000	2.0 G	Johnson and Finley (1980)
		LC_{50}	4	>100,000	1.8 G	Johnson and Finley (1980)
Insect	Backswimmer family	LC_{50}	1	10,400	ADULT	Hussain et al. (1984)
	Insect class	BCF	3 hr	974	NYMPH AND LARVAE	Geen et al. (1981)
		BCF	5 hr	1,141	NYMPH AND LARVAE	Geen et al. (1981)
		MOR-D	5 hr	1,141	NYMPH AND LARVAE	Geen et al. (1981)
	Midge (*Chironomus plumosus*)	EC_{50}IM	2	>50,000	4TH INSTAR, LARVAE	Sanders et al. (1983)
		EC_{50}IM	2	>50,000	4TH INSTAR LARVAE	Sanders et al. (1983)
		LC_{50}	4	1,000,000	4TH INSTAR	Johnson and Finley (1980)
	Stonefly (*Pteronarcella badia*)	LC_{50}	4	6,400	NR	Woodward and Mauck (1980)
		LC_{50}	4	6,400	NAIAD	Johnson and Finley (1980)
		LC_{50}	4	9,500	NR	Woodward and Mauck (1980)
		LC_{50}	4	9,500	NAIAD	Johnson and Finley (1980)
		LC_{50}	4	21,000	NAIAD	Johnson and Finley (1980)
		LC_{50}	4	21,200	NR	Woodward and Mauck (1980)
	Stonefly (*Skwala* sp)	LC_{50}	4	12,000	NAIAD	Johnson and Finley (1980)
		LC_{50}	4	12,000	NAIAD	Johnson and Finley (1980)
	Water boatman family	LC_{50}	1	8,200	ADULT	Hussain et al. (1984)

Table 11. Toxicology of Aquatic and Terrestrial Species: Chemical: Acephate, CAS Number: 30560191 (continued)

Class	Species	Effect	Duration (days)	Concentration (µg L^{-1})	Lifestage	Ref.
Bird	Mallard duck	LD_{50}	14	234 mg kg^{-1}	120–180 D	Hudson et al. (1984)
	Passerine birds	LO SIGNS	15	1.13 kg ha^{-1}		Zinkl et al. (1979)
		LO SIGNS	89	1.13 kg ha^{-1}		Zinkl et al. (1979)
		NO SIGNS	15	0.56 kg ha^{-1}		Zinkl et al. (1979)
Invertebrate	Honey bee	LO MOR	48	1.0 lb ac^{-1}		Robinson and Johnson (1978)
		LO POP	48	1.0 lb ac^{-1}		Robinson and Johnson (1978)

Table 12. Toxicology of Aquatic and Terrestrial Species: Chemical: Atrazine, CAS Number: 1912249

Class	Species	Effect	Duration (days)	Concentration ($\mu g\ L^{-1}$)	Lifestage	Ref.
Algae	Algae, phytoplankton, algal mat	BMS	2	20	NR	DeNoyelles et al. (1982)
		CLR	12	24	STREAM AUFWUCHS COMMUNITY	Krieger et al. (1988)
		CLR	12	24	STREAM AUFWUCHS COMMUNITY	Krieger et al. (1988)
		CLR	12	134	STREAM AUFWUCHS COMMUNITY	Krieger et al. (1988)
		PGR	12–49	80	4 MAJOR GROUPS	Hamilton et al. (1987)
		PGR	35	140	4 MAJOR GROUPS	Hamilton et al. (1987)
		POP	14	100	NR	Hamala and Kollig (1985)
		POP	63	20	NR	DeNoyelles et al. (1982)
		POP-D	NR	10	NR	Kosinski (1984)
		PSE	3	100	NR	Moorhead and Kosinski (1986)
		PSE	10	0.01	NR	Kosinski and Merkle (1984)
		PSE	14	100	NR	Hamala and Kollig (1985)
		PSE	21	1	NR	Kosinski and Merkle (1984)
		PSE	30	25	NR	Lynch et al. (1985)
	Blue-green algae	$EC_{50}GR$	12–14	30	6.5E4 CELLS ML^{-1}	Stratton (1984)
	(*Anabaena inaequalis*)	$EC_{50}GR$	12–14	100	6.5E4 CELLS ML^{-1}	Stratton (1984)
	Blue-green algae	$EC_{50}GR$	12–14	1,200	6.5E4 CELLS ML^{-1}	Stratton (1984)
	(*Anabaena cylindrica*)	$EC_{50}GR$	12–14	3,600	6.5E4 CELLS ML^{-1}	Stratton (1984)
	Blue-green algae	$EC_{50}GR$	12–14	4,000	6.5E4 CELLS ML^{-1}	Stratton (1984)
	(*Anabaena variabilis*)	$EC_{50}GR$	12–14	5,000	6.5E4 CELLS ML^{-1}	Stratton (1984)

Table 12. Toxicology of Aquatic and Terrestrial Species: Chemical: Atrazine, CAS Number: 1912249 (continued)

Class	Species	Effect	Duration (days)	Concen-tration ($\mu g\ L^{-1}$)	Lifestage	Ref.
Algae (continued)	Blue-green algae (*Anabaena inaequalis*)	EC_{50}NF	5 hr	55,000	6.5E4 CELLS ML^{-1}	Stratton (1984)
	Blue-green algae (*Anabaena cylindrica*)	EC_{50}NF	5 hr	>100,000	6.5E4 CELLS ML^{-1}	Stratton (1984)
	Blue-green algae (*Anabaena variabilis*)	EC_{50}PS	3 hr	100	6.5E4 CELLS ML^{-1}	Stratton (1984)
	Blue-green algae (*Anabaena inaequalis*)	EC_{50}PS	3 hr	300	6.5E4 CELLS ML^{-1}	Stratton (1984)
	Blue-green algae (*Anabaena cylindrica*)	EC_{50}PS	3 hr	500	6.5E4 CELLS ML^{-1}	Stratton (1984)
	Blue-green algae (*Anabaena inaequalis*)	EC_{50}PS	NR	40	6,500 CELLS ML^{-1}	Stratton and Corke (1981)
	Blue-green algae (*Anabaena variabilis*)	EC_{50}PS	NR	40	6,500 CELLS ML^{-1}	Stratton and Corke (1981)
	Blue-green algae (*Anabaena inaequalis*)	EC_{50}PS	NR	50	6,500 CELLS ML^{-1}	Stratton and Corke (1981)
		EC_{50}PS	NR	50	6,500 CELLS ML^{-1}	Stratton and Corke (1981)
		EC_{50}PS	NR	50	6,500 CELLS ML^{-1}	Stratton and Corke (1981)
		EC_{50}PS	NR	50	6,500 CELLS ML^{-1}	Stratton and Corke (1981)
	Blue-green algae (*Anabaena variabilis*)	EC_{50}PS	NR	50	6,500 CELLS ML^{-1}	Stratton and Corke (1981)
	Blue-green algae (*Anabaena inaequalis*)	EC_{50}PS	NR	60	6,500 CELLS ML^{-1}	Stratton and Corke (1981)

Blue-green algae	EC_{50}PS	NR	60	6,500 CELLS ML^{-1}	Stratton and Corke (1981)
(*Anabaena variabilis*)	EC_{50}PS	NR	60	6,500 CELLS ML^{-1}	Stratton and Corke (1981)
	EC_{50}PS	NR	100	6,500 CELLS ML^{-1}	Stratton and Corke (1981)
	EC_{50}PS	NR	100	6,500 CELLS ML^{-1}	Stratton and Corke (1981)
Blue-green algae	EC_{50}PS	NR	250	6,500 CELLS ML^{-1}	Stratton and Corke (1981)
(*Anabaena cylindrica*)	EC_{50}PS	NR	290	6,500 CELLS ML^{-1}	Stratton and Corke (1981)
	EC_{50}PS	NR	300	6,500 CELLS ML^{-1}	Stratton and Corke (1981)
	EC_{50}PS	NR	360	6,500 CELLS ML^{-1}	Stratton and Corke (1981)
	EC_{50}PS	NR	370	6,500 CELLS ML^{-1}	Stratton and Corke (1981)
	EC_{50}PS	NR	>500	6,500 CELLS ML^{-1}	Stratton and Corke (1981)
Blue-green algae	MOR	NR	3	NR	Bringmann and Kuhn (1978a)
(*Anacystis aeruginosa*)	PGR	8	3	NR	Bringmann and Kuhn (1978b)
	PSE	NR	100	NR	Braginskii and Migal (1973)
Blue-green algae (*Cyanophyta*)	BMS	34	100	NR	Herman et al. (1986)
Blue-green algae (*Nostoc muscorum*)	PGR	NR	2,000,000	CYANOBACTERIAL LAWN	Mallison and Cannon (1984)
Blue-green algae (*Phormidium* sp)	CEL	1–3 hr	15	NR	Noll and Bauer (1974)
Blue-green algae (*Plectonema boryanum*)	PGR	31	10,000	10 ML	Mallison and Cannon (1984)
Cuvie, tangleweed	CEL	1	1,000,000	ADULTS	Hopkin and Kain (1971)
	GRO	NR	10	SPOROPHYTE	Hopkin and Kain (1978)
	GRO	NR	1,000	GAMETOPHYTE	Hopkin and Kain (1978)
	PGR	28	10	ZOOSPORES	Hopkin and Kain (1971)
	PSE	NR	1,000,000	MATURE FROND TISSUE	Hopkin and Kain (1978)
Diatom (*Achnanthes brevipes*)	EC_{50}OX	1.5 hr	93	NR	Hollister and Walsh (1973)
Diatom (*Amphora exigua*)	EC_{50}OX	1.5 hr	300	NR	Hollister and Walsh (1973)
Diatom	EC_{50}OX	0.86 hr	99	NR	Millie and Hersh (1987)
(*Cyclotella meneghiniana*)	EC_{50}OX	0.86 hr	105	NR	Millie and Hersh (1987)
	EC_{50}OX	0.86 hr	243	NR	Millie and Hersh (1987)

Table 12. Toxicology of Aquatic and Terrestrial Species: Chemical: Atrazine, CAS Number: 1912249 (continued)

Class	Species	Effect	Duration (days)	Concentration ($\mu g\ L^{-1}$)	Lifestage	Ref.
Algae (continued)	Diatom (*Navicula incerta*)	$EC_{50}OX$	1.5 hr	460	NR	Hollister and Walsh (1973)
	Diatom (*Nitzschia sigma*)	$EC_{50}GR$	7	280	NR	Plumley and Davis (1980)
	Diatom (*Nitzschia* sp)	$EC_{50}OX$	1.5 hr	434	NR	Hollister and Walsh (1973)
	Diatom (*Nitzschia sigma*)	CLR	7	220	NR	Plumley and Davis (1980)
		CLR	7	2,200	NR	Plumley and Davis (1980)
		PSE	7	22	NR	Plumley and Davis (1980)
		PSE	7	220	NR	Plumley and Davis (1980)
	Diatom (*Phaeodactylum tricornutum*)	$EC_{50}OX$	1.5 hr	100	LOG PHASE	Walsh (1972)
		$EC_{50}OX$	1.5 hr	100	NR	Hollister and Walsh (1973)
		$EC_{50}OX$	1.5 hr	200	LOG PHASE	Walsh (1972)
		$EC_{50}OX$	1.5 hr	287	NR	Hollister and Walsh (1973)
		GRO	7	15	40,000 CELLS ML^{-1}	Mayasich et al. (1987)
		GRO	7	15	40,000 CELLS ML^{-1}	Mayasich et al. (1987)
		PGR	10	200	LOG PHASE	Walsh (1972)
		PGR	10	200	LOG PHASE	Walsh (1972)
	Diatom (*Skeletonema costatum*)	$EC_{50}PP$	2	265	INITIAL CONC. 40,000 CELLS ML^{-1}	Walsh (1983)
	Diatom (*Stauroneis amphoroides*)	$EC_{50}OX$	1.5 hr	348	NR	Hollister and Walsh (1973)
	Diatom (*Thalassiosira weissflogii*)	$EC_{50}GR$	7	280	NR	Plumley and Davis (1980)
		$EC_{50}GR$	7	960	NR	Plumley and Davis (1980)
	Diatom (*Thalassiosira guillardii*)	$EC_{50}OX$	1.5 hr	84	NR	Hollister and Walsh (1973)

Diatom (*Thalassiosira weissflogii*)	EC_{50}OX	1.5 hr	110	NR	Hollister and Walsh (1973)
	CLR	7	220	NR	Plumley and Davis (1980)
	CLR	7	220	NR	Plumley and Davis (1980)
	CLR	7	2,200	NR	Plumley and Davis (1980)
	CLR	7	2,200	NR	Plumley and Davis (1980)
	PSE	7	22	NR	Plumley and Davis (1980)
	PSE	7	220	NR	Plumley and Davis (1980)
	PSE	7	220	NR	Plumley and Davis (1980)
Diatoms, chrysophyte division	CLR-D	5	2,200	NR	Plumley and Davis (1980)
	PGR-D	5	2,200	NR	Plumley and Davis (1980)
	PGR-D	5	2,200	NR	Plumley and Davis (1980)
	PSE-D	5	2,200	NR	Plumley and Davis (1980)
	PSE-D	5	2,200	NR	Plumley and Davis (1980)
	PSE-D	16	50	NR	Plumley and Davis (1980)
Flagellate euglenoid (*Euglena gracilis*)	PSE	6 hr	1,000	NR	Valentine and Bingham (1976)
Green algae (*Ankistrodesmus braunii*)	EC_{50}GR	11	60	100,000 CELLS ML^{-1}	Burrell et al. (1985)
Green algae (*Ankistrodesmus falcatus*)	EC_{50}PP	NR	130	NR	Tscheu-Schluter (1976)
Green algae (*Chlamydomonas reinhardtii*)	EC_{50}CH	NR	0.00000012	NR	Tellenbach et al. (1983)
Green algae (*Chlamydomonas* sp)	EC_{50}OX	1.5 hr	60	NR	Hollister and Walsh (1973)
Green algae (*Chlamydomonas reinhardtii*)	CLR	2.71	832	NR	Foy and Hiranpradit (1977)
	LET	1–2	216	ACTIVELY GROWING CULTURE	Hersh and Crumpton (1987)
	PGR	1–2	21.6	ACTIVELY GROWING CULTURE	Hersh and Crumpton (1987)
Green algae (*Chlamydomonas eugametos*)	PGR	2	22	INITIAL CONC. 100,000 CELLS ML^{-1}	Hess (1980)

Table 12. Toxicology of Aquatic and Terrestrial Species: Chemical: Atrazine, CAS Number: 1912249 (continued)

Class	Species	Effect	Duration (days)	Concentration ($\mu g\ L^{-1}$)	Lifestage	Ref.
Algae (continued)	Green algae (*Chlamydomonas reinhardtii*)	PGR	2.71	52	NR	Foy and Hiranpradit (1977)
		PGR	3.79	50	NR	Foy and Hiranpradit (1977)
		PSE	6 hr	1,000	NR	Valentine and Bingham (1976)
	Green algae (*Chlorella vulgaris*)	$EC_{50}GR$	11	25	100,000 CELLS ML^{-1}	Burrell et al. (1985)
	Green algae (*Chlorella pyrenoidosa*)	$EC_{50}GR$	12–14	300	10E5 CELLS ML^{-1}	Stratton (1984)
		$EC_{50}GR$	12–14	1,000	10E5 CELLS ML^{-1}	Stratton (1984)
	Green algae (*Chlorococcum* sp)	$EC_{50}OX$	1.5 hr	80	NR	Hollister and Walsh (1973)
		$EC_{50}OX$	1.5 hr	100	LOG PHASE	Walsh (1972)
	Green algae (*Chlorella* sp)	$EC_{50}OX$	1.5 hr	143	NR	Hollister and Walsh (1973)
	Green algae (*Chlorococcum* sp)	$EC_{50}OX$	1.5 hr	400	LOG PHASE	Walsh (1972)
	Green algae (*Chlorella pyrenoidosa*)	$EC_{50}PS$	3 hr	500	10E5 CELLS ML^{-1}	Stratton (1984)
		CLR	3	0.25	INITIAL CONC. 100,000 CELLS ML^{-1}	Gonzalez-Murua et al. (1985)
		CLR	4.6	832	NR	Foy and Hiranpradit (1977)
	Green algae (*Chlorococcum hypnosporum*)	PGR	1	10,000	NR	Virmani et al. (1975)
	Green algae (*Chlorella pyrenoidosa*)	PGR	1	10,000	NR	Virmani et al. (1975)
		PGR	4.6	52	NR	Foy and Hiranpradit (1977)
		PGR	10	0.25	INITIAL CONC. 100,000 CELLS ML^{-1}	Gonzalez-Murua et al. (1985)
	Green algae (*Chlorococcum* sp)	PGR	10	100	LOG PHASE	Walsh (1972)
		PGR	10	100	LOG PHASE	Walsh (1972)
	Green algae (*Chlorella pyrenoidosa*)	PSE	6 hr	1,000	NR	Valentine and Bingham (1976)

Green algae	$EC_{50}OX$	1.5 hr	159	NR	Hollister and Walsh (1973)
(*Dunaliella tertiolecta*)	$EC_{50}OX$	1.5 hr	300	LOG PHASE	Walsh (1972)
	$EC_{50}OX$	1.5 hr	600	LOG PHASE	Walsh (1972)
Green algae	LET	2	100	LOG PHASE	Felix et al. (1988)
(*Dunaliella bioculata*)					
Green algae	MOR	NR	10,000	NR	Portmann (1972)
(*Dunaliella tertiolecta*)					
Green algae	PGR	2	1	LOG PHASE	Felix et al. (1988)
(*Dunaliella bioculata*)					
Green algae	PGR	10	300	LOG PHASE	Walsh (1972)
(*Dunaliella tertiolecta*)	PGR	10	400	LOG PHASE	Walsh (1972)
	PGR	NR	1,000	NR	Portmann (1972)
Green algae (*Gloeotaenium*	LET	4	2,200,000	NR	Prasad and Chowdary (1981)
loitlesbergerianu)	PHY	1.54	220	NR	Prasad and Chowdary (1981)
	PHY	4	220	NR	Prasad and Chowdary (1981)
	PHY	4	2,200	NR	Prasad and Chowdary (1981)
	PHY	>1.54	2,200	NR	Prasad and Chowdary (1981)
Green algae	GRO	7	15	40,000 CELLS ML^{-1}	Mayasich et al. (1987)
(*Nannochloropsis oculata*)	GRO	7	15	40,000 CELLS ML^{-1}	Mayasich et al. (1987)
Green algae (*Neochloris* sp)	$EC_{50}OX$	1.5 hr	82	NR	Hollister and Walsh (1973)
Green algae	BCF	18	3.8	NR	Kearney et al. (1977a)
(*Oedogonium cardiacum*)	BCF	33	4.1	NR	Isensee (1976)
	BCF	33	42.4	NR	Isensee (1976)
	BCF	33	456.8	NR	Isensee (1976)
	BCF	33	4,710	NR	Isensee (1976)
	BCF	33	38,850	NR	Isensee (1976)
Green algae (*Oocystis* sp)	BMS	19	500	NR	DeNoyelles et al. (1982)
Green algae (*Platymonas* sp)	$EC_{50}OX$	1.5 hr	102	NR	Hollister and Walsh (1973)
Green algae	PSE	NR	100	NR	Braginskii and Migal (1973)
(*Rhizoclonium hieroglyphicum*)					

Table 12. Toxicology of Aquatic and Terrestrial Species: Chemical: Atrazine, CAS Number: 1912249 (continued)

Class	Species	Effect	Duration (days)	Concentration ($\mu g\ L^{-1}$)	Lifestage	Ref.
Algae (continued)	Green algae (*Scenedesmus abundans*)	EC_{50}GR	4	110	10E4 CELLS ML^{-1}	Geyer et al. (1985)
	Green algae (*Scenedesmus quadricauda*)	EC_{50}GR	12–14	100	10E5 CELLS ML^{-1}	Stratton (1984)
		EC_{50}GR	12–14	200	10E5 CELLS ML^{-1}	Stratton (1984)
		EC_{50}PS	3 hr	300	10E5 CELLS ML^{-1}	Stratton (1984)
	Green algae (*Scenedesmus obliquus*)	BCF	NR	1	LIVING CELLS	Bohm and Muller (1976)
		BCF	NR	1	DEAD CELLS	Bohm and Muller (1976)
		BCF	NR	10.03	LIVING CELLS	Bohm and Muller (1976)
		BCF	NR	10.03	DEAD CELLS	Bohm and Muller (1976)
		BCF	NR	10.79	NR	Bohm and Muller (1976)
		BCF	NR	100.31	LIVING CELLS	Bohm and Muller (1976)
		BCF	NR	100.31	DEAD CELLS	Bohm and Muller (1976)
		BCF	NR	1,003.10	LIVING CELLS	Bohm and Muller (1976)
		BCF	NR	1,003.10	DEAD CELLS	Bohm and Muller (1976)
	Green algae (*Scenedesmus quadricauda*)	MOR	NR	30	NR	Bringmann and Kuhn (1978a)
		PGR	4	50	NR	Foy and Hiranpradit (1977)
		PGR	8	30	NR	Bringmann and Kuhn (1978b)
		PSE	6 hr	1,000	NR	Valentine and Bingham (1976)
	Green algae (*Scenedesmus obliquus*)	RSD	4.5 hr	10–100	1,000,000 CELLS ML^{-1}	Ellgehausen et al. (1980)
	Green algae (*Scenedesmus quadricauda*)	RSD	6 hr	10–1,000	NR	Valentine and Bingham (1976)

	Green algae (*Selenastrum capricornutum*)	EC_{50}BM	14–21	58.7	LOG PHASE	Turbak et al. (1986)
		EC_{50}BM	14–21	410	LOG PHASE	Turbak et al. (1986)
		EC_{50}OX	NR	69.7	LOG PHASE	Turbak et al. (1986)
		EC_{50}OX	NR	854	LOG PHASE	Turbak et al. (1986)
		BCF	1.17	25	10E5 CELLS ML^{-1}	Mailhot (1987)
		BMS	30	1,000	ACTIVELY GROWING	Johnson (1986)
	Green algae division	BMS	34	100	NR	Herman et al. (1986)
		PGR	14	10	NR	Butler et al. (1975)
	Haptophyte (*Isochrysis galbana*)	EC_{50}OX	1.5 hr	100	LOG PHASE	Walsh (1972)
		EC_{50}OX	1.5 hr	100	NR	Hollister and Walsh (1973)
		EC_{50}OX	1.5 hr	200	LOG PHASE	Walsh (1972)
		PGR	10	100	LOG PHASE	Walsh (1972)
		PGR	10	100	LOG PHASE	Walsh (1972)
	Haptophyte (*Pavlova lutheri*)	EC_{50}OX	1.5 hr	77	NR	Hollister and Walsh (1973)
	Red algae (*Porphyridium cruentum*)	EC_{50}OX	1.5 hr	79	NR	Hollister and Walsh (1973)
Amphibia	American toad	EC_{50}TE	NR	>48,000	EMBRYO	Birge et al. (1980)
		EC_{50}TE	NR	>48,000	EMBRYO	Birge et al. (1980)
	Bullfrog	EC_{50}TE	NR	410	EMBRYO	Birge et al. (1980)
		EC_{50}TE	NR	11,550	EMBRYO	Birge et al. (1980)
		BCF	2	269	TADPOLE	Klaassen and Kadoum (1979)
		BCF	23	263	TADPOLE	Klaassen and Kadoum (1979)
		BCF	51	252	TADPOLE	Klaassen and Kadoum (1979)
		BCF	85	270	TADPOLE	Klaassen and Kadoum (1979)
	Common or european toad	DVP	NR	5,000–8,000	TADPOLE	Paulov (1976)
	Leopard frog (*Rana pipiens*)	EC_{50}TE	NR	7,680	EMBRYO	Birge et al. (1980)
		EC_{50}TE	NR	22,890	EMBRYO	Birge et al. (1980)
	Pickeral frog	EC_{50}TE	NR	17,960	EMBRYO	Birge et al. (1980)
		EC_{50}TE	NR	20,200	EMBRYO	Birge et al. (1980)

Table 12. Toxicology of Aquatic and Terrestrial Species: Chemical: Atrazine, CAS Number: 1912249 (continued)

Class	Species	Effect	Duration (days)	Concentration ($\mu g\ L^{-1}$)	Lifestage	Ref.
Annelida	Leech (*Glossiphonia complanata*)	LC_{50}	28	6,300	7–162 MG	Streit and Peter (1978)
		BCF	2.6 hr	200	3.0–51.7 MG	Streit (1978)
		BCF	3 hr	200	NR	Streit and Schwoerbel (1977)
	Leech (*Helobdella stagnalis*)	LC_{50}	27	9,900	2–12 MG	Streit and Peter (1978)
		BCF	0.53 hr	200	3.0–51.7 MG	Streit (1978)
		BCF	NR	200	NR	Streit and Schwoerbel (1977)
Aquatic invertebrate	Aquatic community	CLR	3–21	3.2	MICROCOSM	Pratt et al. (1988)
		STR	7	>50	NR	Brockway et al. (1984)
		STR	28	50	NR	Brockway et al. (1984)
	Invertebrates	ABD	42–56	500–2,000	NR	Walker (1964)
		BCF	1	309	20 G TOTAL WEIGHT	Klaassen and Kadoum (1979)
		BCF	2	269	20 G TOTAL WEIGHT	Klaassen and Kadoum (1979)
		BCF	22	230	20 G TOTAL WEIGHT	Klaassen and Kadoum (1979)
		BCF	23	263	20 G TOTAL WEIGHT	Klaassen and Kadoum (1979)
		BCF	51	252	20 G TOTAL WEIGHT	Klaassen and Kadoum (1979)
		BCF	55	237	20 G TOTAL WEIGHT	Klaassen and Kadoum (1979)
		BCF	85	270	20 G TOTAL WEIGHT	Klaassen and Kadoum (1979)
		BCF	120	206	20 G TOTAL WEIGHT	Klaassen and Kadoum (1979)
		DRF	1	25	NR	Lynch et al. (1985)
		POP	10	500	NR	DeNoyelles et al. (1982)
		POP	30	25	NR	Lynch et al. (1985)

Cnidaria	Green hydra	PHY	21	5,000	1 SMALL BUD STAGE	Benson and Boush (1983)
		PHY	21	5,000	1 SMALL BUD STAGE	Benson and Boush (1983)
		REP	21	5,000	1 SMALL BUD STAGE	Benson and Boush (1983)
		REP	21	5,000	1 SMALL BUD STAGE	Benson and Boush (1983)
Crustacea	Brown shrimp	MOR	1	1,000	NR	Butler (1965)
		MOR	2	1,000	NR	Butler (1965)
	Calanoid copepod (*Acartia tonsa*)	LC_{50}	4	94	NR	Ward and Ballantine (1985)
	Common shrimp	LC_{50}	2	10,000–33,000	LARVAL	Portmann (1972)
		AVO	NR	10,000	LARVAL	Portmann (1972)
	Fiddler crab (*Uca pugilator*)	LC_{50}	4	>29,000	NR	Ward and Ballantine (1985)
	Fiddler crab (*Uca pugnax*)	LET	8	100,000	LARGE FEMALES, >1.5 CM WIDE	Plumley et al. (1980)
		LET	8	1,000,000	SMALL MALES, <1.5 CM WIDE	Plumley et al. (1980)
		LET	9	1,000,000	SMALL MALES, <1.5 CM WIDE	Plumley et al. (1980)
		MOR	8	1,000	LARGE MALES, >1.5 CM WIDE	Plumley et al. (1980)
		MOR	8	1,000	LARGE FEMALES, >1.5 CM WIDE	Plumley et al. (1980)
		MOR	8	1,000	SMALL MALES, <1.5 CM WIDE	Plumley et al. (1980)
		MOR	8	1,000	SMALL FEMALES, <1.5 CM WIDE	Plumley et al. (1980)
		MOR	8	100,000	LARGE MALES, >1.5 CM WIDE	Plumley et al. (1980)
		MOR	8	100,000	SMALL MALES, <1.5 CM WIDE	Plumley et al. (1980)

Table 12. Toxicology of Aquatic and Terrestrial Species: Chemical: Atrazine, CAS Number: 1912249 (continued)

Class	Species	Effect	Duration (days)	Concentration ($\mu g\ L^{-1}$)	Lifestage	Ref.
Crustacea (continued)	Fiddler crab (*Uca pugnax*)	MOR	8	100,000	SMALL FEMALES, <1.5 CM WIDE	Plumley et al. (1980)
		MOR	8	1,000,000	LARGE MALES, >1.5 CM WIDE	Plumley et al. (1980)
		MOR	8	1,000,000	SMALL FEMALES, <1.5 CM WIDE	Plumley et al. (1980)
		MOR	9	100,000	SMALL MALES, <1.5 CM WIDE	Plumley et al. (1980)
		MOR	9	180,000	SMALL MALES, <1.5 CM WIDE	Plumley et al. (1980)
		MOR	9	320,000	SMALL MALES, <1.5 CM WIDE	Plumley et al. (1980)
		MOR	9	560,000	SMALL MALES, <1.5 CM WIDE	Plumley et al. (1980)
		MOR	9	1,000,000	SMALL MALES, <1.5 CM WIDE	Plumley et al. (1980)
		MOR	30	100,000	SMALL MALES, <1.5 CM WIDE	Plumley et al. (1980)
		MOR	30	180,000	SMALL MALES, <1.5 CM WIDE	Plumley et al. (1980)
		MOR	30	320,000	SMALL MALES, <1.5 CM WIDE	Plumley et al. (1980)
		MOR	30	560,000	SMALL MALES, <1.5 CM WIDE	Plumley et al. (1980)

	MOR	30	1,000,000	SMALL MALES, <1.5 CM WIDE	Plumley et al. (1980)
	MOR	70	1,000,000	NR	Plumley et al. (1980)
	MOR	70	1,000,000	NR	Plumley et al. (1980)
Grass shrimp (*Palaemonetes pugio*)	LC_{50}	4	9,000	NR	Ward and Ballantine (1985)
Opossum shrimp (*Mysidopsis bahia*)	LC_{50}	4	1,000	NR	Ward and Ballantine (1985)
	MOR	28	50	NR	Ward and Ballantine (1985)
	MOR	28	80	NR	Ward and Ballantine (1985)
	MOR	28	190	NR	Ward and Ballantine (1985)
	MOR	28	290	NR	Ward and Ballantine (1985)
	MOR	28	470	NR	Ward and Ballantine (1985)
	REP	28	290	NR	Ward and Ballantine (1985)
	REP	28	470	NR	Ward and Ballantine (1985)
Pink shrimp (*america*)	LC_{50}	4	6,900	NR	Ward and Ballantine (1985)
Scud (*Gammarus fasciatus*)	LC_{50}	2	5,700	1ST INSTAR	Macek et al. (1976)
	GRO	119	60	1–22 D, 2ND GENERATION	Macek et al. (1976)
	GRO	119	140	1–22 D, 2ND GENERATION	Macek et al. (1976)
Water flea (*Ceriodaphnia quadrangula*)	GRO	30–50	1,000	5–8 GENERATIONS	Shcherban (1972b)
	REP	30	1,000	NR	Shcherban (1972a)
	REP	30–50	1,000	5–8 GENERATIONS	Shcherban (1972b)
Water flea (*Daphnia pulex*)	EC_{50}	2	36,500	NR	Hartman and Martin (1985)
	EC_{50}FC	2	1,600	NR	Pott (1980)
Water flea (*Daphnia magna*)	EC_{50}IM	1	>39,000	<24 HR	Marchini et al. (1988)
	EC_{50}IM	2	>39,000	<24 HR	Marchini et al. (1988)
Water flea (*Daphnia pulex*)	EC_{50}RE	2	240	NR	Pott (1980)
Water flea (*Daphnia magna*)	LC_{50}	1.08	3,600	1ST INSTAR, <24 HR	Frear and Boyd (1967)
	LC_{50}	2	6,900	<24 H	Macek et al. (1976)
	LC_{50}	2	54,000	NR	Semov and Iosifov (1973)

Table 12. Toxicology of Aquatic and Terrestrial Species: Chemical: Atrazine, CAS Number: 1912249 (continued)

Class	Species	Effect	Duration (days)	Concentration ($\mu g\ L^{-1}$)	Lifestage	Ref.
Crustacea (continued)	Water flea (*Daphnia pulex*)	LC_{50}	3 hr	>40,000	FEMALE ADULT	Nishiuchi and Hashimoto (1969)
		LC_{50}	3 hr	>40,000	NR	Nishiuchi and Hashimoto (1975)
	Water flea (*Daphnia magna*)	DVP	30–45	1,000	EMBRYO	Shcherban (1973)
		MOR	30	1,000	<24 HR ADULTS	Johnson (1986)
		MOR	42	20	1ST INSTAR LARVAE	Huckins et al. (1986)
	Water flea (*Daphnia pulex*)	MOR	70	10,000	<15 HR	Schober and Lampert (1977)
		REP	28	1,000	<15 HR	Schober and Lampert (1977)
		REP	28	2,000	<15 HR	Schober and Lampert (1977)
	Water flea (*Daphnia magna*)	REP	30–45	1,000	EMBRYO	Shcherban (1973)
		REP	64	140	<24 HR, 3RD GENERATION	Macek et al. (1976)
		REP	64	250	<24 HR, 3RD GENERATION	Macek et al. (1976)
		RSD	2	10–100	1 MG ML^{-1}	Ellgehausen et al. (1980)
	Water flea (*Moina macrocopa*)	LC_{50}	3 hr	>40,000	FEMALE ADULT	Nishiuchi and Hashimoto (1969)
		LC_{50}	3 hr	>40,000	NR	Nishiuchi and Hashimoto (1975)
	Water flea (*Moina rectrirostris*)	DVP	30–45	1,000	EMBRYO	Shcherban (1973)
	Water flea (*Moina macrocopa*)	GRO	30–50	1,000	5–8 GENERATIONS	Shcherban (1972b)
		MOR	30	1,000	NR	Shcherban (1972a)
		REP	30	1,000	NR	Shcherban (1972a)
	Water flea (*Moina rectrirostris*)	REP	30–45	1,000	EMBRYO	Shcherban (1973)

	Water flea (*Moina macrocopa*)	REP	30–50	1,000	5–8 GENERATIONS	Shcherban (1972b)
	Water flea	GRO	30–50	1,000	5–8 GENERATIONS	Shcherban (1972b)
	(*Scapholeberis mucronata*)	REP	30	1,000	NR	Shcherban (1972a)
		REP	30–50	1,000	5–8 GENERATIONS	Shcherban (1972b)
	Water flea	GRO	10	500	JUVENILE, 12 HR	DeNoyelles et al. (1982)
	(*Simocephalus serrulatus*)	GRO	10	500	YOUNG, 12 HR	DeNoyelles et al. (1982)
Fish	Black bullhead	LC_{50}	2	8,000	2–10 CM, 0.5–14 G, 4–12 MONTHS	Bathe et al. (1975a)
		LC_{50}	2	>37,000	2–10 CM, 0.5–14 G, 4–12 MONTHS	Bathe et al. (1975a)
		LC_{50}	4	35,000	2–10 CM, 0.5–14 G, 4–12 MONTHS	Bathe et al. (1975a)
	Bluegill	LC_{50}	2	80,000	2–10 CM, 0.5–14 G	Bathe et al. (1973)
		LC_{50}	2	>21,000	2–10 CM, 0.5–14 G	Bathe et al. (1973)
		LC_{50}	4	15,000	FINGERLING	Klaassen and Kadoum (1979)
		LC_{50}	4	16,000	2–10 CM, 0.5–14 G	Bathe et al. (1973)
		LC_{50}	4	50,000	2–10 CM, 0.5–14 G	Bathe et al. (1973)
		LC_{50}	4	>8,000	6.5 G	Macek et al. (1976)
		LC_{50}	NR	6,700	6.5 G	Macek et al. (1976)
		BCF	1	269	<75 MM	Klaassen and Kadoum (1979)
		BCF	1	269	76–150 MM	Klaassen and Kadoum (1979)
		BCF	1	309	<75 MM	Klaassen and Kadoum (1979)
		BCF	1	309	76–150 MM	Klaassen and Kadoum (1979)
		BCF	1	309	>150 MM	Klaassen and Kadoum (1979)
		BCF	22	230	<75 MM	Klaassen and Kadoum (1979)
		BCF	22	230	76–150 MM	Klaassen and Kadoum (1979)
		BCF	22	230	>150 MM	Klaassen and Kadoum (1979)
		BCF	23	263	<75 MM	Klaassen and Kadoum (1979)
		BCF	23	263	76–150 MM	Klaassen and Kadoum (1979)
		BCF	23	263	>150 MM	Klaassen and Kadoum (1979)

Table 12. Toxicology of Aquatic and Terrestrial Species: Chemical: Atrazine, CAS Number: 1912249 (continued)

Class	Species	Effect	Duration (days)	Concentration ($\mu g\ L^{-1}$)	Lifestage	Ref.
Fish (continued)	Bluegill	BCF	51	252	<75 MM	Klaassen and Kadoum (1979)
		BCF	51	252	76–150 MM	Klaassen and Kadoum (1979)
		BCF	55	237	<75 MM	Klaassen and Kadoum (1979)
		BCF	55	237	76–150 MM	Klaassen and Kadoum (1979)
		BCF	55	237	>150 MM	Klaassen and Kadoum (1979)
		BCF	85	270	<75 MM	Klaassen and Kadoum (1979)
		BCF	85	270	76–150 MM	Klaassen and Kadoum (1979)
		BCF	120	206	<75 MM	Klaassen and Kadoum (1979)
		BCF	120	206	76–150 MM	Klaassen and Kadoum (1979)
		BCF	120	206	>150 MM	Klaassen and Kadoum (1979)
		BCF	546	94	YEARLING, PROGENY	Macek et al. (1976)
		EQU	28	500	15 G	Macek et al. (1976)
		FCR	136	20	85 MM	Kettle et al. (1987)
		MOR	4.67	10,000	FRY, 19 MM	Jones (1962)
		MOR	8	10,000	FRY	Hiltibran (1967)
		MOR	12	5,000	SMALL	Hiltibran (1967)
		MOR	12	10,000	SMALL	Hiltibran (1967)
		REP	136	20	85 MM	Kettle et al. (1987)
		REP	736	95	7–10 CM	Macek et al. (1976)
	Bluntnose minnow	LET	1	80,000	NR	Zora and Paladino (1986)
	Brook trout	LC_{50}	4	4,900	52 G	Macek et al. (1976)
		LC_{50}	4	6,300	52 G	Macek et al. (1976)
		BCF	308	740	YEARLING, PROGENY	Macek et al. (1976)
		GRO	308	120	YEARLING, PROGENY	Macek et al. (1976)
		MOR	308	65	YEARLING, PROGENY	Macek et al. (1976)

Bullhead, catfish	LC_{50}	2	8,000	2–10 CM, 0.5–14 G	Bathe et al. (1973)
	LC_{50}	2	>37,000	2–10 CM, 0.5–14 G	Bathe et al. (1973)
	LC_{50}	4	7,600	2–10 CM, 0.5–14 G	Bathe et al. (1973)
	LC_{50}	4	35,000	2–10 CM, 0.5–14 G	Bathe et al. (1973)
Channel catfish	LC_{50}	4.5	310	EGGS 4 D POSTHATCH	Birge et al. (1979)
	LC_{50}	4.5	340	EGGS 4 D POSTHATCH	Birge et al. (1979)
	LC_{50}	8.5	220	EGGS 4 D POSTHATCH	Birge et al. (1979)
	LC_{50}	8.5	240	EGGS 4 D POSTHATCH	Birge et al. (1979)
	MOR	4	10,000	FRY, YOLK SAC	Jones (1962)
	RSD	4	10–100	1–2 G	Ellgehausen et al. (1980)
Coho salmon, silver salmon	MOR	6	18,000–18,800	YEARLING	Lorz et al. (1979)
	RSD	NR	8,000	6–18 MONTHS, 8.9–25.5 G	Walsh and Ribelin (1973)
Common, mirror, colored, carp	LC_{50}	2	50,000	NR	Svobodova (1980)
	LC_{50}	2	>10,000	4.5 CM, 1.1 G	Nishiuchi and Hashimoto (1969)
	LC_{50}	2	>10,000	NR	Nishiuchi and Hashimoto (1975)
	BIO	0.08 hr	33,000	0.37–1.36 KG	McBride and Richards (1975)
	BIO	6 hr	100	NR	Gluth and Hanke (1984)
	BIO	6 hr	100	NR	Gluth and Hanke (1984)
	BIO	6 hr	100	NR	Gluth and Hanke (1984)
	MOR	6	5,000	32–65 G, 1 YEAR	Antychowicz et al. (1979)
	PHY	6 hr	50	30–50 G	Hanke et al. (1983)
	PHY	6 hr	100	30–50 G	Hanke et al. (1983)
	PHY	6 hr	100	12–17 CM, 40–80 G	Gluth and Hanke (1985)

Table 12. Toxicology of Aquatic and Terrestrial Species: Chemical: Atrazine, CAS Number: 1912249 (continued)

Class	Species	Effect	Duration (days)	Concentration ($\mu g\ L^{-1}$)	Lifestage	Ref.
Fish (continued)	Crucian carp	LC_{50}	2	100,000	2–10 CM, 0.5–14 G	Bathe et al. (1973)
		LC_{50}	2	100,000	2–10 CM, 0.5–14 G, 4–12 MONTHS	Bathe et al. (1975a)
		LC_{50}	2	>100,000	2–10 CM, 0.5–14 G	Bathe et al. (1973)
		LC_{50}	2	>100,000	2–10 CM, 0.5–14 G, 4–12 MONTHS	Bathe et al. (1975a)
		LC_{50}	4	76,000	2–10 CM, 0.5–14 G	Bathe et al. (1973)
		LC_{50}	4	>100,000	2–10 CM, 0.5–14 G	Bathe et al. (1973)
		LC_{50}	4	>100,000	2–10 CM, 0.5–14 G, 4–12 MONTHS	Bathe et al. (1975a)
		LC_{50}	4	>100,000	25–50 G	Bathe et al. (1975b)
	Fathead minnow	LC_{50}	4	15,000	1.8 G	Macek et al. (1976)
		BCF	301	210	YEARLING, PROGENY	Macek et al. (1976)
		MOR	4	520	3–5 D	Macek et al. (1976)
		MOR	301	210	18 D, PROGENY	Macek et al. (1976)
	Goldfish	LC_{50}	2	>10,000	4.01 CM, 1.04 G	Nishiuchi and Hashimoto (1969)
		LC_{50}	2	>10,000	NR	Nishiuchi and Hashimoto (1975)
	Green sunfish	HAT	8	10,000	FERTILIZED EGGS	Hiltibran (1967)
		HAT	8	10,000	FERTILIZED EGGS	Hiltibran (1967)
		MOR	8	10,000	FRY	Hiltibran (1967)

Guppy	LC_{50}	2	38,200	NR	Tscheu-Schluter (1976)
	LC_{50}	2	71,000	NR	Svobodova (1980)
	LC_{50}	2	>10,000	2–10 CM, 0.5–14 G	Bathe et al. (1973)
	LC_{50}	2	>10,000	2–10 CM, 0.5–14 G, 4–12 MONTHS	Bathe et al. (1975a)
	LC_{50}	3	31,600	NR	Tscheu-Schluter (1976)
	LC_{50}	4	4,300	2–10 CM, 0.5–14 G	Bathe et al. (1973)
	LD_{100}	NR	50	NR	Gzhetotskii et al. (1977)
	MOR	NR	500	NR	Gzhetotskii et al. (1977)
Ide, silver or golden orfe	LC_{50}	2	70,000	2–10 CM, 0.5–14 G, 4–12 MONTHS	Bathe et al. (1975a)
	LC_{50}	4	44,000	2–10 CM, 0.5–14 G, 4–12 MONTHS	Bathe et al. (1975a)
	LC_{50}	NR	33,000	NR	Juhnke and Luedemann (1978)
Lake chubsucker	MOR	8	10,000	FRY	Hiltibran (1967)
Lake trout, siscowet	RSD	NR	8,000	6–18 MONTHS, 8.9–25.5 G	Walsh and Ribelin (1973)
Largemouth bass	MOR	2	5,000	FRY, 10–15 MM	Jones (1962)
Medaka, high-eyes	LC_{50}	2	>10,000	2.54 CM, 0.16 G	Nishiuchi and Hashimoto (1969)
	LC_{50}	2	>10,000	NR	Nishiuchi and Hashimoto (1975)
Mosquitofish (*Gambusia affinis*)	BCF	3	4.1	NR	Isensee (1976)
	BCF	3	42.4	NR	Isensee (1976)
	BCF	3	45.68	NR	Isensee (1976)
	BCF	3	4,710	NR	Isensee (1976)
	BCF	3	38,850	NR	Isensee (1976)
	BCF	9	6	NR	Kearney et al. (1977a)
	BCF	18	3.8	NR	Kearney et al. (1977a)
	MOR	2	1	2.5 CM	Darwazeh and Mulla (1974)

Table 12. Toxicology of Aquatic and Terrestrial Species: Chemical: Atrazine, CAS Number: 1912249 (continued)

Class	Species	Effect	Duration (days)	Concentration ($\mu g\ L^{-1}$)	Lifestage	Ref.
Fish (continued)	Mosquitofish (*Gambusia affinis*)	MOR	2	5	2.5 CM	Darwazeh and Mulla (1974)
		MOR	2	10,000	2.5 CM	Darwazeh and Mulla (1974)
	Perch (*Perca* sp)	LC_{50}	2	80,000	2–10 CM, 0.5–14 G, 4–12 MONTHS	Bathe et al. (1975a)
		LC_{50}	2	>21,000	2–10 CM, 0.5–14 G, 4–12 MONTHS	Bathe et al. (1975a)
		LC_{50}	4	50,000	2–10 CM, 0.5–14 G, 4–12 MONTHS	Bathe et al. (1975a)
	Rainbow trout, donaldson trout	LC_{50}	2	10,000	2–10 CM, 0.5–14 G	Bathe et al. (1973)
		LC_{50}	2	10,000	2–10 CM, 0.5–14 G, 4–12 MONTHS	Bathe et al. (1975a)
		LC_{50}	2	26,400	NR	Svobodova (1980)
		LC_{50}	2	30,000	2–10 CM, 0.5–14 G	Bathe et al. (1973)
		LC_{50}	2	30,000	2–10 CM, 0.5–14 G, 4–12 MONTHS	Bathe et al. (1975a)
		LC_{50}	4	4,500	25–50 G	Bathe et al. (1975b)
		LC_{50}	4	8,800	2–10 CM, 0.5–14 G	Bathe et al. (1973)
		LC_{50}	4	17,000	2–10 CM, 0.5–14 G	Bathe et al. (1973)
		LC_{50}	23	920	EGGS 4 D POSTHATCH	Birge et al. (1979)
		LC_{50}	23	1,100	EGGS 4 D POSTHATCH	Birge et al. (1979)
		LC_{50}	27	870	EGGS 4 D POSTHATCH	Birge et al. (1979)
		LC_{50}	27	1,080	EGGS 4 D POSTHATCH	Birge et al. (1979)

Sheepshead minnow	LC_{50}	4	>16,000	NR	Ward and Ballantine (1985)
	GRO	28	3,400	EMBRYO	Ward and Ballantine (1985)
	HAT	NR	3,400	EMBRYO	Ward and Ballantine (1985)
	MOR	28	410	EMBRYO	Ward and Ballantine (1985)
	MOR	28	1,100	EMBRYO	Ward and Ballantine (1985)
	MOR	28	1,800	EMBRYO	Ward and Ballantine (1985)
	MOR	28	1,900	EMBRYO	Ward and Ballantine (1985)
	MOR	28	3,400	EMBRYO	Ward and Ballantine (1985)
Smallmouth bass	LET	3	10,000	FRY	Hiltibran (1967)
Spot	LC_{50}	4	8,500	NR	Ward and Ballantine (1985)
	MOR	2	1,000	JUVENILE	Butler (1965)
Whitefish	EC_{50}IM	1	27,500	STARVED FRY	Gunkel and Kausch (1976)
	EC_{50}IM	1	36,600	STARVED FRY	Gunkel and Kausch (1976)
	EC_{50}IM	2	21,400	STARVED FRY	Gunkel and Kausch (1976)
	EC_{50}IM	2	31,200	STARVED FRY	Gunkel and Kausch (1976)
	EC_{50}IM	3	10,700	STARVED FRY	Gunkel and Kausch (1976)
	EC_{50}IM	3	27,600	STARVED FRY	Gunkel and Kausch (1976)
	EC_{50}IM	4	9,000	STARVED FRY	Gunkel and Kausch (1976)
	EC_{50}IM	4	25,100	STARVED FRY	Gunkel and Kausch (1976)
	EC_{50}IM	5	23,200	STARVED FRY	Gunkel and Kausch (1976)
	EC_{50}IM	5	<3,000	STARVED FRY	Gunkel and Kausch (1976)
	LC_{50}	1	>44,000	STARVED FRY	Gunkel and Kausch (1976)
	LC_{50}	2	33,300	STARVED FRY	Gunkel and Kausch (1976)
	LC_{50}	2	35,500	STARVED FRY	Gunkel and Kausch (1976)
	LC_{50}	3	22,400	STARVED FRY	Gunkel and Kausch (1976)
	LC_{50}	3	29,500	STARVED FRY	Gunkel and Kausch (1976)
	LC_{50}	4	11,200	STARVED FRY	Gunkel and Kausch (1976)
	LC_{50}	4	26,300	STARVED FRY	Gunkel and Kausch (1976)
	LC_{50}	5	6,300	STARVED FRY	Gunkel and Kausch (1976)
	LC_{50}	5	24,700	STARVED FRY	Gunkel and Kausch (1976)

Table 12. Toxicology of Aquatic and Terrestrial Species: Chemical: Atrazine, CAS Number: 1912249 (continued)

Class	Species	Effect	Duration (days)	Concentration ($\mu g\ L^{-1}$)	Lifestage	Ref.
Fish (continued)	Whitefish	BCF	1	50	1 YEAR, 2.5–5.0 G, 7.5–9.0 CM	Gunkel and Streit (1980)
		BCF	1	50	1 YEAR, 2.5–5.0 G, 7.5–9.0 CM	Gunkel and Streit (1980)
		BCF	1	50	1 YEAR, 2.5–5.0 G, 7.5–9.0 CM	Gunkel and Streit (1980)
		BCF	1	50	1 YEAR, 2.5–5.0 G, 7.5–9.0 CM	Gunkel and Streit (1980)
		BCF	1	50	1 YEAR, 2.5–5.0 G, 7.5–9.0 CM	Gunkel and Streit (1980)
		BCF	1	50	1 YEAR, 2.5–5.0 G, 7.5–9.0 CM	Gunkel and Streit (1980)
		BCF	1	50	1 YEAR, 2.5–5.0 G, 7.5–9.0 CM	Gunkel and Streit (1980)
		BCF	1	50	1 YEAR, 2.5–5.0 G, 7.5–9.0 CM	Gunkel and Streit (1980)
		BCF	1	50	1 YEAR, 2.5–5.0 G, 7.5–9.0 CM	Gunkel and Streit (1980)
		BCF	1	50	1 YEAR, 2.5–5.0 G, 7.5–9.0 CM	Gunkel and Streit (1980)

Insect	Insect class	POP	112	20	BENTHIC	Dewey (1986)
	Midge (*Chironomus riparius*)	EC_{50}	2	1,000	4TH INSTAR LARVA	Johnson (1986)
	Midge (*Chironomus tentans*)	LC_{50}	2	720	1ST INSTAR	Macek et al. (1976)
		DVP	NR	230	EGG 2ND GENERATION	Macek et al. (1976)
		GRO	NR	110	EGG 2ND GENERATION	Macek et al. (1976)
	Midge (*Chironomus riparius*)	MOR	42	20	1ST INSTAR LARVAE	Huckins et al. (1986)
	Midge (*Prodiamesa olivacea*)	BCF	0.92 hr	200	3.0–51.7 MG	Streit (1978)
		BCF	NR	200	NR	Streit and Schwoerbel (1977)
	Mosquito (*Aedes aegypti*)	MOR	1	10,000	LARVAE, 3RD INSTAR	Lichtenstein et al. (1973)
Mullusca	American or virginia oyster	EC_{50}DV	2	>30,000	EMBRYO	Ward and Ballantine (1985)
		GRO	4	1,000	SMALL	Butler (1965)
	Bivalve, clam, mussel class	BCF	55	237	NR	Klaassen and Kadoum (1979)
	Cockle (*Cerastoderma edule*)	LC_{50}	2	>100,000	LARVAL	Portmann (1972)
	Pouch snail (*Physa* sp)	BCF	33	4.1	NR	Isensee (1976)
		BCF	33	42.4	NR	Isensee (1976)
		BCF	33	456.8	NR	Isensee (1976)
		BCF	33	4,710	NR	Isensee (1976)
		BCF	33	38,850	NR	Isensee (1976)
	Ramshorn snail (*Helisoma* sp)	BCF	18	3.8	NR	Kearney et al. (1977a)
	River limpet	BCF	0.7 hr	200	3.0–51.7 MG	Streit (1978)
		BCF	1	50	4 MM	Gunkel and Streit (1980)
		BCF	1	200	4 MM	Gunkel and Streit (1980)
		BCF	NR	200	NR	Streit and Schwoerbel (1977)
		BCF	NR	200	3.0–51.7 MG	Streit (1978)
		REP	40	1,000	4.35 MM	Streit and Peter (1978)
		REP	40	16,000	5.97 MM	Streit and Peter (1978)

Table 12. Toxicology of Aquatic and Terrestrial Species: Chemical: Atrazine, CAS Number: 1912249 (continued)

Class	Species	Effect	Duration (days)	Concentration ($\mu g\ L^{-1}$)	Lifestage	Ref.
Protozoa	Amoeba	PGR	6	4,000	6 D CULTURE	Prescott et al. (1977)
		PGR	6	10,000	6 D CULTURE	Prescott et al. (1977)
	Ciliate (*Tetrahymena pyriformis*)	LC_{50}	1	5.83	24 HR, LOG PHASE	Toth and Tomasovicova (1979)
		LC_{50}	7	38.4	24 HR, LOG PHASE	Toth and Tomasovicova (1979)
	Protozoan phylum	CLN	3–21	3.2	AQUATIC COMMUNITY MICROCOSM	Pratt et al. (1988)
Tracheophy	Duckweed (*Lemna minor*)	EC_{50}PS	NR	1,000	10–20 D	Beaumont et al. (1976b)
		BIO	15	1,000	NR	Beaumont et al. (1978)
		BIO	20	1,000	NR	Beaumont et al. (1978)
		BIO	NR	50	15 D	Grenier et al. (1982)
		BIO	NR	100	10 D	Grenier and Beaumont (1983)
		CEL	15	250	NR	Beaumont et al. (1980)
		CYT	15	250	NR	Beaumont et al. (1978)
		GRO	20	20–250	NR	Beaumont et al. (1976a)
		PGR	NR	<100	NR	Hartman and Martin (1985)
		PHY	15	50	NR	Grenier et al. (1979)
		PSE	3–6 hr	20–100	10 D	Beaumont et al. (1976b)
		PSE	5	20–100	20 D	Beaumont et al. (1976b)
		PSE	5–15	20–100	15 D	Beaumont et al. (1976b)
		PSE	NR	100	NR	Braginskii and Migal (1973)

Eelgrass	LC_{50}	21	100	SINGLE SHOOT	Delistraty and Hershner (1984)
	LC_{50}	21	365	SINGLE SHOOT	Delistraty and Hershner (1984)
	LC_{50}	21	367	SINGLE SHOOT	Delistraty and Hershner (1984)
	LC_{50}	21	540	SINGLE SHOOT	Delistraty and Hershner (1984)
	BIO	21	0.1	SINGLE SHOOT	Delistraty and Hershner (1984)
	ENZ	6 hr	10	SINGLE SHOOT	Delistraty and Hershner (1984)
Pondweed	EC_{50}GR	29	474	NR	Davis (1981)
(*Potamogeton perfoliatus*)	EC_{50}GR	29	474	TUBER	Forney (1979)
	EC_{50}GR	29	907	NR	Davis (1981)
	EC_{50}GR	29	907	TUBER	Forney (1979)
	GRO	22	320	TUBER	Forney (1979)
	GRO	25	100	TUBERS	Forney (1979)
	LET	29	320	TUBER	Forney (1979)
	MOR	29	100	TUBER	Forney (1979)
	PSE	4 hr	100	LEAVES	Jones and Estes (1984)
Sago pondweed	GRO	NR	100	TUBERS	Hartman and Martin (1985)
Smooth cordgrass	RSD	2	200	NR	Pillai et al. (1979)
	RSD	2	260	ROOTS	McEnerney and Davis (1979)
	RSD	2	265	ROOTS	Weete et al. (1980)

Table 12. Toxicology of Aquatic and Terrestrial Species: Chemical: Atrazine, CAS Number: 1912249 (continued)

Class	Species	Effect	Duration (days)	Concentration ($\mu g\ L^{-1}$)	Lifestage	Ref.
Tracheophy (continued)	Turtle-grass	EC_{50}PS	1.67	320	LEAVES	Walsh et al. (1982)
	Vascular plants	MOR	>60	500	NR	Walker (1964)
	Water-celery, tapegrass	EC_{50}GR	21	532	TUBER	Forney (1979)
		EC_{50}GR	32	414	TUBER	Forney (1979)
		EC_{50}GR	42	163	NR	Davis (1981)
		EC_{50}GR	42	163	TUBER	Forney (1979)
		EC_{50}GR	42	414	NR	Davis (1981)
		EC_{50}GR	42	532	NR	Davis (1981)
		EC_{50}GR	63	450	SPROUTED TUBERS	Forney (1979)
		EC_{50}GR	63	2,876	SPROUTED TUBERS	Forney (1979)
		GRO	32	320	TUBER	Forney (1979)
		GRO	63	320	SPROUTED TUBERS	Forney (1979)
	Water-milfoil	EC_{50}GR	28	1,104	NR	Davis (1981)
	(*Myriophyllum spicatum*)	EC_{50}GR	28	1,104	7 CM, TERMINAL CUTTING	Forney (1979)
	Waterweed (*Elodea canadensis*)	EC_{50}GR	21	109	7 CM, TERMINAL CUTTING	Forney (1979)
		EC_{50}GR	28	13	7 CM, TERMINAL CUTTING	Forney (1979)
		EC_{50}GR	28	75	ROOTED PLANT	Forney (1979)
		EC_{50}GR	28	80	7 CM, TERMINAL CUTTING	Forney (1979)

		$EC_{50}GR$	28	123	ROOTED PLANT	Forney (1979)
		$EC_{50}GR$	NR	80	NR	Davis (1981)
		$EC_{50}GR$	NR	109	NR	Davis (1981)
		GRO	35	320	7 CM, TERMINAL CUTTING	Forney (1979)
		GRO	36	100	7 CM, TERMINAL CUTTING	Forney (1979)
Invertebrate	Honey bee	LO MOR	NR	40 ppm		Sonnet (1978)
		NO POP	NR	140 ppm		Sonnet (1978)

Table 13. Toxicology of Aquatic and Terrestrial Species: Chemical: Bendiocarb, CAS Number: 22781233

Class	Species	Effect	Duration (days)	Concentration ($\mu g/L^{-1}$)	Lifestage	Ref.
Crustacea	Red swamp crayfish	LC_{50}	4	5,550	IMMATURE, 25–40 MM	Holck and Meek (1987)
		LC_{50}	4	5,550	IMMATURE, 25–40 MM	Holck and Meek (1987)
Insect	Mosquito (*Aedes punctor*)	LC_{50}	1	88.7	LARVAE, 4TH INSTAR	Rettich (1977)
		LC_{50}	1	88.7	LARVAE, 4TH INSTAR	Rettich (1977)
	Mosquito (*Aedes cantans*)	LC_{50}	1	175.1	LARVAE, 4TH INSTAR	Rettich (1977)
		LC_{50}	1	175.1	LARVAE, 4TH INSTAR	Rettich (1977)
	Mosquito (*Anopheles quadrimaculatus*)	LC_{50}	1	81	4TH INSTAR	Holck and Meek (1987)
		LC_{50}	1	81	4TH INSTAR	Holck and Meek (1987)
	Mosquito (*Anopheles stephensi*)	LC_{50}	1	1,700	4TH INSTAR LARVA	Scott and Georghiou (1986)
		LC_{50}	1	1,700	4TH INSTAR LARVA	Scott and Georghiou (1986)
	Mosquito (*Culex pipiens pipiens*)	LC_{50}	1	57	LARVAE, 4TH INSTAR	Rettich (1977)
		LC_{50}	1	57	LARVAE, 4TH INSTAR	Rettich (1977)
	Mosquito (*Culex pipiens molestus*)	LC_{50}	1	96.5	LARVAE, 4TH INSTAR	Rettich (1977)
		LC_{50}	1	96.5	LARVAE, 4TH INSTAR	Rettich (1977)
	Mosquito (*Culiseta annulata*)	LC_{50}	1	178.2	LARVAE, 4TH INSTAR	Rettich (1977)
		LC_{50}	1	178.2	LARVAE, 4TH INSTAR	Rettich (1977)
Mammal	Rat	LD_{50}		40–179 mg kg^{-1}		*Farm Chemicals Handbook* (1990)

Table 14. Toxicology of Aquatic and Terrestrial Species: Chemical: Benefin, CAS Number: 1861401

Class	Species	Effect	Duration (days)	Concentration (µg/L)	Lifestage	Ref.
Amphibia	Frog (*Bufo bufo japonicus*)	LC_{50}	1	11,000	NR	Nishiuchi (1980a)
		LC_{50}	1	11,000	NR	Nishiuchi (1980a)
		LC_{50}	1	11,000	NR	Nishiuchi (1980a)
		LC_{50}	1	11,000	NR	Nishiuchi (1980a)
		LC_{50}	1	11,000	NR	Nishiuchi (1980a)
		LC_{50}	1	11,000	NR	Nishiuchi (1980a)
		LC_{50}	1	11,000	NR	Nishiuchi (1980a)
		LC_{50}	1	11,000	NR	Nishiuchi (1980a)
		LC_{50}	1	11,000	NR	Nishiuchi (1980a)
		LC_{50}	1	11,000	NR	Nishiuchi (1980a)
		LC_{50}	1	11,000	NR	Nishiuchi (1980a)
		LC_{50}	1	11,000	NR	Nishiuchi (1980a)
Crustacea	Scud (*Gammarus fasciatus*)	LC_{50}	1	8,200	EARLY INSTAR	Sanders (1970a)
		LC_{50}	1	8,200	EARLY INSTAR	Sanders (1970a)
		LC_{50}	2	4,000	EARLY INSTAR	Sanders (1970a)
		LC_{50}	2	4,000	EARLY INSTAR	Sanders (1970a)
		LC_{50}	4	1,100	EARLY INSTAR	Sanders (1970a)
		LC_{50}	4	1,100	MATURE	Johnson and Finley (1980)
		LC_{50}	4	1,100	EARLY INSTAR	Sanders (1970a)
		LC_{50}	4	1,100	MATURE	Johnson and Finley (1980)
Fish	Fathead minnow	LC_{50}	4	<1,000	0.9 G	Johnson and Finley (1980)
		LC_{50}	4	<1,000	0.9 G	Johnson and Finley (1980)
	Goldfish	LC_{50}	4	800	1 G	Johnson and Finley (1980)
		LC_{50}	4	800	1 G	Johnson and Finley (1980)

Table 14. Toxicology of Aquatic and Terrestrial Species: Chemical: Benefin, CAS Number: 1861401 (continued)

Class	Species	Effect	Duration (days)	Concentration (µg/L)	Lifestage	Ref.
Tracheophy	Duckweed (*Lemna perpusilla*)	MOR	7	100,000	NR	Nishiuchi (1974)
		MOR	7	100,000	NR	Nishiuchi (1974)
		MOR	7	300,000	NR	Nishiuchi (1974)
		MOR	7	300,000	NR	Nishiuchi (1974)
Mammal	Rat	LD_{50}	NR	790 mg kg^{-1}	1 D	Goldenthal (1971)

Table 15. Toxicology of Aquatic and Terrestrial Species: Chemical: Benomyl, CAS Number: 17804352

Class	Species	Effect	Duration (days)	Concentration (µg/L)	Lifestage	Ref.
Algae	Algae, phytoplankton, algal mat	PSE	4 hr	100,000	NR	Somashekar and Sreenath (1984)
Fish	Bluegill	LC_{50}	4	850	0.9 G	Johnson and Finley (1980)
		LC_{50}	4	1,200	0.6 G	Johnson and Finley (1980)
		LC_{50}	4	1,300	FRY, 0.2 G	Palawski and Knowles (1986)
	Channel catfish	LC_{50}	4	5.6	YOLK SAC FRY	Johnson and Finley (1980)
		LC_{50}	4	6	YOLK SAC FRY	Palawski and Knowles (1986)
		LC_{50}	4	12	SWIM-UP FRY	Johnson and Finley (1980)
		LC_{50}	4	12	SWIM-UP FRY	Palawski and Knowles (1986)
		LC_{50}	4	24	FRY, 0.2 G	Palawski and Knowles (1986)
		LC_{50}	4	28	1.2 G	Johnson and Finley (1980)
		LC_{50}	4	29	1.2 G	Johnson and Finley (1980)
		LC_{50}	4	29	FINGERLING	Johnson and Finley (1980)
		LC_{50}	4	29	FINGERLING, 1.2 G	Palawski and Knowles (1986)
		LC_{50}	4	42	FINGERLING	Palawski and Knowles (1986)
		LC_{50}	4	43	FINGERLING	Palawski and Knowles (1986)
		LC_{50}	4	44	FINGERLING	Palawski and Knowles (1986)
		LC_{50}	4	47	FINGERLING	Palawski and Knowles (1986)
		LC_{50}	4	56	FINGERLING	Palawski and Knowles (1986)
		LC_{50}	4	76	FINGERLING	Palawski and Knowles (1986)
		LC_{50}	4	120	FINGERLING	Palawski and Knowles (1986)
		LC_{50}	4	720	FINGERLING	Palawski and Knowles (1986)
	Fathead minnow	LC_{50}	4	1,900	0.5 G	Johnson and Finley (1980)
		LC_{50}	4	2,200	0.9 G	Johnson and Finley (1980)

Table 15. Toxicology of Aquatic and Terrestrial Species: Chemical: Benomyl, CAS Number: 17804352 (continued)

Class	Species	Effect	Duration (days)	Concentration (μg/L)	Lifestage	Ref.
Fish (continued)	Rainbow trout, donaldson trout	LC_{50}	4	120	FRY, 0.2 G	Palawski and Knowles (1986)
		LC_{50}	4	160	SWIM-UP FRY	Palawski and Knowles (1986)
		LC_{50}	4	160	FINGERLING	Palawski and Knowles (1986)
		LC_{50}	4	170	1.2 G	Johnson and Finley (1980)
		LC50	4	170	FINGERLING, 1.2 G	Palawski and Knowles (1986)
		LC_{50}	4	170	FINGERLING	Palawski and Knowles (1986)
		LC_{50}	4	190	FINGERLING	Palawski and Knowles (1986)
		LC_{50}	4	200	FINGERLING	Palawski and Knowles (1986)
		LC_{50}	4	230	FINGERLING	Palawski and Knowles (1986)
		LC_{50}	4	280	YOLK SAC FRY	Palawski and Knowles (1986)
		LC_{50}	4	280	FINGERLING	Palawski and Knowles (1986)
		LC_{50}	4	310	1 G	Johnson and Finley (1980)
		LC_{50}	4	600	FINGERLING	Palawski and Knowles (1986)
		LC_{50}	4	880	FINGERLING	Palawski and Knowles (1986)
Invertebrate	Earthworm	LO AVO	2	0.87 μg c^{-1}		Stringer and Wright (1973)
		LO AVO	12	0.87 μg		Stringer and Wright (1976)
		LO AVO	21	0.87 μg		Stringer and Wright (1976)
		LO AVO	27	0.00%		Stringer and Wright (1973)
		LO AVO	49	560 g ha^{-1}		Keogh and Whitehead (1975)
		LO AVO	NR	0.80 μg c^{-1}	ADULT	Wright (1977)
		LO MOR	14	1.55 kg ha^{-1}		Stringer and Wright (1973)
		LO MOR	21	0.87 μg		Stringer and Wright (1976)
		LO MOR	27	0.05%		Stringer and Wright (1973)
		LO MOR	NR	0.05%	ADULT	Wright (1977)
		LO MOR	NR	1.55 kg ha^{-1}	ADULT	Wright (1977)

LO POP	21	7.80 kg ha^{-1}		Tomlin and Gore (1974)
LO POP	NR	0.28 kg ha^{-1}		Wright (1977)
LO POP	NR	0.28 kg ha^{-1}		Wright (1977)
LO POP	NR	0.28 kg ha^{-1}		Wright (1977)
LO POP	NR	0.28 kg ha^{-1}		Wright (1977)
LO POP	NR	0.28 kg ha^{-1}		Wright (1977)
LO POP	NR	0.28 kg ha^{-1}		Wright (1977)
LO POP	NR	0.28 kg ha^{-1}		Wright (1977)
LO POP	NR	0.28 kg ha^{-1}		Wright (1977)
LO POP	NR	0.28 kg ha^{-1}	YOUNG	Wright (1977)
LO POP	NR	0.28 kg ha^{-1}	YOUNG	Wright (1977)
LO POP	NR	0.28 kg ha^{-1}		Wright (1977)
LO POP	NR	0.28 kg ha^{-1}		Wright (1977)
LO POP	NR	0.28 kg ha^{-1}		Wright (1977)
LO POP	NR	0.28 kg ha^{-1}		Wright (1977)
LO POP	NR	0.28 kg ha^{-1}		Wright (1977)
LO POP	NR	0.28 kg ha^{-1}		Wright (1977)
LO POP	NR	0.28 kg ha^{-1}		Wright (1977)
LO POP	NR	0.28 kg ha^{-1}		Stringer and Wright (1973)
LO POP	370	0.025%		Cook and Swait (1975)
LO POP	825	18 g sq^{-1}		Black and Neely (1975)
LO SIGNS	27	0.05%		Stringer and Wright (1973)
NO AVO	2	0.43 μg c^{-1}		Stringer and Wright (1973)
NO AVO	2	7 μg c^{-1}		Stringer and Wright (1973)
NO AVO	12	0.43 μg		Stringer and Wright (1976)
NO MOR	27	0.005%		Stringer and Wright (1973)
NO SIGNS	27	0.005%		Stringer and Wright (1973)

Table 16. Toxicology of Aquatic and Terrestrial Species: Chemical: Bensulide, CAS Number: 741582

Class	Species	Effect	Duration (days)	Concentration (μg/L)	Lifestage	Ref.
Algae	Green algae (*Chlamydomona eugametos)*	PGR	2	400	INITIAL CONC. 100,000 CELLS ML^{-1}	Hess (1980)
		PGR	2	4,000	INITIAL CONC. 100,000 CELLS ML^{-1}	Hess (1980)
Crustacea	Scud (*Gammarus fasciatus*)	LC_{50}	1	7,600	EARLY INSTAR	Sanders (1970a)
		LC_{50}	2	3,300	EARLY INSTAR	Sanders (1970a)
		LC_{50}	4	1,400	EARLY INSTAR	Sanders (1970a)
		LC_{50}	4	1,400	MATURE	Johnson and Finley (1980)
Fish	Bluegill	LC_{50}	1	970	1.04 G	Cope (1965)
		LC_{50}	2	810	1.04 G	Cope (1965)
		LC_{50}	4	800	0.2 G	Johnson and Finley (1980)
		LC_{50}	4	810	1.04 G	Cope (1965)
	Channel catfish	LC_{50}	4	379	1 YEAR, 14 G, 12 CM, FINGERLINGS	Mccorkle et al. (1977)
	Rainbow trout, donaldson trout	LC_{50}	1	960	1.62 G	Cope (1965)
		LC_{50}	2	730	1.62 G	Cope (1965)
		LC_{50}	4	700	1.6 G	Johnson and Finley (1980)
		LC_{50}	4	720	1.62 G	Cope (1965)
Tracheophy	Duckweed (*Lemna perpusilla*)	MOR	7	100,000	NR	Nishiuchi (1974)
Mammal	Rat	LD_{50}		241–1,470 mg kg^{-1}		*Farm Chemicals Handbook* (1990)

Table 17. Toxicology of Aquatic and Terrestrial Species: Chemical: Bromoxynil, CAS Number: 1689845

Class	Species	Effect	Duration (days)	Concentration (µg/L)	Lifestage	Ref.
Algae	Diatom (*Skeletonema costatum*)	EC_{50}PP	3	720	4.7E4 CELLS ML^{-1}	Walsh et al. (1977b)
		EC_{50}PP	3	850	4.7E4 CELLS ML^{-1}	Walsh et al. (1977b)
		EC_{50}PP	3	1,000	4.7E4 CELLS ML^{-1}	Walsh et al. (1977b)
		EC_{50}PP	3	1,000	4.7E4 CELLS ML^{-1}	Walsh et al. (1977b)
		EC_{50}PP	3	1,300	4.7E4 CELLS ML^{-1}	Walsh et al. (1977b)
		EC_{50}PP	3	1,400	4.7E4 CELLS ML^{-1}	Walsh et al. (1977b)
		EC_{50}PP	3	2,200	4.7E4 CELLS ML^{-1}	Walsh et al. (1977b)
		EC_{50}PP	3	2,700	4.7E4 CELLS ML^{-1}	Walsh et al. (1977b)
		EC_{50}PP	3	3,200	4.7E4 CELLS ML^{-1}	Walsh et al. (1977b)
		EC_{50}PP	3	3,200	4.7E4 CELLS ML^{-1}	Walsh et al. (1977b)
	Diatom (*Thalassiosira* sp)	EC_{50}PP	3	1,000	4.7E4 CELLS ML^{-1}	Walsh et al. (1977b)
		EC_{50}PP	3	1,400	4.7E4 CELLS ML^{-1}	Walsh et al. (1977b)
		EC_{50}PP	3	1,400	4.7E4 CELLS ML^{-1}	Walsh et al. (1977b)
		EC_{50}PP	3	1,600	4.7E4 CELLS ML^{-1}	Walsh et al. (1977b)
		EC_{50}PP	3	1,900	4.7E4 CELLS ML^{-1}	Walsh et al. (1977b)
		EC_{50}PP	3	2,400	4.7E4 CELLS ML^{-1}	Walsh et al. (1977b)
		EC_{50}PP	3	2,900	4.7E4 CELLS ML^{-1}	Walsh et al. (1977b)
		EC_{50}PP	3	3,500	4.7E4 CELLS ML^{-1}	Walsh et al. (1977b)
		EC_{50}PP	3	3,500	4.7E4 CELLS ML^{-1}	Walsh et al. (1977b)
		EC_{50}PP	3	4,500	4.7E4 CELLS ML^{-1}	Walsh et al. (1977b)
		EC_{50}PP	3	4,600	4.7E4 CELLS ML^{-1}	Walsh et al. (1977b)
		EC_{50}PP	3	5,300	4.7E4 CELLS ML^{-1}	Walsh et al. (1977b)
	Green algae (*Chlamydomonas eugametos*)	PGR	2	28,000	INITIAL CONC. 100,000 CELLS ML^{-1}	Hess (1980)
		PGR	2	280,000	INITIAL CONC. 100,000 CELLS ML^{-1}	Hess (1980)

Table 17. Toxicology of Aquatic and Terrestrial Species: Chemical: Bromoxynil, CAS Number: 1689845 (continued)

Class	Species	Effect	Duration (days)	Concentration (µg/L)	Lifestage	Ref.
Algae (continued)	Green algae (*Chlorella* sp)	EC_{50}PP	4	2,600	4.7E4 CELLS ML^{-1}	Walsh et al. (1977b)
		EC_{50}PP	4	2,900	4.7E4 CELLS ML^{-1}	Walsh et al. (1977b)
		EC_{50}PP	4	3,200	4.7E4 CELLS ML^{-1}	Walsh et al. (1977b)
		EC_{50}PP	4	3,500	4.7E4 CELLS ML^{-1}	Walsh et al. (1977b)
		EC_{50}PP	4	3,500	4.7E4 CELLS ML^{-1}	Walsh et al. (1977b)
		EC_{50}PP	4	4,400	4.7E4 CELLS ML^{-1}	Walsh et al. (1977b)
		EC_{50}PP	4	7,800	4.7E4 CELLS ML^{-1}	Walsh et al. (1977b)
		EC_{50}PP	4	8,700	4.7E4 CELLS ML^{-1}	Walsh et al. (1977b)
		EC_{50}PP	4	9,400	4.7E4 CELLS ML^{-1}	Walsh et al. (1977b)
		EC_{50}PP	4	9,700	4.7E4 CELLS ML^{-1}	Walsh et al. (1977b)
		EC_{50}PP	4	10,000	4.7E4 CELLS ML^{-1}	Walsh et al. (1977b)
		EC_{50}PP	4	10,000	4.7E4 CELLS ML^{-1}	Walsh et al. (1977b)
Fish	Fathead minnow	LC_{50}	4	11,500	29–30 D, 28.0 MM, 0.154 G	Brooke et al. (1984)
		LC_{50}	4	13,800	33 D, 17.5 MM, 0.085 G	Geiger et al. (1988)
Mammal	Rat	LD_{50}		260–779 mg kg^{-1}		*Farm Chemicals Handbook* (1990)

Table 18. Toxicology of Aquatic and Terrestrial Species: Chemical: Carbaryl, CAS Number: 63252

Class	Species	Effect	Duration (days)	Concentration ($\mu g\ L^{-1}$)	Lifestage	Ref.
Algae	Algae, phytoplankton, algal mat	CLR	1	7,500	NR	Vickers and Boyd (1971)
		MOR	50	1,000	PHYTOPLANKTON	Hanazato and Yasuno (1987)
		PSE	1	500	NR	Vickers and Boyd (1971)
		PSE	4 hr	1,000	NR	Butler (1963)
		PSE	14	5,000	NR	Murray and Guthrie (1980)
	Blue-green algae	MOR	NR	30	NR	Bringmann and Kuhn (1978a)
	(*Anacystis aeruginosa*)	PGR	8	33	NR	Bringmann and Kuhn (1978b)
	Blue-green algae (*Anacystis nidulans*)	PGR	NR	100–100,000	NR	Kopecek et al. (1975)
	Blue-green algae (*Cyanophyta*)	ABD	14	5,000	NR	Murray and Guthrie (1980)
	Blue-green algae (*Spirulina platensis*)	PGR	NR	500–25,000	NR	Abd El-Magid (1986)
	Diatom (*Amphiprora* sp)	PGR	2	500	NR	Maly and Ruber (1983)
		PGR	2	500	NR	Maly (1980)
		PGR	2	1,000	NR	Maly and Ruber (1983)
		PGR	2	1,000	NR	Maly (1980)
	Diatom (*Amphora coffeaformis* v. *boreal*)	PGR	4	500	NR	Maly and Ruber (1983)
	Diatom (*Amphora coffeaeformis*)	PGR	4	500	NR	Maly (1980)
	Diatom (*Amphora coffeaformis* v. *boreal*)	PGR	4	1,000	NR	Maly and Ruber (1983)
	Diatom (*Amphora coffeaeformis*)	PGR	4	1,000	NR	Maly (1980)

Table 18. Toxicology of Aquatic and Terrestrial Species: Chemical: Carbaryl, CAS Number: 63252 (continued)

Class	Species	Effect	Duration (days)	Concentration ($\mu g\ L^{-1}$)	Lifestage	Ref.
Algae (continued)	Diatom (*Nitzschia angularis affinis*)	EC_{50}GR	4	1,000	EXPONENTIAL GROWTH PHASE	Walsh and Alexander (1980)
		EC_{50}GR	4	1,500	EXPONENTIAL GROWTH PHASE	Walsh and Alexander (1980)
		EC_{50}GR	12	1,500	EXPONENTIAL GROWTH PHASE	Walsh and Alexander (1980)
		EC_{50}GR	12	1,600	EXPONENTIAL GROWTH PHASE	Walsh and Alexander (1980)
	Diatom (*Phaeodactylum tricornutum*)	LET	10	10,000	250,000 CELLS ML^{-1} INITIAL CONC.	Ukeles (1962)
		MOR	10	100	250,000 CELLS ML^{-1} INITIAL CONC.	Ukeles (1962)
		PGR	2	1,000	NR	Maly and Ruber (1983)
		PGR	2	1,000	NR	Maly (1980)
	Diatom (*Skeletonema costatum*)	EC_{50}GR	4	900	EXPONENTIAL GROWTH PHASE	Walsh and Alexander (1980)
		EC_{50}GR	4	1,700	EXPONENTIAL GROWTH PHASE	Walsh and Alexander (1980)
		EC_{50}GR	12	1,600	EXPONENTIAL GROWTH PHASE	Walsh and Alexander (1980)
		EC_{50}GR	12	1,800	EXPONENTIAL GROWTH PHASE	Walsh and Alexander (1980)
	Diatom (*Thalassiosira guillardii*)	PGR	3	20,000	3E+5 CELLS ML^{-1}	Sikka and Rice (1974)
		PGR	NR	20,000	NR	Sikka et al. (1973)
		PSE	0.25 hr	20,000	NR	Sikka et al. (1973)
		PSE	0.5 hr	20,000	NR	Sikka and Rice (1974)

Diatoms, chrysophyte division	ABD	14	5,000	NR	Murray and Guthrie (1980)
Dinoflagellate (*Gonyaulax* sp)	PGR	4	10,000	NR	Maly and Ruber (1983)
	PGR	4	20,000	NR	Maly (1980)
Green algae	PGR	NR	100–100,000	NR	Kopecek et al. (1975)
(*Ankistrodesmus braunii*)	PSE	NR	100–100,000	NR	Kopecek et al. (1975)
Green algae	PGR	NR	80 μm	NR	Cain and Cain (1984)
(*Chlamydomonas moewusii*)	REP	NR	80 μm	NR	Cain and Cain (1984)
Green algae (*Chlorella* sp)	$EC_{50}GR$	4	600	EXPONENTIAL GROWTH PHASE	Walsh and Alexander (1980)
	$EC_{50}GR$	4	1,000	EXPONENTIAL GROWTH PHASE	Walsh and Alexander (1980)
Green algae (*Chlorococcum* sp)	$EC_{50}GR$	4	1,800	EXPONENTIAL GROWTH PHASE	Walsh and Alexander (1980)
	$EC_{50}GR$	4	2,100	EXPONENTIAL GROWTH PHASE	Walsh and Alexander (1980)
Green algae (*Chlorella* sp)	$EC_{50}GR$	12	1,200	EXPONENTIAL GROWTH PHASE	Walsh and Alexander (1980)
	$EC_{50}GR$	12	1,400	EXPONENTIAL GROWTH PHASE	Walsh and Alexander (1980)
Green algae (*Chlorococcum* sp)	$EC_{50}GR$	12	2,700	EXPONENTIAL GROWTH PHASE	Walsh and Alexander (1980)
	$EC_{50}GR$	12	2,900	EXPONENTIAL GROWTH PHASE	Walsh and Alexander (1980)
Green algae (*Chlorella* sp)	LET	10	10,000	150,000 CELLS ML^{-1} INITIAL CONC.	Ukeles (1962)
Green algae (*Chlorella pyrenoidosa*)	PGR	4	10	7 D, 6,000,000–7,000,000 CELLS ML^{-1}	Christie (1969)
Green algae (*Chlorococcum* sp)	PGR	4	500	NR	Maly and Ruber (1983)
	PGR	4	500	NR	Maly (1980)
	PGR	4	1,000	NR	Maly and Ruber (1983)
	PGR	4	1,000	NR	Maly (1980)

Table 18. Toxicology of Aquatic and Terrestrial Species: Chemical: Carbaryl, CAS Number: 63252 (continued)

Class	Species	Effect	Duration (days)	Concentration ($\mu g\ L^{-1}$)	Lifestage	Ref.
Algae (continued)	Green algae (*Chlorella pyrenoidosa*)	PGR	4	>100	7 D, 6,000,000–7,000,000 CELLS ML^{-1}	Christie (1969)
		PGR	7	100,000	7 D, 6,000,000–7,000,000 CELLS ML^{-1}	Christie (1969)
		PGR	7	100,000	7 D, 6,000,000–7,000,000 CELLS ML^{-1}	Christie (1969)
	Green algae (*Chlorella* sp)	PGR	10	1,000	150,000 CELLS ML^{-1} INIT CONC.	Ukeles (1962)
	Green algae (*Dunaliella tertiolecta*)	ABD	3	20,000	3E+5 CELLS ML^{-1}	Sikka and Rice (1974)
	Green algae (*Dunaliella euchlora*)	LET	10	100,000	250,000 CELLS ML^{-1} INITIAL CONC.	Ukeles (1962)
		MOR	10	10,000	150,000 CELLS ML^{-1} INITIAL CONC.	Ukeles (1962)
		PGR	10	1,000	150,000 CELLS ML^{-1} INITIAL CONC.	Ukeles (1962)
	Green algae (*Dunaliella tertiolecta*)	PGR	NR	20,000	NR	Sikka et al. (1973)
		PSE	0.25 hr	20,000	NR	Sikka et al. (1973)
		PSE	0.5 hr	20,000	NR	Sikka and Rice (1974)
	Green algae (*Oedogonium cardiacum*)	BCF	7	9	0.5–1 G	Fisher and Lohner (1986)
	Green algae (*Phytoconis* sp)	LET	10	100,000	150,000 CELLS ML^{-1} INITIAL CONC.	Ukeles (1962)
		MOR	10	10,000	150,000 CELLS ML^{-1} INITIAL CONC.	Ukeles (1962)
		PGR	10	1,000	150,000 CELLS ML^{-1} INITIAL CONC.	Ukeles (1962)

	Green algae (*Scenedesmus quadricauda*)	CLR	10	100	NR	Lejczak (1977)
		CLR	NR	600–2,900	NR	Bogacka and Groba (1980)
		CLR	NR	4,400–6,600	NR	Bogacka and Groba (1980)
		CLR	NR	50,000	NR	Bogacka and Groba (1980)
		CLR	NR	9,900–14,800	NR	Bogacka and Groba (1980)
		CLR	NR	22,200–33,300	NR	Bogacka and Groba (1980)
		MOR	NR	1,400	NR	Bringmann and Kuhn (1978a)
		PGR	8	1,350	NR	Bringmann and Kuhn (1978b)
		PGR	NR	1,400	NR	Bringmann and Kuhn (1977b)
		PSE	2	100	NR	Stadnyk et al. (1971)
		PSE	4	1,000	NR	Stadnyk et al. (1971)
		PSE	6	1,000	NR	Stadnyk et al. (1971)
	Green algae division	PGR	14	1,000	NR	Butler et al. (1975)
		PGR	14	10,000	NR	Butler et al. (1975)
	Haptophyte (*Pavlova lutheri*)	LET	10	10,000	150,000 CELLS ML^{-1} INITIAL CONC.	Ukeles (1962)
		MOR	10	1,000	150,000 CELLS ML^{-1} INITIAL CONC.	Ukeles (1962)
		PGR	10	100	150,000 CELLS ML^{-1} INITIAL CONC.	Ukeles (1962)
Amphibia	Clawed toad	$EC_{50}AB$	1	110	EMBRYO	Elliott-Feeley and Armstrong (1982)
		LC_{50}	1	4,700	EMBRYO	Elliott-Feeley and Armstrong (1982)
		ABN	1	0.01%	TADPOLE, STAGE 19–20	Rzehak et al. (1977)
		ABN-D	0.42	0.01%	TADPOLE, STAGE 19–20	Rzehak et al. (1977)
		DVP	8	0.001%	TADPOLE, STAGE 19–20	Rzehak et al. (1977)

Table 18. Toxicology of Aquatic and Terrestrial Species: Chemical: Carbaryl, CAS Number: 63252 (continued)

Class	Species	Effect	Duration (days)	Concentration ($\mu g\ L^{-1}$)	Lifestage	Ref.
Amphibia (continued)	Clawed toad	LOC	1	100	TADPOLE	Elliott-Feeley and Armstrong (1982)
		MOR-D	1	100	TADPOLE	Elliott-Feeley and Armstrong (1982)
		MOR-D	1	1,000	TADPOLE	Elliott-Feeley and Armstrong (1982)
		MOR-D	1	10,000	TADPOLE	Elliott-Feeley and Armstrong (1982)
		MOR-D	4 hr	0.05%	TADPOLE, STAGE 19–20	Rzehak et al. (1977)
	Frog (*Bufo bufo japonicus*)	LC_{50}	2	7,200	TADPOLE	Hashimoto and Nishiuchi (1981)
		MOR	10	10,000	TADPOLE, 23 MM 110 MG	Nishiuchi (1976)
	Frog (*Rana hexadactyla*)	LC_{50}	0.5	150,000	TADPOLE, 20(15–25) MM	Khangarot et al. (1985)
		LC_{50}	1	150,000	TADPOLE, 20(15–25) MM	Khangarot et al. (1985)
		LC_{50}	2	107,900	TADPOLE, 20(15–25) MM	Khangarot et al. (1985)
		LC_{50}	3	63,890	TADPOLE, 20(15–25) MM	Khangarot et al. (1985)
		LC_{50}	4	55,340	TADPOLE, 20(15–25) MM	Khangarot et al. (1985)

	Frog (*Rana temporaria*)	ABN	1	0.01%	TADPOLE, STAGE 25	Rzehak et al. (1977)
		LET	0.42–0	0.1%	TADPOLE, STAGE 25	Rzehak et al. (1977)
		LET	0.83–1	0.05%	TADPOLE, STAGE 25	Rzehak et al. (1977)
		MOR	0.5 hr	0.05%	TADPOLE, STAGE 25	Rzehak et al. (1977)
		MOR	7–10	0.001%	TADPOLE, STAGE 25	Rzehak et al. (1977)
		MOR-D	0.25 hr	0.1%	TADPOLE, STAGE 25	Rzehak et al. (1977)
	Tiger frog, indian bullfrog	LC_{50}	1	12,800	0.1 G	Marian et al. (1983)
		LC_{50}	2	8,250	0.1 G	Marian et al. (1983)
		LC_{50}	3	6,700	0.1 G	Marian et al. (1983)
		LC_{50}	4	6,200	0.02 G	Marian et al. (1983)
		LC_{50}	4	6,300	0.1 G	Marian et al. (1983)
		LC_{50}	4	11,700	1.2 G	Marian et al. (1983)
		LT_{50}	4	10,000	0.5 G	Marian et al. (1983)
		LT_{50}	4	30,000	0.5 G	Marian et al. (1983)
		DVP	NR	100	HAT TO FROGLET	Marian et al. (1983)
		FCR	NR	100	HAT TO FROGLET	Marian et al. (1983)
		FCR	NR	500	HAT TO FROGLET	Marian et al. (1983)
		LET	NR	5,000	HAT TO FROGLET	Marian et al. (1983)
		MOR	NR	100	HAT TO FROGLET	Marian et al. (1983)
		MOR	NR	500	HAT TO FROGLET	Marian et al. (1983)
		MOR	NR	1,000	HAT TO FROGLET	Marian et al. (1983)
		MOR	NR	2,000	HAT TO FROGLET	Marian et al. (1983)
Annelida	Asian leech	EC_{50}IM	1	20,000	ADULT, 4–5 CM	Kimura and Keegan (1966)
		EC_{50}IM	1	20,000	ADULT, 4–5 CM	Kimura and Keegan (1966)
		EC_{50}IM	2	5,500	ADULT, 4–5 CM	Kimura and Keegan (1966)
		EC_{50}IM	2	5,500	ADULT, 4–5 CM	Kimura and Keegan (1966)
	Lugworm	LC_{50}	2	7,200	NR	Conti (1987)
		HIS	2	8,000	NR	Conti (1987)

Table 18. Toxicology of Aquatic and Terrestrial Species: Chemical: Carbaryl, CAS Number: 63252 (continued)

Class	Species	Effect	Duration (days)	Concentration (µg L^{-1})	Lifestage	Ref.
Annelida (continued)	Oligochaete (*Branchiura sowerbyi*)	LET	3	4,000	NR	Naqvi (1973)
		LET	3	4,000	NR	Naqvi (1973)
		MOR	3	4,000	NR	Naqvi (1973)
		MOR	90	4,000	NR	Naqvi (1973)
	Oligochaete (*Lumbriculus variegatus*)	LC_{50}	7	13,000	NR	Bailey and Liu (1980)
		LC_{50}	9	8,200	NR	Bailey and Liu (1980)
	Oligochaete family	LD_{50}	1	>100,000	<200 MM, <2 MM DIAMETER	Whitten and Goodnight (1966)
		LD_{50}	2	>50,000	<200 MM, <2 MM DIAMETER	Whitten and Goodnight (1966)
		LD_{50}	4	>50,000	<200 MM, <2 MM DIAMETER	Whitten and Goodnight (1966)
	Polychaete	POP	70	11.1	LARVA	Tagatz et al. (1979)
		POP	70	103	LARVA	Tagatz et al. (1979)
	Polychaete or rag worm	ABD	1	11 kg ha^{-1}	NR	Chambers (1970)
		ABD	1	11 kg ha^{-1}	NR	Chambers (1970)
		ABD	60		NR	Chambers (1970)
		STR	NR	1.2 kg m^{-2}	NR	Chambers (1970)
	Polychaete, annelid class	ABD	30		NR	Armstrong and Millemann (1974c)
	Spionid polychaete	POP	70	1.1	LARVA	Tagatz et al. (1979)
Aquatic	Arthropod phylum	ABD	3	570 g ha	NAIADS, LARVAE, ADULTS	Jamnback and Frempong-Boadu (1966)

Invertebrate	Invertebrates	ABD	30–60	840 g ha^{-1}	NR	Courtemanch and Gibbs (1980)
	'	ABD	30–60	840 g ha^{-1}	NR	Courtemanch and Gibbs (1980)
		ABD	50	1,000	ZOOPLANKTON	Hanazato and Yasuno (1987)
		DRF	13	1.1 kg ha^{-1}	NR	Coutant (1964)
Crustacea	Aquatic sowbug	LC_{50}	1	320	MATURE	Sanders (1972)
	(*Asellus brevicaudus*)	LC_{50}	4	240	MATURE	Sanders (1972)
		LC_{50}	4	280	MATURE	Johnson and Finley (1980)
	Barnacle	POP	70	11.1	LARVA	Tagatz et al. (1979)
		POP	70	103	LARVA	Tagatz et al. (1979)
	Blue crab	EC_{50}EQ	1	550	JUVENILE	Butler (1963)
		EC_{50}EQ	2	550	JUVENILE	Butler (1963)
	Brine shrimp	HAT	2	10,000	DRIED EGG	Kuwabara et al. (1980)
	Brown shrimp	EC_{50}EQ	1	5.5	ADULT	Butler (1963)
		EC_{50}EQ	2	2.5	ADULT	Butler (1963)
	Calanoid copepod	LC_{50}	1	240	ADULT, 2.2–2.8 MM	Kader et al. (1976)
	(*Spicodiaptomus chilospinus*)	LC_{50}	2	130	ADULT, 2.2–2.8 MM	Kader et al. (1976)
	Caridean shrimp	ABD	0.25 hr		NR	Chambers (1970)
		ABD	NR	1.2 kg m^{-2}	NR	Chambers (1970)
		MOR	2.5 hr	5.7 kg ha^{-1}	NR	Armstrong and Millemann (1974c)
		STR	2 hr	120 kg ha^{-1}	NR	Tegelberg and Magoon (1970)
	Crayfish (*Orconectes nais*)	LC_{50}	1	34	3–5 WEEKS	Sanders (1972)
		LC_{50}	1	2,900	MATURE	Sanders (1972)
		LC_{50}	4	8.6	3–5 WEEKS	Sanders (1972)
		LC_{50}	4	1,000	MATURE	Sanders (1972)
	Crayfish (*Orconectes immunis*)	LC_{50}	4	2,870	3.9 G	Phipps and Holcombe (1985)
	Crayfish	LC_{50}	1	2,870	60–70 MM	Chaiyarach et al. (1975)
	(*Procambarus simulans simulans*)	LC_{50}	2	2,660	60–70 MM	Chaiyarach et al. (1975)
		LC_{50}	3	2,430	60–70 MM	Chaiyarach et al. (1975)
	Crayfish (*Procambarus* sp)	LC_{50}	4	1,900	EARLY INSTAR	Johnson and Finley (1980)

Table 18. Toxicology of Aquatic and Terrestrial Species: Chemical: Carbaryl, CAS Number: 63252 (continued)

Class	Species	Effect	Duration (days)	Concentration ($\mu g\ L^{-1}$)	Lifestage	Ref.
Crustacea (continued)	Crayfish (*Procambarus simulans simulans*)	LC_{50}	4	2,430	60–70 MM	Chaiyarach et al. (1975)
	Dungeness or edible crab	EC_{50}DV	1	6	EGGS	Buchanan et al. (1970)
		EC_{50}DV	1	20	EGGS	Buchanan et al. (1970)
		EC_{50}DV	1	30	EGGS	Buchanan et al. (1970)
		EC_{50}IM	1	76	JUVENILE, 2ND STAGE, 9.5 MM WIDE	Buchanan et al. (1970)
		EC_{50}IM	1	320	JUVENILE, 9TH STAGE, 67 MM WIDE	Buchanan et al. (1970)
		EC_{50}IM	1	320	ADULT, FEMALE, 110 MM WIDE	Buchanan et al. (1970)
		EC_{50}IM	1	350	JUVENILE, 9TH STAGE, 67 MM WIDE	Buchanan et al. (1970)
		EC_{50}IM	1	490	ADULT, FEMALE, 110 MM WIDE	Buchanan et al. (1970)
		EC_{50}IM	1 hr	4,300	JUVENILE, 2ND STAGE, 9.5 MM WIDE	Buchanan et al. (1970)
		EC_{50}IM	2	57	JUVENILE, 2ND STAGE, 9.5 MM WIDE	Buchanan et al. (1970)

EC_{50}IM	2	220	JUVENILE, 9TH STAGE, 67 MM WIDE	Buchanan et al. (1970)
EC_{50}IM	2	620	JUVENILE, 9TH STAGE, 67 MM WIDE	Buchanan et al. (1970)
EC_{50}IM	4	180	ADULT, FEMALE, 110 MM WIDE	Buchanan et al. (1970)
EC_{50}IM	4	260	ADULT, FEMALE, 110 MM WIDE	Buchanan et al. (1970)
EC_{50}IM	4	280	JUVENILE, 9TH STAGE, 67 MM WIDE	Buchanan et al. (1970)
EC_{50}IM	4	300	JUVENILE, 9TH STAGE, 67 MM WIDE	Buchanan et al. (1970)
EC_{50}SW	1	3.2	LARVAE, 1 D	Buchanan et al. (1970)
EC_{50}SW	1	3.2	LARVAE, 1 D	Buchanan et al. (1970)
EC_{50}SW	1	6.5	LARVAE, 1 D	Buchanan et al. (1970)
EC_{50}SW	1	12	LARVAE, 1 D	Buchanan et al. (1970)
EC_{50}SW	1	13	LARVAE, 1 D	Buchanan et al. (1970)
EC_{50}SW	1	15	LARVAE, 1 D	Buchanan et al. (1970)
EC_{50}SW	1	16	LARVAE, 1 D	Buchanan et al. (1970)
EC_{50}SW	1	16	LARVAE, 1 D	Buchanan et al. (1970)
EC_{50}SW	1	16	LARVAE, 1 D	Buchanan et al. (1970)
EC_{50}SW	1	16	LARVAE, 1 D	Buchanan et al. (1970)
EC_{50}SW	1	17	LARVAE, 8 D	Buchanan et al. (1970)
EC_{50}SW	1	17	LARVAE, 1 D	Buchanan et al. (1970)
EC_{50}SW	1 hr	56	LARVAE, 1 D	Buchanan et al. (1970)
EC_{50}SW	1 hr	190	LARVAE, 8 D	Buchanan et al. (1970)
EC_{50}SW	1 hr	210	LARVAE, 1 D	Buchanan et al. (1970)

Table 18. Toxicology of Aquatic and Terrestrial Species: Chemical: Carbaryl, CAS Number: 63252 (continued)

Class	Species	Effect	Duration (days)	Concentration (μg L^{-1})	Lifestage	Ref.
Crustacea (continued)	Dungeness or edible crab	EC_{50}SW	1 hr	210	LARVAE, 1 D	Buchanan et al. (1970)
		EC_{50}SW	1 hr	260	LARVAE, 1 D	Buchanan et al. (1970)
		EC_{50}SW	1 hr	560	LARVAE, 1 D	Buchanan et al. (1970)
		EC_{50}SW	1 hr	1,200	LARVAE, 1 D	Buchanan et al. (1970)
		EC_{50}SW	1 hr	1,400	LARVAE, 1 D	Buchanan et al. (1970)
		EC_{50}SW	1 hr	1,600	LARVAE, 1 D	Buchanan et al. (1970)
		EC_{50}SW	1 hr	1,600	LARVAE, 1 D	Buchanan et al. (1970)
		EC_{50}SW	1 hr	1,700	LARVAE, 1 D	Buchanan et al. (1970)
		EC_{50}SW	1 hr	1,700	LARVAE, 1 D	Buchanan et al. (1970)
		EC_{50}SW	5 hr	18	LARVAE, 1 D	Buchanan et al. (1970)
		EC_{50}SW	5 hr	24	LARVAE, 1 D	Buchanan et al. (1970)
		EC_{50}SW	5 hr	32	LARVAE, 1 D	Buchanan et al. (1970)
		EC_{50}SW	5 hr	110	LARVAE, 1 D	Buchanan et al. (1970)
		EC_{50}SW	5 hr	120	LARVAE, 1 D	Buchanan et al. (1970)
		EC_{50}SW	5 hr	130	LARVAE, 1 D	Buchanan et al. (1970)
		EC_{50}SW	5 hr	220	LARVAE, 1 D	Buchanan et al. (1970)
		EC_{50}SW	5 hr	520	LARVAE, 1 D	Buchanan et al. (1970)
		EC_{50}SW	5 hr	550	LARVAE, 1 D	Buchanan et al. (1970)
		EC_{50}SW	5 hr	730	LARVAE, 1 D	Buchanan et al. (1970)
		EC_{50}SW	5 hr	780	LARVAE, 1 D	Buchanan et al. (1970)
		LC_{50}	1	3.2	LARVAE, 1 D	Buchanan et al. (1970)
		LC_{50}	1	3.2	LARVAE, 1 D	Buchanan et al. (1970)
		LC_{50}	1	12	LARVAE, 1 D	Buchanan et al. (1970)
		LC_{50}	1	14	LARVAE, 1 D	Buchanan et al. (1970)
		LC_{50}	1	15	LARVAE, 1 D	Buchanan et al. (1970)
		LC_{50}	1	16	LARVAE, 1 D	Buchanan et al. (1970)

LC_{50}	1	16	LARVAE, 1 D	Buchanan et al. (1970)
LC_{50}	1	16	LARVAE, 1 D	Buchanan et al. (1970)
LC_{50}	1	17	LARVAE, 8 D	Buchanan et al. (1970)
LC_{50}	1	17	LARVAE, 1 D	Buchanan et al. (1970)
LC_{50}	1	19	LARVAE, 8 D	Buchanan et al. (1970)
LC_{50}	1	24	LARVAE, 8 D	Buchanan et al. (1970)
LC_{50}	1	80	LARVAE, 1–2 D	Buchanan et al. (1970)
LC_{50}	1	130	LARVAE, 1 D	Buchanan et al. (1970)
LC_{50}	1 hr	190	LARVAE, 8 D	Buchanan et al. (1970)
LC_{50}	1 hr	210	LARVAE, 1 D	Buchanan et al. (1970)
LC_{50}	1 hr	210	LARVAE, 1 D	Buchanan et al. (1970)
LC_{50}	1 hr	240	LARVAE, 8 D	Buchanan et al. (1970)
LC_{50}	1 hr	260	LARVAE, 1 D	Buchanan et al. (1970)
LC_{50}	1 hr	560	LARVAE, 1 D	Buchanan et al. (1970)
LC_{50}	1 hr	570	LARVAE, 8 D	Buchanan et al. (1970)
LC_{50}	1 hr	1,300	LARVAE, 1 D	Buchanan et al. (1970)
LC_{50}	1 hr	1,600	LARVAE, 1 D	Buchanan et al. (1970)
LC_{50}	1 hr	1,700	LARVAE, 1 D	Buchanan et al. (1970)
LC_{50}	1 hr	1,800	LARVAE, 1 D	Buchanan et al. (1970)
LC_{50}	1 hr	1,900	LARVAE, 1 D	Buchanan et al. (1970)
LC_{50}	1 hr	2,500	LARVAE, 1 D	Buchanan et al. (1970)
LC_{50}	2	5	LARVAE, 1–2 D	Buchanan et al. (1970)
LC_{50}	4	9	LARVAE, 2 D	Buchanan et al. (1970)
LC_{50}	4	10	LARVAE, 1–2 D	Buchanan et al. (1970)
LC_{50}	5 hr	24	LARVAE, 1 D	Buchanan et al. (1970)
LC_{50}	5 hr	32	LARVAE, 1 D	Buchanan et al. (1970)
LC_{50}	5 hr	110	LARVAE, 1 D	Buchanan et al. (1970)
LC_{50}	5 hr	120	LARVAE, 1 D	Buchanan et al. (1970)
LC_{50}	5 hr	180	LARVAE, 1 D	Buchanan et al. (1970)
LC_{50}	5 hr	450	LARVAE, 1 D	Buchanan et al. (1970)
LC_{50}	5 hr	620	LARVAE, 1 D	Buchanan et al. (1970)

Table 18. Toxicology of Aquatic and Terrestrial Species: Chemical: Carbaryl, CAS Number: 63252 (continued)

Class	Species	Effect	Duration (days)	Concentration ($\mu g\ L^{-1}$)	Lifestage	Ref.
Crustacea (continued)	Dungeness or edible crab	LC_{50}	5 hr	1,400	LARVAE, 1 D	Buchanan et al. (1970)
		LC_{50}	5 hr	2,100	LARVAE, 1 D	Buchanan et al. (1970)
		LC_{50}	5 hr	>10,000	LARVAE, 1 D	Buchanan et al. (1970)
		LC_{50}	20	5	LARVAE, 2 D	Buchanan et al. (1970)
		LC_{50}	25	2	LARVAE, 2 D	Buchanan et al. (1970)
		ABD	1	11 kg ha^{-1}	NR	Chambers (1970)
		ABD	1	11 kg ha^{-1}	NR	Chambers (1970)
		ABD	1	11 kg ha^{-1}	NR	Chambers (1970)
		ABD	1	11 kg ha^{-1}	NR	Chambers (1970)
		ABD	1	22 kg ha^{-1}	NR	Chambers (1970)
		GRO	1	32	JUVENILE, 3RD STAGE, 14.5 MM WIDE	Buchanan et al. (1970)
		MOR	1	5.7 kg ha^{-1}	NR	Armstrong and Millemann (1974c)
		MOR	1	120 kg ha^{-1}	85–152 MM	Tegelberg and Magoon (1970)
		MOR	1–4	120 kg ha^{-1}	10–20 CM	Tegelberg and Magoon (1970)
		MOR	1–4	120 kg ha^{-1}	20–40 CM	Tegelberg and Magoon (1970)
		MOR	1–4	120 kg ha^{-1}	40–60 CM	Tegelberg and Magoon (1970)
		MOR	1–4	120 kg ha^{-1}	60–80 CM	Tegelberg and Magoon (1970)
		MOR	1–4	120 kg ha^{-1}	80–100 CM	Tegelberg and Magoon (1970)
		MOR	1–4	120 kg ha^{-1}	100–120 CM	Tegelberg and Magoon (1970)
		MOR	1–4	120 kg ha^{-1}	120–140 CM	Tegelberg and Magoon (1970)
		MOR	1–4	120 kg ha^{-1}	140–160 CM	Tegelberg and Magoon (1970)
		MOR	4	120 kg ha^{-1}	85–152 MM	Tegelberg and Magoon (1970)
		STR	1	3,450	ADULT, 100 MM WIDE	Buchanan et al. (1970)

Ghost shrimp	ABD	1	11 kg ha^{-1}	NR	Chambers (1970)
	ABD	1	11 kg ha^{-1}	NR	Chambers (1970)
	ABD	1	11 kg ha^{-1}	NR	Chambers (1970)
	ABD	1	11 kg ha^{-1}	NR	Chambers (1970)
	ABD	1	22 kg ha^{-1}	NR	Chambers (1970)
	ABD	60		NR	Chambers (1970)
	ABD	NR	1.2 kg m^{-2}	NR	Chambers (1970)
	ABD	330	5.7 kg ha^{-1}	NR	Armstrong and Millemann (1974c)
Grass shrimp (*Palaemonetes kadiakensis*)	LC_{50}	1	42.5	NR	Naqvi and Ferguson (1970)
	LC_{50}	1	64	NR	Naqvi and Ferguson (1970)
	LC_{50}	1	120	LATE INSTAR	Sanders (1972)
	LC_{50}	1	152.5	NR	Naqvi and Ferguson (1970)
	LC_{50}	1	271.8	NR	Naqvi and Ferguson (1970)
	LC_{50}	1	410	25–31 MM	Chaiyarach et al. (1975)
	LC_{50}	2	240	25–31 MM	Chaiyarach et al. (1975)
	LC_{50}	3	140	25–31 MM	Chaiyarach et al. (1975)
	LC_{50}	4	5.6	LATE INSTAR	Sanders (1972)
	LC_{50}	4	5.6	MATURE	Johnson and Finley (1980)
	LC_{50}	4	120	25–31 MM	Chaiyarach et al. (1975)
Grass shrimp (*Palaemonetes pugio*)	AVO	0.5 hr	100	10–40 MM	Hansen et al. (1973)
Hermit crab (*Pagurus* sp)	ABD	1	11 kg ha^{-1}	NR	Chambers (1970)
	ABD	1	11 kg ha^{-1}	NR	Chambers (1970)
	ABD	1	11 kg ha^{-1}	NR	Chambers (1970)
	ABD	1	11 kg ha^{-1}	NR	Chambers (1970)
	ABD	1	22 kg ha^{-1}	NR	Chambers (1970)
Mud crab (*Upogebia pugettensis*)	ABD	1	11 kg ha^{-1}	NR	Chambers (1970)
	ABD	1	11 kg ha^{-1}	NR	Chambers (1970)
	ABD	1	11 kg ha^{-1}	NR	Chambers (1970)
	ABD	1	11 kg ha^{-1}	NR	Chambers (1970)
	ABD	1	22 kg ha^{-1}	NR	Chambers (1970)

Table 18. Toxicology of Aquatic and Terrestrial Species: Chemical: Carbaryl, CAS Number: 63252 (continued)

Class	Species	Effect	Duration (days)	Concentration (μg L^{-1})	Lifestage	Ref.
Crustacea (continued)	Mud crab (*Upogebia pugettensis*)	ABD	60		NR	Chambers (1970)
		ABD	NR	1.2 kg m^{-2}	NR	Chambers (1970)
		MOR	2 hr	120 kg ha^{-1}	NR	Tegelberg and Magoon (1970)
	Ostracod (*Cypretta kawatai*)	EC_{50}IM	1	4,450	ADULT	Hansen and Kawatski (1976)
		EC_{50}IM	2	5,280	ADULT	Hansen and Kawatski (1976)
		EC_{50}IM	3	1,800	ADULT	Hansen and Kawatski (1976)
		EC_{50}IM	4	1,800	ADULT	Hansen and Kawatski (1976)
	Ostracod (*Cypridopsis vidua*)	EC_{50}IM	2	115	MATURE	Johnson and Finley (1980)
	Prawn	LC_{50}	1	33	64 MM, 1.2 G	Omkar and Shukla (1985)
		LC_{50}	1	48.9	60–65 MM, 1.2–1.5 G	Shukla and Omkar (1984)
		LC_{50}	2	27	64 MM, 1.2 G	Omkar and Shukla (1985)
		LC_{50}	2	40.8	60–65 MM, 1.2–1.5 G	Shukla and Omkar (1984)
		LC_{50}	3	24	64 MM, 1.2 G	Omkar and Shukla (1985)
		LC_{50}	3	36.1	60–65 MM, 1.2–1.5 G	Shukla and Omkar (1984)
		LC_{50}	4	19	64 MM, 1.2 G	Omkar and Shukla (1985)
		LC_{50}	4	32.6	60–65 MM, 1.2–1.5 G	Shukla and Omkar (1984)
	Red crab	ABD	1	11 kg ha^{-1}	NR	Chambers (1970)
		ABD	1	11 kg ha^{-1}	NR	Chambers (1970)
	Red swamp crayfish	LC_{50}	1	5,000	4–10 G	Muncy and Oliver (1963)
		LC_{50}	2	3,000	4–10 G	Muncy and Oliver (1963)
		LC_{50}	3	2,000	4–10 G	Muncy and Oliver (1963)
		LC_{50}	4	1,000	MATURE, 15–38 G	Andreu-Moliner et al. (1986)
		ABD	240	0.90	MALE 10.8 G, FEMALE 9.8 G, STOCKED	Hendrick et al. (1966)

	LET	0.53	3,100	3.8 G	Naqvi (1973)
	MOR	NR	1,000	2.6 G	Naqvi (1973)
Sand shrimp	LC_{50}	2.21	20	6.4–8.3 CM, 2.4–4.5 G	McLeese et al. (1979)
Scud (*Corophium acherusicum*)	POP	70	1.1	LARVA	Tagatz et al. (1979)
Scud (*Gammarus pulex*)	LC_{50}	1	35	NR	Bluzat and Seuge (1979)
Scud (*Gammarus lacustris*)	LC_{50}	1	40	2 MONTHS OLD	Sanders (1969)
Scud (*Gammarus fasciatus*)	LC_{50}	1	50	NR	Sanders (1972)
Scud (*Gammarus lacustris*)	LC_{50}	2	22	2 MONTHS OLD	Sanders (1969)
Scud (*Gammarus pulex*)	LC_{50}	2	29	NR	Bluzat and Seuge (1979)
	LC_{50}	3	25	NR	Bluzat and Seuge (1979)
Scud (*Gammarus pseudolimnaeus*)	LC_{50}	4	7	NR	Woodward and Mauck (1980)
	LC_{50}	4	7.2	NR	Woodward and Mauck (1980)
	LC_{50}	4	13	NR	Woodward and Mauck (1980)
	LC_{50}	4	16	MATURE	Schoettger and Mauck (1978)
Scud (*Gammarus lacustris*)	LC_{50}	4	16	2 MONTHS OLD	Sanders (1969)
Scud (*Gammarus pseudolimnaeus*)	LC_{50}	4	16	MATURE	Sanders et al. (1983)
Scud (*Gammarus lacustris*)	LC_{50}	4	22	MATURE	Johnson and Finley (1980)
Scud (*Gammarus fasciatus*)	LC_{50}	4	26	NR	Sanders (1972)
	LC_{50}	4	26	MATURE	Johnson and Finley (1980)
Scud order	MOR	1 hr	840 g ha^{-1}	NR	Gibbs et al. (1984)
	MOR	2	840 g ha^{-1}	NR	Gibbs et al. (1984)
	MOR	4 hr	840 g ha^{-1}	NR	Gibbs et al. (1984)
Shore crab	ABD	0.25 hr		NR	Chambers (1970)
	ABD	NR	1.2 kg m^{-2}	NR	Chambers (1970)
Water flea (*Daphnia magna*)	EC_{50}IM	0.5 hr	12	0–24 HR	Parker et al. (1970)
	EC_{50}IM	1	0.66	2–26 HR, ADULT	Rawash et al. (1975)
Water flea (*Daphnia carinata*)	EC_{50}IM	1	100	ADULT 2–2.5 MM	Santharam et al. (1976)
Water flea (*Daphnia magna*)	EC_{50}IM	2	5.6	1ST INSTAR	Sanders et al. (1983)
Water flea (*Daphnia pulex*)	EC_{50}IM	2	6.4	1ST INSTAR	Sanders and Cope (1966)
	EC_{50}M	2	6.4	1ST INSTAR	Johnson and Finley (1980)
Water flea (*Daphnia carinata*)	EC_{50}M	2	35	ADULT 2–2.5 MM	Santharam et al. (1976)

Table 18. Toxicology of Aquatic and Terrestrial Species: Chemical: Carbaryl, CAS Number: 63252 (continued)

Class	Species	Effect	Duration (days)	Concentration ($\mu g\ L^{-1}$)	Lifestage	Ref.
Crustacea (continued)	Water flea (*Daphnia magna*)	LC_{50}	1	1	NR	Bogacka and Groba (1980)
		LC_{50}	1	1.1	<24 HR	Gaaboub et al. (1975)
		LC_{50}	1	2	NR	Bogacka and Groba (1980)
		LC_{50}	1	22,900	NR	Lejczak (1977)
		LC_{50}	2	1.25	NR	Koval'Chuk et al. (1971)
		LC_{50}	2	16,760	NR	Lejczak (1977)
		LC_{50}	3	9,360	NR	Lejczak (1977)
	Water flea (*Daphnia pulex*)	LC_{50}	3 hr	0.85	NR	Nishiuchi and Hashimoto (1975)
		LC_{50}	3 hr	29	NR	Hashimoto and Nishiuchi (1981)
		LC_{50}	3 hr	30	NR	Hashimoto and Nishiuchi (1981)
		LC_{50}	3 hr	35	NR	Hashimoto and Nishiuchi (1981)
		LC_{50}	3 hr	50	NR	Hashimoto and Nishiuchi (1981)
	Water flea (*Daphnia magna*)	LC_{50}	4	3,280	NR	Lejczak (1977)
	Water flea (*Daphnia pulex*)	RSD	1	1	NR	Tsuge et al. (1980)
	Water flea (*Moina macrocopa*)	LC_{50}	3 hr	1	NR	Nishiuchi and Hashimoto (1975)
		LC_{50}	3 hr	50	NR	Hashimoto and Nishiuchi (1981)
	Water flea	EC_{50}IM	2	7.6	1ST INSTAR	Sanders and Cope (1966)
	(*Simocephalus serrulatus*)	EC_{50}IM	2	7.6	1ST INSTAR	Johnson and Finley (1980)
	White river crayfish	LC_{50}	4	500	0.4 G	Cheah et al. (1980)
		LC_{50}	4	500	0.7 G	Carter and Graves (1972)
Fish	Agohaze, goby	LC_{50}	1	2,000	0.01–0.02 G	Hirose and Kitsukawa (1976)
		LC_{50}	1	3,700	0.01–0.02 G	Hirose and Kitsukawa (1976)
		LC_{50}	2	1,200	0.01–0.02 G	Hirose and Kitsukawa (1976)
		LC_{50}	2	1,800	0.01–0.02 G	Hirose and Kitsukawa (1976)

Atlantic salmon	LC_{50}	4	4,500	0.4 G	Johnson and Finley (1980)
	BEH	1	250–1,000	1 G, 3–4 CM, JUVENILE	Peterson (1976)
	MOR	1	1,000	1 G, 3–4 CM, JUVENILE	Peterson (1976)
Atlantic silverside	BEH	3	100	4–6 CM	Weis and Weis (1974a)
Black bullhead	LC_{50}	4	20,000	1.2 G	Johnson and Finley (1980)
	LC_{50}	4	20,000	NR	Macek and McAllister (1970)
Black crappie	LC_{50}	4	2,600	1 G	Johnson and Finley (1980)
Blennies	ABD	1	11 kg ha^{-1}	NR	Chambers (1970)
	STR	1	11 kg ha^{-1}	NR	Chambers (1970)
	STR	1	11 kg ha^{-1}	NR	Chambers (1970)
	STR	1	22 kg ha^{-1}	NR	Chambers (1970)
Bluegill	LC_{50}	1	3,400	0.87 G	Cope (1965)
	LC_{50}	1	10,100	0.5 G	Phipps and Holcombe (1985)
	LC_{50}	1	11,000	3.8–6.4 CM,1–2 G	Henderson et al.(1959)
	LC_{50}	1	>18,000	3.8–6.4 CM,1–2 G	Henderson et al.(1959)
	LC_{50}	2	2,500	0.87 G	Cope (1965)
	LC_{50}	2	9,510	0.5 G	Phipps and Holcombe (1985)
	LC_{50}	2	11,000	3.8–6.4 CM,1–2 G	Henderson et al.(1959)
	LC_{50}	2	14,000	3.8–6.4 CM,1–2 G	Henderson et al.(1959)
	LC_{50}	3	8,120	0.5 G	Phipps and Holcombe (1985)
	LC_{50}	4	1,000	NR	Sanders et al. (1983)
	LC_{50}	4	1,800	NR	Sanders et al. (1983)
	LC_{50}	4	2,000	0.87 G	Cope (1965)
	LC_{50}	4	2,200	NR	Sanders et al. (1983)
	LC_{50}	4	5,200	NR	Sanders et al. (1983)
	LC_{50}	4	5,300	3.8–6.4 CM,1–2 G	Henderson et al.(1959)
	LC_{50}	4	5,400	NR	Sanders et al. (1983)
	LC_{50}	4	5,500	3.8–6.4 CM,1–2 G	Henderson et al.(1959)
	LC_{50}	4	5,600	3.8–6.4 CM,1–2 G	Henderson et al.(1959)

Table 18. Toxicology of Aquatic and Terrestrial Species: Chemical: Carbaryl, CAS Number: 63252 (continued)

Class	Species	Effect	Duration (days)	Concentration ($\mu g\ L^{-1}$)	Lifestage	Ref.
Fish (continued)	Bluegill	LC_{50}	4	5,900	0.5 G	Carter and Graves (1972)
		LC_{50}	4	6,760	1.2 G	Johnson and Finley (1980)
		LC_{50}	4	6,760	NR	Macek and McAllister (1970)
		LC_{50}	4	6,970	0.5 G	Phipps and Holcombe (1985)
		LC_{50}	4	7,000	NR	Sanders et al. (1983)
		LC_{50}	4	8,200	NR	Sanders et al. (1983)
		LC_{50}	4	11,000	3.8–6.4 CM,1–2 G	Henderson et al.(1959)
		LC_{50}	4	16,000	NR	Sanders et al. (1983)
		LC_{50}	4	39,000	0.7 G	Johnson and Finley (1980)
		BIO	NR		NR	Hiltibran (1974)
		BIO	NR		NR	Hiltibran (1974)
		ENZ	0.42 hr	8,380	NR	Cutkomp et al. (1971b)
	Brook trout	LC_{50}	1	1,830	1.15 G	Post and Schroeder (1971)
		LC_{50}	2	1,500	1.15 G	Post and Schroeder (1971)
		LC_{50}	3	1,150	1.15 G	Post and Schroeder (1971)
		LC_{50}	3	1,640	2.04 G	Post and Schroeder (1971)
		LC_{50}	4	900	NR	Schoettger and Mauck (1978)
		LC_{50}	4	1,070	1.15 G	Post and Schroeder (1971)
		LC_{50}	4	1,100	NR	Schoettger and Mauck (1978)
		LC_{50}	4	1,450	2.04 G	Post and Schroeder (1971)
		LC_{50}	4	2,100	NR	Schoettger and Mauck (1978)
		LC_{50}	4	2,100	NR	Schoettger and Mauck (1978)
		LC_{50}	4	2,100	0.8 G	Johnson and Finley (1980)
		LC_{50}	4	2,500	NR	Schoettger and Mauck (1978)
		LC_{50}	4	3,000	NR	Schoettger and Mauck (1978)
		LC_{50}	4	3,700	NR	Schoettger and Mauck (1978)

	LC_{50}	4	4,500	1.3 G	Johnson and Finley (1980)
	LC_{50}	4	4,600	NR	Schoettger and Mauck (1978)
	LC_{50}	4	5,400	NR	Schoettger and Mauck (1978)
	ENZ	3–7	0.56 kg ha^{-1}	141 (93–184) MM	Haines (1981)
	GRO-D	7	0.27–0.79	NR	Wilder and Stanley (1983)
	MOR	10	0.56 kg ha^{-1}	YOUNG OF YEAR	Haines (1981)
	RSD	1	7–314	131 (126–139) MM, 19.7 (17.4–23.3) G	Sundaram and Szeto (1987)
Brown trout	LC_{50}	4	1,950	NR	Macek and McAllister (1970)
	LC_{50}	4	6,300	0.6 G	Johnson and Finley (1980)
Carp (*Cyprinus carpio var communis*)	EC_{50}HA	NR	1,400	EGGS	Kaur and Toor (1977)
Carp, hawk fish	LC_{50}	4	1,370	72 H LARVAE, 8 MM	Bansal et al. (1980)
	LC_{50}	4	1,940	4.5 MM, 51 MG	Verma et al. (1984a)
	LC_{50}	4	2,500	4–6 CM, 11 G	Rao et al. (1984b)
	LC_{50}	4	5,700	4–6 CM, 11 G	Rao et al. (1984b)
	LC_{50}	4	5,900	4–6 CM, 11 G	Rao et al. (1984b)
	BIO	4	2,000	4–6 CM, 11 G	Rao et al. (1984b)
Catfish	LC_{50}	3	17,500	1.10 G, JUVENILE	Arunachalam et al. (1980)
	FCR	27	5,000	1.10 G, JUVENILE	Arunachalam et al. (1980)
	GRO	27	5,000	1.10 G, JUVENILE	Arunachalam et al. (1980)
	GRO	27	7,500	1.10 G, JUVENILE	Arunachalam et al. (1980)
Catla	LC_{50}	4	1,420	72 H LARVAE, 8 MM	Bansal et al. (1980)
Channel catfish	EC_{50}EZ	0.08 hr		1–1.5 KG	Christensen and Tucker (1976)
	LC_{50}	1	2,270	6 WEEKS, 0.3 G	Brown et al. (1979)
	LC_{50}	1	6,020	6 WEEKS, 0.3 G	Brown et al. (1979)
	LC_{50}	1	6,710	6 WEEKS, 0.3 G	Brown et al. (1979)
	LC_{50}	1	11,500	10 G	Carter and Graves (1972)
	LC_{50}	1	18,600	27.6 G	Phipps and Holcombe (1985)
	LC_{50}	2	1,300	6 WEEKS, 0.3 G	Brown et al. (1979)
	LC_{50}	2	1,530	6 WEEKS, 0.3 G	Brown et al. (1979)

Table 18. Toxicology of Aquatic and Terrestrial Species: Chemical: Carbaryl, CAS Number: 63252 (continued)

Class	Species	Effect	Duration (days)	Concentration (μg L^{-1})	Lifestage	Ref.
Fish (continued)	Channel catfish	LC_{50}	2	2,000	6 WEEKS, 0.3 G	Brown et al. (1979)
		LC_{50}	2	13,600	27.6 G	Phipps and Holcombe (1985)
		LC_{50}	3	12,600	27.6 G	Phipps and Holcombe (1985)
		LC_{50}	4	140	6 WEEKS, 0.3 G	Brown et al. (1979)
		LC_{50}	4	1,300	6 WEEKS, 0.3 G	Brown et al. (1979)
		LC_{50}	4	1,560	6 WEEKS, 0.3 G	Brown et al. (1979)
		LC_{50}	4	12,400	27.6 G	Phipps and Holcombe (1985)
		LC_{50}	4	15,800	NR	Sanders et al. (1983)
		LC_{50}	4	15,800	1.5 G	Johnson and Finley (1980)
		LC_{50}	4	15,800	NR	Macek and McAllister (1970)
		RSD	56	50–250	78 G	Korn (1973)
	Chinook salmon	LC_{50}	4	2,400	FINGERLING	Johnson and Finley (1980)
		MOR	1	10,000	5–10 CM	MacPhee and Ruelle (1969)
	Climbing perch (*Anabas scandens*)	LC_{50}	1	48,500	9.76 CM, 40.6 G	Gouda et al. (1981)
		LC_{50}	4	27,000	9.76 CM, 40.6 G	Gouda et al. (1981)
	Coho salmon, silver salmon	LC_{50}	1	1,330	57–76 MM, 2.7–4.1 G	Katz (1961)
		LC_{50}	1	2,950	1.50 G	Post and Schroeder (1971)
		LC_{50}	2	997	57–76 MM, 2.7–4.1 G	Katz (1961)
		LC_{50}	2	2,700	1.50 G	Post and Schroeder (1971)
		LC_{50}	3	997	57–76 MM, 2.7–4.1 G	Katz (1961)
		LC_{50}	3	1,690	1.50 G	Post and Schroeder (1971)
		LC_{50}	4	764	NR	Macek and McAllister (1970)
		LC_{50}	4	997	57–76 MM, 2.7–4.1 G	Katz (1961)
		LC_{50}	4	1,300	1.50 G	Post and Schroeder (1971)
		LC_{50}	4	4,340	1 G	Johnson and Finley (1980)

	BCF	NR	2.6	6–18 MONTHS, 8.9–25.5 G	Walsh and Ribelin (1973)
	MOR	1	13,000	5–10 CM	MacPhee and Ruelle (1969)
Common, mirror, colored, carp	LC_{50}	1	12,000	0.3813 G, 8 WEEKS, FINGERLING	Chin and Sudderuddin (1979)
	LC_{50}	1	13,510	7–9 CM	Toor and Kaur (1974)
	LC_{50}	1	20,000	4.2 CM, 1.2 G	Nishiuchi and Asano (1981)
	LC_{50}	2	2,500	4.45 CM, 1.10 G	Li and Chen (1981)
	LC_{50}	2	10,000	4.45 CM, 1.10 G	Li and Chen (1981)
	LC_{50}	2	11,740	7–9 CM	Toor and Kaur (1974)
	LC_{50}	2	13,000	NR	Hashimoto and Nishiuchi (1981)
	LC_{50}	2	13,000	6 CM, 3 G	Nishiuchi and Asano (1981)
	LC_{50}	2	15,000	4.2 CM, 1.2 G	Nishiuchi and Asano (1981)
	LC_{50}	2	>10,000	NR	Nishiuchi and Hashimoto (1975)
	LC_{50}	3	10,360	7–9 CM	Toor and Kaur (1974)
	LC_{50}	3	13,000	4.2 CM, 1.2 G	Nishiuchi and Asano (1981)
	LC_{50}	4	1,700	0.3813 G, 8 WEEKS, FINGERLING	Chin and Sudderuddin (1979)
	LC_{50}	4	2,000	72 H LARVAE, 8 MM	Bansal et al. (1980)
	LC_{50}	4	2,000	LARVAE, 8 MM	Verma et al. (1981a)
	LC_{50}	4	5,280	0.6 G	Johnson and Finley (1980)
	LC_{50}	4	5,280	NR	Macek and McAllister (1970)
	LC_{50}	7	7,410	3.8–6 MM, 0.74–4 G	Basak and Konar (1976)
	LD_{50}	NR	>10	NR	Hashimoto and Nishiuchi (1981)
	MOR	5	150–184 mg kg^{-1}	NR	Loeb and Kelly (1963)
	RSD	3–24 hr	50–250 mg	NR	Ishii and Hasimoto (1970)

Table 18. Toxicology of Aquatic and Terrestrial Species: Chemical: Carbaryl, CAS Number: 63252 (continued)

Class	Species	Effect	Duration (days)	Concentration ($\mu g\ L^{-1}$)	Lifestage	Ref.
Fish (continued)	Cutthroat trout	LC_{50}	3	2,000	0.37 G	Post and Schroeder (1971)
		LC_{50}	4	970	NR	Woodward and Mauck (1980)
		LC_{50}	4	1,500	.037 G	Post and Schroeder (1971)
		LC_{50}	4	2,169	1.30 G	Post and Schroeder (1971)
		LC_{50}	4	3,950	NR	Woodward and Mauck (1980)
		LC_{50}	4	3,950	NR	Woodward and Mauck (1980)
		LC_{50}	4	5,000	NR	Woodward and Mauck (1980)
		LC_{50}	4	6,000	NR	Woodward and Mauck (1980)
		LC_{50}	4	6,700	NR	Woodward and Mauck (1980)
		LC_{50}	4	7,100	0.5 G	Johnson and Finley (1980)
	Electric eel	EC_{50}EZ	NR	36	NR	Basol et al. (1980)
		EC_{50}EZ	NR	50	NR	Basol et al. (1980)
	English sole	MOR	2.5 hr	5.7 kg ha^{-1}	NR	Armstrong and Millemann (1974c)
	Fathead minnow	LC_{50}	1	5,940	0.3 G	Phipps and Holcombe (1985)
		LC_{50}	1	10,700	2 MONTHS	Carlson (1971)
		LC_{50}	1	12,000	3.8–6.4 CM, 1–2 G	Henderson et al.(1959)
		LC_{50}	1	56,000	3.8–6.4 CM, 1–2 G	Henderson et al.(1959)
		LC_{50}	1	>32,000	3.8–6.4 CM, 1–2 G	Henderson et al.(1959)
		LC_{50}	2	5,010	0.3 G	Phipps and Holcombe (1985)
		LC_{50}	2	7,100	3.8–6.4 CM, 1–2 G	Henderson et al.(1959)
		LC_{50}	2	9,200	2 MONTHS	Carlson (1971)
		LC_{50}	2	20,000	3.8–6.4 CM, 1–2 G	Henderson et al.(1959)
		LC_{50}	2	42,000	3.8–6.4 CM, 1–2 G	Henderson et al.(1959)
		LC_{50}	3	5,010	0.3 G	Phipps and Holcombe (1985)
		LC_{50}	3	9,000	2 MONTHS	Carlson (1971)

	LC_{50}	4	5,010	0.3 G	Phipps and Holcombe (1985)
	LC_{50}	4	6,670	29 D, 21.9 MM, 0.169 G	Geiger et al. (1988)
	LC_{50}	4	6,700	3.8–6.4 CM, 1–2 G	Henderson et al.(1959)
	LC_{50}	4	7,000	3.8–6.4 CM, 1–2 G	Henderson et al.(1959)
	LC_{50}	4	8,930	28 D, 21.9 MM, 0.169 G	Geiger et al. (1988)
	LC_{50}	4	9,000	2 MONTHS	Carlson (1971)
	LC_{50}	4	9,470	31 D, 16.5 MM, 0.075 G	Geiger et al. (1988)
	LC_{50}	4	10,400	28 D, 17 MM, 0.073 G	Geiger et al. (1988)
	LC_{50}	4	12,000	3.8–6.4 CM, 1–2 G	Henderson et al.(1959)
	LC_{50}	4	13,000	3.8–6.4 CM, 1–2 G	Henderson et al.(1959)
	LC_{50}	4	14,600	NR	Sanders et al. (1983)
	LC_{50}	4	14,600	0.8 G	Johnson and Finley (1980)
	LC_{50}	4	14,600	NR	Macek and McAllister (1970)
	LC_{50}	4	41,000	3.8–6.4 CM, 1–2 G	Henderson et al.(1959)
	MOR	270	210	1 D—PROGENY	Carlson (1971)
	REP	270	680	1 D—PROGENY	Carlson (1971)
Flying barb	STR	NR	250	NR	Mathur (1974)
Goby family	STR	NR	1.2 kg m^{-2}	NR	Chambers (1970)
Goldfish	LC_{50}	1	23,800	14.2 G	Phipps and Holcombe (1985)
	LC_{50}	2	14,000	6.4–7.6 CM	Haynes et al. (1958)
	LC_{50}	2	21,400	14.2 G	Phipps and Holcombe (1985)
	LC_{50}	2	28,000	6.4–7.6 CM	Haynes et al. (1958)
	LC_{50}	2	>10,000	NR	Hashimoto and Nishiuchi (1981)
	LC_{50}	2	>10,000	NR	Nishiuchi and Hashimoto (1975)
	LC_{50}	3	19,300	14.2 G	Phipps and Holcombe (1985)
	LC_{50}	4	13,200	0.9 G	Johnson and Finley (1980)
	LC_{50}	4	13,200	NR	Macek and McAllister (1970)
	LC_{50}	4	16,700	14.2 G	Phipps and Holcombe (1985)
	LET	2	35,000	6.4–7.6 CM	Haynes et al. (1958)
	MOR	2	7,500	6.4–7.6 CM	Haynes et al. (1958)
	MOR	2	15,000	6.4–7.6 CM	Haynes et al. (1958)

Table 18. Toxicology of Aquatic and Terrestrial Species: Chemical: Carbaryl, CAS Number: 63252 (continued)

Class	Species	Effect	Duration (days)	Concentration (μg L^{-1})	Lifestage	Ref.
Fish (continued)	Goldfish	MOR	2	25,000	6.4–7.6 CM	Haynes et al. (1958)
		MOR	2	30,000	6.4–7.6 CM	Haynes et al. (1958)
		MOR	10	1,000	NR	Shea and Berry (1983)
		MOR	10	5,000	NR	Shea and Berry (1983)
		MOR	10	10,000	NR	Shea and Berry (1983)
		VTE	7	5,000	NR	Imada (1976)
		VTE	7	10,000	NR	Imada (1976)
	Green fish	LC_{50}	1	1,800	0.15–0.24 G	Hirose and Kitsukawa (1976)
		LC_{50}	2	900	0.15–0.24 G	Hirose and Kitsukawa (1976)
	Green sunfish	LC_{50}	4	11,200	1.1 G	Johnson and Finley (1980)
	Guppy	EC_{50}IM	0.5	16,900	NR	Lejczak (1977)
		EC_{50}IM	1	12,500	NR	Lejczak (1977)
		EC_{50}IM	2	10,100	NR	Lejczak (1977)
		EC_{50}IM	3	9,740	NR	Lejczak (1977)
		EC_{50}IM	4	9,740	NR	Lejczak (1977)
		LC_{50}	0.5	39,200	NR	Lejczak (1977)
		LC_{50}	1	2,600	7 WEEKS	Chen et al. (1971)
		LC_{50}	1	3,600	7 WEEKS	Chen et al. (1971)
		LC_{50}	1	8,200	NR	Bogacka and Groba (1980)
		LC_{50}	1	17,700	NR	Lejczak (1977)
		LC_{50}	2	11,200	NR	Lejczak (1977)
		LC_{50}	3	10,300	NR	Lejczak (1977)
		LC_{50}	4	4,700	NR	Bogacka and Groba (1980)
		LC_{50}	4	9,740	NR	Lejczak (1977)
		LT_{50}	4	6,200	NR	Lejczak (1977)

	MOR	0.5 hr	7,000	7 WEEKS	Chen et al. (1971)
	RSD	1	1	NR	Tsuge et al. (1980)
Ide, silver or golden orfe	LC_{50}	NR	20,000	NR	Juhnke and Luedemann (1978)
Indian catfish	LC_{50}	1	22,950	23 G	Singh et al. (1984)
	LC_{50}	1	23,010	50–75 MM, 5–10 G	Verma et al. (1982a)
	LC_{50}	1	64,000	12.30 CM, 10.6 G	Gouda et al. (1981)
	LC_{50}	2	20,800	50–75 MM, 5–10 G	Verma et al. (1982a)
	LC_{50}	2	22,300	23 G	Singh et al. (1984)
	LC_{50}	3	19,790	50–75 MM, 5–10 G	Verma et al. (1982a)
	LC_{50}	3	21,450	23 G	Singh et al. (1984)
	LC_{50}	4	19,580	50–75 MM, 5–10 G	Verma et al. (1982a)
	LC_{50}	4	20,100	23 G	Singh et al. (1984)
	LC_{50}	4	50,000	12.30 CM, 10.6 G	Gouda et al. (1981)
	LC_{50}	7	21,230	81–131 MM, 4.2–12.4 G	Basak and Konar (1976)
	LET	4 hr	50,000	NR	Saxena and Aggarwal (1970)
	MOR	4 hr	40,000	NR	Saxena and Aggarwal (1970)
	RSD	NR	8,420	128 MM, 35 G	Verma et al. (1977)
	STR	NR	250	NR	Mathur (1974)
	LC_{50}	1	4,207	25–40 MM STD LENGTH	Jacob et al. (1982)
Killifish	LC_{50}	2	3,747	25–40 MM STD LENGTH	Jacob et al. (1982)
Lake trout, siscowet	LC_{50}	4	690	1.7 G	Johnson and Finley (1980)
	BCF	NR	2.6	6–18 MONTHS, 8.9–25.5 G	Walsh and Ribelin (1973)
Largemouth bass	LC_{50}	4	6,400	0.9 G	Johnson and Finley (1980)
	LC_{50}	4	6,400	NR	Macek and McAllister (1970)
	RES	0.33	9,840	102.7 G, 20.7 CM	Morgan (1975)
	RES	1.08	810	102.7 G, 20.7 CM	Morgan (1975)
	RES	35	90	102.7 G, 20.7 CM	Morgan (1975)

Table 18. Toxicology of Aquatic and Terrestrial Species: Chemical: Carbaryl, CAS Number: 63252 (continued)

Class	Species	Effect	Duration (days)	Concentration ($\mu g\ L^{-1}$)	Lifestage	Ref.
Fish (continued)	Longnose killifish	LC_{50}	1	1,750	JUVENILE	Butler (1963)
		LC_{50}	2	1,750	JUVENILE	Butler (1963)
	Medaka, high-eyes	LC_{50}	2	2,800	NR	Hashimoto and Nishiuchi (1981)
		LC_{50}	2	10,000	NR	Nishiuchi and Hashimoto (1975)
		AVO	NR	246	NR	Hidaka et al. (1984)
		TER	2	5,000	NEWLY HATCHED FRY	Solomon (1977)
		TER	NR	500–30,000	NEWLY FERTILIZED EGGS	Solomon (1977)
		VTE	7	2,000	NR	Imada (1976)
		VTE	7	4,000	NR	Imada (1976)
		VTE	7	6,000	NR	Imada (1976)
	Mosquitofish (*Gambusia affinis*)	EC_{50}IR	1	10,000	NR	Hansen et al. (1972)
		LC_{50}	1	1,300	MALE, 15–30 G	Krieger and Lee (1973)
		LC_{50}	1	40,000	30–40 MM	Chaiyarach et al. (1975)
	Mosquitofish (*Gambusia patruelis*)	LC_{50}	2	736	3.28(3–4) CM, 0.36(0.31–0.51) G	Li and Chen (1981)
	Mosquitofish (*Gambusia affinis*)	LC_{50}	2	35,000	30–40 MM	Chaiyarach et al. (1975)
		LC_{50}	3	32,400	30–40 MM	Chaiyarach et al. (1975)
		LC_{50}	4	1,400	0.5 G	Carter and Graves (1972)
		LC_{50}	4	31,800	30–40 MM	Chaiyarach et al. (1975)
		LC_{50}	4	204,000	ADULT, 0.289 G, 2.76 CM	Naqvi and Hawkins (1988)
		AVO	1 hr	10,000	20–45 MM	Hansen et al. (1972)

	BCF	7	9	3RD TO 4TH INSTAR LARVAE	Fisher and Lohner (1986)
	ENZ	0.17 hr	20,000	NR	Whitmore and Hodges (1978)
Motsuga, stone moroko	BCF	1	575	4–8 CM, 2–5 G	Kanazawa (1975)
	BCF	1–14	5–20	4–8 CM, 2–5 G	Kanazawa (1981)
	ENZ	0.2 hr		4–8 CM, 1–5 G	Kanazawa (1983)
	ENZ	1	1,000	4–8 CM, 1–5 G	Kanazawa (1983)
Mozambique tilapia	LC_{50}	1	13,000	NR	Koundinya and Ramamurthi (1980)
	LC_{50}	2	5,495	NR	Basha et al. (1984)
	LC_{50}	2	5,495	NR	Basha et al. (1983)
	LC_{50}	2	10,000	10 G	Koundinya and Ramamurthi (1979a)
	LC_{50}	2	10,000	NR	Koundinya and Ramamurthi (1980)
	LC_{50}	2	10,000	10 G, 13–18 CM	Koundinya and Ramamurthi (1979b)
	LC_{50}	3	8,000	NR	Koundinya and Ramamurthi (1980)
	LC_{50}	7	3,210	39–57 MM, 1.1–2.7 G	Basak and Konar (1976)
	BIO	NR	3,000	NR	Ramaswamy (1988)
	ENZ	NR	10,000	10 G, 13–18 CM	Koundinya and Ramamurthi (1979b)
	HEM	2	10,000	10 G, 12.7–17.8 CM	Koundinya and Murthi (1979)
	OC	1.5	1,832	NR	Basha et al. (1984)
	PHY	2	1,832	NR	Basha et al. (1984)

Table 18. Toxicology of Aquatic and Terrestrial Species: Chemical: Carbaryl, CAS Number: 63252 (continued)

Class	Species	Effect	Duration (days)	Concentration (μg L^{-1})	Lifestage	Ref.
Fish (continued)	Mummichog	DVP	NR	1,000	EMBRYO, HIGH BLASTULA, 8–16 CELLS	Weis and Weis (1974b)
		DVP	NR	10,000	EMBRYO	Weis and Weis (1974b)
		MOR	10	10,000	NR	Shea and Berry (1983)
		MOR	10	50,000	NR	Shea and Berry (1983)
		RGN	7	10	4–5 CM	Weis and Weis (1975)
	Northern squawfish	MOR	1	10,000	5–10 CM	MacPhee and Ruelle (1969)
	Oriental weatherfish	LC_{50}	2	13,000	NR	Hashimoto and Nishiuchi (1981)
	Paradise fish	LC_{50}	1	14,730	20–28 MM STD LENGTH	Jacob et al. (1982)
		LC_{50}	2	13,910	20–28 MM STD LENGTH	Jacob et al. (1982)
		LC_{50}	4	3,500	JUVENILE, 305 MG	Arunachalam and Palanichamy (1982)
		GRO	26	1,000	JUVENILE, 305 MG	Arunachalam and Palanichamy (1982)
	Pinfish	ENZ	0.63	1,333	46–102 MM	Coppage (1977)
	Rainbow trout, donaldson trout	LC_0	2	30,000	2 YEARS	Lysak and Marcinek (1972)
		LC_{100}	2	100,000	2 YEARS	Lysak and Marcinek (1972)
		LC_{50}	1	860	19.7 G	Phipps and Holcombe (1985)
		LC_{50}	1	1,410	30–80 G, 14–21 CM	Zinkl et al. (1987b)
		LC_{50}	1	2,300	10 CM, 15 G	Smith and Grigoropoulos (1968)
		LC_{50}	2	860	19.7 G	Phipps and Holcombe (1985)
		LC_{50}	2	1,500	10 CM, 15 G	Smith and Grigoropoulos (1968)
		LC_{50}	2	1,600	51–79 MM, 3.2 G	Katz (1961)

LC_{50}	3	860	19.7 G	Phipps and Holcombe (1985)
LC_{50}	3	1,350	51–79 MM, 3.2 G	Katz (1961)
LC_{50}	4	800	NR	Sanders et al. (1983)
LC_{50}	4	800	NR	Sanders et al. (1983)
LC_{50}	4	860	19.7 G	Phipps and Holcombe (1985)
LC_{50}	4	900	NR	Sanders et al. (1983)
LC_{50}	4	935	FRY, 1 G	Marking et al. (1984)
LC_{50}	4	1,000	10 CM, 15 G	Smith and Grigoropoulos (1968)
LC_{50}	4	1,000	FRY, 1 G	Marking et al. (1984)
LC_{50}	4	1,000	FRY, 1 G	Marking et al. (1984)
LC_{50}	4	1,100	NR	Sanders et al. (1983)
LC_{50}	4	1,350	51–79 MM, 3.2 G	Katz (1961)
LC_{50}	4	1,400	FRY, 1 G	Marking et al. (1984)
LC_{50}	4	1,470	1.24 G	Post and Schroeder (1971)
LC_{50}	4	1,500	NR	Sanders et al. (1983)
LC_{50}	4	1,740	FRY, 1 G	Marking et al. (1984)
LC_{50}	4	1,950	1.5 G	Johnson and Finley (1980)
LC_{50}	4	2,200	NR	Sanders et al. (1983)
LC_{50}	4	2,200	NR	Sanders et al. (1983)
LC_{50}	4	2,800	NR	Sanders et al. (1983)
LC_{50}	4	2,830	100 G	McKim et al. (1987)
LC_{50}	4	4,330	JUVENILE	Douglas et al. (1986)
LC_{50}	4	4,340	NR	Macek and McAllister (1970)
LC_{50}	4	5,400	JUVENILE	Douglas et al. (1986)
LC_{50}	5	700	10 CM, 15 G	Smith and Grigoropoulos (1968)
LC_{50}	NR	1,400	NR	Meyer (1981)
LC_{50}	NR	4,000	NR	Meyer (1981)
LT_{50}	0.57	5,167	600–900 G	McKim et al. (1987)
BIO	1	250	30–80 G, 14–21 CM	Zinkl et al. (1987b)
ENZ	2 hr		31–41 CM	Smith and Grigoropoulos (1968)
HEM	7–14	3,200	13–15 CM, 30 G	Groba and Trzcinska (1979)

Table 18. Toxicology of Aquatic and Terrestrial Species: Chemical: Carbaryl, CAS Number: 63252 (continued)

Class	Species	Effect	Duration (days)	Concentration ($\mu g\ L^{-1}$)	Lifestage	Ref.
Fish (continued)	Rainbow trout, donaldson trout	LET	4.25 hr	1,000	150–250 G	Van Hoof (1980)
		MOR	1–5 hr	2,000	JUVENILE, 1–1.5 YEARS	Lunn et al. (1976)
		MOR	2	1,000	FINGERLING	Statham and Lech (1975b)
		MOR	2	1,000	7.6–10.2 CM, FINGERLING	Statham and Lech (1975a)
		PHY	5 hr	2,000	JUVENILE, 1–1.5 YEARS 120 G, 22.8 CM	Lunn et al. (1976)
		PHY	NR	5,167	600–900 G	McKim et al. (1987)
		RES	5 hr	1,000	JUVENILE, 1–1.5 YEARS 120 G, 22.8 CM	Lunn et al. (1976)
		RES	5 hr	2,000	JUVENILE, 1–1.5 YEARS 120 G, 22.8 CM	Lunn et al. (1976)
	Red shiner	LC_{50}	1	13,000	4.6 CM, 1.4 G	Smith and Grigoropoulos (1968)
		LC_{50}	2	12,000	4.6 CM, 1.4 G	Smith and Grigoropoulos (1968)
		LC_{50}	4	9,200	4.6 CM, 1.4 G	Smith and Grigoropoulos (1968)
		LC_{50}	5	7,400	4.6 CM, 1.4 G	Smith and Grigoropoulos (1968)

Redear sunfish	LC_{50}	4	11,200	NR	Macek and McAllister (1970)
Rohu	LC_{50}	4	1,870	72 H LARVAE, 8 MM	Bansal et al. (1980)
	LC_{50}	4	4,600	1–2.5 CM, 0.5 G	Tilak et al. (1980)
	LC_{50}	4	7,750	4–6 CM, 4.5 G	Tilak et al. (1980)
	OC	4	660	15–20 G, 6.2–7.5 CM	Bansal et al. (1979)
	RSD	NR	8,420	110 MM, 20 G	Verma et al. (1977)
Rosy or red barb	LC_{50}	4	2,142	>2 YEARS, 5.66 CM TOTAL LENGTH	Pant and Singh (1983)
	HIS	15	194	5.66 CM	Gill et al. (1988)
	PHY	0.5	2142	>2 YEARS, 5.66 CM TOTAL LENGTH	Pant and Singh (1983)
	PHY	15	194	>2 YEARS, 5.66 CM TOTAL LENGTH	Pant and Singh (1983)
Sheepshead minnow	AVO	0.5 hr	10,000	20–40 MM	Hansen (1969)
	AVO	NR	10,000	20–40 MM	Hansen (1970)
Slender rasbora	STR	NR	250	NR	Mathur (1974)
Slimy sculpin	RSD	1	7–314	70(65–82) MM, 6.1(5.8–6.6) G	Sundaram and Szeto (1987)
Snake-head catfish (*Channa punctatus*)	LC_{50}	1	21,200	24.5 G	Singh et al. (1984)
	LC_{50}	2	8,710	20 G	Rao et al. (1985a)
	LC_{50}	2	8,710	ADULT, 20 G	Rao et al. (1985b)
	LC_{50}	2	20,530	24.5 G	Singh et al. (1984)
	LC_{50}	3	20,020	24.5 G	Singh et al. (1984)
	LC_{50}	4	6,000	2.1 G	Arunachalam et al. (1985)
	LC_{50}	4	19,500	24.5 G	Singh et al. (1984)
	BIO	1 hr	100	14 CM, 50 G	Sastry and Siddiqui (1985)
	BIO	1 hr	1,000	14 CM, 50 G	Sastry and Siddiqui (1985)
	BIO	15	1,050	14–16 CM, 50–60 G	Sastry and Siddiqui (1982)
	ENZ	1 hr	10,000	14 CM, 50 G	Sastry and Siddiqui (1985)

Table 18. Toxicology of Aquatic and Terrestrial Species: Chemical: Carbaryl, CAS Number: 63252 (continued)

Class	Species	Effect	Duration (days)	Concentration (μg L^{-1})	Lifestage	Ref.
Fish (continued)	Snake-head catfish (*Channa striatus*)	ENZ	8	20,000	14 CM	Arora and Kulshrestha (1985)
	Snake-head catfish (*Channa punctatus*)	ENZ	15	1,050	14–16 CM, 50–60 G	Sastry and Siddiqui (1982)
	Snake-head catfish (*Channa striatus*)	ENZ	15	10,000	14 CM	Arora and Kulshrestha (1985)
		HIS	2	10,000	14 CM	Jauhar and Kulshrestha (1985)
		HIS	2	10,000	14 CM	Kulshrestha and Arora (1984b)
		HIS	2	10,000	14 CM, 25 G	Arora and Kulshrestha (1984)
		HIS	2	10,000	15 CM	Kulshrestha and Jauhar (1986)
	Snake-head catfish (*Channa punctatus*)	MOR	60	1,050	14–16 CM, 50–60 G	Sastry and Siddiqui (1982)
		PHY	2	3,000	20 G	Rao et al. (1985a)
		PHY	2	3,000	NR	Rao (1987)
	Snake-head catfish (*Channa striatus*)	PHY	2	10,000	15 CM	Kulshrestha and Arora (1984a)
	Snake-head catfish (*Channa punctatus*)	PHY	NR	5,000	2.1 G	Arunachalam et al. (1985)
		PHY	150	2,000	15–20 CM	Saxena and Garg (1978)
		STR	NR	250	NR	Mathur (1974)
	Spot	MOR	150	100	18 MM	Lowe (1967)
	Striped bass	LC_{50}	4	760	63 D	Palawski et al. (1985)
		LC_{50}	4	1,000	0.42 G JUVENILE	Korn and Earnest (1974)
		LC_{50}	4	2,300	63 D	Palawski et al. (1985)

Three spine stickleback	LC_{50}	2	10,450	22–44 MM, 0.38–0.77 G	Katz (1961)
	LC_{50}	2	16,625	22–44 MM, 0.38–0.77 G	Katz (1961)
	LC_{50}	3	4,940	22–44 MM, 0.38–0.77 G	Katz (1961)
	LC_{50}	3	6,175	22–44 MM, 0.38–0.77 G	Katz (1961)
	LC_{50}	4	399	22–44 MM, 0.38–0.77 G	Katz (1961)
	LC_{50}	4	3,990	22–44 MM, 0.38–0.77 G	Katz (1961)
Tiger fish, crescent perch	LC_{50}	4	2,200	6.9–9.2 CM, 6.5–13.1 G, 6–9 MONTHS	Lingaraja and Venugopalan (1978)
Tilapia (*Tilapia* sp)	LC_{50}	2	1,958.5	4.99(4.6–5.38) CM, 1.46(1.23–2.04) G	Li and Chen (1981)
Two spotted, tic tac toe barb	LC_{50}	1	4,200	50–68 MM, 1.560–5.125 G	Bhatia (1971)
	LC_{50}	2	3,900	50–68 MM, 1.560–5.125 G	Bhatia (1971)
	LC_{50}	3	3,800	50–68 MM, 1.560–5.125 G	Bhatia (1971)
	LC_{50}	4	3,700	50–68 MM, 1.560–5.125 G	Bhatia (1971)
Walking catfish	LC_{50}	1	16,260	65–70 G, 17–18 CM	Tripathi and Shukla (1988)
	LC_{50}	1	61,140	65–70 G, 17–18 CM	Tripathi and Shukla (1988)
	LC_{50}	2	53,650	65–70 G, 17–18 CM	Tripathi and Shukla (1988)
	LC_{50}	2	134,080	65–70 G, 17–18 CM	Tripathi and Shukla (1988)
	LC_{50}	3	48,580	65–70 G, 17–18 CM	Tripathi and Shukla (1988)
	LC_{50}	3	123,360	65–70 G, 17–18 CM	Tripathi and Shukla (1988)
	LC_{50}	4	46,850	65–70 G, 17–18 CM	Tripathi and Shukla (1988)
	LC_{50}	4	107,660	65–70 G, 17–18 CM	Tripathi and Shukla (1988)
White mullet	LC_{50}	1	4,250	JUVENILE	Butler (1963)
	LC_{50}	2	2,500	JUVENILE	Butler (1963)
Yellow perch	LC_{50}	4	745	NR	Macek and McAllister (1970)
	LC_{50}	4	5,100	0.6 G	Johnson and Finley (1980)
Yellowtail	LC_{50}	1	3,500	0.01–0.02 G	Hirose and Kitsukawa (1976)

Table 18. Toxicology of Aquatic and Terrestrial Species: Chemical: Carbaryl, CAS Number: 63252 (continued)

Class	Species	Effect	Duration (days)	Concentration (μg L^{-1})	Lifestage	Ref.
Insect	Alderfly	LC_{50}	1	650	LARVAE	Bluzat and Seuge (1979)
		LC_{50}	3	200	LARVAE	Bluzat and Seuge (1979)
	Backswimmer (*Notonecta undulata*)	LC_{50}	2	360	ADULT, 66 MG	Federle and Collins (1976)
		LC_{50}	3	230	ADULT, 66 MG	Federle and Collins (1976)
		LC_{50}	4	200	ADULT, 66 MG	Federle and Collins (1976)
	Beetle (*Eretes sticticus*)	LC_{50}	1	910	NR	Jeyasingam et al. (1978)
		LC_{50}	2	890	NR	Jeyasingam et al. (1978)
	Beetle (*Peltodytes* sp)	LC_{50}	1	100,000	ADULT, 5 MG	Federle and Collins (1976)
		LC_{50}	2	25,000	ADULT, 5 MG	Federle and Collins (1976)
		LC_{50}	3	6,000	ADULT, 5 MG	Federle and Collins (1976)
		LC_{50}	4	3,300	ADULT, 5 MG	Federle and Collins (1976)
	Blackfly (*Prosimulium magnum*)	ABD	NR	500	LARVAE	Jamnback and Frempong-Boadu (1966)
	Blackfly (*Simulium venustum*)	ABD	3 hr	570 g ha^{-1}	LARVAE	Jamnback and Frempong-Boadu (1966)
		ABD	NR	1,350	LARVAE	Jamnback and Frempong-Boadu (1966)
		ABD	NR	2,700	LARVAE	Jamnback and Frempong-Boadu (1966)
	Blackfly family	$EC_{50}DT$	1	106	MEDIUM TO LARGE LARVAE	Frempong-Boadu (1966)
		DRF	0.08 hr	40	LARVAE, 3RD–6TH INSTAR	Jamnback and Frempong-Boadu (1966)
		DRF	0.08 hr	400	LARVAE, 3RD–6TH INSTAR	Jamnback and Frempong-Boadu (1966)

	DRF	0.08 hr	400	LARVAE, 3RD–6TH INSTAR	Jamnback and Frempong-Boadu (1966)
	DRF	0.08 hr	4,000	LARVAE, 3RD–6TH INSTAR	Jamnback and Frempong-Boadu (1966)
	DRF	0.08 hr	4,000	LARVAE, 3RD–6TH INSTAR	Jamnback and Frempong-Boadu (1966)
	DRF	0.08 hr	4,000	LARVAE, 3RD–6TH INSTAR	Jamnback and Frempong-Boadu (1966)
Caddisfly (*Pycnopsyche* sp)	BEH	NR	840 g ha^{-1}	NR	Courtemanch and Gibbs (1980)
Caddisfly order	ABD	30–60	840 g ha^{-1}	NR	Courtemanch and Gibbs (1980)
	ABD	30–60	840 g ha^{-1}	NR	Courtemanch and Gibbs (1980)
	MOR	1	840 g ha^{-1}	NR	Gibbs et al. (1984)
	MOR	4 hr	840 g ha^{-1}	NR	Gibbs et al. (1984)
Dragonfly	LC_{50}	1	550	NR	Nishiuchi and Asano (1978)
(*Orthetrum albistylum speciosum*)	LC_{50}	1 hr	5,500	NR	Nishiuchi and Asano (1978)
	LC_{50}	2	430	LARVAE	Hashimoto and Nishiuchi (1981)
	LC_{50}	2	430	NR	Nishiuchi and Asano (1978)
	LC_{50}	2 hr	3,700	NR	Nishiuchi and Asano (1978)
	LC_{50}	3 hr	3,000	NR	Nishiuchi and Asano (1978)
	LC_{50}	6 hr	550	NR	Nishiuchi and Asano (1978)
Fly, mosquito, midge order	DRF	2	840 g ha^{-1}	NR	Courtemanch and Gibbs (1980)
Insect class	DRF	2	840 g ha^{-1}	NR	Courtemanch and Gibbs (1980)
	MOR	NR	840 g ha^{-1}	NR	Courtemanch and Gibbs (1980)
	POP	NR	840 g ha^{-1}	NR	Gibbs et al. (1984)
Mayfly (*Cloeon dipterum*)	LC_{50}	2	370	LARVAE	Hashimoto and Nishiuchi (1981)
Mayfly (*Cloeon* sp)	LC_{50}	2	480	LARVAE	Bluzat and Seuge (1979)
	LC_{50}	3	390	LARVAE	Bluzat and Seuge (1979)
Mayfly (*Ephemerella* sp)	ABD	30–60	840 g ha^{-1}	NR	Courtemanch and Gibbs (1980)
Mayfly order	ABD	30–60	840 g ha^{-1}	NR	Courtemanch and Gibbs (1980)
	ABD	30–60	840 g ha^{-1}	NR	Courtemanch and Gibbs (1980)

Table 18. Toxicology of Aquatic and Terrestrial Species: Chemical: Carbaryl, CAS Number: 63252 (continued)

Class	Species	Effect	Duration (days)	Concentration (µg L^{-1})	Lifestage	Ref.
Insect (continued)	Midge (*Chironomus tentans*)	EC_{50}IM	1	1.6	3RD–4TH INSTAR	Karnak and Collins (1974)
		EC_{50}IM	1	4.2	3RD–4TH INSTAR	Karnak and Collins (1974)
		EC_{50}IM	1	7.0	3RD–4TH INSTAR	Karnak and Collins (1974)
	Midge (*Chironomus riparius*)	EC_{50}IM	1	104.5	4TH INSTAR	Estenik and Collins (1979)
	Midge (*Chironomus tentans*)	EC_{50}IM	1	>18,000	LARVAE, 4TH INSTAR	Hansen and Kawatski (1976)
	Midge (*Chironomus plumosus*)	EC_{50}IM	2	10	4TH INSTAR LARVAE	Sanders et al. (1983)
	Midge (*Chironomus tentans*)	EC_{50}IM	2	>18,000	LARVAE, 4TH INSTAR	Hansen and Kawatski (1976)
		EC_{50}IM	3	12,000	LARVAE, 4TH INSTAR	Hansen and Kawatski (1976)
		EC_{50}IM	4	5,900	LARVAE, 4TH INSTAR	Hansen and Kawatski (1976)
	Midge (*Chironomus riparius*)	LC_{50}	1	127	4TH INSTAR LARVAE	Fisher and Lohner (1986)
		BCF	7	9	3RD TO 4TH INSTAR LARVAE	Fisher and Lohner (1986)
	Midge (*Forcipomyia* sp)	MOR	1	16 kg ha^{-1}	NR	Soria and Abreu (1974)
		MOR	1	16 kg ha^{-1}	NR	Soria and Abreu (1974)
	Midge family	ABD	14		LARVAE	Mulla et al. (1971)
	Mosquito (*Aedes excrucians*)	LC_{50}	1	145.5	LARVAE, 4TH INSTAR	Rettich (1977)
	Mosquito (*Aedes communis*)	LC_{50}	1	167.9	LARVAE, 4TH INSTAR	Rettich (1977)
	Mosquito (*Aedes punctor*)	LC_{50}	1	298.3	LARVAE, 4TH INSTAR	Rettich (1977)
	Mosquito (*Aedes vexans*)	LC_{50}	1	322.6	LARVAE, 4TH INSTAR	Rettich (1977)
	Mosquito (*Aedes cantans*)	LC_{50}	1	376.6	LARVAE, 4TH INSTAR	Rettich (1977)

Mosquito (*Aedes aegypti*)	LC_{50}	NR	4,400	LARVAE, 4TH INSTAR	Klassen et al. (1965)
	ABD	90	0.5%	LARVAE	von Windeguth (1971)
	ABD	90	1.25%	LARVAE	von Windeguth (1971)
	ABD	90	2.5%	LARVAE	von Windeguth (1971)
	LET	1	1000	LARVAE, LAST INSTAR	Mitsuhashi et al. (1970)
	MOR	0.08–1	50,000	3RD INSTAR, LARVAE	Lichtenstein et al. (1966)
	MOR	1	1,000	3RD INSTAR, LARVAE	Lichtenstein et al. (1966)
	MOR	1	1,000	3RD INSTAR, LARVAE	Lichtenstein et al. (1966)
	MOR	1	1,000	3RD INSTAR, LARVAE	Lichtenstein et al. (1966)
Mosquito (*Anopheles stephensi*)	LC_{50}	1	720	4TH INSTAR LARVA,	Scott and Georghiou (1986)
Mosquito (*Anopheles albimanus*)	LC_{50}	1	890	4TH INSTAR	Hemingway and Georghiou (1983)
Mosquito (*Culex pipiens*)	EC_{50}IM	NR	75	4TH INSTAR	Rawash et al. (1975)
Mosquito (*Culex pipiens pipiens*)	LC_{50}	1	333.9	LARVAE, 4TH INSTAR	Rettich (1977)
Mosquito (*Culex pipiens quinquefasciata*)	LC_{50}	1	400	3RD–4TH INSTAR LARVAE	Chen et al. (1971)
Mosquito (*Culex pipiens molestus*)	LC_{50}	1	418.8	LARVAE, 4TH INSTAR	Rettich (1977)
Mosquito (*Culex pipiens quinquefasciata*)	LC_{50}	1	1,500	3RD–4TH INSTAR LARVAE	Chen et al. (1971)
Mosquito (*Culex pipiens*)	LC_{50}	NR	170	LARVAE, 72 H, 4TH INSTAR	Gaaboub et al. (1975)
Mosquito (*Culex pipiens pallens*)	RSD	1	1	LARVAE	Tsuge et al. (1980)
Mosquito (*Culiseta annulata*)	LC_{50}	1	179.5	LARVAE, 4TH INSTAR	Rettich (1977)

Table 18. Toxicology of Aquatic and Terrestrial Species: Chemical: Carbaryl, CAS Number: 63252 (continued)

Class	Species	Effect	Duration (days)	Concentration ($\mu g\ L^{-1}$)	Lifestage	Ref.
Insect (continued)	Mosquito (*Wyeomyia smithii*)	DVP	7	1,000	2ND INSTAR LARVAE	Strickman (1985)
		MOR	7	1,000	2ND INSTAR LARVAE	Strickman (1985)
	Phantom midge (*Chaoborus* sp)	LC_{50}	2	290	LARVAE	Bluzat and Seuge (1979)
	Stonefly (*Claassenia sabulosa*)	LC_{50}	1	12	20–25 MM	Sanders and Cope (1968)
		LC_{50}	2	6.8	20–25 MM	Sanders and Cope (1968)
		LC_{50}	4	5.6	20–25 MM	Sanders and Cope (1968)
		LC_{50}	4	5.6	2ND YEAR CLASS	Johnson and Finley (1980)
	Stonefly (*Leuctra* sp)	ABD	30–60	840 g ha^{-1}	NR	Courtemanch and Gibbs (1980)
	Stonefly (*Plecoptera*)	ABD	30–60	840 g ha^{-1}	NR	Courtemanch and Gibbs (1980)
	Stonefly (*Pteronarcella badia*)	LC_{50}	1	5	15–20 MM	Sanders and Cope (1968)
	Stonefly (*Pteronarcys californica*)	LC_{50}	1	30	30–35 MM	Sanders and Cope (1968)
		LC_{50}	1	30	NYMPH	Cope (1965)
	Stonefly (*Pteronarcella badia*)	LC_{50}	2	3.6	15–20 MM	Sanders and Cope (1968)
	Stonefly (*Pteronarcys californica*)	LC_{50}	2	13	30–35 MM	Sanders and Cope (1968)
		LC_{50}	2	15	NYMPH	Cope (1965)
	Stonefly (*Pteronarcella badia*)	LC_{50}	4	1.7	15–20 MM	Sanders and Cope (1968)
		LC_{50}	4	1.7	NAIAD	Johnson and Finley (1980)
	Stonefly (*Pteronarcys californica*)	LC_{50}	4	2	LATE INSTAR NAIAD	Schoettger and Mauck (1978)
		LC_{50}	4	4.8	30–35 MM	Sanders and Cope (1968)
		LC_{50}	4	4.8	NYMPH	Cope (1965)
		LC_{50}	4	4.8	2ND YEAR CLASS	Johnson and Finley (1980)
	Stonefly (*Pteronarcella badia*)	LC_{50}	4	11	NR	Woodward and Mauck (1980)
		LC_{50}	4	13	NR	Woodward and Mauck (1980)
		LC_{50}	4	29	NR	Woodward and Mauck (1980)

	Stonefly (*Skwala* sp)	LC_{50}	4	3.6	NAIAD	Johnson and Finley (1980)
		LC_{50}	4	9.2	1ST YEAR CLASS	Johnson and Finley (1980)
Mullusca	Abalone	LC_{50}	1	8,300	1.4 CM, 1.4 G	Nishiuchi (1977b)
		LC_{50}	2	8,100	1.4 CM, 1.4 G	Nishiuchi (1977b)
		LC_{50}	4	8,100	1.4 CM, 1.4 G	Nishiuchi (1977b)
	American or Virginia oyster	EC_{50}DV	2	3,300	EGGS, 2 CELL STAGE	Davis and Hidu (1979)
		LC_{50}	12	3,000	2 D, LARVAE	Davis and Hidu (1979)
		GRO	4	2,000	2.5–5.1 CM	Butler (1963)
		GRO	4	2,000	2.5–5.1 CM	Butler (1963)
		GRO	12	20	2 D, LARVAE	Davis and Hidu (1979)
	Bent-nosed clam	EC_{50}FR	2	27,500	2.5 G, 40 MM LONG, 30 MM WIDE	Armstrong and Millemann (1974a)
		EC_{50}FR	4	17,000	2.5 G; 40 MM LONG, 30 MM WIDE	Armstrong and Millemann (1974a)
		ABD	NR	1.2 kg m^{-2}	NR	Chambers (1970)
	Bivalve, clam, mussel class	ABD	30	5.7 kg ha^{-1}	NR	Armstrong and Millemann (1974c)
	Bladder snail	LC_{50}	2	27,000	NR	Hashimoto and Nishiuchi (1981)
		LC_{50}	2	27,000	NR	Nishiuchi and Yoshida (1972)
	Cockle (*Clinocardium nuttallii*)	STR	1	100,000	60 MM	Buchanan et al. (1970)
	Common atlantic abra	POP	70	1.1	LARVA	Tagatz et al. (1979)
		POP	70	11.1	LARVA	Tagatz et al. (1979)
	Common bay mussel, blue mussel	EC_{50}AB	1 hr	5,300	0.33 HR, 1ST POLAR BODY	Armstrong and Millemann (1974b)
		EC_{50}AB	1 hr	7,000	2 CELL, 1 HR	Armstrong and Millemann (1974b)
		EC_{50}AB	1 hr	8,300	≥64 CELLS, 4 HR	Armstrong and Millemann (1974b)

Table 18. Toxicology of Aquatic and Terrestrial Species: Chemical: Carbaryl, CAS Number: 63252 (continued)

Class	Species	Effect	Duration (days)	Concentration (µg L^{-1})	Lifestage	Ref.
Mullusca (continued)	Common bay mussel, blue mussel	EC_{50}AB	1 hr	16,000	BLASTULA, 8 HR	Armstrong and Millemann (1974b)
		EC_{50}AB	1 hr	19,000	20 HR TROCOPHORE	Armstrong and Millemann (1974b)
		EC_{50}AB	1 hr	20,700	UNFERTILIZED EGG	Armstrong and Millemann (1974b)
		EC_{50}AB	1 hr	24,000	EARLY VELIGER, 32 HR	Armstrong and Millemann (1974b)
		EC_{50}BA	1	>30,000	0.25 G	Roberts (1975)
		EC_{50}BA	2	>30,000	0.25 G	Roberts (1975)
		EC_{50}BA	4	10,300	ADULT	Liu and Lee (1975)
		EC_{50}DV	2	1,500	EGGS	Liu and Lee (1975)
		LC_{50}	4	22,700	ADULT	Liu and Lee (1975)
		DVP	40	360	29–30 D, LARVAE	Liu and Lee (1975)
		GRO	20	330	LARVAE	Liu and Lee (1975)
		GRO	20	650	LARVAE	Liu and Lee (1975)
	Common rangia or clam	LC_{50}	1	2,000,000	35–50 MM	Chaiyarach et al. (1975)
		LC_{50}	2	1,860,000	35–50 MM	Chaiyarach et al. (1975)
		LC_{50}	3	350,000	35–50 MM	Chaiyarach et al. (1975)
		LC_{50}	4	125,000	35–50 MM	Chaiyarach et al. (1975)

Great pond snail	LC_{50}	2	21,000	NR	Bluzat and Seuge (1979)
	LC_{50}	3	10,700	NR	Bluzat and Seuge (1979)
	GRO	NR	1,000	NR	Seuge and Bluzat (1983a)
	GRO	NR	2,000	NR	Seuge and Bluzat (1983a)
	GRO	NR	2,500	NR	Seuge and Bluzat (1983a)
	MOR	30	1,000	JUVENILE	Seuge and Bluzat (1983a)
	MOR	30	1,000	LARVAE	Seuge and Bluzat (1979a)
	MOR	30	2,000	JUVENILE	Seuge and Bluzat (1983a)
	MOR	30	2,000	LARVAE	Seuge and Bluzat (1979a)
	MOR	30	2,500	JUVENILE	Seuge and Bluzat (1983a)
	MOR	30	3,000	JUVENILE	Seuge and Bluzat (1983a)
	REP	NR	1,000	NR	Seuge and Bluzat (1983a)
	REP	NR	4,000	F1 GENERATION	Seuge and Bluzat (1979b)
	REP	300	2,000	LARVAE	Seuge and Bluzat (1979a)
	REP	300	2,000	LARVAE	Seuge and Bluzat (1979a)
Jackknife clam	GRO	70	1.1	LARVA	Tagatz et al. (1979)
	GRO	70	11.1	LARVA	Tagatz et al. (1979)
	POP	70	11.1	LARVA	Tagatz et al. (1979)
Japanese littleneck clam	LC_{50}	1	18,000	2.3–2.6 CM, 6.5–11 G	Nishiuchi (1977b)
	LC_{50}	2	12,000	2.3–2.6 CM, 6.5–11 G	Nishiuchi (1977b)
	LC_{50}	4	11,000	2.3–2.6 CM, 6.5–11 G	Nishiuchi (1977b)
Manilla clam	STR	NR	1.2 kg m^{-2}	NR	Chambers (1970)
Marsh snail	LC_{50}	2	25,000	NR	Hashimoto and Nishiuchi (1981)
	LC_{50}	2	25,000	NR	Nishiuchi and Yoshida (1972)
Mediterranean mussel	MOR	4	10,000	10–30 MM	Rao (1981)
	MOR	4	10,000	50–70 MM	Rao (1981)
	MOR	7	10,000	10–30 MM	Rao (1981)
Milk moon shell	POP	70	1.1	LARVA	Tagatz et al. (1979)
Mollusc	LC_{50}	4	10,100	1.4–1.8 MM	Almar et al. (1988)
	LC_{50}	4	12,800	1.4–1.8 MM	Almar et al. (1988)
	LC_{50}	4	14,870	1.4–1.8 MM	Almar et al. (1988)

Table 18. Toxicology of Aquatic and Terrestrial Species: Chemical: Carbaryl, CAS Number: 63252 (continued)

Class	Species	Effect	Duration (days)	Concentration ($\mu g\ L^{-1}$)	Lifestage	Ref.
Mullusca (continued)	Mud snail	LC_{50}	2	3,500	48 HR	Nishiuchi and Yoshida (1972)
		LC_{50}	2	15,000	NR	Nishiuchi and Yoshida (1972)
		LC_{50}	2	16,000	NR	Nishiuchi and Yoshida (1972)
		LC_{50}	2	18,000	ADULT	Nishiuchi and Yoshida (1972)
		LC_{50}	2	18,000	NR	Nishiuchi and Yoshida (1972)
		LC_{50}	2	30,000	NR	Nishiuchi and Yoshida (1972)
	Northern quahog or hard clam	$EC_{50}DV$	2	3,820	EGGS, 2 CELL STAGE	Davis and Hidu (1979)
		GRO	10	1,000	2 D, LARVAE	Davis and Hidu (1979)
		MOR	10	1,000	2 D, LARVAE	Davis and Hidu (1979)
		MOR	10	2,500	2 D, LARVAE	Davis and Hidu (1979)
	Opistobranch	MOR	2.5 hr	5.7 kg ha^{-1}	NR	Armstrong and Millemann (1974c)
	Pacific oyster	RSD	1	11 kg ha^{-1}	NR	Sayce (1970)
		RSD	5–60	11 kg ha^{-1}	ADULT	Sayce and Chambers (1970)
	Pond snail (*Lymnaea acuminata*)	LC_{50}	1	15,980	3 MONTHS	Goel and Srivastava (1981)
		LC_{50}	2	14,000	ADULT, 2.6 CM	Singh and Agarwal (1983)
		LC_{50}	3	7,000	ADULT, 2.6 CM	Singh and Agarwal (1983)
		LC_{50}	4	4,400	ADULT, 2.6 CM	Singh and Agarwal (1983)
		LC_{50}	5	2,400	ADULT, 2.6 CM	Singh and Agarwal (1983)
		LC_{50}	6	1,200	ADULT, 2.6 CM	Singh and Agarwal (1983)
		ENZ	2	30	ADULT, 2.6 CM	Singh and Agarwal (1986)
		ENZ	2	5,600	ADULT, 2.6 CM	Singh and Agarwal (1986)
		PHY	2	11,200	ADULT, 2.6 CM	Singh and Agarwal (1986)
	Sand gaper, soft shell clam	ABD	NR	1.2 kg m^{-2}	NR	Chambers (1970)

	Scallop	LC_{50}	1	7,900	3.2 CM, 3.7 G	Nishiuchi (1977b)
		LC_{50}	2	6,800	3.2 CM, 3.7 G	Nishiuchi (1977b)
		LC_{50}	3	5,600	3.2 CM, 3.7 G	Nishiuchi (1977b)
		LC_{50}	4	5,600	3.2 CM, 3.7 G	Nishiuchi (1977b)
	Snail (*Aplexa hypnorum*)	LC_{50}	4	>27,000	ADULT	Phipps and Holcombe (1985)
	Snail (*Biomphalaria glabrata*)	PHY	1	16,100	14–16 MM SHELL DIAMETER	Cocks (1973)
		PHY	1	64,400	14–16 MM SHELL DIAMETER	Cocks (1973)
	Snail (*Indoplanorbis exustus*)	LC_{50}	1	30,000	3 MONTHS	Goel and Srivastava (1981)
		LC_{50}	2	28,000	NR	Hashimoto and Nishiuchi (1981)
		LC_{50}	2	28,000	NR	Nishiuchi and Yoshida (1972)
		RSD	1	1	NR	Tsuge et al. (1980)
	White cockle	STR	1	11 kg ha^{-1}	NR	Chambers (1970)
		STR	1	1[illegible] kg ha^{-1}	NR	Chambers (1970)
		STR	1	22 kg ha^{-1}	NR	Chambers (1970)
Nemertea	Proboscis worm, nemertean (*Cerebratulus* sp)	ABD	30		NR	Armstrong and Millemann (1974c)
	Proboscis worm, nemertean (*Nemertea* sp)	POP	70	11.1	LARVA	Tagatz et al. (1979)
		POP	70	103	LARVA	Tagatz et al. (1979)
Protozoa	Amoeba	PGR	6	10,000	6 D CULTURE	Prescott et al. (1977)
	Ciliate (*Colpidium campylum*)	PGR	1.79	>10,000	>96 HR	Dive et al. (1980)
	Ciliate (*Paramecium multimicronucleatum*)	LC_{50}	0.29	105,000	12 D	Edmiston et al. (1985)
		LC_{50}	0.38	65,000	12 D	Edmiston et al. (1985)
		LC_{50}	0.54	46,000	12 D	Edmiston et al. (1985)
		LC_{50}	0.71	34,000	12 D	Edmiston et al. (1985)
	Ciliate (*Paramecium caudatum*)	LC_{50}	1	7,900	NR	Lejczak (1977)
		LC_{50}	1	10,000	7 D CULT	Edmiston et al. (1984)

Table 18. Toxicology of Aquatic and Terrestrial Species: Chemical: Carbaryl, CAS Number: 63252 (continued)

Class	Species	Effect	Duration (days)	Concentration (µg L^{-1})	Lifestage	Ref.
Protozoa (continued)	Ciliate (*Paramecium multimicronucleatum*)	LC_{50}	1	24,000	7 D CULT	Edmiston et al. (1984)
		LC_{50}	1	28,000	12 D	Edmiston et al. (1985)
	Ciliate (*Paramecium bursaria*)	LC_{50}	1	31,000	7 D CULT	Edmiston et al. (1984)
	Ciliate (*Paramecium aurelia*)	LC_{50}	1	46,000	7 D CULT	Edmiston et al. (1984)
	Ciliate (*Paramecium caudatum*)	LC_{50}	2	12,400	NR	Lejczak (1977)
		LC_{50}	4	10,200	NR	Lejczak (1977)
	Ciliate (*Paramecium multimicronucleatum*)	OC	1	120,000	12 D	Edmiston et al. (1985)
Tracheophy	Water-meal	LET	12–30	1,000,000	NR	Worthley and Schott (1972)
Arachnida	Predatory mite	LO MOR	NR	1.50%	YNG#	Elghar et al. (1971)
		NO MOR	NR	0.50%	YNG#	Elghar et al. (1971)
		NO MOR	NR	4.50%	EMB#	Elghar et al. (1971)
		NO MOR	NR	4.50%	ADULT#	Elghar et al. (1971)
Bird	Bobwhite quail	NO AVO	21	1,235 ppm	ADULT	Robel et al. (1982)
		NO PHY	2–8	90 mg kg^{-1}	ADULT	Solomon and Robel (1980)
		NO PHY	15	35 mg kg^{-1}	ADULT	Rodgers and Robel (1977)
	Japanese quail	LO BEH	21	20 ppm	70 D	Derosa et al. (1976)
		LO BEH	63	20 ppm	70 D	Derosa et al. (1976)
		LO PHY	60	100 ppm	39 D	Cecil et al. (1974)
		LO PHY	140	900 ppm	1 D	Bursian and Edens (1977)
		LO REP	6	1,000 mg kg^{-1}	ADULT	Haegele and Tucker (1974)
		LO REP	21	40 ppm	70 D	Derosa et al. (1976)
		LO SIGNS	6	1,000 mg kg^{-1}	ADULT	Haegele and Tucker (1974)
		NO PHY	140	600 ppm	1 D	Bursian and Edens (1977)
		NO REP	1	160 mg kg^{-1}	ADULT	Hill (1979)

		NO REP	21	[illegible]0 ppm	70 D	Derosa et al. (1976)
		NO REP	140	1,2[illegible]0 ppm	1 D	Bursian and Edens (1977)
	Mallard duck	NO REP	6	1,000 mg kg^{-1}	ADULT	Haegele and Tucker (1974)
	Towhee	NO RSD	3	1 lb ac^{-1}	ADULT	Kurtz and Studholme (1974)
Invertebrate	Earthworm	LO POP	21	2 lb ac^{-1}		Thompson (1971)
		LO POP	365	2.24 kg ha^{-1}		Thompson and Sans (1974)
	Honey bee	LD_{50}	NR	140 mg be^{-1}	ADULT	Stevenson et al. (1978)
		LD_{50}	NR	1,300 mg be^{-1}	ADULT	Stevenson et al. (1978)
		LO MOR	NR	1 ppm		Sonnet (1978)
		LO MOR	NR	[illegible] lb ac^{-1}	MULT	Robinson and Johnson (1978)
		LO POP	NR	[illegible] lb ac^{-1}	MULT	Robinson and Johnson (1978)
		NO POP	NR	5 ppm		Sonnet (1978)
Mammal	Cotton rat	LO PHY	300	[illegible] lb ac^{-1}	ADULT	Pomeroy and Barrett (1975)
		LO POP	300	[illegible] lb ac^{-1}	ADULT	Pomeroy and Barrett (1975)
		LO REP	300	2 lb ac^{-1}	ADULT	Pomeroy and Barrett (1975)
		NO MOR	300	2 lb ac^{-1}	ADULT	Pomeroy and Barrett (1975)
	House mouse	LO POP	300	2 lb ac^{-1}	ADULT	Pomeroy and Barrett (1975)
		LO REP	300	2 lb ac^{-1}	ADULT	Pomeroy and Barrett (1975)
		NO MOR	300	2 lb ac^{-1}	ADULT	Pomeroy and Barrett (1975)
	Mouse (*Mus* sp.)	LD_{50}	1	112 mg kg^{-1}	56 D	Ahdaya et al. (1976)
		LD_{50}	1	263 mg kg^{-1}	56 D	Ahdaya et al. (1976)
		LD_{50}	1	588 mg kg^{-1}	56 D	Ahdaya et al. (1976)
		LD_{50}	14	1,80[illegible] mg kg^{-1}	YNG	Cress and Strother (1974)
		LO MOR	1	5[illegible] mg kg^{-1}	49–56 D	Ahdaya et al. (1976)
		LO MOR	14	25 mmol	YNG	Cress and Strother (1974)
		LO PATH	20	2[illegible] mg kg^{-1}		Jordan et al. (1975)
		LO PHY	1	5[illegible] mg kg^{-1}	49–56 D	Ahdaya et al. (1976)
		LO PHY	14	25 mmol	YNG	Cress and Strother (1974)
		LO PHY	14	25 mmol	YNG	Cress and Strother (1974)
		NO REP	20	2[illegible] mg kg^{-1}		Jordan et al. (1975)

Table 18. Toxicology of Aquatic and Terrestrial Species: Chemical: Carbaryl, CAS Number: 63252 (continued)

Class	Species	Effect	Duration (days)	Concentration ($\mu g\ L^{-1}$)	Lifestage	Ref.
Mammal (continued)	Norway rat	LO PHY	60	100 ppm	21 D	Cecil et al. (1974)
	Old–field mouse	NO MOR	300	2 lb ac^{-1}	ADULT	Pomeroy and Barrett (1975)
		NO POP	300	2 lb ac^{-1}	ADULT	Pomeroy and Barrett (1975)
		NO REP	300	2 lb ac^{-1}	ADULT	Pomeroy and Barrett (1975)
	Rat	LD_{50}		246–283 mg kg^{-1}		*Farm Chemicals Handbook* (1990)
		LO BEH	50	100 ppm	Desi et al. (1974)	
		LO PHY	50	100 ppm	Desi et al. (1974)	

Table 19. Toxicology of Aquatic and Terrestrial Species: Chemical: Chlorothalonil, CAS Number: 1897456

Class	Species	Effect	Duration (days)	Concentration (µg L^{-1})	Lifestage	Ref.
Algae	Algae, phytoplankton, algal mat	BCF	14	20	AUFWUCHS	Davies (1988)
Amphibia	Frog (*Bufo bufo japonicus*)	LC_{50}	2	160	TADPOLE	Hashimoto and Nishiuchi (1981)
Crustacea	Dungeness or edible crab	EC_{50}SW	2	170	FIRST STAGE ZOEAE	Armstrong et al. (1976)
		EC_{50}SW	4	<100	FIRST STAGE ZOEAE	Armstrong et al. (1976)
		LC_{50}	2	560	FIRST STAGE ZOEAE	Armstrong et al. (1976)
		LC_{50}	4	140	FIRST STAGE ZOEAE	Armstrong et al. (1976)
	Water flea (*Daphnia pulex*)	LC_{50}	3 hr	7,800	NR	Hashimoto and Nishiuchi (1981)
		LC_{50}	3 hr	7,800	NR	Nishiuchi and Hashimoto (1975)
	Water flea (*Moina macrocopa*)	LC_{50}	3 hr	>10,000	NR	Hashimoto and Nishiuchi (1981)
		LC_{50}	3 hr	>10,000	NR	Nishiuchi and Hashimoto (1975)
Fish	Common, mirror, colored, carp	LC_{50}	2	110	NR	Hashimoto and Nishiuchi (1981)
		LC_{50}	2	110	NR	Nishiuchi and Hashimoto (1975)
	Golden galaxias	BCF	14	20	NR	Davies (1988)
	Goldfish	LC_{50}	2	170	NR	Hashimoto and Nishiuchi (1981)
		LC_{50}	2	170	NR	Nishiuchi and Hashimoto (1975)
	Medaka, high-eyes	LC_{50}	2	38	NR	Hashimoto and Nishiuchi (1981)
		LC_{50}	2	38	NR	Nishiuchi and Hashimoto (1975)
	Oriental weatherfish	LC_{50}	2	150	NR	Hashimoto and Nishiuchi (1981)
	Rainbow trout, donaldson trout	LC_{50}	4	7.6	9.5 (6–12) G	Davies (1987)
		LC_{50}	4	10.5	9.5 (6–12) G	Davies (1987)
		LC_{50}	4	13.6	9.5 (6–12) G	Davies (1987)
		LC_{50}	4	17.1	9.5 (6–12) G	Davies (1987)
		BIO	2	50	9.5 (6–12) G	Davies (1987)

Table 19. Toxicology of Aquatic and Terrestrial Species: Chemical: Chlorothalonil, CAS Number: 1897456 (continued)

Class	Species	Effect	Duration (days)	Concentration (µg L^{-1})	Lifestage	Ref.
Fish (continued)	Rainbow trout, donaldson trout	HEM	1	>20	9.5 (6–12) G	Davies (1987)
		HIS	24	2	9.5 (6–12) G	Davies (1987)
		RES	2 hr	30	9.5 (6–12) G	Davies (1987)
Insect	Mayfly (*Cloeon dipterum*)	LC_{50}	2	1,800	LARVAE	Hashimoto and Nishiuchi (1981)
Mullusca	Bladder snail	LC_{50}	2	37,000	NR	Hashimoto and Nishiuchi (1981)
		LC_{50}	2	37,000	NR	Nishiuchi and Yoshida (1972)
	Marsh snail	LC_{50}	2	9,000	NR	Hashimoto and Nishiuchi (1981)
		LC_{50}	2	9,000	NR	Nishiuchi and Yoshida (1972)
	Mud snail	LC_{50}	2	30,000	NR	Nishiuchi and Yoshida (1972)
	Snail (*Indoplanorbis exustus*)	LC_{50}	2	15,000	NR	Hashimoto and Nishiuchi (1981)
		LC_{50}	2	15,000	NR	Nishiuchi and Yoshida (1972)
Mammal	Rat	LC_{50}	1 hr	0.31 mg L^{-1}		*Farm Chemicals Handbook* (1990)
		LD_{50}		>10,000 mg kg^{-1}		*Farm Chemicals Handbook* (1990)

Table 20. Toxicology of Aquatic and Terrestrial Species: Chemical: Chlorpyriphos, CAS Number: 2921882

Class	Species	Effect	Duration (days)	Concen-tration (µg L^{-1})	Lifestage	Ref.
Algae	Algae, phytoplankton, algal mat	ABD	30	10	NR	Hughes et al. (1980)
		ABD	74	7.2	NR	Hurlbert et al. (1972)
		ABD	74	72	NR	Hurlbert et al. (1972)
		PGR	NR	4	NR	Butcher et al. (1975)
		POP	80	4	NR	Butcher et al. (1977)
		POP	80	10	NR	Butcher et al. (1977)
		POP	80	1,000	NR	Butcher et al. (1977)
	Blue-green algae (*Anabaena* sp)	ABD	74	72	NR	Hurlbert et al. (1972)
		BCFD	1	1,000	NR	Lal and Singh (1987a)
		BCFD	1	5,000	NR	Lal and Singh (1987a)
		BCFD	1	10,000	NR	Lal and Singh (1987a)
		BCFD	2	1,000	NR	Lal and Singh (1987a)
		BCFD	2	5,000	NR	Lal and Singh (1987a)
		BCFD	2	10,000	NR	Lal and Singh (1987a)
		BCFD	3	1,000	NR	Lal and Singh (1987a)
		BCFD	3	5,000	NR	Lal and Singh (1987a)
		BCFD	3	10,000	NR	Lal and Singh (1987a)
		BCFD	4	1,000	NR	Lal and Singh (1987a)
		BCFD	4	5,000	NR	Lal and Singh (1987a)
		BCFD	4	10,000	NR	Lal and Singh (1987a)
		BCFD	5	1,000	NR	Lal and Singh (1987a)
		BCFD	5	5,000	NR	Lal and Singh (1987a)
		BCFD	5	10,000	NR	Lal and Singh (1987a)

Table 20. Toxicology of Aquatic and Terrestrial Species: Chemical: Chlorpyriphos, CAS Number: 2921882 (continued)

Class	Species	Effect	Duration (days)	Concentration ($\mu g\ L^{-1}$)	Lifestage	Ref.
Algae (continued)	Blue-green algae (*Aulosira fertilissima*)	BCFD	1	1,000	NR	Lal and Singh (1987a)
		BCFD	1	5,000	NR	Lal and Singh (1987a)
		BCFD	1	10,000	NR	Lal and Singh (1987a)
		BCFD	2	1,000	NR	Lal and Singh (1987a)
		BCFD	2	5,000	NR	Lal and Singh (1987a)
		BCFD	2	10,000	NR	Lal and Singh (1987a)
		BCFD	3	1,000	NR	Lal and Singh (1987a)
		BCFD	3	5,000	NR	Lal and Singh (1987a)
		BCFD	3	10,000	NR	Lal and Singh (1987a)
		BCFD	4	1,000	NR	Lal and Singh (1987a)
		BCFD	4	5,000	NR	Lal and Singh (1987a)
		BCFD	4	10,000	NR	Lal and Singh (1987a)
		BCFD	5	1,000	NR	Lal and Singh (1987a)
		BCFD	5	5,000	NR	Lal and Singh (1987a)
		BCFD	5	10,000	NR	Lal and Singh (1987a)
	Chrysophyte	ABD	17	1.2	NR	Brown and Chow (1974)
	Diatom (*Amphiprora* sp)	PGR	4	10,000	NR	Maly and Ruber (1983)
	Diatom (*Skeletonema costatum*)	EC_{50}PP	2	1,200	INITIAL CONC. 40,000 CELLS ML^{-1}	Walsh (1983)
	Diatom class	POP	42	250	NR	Nelson et al. (1976)
		POP	84	500	NR	Nelson et al. (1976)
	Diatoms, chrysophyte division	ABD	74	7.2	NR	Hurlbert et al. (1972)
		PGR	10	400	NR	Roberts and Miller (1970)
	Dinoflagellate (*Ceratium* sp)	ABD	17	240	NR	Brown et al. (1976)
	Dinoflagellate (*Glenodinium* sp)	ABD	17	1.2	NR	Brown and Chow (1974)

	Dinoflagellate (*Gonyaulax* sp)	PGR	4	10,000	NR	Maly and Ruber (1983)
	Green algae (*Ankistrodesmus falcatus*)	ABD	6	200	NR	Brown and Chow (1974)
		ABD	9	1.2	NR	Brown et al. (1976)
		ABD	17	1.2	NR	Brown and Chow (1974)
	Green algae (*Ankistrodesmus spiralis*)	ABD	17	1.2	NR	Brown and Chow (1974)
	Green algae (*Ankistrodesmus falcatus*)	PGR	6	100	NR	Brown and Chow (1974)
	Green algae (*Chlorella vulgaris*)	ABD	6	10	NR	Brown and Chow (1974)
		ABD	6	100	NR	Brown and Chow (1974)
	Green algae (*Chlorococcum* sp)	PGR	4	10,000	NR	Maly and Ruber (1983)
	Green algae (*Chlorella vulgaris*)	PGR	6	1.0	NR	Brown and Chow (1974)
	Green algae (*Schroederia setigera*)	ABD	74	72	NR	Hurlbert et al. (1972)
	Green algae (*Tetraedron* sp)	ABD	9	1.2	NR	Brown et al. (1976)
		ABD	17	1.2	NR	Brown et al. (1976)
Amphibia	Western toad	THL	1	30	19.8 (15–28) MM	Johnson and Prine (1976)
Annelida	Oligochaete (*Branchiura sowerbyi*)	LET	3	1,500	NR	Naqvi (1973)
		LET	3	1,500	NR	Naqvi (1973)
		MOR	3	1,500	NR	Naqvi (1973)
		MOR	90	1,500	NR	Naqvi (1973)
	Oligochaete family	ABD	21	7.4	NR	Ali and Mulla (1978b)
Aquatic invertebrate	Invertebrates	ABD	28	0.16–17.3	MICRO-INVERTEBRATES	Siefert (1987)
		ABD	28	0.16–17.3	MACRO-INVERTEBRATES	Siefert (1987)
Cnidaria	Hydra (*Hydra* sp)	MOR	2	100	NR	Hughes (1977)

Table 20. Toxicology of Aquatic and Terrestrial Species: Chemical: Chlorpyriphos, CAS Number: 2921882 (continued)

Class	Species	Effect	Duration (days)	Concen-tration ($\mu g\ L^{-1}$)	Lifestage	Ref.
Crustacea	Calanoid copepod (*Diaptomus* sp)	ABD	21	7.4	NR	Ali and Mulla (1978b)
	Calanoid copepod	ABD	74	7.2	NAUPLII	Hurlbert et al. (1972)
	(*Diaptomus pallidus*)	ABD	74	7.2	COPEPODIDS	Hurlbert et al. (1972)
		ABD	74	72	COPEPODIDS	Hurlbert et al. (1972)
	Calanoid copepod (*Diaptomus* sp)	MOR	2	100	NR	Hughes (1977)
	Copepod order	LC_{50}	2	2.13	NR	Siefert (1987)
		LC_{50}	3	0.94	NR	Siefert (1987)
		ABD	20	10	NAUPLII	Hughes et al. (1980)
		ABD	74	7.2	NAUPLII	Hurlbert et al. (1972)
		STR	2	6.1–15.9	NR	Siefert (1987)
	Copepod suborder	ABD	40	10	COPEPODIDS	Hughes et al. (1980)
	Crayfish (*Orconectes immunis*)	LC_{50}	4	6	1.8 G	Phipps and Holcombe (1985)
	Crustacean class	ABD	7	0.056–0.112	NR	Frank and Sjogren (1978)
	Cyclopoid copepod	ABD	28	0.011	NR	Hurlbert et al. (1970)
	(*Acanthocyclops vernalis*)	ABD	74	7.2	NAUPLII	Hurlbert et al. (1972)
		ABD	74	7.2	COPEPODIDS	Hurlbert et al. (1972)
		ABD	74	72	COPEPODIDS	Hurlbert et al. (1972)
		MOR	28	0.011	NR	Hurlbert et al. (1970)
	Cyclopoid copepod (*Cyclops* sp)	ABD	21	7.4	NR	Ali and Mulla (1978b)
	Cyclopoid copepod	MOR	1	1	NR	Johnson (1978a)
	(*Macrocyclops albidus*)	MOR	1	5	NR	Johnson (1978a)
		MOR	1	10	NR	Johnson (1978a)
		MOR	1	25	NR	Johnson (1978a)

Fairy shrimp order	LC_{50}	0.4583	0.39	NR	Siefert (1987)
	LC_{50}	5 hr	0.57	NR	Siefert (1987)
Grass shrimp (*Palaemonetes pugio*)	AVO	0.50	1	10–40 MM	Hansen et al. (1973)
Opossum shrimp (*Mysidopsis bahia*)	LC_{50}	4	0.085	NEWLY HATCHED, ≤24 HR	Schimmel et al. (1983)
Ostracod (*Cyprinotus* sp)	ABD	7	3	NR	Ali and Mulla (1978b)
	ABD	28	7.4	NR	Ali and Mulla (1978b)
Ostracod, seed shrimp subclass	LET	2	100	NR	Hughes (1977)
	MOR	2	1	NR	Hughes (1977)
	MOR	4	6.93	NR	Siefert (1987)
Red swamp crayfish	LET	3.5 hr	3,720	10.6 G	Naqvi (1973)
Scud (*Gammarus fasciatus*)	LC_{10}	1	0.8	NR	Thayer and Ruber (1976)
Scud (*Gammarus lacustris*)	LC_{50}	1	0.76	2 MONTHS OLD	Sanders (1969)
Scud (*Gammarus fasciatus*)	LC_{50}	1	5.6	NR	Sanders (1972)
Scud (*Gammarus pseudolimnaeus*)	LC_{50}	2	0.33	NR	Siefert (1987)
Scud (*Gammarus lacustris*)	LC_{50}	2	0.40	2 MONTHS OLD	Sanders (1969)
Scud (*Gammarus pseudolimnaeus*)	LC_{50}	3	0.19	NR	Siefert (1987)
Scud (*Gammarus lacustris*)	LC_{50}	4	0.11	2 MONTHS OLD	Sanders (1969)
	LC_{50}	4	0.11	MATURE	Johnson and Finley (1980)
Scud (*Gammarus pseudolimnaeus*)	LC_{50}	4	0.18	NR	Siefert (1987)
Scud (*Gammarus fasciatus*)	LC_{50}	4	0.32	NR	Sanders (1972)
	DRF	1	0.8	NR	Thayer and Ruber (1976)
	DRF	1	0.8	NR	Thayer and Ruber (1976)
Scud (*Hyalella azteca*)	LC_{50}	1	0.68	NR	Siefert (1987)
	LC_{50}	2	0.29	NR	Siefert (1987)
	LC_{50}	3	0.18	NR	Siefert (1987)
	LC_{50}	4	0.14	NR	Siefert (1987)
	ABD	14	3	NR	Ali and Mulla (1978b)
	ABD	21	7.4	NR	Ali and Mulla (1978b)

Table 20. Toxicology of Aquatic and Terrestrial Species: Chemical: Chlorpyriphos, CAS Number: 2921882 (continued)

Class	Species	Effect	Duration (days)	Concentration (μg L^{-1})	Lifestage	Ref.
Crustacea (continued)	Water flea (*Ceriodaphnia reticulata*)	LET	0.75	2.6	NR	Siefert (1987)
	Water flea (*Daphnia pulex*)	LC_{50}	0.75	0.24	NR	Siefert (1987)
		LC_{50}	1	0.17	NR	Siefert (1987)
	Water flea (*Daphnia magna*)	LC_{50}	1	0.4	FEMALE	Roberts and Miller (1971)
	Water flea (*Daphnia pulex*)	LC_{50}	3	0.12	NR	Siefert (1987)
	Water flea (*Daphnia* sp)	LC_{50}	4 hr	0.88	NR	Siefert (1987)
		ABD	7	7.4	NR	Ali and Mulla (1978b)
	Water flea (*Daphnia magna*)	LET	1	1.3	FEMALE	Roberts and Miller (1971)
	Water flea (*Daphnia* sp)	LET	6 hr	100	NR	Hughes (1977)
	Water flea (*Daphnia magna*)	MOR	1	0.4	FEMALE	Roberts and Miller (1971)
	Water flea (*Daphnia* sp)	MOR	6 hr	1	NR	Hughes (1977)
	Water flea (*Moina rectrirostris*)	ABD	3	0.06	LARVAE	Mulla and Khasawinah (1969)
		ABD	3	0.06	LARVAE	Mulla and Khasawinah (1969)
	Water flea (*Moina micrura*)	ABD	74	7.2	NR	Hurlbert et al. (1972)
		ABD	74	72	NR	Hurlbert et al. (1972)
		LET	28	0.011	NR	Hurlbert et al. (1970)
		MOR	28	0.011	NR	Hurlbert et al. (1970)
	Water flea order	ABD	1	10	NR	Hughes et al. (1980)
		LET	NR	0.88	NR	Siefert (1987)
	White river crayfish	LC_{50}	4	2	0.7 G	Carter and Graves (1972)

Fish	Atlantic salmon	BEH	1	100–250	1 G, 3.0–4.0 CM, JUVENILE	Peterson (1976)
		MOR	1	250	1 G, 3.0–4.0 CM, JUVENILE	Peterson (1976)
	Atlantic silverside	LC_{50}	4	1.7	EMBRYO, >48 HR	Schimmel et al. (1983)
	Banded tetra	LET	2	80	NR	Rodrigues et al. (1980)
		MOR	2	7	NR	Rodrigues et al. (1980)
		MOR	2	14	NR	Rodrigues et al. (1980)
		MOR	2	25	NR	Rodrigues et al. (1980)
		MOR	2	45	NR	Rodrigues et al. (1980)
		MOR	2	50	NR	Rodrigues et al. (1980)
		MOR	2	70	NR	Rodrigues et al. (1980)
	Barb	LC_{50}	1	0.8	NR	Scirocchi and Erme (1980)
	Bluegill	LC_{50}	4	2.4	0.6 G	Johnson and Finley (1980)
		LC_{50}	4	10	0.8 G	Phipps and Holcombe (1985)
		LC_{50}	4	30	0.5 G	Carter and Graves (1972)
		BCF	1	2.33	14.1 G, 8.6 CM	Macek et al. (1972)
		BCF	1	6.4	JUVENILE	Siefert (1987)
		BCF	1	0.995	14.1 G, 8.6 CM	Macek et al. (1972)
		BCF	3	1.73	14.1 G, 8.6 CM	Macek et al. (1972)
		BCF	3	4.5	JUVENILE	Siefert (1987)
		BCF	3	0.305	14.1 G, 8.6 CM	Macek et al. (1972)
		BCF	7	0.20	14.1 G, 8.6 CM	Macek et al. (1972)
		BCF	7	0.39	JUVENILE	Siefert (1987)
		BCF	7	1.08	14.1 G, 8.6 CM	Macek et al. (1972)
		BCF	7	1.75	JUVENILE	Siefert (1987)
		BCF	14	0.23	JUVENILE	Siefert (1987)
		BCF	14	0.77	JUVENILE	Siefert (1987)
		BCF	28	0.15	14.1 G, 8.6 CM	Macek et al. (1972)
		BCF	28	0.165	14.1 G, 8.6 CM	Macek et al. (1972)
		BCF	35	1.66	14.1 G, 8.6 CM	Macek et al. (1972)

Table 20. Toxicology of Aquatic and Terrestrial Species: Chemical: Chlorpyriphos, CAS Number: 2921882 (continued)

Class	Species	Effect	Duration (days)	Concentration ($\mu g\ L^{-1}$)	Lifestage	Ref.
Fish (continued)	Bluegill	BCF	35	0.595	14.1 G, 8.6 CM	Macek et al. (1972)
		BCF	37	0.38	14.1 G, 8.6 CM	Macek et al. (1972)
		BCF	37	1.02	14.1 G, 8.6 CM	Macek et al. (1972)
		BCF	49	0.04	14.1 G, 8.6 CM	Macek et al. (1972)
		BCF	49	0.06	14.1 G, 8.6 CM	Macek et al. (1972)
		STR	28	0.16–17.3	JUVENILE	Siefert (1987)
	Channel catfish	LC_{50}	1	160	10.0 G	Carter and Graves (1972)
		LC_{50}	3	806	7.9 G	Phipps and Holcombe (1985)
		LC_{50}	4	280	0.8 G	Johnson and Finley (1980)
		LC_{50}	4	806	7.9 G	Phipps and Holcombe (1985)
	Common, mirror, colored, carp	LC_{50}	0.5	123	132 MM, 31.5 G	El-Refai et al. (1976)
		LC_{50}	0.5	430	49 MM, 1.75 G	El-Refai et al. (1976)
		LC_{50}	1	81	132 MM, 31.5 G	El-Refai et al. (1976)
		LC_{50}	1	310	49 MM, 1.75 G	El-Refai et al. (1976)
		LC_{50}	2	59	132 MM, 31.5 G	El-Refai et al. (1976)
		LC_{50}	2	280	49 MM, 1.75 G	El-Refai et al. (1976)
	Crucian carp	LC_{50}	NR	14	NR	Jirasek et al. (1978)
	Cutthroat trout	LC_{50}	4	18	1.4 G	Johnson and Finley (1980)
	Fathead minnow	$EC_{50}AB$	4	54.9	<24 HR, HATCHED LARVAE	Jarvinen et al. (1988)
		LC_{50}	1	320	31–32 D, 0.1 G	Holcombe et al. (1982)
		LC_{50}	2	248	31–32 D, 0.1 G	Holcombe et al. (1982)
		LC_{50}	3	220	31–32 D, 0.1 G	Holcombe et al. (1982)
		LC_{50}	4	120	NEWLY HATCHED LARVAE	Jarvinen and Tanner (1982)

LC_{50}	4	122.2	<24 HR, LARVAE	Jarvinen et al. (1988)
LC_{50}	4	130	NEWLY HATCHED LARVAE	Jarvinen and Tanner (1982)
LC_{50}	4	140	NEWLY HATCHED LARVAE	Jarvinen and Tanner (1982)
LC_{50}	4	170	NEWLY HATCHED LARVAE	Jarvinen and Tanner (1982)
LC_{50}	4	200	32 D, 0.0 MM, 0.000 G	Geiger et al. (1988)
LC_{50}	4	203	31–32 D, 0.1 G	Holcombe et al. (1982)
LC_{50}	4	506	44 D, 0.0 MM, 0.000 G	Geiger et al. (1988)
LC_{50}	4	542	0.5 G	Phipps and Holcombe (1985)
ABN	30	1.29	<24 HR, LARVAE	Jarvinen et al. (1988)
ABN	30	2.10	<24 HR, LARVAE	Jarvinen et al. (1988)
ABN	NR	2.68	<24 HR, LARVAE	Jarvinen et al. (1983)
BMS	NR	0.12	<24 HR, LARVAE	Jarvinen et al. (1983)
ENZ	60	0.27	<24 HR, LARVAE	Jarvinen et al. (1983)
GRO	7	3.7	NEWLY HATCHED, <24 HR	Norberg and Mount (1985)
GRO	7	7.4	NEWLY HATCHED, <24 HR	Norberg and Mount (1985)
GRO	30	2.68	<24 HR, LARVAE	Jarvinen et al. (1983)
GRO	30	3.88	<24 HR, LARVAE	Jarvinen et al. (1988)
GRO	30	2.08	<24 HR, LARVAE	Jarvinen et al. (1988)
GRO	32	1.6	EMBRYO	Jarvinen and Tanner (1982)
GRO	32	2.2	EMBRYO	Jarvinen and Tanner (1982)
GRO	32	3.2	EMBRYO	Jarvinen and Tanner (1982)
GRO	32	4.8	EMBRYO	Jarvinen and Tanner (1982)
GRO	60	1.21	<24 HR, LARVAE	Jarvinen et al. (1983)
GRO	NR	0.12	<24 HR, LARVAE	Jarvinen et al. (1983)

Table 20. Toxicology of Aquatic and Terrestrial Species: Chemical: Chlorpyriphos, CAS Number: 2921882 (continued)

Class	Species	Effect	Duration (days)	Concen-tration ($\mu g\ L^{-1}$)	Lifestage	Ref.
Fish (continued)	Fathead minnow	MOR	30	2.68	<24 HR, LARVAE	Jarvinen et al. (1983)
		MOR	30	7.08	<24 HR, LARVAE	Jarvinen et al. (1988)
		MOR	32	2.2	EMBRYO	Jarvinen and Tanner (1982)
		MOR	32	3.2	EMBRYO	Jarvinen and Tanner (1982)
		MOR	32	4.8	EMBRYO	Jarvinen and Tanner (1982)
		MOR	32	5.7	EMBRYO	Jarvinen and Tanner (1982)
		MOR	NR	0.12	<24 HR, LARVAE	Jarvinen et al. (1983)
		MOR	NR	0.27	<24 HR, LARVAE	Jarvinen et al. (1983)
		MOR	NR	0.63	<24 HR, LARVAE	Jarvinen et al. (1983)
		MOR	NR	1.21	<24 HR, LARVAE	Jarvinen et al. (1983)
		MOR	NR	2.68	<24 HR, LARVAE	Jarvinen et al. (1983)
		REP	NR	0.63	<24 HR, LARVAE	Jarvinen et al. (1983)
		REP	NR	2.68	<24 HR, LARVAE	Jarvinen et al. (1983)
		REP	136	0.12	<24 HR, LARVAE	Jarvinen et al. (1983)
	Golden shiner	MOR	6 hr	0.1 kg ha^{-1}	NR	Mount et al. (1970)
	Goldfish	LC_{50}	4	>806	10.7 G	Phipps and Holcombe (1985)
	Green sunfish	LC_{50}	1	110	1.8 G, 5.7 CM	Davey et al. (1976)
		LC_{50}	2	50	1.8 G, 5.7 CM	Davey et al. (1976)
		LC_{50}	3	40	1.8 G, 5.7 CM	Davey et al. (1976)
	Gulf toadfish	LC_{50}	4	520	JUVENILE, 99 (73–116) MG	Hansen et al. (1986)
		BCF	49	1.4	EMBRYO - LARVAE	Hansen et al. (1986)
		BCF	49	3.7	EMBRYO - LARVAE	Hansen et al. (1986)
		BCF	49	8.2	EMBRYO - LARVAE	Hansen et al. (1986)
		BCF	49	9.7	EMBRYO - LARVAE	Hansen et al. (1986)

	BCF	49	18	EMBRYO - LARVAE	Hansen et al. (1986)
	BCF	49	21	EMBRYO - LARVAE	Hansen et al. (1986)
	BCF	49	46	EMBRYO - LARVAE	Hansen et al. (1986)
	BCF	49	46	EMBRYO - LARVAE	Hansen et al. (1986)
	BCF	49	50	EMBRYO - LARVAE	Hansen et al. (1986)
	BCF	49	93	EMBRYO - LARVAE	Hansen et al. (1986)
	BCF	49	150	EMBRYO - LARVAE	Hansen et al. (1986)
	GRO	49	1.4	EMBRYO - LARVAE	Hansen et al. (1986)
	GRO	49	3.7	EMBRYO - LARVAE	Hansen et al. (1986)
	GRO	49	18	EMBRYO - LARVAE	Hansen et al. (1986)
	MOR	49	46	EMBRYO - LARVAE	Hansen et al. (1986)
	MOR	49	93	EMBRYO - LARVAE	Hansen et al. (1986)
	MOR	49	150	EMBRYO - LARVAE	Hansen et al. (1986)
Inland silverside	LC_{50}	4	4.2	72 D	Clark et al. (1985)
Lake trout, siscowet	LC_{50}	4	98	2.3 G	Johnson and Finley (1980)
Largemouth bass	BCF	1	2.38	17.1 G, 10.5 CM	Macek et al. (1972)
	BCF	1	0.995	17.1 G, 10.5 CM	Macek et al. (1972)
	BCF	3	1.73	17.1 G, 10.5 CM	Macek et al. (1972)
	BCF	3	0.305	17.1 G, 10.5 CM	Macek et al. (1972)
	BCF	7	0.20	17.1 G, 10.5 CM	Macek et al. (1972)
	BCF	7	1.08	17.1 G, 10.5 CM	Macek et al. (1972)
	BCF	28	0.15	17.1 G, 10.5 CM	Macek et al. (1972)
	BCF	28	0.165	17.1 G, 10.5 CM	Macek et al. (1972)
	BCF	35	1.66	17.1 G, 10.5 CM	Macek et al. (1972)
	BCF	35	0.595	17.1 G, 10.5 CM	Macek et al. (1972)
	BCF	37	0.38	17.1 G, 10.5 CM	Macek et al. (1972)
	BCF	37	1.02	17.1 G, 10.5 CM	Macek et al. (1972)
	BCF	49	0.04	17.1 G, 10.5 CM	Macek et al. (1972)
	BCF	49	0.06	17.1 G, 10.5 CM	Macek et al. (1972)
Longnose killifish	LC_{50}	4	4.1	NR	Schimmel et al. (1983)
Medaka, high-eyes	LC_{50}	1	310	NR	Shim and Self (1973)

Table 20. Toxicology of Aquatic and Terrestrial Species: Chemical: Chlorpyriphos, CAS Number: 2921882 (continued)

Class	Species	Effect	Duration (days)	Concen- tration ($\mu g\ L^{-1}$)	Lifestage	Ref.
Fish (continued)	Milkfish, salmon-herring	LC_{50}	3	0.15 mg kg^{-1}	FINGERLING	Tsai (1978)
		LC_{50}	3	0.010 mg kg^{-1}	FRY, 6.0–6.5 CM, 2.8–3.2 G	Tsai (1978)
	Mosquitofish (*Gambusia affinis*)	LC_0	2	500	2.54 CM	Darwazeh and Mulla (1974)
		LC_{50}	1	190	NR	Ahmed (1977)
		LC_{50}	1	200	NR	Ahmed (1977)
		LC_{50}	1	200	NR	Ahmed (1977)
		LC_{50}	1	220	NR	Ahmed (1977)
		LC_{50}	1	1,400	MATURE GRAVID FEMALES, 0.6 G, 3.2 CM	Davey et al. (1976)
		LC_{50}	1	4,000	NR	Hansen et al. (1972)
		LC_{50}	2	440	MATURE GRAVID FEMALES, 0.6 G, 3.2 CM	Davey et al. (1976)
		LC_{50}	3	260	MATURE GRAVID FEMALES, 0.6 G, 3.2 CM	Davey et al. (1976)
		LC_{50}	4	280	0.5 G	Carter and Graves (1972)
		LC_{50}	4	570	NR	Scirocchi and Erme (1980)
		AVO	1 hr	100	20–45 MM	Hansen et al. (1972)
		BCF	1	<4	NR	Hurlbert et al. (1970)
		BCF	1	<4	NR	Hurlbert et al. (1970)
		BCF	2	<4	NR	Hurlbert et al. (1970)
		BCF	2	<4	NR	Hurlbert et al. (1970)

	BCF	4 hr	10	NR	Hurlbert et al. (1970)
	BCF	4 hr	10	NR	Hurlbert et al. (1970)
	BCF	4 hr	223	NR	Hurlbert et al. (1970)
	BCF	4 hr	223	NR	Hurlbert et al. (1970)
	LET	4 hr	<4–223	ADULT AND NYMPHS	Hurlbert et al. (1970)
	MOR	1	0.10	2.5 CM	Darwazeh and Mulla (1974)
	MOR	1	0.50	2.5 CM	Darwazeh and Mulla (1974)
	MOR	1	0.50	2.5 CM	Darwazeh and Mulla (1974)
	MOR	1	1,[illegible]00	2.54 CM	Darwazeh and Mulla (1974)
	MOR	2	<4–10	NR	Hurlbert et al. (1970)
	MOR	2	<4–223	NR	Hurlbert et al. (1970)
	MOR	5	0.05	2.5 CM	Darwazeh and Mulla (1974)
	MOR	15	0.011	NR	Hurlbert et al. (1970)
	THL	1	1	NR	Johnson (1978a)
	THL	1	25	18–53 MM	Johnson (1977)
	THL	2	5	NR	Johnson (1978a)
Motsuga, stone moroko	ENZ	0.20	0.000025 *M*	4–8 CM, 1–5 G	Kanazawa (1983)
Mullet	ENZ	NR	3	NR	Fytizas and Vassiliou (1980)
Mummichog	$EC_{50}EZ$	0.50	.44	NR	Thirugnanam and Forgash (1977)
	$EC_{50}EZ$	0.50	25,200	NR	Thirugnanam and Forgash (1977)
	LC_{50}	4	4.65	1.67 G, 5.54 CM	Thirugnanam and Forgash (1977)
	LT_{50}	NR	5.6	1.67 G, 5.54 CM	Thirugnanam and Forgash (1977)
	ENZ	1	1	1.67 G, 5.54 CM	Thirugnanam and Forgash (1977)
Northern pike, pike	LC_{50}	2	3.3	NR	Scirocchi and Erme (1980)
Oikawa	LC_{50}	1	1,700	NR	Shim and Self (1973)

Table 20. Toxicology of Aquatic and Terrestrial Species: Chemical: Chlorpyriphos, CAS Number: 2921882 (continued)

Class	Species	Effect	Duration (days)	Concen- tration ($\mu g\ L^{-1}$)	Lifestage	Ref.
Fish (continued)	Rainbow trout, donaldson trout	LC_{50}	1	53	0.6–1.5 G	Macek et al. (1969)
		LC_{50}	1	110	0.6–1.5 G	Macek et al. (1969)
		LC_{50}	1	550	0.6–1.5 G	Macek et al. (1969)
		LC_{50}	2	11.4	JUVENILE, 1 G	Holcombe et al. (1982)
		LC_{50}	2	240	NR	Scirocchi and Erme (1980)
		LC_{50}	3	8	JUVENILE, 1 G	Holcombe et al. (1982)
		LC_{50}	4	7.1	0.6–1.5 G	Macek et al. (1969)
		LC_{50}	4	7.1	1.4 G	Johnson and Finley (1980)
		LC_{50}	4	8	JUVENILE, 1 G	Holcombe et al. (1982)
		LC_{50}	4	9	3 G	Phipps and Holcombe (1985)
		LC_{50}	4	15	0.6–1.5 G	Macek et al. (1969)
		LC_{50}	4	51	0.6–1.5 G	Macek et al. (1969)
		LC_{50}	NR	7,600	NR	Meyer (1981)
		LC_{50}	NR	8,950	NR	Meyer (1981)
	Roach	LC_{50}	4	120	NR	Scirocchi and Erme (1980)
	Sheepshead minnow	LC_{50}	4	136	NR	Schimmel et al. (1983)
		AVO	0.50	100	20–40 MM	Hansen (1969)
		AVO	NR	100	20–40 MM	Hansen (1970)
	Striped mullet	LC_{50}	4	5.4	NR	Schimmel et al. (1983)
	Tench	LC_{50}	3	45	NR	Scirocchi and Erme (1980)
	Tidewater silverside	LC_{50}	4	1.3	60 D	Clark et al. (1985)

Tilapia (*Tilapia nilotica*)	LC_{50}	0.5	132	1.5 G, 41.5 MM	El-Refai et al. (1976)
	LC_{50}	0.5	140	5.36 G, 64 MM	El-Refai et al. (1976)
	LC_{50}	0.5	180	13.8 G, 89 MM	El-Refai et al. (1976)
	LC_{50}	0.5	200	5.36 G, 64 MM	El-Refai et al. (1976)
	LC_{50}	1	84	1.5 G, 41.5 MM	El-Refai et al. (1976)
	LC_{50}	1	90	5.36 G, 64 MM	El-Refai et al. (1976)
	LC_{50}	1	120	5.36 G, 64 MM	El-Refai et al. (1976)
	LC_{50}	1	139	13.8 G, 89 MM	El-Refai et al. (1976)
Tilapia (*Tilapia aurea*)	LC_{50}	1	418	1.14 G	Herzberg (1987)
Tilapia (*Tilapia nilotica*)	LC_{50}	2	62	1.5 G, 41.5 MM	El-Refai et al. (1976)
	LC_{50}	2	70	5.36 G, 64 MM	El-Refai et al. (1976)
	LC_{50}	2	100	5.36 G, 64 MM	El-Refai et al. (1976)
	LC_{50}	2	114	13.8 G, 89 MM	El-Refai et al. (1976)
	LC_{50}	3	151	19.3 G	Herzberg (1987)
Tilapia (*Tilapia aurea*)	BCF	5	72–1,060	1.14 G	Herzberg (1987)
	BCF	5	72–1,060	1.14 G	Herzberg (1987)
	BCF	5	72–1,060	1.14 G	Herzberg (1987)
	BCF	5	72–1,060	1.14 G	Herzberg (1987)
Tooth carp	LC_{50}	2	0.215 $\mu L\ L^{-1}$	NR	Boumaiza et al. (1979)
	LC_{50}	2	0.2045 $\mu L\ L^{-1}$	NR	Boumaiza et al. (1979)
	LC_{50}	4	0.169 $\mu L\ L^{-1}$	NR	Boumaiza et al. (1979)
	LC_{50}	4	0.1634 $\mu L\ L^{-1}$	NR	Boumaiza et al. (1979)

Table 20. Toxicology of Aquatic and Terrestrial Species: Chemical: Chlorpyriphos, CAS Number: 2921882 (continued)

Class	Species	Effect	Duration (days)	Concentration ($\mu g\ L^{-1}$)	Lifestage	Ref.
Fish (continued)	White sucker	BCF	1	6.4	JUVENILE	Siefert (1987)
		BCF	3	4.5	JUVENILE	Siefert (1987)
		BCF	5	2.4	JUVENILE	Siefert (1987)
		BCF	7	1.76	JUVENILE	Siefert (1987)
		BCF	14	0.77	JUVENILE	Siefert (1987)
		BCF	21	0.39	JUVENILE	Siefert (1987)
		STR	28	0.16–17.3	JUVENILE	Siefert (1987)
Insect	Backswimmer (*Neoplea striola*)	LC_{50}	2	2.42	NR	Siefert (1987)
		LC_{50}	2	3.17	NR	Siefert (1987)
		LC_{50}	3	1.59	NR	Siefert (1987)
		LC_{50}	3	1.76	NR	Siefert (1987)
		LC_{50}	4	1.22	NR	Siefert (1987)
		LC_{50}	4	1.56	NR	Siefert (1987)
		LC_{50}	6	0.97	NR	Siefert (1987)
		STR	1.25	>6.1	NR	Siefert (1987)
	Backswimmer (*Notonecta undulata*)	LC_{50}	1	35.2	ADULT	Roberts and Miller (1971)
	Backswimmer (*Notonecta* sp)	LET	6 hr	9–10	NR	Siefert (1987)
	Beetle (*Berosus styliferus*)	LC_{50}	1	9	ADULT	Ahmed (1977)
	Beetle (*Helophorus* sp)	ABD	65	72	ADULT	Hurlbert et al. (1972)
	Beetle (*Hydrophilus triangularis*)	LC_{50}	1	20	LARVA	Ahmed (1977)
		LC_{50}	1	30	ADULT	Ahmed (1977)
	Beetle (*Hygrotus* sp)	LC_{50}	1	40	ADULT	Ahmed (1977)
	Beetle (*Laccophilus fasciatus*)	LC_{50}	1	2.1	ADULT	Roberts and Miller (1971)

Beetle (*Laccophilus decipiens*)	LC_{50}	1	4.5	ADULT	Ahmed (1977)
Beetle (*Laccophilus fasciatus*)	ABD	14	<0.1–1.5	LARVAE	Roberts and Miller (1971)
Beetle (*Laccophilus* sp)	ABD	65	72	LARVAE	Hurlbert et al. (1972)
	ABD	65	72	ADULTS	Hurlbert et al. (1972)
	ABD	65	72	LARVAE	Hurlbert et al. (1972)
	ABD	65	72	ADULTS	Hurlbert et al. (1972)
Beetle (*Laccophilus fasciatus*)	ABD	77	0.9–2.7	LARVAE	Roberts and Miller (1971)
Beetle (*Peltodytes* sp)	LC_{50}	3	0.9	ADULT, 5 MG	Federle and Collins (1976)
	LC_{50}	4	0.8	ADULT, 5 MG	Federle and Collins (1976)
Beetle (*Thermonectus basillaris*)	LC_{50}	1	6	ADULT	Ahmed (1977)
Beetle (*Tropisternus lateralis*)	LC_{50}	1	8	ADULT	Ahmed (1977)
	LC_{50}	1	2	LARVA	Ahmed (1977)
	ABD	65	72	LARVAE	Hurlbert et al. (1972)
	ABD	65	72	ADULT	Hurlbert et al. (1972)
	ABD	65	72	LARVAE	Hurlbert et al. (1972)
	ABD	65	72	ADULT	Hurlbert et al. (1972)
Blackfly (*Simulium ornatum*)	HAT	1	1	EGG	Muirhead-Thomson and Merryweather (1969)
	HAT	1	1,000	EGG	Muirhead-Thomson and Merryweather (1969)
	MOR	1	100	EGG	Muirhead-Thomson and Merryweather (1969)
Blackfly family	ABD	0.25	100	LARVAE	Wallace et al. (1973)
	ABD	0.25	100	LARVAE	Wallace et al. (1973)
	ABD	0.25	100	LARVAE	Wallace et al. (1973)
	ABD	0.25	100	LARVAE	Wallace et al. (1973)
	ABD	0.25	100	LARVAE	Wallace et al. (1973)
	ABD	0.25	100	LARVAE	Wallace et al. (1973)
	ABD	0.25	100	LARVAE	Wallace et al. (1973)
	ABD	0.25	100	LARVAE	Wallace et al. (1973)
	ABD	0.25	100	LARVAE	Wallace et al. (1973)

Table 20. Toxicology of Aquatic and Terrestrial Species: Chemical: Chlorpyriphos, CAS Number: 2921882 (continued)

Class	Species	Effect	Duration (days)	Concen-tration ($\mu g\ L^{-1}$)	Lifestage	Ref.
Insect (continued)	Blackfly family	ABD	0.25	100	LARVAE	Wallace et al. (1973)
		ABD	0.25	100	LARVAE	Wallace et al. (1973)
		ABD	0.25	100	LARVAE	Wallace et al. (1973)
		ABD	0.25	100	LARVAE	Wallace et al. (1973)
		ABD	0.25	100	LARVAE	Wallace et al. (1973)
		ABD	0.25	100	LARVAE	Wallace et al. (1973)
		ABD	0.25	100	LARVAE	Wallace et al. (1973)
		ABD	0.25	100	LARVAE	Wallace et al. (1973)
		ABD	0.25	100	LARVAE	Wallace et al. (1973)
		DRF	0.08 hr	400	LARVAE, 3RD–6TH INSTAR	Jamnback and Frempong-Boadu (1966)
		DRF	0.08 hr	4,000	LARVAE, 3RD–6TH INSTAR	Jamnback and Frempong-Boadu (1966)
		DRF	0.25	100	LARVAE	Wallace et al. (1973)
		DRF	0.25	100	LARVAE	Wallace et al. (1973)
		DRF	0.25	100	LARVAE	Wallace et al. (1973)
		DRF	0.25	100	LARVAE	Wallace et al. (1973)
		DRF	0.25	100	LARVAE	Wallace et al. (1973)
		DRF	0.25	100	LARVAE	Wallace et al. (1973)
	Caddisfly family	LC_{50}	2	0.87	NR	Siefert (1987)
		LC_{50}	4	0.77	NR	Siefert (1987)
		MOR	2	0.62	NR	Siefert (1987)

Caddisfly order	ABD	0.25	100	LARVAE	Wallace et al. (1973)
	ABD	0.25	100	LARVAE	Wallace et al. (1973)
	ABD	0.25	100	LARVAE	Wallace et al. (1973)
	ABD	0.25	100	LARVAE	Wallace et al. (1973)
	ABD	0.25	100	LARVAE	Wallace et al. (1973)
	ABD	0.25	100	LARVAE	Wallace et al. (1973)
	DRF	0.25	100	LARVAE	Wallace et al. (1973)
	STR	5 hr	8.6–15.9	NR	Siefert (1987)
Dragonfly and damselfly order	ABD	65	7.2	NAIADS	Hurlbert et al. (1972)
Giant water bug	LC_{50}	1	15	ADULT	Ahmed (1977)
Insect class	ABD	65	7.2	LARVAE, NYMPHS, AND ADULTS	Hurlbert et al. (1972)
	ABD	65	7.2	LARVAE, NYMPHS, AND ADULTS	Hurlbert et al. (1972)
	ABD	65	72	LARVAE, NYMPHS, AND ADULTS	Hurlbert et al. (1972)
	ABD	65	72	LARVAE, NYMPHS, AND ADULTS	Hurlbert et al. (1972)
Mayfly (*Ephemerella* sp)	LC_{50}	2	0.40	NR	Siefert (1987)
	LC_{50}	3	0.33	NR	Siefert (1987)
Mayfly family	ABD	65	7.2	NYMPH	Hurlbert et al. (1972)
	ABD	65	72	NYMPH	Hurlbert et al. (1972)
Mayfly order	ABD	0.25	100	NYMPH	Wallace et al. (1973)
	ABD	0.25	100	NYMPH	Wallace et al. (1973)
	ABD	0.25	100	NYMPH	Wallace et al. (1973)
	ABD	0.25	100	NYMPH	Wallace et al. (1973)
	ABD	0.25	100	NYMPH	Wallace et al. (1973)
	ABD	0.25	100	NYMPH	Wallace et al. (1973)
	DRF	0.25	100	NYMPH	Wallace et al. (1973)

Table 20. Toxicology of Aquatic and Terrestrial Species: Chemical: Chlorpyriphos, CAS Number: 2921882 (continued)

Class	Species	Effect	Duration (days)	Concentration ($\mu g\ L^{-1}$)	Lifestage	Ref.
Insect (continued)	Midge (*Chaoborus americanus*)	LC_{50}	0.75	2.36	NR	Siefert (1987)
		LC_{50}	1.75	1.29	NR	Siefert (1987)
		LC_{50}	4.75	0.85	NR	Siefert (1987)
		STR	0.4167	9.5–15.9	NR	Siefert (1987)
	Midge (*Chironomus* sp)	EC_{50}IM	1	0.42	LARVAE, 4TH INSTAR	Mulla and Khasawinah (1969)
	Midge (*Chironomus tentans*)	EC_{50}IM	1	6.4	3RD–4TH INSTAR	Karnak and Collins (1974)
	Midge (*Chironomus* sp)	EC_{50}IM	2	1.1	4TH INSTAR	Norland et al. (1974)
		EC_{50}IM	2	230	4TH INSTAR	Norland et al. (1974)
	Midge (*Chironomus utahensis*)	LC_{50}	1	1.2	4TH INSTAR	Ali and Mulla (1978a)
		LC_{50}	1	1.2	4TH INSTAR	Ali and Mulla (1977)
		LC_{50}	1	5.4	4TH INSTAR	Ali and Mulla (1978a)
	Midge (*Chironomus decorus*)	LC_{50}	1	7	4TH INSTAR	Ali and Mulla (1978a)
		LC_{50}	1	46	4TH INSTAR	Ali and Mulla (1978a)
	Midge (*Chironomus stigmaterus*)	ABD	3	0.06	LARVAE	Mulla and Khasawinah (1969)
		ABD	3	0.06	LARVAE	Mulla and Khasawinah (1969)
		ABD	3	0.11	LARVAE	Mulla and Khasawinah (1969)
	Midge (*Chironomus utahensis*)	ABD	42	0.22 kg ha^{-1}	LARVAE	Ali and Mulla (1977)
	Midge (*Chironomus* sp)	ABD	49	0.14 kg ha^{-1}	LARVAE	Ali and Mulla (1977)
	Midge (*Cricotopus* sp)	LC_{50}	1	6.5	4TH INSTAR	Ali and Mulla (1977)
	Midge (*Goeldichironomus holoprassinus*)	EC_{50}IM	1	0.97	LARVAE, 4TH INSTAR	Mulla and Khasawinah (1969)
		ABD	3	0.06	LARVAE	Mulla and Khasawinah (1969)
		ABD	3	0.06	LARVAE	Mulla and Khasawinah (1969)
		ABD	3	0.28	LARVAE	Mulla and Khasawinah (1969)
	Midge (*Paratanytarsus* sp)	LET	0.75	1.6	NR	Siefert (1987)

Midge (*Procladius* sp)	LC_{50}	1	0.5	4TH INSTAR	Ali and Mulla (1978a)
	LC_{50}	1	0.5	4TH INSTAR	Ali and Mulla (1977)
	LC_{50}	1	71	4TH INSTAR	Ali and Mulla (1978a)
	ABD	42	0.22 kg ha^{-1}	LARVAE	Ali and Mulla (1977)
	ABD	70	0.14 kg ha^{-1}	LARVAE	Ali and Mulla (1977)
Midge (*Tanypus grodhausi*)	EC_{50}IM	1	0.5	LARVAE, 4TH INSTAR	Mulla and Khasawinah (1969)
	EC_{50}IM	1	5.7	LARVAE, 4TH INSTAR	Mulla and Khasawinah (1969)
	ABD	3	0.06	LARVAE	Mulla and Khasawinah (1969)
	ABD	3	0.06	LARVAE	Mulla and Khasawinah (1969)
	ABD	3	0.11	LARVAE	Mulla and Khasawinah (1969)
	ABD	3	0.28	LARVAE	Mulla and Khasawinah (1969)
Midge family	ABD	0.25	100	LARVAE	Wallace et al. (1973)
	ABD	0.25	100	LARVAE	Wallace et al. (1973)
	ABD	0.25	100	LARVAE	Wallace et al. (1973)
	ABD	0.25	100	LARVAE	Wallace et al. (1973)
	ABD	0.25	100	LARVAE	Wallace et al. (1973)
	ABD	0.25	100	LARVAE	Wallace et al. (1973)
	ABD	28	<0.1–1.6	ADULT	Roberts and Miller (1971)
	ABD	84	0.9–2.7	ADULT	Roberts and Miller (1971)
	DRF	0.25	100	LARVAE	Wallace et al. (1973)
Mosquito (*Aedes cantans*)	LC_{100}	3	10	YOUNG, 4TH INSTAR	Rettich (1979)
Mosquito (*Aedes vexans*)	LC_{50}	0.2917	1.70	NR	Siefert (1987)
	LC_{50}	0.2917	2.27	NR	Siefert (1987)
Mosquito (*Aedes triseriatus*)	LC_{50}	1	0.20	1ST INSTAR	Nelson and Evans (1973)
Mosquito (*Aedes aegypti*)	LC_{50}	1	0.30	1ST INSTAR	Nelson and Evans (1973)
Mosquito (*Aedes triseriatus*)	LC_{50}	1	0.36	2ND INSTAR	Nelson and Evans (1973)
Mosquito (*Aedes aegypti*)	LC_{50}	1	0.43	2ND INSTAR	Nelson and Evans (1973)
Mosquito (*Aedes vexans*)	LC_{50}	1	0.49	3RD INSTAR	Nelson and Evans (1973)
Mosquito (*Aedes sticticus*)	LC_{50}	1	0.5	LARVAE, 4TH INSTAR	Rettich (1977)

Table 20. Toxicology of Aquatic and Terrestrial Species: Chemical: Chlorpyriphos, CAS Number: 2921882 (continued)

Class	Species	Effect	Duration (days)	Concentration (μg L^{-1})	Lifestage	Ref.
Insect (continued)	Mosquito (*Aedes triseriatus*)	LC_{50}	1	0.77	3RD INSTAR	Nelson and Evans (1973)
	Mosquito	LC_{50}	1	0.86	4TH INSTAR	Nelson and Evans (1973)
	Mosquito (*Aedes vexans*)	LC_{50}	1	1.0	LARVAE, 4TH INSTAR	Rettich (1977)
	Mosquito (*Aedes cantans*)	LC_{50}	1	1.1	LARVAE, 4TH INSTAR	Rettich (1977)
	Mosquito (*Aedes communis*)	LC_{50}	1	1.2	LARVAE, 4TH INSTAR	Rettich (1977)
	Mosquito (*Aedes aegypti*)	LC_{50}	1	1.26	3RD INSTAR	Nelson and Evans (1973)
	Mosquito (*Aedes triseriatus*)	LC_{50}	1	1.44	4TH INSTAR	Nelson and Evans (1973)
	Mosquito (*Aedes aegypti*)	LC_{50}	1	2.31	4TH INSTAR	Nelson and Evans (1973)
	Mosquito (*Aedes punctor*)	LC_{50}	1	2.7	LARVAE, 4TH INSTAR	Rettich (1977)
	Mosquito (*Aedes excrucians*)	LC_{50}	1	3.3	LARVAE, 4TH INSTAR	Rettich (1977)
	Mosquito (*Aedes flavescens*)	MOR	1	0.28	3RD AND 4TH INSTAR LARVAE	Dixon and Brust (1971)
		MOR	1	0.28	3RD AND 4TH INSTAR LARVAE	Dixon and Brust (1971)
		MOR	1	0.28	3RD AND 4TH INSTAR LARVAE	Dixon and Brust (1971)
	Mosquito (*Anopheles freeborni*)	LC_{50}	1	3	LARVA	Ahmed (1977)
	Mosquito (*Anopheles albimanus*)	LC_{50}	1	8	4TH INSTAR	Hemingway and Georghiou (1983)
	Mosquito (*Culex pipiens molestus*)	LC_{100}	3	2	4TH INSTAR	Rettich (1979)
	Mosquito (*Culex pipiens*)	LC_{50}	1	0.16	1ST INSTAR	Nelson and Evans (1973)
	Mosquito (*Culex tritaeniorhynchus*)	LC_{50}	1	0.32	LARVAE	Shim and Self (1973)
	Mosquito (*Culex pipiens*)	LC_{50}	1	0.34	2ND INSTAR	Nelson and Evans (1973)

Mosquito (*Culex pipiens molestus*)	LC_{50}	1	0.5	LARVAE	Rettich (1979)
Mosquito (*Culex pipiens*)	LC_{50}	1	0.5	LAB REARED LARVAE	Roberts and Miller (1971)
	LC_{50}	1	0.50	3RD INSTAR	Nelson and Evans (1973)
	LC_{50}	1	0.64	4TH INSTAR	Nelson and Evans (1973)
	LC_{50}	1	1	FIELD COLLECTED LARVAE	Roberts and Miller (1971)
Mosquito (*Culex pipiens molestus*)	LC_{50}	1	1.2	LARVAE	Rettich (1979)
	LC_{50}	1	1.2	LARVAE, 4TH INSTAR	Rettich (1977)
	LC_{50}	1	1.6	LARVAE, 4TH INSTAR	Rettich (1977)
Mosquito (*Culex tarsalis*)	LC_{50}	1	2	LARVA	Ahmed (1977)
Mosquito (*Culex quinquefasciatus*)	LC_{50}	1	2.22	3RD INSTAR LARVA	Boike et al. (1985)
	LC_{50}	1	2.[illegible]4	3RD INSTAR LARVA	Boike et al. (1985)
	LC_{50}	1	2.8	3RD OR 4TH INSTAR LARVAE	El-Khatib (1985)
	LC_{50}	1	3.75	3RD INSTAR LARVA	Boike et al. (1985)
	LC_{50}	1	3.96	3RD INSTAR LARVA	Boike et al. (1985)
	LC_{50}	1	4.1	LATE 3RD OR EARLY 4TH INSTAR LARVAE	El-Khatib (1985)
	LC_{50}	1	4.[illegible]2	3RD INSTAR LARVA	Boike et al. (1985)
	LC_{50}	1	4.30	3RD INSTAR LARVA	Boike et al. (1985)
	LC_{50}	1	4.33	3RD INSTAR LARVA	Boike et al. (1985)
	LC_{50}	1	6.21	3RD INSTAR LARVA	Boike et al. (1985)
	LC_{50}	1	8.33	3RD INSTAR LARVA	Boike et al. (1985)
Mosquito (*Culex pipiens molestus*)	LC_{50}	2	[illegible].2	LARVAE	Rettich (1979)
	LC_{50}	2	[illegible].5	LARVAE	Rettich (1979)
Mosquito (*Culex pipiens*)	LC_{50}	NR	[illegible].6	2ND INSTAR LARVAE	Saleh (1988)

Table 20. Toxicology of Aquatic and Terrestrial Species: Chemical: Chlorpyriphos, CAS Number: 2921882 (continued)

Class	Species	Effect	Duration (days)	Concentration ($\mu g\ L^{-1}$)	Lifestage	Ref.
Insect (continued)	Mosquito (*Culex pipiens quinquefasciata*)	LET	1	0.7	3RD–4TH INSTAR LARVAE	Roberts and Miller (1970)
		LET	1	0.1 kg ha^{-1}	4TH INSTAR LARVAE	Mount et al. (1970)
		MOR	1	0.7	3RD–4TH INSTAR LARVAE	Roberts and Miller (1970)
	Mosquito (*Culiseta melanura*)	LC_{50}	1	0.46	1ST INSTAR	Nelson and Evans (1973)
		LC_{50}	1	0.58	2ND INSTAR	Nelson and Evans (1973)
		LC_{50}	1	1.97	3RD INSTAR	Nelson and Evans (1973)
		LC_{50}	1	2.20	4TH INSTAR	Nelson and Evans (1973)
	Mosquito (*Culiseta annulata*)	LC_{50}	1	3.5	LARVAE, 4TH INSTAR	Rettich (1977)
	Mosquito (*Wyeomyia smithii*)	DVP	7	1	2ND INSTAR LARVAE	Strickman (1985)
		MOR	7	1	2ND INSTAR LARVAE	Strickman (1985)
	Mosquito family	LET	1	0.018	LARVAE	Dixon and Brust (1971)
		LET	1	0.028	LARVAE	Dixon and Brust (1971)
		LET	1	0.028	LARVAE	Dixon and Brust (1971)
		MOR	1	0.28	LARVAE	Dixon and Brust (1971)
		MOR	1	0.28	LARVAE	Dixon and Brust (1971)
		MOR	1	0.28	LARVAE	Dixon and Brust (1971)
	Phantom midge (*Chaoborus punctipennis*)	LC_{50}	1	5.4	LARVAE	Roberts and Miller (1971)
		ABD	63	0.9–2.7	LARVAE	Roberts and Miller (1971)
		ABD	84	<0.1–1.6	LARVAE	Roberts and Miller (1971)
	Phantom midge (*Chaoborus* sp)	LET	1	100	LARVAE	Hughes (1977)
	Phantom midge (*Chaoborus* sp)	MOR	2	1	LARVAE	Hughes (1977)

Stonefly (*Claassenia sabulosa*)	LC_{50}	1	8.2	20–25 MM	Sanders and Cope (1968)
	LC_{50}	2	1.8	20–25 MM	Sanders and Cope (1968)
	LC_{50}	4	0.57	2ND YEAR CLASS	Johnson and Finley (1980)
	LC_{50}	4	0.57	20–25 MM	Sanders and Cope (1968)
Stonefly (*Plecoptera*)	ABD	0.25	100	NYMPHS	Wallace et al. (1973)
	ABD	0.25	100	NYMPHS	Wallace et al. (1973)
	ABD	0.25	100	NYMPHS	Wallace et al. (1973)
	ABD	0.25	100	NYMPHS	Wallace et al. (1973)
	ABD	0.25	100	NYMPHS	Wallace et al. (1973)
	ABD	0.25	100	NYMPHS	Wallace et al. (1973)
	DRF	0.25	100	NYMPH	Wallace et al. (1973)
Stonefly (*Pteronarcella badia*)	LC_{50}	1	4.2	15–20 MM	Sanders and Cope (1968)
Stonefly (*Pteronarcys californica*)	LC_{50}	1	50	30–35 MM	Sanders and Cope (1968)
Stonefly (*Pteronarcella badia*)	LC_{50}	2	1.8	15–20 MM	Sanders and Cope (1968)
Stonefly (*Pteronarcys californica*)	LC_{50}	2	18	30–35 MM	Sanders and Cope (1968)
Stonefly (*Pteronarcella badia*)	LC_{50}	4	0.38	15–20 MM	Sanders and Cope (1968)
Stonefly (*Pteronarcys californica*)	LC_{50}	4	10	2ND YEAR CLASS	Johnson and Finley (1980)
	LC_{50}	4	10	30–35 MM	Sanders and Cope (1968)
Water boatman (*Corisella decolor*)	ABD	1	<4–10	ADULT AND NYMPHS	Hurlbert et al. (1970)
	ABD	1	<4–223	ADULT AND NYMPHS	Hurlbert et al. (1970)
Water boatman (*Corisella* sp)	ABD	65	7.2	NYMPH	Hurlbert et al. (1972)
	ABD	65	7.2	ADULTS	Hurlbert et al. (1972)
	ABD	65	72	NYMPH	Hurlbert et al. (1972)
	ABD	65	72	ADULTS	Hurlbert et al. (1972)
Water boatman (*Corisella decolor*)	ABD	70	0.011	ADULT AND NYMPHS	Hurlbert et al. (1970)
Water strider	ABD	56	0.9–2.7	NYMPH AND ADULT	Roberts and Miller (1971)
	ABD	56	<0.1–1.6	NYMPH AND ADULT	Roberts and Miller (1971)
Waterbug order	ABD	65	7.2	NYMPH	Hurlbert et al. (1972)
	ABD	65	7.2	ADULT	Hurlbert et al. (1972)
	ABD	65	72	NYMPH	Hurlbert et al. (1972)
	ABD	65	72	ADULT	Hurlbert et al. (1972)

Table 20. Toxicology of Aquatic and Terrestrial Species: Chemical: Chlorpyriphos, CAS Number: 2921882 (continued)

Class	-Species	Effect	Duration (days)	Concen-tration (μg L^{-1})	Lifestage	Ref.
Insect (continued)	White dotted mosquito	LC_{50}	1	0.32	3RD INSTAR	Nelson and Evans (1973)
Mullusca	Salt marsh snail	POP	90	28 g ha^{-1}	NR	Fitzpatrick and Sutherland (1978)
	Snail (*Aplexa hypnorum*)	LC_{50}	4	>806	ADULT	Phipps and Holcombe (1985)
Platyhelmi	Turbellarian, flatworm	MOR	1	4	1–1.5 CM	Levy and Miller (1978)
		MOR	1	4	1–1.5 CM	Levy and Miller (1978)
Rotifer	Rotifer (*Asplanchna brightwelli*)	ABD	28	<4–223	NR	Hurlbert et al. (1970)
		ABD	74	7.2	NR	Hurlbert et al. (1972)
		ABD	74	72	NR	Hurlbert et al. (1972)
	Rotifer (*Brachionus* sp)	ABD	74	7.2	NR	Hurlbert et al. (1972)
	Rotifer (*Brachionus angularis*)	ABD	74	7.2	NR	Hurlbert et al. (1972)
	Rotifer (*Brachionus* sp)	ABD	74	72	NR	Hurlbert et al. (1972)
	Rotifer (*Brachionus quadridentata*)	ABD	76	7.2	NR	Hurlbert et al. (1972)
	Rotifer (*Brachionus budapestinensis*)	ABD	76	7.2	NR	Hurlbert et al. (1972)
	Rotifer (*Brachionus urceolaris*)	ABD	76	7.2	NR	Hurlbert et al. (1972)
	Rotifer (*Brachionus bidentata*)	ABD	76	7.2	NR	Hurlbert et al. (1972)
	Rotifer (*Brachionus plicatilis*)	ABD	76	7.2	NR	Hurlbert et al. (1972)
	Rotifer (*Brachionus calyciflorus*)	ABD	76	7.2	NR	Hurlbert et al. (1972)
	Rotifer (*Brachionus angularis*)	ABD	76	72	NR	Hurlbert et al. (1972)
	Rotifer (*Brachionus quadridentata*)	ABD	76	72	NR	Hurlbert et al. (1972)

Rotifer (*Brachionus budapestinensis*)	ABD	76	72	NR	Hurlbert et al. (1972)
Rotifer (*Brachionus urceolaris*)	ABD	76	72	NR	Hurlbert et al. (1972)
Rotifer (*Brachionus bidentata*)	ABD	76	72	NR	Hurlbert et al. (1972)
Rotifer (*Brachionus plicatilis*)	ABD	76	72	NR	Hurlbert et al. (1972)
Rotifer (*Brachionus calyciflorus*)	ABD	76	72	NR	Hurlbert et al. (1972)
Rotifer (*Epiphanes brachionus*)	ABD	76	7.2	NR	Hurlbert et al. (1972)
	ABD	76	72	NR	Hurlbert et al. (1972)
Rotifer (*Filinia terminalis*)	ABD	76	7.2	NR	Hurlbert et al. (1972)
	ABD	6	72	NR	Hurlbert et al. (1972)
Rotifer (*Hexarthra intermedia*)	ABD	76	7.2	NR	Hurlbert et al. (1972)
	ABD	76	72	NR	Hurlbert et al. (1972)
Rotifer (*Keratella* sp)	ABD	76	7.2	NR	Hurlbert et al. (1972)
	ABD	76	72	NR	Hurlbert et al. (1972)
Rotifer (*Lecane* sp)	ABD	76	7.2	NR	Hurlbert et al. (1972)
	ABD	76	72	NR	Hurlbert et al. (1972)
Rotifer (*Lepadella* sp)	ABD	76	7.2	NR	Hurlbert et al. (1972)
	ABD	76	72	NR	Hurlbert et al. (1972)
Rotifer (*Monostyla* sp)	ABD	76	7.2	NR	Hurlbert et al. (1972)
Rotifer (*Monostyla quadridentata*)	ABD	76	7.2	NR	Hurlbert et al. (1972)
Rotifer (*Monostyla* sp)	ABD	76	72	NR	Hurlbert et al. (1972)
Rotifer (*Monostyla quadridentata*)	ABD	76	72	NR	Hurlbert et al. (1972)
Rotifer (*Platyias quadricornis*)	ABD	76	7.2	NR	Hurlbert et al. (1972)
	ABD	76	72	NR	Hurlbert et al. (1972)
Rotifer (*Polyarthra trigla*)	ABD	76	7.2	NR	Hurlbert et al. (1972)
	ABD	76	72	NR	Hurlbert et al. (1972)
Rotifer (*Testudinella patina*)	ABD	76	7.2	NR	Hurlbert et al. (1972)
	ABD	76	72	NR	Hurlbert et al. (1972)
Rotifer phylum	ABD	76	7.2	NR	Hurlbert et al. (1972)
	ABD	76	72	NR	Hurlbert et al. (1972)

Table 21. Toxicology of Aquatic and Terrestrial Species: Chemical: DCPA, CAS Number: 1861321

Class	Species	Effect	Duration (days)	Concentration ($\mu g\ L^{-1}$)	Lifestage	Ref.
Algae	Algae, phytoplankton, algal mat	PGR	4 hr	1,000	NR	Butler (1964)
	Green algae (*Chlamydomonas eugametos*)	PGR	2	3,300	INITIAL CONC. 100,000 CELLS ML^{-1}	Hess (1980)
		PGR	2	33,000	INITIAL CONC. 100,000 CELLS ML^{-1}	Hess (1980)
		PGR	4	200,000	2ND D OF LOG PHASE	Vance and Smith (1969)
	Green algae (*Chlorella pyrenoidosa*)	PGR	4	200,000	2ND D OF LOG PHASE	Vance and Smith (1969)
	Green algae (*Scenedesmus quadricauda*)	PGR	4	200,000	2ND D OF LOG PHASE	Vance and Smith (1969)
Annelida	Tubificid worm (*Tubifex tubifex*)	LC_{50}	2	286,000	NR	Voronkin and Loshakov (1973)
Crustacea	Brown shrimp	MOR	2	1,000	NR	Butler (1964)
	Water flea (*Daphnia pulex*)	LC_{50}	3 hr	>40,000	NR	Nishiuchi and Hashimoto (1975)
	Water flea (*Moina macrocopa*)	LC_{50}	3 hr	>40,000	NR	Nishiuchi and Hashimoto (1975)

Fish	Bluegill	LC_{50}	1	1,000,000	NR	Hughes and Davis (1964)
		LC_{50}	2	700,000	NR	Hughes and Davis (1964)
	Common, mirror, colored, carp	LC_{50}	2	>10,000	NR	Nishiuchi and Hashimoto (1975)
	Goldfish	LC_{50}	2	>10,000	NR	Nishiuchi and Hashimoto (1975)
	Medaka, high-eyes	LC_{50}	2	>10,000	NR	Nishiuchi and Hashimoto (1975)
	Sheepshead minnow	MOR	2	1,000	JUVENILE	Butler (1964)
Mullusca	American or Virginia oyster	EC_{50}GR	4	250	JUVENILE	Butler (1964)
Tracheophy	Duckweed (*Lemna perpusilla*)	MOR	7	300,000	NR	Nishiuchi (1974)
Mammal	Rat	LC_{50}	4 hr	>5.7 mg L^{-1}		*Farm Chemicals Handbook* (1990)
		LD_{50}		10,000 mg kg^{-1}		*Farm Chemicals Handbook* (1990)

Table 22. Toxicology of Aquatic and Terrestrial Species: Chemical: Diazinon, CAS Number: 333415

Class	Species	Effect	Duration (days)	Concentration ($\mu g\ L^{-1}$)	Lifestage	Ref.
Algae	Algae, phytoplankton, algal mat	ABD	14	5,000	NR	Murray and Guthrie (1980)
		PSE	4 hr	1,000	NR	Butler (1963)
	Blue-green algae (*Aulosira fertilissima*)	GRO	18	400,000	EXPO GRO	Singh (1973)
	Blue-green algae (*Cylindrospermum* sp)	PGR	18	500,000	EXPO GRO, INITIAL % TRAN = 7	Singh (1973)
	Blue-green algae (*Plectonema boryanum*)	GRO	18	400,000	EXPO GRO	Singh (1973)
	Green algae (*Scenedesmus quadricauda*)	PSE	2	1,000	NR	Stadnyk et al. (1971)
		PSE	4	1,000	NR	Stadnyk et al. (1971)
		PSE	6	1,000	NR	Stadnyk et al. (1971)
		PSE	8	1,000	NR	Stadnyk et al. (1971)
		PSE	10	1,000	NR	Stadnyk et al. (1971)
	Green algae (*Selenastrum capricornutum*)	EC_{50}PP	7	6,400	1.42 E6 CELLS ML^{-1}	Hughes (1988)
	Green algae division	PGR	14	10	NR	Butler et al. (1975)
Amphibia	Frog (*Bufo bufo japonicus*)	LC_{50}	2	14,000	TADPOLE	Hashimoto and Nishiuchi (1981)

Annelida	Asian leech	EC_{50}IM	1	4,500	ADULT, 4–5 CM	Kimura and Keegan (1966)
		EC_{50}IM	1	7,000	ADULT, 4–5 CM	Kimura and Keegan (1966)
		EC_{50}IM	2	1,500	ADULT, 4–5 CM	Kimura and Keegan (1966)
		EC_{50}IM	2	2,400	ADULT, 4–5 CM	Kimura and Keegan (1966)
	Tubificid worm (*Tubifex* sp)	LC_{50}	7	3,160	NR	Chatterjee (1975)
Aquatic invertebrate	Invertebrates	ABD	77	3	NR	Morgan (1976)
		CLN	77	3	NR	Morgan (1976)
Crustacea	Aquatic sowbug (*Asellus communis*)	LC_{50}	4	21	NR	Morgan (1976)
	Brine shrimp	HAT	2	10,000	DRIED EGG	Kuwabara et al. (1980)
	Brown shrimp	EC_{50}EQ	1	44	ADULT	Butler (1963)
	Calanoid copepod (*Acartia tonsa*)	EC_{50}EQ	4	2.57	ADULT	Khattat and Farley (1976)
		BCF	4	0.17	NR	Khattat and Farley (1976)
		BCF	4	0.35	NR	Khattat and Farley (1976)
		BCF	4	0.68	NR	Khattat and Farley (1976)
		BCF	4	1.1	NR	Khattat and Farley (1976)
		BCF	4	6.8	NR	Khattat and Farley (1976)
	Crab (*Portunus trituberculatus*)	LC_{50}	1	4–15	ZOEA IV, MEGALOPA	Hirayama and Tamanoi (1980)
	Crayfish (*Orconectes propinquus*)	LC_{50}	1	2,846	NR	Morgan (1976)
		LC_{50}	2	537	NR	Morgan (1976)
		LC_{50}	7	15	NR	Morgan (1976)
		BEH	7	3	6.8–7.6 CM, FEMALES	Morgan (1976)
		BEH	13	3	6.8–7.6 CM, MALES	Morgan (1976)
		BEH	13	3	6.8–7.6 CM	Morgan (1976)
		BEH	14	1	6.8–7.6 CM	Morgan (1976)
	Cyclopoid copepod (*Cyclops* sp)	LC_{50}	7	2,510	NR	Chatterjee (1975)

Table 22. Toxicology of Aquatic and Terrestrial Species: Chemical: Diazinon, CAS Number: 333415 (continued)

Class	Species	Effect	Duration (days)	Concen-tration ($\mu g\ L^{-1}$)	Lifestage	Ref.
Crustacea (continued)	Kuruma shrimp	LC_{50}	1	20–50	POSTLARVAL	Hirayama and Tamanoi (1980)
		LC_{50}	1	>20,000	NAUPLIUS	Hirayama and Tamanoi (1980)
	Scud (*Gammarus pseudolimnaeus*)	LC_{50}	1	48.7	NR	Morgan (1976)
	Scud (*Gammarus lacustris*)	LC_{50}	1	434	NR	Morgan (1976)
		LC_{50}	1	800	2 MONTHS OLD	Sanders (1969)
	Scud (*Gammarus pseudolimnaeus*)	LC_{50}	2	4	NR	Morgan (1976)
	Scud (*Gammarus lacustris*)	LC_{50}	2	229	NR	Morgan (1976)
		LC_{50}	2	500	2 MONTHS OLD	Sanders (1969)
	Scud (*Gammarus pseudolimnaeus*)	LC_{50}	3	3	NR	Morgan (1976)
	Scud (*Gammarus lacustris*)	LC_{50}	3	196	NR	Morgan (1976)
	Scud (*Gammarus fasciatus*)	LC_{50}	4	0.20	MATURE	Johnson and Finley (1980)
	Scud (*Gammarus pseudolimnaeus*)	LC_{50}	4	2	NR	Morgan (1976)
	Scud (*Gammarus lacustris*)	LC_{50}	4	170	NR	Morgan (1976)
		LC_{50}	4	200	2 MONTHS OLD	Sanders (1969)
	Scud (*Gammarus pseudolimnaeus*)	LC_{50}	5	1	NR	Morgan (1976)
	Scud (*Gammarus lacustris*)	LC_{50}	5	142	NR	Morgan (1976)
	Scud (*Gammarus pseudolimnaeus*)	LC_{50}	6	0.7	NR	Morgan (1976)
		LC_{50}	7	0.5	NR	Morgan (1976)
	Scud (*Gammarus lacustris*)	LC_{50}	7	133	NR	Morgan (1976)
		AVO	1.5 hr	1,300	8–10 MM	Morgan (1976)
		BEH	3	3	5–8 MM	Morgan (1976)
	Scud (*Hyalella azteca*)	LC_{50}	0.25	60	NR	Morgan (1976)
		LC_{50}	2	22	NR	Morgan (1976)

Shrimp	BCF	1	16	1.2 G	Seguchi and Asaka (1981)
	BCF	3	21	1.2 G	Seguchi and Asaka (1981)
	BCF	7	20	1.2 G	Seguchi and Asaka (1981)
	BCF	14	21	1.2 G	Seguchi and Asaka (1981)
Water flea (*Daphnia pulex*)	EC_{50}IM	2	0.90	1ST INSTAR	Sanders and Cope (1966)
Water flea (*Daphnia magna*)	EC_{50}IM	2	0.70	<24 HR	Dortland (1980)
Water flea (*Daphnia pulex*)	EC_{50}IM	2	0.8	1ST INSTAR	Johnson and Finley (1980)
Water flea (*Daphnia magna*)	EC_{50}IM	2	0.80	<24 HR	Dortland (1980)
	EC_{50}IM	2	1.22	<24 HR	Dennis et al. (1979a)
	EC_{50}IM	2	1.22	<24 HR	Dennis et al. (1979b)
	EC_{50}IM	2	1.25	<24 HR	Dennis et al. (1979a)
	EC_{50}IM	2	1.25	<24 HR	Dennis et al. (1979b)
	EC_{50}IM	2	1.5	<24 HR	Dortland (1980)
	EC_{50}IM	21	0.22	<24 HR	Dortland (1980)
	EC_{50}IM	21	0.24	<24 HR	Dortland (1980)
	LC_{50}	1	1.81	<20 HR	Vilkas (1976)
	LC_{50}	2	0.74	<24 HR	Mitchell (1985)
	LC_{50}	2	0.96	<20 HR	Vilkas (1976)
	LC_{50}	2	2	NR	Meier et al. (1979)
Water flea (*Daphnia pulex*)	LC_{50}	3 hr	7.8	NR	Nishiuchi and Hashimoto (1975)
	LC_{50}	3 hr	80	NR	Hashimoto and Nishiuchi (1981)
Water flea (*Daphnia magna*)	LC_{50}	4	0.21	<24 HR	Mitchell (1985)
	MOR	21	0.2	<24 HR	Dortland (1980)
	MOR	21	0.3	<24 HR	Dortland (1980)
	MOT	2.08	4.3	<24 HR	Anderson (1959)
	REP	21	0.2	<24 HR	Dortland (1980)
	REP	21	0.3	<24 HR	Dortland (1980)

Table 22. Toxicology of Aquatic and Terrestrial Species: Chemical: Diazinon, CAS Number: 333415 (continued)

Class	Species	Effect	Duration (days)	Concentration ($\mu g\ L^{-1}$)	Lifestage	Ref.
Crustacea (continued)	Water flea (*Moina macrocopa*)	LC_{50}	3 hr	26	NR	Nishiuchi and Hashimoto (1975)
		LC_{50}	3 hr	50	NR	Hashimoto and Nishiuchi (1981)
	Water flea (*Simocephalus serrulatus*)	EC_{50}IM	2	1.4	1ST INSTAR	Sanders and Cope (1966)
		EC_{50}IM	2	1.4	1ST INSTAR	Johnson and Finley (1980)
		EC_{50}IM	2	1.8	1ST INSTAR	Sanders and Cope (1966)
Fish	Agohaze, goby	LC_{50}	1	2.6	NR	Hirose et al. (1979)
		LC_{50}	1	4.5	NR	Hirose et al. (1979)
		LC_{50}	1	10	NR	Hirose et al. (1979)
		LC_{50}	1	34	0.01–0.02 G	Hirose and Kitsukawa (1976)
		LC_{50}	1	38	NR	Hirose et al. (1979)
		LC_{50}	1	60	0.01–0.02 G	Hirose and Kitsukawa (1976)
		LC_{50}	1	70	NR	Hirose et al. (1979)
		LC_{50}	1	86	NR	Hirose et al. (1979)
		LC_{50}	1	58–61	NR	Hirose et al. (1979)
		LC_{50}	2	2.1	NR	Hirose et al. (1979)
		LC_{50}	2	2.2	NR	Hirose et al. (1979)
		LC_{50}	2	8.6	NR	Hirose et al. (1979)
		LC_{50}	2	26	0.01–0.02 G	Hirose and Kitsukawa (1976)
		LC_{50}	2	30	0.01–0.02 G	Hirose and Kitsukawa (1976)
		LC_{50}	2	31	NR	Hirose et al. (1979)
		LC_{50}	2	64	NR	Hirose et al. (1979)
		LC_{50}	2	82	NR	Hirose et al. (1979)

	LC_{50}	2	26–48	NR	Hirose et al. (1979)
	LC_{50}	3	33	NR	Hirose et al. (1979)
	LC_{50}	3	80	NR	Hirose et al. (1979)
	LC_{50}	3	<10–16	NR	Hirose et al. (1979)
	LC_{50}	4	16	NR	Hirose et al. (1979)
	LC_{50}	4	80	NR	Hirose et al. (1979)
	LC_{50}	4	<10	NR	Hirose et al. (1979)
	VTE	1	2	0.01–0.02 G	Hirose and Kitsukawa (1976)
	VTE	1	10	0.01–0.02 G	Hirose and Kitsukawa (1976)
	VTE	2	1.8	NR	Hirose et al. (1979)
	VTE	2	18	NR	Hirose et al. (1979)
	VTE	2	32	NR	Hirose et al. (1979)
	VTE	2	<5.6	NR	Hirose et al. (1979)
	VTE	2	<10	NR	Hirose et al. (1979)
Ayu	MOR	0.5 hr	10,000	YOUNG, 8–9 CM	Matsuo and Tamura (1970)
	MOR	1 hr	10,000	YOUNG, 8–9 CM	Matsuo and Tamura (1970)
Black bullhead	LC_{50}	2	8,000	2–10 CM, 0.5–14 G, 4–12 MONTHS	Bathe et al. (1975a)
	LC_{50}	4	8,000	2–10 CM, 0.5–14 G, 4–12 MONTHS	Bathe et al. (1975a)
Bluegill	LC_{50}	1	52	0.87 G	Cope (1965)
	LC_{50}	1	250	2.5–5.0 CM	Beliles (1965)
	LC_{50}	2	30	0.87 G	Cope (1965)
	LC_{50}	2	195	2.5–5.0 CM	Beliles (1965)
	LC_{50}	2	100–1,000	10 G, FINGERLING	Posner (1970)
	LC_{50}	3	156	2.5–5.0 CM	Beliles (1965)
	LC_{50}	4	22	0.87 G	Cope (1965)
	LC_{50}	4	22	NR	Ciba-Geigy Corp. (1976)
	LC_{50}	4	120	0.8 G	Dennis et al. (1980)
	LC_{50}	4	120	NR	Meier et al. (1979)
	LC_{50}	4	136	2.5–5.0 CM	Beliles (1965)

Table 22. Toxicology of Aquatic and Terrestrial Species: Chemical: Diazinon, CAS Number: 333415 (continued)

Class	Species	Effect	Duration (days)	Concentration ($\mu g\ L^{-1}$)	Lifestage	Ref.
Fish (continued)	Bluegill	LC_{50}	4	160	0.8 G	Dennis et al. (1980)
		LC_{50}	4	168	1 G	Johnson and Finley (1980)
		LC_{50}	4	170	0.8 G	Dennis et al. (1980)
		LC_{50}	4	170	<1 YEAR	Dennis et al. (1979a)
		LC_{50}	4	170	<1 YEAR	Dennis et al. (1979b)
		LC_{50}	4	440	1 YEAR, 56.6 MM	Allison and Hermanutz (1977)
		LC_{50}	4	480	1 YEAR, 50 MM	Allison and Hermanutz (1977)
		LC_{50}	4	530	0.8 G	Dennis et al. (1980)
		LC_{50}	4	530	<1 YEAR	Dennis et al. (1979a)
		LC_{50}	4	530	<1 YEAR	Dennis et al. (1979b)
		LC_{50}	4	400–800	10 G, FINGERLING	Posner (1970)
		LD_{50}	NR	310	NR	Menzie (1983)
		BIO	NR		NR	Hiltibran (1974)
		BIO	NR		NR	Hiltibran (1974)
		ENZ	15	1	63–90 MM	Weiss and Gakstatter (1964)
		STR	2 hr	5,000	FINGERLING, ≤10 CM	Applegate et al. (1957)
	Brook trout	LC_{50}	2	100–1,000	1.8 G, FINGERLING	Posner (1970)
		LC_{50}	4	450	1 YEAR, 220 MM	Allison and Hermanutz (1977)
		LC_{50}	4	800	1 YEAR, 190 MM	Allison and Hermanutz (1977)
		LC_{50}	4	1,050	1 YEAR, 170 MM	Allison and Hermanutz (1977)

LC_{50}	4	400–800	1.8 G, FINGERLING	Posner (1970)
BCF	NR	9.6	EGGS	Allison and Hermanutz (1977)
BCF	180	1.1	ADULT	Allison and Hermanutz (1977)
BCF	180	4.8	ADULT	Allison and Hermanutz (1977)
BCF	210	0.55	ADULT	Allison and Hermanutz (1977)
BCF	210	1.1	ADULT	Allison and Hermanutz (1977)
BCF	210	2.4	ADULT	Allison and Hermanutz (1977)
BCF	210	4.8	MATURE MALES	Allison and Hermanutz (1977)
BCF	210	4.8	IMMATURE MALES	Allison and Hermanutz (1977)
BCF	210	4.8	SPAWNED FEMALES	Allison and Hermanutz (1977)
BCF	210	9.6	ADULT	Allison and Hermanutz (1977)
BEH	14	56	1 YEAR	Allison and Hermanutz (1977)
GRO	2	11.1	EGGS, PARENTS EXPOSED AT 9.6 μG L^{-1}	Allison and Hermanutz (1977)
GRO	91	4.8	YEARLING	Allison and Hermanutz (1977)
GRO	122	0.8	EGGS, PARENTS EXPOSED AT 0.55 μG L^{-1}	Allison and Hermanutz (1977)

Table 22. Toxicology of Aquatic and Terrestrial Species: Chemical: Diazinon, CAS Number: 333415 (continued)

Class	Species	Effect	Duration (days)	Concentration ($\mu g\ L^{-1}$)	Lifestage	Ref.
Fish (continued)	Brook trout	MOR	60	2.6	1 YEAR	Allison and Hermanutz (1977)
		MOR	173	2.4	YEARLING	Allison and Hermanutz (1977)
		MOR	173	9.6	YEARLING	Allison and Hermanutz (1977)
	Bullhead, catfish	LC_{50}	4	8,000	NR	Ciba-Geigy Corp. (1976)
	Carp (*Cyprinus carpio* var *communis*)	EC_{50}HA	NR	8	EGGS	Kaur and Toor (1977)
	Central mudminnow	CYT	11	5.4E–10	NR	Vigfusson et al. (1983)
		CYT	11	5.4E–11	NR	Vigfusson et al. (1983)
	Channel catfish	EC_{50}EZ	0.08 hr	304,000	1.0–1.5 KG	Christensen and Tucker (1976)
		MOR	2–4	100–1,000	6 G, FINGERLING	Posner (1970)
	Climbing perch (*Anabas testudineus*)	HIS	7–15	3.36	30.7 G, 11.5 CM	Kabir and Ahmed (1979)
		PHY	15	3.36	30.7 G, 11.5 CM	Kabir and Ahmed (1979)
	Common, mirror, colored, carp	LC_{50}	1	1,900	50–60 D, 3.4–4.5 CM, 0.62–1.37 G	Hashimoto et al. (1982)
		LC_{50}	1	2,300	25–33 D, 2.4–2.7 CM, 0.19–0.28 G	Hashimoto et al. (1982)
		LC_{50}	1	2,400	70–80 D, 4.6–5.8 CM, 1.46–3.60 G	Hashimoto et al. (1982)
		LC_{50}	1	2,500	FLOATING FRY, 5–6 D, 0.7–0.8 CM	Hashimoto et al. (1982)

	LC_{50}	1	2,700	9–10 D, 1.0–1.4 CM, 0.007–0.020 G	Hashimoto et al. (1982)
	LC_{50}	1	2,800	17–19 D, 1.4–1.7 CM	Hashimoto et al. (1982)
	LC_{50}	1	3,180	7–9 CM	Toor and Kaur (1974)
	LC_{50}	1	3,800	4.2 CM, 1.2 G	Nishiuchi and Asano (1981)
	LC_{50}	1	6,100	SAC FRY, 1–3 D, 0.5–0.7 CM	Hashimoto et al. (1982)
	LC_{50}	1	7,200	EYED EGG	Hashimoto et al. (1982)
	LC_{50}	2	1,800	4.45 CM, 1.10 G	Li and Chen (1981)
	LC_{50}	2	2,000	4.2 CM, 1.2 G	Nishiuchi and Asano (1981)
	LC_{50}	2	3,140	7–9 CM	Toor and Kaur (1974)
	LC_{50}	2	3,200	4.45 CM, 1.10 G	Li and Chen (1981)
	LC_{50}	2	3,200	NR	Hashimoto and Nishiuchi (1981)
	LC_{50}	2	3,200	NR	Nishiuchi and Hashimoto (1975)
	LC_{50}	2	3,200	6 CM, 3 G	Nishiuchi and Asano (1981)
	LC_{50}	3	2,000	4.2 CM, 1.2 G	Nishiuchi and Asano (1981)
	LC_{50}	3	3,110	7–9 CM	Toor and Kaur (1974)
	BCF	1	18	8 G	Seguchi and Asaka (1981)
	BCF	3	18	8 G	Seguchi and Asaka (1981)
	BCF	7	18	8 G	Seguchi and Asaka (1981)
	BCF	14	18	8 G	Seguchi and Asaka (1981)
	HAT	4	100	EGGS	Malone and Blaylock (1970)
	HAT	4	1,000	EGGS	Malone and Blaylock (1970)
Crucian carp	LC_{50}	2	5,000	2–10 CM, 0.5–14 G, 4–12 MONTHS	Bathe et al. (1975a)
	LC_{50}	4	5,000	2–10 CM, 0.5–14 G, 4–12 MONTHS	Bathe et al. (1975a)
	LC_{50}	4	5,000	NR	Ciba-Geigy Corp. (1976)
	LC_{50}	4	23,400	25–50 G	Bathe et al. (1975b)

Table 22. Toxicology of Aquatic and Terrestrial Species: Chemical: Diazinon, CAS Number: 333415 (continued)

Class	Species	Effect	Duration (days)	Concentration ($\mu g\ L^{-1}$)	Lifestage	Ref.
Fish (continued)	Cutthroat trout	LC_{50}	4	1,700	2 G	Johnson and Finley (1980)
		LC_{50}	4	2,760	2.02 G	Swedburg (1973)
		LC_{50}	4	3,850	0.52 G	Swedburg (1973)
	Fathead minnow	LC_{50}	2	100–1,000	1 G, FINGERLING	Posner (1970)
		LC_{50}	4	3,700	<1 YEAR	Dennis et al. (1979a)
		LC_{50}	4	3,700	<1 YEAR	Dennis et al. (1979b)
		LC_{50}	4	400–800	1 G, FINGERLING	Posner (1970)
		LC_{50}	4	4,300	NEWLY HAT LARVAE	Jarvinen and Tanner (1982)
		LC_{50}	4	6,100	NEWLY HAT LARVAE	Jarvinen and Tanner (1982)
		LC_{50}	4	6,600	20 WEEKS, 30 MM	Allison and Hermanutz (1977)
		LC_{50}	4	6,800	15 WEEKS, 30 MM	Allison and Hermanutz (1977)
		LC_{50}	4	6,900	NEWLY HAT LARVAE	Jarvinen and Tanner (1982)
		LC_{50}	4	9,350	31 D, 21.8 MM, 0.160 G	Geiger et al. (1988)
		LC_{50}	4	10,000	13 WEEKS, 35 MM	Allison and Hermanutz (1977)
		LC_{50}	4	10,300	NR	Meier et al. (1979)
		LC_{50}	4	5,600–10,000	<1 YEAR	Dennis et al. (1979a)
		LC_{50}	4	5,600–10,000	<1 YEAR	Dennis et al. (1979b)
		ENZ	1.33 hr	500	8–10 CM	Weiss (1961)
		GRO	32	40	EMBRYO	Jarvinen and Tanner (1982)

GRO	32	50	EMBRYO	Jarvinen and Tanner (1982)
GRO	32	76	EMBRYO	Jarvinen and Tanner (1982)
GRO	32	90	EMBRYO	Jarvinen and Tanner (1982)
GRO	61	229	4 D	Allison and Hermanutz (1977)
HAT	274	3.2	5 D, FISH	Allison and Hermanutz (1977)
LET	5	100	NR	Hilsenhoff (1959)
MOR	5	0.1	NR	Hilsenhoff (1959)
MOR	5	1	NR	Hilsenhoff (1959)
MOR	5	10	NR	Hilsenhoff (1959)
MOR	30	1,100	4 D	Allison and Hermanutz (1977)
MOR	32	140	EMBRYO	Jarvinen and Tanner (1982)
MOR	32	260	EMBRYO	Jarvinen and Tanner (1982)
MOR	32	290	EMBRYO	Jarvinen and Tanner (1982)
MOR	32	490	EMBRYO	Jarvinen and Tanner (1982)
MOR	60	3.3	EGGS	Allison and Hermanutz (1977)
MOR	60	6.8	EGGS	Allison and Hermanutz (1977)
MOR	60	28	EGGS, PARENTS NOT EXPOSED	Allison and Hermanutz (1977)
MOR	274	28	5 D	Allison and Hermanutz (1977)
MOR	274	60.3	5 D	Allison and Hermanutz (1977)
VTE	91	69	4 D	Allison and Hermanutz (1977)
VTE	168	3.2	5 D	Allison and Hermanutz (1977)

Table 22. Toxicology of Aquatic and Terrestrial Species: Chemical: Diazinon, CAS Number: 333415 (continued)

Class	Species	Effect	Duration (days)	Concentration ($\mu g\ L^{-1}$)	Lifestage	Ref.
Fish (continued)	Flagfish	LC_{50}	4	1,500	6 WEEKS, 18.1 MM	Allison and Hermanutz (1977)
		LC_{50}	4	1,800	7 WEEKS, 17.8 MM	Allison and Hermanutz (1977)
		GRO	90	60	JUVENILE, 29 D PROGENY	Allison (1977)
		GRO	90	1,070	JUVENILE, 29 D PROGENY	Allison (1977)
		GRO	120	14	1 D, PROGENY	Allison (1977)
		GRO	120	54	1 D, PROGENY	Allison (1977)
		GRO	120	88	1 D, PROGENY	Allison (1977)
		MOR	60	380	ADULT, 66 D PROGENY	Allison (1977)
		MOR	60	380	ADULT, 66 D PROGENY	Allison (1977)
		MOR	60	1,170	ADULT, 66 D PROGENY	Allison (1977)
		MOR	60	1,170	ADULT, 66 D PROGENY	Allison (1977)
	Golden shiner	LC_{50}	2	100–1,000	2.6 G, FINGERLING	Posner (1970)
		LC_{50}	4	400–800	2.6 G, FINGERLING	Posner (1970)
		ENZ	15	1	63–90 MM	Weiss and Gakstatter (1964)

Goldfish	LC_{50}	1	18,000	2.5–6.0 CM	Beliles (1965)
	LC_{50}	2	5,100	NR	Hashimoto and Nishiuchi (1981)
	LC_{50}	2	5,100	NR	Nishiuchi and Hashimoto (1975)
	LC_{50}	2	14,000	2.5–6.0 CM	Beliles (1965)
	LC_{50}	3	9,600	2.5–6.0 CM	Beliles (1965)
	LC_{50}	4	9,000	2.5–6.0 CM	Beliles (1965)
	ENZ	1	100	45–65 MM	Weiss (1959)
	ENZ	15	1	63–90 MM	Weiss and Gakstatter (1964)
Green fish	LC_{50}	1	54	0.15–0.24 G	Hirose and Kitsukawa (1976)
	LC_{50}	1	80	0.15–0.24 G	Hirose and Kitsukawa (1976)
	LC_{50}	1	90	NR	Hirose et al. (1979)
	LC_{50}	1	160	NR	Hirose et al. (1979)
	LC_{50}	1	190	NR	Hirose et al. (1979)
	LC_{50}	1	230	NR	Hirose et al. (1979)
	LC_{50}	2	40	0.15–0.24 G	Hirose and Kitsukawa (1976)
	LC_{50}	2	74	0.15–0.24 G	Hirose and Kitsukawa (1976)
	LC_{50}	2	95	NR	Hirose et al. (1979)
	LC_{50}	2	140	NR	Hirose et al. (1979)
	LC_{50}	2	150	NR	Hirose et al. (1979)
	LC_{50}	2	220	NR	Hirose et al. (1979)
	LC_{50}	3	56	NR	Hirose et al. (1979)
	LC_{50}	3	220	NR	Hirose et al. (1979)
	LC_{50}	4	56	NR	Hirose et al. (1979)
	LC_{50}	4	160	NR	Hirose et al. (1979)
	VTE	1	2	0.15–0.24 G	Hirose and Kitsukawa (1976)
	VTE	1	20	0.15–0.24 G	Hirose and Kitsukawa (1976)
	VTE	2	<18	NR	Hirose et al. (1979)
	VTE	2	<56	NR	Hirose et al. (1979)
	VTE	4	32	NR	Hirose et al. (1979)

Table 22. Toxicology of Aquatic and Terrestrial Species: Chemical: Diazinon, CAS Number: 333415 (continued)

Class	Species	Effect	Duration (days)	Concentration (μg L^{-1})	Lifestage	Ref.
Fish (continued)	Green fish	VTE	4	<56	NR	Hirose et al. (1979)
		VTE	NR	10	0.15–0.24 G	Hirose and Kitsukawa (1976)
	Guppy	LC_{50}	1	3,700	7 WEEKS	Chen et al. (1971)
		LC_{50}	1	3,800	7 WEEKS	Chen et al. (1971)
		LC_{50}	2	6,000	2–10 CM, 0.5–14 G, 4–12 MONTHS	Bathe et al. (1975a)
		LC_{50}	4	3,000	2–10 CM, 0.5–14 G, 4–12 MONTHS	Bathe et al. (1975a)
		LC_{50}	4	3,000	NR	Ciba-Geigy Corp. (1976)
		LET	2	4,000	NR	Adlung (1957)
		MOR	0.5 hr	7,000	7 WEEKS	Chen et al. (1971)
	Harlequinfish, red rasbora	LC_{50}	1	1,450	1.3–3 CM	Alabaster (1969)
	Ide, silver or golden orfe	LC_{50}	2	500	2–10 CM, 0.5–14 G, 4–12 MONTHS	Bathe et al. (1975a)
		LC_{50}	4	150	2–10 CM, 0.5–14 G, 4–12 MONTHS	Bathe et al. (1975a)
		LC_{50}	4	150	NR	Ciba-Geigy Corp. (1976)
	Indian catfish	LC_{50}	1	3,030	50–75 MM, 5–10 G	Verma et al. (1982a)
		LC_{50}	2	2,750	50–75 MM, 5–10 G	Verma et al. (1982a)
		LC_{50}	3	2,390	50–75 MM, 5–10 G	Verma et al. (1982a)
		LC_{50}	4	2,270	50–75 MM, 5–10 G	Verma et al. (1982a)
		BEH	3.5 hr	5,000	30.2 G, 17.8 CM	Kabir and Begum (1978)
		BEH	25	2,500	30.2 G, 17.8 CM	Kabir and Begum (1978)
		HIS	25	1,000	SEXUALLY MATURE MALES AND FEMALES	Banu et al. (1984)

	HIS	NR	5,000	30.2 G, 17.8 CM	Kabir and Begum (1978)
	MOR	4	12,000	65 G	Sastry and Malik (1982b)
	MOR	14	3,300	65 G	Sastry and Malik (1982b)
	MOR	25	3,300	65 G	Sastry and Malik (1982b)
	MOR	30	3,300	65 G	Sastry and Malik (1982b)
Lake trout	LC_{50}	4	602	3.22 G	Swedburg (1973)
Lake trout, siscowet	LC_{50}	4	602	3.2 G	Johnson and Finley (1980)
Largemouth bass	LT_{50}	1 hr	500	8–10 CM	Weiss (1961)
	ENZ	1 hr	500	8–10 CM	Weiss (1961)
Medaka, high-eyes	LC_{50}	1	4,800	NR	Asaka et al. (1980)
	LC_{50}	1	11,000	NR	Asaka et al. (1980)
	LC_{50}	1	>20,000	NR	Asaka et al. (1980)
	LC_{50}	2	3,800	NR	Asaka et al. (1980)
	LC_{50}	2	5,300	NR	Hashimoto and Nishiuchi (1981)
	LC_{50}	2	5,300	NR	Nishiuchi and Hashimoto (1975)
	LC_{50}	2	9,200	NR	Asaka et al. (1980)
	LC_{50}	2	20,000	NR	Asaka et al. (1980)
	LC_{50}	3	9,100	NR	Asaka et al. (1980)
	LC_{50}	3	<2,100	NR	Asaka et al. (1980)
	LC_{50}	3	<20,000	NR	Asaka et al. (1980)
	AVO	NR	98.1	NR	Hidaka et al. (1984)
	BEH	2	1,000	LARVAL	Hirose and Kawakami (1977)
	BEH	2	5,000	LARVAL	Hirose and Kawakami (1977)
	HAT	24	100	FERTILIZED EGGS	Hirose and Kawakami (1977)
	HAT	24	5,000	FERTILIZED EGGS	Hirose and Kawakami (1977)
Minnow, tanago	LC_{50}	2	3,200	NR	Nishiuchi (1977a)
Molly	LC_{50}	7	1,600	NR	Chatterjee (1975)
Mosquitofish (*Gambusia patruelis*)	LC_{50}	2	1,273	3.28 (3–4) CM, 0.36 (0.31–0.51) G	Li and Chen (1981)

Table 22. Toxicology of Aquatic and Terrestrial Species: Chemical: Diazinon, CAS Number: 333415 (continued)

Class	Species	Effect	Duration (days)	Concentration (μg L^{-1})	Lifestage	Ref.
Fish (continued)	Mosquitofish (*Gambusia affinis*)	ENZ	0.17 hr	30,000	NR	Whitmore and Hodges (1978)
		ENZ	0.17 hr	30,000	NR	Whitmore and Hodges (1978)
		LET	1	0.3	NR	Mulla et al. (1963)
	Motsuga, stone moroko	BCF	1–14	5–20	4–8 CM, 2–5 G	Kanazawa (1981)
		BCF	30	270	4–8 CM, 2–5 G	Kanazawa (1975)
		ENZ	0.2 hr	0.0000075	4–8 CM, 1–5 G	Kanazawa (1983)
		ENZ	1	10	4–8 CM, 1–5 G	Kanazawa (1983)
		ENZ	14	50	4–8 CM, 1–5 G	Kanazawa (1983)
	Mozambique tilapia	GRO	NR	463	NR	Chatterjee and Konar (1984)
		MOR	NR	2	FINGERLING	Kok (1972)
	Mummichog	BCF	3 hr	120	NR	Miller et al. (1966)
		LET	1	120	NR	Miller et al. (1966)
	Oriental weatherfish	LC_{50}	2	500	NR	Hashimoto and Nishiuchi (1981)
		BCF	1	16	2.6 G	Seguchi and Asaka (1981)
		BCF	3	14	2.6 G	Seguchi and Asaka (1981
		BCF	7	15	2.6 G	Seguchi and Asaka (1981)
		BCF	14	14	2.6 G	Seguchi and Asaka (1981)
	Pink salmon	LET	2	500,000	FRY, 1.3 G	Kuroda (1975)
	Rainbow trout, donaldson trout	LC_{50}	1	380	1.20 G	Cope (1965)
		LC_{50}	1	2,900	3–7 CM	Beliles (1965)
		LC_{50}	2	170	1.20 G	Cope (1965)
		LC_{50}	2	1,000	3–7 CM	Beliles (1965)
		LC_{50}	2	8,000	2–10 CM, 0.5–14 G, 4–12 MONTHS	Bathe et al. (1975a)

	LC_{50}	3	420	3–7 CM	Beliles (1965)
	LC_{50}	4	20	FRY	Mitchell (1985)
	LC_{50}	4	90	1.2 G	Johnson and Finley (1980)
	LC_{50}	4	90	1.20 G	Cope (1965)
	LC_{50}	4	90	NR	Ciba-Geigy Corp. (1976)
	LC_{50}	4	400	3–7 CM	Beliles (1965)
	LC_{50}	4	1,350	NR	Meier et al. (1979)
	LC_{50}	4	3,200	25–50 G	Bathe et al. (1975b)
	BCF	1	14	16 G	Seguchi and Asaka (1981)
	BCF	3	12	16 G	Seguchi and Asaka (1981)
	BCF	7	12	16 G	Seguchi and Asaka (1981)
	BCF	14	15	16 G	Seguchi and Asaka (1981)
	MOR	0.5 hr	5,000	YOUNG, 9–11 CM	Matsuo and Tamura (1970)
	MOR	0.5 hr	15,000	YOUNG, 9–11 CM	Matsuo and Tamura (1970)
	MOR	1 hr	15,000	YOUNG, 9–11 CM	Matsuo and Tamura (1970)
	MOR	1 hr	15,000	YOUNG, 9–11 CM	Matsuo and Tamura (1970)
	STR	3 hr	5,000	FINGERLING, ≤10 CM	Applegate et al. (1957)
Sea lamprey	STR	0.5	5,000	LARVAE, 8–13 CM	Applegate et al. (1957)
Sheepshead minnow	LC_{50}	4	1,470	14 MM	Goodman et al. (1979)
	BCF	108	1.8	14 MM	Goodman et al. (1979)
	BCF	108	3.5	14 MM	Goodman et al. (1979)
	BCF	108	6.5	14 MM	Goodman et al. (1979)
	REP	5–7	0.47	14 MM	Goodman et al. (1979)
	REP	5–7	0.47	14 MM	Goodman et al. (1979)
Snake-head catfish	LC_{50}	1	590	60 G, 15 CM	Anees (1975)
(*Channa punctatus*)	LC_{50}	1	590	10–15 CM	Anees (1974a)
	LC_{50}	2	530	60 G, 15 CM	Anees (1975)
	LC_{50}	2	530	10–15 CM	Anees (1974a)
	LC_{50}	3	500	60 G, 15 CM	Anees (1975)
	LC_{50}	3	500	10–15 CM	Anees (1974a)
	LC_{50}	4	455	60 G, 15 CM	Anees (1975)

Table 22. Toxicology of Aquatic and Terrestrial Species: Chemical: Diazinon, CAS Number: 333415 (continued)

Class	Species	Effect	Duration (days)	Concentration ($\mu g\ L^{-1}$)	Lifestage	Ref.
Fish (continued)	Snake-head catfish (*Channa punctatus*)	LC_{50}	4	455	10–15 CM	Anees (1974a)
		LC_{50}	4	3,100	50–60 G, 14–16 CM	Sastry and Sharma (1980)
		ENZ	4	3,100	50–60 G, 14–16 CM	Sastry and Sharma (1980)
		ENZ	4	3,100	60 G	Sastry and Malik (1982a)
		ENZ	15	310	50–60 G, 14–16 CM	Sastry and Sharma (1980)
		ENZ	15	310	60 G	Sastry and Malik (1982a)
		ENZ	30	310	50–60 G, 14–16 CM	Sastry and Sharma (1980)
		ENZ	30	310	60 G	Sastry and Malik (1982a)
		HEM	1	370	MALE	Anees (1974b)
		HEM	4	280	MALE	Anees (1974b)
		HEM	14	150	MALE	Anees (1974b)
		HEM	14	150	ADULT, 15 CM, 60 G	Anees (1978b)
		HEM	30	200	14 CM, 50 G	Sastry and Sharma (1981)
		HIS	1	370	NR	Anees (1976)
		HIS	1	370	NR	Anees (1978a)
		MOR	30	310	50–60 G, 14–16 CM	Sastry and Sharma (1980)
		PHY	1	370	10–15 CM	Anees (1974a)
		PHY	4	280	10–15 CM	Anees (1974a)
		PHY	14	150	10–15 CM	Anees (1974a)
	Tilapia (*Tilapia* sp)	LC_{50}	2	1,492.5	4.99 (4.6–5.38) CM, 1.46 (1.23–2.04) G	Li and Chen (1981)
	Tooth carp	LC_{50}	2	0.189	NR	Boumaiza et al. (1979)
		LC_{50}	4	0.151	NR	Boumaiza et al. (1979)
	Trout family	LC_{50}	4	8,000	NR	Ciba-Geigy Corp. (1976)

	White mullet	LC_{50}	1	250	JUVENILE	Butler (1963)
		LC_{50}	2	250	JUVENILE	Butler (1963)
	Yellowtail	LC_{50}	1	70	0.01–0.02 G	Hirose and Kitsukawa (1976)
		LC_{50}	2	40	0.01–0.02 G	Hirose and Kitsukawa (1976)
	Zebra danio, zebrafish	LC_{50}	1	2,300	NR	Ansari et al. (1987)
		LC_{50}	2	2,180	NR	Ansari et al. (1987)
		LC_{50}	3	2,190	NR	Ansari et al. (1987)
		LC_{50}	4	2,120	NR	Ansari et al. (1987)
Insect	Beetle order	DRF	1	3	NR	Morgan (1976)
		DRF	3 hr	3	NR	Morgan (1976)
		DRF	33	3	NR	Morgan (1976)
		DRF	33	3	NR	Morgan (1976)
	Blackfly (*Simulium* sp)	MOR-D	0.02 hr	1,000	LARVAE	Matsuo and Tamura (1970)
		MOR-D	0.02 hr	10,000	LARVAE	Matsuo and Tamura (1970)
		MOR-D	0.17 hr	100	LARVAE	Matsuo and Tamura (1970)
		MOR-D	0.17 hr	1,000	LARVAE	Matsuo and Tamura (1970)
	Blackfly family	DRF-D	0.08 hr	40	LARVAE, 3RD–6TH INSTAR	Jamnback and Frempong-Boadu (1966)
		DRF-D	0.08 hr	400	LARVAE, 3RD–6TH INSTAR	Jamnback and Frempong-Boadu (1966)
		DRF-D	0.08 hr	400	LARVAE, 3RD–6TH INSTAR	Jamnback and Frempong-Boadu (1966)
		DRF-D	0.08 hr	400	LARVAE, 3RD–6TH INSTAR	Jamnback and Frempong-Boadu (1966)
		DRF-D	0.08 hr	4,000	LARVAE, 3RD–6TH INSTAR	Jamnback and Frempong-Boadu (1966)
		DRF-D	0.08 hr	4,000	LARVAE, 3RD–6TH INSTAR	Jamnback and Frempong-Boadu (1966)
		DRF-D	0.08 hr	4,000	LARVAE, 3RD–6TH INSTAR	Jamnback and Frempong-Boadu (1966)

Table 22. Toxicology of Aquatic and Terrestrial Species: Chemical: Diazinon, CAS Number: 333415 (continued)

Class	Species	Effect	Duration (days)	Concen-tration (µg L^{-1})	Lifestage	Ref.
Insect (continued)	Blackfly family	DRF-D	0.08 hr	4,000	LARVAE, 3RD–6TH INSTAR	Jamnback and Frempong-Boadu (1966)
	Caddisfly (*Cheumatopsyche oxa*)	LC_{50}	1	30	LARVAE	Morgan (1976)
		LC_{50}	1	190	LARVAE	Morgan (1976)
	Caddisfly (*Cheumatopsyche* sp)	DRF	1	3	NR	Morgan (1976)
	Caddisfly (*Hydropsyche morosa*)	EC_{50}IM	0.25	<500	LARVAE, 25 MG	Fredeen (1972)
		EC_{50}IM	0.25	<500	LARVAE, 25 MG	Fredeen (1972)
	Caddisfly (*Hydropsyche recurvata*)	EC_{50}IM	0.25	<500	LARVAE, 25 MG	Fredeen (1972)
		EC_{50}IM	0.25	<500	LARVAE, 25 MG	Fredeen (1972)
	Caddisfly (*Hydropsyche morosa*)	EC_{50}IM	3 hr	2,500	LARVAE, 25 MG	Fredeen (1972)
	Caddisfly (*Hydropsyche recurvata*)	EC_{50}IM	3 hr	>500	LARVAE, 25 MG	Fredeen (1972)
	Caddisfly (*Hydropsyche morosa*)	EC_{50}IM	3 hr	>2,500	LARVAE, 25 MG	Fredeen (1972)
	Caddisfly (*Hydropsyche recurvata*)	EC_{50}IM	3 hr	>2,500	LARVAE, 25 MG	Fredeen (1972)
	Caddisfly (*Hydropsyche morosa*)	EC_{50}IM	6 hr	500	LARVAE, 25 MG	Fredeen (1972)
		EC_{50}IM	6 hr	2,500	LARVAE, 25 MG	Fredeen (1972)
	Caddisfly (*Hydropsyche recurvata*)	EC_{50}IM	6 hr	<500	LARVAE, 25 MG	Fredeen (1972)
		EC_{50}IM	6 hr	>500	LARVAE, 25 MG	Fredeen (1972)
	Caddisfly (*Hydropsyche sparna*)	LC_{50}	1	90	LARVAE	Morgan (1976)
		LC_{50}	1	220	LARVAE	Morgan (1976)
	Caddisfly (*Hydropsyche* sp)	DRF	1	3	NR	Morgan (1976)
	Caddisfly (*Leptocella albida*)	LC_{50}	1	220	LARVAE	Morgan (1976)

Caddisfly order	DRF	3 hr	3	ADULTS	Morgan (1976)
	DR	33	3	NR	Morgan (1976)
	DRF	33	3	NR	Morgan (1976)
Damselfly (*Lestes congener*)	LC_{50}	4	50	NYMPHS, 44 MG	Federle and Collins (1976)
Dragonfly (*Orthetrum albistylum speciosum*)	LC_{50}	1	150	NR	Nishiuchi and Asano (1978)
	LC_{50}	1 hr	6,700	NR	Nishiuchi and Asano (1978)
	LC_{50}	2	140	LARVAE	Hashimoto and Nishiuchi (1981)
	LC_{50}	2	140	NR	Nishiuchi and Asano (1978)
	LC_{50}	2 hr	3,700	NR	Nishiuchi and Asano (1978)
	LC_{50}	3 hr	2,300	NR	Nishiuchi and Asano (1978)
	LC_{50}	6 hr	550	NR	Nishiuchi and Asano (1978)
Fly, mosquito, midge order	DRF	3 hr	3	LARVAE, PUPAE, ADULTS	Morgan (1976)
	DRF	33	3	LARVAE and ADULTS	Morgan (1976)
Mayfly (*Baetis intermedius*)	LC_{50}	1	358	LARVAE	Morgan (1976)
	LC_{50}	2	55	LARVAE	Morgan (1976)
	LC_{50}	4	24	LARVAE	Morgan (1976)
Mayfly (*Cloeon dipterum*)	LC_{50}	1	13	YOUNGER NYMPH, 9.3 MM, 5.6 MG	Nishiuchi and Asano (1979)
	LC_{50}	2	7.8	LARVAE	Hashimoto and Nishiuchi (1981)
	LC_{50}	2	7.8	YOUNGER NYMPH, 9.3 MM, 5.6 MG	Nishiuchi and Asano (1979)
	LC_{50}	3 hr	120	YOUNGER NYMPH, 9.3 MM, 5.6 MG	Nishiuchi and Asano (1979)
	LC_{50}	6 hr	52	YOUNGER NYMPH, 9.3 MM, 5.6 MG	Nishiuchi and Asano (1979)
Mayfly (*Ephemerella* sp)	DRF	1	3	NR	Morgan (1976)

Table 22. Toxicology of Aquatic and Terrestrial Species: Chemical: Diazinon, CAS Number: 333415 (continued)

Class	Species	Effect	Duration (days)	Concentration ($\mu g\ L^{-1}$)	Lifestage	Ref.
Insect (continued)	Mayfly (*Paraleptophlebia pallipes*)	LC_{50}	0.5	440	NYMPH	Morgan (1976)
		LC_{50}	1	243	NYMPH	Morgan (1976)
		LC_{50}	2	134	NYMPH	Morgan (1976)
		LC_{50}	3	85	NYMPH	Morgan (1976)
		LC_{50}	4	44	NYMPH	Morgan (1976)
		LC_{50}	6	43	NYMPH	Morgan (1976)
		LC_{50}	7	32	NYMPH	Morgan (1976)
	Mayfly order	DRF	3 hr	3	NR	Morgan (1976)
		DRF	33	3	NR	Morgan (1976)
		DRF	33	3	NR	Morgan (1976)
	Midge (*Chironomus tentans*)	LC_{50}	1	0.4	LARVAE	Morgan (1976)
		LC_{50}	2	0.1	LARVAE	Morgan (1976)
		LC_{50}	3	0.07	LARVAE	Morgan (1976)
		LC_{50}	4	0.03	LARVAE	Morgan (1976)
		LC_{50}	7	0.027	LARVAE	Morgan (1976)
		DVP	6.6	0.0006	EGG	Morgan (1976)
		DVP	80	0.0006	LARVAE	Morgan (1976)
		DVP	94	0.003	LARVAE	Morgan (1976)
		DVP	108.9	0.003	LARVAE	Morgan (1976)
	Midge (*Chironomus* sp)	MOR	2	10	4TH INSTAR	Norland et al. (1974)
		MOR	2	1,000	4TH INSTAR	Norland et al. (1974)
	Midge (*Chironomus plumosus*)	MOR	5		LARVAE, 4TH INSTAR	Hilsenhoff (1959)
		MOR	5	0.1	LARVAE, 4TH INSTAR	Hilsenhoff (1959)
		MOR	5	1	LARVAE, 4TH INSTAR	Hilsenhoff (1959)
		MOR	5	10	LARVAE, 4TH INSTAR	Hilsenhoff (1959)

Midge family	DRF	1	3	LARVAE	Morgan (1976)
	DRF	33	3	LARVAE	Morgan (1976)
Mosquito (*Aedes aegypti*)	$ET_{50}IM$	0.13 hr	10,000	LARVAE, 2ND INSTAR	Burchfield and Storrs (1954)
	$ET_{50}IM$	0.5 hr	1,000	LARVAE, 2ND INSTAR	Burchfield and Storrs (1954)
	$ET_{50}IM$	>5 hr	100	LARVAE, 2ND INSTAR	Burchfield and Storrs (1954)
Mosquito (*Aedes cantans*)	LC_{100}	3	100	YOUNG 4TH INSTAR	Rettich (1979)
	LC_{50}	1	35.6	LARVAE, 4TH INSTAR	Rettich (1977)
Mosquito (*Aedes vexans*)	LC_{50}	1	37.9	LARVAE, 4TH INSTAR	Rettich (1977)
Mosquito (*Aedes punctor*)	LC_{50}	1	69.5	LARVAE, 4TH INSTAR	Rettich (1977)
Mosquito (*Aedes aegypti*)	LC_{50}	NR	350	LARVAE, 4TH INSTAR	Klassen et al. (1965)
	LET	1	1,000	LARVAE, LAST INSTAR	Mitsuhashi et al. (1970)
Mosquito (*Anopheles quadrimaculatus*)	LET	2	100	4TH INSTAR LARVAE	Labrecque et al. (1956)
Mosquito (*Anopheles* sp)	MOR	1		LARVAE	Labrecque et al. (1956)
	MOR	1		LARVAE	Labrecque et al. (1956)
Mosquito (*Anopheles quadrimaculatus*)	MOR	2	10	4TH INSTAR LARVAE	Labrecque et al. (1956)
Mosquito (*Culex pipiens molestus*)	LC_{100}	3	50	4TH INSTAR	Rettich (1979)
Mosquito (*Culex pipiens pipiens*)	LC_{50}	1	24.3	LARVAE, 4TH INSTAR	Rettich (1977)
Mosquito (*Culex pipiens molestus*)	LC_{50}	1	30.8	LARVAE, 4TH INSTAR	Rettich (1977)
Mosquito (*Culex pipiens*)	LC_{50}	1	45	4TH INSTAR LARVAE	Yasuno et al. (1965)
Mosquito (*Culex pipiens quinquefasciata*)	LC_{50}	1	61	3RD–4TH INSTAR LARVAE	Chen et al. (1971)
	LC_{50}	1	80	3RD–4TH INSTAR LARVAE	Chen et al. (1971)
Mosquito (*Culiseta annulata*)	LC_{50}	1	2.3	LARVAE, 4TH INSTAR	Rettich (1977)
Mosquito family	DRF	33	3	ADULT	Morgan (1976)

Table 22. Toxicology of Aquatic and Terrestrial Species: Chemical: Diazinon, CAS Number: 333415 (continued)

Class	Species	Effect	Duration (days)	Concentration ($\mu g\ L^{-1}$)	Lifestage	Ref.
Insect (continued)	Stonefly (*Acroneuria ruralis*)	LC_{50}	1.5	363	NR	Morgan (1976)
		LC_{50}	2	294	NR	Morgan (1976)
		LC_{50}	2.5	152	NR	Morgan (1976)
		LC_{50}	3	68	NR	Morgan (1976)
		LC_{50}	4	16	NR	Morgan (1976)
	Stonefly (*Pteronarcys californica*)	LC_{50}	1	150	NYMPH	Cope (1965)
		LC_{50}	1	155	30–35 MM	Sanders and Cope (1968)
		LC_{50}	2	60	30–35 MM	Sanders and Cope (1968)
		LC_{50}	2	74	NYMPH	Cope (1965)
		LC_{50}	4	25	30–35 MM	Sanders and Cope (1968)
		LC_{50}	4	25	2ND YEAR CLASS	Johnson and Finley (1980)
		LC_{50}	4	25	NYMPH	Cope (1965)
	Waterbug order	DRF	3 hr	3	ADULTS	Morgan (1976)
		DRF	33	3	NR	Morgan (1976)
Mullusca	American or Virginia oyster	GRO	4	1,000	2.5–5.1 CM	Butler (1963)
	Bladder snail	LC_{50}	2	4,800	NR	Hashimoto and Nishiuchi (1981)
		LC_{50}	2	4,800	NR	Nishiuchi and Yoshida (1972)
	Marsh snail	LC_{50}	2	9,500	NR	Hashimoto and Nishiuchi (1981)
		LC_{50}	2	9,500	NR	Nishiuchi and Yoshida (1972)
	Mud snail	LC_{50}	2	16,000	NR	Nishiuchi and Yoshida (1972)
	Mussel, eastern elliptio	BCF	2	20	NR	Miller et al. (1966)
		BCF	6	10	NR	Miller et al. (1966)

	Pouch snail (*Physa gyrina*)	LC_{50}	4	48	NR	Morgan (1976)
	Ramshorn snail (*Helisoma trivolvis*)	LC_{50}	7	528	NR	Morgan (1976)
	Snail (*Indoplanorbis exustus*)	LC_{50}	2	20,000	NR	Hashimoto and Nishiuchi (1981)
		LC_{50}	2	20,000	NR	Nishiuchi and Yoshida (1972)
Tracheophy	Water-meal	LET	12–30	100,000	NR	Worthley and Schott (1972)
Bird	Bobwhite quail	LC_{50}	8	245 ppm	14–21 D	Heath et al. (1972)
	Japanese quail	LC_{50}	8	47 ppm	14–21 D	Heath et al. (1972)
	Mallard duck	LC_{50}	8	191 ppm	14–21 D	Heath et al. (1972)
		LD_{50}	14	3.54 mg kg^{-1}	90–120 D	Hudson et al. (1984)
	Pheasant	LO AVO	42	2 mg	365 D	Stromborg (1977)
		LO REP	42	2 mg	365 D	Stromborg (1977)
	Ring-necked pheasant	LC_{50}	8	244 ppm	14–21 D	Heath et al. (1972)
		LD_{50}	14	4.33 mg kg^{-1}	90–120 D	Hudson et al. (1984)
		LO AVO	8	2,200 ppm	365 D	Bennett and Prince (1981)
		LO MOR	8	2,200 ppm	365 D	Bennett and Prince (1981)
Invertebrate	Honey bee	LD_{50}	NR	200 mg bee^{-1}	ADULT	Stevenson et al. (1978)
		LD_{50}	NR	220 mg bee^{-1}	ADULT	Stevenson et al. (1978)
		LO MOR	NR	3 ppm		Sonnet (1978)
		NO POP	NR	5 ppm		Sonnet (1978)
Mammal	Rodent	NO MOR	825	3.46 lb ac^{-1}		Robel et al. (1972)
		NO POP	825	3.46 lb ac^{-1}		Robel et al: (1972)

Table 23. Toxicology of Aquatic and Terrestrial Species: Chemical: Dicamba, CAS Number: 1918009

Class	Species	Effect	Duration (days)	Concentration ($\mu g\ L^{-1}$)	Lifestage	Ref.
Amphibia	Brown frog	LC_{50}	1	205,000	TADPOLE,1–2 WEEKS	Johnson (1976)
		LC_{50}	2	166,000	TADPOLE,1–2 WEEKS	Johnson (1976)
		LC_{50}	4	106,000	1–2 WEEKS	Johnson (1976)
	Tusked frog	LC_{50}	1	220,000	TADPOLE,1–2 WEEKS	Johnson (1976)
		LC_{50}	2	202,000	TADPOLE,1–2 WEEKS	Johnson (1976)
		LC_{50}	4	185,000	TADPOLE,1–2 WEEKS	Johnson (1976)
Crustacea	Aquatic sowbug	LC_{50}	2	>100,000	EARLY INSTAR	Sanders (1970a)
	(*Asellus brevicaudus*)	LC_{50}	4	>100,000	MATURE	Johnson and Finley (1980)
	Crayfish (*Orconectes nais*)	LC_{50}	2	>100,000	EARLY INSTAR	Sanders (1970a)
	Grass shrimp	LC_{50}	2	>100,000	EARLY INSTAR	Sanders (1970a)
	(*Palaemonetes kadiakensis*)	LC_{50}	4	>56,000	MATURE	Johnson and Finley (1980)
	Ostracod (*Cypridopsis vidua*)	EC_{50}IM	2	>100,000	EARLY INSTAR	Sanders (1970a)
	Scud (*Gammarus lacustris*)	LC_{50}	1	10,000	2 MONTHS	Sanders (1969)
		LC_{50}	2	5,800	2 MONTHS	Sanders (1969)
	Scud (*Gammarus fasciatus*)	LC_{50}	2	>100,000	EARLY INSTAR	Sanders (1970a)
	Scud (*Gammarus lacustris*)	LC_{50}	4	3,900	2 MONTHS	Sanders (1969)
	Scud (*Gammarus fasciatus*)	LC_{50}	4	>100,000	MATURE	Johnson and Finley (1980)
	Water flea (*Daphnia magna*)	EC_{50}IM	2	>100,000	EARLY INSTAR	Sanders (1970a)
		EC_{50}IM	2	>100,000	1ST INSTAR	Johnson and Finley (1980)
	Water flea (*Daphnia pulex*)	LC_{50}	3 hr	>40,000	NR	Nishiuchi and Hashimoto (1975)
	Water flea (*Moina macrocopa*)	LC_{50}	3 hr	>40,000	NR	Nishiuchi and Hashimoto (1975)

Fish	Bluegill	LC_{50}	4	>50,000	0.9 G	Johnson and Finley (1980)
	Coho salmon, silver salmon	MOR	6	09,900	YEARLING	Lorz et al. (1979)
	Common, mirror, colored, carp	LC_{50}	2	>40,000	NR	Nishiuchi and Hashimoto (1975)
	Cutthroat trout	LC_{50}	4	>50,000	FINGERLING, 0.4–0.8 G	Woodward (1982)
	Goldfish	LC_{50}	2	>40,000	NR	Nishiuchi and Hashimoto (1975)
	Medaka, high-eyes	LC_{50}	2	>40,000	NR	Nishiuchi and Hashimoto (1975)
	Mosquitofish (*Gambusia affinis*)	LC_{50}	1	516,000	NR	Johnson (1978a)
		LC_{50}	2	510,000	NR	Johnson (1978a)
		LC_{50}	4	465,000	NR	Johnson (1978a)
	Rainbow trout, donaldson trout	LC_{50}	4	28,000	0.8 G	Johnson and Finley (1980)
Bird	Bobwhite quail	LC_{50}	NR	>4,640		EPA (1983)
		LC_{50}	NR	>5,620		EPA (1983)
		LC_{50}	NR	>5,620		EPA (1983)
		LC_{50}	NR	>10,000		EPA (1983)
	Mallard duck	LC_{50}	NR	>4,640		EPA (1983)
		LC_{50}	NR	>5,620		EPA (1983)
		LC_{50}	NR	>5,620		EPA (1983)
		LC_{50}	NR	>10,000		EPA (1983)
		LD_{50}	NR	>2,510		EPA (1983)
		LD_{50}	NR	>2,510		EPA (1983)
		LD_{50}	NR	>2,510		EPA (1983)
Mammal	Rat	LD_{50}	NR	757		USDA Forest Service (1989)

Table 24. Toxicology of Aquatic and Terrestrial Species: Chemical: Dicamba, Sodium Salt, CAS Number: 62610393

Class	Species	Effect	Duration (days)	Concentration ($\mu g\ L^{-1}$)	Lifestage	Ref.
Crustacea	Water flea (*Daphnia magna*)	ABN	75	11,500–45,000	GENERATION 3 and 4	Trofimova (1979)
		DVP	75	23,000–45,000	4 GENERATIONS	Trofimova (1979)
		MOR	75	67,000	4 GENERATIONS	Trofimova (1979)
Fish	Bluegill	LC_{50}	1	20,000	NR	Hughes and Davis (1962)
		LC_{50}	1	67,500	NR	Hughes and Davis (1962)
		LC_{50}	1	600,000	NR	Hughes and Davis (1962)
		LC_{50}	2	20,000	NR	Hughes and Davis (1962)
		LC_{50}	2	67,500	NR	Hughes and Davis (1962)
		LC_{50}	2	410,000	NR	Hughes and Davis (1962)

Table 25. Toxicology of Aquatic and Terrestrial Species: Chemical: Dichloropropene, CAS Number: 542756

Class	Species	Effect	Duration (days)	Concentration ($\mu g\ L^{-1}$)	Lifestage	Ref.
Crustacea	Water flea (*Daphnia magna*)	EC_{50}IM	2	90	1ST INSTAR	Johnson and Finley (1980)
		LC_{50}	1	7,200	<24 HR	LeBlanc (1980)
		LC_{50}	1	7,200	<24 HR	LeBlanc (1980)
		LC_{50}	2	6,200	<24 HR	LeBlanc (1980)
		MOR	2	410	<24 HR	LeBlanc (1980)
Fish	Bluegill	LC_{50}	1	6,800	JUVENILE, 0.32–1.2 G	Buccafusco et al. (1981)
		LC_{50}	4	6,100	JUVENILE, 0.32–1.2 G	Buccafusco et al. (1981)
	Fathead minnow	LC_{50}	4	4,100	0.9 G	Johnson and Finley (1980)
	Largemouth bass	LC_{50}	4	3,650	1 G	Johnson and Finley (1980)
	Walleye	LC_{50}	4	1,080	1.3 G	Johnson and Finley (1980)
Mammal	Rat	LC_{50}		904–1,035 mg kg^{-1}		*Farm Chemicals Handbook* (1990)
		LD_{50}		224–300 mg kg^{-1}		*Farm Chemicals Handbook* (1990)

Table 26. Toxicology of Aquatic and Terrestrial Species: Chemical: Dicofol, CAS Number: 115322

Class	Species	Effect	Duration (days)	Concentration (μg L^{-1})	Lifestage	Ref.
Algae	Green algae (*Scenedesmus obliquus*)	GRO	5	500	NR	Krishnakumari (1977)
Annelida	Oligochaete (*Branchiura sowerbyi*)	LET	3	1,500	NR	Naqvi (1973)
		LET	3	1,500	NR	Naqvi (1973)
		LET	3	1,500	NR	Naqvi (1973)
Crustacea	American lobster	BIO	0.17 hr	1,110	NR	Doherty and Matsumura (1974)
		BIO	0.17 hr	3,700	NR	Doherty and Matsumura (1974)
	Bay shrimp	LC_{50}	1	1,290	35–65 MM	Khorram and Knight (1977)
		LC_{50}	2	603	35–65 MM	Khorram and Knight (1977)
		LC_{50}	3	1,380	0.548–1.791 G	Sharp et al. (1979)
		LC_{50}	4	508	0.548–1.791 G	Sharp et al. (1979)
		DVP	13	14	0.548–1.791 G	Sharp et al. (1979)
		FCR	13	14	0.548–1.791 G	Sharp et al. (1979)
		MOR	1	89	ADULT, 42–76 MM	Khorram and Knight (1976)
		MOR	1	89	ADULT, 42–76 MM	Khorram and Knight (1976)
		MOR	1	130	ADULT, 42–76 MM	Khorram and Knight (1976)
		MOR	1	130	ADULT, 42–76 MM	Khorram and Knight (1976)
		MOR	1	208	ADULT, 42–76 MM	Khorram and Knight (1976)
		MOR	1	208	ADULT, 42–76 MM	Khorram and Knight (1976)
		MOR	1	475	ADULT, 42–76 MM	Khorram and Knight (1976)
		MOR	1	475	ADULT, 42–76 MM	Khorram and Knight (1976)

		MOR	2	89	ADULT, 42–76 MM	Khorram and Knight (1976)
		MOR	2	89	ADULT, 42–76 MM	Khorram and Knight (1976)
		MOR	2	89	ADULT, 42–76 MM	Khorram and Knight (1976)
		MOR	2	130	ADULT, 42–76 MM	Khorram and Knight (1976)
		MOR	2	130	ADULT, 42–76 MM	Khorram and Knight (1976)
		MOR	2	130	ADULT, 42–76 MM	Khorram and Knight (1976)
		MOR	2	475	ADULT, 42–76 MM	Khorram and Knight (1976)
		MOR	3	89	ADULT, 42–76 MM	Khorram and Knight (1976)
		MOR	3	89	ADULT, 42–76 MM	Khorram and Knight (1976)
		MOR	3	89	ADULT, 42–76 MM	Khorram and Knight (1976)
		MOR	3	130	ADULT, 42–76 MM	Khorram and Knight (1976)
		MOR	3	208	ADULT, 42–76 MM	Khorram and Knight (1976)
		OC	4	89	0.710–2.052 G, ADULT	Sharp et al. (1978)
		OC	13	14	0.548–1.791 G	Sharp et al. (1979)
		OC	13	33	0.548–1.791 G	Sharp et al. (1979)
	Water flea (*Daphnia magna*)	EC_{50}IM	1	80	2–26 HR, ADULT	Rawash et al. (1975)
		LC_{50}	1.08	390	1ST INSTAR, <24 HR	Frear and Boyd (1967)
	Water flea (*Daphnia pulex*)	LC_{50}	3 hr	3,800	NR	Nishiuchi and Hashimoto (1975)
	Water flea (*Moina macrocopa*)	LC_{50}	3 hr	8,200	NR	Nishiuchi and Hashimoto (1975)
Fish	Bluegill	EC_{50}EZ	NR	300	NR	Cutkomp et al. (1971a)
		LC_{50}	4	520	1 G	Johnson and Finley (1980)
		ENZ	0.42 hr	960	NR	Cutkomp et al. (1971b)
		ENZ	0.42 hr	1,900	NR	Cutkomp et al. (1971b)
	Brook trout	ENZ	NR	3,700	NR	Jackson and Gardner (1973)
	Channel catfish	LC_{50}	4	360	0.8 G	Johnson and Finley (1980)
	Chinook salmon	EQU	1	10,000	5–10 CM	MacPhee and Ruelle (1969)
		MOR	1	10,000	5–10 CM	MacPhee and Ruelle (1969)

Table 26. Toxicology of Aquatic and Terrestrial Species: Chemical: Dicofol, CAS Number: 115322 (continued)

Class	Species	Effect	Duration (days)	Concentration ($\mu g\ L^{-1}$)	Lifestage	Ref.
Fish (continued)	Coho salmon,silver salmon	MOR	1	10,000	5–10 CM	MacPhee and Ruelle (1969)
	Common, mirror, colored, carp	LC_{50}	2	360	NR	Nishiuchi and Hashimoto (1975)
	Cutthroat trout	LC_{50}	4	53	0.1 G	Johnson and Finley (1980)
	Fathead minnow	LC_{50}	4	510	31–32 D, 0.2 G	Holcombe et al. (1982)
		LC_{50}	4	603	32 D, 0.0 MM, 0.000 G	Geiger et al. (1988)
	Goldfish	LC_{50}	2	620	NR	Nishiuchi and Hashimoto (1975)
	Guppy	LET	2	500	NR	Adlung (1957)
	Lake trout, siscowet	LC_{50}	4	87	1.2 G	Johnson and Finley (1980)
	Largemouth bass	LC_{50}	4	395	0.8 G	Johnson and Finley (1980)
	Medaka, high-eyes	LC_{50}	2	560	NR	Nishiuchi and Hashimoto (1975)
	Mosquitofish (*Gambusia affinis*)	LC_{50}	1	1,900	NR	Boyd and Ferguson (1964)
	Northern squawfish	EQU	1	10,000	5–10 CM	MacPhee and Ruelle (1969)
		MOR	1	10,000	5–10 CM	MacPhee and Ruelle (1969)
	Rainbow trout, donaldson trout	EC_{50}EZ	0.25 hr	2,000	YEARLING	Davis and Wedemeyer (1971)
		LC_{50}	1	335	1.9 G, JUVENILE	Holcombe et al. (1982)
		LC_{50}	2	309	1.9 G, JUVENILE	Holcombe et al. (1982)
		LC_{50}	3	262	1.9 G, JUVENILE	Holcombe et al. (1982)
		LC_{50}	4	210	1.9 G, JUVENILE	Holcombe et al. (1982)

Insect	Mosquito (*Aedes aegypti*)	LC_{50}	NR	[illegible],600	LARVAE, 4TH INSTAR	Klassen et al. (1965)
		LC_{50}	NR	1[illegible],800	LARVAE, 4TH INSTAR	Klassen et al. (1965)
		LC_{50}	NR	3[illegible],000	LARVAE, 4TH INSTAR	Klassen et al. (1965)
	Mosquito (*Culex pipiens*)	EC_{50}IM	NR	200	4TH INSTAR	Rawash et al. (1975)
	Stonefly (*Pteronarcys californica*)	LC_{50}	4	650	2ND YEAR CLASS	Johnson and Finley (1980)
Mullusca	Bladder snail	LC_{50}	2	1,600	NR	Nishiuchi and Yoshida (1972)
	Marsh snail	LC_{50}	2	1,400	NR	Nishiuchi and Yoshida (1972)
	Mud snail	LC_{50}	2	5,600	NR	Nishiuchi and Yoshida (1972)
	Snail (*Indoplanorbis exustus*)	LC_{50}	2	3,000	NR	Nishiuchi and Yoshida (1972)
Protozoa	Ciliate (*Paramecium aurelia*)	LET	1.5 hr	1,000	10,000 CELLS L^{-1}	Joshi and Misra (1986)
Bird	Bobwhite quail	LC_{50}	8	3,01[illegible] ppm	14–21 D	Heath et al. (1972)
	Japanese quail	LC_{50}	8	1,413 ppm	14–21 D	Heath et al. (1972)
	Mallard duck	LC_{50}	8	1,65[illegible] ppm	14–21 D	Heath et al. (1972)
	Ring-necked pheasant	LC_{50}	8	2,125 ppm	14–21 D	Heath et al. (1972)
		LD_{50}	14	265 mg kg^{-1}	120 D	Hudson et al. (1984)

Table 27. Toxicology of Aquatic and Terrestrial Species: Chemical: Diesel Fuel

Class	Species	Effect	Duration (days)	Concentration ($\mu g\ L^{-1}$)	Lifestage	Ref.
Bird	Mallard duck	LD_{50}		16,400		USDA Forest Service (1989)
Mammal	Rat	LD_{50}		>7,380		USDA Forest Service (1989)

Table 28. Toxicology of Aquatic and Terrestrial Species: Chemical: DSMA, CAS Number: 144218

Class	Species	Effect	Duration (days)	Concentration ($\mu g\ L^{-1}$)	Lifestage	Ref.
Amphibia	Brown frog	LC_{50}	1	324,000	TADPOLE, 1–2 WEEKS	Johnson (1976)
		LC_{50}	2	310,000	TADPOLE, 1–2 WEEKS	Johnson (1976)
		LC_{50}	4	271,000	TADPOLE, 1–2 WEEKS	Johnson (1976)
	Tusked frog	LC_{50}	1	600,000	TADPOLE, 1–2 WEEKS	Johnson (1976)
		LC_{50}	2	525,000	TADPOLE, 1–2 WEEKS	Johnson (1976)
		LC_{50}	4	453,000	TADPOLE, 1–2 WEEKS	Johnson (1976)
Crustacea	Scud (*Gammarus pseudolimnaeus*)	MOR	14	970	0.004-0.02 G	Spehar et al. (1980)
	Water flea (*Daphnia pulex*)	LC_{50}	3 hr	>40,000	NR	Nishiuchi and Hashimoto (1975)
	Water flea (*Daphnia Magna*)	MOR	14	833	NR	Spehar et al. (1980)
		REP	14	833	NR	Spehar et al. (1980)
	Water flea (*Moina macrocopa*)	LC_{50}	3 hr	>40,000	NR	Nishiuchi and Hashimoto (1975)

Fish	Channel catfish	MOR	2	10,000	1 YEAR, 14 G, 12 CM, FINGERLINGS	Mccorkle et al. (1977)
	Common, mirror, colored, carp	LC_{50}	2	>10,000	NR	Nishiuchi and Hashimoto (1975)
	Goldfish	LC_{50}	2	>10,000	NR	Nishiuchi and Hashimoto (1975)
	Medaka, high-eyes	LC_{50}	2	>10,000	NR	Nishiuchi and Hashimoto (1975)
	Mosquitofish	LC_{50}	1	1,500,000	NR	Johnson (1978a)
	(*Gambusia affinis*)	LC_{50}	2	1,400,000	NR	Johnson (1978a)
		LC_{50}	4	1,300,000	NR	Johnson (1978a)
	Rainbow trout, donaldson trout	MOR	28	970	9.5–21.9 G	Spehar et al. (1980)
Insect	Stonefly (*Pteronarcys dorsata*)	BCF	28	970	0.1–0.6 G	Spehar et al. (1980)
		MOR	28	970	0.1–0.6 G	Spehar et al. (1980)
Mullusca	Pond snail (*Lymnaea emarginata*)	BCF	28	970	0.1–0.4 G	Spehar et al. (1980)
	Ramshorn snail	MOR	28	970	0.1–0.4 G	Spehar et al. (1980)
	(*Helisoma campanulata*)	MOR	28	970	0.1–0.4 G	Spehar et al. (1980)
Mammal	Rat	LC_{50}	4 hr	>22.1 mg L^{-1}		*Farm Chemicals Handbook* (1990)
		LD_{50}		1,800 mg kg^{-1}		*Farm Chemicals Handbook* (1990)

Table 29. Toxicology of Aquatic and Terrestrial Species: Chemical: Dyrene (Anilazine), CAS Number: 101053

Class	Species	Effect	Duration (days)	Concentration ($\mu g\ L^{-1}$)	Lifestage	Ref.
Algae	Algae, phytoplankton, algal mat	PSE	4 hr	1,000	NR	Butler (1963)
Crustacea	Brown shrimp	MOR	1	100	ADULT	Butler (1964)
		MOR	2	100	ADULT	Butler (1964)
	Scud (*Gammarus fasciatus*)	LC_{50}	4	0.27	MATURE	Johnson and Finley (1980)
	Water flea (*Daphnia magna*)	LC_{50}	1.0833	490	1ST INSTAR, <24 HR	Frear and Boyd (1967)
	Water flea (*Daphnia pulex*)	LC_{50}	3 hr	4,500	NR	Nishiuchi and Hashimoto (1975)
		LC_{50}	6 hr	1,200	NR	Hashimoto and Nishiuchi (1981)
		LC_{50}	6 hr	5,300	NR	Hashimoto and Nishiuchi (1981)
		LC_{50}	6 hr	5,400	NR	Hashimoto and Nishiuchi (1981)
	Water flea (*Moina macrocopa*)	LC_{50}	3 hr	5,500	NR	Nishiuchi and Hashimoto (1975)
Fish	Bluegill	LC_{50}	1	196	1.365–1.827 G	McCann and Jasper (1972)
		LC_{50}	1	675	0.844 G	McCann and Jasper (1972)
		LC_{50}	4	320	1.1 G	Johnson and Finley (1980)
	Channel catfish	LC_{50}	4	240	0.7 G	Johnson and Finley (1980)
	Common, mirror, colored, carp	LC_{50}	2	95	NR	Nishiuchi and Hashimoto (1975)
	Goldfish	LC_{50}	2	160	NR	Nishiuchi and Hashimoto (1975)

	Medaka, high-eyes	LC_{50}	2	85	NR	Nishiuchi and Hashimoto (1975)
	Rainbow trout, donaldson trout	LC_{50}	4	140	1.5 G	Johnson and Finley (1980)
	Redear sunfish	LC_{50}	4	<140	2.5 G	Johnson and Finley (1980)
	Spot	LC_{50}	2	8.5	JUVENILE	Butler (1964)
	Striped mullet	MOR	1	100	JUVENILE	Butler (1964)
		MOR	2	100	JUVENILE	Butler (1964)
Mullusca	American or Virginia oyster	$EC_{50}GR$	4	46	JUVENILE	Butler (1964)
		$EC_{50}GR$	4	64	JUVENILE	Butler (1964)
	Bladder snail	LC_{50}	2	1,000	NR	Nishiuchi and Yoshida (1972)
	Marsh snail	LC_{50}	2	580	NR	Nishiuchi and Yoshida (1972)
	Mud snail	LC_{50}	2	1,800	NR	Nishiuchi and Yoshida (1972)
	Snail (*Indoplanorbis exustus*)	LC_{50}	2	850	NR	Nishiuchi and Yoshida (1972)

Table 30. Toxicology of Aquatic and Terrestrial Species: Chemical: Endothall, CAS Number: 145733

Class	Species	Effect	Duration (days)	Concentration (µg L^{-1})	Lifestage	Ref.
Algae	Blue-green algae (*Lyngbya* sp)	PGR	30	4,000	NR	Berry (1976)
	Diatom (*Phaeodactylum tricornutum*)	EC_{50}OX	1.5 hr	75,000	LOG PHASE	Walsh (1972)
		PGR	10	15,000	LOG PHASE	Walsh (1972)
	Green algae (*Chlorococcum* sp)	EC_{50}OX	1.5 hr	100,000	LOG PHASE	Walsh (1972)
		PGR	10	50,000	LOG PHASE	Walsh (1972)
	Green algae (*Dunaliella tertiolecta*)	EC_{50}OX	1.5 hr	425,000	LOG PHASE	Walsh (1972)
		PGR	10	50,000	LOG PHASE	Walsh (1972)
	Green algae (*Oedogonium cardiacum*)	BCF	33	740	NR	Isensee (1976)
	Haptophyte (*Isochrysis galbana*)	EC_{50}OX	1.5 hr	60,000	LOG PHASE	Walsh (1972)
		PGR	10	25,000	LOG PHASE	Walsh (1972)
Crustacea	Grass shrimp (*Palaemonetes kadiakensis*)	LC_{50}	4	50	MATURE	Johnson and Finley (1980)
	Scud (*Gammarus lacustris*)	LC_{50}	4	500	MATURE	Johnson and Finley (1980)
	Scud (*Gammarus fasciatus*)	LC_{50}	4	510	MATURE	Johnson and Finley (1980)
	Water flea (*Daphnia magna*)	EC_{50}IM	1.08	46,000	1ST INSTAR	Crosby and Tucker (1966)
		BCF	33	740	NR	Isensee (1976)
		LET	1.08	25,000	1ST INSTAR	Crosby and Tucker (1966)
		LET	1.08	84,000	1ST INSTAR	Crosby and Tucker (1966)

Fish	Bluegill	LC_{50}	4	940	0.5 G	Johnson and Finley (1980)
		LC_{50}	4	1,200	0.5 G	Johnson and Finley (1980)
		LC_{50}	4	77,700	33–46 MM	Berry (1976)
		LC_{50}	4	343,000	1.3 G	Johnson and Finley (1980)
		LC_{50}	4	1,740,000	NR	Johnson and Finley (1980)
		MOR	8	10,000	FRY	Hiltibran (1967)
		MOR	8	25,000	FRY	Hiltibran (1967)
		MOR	12	50,000	SMALL	Hiltibran (1967)
		MOR	12	100,000	SMALL	Hiltibran (1967)
		RES	1	450	120–144 MM	Berry (1976)
		RES	6 hr	450,000	120–144 MM	Berry (1976)
		RSD	0.17–4	2,000	3–5 G, 8–10 CM	Sikka et al. (1975)
		RSD	2	10 mg kg^{-1}	3–5 G, 8–10 CM	Sikka et al. (1975)
	Channel catfish	LC_{50}	4	490	0.3 G	Johnson and Finley (1980)
		LC_{50}	4	>150,000	0.4 G	Johnson and Finley (1980)
	Coho salmon, silver salmon	LC_{50}	4	>100,000	1.4 G	Johnson and Finley (1980)
	Creek chub	LC_{0}	1	1,600,000	8–10 CM	Gillette et al. (1952)
		LC_{100}	1	3,200,000	8–10 CM	Gillette et al. (1952)
	Cutthroat trout	LC_{50}	4	180	1.0 G	Johnson and Finley (1980)
	Fathead minnow	LC_{50}	4	750	0.6 G	Johnson and Finley (1980)
	Goldfish	LC_{50}	4	445,000	51–72 MM	Berry (1976)
		AVO	0.17 hr	1,700	39–60 MM	Berry (1976)
		AVO	0.17 hr	17,000	39–60 MM	Berry (1976)
	Green sunfish	HAT	8	10,000	FERTILIZED EGGS	Hiltibran (1967)
		HAT	8	25,000	FERTILIZED EGGS	Hiltibran (1967)
		MOR	8	10,000	FRY	Hiltibran (1967)
		MOR	8	25,000	FRY	Hiltibran (1967)
	Guppy	LC_{50}	4	181,900	8–15 MM	Berry (1976)
	Lake chubsucker	MOR	8	10,000	FRY	Hiltibran (1967)
		MOR	8	25,000	FRY	Hiltibran (1967)
	Largemouth bass	LC_{50}	4	255,000	25–42 MM	Berry (1976)

Table 30. Toxicology of Aquatic and Terrestrial Species: Chemical: Endothall, CAS Number: 145733 (continued)

Class	Species	Effect	Duration (days)	Concentration (μg L^{-1})	Lifestage	Ref.
Fish (continued)	Mosquitofish (*Gambusia affinis*)	LC_{50}	1	>10,000	NR	Ahmed (1977)
		BCF	3	740	NR	Isensee (1976)
	Rainbow trout, donaldson trout	LC_{50}	4	560	1.2 G	Johnson and Finley (1980)
		LC_{50}	4	980	1.2 G	Johnson and Finley (1980)
		LC_{50}	4	230,000	1.2 G	Johnson and Finley (1980)
	Smallmouth bass	MOR	8	10,000	FRY	Hiltibran (1967)
		MOR	8	25,000	FRY	Hiltibran (1967)
Insect	Stonefly (*Pteronarcys californica*)	LC_{50}	2	3,250	2ND YEAR CLASS	Johnson and Finley (1980)
Mullusca	Pouch snail (*Physa* sp)	BCF	33	740	NR	Isensee (1976)
Tracheophy	American frog's-bit	GRO	30	500	NR	Berry (1976)
		PGR	30	250	NR	Berry (1976)
	Coon-tail	PGR	30	4,000	NR	Berry (1976)
	Duckweed (*Lemna minor*)	GRO	30	4,000	NR	Berry (1976)
	Waterweed (*Elodea* sp)	GRO	30	1,000	NR	Berry (1976)
		PGR	30	500	NR	Berry (1976)
Bird	Mallard duck	LD_{50}	14	229.00 mg kg^{-1}	90 D	Hudson et al. (1984)

Table 31. Toxicology of Aquatic and Terrestrial Species: Chemical: Endothall, Disodium Salt, CAS Number: 129679

Class	Species	Effect	Duration (days)	Concentration (μg L^{-1})	Lifestage	Ref.
Fish	Bluegill	LC_{50}	1	249,000	33 MM, 0.78 G	Inglis and Davis (1972)
		LC_{50}	1	277,000	33 MM, 0.78 G	Inglis and Davis (1972)
		LC_{50}	1	280,000	33 MM, 0.78 G	Inglis and Davis (1972)
		LC_{50}	1	390,000	46 MM	Surber and Pickering (1962)
		LC_{50}	1	450,000	46 MM	Surber and Pickering (1962)
		LC_{50}	1	450,000	NR	Hughes and Davis (1962)
		LC_{50}	1	650,000	NR	Hughes and Davis (1962)
		LC_{50}	2	181,000	33 MM, 0.78 G	Inglis and Davis (1972)
		LC_{50}	2	195,000	33 MM, 0.78 G	Inglis and Davis (1972)
		LC_{50}	2	219,000	33 MM, 0.78 G	Inglis and Davis (1972)
		LC_{50}	2	240,000	46 MM	Surber and Pickering (1962)
		LC_{50}	2	280,000	NR	Hughes and Davis (1962)
		LC_{50}	2	280,000	NR	Hughes and Davis (1962)
		LC_{50}	2	320,000	46 MM	Surber and Pickering (1962)
		LC_{50}	3	123,000	33 MM, 0.78 G	Inglis and Davis (1972)
		LC_{50}	3	123,000	33 MM, 0.78 G	Inglis and Davis (1972)
		LC_{50}	3	179,000	33 MM, 0.78 G	Inglis and Davis (1972)
		LC_{50}	4	102,000	33 MM, 0.78 G	Inglis and Davis (1972)
		LC_{50}	4	105,000	33 MM, 0.78 G	Inglis and Davis (1972)
		LC_{50}	4	140,000	33 MM, 0.78 G	Inglis and Davis (1972)
		LC_{50}	4	160,000	46 MM	Surber and Pickering (1962)
		LC_{50}	4	180,000	46 MM	Surber and Pickering (1962)
		MOR	4	2,000	FRY, 17.8 MM	Jones (1962)
	Central stoneroller	HAT	8	50,000	FERTILIZED EGGS	Hiltibran (1967)
	Channel catfish	MOR	4	50,000	FRY, YOLK SAC STAGE	Jones (1962)
	Chinook salmon	LC_{50}	1	155,000	FINGERLING, 58–96 MM	Bond et al. (1959)

Table 31. Toxicology of Aquatic and Terrestrial Species: Chemical: Endothall, Disodium Salt, CAS Number: 129679 (continued)

Class	Species	Effect	Duration (days)	Concentration ($\mu g\ L^{-1}$)	Lifestage	Ref.
Fish (continued)	Chinook salmon	LC_{50}	2	136,000	FINGERLING, 58–96 MM	Bond et al. (1959)
	Fathead minnow	LC_{50}	1	680,000	47 MM	Surber and Pickering (1962)
		LC_{50}	1	>560,000	45 MM	Surber and Pickering (1962)
		LC_{50}	2	480,000	45 MM	Surber and Pickering (1962)
		LC_{50}	2	660,000	47 MM	Surber and Pickering (1962)
		LC_{50}	4	320,000	45 MM	Surber and Pickering (1962)
		LC_{50}	4	610,000	47 MM	Surber and Pickering (1962)
	Green sunfish	MOR	12	300	NR	Yeo (1970)
	Largemouth bass	LC_{50}	1	200,000	FINGERLING, 55–68 MM	Bond et al. (1959)
		LC_{50}	1	>560,000	63 MM	Surber and Pickering (1962)
		LC_{50}	2	200,000	FINGERLING, 55–68 MM	Bond et al. (1959)
		LC_{50}	2	320,000	63 MM	Surber and Pickering (1962)
		LC_{50}	4	200,000	63 MM	Surber and Pickering (1962)
		MOR	4	2,000	FRY, 18.7 MM	Jones (1962)
	Mosquitofish (*Gambusia affinis*)	MOR	12	300	NR	Yeo (1970)
	Rainbow trout, donaldson trout	LC_{50}	4	310	2 G	Johnson and Finley (1980)
	Smallmouth bass	MOR	12	300	NR	Yeo (1970)
Mullusca	American or Virginia oyster	EC_{50}DV	2	28,220	EGGS, 2 CELL STAGE	Davis and Hidu (1979)
		LC_{50}	12	48,080	2 D LARVAE	Davis and Hidu (1979)
		GRO	12	250	2 D LARVAE	Davis and Hidu (1979)
	Northern quahog or hard clam	EC_{50}DV	2	51,020	EGGS, 2 CELL STAGE	Davis and Hidu (1979)
		LC_{50}	10	12,500	2 D LARVAE	Davis and Hidu (1979)
		GRO	10	250	2 D LARVAE	Davis and Hidu (1979)
Invertebrate	Honey bee	LO MOR	NR	1 lb ac^{-1}		Moffett and Morton (1972)

Table 32. Toxicology of Aquatic and Terrestrial Species: Chemical: Ethoprop, CAS Number: 13194484

Class	Species	Effect	Duration (days)	Concentration ($\mu g\ L^{-1}$)	Lifestage	Ref.
Algae	Diatom (*Skeletonema costatum*)	$EC_{50}GR$	4	3,800	EXPONENTIAL GROWTH PHASE	Walsh and Alexander (1980)
		$EC_{50}GR$	4	8,400	EXPONENTIAL GROWTH PHASE	Walsh and Alexander (1980)
		$EC_{50}GR$	4	8,400	EXPONENTIAL GROWTH PHASE	Walsh and Alexander (1980)
Bird	Mallard duck	LD_{50}	14	12.60 mg kg^{-1}	90 D	Hudson et al. (1984)
		LD_{50}	30	10.60 mg kg^{-1}	365 D	Hudson et al. (1984)
	Ring-necked pheasant	LD_{50}	14	4.21 mg kg^{-1}	90–120 D	Hudson et al. (1984)

Table 33. Toxicology of Aquatic and Terrestrial Species: Chemical: Etridiozole, CAS Number: 2593159

Class	Species	Effect	Duration (days)	Concentration ($\mu g\ L^{-1}$)	Lifestage	Ref.
Amphibia	Frog (*Bufo bufo japonicus*)	LC_{50}	1	12,000	25–35 D, 0.34 G, TADPOLE	Nishiuchi and Yoshida (1974)
		LC_{50}	2	12,000	25–35 D, 0.34 G, TADPOLE	Nishiuchi and Yoshida (1974)
		LC_{50}	3	12,000	25–35 D, 0.34 G, TADPOLE	Nishiuchi and Yoshida 1974)
		LC_{50}	4	12,000	25–35 D, 0.34 G, TADPOLE	Nishiuchi and Yoshida (1974)
Mammal	Rat	LD_{50}		1,077–4,000 mg kg^{-1}		*Farm Chemicals Handbook* (1990)

Table 34. Toxicology of Aquatic and Terrestrial Species: Chemical: Fenoxaprop-ethyl, CAS Number: 1330207

Class	Species	Effect	Duration (days)	Concen-tration ($\mu g\ L^{-1}$)	Lifestage	Ref.
Algae	Algae, phytoplankton, algal mat	EC_{50}PS	0.33	3,000	NR	Brooks et al. (1977)
	Blue-green algae (*Anacystis aeruginosa*)	MOR	NR	>200,000	NR	Bringmann and Kuhn (1978a)
	Coccolithophorid (*Cricosphaera carterae*)	GRO	2–3	10,000	NR	Dunstan et al. (1975)
	Cryptomonad	PGR	2	>80,000	NR	Bringmann et al. (1980)
	Diatom (*Skeletonema costatum*)	GRO	2–3	10,000	NR	Dunstan et al. (1975)
	Dinoflagellate (*Amphidinium carterae*)	GRO	2–3	10,000	NR	Dunstan et al. (1975)
	Green algae (*Dunaliella tertiolecta*)	GRO	2–3	100,000	NR	Dunstan et al. (1975)
	Green algae (*Scenedesmus quadricauda*)	MOR	NR	>200,000	NR	Bringmann and Kuhn (1978a)
		PGR	NR	>200,000	NR	Bringmann and Kuhn (1977b)
Crustacea	Calanoid copepod (*Diaptomus forbesi*)	LC_{50}	4	99,500	NR	Saha and Konar (1983)
	Crab (*Paratelphusa jacuemontii*)	OC	1 hr	350,000	INTERMOULT	Kulkarni and Kamath (1980)
	Grass shrimp (*Palaemonetes pugio*)	LC_{50}	1	14,000	NR	Tatem et al. (1978)
		LC_{50}	2	8,500	NR	Tatem et al. (1978)
		LC_{50}	4	7,400	NR	Tatem et al. (1978)
	Scud (*Gammarus lacustris*)	LC_{50}	1	800	2 MONTHS	Sanders (1969)
		LC_{50}	2	600	2 MONTHS	Sanders (1969)
		LC_{50}	4	350	2 MONTHS	Sanders (1969)

	Water flea (*Daphnia magna*)	LC_{50}	1	150,000	24 HR	Bringmann and Kuhn (1977a)
		LC_{50}	1	100,000–1,000,000	NR	Dowden and Bennett (1965)
Fish	Bluegill	LC_{50}	1	24,000	3.8–6.4 CM, 1–2 G	Pickering and Henderson (1966)
		LC_{50}	1	36,000	0.80 G	Cope (1965)
		LC_{50}	2	19,000	0.80 G	Cope (1965)
		LC_{50}	2	24,000	3.8–6.4 CM, 1–2 G	Pickering and Henderson (1966)
		LC_{50}	4	13,500	0.9 G	Johnson and Finley (1980)
		LC_{50}	4	19,000	0.80 G	Cope (1965)
		LC_{50}	4	20,870	3.8–6.4 CM, 1–2 G	Pickering and Henderson 1966)
	Coho salmon, silver salmon	LC_{100}	1	100,000	5–40 G	Morrow et al. (1975)
		HEM	1	30,000	5–40 G	Morrow et al. (1975)
		HEM	1 hr	30,000	5–40 G	Morrow et al. (1975)
		HEM	1 hr	30,000	5–40 G	Morrow et al. (1975)
		HEM	1 hr	30,000	5–40 G	Morrow et al. (1975)
		HEM	2 hr	30,000	5–40 G	Morrow et al. (1975)
		HEM	2 hr	30,000	5–40 G	Morrow et al. (1975)
		HEM	2 hr	30,000	5–40 G	Morrow et al. (1975)
		HEM	3 hr	30,000	5–40 G	Morrow et al. (1975)
		HEM	3 hr	30,000	5–40 G	Morrow et al. (1975)
		HEM	3 hr	30,000	5–40 G	Morrow et al. (1975)
		HEM	4 hr	30,000	5–40 G	Morrow et al. (1975)
	Common, mirror, colored, carp	LC_{50}	1	1,080,000	4–5 CM	Rao et al. (1975)
		LC_{50}	2	950,000	4–5 CM	Rao et al. (1975)
		LC_{50}	4	780,000	4–5 CM	Rao et al. (1975)

Table 34. Toxicology of Aquatic and Terrestrial Species: Chemical: Fenoxaprop-ethyl, CAS Number: 1330207 (continued)

Class	Species	Effect	Duration (days)	Concentration ($\mu g\ L^{-1}$)	Lifestage	Ref.
Fish (continued)	Fathead minnow	LC_{50}	1	28,770	3.8–6.4 CM, 1–2 G	Pickering and Henderson (1966)
		LC_{50}	1	28,770	3.8–6.4 CM, 1–2 G	Pickering and Henderson (1966)
		LC_{50}	1	42,000	JUVENILE, 4–8 WEEKS, 1.1–3.1 CM	Mattson et al. (1976)
		LC_{50}	1 hr	46,000	JUVENILE, 4–8 WEEKS, 1.1–3.1 CM	Mattson et al. (1976)
		LC_{50}	2	27,710	3.8–6.4 CM, 1–2 G	Pickering and Henderson (1966)
		LC_{50}	2	28,770	3.8–6.4 CM, 1–2 G	Pickering and Henderson (1966)
		LC_{50}	2	42,000	JUVENILE, 4–8 WEEKS, 1.1–3.1 CM	Mattson et al. (1976)
		LC_{50}	3	42,000	JUVENILE, 4–8 WEEKS, 1.1–3.1 CM	Mattson et al. (1976)
		LC_{50}	4	26,700	3.8–6.4 CM, 1–2 G	Pickering and Henderson (1966)
		LC_{50}	4	28,770	3.8–6.4 CM, 1–2 G	Pickering and Henderson (1966)
		LC_{50}	4	42,000	JUVENILE, 4–8 WEEKS, 1.1–3.1 CM	Mattson et al. (1976)

Goldfish	LC_{50}	1	30,550	20–80 G, 1–1.5 YEARS, 13–20 CM	Brenniman et al. (1976)
	LC_{50}	1	36,810	3.8–6.4 CM, 1–2 G	Pickering and Henderson (1966)
	LC_{50}	1	75,000	NR	Jensen (1978)
	LC_{50}	2	25,100	20–80 G, 1–1.5 YEARS, 13–20 CM	Brenniman et al. (1976)
	LC_{50}	2	36,810	3.8–6.4 CM, 1–2 G	Pickering and Henderson (1966)
	LC_{50}	3	20,720	20–80 G, 1–1.5 YEARS, 13–20 CM	Brenniman et al. (1976)
	LC_{50}	4	16,940	20–80 G, 1–1.5 YEARS, 13–20 CM	Brenniman et al. (1976)
	LC_{50}	4	36,810	3.8–6.4 CM, 1–2 G	Pickering and Henderson (1966)
Guppy	LC_{50}	1	34,730	6 MONTHS, 1.9–2.5 CM, 0.1–0.2 G	Pickering and Henderson (1966)
	LC_{50}	2	34,730	6 MONTHS, 1.9–2.5 CM, 0.1–0.2 G	Pickering and Henderson (1966)
	LC_{50}	4	34,730	6 MONTHS, 1.9–2.5 CM, 0.1–0.2 G	Pickering and Henderson (1966)
Ide, silver or golden orfe	LC_{50}	NR	86,000	NR	Juhnke and Luedemann (1978)
	LC_{50}	NR	264,000	NR	Juhnke and Luedemann (1978)
Orangespotted sunfish	LC_{100}	1 hr	47,000–48,000	4-6 G	Shelford (1917)

Table 34. Toxicology of Aquatic and Terrestrial Species: Chemical: Fenoxaprop-ethyl, CAS Number: 1330207 (continued)

Class	Species	Effect	Duration (days)	Concentration ($\mu g\ L^{-1}$)	Lifestage	Ref.
Fish (continued)	Rainbow trout, donaldson trout	LC_0	2 hr	7,100	200 G	Walsh et al. (1977a)
		LC_{100}	2 hr	16,100	200 G	Walsh et al. (1977a)
		LC_{50}	1	13,500	0.9 G	Walsh et al. (1977a)
		LC_{50}	1	17,300	0.9 G	Walsh et al. (1977a)
		LC_{50}	4	8,200	0.6 G	Johnson and Finley (1980)
		LC_{50}	4	13,500	0.9 G	Walsh et al. (1977a)
		LC_{50}	4	17,300	0.9 G	Walsh et al. (1977a)
		AVO	4	100	FRY	Folmar (1976)
		BCF	1	360	200 G	Walsh et al. (1977a)
		BCF	1	740	200 G	Walsh et al. (1977a)
		BCF	1	1,300	200 G	Walsh et al. (1977a)
		BCF	7	360	200 G	Walsh et al. (1977a)
		BCF	7	740	200 G	Walsh et al. (1977a)
		BCF	7	1,300	200 G	Walsh et al. (1977a)
		BCF	56	360	200 G	Walsh et al. (1977a)
		BCF	56	740	200 G	Walsh et al. (1977a)
		BCF	56	1,300	200 G	Walsh et al. (1977a)
		BCF	NR	640	200 G	Walsh et al. (1977a)
		BCF	NR	3,600	200 G	Walsh et al. (1977a)
		BCF	NR	7,100	200 G	Walsh et al. (1977a)
		EQU	1.5 hr	7,100	200 G	Walsh et al. (1977a)
		MOR	56	1,300	200 G	Walsh et al. (1977a)
		RES	1	2,000	NR	Slooff (1979)
	Zebra danio, zebrafish	LC_{50}	2	20,000	NR	Slooff (1979)

Class	Species	Effect	Duration (days)	Concentration (µg L^{-1})	Lifestage	Ref.
Insect	Mayfly (*Ephemerella walkerii*)	AVO	1 hr	10,000	LATE INSTAR NYMPHS	Folmar (1978)
	Mosquito (*Aedes aegypti*)	EC_{50}IM	1	13,940	4TH INSTAR LARVAE	Berry and Brammer (1977)
		MOR	1	7,920	4TH INSTAR LARVAE	Berry and Brammer (1977)
Mullusca	Zebra mussel	BEH	NR	11,900	3.0–3.5 CM	Sloof et al. (1983)
		BEH	NR	17,200	3.0–3.5 CM	Sloof et al. (1983)
		BEH	NR	19,400	3.0–3.5 CM	Sloof et al. (1983)

Table 35. Toxicology of Aquatic and Terrestrial Species: Chemical: Fenoxaprop-ethyl, CAS Number: 66441234

Class	Species	Effect	Duration (days)	Concentration (µg L^{-1})	Lifestage	Ref.
Mammal	Rabbit	LC_{50}	3–4 hr	3,920 mg m^{-1}		*Farm Chemicals Handbook* (1990)
	Rat	LD_{50}		3,310–3,400 mg kg^{-1}		*Farm Chemicals Handbook* (1990)

Table 36. Toxicology of Aquatic and Terrestrial Species: Chemical: Glyphosate, Rodeo

Class	Species	Effect	Duration (days)	Concentration (µg L^{-1})	Lifestage	Ref.
Mammal	Rabbit	LD_{50}	NR	>5,000		Worthington (1979)
	Rat	LD_{50}	NR	5,400		WSSA (1983)
		LD_{50}	NR	>5,000		Monsanto (1983)

Table 37. Toxicology of Aquatic and Terrestrial Species: Chemical: Glyphosate, CAS Number: 1071836

Class	Species	Effect	Duration (days)	Concen- tration (µg L^{-1})	Lifestage	Ref.
Algae	Algae, phytoplankton, algal mat	EC_{100}GR	14	11,000	LOG PHASE	Blanck et al. (1984)
	Flagellate euglenoid	CLR	6	203,000	LOGARITHMIC PHASE	Richardson et al. (1979)
	(*Euglena gracilis*)	PGR	6	203,000	LOGARITHMIC PHASE	Richardson et al. (1979)
		PSE	1.67 hr	5,100	LOGARITHMIC PHASE	Richardson et al. (1979)
		PSE	1.67 hr	510,000	LOGARITHMIC PHASE	Richardson et al. (1979)
		PSE	4	10,100	LOGARITHMIC PHASE	Richardson et al. (1979)
	Green algae (*Chlamydomonas eugametos*)	PGR	2	17,000	INITIAL CONC. 100,000 CELLS ML^{-1}	Hess (1980)
		PGR	2	170,000	INITIAL CONC. 100,000 CELLS ML^{-1}	Hess (1980)
	Green algae	PGR	2	100	LOG PHASE	Felix et al. (1988)
	(*Dunaliella bioculata*)	PGR	2	100	LOG PHASE	Felix et al. (1988)
	Green algae	EC_{50}BM	14–21	3.83	LOG PHASE	Turbak et al. (1986)
	(*Selenastrum capricornutum*)	EC_{50}OX	NR	10	LOG PHASE	Turbak et al. (1986)
Crustacea	Harpacticoid copepod	LC_{50}	4	22,000	0.6–0.8 MM	Linden et al. (1979)
	Red swamp crayfish	LC_{50}	4	47,310	IMMATURE, 25–40 MM	Holck and Meek (1987)

Fish	Bleak	LC_{50}	4	16,000	8 CM	Linden et al. (1979)
	Bluegill	LC_{50}	1	150,000	0.5–2.2 G	Folmar et al. (1979)
		LC_{50}	1	230,000	0.5–2.2 G	Folmar et al. (1979)
		LC_{50}	1	240,000	0.5–2.2 G	Folmar et al. (1979)
		LC_{50}	4	135,000	0.9 G	Johnson and Finley (1980)
		LC_{50}	4	140,000	0.5–2.2 G	Folmar et al. (1979)
		LC_{50}	4	140,000	0.5–2.2 G	Folmar et al. (1979)
		LC_{50}	4	220,000	0.5–2.2 G	Folmar et al. (1979)
	Channel catfish	LC_{50}	1	130,000	0.5–2.2 G	Folmar et al. (1979)
		LC_{50}	4	130,000	0.5–2.2 G	Folmar et al. (1979)
		LC_{50}	4	130,000	2.2 G	Johnson and Finley (1980)
	Common, mirror, colored, carp	LC_{50}	2	12,800	4.45 CM, 1.10 G	Li and Chen (1981)
		LC_{50}	2	115,000	4.45 CM, 1.10 G	Li and Chen (1981)
	Fathead minnow	LC_{50}	1	97,000	0.5–2.2 G	Folmar et al. (1979)
		LC_{50}	4	97,000	0.5–2.2 G	Folmar et al. (1979)
		LC_{50}	4	97,000	0.6 G	Johnson and Finley (1980)
	Flagfish	LC_{20}-D	2 hr	2,940	JUVENILE, 8 D POSTHATCH	Holdway and Dixon (1988)
		LC_{20}-D	2 hr	29,600	JUVENILE, 8 D POSTHATCH	Holdway and Dixon (1988)
		MOR-D	2 hr	30,000	JUVENILE, 2 D POSTHATCH	Holdway and Dixon (1988)
		MOR-D	2 hr	30,000	JUVENILE, 4 D POSTHATCH	Holdway and Dixon (1988)
	Grass carp, white amur	LC_{50}	1	26,000	9.5 CM, 15.8 G, 1+ YEAR	Tooby et al. (1980)
		LC_{50}	2	24,000	9.5 CM, 15.8 G, 1+ YEAR	Tooby et al. (1980)
		LC_{50}	4	15,000	9.5 CM, 15.8 G, 1+ YEAR	Tooby et al. (1980)

Table 37. Toxicology of Aquatic and Terrestrial Species: Chemical: Glyphosate, CAS Number: 1071836 (continued)

Class	Species	Effect	Duration (days)	Concentration (µg L^{-1})	Lifestage	Ref.
Fish (continued)	Rainbow trout, donaldson trout	LC_{50}	1	140,000	0.5–2.2 G	Folmar et al. (1979)
		LC_{50}	1	240,000	0.5–2.2 G	Folmar et al. (1979)
		LC_{50}	1	240,000	0.5–2.2 G	Folmar et al. (1979)
		LC_{50}	4	130,000	0.8 G	Johnson and Finley (1980)
		LC_{50}	4	140,000	0.5–2.2 G	Folmar et al. (1979)
		LC_{50}	4	140,000	0.5–2.2 G	Folmar et al. (1979)
		LC_{50}	4	240,000	0.5–2.2 G	Folmar et al. (1979)
		LC_{50}	4	240,000	0.5–2.2 G	Folmar et al. (1979)
		AVO	4	10,000	FRY	Folmar (1976)
Insect	Midge (*Chironomus plumosus*)	EC_{50}IM	2	55,000	EARLY 4TH INSTAR LARVAE	Folmar et al. (1979)
	Mosquito (*Anopheles quadrimaculatus*)	LC_{50}	1	673,430	4TH INSTAR	Holck and Meek (1987)
Tracheophy	Duckweed (*Lemna minor*)	PHY	0.5	1,690	FROND	O'Brien and Prendeville (1979)
		PHY	1	1,690	FROND	O'Brien and Prendeville (1979)
	Water-milfoil (*Myriophyllum spicatum*)	GRO	28	320	NR	Davis (1981)
	Waterweed (*Elodea canadensis*)	GRO	21	320	NR	Davis (1981)

Bird	Bobwhite quail	LD_{50}	NR	>2,000	USDA Forest Service (1989)
	Duck	LC_{50}	8	>4,640	WSSA (1983)
	Harlequin	LC_{50}	4	163	Monsanto (1982)
	Quail	LC_{50}	8	>4,640	WSSA (1983)
		LD_{50}	NR	>3,800	WSSA (1983)
Mammal	Rabbit	EC_{50}RE	6–27	>175	EPA (1982a)
		LD_{50}	NR	3,800	EPA (1982b)
		LD_{50}	NR	>7,900	Worthington (1979)
		BMS	NR	>350	EPA (1982a)
		MOR	NR	>350	EPA (1982a)
	Rat	LD_{50}	NR	4,300	EPA (1982b)
		LD_{50}	NR	4,320	USDA Forest Service (1989)
		LD_{50}	NR	4,900	EPA (1982b)
		LD_{50}	NR	5,640	WSSA (1983)
		BEH	90	>2,000	WSSA (1983)
		BEH	3 generations	>10	EPA (1982a)
		BMS	90	>2,000	WSSA (1983)
		BMS	3 generations	>10	EPA (1982a)
		FCR	90	>2,000	WSSA (1983)
		FCR	730	>300	WSSA (1983)
		HEM	90	>2,000	WSSA (1983)
		HIST	90	>2,000	WSSA (1983)
		MOR	90	>2,000	WSSA (1983)
		MOR	3 generations	>10	EPA (1982a)
		PHY	90	>2,000	WSSA (1983)
		REP	NR	>100	EPA (1982a)
		TER	6–19	>3.5 G K^{-1} D^{-1}	EPA (1982a)

Table 38. Toxicology of Aquatic and Terrestrial Species: Chemical: Glyphosate, Roundup, CAS Number: 38641940

Class	Species	Effect	Duration (days)	Concentration ($\mu g\ L^{-1}$)	Lifestage	Ref.
Algae	Algae, phytoplankton, algal mat	EC_{50}PS	4 hr	35,400	PERIPHYTIC ALGAL COMMUNITY	Goldsborough and Brown (1988)
		EC_{50}PS	4 hr	44,400	PERIPHYTIC ALGAL COMMUNITY	Goldsborough and Brown (1988)
		EC_{50}PS	4 hr	69,700	PERIPHYTIC ALGAL COMMUNITY	Goldsborough and Brown (1988)
		EC_{50}PS	4 hr	8,900–89,000	PERIPHYTIC ALGAL COMMUNITY	Goldsborough and Brown (1988)
	Green algae (*Chlorella sorokiniana*)	LET	NR	5,000	NR	Christy et al. (1981)
		PGR	NR	2,000	NR	Christy et al. (1981)
		PGR	NR	3,000	NR	Christy et al. (1981)
Crustacea	Scud (*Gammarus pseudolimnaeus*)	LC_{50}	2	62,000	MATURE	Folmar et al. (1979)
		LC_{50}	4	43,000	MATURE	Johnson and Finley (1980)
		LC_{50}	4	43,000	MATURE	Folmar et al. (1979)
	Water flea (*Daphnia magna*)	EC_{50}IM	2	3000	1ST INSTAR	Johnson and Finley (1980)
		EC_{50}IM	2	3,000	1ST INSTAR	Folmar et al. (1979)
		EC_{50}IM	2	7,900	MATURE	Hartman and Martin (1984)
		EC_{50}IM	4	25,500	<24 HR	Servizi et al. (1987)
		MOR	2	2.2	2 D	Hildebrand et al. (1980)
		MOR	2	22	2 D	Hildebrand et al. (1980)
		MOR	2	220	2 D	Hildebrand et al. (1980)
		POP	7	1,000	MATURE	Hartman and Martin (1984)

Fish	Bluegill	LC_{50}	1	2,400	0.5–2.2 G	Folmar et al. (1979)
		LC_{50}	1	3,900	0.5–2.2 G	Folmar et al. (1979)
		LC_{50}	1	4,000	0.5–2.2 G	Folmar et al. (1979)
		LC_{50}	1	4,300	0.5–2.2 G	Folmar et al. (1979)
		LC50	1	4,300	0.5–2.2 G	Folmar et al. (1979)
		LC_{50}	1	6,200	0.5–2.2 G	Folmar et al. (1979)
		LC_{50}	1	6,400	0.5–2.2 G	Folmar et al. (1979)
		LC_{50}	1	6,400	0.5–2.2 G	Folmar et al. (1979)
		LC_{50}	1	6,600	0.5–2.2 G	Folmar et al. (1979)
		LC_{50}	1	7,600	0.5–2.2 G	Folmar et al. (1979)
		LC_{50}	1	8,000	0.5–2.2 G	Folmar et al. (1979)
		LC_{50}	1	9,600	0.5–2.2 G	Folmar et al. (1979)
		LC_{50}	4	1,800	0.5–2.2 G	Folmar et al. (1979)
		LC_{50}	4	2,400	0.5–2.2 G	Folmar et al. (1979)
		LC_{50}	4	2,400	0.5–2.2 G	Folmar et al. (1979)
		LC_{50}	4	4,000	0.5–2.2 G	Folmar et al. (1979)
		LC_{50}	4	4,000	0.5–2.2 G	Folmar et al. (1979)
		LC_{50}	4	4,200	0.5–2.2 G	Folmar et al. (1979)
		LC_{50}	4	5,000	0.5–2.2 G	Folmar et al. (1979)
		LC_{50}	4	5,000	0.5–2.2 G	Folmar et al. (1979)
		LC_{50}	4	5,600	0.7 G	Johnson and Finley (1980)
		LC_{50}	4	5,600	0.5–2.2 G	Folmar et al. (1979)
		LC_{50}	4	6,000	0.5–2.2 G	Folmar et al. (1979)
		LC_{50}	4	7,000	0.5–2.2 G	Folmar et al. (1979)
		LC_{50}	4	7,500	0.5–2.2 G	Folmar et al. (1979)
	Carp, hawk fish	LET	0.16 hr	100,000	4–5 CM, FINGERLING	Singh and Yadav (1978)
	Channel catfish	LC_{50}	1	3,700	SWIM-UP FRY	Folmar et al. (1979)
		LC_{50}	1	4,300	SAC FRY	Folmar et al. (1979)
		LC_{50}	1	13,000	0.5–2.2 G	Folmar et al. (1979)
		LC_{50}	1	13,000	FINGERLING, 2.2 G	Folmar et al. (1979)
		LC_{50}	1	43,000	EYED EGGS	Folmar et al. (1979)

Table 38. Toxicology of Aquatic and Terrestrial Species: Chemical: Glyphosate, Roundup, CAS Number: 38641940 (continued)

Class	Species	Effect	Duration (days)	Concentration ($\mu g\ L^{-1}$)	Lifestage	Ref.
Fish (continued)	Channel catfish	LC_{50}	4	3,300	SWIM-UP FRY	Folmar et al. (1979)
		LC_{50}	4	4,300	SAC FRY	Folmar et al. (1979)
		LC_{50}	4	13,000	0.6 G	Johnson and Finley (1980)
		LC_{50}	4	13,000	0.5–2.2 G	Folmar et al. (1979)
		LC_{50}	4	13,000	FINGERLING, 2.2 G	Folmar et al. (1979)
	Chinook salmon	LC_{50}	4	20,000	4.6 G	Mitchell (1987b)
	Coho salmon, silver salmon	LC_{50}	4	22,000	11.8 G	Mitchell (1987b)
		LC_{50}	4	42,000	FRY, 0.30 G, 3.5 CM	Servizi et al. (1987)
		BEH	10	29	SMOLT, 12.39 G	Mitchell et al. (1987a)
	Fathead minnow	LC_{50}	1	2,400	0.5–2.2 G	Folmar et al. (1979)
		LC_{50}	4	2,300	0.6 G	Johnson and Finley (1980)
		LC_{50}	4	2,300	0.5–2.2 G	Folmar et al. (1979)
	Rainbow trout, donaldson trout	LC_{50}	1	2,200	FINGERLING, 1 G	Folmar et al. (1979)
		LC_{50}	1	2,400	0.5–2.2 G	Folmar et al. (1979)
		LC_{50}	1	2,400	0.5–2.2 G	Folmar et al. (1979)
		LC_{50}	1	2,400	0.5–2.2 G	Folmar et al. (1979)
		LC_{50}	1	7,500	0.5–2.2 G	Folmar et al. (1979)
		LC_{50}	1	8,300	0.5–2.2 G	Folmar et al. (1979)
		LC_{50}	1	8,300	FINGERLING, 2 G	Folmar et al. (1979)
		LC_{50}	1	11,000	SAC FRY	Folmar et al. (1979)
		LC_{50}	1	14,000	0.5–2.2 G	Folmar et al. (1979)
		LC_{50}	1	14,000	0.5–2.2 G	Folmar et al. (1979)
		LC_{50}	1	14,000	0.5–2.2 G	Folmar et al. (1979)
		LC_{50}	1	14,000	0.5–2.2 G	Folmar et al. (1979)
		LC_{50}	1	14,000	0.5–2.2 G	Folmar et al. (1979)

LC_{50}	1	14,000	0.5–2.2 G	Folmar et al. (1979)
LC_{50}	1	19,000	0.5–2.2 G	Folmar et al. (1979)
LC_{50}	1	46,000	EYED EGGS	Folmar et al. (1979)
LC_{50}	4	1,300	FINGERLING, 1 G	Folmar et al. (1979)
LC_{50}	4	1,400	0.5–2.2 G	Folmar et al. (1979)
LC_{50}	4	1,400	0.5–2.2 G	Folmar et al. (1979)
LC_{50}	4	1,600	0.5–2.2 G	Folmar et al. (1979)
LC_{50}	4	2,400	SAC FRY	Folmar et al. (1979)
LC_{50}	4	7,400	0.5–2.2 G	Folmar et al. (1979)
LC_{50}	4	7,500	0.5–2.2 G	Folmar et al. (1979)
LC_{50}	4	7,600	0.5–2.2 G	Folmar et al. (1979)
LC_{50}	4	7,600	0.5–2.2 G	Folmar et al. (1979)
LC_{50}	4	7,600	0.5–2.2 G	Folmar et al. (1979)
LC_{50}	4	7,600	0.5–2.2 G	Folmar et al. (1979)
LC_{50}	4	8,300	1 G	Johnson and Finley (1980)
LC_{50}	4	8,300	0.5–2.2 G	Folmar et al. (1979)
LC_{50}	4	8,300	FINGERLING, 2 G	Folmar et al. (1979)
LC_{50}	4	9,000	0.5–2.2 G	Folmar et al. (1979)
LC_{50}	4	14,000	0.5–2.2 G	Folmar et al. (1979)
LC_{50}	4	15,000	0.37 G	Mitchell (1987b)
LC_{50}	4	16,000	EYED EGGS	Folmar et al. (1979)
LC_{50}	4	22,000	0.37 G	Mitchell (1987b)
LC_{50}	4	25,500	FRY, 0.60 G, 3.9 CM	Servizi et al. (1987)
LC_{50}	4	26,000	0.37 G	Mitchell (1987b)
LC_{50}	4	28,000	FRY, 0.33 G, 3.4 CM	Servizi et al. (1987)
LC_{50}	4	52,000	1.6 G, 5 CM	Hildebrand et al. (1982)
LC_{50}	4	54,800	1.4 G, 4.9 CM, FINGERLING	Hildebrand et al. (1982)
AVO	0.33 hr	50,000	FINGERLING	Hildebrand et al. (1982)
AVO	1 hr	10,000	5 WEEKS AFTER SWIM-UP	Folmar (1976)

Table 38. Toxicology of Aquatic and Terrestrial Species: Chemical: Glyphosate, Roundup, CAS Number: 38641940 (continued)

Class	Species	Effect	Duration (days)	Concentration ($\mu g\ L^{-1}$)	Lifestage	Ref.
Fish (continued)	Rainbow trout, donaldson trout	AVO	1 hr	10,000	FRY	Folmar et al. (1979)
		AVO	1 hr	40,000	FINGERLING	Hildebrand et al. (1982)
		AVO	1 hr	<10,000	FRY	Folmar et al. (1979)
		HAT	4 hr	5,000	EYED EGGS	Folmar et al. (1979)
		HAT	4 hr	10,000	EYED EGGS	Folmar et al. (1979)
		MOR	4 hr	2,000	SAC FRY TO SWIM-UP FRY	Folmar et al. (1979)
		MOR	4 hr	5,000	SAC FRY TO SWIM-UP FRY	Folmar et al. (1979)
		MOR-D	NR		2.1 G, 5.8 CM, FINGERLING	Hildebrand et al. (1982)
		MOR-D	NR		2.3 G, 5.5 CM, FINGERLING	Hildebrand et al. (1982)
		REP-D	0.5	2,000	NR	Folmar et al. (1979)
		REP-D	0.5	2,000	NR	Folmar et al. (1979)
	Sockeye salmon	LC_{50}	4	26,700	FINGERLING, 3.80 G, 6.5 CM	Servizi et al. (1987)
		LC_{50}	4	27,700	FINGERLING, 3.70 G, 6.5 CM	Servizi et al. (1987)
		LC_{50}	4	28,800	FRY, 0.25 G, 3 CM	Servizi et al. (1987)
Insect	Mayfly (*Ephemerella walkerii*)	AVO	1 hr	1,000	LATE INSTAR NYMPH	Folmar et al. (1979)
		AVO	1 hr	10,000	LATE INSTAR NYMPH	Folmar (1978)
		AVO	1 hr	10,000	LATE INSTAR NYMPH	Folmar et al. (1979)

Class	Species	Effect	Duration (days)	Concentration (µg L^{-1})	Lifestage	Ref.
	Midge (*Chironomus plumosus*)	EC_{50}IM	2	18,000	4TH INSTAR LARVAE	Folmar et al. (1979)
		EC_{50}IM	2	55,000	4TH INSTAR	Johnson and Finley (1980)
		LC_{50}	2	30,000	4TH INSTAR LARVAE	Folmar et al. (1979)
		LC_{50}	2	34,000	4TH INSTAR LARVAE	Folmar et al. (1979)
		LC_{50}	2	34,000	4TH INSTAR LARVAE	Folmar et al. (1979)
		LC_{50}	2	43,000	4TH INSTAR LARVAE	Folmar et al. (1979)
		DRF	7	200	LARVAE	Folmar et al. (1979)
		DRF	7	2,000	LARVAE	Folmar et al. (1979)
		DRF	7	2,000	LARVAE	Folmar et al. (1979)
Tracheophy	Duckweed (*Lemna minor*)	EC_{50}GR	14	2,000	2 FROND	Hartman and Martin (1984)
	Sago pondweed	GRO	NR	<10,000	TUBERS	Hartman and Martin (1985)

Table 39. Toxicology of Aquatic and Terrestrial Species: Chemical: Kerosene

Class	Species	Effect	Duration (days)	Concentration (µg L^{-1})	Lifestage	Ref.
Mammal	Rat	LD_{50}		>28,000		USDA Forest Service (1989)

Table 40. Toxicology of Aquatic and Terrestrial Species: Chemical: Malathion, CAS Number: 121755

Class	Species	Effect	Duration (days)	Concentration ($\mu g\ L^{-1}$)	Lifestage	Ref.
Algae	Algae, phytoplankton, algal mat	ABD	4	10	NR	Taub et al. (1983)
		ABD	14	5,000	NR	Murray and Guthrie (1980)
		POP	11	10	NR	Taub et al. (1983)
		POP	NR	1,270–5,070	NR	Mani and Konar (1985a)
		PSE	1	10,000	NR	Vickers and Boyd (1971)
		PSE	4	10	NR	Taub et al. (1983)
		PSE	4 hr	1,000	NR	Butler (1963)
	Blue-green algae (*Anabaena cylindrica*)	NFX	1 hr	10,000	1 ML EXPO GROWING CULTURE	DaSilva et al. (1975)
	Blue-green algae (*Anabaena* sp)	NFX	4	100,000	NR	Tandon et al. (1988)
		PGR	5–20	50,000	NR	Tandon et al. (1988)
		PGR	18	10	NR	Taub et al. (1983)
		PGR	30	10,000	NR	Tandon et al. (1988)
		PSE	1.5 hr	100,000	EXPO GRO PHASE	Tandon et al. (1988)
	Blue-green algae (*Aulosira* sp)	NFX	1 hr	10,000	1 ML EXPO GROWING CULTURE	DaSilva et al. (1975)
	Blue-green algae (*Aulosira fertilissima*)	NFX	4	10,000	NR	Tandon et al. (1988)
		NFX	4	50,000	NR	Tandon et al. (1988)
		PGR	10	50,000	NR	Tandon et al. (1988)
		PGR	30	10,000	NR	Tandon et al. (1988)
		PSE	1.5 hr	10,000	EXPO GRO PHASE	Tandon et al. (1988)
		PSE	1.5 hr	50,000	EXPO GRO PHASE	Tandon et al. (1988)

Blue-green algae (*Calothrix elenkinii*)	NFX	1 hr	10,000	1 ML EXPO GROWING CULTURE	DaSilva et al. (1975)
	NFX	1 hr	10,000	1 ML EXPO GROWING CULTURE	DaSilva et al. (1975)
Blue-green algae (*Chlorogloea fritschii*)	NFX	1 hr	10,000	1 ML EXPO GROWING CULTURE	DaSilva et al. (1975)
Blue-green algae (*Cylindrospermum muscicola*)	NFX	1 hr	10,000	1 ML EXPO GROWING CULTURE	DaSilva et al. (1975)
Blue-green algae (*Lyngbya* sp)	PGR	28	10	NR	Taub et al. (1983)
Blue-green algae (*Nostoc* sp)	NFX	1 hr	10,000	1 ML EXPO GROWING CULTURE	DaSilva et al. (1975)
Blue-green algae (*Nostoc muscorum*)	NFX	1 hr	10,000	1 ML EXPO GROWING CULTURE	DaSilva et al. (1975)
	PGR	NR	2,000,000	CYANOBACTERIAL LAWN	Mallison and Cannon (1984)
Blue-green algae (*Plectonema boryanum*)	PGR	31	10,000	10 ML	Mallison and Cannon (1984)
Blue-green algae (*Spirulina platensis*)	PGR	NR	500–25,000	NR	Abd El-Magid (1986)
Blue-green algae (*Tolypothrix tenuis*)	NFX	1 hr	10,000	1 ML EXPO GROWING CULTURE	DaSilva et al. (1975)
Diatom (*Amphiprora* sp)	PGR	4	10,000	NR	Maly and Ruber (1983)
Diatom (*Gomphonema parvulum*)	PGR	4	10	NR	Taub et al. (1983)
Diatom (*Nitzschia kuetzingianum*)	PGR	11	10	NR	Taub et al. (1983)
Flagellate euglenoid (*Euglena gracilis*)	PGR	1	100,000	1 WEEK	Poorman (1973)
	PGR	5	7,250	NR	Moore (1970)
	PGR	7	100,000	1 WEEK	Poorman (1973)
	PGR	9	7,250	NR	Moore (1970)
Green algae (*Ankistrodesmus* sp)	PGR	4	10	NR	Taub et al. (1983)

Table 40. Toxicology of Aquatic and Terrestrial Species: Chemical: Malathion, CAS Number: 121755 (continued)

Class	Species	Effect	Duration (days)	Concen-tration (μg L^{-1})	Lifestage	Ref.
Algae (continued)	Green algae (*Chlamydomonas reinhardtii*)	PGR	1	10	NR	Taub et al. (1983)
		REP-D	2 hr	0.001	NR	Netrawali et al. (1986)
	Green algae (*Chlorella pyrenoidosa*)	PGR	4	0.01–10	7 D, 6,000,000–7,000,000 CELLS ML^{-1}	Christie (1969)
		PGR	7	100,000	7 D, 6,000,000–7,000,000 CELLS ML^{-1}	Christie (1969)
		PGR	7	100,000	7 D, 6,000,000–7,000,000 CELLS ML^{-1}	Christie (1969)
	Green algae (*Chlorella vulgaris*)	PGR	14	10	NR	Taub et al. (1983)
	Green algae (*Scenedesmus obliquus*)	PGR	4	10	NR	Taub et al. (1983)
	Green algae (*Selenastrum capricornutum*)	PGR	4	10	NR	Taub et al. (1983)
	Green algae (*Stigeoclonium* sp)	PGR	11	10	NR	Taub et al. (1983)
	Green algae (*Ulothrix* sp)	PGR	14	10	NR	Taub et al. (1983)
Amphibia	Bullfrog	MOR	4	5,000	TADPOLES	Hall and Kolbe (1980)
	Fowler's toad	LC_{50}	1	1,900	TADPOLE, 28–35 D	Sanders (1970b)
		LC_{50}	2	500	TADPOLE, 28–35 D	Sanders (1970b)
		LC_{50}	4	420	TADPOLE, 28–35 D	Sanders (1970b)

Frog (*Microhyla ornata*)	ABN	1	1,000	EMBRYO, YOLK-PLUG STAGE	Pawar et al. (1983)
	ABN	1	5,000	EMBRYO, YOLK-PLUG STAGE	Pawar et al. (1983)
	LET	2	20,000	EMBRYO, YOLK-PLUG STAGE	Pawar et al. (1983)
	LET	3	15,000	EMBRYO, YOLK-PLUG STAGE	Pawar et al. (1983)
	MOR	1	20,000	EMBRYO, YOLK-PLUG STAGE	Pawar et al. (1983)
	MOR	2	5,000	EMBRYO, YOLK-PLUG STAGE	Pawar et al. (1983)
	MOR	2	10,000	EMBRYO, YOLK-PLUG STAGE	Pawar et al. (1983)
	MOR	2	15,000	EMBRYO, YOLK-PLUG STAGE	Pawar et al. (1983)
	MOR	3	5,000	EMBRYO, YOLK-PLUG STAGE	Pawar et al. (1983)
	MOR	3	10,000	EMBRYO, YOLK-PLUG STAGE	Pawar et al. (1983)
	MOR	4	5,000	EMBRYO, YOLK-PLUG STAGE	Pawar et al. (1983)
	MOR	4	10,000	EMBRYO, YOLK-PLUG STAGE	Pawar et al. (1983)

Table 40. Toxicology of Aquatic and Terrestrial Species: Chemical: Malathion, CAS Number: 121755 (continued)

Class	Species	Effect	Duration (days)	Concen-tration ($\mu g\ L^{-1}$)	Lifestage	Ref.
Amphibia (continued)	Frog (*Rana hexadactyla*)	LC_{50}	0.5	3.54	TADPOLE, 20 (15–25) MM	Khangarot et al. (1985)
		LC_{50}	1	0.846	TADPOLE, 20 (15–25) MM	Khangarot et al. (1985)
		LC_{50}	2	0.613	TADPOLE, 20 (15–25) MM	Khangarot et al. (1985)
		LC_{50}	3	0.613	TADPOLE, 20 (15–25) MM	Khangarot et al. (1985)
		LC_{50}	4	0.59	TADPOLE, 20 (15–25) MM	Khangarot et al. (1985)
	Tiger frog, indian bullfrog	LT_{50}	3	30,000	EGG	Mohanty-Hejmadi and Dutta (1981)
		LT_{50}	7	40,000	FEEDING STAGE TADPOLE	Mohanty-Hejmadi and Dutta (1981)
		LT_{50}	NR	50,000	LIMB BUD STAGE TADPOLE	Mohanty-Hejmadi and Dutta (1981)
		LET	<1	40,000	EGG	Mohanty-Hejmadi and Dutta (1981)
		LET	<1 hr	50,000	FEEDING STAGE TADPOLE	Mohanty-Hejmadi and Dutta (1981)
		LET	<1 hr	60,000	LIMB BUD STAGE TADPOLE	Mohanty-Hejmadi and Dutta (1981)
	Toad	ENZ	25	4.7	FERTILIZED EMBRYO	De Llamas et al. (1985)
		LET	5	47,300	FERTILIZED EMBRYO	De Llamas et al. (1985)

	Western chorus frog	LC_{50}	1	560	TADPOLE, 7 D	Sanders (1970b)
		LC_{50}	2	320	TADPOLE, 7 D	Sanders (1970b)
		LC_{50}	4	200	TADPOLE, 7 D	Sanders (1970b)
Annelida	Asian leech	EC_{50}IM	1	45,000	ADULT, 4–5 CM	Kimura and Keegan (1966)
		EC_{50}IM	1	45,000	ADULT, 4–5 CM	Kimura and Keegan (1966)
		EC_{50}IM	2	17,000	ADULT, 4–5 CM	Kimura and Keegan (1966)
		EC_{50}IM	2	17,000	ADULT, 4–5 CM	Kimura and Keegan (1966)
	Oligochaete	LC_{5}	NR	50–660	NR	Mani and Konar (1985b)
	(*Branchiura sowerbyi*)	LET	3	4,000	NR	Naqvi (1973)
		LET	3	4,000	NR	Naqvi (1973)
		MOR	3	4,000	NR	Naqvi (1973)
	Oligochaete	EC_{50}IM	1	20,000	NR	Dad et al. (1982)
	(*Limnodrilus hoffmeisteri*)	EC_{50}IM	2	18,000	NR	Dad et al. (1982)
		EC_{50}IM	4	16,000	NR	Dad et al. (1982)
	Oligochaete	LC_{50}	7	30,900	NR	Bailey and Liu (1980)
	(*Lumbriculus variegatus*)	LC_{50}	9	20,500	NR	Bailey and Liu (1980)
	Oligochaete family	LC_{50}	1	26,500	<200 MM, <2 MM DIAMETER	Whitten and Goodnight (1966)
		LC_{50}	1	96,140	NR	Ardo (1974)
		LC_{50}	2	20,700	<200 MM, <2 MM DIAMETER	Whitten and Goodnight (1966)
		LC_{50}	2	35,080	NR	Ardo (1974)
		LC_{50}	4	16,700	<200 MM, <2 MM DIAMETER	Whitten and Goodnight (1966)
	Tubificid worm (*Tubifex tubifex*)	EC_{50}IM	1	36,000	NR	Dad et al. (1982)
		EC_{50}IM	2	32,000	NR	Dad et al. (1982)
		EC_{50}IM	4	28,000	NR	Dad et al. (1982)

Table 40. Toxicology of Aquatic and Terrestrial Species: Chemical: Malathion, CAS Number: 121755 (continued)

Class	Species	Effect	Duration (days)	Concen-tration ($\mu g\ L^{-1}$)	Lifestage	Ref.
Aquatic invertebrate	Aquatic community	ABD	3	12,540	NR	Ranke-Rybicaka and Stanislawska (1972)
		ABD	3	12,540	NR	Ranke-Rybicaka and Stanislawska (1972)
	Invertebrates	ABD	78	2	NR	Kennedy and Walsh (1970)
		ABD	78	20	NR	Kennedy and Walsh (1970)
Chondricht	Spiny dogfish	MOR	3	10 mg kg^{-1}	NR	Guarino et al. (1976)
Crustacea	Aquatic sowbug (*Asellus brevicaudus*)	LC_{50}	1	6,000	MATURE	Sanders (1972)
		LC_{50}	4	3,000	MATURE	Sanders (1972)
		LC_{50}	4	3,000	MATURE	Johnson and Finley (1980)
	Blue crab	$EC_{50}IR$	1	1,000	JUVENILE	Butler (1963)
		$EC_{50}IR$	2	1,000	JUVENILE	Butler (1963)
		DVP	NR	20	LARVAE	Bookhout and Monroe (1977)
		MOR	1	57 g ha^{-1}	15–25 MM	Tagatz et al. (1974)
		MOR	28	420 g ha^{-1}	JUVENILE, 83.2 (44–115) MM	Tagatz et al. (1974)
		MOR	NR	20	LARVAE	Bookhout and Monroe (1977)
		MOR	NR	20	LARVAE	Bookhout and Monroe (1977)
		MOR	NR	20	LARVAE	Bookhout and Monroe (1977)
		MOR	NR	20	LARVAE	Bookhout and Monroe (1977)
		MOR	NR	20	LARVAE	Bookhout and Monroe (1977)
		MOR	NR	20	LARVAE	Bookhout and Monroe (1977)

MOR	NR	20	LARVAE	Bookhout and Monroe (1977)
MOR	NR	20	LARVAE	Bookhout and Monroe (1977)
MOR	NR	20	LARVAE	Bookhout and Monroe (1977)
MOR	NR	50	LARVAE	Bookhout and Monroe (1977)
MOR	NR	50	LARVAE	Bookhout and Monroe (1977)
MOR	NR	50	LARVAE	Bookhout and Monroe (1977)
MOR	NR	50	LARVAE	Bookhout and Monroe (1977)
MOR	NR	50	LARVAE	Bookhout and Monroe (1977)
MOR	NR	50	LARVAE	Bookhout and Monroe (1977)
MOR	NR	50	LARVAE	Bookhout and Monroe (1977)
MOR	NR	50	LARVAE	Bookhout and Monroe (1977)
MOR	NR	50	LARVAE	Bookhout and Monroe (1977)
MOR	NR	80	LARVAE	Bookhout and Monroe (1977)
MOR	NR	80	LARVAE	Bookhout and Monroe (1977)
MOR	NR	80	LARVAE	Bookhout and Monroe (1977)
MOR	NR	80	LARVAE	Bookhout and Monroe (1977)
MOR	NR	80	LARVAE	Bookhout and Monroe (1977)
MOR	NR	80	LARVAE	Bookhout and Monroe (1977)
MOR	NR	80	LARVAE	Bookhout and Monroe (1977)
MOR	NR	80	LARVAE	Bookhout and Monroe (1977)
MOR	NR	80	LARVAE	Bookhout and Monroe (1977)
MOR	NR	110	LARVAE	Bookhout and Monroe (1977)
MOR	NR	110	LARVAE	Bookhout and Monroe (1977)
MOR	NR	110	LARVAE	Bookhout and Monroe (1977)
MOR	NR	110	LARVAE	Bookhout and Monroe (1977)
MOR	NR	110	LARVAE	Bookhout and Monroe (1977)
MOR	NR	110	LARVAE	Bookhout and Monroe (1977)
MOR	NR	110	LARVAE	Bookhout and Monroe (1977)
MOR	NR	110	LARVAE	Bookhout and Monroe (1977)
MOR	NR	110	LARVAE	Bookhout and Monroe (1977)

Table 40. Toxicology of Aquatic and Terrestrial Species: Chemical: Malathion, CAS Number: 121755 (continued)

Class	Species	Effect	Duration (days)	Concen-tration ($\mu g\ L^{-1}$)	Lifestage	Ref.
Crustacea (continued)	Brine shrimp	HAT	2	10,000	DRIED EGG	Kuwabara et al. (1980)
	Brown shrimp	BCF	0.38	3	NR	Conte and Parker (1975)
		BCF	0.38	3	NR	Conte and Parker (1975)
		BCF	0.38	3	NR	Conte and Parker (1975)
		BCF	0.38	3	NR	Conte and Parker (1975)
		BCF	1	1.50	NR	Conte and Parker (1975)
		BCF	1	1.50	NR	Conte and Parker (1975)
		BCF	1	1.50	NR	Conte and Parker (1975)
		BCF	1	1.50	NR	Conte and Parker (1975)
		BCF	1.38	1	NR	Conte and Parker (1975)
		BCF	1.38	1	NR	Conte and Parker (1975)
		BCF	1.38	1	NR	Conte and Parker (1975)
		BCF	1.38	1	NR	Conte and Parker (1975)
		BCF	2	0.80	NR	Conte and Parker (1975)
		BCF	2	0.80	NR	Conte and Parker (1975)
		BCF	2	0.80	NR	Conte and Parker (1975)
		MOR	2	85.7 g ha^{-1}	60–85 MM	Conte and Parker (1975)
		MOR	2.04	85.7 g ha^{-1}	60–85 MM	Conte and Parker (1975)
	Calanoid copepod (*Diaptomus forbesi*)	LC_{50}	NR	10–540	NR	Mani and Konar (1985b)
	Common shrimp	LC_{50}	2	330–1,000	LARVAL	Portmann (1972)

Crab (*Barytelphusa cunicularis*)	LC_{50}	1	8,920	NR	Rao and Nagabhushanam (1987)
	LC_{50}	2	6,480	NR	Rao and Nagabhushanam (1987)
	LC_{50}	3	5,640	NR	Rao and Nagabhushanam (1987)
	LC_{50}	4	3,780	NR	Rao and Nagabhushanam (1987)
Crab (*Oziotelphusa senex senex*)	BIO	10	2,000	30 G, MALE, INTERMOLT, STAGE C4	Reddy et al. (1986)
	DVP	55	100	INTERMOLT MALES, 30–32 G, 35–40 MM	Reddy et al. (1985)
Crab (*Paratelphusa jacuemontii*)	OC	1 hr	1,000	INTERMOLT	Kulkarni and Kamath (1980)
Crayfish (*Orconectes nais*)	LC_{50}	1	290	3–5 WEEKS	Sanders (1972)
	LC_{50}	4	50	3–5 WEEKS	Sanders (1972)
	LC_{50}	4	180	EARLY INSTAR	Johnson and Finley (1980)
	MOR	4	100,000	MATURE	Sanders (1972)
Cyclopoid copepod (*Macrocyclops albidus*)	MOR	1	1	NR	Johnson (1978c)
	MOR	1	5	NR	Johnson (1978c)
	MOR	1	10	NR	Johnson (1978c)
	MOR	1	25	NR	Johnson (1978c)
	MOR	1	50	NR	Johnson (1978c)
Grass shrimp (*Palaemonetes vulgaris*)	LC_{50}	1	131	31 MM, 0.47 G	Eisler (1969)
Grass shrimp (*Palaemonetes kadiakensis*)	LC_{50}	1	150	LATE INSTAR	Sanders (1972)
	LC_{50}	1	320	LATE INSTAR	Sanders (1972)
	LC_{50}	2	25	LATE INSTAR	Sanders (1972)
Grass shrimp (*Palaemonetes vulgaris*)	LC_{50}	2	90	31 MM, 0.47 G	Eisler (1969)

Table 40. Toxicology of Aquatic and Terrestrial Species: Chemical: Malathion, CAS Number: 121755 (continued)

Class	Species	Effect	Duration (days)	Concen-tration ($\mu g\ L^{-1}$)	Lifestage	Ref.
Crustacea	Grass shrimp	LC_{50}	2	100	LATE INSTAR	Sanders (1972)
(continued)	(*Palaemonetes kadiakensis*)	LC_{50}	4	12	LATE INSTAR	Sanders (1972)
	Grass shrimp	LC_{50}	4	82	31 MM, 0.47 G	Eisler (1969)
	(*Palaemonetes vulgaris*)					
	Grass shrimp	LC_{50}	4	90	LATE INSTAR	Sanders (1972)
	(*Palaemonetes kadiakensis*)	LC_{50}	4	90	MATURE	Johnson and Finley (1980)
		LC_{50}	5	9	LATE INSTAR	Sanders (1972)
		LC_{50}	5	60	LATE INSTAR	Sanders (1972)
	Grass shrimp	AVO	0.5 hr	1,000	10–40 MM	Hansen et al. (1973)
	(*Palaemonetes pugio*)	MOR	1	57 g ha^{-1}	ADULT	Tagatz et al. (1974)
	Grass shrimp	MOR	7	420 g ha^{-1}	ADULT	Tagatz et al. (1974)
	(*Palaemonetes vulgaris*)					
	Hermit crab	LC_{50}	1	118	3.5 MM, 0.28 G	Eisler (1969)
	(*Pagurus longicarpus*)	LC_{50}	2	100	3.5 MM, 0.28 G	Eisler (1969)
		LC_{50}	4	83	3.5 MM, 0.28 G	Eisler (1969)
	Indian prawn	LC_{50}	1	384	INTERMOLT STAGE	Rao et al. (1988)
		LC_{50}	2	320	INTERMOLT STAGE	Rao et al. (1988)
		LC_{50}	3	278	INTERMOLT STAGE	Rao et al. (1988)
		LC_{50}	4	214	INTERMOLT STAGE	Rao et al. (1988)
	Jumbo tiger prawn	LC_{50}	1	649	INTERMOLT STAGE	Rao et al. (1988)
		LC_{50}	2	521	INTERMOLT STAGE	Rao et al. (1988)
		LC_{50}	3	478	INTERMOLT STAGE	Rao et al. (1988)
		LC_{50}	4	392	INTERMOLT STAGE	Rao et al. (1988)

Mud crab	DVP	NR	11	LARVAE	Bookhout and Monroe (1977)
(*Rhithropanopeus harrisii*)	LET	NR	50	LARVAE	Bookhout and Monroe (1977)
	MOR	NR	11	LARVAE	Bookhout and Monroe (1977)
	MOR	NR	11	LARVAE	Bookhout and Monroe (1977)
	MOR	NR	11	LARVAE	Bookhout and Monroe (1977)
	MOR	NR	11	LARVAE	Bookhout and Monroe (1977)
	MOR	NR	11	LARVAE	Bookhout and Monroe (1977)
	MOR	NR	14	LARVAE	Bookhout and Monroe (1977)
	MOR	NR	14	LARVAE	Bookhout and Monroe (1977)
	MOR	NR	14	LARVAE	Bookhout and Monroe (1977)
	MOR	NR	14	LARVAE	Bookhout and Monroe (1977)
	MOR	NR	14	LARVAE	Bookhout and Monroe (1977)
	MOR	NR	17	LARVAE	Bookhout and Monroe (1977)
	MOR	NR	17	LARVAE	Bookhout and Monroe (1977)
	MOR	NR	17	LARVAE	Bookhout and Monroe (1977)
	MOR	NR	17	LARVAE	Bookhout and Monroe (1977)
	MOR	NR	17	LARVAE	Bookhout and Monroe (1977)
	MOR	NR	20	LARVAE	Bookhout and Monroe (1977)
	MOR	NR	20	LARVAE	Bookhout and Monroe (1977)
	MOR	NR	20	LARVAE	Bookhout and Monroe (1977)
	MOR	NR	20	LARVAE	Bookhout and Monroe (1977)
	MOR	NR	20	LARVAE	Bookhout and Monroe (1977)
	MOR	NR	50	LARVAE	Bookhout and Monroe (1977)
Ostracod (*Cypretta kawatai*)	EC_{50}IM	1	459	ADULT	Hansen and Kawatski (1976)
	EC_{50}IM	3	86	ADULT	Hansen and Kawatski (1976)
	EC_{50}IM-D	1	344	ADULT	Hansen and Kawatski (1976)
	EC_{50}IM-D	3	51	ADULT	Hansen and Kawatski (1976)
Ostracod (*Cypridopsis vidua*)	EC_{50}IM	2	47	MATURE	Johnson and Finley (1980)
Ostracod, seed shrimp subclass	ABD	25	10	NR	Taub et al. (1983)

Table 40. Toxicology of Aquatic and Terrestrial Species: Chemical: Malathion, CAS Number: 121755 (continued)

Class	Species	Effect	Duration (days)	Concentration (μg L^{-1})	Lifestage	Ref.
Crustacea (continued)	Pink shrimp	EC_{50}EQ	1	820	ADULT	Butler (1963)
		EC_{50}EQ	2	500	ADULT	Butler (1963)
		LC_{50}	2	12.5	YOUNG ADULTS	Bahner and Nimmo (1975)
		ENZ	2	1,000	78–122 MM	Coppage and Matthews (1974)
		MOR	3	420 g ha^{-1}	JUVENILE	Tagatz et al. (1974)
	Prawn	LC_{50}	1	2,907	50–65 MM, 1.0–1.5 G	Shukla and Omkar (1983b)
		LC_{50}	2	1,687	50–65 MM, 1.0–1.5 G	Shukla and Omkar (1983b)
		LC_{50}	3	1,454	50–65 MM, 1.0–1.5 G	Shukla and Omkar (1983b)
		LC_{50}	4	1,261	50–65 MM, 1.0–1.5 G	Shukla and Omkar (1983b)
	Red swamp crayfish	LC_{50}	4	49,170	IMMATURE, 25–40 MM	Holck and Meek (1987)
		MOR	3	20,000	4–10 G	Muncy and Oliver (1963)
		MOR	4	1,600	MATURE, 15–38 G	Andreu-Moliner et al. (1986)
		MOR	4	1,600	MATURE, 15–38 G	Andreu-Moliner et al. (1986)
	Sand shrimp	LC_{50}	1	246	26 MM, 0.25 G	Eisler (1969)
		LC_{50}	2	210	26 MM, 0.25 G	Eisler (1969)
		LC_{50}	4	33	26 MM, 0.25 G	Eisler (1969)
	Scud (*Gammarus fasciatus*)	LC_{50}	1	1.2	NR	Sanders (1972)
		LC_{50}	1	3.2	NR	Sanders (1972)
	Scud (*Gammarus lacustris*)	LC_{50}	1	3.8	2 MONTHS OLD	Sanders (1969)
	Scud (*Gammarus fasciatus*)	LC_{50}	1	3.8	NR	Sanders (1972)
		LC_{50}	2	0.50	NR	Sanders (1972)
	Scud (*Gammarus lacustris*)	LC_{50}	2	1.8	2 MONTHS OLD	Sanders (1969)

Scud (*Gammarus fasciatus*)	LC_{50}	2	2	NR	Sanders (1972)
	LC_{50}	4	0.50	NR	Sanders (1972)
	LC_{50}	4	0.76	NR	Sanders (1972)
	LC_{50}	4	0.76	MATURE	Johnson and Finley (1980)
	LC_{50}	4	0.90	NR	Sanders (1972)
Scud (*Gammarus lacustris*)	LC_{50}	4	1.62	NR	Gaufin et al. (1965)
	LC_{50}	4	1.8	2 MONTHS OLD	Sanders (1969)
Scud (*Gammarus fasciatus*)	LC_{50}	5	0.48	NR	Sanders (1972)
	LC_{50}	5	0.50	NR	Sanders (1972)
Scud order	ABD	4	10	NR	Taub et al. (1983)
Water flea	LC_{100}	0.92	50	NR	Prokopenko et al. (1976b)
(*Ceriodaphnia reticulata*)	HAT	10	1	NR	Prokopenko et al. (1976b)
Water flea (*Daphnia magna*)	EC_{50}IM	1	0.098	ADULT, 2–26 HR	Rawash et al. (1975)
Water flea (*Daphnia carinata*)	EC_{50}IM	1	200	ADULT, 2–2.5 MM	Santharam et al. (1976)
Water flea (*Daphnia magna*)	EC_{50}IM	1.08	0.90	NR	Crosby et al. (1966)
	EC_{50}IM	1.08	4	NR	Crosby et al. (1966)
	EC_{50}IM	2	1	1ST INSTAR	Johnson and Finley (1980)
	EC_{50}IM	2	1.6	<24 HR	Dortland (1980)
	EC_{50}IM	2	1.7	<24 HR	Dortland (1980)
Water flea (*Daphnia pulex*)	EC_{50}IM	2	1.8	1ST INSTAR	Sanders and Cope (1966)
	EC_{50}IM	2	1.8	1ST INSTAR	Johnson and Finley (1980)
Water flea (*Daphnia magna*)	EC50IM	2	2.1	<24 HR	Dortland (1980)
	EC_{50}IM	2	2.2	<24 HR	Dortland (1980)
Water flea (*Daphnia carinata*)	EC_{50}IM	2	100	ADULT, 2–2.5 MM	Santharam et al. (1976)
Water flea (*Daphnia magna*)	EC_{50}IM	21	0.34	<24 HR	Dortland (1980)
	EC_{50}IM	21	0.38	<24 HR	Dortland (1980)
	EC_{50}RE	16	0.36	NR	Hermens (1984b)
	EC_{50}RE	16	0.36	<1 D	Hermens et al. (1984a)
	EC_{50}RE	16	0.56	<1 D	Hermens et al. (1984a)
	LC_{50}	1	0.27	<24 HR	Gaaboub et al. (1975)
	LC_{50}	1	12	NR	Ardo (1974)

Table 40. Toxicology of Aquatic and Terrestrial Species: Chemical: Malathion, CAS Number: 121755 (continued)

Class	Species	Effect	Duration (days)	Concentration ($\mu g\ L^{-1}$)	Lifestage	Ref.
Crustacea (continued)	Water flea (*Daphnia carinata*)	LC_{50}	1	27	NR	Nishiuchi (1979)
		LC_{50}	1 hr	510	NR	Nishiuchi (1979)
	Water flea (*Daphnia magna*)	LC_{50}	1.08	0.90	1ST INSTAR, <24 HR	Frear and Boyd (1967)
		LC_{50}	2	9	NR	Ardo (1974)
	Water flea (*Daphnia carinata*)	LC_{50}	2	13	NR	Nishiuchi (1979)
	Water flea (*Daphnia magna*)	LC_{50}	2	33	NR	Hermens (1984b)
		LC_{50}	2	33	<1 D	Hermens et al. (1984a)
		LC_{50}	2	52	<1 D	Hermens et al. (1984a)
	Water flea (*Daphnia pulex*)	LC_{50}	3 hr	5.9	NR	Hashimoto and Nishiuchi (1981)
		LC_{50}	3 hr	13	FEMALE ADULT	Nishiuchi and Hashimoto (1969)
		LC_{50}	3 hr	13	NR	Nishiuchi and Hashimoto (1975)
		LC_{50}	3 hr	36	NR	Hashimoto and Nishiuchi (1981)
		LC_{50}	3 hr	120	NR	Hashimoto and Nishiuchi (1981)
	Water flea (*Daphnia carinata*)	LC_{50}	3 hr	150	NR	Nishiuchi (1979)
	Water flea (*Daphnia pulex*)	LC_{50}	3 hr	260	NR	Hashimoto and Nishiuchi (1981)
	Water flea (*Daphnia carinata*)	LC_{50}	6 hr	110	NR	Nishiuchi (1979)

Water flea (*Daphnia magna*)	ABD	4	10	NR	Taub et al. (1983)
	LET	7	10	NR	Desi et al. (1975)
	MOR	7	1	NR	Desi et al. (1975)
	MOR	21	0.15	<24 HR	Dortland (1980)
	MOR	21	0.3	<24 HR	Dortland (1980)
	MOT	2.08	0.9	<24 HR	Anderson (1959)
	REP	21	0.3	<24 HR	Dortland (1980)
	REP	21	0.6	<24 HR	Dortland (1980)
Water flea (*Moina macrocopa*)	LC_{50}	3 hr	58	FEMALE ADULT	Nishiuchi and Hashimoto (1969)
	LC_{50}	3 hr	58	NR	Nishiuchi and Hashimoto (1975)
Water flea (*Moina rectrirostris*)	ABD	3–4		LARVAE	Mulla and Khasawinah (1969)
Water flea (*Simocephalus serrulatus*)	EC_{50}IM	2	3.5	1ST INSTAR	Sanders and Cope (1966)
	EC_{50}IM	2	3.5	1ST INSTAR	Johnson and Finley (1980)
	EC_{50}IM	2	6.2	1ST INSTAR	Sanders and Cope (1966)
Water flea (*Simocephalus vetulus*)	LC_{100}	0.83	10	NR	Prokopenko et al. (1976b)
	MOT	2 hr	50	NR	Prokopenko et al. (1976b)
White river crayfish	LC_{50}	4	50,000	0.4 G	Cheah et al. (1980)
	LC_{50}	4	50,000	0.7 G	Carter and Graves (1972)
White shrimp	BCF	0.33	1.50	NR	Conte and Parker (1975)
	BCF	0.33	1.50	NR	Conte and Parker (1975)
	BCF	0.33	2.40	NR	Conte and Parker (1975)
	BCF	0.33	2.40	NR	Conte and Parker (1975)
	BCF	1	1.20	NR	Conte and Parker (1975)
	BCF	1	1.20	NR	Conte and Parker (1975)
	BCF	1	1.20	NR	Conte and Parker (1975)
	BCF	1	1.20	NR	Conte and Parker (1975)
	BCF	1	2.20	NR	Conte and Parker (1975)
	BCF	1 hr	2	NR	Conte and Parker (1975)
	BCF	1 hr	2	NR	Conte and Parker (1975)

Table 40. Toxicology of Aquatic and Terrestrial Species: Chemical: Malathion, CAS Number: 121755 (continued)

Class	Species	Effect	Duration (days)	Concentration ($\mu g\ L^{-1}$)	Lifestage	Ref.
Crustacea (continued)	White shrimp	BCF	3 hr	2	NR	Conte and Parker (1975)
		BCF	3 hr	2	NR	Conte and Parker (1975)
		BCF	3 hr	2	NR	Conte and Parker (1975)
		BCF	3 hr	3.20	NR	Conte and Parker (1975)
		BCF	3 hr	3.20	NR	Conte and Parker (1975)
		MOR	1	85.7 g ha^{-1}	60–85 MM	Conte and Parker (1975)
		MOR	2.04	85.7 g ha^{-1}	60–85 MM	Conte and Parker (1975)
Fish	American eel	LC_{50}	1	82	57 MM, 0.16 G	Eisler (1970a)
		LC_{50}	1	1,600	YOUNG OF THE YEAR	Rehwoldt et al. (1977)
		LC_{50}	2	82	57 MM, 0.16 G	Eisler (1970a)
		LC_{50}	2	710	YOUNG OF THE YEAR	Rehwoldt et al. (1977)
		LC_{50}	4	82	57 MM, 0.16 G	Eisler (1970a)
		LC_{50}	4	500	YOUNG OF THE YEAR	Rehwoldt et al. (1977)
	Atlantic croaker	ENZ	1	1,000	85–150 MM	Coppage and Matthews (1974)
	Atlantic herring	ENZ	1	26.4–6,600	NR	Kalac (1974)
	Atlantic salmon	MOR	1	50	1 G, 3–4 CM, JUVENILE	Peterson (1976)
		THL	1	25–50	1 G, 3–4 CM, JUVENILE	Peterson (1976)
	Atlantic silverside	LC_{50}	1	315	50 MM, 0.8 G	Eisler (1970a)
		LC_{50}	2	315	50 MM, 0.8 G	Eisler (1970a)
		LC_{50}	4	125	50 MM, 0.8 G	Eisler (1970a)
	Ayu	MOR	1 hr	15,000	YOUNG, 8–9 CM	Matsuo and Tamura (1970)

Banded killifish	LC_{50}	1	380	YOUNG OF THE YEAR	Rehwoldt et al. (1977)
	LC_{50}	2	290	YOUNG OF THE YEAR	Rehwoldt et al. (1977)
	LC_{50}	4	240	YOUNG OF THE YEAR	Rehwoldt et al. (1977)
Barb	LC_{50}	1	6,600	NR	Scirocchi and Erme (1980)
Beluga	LET	NR	500	YEARLING	Prokopenko et al. (1976a)
Black bullhead	LC_{50}	4	12,900	1.2 G	Johnson and Finley (1980)
	LC_{50}	4	12,900	NR	Macek and McAllister (1970)
	ENZ	2 hr	100	40–80 G	Murphy et al. (1968)
	MOR	2 hr	100	40–80 G	Murphy et al. (1968)
Bluegill	LC_{50}	1	110	WEIGHT 0.6–1.5 G	Macek et al. (1969)
	LC_{50}	1	140	WEIGHT 0.6–1.5 G	Macek et al. (1969)
	LC_{50}	1	140	3.8–6.4 CM, 1–2 G	Pickering et al. (1962)
	LC_{50}	1	190	3.8–6.4 CM, 1–2 G	Pickering et al. (1962)
	LC_{50}	1	220	WEIGHT 0.6–1.5 G	Macek et al. (1969)
	LC_{50}	1	600	3.8–6.4 CM, 1–2 G	Pickering et al. (1962)
	LC_{50}	1	1,700	10 G, 9 CM	Pickering et al. (1962)
	LC_{50}	2	110	3.8–6.4 CM, 1–2 G	Pickering et al. (1962)
	LC_{50}	2	120	3.8–6.4 CM, 1–2 G	Pickering et al. (1962)
	LC_{50}	2	550	3.8–6.4 CM, 1–2 G	Pickering et al. (1962)
	LC_{50}	2	1,300	10 G, 9 CM	Pickering et al. (1962)
	LC_{50}	4	46	WEIGHT 0.6–1.5 G	Macek et al. (1969)
	LC_{50}	4	55	WEIGHT 0.6–1.5 G	Macek et al. (1969)
	LC_{50}	4	88	3.8–6.4 CM, 1–2 G	Pickering et al. (1962)
	LC_{50}	4	90	3.8–6.4 CM, 1–2 G	Pickering et al. (1962)
	LC_{50}	4	103	1.5 G	Johnson and Finley (1980)
	LC_{50}	4	103	NR	Macek and McAllister (1970)
	LC_{50}	4	120	WEIGHT 0.6–1.5 G	Macek et al. (1969)
	LC_{50}	4	120	0.5 G	Carter and Graves (1972)
	LC_{50}	4	550	3.8–6.4 CM, 1–2 G	Pickering et al. (1962)
	LC_{50}	4	1,200	10 G, 9 CM	Pickering et al. (1962)

Table 40. Toxicology of Aquatic and Terrestrial Species: Chemical: Malathion, CAS Number: 121755 (continued)

Class	Species	Effect	Duration (days)	Concentration ($\mu g\ L^{-1}$)	Lifestage	Ref.
Fish (continued)	Bluegill	LD_{50}	NR	90	NR	Whitten and Goodnight (1966)
		BIO	NR		NR	Hiltibran (1974)
		BIO	NR		NR	Hiltibran (1974)
		ENZ	0.42 hr	13,100	NR	Cutkomp et al. (1971b)
		GRO	78	20	69.6 MM FORK LENGTH, 5.5 G	Kennedy and Walsh (1970)
		HIS	1	50	NR	Richmonds and Dutta (1988)
		MOR	78	20	69.6 MM FORK LENGTH, 5.5 G	Kennedy and Walsh (1970)
		STR	0.42	5,000	FINGERLING, ≤10 CM	Applegate et al. (1957)
		STR	1 hr	5,000	FINGERLING, ≤10 CM	Applegate et al. (1957)
		STR	1 hr	5,000	FINGERLING, ≤10 CM	Applegate et al. (1957)
	Bluehead wrasse	LC_{50}	1	33	80 MM, 5.4 G	Eisler (1970a)
		LC_{50}	2	27	80 MM, 5.4 G	Eisler (1970a)
		LC_{50}	4	27	80 MM, 5.4 G	Eisler (1970a)
	Brook trout	LC_{50}	3	150	2.13 G	Post and Schroeder (1971)
		LC_{50}	3	160	1.15 G	Post and Schroeder (1971)
		LC_{50}	4	120	2.13 G	Post and Schroeder (1971)
		LC_{50}	4	130	1.15 G	Post and Schroeder (1971)
		ENZ	7	40	53 G	Post and Leasure (1974)
	Brown trout	LC_{50}	4	101	1.1 G	Johnson and Finley (1980)
		LC_{50}	4	200	NR	Macek and McAllister (1970)
	Capelin	ENZ	1	26.4–6,600	NR	Kalac (1974)

Carp (*Cyprinus carpio* var *communis*)	EC_{50}HA	NR	4,570	EGGS	Kaur and Toor (1977)
	LET	1	20,000	7.5–10.5 CM, 12–27 G	Toor et al. (1973)
	MOR	3	17,000	7.5–10.5 CM, 12–27 G	Toor et al. (1973)
Carp, hawk fish	LC_{50}	1	7,150	2.5–6 CM	Arora et al. (1971b)
	LC_{50}	2	5,397	51 G	Roy and Munshi (1988)
	LC_{50}	2	6,350	2.5–6 CM	Arora et al. (1971b)
	LC_{50}	2	7,000	1–2 CM	Sreenivasan and Swaminathan (1967)
	LC_{50}	3	2,520	2.5–6 CM	Arora et al. (1971b)
	LC_{50}	4	880	4.5 MM, 51 MG	Verma et al. (1984a)
	LC_{50}	4	2,250	2.5–6 CM	Arora et al. (1971b)
	ENZ	3	2,707.5	145–195 MM, 88–109 G	Verma et al. (1979)
	ENZ	3	2,707.5	145–195 MM, 88–109 G	Verma et al. (1979)
	ENZ	3	2,707.5	145–195 MM, 88–109G	Verma et al. (1979)
	ENZ	7	2,707.5	145–195 MM, 88–109 G	Verma et al. (1979)
	ENZ	7	2,707.5	145–195 MM, 88–109 G	Verma et al. (1979)
	ENZ	7	2,707.5	145–195 MM, 88–109 G	Verma et al. (1979)
	OC	1	1,000	51 G	Roy and Munshi (1988)
Catfish	HIS	84	2,500	FEMALE, 12 G, OVARY	Haider and Upadhyaya (1985)
Catla	LC_{50}	1	6,620	2.5–6 CM	Arora et al. (1971b)
	LC_{50}	2	3,300	2.5–6 CM	Arora et al. (1971b)
	LC_{50}	3	2,350	2.5–6 CM	Arora et al. (1971b)

Table 40. Toxicology of Aquatic and Terrestrial Species: Chemical: Malathion, CAS Number: 121755 (continued)

Class	Species	Effect	Duration (days)	Concentration ($\mu g\ L^{-1}$)	Lifestage	Ref.
Fish (continued)	Channel catfish	LC_{50}	1	10,000	10 G	Carter and Graves (1972)
		LC_{50}	4	8,970	1.5 G	Johnson and Finley (1980)
		LC_{50}	4	8,970	NR	Macek and McAllister (1970)
		GRO	78	20	68.8 MM FORK LENGTH, 4.1 G	Kennedy and Walsh (1970)
		MOR	78	20	68.8 MM FORK LENGTH, 4.1 G	Kennedy and Walsh (1970)
	Chinook salmon	LC_{50}	1	24.5	51–114 MM, 1.45–5 G	Katz (1961)
		LC_{50}	1	170	FINGERLING, 3.8 CM	Parkhurst and Johnson (1955)
		LC_{50}	2	23.9	51–114 MM, 1.45–5 G	Katz (1961)
		LC_{50}	2	150	FINGERLING, 3.8 CM	Parkhurst and Johnson (1955)
		LC_{50}	3	23.6	51–114 MM, 1.45–5 G	Katz (1961)
		LC_{50}	4	23	51–114 MM, 1.45–5 G	Katz (1961)
		LC_{50}	4	120	FINGERLING, 3.8 CM	Parkhurst and Johnson (1955)
		MOR	1	10,000	5–10 CM	MacPhee and Ruelle (1969)
		MOR	1	10,000	5–10 CM	MacPhee and Ruelle (1969)
	Climbing perch (*Anabas testudineus*)	HIS	2	800	ADULT, 14 CM, 25 G	Ray and Bhattacharya (1984)

Coho salmon, silver salmon	LC_{50}	1	300	1.70 G	Post and Schroeder (1971)
	LC_{50}	4	101	NR	Macek and McAllister (1970)
	LC_{50}	4	170	0.9 G	Johnson and Finley (1980)
	LC_{50}	4	265	1.70 G	Post and Schroeder (1971)
	BCF	NR	178	6–18 MONTHS, 8.9–25.5 G	Walsh and Ribelin (1973)
	ENZ	9	100	11 G	Post and Leasure (1974)
	MOR	1	10,000	5–10 CM	MacPhee and Ruelle (1969)
	MOR	1	10,000	5–10 CM	MacPhee and Ruelle (1969)
Common, mirror, colored, carp	LC_{50}	1	2,600	YOUNG OF THE YEAR	Rehwoldt et al. (1977)
	LC_{50}	1	6,000	2.5–6 CM	Arora et al. (1971b)
	LC_{50}	1	18,000	4.2 CM, 1.2 G	Nishiuchi and Asano (1981)
	LC_{50}	2	88	EGGS	Prokopenko et al. (1976b)
	LC_{50}	2	165	YEARLINGS, 20–25 G	Prokopenko et al. (1976b)
	LC_{50}	2	750	LARVAE	Prokopenko et al. (1976b)
	LC_{50}	2	2,100	YOUNG OF THE YEAR	Rehwoldt et al. (1977)
	LC_{50}	2	4,500	4.5 CM, 1.1 G	Nishiuchi and Hashimoto (1969)
	LC_{50}	2	4,500	NR	Nishiuchi and Hashimoto (1975)
	LC_{50}	2	4,650	2.5–6 CM	Arora et al. (1971b)
	LC_{50}	2	5,600	4.2 CM, 1.2 G	Nishiuchi and Asano (1981)
	LC_{50}	2	8,500	2.5–6 CM	Sreenivasan and Swaminathan (1967)
	LC_{50}	2	10,000	2.5–6 CM	Sreenivasan and Swaminathan (1967)
	LC_{50}	2	23,000	6 CM, 3 G	Nishiuchi and Asano (1981)
	LC_{50}	2	50,000	NR	Kozlovskaya and Mayer (1984)
	LC_{50}	2	9,800–10,400	10 G	Reddy and Bashamohideen (1988)

Table 40. Toxicology of Aquatic and Terrestrial Species: Chemical: Malathion, CAS Number: 121755 (continued)

Class	Species	Effect	Duration (days)	Concen-tration ($\mu g\ L^{-1}$)	Lifestage	Ref.
Fish (continued)	Common, mirror, colored, carp	LC_{50}	3	4,250	2.5–6 CM	Arora et al. (1971b)
		LC_{50}	3	4,400	4.2 CM, 1.2 G	Nishiuchi and Asano (1981)
		LC_{50}	3	11,000	NR	Kobayashi (1978)
		LC_{50}	4	85	LARVAE, 8 MM	Verma et al. (1981a)
		LC_{50}	4	1,900	YOUNG OF THE YEAR	Rehwoldt et al. (1977)
		LC_{50}	4	3,150	2.5–6 CM	Arora et al. (1971b)
		LC_{50}	4	6,590	0.6 G	Johnson and Finley (1980)
		LC_{50}	4	6,590	NR	Macek and McAllister (1970)
		LC_{50}	NR	88	SPAWNS	Nikulina and Sokolskaya (1975)
		BIO	4	10,000	15–20 G, FRY	Dragomirescu et al. (1975)
		BIO	NR	10	NR	Dragomirescu et al. (1975)
		ENZ	4	10,000	15–20 G, FRY	Dragomirescu et al. (1975)
		ENZ	20	30,000	NR	Kozlovskaya and Mayer (1984)
		ENZ	NR	10	NR	Dragomirescu et al. (1975)
		ENZ	NR	750	YEARLING	Gantverg and Perevoznikov (1983)
		LET	NR	100	SPAWNS	Nikulina and Sokolskaya (1975)
		LET	NR	500	YEARLING	Prokopenko et al. (1976a)
		MOR	0.79	75–116 mg kg^{-1}	NR	Loeb and Kelly (1963)
		MOR	0.88	40–218 mg kg^{-1}	NR	Loeb and Kelly (1963)

	MOR	NR	10	SPAWNS	Nikulina and Sokolskaya (1975)
	RSD	4	5,000	NR	Bender (1969)
Cutthroat trout	LC_{50}	3	200	0.33 G	Post and Schroeder (1971)
	LC_{50}	4	150	0.33 G	Post and Schroeder (1971)
	LC_{50}	4	201	1.25 G	Post and Schroeder (1971)
	LC_{50}	4	280	1 G	Johnson and Finley (1980)
	ENZ	30	4 mg kg^{-1}	ADULT	Cope (1965)
Cyprinid	LC_{50}	1	11,610	NR	Ardo (1974)
	LC_{50}	2	7,640	NR	Ardo (1974)
Danio	LC_{50}	2	13,500	EARLY FRY	Sreenivasan and Swaminathan (1967)
Eastern mudminnow	LC_{50}	4	240	3–5 CM	Bender and Westman (1976)
	LC_{50}	14	140	3–5 CM	Bender and Westman (1976)
Fathead minnow	LC_{50}	1	14,250	1–1.5 G, 5–6.5 CM	Henderson and Pickering (1958)
	LC_{50}	1	14,800	1–1.5 G, 5–6.5 CM	Henderson and Pickering (1958)
	LC_{50}	1	23,000	3.8–6.4 CM, 1–2 G	Pickering et al. (1962)
	LC_{50}	1	25,000	3.8–6.4 CM, 1–2 G	Pickering et al. (1962)
	LC_{50}	1	26,000	3.8–6.4 CM, 1–2 G	Pickering et al. (1962)
	LC_{50}	2	13,100	1–1.5 G, 5–6.5 CM	Henderson and Pickering (1958)
	LC_{50}	2	13,100	1–1.5 G, 5–6.5 CM	Henderson and Pickering (1958)
	LC_{50}	2	18,000	3.8–6.4 CM, 1–2 G	Pickering et al. (1962)
	LC_{50}	2	24,000	3.8–6.4 CM, 1–2 G	Pickering et al. (1962)
	LC_{50}	2	25,000	3.8–6.4 CM, 1–2 G	Pickering et al. (1962)
	LC_{50}	4	8,650	0.9 G	Johnson and Finley (1980)
	LC_{50}	4	8,650	NR	Macek and McAllister (1970)

Table 40. Toxicology of Aquatic and Terrestrial Species: Chemical: Malathion, CAS Number: 121755 (continued)

Class	Species	Effect	Duration (days)	Concentration (μg L⁻¹)	Lifestage	Ref.
Fish (continued)	Fathead minnow	LC_{50}	4	12,500	1–1.5 G, 5–6.5 CM	Henderson and Pickering (1958)
		LC_{50}	4	12,500	1–1.5 G, 5–6.5 CM	Henderson and Pickering (1958)
		LC_{50}	4	14,100	29–30 D, 17.3 MM, 0.069 G	Geiger et al. (1988)
		LC_{50}	4	16,000	3.8–6.4 CM, 1–2 G	Pickering et al. (1962)
		LC_{50}	4	23,000	3.8–6.4 CM, 1–2 G	Pickering et al. (1962)
		LC_{50}	4	25,000	3.8–6.4 CM, 1–2 G	Pickering et al. (1962)
		LD_{50}	NR	16,000	NR	Whitten and Goodnight (1966)
		BEH	300	<200	2.5 CM	Mount and Stephan (1967)
		ENZ	3	500	8–10 CM	Weiss (1961)
		MOR	49	200	2.5 CM	Mount and Stephan (1967)
		MOR	49	580	2.5 CM	Mount and Stephan (1967)
		MOR	5	0.1 kg ha^{-1}	NR	Hilsenhoff (1959)
		MOR	5	1 kg ha^{-1}	NR	Hilsenhoff (1959)
		MOR	5	10 kg ha^{-1}	NR	Hilsenhoff (1959)
		MOR	5	100 kg ha^{-1}	NR	Hilsenhoff (1959)
		MOR	7	60 kg ha^{-1}	NR	Hilsenhoff (1959)
		MOR	7	100 kg ha^{-1}	NR	Hilsenhoff (1959)
		MOR	14	1,000	EGG TO FRY	Mount and Stephan (1967)
		MOR	120	580	2.5 CM	Mount and Stephan (1967)
		MOR	300	580	2.5 CM	Mount and Stephan (1967)

Filament or featherfin barb	LC_{50}	2	800	EARLY FRY	Sreenivasan and Swaminathan (1967)
Fish (*Labeo fimbriatus*)	LC_{50}	2	8,500	1–2 CM	Sreenivasan and Swaminathan (1967)
Flagfish	LC_{50}	1	320	37 D	Hermanutz et al. (1985)
	LC_{50}	2	280	37 D	Hermanutz et al. (1985)
	GRO	30	18.5	2–3 D	Hermanutz et al. (1985)
	MOR	30	13.8	2–3 D	Hermanutz et al. (1985)
	MOR	30	18.5	2–3 D	Hermanutz et al. (1985)
	MOR	30	23.1	2–3 D	Hermanutz et al. (1985)
	REP	30	23.1	2–3 D	Hermanutz et al. (1985)
Giant gourami	LC_{50}	1	3,300	34.6–36 MM, 0.802–0.882 G	Panwar et al. (1976)
	LC_{50}	4	2,150	34.6–36 MM, 0.802–0.882 G	Panwar et al. (1976)
Golden shiner	MOR	6 hr	0.2 kg ha^{-1}	NR	Mount et al. (1970)
Goldfish	LC_{50}	1	790	3.8–6.4 CM, 1–2 G	Pickering et al. (1962)
	LC_{50}	2	790	3.8–6.4 CM, 1–2 G	Pickering et al. (1962)
	LC_{50}	2	7,800	4.01 CM, 1.04 G	Nishiuchi and Hashimoto (1969)
	LC_{50}	2	7,800	NR	Nishiuchi and Hashimoto (1975)
	LC_{50}	4	790	3.8–6.4 CM, 1–2 G	Pickering et al. (1962)
	LC_{50}	4	2,610	EGGS-4 D POSTHATCH	Birge et al. (1979)
	LC_{50}	4	3,150	EGGS-4 D POSTHATCH	Birge et al. (1979)
	LC_{50}	4	10,700	0.9 G	Johnson and Finley (1980)
	LC_{50}	4	10,700	NR	Macek and McAllister (1970)
	LC_{50}	8	1,200	EGGS-4 D POSTHATCH	Birge et al. (1979)
	LC_{50}	8	1,650	EGGS-4 D POSTHATCH	Birge et al. (1979)
	ENZ	1	100	45–65 MM	Weiss (1959)

Table 40. Toxicology of Aquatic and Terrestrial Species: Chemical: Malathion, CAS Number: 121755 (continued)

Class	Species	Effect	Duration (days)	Concen-tration (μg L^{-1})	Lifestage	Ref.
Fish (continued)	Goldfish	MOR	34		5–7.6 CM	Joseph et al. (1972)
		MOR	34		5–7.6 CM	Joseph et al. (1972)
		MOR	35		5–7.6 CM	Joseph et al. (1972)
		MOR	35		5–7.6 CM	Joseph et al. (1972)
	Green sunfish	LC_{50}	1	1,200	15 G, 10 CM	Pickering et al. (1962)
		LC_{50}	2	700	15 G, 10 CM	Pickering et al. (1962)
		LC_{50}	4	175	1.1 G	Johnson and Finley (1980)
		LC_{50}	4	600	15 G, 10 CM	Pickering et al. (1962)
	Guppy	LC_{50}	1	930	6 MONTHS, 0.1–0.2 G, 2–2.5 CM	Pickering et al. (1962)
		LC_{50}	1	1,050	7 WEEKS	Chen et al. (1971)
		LC_{50}	1	1,360	7 WEEKS	Chen et al. (1971)
		LC_{50}	1	2,200	NR	Rehwoldt et al. (1977)
		LC_{50}	2	880	6 MONTHS, 0.1–0.2 G, 2–2.5 CM	Pickering et al. (1962)
		LC_{50}	2	1,800	NR	Rehwoldt et al. (1977)
		LC_{50}	4	840	6 MONTHS, 0.1–0.2 G, 2–2.5 CM	Pickering et al. (1962)
		LC_{50}	4	1,200	NR	Rehwoldt et al. (1977)
		LC_{50}	7	819	MALE	Desi et al. (1975)
		LET	1	2,000	NR	Adlung (1957)
		LET	1	2,000	NR	Adlung (1957)
		MOR	0.5 hr	7,000	7 WEEKS	Chen et al. (1971)
		MOR	7	100	MALE	Desi et al. (1975)
		OC	0.5 hr	2,600	NR	Ranke-Rybicka (1975)

Indian catfish	LC_{50}	1	5,800	23 G	Singh et al. (1984)
	LC_{50}	1	9,500	37.9 G, 14.2 CM	Mishra and Srivastava (1983)
	LC_{50}	1	13,490	50–75 MM, 5–10 G	Verma et al. (1982a)
	LC_{50}	1	45,300	NR	Mishra et al. (1986)
	LC_{50}	2	3,350	23.0 G	Singh et al. (1984)
	LC_{50}	2	9,250	37.9 G, 14.2 CM	Mishra and Srivastava (1983)
	LC_{50}	2	17,180	50–75 MM, 5–10 G	Verma et al. (1982a)
	LC_{50}	3	8,620	37.9 G, 14.2 CM	Mishra and Srivastava (1983)
	LC_{50}	3	16,180	50–75 MM, 5–10 G	Verma et al. (1982a)
	LC_{50}	4	7,300	34.20 G, 13.5 CM, MALES AND FEMALES	Srivastava and Srivastava (1988)
	LC_{50}	4	8,500	37.9 G, 14.2 CM	Mishra and Srivastava (1983)
	LC_{50}	4	15,000	50–75 MM, 5–10 G	Verma et al. (1982a)
	LC_{50}	4	38,000	ADULT, 35–42 G, 19–21 CM	Singh and Singh (1980a)
	LC_{50}	4	42,000	ADULT, 35–42 G, 19–21 CM	Singh and Singh (1980a)
	LC_{50}	4	45,000	ADULT, 35–42 G, 19–21 CM	Singh and Singh (1980a)
	LC_{50}	4	45,000	FEMALE, 45–55 G, 19–20 CM	Yadav and Singh (1986)
	BEH	1	>30,000	NR	Mishra et al. (1986)
	BIO	2 hr	7,600	37.9 G, 14.2 CM	Mishra and Srivastava (1983)
	BIO	4	7,600	37.9 G, 14.2 CM	Mishra and Srivastava (1983)
	BIO	4	8,000	ADULT MALE, 40 (38–42) G, 16 CM	Lal et al. (1986)
	BIO	4	9,000	ADULT FEMALE, 35–42 G, 19–21 CM	Singh and Singh (1980a)
	BIO	28	9,000	38 G, 19.5 CM, ADULT	Singh and Singh (1980f)
	BIO	28	9,000	MALE, 22 CM, 70 G	Singh and Singh (1980b)

Table 40. Toxicology of Aquatic and Terrestrial Species: Chemical: Malathion, CAS Number: 121755 (continued)

Class	Species	Effect	Duration (days)	Concentration (µg L^{-1})	Lifestage	Ref.
Fish (continued)	Indian catfish	BIO	28	35,000	ADULT, FEMALE, 35.5 G, 22.5 CM	Singh and Singh (1980c)
		ENZ	4	10,000	FEMALE, 45–60 G, 19–20 CM	Yadav and Singh (1987a)
		HEM	2 hr	7,600	37.9 G, 14.2 CM	Mishra and Srivastava (1983)
		HEM	4	5,840	34.20 G, 13.5 CM, MALES AND FEMALES	Srivastava and Srivastava (1988)
		HEM	4	7,600	37.9 G, 14.2 CM	Mishra and Srivastava (1983)
		HEM	4	8,000	ADULT MALE, 40 (38–42) G, 16 CM	Lal et al. (1986)
		HEM	15	0.011%	30–50 G	Goel et al. (1982)
		HEM	15	5,200	19–22 CM, 50–70 G	Verma et al. (1981b)
		HEM	28	10,000	FEMALE, 45–55 G, 19–20 CM	Yadav and Singh (1986)
		HEM	30	5,200	19–22 CM, 50–70 G	Verma et al. (1981b)
		HRM	4	9,000	ADULT FEMALE, 35–42 G, 19–21 CM	Singh and Singh (1980a)
		HRM	4	10,000	FEMALE, 45–60 G, 19–20 CM	Yadav and Singh (1987a)
		HRM	28	10,000	45–60 G, 19–20 CM, FEMALE	Yadav and Singh (1987b)
		PHY	4	900	35–52 G, 19–21 CM	Singh and Singh (1980e)
		PHY	4	9,000	FEMALES, ADULT, 35–52 G, 19–21 CM	Singh and Singh (1980e)

	PHY	4	45,000	FEMALES, ADULT, 35–52 G, 19–21 CM	Singh and Singh (1980e)
	PHY	28	9,000	35.5 G, ADULT FEMALE, 22.5 CM	Singh and Singh (1980d)
	PHY	28	9,000	35.5 G, ADULT FEMALE, 22.5 CM	Singh and Singh (1980d)
	PHY	28	35,000	35.5 G, ADULT FEMALE, 22.5 CM	Singh and Singh (1980d)
	PHY	28	35,000	35.5 G, ADULT FEMALE, 22.5 CM	Singh and Singh (1980d)
	PHY	42	12,400	MALE, 50–60 G	Pandey et al. (1988)
	PHY	NR	35,000	42.5–48 G, 21.5–23 CM	Singh and Singh (1982)
Killifish	LC_{50}	1	1,150	25–40 MM STANDARD LENGTH	Jacob et al. (1982)
	LC_{50}	2	975	25–40 MM STANDARD LENGTH	Jacob et al. (1982)
Lake trout, siscowet	LC_{50}	4	76	0.3 G	Johnson and Finley (1980)
	BCF	NR	178	6–18 MONTHS, 8.9–25.5 G	Walsh and Ribelin (1973)
Largemouth bass	LC_{50}	1	420	5 G, 8 CM	Pickering et al. (1962)
	LC_{50}	2	280	5 G, 8 CM	Pickering et al. (1962)
	LC_{50}	4	250	5 G, 8 CM	Pickering et al. (1962)
	LC_{50}	4	285	0.9 G	Johnson and Finley (1980)
	LC_{50}	4	285	NR	Macek and McAllister (1970)
	LT_{50}	2.67 hr	500	8–10 CM	Weiss (1961)
	ENZ	2.67 hr	500	8–10 CM	Weiss (1961)
Loach (*Lepidocephalus thermalis*)	LC_{50}	1	12,500	35 MM JUVENILE	Kumari and Nair (1978)
	LC_{50}	1	22,690	50 MM ADULT	Kumari and Nair (1978)
	LC_{50}	2	7,750	35 MM JUVENILE	Kumari and Nair (1978)
	LC_{50}	2	20,610	50 MM ADULT	Kumari and Nair (1978)

Table 40. Toxicology of Aquatic and Terrestrial Species: Chemical: Malathion, CAS Number: 121755 (continued)

Class	Species	Effect	Duration (days)	Concentration ($\mu g\ L^{-1}$)	Lifestage	Ref.
Fish (continued)	Medaka, high-eyes	LC_{50}	1	460	NR	Kobayashi (1978)
		LC_{50}	1	2,800	NR	Shim and Self (1973)
		LC_{50}	2	750	2.54 CM, 0.16 G	Nishiuchi and Hashimoto (1969)
		LC_{50}	2	750	NR	Nishiuchi and Hashimoto (1975)
		MOR	3	0.35	NR	Shim and Self (1973)
		MOR	3	0.70	NR	Shim and Self (1973)
		TER	2	3,000–5,000	NEWLY HATCHED FRY	Solomon (1977)
		TER	NR	5,000–40,000	NEWLY FERTILIZED EGGS	Solomon (1977)
	Mosquitofish (*Gambusia affinis*)	LC_{50}	1	2,000	NR	Hansen et al. (1972)
		LC_{50}	4	200	ADULT, 0.289 G, 2.76 CM	Naqvi and Hawkins (1988)
		LC_{50}	4	660	NR	Scirocchi and Erme (1980)
		AVO	1 hr	50	20–45 MM	Hansen et al. (1972)
		BEH	1	1,000	18–55 MM	Hansen (1970)
		ENZ	0.17 hr	30,000	NR	Whitmore and Hodges (1978)
		ENZ	0.17 hr	30,000	NR	Whitmore and Hodges (1978)
		ENZ	0.17 hr	30,000	NR	Whitmore and Hodges (1978)
		LET	1	500	18–53 MM	Johnson (1977)
		MOR	1	0.5	NR	Mulla and Isaak (1961)
		MOR	1	0.5	NR	Mulla and Isaak (1961)
		MOR	1	2	NR	Mulla and Isaak (1961)

	MOR	1	2	NR	Mulla and Isaak (1961)
	MOR	1	25	18–53 MM	Johnson (1977)
	MOR	1	50	2.5–3 CM	Lewallen (1959)
	MOR	1	50	2.5–3 CM	Lewallen (1959)
	MOR	1	50	2.5–3 CM	Lewallen (1959)
	MOR	1	100	1.8–4.8 CM	Hansen (1972)
	MOR	1	1,000	1.8–4.8 CM	Hansen (1972)
	MOR	2	0.5	NR	Mulla and Isaak (1961)
	MOR	2	0.5	NR	Mulla and Isaak (1961)
	MOR	2	2	NR	Mulla and Isaak (1961)
	MOR	2	2	NR	Mulla and Isaak (1961)
	MOR	3	0.5	NR	Mulla and Isaak (1961)
	MOR	3	2	NR	Mulla and Isaak (1961)
	MOR	4	0.5	NR	Mulla and Isaak (1961)
	MOR	4	2	NR	Mulla and Isaak (1961)
	THL	1	10	NR	Johnson (1978b)
	THL	1	50	18–53 MM	Johnson (1977)
Motsuga, stone moroko	BCF	3	1,200	4–8 CM, 2–5 G	Kanazawa (1975)
	BCF	7	<1	4–8 CM, 2–5 G	Kanazawa (1975)
	ENZ	0.2 hr	0.000025 M	4–8 CM, 1–5 G	Kanazawa (1983)
Mozambique tilapia	LC_{50}	NR	9,200–16,700	NR	Mani and Konar (1985b)
	LC_{50}	1	5,500	12 G	Rao et al. (1984a)
	LC_{50}	2	358.9	8 G	Sailatha et al. (1981)
	LC_{50}	2	367	NR	Basha et al. (1984)
	LC_{50}	2	367	NR	Basha et al. (1983)
	LC_{50}	2	5,495	8 G	Sailatha et al. (1981)
	LC_{50}	2	5,600	8 G	Sahib and Rao (1980a)
	LC_{50}	2	8,300	2.5–6 CM	Sreenivasan and Swaminathan (1967)
	LC_{50}	4	290.1	6–11 CM	Lui et al. (1983)

Table 40. Toxicology of Aquatic and Terrestrial Species: Chemical: Malathion, CAS Number: 121755 (continued)

Class	Species	Effect	Duration (days)	Concen-tration (µg L^{-1})	Lifestage	Ref.
Fish (continued)	Mozambique tilapia	LC_{50}	4	778.7	6–11 CM	Lui et al. (1983)
		BEH	NR	1,270–5,070	NR	Mani and Konar (1986)
		BIO	2	2,000	8 G	Sahib et al. (1980)
		BIO	2	2,000	8 G, 6 CM	Sahib et al. (1981)
		BIO	2	2,000	8 G	Sahib (1984b)
		BIO	2	2,000	8 G	Sahib et al. (1984a)
		BIO	7	950	15–20 G	Ramalingam and Ramalingam (1982)
		ENZ	2	2,000	NR	Sahib and Rao (1980b)
		ENZ	2	2,000	8 G	Sahib et al. (1980)
		ENZ	2	2,000	8 G	Sahib (1984b)
		GRO	2	2,000	8 G, 6 CM	Sahib et al. (1981)
		HEM	15	1,100	12 G	Rao et al. (1984a)
		HIS	7	1,100	12 G	Rao et al. (1983)
		HIS	10	2,000	26 G, 5.8 CM	Pandey and Shukla (1982)
		HIS	20	200	ADULT	Shulka et al. (1984)
		HIS	20	2,000	NR	Pandey and Shukla (1983)
		OC	1	1,000	8 G	Bashamohideen et al. (1988)
		OC	1.5	122	NR	Basha et al. (1984)
		OC	2	2,000	8 G, 6 CM	Sahib et al. (1981)
		PHY	2	122	NR	Basha et al. (1984)
		PHY	2	2,000	8 G, 6 CM	Sahib et al. (1981)
		REP	NR	3,800	NR	Mani and Konar (1986)
		REP	NR	5,070	NR	Mani and Konar (1986)
		RES	NR	2,530–5,070	NR	Mani and Konar (1986)

Mummichog	LC_{50}	1	130	56 MM, 2.5 G	Eisler (1970a)
	LC_{50}	1	810	55 MM, 1.8 G	Eisler (1970a)
	LC_{50}	2	80	56 MM, 2.5 G	Eisler (1970a)
	LC_{50}	2	440	55 MM, 1.8 G	Eisler (1970a)
	LC_{50}	4	70	42 (36–45) MM	Eisler (1970b)
	LC_{50}	4	80	56 MM, 2.5 G	Eisler (1970a)
	LC_{50}	4	400	55 MM, 1.8 G	Eisler (1970a)
	LC_{50}	10	70	42 (36–45) MM	Eisler (1970b)
	DVP	31	10,000	EMBRYO	Weis and Weis (1974b)
	RGN	7	10	4–5 CM	Weis and Weis (1975)
Northern puffer	LC_{50}	1	9,000	183 MM, 126 G	Eisler (1970a)
	LC_{50}	2	6,000	183 MM, 126 G	Eisler (1970a)
	LC_{50}	4	3,250	183 MM, 126 G	Eisler (1970a)
Northern squawfish	MOR	1	10,000	5–10 CM	MacPhee and Ruelle (1969)
	MOR	1	10,000	5–10 CM	MacPhee and Ruelle (1969)
Ocellated killifish	MOR	4 hr		49.9 MM	Darsie and Corriden (1959)
	MOR	4 hr		40.1–63.1 MM	Darsie and Corriden (1959)
Oikawa	LC_{50}	1	9,700	NR	Shim and Self (1973)
	MOR	3	0.70	NR	Shim and Self (1973)
	MOR	3	1.4	NR	Shim and Self (1973)
Paradise fish	LC_{50}	1	4,952	20–28 MM STD LENGTH	Jacob et al. (1982)
	LC_{50}	2	4,594	20–28 MM STD LENGTH	Jacob et al. (1982)
Pearl spot	DVP	5.17	400	LARVAE	Nair and Geevarghese (1984)
	GRO	5.17	400	LARVAE	Nair and Geevarghese (1984)
Perch (*Perca fluviatilis*)	ENZ	NR	5	YEARLING	Gantverg and Perevoznikov (1983)

Table 40. Toxicology of Aquatic and Terrestrial Species: Chemical: Malathion, CAS Number: 121755 (continued)

Class	Species	Effect	Duration (days)	Concentration ($\mu g\ L^{-1}$)	Lifestage	Ref.
Fish (continued)	Pinfish	ENZ	0.33	26.9	52–101 MM	Cook et al. (1976)
		ENZ	0.33	26.9	52–101 MM	Cook et al. (1976)
		ENZ	1	19.8	52–101 MM	Cook et al. (1976)
		ENZ	1	19.8	52–101 MM	Cook et al. (1976)
		ENZ	1	1,000	65–125 MM	Coppage and Matthews (1974)
		ENZ	3	31	NR	Coppage et al. (1975)
		ENZ	4 hr	26.9	52–101 MM	Cook et al. (1976)
		ENZ	6 hr	19.8	52–101 MM	Cook et al. (1976)
		MOR	3	58	NR	Coppage et al. (1975)
		RSD	1	75	5–7 CM	Cook and Moore (1976)
	Pumpkinseed	LC_{50}	1	920	YOUNG OF THE YEAR	Rehwoldt et al. (1977)
		LC_{50}	2	600	YOUNG OF THE YEAR	Rehwoldt et al. (1977)
		LC_{50}	4	480	YOUNG OF THE YEAR	Rehwoldt et al. (1977)
		ENZ	2 hr	200	40–80 G	Murphy et al. (1968)
		ENZ	2 hr	200	40–80 G	Murphy et al. (1968)
		MOR	2 hr	200	40–80 G	Murphy et al. (1968)
		MOR	2 hr	200	40–80 G	Murphy et al. (1968)
	Rainbow trout, donaldson trout	LC_{50}	1	5	10 CM, 15 G	Smith and Grigoropoulos (1968)
		LC_{50}	1	85	1 G	Cope (1965)
		LC_{50}	1	100	1 G	Cope (1965)
		LC_{50}	1	130	1 G	Cope (1965)
		LC_{50}	1	240	0.41 G	Post and Schroeder (1971)
		LC_{50}	2	4.6	10 CM, 15 G	Smith and Grigoropoulos (1968)

LC_{50}	2	17	NR	Scirocchi and Erme (1980)
LC_{50}	2	70	1 G	Cope (1965)
LC_{50}	2	79	1 G	Cope (1965)
LC_{50}	2	120	1 G	Cope (1965)
LC_{50}	2	196	0.41 G	Post and Schroeder (1971)
LC_{50}	3	175	0.41 G	Post and Schroeder (1971)
LC_{50}	4	2.8	10 CM, 15 G	Smith and Grigoropoulos (1968)
LC_{50}	4	68	1 G	Cope (1965)
LC_{50}	4	70	NR	Marking and Dawson (1975)
LC_{50}	4	77	1 G	Cope (1965)
LC_{50}	4	110	1 G	Cope (1965)
LC_{50}	4	111	FRY, 1 G	Marking et al. (1984)
LC_{50}	4	115	JUVENILE	Douglas et al. (1986)
LC_{50}	4	122	0.41 G	Post and Schroeder (1971)
LC_{50}	4	160	100 G	McKim et al. (1987)
LC_{50}	4	161	JUVENILE	Douglas et al. (1986)
LC_{50}	4	170	NR	Macek and McAllister (1970)
LC_{50}	4	190	FRY, 1 G	Marking et al. (1984)
LC_{50}	4	191	FRY, 1 G	Marking et al. (1984)
LC_{50}	4	200	1.4 G	Johnson and Finley (1980)
LC_{50}	4	200	FRY, 1 G	Marking et al. (1984)
LC_{50}	4	234	FRY, 1 G	Marking et al. (1984)
LC_{50}	5	2.3	10 CM, 15 G	Smith and Grigoropoulos (1968)
LC_{50}	NR	32.8	NR	Meyer (1981)
LC_{50}	NR	115	NR	Meyer (1981)
LT_{50}	1.67	296.1	600–900 G	McKim et al. (1987)
CYT	3	0.1 $\mu L\ L^{-1}$	10–25 G, 7–12 CM	Al-Sabti (1985)
ENZ	2 hr	7,200 $\mu g\ mg^{-1}$	31–41 CM	Smith and Grigoropoulos (1968)

Table 40. Toxicology of Aquatic and Terrestrial Species: Chemical: Malathion, CAS Number: 121755 (continued)

Class	Species	Effect	Duration (days)	Concentration ($\mu g\ L^{-1}$)	Lifestage	Ref.
Fish (continued)	Rainbow trout, donaldson trout	ENZ	10	55	10 G	Post and Leasure (1974)
		LET	1	1,000	FEEDING FRY, 1 MONTH	Lewallen and Wilder (1962)
		LET	1	10,000	SAC FRY, 1 WEEK	Lewallen and Wilder (1962)
		LET	1.3 hr	5,000	150–250 G	Van Hoof (1980)
		LET	3 hr	2,000	150–250 G	Van Hoof (1980)
		MOR	1	1,000	SAC FRY, 1 WEEK	Lewallen and Wilder (1962)
		PHY	NR	296.1	600–900 G	McKim et al. (1987)
		PHY	NR	5,000	NR	Fromm et al. (1971)
		STR	1 hr	5,000	FINGERLING, ≤10 CM	Applegate et al. (1957)
		STR	1 hr	5,000	FINGERLING, ≤10 CM	Applegate et al. (1957)
		STR	1 hr	5,000	FINGERLING, ≤10 CM	Applegate et al. (1957)
	Red shiner	LC_{50}	1	40	4.8 CM, 1.9 G	Smith and Grigoropoulos (1968)
		LC_{50}	2	36	4.8 CM, 1.9 G	Smith and Grigoropoulos (1968)
		LC_{50}	4	25	4.8 CM, 1.9 G	Smith and Grigoropoulos (1968)
		LC_{50}	5	23	4.8 CM, 1.9 G	Smith and Grigoropoulos (1968)
	Redear sunfish	LC_{50}	4	62	3.2 G	Johnson and Finley (1980)
		LC_{50}	4	170	NR	Macek and McAllister (1970)
	Roach	LC_{50}	4	12,000	NR	Scirocchi and Erme (1980)

Rohu	LC_{50}	1	7,150	2.5–6 CM	Arora et al. (1971b)
	LC_{50}	2	6,320	2.5–6 CM	Arora et al. (1971b)
	LC_{50}	3	6,320	2.5–6 CM	Arora et al. (1971b)
	LC_{50}	4	4,980	2.5–6 CM	Arora et al. (1971b)
Sea lamprey	STR	1	5,000	LARVAE, 8–13 CM	Applegate et al. (1957)
	STR	1	5,000	LARVAE, 8–13 CM	Applegate et al. (1957)
	STR	1	5,000	LARVAE, 8–13 CM	Applegate et al. (1957)
Sheepshead minnow	LC_{50}	4	51	JUVENILE	Hansen and Parrish (1977)
	AVO	0.5 hr	1,000	20–40 MM	Hansen (1969)
	AVO	NR	1,000	20–40 MM	Hansen (1970)
	ENZ	1	57 g ha^{-1}	25–40 MM TOTAL LENGTH	Tagatz et al. (1974)
	ENZ	1	200	45–70 MM	Coppage and Matthews (1974)
	MOR	1	57 g ha^{-1}	25–40 MM TOTAL LENGTH	Tagatz et al. (1974)
	MOR	28	420 g ha^{-1}	ADULT AND SUBADULT	Tagatz et al. (1974)
	MOR	140	4	FRY	Hansen and Parrish (1977)
	MOR	140	9	FRY	Hansen and Parrish (1977)
	MOR	140	18	ADULT	Hansen and Parrish (1977)
	MOR	140	37	ADULT	Hansen and Parrish (1977)
Silver carp	LET	NR	500	YEARLING	Prokopenko et al. (1976a)
Slender rasbora	LC_{50}	4	6,000	10–12 CM, 2–4 G	Singh and Sahai (1984)
Snake-head catfish (*Channa striatus*)	LC_{30}	4	350	NR	Choudhuri et al. (1984)
Snake-head catfish (*Channa punctatus*)	LC_{50}	1	1,420	60 G, 15 CM	Anees (1975)
	LC_{50}	1	1,420	10–15 CM	Anees (1974a)
	LC_{50}	1	4,200	24.5 G	Singh et al. (1984)
Snake-head catfish (*Channa gachua*)	LC_{50}	1	8,700	140 MM	Dalela et al. (1978)
	LC_{50}	1	8,750	110 MM	Dalela et al. (1978)
	LC_{50}	1	8,800	140 MM	Dalela et al. (1978)

Table 40. Toxicology of Aquatic and Terrestrial Species: Chemical: Malathion, CAS Number: 121755 (continued)

Class	Species	Effect	Duration (days)	Concentration ($\mu g\ L^{-1}$)	Lifestage	Ref.
Fish (continued)	Snake-head catfish (*Channa punctatus*)	LC_{50}	1	9,160	ADULT, 59.8 G, 19 CM	Haider and Inbaraj (1986)
	Snake-head catfish (*Channa gachua*)	LC_{50}	1	9,200	140 MM	Dalela et al. (1978)
	Snake-head catfish (*Channa punctatus*)	LC_{50}	1	9,930	ADULT, 59.8 G, 19 CM	Haider and Inbaraj (1986)
		LC_{50}	1	10,950	FINGERLING	Shukla et al. (1987)
		LC_{50}	2	1,310	60 G, 15 CM	Anees (1975)
		LC_{50}	2	1,310	10–15 CM	Anees (1974a)
		LC_{50}	2	3,700	24.5 G	Singh et al. (1984)
		LC_{50}	2	6,590	ADULT, 59.8 G, 19 CM	Haider and Inbaraj (1986)
	Snake-head catfish (*Channa gachua*)	LC_{50}	2	7,850	140 MM	Dalela et al. (1978)
		LC_{50}	2	7,950	140 MM	Dalela et al. (1978)
		LC_{50}	2	8,000	110 MM	Dalela et al. (1978)
		LC_{50}	2	8,100	140 MM	Dalela et al. (1978)
	Snake-head catfish (*Channa punctatus*)	LC_{50}	2	8,210	ADULT, 59.8 G, 19 CM	Haider and Inbaraj (1986)
		LC_{50}	2	8,360	FINGERLING	Shukla et al. (1987)
		LC_{50}	3	1,200	60 G, 15 CM	Anees (1975)
		LC_{50}	3	1,200	10–15 CM	Anees (1974a)
		LC_{50}	3	3,320	24.5 G	Singh et al. (1984)
		LC_{50}	3	4,920	FINGERLING	Shukla et al. (1987)
		LC_{50}	3	5,100	ADULT, 59.8 G, 19 CM	Haider and Inbaraj (1986)
		LC_{50}	3	5,940	ADULT, 59.8 G, 19 CM	Haider and Inbaraj (1986)

Snake-head catfish (*Channa gachua*)	LC_{50}	3	7,300	140 MM	Dalela et al. (1978)
	LC_{50}	3	7,600	140 MM	Dalela et al. (1978)
	LC_{50}	3	7,650	110 MM	Dalela et al. (1978)
	LC_{50}	3	7,900	140 MM	Dalela et al. (1978)
Snake-head catfish (*Channa punctatus*)	LC_{50}	4	920	60 G, 15 CM	Anees (1975)\
	LC_{50}	4	920	10–15 CM	Anees (1974a)
	LC_{50}	4	2,900	24.5 G	Singh et al. (1984)
	LC_{50}	4	3,220	FINGERLING	Shukla et al. (1987)
	LC_{50}	4	3,890	ADULT, 59.8 G, 19 CM	Haider and Inbaraj (1986)
	LC_{50}	4	4,510	ADULT, 59.8 G, 19 CM	Haider and Inbaraj (1986)
Snake-head catfish (*Channa gachua*)	LC_{50}	4	6,950	110 MM	Dalela et al. (1978)
	LC_{50}	4	7,050	140 MM	Dalela et al. (1978)
	LC_{50}	4	7,350	140 MM	Dalela et al. (1978)
	LC_{50}	4	7,600	140 MM	Dalela et al. (1978)
Snake-head catfish (*Channa punctatus*)	BIO	180	2,000	56 G, 15 CM, ADULT FEMALES AND MALES	Ram and Sathyanesan (1987)
Snake-head catfish (*Channa gachua*)	ENZ	3	3,800	120–167 MM, 75–102 G	Verma et al. (1979)
	ENZ	3	3,800	120–167 MM, 75–102 G	Verma et al. (1979)
	ENZ	3	3,800	120–167 MM, 75–102 G	Verma et al. (1979)
	ENZ	7	3,800	120–167 MM, 75–102 G	Verma et al. (1979)
	ENZ	7	3,800	120–167 MM, 75–102 G	Verma et al. (1979)
	ENZ	7	3,800	120–167 MM, 75–102 G	Verma et al. (1979)
Snake-head catfish (*Channa striatus*)	ENZ	10	100	80 G	Sadhu et al. (1985)

Table 40. Toxicology of Aquatic and Terrestrial Species: Chemical: Malathion, CAS Number: 121755 (continued)

Class	Species	Effect	Duration (days)	Concen-tration (μg L^{-1})	Lifestage	Ref.
Fish (continued)	Snake-head catfish (*Channa punctatus*)	ENZ	120	190	36 G	Inbaraj and Haider (1988)
		GRO	13	2,500	FINGERLING	Shukla et al. (1987)
		GRO	19	2,500	FINGERLING	Shukla et al. (1987)
		HEM	1	50	10–17 CM	Shakoori et al. (1976)
		HEM	1	50	10–17 CM	Shakoori et al. (1976)
		HEM	1	50	10–17 CM	Shakoori et al. (1976)
		HEM	1	50	10–17 CM	Shakoori et al. (1976)
		HEM	1	50	10–17 CM	Shakoori et al. (1976)
		HEM	1	50	10–17 CM	Shakoori et al. (1976)
		HEM	4	1.5	10–17 CM	Shakoori et al. (1976)
		HEM	4	1.5	10–17 CM	Shakoori et al. (1976)
		HEM	4	1.5	10–17 CM	Shakoori et al. (1976)
		HEM	4	25	NR	Lone and Javaid (1976)
		HIS	7	1,000	12–15 CM ADULT	Dubale and Shah (1984)
		LET	12	2,000	12–15 CM ADULT	Dubale and Shah (1984)
		MOR	1	50	10–17 CM	Shakoori et al. (1976)
		MOR	4	1.5	10–17 CM	Shakoori et al. (1976)
		REP	120	190	36 G	Inbaraj and Haider (1988)
	Spot	LC_{50}	1	550	JUVENILE	Butler (1964)
		LC_{50}	2	550	JUVENILE	Butler (1964)
		ENZ	1	1,250	65–150 MM	Coppage and Matthews (1974)
		ENZ	182	10	JUVENILE, 39 MM	Holland and Lowe (1966)

Striped bass	LC_{50}	1	91	YOUNG OF THE YEAR	Rehwoldt et al. (1977)
	LC_{50}	1	790	FINGERLINGS, 0.93 G, 46 MM	Wellborn (1971)
	LC_{50}	2	70	YOUNG OF THE YEAR	Rehwoldt et al. (1977)
	LC_{50}	2	460	FINGERLINGS, 0.93 G, 46 MM	Wellborn (1971)
	LC_{50}	4	17.8	JUVENILE 0.4 G	Korn and Earnest (1974)
	LC_{50}	4	24.5	56 D	Palawski et al. (1985)
	LC_{50}	4	39	YOUNG OF THE YEAR	Rehwoldt et al. (1977)
	LC_{50}	4	65	56 D	Palawski et al. (1985)
	LC_{50}	4	240	FINGERLINGS, 0.93 G, 46 MM	Wellborn (1971)
Striped killifish	LC_{50}	1	280	84 MM, 6.5 G	Eisler (1970a)
	LC_{50}	2	250	84 MM, 6.5 G	Eisler (1970a)
	LC_{50}	4	250	84 MM, 6.5 G	Eisler (1970a)
Striped mullet	LC_{50}	1	>960	48 MM, 0.78 G	Eisler (1970a)
	LC_{50}	2	550	48 MM, 0.78 G	Eisler (1970a)
	LC_{50}	4	550	48 MM, 0.78 G	Eisler (1970a)
Tench	LC_{50}	3	16,200	NR	Scirocchi and Erme (1980)
Three spine stickleback	LC_{50}	1	76.9	22–44 MM, 0.38–0.77 G	Katz (1961)
	LC_{50}	1	96.9	22–44 MM, 0.38–0.77 G	Katz (1961)
	LC_{50}	2	76.9	22–44 MM, 0.38–0.77 G	Katz (1961)
	LC_{50}	2	94	22–44 MM, 0.38–0.77 G	Katz (1961)
	LC_{50}	3	76.9	22–44 MM, 0.38–0.77 G	Katz (1961)
	LC_{50}	3	94	22–44 MM, 0.38–0.77 G	Katz (1961)

Table 40. Toxicology of Aquatic and Terrestrial Species: Chemical: Malathion, CAS Number: 121755 (continued)

Class	Species	Effect	Duration (days)	Concentration (μg L^{-1})	Lifestage	Ref.
Fish (continued)	Three spine stickleback	LC_{50}	4	76.9	22–44 MM, 0.38–0.77 G	Katz (1961)
		LC_{50}	4	94	22–44 MM, 0.38–0.77 G	Katz (1961)
	Tooth carp	LC_{50}	2	348	NR	Boumaiza et al. (1979)
		LC_{50}	4	314	NR	Boumaiza et al. (1979)
	Two spot african barb	LC_{50}	1	6,400	6–8.5 CM	Rao et al. (1967)
		LC_{50}	2	4,600	6–8.5 CM	Rao et al. (1967)
		LC_{50}	4	3,700	6–8.5 CM	Rao et al. (1967)
	Two spot barb, dotted barb	LC_{50}	0.96	4,750	2.54–7.62 CM	Arora et al. (1971a)
		LC_{50}	2	3,500	2.54–7.62 CM	Arora et al. (1971a)
		LC_{50}	4	3,200	2.54–7.62 CM	Arora et al. (1971a)
	Two spotted, tic tac toe barb	LC_{50}	1	16.3	50–68 MM, 1.560–5.125 G	Bhatia (1971)
		LC_{50}	2	13	50–68 MM, 1.560–5.125 G	Bhatia (1971)
		LC_{50}	3	13	50–68 MM, 1.560–5.125 G	Bhatia (1971)
		LC_{50}	4	8.9	50–68 MM, 1.560–5.125 G	Bhatia (1971)
		LC_{50}	4	4,000	10–12 CM, 2–4 G	Singh and Sahai (1984)
		RSD	15	6,000	2–4 G, 6–8 CM	Singh and Sahai (1986)

Walking catfish	LC_{50}	4	12,000	18–20 CM, 65–75 G	Singh and Singh (1987b)
	BIO	28	1 $\mu L\ L^{-1}$	65 (60–70) G, 18.5 (18–19) CM	Lal and Singh (1987b)
	BIO	28	1,000	60–70 G, 18–19 CM	Lal and Singh (1987c)
	BIO	28	1,000	60–70 G, 18–19 CM	Lal and Singh (1987c)
	BIO	28	1,000	60–70 G, 18–19 CM	Lal and Singh (1987c)
	BIO	28	1,000	60–70 G, 18–19 CM	Lal and Singh (1987c)
	BIO	28	4,000	65 (60–70) G, 18.5 (18–19) CM	Lal and Singh (1986)
	BIO	28	4,000	65 (60–70) G, 18.5 (18–19) CM	Lal and Singh (1986)
	BIO	28	4,000	65 (60–70) G, 18.5 (18–19) CM	Lal and Singh (1986)
	BIO	28	4,000	65 (60–70) G, 18.5 (18–19) CM	Lal and Singh (1986)
	BIO	40	500	30.0 G	Mukhopadhyay and Dehadrai (1980b)
	ENZ	4	4,800	150–170 G, 20.5–22.0 CM	Tandon and Dubey (1983)
	ENZ	30	1,000	30 G	Mukhopadhyay and Dehadrai (1980b)
	ENZ	40	.500	30 G	Mukhopadhyay and Dehadrai (1980a)
	ENZ	40	500	30 G	Mukhopadhyay and Dehadrai (1980b)
	HEM	15	4	45 G, 16 CM	Mandal and Kulshrestha (1983)
	HIS	30	500	50 G	Sadhu and Mukhopadhyay (1985)
	HRM	28	1,000	18–20 CM, 65–75 G	Singh and Singh (1987b)

Table 40. Toxicology of Aquatic and Terrestrial Species: Chemical: Malathion, CAS Number: 121755 (continued)

Class	Species	Effect	Duration (days)	Concen-tration (µg L^{-1})	Lifestage	Ref.
Fish (continued)	Walking catfish	HRM	28	1,000	18–20 CM, 65–75 G, FEMALE	Singh and Singh (1987a)
		LET	3	1,750	15–16.5 CM	Sharma et al. (1983)
		LET	4	5,600	150–170 G, 20.5–22 CM	Tandon and Dubey (1983)
		LET	4	6,200	150–170 G, 20.5–22 CM	Tandon and Dubey (1983)
		MOR	1	5,000	150–170 G, 20.5–22 CM	Tandon and Dubey (1983)
		MOR	1	5,600	150–170 G, 20.5–22 CM	Tandon and Dubey (1983)
		MOR	1	6,200	150–170 G, 20.5–22 CM	Tandon and Dubey (1983)
		MOR	2	5,000	150–170 G, 20.5–22 CM	Tandon and Dubey (1983)
		MOR	2	5,600	150–170 G, 20.5–22 CM	Tandon and Dubey (1983)
		MOR	2	6,200	150–170 G, 20.5–22 CM	Tandon and Dubey (1983)
		MOR	3	750	15–16.5 CM	Sharma et al. (1983)
		MOR	3	1,000	15–16.5 CM	Sharma et al. (1983)
		MOR	3	1,250	15–16.5 CM	Sharma et al. (1983)
		MOR	3	1,500	15–16.5 CM	Sharma et al. (1983)
		MOR	3	4,800	150–170 G, 20.5–22 CM	Tandon and Dubey (1983)

	MOR	3	5,000	150–170 G, 20.5–22 CM	Tandon and Dubey (1983)
	MOR	3	5,600	150–170 G, 20.5–22 CM	Tandon and Dubey (1983)
	MOR	3	6,200	150–170 G, 20.5–22 CM	Tandon and Dubey (1983)
	MOR	4	4,800	150–170 G, 20.5–22 CM	Tandon and Dubey (1983)
	MOR	4	5,000	150–170 G, 20.5–22 CM	Tandon and Dubey (1983)
	MOR	5	4,800	150–170 G, 20.5–22 CM	Tandon and Dubey (1983)
	MOR	5	5,000	150–170 G, 20.5–22 CM	Tandon and Dubey (1983)
	PHY	28	1,000	65 (60–70) G, 18.5 (18–19) CM	Lal and Singh (1987a)
	PHY	40	500	NR	Mukhopadhyay et al. (1984)
Walleye	LC_{50}	4	64	1.3 G	Johnson and Finley (1980)
	ENZ	0.5	2[illegible]0 g ha^{-1}	JUVENILE, >1 MONTH	Lockhart et al. (1985)
	MOR	1		7.2–15.2 CM, FINGERLING	Hilsenhoff (1962)
	MOR	1		7.2–15.2 CM, FINGERLING	Hilsenhoff (1962)
White mullet	LC_{50}	1	950	JUVENILE	Butler (1963)
	LC_{50}	2	570	JUVENILE	Butler (1963)
White perch	LC_{50}	1	2,100	YOUNG OF THE YEAR	Rehwoldt et al. (1977)
	LC_{50}	2	1,900	YOUNG OF THE YEAR	Rehwoldt et al. (1977)
	LC_{50}	4	1,100	YOUNG OF THE YEAR	Rehwoldt et al. (1977)
Yellow perch	LC_{50}	4	263	1.4 G	Johnson and Finley (1980)
	LC_{50}	4	263	NR	Macek and McAllister (1970)

Table 40. Toxicology of Aquatic and Terrestrial Species: Chemical: Malathion, CAS Number: 121755 (continued)

Class	Species	Effect	Duration (days)	Concentration ($\mu g\ L^{-1}$)	Lifestage	Ref.
Fish (continued)	Zebra danio, zebrafish	LC_{50}	1	1,450	NR	Kumar and Ansari (1984)
		LC_{50}	2	1,250	NR	Kumar and Ansari (1984)
		LC_{50}	3	1,150	NR	Kumar and Ansari (1984)
		LC_{50}	4	1,050	NR	Kumar and Ansari (1984)
		BIO	7	500	4 MONTHS, ADULT FEMALE	Ansari and Kumar (1987)
		BIO	120	500	4 MONTHS, ADULT FEMALE	Ansari and Kumar (1987)
		ENZ	7	500	4 MONTHS, ADULT	Ansari and Kumar (1984)
		REP	120	500	NR	Kumar and Ansari (1984)
Insect	Backswimmer (*Notonecta undulata*)	LC_{50}	1	220	ADULT, 66 MG	Federle and Collins (1976)
		LC_{50}	2	110	ADULT, 66 MG	Federle and Collins (1976)
		LC_{50}	3	80	ADULT, 66 MG	Federle and Collins (1976)
		LC_{50}	4	80	ADULT, 66 MG	Federle and Collins (1976)
	Beetle (*Eretes sticticus*)	LC_{50}	1	520	NR	Jeyasingam et al. (1978)
		LC_{50}	2	430	NR	Jeyasingam et al. (1978)
	Beetle (*Peltodytes* sp)	LC_{50}	1	6,800	ADULT, 5 MG	Federle and Collins (1976)
		LC_{50}	2	1,500	ADULT, 5 MG	Federle and Collins (1976)
		LC_{50}	3	1,200	ADULT, 5 MG	Federle and Collins (1976)
		LC_{50}	4	1,000	ADULT, 5 MG	Federle and Collins (1976)
	Beetle (*Tropisternus lateralis*)	LET	1	60	LARVAE, 1–1.5 CM	Lewallen (1962)

Blackfly (*Simulium* sp)	MOR-D	0.02 hr	1,000	LARVAE	Matsuo and Tamura (1970)
	MOR-D	0.02 hr	10,000	LARVAE	Matsuo and Tamura (1970)
	MOR-D	0.17 hr	100	LARVAE	Matsuo and Tamura (1970)
	MOR-D	0.17 hr	1,000	LARVAE	Matsuo and Tamura (1970)
Caddisfly (*Arctopsyche grandis*)	LC_{50}	4	32	LARVAE	Gaufin et al. (1965)
	LC_{50}	4	32	2–5 CM	Gaufin et al. (1961)
Caddisfly (*Hydropsyche* sp)	LC_{50}	1	12.3	LARVAE	Carlson (1966)
Caddisfly (*Hydropsyche morosa*)	LC_{50}	3 hr	500	LARVAE, 25 MG	Fredeen (1972)
Caddisfly (*Hydropsyche recurvata*)	LC_{50}	3 hr	<500	LARVAE, 25 MG	Fredeen (1972)
Caddisfly (*Hydropsyche morosa*)	LC_{50}	3 hr	>500	LARVAE, 25 MG	Fredeen (1972)
Caddisfly (*Hydropsyche recurvata*)	LC_{50}	3 hr	>500	LARVAE, 25 MG	Fredeen (1972)
Caddisfly (*Hydropsyche* sp)	LC_{50}	4	5	JUVENILE	Johnson and Finley (1980)
Caddisfly	LC_{50}	4	22.5	LARVAE	Gaufin et al. (1965)
(*Hydropsyche californica*)	LC_{50}	4	22.5	2–5 CM	Gaufin et al. (1961)
Caddisfly (*Hydropsyche morosa*)	LC_{50}	6 hr	500	LARVAE, 25 MG	Fredeen (1972)
Caddisfly (*Hydropsyche recurvata*)	LC_{50}	6 hr	500	LARVAE, 25 MG	Fredeen (1972)
Caddisfly (*Hydropsyche morosa*)	LC_{50}	6 hr	<500	LARVAE, 25 MG	Fredeen (1972)
Caddisfly (*Hydropsyche recurvata*)	LC_{50}	6 hr	<500	LARVAE, 25 MG	Fredeen (1972)
Caddisfly (*Hydropsyche morosa*)	LC_{50}-D	6 hr	<500	LARVAE, 25 MG	Fredeen (1972)
	LC_{50}-D	6 hr	<500	LARVAE, 25 MG	Fredeen (1972)
Caddisfly	LC_{50}-D	6 hr	<500	LARVAE, 25 MG	Fredeen (1972)
(*Hydropsyche recurvata*)	LC_{50}-D	6 hr	<500	LARVAE, 25 MG	Fredeen (1972)
Caddisfly (*Hydropsyche* sp)	DRF	NR	<50	NR	Peterson (1972)
Caddisfly (*Leptocella* sp)	DRF	NR	<50	NR	Peterson (1972)
Caddisfly (*Limnephilus* sp)	LC_{50}	4	1.3	JUVENILE	Johnson and Finley (1980)
Caddisfly (*Potamyia flava*)	DRF	NR	<50	NR	Peterson (1972)

Table 40. Toxicology of Aquatic and Terrestrial Species: Chemical: Malathion, CAS Number: 121755 (continued)

Class	Species	Effect	Duration (days)	Concen-tration ($\mu g\ L^{-1}$)	Lifestage	Ref.
Insect (continued)	Damselfly (*Lestes congener*)	LC_{50}	1	300	NYMPHS, 44 MG	Federle and Collins (1976)
		LC_{50}	4	10	JUVENILE	Johnson and Finley (1980)
	Dragonfly (*Orthetrum albistylum speciosum*)	LC_{50}	1	2,400	NR	Nishiuchi and Asano (1978)
		LC_{50}	1 hr	>10,000	NR	Nishiuchi and Asano (1978)
		LC_{50}	2	730	NR	Nishiuchi and Asano (1978)
		LC_{50}	2 hr	>10,000	NR	Nishiuchi and Asano (1978)
		LC_{50}	3 hr	10,000	NR	Nishiuchi and Asano (1978)
		LC_{50}	6 hr	5,100	NR	Nishiuchi and Asano (1978)
	Mayfly (*Caenis* sp)	DRF	NR	<50	NR	Peterson (1972)
	Mayfly (*Centroptilum* sp)	DRF	NR	<50	NR	Peterson (1972)
	Mayfly (*Cloeon dipterum*)	LC_{50}	1	28	YOUNGER NYMPH, 9.3 MM, 5.6 MG	Nishiuchi and Asano (1979)
		LC_{50}	2	18	YOUNGER NYMPH, 9.3 MM, 5.6 MG	Nishiuchi and Asano (1979)
		LC_{50}	3 hr	280	YOUNGER NYMPH, 9.3 MM, 5.6 MG	Nishiuchi and Asano (1979)
		LC_{50}	6 hr	75	YOUNGER NYMPH, 9.3 MM, 5.6 MG	Nishiuchi and Asano (1979)
	Mayfly (*Ephemerella grandis*)	LC_{50}	4	100	NYMPH	Gaufin et al. (1965)
	Mayfly (*Heptagenia diabasia*)	DRF	NR	<50	NR	Peterson (1972)
	Mayfly (*Hexagenia* sp)	LC_{50}	1	631	NAIAD	Carlson (1966)
	Mayfly (*Isonychia* rufa)	DRF	NR	<50	NR	Peterson (1972)
	Mayfly (*Tricorythodes* sp)	DRF	NR	<50	NR	Peterson (1972)
	Midge (*Chironomus riparius*)	EC_{50}IM	1	1.9	4TH INSTAR	Estenik and Collins (1979)
	Midge (*Chironomus tentans*)	EC_{50}IM	1	2	3RD–4TH INSTAR	Karnak and Collins (1974)

Midge (*Chironomus* sp)	EC_{50}IM	1	2.1	LARVAE, 4TH INSTAR	Mulla and Khasawinah (1969)
Midge (*Chironomus tentans*)	EC_{50}IM	1	30,000	4TH INSTAR LARVAE	Hansen and Kawatski (1976)
Midge (*Chironomus* sp)	EC_{50}IM	2	1.2	4TH INSTAR	Norland et al. (1974)
	EC_{50}IM	2	3	4TH INSTAR	Norland et al. (1974)
Midge (*Chironomus tentans*)	EC_{50}IM	3	2,500	4TH INSTAR LARVAE	Hansen and Kawatski (1976)
	EC_{50}IM-D	1	36,000	4TH INSTAR LARVAE	Hansen and Kawatski (1976)
	EC_{50}IM-D	3	620	4TH INSTAR LARVAE	Hansen and Kawatski (1976)
Midge (*Chironomus utahensis*)	LC_{50}	1	3	4TH INSTAR	Ali and Mulla (1978a)
	LC_{50}	1	3	4TH INSTAR	Ali and Mulla (1977)
	LC_{50}	1	5	4TH INSTAR	Ali and Mulla (1978a)
Midge (*Chironomus decorus*)	LC_{50}	1	29	4TH INSTAR	Ali and Mulla (1978a)
	LC_{50}	1	81	4TH INSTAR	Ali and Mulla (1978a)
Midge (*Chironomus stigmaterus*)	ABD	3–4		LARVAE	Mulla and Khasawinah (1969)
	ABD	3–4		LARVAE	Mulla and Khasawinah (1969)
Midge (*Chironomus plumosus*)	MOR	5		LARVAE, 4TH INSTAR	Hilsenhoff (1959)
	MOR	5	0.1 kg ha^{-1}	LARVAE, 4TH INSTAR	Hilsenhoff (1959)
	MOR	5	1 kg ha^{-1}	LARVAE, 4TH INSTAR	Hilsenhoff (1959)
	MOR	5	10 kg ha^{-1}	LARVAE, 4TH INSTAR	Hilsenhoff (1959)
	MOR	7		LARVAE, 4TH INSTAR	Hilsenhoff (1959)
	MOR	7		LARVAE, 4TH INSTAR	Hilsenhoff (1959)
Midge (*Cricotopus* sp)	LC_{50}	1	25	4TH INSTAR	Ali and Mulla (1977)
Midge (*Goeldichironomus holoprassinus*)	ABD	3–4		LARVAE	Mulla and Khasawinah (1969)
	ABD	3–4		LARVAE	Mulla and Khasawinah (1969)
Midge (*Paratanytarsus parthenogenetic*)	DVP	NR	3,500	EGG	Anderson and Shubat (1983)
	HAT	NR	3,500	EGG	Anderson and Shubat (1983)

Table 40. Toxicology of Aquatic and Terrestrial Species: Chemical: Malathion, CAS Number: 121755 (continued)

Class	Species	Effect	Duration (days)	Concen-tration ($\mu g\ L^{-1}$)	Lifestage	Ref.
Insect (continued)	Midge (*Procladius* sp)	LC_{50}	1	5	4TH INSTAR	Ali and Mulla (1978a)
		LC_{50}	1	5	4TH INSTAR	Ali and Mulla (1977)
		LC_{50}	1	43	4TH INSTAR	Ali and Mulla (1978a)
	Midge (*Tanypus grodhausi*)	ABD	3–4		LARVAE	Mulla and Khasawinah (1969)
	Midge family	LC_{50}	1	37.63	LARVAE, 9–11 MM	Joshi et al. (1975)
		DRF	NR	<50	NR	Peterson (1972)
		POP	NR	2,530–5,070	LARVAE	Mani and Konar (1985a)
	Mosquito (*Aedes cantans*)	LC_{50}	3	200	YOUNG, 4TH INSTAR	Rettich (1979)
	Mosquito (*Aedes sticticus*)	LC_{50}	1	15.5	LARVAE, 4TH INSTAR	Rettich (1977)
	Mosquito (*Aedes vexans*)	LC_{50}	1	26.1	LARVAE, 4TH INSTAR	Rettich (1977)
	Mosquito (*Aedes excrucians*)	LC_{50}	1	30.3	LARVAE, 4TH INSTAR	Rettich (1977)
	Mosquito (*Aedes communis*)	LC_{50}	1	38.2	LARVAE, 4TH INSTAR	Rettich (1977)
	Mosquito (*Aedes punctor*)	LC_{50}	1	44.1	LARVAE, 4TH INSTAR	Rettich (1977)
	Mosquito (*Aedes cantans*)	LC_{50}	1	48.8	LARVAE, 4TH INSTAR	Rettich (1977)
	Mosquito (*Aedes caspius*)	LC_{50}	NR	27.9	4TH INSTAR LARVAE	Grandes and Sagrado (1988)
	Mosquito (*Aedes quasirusticus*)	LC_{50}	NR	34.5	4TH INSTAR LARVAE	Grandes and Sagrado (1988)
	Mosquito (*Aedes punctor*)	LC_{50}	NR	51.2	4TH INSTAR LARVAE	Grandes and Sagrado (1988)
	Mosquito (*Aedes vexans*)	LC_{50}	NR	54.7	4TH INSTAR LARVAE	Grandes and Sagrado (1988)
	Mosquito (*Aedes excrucians*)	LC_{50}	NR	68.1	4TH INSTAR LARVAE	Grandes and Sagrado (1988)
	Mosquito (*Aedes rusticus*)	LC_{50}	NR	86.3	4TH INSTAR LARVAE	Grandes and Sagrado (1988)
	Mosquito (*Aedes aegypti*)	LC_{50}	NR	250	LARVAE, 4TH INSTAR	Klassen et al. (1965)
		LC_{50}-D	1 hr	1,530	3RD INSTAR LARVAE	Chadwick et al. (1984)
		ABD	30	2.5%	LARVAE	von Windeguth (1971)
		LET	1	1,000	LARVAE, LAST INSTAR	Mitsuhashi et al. (1970)

Mosquito (*Aedes taeniorhynchus*)	LET	NR	57 g ha^{-1}	NR	Tagatz et al. (1974)
Mosquito (*Aedes flavescens*)	MOR-D	1	5 lb ac^{-1}	3RD AND 4TH INSTAR LARVAE	Dixon and Brust (1971)
Mosquito (*Anopheles quadrimaculatus*)	LC_{50}	1	69	4TH INSTAR	Holck and Meek (1987)
Mosquito (*Anopheles stephensi*)	LC_{50}	1	180	4TH INSTAR LARVA	Scott and Georghiou (1986)
Mosquito (*Anopheles albimanus*)	LC_{50}	1	350	4TH INSTAR	Hemingway and Georghiou (1983)
Mosquito (*Anopheles atroparvus*)	LC_{50}	NR	83.3	4TH INSTAR LARVAE	Grandes and Sagrado (1988)
	LC_{50}	NR	151	4TH INSTAR LARVAE	Grandes and Sagrado (1988)
Mosquito (*Culex pipiens*)	EC_{50}IM	NR	3.4	4TH INSTAR	Rawash et al. (1975)
Mosquito (*Culex pipiens quinquefasciata*)	LC_{100}	1	2,000	LARVAE, 4TH INSTAR	Mulla (1963)
Mosquito (*Culex pipiens molestus*)	LC_{100}	3	50	4TH INSTAR	Rettich (1979)
Mosquito (*Culex pipiens*)	LC_{50}	1	20	4TH INSTAR LARVAE	Yasuno et al. (1965)
Mosquito (*Culex tritaeniorhynchus*)	LC_{50}	1	21	LARVAE	Shim and Self (1973)
Mosquito (*Culex pipiens molestus*)	LC_{50}	1	30.3	LARVAE	Rettich (1979)
Mosquito (*Culex pipiens pipiens*)	LC_{50}	1	32.2	LARVAE, 4TH INSTAR	Rettich (1977)
Mosquito (*Culex pipiens molestus*)	LC_{50}	1	34.2	LARVAE, 4TH INSTAR	Rettich (1977)
Mosquito (*Culex pipiens quinquefasciata*)	LC_{50}	1	46	3RD–4TH INSTAR LARVAE	Chen et al. (1971)
	LC_{50}	1	67	3RD–4TH INSTAR LARVAE	Chen et al. (1971)

Table 40. Toxicology of Aquatic and Terrestrial Species: Chemical: Malathion, CAS Number: 121755 (continued)

Class	Species	Effect	Duration (days)	Concen-tration ($\mu g\ L^{-1}$)	Lifestage	Ref.
Insect (continued)	Mosquito (*Culex quinquefasciatus*)	LC_{50}	1	76	LATE 3RD OR EARLY 4TH INSTAR LARVAE	El-Khatib (1985)
		LC_{50}	1	76	3RD OR 4TH INSTAR LARVAE	El-Khatib and Georghiou (1985)
		LC_{50}	1	148	3RD INSTAR LARVA	Boike et al. (1985)
		LC_{50}	1	203	3RD INSTAR LARVA	Boike et al. (1985)
		LC_{50}	1	293	3RD INSTAR LARVA	Boike et al. (1985)
		LC_{50}	1	310	3RD INSTAR LARVA	Boike et al. (1985)
		LC_{50}	1	323	3RD INSTAR LARVA	Boike et al. (1985)
		LC_{50}	1	451	3RD INSTAR LARVA	Boike et al. (1985)
		LC_{50}	1	466	3RD INSTAR LARVA	Boike et al. (1985)
		LC_{50}	1	700	3RD INSTAR LARVA	Boike et al. (1985)
		LC_{50}	1	1,133	3RD INSTAR LARVA	Boike et al. (1985)
	Mosquito (*Culex pipiens molestus*)	LC_{50}	2	24	LARVAE	Rettich (1979)
	Mosquito (*Culex pipiens*)	LC_{50}	NR	1.5	LARVAE, 72 HR, 4TH INSTAR	Gaaboub et al. (1975)
	Mosquito (*Culex theileri*)	LC_{50}	NR	48.5	4TH INSTAR LARVAE	Grandes and Sagrado (1988)
	Mosquito (*Culex quinquefasciatus*)	LC_{50}	NR	65	4TH INSTAR LARVAE	Hemingway and Georghiou (1984)
	Mosquito (*Culex pipiens*)	LC_{50}	NR	84.5	4TH INSTAR LARVAE	Grandes and Sagrado (1988)
	Mosquito (*Culex theileri*)	LC_{50}	NR	154	4TH INSTAR LARVAE	Grandes and Sagrado (1988)
	Mosquito (*Culex pipiens quinquefasciata*)	MOR	1	0.2 kg ha^{-1}	4TH INSTAR LARVAE	Mount et al. (1970)
	Mosquito (*Culex nigripalpus*)	MOR	NR	57 g ha^{-1}	NR	Tagatz et al. (1974)

Mosquito (*Culex tarsalis*)	STR	1	100	4TH INSTAR LARVAE	McDonald and Asman (1982)
Mosquito (*Culiseta annulata*)	LC_{50}	1	24.5	LARVAE, 4TH INSTAR	Rettich (1977)
Mosquito (*Wyeomyia smithii*)	DVP	7	100	2ND INSTAR LARVAE	Strickman (1985)
	MOR	7	100	2ND INSTAR LARVAE	Strickman (1985)
Mosquito family	LET	1	0.5 lb	LARVAE	Dixon and Brust (1971)
	MOR	<7	5 lb ac^{-1}	LARVAE	Dixon and Brust (1971)
Snipefly	LC_{50}	4	385	JUVENILE	Johnson and Finley (1980)
Stonefly (*Acroneuria pacifica*)	LC_{50}	2	12	2–2.5 CM, NAIADS	Jensen and Gaufin (1964b)
	LC_{50}	3	16	2–2.5 CM, NAIADS	Jensen and Gaufin (1964b)
	LC_{50}	4	7	2–2.5 CM, NAIADS	Jensen and Gaufin (1964b)
	LC_{50}	4	7	NAIAD, 2–2.5 CM	Gaufin et al. (1965)
	LC_{50}	4	7.2	2–5 CM	Gaufin et al. (1961)
	LC_{50}	4	7.7	NAIADS	Jensen and Gaufin (1964a)
	LC_{50}	5	7.70	NAIADS	Jensen and Gaufin (1964a)
	LC_{50}	10	5.10	NAIADS	Jensen and Gaufin (1964a)
	LC_{50}	15	3.30	NAIADS	Jensen and Gaufin (1964a)
	LC_{50}	20	3.20	NAIADS	Jensen and Gaufin (1964a)
	LC_{50}	25	2.40	NAIADS	Jensen and Gaufin (1964a)
	LC_{50}	30	0.78	NAIADS	Jensen and Gaufin (1964a)
Stonefly (*Claassenia sabulosa*)	LC_{50}	1	13	20–25 MM	Sanders and Cope (1968)
	LC_{50}	2	6	20–25 MM	Sanders and Cope (1968)
	LC_{50}	4	2.8	20–25 MM	Sanders and Cope (1968)
	LC_{50}	4	2.8	2ND YEAR CLASS	Johnson and Finley (1980)
	LC_{50}	4	56	2–5 CM	Gaufin et al. (1961)
Stonefly (*Isoperla* sp)	LC_{50}	4	0.69	1ST YEAR CLASS	Johnson and Finley (1980)
Stonefly (*Pteronarcella badia*)	LC_{50}	1	10	15–20 MM	Sanders and Cope (1968)
Stonefly (*Pteronarcys californica*)	LC_{50}	1	35	30–35 MM	Sanders and Cope (1968)
Stonefly (*Pteronarcella badia*)	LC_{50}	2	6	15–20 MM	Sanders and Cope (1968)
Stonefly (*Pteronarcys californica*)	LC_{50}	2	20	30–35 MM	Sanders and Cope (1968)
	LC_{50}	2	180	4–6 CM, NAIADS	Jensen and Gaufin (1964b)
	LC_{50}	3	72.5	4–6 CM, NAIADS	Jensen and Gaufin (1964b)

Table 40. Toxicology of Aquatic and Terrestrial Species: Chemical: Malathion, CAS Number: 121755 (continued)

Class	Species	Effect	Duration (days)	Concentration ($\mu g\ L^{-1}$)	Lifestage	Ref.
Insect (continued)	Stonefly (*Pteronarcella badia*)	LC_{50}	4	1.1	15–20 MM	Sanders and Cope (1968)
		LC_{50}	4	1.1	NAIAD	Johnson and Finley (1980)
	Stonefly (*Pteronarcys californica*)	LC_{50}	4	10	30–35 MM	Sanders and Cope (1968)
		LC_{50}	4	10	2ND YEAR CLASS	Johnson and Finley (1980)
		LC_{50}	4	50	NAIAD, 4–6 CM	Gaufin et al. (1965)
		LC_{50}	4	50	4–6 CM, NAIADS	Jensen and Gaufin (1964b)
		LC_{50}	4	100	2–5 CM	Gaufin et al. (1961)
		LC_{50}	15	45	NAIADS	Jensen and Gaufin (1964a)
		LC_{50}	20	24	NAIADS	Jensen and Gaufin (1964a)
		LC_{50}	25	15.50	NAIADS	Jensen and Gaufin (1964a)
		LC_{50}	30	8.80	NAIADS	Jensen and Gaufin (1964a)
Mullusca	American or Virginia oyster	EC_{50}DV	2	9,070	EGGS, 2 CELL STAGE	Davis and Hidu (1979)
		LC_{50}	12	2,660	2 D LARVAE	Davis and Hidu (1979)
		GRO	4	1,000	2.5–5.1 CM	Butler (1963)
		GRO	4	1,000	2.5–5.1 CM	Butler (1963)
		GRO	12	250	2 D LARVAE	Davis and Hidu (1979)
	Apple snail	LC_{50}	2	10,000	NR	Ahamad (1978b)
		BIO	2	5,000	NR	Ahamad et al. (1979)
		BIO	2	5,000	20 G	Sivaiah and Rao (1978)
		BIO	2	5,000	NR	Ahamad (1978b)
		ENZ	2	5,000	NR	Ahamad (1978b)
		ENZ	2	5,000	NR	Sahib and Rao (1988)

Bivalve	LC_{50}	4	118.55	ADULT, 70–75 MM SHELL LENGTH	Mane and Muley (1987)
	LC_{50}	4	120	70–75 MM	Mane et al. (1984)
	LC_{50}	4	248.97	ADULT, 70–75 MM SHELL LENGTH	Mane and Muley (1987)
	LC_{50}	4	284.11	ADULT, 70–75 MM SHELL LENGTH	Mane and Muley (1987)
	OC	4	60–140	ADULT, 70–75 MM SHELL LENGTH	Muley and Mane (1987)
	OC	4	120–240	ADULT, 70–75 MM SHELL LENGTH	Muley and Mane (1987)
Bivalve, clam, mussel class	BEH	5	0.1	GLOCHIDIUM	Desi et al. (1975)
	BEH	5	1	GLOCHIDIUM	Desi et al. (1975)
Clam	LC_{50}	2	35,000 $\mu L\ L^{-1}$	ADULT, 75 MM, 18 G	Varanka (1987)
	LC_{50}	3	14,250 $\mu L\ L^{-1}$	ADULT, 75 MM, 18 G	Varanka (1987)
	LC_{50}	4	5,000 $\mu L\ L^{-1}$	ADULT, 75 MM, 18 G	Varanka (1987)
	LC_{50}	7	500 $\mu L\ L^{-1}$	ADULT, 75 MM, 18 G	Varanka (1987)
	BEH	5	0.1 $\mu L\ L^{-1}$	ADULT, 75 MM, 18 G, SOFT PART	Varanka (1987)
Cockle (*Cerastoderma edule*)	LC_{50}	2	3,300–10,000	LARVAL	Portmann (1972)
Common bay mussel, blue mussel	$EC_{50}DV$	2	13,400	EGGS	Liu and Lee (1975)
	LC_0	4	39,400	ADULT	Liu and Lee (1975)
	LC_{100}	4	6,900	ADULT	Liu and Lee (1975)
	DVP	40	6,050	LARVAE, 29–30 D	Liu and Lee (1975)
	DVP	40	12,100	LARVAE, 29–30 D	Liu and Lee (1975)
	GRO	20	3,100	LARVAE	Liu and Lee (1975)
	GRO	20	6,200	LARVAE	Liu and Lee (1975)

Table 40. Toxicology of Aquatic and Terrestrial Species: Chemical: Malathion, CAS Number: 121755 (continued)

Class	Species	Effect	Duration (days)	Concen-tration (µg L^{-1})	Lifestage	Ref.
Mullusca (continued)	Eastern mud snail	MOR	4	25,000	15 MM, 0.16 G SHELL	Eisler (1970c)
	Marine bivalve (*Donax cuneatus*)	FLT	3	1,000	ADULT, 20–25 MM	Mane et al. (1979)
		FLT	3	1,000	ADULT, 20–25 MM	Mane et al. (1979)
		MOR	9	1,000	ADULT, 20–25 MM	Mane et al. (1979)
		OC	8	1,000	ADULT, 20–25 MM	Mane et al. (1979)
	Marine bivalve (*Katelysia opima*)	FLT	3	1,000	ADULT, 25–30 MM	Mane et al. (1979)
		FLT	3	1,000	ADULT, 25–30 MM	Mane et al. (1979)
		LET	3.33	1,000	ADULT, 25–30 MM	Mane et al. (1979)
		OC	3	1,000	ADULT, 25–30 MM	Mane et al. (1979)
	Mediterranean mussel	LC_{50}	7	6,000	10–70 MM	Rao and Mane (1978)
		OC	6	1,000	10–30 MM	Rao and Mane (1978)
		RES	6	1,000	50–70 MM	Rao and Mane (1978)
	Mussel (*Lamellidens marginalis*)	LC_{50}	4	55.63	ADULT, 65–70 MM SHELL LENGTH	Mane and Muley (1987)
		LC_{50}	4	100	65–70 MM	Mane et al. (1984)
		LC_{50}	4	168.36	ADULT, 65–70 MM SHELL LENGTH	Mane and Muley (1987)
		LC_{50}	4	797.78	ADULT, 65–70 MM SHELL LENGTH	Mane and Muley (1987)
		BIO	2	5,000	24.4 G	Ahamad et al. (1978a)
		OC	4	20–100	ADULT, 65–70 MM SHELL LENGTH	Muley and Mane (1987)
		OC	4	60–200	ADULT, 65–70 MM SHELL LENGTH	Muley and Mane (1987)

Mussel (*Unio pictorum*)	LC_{50}	7	31,250 µL L^{-1}	ADULT, 70 MM, 10.1 G	Varanka (1987)
	BEH	5	0.1 µL L^{-1}	ADULT, 70 MM, 10.1 G, SOFT PART	Varanka (1987)
Northern quahog or hard clam	MOR	4	25,000	ADULT, 82 MM, 21 G SHELL	Eisler (1970c)
Pond snail (*Lymnaea acuminata*)	LC_{50}	1	38.1	NR	Chaudhari et al. (1988)
	LC_{50}	2	16.9	NR	Chaudhari et al. (1988)
Snail (*Biomphalaria glabrata*)	PHY	1	80,000	14–16 MM SHELL DIAMETER	Cocks (1973)
	PHY	1	124,000	14–16 MM SHELL DIAMETER	Cocks (1973)
	PHY	1	161,000	14–16 MM SHELL DIAMETER	Cocks (1973)
Snail (*Thiara scabra*)	LC_{50}	1	48.9	NR	Chaudhari et al. (1988)
Snail (*Thiara lineata*)	LC_{50}	1	48.9	NR	Chaudhari et al. (1988)
	LC_{50}	2	31.1	NR	Chaudhari et al. (1988)
Snail (*Thiara scabra*)	LC_{50}	2	37.4	NR	Chaudhari et al. (1988)
Swan mussel	EC_{50}PH	0.5 hr	0.0018	LARVAE	Varanka (1979)
	EC_{50}PH	1	0.000075	LARVAE	Varanka (1979)
	LC_{50}	1	25,000 µL L^{-1}	ADULT, 92 MM, 22.8 G	Varanka (1987)
	LC_{50}	2	14,500 µL L^{-1}	ADULT, 92 MM, 22.8 G	Varanka (1987)
	LC_{50}	3	6,350 µL L^{-1}	ADULT, 92 MM, 22.8 G	Varanka (1987)
	LC_{50}	4	975 µL L^{-1}	ADULT, 92 MM, 22.8 G	Varanka (1987)
	LC_{50}	7	225 µL L^{-1}	ADULT, 92 MM, 22.8 G	Varanka (1987)
	BEH	2	1,000	NR	Desi et al. (1975)

Table 40. Toxicology of Aquatic and Terrestrial Species: Chemical: Malathion, CAS Number: 121755 (continued)

Class	Species	Effect	Duration (days)	Concen-tration (μg L^{-1})	Lifestage	Ref.
Mullusca (continued)	Swan mussel	BEH	2	10,000	NR	Desi et al. (1975)
		BEH	5	0.1 μL L^{-1}	ADULT, 92 MM, 22.8 G, SOFT PART	Varanka (1987)
		BEH	5	1 μL L^{-1}	ADULT, 92 MM, 22.8 G, SOFT PART	Varanka (1987)
	Unionid clam	LC_{50}	4	12	55–60 MM	Mane et al. (1984)
		LC_{50}	4	14	55–60 MM	Mane et al. (1984)
		LT_{50}	3.54	1,000	ADULT, 55–65 MM	Mane et al. (1979)
Platyhelmi	Turbellarian, flatworm	MOR	1	100	1–1.5 CM	Levy and Miller (1978)
		MOR	1	100	1–1.5 CM	Levy and Miller (1978)
Protozoa	Ciliate (*Colpidium campylum*)	PGR	1.79	>10,000	>96 HR	Dive et al. (1980)
	Ciliate (*Paramecium aurelia*)	GRO	1	10,000	35 ML^{-1}	Tandon et al. (1987)
		GRO	1	20,000	35 ML^{-1}	Tandon et al. (1987)
		GRO	5	1,000	35 ML^{-1}	Tandon et al. (1987)
		MOR	1.5 hr	5,000	10,000 CELLS L^{-1}	Joshi and Misra (1986)
	Protozoan phylum	ABD	11	10	NR	Taub et al. (1983)
Rotifera	Rotifer (*Philodina* sp)	ABD	39	10	NR	Taub et al. (1983)
	Rotifer phylum	ABD	25	10	NR	Taub et al. (1983)
Tracheophy	Bog-rush	BCF	1	<0.10	NR	Tagatz et al. (1974)
		BCF	3	<0.10	NR	Tagatz et al. (1974)
	Water-meal	LET	12–30	100,000	NR	Worthley and Schott (1972)

Bird	Bobwhite quail	LC_{50}	8	3,497 ppm	14–21 D	Heath et al. (1972)
	European starling	NO MOR	84	160 ppm		Dieter (1975)
	Horned lark	LD_{50}	14	403 mg kg^{-1}	ADULT	Hudson et al. (1984)
	Japanese quail	LC_{50}	8	2,123 ppm	14–21 D	Heath et al. (1972)
		LO BEH	51	20 mg kg^{-1}	3 D	Meydani and Post (1979)
		LO PHY	60	100 ppm	39 D	Cecil et al. (1974)
	Mallard duck	LD_{50}	14	1,435 mg kg^{-1}	90 D	Hudson et al. (1984)
	Ring-necked pheasant	LC_{50}	8	4,320 ppm	14–21 D	Heath et al. (1972)
		LD_{50}	14	167 mg kg^{-1}	90 D	Hudson et al. (1984)
Invertebrate	Honey bee	LD_{50}	NR	270 mg bee^{-1}	ADULT	Stevenson et al. (1978)
		LD_{50}	NR	380 mg bee^{-1}	ADULT	Stevenson et al. (1978)
		LO MOR	NR	1 kg ha^{-1}	ADULT	Korpela and Tulisalo (1974)
		LO MOR	NR	3 ppm		Sonnet (1978)
		LO MOR	NR	3 ppm		Sonnet (1978)
		LO MOR	NR	3 ppm		Sonnet (1978)
		LO MOR	NR	3 ppm		Sonnet (1978)
		NO POP	NR	3 ppm		Sonnet (1978)
		NO POP	NR	3 ppm		Sonnet (1978)
		NO POP	NR	3 ppm		Sonnet (1978)
		NO POP	NR	5 ppm		Sonnet (1978)
Mammal	Mouse (*Mus* sp)	LD_{50}	14	1,680 mg kg^{-1}	ADULT	Berteau and Deen (1978)
		NO POP	60–90	0.585 L ha^{-1}	MULT	Erwin and Sharpe (1978)
	Norway rat	LO PHY	60	100 ppm	21 D	Cecil et al. (1974)
	Prairie mole	NO POP	60–90	0.585 L ha^{-1}	MULT	Erwin and Sharpe (1978)

Table 40. Toxicology of Aquatic and Terrestrial Species: Chemical: Malathion, CAS Number: 121755 (continued)

Class	Species	Effect	Duration (days)	Concen-tration (µg L^{-1})	Lifestage	Ref.
Mammal (continued)	Rat	LD_{50}	1	209 mg kg^{-1}	1 D	Mendoza (1976)
		LD_{50}	1	469 mg kg^{-1}	6 D	Mendoza (1976)
		LD_{50}	1	707 mg kg^{-1}	6 D	Mendoza (1976)
		LD_{50}	1	1,085 mg kg^{-1}	12 D	Mendoza (1976)
		LD_{50}	1	1,806 mg kg^{-1}	17 D	Mendoza (1976)
		LO PHY	50	20 mg kg^{-1}	4 D	Krause et al. (1976)
	Short-tailed shrew	NO POP	60–90	0.585 L ha^{-1}	MULT	Erwin and Sharpe (1978)

Table 41. Toxicology of Aquatic and Terrestrial Species: Chemical: Mancozeb, CAS Number: 8018017

Class	Species	Effect	Duration (days)	Concentration ($\mu g\ L^{-1}$)	Lifestage	Ref.
Algae	Green algae (*Chlorella pyrenoidosa*)	EC_{50}GR	4	1,100	10E+8 CELLS 100 ML^{-1}	Van Leeuwen et al. (1985a)
Crustacea	Water flea (*Daphnia magna*)	LC_{50}	2	1,300	NR	Van Leeuwen et al. (1985a)
Fish	Common, mirror, colored, carp	MOR	6	5,000	32–65 G, 1 YEAR	Antychowicz et al. (1979)
	Cyprinid fish (*Barilius bendelisis*)	LC_{50}	NR	400	NR	Deoray and Wagh (1987)
	Guppy	LC_{50}	4	2,600	NR	Van Leeuwen et al. (1985a)
	Pink salmon	LET	2	500	FRY, 1.3 G	Kuroda (1975)
	Trout family	LET	2	300–1,000	FRY, 2.8–5.4 CM, 0.25–0.56 G	Kuroda (1974)
		LET	2	4,000–7,000	FRY, 2.8–5.4 CM, 0.25–0.56 G	Kuroda (1974)
Protozoa	Ciliate (*Paramecium aurelia*)	LET	1.5 hr	5,000	10,000 CELLS ML^{-1}	Joshi and Misra (1986)
		MOR	1.5 hr	1,000	10,000 CELLS ML^{-1}	Joshi and Misra (1986)
Mammal	Rat	LD_{50}		11,200 mg kg^{-1}		*Farm Chemicals Handbook* (1990)

Table 42. Toxicology of Aquatic and Terrestrial Species: Chemical: Maneb, CAS Number: 12427382

Class	Species	Effect	Duration (days)	Concentration ($\mu g\ L^{-1}$)	Lifestage	Ref.
Algae	Green algae (*Chlorella pyrenoidosa*)	EC_{50}GR	4	3,200	10E+8 CELLS 100 ML^{-1}	Van Leeuwen et al. (1985a)
		EC_{50}GR	4 hr	3,200	5E+8 CELLS L^{-1}	Van Leeuwen et al. (1985a)
Amphibia	Clawed toad	PHY	1–10	2,500	11 HR, STAGES 10–11, EMBRYO	Bancroft and Prahlad (1973)
	Crested newt	LC_{50}	0.37	125,000	ADULT MALE 7.2–7.5 CM, 8 G	Zaffaroni et al. (1978)
		LC_{50}	0.67	125,000	ADULT FEMALE 7.6–8.8 M, 12 G	Zaffaroni et al. (1978)
		LC_{50}	0.79	75,000	ADULT MALE 7.2–7.5 CM, 8 G	Zaffaroni et al. (1978)
		LC_{50}	0.81	100,000	ADULT FEMALE 7.6–8.8 M, 12 G	Zaffaroni et al. (1978)
		LC_{50}	1.06	75,000	ADULT FEMALE 7.6–8.8 M, 12 G	Zaffaroni et al. (1978)
		LC_{50}	1.17	100,000	ADULT MALE 7.2–7.5 CM, 8 G	Zaffaroni et al. (1978)
		LC_{50}	3.17	50,000	ADULT MALE 7.2–7.5 CM, 8 G	Zaffaroni et al. (1978)
		LC_{50}	7	50,000	ADULT FEMALE 7.6–8.8 M, 12 G	Zaffaroni et al. (1978)
		LC_{50}	10	25,000	ADULT MALE 7.2–7.5 CM, 8 G	Zaffaroni et al. (1978)

LT_{50}	0.35	125,000	ADULT MALE, 7.4 CM, 8 G	Zaffaroni et al. (1978)
LT_{50}	0.37	125,000	ADULT MALE, 7.1 CM, 6.5 G	Zaffaroni et al. (1978)
LT_{50}	0.44	125,000	ADULT MALE, 7.4 CM, 8 G	Zaffaroni et al. (1978)
LT_{50}	0.46	100,000	ADULT MALE, 7.1 CM, 6.5 G	Zaffaroni et al. (1978)
LT_{50}	0.67	100,000	ADULT FEMALE, 7.5 CM, 8.5 G	Zaffaroni et al. (1978)
LT_{50}	0.67	125,000	ADULT FEMALE, 7.5 CM, 8.5 G	Zaffaroni et al. (1978)
LT_{50}	0.79	75,000	ADULT MALE, 7.4 CM, 8 G	Zaffaroni et al. (1978)
LT_{50}	0.81	100,000	ADULT FEMALE, 7.8 CM, 12 G	Zaffaroni et al. (1978)
LT_{50}	1.06	75,000	ADULT FEMALE, 7.8 CM, 12 G	Zaffaroni et al. (1978)
LT_{50}	1.17	100,000	ADULT MALE, 7.4 CM, 8 G	Zaffaroni et al. (1978)
LT_{50}	1.19	125,000	ADULT FEMALE, 7.8 CM, 12 G	Zaffaroni et al. (1978)
LT_{50}	3.17	50,000	ADULT MALE, 7.4 CM, 8 G	Zaffaroni et al. (1978)
LT_{50}	7	50,000	ADULT FEMALE, 7.8 CM, 12 G	Zaffaroni et al. (1978)

Table 42. Toxicology of Aquatic and Terrestrial Species: Chemical: Maneb, CAS Number: 12427382 (continued)

Class	Species	Effect	Duration (days)	Concentration ($\mu g\ L^{-1}$)	Lifestage	Ref.
Amphibia (continued)	European crested newt	HIS	105	500	ADULT, 4 YR, 7.4–7.9 CM	Zavanella et al. (1979)
		HIS	105	2,500	ADULT, 4 YR, 7.4–7.9 CM	Zavanella et al. (1979)
		RGN	85	5,000	ADULT FEMALE	Arias and Zavanella (1979)
		RGN	85	5,000	ADULT MALE	Arias and Zavanella (1979)
		RGN	85	5,000	ADULT FEMALE	Arias and Zavanella (1979)
		RGN	85	5,000	ADULT MALE	Arias and Zavanella (1979)
		TMR	259	5,000	6.1 CM MALE, 6.7 CM FEMALE	Zavanella et al. (1980)
		TMR	259	5,000	6.6 CM MALE, 7 CM FEMALE	Zavanella et al. (1980)
Crustacea	Common shrimp	LC_{50}	2	3,300–10,000	LARVAL	Portmann (1972)
	Harpacticoid copepod	LC_{50}	4	110	0.6 – 0.8 MM	Linden et al. (1979)
	Water flea (*Daphnia magna*)	EC_{50}	NR	60	EXPO GRO CULTURE OF VARIOUS AGE	Van Leeuwen et al. (1987)
		LC_{50}	2	1,000	NR	Van Leeuwen et al. (1985a)
		LC_{50}	21	110	0–24 HR	Van Leeuwen et al. (1987)
		LC_{50}	21	111	<24 HR	Van Leeuwen et al. (1985b)

Fish	Bleak	LC_{50}	4	520	8 CM	Linden et al. (1979)
	Guppy	LC_{50}	4	3,700	NR	Van Leeuwen et al. (1985a)
	Hooknose or pogge	LC_{50}	2	330–1,000	NR	Portmann (1972)
	Rainbow trout, donaldson trout	EC_{50}	60	148	FERTILIZED EGG — EARLY FRY STAGE	Van Leeuwen et al. (1986a)
		LC_{50}	4	320	SAC FRY, 42 D	Van Leeuwen et al. (1985c)
		LC_{50}	4	340	EARLY FRY, 77 D	Van Leeuwen et al. (1985c)
		LC_{50}	4	1,300	LATE-EYED EGG, 28 D	Van Leeuwen et al. (1985c)
		LC_{50}	4	1,800	EARLY EYED EGG, 14 D	Van Leeuwen et al. (1985c)
		LC_{50}	4	5,600	FERTILIZED EGG, 24 HR	Van Leeuwen et al. (1985c)
		LC_{50}	4	6,000	FERTILIZED EGG, 0 HR	Van Leeuwen et al. (1985c)
		LC_{50}	60	165	FERTILIZED EGG — EARLY FRY STAGE	Van Leeuwen et al. (1986a)
Mullusca	Cockle (*Cerastoderma edule*)	LC_{50}	2	100,000–330,000	LARVAL	Portmann (1972)
Mammal	Rat	LD_{50}		7,990 mg kg^{-1}		*Farm Chemicals Handbook* (1990)

Table 43. Toxicology of Aquatic and Terrestrial Species: Chemical: MCPA, CAS Number: 94746

Class	Species	Effect	Duration (days)	Concentration ($\mu g\ L^{-1}$)	Lifestage	Ref.
Algae	Blue-green algae (*Anabaena cylindrica*)	NFX	1 hr	20,000	EXPO GROWING CULTURE	DaSilva et al. (1975)
	Blue-green algae (*Anacystis aeruginosa*)	PGR	21	2,000	INITIAL CONC. 125,000 CELLS ML^{-1}	Palmer and Maloney (1955)
	Blue-green algae (*Aulosira* sp)	NFX	1 hr	20,000	EXPO GROWING CULTURE	DaSilva et al. (1975)
	Blue-green algae (*Calothrix elenkinii*)	NFX	1 hr	20,000	EXPO GROWING CULTURE	DaSilva et al. (1975)
	Blue-green algae (*Chlorogloea fritschii*)	NFX	1 hr	20,000	EXPO GROWING CULTURE	DaSilva et al. (1975)
	Blue-green algae (*Cylindrospermum muscicola*)	NFX	1 hr	20,000	EXPO GROWING CULTURE	DaSilva et al. (1975)
	Blue-green algae (*Cylindrospermum licheniforme*)	PGR	3	2,000	INITIAL CONC. 125,000 CELLS ML^{-1}	Palmer and Maloney (1955)
	Blue-green algae (*Nostoc* sp)	NFX	1 hr	20,000	EXPO GROWING CULTURE	DaSilva et al. (1975)
	Blue-green algae (*Nostoc muscorum*)	NFX	1 hr	20,000	EXPO GROWING CULTURE	DaSilva et al. (1975)
	Blue-green algae (*Phormidium* sp)	CEL	1–3 hr	9.0 μg	NR	Noll and Bauer (1974)
	Blue-green algae (*Tolypothrix tenuis*)	NFX	1 hr	20,000	EXPO GROWING CULTURE	DaSilva et al. (1975)
	Blue-green algae (*Westiellopsis* sp)	NFX	1 hr	20,000	EXPO GROWING CULTURE	DaSilva et al. (1975)
	Cuvie, tangleweed	PGR	28	1,000	ZOOSPORES	Hopkins and Kain (1971)

Diatom (*Gomphonema parvulum*)	PGR	3	2,000	INITIAL CONC. 125,000 CELLS ML^{-1}	Palmer and Maloney (1955)
Diatom (*Nitzschia palea*)	PGR	3	2,000	INITIAL CONC. 125,000 CELLS ML^{-1}	Palmer and Maloney (1955)
Green algae (*Chlamydomonas globosa*)	PGR	24	500,000	NR	Kirkwood and Fletcher (1970)
	PGR	24	500,000	NR	Kirkwood and Fletcher (1970)
	PGR	24	500,000	NR	Kirkwood and Fletcher (1970)
Green algae (*Chlorella variegata*)	PGR	21	2,000	INITIAL CONC. 125,000 CELLS ML^{-1}	Palmer and Maloney (1955)
Green algae (*Chlorella pyrenoidosa*)	PGR	24	500,000	NR	Kirkwood and Fletcher (1970)
	PGR	24	500,000	NR	Kirkwood and Fletcher (1970)
	PGR	24	500,000	NR	Kirkwood and Fletcher (1970)
Green algae (*Scenedesmus obliquus*)	PGR	21	2,000	INITIAL CONC. 125,000 CELLS ML^{-1}	Palmer and Maloney (1955)
Green algae (*Stichococcus bacillaris*)	PGR	24	500,000	NR	Kirkwood and Fletcher (1970)
	PGR	24	500,000	NR	Kirkwood and Fletcher (1970)
	PGR	24	500,000	NR	Kirkwood and Fletcher (1970)

Table 43. Toxicology of Aquatic and Terrestrial Species: Chemical: MCPA, CAS Number: 94746 (continued)

Class	Species	Effect	Duration (days)	Concentration ($\mu g\ L^{-1}$)	Lifestage	Ref.
Crustacea	Calanoid copepod (*Spicodiaptomus chilospinus*)	MOR	2	4,000	ADULT, 2.2–2.8 MM	Kader et al. (1976)
	Water flea (*Daphnia magna*)	EC_{50}IM	1.08	>100,000	1ST INSTAR	Crosby and Tucker (1966)
	Water flea (*Daphnia pulex*)	LC_{50}	3 hr	>40,000	FEMALE, ADULT	Nishiuchi and Hashimoto (1969)
		LC_{50}	3 hr	>40,000	NR	Nishiuchi and Hashimoto (1975)
	Water flea (*Daphnia magna*)	LC_{50}	4	11,000	NR	Knapek and Lakota (1974)
	Water flea (*Moina macrocopa*)	LC_{50}	3 hr	>40,000	FEMALE, ADULT	Nishiuchi and Hashimoto (1969)
		LC_{50}	3 hr	>40,000	NR	Nishiuchi and Hashimoto (1975)
Fish	Bluegill	LC_{50}	1	163,500	2.5–7.6 CM	Davis and Hughes (1963)
		LC_{50}	2	163,500	2.5–7.6 CM	Davis and Hughes (1963)
		LC_{50}	4	>10,000	FINGERLING	Johnson and Finley (1980)
		STR	1	5,000	FINGERLING, ≤10 CM	Applegate et al. (1957)
	Carp (*Carassius* sp)	LC_{50}	4	45,000	FRY	Knapek and Lakota (1974)
	Common, mirror, colored, carp	LC_{50}	2	>40,000	4.5 CM, 1.1 G	Nishiuchi and Hashimoto (1969)
		LC_{50}	2	>40,000	NR	Nishiuchi and Hashimoto (1975)
		LC_{50}	4	59,000	FRY	Knapek and Lakota (1974)

	Goldfish	LC_{50}	2	>40,000	4.01 CM, 1.04 G	Nishiuchi and Hashimoto (1969)
		LC_{50}	2	>40,000	NR	Nishiuchi and Hashimoto (1975)
	Medaka, high-eyes	LC_{50}	2	>40,000	2.54 CM, 0.16 G	Nishiuchi and Hashimoto (1969)
		LC_{50}	2	>40,000	NR	Nishiuchi and Hashimoto (1975)
	Mosquitofish (*Gambusia affinis*)	LC_{50}	1	>10,000	NR	Ahmed (1977)
	Rainbow trout, donaldson trout	LC_{0}	2	20,000	2 YEARS	Lysak and Marcinek (1972)
		LC_{100}	2	75,000–100,000	2 YEARS	Lysak and Marcinek (1972)
		STR	1	5,000	FINGERLING, ≤10 CM	Applegate et al. (1957)
	Sea lamprey	STR	1	5,000	LARVAE, 8–13 CM	Applegate et al. (1957)
	Tench	LC_{50}	4	45,000	FRY	Knapek and Lakota (1974)
	Trout family	LC_{50}	4	25,000	FRY	Knapek and Lakota (1974)
Insect	Mosquito (*Aedes aegypti*)	LC_{50}	4	335,000	LARVAE	Knapek and Lakota (1974)
Protozoa	Ciliate (*Colpidium campylum*)	PGR	1.79	>10,000	>96 HR	Dive et al. (1980)
Bird	Bobwhite quail	LC_{50}	8	>5,000		Heath et al. (1972)
	Japanese quail	LC_{50}	8	>5,000		Heath et al. (1972)
	Ring-necked pheasant	LC_{50}	8	>5,000		Heath et al. (1972)

Table 44. Toxicology of Aquatic and Terrestrial Species: Chemical: Methomyl, CAS Number: 16752775

Class	Species	Effect	Duration (days)	Concentration ($\mu g\ L^{-1}$)	Lifestage	Ref.
Algae	Diatom (*Skeletonema costatum*)	$EC_{50}PS$	2 hr	524,000	LOG GRO PHASE	Roberts et al. (1982)
	Dinoflagellate (*Prorocentrum minimum*)	$EC_{50}PS$	2 hr	388,000	LOG GRO PHASE	Roberts et al. (1982)
	Microflagellate	$EC_{50}PS$	2 hr	580,000	LOG GRO PHASE	Roberts et al. (1982)
Amphibia	Frog (*Bufo vulgaris formosus*)	LC_{50}	1	34,000	NR	Nishiuchi (1980b)
		LC_{50}	1	23,000–26,000	NR	Nishiuchi (1980b)
		LC_{50}	1	>42,000	NR	Nishiuchi (1980b)
	Frog (*Bufo bufo japonicus*)	LC_{50}	2	>40,000	TADPOLE	Hashimoto and Nishiuchi (1981)
Crustacea	Calanoid copepod (*Acartia tonsa*)	LC_{50}	4	410	NAUPLII	Roberts et al. (1982)
	Calanoid copepod (*Eurytemora affinis*)	LC_{50}	4	290	NAUPLII	Roberts et al. (1982)
	Fiddler crab (*Uca pugilator*)	LC_{50}	4	2,380	NR	Kaplan and Sherman (1977)
	Grass shrimp (*Palaemonetes vulgaris*)	LC_{50}	4	49	NR	Kaplan and Sherman (1977)
	Mud crab (*Neopanope texana*)	LC_{50}	4	410	NR	Kaplan and Sherman (1977)
	Opossum shrimp	LC_{50}	4	50	ADULT	Roberts et al. (1982)
	(*Mysidopsis bahia*)	LC_{50}	4	64	ADULT	Roberts et al. (1982)
	Opossum shrimp	LC_{50}	4	30	ADULT	Roberts et al. (1982)
	(*Neomysis americana*)	LC_{50}	4	34	ADULT	Roberts et al. (1982)
	Pink shrimp	LC_{50}	4	19	NR	Kaplan and Sherman (1977)

	Scud	LC_{50}	4	720	MATURE	Sanders et al. (1983)
	(*Gammarus pseudolimnaeus*)	LC_{50}	4	920	MATURE	Sanders et al. (1983)
		LC_{50}	4	1,050	MATURE	Johnson and Finley (1980)
	Water flea (*Daphnia magna*)	EC_{50}IM	2	7.6	1ST INSTAR	Sanders et al. (1983)
		EC_{50}IM	2	8.8	1ST INSTAR	Sanders et al. (1983)
		EC_{50}IM	2	8.8	1ST INSTAR	Johnson and Finley (1980)
		EC_{50}IM	2	3,200	1ST INSTAR	Johnson and Finley (1980)
	Water flea (*Daphnia pulex*)	LC_{50}	3 hr	37	NR	Hashimoto and Nishiuchi (1981)
		LC_{50}	3 hr	45	NR	Hashimoto and Nishiuchi (1981)
		LC_{50}	3 hr	49	NR	Hashimoto and Nishiuchi (1981)
		LC_{50}	3 hr	64	NR	Hashimoto and Nishiuchi (1981)
		LC_{50}	3 hr	69	NR	Hashimoto and Nishiuchi (1981)
	Water flea (*Moina macrocopa*)	LC_{50}	3 hr	25	NR	Hashimoto and Nishiuchi (1981)
	White river crayfish	LC_{50}	4	1,000	0.7 G	Carter and Graves (1972)
Fish	Atlantic salmon	LC_{50}	4	1,120	0.5 G	Johnson and Finley (1980)
		LC_{50}	4	1,200	0.3 G	Johnson and Finley (1980)
		LC_{50}	4	1,400	0.3 G	Johnson and Finley (1980)
	Atlantic silverside	LC_{50}	4	340	59.4 MM, 2.15 G	Roberts et al. (1982)
	Bluegill	LC_{50}	4	480	NR	Sanders et al. (1983)
		LC_{50}	4	600	NR	Sanders et al. (1983)
		LC_{50}	4	620	NR	Sanders et al. (1983)
		LC_{50}	4	660	NR	Sanders et al. (1983)
		LC_{50}	4	670	0.8 G	Johnson and Finley (1980)
		LC_{50}	4	710	0.9 G	Johnson and Finley (1980)

Table 44. Toxicology of Aquatic and Terrestrial Species: Chemical: Methomyl, CAS Number: 16752775 (continued)

Class	Species	Effect	Duration (days)	Concentration ($\mu g\ L^{-1}$)	Lifestage	Ref.
Fish (continued)	Bluegill	LC_{50}	4	840	NR	Sanders et al. (1983)
		LC_{50}	4	860	NR	Sanders et al. (1983)
		LC_{50}	4	875	NR	Kaplan and Sherman (1977)
		LC_{50}	4	1,050	0.9 G	Johnson and Finley (1980)
		LC_{50}	4	1,200	NR	Sanders et al. (1983)
		LC_{50}	4	1,200	NR	Sanders et al. (1983)
		LC_{50}	4	1,200	NR	Sanders et al. (1983)
		LC_{50}	4	2,000	0.5 G	Carter and Graves (1972)
		LC_{50}	4	2,000	NR	Sanders et al. (1983)
	Brook trout	LC_{50}	4	1,220	1.2 G	Johnson and Finley (1980)
		LC_{50}	4	1,500	1.2 G	Johnson and Finley (1980)
		LC_{50}	4	2,200	1.2 G	Johnson and Finley (1980)
	Channel catfish	LC_{50}	1	920	10 G	Carter and Graves (1972)
		LC_{50}	4	300	NR	Sanders et al. (1983)
		LC_{50}	4	300	0.5 G	Johnson and Finley (1980)
		LC_{50}	4	320	NR	Sanders et al. (1983)
		LC_{50}	4	320	0.8 G	Johnson and Finley (1980)

	LC_{50}	2	2,800	NR	Hashimoto and Nishiuchi (1981)
	LC_{50}	2	2,960	49 MM, 1.75 G	El-Refai et al. (1976)
Cutthroat trout	EC_{50}IM	2	6,800	1 G	Johnson and Finley (1980)
Fathead minnow	LC_{50}	4	1,500	NR	Sanders et al. (1983)
	LC_{50}	4	1,500	0.8 G	Johnson and Finley (1980)
	LC_{50}	4	1,800	NR	Sanders et al. (1983)
	LC_{50}	4	1,800	0.2 G	Johnson and Finley (1980)
	LC_{50}	4	2,110	38 D, 18.8 MM, 0.100 G	Geiger et al. (1988)
	LC_{50}	4	2,800	NR	Sanders et al. (1983)
	LC_{50}	4	2,800	0.8 G	Johnson and Finley (1980)
Goldfish	LC_{50}	2	2,700	NR	Hashimoto and Nishiuchi (1981)
	LC_{50}	4	>100	NR	Kaplan and Sherman (1977)
Largemouth bass	LC_{50}	4	760	3 G	Johnson and Finley (1980)
	LC_{50}	4	1,250	3 G	Johnson and Finley (1980)
Medaka, high-eyes	LC_{50}	2	870	NR	Hashimoto and Nishiuchi (1981)
Motsuga, stone moroko	ENZ	0.2 hr		4–8 CM, 1–5 G	Kanazawa (1983)
	ENZ	1	100	4–8 CM, 1–5 G	Kanazawa (1983)
Oriental weatherfish	LC_{50}	2	1,500	NR	Hashimoto and Nishiuchi (1981)
Pinfish	ENZ	0.79	500	46–102 MM	Coppage (1977)
Rainbow trout, donaldson trout	EC_{50}IM	2	1,600	1.1 G	Johnson and Finley (1980)
	LC_{50}	4	1,000	NR	Sanders et al. (1983)
	LC_{50}	4	1,100	NR	Sanders et al. (1983)
	LC_{50}	4	1,200	NR	Sanders et al. (1983)
	LC_{50}	4	1,200	NR	Sanders et al. (1983)
	LC_{50}	4	1,200	NR	Sanders et al. (1983)
	LC_{50}	4	1,200	1 G	Johnson and Finley (1980)
	LC_{50}	4	1,200	0.6 G	Johnson and Finley (1980)

Table 44. Toxicology of Aquatic and Terrestrial Species: Chemical: Methomyl, CAS Number: 16752775 (continued)

Class	Species	Effect	Duration (days)	Concentration ($\mu g\ L^{-1}$)	Lifestage	Ref.
Fish (continued)	Rainbow trout, donaldson trout	LC_{50}	4	1,400	NR	Sanders et al. (1983)
		LC_{50}	4	1,500	NR	Sanders et al. (1983)
		LC_{50}	4	1,600	NR	Sanders et al. (1983)
		LC_{50}	4	1,700	NR	Sanders et al. (1983)
		LC_{50}	4	2,000	NR	Sanders et al. (1983)
		LC_{50}	4	3,200	YOLK SAC FRY	Johnson and Finley (1980)
		LC_{50}	4	3,400	NR	Kaplan and Sherman (1977)
		LC_{50}	4	32,000	EYED EGG	Johnson and Finley (1980)
		BCF	28	750	NR	Kaplan and Sherman (1977)
	Sheepshead minnow	LC_{50}	4	960	25.8 MM, 0.70 G	Roberts et al. (1982)
	Tilapia (*Tilapia nilotica*)	LC_{50}	0.5	1,291	1.5 G, 41.5 MM	El-Refai et al. (1976)
		LC_{50}	0.5	1,300	5.36 G, 64 MM	El-Refai et al. (1976)
		LC_{50}	0.5	1,300	5.36 G, 64 MM	El-Refai et al. (1976)
		LC_{50}	0.5	1,560	13.9 G, 89 MM	El-Refai et al. (1976)
		LC_{50}	0.5	15,000	5.36 G, 64 MM	El-Refai et al. (1976)
		LC_{50}	1	1,000	5.36 G, 64 MM	El-Refai et al. (1976)
		LC_{50}	1	1,054	1.5 G, 41.5 MM	El-Refai et al. (1976)
		LC_{50}	1	1,100	5.36 G, 64 MM	El-Refai et al. (1976)
		LC_{50}	1	1,100	5.36 G, 64 MM	El-Refai et al. (1976)
		LC_{50}	1	1,301	13.8 G, 89 MM	El-Refai et al. (1976)
		LC_{50}	2	850	5.36 G, 64 MM	El-Refai et al. (1976)
		LC_{50}	2	884	13.8 G, 89 MM	El-Refai et al. (1976)
		LC_{50}	2	916	1.5 G, 41.5 MM	El-Refai et al. (1976)
		LC_{50}	2	1,100	5.36 G, 64 MM	El-Refai et al. (1976)
		LC_{50}	2	1,800	5.36 G, 64 MM	El-Refai et al. (1976)

Insect	Mayfly (*Cloeon dipterum*)	LC_{50}	3 hr	100	LARVAE	Hashimoto and Nishiuchi (1981)
	Midge (*Chironomus plumosus*)	EC_{50}IM	2	32	4TH INSTAR LARVAE	Sanders et al. (1983)
		EC_{50}IM	2	88	4TH INSTAR LARVAE	Sanders et al. (1983)
		LC_{50}	2	32	4TH INSTAR	Johnson and Finley (1980)
	Midge (*Chironomus* sp)	MOR	2	1,000	4TH INSTAR	Norland et al. (1974)
	Mosquito (*Wyeomyia smithii*)	DVP	7	1,000	2ND INSTAR LARVAE	Strickman (1985)
		MOR	7	5,000	2ND INSTAR LARVAE	Strickman (1985)
	Stonefly (*Pteronarcella badia*)	EC_{50}IM	2	69	NAIAD	Johnson and Finley (1980)
		LC_{50}	4	60	NAIAD	Johnson and Finley (1980)
	Stonefly (*Skwala* sp)	EC_{50}IM	2	34	NAIAD	Johnson and Finley (1980)
		LC_{50}	4	29	NAIAD	Johnson and Finley (1980)
Mullusca	Bladder snail	LC_{50}	2	18,000	NR	Nishiuchi and Yoshida (1972)
		LC_{50}	2	18,000	NR	Hashimoto and Nishiuchi (1981)
	Marsh snail	LC_{50}	2	12,000	NR	Nishiuchi and Yoshida (1972)
		LC_{50}	3 hr	12,000	NR	Hashimoto and Nishiuchi (1981)
	Mud snail	LC_{50}	2	25,000	NR	Nishiuchi and Yoshida (1972)
	Snail (*Indoplanorbis exustus*)	LC_{50}	2	6,600	NR	Nishiuchi and Yoshida (1972)
		LC_{50}	3 hr	6,600	NR	Hashimoto and Nishiuchi (1981)
Bird	Mallard duck	LD_{50}	14	15.90 mg kg^{-1}	240–720 D	Hudson et al. (1984)
	Ring-necked pheasant	LD_{50}	14	15 mg kg^{-1}	90–120 D	Hudson et al. (1984)
Invertebrate	Earthworm	LO POP	21	3.40 kg ha^{-1}		Tomlin and Gore (1974)

Table 45. Toxicology of Aquatic and Terrestrial Species: Chemical: MSMA, CAS Number: 6484522

Class	Species	Effect	Duration (days)	Concentration ($\mu g\ L^{-1}$)	Lifestage	Ref.
Crustacea	Cape spiney lobster	BEH	1–1.5 hr	18,000	ADULT	Brown (1974)
		BEH	1–1.5 hr	130,000	ADULT	Brown (1974)
	Isopod	LT_{50}	0.97 hr	130,000	LARGE ADULT	Brown (1974)
		BEH	1–1.5 hr	1,800	NR	Brown (1974)
		BEH	1–1.5 hr	18,000	NR	Brown (1974)
	Marine isopod (*Exosphaeroma truncatitelson*)	LT_{50}	5.5 hr	130,000	LARGE ADULT	Brown (1974)
		BEH	1–1.5 hr	18,000	NR	Brown (1974)
		BEH	1–1.5 hr	53,000	NR	Brown (1974)
	Marine isopod (*Pontogeloides latipes*)	BEH	1.5 hr	130,000	LARGE ADULT	Brown (1974)
	Mysid	LT_{50}	0.35 hr	130,000	LARGE ADULT	Brown (1974)
		BEH	1 hr	1,800	NR	Brown (1974)
		BEH	1 hr	18,000	NR	Brown (1974)
Echinoderm	Sea urchin	REP	0.17 hr	140	EGG	Greenwood and Brown (1974)
		REP	0.17 hr	1,400	EGG	Greenwood and Brown (1974)
Fish	Common, mirror, colored, carp	LC_{50}	2	74,000	FRY, ≥AGE 0, 33.6 G	Dabrowska and Sikora (1987)
		LC_{50}	2	79,000	FRY, ≥AGE 0, 32.6 G	Dabrowska and Sikora (1987)

Mullusca	Gastropod	BEH	7	4,400	ADULT	Brown and Currie (1973)
		BEH	7	8,750	ADULT	Brown and Currie (1973)
		MOR	7	52,500	ADULT	Brown and Currie (1973)
		OC	2	175,000	ADULT	Brown and Currie (1973)
		STR	7	52,500	ADULT	Brown and Currie (1973)
Nemertea	Nemertean	LT_{50}	1.77 hr	130,000	LARGE ADULT	Brown (1974)
		BEH	1–1.5 hr	18,000	NR	Brown (1974)
		BEH	1–1.5 hr	53,000	NR	Brown (1974)
Urochordat	Sea squirt	PHY	0.17 hr	70,000	NR	Brown (1974)

Table 46. Toxicology of Aquatic and Terrestrial Species: Chemical: MSMA + Surfactant, CAS Number: 2163806

Class	Species	Effect	Duration (days)	Concentration ($\mu g\ L^{-1}$)	Lifestage	Ref.
Algae	Algae, phytoplankton, algal mat	POP-D	NR	10 mg kg^{-1}	NR	Kosinski (1984)
		PSE	21	10 mg kg^{-1}	NR	Kosinski and Merkle (1984)
		PSE	28	0.01 mg kg^{-1}	NR	Kosinski and Merkle (1984)
	Flagellate euglenoid (*Euglena gracilis*)	CLR	6	97,000	LOGARITHMIC PHASE	Richardson et al. (1979)
		PGR	6	97,000	LOGARITHMIC PHASE	Richardson et al. (1979)
		PSE	1.67 hr	81,000	LOGARITHMIC PHASE	Richardson et al. (1979)
		PSE	1.67 hr	970,000	LOGARITHMIC PHASE	Richardson et al. (1979)
		PSE	4	97,000	LOGARITHMIC PHASE	Richardson et al. (1979)
	Green algae (*Chlorella pyrenoidosa*)	PGR	5.92	3,000,000	NR	Blythe et al. (1979)
Amphibia	Brown tree frog	MOR	4	400,000	ADULT	Johnson (1976)
	Couch's spadefoot toad	MOR	5	1,000,000	14–19 MM, JUVENILE	Judd (1977)
		MOR	7	1,000,000	14–19 MM, JUVENILE	Judd (1977)
		MOR	8	1,000,000	14–19 MM, JUVENILE	Judd (1977)
		MOR	NR	100,000	14–19 MM, JUVENILE	Judd (1977)
	Frog (*Limnodynastes tasmaniensis*)	MOR	4	520,000	ADULT	Johnson (1976)

Crustacea	Calanoid copepod (*Diaptomus* sp)	LC_{50}	2	37,600	NR	Naqvi et al. (1985)
	Cladoceran	LC_{50}	2	39,300	NR	Naqvi et al. (1985)
	Crayfish (*Procambarus* sp)	LC_{50}	2	5,100,000	NR	Anderson et al. (1975)
		LC_{50}	4	1,100,000	NR	Anderson et al. (1975)
		BCF	56	1.1	MIXED SEX AND SIZE	Abdelghani et al. (1980b)
		BCF	56	1.1	MIXED SEX AND SIZE	Abdelghani et al. (1980b)
		BCF	56	11	MIXED SEX AND SIZE	Abdelghani et al. (1980b)
		BCF	56	11	MIXED SEX AND SIZE	Abdelghani et al. (1980b)
		BCF	56	200	NR	Abdelghani et al. (1976)
		BCF	56	200	NR	Abdelghani et al. (1976)
		BCF	56	200	NR	Abdelghani et al. (1976)
		BCF	56	200	NR	Abdelghani et al. (1976)
		BCF	56	200	NR	Abdelghani et al. (1976)
		BCF	56	200	NR	Abdelghani et al. (1976)
		BCF	56	510	24.2–31 G	Abdelghani (1978)
		BCF	56	510	24.2–31 G	Abdelghani (1978)
		BCF	56	510	24.2–31 G	Abdelghani (1978)
		BCF	56	510	24.2–31 G	Abdelghani (1978)
		BCF	56	510	24.2–31 G	Abdelghani (1978)
		BCF	56	510	24.2–31 G	Abdelghani (1978)
		BCF	56	2,000	NR	Abdelghani et al. (1976)
		BCF	56	2,000	NR	Abdelghani et al. (1976)
		BCF	56	2,000	NR	Abdelghani et al. (1976)
		BCF	56	2,000	NR	Abdelghani et al. (1976)
		BCF	56	2,000	NR	Abdelghani et al. (1976)
		BCF	56	2,000	NR	Abdelghani et al. (1976)
		BCF	56	500–5,000	NR	Mason et al. (1976)
		BCF	56	500–5,000	NR	Mason et al. (1976)
		BCF	56	500–5,000	NR	Mason et al. (1976)
		BCF	56	500–5,000	NR	Mason et al. (1976)

Table 46. Toxicology of Aquatic and Terrestrial Species: Chemical: MSMA + Surfactant, CAS Number: 2163806 (continued)

Class	Species	Effect	Duration (days)	Concentration ($\mu g\ L^{-1}$)	Lifestage	Ref.
Crustacea (continued)	Crayfish (*Procambarus* sp)	BCF	56	500–5,000	NR	Mason et al. (1976)
		BCF	56	500–5,000	NR	Mason et al. (1976)
		BCF	56	5,100	24.2–31 G	Abdelghani (1978)
		BCF	56	5,100	24.2–31 G	Abdelghani (1978)
		BCF	56	5,100	24.2–31 G	Abdelghani (1978)
		BCF	56	5,100	24.2–31 G	Abdelghani (1978)
		BCF	56	5,100	24.2–31 G	Abdelghani (1978)
		BCF	56	5,100	24.2–31 G	Abdelghani (1978)
		MOR	56	5,100	24.2–31 G	Abdelghani (1978)
	Cyclopoid copepod (*Eucyclops* sp)	LC_{50}	2	96,300	NR	Naqvi et al. (1985)
	Ostracod (*Cypria* sp)	LC_{50}	2	98,200	NR	Naqvi et al. (1985)
	Red swamp crayfish	LC_{50}	4	101,000	JUVENILE, 3–3.4 CM, 1.1–1.5 G	Naqvi et al. (1987)
		LC_{50}	4	1,019,000	ADULT, 9–10 CM, 25–32 G	Naqvi et al. (1987)
	Scud (*Gammarus fasciatus*)	LC_{50}	4	>100,000	MATURE	Johnson and Finley (1980)
Fish	Bluegill	LC_{50}	4	1,200	0.9 G	Johnson and Finley (1980)
		LC_{50}	4	49,200	1 G	Johnson and Finley (1980)
	Channel catfish	LC_{50}	2	4,700,000	NR	Anderson et al. (1975)
		LC_{50}	4	26,800	2.1 G	Johnson and Finley (1980)
		LC_{50}	4	3,050,000	NR	Anderson et al. (1975)
		MOR	2	10,000	1 YEAR, 14 G, 12 CM, FINGERLINGS	Mccorkle et al. (1977)

	Cutthroat trout	LC_{50}	4	>100,000	0.6 G	Johnson and Finley (1980)
	Fathead minnow	LC_{50}	4	13,300	0.9 G	Johnson and Finley (1980)
	Goldfish	LC_{50}	4	31,100	0.9 G	Johnson and Finley (1980)
	Mosquitofish (*Gambusia affinis*)	LC_{50}	1	207,000	NR	Johnson (1978a)
		LC_{50}	2	136,000	NR	Johnson (1978a)
		LC_{50}	4	132,000	NR	Johnson (1978a)
	Sailfin molly	LC_{50}	2	1,600,000	ADULT	Abdelghani et al. (1980a)
		LC_{50}	4	1,300,000	ADULT	Abdelghani et al. (1980a)
	Smallmouth bass	LC_{50}	2	1,650,000	FINGERLING	Anderson et al. (1975)
		LC_{50}	4	900,000	FINGERLING	Anderson et al. (1975)
Mullusca	Marsh periwinkle	ABD	120	100	NR	Edwards and Davis (1975)
		ABD	120	1,000	NR	Edwards and Davis (1975)
		ABD	120	10,000	NR	Edwards and Davis (1975)
	Ribbed mussel (*Modiolus demissus*)	MOR	120	100	NR	Edwards and Davis (1975)
Tracheophy	Alligator-weed	BCF	42	0.024	NR	Anderson et al. (1980)
		BCF	NR	9,800	NR	Anderson et al. (1981)
	Aquatic plant	BCF	42	0.024	NR	Anderson et al. (1980)
		BCF	NR	9,800	NR	Anderson et al. (1981)
	Coon-tail	BCF	21	9,800	NR	Anderson et al. (1981)
	Duckweed (*Lemna minor*)	BCF	42	0.024	NR	Anderson et al. (1980)
	Smooth cordgrass	BMS	120	100	NR	Edwards and Davis (1975)
		BMS	120	100	NR	Edwards and Davis (1975)
		BMS	120	10,000	NR	Edwards and Davis (1975)
		MOR	120	90,000	NR	Edwards and Davis (1975)
		REP	120	1,000	NR	Edwards and Davis (1975)
	Water-hyacinth	BCF	42	0.024	NR	Anderson et al. (1980)
		BCF	NR	9,800	NR	Anderson et al. (1981)

Table 46. Toxicology of Aquatic and Terrestrial Species: Chemical: MSMA + Surfactant, CAS Number: 2163806 (continued)

Class	Species	Effect	Duration (days)	Concentration ($\mu g\ L^{-1}$)	Lifestage	Ref.
Amphibian	Spadefoot toad	LO MOR	8	1,000 ppm	YOUNG, ADULT	Judd (1977)
		NO MOR	8	100 ppm	YOUNG, ADULT	Judd (1977)
Mammal	Rat	LC_{50}		>20 mg L^{-1}		*Farm Chemicals Handbook* (1990)
		LD_{50}		700–1,738 mg kg^{-1}		*Farm Chemicals Handbook* (1990)
	White footed mouse	LD_{50}	7	300 mg kg^{-1}	ADULT	Judd (1979)
		LO AVO	21	3,000 ppm	ADULT	Judd (1979)
		LO BEH	14	477 ppm	ADULT	Lopez and Judd (1979)
		LO MOR	21	3,000 ppm	ADULT	Judd (1979)
		LO PHY	7	200 mg	ADULT	Judd (1979)
		LO PHY	21	3,000 ppm	ADULT	Judd (1979)
		LO PHY	30	1,000 ppm	ADULT	Judd (1979)
		NO AVO	21	1,000 ppm	ADULT	Judd (1979)
		NO MOR	21	1,000 ppm	ADULT	Judd (1979)
		NO PHY	7	150 *M*	ADULT	Judd (1979)
		NO PHY	21	1,000 ppm	ADULT	Judd (1979)
		NO PHY	60	1,000 ppm	ADULT	Judd (1979)

Table 47. Toxicology of Aquatic and Terrestrial Species: Chemical: Oxadiazon, CAS Number: 19666309

Class	Species	Effect	Duration (days)	Concentration ($\mu g\ L^{-1}$)	Lifestage	Ref.
Algae	Green algae (*Oedogonium cardiacum*)	BCF	48	2.7	NR	Ambrosi et al. (1978)
		BCF	48	26.6	1 G	Ambrosi et al. (1978)
Amphibia	Frog (*Bufo bufo japonicus*)	LC_{50}	1	2,800	25–35 D, 0.34 G, TADPOLE	Nishiuchi and Yoshida (1974)
		LC_{50}	2	2,500	25–35 D, 0.34 G, TADPOLE	Nishiuchi and Yoshida (1974
		LC_{50}	3	1,300	25–35 D, 0.34 G, TADPOLE	Nishiuchi and Yoshida (1974)
		LC_{50}	4	1,300	25–35 D, 0.34 G, TADPOLE	Nishiuchi and Yoshida (1974)
Arachnida	Water mite (*Hydracarina*)	ABD	106	0.8 kg ha^{-1}	NR	Ishibashi et al. (1983)
Crustacea	Crustacean class	ABD	105	0.8 kg ha^{-1}	NR	Ishibashi et al. (1983)
		ABD	106	0.8 kg ha^{-1}	NR	Ishibashi et al. (1983)
	Water flea (*Daphnia magna*)	BCF	48	2.7	NR	Ambrosi et al. (1978)
		BCF	48	26.6	NR	Ambrosi et al. (1978)
Fish	Common, mirror, colored, carp	LC_{50}	2	3,200	NR	Otsuka (1975)
	Mosquitofish (*Gambusia patruelis*)	LC50	2	3,189	3.28 (3–4) CM, 0.36 (0.31–0.51) G	Li and Chen (1981)
	Mosquitofish (*Gambusia affinis*)	BCF	48	26.6	NR	Ambrosi et al. (1978)
	Tilapia (*Tilapia* sp)	LC_{50}	2	1,459.8	4.99 (4.6–5.38) CM, 1.46 (1.23–2.04) G	Li and Chen (1981)

Table 47. Toxicology of Aquatic and Terrestrial Species: Chemical: Oxadiazon, CAS Number: 19666309 (continued)

Class	Species	Effect	Duration (days)	Concen-tration ($\mu g\ L^{-1}$)	Lifestage	Ref.
Mullusca	Ramshorn snail (*Helisoma* sp)	BCF	48	2.7	NR	Ambrosi et al. (1978)
		BCF	48	26.6	NR	Ambrosi et al. (1978)
	Snails, limpets class	ABD	105	0.8 kg ha^{-1}	NR	Ishibashi et al. (1983)
Tracheophy	Duckweed (*Lemna perpusilla*)	MOR	7	100,000	NR	Nishiuchi (1974)
		MOR	7	300,000	NR	Nishiuchi (1974)
		MOR	7	300,000	NR	Nishiuchi (1974)

Table 48. Toxicology of Aquatic and Terrestrial Species: Chemical: PCNB, CAS Number: 82688

Class	Species	Effect	Duration (days)	Concentration ($\mu g\ L^{-1}$)	Lifestage	Ref.
Algae	Algae, phytoplankton, algal mat	PSE	4 hr	100,000	NR	Somashekar and Sreenath (1984)
	Green algae (*Chlorella fusca*)	BCF	1	49	NR	Korte et al. (1978)
		BCF	1	49	NR	Korte et al. (1978)
Fish	Chinook salmon	MOR	1	10,000	5–10 CM	MacPhee and Ruelle (1969)
	Coho salmon, silver salmon	MOR	1	10,000	5–10 CM	MacPhee and Ruelle (1969)
	Common, mirror, colored, carp	LC_{50}	1	1,000	NR	Hashimoto and Nishiuchi (1981)
		LC_{50}	1	3,800	NR	Hashimoto and Nishiuchi (1981)
		LC_{50}	1	15,000	NR	Hashimoto and Nishiuchi (1981)
		LC_{50}	1	20,000	NR	Hashimoto and Nishiuchi (1981)
		LC_{50}	1	40,000	NR	Hashimoto and Nishiuchi (1981)
		MOR	1.75	140–325 mg kg^{-1}	NR	Loeb and Kelly (1963)
	Ide, silver or golden orfe	BCF	1	0.4	2–5 G	Korte et al. (1978)
		BCF	1	6	2–5 G	Korte et al. (1978)
		BCF	1	33	2–5 G	Korte et al. (1978)
		BCF	2	33	2–5 G	Korte et al. (1978)
		BCF	3	0.4	2–5 G	Korte et al. (1978)
		BCF	3	6	2–5 G	Korte et al. (1978)

Table 48. Toxicology of Aquatic and Terrestrial Species: Chemical: PCNB, CAS Number: 82688 (continued)

Class	Species	Effect	Duration (days)	Concentration ($\mu g\ L^{-1}$)	Lifestage	Ref.
	Motsuga, stone moroko	BCF	1–14	5–20	4–8 CM, 2–5 G	Kanazawa (1981)
	Northern squawfish	MOR	1	10,000	5–10 CM	MacPhee and Ruelle (1969)
	Pink salmon	LET	2	90,000	FRY, 1.3 G	Kuroda (1975)
	Trout family	LET	2	75,000–300,000	FRY, 2.8–5.4 CM, 0.25–0.56 G	Kuroda (1974)
		LET	2	200,000–250,000	FRY, 2.8–5.4 CM, 0.25–0.56 G	Kuroda (1974)
Protozoa	Ciliate (*Colpidium campylum*)	PGR	1.79	>10,000	>96 HR	Dive et al. (1980)
Mammal	Rat	LD_{50}		1,700 mg kg^{-1}		*Farm Chemicals Handbook* (1990)

Table 49. Toxicology of Aquatic and Terrestrial Species: Chemical: Pronamide, Propyzamide, CAS Number: 23950585

Class	Species	Effect	Duration (days)	Concentration ($\mu g\ L^{-1}$)	Lifestage	Ref.
Amphibia	Frog (*Bufo bufo japonicus*)	LC_{50}	1	>40,000	25–35 D, 0.34 G, TADPOLE	Nishiuchi and Yoshida (1974)
		LC_{50}	2	>40,000	25–35 D, 0.34 G, TADPOLE	Nishiuchi and Yoshida (1974)
		LC_{50}	3	>40,000	25–35 D, 0.34 G, TADPOLE	Nishiuchi and Yoshida (1974)
		LC_{50}	4	>40,000	25–35 D, 0.34 G, TADPOLE	Nishiuchi and Yoshida (1974)
Mammal	Dog	LD_{50}		10,000 mg kg^{-1}		*Farm Chemicals Handbook* (1990)
	Rat	LD_{50}		5,620–8,350 mg kg^{-1}		*Farm Chemicals Handbook* (1990)

Table 50. Toxicology of Aquatic and Terrestrial Species: Chemical: Simazine, CAS Number: 122349

Class	Species	Effect	Duration (days)	Concentration ($\mu g\ L^{-1}$)	Lifestage	Ref.
Algae	Algae, phytoplankton, algal mat	$EC_{50}CH$	18	100–1,000	NR	Goldsborough and Robinson (1983)
	Blue-green algae (*Anabaena* sp)	MOR	21	260	NR	Patnaik and Ramachandran (1976)
	Blue-green algae (*Anabaena variabilis*)	RSD	30	2,000	NR	Kruglor and Mikhaylova (1975)
	Blue-green algae (*Anacystis nidulans*)	CLR	0–3	0.00010	10 D	Mehta and Hawxby (1983)
	Blue-green algae (*Anacystis aeruginosa*)	MOR	21	260	NR	Patnaik and Ramachandran (1976)
	Blue-green algae (*Anacystis nidulans*)	PHY	3	2,020	10 D CULTURE	Mehta and Hawxby (1979)
	Blue-green algae (*Chloridella* sp)	RSD	30	2,000	NR	Kruglor and Mikhaylova (1975)
	Blue-green algae (*Phormidium* sp)	CEL	1–3 hr	3	NR	Noll and Bauer (1974)
	Blue-green algae (*Phormidium foveolarum*)	RSD	30	2,000	NR	Kruglor and Mikhaylova (1975)
	Diatom (*Phaeodactylum tricornutum*)	$EC_{50}OX$	1.5 hr	600	LOG PHASE	Walsh (1972)
		$EC_{50}OX$	1.5 hr	10,000	LOG PHASE	Walsh (1972)
		PGR	10	500	LOG PHASE	Walsh (1972)
		PGR	10	2,000	LOG PHASE	Walsh (1972)
	Diatom (*Synedra* sp)	MOR	21	260	NR	Patnaik and Ramachandran (1976)

Green algae (*Actinastrum* sp)	MOR	21	260	NR	Patnaik and Ramachandran (1976)
Green algae (*Ankistrodesmus falcatus*)	EC_{50}PP	NR	640	NR	Tscheu-Schluter (1976)
Green algae (*Ankistrodesmus braunii*)	EC_{50}PS	NR	4.7	NR	O'Neal and Lembi (1983)
Green algae	CLR	2.71	832	NR	Foy and Hiranpradit (1977)
(*Chlamydomonas reinhardtii*)	PGR	2.71	52	NR	Foy and Hiranpradit (1977)
	PGR	3.79	50	NR	Foy and Hiranpradit (1977)
Green algae (*Chlamydomonas eugametos*)	PGR	4	200,000	2ND DAY OF LOG PHASE	Vance and Smith (1969)
Green algae (*Chlorococcum* sp)	EC_{50}OX	1.5 hr	2,500	LOG PHASE	Walsh (1972)
	EC_{50}OX	1.5 hr	50,000	LOG PHASE	Walsh (1972)
Green algae (*Chlorella pyrenoidosa*)	CLR	4.6	52	NR	Foy and Hiranpradit (1977)
Green algae (*Chlorococcum hypnosporum*)	PGR	1	5,000	NR	Virmani et al. (1975)
Green algae	PGR	1	5,000	NR	Virmani et al. (1975)
(*Chlorella pyrenoidosa*)	PGR	4	200,000	2ND DAY OF LOG PHASE	Vance and Smith (1969)
	PGR	4.6	52	NR	Foy and Hiranpradit (1977)
	PGR	4.6	104	NR	Foy and Hiranpradit (1977)
Green algae (*Chlorococcum* sp)	PGR	10	2,000	LOG PHASE	Walsh (1972)
	PGR	10	2,500	LOG PHASE	Walsh (1972)
Green algae (*Chlorella vulgaris*)	RSD	30	2,000	NR	Kruglor and Mikhaylova (1975)
Green algae	EC_{50}PS	NR	3.8	NR	O'Neal and Lembi (1983)
(*Cladophora glomerata*)	PGR	45	5	FILAMENT CLUMPS	O'Neal and Lembi (1983)

Table 50. Toxicology of Aquatic and Terrestrial Species: Chemical: Simazine, CAS Number: 122349 (continued)

Class	Species	Effect	Duration (days)	Concentration (μg L^{-1})	Lifestage	Ref.
Algae (continued)	Green algae	EC_{50}OX	1.5 hr	4,000	LOG PHASE	Walsh (1972)
	(*Dunaliella tertiolecta*)	EC_{50}OX	1.5 hr	50,000	LOG PHASE	Walsh (1972)
		MOR	NR	10,000	NR	Portmann (1972)
		PGR	10	5,000	LOG PHASE	Walsh (1972)
		PGR	10	20,000	LOG PHASE	Walsh (1972)
		PGR	NR	1,000	NR	Portmann (1972)
	Green algae	EC_{50}PS	NR	3	NR	O'Neal and Lembi (1983)
	(*Pithophora oedogonia*)	PGR	45	5	FILAMENT CLUMPS	O'Neal and Lembi (1983)
	Green algae (*Scenedesmus* sp)	MOR	21	260	NR	Patnaik and Ramachandran (1976)
	Green algae	PGR	4	50	NR	Foy and Hiranpradit (1977)
	(*Scenedesmus quadricauda*)	PGR	4	200,000	2ND DAY OF LOG PHASE	Vance and Smith (1969)
	Green algae	EC_{50}BM	14–21	0.614	LOG PHASE	Turbak et al. (1986)
	(*Selenastrum capricornutum*)	EC_{50}OX	NR	2.24	LOG PHASE	Turbak et al. (1986)
	Green algae	EC_{50}PS	NR	1.1	NR	O'Neal and Lembi (1983)
	(*Spirogyra juergensii*)	PGR	45	5	FILAMENT CLUMPS	O'Neal and Lembi (1983)
	Haptophyte (*Isochrysis galbana*)	EC_{50}OX	1.5 hr	600	LOG PHASE	Walsh (1972)
		EC_{50}OX	1.5 hr	10,000	LOG PHASE	Walsh (1972)
		PGR	10	500	LOG PHASE	Walsh (1972)
		PGR	10	5,000	LOG PHASE	Walsh (1972)
Amphibia	Frog (*Bufo bufo japonicus*)	LC_{50}	2	>100,000	TADPOLE	Hashimoto and Nishiuchi (1981)

Annelida	Annelid worm class	LET	42	250,000	LARVAE	Walker (1964)
Aquatic invertebrate	Invertebrates	ABD	450	250–500	NR	Harman (1977)
		BCF	3	230	NYMPH AND LARVAE	Mauck et al. (1976)
		BCF	3	420	NYMPH AND LARVAE	Mauck et al. (1976)
		BCF	8	370	NYMPH AND LARVAE	Mauck et al. (1976)
		BCF	29	50	NYMPH AND LARVAE	Mauck et al. (1976)
		BCF	29	170	NYMPH AND LARVAE	Mauck et al. (1976)
		BCF	83	200	NYMPHS AND LARVAE	Mauck et al. (1976)
		BCF	86	20	NYMPH AND LARVAE	Mauck et al. (1976)
		BCF	86	50	NYMPH AND LARVAE	Mauck et al. (1976)
		BCF	110	70	NYMPHS AND LARVAE	Mauck et al. (1976)
		BCF	110	210	NYMPHS AND LARVAE	Mauck et al. (1976)
		BCF	114	20	NYMPH AND LARVAE	Mauck et al. (1976)
		BCF	114	110	NYMPH AND LARVAE	Mauck et al. (1976)
		BCF	138	200	NYMPH AND LARVAE	Mauck et al. (1976)
		BCF	142	40	NYMPH AND LARVAE	Mauck et al. (1976)
		BCF	176	80	NYMPH AND LARVAE	Mauck et al. (1976)
		BCF	197	40	NYMPH AND LARVAE	Mauck et al. (1976)
		BCF	367	30	NYMPH AND LARVAE	Mauck et al. (1976)
		BCF	456	10	NYMPHS AND LARVAE	Mauck et al. (1976)
		BCF	456	70	NYMPH AND LARVAE	Mauck et al. (1976)
		BMS	3	500	NR	Harman (1977)
		BMS	4	250	NR	Harman (1977)
		BMS	330	500–10,000	NR	Walker (1964)
		MOR	21	260	NR	Patnaik and Ramachandran (1976)

Table 50. Toxicology of Aquatic and Terrestrial Species: Chemical: Simazine, CAS Number: 122349 (continued)

Class	Species	Effect	Duration (days)	Concentration ($\mu g\ L^{-1}$)	Lifestage	Ref.
Crustacea	Aquatic sowbug (*Asellus brevicaudus*)	LC_{50}	2	>100,000	EARLY INSTAR	Sanders (1970a)
	Calanoid copepod (*Heliodiaptomus viduus*)	LT_{50}	0.46	50,000	NR	George et al. (1982)
		LT_{50}	0.54	10,000	NR	George et al. (1982)
		LT_{50}	1	1,000	NR	George et al. (1982)
		LT_{50}	5 hr	100,000	NR	George et al. (1982)
		LET	0.5	100,000	NR	George et al. (1982)
		LET	0.67	50,000	NR	George et al. (1982)
		LET	1.42	10,000	NR	George et al. (1982)
		LET	2.13	1,000	NR	George et al. (1982)
	Common shrimp	LC_{50}	2	>100,000	LARVAL	Portmann (1972)
	Crayfish (*Orconectes nais*)	LC_{50}	2	>100,000	EARLY INSTAR	Sanders (1970a)
	Grass shrimp (*Palaemonetes kadiakensis*)	LC_{50}	2	>100,000	EARLY INSTAR	Sanders (1970a)
	Ostracod (*Cypridopsis vidua*)	EC_{50}IM	2	3,200	EARLY INSTAR	Sanders (1970a)
		EC_{50}IM	2	3,700	MATURE	Johnson and Finley (1980)
	Scud (*Gammarus lacustris*)	LC_{50}	1	30,000	2 MONTHS OLD	Sanders (1969)
		LC_{50}	2	21,000	2 MONTHS OLD	Sanders (1969)
	Scud (*Gammarus fasciatus*)	LC_{50}	2	>100,000	EARLY INSTAR	Sanders (1970a)
	Scud (*Gammarus lacustris*)	LC_{50}	4	13,000	2 MONTHS OLD	Sanders (1969)
	Scud (*Gammarus fasciatus*)	LC_{50}	4	>100,000	MATURE	Johnson and Finley (1980)
	Water flea (*Daphnia magna*)	EC_{50}IM	1	>3,500	<24 HR	Marchini et al. (1988)
		EC_{50}IM	2	1,000	EARLY INSTAR	Sanders (1970a)
		EC_{50}IM	2	1,100	1ST INSTAR	Johnson and Finley (1980)
		EC_{50}IM	2	>3,500	<24 HR	Marchini et al. (1988)

	Water flea (*Daphnia pulex*)	LC_{50}	2	52,100	24 HR, EARLY INSTAR	Fitzmayer et al. (1982)
	Water flea (*Daphnia magna*)	LC_{50}	2	54,000	NR	Semov and Iosifov (1973)
	Water flea (*Daphnia pulex*)	LC_{50}	2	424,000	24 HR, EARLY INSTAR	Fitzmayer et al. (1982)
		LC_{50}	3 hr	>40,000	NR	Hashimoto and Nishiuchi (1981)
		LC_{50}	3 hr	>40,000	FEMALE ADULT	Nishiuchi and Hashimoto (1969)
		LC_{50}	3 hr	>40,000	NR	Nishiuchi and Hashimoto (1975)
	Water flea (*Moina macrocopa*)	LC_{50}	3 hr	>40,000	NR	Hashimoto and Nishiuchi (1981)
		LC_{50}	3 hr	>40,000	FEMALE ADULT	Nishiuchi and Hashimoto (1969)
		LC_{50}	3 hr	>40,000	NR	Nishiuchi and Hashimoto (1975)
Fish	Black bullhead	LC_{50}	2	30,000	2–10 CM, 0.5–14 G, 4–12 MONTHS	Bathe et al. (1975a)
		LC_{50}	4	55,000	2–10 CM, 0.5–14 G, 4–12 MONTHS	Bathe et al. (1975a)
	Bluegill	LC_{50}	1	130,000	1.04 G	Cope (1965)
		LC_{50}	2	118,000	1.04 G	Cope (1965)
		LC_{50}	2	>100,000	2–10 CM, 0.5–14 G	Bathe et al. (1973)
		LC_{50}	4	90,000	2–10 CM, 0.5–14 G	Bathe et al. (1973)
		LC_{50}	4	100,000	1 G	Johnson and Finley (1980)
		LC_{50}	4	118,000	1.04 G	Cope (1965)
		LC_{50}	4	8,100,000	1.2–2.8 G, 40–61 MM	Watkins et al. (1985)
		BCF	1	20	SUB ADULT	Mauck et al. (1976)
		BCF	1	100	SUB ADULT	Mauck et al. (1976)
		BCF	1	450	SUB ADULT	Mauck et al. (1976)
		BCF	1	870	SUB ADULT	Mauck et al. (1976)

Table 50. Toxicology of Aquatic and Terrestrial Species: Chemical: Simazine, CAS Number: 122349 (continued)

Class	Species	Effect	Duration (days)	Concentration ($\mu g\ L^{-1}$)	Lifestage	Ref.
Fish (continued)	Bluegill	BCF	3	50	SUB ADULT	Mauck et al. (1976)
		BCF	3	420	SUB ADULT	Mauck et al. (1976)
		BCF	3	580	SUB ADULT	Mauck et al. (1976)
		BCF	8	70	SUB ADULT	Mauck et al. (1976)
		BCF	8	180	SUB ADULT	Mauck et al. (1976)
		BCF	8	370	SUB ADULT	Mauck et al. (1976)
		BCF	8	650	SUB ADULT	Mauck et al. (1976)
		BCF	15	70	SUB ADULT	Mauck et al. (1976)
		BCF	15	180	SUB ADULT	Mauck et al. (1976)
		BCF	15	370	SUB ADULT	Mauck et al. (1976)
		BCF	15	650	SUB ADULT	Mauck et al. (1976)
		BCF	29	50	SUB ADULT	Mauck et al. (1976)
		BCF	29	170	SUB ADULT	Mauck et al. (1976)
		BCF	29	480	SUB ADULT	Mauck et al. (1976)
		BCF	29	860	SUB ADULT	Mauck et al. (1976)
		BCF	31	360	SUB ADULT	Mauck et al. (1976)
		BCF	31	1,400	SUB ADULT	Mauck et al. (1976)
		BCF	31	2,500	SUB ADULT	Mauck et al. (1976)
		BCF	72	230	SUB ADULT	Mauck et al. (1976)
		BCF	86	50	SUB ADULT	Mauck et al. (1976)
		BCF	86	320	SUB ADULT	Mauck et al. (1976)
		BCF	86	620	SUB ADULT	Mauck et al. (1976)
		BCF	110	1,000	SUB ADULT	Mauck et al. (1976)
		BCF	110	2,600	SUB ADULT	Mauck et al. (1976)
		BCF	114	300	SUB ADULT	Mauck et al. (1976)

	BCF	114	500	SUB ADULT	Mauck et al. (1976)
	BCF	142	40	SUB ADULT	Mauck et al. (1976)
	BCF	142	320	SUB ADULT	Mauck et al. (1976)
	BCF	142	1,000	SUB ADULT	Mauck et al. (1976)
	BCF	143	820	SUB ADULT	Mauck et al. (1976)
	BCF	143	2,000	SUB ADULT	Mauck et al. (1976)
	BCF	176	80	SUB ADULT	Mauck et al. (1976)
	BCF	176	300	SUB ADULT	Mauck et al. (1976)
	BCF	176	1,200	SUB ADULT	Mauck et al. (1976)
	BCF	197	40	SUB ADULT	Mauck et al. (1976)
	BCF	197	190	SUB ADULT	Mauck et al. (1976)
	BCF	197	450	SUB ADULT	Mauck et al. (1976)
	BCF	367	30	SUB ADULT	Mauck et al. (1976)
	BCF	367	90	SUB ADULT	Mauck et al. (1976)
	BCF	367	140	SUB ADULT	Mauck et al. (1976)
	BCF	367	<10	SUB ADULT	Mauck et al. (1976)
	BCF	456	50	SUB ADULT	Mauck et al. (1976)
	BCF	456	420	SUB ADULT	Mauck et al. (1976)
	MOR	4	100,000	FRY, 19.7 MM	Jones (1962)
	MOR	8	10,000	FRY	Hiltibran (1967)
	MOR	12	10,000	SMALL	Hiltibran (1967)
	MOR	12	30,000	SMALL	Hiltibran (1967)
	RSD-D	2	2,000	NR	Funderburk (1963)
Bullhead, catfish	LC_{50}	2	80,000	2–10 CM, 0.5–14 G	Bathe et al. (1973)
	LC_{50}	4	65,000	2–10 CM, 0.5–14 G	Bathe et al. (1973)
Carp, hawk fish	LET	0.75 hr	100,000	4–5 CM, FINGERLING	Singh and Yadav (1978)
	MOR	21	750	NR	Patnaik and Ramachandran (1976)
	MOR	24	2,500	4–5 CM, FINGERLING	Singh and Yadav (1978)
	MOR	24	5,000	4–5 CM, FINGERLING	Singh and Yadav (1978)
	MOR	24	7,500	4–5 CM, FINGERLING	Singh and Yadav (1978)

Table 50. Toxicology of Aquatic and Terrestrial Species: Chemical: Simazine, CAS Number: 122349 (continued)

Class	Species	Effect	Duration (days)	Concentration ($\mu g\ L^{-1}$)	Lifestage	Ref.
Fish (continued)	Catla	MOR	21	750	NR	Patnaik and Ramachandran (1976)
	Channel catfish	MOR	4	100,000	FRY, 15–18 MM	Jones (1962)
	Chinook salmon	LC_{50}	1	7,000	FINGERLING, 58–96 MM	Bond et al. (1959)
		LC_{50}	2	6,600	FINGERLING, 58–96 MM	Bond et al. (1959)
	Common, mirror, colored, carp	LC_{50}	2	>40,000	NR	Hashimoto and Nishiuchi (1981)
		LC_{50}	2	>40,000	4.5 CM, 1.1 G	Nishiuchi and Hashimoto (1969)
		LC_{50}	2	>40,000	NR	Nishiuchi and Hashimoto (1975)
		MOR	1.75	161–202	NR	Loeb and Kelly (1963)
	Crucian carp	LC_{50}	2	>100,000	2–10 CM, 0.5–14 G	Bathe et al. (1973)
		LC_{50}	2	>100,000	2–10 CM, 0.5–14 G, 4–12 MONTHS	Bathe et al. (1975a)
		LC_{50}	4	>100,000	2–10 CM, 0.5–14 G	Bathe et al. (1973)
		LC_{50}	4	>100,000	2–10 CM, 0.5–14 G, 4–12 MONTHS	Bathe et al. (1975a)
	Danio	EC_{50}IM	4	12,600	3.2 G	Rao and Dad (1979)
	Fathead minnow	LC_{50}	4	>100,000	0.7 G	Johnson and Finley (1980)

Goldfish	LC_{50}	2	>40,000	NR	Hashimoto and Nishiuchi (1981)
	LC_{50}	2	>40,000	4.01 CM, 1.04 G	Nishiuchi and Hashimoto (1969)
	LC_{50}	2	>40,000	NR	Nishiuchi and Hashimoto (1975)
Green sunfish	BCF	1	660	34 G	Rodgers (1970)
	BCF	1	1,000	34 G	Rodgers (1970)
	BCF	1	1,000	34 G	Rodgers (1970)
	BCF	1	1,030	34 G	Rodgers (1970)
	BCF	1	2,200	34 G	Rodgers (1970)
	BCF	1	2,200	34 G	Rodgers (1970)
	BCF	3	700	34 G	Rodgers (1970)
	BCF	3	700	34 G	Rodgers (1970)
	BCF	3	740	34 G	Rodgers (1970)
	BCF	3	1,380	34 G	Rodgers (1970)
	BCF	3	1,900	34 G	Rodgers (1970)
	BCF	3	1,900	34 G	Rodgers (1970)
	BCF	7	800	34 G	Rodgers (1970)
	BCF	7	800	34 G	Rodgers (1970)
	BCF	7	980	34 G	Rodgers (1970)
	BCF	7	1,800	34 G	Rodgers (1970)
	BCF	7	2,200	34 G	Rodgers (1970)
	BCF	7	2,200	34 G	Rodgers (1970)
	BCF	10	650	34 G	Rodgers (1970)
	BCF	10	2,240	34 G	Rodgers (1970)
	BCF	14	700	34 G	Rodgers (1970)
	BCF	14	700	34 G	Rodgers (1970)
	BCF	14	860	34 G	Rodgers (1970)
	BCF	14	2,100	34 G	Rodgers (1970)
	BCF	14	2,100	34 G	Rodgers (1970)

Table 50. Toxicology of Aquatic and Terrestrial Species: Chemical: Simazine, CAS Number: 122349 (continued)

Class	Species	Effect	Duration (days)	Concentration (µg L^{-1})	Lifestage	Ref.
Fish (continued)	Green sunfish	BCF	14	2,160	34 G	Rodgers (1970)
		BCF	21	24	34 G	Rodgers (1970)
		BCF	21	800	34 G	Rodgers (1970)
		BCF	21	800	34 G	Rodgers (1970)
		BCF	21	810	34 G	Rodgers (1970)
		BCF	21	1,900	34 G	Rodgers (1970)
		BCF	21	1,900	34 G	Rodgers (1970)
		HAT	8	10,000	FERTILIZED EGGS	Hiltibran (1967)
		HAT	8	10,000	FERTILIZED EGGS	Hiltibran (1967)
		LET	5	10,000	FRY	Hiltibran (1967)
		LET	7	10,000	FRY	Hiltibran (1967)
	Guppy	LC_{50}	2	3,900	NR	Tscheu-Schluter (1976)
		LC_{50}	2	>49,000	2–10 CM, 0.5–14 G	Bathe et al. (1973)
		LC_{50}	2	>49,000	2–10 CM, 0.5–14 G, 4–12 MONTHS	Bathe et al. (1975a)
		LC_{50}	3	3,000	NR	Tscheu-Schluter (1976)
		LC_{50}	4	49,000	2–10 CM, 0.5–14 G	Bathe et al. (1973)
		LC_{50}	4	49,000	2–10 CM, 0.5–14 G, 4–12 MONTHS	Bathe et al. (1975a)
	Indian catfish	LC_{100}	19	10,000	NR	Upadhyaya and Rao (1980)
		LC_{50}	15	10,000	NR	Upadhyaya and Rao (1980)
	Lake chubsucker	MOR	8	10,000	FRY	Hiltibran (1967)
		MOR	8	10,000	FRY	Hiltibran (1967)
	Largemouth bass	MOR	3.88	2,000	FRY, 10–15 MM	Jones (1962)

Medaka, high-eyes	LC_{50}	2	>10,000	NR	Hashimoto and Nishiuchi (1981)
	LC_{50}	2	>10,000	2.54 CM, 0.16 G	Nishiuchi and Hashimoto (1969)
	LC_{50}	2	>40,000	NR	Nishiuchi and Hashimoto (1975)
Mosquitofish (*Gambusia affinis*)	MOR	2	1	2.5 CM	Darwazeh and Mulla (1974)
	MOR	2	5	2.5 CM	Darwazeh and Mulla (1974)
Mozambique tilapia	EC_{50}IM	4	3,100	5 G	Rao and Dad (1979)
	LC_{100}	8	10,000	NR	Upadhyaya and Rao (1980)
	LC_{50}	11	5,000	NR	Upadhyaya and Rao (1980)
	LC_{50}	26	1,000	NR	Upadhyaya and Rao (1980)
	LD_{100}	16	5,000	NR	Upadhyaya and Rao (1980)
Muskellunge	MOR	4	1,500	FRY, 4 CM	Shaw and Hopke (1975)
Oriental weatherfish	LC_{50}	2	>40,000	NR	Hashimoto and Nishiuchi (1981)
Perch (*Perca* sp)	LC_{50}	2	>100,000	2–10 CM, 0.5–14 G, 4–12 MONTHS	Bathe et al. (1975a)
	LC_{50}	4	90	2–10 CM, 0.5–14 G, 4–12 MONTHS	Bathe et al. (1975a)
Rainbow trout, donaldson trout	LC_{50}	1	68,000	1.20 G	Cope (1965)
	LC_{50}	1	95,000	UNDER-YEARLING	Alabaster (1969)
	LC_{50}	1	2,200,000	UNDER-YEARLING	Alabaster (1969)
	LC_{50}	2	60,000	1.20 G	Cope (1965)
	LC_{50}	2	85,000	UNDER-YEARLING	Alabaster (1969)
	LC_{50}	2	>100,000	2–10 CM, 0.5–14 G	Bathe et al. (1973)
	LC_{50}	2	>100,000	2–10 CM, 0.5–14 G, 4–12 MONTHS	Bathe et al. (1975a)
	LC_{50}	4	56,000	1.20 G	Cope (1965)
	LC_{50}	4	>100,000	2–10 CM, 0.5–14 G	Bathe et al. (1973)

Table 50. Toxicology of Aquatic and Terrestrial Species: Chemical: Simazine, CAS Number: 122349 (continued)

Class	Species	Effect	Duration (days)	Concen-tration ($\mu g\ L^{-1}$)	Lifestage	Ref.
Fish (continued)	Rainbow trout, donaldson trout	LC_{50}	4	>100,000	2–10 CM, 0.5–14 G, 4–12 MONTHS	Bathe et al. (1975a)
		LC_{50}	4	>100,000	1.2 G	Johnson and Finley (1980)
		BCF	1	11,600	1 YEAR	Dodson and Mayfield (1979b)
		BCF	1	11,800	1 YEAR	Dodson and Mayfield (1979b)
		BEH	1	1,000	1 YEAR	Dodson and Mayfield (1979b)
		BEH	1	1,000	1 YEAR	Dodson and Mayfield (1979b)
		BEH	1	12,500	1 YEAR	Dodson and Mayfield (1979b)
		HIS	28	2,500	25–40 G	Bathe et al. (1975a)
		LOC	1	1,000	1 YEAR	Dodson and Mayfield (1979b)
		LOC	1	12,500	1 YEAR	Dodson and Mayfield (1979b)
		LOC	1	12,500	1 YEAR	Dodson and Mayfield (1979b)
		RSD	1	1,000–4,000	1 YEAR	Dodson and Mayfield (1979b)
		RSD	1	1,000–4,000	1 YEAR	Dodson and Mayfield (1979b)
	Rohu	MOR	21	750	NR	Patnaik and Ramachandran (1976)
	Smallmouth bass	LET	3	10,000	FRY	Hiltibran (1967)
		MOR	8	10,000	FRY	Hiltibran (1967)

Striped bass	LC_{50}	1	600	60 MM, 2.7 G	Wellborn (1969)
	LC_{50}	2	440	60 MM, 2.7 G	Wellborn (1969)
	LC_{50}	2	15,900	3 D, LARVAE	Fitzmayer et al. (1982)
	LC_{50}	2	18,400	3 D, LARVAE	Fitzmayer et al. (1982)
	LC_{50}	2	127,000	7 D, LARVAE	Fitzmayer et al. (1982)
	LC_{50}	2	157,200	7 D, LARVAE	Fitzmayer et al. (1982)
	LC_{50}	4	250	60 MM, 2.7 G	Wellborn (1969)
	LC_{50}	4	>180,000	0.71 G, 4.16 CM, FINGERLING	McCann and Hitch (1980)
	LC_{50}	4	>180,000	0.71 G, 4.16 CM, FINGERLING	McCann and Hitch (1980)
Sunfish family	LC_{50}	4	[illegible]4,300	NR	Walker (1964)
	LC_{50}	4	[illegible]5,000	NR	Walker (1964)
	LC_{50}	4	56,000	NR	Walker (1964)
	LC_{50}	4	[illegible]6,000	NR	Walker (1964)
	LC_{50}	4	6[illegible]5,000	NR	Walker (1964)
Tilapia (*Tilapia nilotica*)	GRO	42	1,000	SWIM-UP FRY, <12 MM	McGinty (1984)
	MOR	126	1,000	SWIM-UP FRY, <12 MM	McGinty (1984)
Two spotted, tic tac toe barb	EC_{50}IM	4	24,500	2.9 G	Rao and Dad (1979)
	LC_{100}	8	10,000	NR	Upadhyaya and Rao (1980)
	LC_{100}	20	5,000	NR	Upadhyaya and Rao (1980)
	LC_{50}	16	5,000	NR	Upadhyaya and Rao (1980)
	LC_{50}	30	1,000	NR	Upadhyaya and Rao (1980)

Table 50. Toxicology of Aquatic and Terrestrial Species: Chemical: Simazine, CAS Number: 122349 (continued)

Class	Species	Effect	Duration (days)	Concen-tration (μg L^{-1})	Lifestage	Ref.
Insect	Mayfly (*Cloeon dipterum*)	LC_{50}	2	>40,000	LARVAE	Hashimoto and Nishiuchi (1981)
	Mayfly (*Hexagenia* sp)	BCF	3	580	NR	Mauck et al. (1976)
		BCF	8	650	NR	Mauck et al. (1976)
		BCF	15	650	NR	Mauck et al. (1976)
		BCF	29	480	NR	Mauck et al. (1976)
		BCF	29	860	NR	Mauck et al. (1976)
		BCF	40	1,500	NR	Mauck et al. (1976)
		BCF	40	2,200	NR	Mauck et al. (1976)
		BCF	55	1,500	NR	Mauck et al. (1976)
		BCF	55	3,300	NR	Mauck et al. (1976)
		BCF	83	940	NR	Mauck et al. (1976)
		BCF	83	1,200	NR	Mauck et al. (1976)
		BCF	86	320	NR	Mauck et al. (1976)
		BCF	86	620	NR	Mauck et al. (1976)
		BCF	110	1,000	NR	Mauck et al. (1976)
		BCF	110	2,600	NR	Mauck et al. (1976)
		BCF	114	500	NR	Mauck et al. (1976)
		BCF	123	650	NR	Mauck et al. (1976)
		BCF	123	1,700	NR	Mauck et al. (1976)
		BCF	138	820	NR	Mauck et al. (1976)
		BCF	138	2,200	NR	Mauck et al. (1976)
		BCF	142	320	NR	Mauck et al. (1976)
		BCF	142	1,000	NR	Mauck et al. (1976)
		BCF	176	300	NR	Mauck et al. (1976)

		BCF	176	1,200	NR	Mauck et al. (1976)
		BCF	193	1,200	NR	Mauck et al. (1976)
		BCF	193	2,500	NR	Mauck et al. (1976)
		BCF	197	190	NR	Mauck et al. (1976)
		BCF	197	450	NR	Mauck et al. (1976)
		BCF	367	90	NR	Mauck et al. (1976)
		BCF	456	420	NR	Mauck et al. (1976)
		BCF	456	500	NR	Mauck et al. (1976)
	Midge family	LC_{50}	42	28,000	LARVAE	Walker (1964)
	Mosquito (*Aedes aegypti*)	MOR	1	10,000	LARVAE, 3RD INSTAR	Lichtenstein et al. (1973)
	Stonefly (*Pteronarcys californica*)	LC_{50}	4	1,900	2ND YEAR CLASS	Johnson and Finley (1980)
Mullusca	Banded mystery snail	LET	NR	500	IMMATURE	Harman (1977)
		MOR	>4	500	NR	Harman et al. (1978)
	Bladder snail	LC_{50}	2	>100,000	NR	Hashimoto and Nishiuchi (1981)
	Cockle (*Cerastoderma edule*)	LC_{50}	2	>100,000	LARVAL	Portmann (1972)
	Great pond snail	EC_{50}	20	<0.000001	EGG MASS <24 HR	Kosanke et al. (1988)
	Marsh snail	LC_{50}	2	>100,000	NR	Hashimoto and Nishiuchi (1981)
	River snail	BMS	3	500	NR	Harman (1977)
		MOR	>4	500	NR	Harman et al. (1978)
	Snail (*Indoplanorbis exustus*)	LC_{50}	2	>100,000	NR	Hashimoto and Nishiuchi (1981)
Protozoa	Amoeba	PGR	6	4,000	6 D CULTURE	Prescott et al. (1977)

Table 50. Toxicology of Aquatic and Terrestrial Species: Chemical: Simazine, CAS Number: 122349 (continued)

Class	Species	Effect	Duration (days)	Concentration ($\mu g\ L^{-1}$)	Lifestage	Ref.
Tracheophy	Creeping bentgrass	MOR	NR	500	NR	Hiltibran and Turgeon (1977)
	Duckweed (*Lemna perpusilla*)	MOR	7	30,000	NR	Nishiuchi (1974)
	Duckweed (*Lemna minor*)	PHY	2	20	FROND	O'Brien and Prendeville (1979)
		PHY	3	20	FROND	O'Brien and Prendeville (1979)
	Vascular plants	MOR	>60	45–134	NR	Walker (1964)
		MOR	>60	10,000	NR	Walker (1964)
		MOR	>60	11.2–22.4	NR	Walker (1964)
		MOR	>60	120,000	NR	Walker (1964)
Bird	Bobwhite quail	LC_{50}	8	8,800		Ciba-Geigy Corp. (1988)
	Mallard duck	LC_{50}	8	51,200		Ciba-Geigy Corp. (1988)
Mammal	Rat	LD_{50}		>4,000		Ciba-Geigy Corp. (1988)

Table 51. Toxicology of Aquatic and Terrestrial Species: Chemical: Thiophanate-methyl, CAS Number: 23564058

Class	Species	Effect	Duration (days)	Concentration (µg L^{-1})	Lifestage	Ref.
Algae	Green algae (*Chlorella pyrenoidosa*)	EC_{50}GR	2	8,500	CULTURE IN LOGPHASE	Canton (1976)
Crustacea	Water flea (*Daphnia magna*)	LC_{50}	2	16,000	<1 D	Canton (1976)
Fish	Common, mirror, colored, carp	LC_{50}	3	>75,000	FRY	Hashimoto et al. (1972)
		LC_{50}	3	>75,000	FRY	Hashimoto et al. (1972)
	Guppy	LC_{50}	2	30,000	3 WEEKS	Canton (1976)
	Rainbow trout, donaldson trout	LC_{50}	1	3,200	FRY	Hashimoto et al. (1972)
		LC_{50}	1	15,700	FRY	Hashimoto et al. (1972)
		LC_{50}	2	7,800	3 MONTHS	Canton (1976)
		LC_{50}	2	8,800	FRY	Hashimoto et al. (1972)
		LC_{50}	2	14,600	FRY	Hashimoto et al. (1972)
		LC_{50}	3	9,400	FRY	Hashimoto et al. (1972)
		LC_{50}	3	13,400	FRY	Hashimoto et al. (1972)
Bird	Japanese quail	NO SIGNS	14	5 g kg^{-1}		Hashimoto et al. (1972)
Invertebrate	Earthworm	LO AVO	2	0.87 µg c^{-1}		Stringer and Wright (1973)
		LO AVO	NR	0.80 µg c^{-1}	ADULT	Wright (1977)
		LO MOR	NR	1.55 kg ha^{-1}	ADULT	Wright (1977)
		LO POP	NR	0.78 kg ha^{-1}		Wright (1977)
		LO POP	NR	0.78 kg ha^{-1}		Wright (1977)
		LO POP	NR	0.78 kg ha^{-1}		Wright (1977)
		LO POP	NR	0.78 kg ha^{-1}		Wright (1977)
		LO POP	NR	0.78 kg ha^{-1}	YOUNG	Wright (1977)

Table 51. Toxicology of Aquatic and Terrestrial Species: Chemical: Thiophanate-methyl, CAS Number: 23564058 (continued)

Class	Species	Effect	Duration (days)	Concentration (μg L^{-1})	Lifestage	Ref.
		LO POP	NR	0.78 kg ha^{-1}		Wright (1977)
		LO POP	NR	0.78 kg ha^{-1}		Wright (1977)
		LO POP	NR	0.78 kg ha^{-1}		Wright (1977)
		LO POP	NR	0.78 kg ha^{-1}		Wright (1977)
		LO POP	NR	0.78 kg ha^{-1}		Wright (1977)
		LO POP	NR	0.78 kg ha^{-1}	YOUNG	Wright (1977)
		LO POP	NR	0.78 kg ha^{-1}		Stringer and Wright (1973)
		NO AVO	2	0.43 μg c^{-1}		Stringer and Wright (1973)
		NO AVO	2	7 μg c^{-1}		Stringer and Wright (1973)
Mammal	Dog	LD_{50}	14	4,000 mg kg^{-1}		Hashimoto et al. (1972)
		NO SIGNS	14	10 g kg^{-1}		Hashimoto et al. (1972)
	Guinea pig	LD_{50}	14	3,640 mg kg^{-1}		Hashimoto et al. (1972)
		NO SIGNS	14	10 g kg^{-1}		Hashimoto et al. (1972)
		NO SIGNS	NR	1%	ADULT	Hashimoto et al. (1972)
		NO SIGNS	NR	2%	ADULT	Hashimoto et al. (1972)
		NO SIGNS	NR	4%	ADULT	Hashimoto et al. (1972)
	Mouse (*Mus* sp)	LD_{50}	14	790 mg kg^{-1}		Hashimoto et al. (1972)
		LD_{50}	14	1,110 mg kg^{-1}		Hashimoto et al. (1972)
		LD_{50}	14	3,400 mg kg^{-1}		Hashimoto et al. (1972)
		LD_{50}	14	3,510 mg kg^{-1}		Hashimoto et al. (1972)
		NO SIGNS	14	10 g kg^{-1}		Hashimoto et al. (1972)

Rabbit	LD_{50}	14	2,270 mg kg^{-1}		Hashimoto et al. (1972)
	LD_{50}	14	2,500 mg kg^{-1}		Hashimoto et al. (1972)
	LO SIGNS	22	10%	ADULT	Hashimoto et al. (1972)
	NO SIGNS	14	10 g kg^{-1}		Hashimoto et al. (1972)
	NO SIGNS	22	0.10%	ADULT	Hashimoto et al. (1972)
Rat	LD_{50}	14	1,140 mg kg^{-1}		Hashimoto et al. (1972)
	LD_{50}	14	1,640 mg kg^{-1}		Hashimoto et al. (1972)
	LD_{50}	14	6,640 mg kg^{-1}		Hashimoto et al. (1972)
	LD_{50}	14	7,500 mg kg^{-1}		Hashimoto et al. (1972)
	NO SIGNS	14	10 g kg^{-1}		Hashimoto et al. (1972)

Table 52. Toxicology of Aquatic and Terrestrial Species: Chemical: Thiram, CAS Number: 137268

Class	Species	Effect	Duration (days)	Concentration ($\mu g\ L^{-1}$)	Lifestage	Ref.
Algae	Green algae (*Chlorella vulgaris*)	EC_{50}GR	3	5,500	EXPO GRO, 15E6 CELLS ML^{-1}	Jouany et al. (1985)
	Green algae	EC_{50}GR	4	1,000	10E+8 CELLS 100 ML^{-1}	Van Leeuwen et al. (1985a)
	(*Chlorella pyrenoidosa*)	EC_{50}GR	4 hr	1,000	5E+8 CELLS L^{-1}	Van Leeuwen et al. (1985a)
	Green algae	GRO	5	500	NR	Krishnakumari (1977)
	(*Scenedesmus obliquus*)	LET	3	10,000	NR	Krishnakumari (1977)
Amphibia	Clawed toad	LC_{50}	1	17	STAGE 47	Seuge et al. (1983)
		LC_{50}	1	25	STAGE 53	Seuge et al. (1983)
		LC_{50}	2	14	STAGE 47	Seuge et al. (1983)
		LC_{50}	2	22	STAGE 53	Seuge et al. (1983)
		LC_{50}	3	13	STAGE 47	Seuge et al. (1983)
		LC_{50}	3	21	STAGE 53	Seuge et al. (1983)
		LC_{50}	4	13	STAGE 47	Seuge et al. (1983)
		LC_{50}	4	21	STAGE 53	Seuge et al. (1983)
Annelida	Tubificid worm (*Tubifex tubifex*)	LC_{50}	2	670	NR	Voronkin and Loshakov (1973)
Crustacea	Aquatic sowbug	LC_{50}	1	1,882,000	NR	Seuge et al. (1983)
	(*Asellus aquaticus*)	LC_{50}	2	688,000	NR	Seuge et al. (1983)
		LC_{50}	3	161,000	NR	Seuge et al. (1983)
		LC_{50}	4	61,000	NR	Seuge et al. (1983)

Red swamp crayfish	LC_{50}	4	4,300	0.4 G, 25–35 MM, JUVENILE	Cheah et al. (1980)
	LC_{50}	4	4,300	0.4 G, 25–35 MM, JUVENILE	Cheah (1980b)
	LC_{50}	4	4,300	25–35 MM, IMMATURE	Cheah et al. (1978a)
Scud (*Gammarus pulex*)	LC_{50}	1	1,370	NR	Bluzat et al. (1982a)
	LC_{50}	1	3,260	NR	Bluzat et al. (1982a)
	LC_{50}	1	4,770	NR	Bluzat et al. (1982a)
	LC_{50}	1	13,990	NR	Bluzat et al. (1982a)
	LC_{50}	2	360	NR	Bluzat et al. (1982a)
	LC_{50}	2	480	NR	Bluzat et al. (1982a)
	LC_{50}	2	870	NR	Bluzat et al. (1982a)
	LC_{50}	2	1,210	NR	Bluzat et al. (1982a)
	LC_{50}	3	130	NR	Bluzat et al. (1982a)
	LC_{50}	3	200	NR	Bluzat et al. (1982a)
	LC_{50}	3	400	NR	Bluzat et al. (1982a)
	LC_{50}	3	410	NR	Bluzat et al. (1982a)
	LC_{50}	4	60	NR	Bluzat et al. (1982a)
	LC_{50}	4	130	NR	Bluzat et al. (1982a)
	LC_{50}	4	195	NR	Bluzat et al. (1982a)
	LC_{50}	4	225	NR	Bluzat et al. (1982a)
	LET	1 hr	50,000	NR	Bluzat (1982b)
	MOR	1	700	NR	Bluzat (1982b)
	MOR	1 hr	1,000	NR	Bluzat (1982b)
	MOR	1 hr	2,000	NR	Bluzat (1982b)
	MOR	1 hr	5,000	NR	Bluzat (1982b)
	MOR	1 hr	10,000	NR	Bluzat (1982b)
	MOR	2	700	NR	Bluzat (1982b)
	MOR	3	700	NR	Bluzat (1982b)
	MOR	4	700	NR	Bluzat (1982b)

Table 52. Toxicology of Aquatic and Terrestrial Species: Chemical: Thiram, CAS Number: 137268 (continued)

Class	Species	Effect	Duration (days)	Concentration ($\mu g\ L^{-1}$)	Lifestage	Ref.
Fish (continued)	Water flea (*Daphnia magna*)	LC_{50}	1.0833	1,300	1ST INSTAR, <24 HR	Frear and Boyd (1967)
		LC_{50}	2	210	NR	Van Leeuwen et al. (1985a)
		LC_{50}	21	8	<24 HR	Van Leeuwen et al. (1985a)
		LET	1	1,000	7 D	Jouany et al. (1985)
		MOR	1	100	7 D	Jouany et al. (1985)
Fish	Catfish	LC_{50}	1	1.2	80–100 MM, 6–10 G	Verma et al. (1981c)
		LC_{50}	2	0.9	80–100 MM, 6–10 G	Verma et al. (1981c)
		LC_{50}	3	0.8	80–100 MM, 6–10 G	Verma et al. (1981c)
		LC_{50}	4	0.67	80–100 MM, 6–10 G	Verma et al. (1980)
		LC_{50}	4	0.67	80–100 MM, 6–10 G	Verma et al. (1981c)
		LC_{50}	4	0.67	8–12 G, 90–110 MM, STD LENGTH	Verma et al. (1983)
		BIO	30	0.168	8–12 G, 90–110 MM, STD LENGTH	Verma et al. (1983)
		ENZ	30	0.045	80–110 MM, 8–12 G	Verma et al. (1981d)
		ENZ	30	0.067	80–110 MM, 8–12 G	Verma et al. (1981d)
		HEM	30	0.13	80–110 MM, 8–12 G	Verma et al. (1982b)
		HEM	30	0.168	8–12 G, 90–110 MM, STD LENGTH	Verma et al. (1983)
	Common, mirror, colored, carp	LC_{50}	4	0.3	LARVAE, 8 MM	Verma et al. (1981a)
		MOR	1	1,000	30 G FRESH WT PER SET	Jouany et al. (1985)
		MOR	2.875	121 $mg\ kg^{-1}$	NR	Loeb and Kelly (1963)
		STR	2.875	93 $mg\ kg^{-1}$	NR	Loeb and Kelly (1963)

Guppy	LC_{50}	4	270	NR	Van Leeuwen et al. (1985a)
Indian catfish	LC_{50}	1	8.1	50–75 MM, 5–10 G	Verma et al. (1982a)
	LC_{50}	2	7.7	50–75 MM, 5–10 G	Verma et al. (1982a)
	LC_{50}	3	7.2	50–75 MM, 5–10 G	Verma et al. (1982a)
	LC_{50}	4	6.3	28–35 G, 160–200 MM, STD LENGTH	Verma et al. (1983)
	BIO	30	1.58	28–35 G, 160–200 MM, STD LENGTH	Verma et al. (1983)
	HEM	30	1.58	28–35 G, 160–200 MM, STD LENGTH	Verma et al. (1983)
	HEM	30	2	19–22 CM, 50–70 G	Verma et al. (1981a)
Rainbow trout, donaldson trout	EC_{50}	60	0.64	FERTILIZED EGG — EARLY FRY STAGE	Van Leeuwen et al. (1986a)
	LC_{50}	1	260	34 G, 15.3 CM	Van Leeuwen et al. (1986c)
	LC_{50}	1	300	46.7 G, 16.8 CM	Van Leeuwen et al. (1986c)
	LC_{50}	60	1.1	FERTILIZED EGG — EARLY FRY STAGE	Van Leeuwen et al. (1986a)
	HIS	21	5	JUVENILE, 5 CM	Van Leeuwen et al. (1986b)
	PHY	1	180	34–46.7 G, 15.3–16.8 CM	Van Leeuwen et al. (1986c)
Snake-head catfish	LC_{50}	1	430	90–110 MM, 40–55 G	Verma et al. (1981c)
(*Channa punctatus*)	LC_{50}	2	360	90–110 MM, 40–55 G	Verma et al. (1981c)
	LC_{50}	3	270	90–110 MM, 40–55 G	Verma et al. (1981c)
	LC_{50}	4	220	90–110 MM, 40–55 G	Verma et al. (1981c)
Walking catfish	LC_{50}	4	9.2	26–31 G, 150–180 MM, STD LENGTH	Verma et al. (1983)
	BIO	30	2.3	26–31 G, 150–180 MM, STD LENGTH	Verma et al. (1983)
	HEM	30	2.3	26–31 G, 150–180 MM, STD LENGTH	Verma et al. (1983)

Table 52. Toxicology of Aquatic and Terrestrial Species: Chemical: Thiram, CAS Number: 137268 (continued)

Class	Species	Effect	Duration (days)	Concen-tration ($\mu g\ L^{-1}$)	Lifestage	Ref.
Fish (continued)	Zebra danio, zebrafish	LET	1	1,000	1.5 G FRESH WT PER SET	Jouany et al. (1985)
		MOR	1	100	1.5 G FRESH WT PER SET	Jouany et al. (1985)
Insect	Mayfly (*Cloeon dipterum*)	LC_{50}	1	1,030	LARVAE	Seuge and Bluzat (1983b)
		LC_{50}	1	1,920	LARVAE	Seuge and Bluzat (1983b)
		LC_{50}	2	800	LARVAE	Seuge and Bluzat (1983b)
		LC_{50}	2	1,300	LARVAE	Seuge and Bluzat (1983b)
		LC_{50}	3	560	LARVAE	Seuge and Bluzat (1983b)
		LC_{50}	3	1,080	LARVAE	Seuge and Bluzat (1983b)
		LC_{50}	4	390	LARVAE	Seuge and Bluzat (1983b)
		LC_{50}	4	1,010	LARVAE	Seuge and Bluzat (1983b)
		MOR	1	1,000	LARVAE	Seuge and Bluzat (1983b)
		MOR	1	1,000	LARVAE	Seuge and Bluzat (1983b)
		MOR	1	1,000	LARVAE	Seuge and Bluzat (1983b)
		MOR	1	1,000	LARVAE	Seuge and Bluzat (1983b)
		MOR	1	1,000	LARVAE	Seuge and Bluzat (1983b)
		MOR	1	1,000	LARVAE	Seuge and Bluzat (1983b)
		MOR	2	1,000	LARVAE	Seuge and Bluzat (1983b)
		MOR	2	1,000	LARVAE	Seuge and Bluzat (1983b)
		MOR	2	1,000	LARVAE	Seuge and Bluzat (1983b)
		MOR	2	1,000	LARVAE	Seuge and Bluzat (1983b)
		MOR	2	1,000	LARVAE	Seuge and Bluzat (1983b)
		MOR	2	1,000	LARVAE	Seuge and Bluzat (1983b)

		MOR	3	1,000	LARVAE	Seuge and Bluzat (1983b)
		MOR	3	1,000	LARVAE	Seuge and Bluzat (1983b)
		MOR	3	1,000	LARVAE	Seuge and Bluzat (1983b)
		MOR	3	1,000	LARVAE	Seuge and Bluzat (1983b)
		MOR	3	1,000	LARVAE	Seuge and Bluzat (1983b)
		MOR	3	1,000	LARVAE	Seuge and Bluzat (1983b)
		MOR	4	1,000	LARVAE	Seuge and Bluzat (1983b)
		MOR	4	1,000	LARVAE	Seuge and Bluzat (1983b)
		MOR	4	1,000	LARVAE	Seuge and Bluzat (1983b)
		MOR	4	1,000	LARVAE	Seuge and Bluzat (1983b)
		MOR	4	1,000	LARVAE	Seuge and Bluzat (1983b)
		MOR	4	1,000	LARVAE	Seuge and Bluzat (1983b)
Platyhelmi	*Turbellarian, planarian*	LC_{50}	1	530	NR	Seuge et al. (1983)
		LC_{50}	2	260	NR	Seuge et al. (1983)
		LC_{50}	3	88	NR	Seuge et al. (1983)
		LC_{50}	4	48	NR	Seuge et al. (1983)
Protozoa	Ciliate (*Colpidium campylum*)	PGR	1.7917	300	>96 HR	Dive et al. (1980)

Table 53. Toxicology of Aquatic and Terrestrial Species: Chemical: Trichlorfon, CAS Number: 52686

Class	Species	Effect	Duration (days)	Concentration (μg L^{-1})	Lifestage	Ref.
Invertebrate	Honey bee	LO MOR	NR	2 kg ha^{-1}	ADULT	Korpela and Tulisalo (1974)
Mammal	Mouse (*Mus* sp)	$LD_{0.1}$	7	362.18 mg kg^{-1}	ADULT	Haley et al. (1975)
		LD_1	7	430.23 mg kg^{-1}	ADULT	Haley et al. (1975)
		LD_1	7	458.28	ADULT	Haley et al. (1975)
		LD_{16}	7	580.97 mg kg^{-1}	ADULT	Haley et al. (1975)
		LD_{16}	7	659.87	ADULT	Haley et al. (1975)
		LD_{50}	7	726.97 mg kg^{-1}	ADULT	Haley et al. (1975)
		LD_{50}	7	866.23	ADULT	Haley et al. (1975)
		LO PHY	7	450 mg kg^{-1}	ADULT	Haley et al. (1975)
		LO PHY	7	450 mg kg^{-1}	ADULT	Haley et al. (1975)
		LO SIGNS	7	450 mg kg^{-1}	ADULT	Haley et al. (1975)
		LO SIGNS	7	450 mg kg^{-1}	ADULT	Haley et al. (1975)
	Rat	LO PATH	30	50 mg kg^{-1}	120–150 D	Finkiewicz-Murawiejska (1978)
		LO PHY	16	50 mg kg^{-1}	ADULT	Staples et al. (1976)
		LO PHY	16	76 mg kg^{-1}	ADULT	Staples et al. (1976)
		LO REP	16	75 mg kg^{-1}	ADULT	Staples et al. (1976)
		LO REP	16	432 mg kg^{-1}	ADULT	Staples et al. (1976)
		LO SIGNS	16	150 mg kg^{-1}	ADULT	Staples et al. (1976)
		NO REP	16	50 mg kg^{-1}	ADULT	Staples et al. (1976)
		NO REP	16	375 mg kg^{-1}	ADULT	Staples et al. (1976)
		NO SIGNS	16	75 mg kg^{-1}	ADULT	Staples et al. (1976)

Table 54. Toxicology of Aquatic and Terrestrial Species: Chemical: Triclopyr, Ester

Class	Species	Effect	Duration (days)	Concentration (µg L^{-1})	Lifestage	Ref.
Bird	Bobwhite quail	LC_{50}	NR	9,026		Dow Chemical Company (undated)
	Mallard duck	LC_{50}	NR	>10,000		Dow Chemical Company (undated)
		LD_{50}	NR	>4,640		USDA Forest Service (1989)
Mammal	Rat	LD_{50}	NR	2,140		Dow Chemical Company (undated)
		LD_{50}	NR	2,460		Dow Chemical Company (undated)

Table 55. Toxicology of Aquatic and Terrestrial Species: Chemical: Triclopyr, Acid, CAS Number: 55335063

Class	Species	Effect	Duration (days)	Concentration ($\mu g\ L^{-1}$)	Lifestage	Ref.
Crustacea	Water flea (*Daphnia pulex*)	EC_{50}IM	4	1,200	<24 HR	Servizi et al. (1987)
Fish	Bluegill	LC_{50}	4	>100,000	0.8 G	Johnson and Finley (1980)
		BCF	1–4	2,510	4–6 CM, 1–2.5 G	Lickly and Murphy (1987)
		BCF	1–4	2,510	4–6 CM, 1–2.5 G	Lickly and Murphy (1987)
	Chinook salmon	LC_{50}	1	9,700	6.8 (5.8–7.5) CM, 2.7 (1.4–3.8) G	Wan et al. (1987)
		LC_{50}	2	9,700	6.8 (5.8–7.5) CM, 2.7 (1.4–3.8) G	Wan et al. (1987)
		LC_{50}	3	9,700	6.8 (5.8–7.5) CM, 2.7 (1.4–3.8) G	Wan et al. (1987)
		LC_{50}	4	9,700	6.8 (5.8–7.5) CM, 2.7 (1.4–3.8) G	Wan et al. (1987)
	Chum salmon	LC_{50}	1	7,900	4.5 (3.9–5) CM, 0.5 (0.3–0.8) G	Wan et al. (1987)
		LC_{50}	2	7,500	4.5 (3.9–5) CM, 0.5 (0.3–0.8) G	Wan et al. (1987)
		LC_{50}	3	7,500	4.5 (3.9–5) CM, 0.5 (0.3–0.8) G	Wan et al. (1987)
		LC_{50}	4	7,500	4.5 (3.9–5) CM, 0.5 (0.3–0.8) G	Wan et al. (1987)

Coho salmon, silver salmon	LC_{50}	1	9,900	4 (3.5–4.5) CM, 0.5 (0.3–0.9) G	Wan et al. (1987)
	LC_{50}	2	9,600	4 (3.5–4.5) CM, 0.5 (0.3–0.9) G	Wan et al. (1987)
	LC_{50}	3	9,600	4 (3.5–4.5) CM, 0.5 (0.3–0.9) G	Wan et al. (1987)
	LC_{50}	4	2,200	FRY, 0.29 G, 3.4 CM	Servizi et al. (1987)
	LC_{50}	4	9,600	4 (3.5–4.5) CM, 0.5 (0.3–0.9) G	Wan et al. (1987)
Pink salmon	LC_{50}	1	13,300	3.5 (3.4–3.7) CM, 0.2 (0.2–0.2) G	Wan et al. (1987)
	LC_{50}	2	8,800	3.5 (3.4–3.7) CM, 0.2 (0.2–0.2) G	Wan et al. (1987)
	LC_{50}	3	6,100	3.5 (3.4–3.7) CM, 0.2 (0.2–0.2) G	Wan et al. (1987)
	LC_{50}	4	5,300	3.5 (3.4–3.7) CM, 0.2 (0.2–0.2) G	Wan et al. (1987)
Rainbow trout, donaldson trout	LC_{50}	1	8,400	4.1 (3.7–4.5) CM, 0.7 (0.4–0.9) G	Wan et al. (1987)
	LC_{50}	2	7,800	4.1 (3.7–4.5) CM, 0.7 (0.4–0.9) G	Wan et al. (1987)
	LC_{50}	3	7,600	4.1 (3.7–4.5) CM, 0.7 (0.4–0.9) G	Wan et al. (1987)
	LC_{50}	4	2,200	FRY, 0.33 G, 3.4 CM	Servizi et al. (1987)
	LC_{50}	4	7,500	4.1 (3.7–4.5) CM, 0.7 (0.4–0.9) G	Wan et al. (1987)
	LC_{50}	4	>100,000	0.9 G	Johnson and Finley (1980)

Table 55. Toxicology of Aquatic and Terrestrial Species: Chemical: Triclopyr, Acid, CAS Number: 55335063 (continued)

Class	Species	Effect	Duration (days)	Concen-tration ($\mu g\ L^{-1}$)	Lifestage	Ref.
Fish (continued)	Sockeye salmon	LC_{50}	1	7,800	3.9 (3.5–4.3) CM, 0.5 (0.3–0.6) G	Wan et al. (1987)
		LC_{50}	2	7,500	3.9 (3.5–4.3) CM, 0.5 (0.3–0.6) G	Wan et al. (1987)
		LC_{50}	3	7,500	3.9 (3.5–4.3) CM, 0.5 (0.3–0.6) G	Wan et al. (1987)
		LC_{50}	4	1,200	FRY, 0.22 G, 2.9 CM	Servizi et al. (1987)
		LC_{50}	4	1,400	FINGERLING, 4.50 G, 7.1 CM	Servizi et al. (1987)
		LC_{50}	4	7,500	3.9 (3.5–4.3) CM, 0.5 (0.3–0.6) G	Wan et al. (1987)
Bird	Bobwhite quail	LC_{50}	NR	2,935		Kenaga (1979)
	Japanese quail	LC_{50}	NR	3,278		USDA Forest Service (1989)
	Mallard duck	LC_{50}	NR	>5,640		Dow Chemical Company (1987)
		LD_{50}	NR	1,698		Kenaga (1979)
Mammal	Guinea pig	LD_{50}	NR	310		EPA (1985b)
	Mouse	LD_{50}	NR	471		EPA (1985b)
	Rabbit	LD_{50}	NR	550		EPA (1985b)
	Rat	LD_{50}	NR	630		Dow Chemical Company (undated)
		LD_{50}	NR	729		Dow Chemical Company (undated)

Table 56. Toxicology of Aquatic and Terrestrial Species: Chemical: Triclopyr, Amine, CAS Number: 5721369

Class	Species	Effect	Duration (days)	Concentration ($\mu g\ L^{-1}$)	Lifestage	Ref.
Crustacea	Water flea (*Daphnia magna*)	LC_{50}	2	1,170,000	NEONATE	Gersich et al. (1984)
		LC_{50}	21	1,140,000	NEONATE	Gersich et al. (1984)
		REP	21	80,700	NEONATE	Gersich et al. (1984)
		REP	21	149,000	NEONATE	Gersich et al. (1984)
Fish	Chinook salmon	LC_{50}	1	472,000	6.8 (5.8–7.5) CM, 2.7 (1.4–3.8) G	Wan et al. (1987)
		LC_{50}	2	312,000	6.8 (5.8–7.5) CM, 2.7 (1.4–3.8) G	Wan et al. (1987)
		LC_{50}	3	283,000	6.8 (5.8–7.5) CM, 2.7 (1.4–3.8) G	Wan et al. (1987)
		LC_{50}	4	275,000	6.8 (5.8–7.5) CM, 2.7 (1.4–3.8) G	Wan et al. (1987)
	Chum salmon	LC_{50}	1	316,000	4.5 (3.9–5.0) CM, 0.5 (0.3–0.8) G	Wan et al. (1987)
		LC_{50}	2	290,000	4.5 (3.9–5.0) CM, 0.5 (0.3–0.8) G	Wan et al. (1987)
		LC_{50}	3	275,000	4.5 (3.9–5.0) CM, 0.5 (0.3–0.8) G	Wan et al. (1987)
		LC_{50}	4	267,000	4.5 (3.9–5.0) CM, 0.5 (0.3–0.8) G	Wan et al. (1987)

Table 56. Toxicology of Aquatic and Terrestrial Species: Chemical: Triclopyr, Amine, CAS Number: 5721369 (continued)

Class	Species	Effect	Duration (days)	Concen-tration (µg L^{-1})	Lifestage	Ref.
Fish (continued)	Coho salmon, silver salmon	LC_{50}	1	498,000	4 (3.5–4.5) CM, 0.5 (0.3–0.9) G	Wan et al. (1987)
		LC_{50}	2	476,000	4 (3.5–4.5) CM, 0.5 (0.3–0.9) G	Wan et al. (1987)
		LC_{50}	3	476,000	4 (3.5–4.5) CM, 0.5 (0.3–0.9) G	Wan et al. (1987)
		LC_{50}	4	463,000	4 (3.5–4.5) CM, 0.5 (0.3–0.9) G	Wan et al. (1987)
	Fathead minnow	LC_{50}	4	120,000	0.22 G, 0.9–1.3 CM, 36 D	Mayes et al. (1984)
		LC_{50}	4	245,000	0.22 G, 1.6–3.1 CM	Mayes et al. (1984)
		LC_{50}	8	101,000	0.22 G, 0.9–1.3 CM, 36 D	Mayes et al. (1984)
		ABN	NR	114,000	EMBRYO, <24 HR	Mayes et al. (1984)
		GRO	31	72,700	EMBRYO, <24 HR	Mayes et al. (1984)
		HAT	NR	114,000	EMBRYO, <24 HR	Mayes et al. (1984)
		MOR	NR	19,300	EMBRYO, <24 HR	Mayes et al. (1984)
		MOR	NR	29,200	EMBRYO, <24 HR	Mayes et al. (1984)
		MOR	NR	46,700	EMBRYO, <24 HR	Mayes et al. (1984)
		MOR	NR	72,700	EMBRYO, <24 HR	Mayes et al. (1984)
		MOR	NR	114,000	EMBRYO, <24 HR	Mayes et al. (1984)
		MOR	NR	117,000	EMBRYO, <24 HR	Mayes et al. (1984)

	Rainbow trout, donaldson trout	LC_{50}	1	457,000	4.1 (3.7–4.5) CM, 0.7 (0.4–0.9) G	Wan et al. (1987)
		LC_{50}	2	435,000	4.1 (3.7–4.5) CM, 0.7 (0.4–0.9) G	Wan et al. (1987)
		LC_{50}	3	420,000	4.1 (3.7–4.5) CM, 0.7 (0.4–0.9) G	Wan et al. (1987)
		LC_{50}	4	420,000	4.1 (3.7–4.5) CM, 0.7 (0.4–0.9) G	Wan et al. (1987)
	Sockeye salmon	LC_{50}	1	353,000	3.9 (3.5–4.3) CM, 0.5 (0.3–0.6) G	Wan et al. (1987)
		LC_{50}	2	311,000	3.9 (3.5–4.3) CM, 0.5 (0.3–0.6) G	Wan et al. (1987)
		LC_{50}	3	311,000	3.9 (3.5–4.3) CM, 0.5 (0.3–0.6) G	Wan et al. (1987)
		LC_{50}	4	311,000	3.9 (3.5–4.3) CM, 0.5 (0.3–0.6) G	Wan et al. (1987)
Bird	Bobwhite quail	LC_{50}	NR	11,622		Dow Chemical Company (undated)
	Mallard duck	LC_{50}	NR	>10,000		Dow Chemical Company (undated)
Mammal	Rat	LD_{50}	NR	2,140		Dow Chemical Company (undated)
		LD_{50}	NR	2,830		Dow Chemical Company (undated)

Table 57. Toxicology of Aquatic and Terrestrial Species: Chemical: Trifluralin, CAS Number: 1582098

Class	Species	Effect	Duration (days)	Concentration ($\mu g\ L^{-1}$)	Lifestage	Ref.
Algae	Algae, phytoplankton, algal mat	POP	NR	10 mg kg^{-1}	NR	Kosinski (1984)
		PSE	21	10 mg kg^{-1}	NR	Kosinski (1984)
		PSE	28	0.01 mg kg^{-1}	NR	Kosinski (1984)
	Blue-green algae (*Anacystis nidulans*)	CLR	0	0.00010 mg kg^{-1}	10 D	Mehta and Hawxby (1983)
	Diatom	$EC_{50}OX$	1.5 hr	>100,000	LOG PHASE	Walsh (1972)
	(*Phaeodactylum tricornutum*)	$EC_{50}OX$	1.5 hr	>500,000	LOG PHASE	Walsh (1972)
		PGR	10	2,500	LOG PHASE	Walsh (1972)
		PGR	10	5,000	LOG PHASE	Walsh (1972)
	Green algae (*Chlamydomonas eugametos*)	PGR	2	33	INITIAL CONC. 100,000 CELLS ML^{-1}	Hess (1980)
		PGR	2	330	INITIAL CONC. 100,000 CELLS ML^{-1}	Hess (1980)
		PGR	2	0.1–≥1 µm	NR	Hess (1979)
		RGN	2 hr	0.01–10 µm	NR	Hess (1979)
	Green algae (*Chlorococcum* sp)	$EC_{50}OX$	1.5 hr	>100,000	LOG PHASE	Walsh (1972)
		$EC_{50}OX$	1.5 hr	>500,000	LOG PHASE	Walsh (1972)
		PGR	10	2,500	LOG PHASE	Walsh (1972)
		PGR	10	2,500	LOG PHASE	Walsh (1972)
	Green algae	$EC_{50}OX$	1.5 hr	>100,000	LOG PHASE	Walsh (1972)
	(*Dunaliella tertiolecta*)	$EC_{50}OX$	1.5 hr	>500,000	LOG PHASE	Walsh (1972)
	Green algae	LET	2	100 µm	LOG PHASE	Felix et al. (1988)
	(*Dunaliella bioculata*)	PGR	2	100 µm	LOG PHASE	Felix et al. (1988)
	Green algae	PGR	10	2,500	LOG PHASE	Walsh (1972)
	(*Dunaliella tertiolecta*)	PGR	10	5,000	LOG PHASE	Walsh (1972)

Green algae (*Oedogonium cardiacum*)	BCF	1	0.2	NR	Yockim et al. (1980)
	BCF	1	0.5	NR	Yockim et al. (1980)
	BCF	1	3.4	NR	Yockim et al. (1980)
	BCF	1	9.3	NR	Yockim et al. (1980)
	BCF	1	36.9	NR	Yockim et al. (1980)
	BCF	1	200	NR	Yockim et al. (1980)
	BCF	1	3,400	NR	Yockim et al. (1980)
	BCF	1	36,900	NR	Yockim et al. (1980)
	BCF	3	0.2	NR	Yockim et al. (1980)
	BCF	3	0.2	NR	Yockim et al. (1980)
	BCF	3	0.7	NR	Yockim et al. (1980)
	BCF	3	0.9	NR	Yockim et al. (1980)
	BCF	3	2.5	NR	Yockim et al. (1980)
	BCF	3	16.2	NR	Yockim et al. (1980)
	BCF	3	31.3	NR	Yockim et al. (1980)
	BCF	3	700	NR	Yockim et al. (1980)
	BCF	3	2,500	NR	Yockim et al. (1980)
	BCF	3	31,300	NR	Yockim et al. (1980)
	BCF	7	0.2	NR	Yockim et al. (1980)
	BCF	7	0.2	NR	Yockim et al. (1980)
	BCF	7	1.4	NR	Yockim et al. (1980)
	BCF	7	2.9	NR	Yockim et al. (1980)
	BCF	7	19.6	NR	Yockim et al. (1980)
	BCF	7	55.1	NR	Yockim et al. (1980)
	BCF	7	2,900	NR	Yockim et al. (1980)
	BCF	7	55,100	NR	Yockim et al. (1980)
	BCF	15	0.5	NR	Yockim et al. (1980)
	BCF	15	0.8	NR	Yockim et al. (1980)
	BCF	15	2.6	NR	Yockim et al. (1980)
	BCF	15	8.4	NR	Yockim et al. (1980)
	BCF	15	29.8	NR	Yockim et al. (1980)

Table 57. Toxicology of Aquatic and Terrestrial Species: Chemical: Trifluralin, CAS Number: 1582098 (continued)

Class	Species	Effect	Duration (days)	Concentration ($\mu g\ L^{-1}$)	Lifestage	Ref.
Algae (continued)	Green algae (*Oedogonium cardiacum*)	BCF	15	148.8	NR	Yockim et al. (1980)
		BCF	15	800	NR	Yockim et al. (1980)
		BCF	15	8,400	NR	Yockim et al. (1980)
		BCF	15	148,800	NR	Yockim et al. (1980)
		BCF	30	0.8	NR	Yockim et al. (1980)
		BCF	30	2.5	NR	Yockim et al. (1980)
		BCF	30	9.1	NR	Yockim et al. (1980)
		BCF	30	21.6	NR	Yockim et al. (1980)
		BCF	30	160.1	NR	Yockim et al. (1980)
		BCF	30	9,100	NR	Yockim et al. (1980)
		BCF	30	160,100	NR	Yockim et al. (1980)
		BCF	33	7.5	NR	Kearney et al. (1977b)
	Green algae (*Selenastrum capricornutum*)	BMS	30	1,000	ACTIVELY GROWING	Johnson (1986)
	Haptophyte (*Isochrysis galbana*)	$EC_{50}OX$	1.5 hr	>100,000	LOG PHASE	Walsh (1972)
		$EC_{50}OX$	1.5 hr	>500,000	LOG PHASE	Walsh (1972)
		PGR	10	2,500	LOG PHASE	Walsh (1972)
		PGR	10	2,500	LOG PHASE	Walsh (1972)
Amphibia	Fowler's toad	LC_{50}	1	180	TADPOLE, 28–35 D	Sanders (1970a)
		LC_{50}	2	170	TADPOLE, 28–35 D	Sanders (1970a)
		LC_{50}	4	100	TADPOLE, 28–35 D	Sanders (1970a)
	Frog (*Bufo bufo japonicus*)	LC_{50}	2	14,000	TADPOLE	Hashimoto and Nishiuchi (1981)

Annelida	Oligochaete (*Lumbriculus variegatus*)	LC_{50}	2	>300	NR	Bailey and Liu (1980)
		LC_{50}	4	>300	NR	Bailey and Liu (1980)
Crustacea	Aquatic sowbug (*Asellus brevicaudus*)	LC_{50}	2	2,000	EARLY INSTAR	Sanders (1970a)
	Calanoid copepod (*Diaptomus* sp)	LC_{50}	2	80	NR	Naqvi et al. (1985)
	Cladoceran	LC_{50}	2	60	NR	Naqvi et al. (1985)
	Crayfish (*Orconectes* nais)	LC_{50}	2	50,000	EARLY INSTAR	Sanders (1970a)
	Cyclopoid copepod (*Eucyclops* sp)	LC_{50}	2	50	NR	Naqvi et al. (1985)
	Dungeness or edible crab	EC_{50}SW	2	170	1ST STAGE ZOEAE	Armstrong et al. (1976)
		EC_{50}SW	4	150	1ST STAGE ZOEAE	Armstrong et al. (1976)
		LC_{50}	2	>320	1ST STAGE ZOEAE	Armstrong et al. (1976)
		LC_{50}	4	300	1ST STAGE ZOEAE	Armstrong et al. (1976)
		LET	50	220	1ST STAGE ZOEAE	Caldwell et al. (1979)
		MOR	50	26	1ST STAGE ZOEAE	Caldwell et al. (1979)
		MOR	80	590	JUVENILE, 3RD INSTAR	Caldwell et al. (1979)
		MOR	85	300	ADULT, 80–100 MM	Caldwell et al. (1979)
	Grass shrimp (*Palaemonetes kadiakensis*)	LC_{50}	2	1,200	EARLY INSTAR	Sanders (1970a)
	Ostracod (*Cypridopsis vidua*)	EC_{50}IM	2	250	EARLY INSTAR	Sanders (1970a)
	Ostracod (*Cypria* sp)	LC_{50}	2	60	NR	Naqvi et al. (1985)
	Red swamp crayfish	LC_{50}	1	13,000	JUVENILE, 3–4 CM	Naqvi and Leung (1983)
		LC_{50}	4	12,000	JUVENILE, 3–4 CM	Naqvi and Leung (1983)
		LC_{50}	4	13,000	JUVENILE, 3–3.4 CM, 1.1–1.5 G	Naqvi et al. (1987)
		LC_{50}	4	26,000	ADULT, 9–10 CM, 25–32 G	Naqvi et al. (1987)

Table 57. Toxicology of Aquatic and Terrestrial Species: Chemical: Trifluralin, CAS Number: 1582098 (continued)

Class	Species	Effect	Duration (days)	Concentration ($\mu g\ L^{-1}$)	Lifestage	Ref.
Crustacea (continued)	Scud (*Gammarus fasciatus*)	LC_{50}	1	3,200	EARLY INSTAR	Sanders (1970a)
	Scud (*Gammarus lacustris*)	LC_{50}	1	8,800	2 MONTHS OLD	Sanders (1969)
	Scud (*Gammarus fasciatus*)	LC_{50}	2	1,800	EARLY INSTAR	Sanders (1970a)
		LC_{50}	2	1,800	EARLY INSTAR	Sanders (1970a)
	Scud (*Gammarus lacustris*)	LC_{50}	2	5,600	2 MONTHS OLD	Sanders (1969)
	Scud (*Gammarus fasciatus*)	LC_{50}	4	1,000	EARLY INSTAR	Sanders (1970a)
	Scud (*Gammarus lacustris*)	LC_{50}	4	2,200	2 MONTHS OLD	Sanders (1969)
	Scud (*Gammarus fasciatus*)	LC_{50}	4	2,200	MATURE	Johnson and Finley (1980)
	Water flea (*Daphnia pulex*)	EC_{50}IM	2	240	1ST INSTAR	Sanders and Cope (1966)
	Water flea (*Daphnia magna*)	EC_{50}IM	2	560	EARLY INSTAR	Sanders (1970a)
		EC_{50}IM	2	560	1ST INSTAR	Johnson and Finley (1980)
	Water flea (*Daphnia pulex*)	EC_{50}IM	2	625	1ST INSTAR	Johnson and Finley (1980)
	Water flea (*Daphnia magna*)	LC_{50}	2	193	NR	Macek et al. (1976)
	Water flea (*Daphnia pulex*)	LC_{50}	3 hr	>40,000	NR	Hashimoto and Nishiuchi (1981)
	Water flea (*Daphnia magna*)	BCF	1	0.2	NR	Yockim et al. (1980)
		BCF	1	3.4	NR	Yockim et al. (1980)
		BCF	1	36.9	NR	Yockim et al. (1980)
		BCF	1	200	NR	Yockim et al. (1980)
		BCF	1	3,400	NR	Yockim et al. (1980)
		BCF	1	36,900	NR	Yockim et al. (1980)
		BCF	3	0.7	NR	Yockim et al. (1980)
		BCF	3	2.5	NR	Yockim et al. (1980)
		BCF	3	31.3	NR	Yockim et al. (1980)
		BCF	3	700	NR	Yockim et al. (1980)

	BCF	3	2,500	NR	Yockim et al. (1980)
	BCF	3	31,300	NR	Yockim et al. (1980)
	BCF	7	0.4	NR	Yockim et al. (1980)
	BCF	7	2.9	NR	Yockim et al. (1980)
	BCF	7	55.1	NR	Yockim et al. (1980)
	BCF	7	400	NR	Yockim et al. (1980)
	BCF	7	2,900	NR	Yockim et al. (1980)
	BCF	7	55,100	NR	Yockim et al. (1980)
	BCF	15	8.4	NR	Yockim et al. (1980)
	BCF	15	148.8	NR	Yockim et al. (1980)
	BCF	15	8,400	NR	Yockim et al. (1980)
	BCF	15	148,800	NR	Yockim et al. (1980)
	BCF	30	0.9	NR	Yockim et al. (1980)
	BCF	30	7.5	NR	Kearney et al. (1977b)
	BCF	30	9.1	NR	Yockim et al. (1980)
	BCF	30	160.1	NR	Yockim et al. (1980)
	BCF	30	900	NR	Yockim et al. (1980)
	BCF	30	9,100	NR	Yockim et al. (1980)
	BCF	30	160,100	NR	Yockim et al. (1980)
	LET	64	7.2	3RD GENERATION	Macek et al. (1976)
	MOR	42	4	1ST INSTAR LARVAE	Huckins et al. (1986)
	MOR	64	2.4	3RD GENERATION	Macek et al. (1976)
	REP	64	7.2	3RD GENERATION	Macek et al. (1976)
	REP	64	14	3RD GENERATION	Macek et al. (1976)
Water flea (*Moina macrocopa*)	LC_{50}	3 hr	>40,000	NR	Hashimoto and Nishiuchi (1981)
Water flea (*Simocephalus serrulatus*)	EC_{50}IM	2	450	1ST INSTAR	Sanders and Cope (1966)
	EC_{50}IM	2	900	1ST INSTAR	Johnson and Finley (1980)

Table 57. Toxicology of Aquatic and Terrestrial Species: Chemical: Trifluralin, CAS Number: 1582098 (continued)

Class	Species	Effect	Duration (days)	Concentration ($\mu g\ L^{-1}$)	Lifestage	Ref.
Fish	Bluegill	LC_{50}	1	23	1.20 G	Cope (1965)
		LC_{50}	1	100	0.97 G	Cope (1965)
		LC_{50}	1	130	WEIGHT 0.6–1.5 G	Macek et al. (1969)
		LC_{50}	1	360	WEIGHT 0.6–1.5 G	Macek et al. (1969)
		LC_{50}	1	540	WEIGHT 0.6–1.5 G	Macek et al. (1969)
		LC_{50}	2	20	1.20 G	Cope (1965)
		LC_{50}	2	96	0.97 G	Cope (1965)
		LC_{50}	4	18	1.20 G	Cope (1965)
		LC_{50}	4	47	0.6–1.5 G	Macek et al. (1969)
		LC_{50}	4	58	0.8 G	Johnson and Finley (1980)
		LC_{50}	4	68	0.97 G	Cope (1965)
		LC_{50}	4	120	WEIGHT 0.6–1.5 G	Macek et al. (1969)
		LC_{50}	4	190	WEIGHT 0.6–1.5 G	Macek et al. (1969)
	Channel catfish	LC_{50}	4	417	1 YEAR, 14 G, 12 CM, FINGERLINGS	Mccorkle et al. (1977)
		LC_{50}	4	2,200	0.8 G	Johnson and Finley (1980)
		RSD	NR	0.00043–548	703 G	Spacie and Hamelink (1979)
	Common, mirror, colored, carp	LC_{50}	2	1,000	NR	Hashimoto and Nishiuchi (1981)
		BIO	0.08 hr	500	0.37–1.36 KG	McBride and Richards (1975)
		RSD	NR	0.00043–548	1591 G	Spacie and Hamelink (1979)

Fathead minnow	LC_{50}	4	105	0.8 G	Johnson and Finley (1980)
	LC_{50}	12	115	44 D	Macek et al. (1976)
	BCF	5 hr	20	0.1–1.6 G	Spacie and Hamelink (1979)
	BCF	425	1.5	2ND GENERATION	Macek et al. (1976)
	BCF	425	1.9	2ND GENERATION	Macek et al. (1976)
	BCF	425	5.1	2ND GENERATION	Macek et al. (1976)
	LET	60	16.5	26 MM	Macek et al. (1976)
	MOR	30	16.5	19–20 MM	Macek et al. (1976)
	MOR	60	8.2	NR	Macek et al. (1976)
	MOR	427	5.1	2ND GENERATION	Macek et al. (1976)
	REP	427	1.95	2ND GENERATION	Macek et al. (1976)
Freshwater drum	RSD	NR	0.00043–548	713 G	Spacie and Hamelink (1979)
Golden redhorse	BCF	NR	1.8 mg kg^{-1}	794 G	Spacie and Hamelink (1979)
Goldfish	LC_{50}	2	850	NR	Hashimoto and Nishiuchi (1981)
	LC_{50}	4	145	1 G	Johnson and Finley (1980)
Green sunfish	BCF	4	4.4	0.30–0.95 G	Reinbold and Metcalf (1976)
	BCF	16	4.2	0.30–0.95 G	Reinbold (1974)
	BCF	16	0.0068	0.30–0.95 G	Reinbold and Metcalf (1976)
	LET	8	100	5–10 G	Reinbold (1974)
	MOR	30	100	15–30 G	Reinbold (1974)
Harlequinfish, red rasbora	LC_{50}	1	600	1.3–3 CM	Alabaster (1969)
	LC_{50}	1	1,000	1.3–3 CM	Alabaster (1969)
	LC_{50}	2	600	1.3–3 CM	Alabaster (1969)
Largemouth bass	LC_{50}	4	75	0.7 G	Johnson and Finley (1980)
Medaka, high-eyes	LC_{50}	2	430	NR	Hashimoto and Nishiuchi (1981)
	PHY	5	100	NR	Kubota and Ochiai (1979)
Mooneye	RSD	NR	0.00043–548	273 G	Spacie and Hamelink (1979)

Table 57. Toxicology of Aquatic and Terrestrial Species: Chemical: Trifluralin, CAS Number: 1582098 (continued)

Class	Species	Effect	Duration (days)	Concen-tration (μg L^{-1})	Lifestage	Ref.
Fish (continued)	Mosquitofish (*Gambusia affinis*)	LC_{50}	1	28,000	ADULT, 2–2.5 CM	Naqvi and Leung (1983)
		LC_{50}	4	12,000	ADULT, 2–2.5 CM	Naqvi and Leung (1983)
		LC_{50}	NR	2,000	NR	Fabacher and Chambers (1974)
		LC_{50}	NR	4,100	NR	Fabacher and Chambers (1974)
		BCF	1	0.1	NR	Yockim et al. (1980)
		BCF	1	0.1	NR	Yockim et al. (1980)
		BCF	1	0.2	NR	Yockim et al. (1980)
		BCF	1	0.5	NR	Yockim et al. (1980)
		BCF	1	0.5	NR	Yockim et al. (1980)
		BCF	1	3.4	NR	Yockim et al. (1980)
		BCF	1	9.3	NR	Yockim et al. (1980)
		BCF	1	9.3	NR	Yockim et al. (1980)
		BCF	1	36.9	NR	Yockim et al. (1980)
		BCF	1	200	NR	Yockim et al. (1980)
		BCF	1	3,400	NR	Yockim et al. (1980)
		BCF	1	36,900	NR	Yockim et al. (1980)
		BCF	3	0.2	NR	Yockim et al. (1980)
		BCF	3	0.2	NR	Yockim et al. (1980)
		BCF	3	0.7	NR	Yockim et al. (1980)
		BCF	3	0.9	NR	Yockim et al. (1980)
		BCF	3	0.9	NR	Yockim et al. (1980)
		BCF	3	2.5	NR	Yockim et al. (1980)
		BCF	3	7.5	NR	Kearney et al. (1977b)

BCF	3	7.5	NR	Kearney et al. (1977b)
BCF	3	7.5	NR	Kearney et al. (1977b)
BCF	3	7.5	NR	Kearney et al. (1977b)
BCF	3	0.184	NR	Sanborn (1974)
BCF	3	0.282	NR	Sanborn (1974)
BCF	3	16.2	NR	Yockim et al. (1980)
BCF	3	16.2	NR	Yockim et al. (1980)
BCF	3	31.3	NR	Yockim et al. (1980)
BCF	3	700	NR	Yockim et al. (1980)
BCF	3	2,500	NR	Yockim et al. (1980)
BCF	3	31,300	NR	Yockim et al. (1980)
BCF	7	0.2	NR	Yockim et al. (1980)
BCF	7	0.2	NR	Yockim et al. (1980)
BCF	7	0.4	NR	Yockim et al. (1980)
BCF	7	1.4	NR	Yockim et al. (1980)
BCF	7	1.4	NR	Yockim et al. (1980)
BCF	7	2.9	NR	Yockim et al. (1980)
BCF	7	19.6	NR	Yockim et al. (1980)
BCF	7	19.6	NR	Yockim et al. (1980)
BCF	7	55.1	NR	Yockim et al. (1980)
BCF	7	400	NR	Yockim et al. (1980)
BCF	7	2,900	NR	Yockim et al. (1980)
BCF	7	55,100	NR	Yockim et al. (1980)
BCF	15	0.5	NR	Yockim et al. (1980)
BCF	15	0.5	NR	Yockim et al. (1980)
BCF	15	0.8	NR	Yockim et al. (1980)
BCF	15	2.6	NR	Yockim et al. (1980)
BCF	15	2.6	NR	Yockim et al. (1980)
BCF	15	8.4	NR	Yockim et al. (1980)
BCF	15	29.8	NR	Yockim et al. (1980)
BCF	15	29.8	NR	Yockim et al. (1980)

Table 57. Toxicology of Aquatic and Terrestrial Species: Chemical: Trifluralin, CAS Number: 1582098 (continued)

Class	Species	Effect	Duration (days)	Concen-tration ($\mu g\ L^{-1}$)	Lifestage	Ref.
Fish (continued)	Mosquitofish (*Gambusia affinis*)	BCF	15	148.8	NR	Yockim et al. (1980)
		BCF	15	800	NR	Yockim et al. (1980)
		BCF	15	8,400	NR	Yockim et al. (1980)
		BCF	15	148,800	NR	Yockim et al. (1980)
		BCF	30	0.8	NR	Yockim et al. (1980)
		BCF	30	0.8	NR	Yockim et al. (1980)
		BCF	30	0.9	NR	Yockim et al. (1980)
		BCF	30	2.5	NR	Yockim et al. (1980)
		BCF	30	2.5	NR	Yockim et al. (1980)
		BCF	30	9.1	NR	Yockim et al. (1980)
		BCF	30	21.6	NR	Yockim et al. (1980)
		BCF	30	21.6	NR	Yockim et al. (1980)
		BCF	30	160.1	NR	Yockim et al. (1980)
		BCF	30	900	NR	Yockim et al. (1980)
		BCF	30	9,100	NR	Yockim et al. (1980)
		BCF	30	160,100	NR	Yockim et al. (1980)
	Motsuga, stone moroko	BCF	1	5–20	4–8 CM, 2–5 G	Kanazawa (1981)
	Oriental weatherfish	LC_{50}	2	350	NR	Hashimoto and Nishiuchi (1981)
	Quillback	RSD	NR	0.00043–548	655 G	Spacie and Hamelink (1979)
	Rainbow trout, donaldson trout	LC_{50}	1	14	1.32 G	Cope (1965)
		LC_{50}	1	98	0.6–1.5 G	Macek et al. (1969)
		LC_{50}	1	210	3.52 G	Cope (1965)
		LC_{50}	1	239	WEIGHT 0.6–1.5 G	Macek et al. (1969)
		LC_{50}	1	308	WEIGHT 0.6–1.5 G	Macek et al. (1969)

		LC_{50}	2	11	1.32 G	Cope (1965)
		LC_{50}	2	130	3.52 G	Cope (1965)
		LC_{50}	4	10	1.32 G	Cope (1965)
		LC_{50}	4	41	0.8 G	Johnson and Finley (1980)
		LC_{50}	4	42	WEIGHT 0.6–1.5 G	Macek et al. (1969)
		LC_{50}	4	86	3.52 G	Cope (1965)
		LC_{50}	4	152	WEIGHT 0.6–1.5 G	Macek et al. (1969)
		LC_{50}	4	210	0.6–1.5 G	Macek et al. (1969)
	River carpsucker	RSD	NR	0.00043–548	669 G	Spacie and Hamelink (1979)
	Sauger	BCF	NR	1.8 mg kg^{-1}	629 G	Spacie and Hamelink (1979)
	Sheepshead minnow	HEM	4	16.6	35–60 MM, ADULT	Couch et al. (1979)
		VTE	28	5.5	ZYGOTE TO 28 D	Couch et al. (1979)
		VTE	51	16.6	ZYGOTE TO 51 D	Couch et al. (1979)
	Shorthead redhorse	BCF	NR	1.8 mg kg^{-1}	784 G	Spacie and Hamelink (1979)
	Smallmouth buffalo	RSD	NR	0.00043–548	1,747 G	Spacie and Hamelink (1979)
	White sucker	RSD	NR	0.00043–548	469 G	Spacie and Hamelink (1979)
Insect	Mayfly (*Cloeon dipterum*)	LC_{50}	2	>40,000	LARVAE	Hashimoto and Nishiuchi (1981)
	Midge (*Chironomus riparius*)	EC_{50}	2	1,000	4TH INSTAR LARVA	Johnson (1986)
		MOR	42	4	1ST INSTAR LARVAE	Huckins et al. (1986)
	Stonefly	LC_{50}	1	13,000	30–35 MM	Sanders and Cope (1968)
	(*Pteronarcys californica*)	LC_{50}	1	13,000	NYMPH	Cope (1965)
		LC_{50}	2	4,200	30–35 MM	Sanders and Cope (1968)
		LC_{50}	2	4,200	NYMPH	Cope (1965)
		LC_{50}	4	2,800	2ND YEAR CLASS	Johnson and Finley (1980)
		LC_{50}	4	3,000	30–35 MM	Sanders and Cope (1968)
		LC_{50}	4	3,000	NYMPH	Cope (1965)

Table 57. Toxicology of Aquatic and Terrestrial Species: Chemical: Trifluralin, CAS Number: 1582098 (continued)

Class	Species	Effect	Duration (days)	Concen-tration (µg L^{-1})	Lifestage	Ref.
Mullusca	Bladder snail	LC_{50}	2	30,000	NR	Nishiuchi and Yoshida (1972)
		LC_{50}	2	30,000	NR	Hashimoto and Nishiuchi (1981)
	Common bay mussel, blue mussel	EC_{50}BA	4	350	ADULT	Liu and Lee (1975)
		LT_{50}	4	240	ADULT	Liu and Lee (1975)
		DVP	2	120	EGGS	Liu and Lee (1975)
		GRO	20	48	LARVAE	Liu and Lee (1975)
		GRO	20	96	LARVAE	Liu and Lee (1975)
	Marsh snail	LC_{50}	2	8,000	NR	Nishiuchi and Yoshida (1972)
		LC_{50}	2	8,000	NR	Hashimoto and Nishiuchi (1981)
	Mud snail	LC_{50}	2	35,000	NR	Nishiuchi and Yoshida (1972)
	Pouch snail (*Physa* sp)	BCF	33	0.184	NR	Sanborn (1974)
		BCF	33	0.282	NR	Sanborn (1974)
	Ramshorn snail (*Helisoma* sp)	BCF	0.2917	1.4	NR	Yockim et al. (1980)
		BCF	1	0.1	NR	Yockim et al. (1980)
		BCF	1	0.1	NR	Yockim et al. (1980)
		BCF	1	0.2	NR	Yockim et al. (1980)
		BCF	1	0.5	NR	Yockim et al. (1980)
		BCF	1	0.5	NR	Yockim et al. (1980)
		BCF	1	3.4	NR	Yockim et al. (1980)
		BCF	1	9.3	NR	Yockim et al. (1980)
		BCF	1	9.3	NR	Yockim et al. (1980)
		BCF	1	36.9	NR	Yockim et al. (1980)
		BCF	1	200	NR	Yockim et al. (1980)

BCF	1	3,400	NR	Yockim et al. (1980)
BCF	1	36,900	NR	Yockim et al. (1980)
BCF	3	0.2	NR	Yockim et al. (1980)
BCF	3	0.2	NR	Yockim et al. (1980)
BCF	3	0.7	NR	Yockim et al. (1980)
BCF	3	0.9	NR	Yockim et al. (1980)
BCF	3	0.9	NR	Yockim et al. (1980)
BCF	3	2.5	NR	Yockim et al. (1980)
BCF	3	16.2	NR	Yockim et al. (1980)
BCF	3	16.2	NR	Yockim et al. (1980)
BCF	3	31.3	NR	Yockim et al. (1980)
BCF	3	700	NR	Yockim et al. (1980)
BCF	3	2,500	NR	Yockim et al. (1980)
BCF	3	31,300	NR	Yockim et al. (1980)
BCF	7	0.2	NR	Yockim et al. (1980)
BCF	7	0.2	NR	Yockim et al. (1980)
BCF	7	1.4	NR	Yockim et al. (1980)
BCF	7	19.6	NR	Yockim et al. (1980)
BCF	7	19.6	NR	Yockim et al. (1980)
BCF	7	55.1	NR	Yockim et al. (1980)
BCF	7	55,100	NR	Yockim et al. (1980)
BCF	15	0.5	NR	Yockim et al. (1980)
BCF	15	0.5	NR	Yockim et al. (1980)
BCF	15	2.6	NR	Yockim et al. (1980)
BCF	15	2.6	NR	Yockim et al. (1980)
BCF	15	2.9	NR	Yockim et al. (1980)
BCF	15	29.8	NR	Yockim et al. (1980)
BCF	15	29.8	NR	Yockim et al. (1980)
BCF	15	148.8	NR	Yockim et al. (1980)
BCF	15	8,400	NR	Yockim et al. (1980)
BCF	15	148,800	NR	Yockim et al. (1980)

Table 57. Toxicology of Aquatic and Terrestrial Species: Chemical: Trifluralin, CAS Number: 1582098 (continued)

Class	Species	Effect	Duration (days)	Concentration ($\mu g\ L^{-1}$)	Lifestage	Ref.
Mullusca	Ramshorn snail (*Helisoma* sp)	BCF	30	0.8	NR	Yockim et al. (1980)
		BCF	30	0.8	NR	Yockim et al. (1980)
		BCF	30	2.5	NR	Yockim et al. (1980)
		BCF	30	2.5	NR	Yockim et al. (1980)
		BCF	30	9.1	NR	Yockim et al. (1980)
		BCF	30	21.6	NR	Yockim et al. (1980)
		BCF	30	21.6	NR	Yockim et al. (1980)
		BCF	30	160.1	NR	Yockim et al. (1980)
		BCF	30	9,100	NR	Yockim et al. (1980)
		BCF	30	160,100	NR	Yockim et al. (1980)
		BCF	33	7.5	NR	Kearney et al. (1977b)
		BCF	33	7.5	NR	Kearney et al. (1977b)
	Snail (*Indoplanorbis exustus*)	LC_{50}	2	30,000	NR	Nishiuchi and Yoshida (1972)
		LC_{50}	2	30,000	NR	Hashimoto and Nishiuchi (1981)
Tracheophy	Duckweed (*Lemna perpusilla*)	MOR	7	100,000	NR	Nishiuchi and Yoshida (1974)

REFERENCES

Abd El-Magid, M. M. 1986. Effect of some pesticides on the growth of blue-green alga *Spirulina platensis. C.A. Sel. Environ. Pollut.* 26:105–220575T; *Egypt. J. Food Sci.* 14(1): 67–74.

Abdelghani, A. A. 1978. Uptake, Distribution and Loss of Monosodium Methane Arsonate by Crawfish *Procambarus* sp. Ph.D. Thesis, Tulane University, New Orleans, LA, 208 pp.; *Diss. Abstr. Int. B* 39(9):224.

Abdelghani, A. A., Anderson, A. C. and McDonell, D. B. 1980a. Toxicity of three arsenical compounds. *Can. Res.* 13(7):31–32.

Abdelghani, A. A., Anderson, A. C., Hughes, J. and Englande, A. J. 1980b. Uptake, Distribution and Excretion of Monosodium Methane Arsonate by Crawfish (*Procambarus* sp.). in Proc. Int. Symp. Arsenic Nickel, July 7–11, 1980, Jena, East Germany, pp. 147–153.

Abdelghani, A. A., Mason, J. W., Anderson, A. C., Englande, A. J. and Diem, J. E. 1976. Bioconcentration of MSMA in crayfish (*Procambarus* sp.). *Trace Subst. Environ. Health* 10:235–245.

Adlung, K. G. 1957. Zur toxizitat insektizider und akarizider wirkstoffe fur fische. (The toxicity of insecticidal and acaricidal agents to fish). *Die Naturwiss.* 44:471–472.

Ahamad, I. K., Begum, M. R., Sivaiah, S. and Rao, K. V. R. 1978a. Effect of malathion on free amino acids, total proteins, glycogen and some enzymes of *Pelecypod lamellidens marginalis* (Lamarck). *Proc. Indian Acad. Sci.* 87B(12):377–380.

Ahamad, I. K., Rao, K. V. R. and Swami, K. S. 1978b. Effect of malathion on enzyme activity in foot, mantle and hepatopancreas of snail *Pila globosa. Indian J. Exp. Biol.* 16:258–260.

Ahamad, I. K., Sivaiah, S. and Rao, K. V. R. 1979. On the possible significance of the changes in organic constituents in selected tissues of malathion exposed snail, *Pila globosa* (Swainson). *Comp. Physiol. Ecol.* 4(2):81–83.

Ahdaya, S. M., Shah, P. V. and Guthrie, F. E. 1976. Thermoregulation in mice treated with parathion, carbaryl, or DDT. *Toxicol. Appl. Pharmacol.* 35:575–580.

Ahmed, W. 1977. The Effectiveness of Predators of Rice Field Mosquitoes in Relation to Pesticide Use in Rice Culture. Ph.D. Thesis, University of California, Davis, CA, 56 pp.; *Dis. Abstr. Int. B* 37(9):430B.

Al-Sabti, K. 1985. Frequency of chromosomal aberrations in the rainbow trout, *Salmo gairdneri* Rich., exposed to five pollutants. *J. Fish Biol.* 26(1):13–19.

Alabaster, J. S. 1969. Survival of fish in 164 herbicides, insecticides, fungicides, wetting agents and miscellaneous substances. *Int. Pest Control* 11(2):29–35.

Alexander, H. C., Gersich, F. M. and Mayes, M. A. 1985. Acute toxicity of four phenoxy herbicides to aquatic organisms. *Bull. Environ. Contam. Toxicol.* 35(3):314–321.

Ali, A. and Mulla, M. S. 1977. The IGR diflubenzuron and organophosphorus insecticides against nuisance midges in man-made residential-recreational lakes. *J. Econ. Entomol.* 70(5):571–577.

Ali, A. and Mulla, M. S. 1978a. Declining field efficacy of chlorpyrifos against chironomid midges and laboratory evaluation of substitute larvicides. *J. Econ. Entomol.* 71(5):778–782.

Ali, A. and Mulla, M. S. 1978b. Effects of chironomid larvicides and diflubenzuron on nontarget invertebrates in residential-recreational lakes. *Environ. Entomol.* 7(1):21–27.

Allison, D. T. 1977. Use of Exposure Units for Estimating Aquatic Toxicity of Organophosphate Pesticides. EPA-600/3-77-077. U.S. EPA, Duluth, MN; U.S. NTIS PB–272 796. 25 pp.

Allison, D. T. and Hermanutz, R. O. 1977. Toxicity of Diazinon to Brook Trout and Fathead Minnows. Ecol. Res. Ser., EPA-600/3-77-060. Environ. Res. Lab., U.S. EPA, Duluth, MN, 69 pp.

Almar, M. M., Ferrando, M. M. D., Alarcon, V., Soler, C. and Andreu, E. 1988. Influence of temperature on several pesticides toxicity to *Melanopsis dufouri* under laboratory conditions. *J. Environ. Biol.* 9(2):183–190.

Ambrosi, D., Isensee, A. R. and Macchia, J. A. 1978. Distribution of oxadiazon and phosalone in an aquatic model ecosystem. *J. Agric. Food Chem.* 26(1):50–53.

Anderson, A. C., Abdelghani, A. A. and McDonnell, D. 1980. Screening of four vascular aquatic plants for uptake of monosodium methanearsonate (MSMA). *Sci. Total Environ.* 16(2):95–98.

Anderson, A. C., Abdelghani, A. A., McDonnell, D. and Craig, L. 1981. Uptake of monosodium methanearsonate (MSMA) by vascular aquatic plants. *J. Plant Nutr.* 3(1–4): 193–201.

Anderson, A. C., Abdelghani, A. A., Smith, P. M., Mason, J. W. and Englande, A. J., Jr. 1975. The acute toxicity of MSMA to black bass (*Micropterus dolomieu*), crayfish (*Procambarus* sp.) and channel catfish (*Ictalurus lacustris*). *Bull. Environ. Contam. Toxicol.* 14(3): 330–333.

Anderson, B. G. 1959. The Toxicity of Organic Insecticides to Daphnia. in Trans. 2nd Sem. Biol. Probl. Water Pollut., U.S. Public Health, Cincinnati, OH, pp. 94–95.

Anderson, R. L. and Shubat, P. 1983. Insecticide effects on normal development and hatch of embryos of *Paratanytarsus parthenogeneticus* (*Diptera: Chironomidae*). *Great Lakes Entomol.* 16(4):177–181.

Andreu-Moliner, E. S., Almar, M. M., Legarra, I. and Nunez, A. 1986. Toxicity of some ricefield pesticides to the crayfish *P. clarkii*, under laboratory and field conditions in Lake Albufera (Spain). *J. Environ. Sci. Health* B21(6):529–537.

Anees, M. A. 1974a. Susceptibility of a freshwater teleost *Channa punctatus* to acute, sublethal and chronic levels of organophosphorus insecticides. Masters Thesis, University of Punjab, Pakistan, Xerox University Microfilms M-6941; *Masters Abstr.* 13(02):103.

Anees, M. A. 1974b. Changes in starch-gel electrophoretic pattern of serum proteins of a freshwater teleost *Channa punctatus* (Bloch), exposed to sublethal and chronic levels of three organophosphorous insecticides. *Ceylon J. Sci. Biol. Sci.* 11(1):53–58.

Anees, M. A. 1975. Acute toxicity of four organophosphorus insecticides to a freshwater teleost *Channa punctatus* (Bloch.). *Pak. J. Zool.* 7(2):135–141.

Anees, M. A. 1976. Intestinal pathology in a freshwater teleost, *Channa punctatus* (Bloch) exposed to sub-lethal and chronic levels of three organophosphorus insecticides. *Acta Physiol. Lat. Am.* 26(1):63–67.

Anees, M. A. 1978a. Hepatic pathology in a fresh-water teleost *Channa punctatus* (Bloch) exposed to sub-lethal and chronic levels of three organophosphorus insecticides. *Bull. Environ. Contam. Toxicol.* 19(5):524–527.

Anees, M. A. 1978b. Haematological abnormalities in a freshwater teleost, *Channa punctatus* (Bloch), exposed to sublethal and chronic levels of three organophosphorus insecticides. *Int. J. Ecol. Environ. Sci.* 4(1–3):53–60.

Ansari, B. A. and Kumar, K. 1984. Malathion toxicity: in vivo inhibition of acetylcholinesterase in the fish *Brachydanio rerio* (Cyprinidae). *Toxicol. Lett.* 20:283–287.

Ansari, B. A. and Kumar, K. 1987. Malathion toxicity: Effect on the ovary of the zebra fish *Brachydanio rerio* (Cyprinidae). *Int. Rev. Gesamten Hydrobiol.* 72(4):517–528.

Ansari, B. A., Aslam, M. and Kumar, K. 1987. Diazinon toxicity: activities of acetylcholinesterase and phosphatases in the nervous tissue of zebra fish, *Brachydanio rerio* (Cyprinidae). *Acta Hydrochim. Hydrobiol.* 15(3):301–306.

Antychowicz, J., Szymbor, E. and Roszkowski, J. 1979. Investigations upon the effects of some pesticides on carp (*Cyprinus carpio*). *Bull. Vet. Inst. Pulawy* 23(3–4):124–130.

Applegate, V. C., Howell, J. H., Hall, A. E., Jr. and Smith, M. A. 1957. Toxicity of 4,346 Chemicals to Larval Lampreys and Fishes. Spec. Sci. Rep.-Fish. No. 207, Fish Wildl. Serv., U.S. Dept. Int., Washington, D.C., 157 P.

Ardo, J. 1974. Kotazke akutnej toxicity niektorych pesticidov. (Acute toxicity of some pesticides). *Vodni Hospod.* 24(8):222–224.

Arias, E. and Zavanella, T. 1979. Teratogenic effects of manganese ethylenebisdithio-carbamate (maneb) on forelimb regeneration in the adult newt, *Triturus cristatus* Carnifex. *Bull. Environ. Contam. Toxicol.* 22(3):297–304.

Armstrong, D. A. and Millemann, R. E. 1974a. Pathology of acute poisoning with the insecticide Sevin in the bent-nosed clam, *Macoma nasuta*. *J. Invertebr. Pathol.* 24(2): 201–212.

Armstrong, D. A. and Millemann, R. E. 1974b. Effects of the insecticide Sevin and its first hydrolytic product, 1-naphthol, on some early developmental stages of the bay mussel *Mytilus edulis*. *Mar. Biol.* 28:11–15; U.S. NTIS COM-75-10967.

Armstrong, D. A. and Millemann, R. E. 1974c. Effects of the insecticide carbaryl on clams and some other intertidal mud flat animals. *J. Fish. Res. Board Can.* 31(4):466–470.

Armstrong, D. A., Buchanan, D. V. and Caldwell, R. S. 1976. A mycosis caused by *Lagneidium* sp. in laboratory-reared larvae of the dungeness crab, *Cancer magister*, and possible chemical treatments. *J. Invertebr. Pathol.* 28:329–336.

Arora, H. C., Shrivastava, S. K. and Seth, A. K. 1971a. Bioassay studies of some commercial organic insecticides. Part I. Studies with exotic carp *Puntius sophore* (Ham.). *Indian J. Environ. Health* 13(3):226–233.

Arora, H. C., Shrivastava, S. K. and Seth, A. K. 1971b. Bioassay studies of some commercial organic insecticides. Part II. Trials of malathion with exotic and indigenous carps. *Indian J. Environ. Health* 13(4):300–306.

Arora, N. and Kulshrestha, S. K. 1984. Comparison of the toxic effects of two pesticides on the testes of a fresh water teleost *Channa striatus* Bl. *Acta Hydrochim. Hydrobiol.* 12(4): 435–441.

Arora, N. and Kulshrestha, S. K. 1985. Effects of chronic exposure to sublethal doses of two pesticides on alkaline and acid phosphatase activities in the intestine of a fresh water teleost, *Channa striatus* Bl. (channidae). *Acta Hydrochim. Hydrobiol.* 13(5):619–624.

Arunachalam, S. and Palanichamy, S. 1982. Sublethal effects of carbaryl on surfacing behavior and food utilization in the air-breathing fish, *Macropodus cupanus*. *Physiol. Behav.* 29(1): 23–27.

Arunachalam, S., Jeyalakshmi, K. and Aboobucker, S. 1980. Toxic and sublethal effects of carbaryl on a freshwater catfish, *Mystus vittatus* (Bloch). *Arch. Environ. Contam. Toxicol.* 9(3):307–316.

Arunachalam, S., Palanichamy, S. and Balasubramanian, M. P. 1985. Sublethal effects of carbaryl on food utilization and oxygen consumption in the air-breathing fish, *Channa punctatus* (Bloch). *J. Environ. Biol.* 6(4):279–286.

Asaka, A., Sakai, M. and Tan, N. 1980. Influences of certain environmental factors on fish toxicity of cartap. *J. Takeda Res. Lab.* 39(1–2):28–33.

Atkinson, D. 1985. Toxicological properties of glyphosate - A summary. in Grossbard, E. and Atkinson, D. (Eds.). *The Herbicide Glyphosate*. Butterworth & Co., Boston, MA, pp. 127–133.

Bahner, L. H. and Nimmo, D. R. 1975. Methods to assess effects of combinations of toxicants, salinity and temperature on estuarine animals. *Trace Subst. Environ. Health* 9:169–177.

Bailey, H. C. and Liu, D. H. W. 1980. *Lumbriculus variegatus*, a benthic oligochaete, as a bioassay organism. in Eaton, J. C., Parrish, P. R. and Hendricks, A. C. (Eds.). *Aquatic Toxicology*, ASTM STP 707, pp. 205–215.

Balogh, J. C. and Anderson, J. L. 1992. Environmental Impacts of Turfgrass Pesticides. in Balogh, J. C. and Walker, W. J. (Eds.). *Golf Course Management & Construction: Environmental Issues.* Lewis Publishers, Chelsea, MI, Chapter 4.

Balogh, J. C., Gordon, G. A., Murphy, S. R. and Tietge, R. M. 1990. The Use and Potential Impacts of Forestry Herbicides in Maine. 89-12070101. Final Report, Maine Dept. Conserv. 237 pp.

Bancroft, R. and Prahlad, K. V. 1973. Effect of ethylenebis [dithiocarbamic acid] disodium salt (nabam) and ethylenebis [dithiocarbamate] manganese (maneb) on *Xenopus laevis* development. *Teratology* 7(2):143–150.

Bansal, S. K., Verma, S. R., Gupta, A. K. and Dalela, R. C. 1980. Predicting long-term toxicity by subacute screening of pesticides with larvae and early juveniles of four species of freshwater major carp. *Ecotoxicol. Environ. Saf.* 4:224–231.

Bansal, S. K., Verma, S. R., Gupta, A. K., Rani, S. and Dalela, R. C. 1979. Pesticide-induced alterations in the oxygen uptake rate of a freshwater major carp *Labeo rohita. Ecotoxicol. Environ. Saf.* 3(4):374–382.

Banu, N., Mustafa, G., Khan, A. M. and Ahmed, M. 1984. Histopathological effects of diazinon on the gonads of freshwater catfish, *Heteropneustes fossilis* (Bloch). *Dhaka Univ. Stud.* 32B(1):17–23.

Basak, P. K. and Konar, S. K. 1976. Toxicity of six insecticides to fish. *Geobios* 3(6):209–210.

Basha, S. M., Rao, K. S. P, Rao, K. R. S. S. and Rao, K. V. R. 1984. Respiratory potentials of the fish (*Tilapia mossambica*) under malathion, carbaryl and lindane intoxication. *Bull. Environ. Contam. Toxicol.* 32(5):570–574.

Basha, S. M., Rao, K. S. P., Rao, K. R. S. S. and Rao, K. V. R. 1983. Differential toxicity of malathion, BHC, and carbaryl to the freshwater fish, *Tilapia mossambica* (Peters). *Bull. Environ. Contam. Toxicol.* 31(5):543–546.

Bashamohideen, M., Obilesu, K. and Reddy, P. M. 1988. Effects of malathion and methyl parathion on the metabolic changes in the fish *Tilapia mossambica. Environ. Ecol.* 6(2): 481–483.

Basol, M. S., Eren, S. and Sadar, M. H. 1980. Comparative toxicity of some pesticides on human health and some aquatic species. *J. Environ. Sci. Health* B15(6):993–1004.

Bathe, R., Sachsse, K., Ullmann, L., Hormann, W. D., Zak, F. and Hess, R. 1975a. The evaluation of fish toxicity in the laboratory. *Proc. Eur. Soc. Toxicol.* 16:113–124.

Bathe, R., Ullmann, L., Sachsse, K. and Hess, R. 1975b. Relationship Between Toxicity to Fish and to Mammals: A Comparative Study Under Defined Laboratory Conditions. U.S. EPA-OPP Registration Standard.

Bathe, R., Ullmann, L. and Sachsse, K. 1973. Determination of pesticide toxicity to fish. *Schriftenr. Ver. Wasser-Boden-Lufthyg. Berlin-Dahlem* 37:241–256.

Beaumont, G., Bastin, R. and Therrien, H. P. 1976a. Effets physiologiques de l'atrazine a doses subletales sur *Lemna minor* L. I. Influence sur la croissance, la teneur en chlorophylle, en proteines et en azote soluble et total. *Nat. Can.* 103(6):527–533.

Beaumont, G., Bastin, R. and Therrien, H. P. 1976b. Effets physiologiques de l'atrazine a doses subletales sur *Lemna minor* L. II. Influence sur la photosynthese et sur la respiration. *Nat. Can.* 103(6):535–541.

Beaumont, G., Bastin, R. and Therrien, H. P. 1978. Effets physiologiques de l'atrazine a doses subletales sur *Lemna minor* L. III. Influence sur les proteines solubles et les acides nucleiques. *Nat. Can.* 105(2):103–113.

Beaumont, G., Lord, A. and Grenier, G. 1980. Effets physiologiques de l'atrazine a doses subletales sur *Lemna minor*. V. Influence sur l'ultrastructure des chloroplastes. *Can. J. Bot.* 58(14):1571–1577.

Beliles, R. 1965. Diazinon Safety Evaluation on Fish and Wildlife: Bobwhite Quail, Goldfish, Sunfish, and Rainbow Trout. U.S. EPA-OPP Registration Standard.

Bender, M. E. 1969. Uptake and retention of malathion by the carp. *Prog. Fish Cult.* 31(3): 155–159.

Bender, M. E. and Westman, J. R. 1976. The toxicity of malathion and its hydrolysis products to the eastern mudminnow, *Umbra pygmea* (Dekay). *Chesapeake Sci.* 17(2):125–128.

Benijts-Claus, C. and Persoone, G. 1975. La toxicite de trois herbicides sur l'ecosysteme aquatique. *La Tribune Du Ceredeau* 28(383):340–346.

Bennett, R. S. and Prince, H. H. 1981. Influence of agricultural pesticides on food preference and consumption by ring-necked pheasants. *J. Wildl. Manage.* 45(1):74–82.

Benson, B. and Boush, G. M. 1983. Effect of pesticides and PCBs on budding rates of green hydra. *Bull. Environ. Contam. Toxicol.* 30(3):344–350.

Berry, C. R., Jr. 1976. The Effects of Herbicide Treatment on a Reservoir Ecosystem. Ph.D. Thesis, Virginia Polytechnic Institute and State University, Blacksburg, VA, 212 pp.

Berry, W. O. and Brammer, J. D. 1977. Toxicity of water-soluble gasoline fractions to fourth-instar larvae of the mosquito, *Aedes aegypti L. Environ. Pollut.* 13(3): 229–234.

Bertagnolli, B. L. and Nadakavukaren, M. J. 1970. Effect of 2,4-dichlorophenoxyacetic acid on the fine structure of *Chlorella pyrenoidosa* Chick. *J. Phycol.* 6:98–100.

Bertagnolli, B. L. and Nadakavukaren, M. J. 1974. Some physiological responses of *Chlorella pyrenoidosa* to 2,4-dichlorophenoxyacetic acid. *J. Exp. Bot.* 25(84):180–188.

Berteau, P. E. and Deen, W. A. 1978. A comparison of oral and inhalation toxicities of four insecticides to mice and rats. *Bull. Environ. Contam. Toxicol.* 19(1):113–120.

Bhatia, H. L. 1971. Toxicity of some pesticides to *Puntius ticto* (Hamilton). *Sci. Cult.* 37(3):160–161.

Birge, W. J., Black, J. A. and Bruser, D. M. 1979. Toxicity of Organic Chemicals to Embryo-Larval Stages of Fish. Ecol. Res. Ser. EPA-560/11-79-007. Office of Toxic Substances, U.S. Environmental Protection Agency, Washington, D.C., 60 pp.

Birge, W. J., Black, J. A. and Kuehne, R. A. 1980. Effects of Organic Compounds on Amphibian Reproduction. Project No. A-074-KY. Res. Rep. No. 121, U.S. NTIS PB80-147 523. Water Resourc. Res. Inst., University of Kentucky, Lexington, KY, 39 pp.

Biro, P. 1979. Acute effects of the sodium salt of 2,4-D on the early developmental stages of bleak, *Alburnus alburnus*. *J. Fish Biol.* 14(1):101–109.

Black, W. M. and Neely, D. 1975. Effect of soil-injected benomyl on resident earthworm populations. *Pestic. Sci.* 6:543–545.

Blanck, H., Wallin, G. and Wangberg, S. A. 1984. Species-dependent variation in algal sensitivity to chemical compounds. *Ecotoxicol. Environ. Saf.* 8:339–351.

Bluzat, R. and Seuge, J. 1979. Effets de trois insecticides (lindane, fenthion et carbaryl): Toxicite aigue sur quatre especes d'invertebres limniques; toxicite chronique chez le mollusque pulmone lymnea. *Environ. Pollut.* 18(1):51–70.

Bluzat, R., Jonot, O. and Seuge, J. 1982a. Acute toxicity of a fungicide, thiram (dithiocarbamate) in the freshwater amphipodal crustacean *Gammarus pulex*. *Environ. Pollut. Ser. A Ecol. Biol.* 29(3):225–233.

Bluzat, R., Jonot, O. and Seuge, J. 1982b. Acute toxicity of thiram in *Gammarus pulex*: Effect of a one-hour contamination and degradation of an aqueous suspension. *Bull. Environ. Contam. Toxicol.* 29(2):248–252.

Blythe, T. O., Grooms, S. M. and Frans, R. E. 1979. Determination and characterization of the effects of fluometuron and MSMA on *Chlorella. Weed Sci.* 27(3):294–299.

Bogacka, T. and Groba, J. 1980. Toksycznosc i biodegradacja chlorfenwinfosu, karbarylu oraz propoksuru w srodowisku wodnym. *Bromatol. Chem. Toksykol.* 13(2):151–158; Pestab:0175 (1981).

Bohm, H. H. and Muller, H. 1976. Model studies on the accumulation of herbicides by microalgae. *Naturwissenschaften* 63(6):296.

Boike, A. H., Jr., Rathburn, C. B., Jr., Lang, K. L., Masters, H. M. and Floore, T. G. 1985. Current status on the Florida abate monitoring program — Susceptibility levels of three species of mosquitoes during 1984. *J. Am. Mosq. Control Assoc.* 1(4): 498–501.

Bond, C. E., Lewis, R. H. and Fryer, J. L. 1959. Toxicity of Various Herbicidal Materials to Fishes. in Second Seminar Biol. Problems Water Pollut., R.A. Taft Sanit. Eng. Center Tech. Rept. W60-3. pp. 96–101.

Bookhout, C. G. and Monroe, R. J. 1977. Effects of malathion on the development of crabs. in Vernberg, F. J., Calabrese, A., Thurberg, F. P. and Vernberg, W. P. (Eds.). *Physiological Responses of Marine Biota to Pollutants*, Academic Press, New York, pp. 3–19.

Boumaiza, M., Ktari, M. H. and Vitiello, P. 1979. Toxicite de divers pesticides utilises en tunisie pour *Aphanius fasciatus Nardo*, 1827 (*Pisces, Cyprinodontidae*). *Arch. Inst. Pasteur Tunis* 56(3):307–342.

Bovey, R. W. and Young, A. L. 1980. *The Science of 2,4,5-T and Associated Phenoxy Herbicides*. John Wiley & Sons, New York, 462 pp.

Boyd, C. E. and Ferguson, D. E. 1964. Spectrum of cross-resistance to insecticides in the mosquito fish, *Gambusia affinis*. *Mosq. News* 24(1):19–21.

Braginskii, L. P. and Migal, A. K. 1973. Effect of atrazine on the vital activity of some aquatic plants. *Eksp. Vodn. Toksikol.* 5:179–187.

Brenniman, G., Hartung, R. and Weber, W. J., Jr. 1976. A continuous flow bioassay method to evaluate the effect of outboard motor exhausts and selected aromatic toxicants on fish. *Water Res.* 10(2):165–169.

Bringmann, G. and Kuhn, R. 1978a. Limiting values for the noxious effects of water pollutant material to blue algae (*Microcystis aeruginosa*) and green algae *Scenedesmus quadricauda*. *Vom Wasser* 50:45–60.

Bringmann, G. and Kuhn, R. 1977a. Results of the damaging effect of water pollutants on *Daphnia magna*. *Z. Wasser Abwasser Forsch.* 10(5):161–166.

Bringmann, G. and Kuhn, R. 1977b. Limiting values for the damaging action of water pollutants to bacteria (*Pseudomonas putida*) and green algae (*Scenedesmus quadricauda*) in the cell multiplication inhibition test. *Z. Wasser Abwasser Forsch.* 10(3–4):87–98.

Bringmann, G. and Kuhn, R. 1978b. Testing of substances for their toxicity threshold: Model organisms *Microcystis (diplocystis) aeruginosa* and *Scenedesmus quadricauda*. *Mitt. Int. Ver. Theor. Angew. Limnol.* 21:275–284.

Bringmann, G., Kuhn, R. and Winter, A. 1980. Determination of biological damage from water pollutants to protozoa. III. Saprozoic flagellates. *Z. Wasser Abwasser Forsch.* 13(5): 170–173.

Brockway, D. L., Smith, P. D. and Stancil, F. E. 1984. Fate and effects of atrazine in small aquatic microcosms. *Bull. Environ. Contam. Toxicol.* 32:345–353.

Brooke, L. T., Call, D. J., Geiger, D. L. and Northcott, C. E. 1984. Acute Toxicities of Organic Chemicals to Fathead Minnows (*Pimephales promelas*), Vol. 1. Center for Lake Superior Environmental Studies, University of Wisconsin, Superior, WI, 414 pp.

Brooks, J. M., Fryxell, G. A., Reid, D. F. and Sackett, W. M. 1977. Gulf underwater flare experiment (gufex): Effects of hydrocarbons on phytoplankton. in Giam, C. S. (Ed.). *Proc. Pollution Effects Marine Organisms*, D.C. Heath Co., Lexington, MA, pp. 45–75.

Brown, A. C. 1974. Observations on the effect of ammonium nitrate solutions on some common marine animals from table bay. *Trans. R. Soc. S. Afr.* 41(2):217–223.

Brown, A. C. and Currie, A. B. 1973. Tolerance of *Bullia digitalis* (Prosobranchiata) to solutions of ammonium nitrate in natural sea water. *S. Afr. J. Sci.* 69(7):219–220.

Brown, J. R. and Chow, L. Y. 1974. The effect of Dursban on micro-flora in non-saline waters. in Coulston, P., Frederick, P. and Korte, F. (Eds.). *Environmental Quality and Safety Supplement.* Vol. 3. Pesticides, Intl. Union of Pure and Applied Chemistry, Third Int. Congress, Helsinki, Finland, pp. 774–779.

Brown, J. R., Chow, L. Y. and Deng, C. B. 1976. The effect of Dursban upon fresh water phytoplankton. *Bull. Environ. Contam. Toxicol.* 15(4):437–441.

Brown, K. W., Anderson, D. C., Jones, S. G., Deuel, L. E. and Price, J. D. 1979. The relative toxicity of four pesticides in tap water and water from flooded rice paddies. *Int. J. Environ. Stud.* 14(1):49–54.

Buccafusco, R. J., Ells, S. J. and LeBlanc, G. A. 1981. Acute toxicity of priority pollutants to bluegill (*Lepomis macrochirus*). *Bull. Environ. Contam. Toxicol.* 26(4):446–452.

Buchanan, D. V., Millemann, R. E. and Stewart, N. E., 1970. Effects of the insecticide Sevin on various stages of the dungeness crab, *Cancer magister. J. Fish. Res. Board Can.* 27(1):93–104.

Burchfield, H. P. and Storrs, E. E. 1954. Kinetics of insecticidal action based on the photomigration of larvae of *Aedes aegypti* (L.). *Contrib. Boyce Thompson Inst.* 17:439–452.

Burrell, R. E., Inniss, W. E. and Mayfield, C. I. 1985. Detection and analysis of interactions between atrazine and sodium pentachlorophenate with single and multiple algal-bacterial populations. *Arch. Environ. Contam. Toxicol.* 14:167–177.

Bursian, S. J. and Edens, F. W. 1977. The prolonged exposure of Japanese quail to carbaryl and its effects on growth and reproductive parameters. *Bull. Environ. Contam. Toxicol.* 17(3):360–368.

Buslovich, S. Yu. and Borushko, N. V. 1976. Chloroderivatives of phenoxyacetic acid as antagonists of thyroid hormones. *Farmakol. Toksikol. (Moscon.)* (4):481–483.

Butcher, J. E., Boyer, M. G. and Fowle, C. D. 1977. Some changes in pond chemistry and photosynthetic activity following treatment with increasing concentrations of chlorpyrifos. *Bull. Environ. Contam. Toxicol.* 17(6):752–758.

Butcher, J., Boyer, M. and Fowle, C. D. 1975. Impact of Dursban and Abate on microbial numbers and some chemical properties of standing ponds. *Water Pollut. Res. Can.* 10: 33–41.

Butler, G. L., Deason, T. R. and O'Kelley, J. C. 1975. The effect of atrazine, 2,4-D, methoxychlor, carbaryl and diazinon on the growth of planktonic algae. *Br. Phycol. J.* 10(4):371–376.

Butler, P. A. 1963. A Review of Fish and Wildlife Service Investigations During 1961 and 1962. Circular No. 7. U.S. Dept. Int. Fish Wildl. Serv., Washington, D.C., 25 pp.

Butler, P. A. 1964. Commercial fishery investigations. in Pesticide-Wildlife Studies, 1963. U.S. Dept. Int. Fish Wildl. Serv. Circular 199. 28 pp.

Butler, P. A. 1965. Commercial Fishery Investigations. U.S. Fish Wildl. Serv., Pestic. Wildl. Stud., Circ. 226, Washington, D.C., pp. 65–77.

Cain, J. R. and Cain, R. K. 1984. Effects of five insecticides on zygospore germination and growth of the green alga *Chlamydomonas moewusii*. *Bull. Environ. Contam. Toxicol.* 33(5):571–574.

Caldwell, R. S., Buchanan, D. V., Armstrong, D. A., Mallon, M. H. and Millemann, R. E. 1979. Toxicity of the herbicides 2,4-D, DEF, propanil and trifluralin to the dungeness crab, *Cancer magister*. *Arch. Environ. Contam. Toxicol.* 8(4):383–396.

Canton, J. H. 1976. The toxicity of benomyl, thiophanate-methyl, and bcm to four freshwater organisms. *Bull. Environ. Contam. Toxicol.* 16(2):214–218.

Carlson, A. R. 1971. Effects of long-term exposure to carbaryl (Sevin) on survival, growth, and reproduction of the fathead minnow (*Pimephales promelas*). *J. Fish. Res. Board Can.* 29:583–587.

Carlson, C. A. 1966. Effects of three organophosphorus insecticides on immature hexagenia and hydropsyche of the upper Mississippi River. *Trans. Am. Fish. Soc.* 95(1):1–5.

Carpenter, L. A. and Eaton, D. L. 1983. The disposition of 2,4,-dichlorophenoxyacetic acid in rainbow trout. *Arch. Environ. Contam. Toxicol.* 12(2):169–173.

Carpentier, A. G., Mackenzie, D. L. and Frank, R. 1988. Residues and efficacy of two formulations of 2,4-D on aquatic macrophytes in Buckhorn Lake, Ontario. *J. Aquat. Plant Manage.* 26:29–37.

Carter, F. L. and Graves, J. B. 1972. Measuring effects of insecticides on aquatic animals. *La. Agric.* 16(2):14–15.

Cecil, H. C., Harris, S. J. and Bitman, J. 1974. Effects of nonpersistent pesticides on liver weight, lipids and vitamin A of rats and quail. *Bull. Environ. Contam. Toxicol.* 11(6): 496–499.

Chadwick, P. R., Slatter, R. and Bowron, M. J. 1984. Cross-resistance to pyrethroids and other insecticides in *Aedes aegypti*. *Pestic. Sci.* 15(2):112–120.

Chaiyarach, S., Ratananun, V. and Harrel, R. C. 1975. Acute toxicity of the insecticides toxaphene and carbaryl and the herbicides propanil and molinate to four species of aquatic organisms. *Bull. Environ. Contam. Toxicol.* 14(3):281–284.

Chambers, J. S. 1970. Investigation of Chemical Control of Ghost Shrimp on Oyster Grounds 1960–1963. Washington (State) Dept. Fish., Tech. Rep. No. 1:25–62.

Chatterjee, K. 1975. Toxicity of diazinon to fish and fish food organisms. *Indian Sci. Congr. Assoc. Proc.* 62:166–167.

Chatterjee, K. and Konar, S. K. 1984. Effects of the pesticide diazinon at various ph and turbidity on fish and aquatic ecosystem. *Environ. Ecol.* 2(1):49–53.

Chaudhari, T. R., Jadhav, M. L. and Lomte, V. S. 1988. Acute toxicity of organophosphates to fresh water snails from Panzara River at Dhule, MS. *Environ. Ecol.* 6(1):244–246.

Cheah, M. L., Avault, J. W., Jr. and Graves, J. B. 1978a. Some effects of thirteen rice pesticides on crawfish *Procambarus clarkii* and *P. acutus acutus*. in Laurent, P. J. (Ed.). *Proc. of the Int. Symp. Freshwater Crayfish*, Vol. 4. Thonon-Lesbains, France, pp. 349–361.

Cheah, M. L., Avault, J. W., Jr. and Graves, J. B. 1980. Some effects of rice pesticides on crawfish. *La. Agric.* 23(2):8, 9, 11.

Cheah, M. L., Avault, J. W., Jr. and Graves, J. B. 1980b. Acute toxicity of selected rice pesticides to crayfish *Procambarus clarkii*. *Prog. Fish Cult.* 42(3):169–172.

Chen, P. S., Lin, Y. N. and Chung, C. L. 1971. Laboratory studies on the susceptibility of mosquito-eating fish, *Lebistes reticulatus* and the larvae of *Culex pipiens fatigans* to insecticides. *Tai-Wan I. Hsueh Hui Tsa Chih* 70(1):28–35.

Chin, Y. N. and Sudderuddin, K. I. 1979. Effect of methamidophos on the growth rate and esterase activity of the common carp *Cyprinus carpio* L. *Environ. Pollut.* 18(3): 213–220.

Chinnaswamy, R. and Patel, R. J. 1983. Effect of pesticide mixtures on the blue-green alga *Anabaena flos-aquae*. *Microbios Lett.* 24:141–143.

Choudhuri, D. K., Sadhu, A. K. and Mukhopadhyay, P. K. 1984. Toxicity of two organophosphorus insecticides — malathion and phosphamidon to the fish *Channa striatus*. *Aquat. Sci. Fish. Abstr.* 14:264; *Environ. Ecol.* 2(2):143–148.

Chouhan, M. S., Verma, D. and Pandey, A. K. 1983. Herbicide induced haematological changes and their recovery in a freshwater fish, *Puntius ticto* (Ham.). *Comp. Physiol. Ecol.* 8(4):249–251.

Christensen, G. M. and Tucker, J. H. 1976. Effects of selected water toxicants on the in vitro activity of fish carbonic anhydrase. *Chem. Biol. Interact.* 13:181–192.

Christie, A. E. 1969. Effects of insecticides on algae. *Water Sewage Works* 116(5):172–176.

Christy, S. L., Karlander, E. P. and Parochetti, J. V. 1981. Effects of glyphosate on the growth rate of *Chlorella*. *Weed Sci.* 29(1):5–7.

Ciba-Geigy Corporation. 1976. Reports of Investigations Made with Respect to Fish and Wildlife Requirements for Diazinon and Its Formulated Products. U.S. EPA-OPP Registration Standard.

Ciba-Geigy Corporation. 1988. Princep Herbicide. Technical bulletin. Agricultural Division, Greensboro, NC, 9 pp.

Clark, J. R., Patrick, J. M., Jr., Middaugh, D. P. and Moore, J. C. 1985. Relative sensitivity of six estuarine fishes to carbophenothion, chlorpyrifos,and fenvalerate. *Ecotoxicol. Environ. Saf.* 10(3):382–390.

Cocks, J. A. 1973. The effect of aldrin on water balance in the freshwater pulmonate gastropod (*Biomphalaria glabrata*). *Environ. Pollut.* 5(2):149–151.

Conte, F. S. and Parker, J. C. 1975. Effect of aerially-applied malathion on juvenile brown and white shrimp *Penaeus aztecus* and *P. setiferus*. *Trans. Am. Fish. Soc.* 104(4):793–799.

Conti, E. 1987. Acute toxicity of three detergents and two insecticides in the lugworm, *Arenicola marina* (L.): A histological and scanning electron microscope study. *Aquat. Toxicol.* 10:325–334.

Cook, G. H. and Moore, J. C. 1976. Determination of malathion, malaoxon, and mono- and dicarboxylic acids of malathion in fish, oyster, and shrimp tissue. *J. Agric. Food Chem.* 24(3):631–634.

Cook, G. H., Moore, J. C. and Coppage, D. L. 1976. The relationship of malathion and its metabolities to fish poisoning. *Bull. Environ. Contam. Toxicol.* 16(3):283–290.

Cook, M. E. and Swait, S.A.J. 1975. Effects of some fungicide treatments on earthworm populations and leaf removal in apple orchards. *J. Hort. Sci.* 50:495–499.

Cooke, A. S. 1972. The effects of DDT, dieldrin and 2,4-D on amphibian spawn and tadpoles. *Environ. Pollut.* 3:51–68.

Cope, O. B. 1965. Sport Fishery Investigations. U.S. Fish Wildl. Serv. Circ. 226, pp. 51–63.

Cope, O. B., Wood, E. M. and Wallen, G. H. 1970. Some chronic effects of 2,4-D on the bluegill (*Lepomis macrochirus*). *Trans. Am. Fish. Soc.* 99(1):1–12.

Coppage, D. L. 1977. Anticholinesterase action of pesticidal carbamates in the central nervous system of poisoned fishes. in Vernberg, J. F. (Ed.). *Symp. Physiological Responses of Marine Biota to Pollutants*, Academic Press, New York, pp. 93–102.

Coppage, D. L. and Matthews, E. 1974. Short-term effects of organophosphate pesticides on cholinesterases of estuarine fishes and pink shrimp. *Bull. Environ. Contam. Toxicol.* 11(5):483–488.

Coppage, D. L., Matthews, E., Cook, G. H. and Knight, J. 1975. Brain acetylcholinesterase inhibition in fish as a diagnosis of environmental poisoning by malathion, o,o-dimethyl s-(1,2-dicarbethoxyethyl) phosphorodithioate. *Pestic. Biochem. Physiol.* 5(6):536–542.

Couch, J. A., Winstead, J. T., Hansen, D. J. and Goodman, L. R. 1979. Vertebral dysplasia in young fish exposed to the herbicide trifluralin. *J. Fish Dis.* 2:35–42; EPA-600/J-79-072. U.S. NTIS PB80–177751. Environ. Res. Lab., U.S. EPA, Gulf Breeze, FL, 11 pp.

Courtemanch, D. L. and Gibbs, K. E. 1980. Short- and long-term effects of forest spraying of carbaryl (Sevin-4-oil) on stream invertebrates. *Can. Entomol.* 112(3):271–276.

Coutant, C. C. 1964. Insecticide Sevin: Effect of aerial spraying on drift of stream insects. *Science* 146:420–421.

Cress, C. R. and Strother, A. 1974. Effects on drug metabolism of carbaryl and 1-naphthol in the mouse. *Life Sci.* 14:861–872.

Crosby, D. G. and Tucker, R. K. 1966. Toxicity of aquatic herbicides to *Daphnia magna*. *Science* 154:289–290.

Crosby, D. G., Tucker, R. K. and Aharonson, N. 1966. The detection of acute toxicity with *Daphnia magna*. *Food Cosmet. Toxicol.* 4:503–514.

Cutkomp, L. K., Yap, H. H., Cheng, E. Y. and Koch, R. B. 1971b. ATPase activity in fish tissue homogenates and inhibitory effects of DDT and related compounds. *Chem. Biol. Interact.* 3:439–447.

Cutkomp, L. K., Yap, H. H., Vea, E. V. and Koch, R. B. 1971a. Inhibition of oligomycin-sensitive (mitochondrial) Mg2+ ATPase by DDT and selected analogs in fish and insect tissue. *Life Sci.* 10(2):1201–1209.

Dabrowska, H. and Sikora, H. 1986. Acute toxicity of ammonia to common carp (*Cyprinus carpio* L.). *Pol. Arch. Hydrobiol.* 33(1):121–128.

Dad, N. K., Qureshi, S. A. and Pandya, V. K. 1982. Acute toxicity of two insecticides to tubificid worms, *Tubifex tubifex* and *Limnodrilus hoffmeisteri*. *Environ. Int.* 7(5): 361–363.

Dahlgren, R. B. and Linder, R. L. 1971. Effects of polychlorinated biphenyls on pheasant reproduction, behavior, and survival. *J. Wildl. Manage.* 35(2):315–318.

Dalela, R. C., Verma, S. R. and Bhatnagar, M. C. 1978. Biocides in relation to water pollution. Part I: Bioassay studies on the effects of a few biocides on fresh water fish, *Channa gachua*. *Acta Hydrochim. Hydrobiol.* 6(1):15–25.

Darsie, R. F., Jr. and Corriden, F. E. 1959. The toxicity of malathion to killifish (Cyprinodontidae) in Delaware. *J. Econ. Entomol.* 52(4):696–700.

Darwazeh, H. A. and Mulla, M. S. 1974. Toxicity of herbicides and mosquitoe larvicides to the mosquito fish *Gambusia affinis*. *Mosq. News* 34(2):214–219.

Das, B. and Singh, P. K. 1977. Mutagenecity of pesticides in blue-green algae. *Microbios Lett.* 5(19–20):103–107.

DaSilva, E. J., Henriksson, L. E. and Henriksson, E. 1975. Effect of pesticides on blue-green algae and nitrogen-fixation. *Arch. Environ. Contam. Toxicol.* 3(2):193–204.

Davey, R. B., Meisch, M. V. and Carter, F. L. 1976. Toxicity of five ricefield pesticides to the mosquitofish, *Gambusia affinis*, and green sunfish, *Lepomis cyanellus*, under laboratory and field conditions in Arkansas. *Environ. Entomol.* 5(6):1053–1056.

Davies, P. E. 1987. Physiological, anatomic and behavioral changes in the respiratory system of *Salmo gairdneri* Rich. on acute and chronic exposure to chlorothalonil. *Comp. Biochem. Physiol.* 87C(1):113–119.

Davies, P. E. 1988. Disappearance rates of chlorothalonil (TCIN) in the aquatic environment. *Bull. Environ. Contam. Toxicol.* 40(3):405–409.

Davis, D. E. 1981. Effects of Herbicides on Submerged Seed Plants. Govt. Reports Announce. Index 1, U.S. NTIS PB81-103103. 19 pp.

Davis, H. C. and Hidu, H. 1979. Effects of pesticides on embryonic development of clams and oysters and on survival and growth of the larvae. *Fish. Bull.* 67(2):393–404.

Davis, J. T. and Hughes, J. S. 1963. Further observations on the toxicity of commercial herbicides to bluegill sunfish. *Proc. South. Weed Conf.* 16:337–340.

Davis, P. W. and Wedemeyer, G. A. 1971. Na+, K+-activated-ATPase inhibition in rainbow trout: a site for organochlorine pesticide toxicity. *Comp. Biochem. Physiol.* B 40(3): 823–827.

Davis, P. W., Friedhoff, J. M. and Wedemeyer, G. A. 1972. Organochlorine insecticide, herbicide and polychlorinated biphenyl (PCB) inhibition of NAK-Atpase in rainbow trout. *Bull. Environ. Contam. Toxicol.* 8(2):69–72.

De Llamas, M. C., De Castro, A. C. and De d'Angelo, A. M. P. 1985. Cholinesterase activities in developing amphibian embryos following exposure to the insecticides dieldrin and malathion. *Arch. Environ. Contam. Toxicol.* 14(2):161–166.

Delistraty, D. A. and Hershner. C. 1984. Effects of the herbicide atrazine on adenine nucleotide levels in *Zostera marina* L. (eelgrass). *Aquat. Bot.* 18(4):353–369.

Dennis, W. H., Jr., Rosencrance, A. B. and Randall, W. F. 1980. Acid hydrolysis of military standard formulations of diazinon. *J. Environ. Sci. Health* B15(1):47–60.

Dennis, W. H., Jr., Meier, E. P. Randall, W. F., Rosencrance, A. B. and Rosenblatt, D. H. 1979a. Degradation of diazinon by sodium hypochlorite. Chemistry and aquatic toxicity. *Environ. Sci. Tech.* 13(5):594–598.

Dennis, W. H., Jr., Meier, E. P., Rosencrance, A. B., Randall, W. F., Reagan, M. T. and Rosenblatt, D. 1979b. Chemical Degradation of Military Standard Formulations of Organophosphorus and Carbamate Pesticides. II. Degradation of diazinon by sodium hypochlorite. U.S. Army Med. Bioeng. Res. Dev. Lab., Tech. Rep. No. 7904. U.S. NTIS AD-AO81 098/6. Fort Detrick, MD, 40 pp.

Dennis, W. H., Jr., Rosencrance, A. B. and Randall, W. F. 1980. Acid hydrolysis of military standard formulations of diazinon. *J. Environ. Sci. Health* B15(1):47–60.

DeNoyelles, F., Kettle, W. D. and Sinn, D. E. 1982. The responses of plankton communities in experimental ponds to atrazine, the most heavily used pesticide in the United States. *Ecology* 63(5):1285–1293.

Deoray, B. M. and Wagh, S. B. 1987. Acute toxicity of thiodan, Nuvan and Dithane m-45 to the freshwater fish, *Barilius bendelisis* (Ham.). *C.A. Sel. Environ. Pollut.* 24:107–192596W; *Geobios* 14(4):151–153.

Derosa, C. T., Taylor, D. H., Farrell, M. P. and Seilkop, S. K. 1976. Effects of Sevin on the reproductive biology of the *Coturnix*. *Poult. Sci.* 55:2133–2141.

Desi, I., Dura, G., Gonczi, L., Kneffel, Z., Strohmayer, A. and Szabo, Z. 1975. Toxicity of malathion to mammals, aquatic organisms and tissue culture cells. *Arch. Environ. Contam. Toxicol.* 3(4):410–425.

Desi, I., Gonczi, L., Simon, G., Farkas, I. and Kneffel, Z. 1974. Neurotoxicologic studies of two carbamate pesticides in subacute animal experiments. *Toxicol. Appl. Pharmacol.* 27: 465–476.

Dewey, S. L. 1986. Effects of the herbicide atrazine on aquatic insect community structure and emergence. *Ecology* 67(1):148–162.

Dieter, M. P. 1975. Further studies on the use of enzyme profiles to monitor residue accumulation in wildlife: plasma enzymes in starlings. *Arch. Environ. Contam. Toxicol.* 3(2):142–150.

Dive, D., Leclerc, H. and Persoone, G. 1980. Pesticide toxicity on the ciliate protozoan *Colpidium campylum*: Possible consequences of the effect of pesticides in the aquatic environment. *Ecotoxicol. Environ. Saf.* 4:129–133.

Dixon, R. D. and Brust, R. A. 1971. Field testing of insecticides used in mosquito control, and a description of the bioassay technique used in temporary pools. *J. Econ. Entomol.* 64(1): 11–14.

Dodson, J. J. and Mayfield, C. I. 1979a. The dynamics and behavioral toxicology of aquakleen (2,4-D butoxyethanol ester) as revealed by the modification of rheotropism in rainbow trout. *Trans. Am. Fish. Soc.* 108(6):632–640.

Dodson, J. J. and Mayfield, C. I. 1979b. Modification of the rheotropic response of rainbow trout (*Salmo gairdneri*) by sublethal doses of the aquatic herbicides diquat and simazine. *Environ. Pollut.* 18(2):147–157.

Doe, K. G., Ernst, W. R., Parker, W. R., Julien, G. R. J. and Hennigar, P. A. 1988. Influence of pH on the acute lethality of fenitrothion, 2,4-D, and aminocarb and some pH-altered sublethal effects of aminocarb on aminocarb. *Can. J. Fish. Aquat. Sci.* 45(2):287–293.

Doherty, J. D. and Matsumura, F. 1974. DDT effect on 32P incorporation from gamma-labelled ATP into proteins from lobster nerve. *J. Neurochem.* 22:765–772.

Dortland, R. J. 1980. Toxicological evaluation of parathion and azinphosmethyl in freshwater model ecosystems. *Versl. Landbouwkd. Onderz.* 898:1–112.

Douglas, M. T., Chanter, D. O., Pell, I. B. and Burney, G. M. 1986. A proposal for the reduction of animal numbers required for the acute toxicity to fish test (LC_{50} determination). *Aquat. Toxicol.* 8(4):243–249.

Dow Chemical Company. 1987. Supplemental Information Concerning Triclopyr and Picloram (with Additional 2,4-D Comments) for Use in Development of a Vegetation Management Environmental Impact Statement for the Southern Region. March 1987 submitted to U.S. For. Serv. and Labat-Anderson.

Dow Chemical Company. Undated. Technical Information on Triclopyr, the Active Ingredient of Garlon Herbicides. Agricultural Productions Department, Midland, MT.

Dowden, B. F. and Bennett, H. J. 1965. Toxicity of selected chemicals to certain animals. *J. Water Pollut. Control Fed.* 37(9):1308–1316.

Dragomirescu, A., Raileanu, L. and Ababei, L. 1975. The effect of carbetox on glycolysis and the activity of some enzymes in carbohydrate metabolism in the fish and rat liver. *Water Res.* 9(2):205–209.

Duangsawasdi, M. 1977. Organophosphate insecticide toxicity in rainbow trout (*Salmo gairdneri*). Effects of Temperature and Investigations on the Sites of Action. Ph.D. Thesis, University of Manitoba, Manitoba, Canada, 138 pp. *Diss. Abstr. Int. B* 38(11):5228 (1978).

Duangsawasdi, M. and Klaverkamp, J. F. 1979. Acephate and fenitrothion toxicity in rainbow trout: Effects of temperature stress and investigations on the sites of action. in Marking, L. L. and Kimerle, R. A. (Eds.). *Aquatic Toxicology*, ASTM STP 667, Philadelphia, PA, pp. 35–51.

Dubale, M. S. and Shah, P. 1984. Toxic effect of malathion on the kidney of a fresh water teleost *Channa punctatus. Comp. Physiol. Ecol.* 9(3):238–244.

Dunstan, W. M., Atkinson, L. P. and Natoli, J. 1975. Stimulation and inhibition of phytoplankton growth by low molecular weight hydrocarbons. *Mar. Biol.* 31(4):305–310.

Edmiston, C. E., Jr., Goheen, M. and Malaney, G. W. 1984. Environmental assessment of carbamate toxicity: Utilization of the coomassie blue g soluble protein assay as an index of environmental toxicity. *Hazard. Waste* 1(2):205–215.

Edmiston, C. E., Jr., Goheen, M., Malaney, G. W. and Mills, W. L. 1985. Evaluation of carbamate toxicity: Acute toxicity in a culture of *paramecium multimicronucleatum* upon exposure to aldicarb, carbaryl, and mexacarbate as measured by warburg respirometry and acute plate assay. *Environ. Res.* 36(2):338–350.

Edwards, A. C. and Davis, D. E. 1975. Effects of an organic arsenical herbicide on a salt marsh ecosystem. *J. Environ. Qual.* 4(2):215–219.

Eisler, R. 1969. Acute toxicities of insecticides to marine decapod crustaceans. *Crustaceana* 16(3):302–310.

Eisler, R. 1970a. Acute Toxicities of Organochlorine and Organophosphorus Insecticides to Estuarine Fishes. Tech. Paper No. 46. Bur. Sport Fish. Wildl., U.S. Dept. Int., Washington, D.C., 12 pp.

Eisler, R. 1970b. Factors Affecting Pesticide-Induced Toxicity in an Estuarine Fish. Technical Paper No. 45. Bureau Sport Fish. Wildl., Fish Wildl. Dept., U.S. Dept. Int., Washington, D.C., 20 pp.

Eisler, R. 1970c. Latent effects of insecticide intoxication to marine molluscs. *Hydrobiology.* 36(3–4):345–352.

El-Ayouty, E. Y., Khalil, A. I., Ishak, M. M. and Ibrahim, E. A. 1978. Effect of 2,4-D, iron, mercury and bayluscide on the cellular structure of microcystis *Flos-aquae, Oscillatoria amphibia, Coelastrum microporum* and *Scenedesmus opaliensis. Egypt. J. Bot.* 21(2): 121–130.

El-Khatib, Z. I. and Georghiou, G. P. 1985. Geographic variation of resistance to organophosphates, propoxur and ddt in the southern house mosquito, *Culex quinquefasciatis,* in California. *J. Am. Mosq. Control Assoc.* 1(3):279–283.

El-Khatib, Z. I. 1985. Isolation of an organophosphate susceptible strain of *Culex quinquefasciatus* from a resistant field population by discrimination against esterase-2 phenotypes. *J. Am. Mosq. Control Assoc.* 1(1):105–107.

El-Refai, A., Fahmy, F. A., Abdel-Lateef, M. F. A. and Imam, A. K. E. 1976. Toxicity of three insecticides to two species of fish. *Int. Pest Control* 18(6):4–8.

Elder, J. H., Lembi, C. A. and Morre, D. J. 1970. Toxicity of 2,4-D and Picloram to Fresh Water Algae. Project C-36-48C. Dept. of Botany and Plant Pathology, Purdue University, U.S. NTIS PB-199 114. 10 pp.

Elghar, M. R. A., Elbadry, E. A., Hassan, S. M. and Kilany. S. M. 1971. Effect of some pesticides on the predatory mite *Agestemus exsertus. J. Econ. Entomol.* 64(1):26–27.

Ellgehausen, H., Guth, J. A. and Esser, H. O. 1980. Factors determining the bioaccumulation potential of pesticides in the individual compartments of aquatic food chains. *Ecotoxicol. Environ. Saf.* 4(2):134–157.

Elliott-Feeley, E. and Armstrong, J. B. 1982. Effects of fenitrothion and carbaryl on *Xenopus laevis* development. *Toxicology* 22(2):319–335.

Environmental Protection Agency. 1982a. Glyphosate; tolerances and exemptions from tolerances for pesticide chemicals in or on raw agricultural commodities. *Fed. Reg.*. 47(241):56136–56138.

Environmental Protection Agency. 1982b. Code of Federal Regulations: Protection of Environment. 40 CFR:180–364.

Environmental Protection Agency. 1983. Dicamba Science Chapters. Office of Pesticides and Toxic Substances, Washington, D.C.

Environmental Protection Agency. 1985a. Hazard Evaluation Division, Standard Evaluation Procedure, Acute Toxicity Test for Freshwater Fish. EPA-540/9-85-006. U.S. Environmental Protection Agency, Washington, D.C.

Environmental Protection Agency. 1985b. Summary of Results of Studies Submitted in Support of the Registration of Triclopyr. Office of Pesticides and Toxic Substances, Washington, D.C.

Erickson, L. C., Wedding, R. T. and Brannaman, B. L. 1955. Influence of pH on 2,4-dichlorophenoxyacetic and acetic acid activity in *Chlorella. Plant Physiol.* 30:69–74.

Erwin, W. J. and Sharpe, R. S. 1978. Effect of wide area ultra low volume application of malathionon small mammal populations. *Trans. Nebr. Acad. Sci.* 5:25–28.

Estenik, J. F. and Collins, W. J. 1979. in vivo and in vitro studies of mixed-function oxidase in an aquatic insect, *Chironomus riparius. ACS Symp. Ser.* 99(21):349–370.

Fabacher, D. L. and Chambers, H. 1974. Resistance to herbicides in insecticide-resistant mosquitofish, *Gambusia affinis*. *Environ. Lett.* 7(1):15–20.

Farm Chemicals Handbook. 1991. *Pesticide and Fertilizer Dictionary.* Meister Publishing Company, Willoughby, OH, 770 pp.

Federle, P. F. and Collins, W. J. 1976. Insecticide toxicity to three insects from Ohio ponds. *Ohio J. Sci.* 76(1):19–24.

Felix, H. R., Chollet, R. and Harr, J. 1988. Use of the cell wall-less alga *Dunaliella bioculata* in herbicide screening tests. *Ann. Appl. Biol.* 113(1):55–60.

Finkiewicz-Murawiejska, L. 1978. The influence of subacute poisoning with the organophosphate insecticide trichlorfon on morphology of the spinal cord. *Folia Morphol. (Warszaw)* 37(2):135–150.

Finlayson, B. J. and Verrue, K. M. 1985. Toxicities of butoxyethanol ester and propylene glycol butyl ether ester formulations of 2,4-dichlorophenoxy acetic acid (2,4-D) to juvenile salmonids. *Arch. Environ. Contam. Toxicol.* 14(2):153–160.

Fisher, S. W. and Lohner, T. W. 1986. Studies on the environmental fate of carbaryl as a function of pH. *Arch. Environ. Contam. Toxicol.* 15(6):661–667.

Fitzgerald, G. P., Gerloff, G. C. and Skoog, F. 1952. Stream pollution: Studies on chemicals with selective toxicity to blue-green algae. *Sewage Ind. Wastes* 24(7):888–896.

Fitzmayer, K. M., Geiger, J. G. and Van Den Avyle, M. J. 1982. Acute toxicity effects of simazine on *Daphnia pulex* and larval striped bass. *Proc. Ann. Conf. Southeast. Assoc. Fish. Wildl. Agencies* 36:146–156.

Fitzpatrick, G. and Sutherland, D. J. 1978. Effects of the organophosphorous insecticides temephos (Abate) and chlorpyrifos (Dursban) on populations of the salt-marsh snail *Melampus bidentatus*. *Mar. Biol.* 46(1):23–28.

Folmar, L. C. 1976. Overt avoidance reaction of rainbow trout fry to nine herbicides. *Bull. Environ. Contam. Toxicol.* 15(5):509–514.

Folmar, L. C. 1978. Avoidance chamber responses of mayfly nymphs exposed to eight herbicides. *Bull. Environ. Contam. Toxicol.* 19(3):312–318.

Folmar, L. C., Sanders, H. O. and Julin, A. M. 1979. Toxicity of the herbicide glyphosate and several of its formulations to fish and aquatic invertebrates. *Arch. Environ. Contam. Toxicol.* 8(3):269–278.

Forney, D. R. 1979. Effects of Atrazine on Chesapeake Bay Aquatic Plants. U.S. NTIS PB81-115560. Office of Water Res. Technol., Washington, D.C., 88 pp.

Foy, C. L. and Hiranpradit, H. 1977. Herbicide Movement with Water and Effects of Contaminant Levels on Non-Target Organisms. OWRT Project A-046-VA. U.S. NTIS PB–263 285. Virginia Water Resourc. Res. Center, Virginia Polytechnic Instit. State University, Blacksburg, VA, 89 pp.

Frank, A. M. and Sjogren, R. D. 1978. Effect of temephos and chlorpyrifos on crustacea. *Mosq. News* 38(1):138–139.

Frank, R., Carpentier, A. G. and Mackenzie, D. L. 1987. Monitoring for 2,4-D residues in fish species resident in treated lakes in East-central Ontario 1977–80. *Environ. Monit. Assess.* 9:71–82.

Frear, D. E. H. and Boyd, J. E. 1967. Use of *Daphnia magna* for the microbioassay of pesticides. I. Development of standardized techniques for rearing *Daphnia* and preparation of dosage-mortality curves for pesticides. *J. Econ. Entomol.* 60(5):1228–1236.

Fredeen, F. J. H. 1972. Reactions of the larvae of three rheophilic species of *Trichoptera* to selected insecticides. *Can. Entomol.* 104:945–953.

Frempong-Boadu, J. 1966. A laboratory study of the effectiveness of methoxychlor, fenthion and carbaryl against blackfly larvae (*Diptera: Simuliidae*). *Mosq. News* 26(4): 562–564.

Fromm, P. O., Richards, B. D. and Hunter, R. C. 1971. Effects of some insecticides and MS-222 on isolated-perfused gills of trout. *Prog. Fish Cult.* 33(3):138–140.

Funderburk, H. H. 1963. Distribution of C14 Labeled Herbicides in Bluegills and Shellcrackers. U.S. EPA-OPP Registration Standard.

Fytizas, R. and Vassiliou, G. 1980. L'inhibition de la cholinesterase du cerveau de quelques especes de poisson critere de la pollution de la mer par les esters organophosphores. *Meded. Rijksfac. Landbouwwet. Gent.* 45(4):923–927.

Gaaboub, I. A., El-Gayar, F. M. and Helal, E. M. 1975. Comparative bioassay studies on larvae of *Culex pipiens* and the microcrustacean *Daphnia magna*. *Bull. Entomol. Soc. Egypt, Econ. Ser.* 9:77–84.

Gantverg, A. N. and Perevoznikov, M. A. 1983. Inhibition of cholinesterase in the brain of perch, *Perca fluviatilis* (Percidae), and common carp, *Cyprinus carpio* (Cyprinidae), under the action of carbophos. *J. Ichthyol.* 23(4):174–175.

Gaufin, A. R., Jensen, L. and Nelson, T. 1961. Bioassays to determine pesticide toxicity to aquatic invertebrates. *Water Sewage Works* 108:355–359.

Gaufin, A. R., Jensen, L. D., Nebeker, A. V., Nelson, T. and Teel, R. W. 1965. The toxicity of ten organic insecticides to various aquatic invertebrates. *Water Sewage Works* 12: 276–279.

Geen, G. H., Hussain, M. A., Oloffs, P. C. and McKeown, B. A. 1981. Fate and toxicity of acephate (Orthene) added to a coastal B. C. stream. *J. Environ. Sci. Health* B16(3):253–271.

Geen, G. H., McKeown, B. A., Watson, T. A. and Parker, D. B. 1984b. Effects of acephate (Orthene) on development and survival of the salamander, *Ambystoma gracile* (Baird). *J. Environ. Sci. Health* 19B(2):157–170.

Geen, G. H., McKeown, B. A. and Oloffs, P. C. 1984a. Acephate in rainbow trout (*Salmo gairdneri*): acute toxicity, uptake, elimination. *J. Environ. Sci. Health* 19B(2):131–155.

Geiger, D. L., Call, D. J. and Brooke, L. T. 1988. Acute Toxicities of Organic Chemicals to Fathead Minnows (*Pimephales promelas*), Vol. 4. Center for Lake Superior Environmental Studies, University of Wisconsin, Superior, WI, 355 pp.

George, J. P., Hingorani, H. G. and Rao, K. S. 1982. Herbicide toxicity to fish-food organisms. *Environ. Pollut. Ser. A Ecol. Biol.* 28(3):183–188.

Gersich, F. M., Mendoza, C. G., Hopkins, D. L. and Bodner, K. M. 1984. Acute and chronic toxicity of triclopyr triethylamine salt to *Daphnia magna* Straus. *Bull. Environ. Contam. Toxicol.* 32:497–502.

Geyer, H., Scheunert, I. and Korte, F. 1985. The effects of organic environmental chemicals on the growth of the alga *Scenedesmus subspicatus*: A contribution to environmental biology. *Chemosphere* 14(9):1355–1369.

Gibbs, K. E., Mingo, T. M. and Courtemanch, D. L. 1984. Persistence of carbaryl (Sevin-4-oil) in woodland ponds and its effects on pond macroinvertebrates following forest spraying. *Can. Entomol.* 116:203–213.

Gill, T. S., Pant, J. C. and Pant, J. 1988. Gill, liver, and kidney lesions associated with experimental exposures to carbaryl and dimethoate in the fish (*Puntius conchonius* Ham.). *Bull. Environ. Contam. Toxicol.* 41(1):71–78.

Gillette, L. A., Miller, D. L. and Redman, H. E. 1952. Appraisal of a chemical waste problem by fish toxicity tests. *Sewage Ind. Wastes* 24:1397–1401.

Gluth, G. and Hanke, W. 1984. A comparison of physiological changes in carp, *Cyprinus carpio*, induced by several pollutants at sublethal concentration — II. The dependency on the temperature. *Comp. Biochem. Physiol.* 79C(1):39–45.

Gluth, G. and Hanke, W. 1985. A comparison of physiological changes in carp, *Cyprinus carpio*, induced by several pollutants at sublethal concentrations. *Ecotoxicol. Environ. Saf.* 9(2):179–188.

Goel, H. C. and Srivastava, C. P. 1981. Laboratory evaluation of some molluscicides against fresh water snails, *Indoplanorbis* and *Lymnaea* species. *J. Commun. Dis.* 13(2):121–127.

Goel, K. A., Tyagi, S. K. and Awasthi, A. K. 1982. Effect of malathion on some haematological values in *Heteropneustes fossilis. Comp. Physiol. Ecol.* 7(4):259–261.

Goldenthal, E. I. 1971. A compilation of LD50 values in newborn and adult animals. *Toxicol. Appl. Pharmacol.* 18:185–207.

Goldsborough, L. G. and Brown, D. J. 1988. Effect of glyphosate (Roundup formulation) on periphytic algal photosynthesis. *Bull. Environ. Contam. Toxicol.* 41(2):253–260.

Goldsborough, L. G. and Robinson, G. G. C. 1983. The effect of two triazine herbicides on the productivity of freshwater marsh periphyton. *Aquat. Toxicol.* 4(2):95–112.

Gonzalez-Murua, C., Munoz-Rueda, A., Hernando, F. and Sanchez-Diaz, M. 1985. Effect of atrazine and methabenzthiazuron on oxygen evolution and cell growth of Chlorella pyrenoidosa. *Weed Res.* 25(1):61–66.

Goodman, L. R., Hansen, D. J., Coppage, D. L., Moore, J. C. and Matthews, E. 1979. Diazinon: chronic toxicity to, and brain acetylcholinesterase inhibition in, the sheepshead minnow, Cyprinodon variegatus. *Trans. Am. Fish. Soc.* 108(5):479–488.

Gouda, R. K., Tripathy, N. K. and Das, C. C. 1981. Toxicity of dimecron, Sevin and lindex to *Anabas scandens* and *Heteropneustes fossilis. Comp. Physiol. Ecol.* 6(3):170–172.

Grandes, A. E. and Sagrado, E. A. 1988. The susceptibility of mosquitoes to insecticides in Salamanca Province, Spain. *J. Am. Mosq. Control Assoc.* 4(2):168–171.

Greenwood, P. J. and Brown, A. C. 1974. Effect of ammonium nitrate solutions on fertilization and development of the sea urchin, *Parechinus angulosus. Zool. Afr.* 9(2):205–209.

Grenier, G. and Beaumont, G. 1983. Physiological effects of sublethal doses of atrazine on *Lemna minor*. VII. 1,2–[14C] acetate incorporation into the groups of lipids and their fatty acids. *Physiol. Plant* 57:477–484.

Grenier, G., Marier, J. P. and Beaumont, G. 1979. Effets physiologiques de l'atrazine a doses subletales sur *Lemna minor* L. IV. Influence sur la composition lipidique. *Can. J. Bot.* 57(9):1015–1020.

Grenier, G., Marier, J. P. and Beaumont, G. 1982. Effets physiologiques de l'atrazine a doses subletales sur *Lemna minor*. VI. Influence sur les classes importantes de phospholipides. *Physiol. Veg.* 20(2):179–185.

Groba, J. and Trzcinska, B. 1979. Wplyw wybranych insektycydow fosforoorganicznych i karbaminianowych na pstraga teczowego *Salmo gairdneri* R. Bromatol. *Chem. Toksykol.* 12(1):33–38; Pestab 0413 (1980).

Guarino, A. M., Rieck, G., Arnold, S., Fenstermacher, P., Bend, J., Knutson, M. J. and Anderson, J. B. 1976. Distribution and toxicity of selected water pollutants in the spiny dogfish, *Squalus acanthias. Bull. Mt. Desert Isl. Biol. Lab.* 16:50–53.

Gunkel, G. and Kausch, H. 1976. Die akute toxizitat von atrazin (s-triazin) auf sandfelchen (*Coregonus fera* Jurine) im hunger. (Acute toxicity of atražine (s-triazine) on *Coregonus fera* Jurine under starvation conditions. *Arch. Hydrobiol. Suppl.* 48(2):207–234.

Gunkel, G. and Streit, B. 1980. Mechanisms of bioaccumulation of a herbicide (atrazine, s-triazine) in a freshwater mollusc (*Ancylus fluviatilis* Mull.) and a fish (*Coregonus fera* jurine). *Water Res.* 14(11):1573–1584.

Gzhetotskii, M. I., Shkliaruk, L. V. and Dychok, L. A. 1977. Toxicological characteristics of the herbicide Zeazin. *Vrach. Delo* 5:133–136 (RUS); *Pestab*:2826.

Haegele, M. A. and Tucker. R. K. 1974. Effects of 15 common environmental pollutants on eggshell thickness in mallards and *Coturnix*. *Bull. Environ. Contam. Toxicol.* 11(1): 98–102.

Haider, S. and Inbaraj, R. M. 1986. Relative toxicity of technical material and commercial formulation of malathion and endosulfan to a freshwater fish, *Channa punctàtus* (Bloch). *Ecotoxicol. Environ. Saf.* 11(3):347–351.

Haider, S. and Upadhyaya, N. 1985. Effect of commercial formulation of four organophosphorus insecticides on the ovaries of a freshwater teleost, *Mystus vittatus* (Bloch) — a histologic histological and histochemical study. *J. Environ. Sci. Health* 20B(3):321–340.

Haines, T. A. 1981. Effect of an aerial application of carbaryl on brook trout (*Salvelinus fontinalis*). *Bull. Environ. Contam. Toxicol.* 27:534–542.

Haley, T. J., Farmer, J. H., Harmon, J. R. and Dooley, K. L. 1975. Estimation of the LD1 and extrapolation of the LD0.1 for five organophosphate pesticides. *Arch. Toxicol.* 34: 103–109.

Hall, R. J. and Kolbe, E. 1980. Bioconcentration of organophosphorus pesticides to hazardous levels by amphibians. *J. Toxicol. Environ. Health* 6(4):853–860.

Hamala, J. A. and Kollig, H. P. 1985. The effects of atrazine on periphyton communities in controlled laboratory ecosystems. *Chemosphere* 14(9):1391–1408.

Hamilton, P. B., Jackson, G. S., Kaushik, N. K. and Solomon, K. R. 1987. The impact of atrazine on lake periphyton communities, including carbon uptake dynamics using track autoradiography. *Environ. Pollut.* 46(2):83–103.

Hanazato, T. and Yasuno, M. 1987. Effects of a carbamate insecticide, carbaryl, on the summer phyto- and zooplankton communities in ponds. *Environ. Pollut.* 48(2):145–159.

Hanke, W., Gluth, G., Bubel, H. and Muller, R. 1983. Physiological changes in carps induced by pollution. *Ecotoxicol. Environ. Saf.* 7(2):229–241.

Hansen, C. R., Jr. and Kawatski, J. A. 1976. Application of 24-hour postexposure observation to acute toxicity studies with invertebrates. *J. Fish. Res. Board Can.* 33(5):1198–1201.

Hansen, D. J. 1969. Avoidance of pesticides by untrained sheepshead minnows. *Trans. Am. Fish. Soc.* 98(3):426–429.

Hansen, D. J. 1970. Behavior of Estuarine Organisms. U.S. Fish Wildl. Serv., Circ. 335, Washington, D.C., pp. 23–28.

Hansen, D. J. 1972. DDT and malathion: effect on salinity selection by mosquitofish. *Trans. Am. Fish. Soc.* 101(2):346–350.

Hansen, D. J. and Parrish, P. R. 1977. Suitability of sheepshead minnows (*Cyprinodon variegatus*) for life-cycle toxicity tests. in Mayer, F. L. and Hamelink, J. L. (Eds.). *Aquatic Toxicology and Hazard Evaluation*, ASTM STP 634. pp. 117–126.

Hansen, D. J., Goodman, L. R., Cripe, G. M. and MaCauley, S. F. 1986. Early life-stage toxicity test methods for Gulf toadfish (*Opsanus beta*) and results using chlorpyrifos. *Ecotoxicol. Environ. Saf.* 11(1):15–22.

Hansen, D. J., Matthews, E., Nall, S. L. and Dumas, D. P. 1972. Avoidance of pesticides by untrained mosquitofish, *Gambusia affinis*. *Bull. Environ. Contam. Toxicol.* 8(1):46–51.

Hansen, D. J., Schimmel, S. C. and Keltner, J. M., Jr. 1973. Avoidance of pesticides by grass shrimp (*Palaemonetes pugio*). *Bull. Environ. Contam. Toxicol.* 9(3):129–133.

Harman, W. N. 1977. The effects of simazine treatments on the benthic fauna of Moriane Lake, Madison County, New York. *Proc. Northeast. Weed Sci. Soc.* 31:122–137.

Harrisson, J. W. E. and Rees, E. W. 1946. 2,4-D toxicity—I: Toxicity towards certain species of fish. *Am. J. Pharm.* (Dec.):422–425.

Hartman, W. A. and Martin, D. B. 1984. Effect of suspended bentonite clay on the acute toxicity of glyphosate to *Daphnia pulex* and *Lemna minor*. *Bull. Environ. Contam. Toxicol.* 33:355–361.

Hartman, W. A. and Martin, D. B. 1985. Effects of four agricultural pesticides on *Daphnia pulex*, *Lemna minor*, and *Potamogeton pectinatus*. *Bull. Environ. Contam. Toxicol.* 35(5): 646–651.

Hashimoto, Y. and Nishiuchi, Y. 1981. Establishment of bioassay methods for the evaluation of acute toxicity of pesticides to aquatic organisms. *J. Pestic. Sci.* 6(2):257–264.

Hashimoto, Y., Makita, T., Ohnuma, N. and Noguchi, T. 1972. Acute toxicity on dimethyl 4,4′-o-phenylene bis (3-thioallophanate), thiophanate-methyl fungicide. *Toxicol. Appl. Pharmacol.* 23(4):606–615.

Hashimoto, Y., Okubo, E., Ito, T., Yamaguchi, M. and Tanaka, S. 1982. Changes in susceptibility of carp to several pesticides with growth. *J. Pestic. Sci.* 7(4):457–461.

Haven, D. 1963. Mass treatment with 2,4-D on milfoil in tidal creeks in VA. *Proc. South. Weed Conf.* 16:345–350.

Hawxby, K., Tubea, B., Ownby, J. and Basler, E. 1977. Effects of various classes of herbicides on four species of algae. *Pestic. Biochem. Physiol.* 7(3):203–209.

Haynes, H. L., Moorefield, H. H., Borash, A. J. and Keays, J. W. 1958. The toxicity of Sevin to goldfish. *J. Econ. Entomol.* 51(4):540.

Heath, R. G., Spann, J. W., Hill, E. F. and Kreitzer, J. F. 1972. Comparative Dietary Toxicities of Pesticides to Birds. U.S. Bureau of Sport Fisheries and Wildlife, Special Scientific Report-Wildlife No. 152:1–57.

Hemingway, J. and Georghiou, G. P. 1983. Studies on the acetylcholinesterase of *Anopheles abimanus* resistant and susceptible to organophosphate and carbamate insecticides. *Pestic. Biochem. Physiol.* 19(2):167–171.

Hemingway, J. and Georghiou, G. P. 1984. Differential suppression of organophosphorus resistance in *Culex quinquefasciatus* by the synergists IBP, DEF, and TPP. *Pestic. Biochem. Physiol.* 21(1):1–9.

Henderson, C. and Pickering, Q. H. 1958. Toxicity of organic phosphorus insecticides to fish. *Trans. Am. Fish. Soc.* 87:39–51.

Henderson, C., Pickering, Q. H. and Tarzwell, C. M. 1959. The Toxicity of Organic Phosphorus and Chlorinated Hydrocarbon Insecticides to Fish. Trans. Second Sem. Biol. Problems Water Pollut., U.S. Public Health Serv., Robert A. Taft Sanit. Eng. Center, Cincinnati, OH, 13 pp.

Hendrick, R. D., Everett, T. R. and Caffey, H. R. 1966. Effects of some insecticides on the survival, reproduction, and growth of the Louisiana red crawfish. *J. Econ. Entomol.* 59(1): 188–192.

Herman, D., Kaushik, N. K. and Solomon, K. R. 1986. Impact of atrazine on periphyton in freshwater enclosures and some ecological consequences. *Can. J. Fish. Aquat. Sci.* 43(10):1917–1925.

Hermanutz, R. O., Eaton, J. G. and Mueller, L. H. 1985. Toxicity of endrin and malathion mixtures to flagfish (*Jordanella floridae*). *Arch. Environ. Contam. Toxicol.* 14:307–314.

Hermens, J., Canton, H., Steyger, N. and Wegman, R. 1984a. Joint effects of a mixture of 14 chemicals on mortality and inhibition of reproduction of *Daphnia magna*. *Aquat. Toxicol.* 5(4):315–322.

Hermens, J., Canton, H., Steyger, N. and Wegman, R. 1984b. Joint effects of a mixture of 14 chemicals on mortality and inhibition of reproduction of *Daphnia magna*. *Aquat. Toxicol.* 5(4):315–322.

Hersh, C. M. and Crumpton, W. G. 1987. Determination of growth rate depression of some green algae by atrazine. *Bull. Environ. Contam. Toxicol.* 39:1041–1048.

Herzberg, A. M. 1987. Toxicity of chlorpyrifos (Dursban) in *Oreochromis aureus* and *O. niloticus* and data on its residues in *O. aureus. Bamidgeh* 39(1):13–20.

Hess, F. D. 1979. The influence of the herbicide trifluralin on flagellar regeneration in *Chlamydomonas. Exp. Cell Res.* 119(1):99–109.

Hess, F. D. 1980. A chlamydomonas algal bioassay for detecting growth inhibitor herbicides. *Weed Sci.* 28(5):515–520.

Hidaka, H., Hattanda, M. and Tatsukawa, R. 1984. Avoidance of pesticides with medakas (*Oryzias latipes*). *J. Agric. Chem. Soc. Jpn.* (Nippon Nogeikagaku Kaishi) 58(2):145–151.

Hildebrand, L. D., Sullivan, D. S. and Sullivan, T. P. 1980. Effects of Roundup herbicide on populations of *Daphnia magna* in a forest pond. *Bull. Environ. Contam. Toxicol.* 25: 353–357.

Hildebrand, L. D., Sullivan, D. S. and Sullivan, T. P. 1982. Experimental studies of rainbow trout populations exposed to field applications of Roundup herbicide. *Arch. Environ. Contam. Toxicol.* 11(1):93–98.

Hill, E. F. 1979. Cholinesterase activity in Japanese quail dusted with carbaryl. *Lab. Anim. Sci.* 29:349–352.

Hilsenhoff, W. L. 1959. The evaluation of insecticides for the control of *Tendipes plumosus* (Linnaeus). *J. Econ. Entomol.* 52(2):331–332.

Hilsenhoff, W. L. 1962. Toxicity of granular malathion to walleyed pike fingerlings. *Mosq. News* 22(1):14–15.

Hiltibran, R. C. 1967. Effects of some herbicides on fertilized fish eggs and fry. *Trans. Am. Fish. Soc.* 96:414–416.

Hiltibran, R. C. 1969. Oxygen and phosphate metabolism of bluegill liver mitochondria in the presence of 2,4-dichlorophenoxyacetic acid derivatives. *Trans. Ill. State Acad. Sci.* 62(2): 175–180.

Hiltibran, R. C. 1974. Oxygen and phosphate metabolism of bluegill liver mitochondria in the presence of some insecticides. *Trans. Ill. State Acad. Sci.* 67(2):228–237.

Hiltibran, R. C. and Turgeon, A. J. 1977. Creeping bentgrass response to aquatic herbicides in irrigation water. *J. Environ. Qual.* 6(3):263–267.

Hirayama, K. and Tamanoi, S. 1980. Acute toxicity of MEP and Diazinon (pesticide) to larvae of Kuruma prawn, *Penaeus japonicus* and of swimming crab *Portunus trituberculatus*. *Bull. Jpn. Soc. Sci. Fish.* 46(2):117–123.

Hirose, K. and Kawakami, K. 1977. Effects of insecticides, oil dispersants and synthetic detergent on the embryonic development in Medaka, *Oryzias latipes*. *Bull. Tokai Reg. Fish. Res. Lab.* 91:9–17.

Hirose, K. and Kitsukawa, M. 1976. Acute toxicity of agricultural chemicals to seawater teleosts, with special respect to TLM and the vertebral abnormality. *Bull. Tokai Reg. Fish. Res. Lab.* 84:11–20.

Hirose, K., Yamazaki, M. and Ishikawa, A. 1979. Effects of water temperature on median lethal concentrations (LC50) of a few pesticides to seawater teleosts. *Bull. Tokai Reg. Fish. Res. Lab.* 98:45–53.

Holck, A. R. and Meek, C. L. 1987. Dose-mortality responses of crawfish and mosquitoes to selected pesticides. *J. Am. Mosq. Control Assoc.* 3(3):407–411.

Holcombe, G. W., Phipps, G. L. and Tanner, D. K. 1982. The acute toxicity of Kelthane, Dursban, disulfoton, pydrin, and permethrin to fathead minnows *Pimephales promelas* and rainbow trout *Salmo gairdneri*. *Environ. Pollut. Ser. A Ecol. Biol.* 29(3):167–178.

Holdway, D. A. and Dixon, D. G. 1988. Acute toxicity of permethrin or glyphosate pulse exposure to larval white sucker (*Catostomus commersoni*) and juvenile flagfish (*Jordanella Floridae*) as modified by age and ration level. *Environ. Toxicol. Chem.* 7(1):63–68.

Holland, H. T. and Lowe, J. I. 1966. Malathion: Chronic effects on estuarine fish. *Mosq. News* 26(3):383–385.

Hollister, T. A. and Walsh, G. E. 1973. Differential responses of marine phytoplankton to herbicides: Oxygen evolution. *Bull. Environ. Contam. Toxicol.* 9(5):291–295.

Hooper, F. N. 1953. The effect of applications of pelleted 2,4-D upon the bottom fauna of Kent Lake, Oakland County, Michigan. *Proc. N.C. Weed Control Conf.* 15:41.

Hopkin, R. and Kain, J. M. 1978. The effects of some pollutants on the survival, growth and respiration of *Laminaria hyperborea. Estuarine Coastal Mar. Sci.* 7(6):531–553.

Hopkin, R. and Kain, J. M. 1971. The effect of marine pollutants on *Laminarea hyperboria. Mar. Pollut. Bull.* 2(5):75–77.

Huang, J. C. and Gloyna, E. F. 1968. Effect of organic compounds on photosynthetic oxygenation-I. Chlorophyll destruction and suppression of photosynthetic oxygen production. *Water Res.* 2:347–366.

Huckins, J. N., Petty, J. D. and England, D. C. 1986. Distribution and impact of trifluralin, atrazine, and fonofos residues in microcosms simulating a northern prairie wetland. *Chemosphere* 15(5):563–588.

Hudson, R. H., Tucker, R. K. and Haegle, M. A. 1984. *Handbook of Toxicity to Wildlife*. U.S. Dept. Int. Fish Wildl. Serv. Resourc. Publ. No. 153. 2nd ed. Washington, D.C., 90 pp.

Hughes, D. N. 1977. The Effects of Three Organophosphorus Insecticides on Zooplankton and Other Invertebrates in Natural and Artificial Ponds. M.S. Thesis, York University, Toronto, Canada. 100 pp.

Hughes, D. N., Boyer, M. G., Papst, M. H., Fowle, C. D., Rees, G. A. V. and Baulu, P. 1980. Persistence of three organophosphorus insecticides in artificial ponds and some biological implications. *Arch. Environ. Contam. Toxicol.* 9(3):269–279.

Hughes, J. 1988. The Toxicity of Diazinon Technical to *Selenastrum capricornutum*. USEPA-OPP Registration Standard.

Hughes, J. S. and Davis, J. T. 1962. Comparative Toxicity to Bluegill Sunfish of Granular and Liquid Herbicides. in Proc. 16th Ann. Conf. S.E. Game Fish Commissioners. pp. 319–323.

Hughes, J. S. and Davis, J. T. 1963. Variations in toxicity to bluegill sunfish of phenoxy herbicides. *Weeds* 2(1):50–53.

Hughes, J. S. and Davis, J. T. 1964. Effects of Selected Herbicides on Bluegill Sunfish. in Proc. 18th Ann. Conf. S.E. Game Fish Commissioners. pp. 480–482.

Hurlbert, S. H., Mulla, M. S. and Wilson, H. R. 1972. Effects of an organophosphorus insecticide on the phytoplankton, zooplankton, and insect populations of freshwater ponds. *Ecol. Monogr.* 42(3):269–299.

Hurlbert, S. H., Mulla, M. S., Keith, J. O., Westlake, W. E. and Dusch, M. E. 1970. Biological effects and persistence of Dursban in freshwater ponds. *J. Econ. Entomol.* 63:43–62.

Hussain, M. A., Mohamad, R. B. and Oloffs, P. C. 1984. Toxicity and metabolism of acephate in adult and larval insects. *J. Environ. Sci. Health* 19B(3):355–377.

Imada, K. 1976. Studies on the vertebral malformation of fishes. III. Vertebral deformation of goldfish (*Carassius auratus*) and medakafish (*Olyzias latipes*) exposed to Carbamate insectides. *Hokkaidoritsu Suisan Fukajo Kenkyu Hokoku* 31:43–65.

Inabinet, J. R. 1979. Environmental Impact of the South Carolina Public Service Authority's FY'76 Aquatic Weed Control Program in Lake Marion. South Carolina Dept. of Health and Environ. Control, Govt. Announce. Index 21. U.S. NTIS AD-A069 572. 60 pp.

Inbaraj, R. M. and Haider, S. 1988. Effect of malathion and endosulfan on brain acetylcholinesterase and ovarian steroidogenesis of *Channa punctatus* (Bloch). *Ecotoxicol. Environ. Saf.* 16(2):123–128.

Inglis, A. and Davis, E. L. 1972. Effects of Water Hardness on the Toxicity of Several Organic and Inorganic Herbicides to Fish. Bur. Sport Fish. Wildl. Tech. Paper No. 67. U.S.D.I. 22 pp.

Isensee, A. R. 1976. Variability of aquatic model ecosystem-derived data. *Int. J. Environ. Stud.* 10:35–41.

Ishibashi, N., Kondo, E. and Ito, S. 1983. Effects of application of certain herbicides on soil nematodes and aquatic invertebrates in rice paddy fields in Japan. *Crop Prot.* 2(3): 289–304.

Ishii, Y. and Hasimoto, Y. 1970. Metabolic fate of carbaryl (1-naphthyl n-methyl carbamate) orally administered to carp, *Cyprinus carpio*. *Bull. Agric. Chem. Insp. Stn.* 10:48–50.

Jackson, D. A. and Gardner, D. R. 1973. The effects of some organochlorine pesticide analogs on Salmonid brain ATPases. *Pestic. Biochem. Physiol.* 2:377–382.

Jacob, S. S., Nair, N. B. and Balasubramanian, N. K. 1982. Toxicity of certain pesticides found in the habitat to the larvivorous fishes *Aplocheilus lineatus* (Cuv. & Val.) and *Macropodus cupanus* (Cuv. & Val.). *Proc. Indian Acad. Sci. Anim. Sci.* 91(3): 323–328.

Jamnback, H. and Frempong-Boadu, J. 1966. Testing blackfly larvicides in the laboratory and in streams. *Bull. W.H.O.* 34:405–421.

Jarvinen, A. W. and Tanner, D. K. 1982. Toxicity of selected controlled release and corresponding unformulated technical grade pesticides to the fathead minnow *Pimephales promelas*. *Environ. Pollut. Ser. A* 27(3):179–195.

Jarvinen, A. W., Nordling, B. R. and Henry, M. E. 1983. Chronic toxicity of Dursban (chlorpyrifos) to the fathead minnow (*Pimephales promelas*) and the resultant acetylcholinesterase inhibition. *Ecotoxicol. Environ. Saf.* 7(4):423–434.

Jarvinen, A. W., Tanner, D. K. and Kline, E. R. 1988. Toxicity of chlorpyrifos, endrin, or fenvalerate to fathead minnows following episodic or continuous exposure. *Ecotoxicol. Environ. Saf.* 15(1):78–95.

Jauhar, L. and Kulshrestha, S. K. 1985. Histopathological effects induced by sublethal doses of Sevin and thiodan on the gills of *Channa striatus* Bl. (*Pisces: Channidae*). *Acta Hydrochim. Hydrobiol.* 13(3):395–400.

Jensen, L. D. and Gaufin, A. R. 1964a. Long-term effects of organic insecticides on two species of stonefly naiads. *Trans. Am. Fish. Soc.* 93(4):357–363.

Jensen, L. D. and Gaufin, A. R. 1964b. Effects of ten organic insecticides on two species of stonefly naiads. *Trans. Am. Fish. Soc.* 93(1):27–34.

Jensen, R. A. 1978. A simplified bioassay using finfish for estimating potential spill damage. in *Proc. Control of Hazardous Material Spills*, Rockville, MD, pp. 104–108.

Jenssen, D. and Renberg, L. 1976. Distribution and cytogenetic test of 2,4-D and 2,4,5-T phenoxyacetic acids in mouse blood tissues. *Chem. Biol. Interact.* 14:291–299.

Jeyasingam, D. N. T., Thayumanavan, B. and Krishnaswamy, S. 1978. The relative toxicities of insecticides on aquatic insect *Eretes sticticus* (Linn.) (*Coleoptera: Dytiscidae*). *J. Madurai Univ.* 7(1):85–87.

Jirasek, J., Adamek, Z., Nguyen, X. T. and Holcman, O. 1978. Estimation of the acute toxicity of the insecticide Dursban for fish. (Stanoveni akutni toxicity insecticidu Dursban pro ryby.). *Acta Univ. Agric. Brno Fac. Agron.* 26(3):51–56 (CZE) (ENG-ABS); Pestab:3134 (1980).

Johnson, B. T. 1986. Potential impact of selected agricultural chemical contaminants on a northern prairie wetland: a microcosm evaluation. *Environ. Toxicol. Chem.* 5(5): 473–485.

Johnson, C. R. 1976. Herbicide toxicities in some Australian anurans and the effect of subacute dosages on temperature tolerance. *Zool. J. Linn. Soc.* 59(1):79–83.

Johnson, C. R. 1977. The effects of field applied rates of five organophosphorus insecticides on thermal tolerance, orientation, and survival in *Gambusia affinis affinis (Pisces: Poeciliidae). Proc. Pap. Annu. Conf. Calif. Mosq. Vector Control Assoc.* 45:56–58.

Johnson, C. R. 1978a. Herbicide toxicities in the mosquito fish, *Gambusia affinis. Proc. R. Soc. Queensl.* 89:25–27.

Johnson, C. R. 1978b. The effects of sublethal concentrations of five organophosphorus insecticides on temperature tolerance, reflexes, and orientation in *Gambusia affinis affinis (Pisces: Poeciliidae). Zool. J. Linn. Soc.* 64(1):63–70.

Johnson, C. R. 1978c. The effect of five organophosphorus insecticides on survival and temperature tolerance in the copepod, *Macrocyclops albidus (Copepoda: Cyclopidae). Zool. J. Linn. Soc.* 64(1):59–62.

Johnson, C. R. and Prine, J. E. 1976. The effects of sublethal concentrations of organophosphorus insecticides and an insect growth regulator on temperature tolerance in hydrated and dehydrated juvenile western toads, *Bufo Boreas. Comp. Biochem. Physiol.* 53(2A): 147–149.

Johnson, D. R. and Hansen, R. M. 1969. Effects of range treatment with 2,4-D on rodent populations. *J. Wildl. Manage.* 33(1):125–132.

Johnson, W. W. and Finley, M. T. 1980. Handbook of Acute Toxicity of Chemicals to Fish and Aquatic Invertebrates. Resour. Publ. 137. Fish Wildl. Serv., U.S. Dept. Int., Washington, D.C., 98 pp.

Jones, R. O. 1962. Tolerance of the Fry of Common Warm-Water Fishes to Some Chemicals Employed in Fish Culture. Proc. 16th Ann. Conf. S.E. Assoc. Game Fish Comm. pp. 436–445.

Jones, T. W. and Estes, P. S. 1984. Uptake and phytotoxicity of soil-sorbed atrazine for the submerged aquatic plant, *Potamogeton perfoliatus* L. *Arch. Environ. Contam. Toxicol.* 13(2):237–241.

Jordan, M., Srebro, Z., Pierscinska, E. and Wozny, L. 1975. Preliminary observations on the effects on the organs of mice of administering some carbamate pesticides. *Bull. Environ. Contam. Toxicol.* 14(2):205–208.

Joseph, S. R., Mallack, J. and George, L. F. 1972. Field applications of ultra low volume malathion to three animal species. *Mosq. News* 32(4):504–506.

Joshi, H. C., Kapoor, D., Panwar, R. S. and Gupta, R. A. 1975. Toxicity of some insecticides to chironomid larvae. *Indian J. Environ. Health* 17(3):238–241.

Joshi, P. C. and Misra, R. B. 1986. Evaluation of chemically-induced phototoxicity to aquatic organism using paramecium as a model. *Biochem. Biophys. Res. Comm.* 139(1):79–84.

Jouany, J. M., Truhaut, R., Vasseur, P., Klein, D., Ferard, J. F. and Deschamps, P. 1985. An example of interaction between environmental pollutants: Modification of thiram toxicity to freshwater organisms by nitrites or nitrates in relation to nitrosamine synthesis. *Ecotoxicol. Environ. Saf.* 9(3):327–338.

Judd, F. W. 1977. Toxicity of monosodium methanearsonate herbicide to Couch's spadefoot toad, *Scaphiopus couchi. Herpetologica* 33(1):44–46.

Judd, F. W. 1979. Acute toxicity and effects of sublethal dietary exposure of monosodium methanearsonate herbicide to *Peromyscus leucopus. Bull. Environ. Contam. Toxicol.* 22: 143–150.

Juhnke, I. and Luedemann, D. 1978. Results of the investigation of 200 chemical compounds for acute fish toxicity with the golden orfe test. *Z. Wasser Abwasser Forsch.* 11(5):161–164.

Kabir, S. M. H. and Ahmed, N. 1979. Histopathological changes in climbing perch, *Anabas testudineus* (Bloch) (*Anabantidae: Perciformes*) due to three granular insecticides. *Bangladesh J. Zool.* 7(1):21–29.

Kabir, S. M. H. and Begum, R. 1978. Toxicity of three organophosphorus insecticides to shinghi fish, *Heteropneustes fossilis* (Bloch). *Dacca. Univ. Stud.* 26(1):115–122.

Kader, H. A., Thayumanavan, B. and Krishnaswamy, S. 1976. The relative toxicities of ten biocides on *Spicodiaptomus chelospinus* Rajendran (1973) [*Copepoda: Calanoida*]. *Comp. Physiol. Ecol.* 1(3):78–82.

Kalac, J. 1974. Ucinky organofosforovych zlucenin na aktivitu niektorych enzymov Zazivacieho traktu. (The effect of organo phosphorus compounds on the activity of some digestive enzymes). *Cesk. Hyg.* 19(8):355–362.

Kamler, E. 1972. Bioenergetical aspects of the influence of 2,4-D-Na on the early development stages in carp (*Cyprinus carpio* L.). *Pol. Arch. Hydrobiol.* 19(4):451–474.

Kamler, E. 1973. Effect of 2,4-D-Na on the respiration and calorific value in embryonic and larval development of carp (*Cyprinus carpio* L.). *Eksp. Vodn. Toksikol.* 5:45–60.

Kamler, E., Matlak, O. and Srokosz, K. 1974. Further observations on the effect of sodium salt of 2,4-D on early developmental stages of carp (*Cyprinus carpio* L.). *Pol. Arch. Hydrobiol.* 21(34):481–502.

Kanazawa, J. 1975. Uptake and excretion of organophosphorus and carbamate insecticides by fresh water fish, motsugo, *Pseudorasbora parva*. *Bull. Environ. Contam. Toxicol.* 14(3): 346–352.

Kanazawa, J. 1981. Measurement of the bioconcentration factors of pesticides by freshwater fish and their correlation with physicochemical properties or acute toxicities. *Pestic. Sci.* 12(4):417–424.

Kanazawa, J. 1983. in vitro and in vivo effects of organophosphorus and carbamate insecticides on brain acetylcholinesterase activity of fresh-water fish, Topmouth gudgeon. *Bull. Nat. Inst. Agric. Sci.* 37C:19–30.

Kaniewska-Prus, M. 1975. The effect of 2,4D-Na herbicide on oxygen consumption and survival of Simocephalus vetulus o.f. muller (Cladocera). *Pol. Arch. Hydrobiol.* 22(4): 593–599.

Kaplan, A. M. and Sherman, H. 1977. Toxicity studies with methyl n-(((methylamino)carbonyl)oxy)-ethanimidothioate. *Toxicol. Appl. Pharmacol.* 40(1):1–17.

Karnak, R. E. and Collins, W. J. 1974. The susceptibility to selected insecticides and acetylcholinesterase activity in a laboratory colony of midge larvae, *Chironomus tentans* (*Dipteria: Chironomidae*). *Bull. Environ. Contam. Toxicol.* 12(1):62–69.

Katz, M. 1961. Acute toxicity of some organic insecticides to three species of salmonids and to the threespine stickleback. *Trans. Am. Fish. Soc.* 90(3):264–268.

Kaur, K. and Toor, H. S. 1977. Toxicity of pesticides to embryonic stages of *Cyprinus carpio communis* Linn. *Indian J. Exp. Biol.* 15:193–196.

Kearney, P. C., Isensee, A. R. and Kontson, A. 1977b. Distribution and degradation of dinitroaniline herbicides in an aquatic ecosystem. *Pestic. Biochem. Physiol.* 7(3): 242–248.

Kearney, P. C., Oliver, J. E., Helling, C. S., Isensee, A. R. and Kontson, A. 1977a. Distribution, movement, persistence, and metabolism of n-nitrosoatrazine in soils and a model aquatic ecosystem. *J. Agric. Food Chem.* 25(5):1177–1181.

Kenaga, E. E. 1975. Acute and chronic toxicity of 75 pesticides to various animal species. *Down Earth* 25(1):5–9.

Kennedy, H. D. and Walsh, D. F. 1970. Effects of Malathion on Two Warmwater Fishes and Aquatic Invertebrates in Ponds. Tech. Paper No. 55. Bureau of Sport Fish. Wildl., Fish Wildl. Service, U.S. Dept. Int., Washington, D.C., 13 pp.

Keogh, R. G. and Whitehead, P. H. 1975. Observations on some effects of pasture spraying with benomyl and carbendazim on earthworm activity and litter removal *F. N. Z. J. Exp. Agric.* 3:103–104.

Kettle, W. D., Denoyelles, F., Jr., Heacock, B. D. and Kadoum, A. M. 1987. Diet and reproductive success of bluegill recovered from experimental ponds treated with atrazine. *Bull. Environ. Contam. Toxicol.* 38(1):47–52.

Khangarot, B. S., Sehgal, A. and Bhasin, M. K. 1985. "Man and biosphere"-studies on the Sikkim Himalayas. Part 6: Toxicity of selected pesticides to frog tadpole *Rana hexadactyla* (Lesson). *Acta Hydrochim. Hydrobiol.* 13(3):391–394.

Khattat, F. and Farley, S. 1976. Acute toxicity of Certain Pesticides to *Acartia tonsa* Dana. U.S. EPA-OPP Registration Standard.

Khera, K. S. and Mckinley, W. P. 1972. Pre- and postnatal studies on 2,4,5-trichlorophenoxyacetic acid, 2,4-dichlorophenoxyacetic acid and their derivatives in. *Toxicol. Appl. Pharmacol.* 22:14–28.

Khorram, S. and Knight, A. W. 1976. Effects of temperature and kelthane on grass shrimp. *J. Environ. Eng. Div. Am. Soc. Civ. Eng.* 102(5):1043–1053.

Khorram, S. and Knight, A. W. 1977. The toxicity of kelthane to the grass shrimp (*Crangon franciscorum*). *Bull. Environ. Contam. Toxicol.* 18(6):674–682.

Kimura, T. and Keegan, H. L. 1966. Toxicity of some insecticides and molluscicides for the asian blood sucking leech, *Hirudo nipponia* Whitman. *Am. J. Trop. Med. Hyg.* 15(1): 113–115.

King, J. E. and Penfound, W. T. 1946. Effects of two of the new formagenic herbicides on bream and largemouth bass. *Ecology* 27(4):372–374.

Kirkwood, R. C. and Fletcher, W. W. 1970. Factors influencing the herbicidal efficiency of MCPA and MCPB in three species of micro-algae. *Weed Res.* 10(1):3–10.

Klaassen, H. E. and Kadoum, D. M. 1979. Distribution and retention of atrazine and carbofuran in farm pond ecosystems. *Arch. Environ. Contam. Toxicol.* 8(3):345–353.

Klassen, W., Keppler, W. J. and Kitzmiller, J. B. 1965. Toxicities of certain larvicides to resistant and susceptible *Aedes aegypti* (L.). *Bull. W.H.O.* 33:117–122.

Klaverkamp, J. F. and Hobden, B. R. 1980. Brain acetylcholinesterase inhibition and hepatic activation of acephate and fenitrothion in rainbow trout (*Salmo gairdneri*). *Can. J. Fish. Aquat. Sci.* 37(9):1450–1453.

Klaverkamp, J. F., Hobden, B. R. and Harrison, S. E. 1975. Acute lethality and in vitro brain cholinesterase inhibition of acephate and fenitrothion in rainbow trout. *Proc. West. Pharmacol. Soc.* 18:358–361.

Klekowski, R. Z. and Zvirgzds, J. 1971. The influence of herbicide 2,4-D-Na on respiration and survival of *Simocephalus vetulus* (O.F. Muller) (Cladocera). *Pol. Arch. Hydrobiol.* 18(4):393–400.

Klekowski, R. Z., Korde, B. and Kaniewska-Prus, M. 1977. The effect of sodium salt of 2,4-D on oxygen consumption of *Misgurnus fossilis* L. During early embryonal development. *Pol. Arch. Hydrobiol.* 24(3):413–421.

Kline, E. R., Mattson, V. R., Pickering, Q. H., Spehar, D. L. and Stephan, C. E. 1987. Effects of pollution on freshwater organisms. *J. Water Pollut. Control Fed.* 59(6):539–572.

Knapek, R. and Lakota, S. 1974. Einige biotests zur untersuchung der toxischen wirkung von pestiziden im wasser. (Biological testing to determine toxic effects of pesticides in water). *Tagungsber. Akad. Landwirtschaftswiss. D.D.R.* 126:105–109; Pestab: 0175 (1977).

Kobayashi, S. 1978. Synergism in pesticide toxicity 2. Acute oral toxicity of anti-che pesticide in mice. *Toho Igakkai Zasshi (J. Med. Soc. Toho, Jpn.)* 25(4):635–649.

Kok, L. T. 1972. Toxicity of Insecticides Used for Asiatic Rice Borer Control to Tropical Fish in Rice Paddies. in The Careless Technol. Conf. Ecological Aspects Intl. Development, Papers. pp. 489–498.

Kopecek, K., Fuller, F., Ratzmann, W. and Simonis, W. 1975. Lichtabhangige insektizidwirkungen auf einzellige algen. (The light dependent effect of insecticides on unicellular algae). *Ber. Dtsch. Bot. Ges.* 88(2):269–281.

Korn, S. 1973. The uptake and persistence of carbaryl in channel catfish. *Trans. Am. Fish. Soc.* 102(1):137–139.

Korn, S. and Earnest, R. 1974. Acute toxicity of twenty insecticides to striped bass, *Morone saxatilis. Calif. Fish Game* 60(3):128–131.

Korpela, S. and Tulisalo, U. 1974. The residual contact toxicity to honey bees of eight organophosphorous insecticides on rape flowers. *Ann. Ent. Fenn.* 40(1):1–9.

Korte, F., Freitag, D., Geyer, H., Klein, W., Kraus, A. G. and Lahaniatis, E. 1978. A concept for establishing ecotoxicologic priority lists for chemicals. *Chemosphere* 7(1): 79–102.

Kosanke, G. J., Schwippert, W. W. and Beneke, T. W. 1988. The impairment of mobility and development in freshwater snails (*Physa fontinalis* and *Lymnaea stagnalis*) caused by herbicides. *Comp. Biochem. Physiol.* 90C(2):373–379.

Kosinski, R. J. 1984. The effect of terrestrial herbicides on the community structure of Stream periphyton. *Environ. Pollut. Ser. A Ecol. Biol.* 36(2):165–189.

Kosinski, R. J. and Merkle, M. G. 1984. The effect of four terrestrial herbicides on the productivity of artificial stream algal communities. *J. Environ. Qual.* 13(1):75–82.

Koundinya, P. R. and Murthi, R. R. 1979. Haematological studies in *Sarotherodon* (*tilapia*) *mossambica* (Peters) exposed to lethal (LC50/48 hrs) concentration of sumithion and Sevin. *Curr. Sci.* 48(19):877–879.

Koundinya, P. R. and Ramamurthi, R. 1979a. Tissue respiration in *Tilapia mossambica* exposed to lethal (LC50) concentration of sumithion and Sevin. *Indian J. Environ. Health* 20(4):426–428.

Koundinya, P. R. and Ramamurthi, R. 1979b. Comparative study of inhibition of acetylcholinesterase activity in the freshwater teleost *Sarotherodon (tilapia) mossambica* (Peters) by Sevin (carbamate). *Curr. Sci.* 48(18):832–833.

Koundinya, P. R. and Ramamurthi, R. 1980. Toxicity of sumithion and Sevin to the freshwater fish, *Sarotherodon mossambicus* (Peters). *Curr. Sci.* 49(22):875–876.

Koval'Chuk, L. Ya., Perevozchenko, I. I. and Braginskii, L. P. 1971. Acute toxicity of Yalan, Eptam and Sevin for *Daphnia magna. Eksp. Vodn. Toksikol.* 2:56–64.

Kozlovskaya, V. I. and Mayer, F. L., Jr. 1984. Brain acetylcholinesterase and backbone collagen in fish intoxicated with organophosphate pesticides. *J. Great Lakes Res.* 10(3): 261–266.

Krause, W., Hamm, K. and Weissmuller, J. 1976. Damage to spermatogenesis in juvenile rat treated with DDVP and malathion. *Bull. Environ. Contam. Toxicol.* 15(4):458–462.

Krieger, K. A., Baker, D. B. and Kramer, J. W. 1988. Effects of herbicides on stream aufwuchs productivity and nutrient uptake. *Arch. Environ. Contam. Toxicol.* 17(3):299–306.

Krieger, R. I. and Lee, P. W. 1973. Inhibition of in vivo and in vitro epoxidation of aldrin, and potentiation of toxicity of various insecticide chemicals by diquat in two species of fish. *Arch. Environ. Contam. Toxicol.* 1(2):112–121.

Krishnakumari, M. K. 1977. Sensitivity of the alga *Scenedesmus acutus* to some pesticides. *Life Sci.* 20:1525–1532.

Kruglor, Y. and Mikhaylova, E. I. 1975. Decomposition of herbicide simazine in cultures of algae. *Mikrobiologiya* 44(4):732–735.

Kubota, S. S. and Ochiai, T. 1979. Studies on toxic effects of pesticides on fish — III. Effect of benthiocarb on the toxicity of the other pesticides. *Bull. Fac. Fish. Mie Univ.* 6:119–128.

Kudla, A. J. 1984. Hydra reaggregation: A rapid assay to predict teratogenic hazards induced by environmental toxicity. *J. Wash. Acad. Sci.* 74(4):102–107.

Kulkarni, K. M. and Kamath, S. V. 1980. The metabolic response of *Paratelphusa jacquemontii* to some pollutants. *Geobios* 7(2):70–73.

Kulshrestha, S. K. and Arora, N. 1984a. Impairments induced by sublethal doses of two pesticides in the ovaries of a freshwater teleost *Channa striatus* Bloch. *Toxicol. Lett.* 20(1):93–98.

Kulshrestha, S. K. and Arora, N. 1984b. Effect of sublethal doses of carbaryl and endosulfan on the skin of *Channa striatus* Bl. *J. Environ. Biol.* 5(3):141–147.

Kulshrestha, S. K. and Jauhar, L. 1986. Impairments induced by sublethal doses of Sevin and thiodan on the brain of a freshwater teleost *Channa striatus* Bl. (Channidae). *Acta Hydrochim. Hydrobiol.* 14(4):429–432.

Kumar, K. and Ansari, B. A. 1984. Malathion toxicity: Skeletal deformities in zebrafish (*Brachydanio rerio*, Cyprinidae). *Pestic. Sci.* 15:107–111.

Kumari, S. D. R. and Nair, N. B. 1978. Toxicity of some insecticides to *Lepidocephalus thermalis* (Cuv. and Val.). *Proc. Indian Natl. Sci. Acad.* 44(3):122–132.

Kuroda, K. 1974. Survival abilities of salmon fry in the presence of pesticides. *Mizu Shori Gijutsa (Water Treat. Technol.)* 15(8): 783–793.

Kuroda, K. 1975. Lethal effect of pesticides on saghalien trout fry. *Mizu Shori Gijutsu (Water Purification Liquid Wastes Treatment)* 16(5):441–448.

Kurtz, D. A. and Studholme., C. R. 1974. Recovery of trichlorfon (Dylox) and carbaryl (Sevin) in songbirds following spraying of forest for gyspy moth. *Bull. Environ. Contam. Toxicol.* 11(1):78–84.

Kuwabara, K., Nakamura, A. and Kashimoto, T. 1980. Effect of petroleum oil, pesticides, pcbs and other environmental contaminants on the hatchability of *Artemia salina* dry eggs. *Bull. Environ. Contam. Toxicol.* 25(1):69–74.

Labrecque, G. C., Noe, J. R. and Gahan, J. B. 1956. Effectiveness of insecticides on granular clay carriers against mosquito larvae. *Mosq. News* 16:1–3.

Lal, B. and Singh, T. P. 1986. in vivo modification of fatty acids and glycerides metabolism in response to 1,2,3,4,5,6 hexachlorocyclohexane and cythion exposure in the catfish, *Clarias batrachus. Ecotoxicol. Environ. Saf.* 11(3):295–307.

Lal, B. and Singh, T. P. 1987a. Impact of pesticides on lipid metabolism in the freshwater catfish, *Clarias batrachus*, during the vitellogenic phase of its annual reproductive cycle. *Ecotoxicol. Environ. Saf.* 13(1):13–23.

Lal, B. and Singh, T. P. 1987b. The effect of malathion and gamma-BHC on the lipid metabolism in relation to reproduction in the tropical teleost, *Clarias batrachus. Environ. Pollut.* 48(1):37–47.

Lal, B. and Singh, T. P. 1987c. Gamma-BHC- and cythion-induced alterations in lipid metabolism in a freshwater catfish, *Clarias batrachus*, during different phases of its annual reproductive cycle. *Ecotoxicol. Environ. Saf.* 14(1):38–47.

Lal, B., Singh, A., Kumari, A. and Sinha, N. 1986. Biochemical and haematological changes following malathion treatment in the freshwater catfish *Heteropneustes fossilis* (Bloch). *Environ. Pollut. Ser. A Ecol. Biol.* 42(2):151–156.

LeBlanc, G. A. 1980. Acute toxicity of priority pollutants to water flea (*Daphnia magna*). *Bull. Environ. Contam. Toxicol.* 24(5):684–691.

Lejczak, B. 1977. Effect of insecticides: Chlorphenvinphos, carbaryl and propoxur on aquatic organisms. *Pol. Arch. Hydrobiol.* 24(4):583–591.

Lembi, C. A. and Coleridge, S. E. 1975. Selective Toxicity of Detergents and Herbicides to Phytoplankton. Water Resourc. Res. Center. NTIS PB-250 497. Purdue University, West Lafayette, IN, 71 pp.

Levy, R. and Miller, T. W., Jr. 1978. Tolerance of the planarian *Dugesia dorotocephala* to high concentrations of pesticides and growth regulators. *Entomophaga* 23(1):31–34.

Lewallen, L. L. 1959. Toxicity of several organophosphorus insecticides to *Gambusia affinis* (Baird and Girard) in laboratory tests. *Mosq. News* 19(1):1–2.

Lewallen, L. L. 1962. Toxicity of certain insecticides to hydrophilid larvae. *Mosq. News* 22(2):112–113.

Lewallen, L. L. and Wilder, W. H. 1962. Toxicity of certain organophosphorus and carbamate insecticides to rainbow trout. *Mosq. News* 22(4):369–372.

Li, G. C. and Chen, C. Y. 1981. Study on the acute toxicities of commonly used pesticides to two kinds of fish. *K'O Hsueh Fa Chan Yueh K'an* 9(2):146–152.

Lichtenstein, E. P., Liang, T. T. and Anderegg, B. N. 1973. Synergism of insecticides by herbicides. *Science* 181:847–849.

Lichtenstein, E. P., Schulz, K. R., Skrentny, R. F. and Tsukano, Y. 1966. Toxicity and fate of insecticide residues in water. *Arch. Environ. Health* 12:199–212.

Lickly, T. D. and Murphy, P. G. 1987. The amount and identity of [14C] residues in bluegills (*Lepomis macrochirus*) exposed to [14C] triclopyr. *Environ. Int.* 13:213–218.

Linden, E., Bengtsson, B. E., Svanberg, O. and Sundstrom, G. 1979. The acute toxicity of 78 chemicals and pesticide formulations against two brackish water organisms, the bleak (*Alburnus alburnus*) and the *Harpacticoid ni. Chemosphere* 8(11/12):843–851.

Lingaraja, T. and Venugopalan, V. K. 1978. Pesticide induced physiological and behavioural changes in an estuarine teleost *Therapon jarbua* (Forsk). *Fish. Technol.* 15(2):115–119.

Liu, D. H. W. and Lee, J. M. 1975. Toxicity of Selected Pesticides to the Bay Mussel (*Mytilus edulis*). Ecol. Res. Ser. EPA-660/3-75-016. U.S. NTIS, PB-243 221. Natl. Environ. Res. Center, U.S. EPA, Corvallis, OR, 102 pp.

Lockhart, W. L., Metner, D. A., Ward, F. J. and Swanson, G. M. 1985. Population and cholinesterase responses in fish exposed to malathion sprays. *Pestic. Biochem. Physiol.* 24:12–18.

Loeb, H. A. and Kelly, W. H. 1963. Acute Oral Toxicity of 1,496 Chemicals Force-Fed to Carp. U.S. Fish. Wildl. Serv., Sp. Sci. Rep. Fish. No. 471. Washington, D.C., 124 pp.

Lone, K. P. and Javaid, M. Y. 1976. Effect of sublethal doses of three organophosphorus insecticides on the haematology of *Channa punctatus* (Bloch). *Pak. J. Zool.* 8(1):77–84.

Lopez, J. F. and Judd, F. W. 1979. Effect of sublethal dietary exposure of monosodium methanearsonate herbicide on the nest-building behavior of the white-footed mouse, *Peromyscus leucopus*. *Bull. Environ. Contam. Toxicol.* 23:030–032.

Lorz, H. W., Glenn, S. W., Williams, R. H., Kunkel, C. M., Norris, L. A. and Loper, B. R. 1979. Effects of Selected Herbicides on Smolting of Coho Salmon. Ecol. Res. Ser. EPA-600/3-79-071. Corvallis Environ. Res. Lab., U.S. EPA, Corvallis, OR, 102 p.

Lowe, J. I. 1967. Effects of prolonged exposure to Sevin on an estuarine fish, *Leiostomus xanthurus* Lacepede. *Bull. Environ. Contam. Toxicol.* 2(3):147–155.

Lui, O. S., Ambak, M. A. and Mohsin, A. K. M. 1983. A comparison of tolerance level of tilapia to malathion on clear and muddy bottom. *Malays. Appl. Biol.* 12(2):25–29.

Lunn, C. R., Toews, D. P. and Pree, D. J. 1976. Effects of three pesticides on respiration, coughing, and heart rates of rainbow trout (*Salmo gairdneri* Richardson). *Can. J. Zool.* 54(2):214–219.

Lynch, T. R., Johnson, H. E. and Adams, W. J. 1985. Impact of atrazine and hexachlorobiphenyl on the structure and function of model stream ecosystems. *Environ. Toxicol. Chem.* 4(3): 399–413.

Lysak, A. and Marcinek, J. 1972. Multiple toxic effect of simultaneous action of some chemical substances on fish. *Rocz. Nauk Roln. Ser. H Rybactivo* 94(3):53–63.

Macek, K. J. and McAllister, W. A. 1970. Insecticide susceptibility of some common fish family representatives. *Trans. Am. Fish. Soc.* 99(1):20–27.

Macek, K. J., Buxton, K. S., Sauter, S., Gnilka, S. and Dean, J. W. 1976. Chronic toxicity of atrazine to selected aquatic invertebrates and fishes. Ecol. Res. Ser. EPA-600/3-76-047. Environ. Res. Lab., U.S. EPA, Duluth, MN, 50 pp.

Macek, K. J., Hutchinson, C. and Cope, O. B. 1969. The effects of temperature on the susceptibility of bluegills and rainbow trout to selected pesticides. *Bull. Environ. Contam. Toxicol.* 4(3):174–183.

Macek, K. J., Walsh, D. F., Hogan, J. W. and Holz, D. D. 1972. Toxicity of the insecticide Dursban to fish and aquatic invertebrates in ponds. *Trans. Am. Fish. Soc.* 101(3):420–427.

MacPhee, C. and Ruelle, R. 1969. Lethal Effects of 1888 Chemicals Upon Four Species of Fish From Western North America. Univ. of Idaho Forest, Wildl. Range Exp. Station Bull. No. 3, Moscow, ID, 112 pp.

Mailhot, H. 1987. Prediction of algal bioaccumulation and uptake rate of nine organic compounds by ten physicochemical properties. *Environ. Sci. Technol.* 21(10):1009–1013.

Mallison, S. M., III and Cannon, R. E. 1984. Effects of pesticides on *Cyanobacterium plectonema* Boryanum and Cyanophage lpp-1. *Appl. Environ. Microbiol.* 47(5):910–914.

Malone, C. R. and Blaylock, B. G. 1970. Toxicity of insecticide formulations to carp embryos reared in vitro. *J. Wildl. Manage.* 34(2):460–463.

Maly, M. and Ruber, E. 1983. Effects of pesticides on pure and mixed species cultures of salt marsh pool algae. *Bull. Environ. Contam. Toxicol.* 30(4):464–472.

Maly, M. P. 1980. A Study of the Effects of Pesticides on Single and Mixed Species Cultures of Algae. Ph.D. Thesis, Northeastern University, Boston, MA, 261 pp.; *Diss. Abstr. Int. B* 41(4):1227.

Mandal, P. K. and Kulshrestha, A. K. 1983. Chronic effects of malathion on erythropoiesis in catfish *Clarias batrachus* (Linn). *Natl. Acad. Sci. Lett.* 6(9):311–313.

Mane, U. H. and Muley, D. V. 1987. Seasonal variations in the toxicity of cythion-malathion to two freshwater bivalve molluscs. *Comp. Physiol. Ecol.* 12(1):25–31.

Mane, U. H., Akarte, S. R. and Muley, D. V. 1984. Effect of cythion-malathion on respiration in three freshwater bivalve molluscs from Godavari River near Paithan. *J. Environ. Biol.* 5(2):71–80.

Mane, U. H., Kachole, M. S. and Pawar, S. S. 1979. Effect of pesticides and narcotants on bivalve molluscs. *Malacologia* 18:347–360.

Mani, V. G. T. and Konar, S. K. 1985b. Acute toxicity of malathion to fish, plankton and worm. *Aquat. Sci. Fish. Abstr. Part 1*, 15(2):287; *Environ. Ecol.* 2(4):248–250 (1984).

Mani, V. G. T. and Konar, S. K. 1986. Chronic effects of malathion on feeding behavior, survival, growth and reproduction of fish. *Aquat. Sci. Fish. Abstr. Part 1*, 16(8):179; *Environ. Ecol.* 3(3):348–350 (1985).

Mani, V. G. T. and Konar, S. K. 1985a. Influence of the organophosphorus insecticide malathion on aquatic ecosystem. *Aquat. Sci. Fish. Abstr.* 16(4):7154–1Q16; *Environ. Ecol.* 3(4):493–495 (1985).

Marchini, S., Passerini, L., Cesareo, D. and Tosato, M. L. 1988. Herbicidal triazines: Acute toxicity on daphnia, fish, and plants and analysis of its relationships with structural factors. *Ecotoxicol. Environ. Saf.* 16(2):148–157.

Marian, M. P., Arul, V. and Pandian, T. J. 1983. Acute and chronic effects of carbaryl on survival, growth, and metamorphosis in the bullfrog (*Rana tigrina*). *Arch. Environ. Contam. Toxicol.* 12(3):271–275.

Marking, L. L. and Dawson, V. K. 1975. Method for Assessment of Toxicity or Efficacy of Mixtures of Chemicals. Invest. Fish Control No. 67. Fish Wildl. Serv., U.S. Dept. Int. Washington, D.C., 8 pp.

Marking, L. L., Bills, T. D. and Crowther, J. R. 1984. Effects of five diets on sensitivity of rainbow trout to eleven chemicals. *Prog. Fish Cult.* 46(1):1–5.

Marshall, C. D. and Rutschsky, C. W., III. 1974. Single herbicide treatment: Effect on the diversity of aquatic insects in Stone Valley Lake, Huntingdon County, PA. *Proc. Pa. Acad. Sci.* 48:127–131.

Mason, J. W., Abdelghani, A. A., Anderson, A. C. and Englande, A. J. 1976. A Study of the Distribution and Fate of MSMA in Crayfish and Blackberries. U.S. NTIS PB-275-794. 86 pp.

Mathur, D. S. 1974. Toxicity of Sevin to certain fishes. *J. Inl. Fish. Soc. India* 6.

Matlak, O. 1972. Sodium salt (2,4-D) related to embryo and larval development to carp. *Pol. Arch. Hydrobiol.* 19(4):437–449.

Matsuo, K. and Tamura, T. 1970. Laboratory experiments on the effect of insecticides against blackfly larvae (*Diptera: Simuliidae*) and fishes. *Bochu-Kagaku* 35(4):125–130.

Mattson, V. R., Arthur, J. W. and Walbridge, C. T. 1976. Acute Toxicity of Selected Organic Compounds to Fathead Minnows. Ecol. Res. Ser. EPA-600/3-76-097. Environ. Res. Lab., U.S. EPA, Duluth, MN, 12 pp.

Mauck, W. L., Mayer, F. L., Jr. and Holz, D. D. 1976. Simazine residue dynamics in small ponds. *Bull. Environ. Contam. Toxicol.* 16(1):1–8.

Mayasich, J. M., Karlander, E. P. and Terlizzi, D. E., Jr. 1987. Growth responses of *Nannochloris oculata* Droop and *Phaeodactylum tricornutum* Bohlin to the herbicide atrazine as influenced by light intensity and temperature in unialgal and bialgal assemblage. *Aquat. Toxicol.* 10(4):187–197.

Mayes, M. A., Dill, D. C., Bodner, K. M. and Mendoza, C. G. 1984. Triclopyr triethylamine salt toxicity to life stages of the fathead minnow (*Pimephales promelas* Rafinesque). *Bull. Environ. Contam. Toxicol.* 33(3):339–347.

McBride, R. K. and Richards, B. D. 1975. The effects of some herbicides and pesticides on sodium uptake by isolated perfused gills from the carp *Cyprinus carpio*. *Comp. Biochem. Physiol.* C 51(1):105–109.

McCann, J. A. and Hitch, R. K. 1980. Simazine toxicty to fingerling striped bass. *Prog. Fish Cult.* 42(3):180–181.

McCann, J. A. and Jasper, R. L. 1972. Vertebral damage to bluegills exposed to acutely toxic levels of pesticides. *Trans. Am. Fish. Soc.* 101(2):317–322.

Mccorkle, F. M., Chambers, J. E. and Yarbrough, J. D. 1977. Acute toxicities of selected herbicides to fingerling channel catfish, *Ictalurus punctatus*. *Bull. Environ. Contam. Toxicol.* 18(3):267–270.

McDonald, P. T. and Asman, S. M. 1982. A genetic-sexing strain based on malathion resistance for *Culex tarsalis*. *Mosq. News* 42(4):531–536.

McEnerney, J. T. and Davis, D. E. 1979. Metabolic fate of atrazine in the *Spartina alterniflora*-detritus-*Uca pugnax* food chain. *J. Environ. Qual.* 8(3):335–338.

McGinty, A. S. 1984. Effects of periodic applications of simazine on the production of *Tilapia nilotica* fingerlings. *J. Agric. Univ. P.R.* 68(4):467–469.

McKim, J. M., Schmieder, P. K., Niemi, G. J., Carlson, R. W. and Henry, T. R. 1987. Use of respiratory-cardiovascular responses of rainbow trout (*Salmo gairdneri*) in identifying acute toxicity syndromes in fish. Part 2. Malathion, carbaryl, acrolein and benzaldehyde. *Environ. Toxicol. Chem.* 6:313–328.

McLeese, D. W., Zitko, V. and Peterson, M. R. 1979. Structure-lethality relationships for phenols, anilines and other aromatic compounds in shrimp and clams. *Chemosphere* 8(2): 53–57.

Meehan, W. R., Norris, L. A. and Sears, H. S. 1974. Toxicity of various formulations of 2,4-D to salmonids in southeast alaska. *J. Fish. Res. Board Can.* 31(4):480–484.

Mehta, R. S. and Hawxby, K. W. 1979. Effects of simazine on the blue-green alga *Anacystis nidulans*. *Bull. Environ. Contam. Toxicol.* 23(3):319–326.

Mehta, R. S. and Hawxby, K. W. 1983. Physiological response of *Anacystis nidulans* (Cyanophyceae) to various herbicides. *J. Phycol.* 19(2):15.

Meier, E. P., Dennis, W. H., Rosencrance, A. B., Randall, W. F., Cooper, W. J. and Warner, M. C. 1979. Sulfotepp, a toxic impurity in formulations of diazinon. *Bull. Environ. Contam. Toxicol.* 23(1/2):158–164.

Mendoza, C. E. 1976. Toxicity and effects of malathion on esterases of suckling albino rats. *Toxicol. Appl. Pharmacol.* 35:229–238.

Menzie, C. 1983. Acute Toxicity of Some Organophosphorus Pesticides Against Fish and Aquatics: Sumithion. U.S. EPA-OPP Registration Standard.

Meydani, M. and Post, G. 1979. Effect of sublethal concentrations of malathion on Coturnix quail. *Bull. Environ. Contam. Toxicol.* 21:661–667.

Meyer, F. P. 1981. Quarterly Report of Progress. Natl. Fish. Res. Lab., Lacrosse, WI and S.E. Fish Control Lab, Warm Springs, GA. Fish Wildl. Serv., U.S. Dept. Int. 34 pp.

Meyers, S. M. and Schiller, S. M. 1985. TERRE-TOX: a data base for effects of anthropogenic substances on terrestrial animals. *J. Chem. Inf. Comput. Sci.* 26(1):33–36.

Miller, C. W., Zuckerman, B. M. and Charig, A. J. 1966. Water translocation of diazinon-c14 and parathion-s35 off a model cranberry bog and subsequent occurrence in fish and mussels. *Trans. Am. Fish. Soc.* 95(4):345–349.

Miller, W. E., Peterson, S. A., Greene, J. C. and Callahan, C. A. 1985. Comparative toxicology of laboratory organisms for assessing hazardous waste sites. *J. Environ. Qual.* 14(4):569–574.

Millie, D. F. and Hersh, C. M. 1987. Statistical characterizations of the atrazine-induced photosynthetic inhibition of *Cyclotella meneghiniana* (Bacillariophyta). *Aquat. Toxicol.* 10(4):239–249.

Mishra, A., Dwivedi, P. P. and Dutta, K. K. 1986. Behavior of freshwater air breathing teleost, *Heteropneustes fossilis* (bi.) exposed to different concentrations of malathion. *C.A. Sel. Environ. Pollut.* 15:3; *J. Recent Adv. Appl. Sci.* 1(1):13–16.

Mishra, J. and Srivastava, A. K. 1983. Malathion induced hematological and biochemical changes in the Indian catfish *Heteropneustes fossilis*. *Environ. Res.* 30(2):393–398.

Mitchell, D. 1985. Bioassay Testing of Herbicide H2 and Insecticidal Soap/Diazinon with Rainbow Trout and Daphnia. U.S. EPA-OPP Registration Standard.

Mitchell, D. G., Chapman, P. M. and Long, T. J. 1987a. Seawater challenge testing of coho salmon smolts following exposure to Roundup herbicide. *Environ. Toxicol. Chem.* 6(11): 875–878.

Mitchell, D. G., Chapman, P. M. and Long, T. J. 1987b. Acute toxicity of Roundup and Rodeo herbicides to rainbow trout, chinook, and coho salmon. *Bull. Environ. Contam. Toxicol.* 39:1028–1035.

Mitsuhashi, J., Grace, T. D. C. and Waterhouse, D. F. 1970. Effects of insecticides on cultures of insect cells. *Entomol. Exp. Appl.* 13:327–341.

Moffett, J. O. and Morton, H. L. 1972. Toxicity of some herbicidal sprays to honey bees. *J. Econ. Entomol.* 65(1):32–36.

Mohanty-Hejmadi, P. and Dutta, S. K. 1981. Effects of some pesticides on the development of the Indian bull frog *Rana tigerina*. *Environ. Pollut. Ser. A* 24(2):145–161.

Monsanto. 1982. Glyphosate Technical. Monsanto Material Safety Sheet. St. Louis, MO, pp. 1–4.

Monsanto. 1983. Rodeo Herbicide: Toxicological and Environmental Properties. Bull. No. 1. EP-005. Monsanto Agric. Prod. Co., St. Louis, MO, 4 pp.

Moore, R. B. 1970. Effects of pesticides on growth and survival of *Euglena gracilis* Z. *Bull. Environ. Contam. Toxicol.* 5(3):226–230.

Moorhead, D. L. and Kosinski, R. J. 1986. Effect of atrazine on the productivity of artificial stream algal communities. *Bull. Environ. Contam. Toxicol.* 37(3):330–336.

Morgan, H. G. 1976. Sublethal Effects of Diazinon on Stream Invertebrates. Ph.D. Thesis, University of Guelph, Guelph, Ontario, Canada 157 p.; *Diss. Abstr. Int. B* 38(1):125 (1977).

Morgan, W. S. G. 1975. Monitoring pesticides by means of changes in electric potential caused by fish opercular rhythms. *Prog. Water Technol.* 7(2):33–40.

Morrow, J. E., Gritz, R. L. and Kirton, M. P. 1975. Effects of some components of crude oil on young coho salmon. *Copeia* 2:326–331.

Mount, D. I. and Stephan, C. E. 1967. A method for establishing acceptable toxicant limits for fish — malathion and the butoxyethanol ester of 2,4-D. *Trans. Am. Fish. Soc.* 96(2): 185–193.

Mount, G. A., Lowe, R. F., Baldwin, K. F., Pierce, N. W. and Savage, K. E. 1970. Ultra-low volume aerial sprays of promising insecticides for mosquito control. *Mosq. News* 30(3): 342–346.

Muirhead-Thomson, R. C. and Merryweather, J. 1969. Effect of larvicides on Simulium eggs. *Nature* 221:858–859.

Mukhopadhyay, P. K. and Dehadrai, P. V. 1980a. Biochemical changes in the air-breathing catfish *Clarias batrachus* (Linn.) exposed to malathion. *Environ. Pollut. Ser. A* 22(2): 149–158.

Mukhopadhyay, P. K. and Dehadrai, P. V. 1980b. Studies on air-breathing catfish *Clarias batrachus* (Linn.) under sublethal malathion exposure. *Indian J. Exp. Biol.* 18(4):400–404.

Mukhopadhyay, P. K., Mukherji, A. P. and Dehadrai, P. V. 1984. Phospholipids and fatty acids in the liver of catfish, *Clarias batrachus* exposed to sublethal malathion. *J. Environ. Biol.* 5(4):221–229.

Muley, D. V. and Mane, U. H. 1987. Malathion induced changes in oxygen consumption in two species of freshwater *Lamellibranch* molluscs from Godavari River, Maharashtra state, *Indian J. Environ. Biol.* 8(3):267–275.

Mulla, M. S. 1963. Persistence of mosquito larvicides in water. *Mosq. News* 23(3):234–237.

Mulla, M. S. and Isaak, L. W. 1961. Field studies on the toxicity of insecticides to the mosquito fish, *Gambusia affinis*. *J. Econ. Entomol.* 54(6):1237–1242.

Mulla, M. S. and Khasawinah, A. M. 1969. Laboratory and field evaluation of larvicides against chironomid midges. *J. Econ. Entomol.* 62(1):37–41.

Mulla, M. S., Isaak, L. W. and Axelrod, H. 1963. Field studies on the effects of insecticides on some aquatic wildlife species. *J. Econ. Entomol.* 56(2):184–188.

Mulla, M. S., Norland, R. L., Fanara, D. M., Darwazeh, H. A. and McKean, D. W. 1971. Control of chironomid midges in recreational lakes. *J. Econ. Entomol.* 64(1):300–307.

Muncy, R. J. and Oliver, A. D., Jr. 1963. Toxicity of ten insecticides to the red crawfish, *Procambarus clarki* (Girard). *Trans. Am. Fish. Soc.* 92(4):428–431.

Murphy, S. D., Lauwerys, R. R. and Cheever, K. L. 1968. Comparative anticholinesterase action of organophosphorus insecticides in vertebrates. *Toxicol. Appl. Pharmacol.* 12: 22–35.

Murphy, S. R. 1990. Aquatic Toxicity Information Retrieval (AQUIRE): A Microcomputer Database. System documentation. Spectrum Research, Inc., Duluth, MN, 37 pp.

Murray, H. E. and Guthrie, R. K. 1980. Effects of carbaryl, diazinon and malathion on native aquatic populations of microorganisms. *Bull. Environ. Contam. Toxicol.* 24(4):535–542.

Murty, A. S. 1986. *Toxicity of Pesticides to Fish.* Vol. 1. CRC Press, Boca Raton, FL, 178 pp.

Nair, J. R. and Geevarghese, C. 1984. Growth of Etroplus suratensis (Bloch) larvae hatched and reared in sublethal levels of malathion. *J. Environ. Biol.* 5(4):261–267.

Naqvi, S. M. and Ferguson, D. E. 1970. Levels of insecticide resistance in fresh-water shrimp, *Palaemonetes kadiakensis*. *Trans. Am. Fish. Soc.* 99(4):696–699.

Naqvi, S. M. and Hawkins, R. 1988. Toxicity of selected insecticides (Thiodan, Security, Spartan, and Sevin) to mosquitofish, *Gambusia affinis*. *Bull. Environ. Contam. Toxicol.* 40(5):779–784.

Naqvi, S. M. and Leung, T. S. 1983. Trifluralin and oryzalin herbicides toxicities to juvenile crawfish (*Procambarus clarkii*) and mosquitofish (*Gambusia affinis*). *Bull. Environ. Contam. Toxicol.* 31(3):304–308.

Naqvi, S. M. Z. 1973. Toxicity of twenty-three insecticides to a tubificid worm *Branchiura sowerbyi* from the Mississippi Delta. *J. Econ. Entomol.* 66(1):70–74.

Naqvi, S. M., Davis, V. O. and Hawkins, R. M. 1985. Percent mortalities and LC_{50} values for selected microcrustaceans exposed to Treflan, Cutrine-plus, and MSMA herbicides. *Bull. Environ. Contam. Toxicol.* 35(1):127–132.

Naqvi, S. M., Hawkins, R. and Naqvi, N. H. 1987. Mortality response and LC_{50} values for juvenile and adult crayfish, *Procambarus clarkii* exposed to Thiodan (insecticide), Treflan, MSMA, Oust (herbicides) and Cutrine (algicide) Plus. *Environ. Pollut.* 48:275–283.

Nelson, J. H. and Evans, E. S., Jr. 1973. Field Evaluation of the Larvicidal Effectiveness, Effects on Nontarget Species and Environmental Residues of a Slow-Release Polymer Formulation of Chlorpyrifos March-October 1973. U.S. NTIS AD/A-002 054. 188 pp.

Nelson, J. H., Stoneburner, D. L., Evans, E. S., Jr., Pennington, N. E. and Meisch, M. V. 1976. Diatom diversity as a function of insecticidal treatment with a controlled-release formulation of chlorpyrifos. *Bull. Environ. Contam. Toxicol.* 15(5):630–634.

Netrawali, M. S., Gandhi, S. R. and Pednekar, M. D. 1986. Effect of endosulfan, malathion, and permethrin on sexual life cycle of *Chlamydomonas reinhardtii*. *Bull. Environ. Contam. Toxicol.* 36(3):412–420.

Nikulina, S. S. and Sokolskaya, N. P. 1975. Influence of some pesticides on ontogenesis of carp. *Veterinariya (Moscow)* 7:94–95; Pestab:0271 (1976).

Nishiuchi, Y. 1974. Control effect of pesticide to duckweed. *Bull. Agric. Chem. Insp. Stn.* (14):69–72.

Nishiuchi, Y. 1976. Toxicity of formulated pesticides to some fresh water organisms. XXXIX. *Suisan Zoshoku* 24(3):102–105.

Nishiuchi, Y. 1977a. Toxicity of formulated pesticides to some fresh water organisms. XXXXI. *Suisan Zoshoku* 24(4):146–150.

Nishiuchi, Y. 1977b. Toxicity of formulated pesticides to some freshwater organisms. XXXXV. *Suisan Zoshoku* 25(3):105–107.

Nishiuchi, Y. 1979. Toxicity of pesticides to animals in freshwater. LXII. *Suisan Zoshoku* 27(2):119–124.

Nishiuchi, Y. 1980a. Toxicity of formulated pesticides to fresh water organisms. LXXIV. *Suisan Zoshoku* 28(2):107–112.

Nishiuchi, Y. 1980b. Toxicity of formulated pesticides to fresh water organisms LXXII. *Suisan Zoshoku* 27(4):238–244.

Nishiuchi, Y. and Asano, K. 1978. Toxicity of formulated agrochemicals to fresh water organisms. III. *Suisan Zoshoku* 26(1):26–30.

Nishiuchi, Y. and Asano, K. 1979. Toxicity of pesticides to some fresh water organisms. LIX. *Suisan Zoshoku* 27(1):48–55.

Nishiuchi, Y. and Asano, K. 1981. Comparison of pesticide susceptibility of colored carp with Japanese common carp. *Bull. Agric. Chem. Insp. Stn.* (21):61–63.

Nishiuchi, Y. and Hashimoto, Y. 1967. Toxicity of pesticide ingredients to some fresh water organisms. *Botyu-Kagaku (Sci. Pest Control)* 32(1):5–11.

Nishiuchi, Y. and Hashimoto, Y. 1969. Toxicity of pesticides to some fresh water organisms. *Rev. Plant Protec. Res.* 2:137–139.

Nishiuchi, Y. and Yoshida, K. 1972. Toxicities of pesticides to some fresh water snails. *Bull. Agric. Chem. Insp. Stn.* (12):86–92.

Nishiuchi, Y. and Yoshida, K. 1974. Effects of pesticides on tadpoles. Part 3. *Noyaku Kensasho Hokoku* 14:66–68; Pestab:1714 (1975).

Noll, M. and Bauer, U. 1974. Phormidium Autumnale as Indicator Organism for Algicidal Substances in Water. U.S. EPA-OPP Registration Standard.

Norberg, T. J. and Mount, D. I. 1985. A new fathead minnow (*Pimephales promelas*) subchronic toxicity test. *Environ. Toxicol. Chem.* 4(5):711–718.

Norland, R. L., Mulla, M. S., Pelsue, F. W. and Ikeshoji, T. 1974. Conventional and new insecticides for the control of chironomid midges. *Proc. Ann. Conf. Calif. Mosq. Control Assoc.* 42:181–183.

O'Brien, M. C. and Prendeville, G. N. 1979. Effect of herbicides on cell membrane permeability in Lemna minor. *Weed Res.* 19(6):331–334.

O'Neal, S. W. and Lembi, C. A. 1983. Effect of simazine on photosynthesis and growth of filamentous algae. *Weed Sci.* 31(6):899–903.

OMKAR and Shukla, G. S. 1984. Toxicity of the herbicide 2,4-D-Na to two species of freshwater prawn of the genus *Macrobrachium*. *Acta Hydrochim. Hydrobiol.* 12(3): 285–289.

OMKAR and Shukla, G. S. 1985. Toxicity of insecticides to *Macrobrachium lamarrei* (H. milne Edwards) (*Decapoda: Palaemonidae*). *Crustaceana* 48(1):1–5.

Otsuka, S. 1975. Plant protection and pesticide news. Pesticides: amitoraz ec, orthobencarb herbicide, and oxadiazon combined fertilizer. *Nogyo Oxobi Engei* 50(9):1171–1172 (JPN); *Pestab*:2701 (1975).

Palawski, D. U. and Knowles, C. O. 1986. Toxicological studies of benomyl and carbendazim in rainbow trout, channel catfish and bluegills. *Environ. Toxicol. Chem.* 5(12):1039–1046.

Palawski, D., Hunn, J. B. and Dwyer, F. J. 1985. Sensitivity of young striped bass to organic and inorganic contaminants in fresh and saline waters. *Trans. Am. Fish.* Soc. 114:748–753.

Palmer, C. M. and Maloney, T. E. 1955. Preliminary screening for potential algicides. *Ohio J. Sci.* 55(1):1–8.

Pandey, A. K. and Shukla, L. 1982. Effect of an organophosphorus insecticide malathion on the testicular histophysiology in *Sarotherodon mossambicus*. *Natl. Acad. Sci. Lett.* 5(4): 141–142.

Pandey, A. K. and Shukla, L. 1983. Antithyroidal effects of an organophosphorus insecticide malathion in *Sarotherodon mossambicus*. *Int. J. Environ. Stud.* 22(1):49–57.

Pandey, P. K., Thakur, G. K. and Choudhary, B. P. 1988. Effect of malathion and metacercarial infection on liver lipofuscin accumulation of the air-breathing fish *Heteropneustes fossilis*. *Environ. Ecol.* 6(1):234–236.

Pant, J. C. and Singh, T. 1983. Inducement of metabolic dysfunction by carbamate and organophosphorus compounds in a fish, *Puntius conchonius*. *Pestic. Biochem. Physiol.* 20(3):294–298.

Panwar, R. S., Kapoor, D., Joshi, H. C. and Gupta, R. A. 1976. Toxicity of some insecticides to the weed fish, *Trichogaster fasciatus* (Bloch and Schneider). *J. Inl. Fish. Soc. India* 8:129–130.

Parker, B. L., Dewey, J. E. and Bache, C. A. 1970. Carbamate bioassay using *Daphnia magna*. *J. Econ. Entomol.* 63(3):710–714.

Parkhurst, Z. E. and Johnson, H. E. 1955. Toxicity of malathion 500 to fall chinook salmon fingerlings. *Prog. Fish Cult.* 17(3):113–116.

Patnaik, S. and Ramachandran, V. 1976. Control of blue-green algal blooms with simazine in fish ponds. in Varshney, C. K. and Rzoska, J. (Eds.). Regional Seminar on Noxious Aquatic Vegetation. Aquatic Weeds in Southeast Asia, New Delhi, India, pp. 285–291.

Paulov, S. 1976. The effects of the herbicides burex, zeazin-50, and aminex on the development of amphibia. (Ucinky herbicidov burex, zeazin-50 a aminex na vyvoj obojzivelnikov). *Agrochemica* 16(4):102–103.

Pawar, K. R., Ghate, H. V. and Katdare, M. 1983. Effect of malathion on embryonic development of the frog *Microhyla ornata* (Dumeril and Bibron). *Bull. Environ. Contam. Toxicol.* 31(2):170–176.

Peterson, J. L. 1972. Biology and Drift of the Aquatic Insects in Seasonal Irrigation Waters and Their Susceptibility to ULV Malathion. Ph.D. Thesis, University of Nebraska, Lincoln, NE, 69 pp.; *Diss. Abstr. Int. B* 33(04):1596 (1972).

Peterson, R. H. 1976. Temperature selection of juvenile atlantic salmon (*Salmo salar*) as influenced by various toxic substances. *J. Fish. Res. Board Can.* 33(8):1722–1730.

Phipps, G. L. and Holcombe, G. W. 1985. A method for aquatic multiple species toxicant testing: Acute toxicity of 10 chemicals to 5 vertebrates and 2 invertebrates. *Environ. Pollut. Ser. A Ecol. Biol.* 38(2):141–157.

Pickering, Q. H. and Henderson, C. 1966. Acute toxicity of some important petrochemicals to fish. *J. Water Pollut. Control Fed.* 38(9):1419–1429.

Pickering, Q., Carle, D. O., Pilli, A., Willingham, T. and Lazorchak, J. M. 1989. Effects of pollution on freshwater organisms. *J. Water Pollut. Control Fed.* 61(6):998–1042.

Pickering, Q. H., Henderson, C. and Lemke, A. E. 1962. The toxicity of organic phosphorus insecticides to different species of warmwater fishes. *Trans. Am. Fish. Soc.* 91: 175–184.

Pillai, P., Weete, J. D., Diner, A. M. and Davis, D. E. 1979. Atrazine metabolism in box crabs. *J. Environ. Qual.* 8(3):277–280.

Pilli, A., Carle, D. O., Kline, E., Pickering, Q. and Lazorchak, J. 1988. Effects of pollution on freshwater organisms. *J. Water Pollut. Control Fed.* 60(6):994–1065.

Pilli, A., Carle, D. O. and Sheedy, B. A. 1989. AQUIRE: Aquatic Toxicity and Retrieval Data Base. Contract No. 68-01-7176. U.S. Environmental Protection Agency, ERL-Duluth, MN, 18 pp.

Plumley, F. G. and Davis, D. E. 1980. The effects of a photosynthesis inhibitor atrazine, on salt marsh edaphic algae, in culture microecosystems, and in the field. *Estuaries* 3(4): 271–277.

Plumley, F. G., Davis, D. E., McEnernery, J. T. and Everest, J. W. 1980. Effects of a photosynthesis inhibitor, atrazine, on the salt-marsh fiddler crab, *Uca pugnax* (Smith). *Estuaries* 3(3):217–223.

Pomeroy, S. E. and Barrett, G. W. 1975. Dynamics of enclosed small mammal populations in relation to an experimental pesticide application. *Am. Midl. Nat.* 93(1):91–106.

Poorman, A. E. 1973. Effects of pesticides on *Euglena gracilis*. I. Growth studies. *Bull. Environ. Contam. Toxicol.* 10(1):25–28.

Portmann, J. E. 1972. Results of acute toxicity tests with marine organisms, using a standard method. in *Marine Pollution and Sea Life*. Fishing News (Books) Ltd., London, England, pp. 212–217.

Posner, S. and Reimer, S. 1970. The Determination of TLM Values of Diazinon on Fingerling Fish. U.S. EPA-OPP Registration Standard.

Post, G. and Leasure, R. A. 1974. Sublethal effect of malathion to three salmonid species. *Bull. Environ. Contam. Toxicol.* 12(3):312–319.

Post, G. and Schroeder, T. R. 1971. Toxicity of four insecticides to four salmonid species. *Bull. Environ. Contam. Toxicol.* 6(2):144–155.

Pott, E. 1980. Die hemmung der futteraufnahme von *Daphnia pulex* — Eine neue limnotoxikologische mebgrobe. (The reduction of food uptake of *Daphnia pulex* — a new indicator of sub-lethal toxic stress). *Z. Wasser Abwasser Forsch.* 13(2):52–54.

Prasad, P. V. D. and Chowdary, Y. B. K. 1981. Effects of metabolic inhibitors on the calcification of a freshwater green alga, *Gloeotaenium ioitlesbergarianum* Hansgirg. I. Effects of some photosynthetic and respiratory inhibitors. *Ann. Bot.* 47(4):451–459.

Pratt, J. R., Bowers, N. J., Niederlehner, B. R. and Cairns, J., Jr. 1988. Effects of atrazine on freshwater microbial communities. *Arch. Environ. Contam. Toxicol.* 17(4):449–457.

Prescott, L. M., Kubovec, M. K. and Tryggestad, D. 1977. The effects of pesticides, polychlorinated biphenyls and metals on the growth and reproduction of *Acanthamoeba castellanii*. *Bull. Environ. Contam. Toxicol.* 18(1):29–34.

Presing, M. and Ponyi, J. E. 1986. Studies on the acute and chronic effect of a 2,4-D-containing herbicide (Dikonirt) on *Eudiaptomus gracilis* (G.O. Sars) (Crustacea, Copepoda). *Arch. Hydrobiol.* 106(2):275–286.

Pritchard, J. B. and James, M. O. 1979. Determinants of the renal handling of 2,4-dichlorophenoxyacetic acid by winter flounder. *J. Pharmacol. Exp. Ther.* 208:280–286.

Prokopenko, V. A., Zhiteneva, L. D., Sokol'Skaya, N. P., Kalyuzhnaya, T. I., Zavgorodnyaya, V. P. and Isaeva, L. N. 1976a. Comparative evaluation of the effect of malathion on yearlings of some fish species. *Tr. Vses. Naucho-Issled. Inst. Morsk. Rybn. Khoz. Okeanogr.* 109:199–209.

Prokopenko, V. A., Zhiteneva, L. D., Sokol'Skaya, N. P., Kalyuzhnaya, T. I., Zavgorodnyaya, V. P., Isaeva, L. N. and Kopylova, Z. N. 1976b. Toxicity of carbophos for certain aquatic organisms. *Hydrobiol. J.* 12(5):36–40.

Quinn, S. A., Cardarelli, N. F. and Gangstad, E. O. 1977. Aquatic herbicide chronicity. *J. Aquat. Plant Manage.* 15:74–76.

Rabeni, C. F. and Stanley, J. G. 1979. Operational spraying of acephate to suppress spruce budworm has minor effects on stream fishes and invertebrates. *Bull. Environ. Contam. Toxicol.* 23(3):327–334.

Ram, R. N. and Sathyanesan, A. G. 1987. Effects of long-term exposure to cythion on the reproduction of the teleost fish, *Channa punctatus* (Bloch). *Environ. Pollut.* 44(1):49–60.

Ramalingam, K. and Ramalingam, K. 1982. Effects of sublethal levels of DDT, malathion and mercury on tissue proteins of *Sarotherodon mossambicus* (Peters). *Proc. Indian Acad. Sci. (Anim. Sci.)* 91(6):501–505.

Ramaswamy, M. 1988. Effects of Sevin on blood free amino acid levels of the fish *Sarotherodon mossambicus. Aquat. Sci. Fish. Abstr.* 18(4):5466–1Q18; *Environ. Biol.* 5(4):633–637 (1987).

Ranke-Rybicaka, B. and Stanislawska, J. 1972. Zmiany w zespolach perifitonowych wywolane dzialaniem pestycydow fosforoorganicznych (malathion, foschlor). *Rocz. Panstu. Zakl. Hig.* 23(2):137–146.

Ranke-Rybicka, B. 1975. The effect of organophosphate pesticides (malathion, foschlor, dichlorvos) on aquatic organisms. *Rocz. Panstu. Zakl. Hig.* 26(3):393–399.

Rao, D. M., Murty, A. S. and Swarup, P. A. 1984b. Relative toxicity of technical grade and formulated carbaryl and 1-naphthol to, and carbaryl-induced biochemical changes in, the fish *Cirrhinus mrigala. Environ. Pollut. Ser. A Ecol. Biol.* 34(1):47–54.

Rao, K. J., Madhu, C. and Murthy, V. S. R. 1983. Histopathology of malathion on gills of a freshwater teleost, *Tilapia mossambica* (Peters). *J. Environ. Biol.* 4(1):9–13.

Rao, K. J., Madhu, C., Rao, V. A. and Ramamurthy, K. 1984a. Hyperglycemia induced by insecticides in *Oreochromis mossambicus* (Trewavas). *J. Curr. Biosci.* 1(3):115–116.

Rao, K. R. S. S. 1987. Variations in the nitrogen products of *Channa punctatus* augmented by interaction of carbaryl and phenthoate in the media. *J. Environ. Biol.* 8(2 Suppl.):173–177.

Rao, K. R. S. S., Rao, K. S. P., Sahib, I. K. A. and Rao, K. V. R. 1985b. Combined action of carbaryl and phenthoate on a freshwater fish (*Channa punctatus* Bloch). *Ecotoxicol. Environ. Saf.* 10(2):209–217.

Rao, K. S. and Dad, N. K. 1979. Studies of herbicide toxicity in some freshwater fishes and ectoprocta. *J. Fish Biol.* 14(6):517–522.

Rao, K. S. and Nagabhushanam, R. 1987. Toxicity of DDT and malathion to the freshwater crab *Barytelphusa cunicularis. Aquat. Sci. Fish. Abstr. Part 1* 17(8):11842–1Q17; *Environ. Ecol.* 5(1):203–204.

Rao, K. S. P., Rao, K. R. S. S., Sahib, I. K. A. and Rao, K. V. R. 1985a. Combined action of carbaryl and phenthoate on tissue lipid derivatives of murrel, *Channa punctatus* (Bloch). *Ecotoxicol. Environ. Saf.* 9(1):107–111.

Rao, K. S., Khan, A. K., Alam, S. and Nagabhushanam, R. 1988. Toxicity of three pesticides to the marine prawn *Penaeus monodon* and *Penaeus indicus. Environ. Ecol.* 6(2):479–480.

Rao, M. B. 1981. Effect of gamma-hexachloran and Sevin on the survival of the black sea mussel, *Mytilus galloprovincialis* Lam. *Hydrobiologia* 78(1):33–37.

Rao, M. B. and Mane, U. K. 1978. The effect of carbofos on the survival rate and respiration of black sea mussels (*Mytilus galloprovincialis*). *Gidrobiol. Zh. (Hydrobiol. J.)* 14(6): 90–94.

Rao, T. S., Dutt, S. and Mangaiah, K. 1967. Tlm values of some modern pesticides to the freshwater fish — *Puntius puckelli. Environ. Health* 9:103–109.

Rao, T. S., Rao, M. S. and Prasad, S. B. S. K. 1975. Median tolerance limits of some chemicals to the fresh water fish "*Cyprinus carpio*". *Indian J. Environ. Health* 17(2):140–146.

Rawash, I. A., Gaaboub, I. A., El-Gayar, F. M. and El-Shazli, A. Y. 1975. Standard curves for nuvacron, malathion, Sevin, DDT and kelthane tested against the mosquito *Vulex pipiens* L. and the microcrustacean *Daphnia magna* Straus. *Toxicology* 4(2):133–144.

Rawls, C. K. 1977. Field studies of shell regrowth as a bioindicator of eastern oyster (*Crassostrea virginica* Gmelin) response to 2,4-D BEE in Maryland tidewaters. *Chesapeake Sci.* 18(3): 266–271.

Ray, A. K. and Bhattacharya, S. 1984. Histopathological changes in the hepatopancreas of the freshwater airbreathing teleost *Anabas testudineus* (Bloch) exposed to acute and chronic levels of cythion. *J. Curr. Biosci.* 1(4):170–174.

Reddy, P. S., Bhagyalakshmi, A. and Ramamurthi, R. 1985. Molt-inhibition in the crab *Oziotelphusa senex* Senex following exposure to malathion and methyl parathion. *Bull. Environ. Contam. Toxicol.* 35(1):92–97.

Reddy, P. S., Bhagyalakshmi, A. and Ramamurthi, R. 1986. Chronic malathion toxicity: Effect on carbohydrate metabolism of *Oziotelphusa senex* Senex, the Indian rice field crab. *Bull. Environ. Contam. Toxicol.* 37(6):816–822.

Reddy, P. M. and Bashamohideen, M. 1988. Toxicity of malathion to the fish *Cyprinus carpio. Environ. Ecol.* 6(2):488–490.

Rehwoldt, R. E., Kelley, E. and Mahoney, M. 1977. Investigations into the acute toxicity and some chronic effects of selected herbicides and pesticides on several fresh water fish species. *Bull. Environ. Contam. Toxicol.* 18(3):361–365.

Reinbold, K. A. 1974. Effects of the Synergist Piperonyl Butoxide on Toxicity and Metabolism of Pesticides in Green Sunfish (*Lepomis cyanellus rafinesque*). Ph.D. Thesis, University of Illinois, Urbana-Champaign, IL, 79 pp.; *Dis. Abstr. Int.* B 35(11):5254.

Reinbold, K. A. and Metcalf, R. L. 1976. Effects of the synergist piperonyl butoxide on metabolism of pesticides in green sunfish. *Pestic. Biochem. Physiol.* 6:401–412.

Rettich, F. 1977. The susceptibility of mosquito larvae to eighteen insecticides in Czechoslovakia. *Mosq. News* 37(2):252–257.

Rettich, F. 1979. Laboratory and field investigations in Czechoslovakia with fenitrothion, pirimiphos-methyl, temephos and other organophosphorous larvicides applied as sprays for control of *culex-pipiens-molestus* and ae des-cantans. *Mosq. News* 39(2):320–328.

Richardson, J. T., Frans, R. E. and Talbert, R. E. 1979. Reactions of *Euglena gracilis* to fluometuron, MSMA, metribuzin, and glyphosate. *Weed Sci.* 27(6):619–624.

Richmonds, C. and Dutta, H. M. 1988. Action of malathion on the gills of bluegill sunfish, *Lepomis macrochirus. Ohio J. Sci.* 88(2):7.

Robel, R. J., Felthousen, R. W. and Dayton, A. D. 1982. Effects of carbamates on bobwhite food intake, body weight, and locomotor activity. *Arch. Environ. Contam. Toxicol.* 11: 611–615.

Robel, R. J., Stalling, C. D., Westfahl, M. E. and Kadoum, A. M. 1972. Effects of insecticides on populations of rodents in Kansas 1965–69. *Pestic. Monit. J.* 6(2):115–121.

Roberts, D. 1975. The effect of pesticides on byssus formation in the common mussel, *Mytilus edulis. Environ. Pollut.* 8(4):241–254.

Roberts, D. and Miller, T. A. 1970. The Effects of Diatoms on the Larvicidal Activity of Dursban November 1969–March 1970. Entomological Special Study No. 31-002-71. U.S. Army Environ. Hygiene Agency, U.S. Dept. of the Army. 14 pp.

Roberts, D. R. and Miller, T. A. 1971. Effects of Polymer Formulations of Dursban and Abate on Non-Target Organism Populations April–October 1970. Entomological Spec. Study No. 31-004-71. U.S. NTIS AD-729 342. U.S. Army Environmental Hygiene Agency, Edgewood Arsenal, MD, 24 pp.

Roberts, M. H., Jr., Warinner, J. E., Tsai, C. F., Wright, D. and Cronin, L. E. 1982. Comparison of estuarine species sensitivities to three toxicants. *Arch. Environ. Contam. Toxicol.* 11(6): 681–692.

Robertson, E. B. and Bunting, D. L. 1976. The acute toxicity of four herbicides to 0–4 hour nauplii of *Cyclops vernalis* Fisher (Copepoda, Cyclopoida). *Bull. Environ. Contam. Toxicol.* 16(6):682–688.

Robinson, W. S. and Johansen. C. A. 1978. Effects of control chemicals for douglas-fir tussock moth *Orgyia pseudotsugata* (Mcdonnough) on forest pollination Lepid. *Melanderia* 30: 9–56.

Rodgers, C. A. 1970. Uptake and elimination of simazine by green sunfish (*Lepomis cyanellus* Raf.). *Weed Sci.* 18(1):134–136.

Rodgers, C. A. and Stalling, D. L. 1972. Dynamics of an ester of 2,4-D in organs of three fish species. *Weed Sci.* 20(1):101–105.

Rodgers, R. D. and Robel, R. J. 1977. Effects of carbaryl on body weight and fat reserves of dietetically-stressed bobwhites. *Bull. Environ. Contam. Toxicol.* 17:184–189.

Rodrigues, J. D., Pedras, J. F., Rodrigues, S. D., Silva, J. A. and Klar, A. E. 1980. Tratamento de plantas aquaticas com herbicidas e insecticidas. I. Efeito sobre algumas especies de peixes das bacias dos rios piracicaba e tiete. *Rev. Agric. (Piracicaba)* 55(1/2):5–12.

Roy, P. K. and Munshi, J. S. D. 1988. Oxygen consumption and ventilation rate of a freshwater major carp, *Cirrhinus mrigala* (Ham.) in fresh and malathion treated waters. *J. Environ. Biol.* 9(1):5–13.

Rzehak, K., Maryanska-Nadachowska, A. and Jordan, M. 1977. The effect of karbatox 75, a carbaryl insecticide, upon the development of tadpoles of *Rana temporaria* and *Xenopus laevis*. *Folia Biol. (Krakow)* 24(4):391–399.

Sadhu, A. K. and Mukhopadhyay, P. K. 1985. Comparative effect of two pesticides malathion and carbofuran on testes of *Clarias batrachus* (linn.). *J. Environ. Biol.* 6(3): 217–222.

Sadhu, A. K., Chowdhury, D. K. and Mukhopadhyay, P. K. 1985. Relationship between serum enzymes, histological features and enzymes in hepatopancreas after sublethal exposure to malathion and phosphamidon in the murrel *Channa striatus* (Bl.). *Int. J. Environ. Stud.* 24(1):35–41.

Saha, M. K. and Konar, S. K. 1983. Acute toxicity of some petroleum pollutants to plankton and fish. *Environ. Ecol.* 1(1):117–119.

Sahib, I. K. A. and Rao, K. R. S. S. 1988. Studies on some kinetic parameters of aminotransferases in tissues of the snail, *Pila globosa* (Swainson) during malathion intoxication. *Acta Physiol. Hung.* 71(1):31–34.

Sahib, I. K. A. and Rao, K. V. 1980a. Toxicity of malathion to the freshwater fish *Tilapia mossambica*. *Bull. Environ. Contam. Toxicol.* 24(6):870–874.

Sahib, I. K. A. and Rao, K. V. R. 1980b. Correlation between subacute toxicity of malathion and acetylcholinesterase inhibition in the tissues of the teleost *Tilapia mossambica*. *Bull. Environ. Contam. Toxicol.* 24(5):711–718.

Sahib, I. K. A., Rao, K. R. S. S. and Rao, K. V. R. 1984a. Effect of malathion on protein synthetic potentiality of the tissues of the teleost, *Tilapia mossambica* (Peters), as measured through incorporation of [^{14}C] amino acids. *Toxicol. Lett.* 20(1):63–67.

Sahib, I. K. A., Rao, K. S. J. and Rao, K. V. R. 1981. Effect of malathion exposure on some physical parameters of whole body and on tissue cations of teleost, *Tilapia mossambica* (Peters). *J. Biosci.* 3(1):17–21.

Sahib, I. K. A., Rao, K. S. P., Rao, K. R. S. S. and Rao, K. V. R. 1984b. Sublethal toxicity of malathion on the proteases and free amino acid composition in the liver of the teleost, *Tilapia mossambica* (Peters). *Toxicol. Lett.* 20(1):59–62.

Sahib, I. K. A., Swami, K. S. and Rao, K.V.R. 1980. Regulation of glutamate dehydrogenase, ammonia and free amino acids in the tissues of the teleost, *Tilapia mossambica* (Peters) consequent to sublethal malathion exposure — a time course study. *Curr. Sci.* 49(20): 779–782.

Sailatha, D., Sahib, I. K. A. and Rao, K. V. R. 1981. Toxicity of technical and commercial grade malathion to the fish, *Tilapia mossambica* (Peters). *Proc. Indian Acad. Sci.* 90(1): 87–92.

Saleh, M. S. 1988. Use of plastic formulations of chlorpyrifos and sumithion as mosquito larvicides and their delayed effects on the basal follicle numbers developed. *Anz. Schaedlingskd. Pflanzenschutz Umweltschutz* 61(1):14–17.

Sanborn, J. R. 1974. The Fate of Select Pesticides in the Aquatic Environment. Ecol. Res. Ser., EPA-660/3-74-025. U.S. EPA, Corvallis, OR, 83 pp.

Sanders, H. O. 1969. Toxicity of Pesticides to the Crustacean Gammarus Lacustris. Tech. Paper No. 25. Bur. Sports Fish. Wildl., Fish Wildl. Serv., U.S. Dept. Int. 18 pp.

Sanders, H. O. 1970a. Toxicities of some herbicides to six species of freshwater crustaceans. *J. Water Pollut. Control Fed.* 24(8):1544–1550.

Sanders, H. O. 1970b. Pesticide toxicities to tadpoles of the western chorus frog *Pseudacris triseriata* and fowler's toad *Bufo woodhousii* Fowleri. *Copeia* 2:246–251.

Sanders, H. O. 1972. Toxicity of Some Insecticides to Four Species of Malacostracan Crustaceans. Tech. Paper No. 66. Bur. Sports Fish. Wildl., Fish Wildl. Serv., U.S. Dept. Int. 19 pp.

Sanders, H. O. and Cope, O. B. 1966. Toxicities of several pesticides to two species of Cladocerans. *Trans. Am. Fish. Soc.* 95(2):165–169.

Sanders, H. O. and Cope, O. B. 1968. The relative toxicities of several pesticides to naiads of three species of stoneflies. *Limnol. Oceanogr.* 13(1):112–117.

Sanders, H. O., Finley, M. T. and Hunn, J. B. 1983. Acute Toxicity of Six Forest Insecticides to Three Aquatic Invertebrates and Four Fishes. Tech. Pap. No. 110. U.S. Fish Wildl. Serv., Washington, D.C., pp. 1–5.

Santharam, K. R., Thayumanavan, B. and Krishnaswamy, S. 1976. Toxicity of some insecticides to *Daphnia Carinata* king, an important link in the food chain in the freshwater ecosystems. *Indian J. Ecol.* 3(1):70–73.

Sastry, K. V. and Malik, P. V. 1982a. Acute and chronic effects of diazinon on the activities of three dehydrogenases in the digestive system of a freshwater teleost fish *Channa punctatus*. *Toxicol. Lett.* 10(1):55–59.

Sastry, K. V. and Malik, P. V. 1982b. Histopathological and enzymological alterations in the digestive system of a freshwater teleost fish, Heteropneustes fossilis, exposed acutely and chronically to diazinon. *Ecotoxicol. Environ. Saf.* 6(3):223–235.

Sastry, K. V. and Sharma, K. 1980. Diazinon effect on the activities of brain enzymes from *Ophiocephalus (channa) punctatus*. *Bull. Environ. Contam. Toxicol.* 24(3):326–332.

Sastry, K. V. and Sharma, K. 1981. Diazinon-induced histopathological and hematological alterations in a freshwater teleost, *Ophiocephalus punctatus*. *Ecotoxicol. Environ. Saf.* 5(3):329–340.

Sastry, K. V. and Siddiqui, A. A. 1982. Chronic toxic effects of the carbamate pesticide Sevin on carbohydrate metabolism in a freshwater snakehead fish, *Channa punctatus*. *Toxicol. Lett.* 14(1/2):123–130.

Sastry, K. V. and Siddiqui, A. A. 1985. Effect of the carbamate pesticide Sevin on the intestinal absorption of some nutrients in the teleost fish, *Channa punctatus*. *Water Air Soil Pollut.* 24(3):247–252.

Saxena, P. K. and Aggarwal, S. 1970. Toxicity of some insecticides to the Indian catfish, *Heteropneustes fossilis* (Bloch). *Anat. Anz.* 127:502–503.

Saxena, P. K. and Garg, M. 1978. Effect of insecticidal pollution on ovarian recrudescence in the fresh water teleost *Channa punctatus*. *Indian J. Exp. Biol.* 16:689–691.

Sayce, C. S. 1970. The Uptake of Sevin by Pacific Oysters and Bottom Muds. Washington (State) Dept. Fish., Tech. Rep. No. 1:9–17.

Sayce, C. S. and Chambers, J. S. 1970. Observations on Potential Uptake of Sevin by Pacific Oysters. Washington (State) Dept. Fish., Tech. Rep. No. 1:18–24.

Schimmel, S. C., Garnas, R. L., Patrick, J. M., Jr. and Moore, J. C. 1983. Acute toxicity, bioconcentration, and persistence of AC 222,705, benthiocarb, chlorpyrifos, and fenvalerate, methyl parathion, and permethrin in in the Estuarine Environment. *J. Agric. Food Chem.* 31(1):104–113.

Schober, U. and Lampert, W. 1977. Effects of sublethal concentrations of the herbicide atrazine on growth and reproduction of *Daphnia pulex. Bull. Environ. Contam. Toxicol.* 17(3):269–277.

Schoettger, R. A. and Mauck, W. L. 1978. Toxicity of Experimental Forest Insecticides to Fish and Aquatic Invertebrates. in Mount, D. I., Swain, W. R. and Ivanikiw, N. K. (Eds.). Proc. First Second USA-USSR Symp. Effects Pollut. Aquatic Ecosystems, Vol. 1. USA Symp. Oct. 21–23, 1975; Vol. 2. USSR Symp. June 22–26, 1976. U.S. NTIS PB-287 219. Duluth, MN, pp. 250–266.

Schott, C. D. and Worthley, E. C. 1974. The Toxicity of TNT and Related Wastes to an Aquatic Flowering Plant, '*Lemna perpusilla*' Torr. Edgewood Arsenal Tech. Rep. EB-TR-74016. U.S. NTIS, AD-778 158. Edgewood Arsenal, Aberdeen Proving Ground, MD, 18 pp.

Schultz, D. P. and Harman, P. D. 1974. Residues of 2,4-D in pond waters, mud, and fish, 1971. *Pestic. Monit. J.* 8(3):173–179.

Scirocchi, A. and Erme, A. D. 1980. Toxicity of seven insecticides on some species of fresh water fishes. *Riv. Parassitol.* 41(1):113–121.

Scott, J. G. and Georghiou, G. P. 1986. Malathion-specific resistance in *Anopheles stephensi* from Pakistan. *J. Am. Mosq. Control Assoc.* 2(1):29–32.

Sears, H. S. and Meehan, W. R. 1971. Residues in fish, wildlife, and estuaries. *Pestic. Monit. J.* 5(2):213–217.

Seguchi, K. and Asaka, S. 1981. Intake and excretion of diazinon in freshwater fishes. *Bull. Environ. Contam. Toxicol.* 27(2):244–249.

Semov, V. and Iosifov, D. 1973. Toxicity of some bulgarian pesticides studied with the test organism *Daphnia magna. Tr. Nauchnoizsled. Inst. Vodosnabdyavane, Kanaliz. Sanit. Tekh.* 9(2):159–167.

Sergeant, M., Blazek, D., Elder, J. H., Lembi, C. A. and Morre, D. J. 1970. The Toxicity of 2,4-D and Picloram Herbicides to Fish. U.S. NTIS PB-201 099. Purdue University and Indiana State Highway Comm. JHRP, Publ. No. 24. Lafayette, IN, 22 pp.

Sergeant, M., Blazek, D., Elder, J. H., Lembi, C. A. and Morre, D. J. 1971. The toxicity of 2,4-D and picloram herbicides to fish. *Proc. Indiana Acad. Sci.* 80:114–123.

Servizi, J. A., Gordon, R. W. and Martens, D. W. 1987. Acute toxicity of Garlon 4 and Roundup herbicides to salmon, daphnia, and trout. *Bull. Environ. Contam. Toxicol.* 39(1): 15–22.

Seuge, J. and Bluzat, R. 1979a. Toxicite chronique du carbaryl et du lindane chez le mollusque d'eau douce *Lymnea stagnalis* L. *Water Res.* 13(3):285–293.

Seuge, J. and Bluzat, R. 1979b. Etude de la toxicite chronique de deux insecticides (carbaryl et lindane) a la generation F1 de *Lymnea stagnalis* L. (Mollusque gasteropode Pulmone). *Hydrobiologia* 66(1):25–31.

Seuge, J. and Bluzat, R. 1983a. Chronic toxicity of three insecticides (carbaryl, fenthion and lindane) in the freshwater snail *Lymnaea stagnalis. Hydrobiologia* 106(1):65–72.

Seuge, J. and Bluzat, R. 1983b. Toxicite aique d'un fongicide dithiocarbamate, le thirame, chez des larves de l'ephemere cloeon diptherum; effets de divers parameters. *Hydrobiologia* 101(3):215–221.

Seuge, J., Marchal-Segault, D. and Bluzat, R. 1983. Toxicite aegue d'un fongicide dithiocarbamete, le thirame, vis a vis de plusieurs especes animales d'eau doucc. *Environ. Pollut. Ser. A* 31(3):177–189.

Shakoori, A. R., Zaheer, S. A. and Ahmad, M. S. 1976. Effect of malathion, dieldrin and endrin on blood serum proteins and free amino acids pool of *Channa punctatus* (Bloch). *Pak. J. Zool.* 8(2):125–134.

Sharma, R. K., Shandilya, S. and Sharma, S. 1983. Observations on the effect of malathion on the mortality of fish, *Clarias batrachus* (Linn.). *Comp. Physiol. Ecol.* 8(2):155–156.

Sharp, J. W., Sitts, R. M. and Knight, A. W. 1978. Effects of kelthane and temperature on the respiration of *Crangon franciscorum. Comp. Biochem. Physiol.* 59C:75–79.

Sharp, J. W., Sitts, R. M. and Knight, A. W. 1979. Effects of kelthane on the estuarine shrimp *Crangon franciscorum. Mar. Biol.* 50(4):367–374.

Shaw, B. and Hopke, P. K. 1975. The dynamics of diquat in a model eco-system. *Environ. Lett.* 8(4):325–335.

Shcherban, E. P. 1972a. Effect of low concentrations of atrazine and diuron on the productivity of *Cladocera. Hydrobiol. J. (Engl. Transl. Gidrobiol. Zh.)* 8(2): 54–58.

Shcherban, E. P. 1972b. The effect of low concentrations of pesticides on the development of some *Cladocera* and the abundance of their progeny. *Hydrobiol. J. (Engl. Trans. Gidrobiol. Zh.)* 6(6):85–89; *Gidrobiol Zh. (Kiev)*, 6(6):101–105.

Shcherban, E. P. 1973. Effect of atrazinc on biological parameters and potential productivity of *Daphnia magna* and *Moina rectirostris. Eksp. Vodn. Toksikol.* 4:80–86.

Shca, T. B. and Berry, E. S. 1983. Toxicity of carbaryl and 1-naphthol to goldfish (*Carassius auratus*) and killifish (*Fundulus heteroclitus*). *Bull. Environ. Contam. Toxicol.* 31(5): 526–529.

Sheedy, B. A., Lazorchak, J. M., Grunwald, D. J., Pickering, Q. H., Pilli, A., Hall, D. and Webb, R. 1991. Effects of pollution on freshwater organisms. *J. Water Pollut. Control Fed.* 63(4):619–696.

Shelford, V. E. 1917. Article VI. An experimental study of the effects of gas waste upon fishes, with especial reference to stream pollution. *Bull. Ill. State Lab. Nat. Hist.* 11(6): 381–410.

Shim, J. C. and Self, L. S. 1973. Toxicity of agricultural chemicals to larvivorous fish in Korean rice fields. *Trop. Med.* 15(3):123–130.

Shukla, G. S. and OMKAR, 1983a. Toxicity of 2,4-D-Na salt to fresh water prawn, *Macrobrachium lamarrei* (M. Edwards). *Comp. Physiol. Ecol.* 8(4):282–284.

Shukla, G. S. and OMKAR, 1983b. Acute toxicity of insecticides to a freshwater prawn *Macrobrachium lamarrei* (M. Edwards). *Indian J. Environ. Health* 25(1):61–63.

Shukla, G. S. and OMKAR, 1984. Insecticide toxicity to *Macrobrachium lamarrei* (H. Milne Edwards) (*Decapoda, Palaemonidae*). *Crustaceana* 46(3):283–287.

Shukla, J. P., Banerjee, M. and Pandey, K. 1987. Deleterious effects of malathion on survivality and growth of the fingerlings of *Channa punctatus* (Bloch), a freshwater murrel. *Acta Hydrochim. Hydrobiol.* 15(6):653–657.

Shukla, L., Shrivastava, A., Merwani, D. and Pandey, A. K. 1984. Effect of sublethal malathion on ovarian histophysiology in *Sarotherodon mossambicus. Comp. Physiol. Ecol.* 9(1): 13–17.

Siefert, R. E. 1987. Effects of Dursban (chlorpyrifos) on Aquatic Organisms in Enclosures in a Natural Pond — Final report. U.S. EPA, Duluth, MN, 214 pp.

Sigmon, C. 1979a. Oxygen consumption in *Lepomis macrochirus* exposed to 2,4-D or 2,4,5-T. *Bull. Environ. Contam. Toxicol.* 21:826–830.

Sigmon, C. 1979b. Oxygen consumption in *Daphnia pulex* exposed to 2,4-D or 2,4,5-T. *Bull. Environ. Contam. Toxicol.* 21(6):822–825.

Sigmon, C. F. 1979c. Influence of 2,4-d and 2,4,5-t on life history characteristics of *Chironomus (Diptera: Chironomidae). Bull. Environ. Contam. Toxicol.* 21:596–599.

Sikka, H. C. and Rice, C. P. 1974. Interaction of Selected Pesticides with Marine Microorganisms. U.S. NTIS AD-785 079. Office of Naval Res., Arlington, VA, 78 pp.

Sikka, H. C., Butler, G. L. and Rice, C. P. 1976. Effects, Uptake, and Metabolism of Methoxychlor, Mirex, and 2,4-D in Seaweeds. EPA-600/3-76-048. Environ. Res. Lab., U.S. EPA, Gulf Breeze, FL, 39 pp.

Sikka, H. C., Ford, D. and Lynch, R. S. 1975. Uptake, distribution, and metabolism of endothall in fish. *J. Agric. Food Chem.* 23(5):849–851.

Sikka, H. C., Miyazaki, S. and Rice, C. P. 1973. Metabolism of Selected Pesticides by Marine Microorganisms. Rept. No. SURC-TR-73-520. U.S. NTIS AD-763 410. Life Sci. Div., Syracuse University Res. Corp., Syracuse, NY, 16 pp.

Singh, D. K. and Agarwal, R. A. 1983. in vivo and in vitro studies on synergism with anticholinesterase pesticides in the snail *Lymnaea acuminata. Arch. Environ. Contam. Toxicol.* 12(4):483–487.

Singh, D. K. and Agarwal, R. A. 1986. Synergistic effect of sulfoxide with carbaryl on the in vivo acetylcholinesterase activity and carbohydrate metabolism of the snail *Lymnaea acuminata. Acta Hydrochim. Hydrobiol.* 14(4):421–427.

Singh, H. and Singh, T. P. 1980a. Short-term effect of 2 pesticides on the survival, ovarian 32p uptake and gonadotrophic potency in a freshwater catfish, *Heteropneustes fossilis* (Bloch). *J. Endocrinol.* 85:193–199.

Singh, H. and Singh, T. P. 1980b. Effects of two pesticides on testicular 32P uptake, gonadotrophic potency, lipid and cholesterol content of testis, liver and blood serum during spawing phase in *Heteropneustes fossilis* (Bloch). *Endokrinologie* 76(3):288–296.

Singh, H. and Singh, T. P. 1980c. Effect of two pesticides on ovarian 32P uptake and gonadotrophin concentration during different phases of annual reproductive cycle in the freshwater catfish, *Heteropneustes fossilis* (Bloch). *Environ. Res.* 22(1):190–200.

Singh, H. and Singh, T. P. 1980d. Effect of two pesticides on total lipid and cholesterol contents of ovary, liver and blood serum during different phases of the annual reproductive cycle. *Environ. Pollut. Ser. A* 23(1):9–18.

Singh, H. and Singh, T. P. 1980e. Short-term effect of two pesticides on lipid and cholesterol content of liver, ovary and blood serum during the pre-spawning phase in the freshwater teleost, *Heteropneustes fossilis* (Bloch). *Environ. Pollut. Ser. A* 22(2):85–90.

Singh, H. and Singh, T. P. 1980f. Thyroid activity and tsh potency of the pituitary gland and blood serum in response to cythion and hexadrin treatment in the freshwater catfish, *Heteropneustes fossilis* (Bloch). *Environ. Res.* 22(1):184–189.

Singh, H. and Singh, T. P. 1982. Effect of some pesticides on hypothalamo-hypophyseal-ovarian axis in the freshwater catfish *Heteropneustes fossilis* (Bloch). *Environ. Pollut. Ser. A Ecol. Biol.* 27(4):283–288.

Singh, P. K. 1973. Effect of pesticides on blue-green algae. *Arch. Mikrobiol.* 89(4):317–320.

Singh, P. K. 1974. Algicidal effect of 2,4-dichlorophenoxy acetic acid on blue-green alga *Cylindrospermum* sp. *Arch. Microbiol.* 97(1):69–72.

Singh, S. and Sahai, S. 1984. Effect of malathion on the mortality and behaviour of two fresh water teleosts. *J. Environ. Biol.* 5(1):23–28.

Singh, S. and Sahai, S. 1986. Accumulation of malathion in the liver, kidney and gills of *Puntius ticto* (Ham.) as assessed by thin layer chromatography (TLC). *J. Environ. Biol.* 7(2):107–112.

Singh, S. and Singh, T. P. 1987a. Impact of malathion and hexachlorocyclohexane on plasma profiles of three sex hormones during different phases of the reproductive cycle in *Clarias batrachus. Pestic. Biochem. Physiol.* 27(3):301–308.

Singh, S. and Singh, T. P. 1987b. Evaluation of toxicity limit and sex hormone production in response to cythion and BHC in the vitellogenic catfish *Clarias batrachus. Environ. Res.* 42(2):482–488.

Singh, S. P. and Yadav, N. K. 1978. Toxicity of some herbicides to major carp fingerlings. *Indian J. Ecol.* 5(2):141–147.

Singh, V. P., Gupta, S. and Saxena, P. K. 1984. Evaluation of acute toxicity of carbaryl and malathion to freshwater teleosts, *Channa punctatus* (Bloch) and *Heteropneustes fossilis* (Bloch). *Toxicol. Lett.* 20(3):271–276.

Sivaiah, S. and Rao, K. V. R. 1978. Effect of malathion on the excretory pattern of the snail, *Pila globosa* (Swainson). *Curr. Sci.* 47(23):894–895.

Slooff, W. 1979. Detection limits of a biological monitoring system based on fish respiration. *Bull. Environ. Contam. Toxicol.* 23(4–5):517–523.

Slooff, W., De Zwart, D. and Marquenie, J. M. 1983. Detection limits of a biological monitoring system for chemical water pollution based on mussel activity. *Bull. Environ. Contam. Toxicol.* 30(4):400–405.

Smith, G. E. and Isom, B. G. 1967. Investigation of effects of large-scale applications of 2,4-D on aquatic fauna and water quality. *Pestic. Monit. J.* 1(3):16–21.

Smith, J. W. and Grigoropoulos, S. G. 1968. Toxic effects of odorous trace organics. *Am. Water Works Assoc. J.* 60:969–979.

Solomon, K. E. and Robel. R. J. 1980. Effects of carbaryl and carbofuran on bobwhite energetics. *J. Wildl. Manage.* 44(3):682–686.

Solomon, K. E., Dahlgren, R. B. and Linder, R. L. 1973. Abnormal embryos in eggs of pheasants given 2,4-D or PCB. *Proc. S. D. Acad. Sci.* 52:95–99.

Solomon, H. M. 1977. The teratogenic effects of the insecticides DDT, carbaryl, malathion and parathion on developing Medaka eggs (*Oryzias latipes*). *Diss. Abst. Int. B.* 39(5): 2176–2177.

Somashekar, R. K. and Sreenath, K. P. 1984. Effect of fungicides on primary productivity. *Indian J. Environ. Health* 26(4):355–357.

Sonnet, P. E. 1978. Effects of selected herbicides on the toxicity of several insecticides to honey bees. *Environ. Entomol.* 7(2):254–256.

Soria, S. De J. and Abreu, J. M. 1974. Mortalidade dos polinizadores *Forcipomyia* spp. (Diptera, Ceratopogonidae) causada pela aplicacao de insecticidas nos cacauais baianos. *Rev. Theobroma* 4(3):13–25.

Spacie, A. and Hamelink, J. L. 1979. Dynamics of trifluralin accumulation in river fishes. *Environ. Sci. Technol.* 13(7):817–822.

Spehar, R. L., Fiandt, J. T., Anderson, R. L. and Defoe, D. L. 1980. Comparative toxicity of arsenic compounds and their accumulation in invertebrates and fish. *Arch. Environ. Contam. Toxicol.* 9(1):53–63.

Sreenivasan, A. and Swaminathan, G. K. 1967. Toxicity of six organophosphorus insecticides to fish. *Curr. Sci.* 36(15):397–398.

Srivastava, A. K. and Srivastava, A. K. 1988. Effects of aldrin and malathion on blood chloride in the Indian catfish *Heteropneustes fossilis. J. Environ. Biol.* 9(Suppl. 1):91–95.

Stadnyk, L., Campbell, R. S. and Johnson, B. T. 1971. Pesticide effect on growth and C14 assimilation in a freshwater alga. *Bull. Environ. Contam. Toxicol.* 6(1):1–8.

Staples, R. E., Kellam, R. G. and Haseman, J. K. 1976. Developmental toxicity in the rat after ingestion or gavage of organophosphate pesticides (Dipterex, Imidan) during pregnancy. *Environ. Health Persp.* 13:133–140.

Statham, C. N. and Lech, J. J. 1975a. Potentiation of the acute toxicity of several pesticides and herbicides in trout by carbaryl. *Toxicol. Appl. Pharmacol.* 34(1):83–87.

Statham, C. N. and Lech, J. J. 1975b. Synergism of the acute toxic effects of 2,4-D butyl ester, dieldrin, rotenone, and pentachlorophenol in rainbow trout by carbaryl. *Toxicol. Appl. Pharmacol.* 33(1):188; in Abstracts: 14th Annual Meeting of Soc. of Toxicol., Williamsburg, VA, p. 133.

Stephan, C. E., Spehar, D. L., Roush, T. H., Phipps, G. L. and Pickering, Q. H. 1986. Effects of pollution on freshwater organisms. *J. Water Pollut. Control Fed.* 58(6):645–671.

Stephenson, M. and Mackie, G. L. 1986. Effects of 2,4-D treatment on natural benthic macroinvertebrate communities in replicate artificial ponds. *Aquat. Toxicol.* 9(4–5): 243–251.

Stevenson, J. H., Needham, P. H. and Walker, J. 1978. Poisoning of honeybees by pesticides: investigations of the changing pattern in Britain over 20 years. *Rothamsted Exp. Stn. Rep.* PT2:55–72.

Stratton, G. W. 1984. Effects of the herbicide atrazine and its degradation products, alone and in combination, on phototrophic microorganisms. *Arch. Environ. Contam. Toxicol.* 13(1): 35–42.

Stratton, G. W. and Corke, C. T. 1981. Effect of acetone on the toxicity of atrazine towards photosynthesis in Anabaena. *J. Environ. Sci. Health B* 16(1):21–33.

Streit, B. 1978. Aufnahme, anreicherung und freisetzung organischer pestizide bei benthischen susswasserinvertebraten. I. Reversible anreicherung von atrazin aus der wasserigen phase. (uptake, accumulation, and release of organic pesticides by benthic freshwater invertebrates). I. Reversible accumulation of atrazine from aqueous solution. *Arch. Hydrobiol. Suppl.* 55(1):1–23.

Streit, B. and Peter, H. M. 1978. Long-term effects of atrazine to selected freshwater invertebrates. *Arch. Hydrobiol. Suppl.* 55(1):62–77.

Streit, B. and Schwoerbel, J. 1977. Experimentelle untersuchungen uber die akkumulation und wirkung von herbiziden bei benthischen susswassertieren. *Verh. Ges. Oekol. Jahresversamml.* 77:371–383.

Strickman, D. 1985. Aquatic bioassay of 11 pesticides using larvae of the mosquito, *Wyeomyia smithii (Diptera: Culicidae)*. *Bull. Environ. Contam. Toxicol.* 35:133–142.

Stringer, A. and Wright, M. A. 1973. The effect of benomyl and some related compounds on *Lumbricus terrestris* and other earthworms. *Pestic. Sci.* 4:165–170.

Stringer, A. and Wright, M. A. 1976. The toxicity of benomyl and some related 2-substituted benzimidazoles to the earthworm *Lumbricus terrestris*. *Pestic. Sci.* 7:459–464.

Stromborg, K. L. 1977. Seed treatment pesticide effects on pheasant reproduction at sublethal doses. *J. Wildl. Manage.* 41(4):632–642.

Sundaram, K. M. S. and Szeto, S. Y. 1987. Distribution and persistence of carbaryl in some terrestrial and aquatic components of a forest environment. *J. Environ. Sci. Health* B22(5):579–599.

Surber, E. W. and Pickering, Q. H. 1962. Acute toxicity of endothal, diquat, hyamine, dalapon, and silvex to fish. *Prog. Fish Cult.* 24(4):164–171.

Sutton, D. L. and Bingham, S. W. 1970. Uptake and translocation of 2,4-D-1-14C in parrot-feather. *Weed Sci.* 18(2):193–196.

Svobodova, Z. 1980. Akutni toxicita pesticidu pro ryby. (Acute toxicity of pesticides to fish). *Agrochemia* 20(11):328–332.

Swedburg, D. 1973. Diazinon Toxicity to Specified Fish. U.S. EPA-OPP Registration Standard.

Tagatz, M. E., Borthwick, P. W., Cook, G. H. and Coppage, D. L. 1974. Effects of ground applications of malathion on salt-marsh environments in northwestern Florida. *Mosq. News* 34(3):309–315.

Tagatz, M. E., Ivey, J. M. and Lehman, H. K. 1979. Effects of Sevin on development of experimental estuarine communities. *J. Toxicol. Environ. Health* 5(4):643–651.

Tandon, R. S. and Dubey, A. 1983. Toxic effects of two organophosphorus pesticides on fructose-1,6-diphosphate aldolase activity of liver, brain and gills of the freshwater fish *Clarias batrachus*. *Environ. Pollut. Ser. A* 31(1):1–7.

Tandon, R. S., Lal, R. and Rao, V. V. S. N. 1987. Effects of malathion and endosulfan on the growth of *Paramecium aurelia*. *Acta Protozool.* 26(4):325–328.

Tandon, R. S., Lal, R. and Rao, V. V. S. N. 1988. Interaction of endosulfan and malathion with blue-green algae *Anabaena* and *Aulosira fertilissima*. *Environ. Pollut.* 52(1):1–9.

Tatem, H. E., Cox, B. A. and Anderson, J. W. 1978. The toxicity of oils and petroleum hydrocarbons to estuarine crustaceans. *Estuarine Coastal Mar. Sci.* 6(4):365–373.

Taub, F.B., Read, P.L., Kindig, A.C., Harrass, M.C., Hartmann, H.J., Conquest, L.L., Hardy, F.J. and Munro, P.T. 1983. Demonstration of the ecological effects of streptomycin and malathion on synthetic aquatic microcosms. in Bishop, W. E., Cardwell, R. D. and Heidolph, B. B. (Eds.). *Aquatic Toxicology and Hazard Assessment*. 6th Symp., ASTM STP 802, Philadelphia, PA, pp. 5–25.

Tegelberg, H. and Magoon, D. 1970. Sevin Treatment of a Subtidal Oyster Bed in Grays Harbor. Washington (State) Dept. Fish., Tech. Rep. No. 1:1–8.

Tellenbach, M., Gerber, A. and Boschetti, A. 1983. Herbicide-binding to thylakoid membranes of a DCMU-resistant mutant of *Chlamydomonas reinhardii*. *Fed. Eur. Biochem. Soc. Lett.* 158(1):147–150.

Thayer, A. and Ruber, E. 1976. Previous feeding history as a factor in the effects of temephos and chlorpyrifos on migration of *Gammarus fasciatus* (*Amphipoda, Crustacea*). *Mosq. News* 36(4):429–432.

Thirugnanam, M. and Forgash, A. J. 1977. Environmental impact of mosquito pesticides: Toxicity and anticholinesterase activity of chlorpyrifos to fish in a salt marsh habitat. *Arch. Environ. Contam. Toxicol.* 5(4):415–425.

Thompson, A. R. 1971. Effects of nine insecticides on the numbers and biomass of earthworms in pasture. *Bull. Environ. Contam. Toxicol.* 5(6):577–586.

Thompson, A. R. and Sans, W. W. 1974. Effects of soil insecticides in southwestern Ontario on non-target invertebrates. Earthworms in pasture. *Environ. Entomol.* 3(2):305–308.

Thursby, G. B., Steele, R. L. and Kane, M. E. 1985. Effect of organic chemicals on growth and reproduction in the marine red alga *Champia parvula*. *Environ. Toxicol. Chem.* 4(6): 797–805.

Tietge, R. M. 1992. Wildlife and golf courses. in Balogh, J. C. and Walker, W. J. (Eds.). *Golf Course Management and Construction: Environmental Issues*. Lewis Publishers, Chelsea, MI, Chapter 6.

Tilak, K. S., Rao, D. M. R., Devi, A. P. and Murty, A. S. 1980. Toxicity of carbaryl and 1-naphthol to the freshwater fish *Labeo rohita*. *Indian J. Exp. Biol.* 18:75–76.

Titova, A. A. 1978. Concentration of the herbicide 2,4-D by some higher water plants. *Hydrobiol. J.* 14(4):96–97.

Tomisek, A., Reid, M. R., Short, W. A. and Skipper, H. E. 1957. Studies on the photosynthetic reaction. III. The effects of various inhibitors upon growth and carbonate-fixation in *Chlorella pyrenoidosa. Plant Physiol.* 32:7–10.

Tomlin, A. D. and Gore, F. L. 1974. Effects of six insecticides and a fungicide on the numbers and biomass of earthworms in pasture. *Bull. Environ. Contam. Toxicol.* 12(4): 487–492.

Tooby, T. E., Lucey, J. and Stott, B. 1980. The tolerance of grass carp, *Ctenopharyngodon idella* Val., to aquatic herbicides. *J. Fish Biol.* 16(5):591–597.

Toor, H. S. and Kaur, K. 1974. Toxicity of pesticides to the fish, *Cyprinus carpio communis* Linn. *Indian J. Exp. Biol.* 12(4):334–336.

Toor, H. S., Mehta, K. and Chhina, S. 1973. Toxicity of insecticides (commercial formulations) to the exotic fish, common carp *Cyprinus carpio communis* Linnaeus. *J. Res. Punjab. Agric. Univ.* 10(3):341–345.

Toth, D. and Tomasovicova, D. 1979. Effect of pesticides on survival of *Tetrahymena pyriformis* in Danube waters. *Biologia (Bratislava)* 34(3):233–239.

Tripathi, G. and Shukla, S. P. 1988. Toxicity bioassay of technical and commercial formulations of carbaryl to the freshwater catfish, *Clarias batrachus. Ecotoxicol. Environ. Saf.* 15(3): 277–281.

Trofimova, M. G. 1979. Chronic effect of the herbicide Banvel-d on *Daphnia magna* Straus. *Byull. Mosk. Ova. Ispyt. Prir. Otd. Biol.* 84(3):70–77.

Tsai, S. C. 1978. Control of chironomids in milkfish (*Chanos chanos*) ponds with Abate (temephos) insecticide. *Trans. Am. Fish. Soc.* 107(3):493–499.

Tscheu-Schluter, M. 1972. Zur bestimmung der akuten toxizitat von herbiziden gegenuber phytoplankton. (Determination of the acute toxicity of herbicides to phytoplankton). *Fortschr. Wasserchem. Ihrer Grenzgeb.* 14:159–168.

Tscheu-Schluter, M. 1976. Zur akuten toxizitat von herbiziden gegenuber ausgewahlten wasserorganismen teil 2: Triazinherbizide und amitrol. *Acta Hydrochim. Hydrobiol.* 4(2): 153–170.

Tsuge, S., Nishimura, T., Kazano, H. and Tomizawa, C. 1980. Uptake of pesticides from aquarium tank water by aquatic organisms. *J. Pestic. Sci. (Nihon Noyakugaku Kaishi)* 5(4):585–593.

Tucker, R. K. and Crabtree, D. G. 1970. Handbook of Toxicity of Pesticides to Wildlife. USDI Fish Wildl. Ser. Resource Publ. No. 84. Washington, D.C., 131 pp.

Turbak, S. C., Olson, S. B. and Mcfeters, G. A. 1986. Comparison of algal assay systems for detecting waterborne herbicides and metals. *Water Res.* 20(1):91–96.

Ukeles, R. 1962. Growth of pure cultures of marine phytoplankton in the presence of toxicants. *Appl. Microbiol.* 10:532–537.

Upadhyaya, A. and Rao, K. S. 1980. Acute toxicity of tafazine to fish. *Int. J. Environ. Stud.* 15(3):236–238.

USDA. 1969. Agricultural Handbook No. 332. United States Department of Agriculture. U.S. Government Printing Office, Washington, D.C.

USDA Forest Service. 1984. Pesticide Background Statements, Vol. 1. Herbicides. Agriculture Handbook No. 633. U.S. Government Printing Office, Washington, D.C.

USDA Forest Service. 1989. Vegetation management in the Appalachian Mountains. Appendices. Vol. 2. Section 6. p. 2.

Valentine, J. P. and Bingham, S. W. 1974. Influence of several algae on 2,4-D residues in water. *Weed Sci.* 22(4):358–363.

Valentine, J. P. and Bingham, S. W. 1976. Influence of algae on amitrole and atrazine residues in water. *Can. J. Bot.* 54(18):2100–2107.

Van Hoof, F. 1980. Evaluation of an automatic system for detection of toxic substances in surface water using trout. *Bull. Environ. Contam. Toxicol.* 25(2):221–225.

Van Leeuwen, C. J., Espeldoorn, A. and Mol, F. 1986a. Aquatic toxicological aspects of dithiocarbamates and related compounds. III. Embryolarval studies with rainbow trout (*Salmo gairdneri*). *Aquat. Toxicol.* 9(2–3):129–145.

Van Leeuwen, C. J., Griffioen, P. S., Vergouw, W. H. A. and Maas-Diepeveen, J. L. 1985c. Differences in susceptibility of early life stages of rainbow trout (*Salmo gairdneri*) to environmental pollutants. *Aquat. Toxicol.* 7(1–2):59–78.

Van Leeuwen, C. J., Helder, T. and Seinen, W. 1986b. Aquatic toxicological aspects of dithiocarbamates and related compounds. IV. Teratogenicity and histopathology in rainbow trout (*Salmo gairdneri*). *Aquat. Toxicol.* 9(2–3):147–159.

Van Leeuwen, C. J., Maas-Diepeveen, J. L. and Overbeck, H. C. M. 1986c. Sublethal effects of tetramethylthiuram disulfide (thiram) in rainbow trout (*Salmo gairdneri*). *Aquat. Toxicol.* 9(1):13–19.

Van Leeuwen, C. J., Moberts, F. and Niebeek, G. 1985b. Aquatic toxicological aspects of dithiocarbamates and related compounds. II. Effects on survival, reproduction and growth of *Daphnia magna*. *Aquat. Toxicol.* 7(3):165–175.

Van Leeuwen, C. J., Niebeek, G. and Rijkeboer, M. 1987. Effects of chemical stress on the population dynamics of *Daphnia magna*: A comparison of two test procedures. *Ecotoxicol. Environ. Saf.* 14(1):1–11.

Van Leeuwen, C. J., Maas-Diepeveen, J. L., Niebeek, G., Vergouw, W. H. A., Griffioen, P. S. and Luijken, M. W. 1985a. Aquatic toxicological aspects of dithiocarbamates and related compounds. I. Short-term toxicity tests. *Aquat. Toxicol.* 7(3):145–164.

Vance, B. D. and Smith, D. L. 1969. Effects of five herbicides on three green algae. *Tex. J. Sci.* 20(4):329–337.

Varanka, I. 1979. Effect of some pesticides on the rhythmic adductor muscle activity of fresh-water mussel larvae. *Symp. Biol. Hung.* 19:177–196.

Varanka, I. 1987. Effect of mosquito killer insecticides on freshwater mussels. *Comp. Biochem. Physiol.* 86C(1):157–162.

Vardia, H. K. and Durve, V. S. 1981a. Bioassay study on some freshwater fishes exposed to 2,4-dichlorophenoxyacetic acid. *Acta Hydrochim. Hydrobiol.* 9(2):219–223.

Vardia, H. K. and Durve, V. S. 1981b. The toxicity of 2,4-D to *Cyprinus carpio* var. *communis* in relation to the seasonal variation in the temperature. *Hydrobiologia* 77(2):155–159.

Vardia, H. K. and Durve, V. S. 1984. Relative toxicity of phenoxy herbicides on *Lebistes* (Poecilia) *reticulatus* (Peters). *Proc. Indian Acad. Sci. Anim. Sci.* 93(7):691–695.

Vardia, H. K., Rao, P. S. and Durve, V. S. 1984. Sensitivity of toad larvae to 2,4-D and endosulfan pesticides. *Arch. Hydrobiol.* 100(3):395–400.

Venkataraman, G. S. and Rajyalakshmi, B. 1971. Tolerance of blue-green algae to pesticides. *Curr. Sci.* 40(6):143–144.

Venkataraman,G.S. and Rajyalakshmi, B. 1972. Relative tolerance of nitrogen-fixing blue-green algae to pesticides. *Indian J. Agric. Sci.* 42(2):119–121.

Verma, D., Chouhan, M. S. and Pandey, A. K. 1984b. Effect of Weedone (a selective 2,4-D herbicide) on neuroendocrine complex of *Puntius ticto* (Ham.). *J. Environ. Biol.* 5(4): 249–254.

Verma, S. R., Bansal, S. K. and Dalela, R. C. 1977. Quantitative estimation of biocide residues in a few tissues of *Labeo rohita* and *Saccobranchus fossilis*. *Indian J. Environ. Health* 19(3):189–198.

Verma, S. R., Rani, S. and Dalela, R. C. 1981d. Isolated and combined effects of pesticides on serum transaminases in *Mystus vittatus* (African catfish). *Toxicol. Lett.* 8: 67–71.

Verma, S. R., Rani, S. and Dalela, R. C. 1982b. Indicators of stress induced by pesticides in *Hystus vittatus*: Haematological parameters. *Indian J. Environ. Health* 24(1):58–64.

Verma, S. R., Rani, S., Bansal, S. K. and Dalela, R. C. 1980. Effects of the pesticides thiotox, dichlorvos, and carbofuran on the test fish *Mystus vittatus*. *Water Air Soil Pollut.* 13: 229–234.

Verma, S. R., Rani, S., Bansal, S. K. and Dalela, R. C. 1981c. Evaluation of the comparative toxicity of thiotox, dichlorvos and carbofuran to two fresh water teleosts *Ophiocephalus punctatus* and *Mystus vittatus*. *Acta Hydrochim. Hydrobiol.* 9(2):119–129.

Verma, S. R., Rani, S., Tonk, I. P. and Dalela, R. C. 1983. Pesticide-induced dysfunction in carbohydrate metabolism in three freshwater fishes. *Environ. Res.* 32(1):127–133.

Verma, S. R., Tonk, I. P. and Dalela, R. C. 1981a. Determination of the maximum acceptable toxicant concentration (MATC) and the safe concentration for certain aquatic pollutants. *Acta Hydrochim. Hydrobiol.* 9(3):247–254.

Verma, S. R., Tonk, I. P., Gupta, A. K. and Dalela, R. C. 1981b. Role of ascorbic acid in the toxicity of pesticides in a fresh water teleost. *Water Air Soil Pollut.* 16(1):107–114.

Verma, S. R., Tonk, I. P., Gupta, A. K. and Saxena, M. 1984a. Evaluation of an application factor for determining the safe concentration of agricultural and industrial chemicals. *Water Res.* 18(1):111–115.

Verma, S. R., Tyagi, A. K., Bhatnagar, M. C. and Dalela, R. C. 1979. Organophosphate poisoning to some fresh water teleosts – acetylcholinesterase inhibition. *Bull. Environ. Contam. Toxicol.* 21(4–5):502–506.

Verma, S. R., Bansal, S. K., Gupta, A. K., Pal, N., Tyagi, A. K., Bhatnagar, M. C., Kumar, V. and Dalela, R. C. 1982a. Bioassay trials with twenty three pesticides to a fresh water teleost, *Saccobranchus fossilis*. *Water Res.* 16(5):525–529.

Vickers, D. H. and Boyd, C. E. 1971. Effects of Organic Insecticides Upon Carbon-14 Uptake by Freshwater Phytoplankton. CONF-710501-PL. Proc. 3rd Natl. Symp. Radioecology, Oak Ridge, TN, pp. 492–496.

Vigfusson, N. V., Vyse, E. R., Pernsteiner, C. A. and Dawson, R. J. 1983. in vivo induction of sister-chromatid exchange in *Umbra limi* by the insecticides endrin, chlordane, diazinon and guthion. *Mutat. Res.* 118:61–68.

Vilkas, A. 1976. Acute Toxicity of Diazinon Technical to the Water Flea, *Daphnia magna* Straus. U.S. EPA-OPP Registration Standard.

Virmani, M., Evans, J. O. and Lynn, R. I. 1975. Preliminary studies of the effects of s-triazine, carbamate, urea, and karbutilate herbicides on growth of fresh water algae. *Chemosphere* 4(2):65–71.

Voight, R. A. and Lynch, D. L. 1974. Effects of 2,4-D and dmso on procaryotic and eucaryotic cells. *Bull. Environ. Contam. Toxicol.* 12(4):400–404.

von Windeguth, D. L., Eliason, D. A. and Schoof, H. F. 1971. The efficacy of carbaryl, propoxur, abate and methoxychlor as larvicides against field infestations of *Aedes aegypti*. *Mosq. News* 31(1):91–95.

Voronkin, A. S. and Loshakov, Y. T. 1973. Toxic effect of pesticides on *Tubifex tubifex*. *Eksp. Vodn. Toksikol.* 5:169–178.

Walker, C. R. 1964. Simazine and other s-triazine compounds as aquatic herbicides in fish habitats. *Weeds* 12(2):134–139.

Wallace, R. R., West, A. S., Downe, A. E. R. and Hynes, H. B. N. 1973. The effects of experimental blackfly (*Diptera: Simuliidae*) larviciding with Abate, Dursban, and methoxychlor on stream invertebrates. *Can. Entomol.* 105(6):817–831.

Walsh, A. H. and Ribelin, W. E. 1973. The pathology of pesticide poisoning. in Ribelin, W. E. and Migaki, G. (Eds.). *The Pathology of Fishes.* University of Wisconsin Press, Madison, WI, pp. 515–541; *Toxicol. Appl. Pharmacol.* 25(3):485.

Walsh, D. F., Armstrong, J. G., Bartley, T. R., Salman, H. J. A. and Frank, P. A. 1977a. Residues of Emulsified Xylene in Aquatic Weed Control and Their Impact on Rainbow Trout *Salmo gairdneri.* REC-ERC-76-11. Appl. Sci. Branch, Eng. Res. Cent., Denver, CO, 15 pp.

Walsh, G. E. 1972. Effects of herbicides on photosynthesis and growth of marine unicellular algae. *Hyacinth Control J.* 10:45–48.

Walsh, G. E. 1983. Cell death and inhibition of population growth of marine unicellular algae by pesticides. *Aquat. Toxicol.* 3(3):209–214.

Walsh, G. E. and Alexander, S. V. 1980. A marine algal bioassay method: results with pesticides and industrial wastes. *Water Air Soil Pollut.* 13(1):45–55.

Walsh, G. E., Hansen, D. L. and Lawrence, D. A. 1982. A flow-through system for exposure of seagrass to pollutants. *Mar. Environ. Res.* 7(1):1–11.

Walsh, G. E., Keltner, J. M., Jr. and Matthews, E. 1970. Effects of Herbicides on Marine Algae. U.S. Fish Wildl. Serv., Circ. 335, Washington, D.C., pp. 10–12.

Walsh, G. E., Yoder, M. J., McLaughlin, L. L. and Lores, E. M. 1987b. Responses of marine unicellular algae to brominated organic compounds in six growth media. *Ecotoxicol. Environ. Saf.* 14:215–222.

Wan, M. T., Moul, D. J. and Watts, R. G. 1987. Acute toxicity to juvenile pacific salmonids of Garlon 3a, Garlon 4, triclopyr, triclopyr ester, and their transformation products: 3,5,6-trichloro-2-pyridinol and 2-methoxy-3,5,6-trichloropyridine. *Environ. Contam. Toxicol.* 39(4):721–728.

Ward, G. S. and Ballantine, L. 1985. Acute and chronic toxicity of atrazine to estuarine fauna. *Estuaries* 8(1):22–27.

Watkins, C. E., II, Thayer, D. D. and Haller, W. T. 1985. Toxicity of adjuvants to bluegill. *Bull. Environ. Contam. Toxicol.* 34(1):138–142.

Wedding, R. T., Erickson, L. C. and Brannaman, B. L. 1954. Effect of 2,4-dichlorophenoxyacetic acid on photosynthesis and respiration. *Plant Physiol.* 29:64–69.

Weete, J. D., Pillai, P. and Davis, D. D. 1980. Metabolism of atrazine by *Spartina alterniflora.* 2. Water-soluble metabolites. *J. Agric. Food Chem.* 28(3):636–640.

Weis, J. S. and Weis, P. 1975. Retardation of fin regeneration in *Fundulus* by several insecticides. *Trans. Am. Fish. Soc.* 104(1):135–137.

Weis, P. and Weis, J. S. 1974a. Schooling behavior of *Menidia menidia* in the presence of the insecticide Sevin (carbaryl). *Mar. Biol.* 28(4):261–263.

Weis, P. and Weis, J. S. 1974b. Cardiac malformations and other effects due to insecticides in embryos of the killifish, *Fundulus heteroclitus. Teratology* 10(3):263–267.

Weiss, C. M. 1959. Stream pollution: Response of fish to sub-lethal exposures of organic phosphorus insecticides. *Sewage Ind. Wastes* 31(5):580–593.

Weiss, C. M. 1961. Physiological effect of organic phosphorus insecticides on several species of fish. *Trans. Am. Fish. Soc.* 90(2):143–152.

Weiss, C. M. and Gakstatter, J. H. 1964. Detection of pesticides in water by biochemical assay. *J. Water Pollut. Control Fed.* 36(2):240–253.

Wellborn, T. L., Jr. 1969. The toxicity of nine therapeutic and herbicidal compounds to striped bass. *Prog. Fish Cult.* 31(1):27–32.

Wellborn, T. L., Jr. 1971. Toxicity of some compounds to striped bass fingerlings. *Prog. Fish Cult.* 33(1):32–36.

Whitmore, D. H., Jr. and Hodges, D. H., Jr. 1978. In vitro pesticide inhibition of muscle esterases of the mosquitofish, *Gambusia affinis. Comp. Biochem. Physiol.* C 59(2):145–149.

Whitten, B. K. and Goodnight, C. J. 1966. Toxicity of some common insecticides to *Tubificids. J. Water Pollut. Control Fed.* 38(2):227–235.

Wierzbicka, M. 1974. Influence of 2,4-D sodium salt on the survival of some copepoda species. *Pol. Arch. Hydrobiol.* 21(2):275–282.

Wilder, I. B. and Stanley, J. G. 1983. RNA-DNA ratio as an index to growth in salmonid fishes in the laboratory and in streams contaminated by carbaryl. *J. Fish Biol.* 22(2):165–172.

Woodward, D. F. 1982. Acute toxicity of mixtures of range management herbicides to cutthroat trout. *J. Range Manage.* 35(4):539–540.

Woodward, D. F. and Mauck, W. L. 1980. Toxicity of five forest insecticides to cutthroat trout and two species of aquatic invertebrates. *Bull. Environ. Contam. Toxicol.* 25(6):846–853.

Woodward, D. F. and Mayer, F. L., Jr. 1978. Toxicity of three herbicides (butyl, isooctyl, and propylene glycol butyl ether esters of 2,4-D) to cutthroat trout and lake trout. *U.S. Fish Wildl. Serv. Tech. Paper* 97:1–6.

Worthing, C. R. (Ed.). 1979. Glyphosate. in *The Pesticide Manual.* The British Crop Protection Council, London, p. 292.

Worthley, E. G. and Schott, C. D. 1972. The Comparative Effects of CS and Various Pollutants on Fresh Water Phytoplankton Colonies of Wolffia Papulifera Thompson. Edgewater Arsenal Technical Rept. EATR 4595. U.S. NTIS, AD 736336. 29 pp. Govt. Rept. Announc. Index:72(6).

Wright, M. A. 1977. Effects of benomyl and some other systemic fungicides on earthworms. *Ann. Appl. Biol.* 87(3):520–524.

WSSA. 1983. Glyphosate, n-(phosphonomethyl)glycine. *Herbicide Handbook of the Weed Science Society of America.* Weed Science Society of America, Champaign, IL.

Yadav, A. K. and Singh, T. P. 1986. Effect of pesticide on circulating thyroid hormone levels in the freshwater catfish, *Heteropneustes fossilis* (Bloch). *Environ. Res.* 39(1):136–142.

Yadav, A. K. and Singh, T. P. 1987a. Pesticide-induced impairment of thyroid physiology in the freshwater catfish, *Heteropneustes fossilis. Environ. Pollut.* 43(1):29–38.

Yadav, A. K. and Singh, T. P. 1987b. Pesticide-induced changes in peripheral thyroid hormone levels during different reproductive phases in *Heteropneustes fossilis. Ecotoxicol. Environ. Saf.* 13(1):97–103.

Yasuno, M., Hirakoso, S., Sasa, M. and Uchida, M. 1965. Inactivation of some organophosphorous insecticides by bacteria in polluted water. *Jpn. J. Exp. Med.* 35(6): 545–563.

Yeo, R. R. 1970. Dissipation of endothall and effects on aquatic weeds and fish. *Weed Sci.* 18(2):282–284.

Yockim, R. S., Isensee, A. R. and Walker, E. A. 1980. Behavior of trifluralin in aquatic model ecosystems. *Bull. Environ. Contam. Toxicol.* 24(1):134–141.

Zaffaroni, N. P., Arias, E., Capodanno, G. and Zavanella, T. 1978. The toxicity of manganese ethylenebisdithiocarbamate to the adult newt, *Triturus cristatus. Bull. Environ. Contam. Toxicol.* 20(2):261–267.

Zavanella, T., Arias, E. and Zaffaroni, N. P. 1979. Preliminary study on the carcinogenic activity of the fungicide manganese ethylenebisdithiocarbamate in the adult newt, *Triturus cristatus Carnifex. Tumori* 65(2):163–167.

Zavanella, T., Zaffaroni, N. P. and Arias, E. 1980. Testing of the fungicide maneb for carcinogenicity in two populations of the European crested newt. *Cancer Lett.* 10(2): 109–116.

Zimakowska-Gnoinska, D. 1977. Toxicological and physiological aspects of the action of herbicide, sodium salt of 2,4-d on *Asellus aquaticus* L. (Isopoda). *Pol. Arch. Hydrobiol.* 24(3):389–411.

Zinkl, J. G., Henny, C. J. and Deweese, L. R. 1977. Brain cholinesterase activities of birds from forests sprayed with trichlorfon (Dylox) and carbaryl (Sevin-4-oil). *Bull. Environ. Contam. Toxicol.* 17(4):379–386.

Zinkl, J. G., Henny, C. J. and Shea, P. J. 1979. *Brain Cholinesterase Activities of Passerine Birds in Forests Sprayed with Cholinesterase Inhibiting Insecticides.* National Academy of Sciences, Washington, D.C., pp. 356–365.

Zinkl, J. G., Shea, P. J., Nakamoto, R. J. and Callman, J. 1987a. Effects of cholinesterases of rainbow trout exposed to acephate and methamidophos. *Bull. Environ. Contam. Toxicol.* 38(1):22–28.

Zinkl, J. G., Shea, P. J., Nakamoto, R. J. and Callman, J. 1987b. Brain cholinesterase activity of rainbow trout poisoned by carbaryl. *Bull. Environ. Contam. Toxicol.* 38(1):29–35.

Zora, K. and Paladino, F. 1986. Combined toxicity of atrazine and acid exposure to bluntnose minnows, *Pimephales notatus*. *Fed. Proc.* 45(4):916.

INDEX

A

B

C

D

E

M

N

O

P

Q

R

S